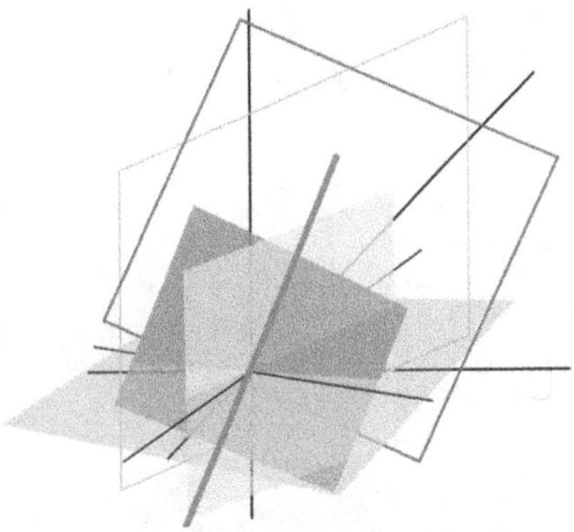

Linear Algebra Tools for Data Mining

Second Edition

Linear Algebra Tools for Data Mining

Second Edition

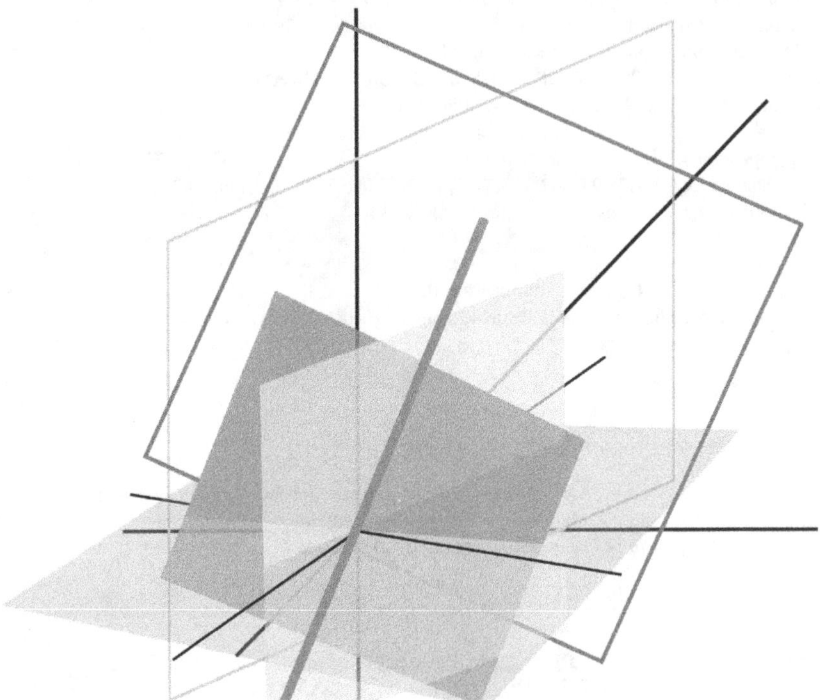

Dan A Simovici

University of Massachusetts Boston, USA
Dana-Farber Cancer Institute, USA

World Scientific

NEW JERSEY • LONDON • SINGAPORE • BEIJING • SHANGHAI • HONG KONG • TAIPEI • CHENNAI

Published by

World Scientific Publishing Co. Pte. Ltd.
5 Toh Tuck Link, Singapore 596224
USA office: 27 Warren Street, Suite 401-402, Hackensack, NJ 07601
UK office: 57 Shelton Street, Covent Garden, London WC2H 9HE

Library of Congress Cataloging-in-Publication Data
Names: Simovici, Dan A., author.
Title: Linear algebra tools for data mining / Dan A. Simovici, University of Massachusetts Boston, USA, Dana-Farber Cancer Institute, USA.
Description: Second edition. | New Jersey : World Scientific Publishing Co. Pte. Ltd., [2023] | Includes bibliographical references and index.
Identifiers: LCCN 2022062233 | ISBN 9789811270338 (hardcover) | ISBN 9789811270345 (ebook for institutions) | ISBN 9789811270352 (ebook for individuals)
Subjects: LCSH: Data mining. | Parallel processing (Electronic computers) | Computer algorithms. | Linear programming.
Classification: LCC QA76.9.D343 S5947 2023 | DDC 006.3/12--dc23/eng/20230210
LC record available at https://lccn.loc.gov/2022062233

British Library Cataloguing-in-Publication Data
A catalogue record for this book is available from the British Library.

For any available supplementary material, please visit
https://www.worldscientific.com/worldscibooks/10.1142/13248#t=suppl

Desk Editors: Balasubramanian/Steven Patt

Typeset by Stallion Press
Email: enquiries@stallionpress.com

To my wife, Doina,
and
to the memory of my brother,
Dr. George Simovici

Preface

Linear algebra plays an increasingly important role in data mining and pattern recognition research either directly, or through the applications of linear algebra in graph theory and optimization.

Linear algebra-based algorithms are elegant and fast, are based on a common mathematical doctrine with its collection of basic ideas and techniques, and are easy to implement; they are especially suitable for parallel and distributed computation to approach large-scale challenging problems such as searching and extracting patterns from the entire web. Thus, the application of linear algebra-based techniques in data mining and machine learning research constitute an increasingly attractive area. Many linear algebra results are important for their applications in biology, chemistry, psychology, and sociology.

The standard undergraduate education of a computer scientist includes one or, rarely, two semesters of linear algebra, which is woefully inadequate for a researcher in data mining or pattern recognition. Even a casual review of publications in these disciplines convincingly demonstrates the use of quite sophisticated tools from linear algebra, optimization, probabilities, functional analysis, and other areas. Linear algebra and its field of applications are constantly growing, and this volume is a mere introduction to a life-long study.

A mathematical background is essential to understand current data mining and pattern recognition research and in conducting research in these disciplines. Therefore, this book was constructed to provide this background and to present a volume of applications that will attract the reader to the study of their mathematical basis. We do not focus on the numerical aspects of the algorithms, particularly error sensitivity, because this extremely important topic has been treated in a vast body of literature in numerical analysis and is not specific to data mining applications.

Among the data mining applications we discuss are the k-means algorithm and several of its relaxations, principal component analysis and singular value decomposition for data dimension reduction, biplots, non-negative matrix factorization for unsupervised and semi-supervised learning, and latent semantic indexing.

Preparing the second edition of this volume involved correcting the existing text, considerable rewriting, and introducing new major topics: tensors, exterior algebra, and multidimensional arrays.

The intended readership consists of graduate students and researchers who work in data mining and pattern recognition. I strived to make this volume as self-contained as possible. The reader interested in applications will find in this volume most of the mathematical background that is currently needed. There are few routine exercises, most of those support the material presented in the main sections of each chapter, and there are more than 600 exercises and supplements.

Special thanks are due to the librarians of the *Joseph Healy Library* at the University of Massachusetts Boston whose dedicated and timely help was essential in completing this project.

I gratefully acknowledge the support provided by MathWorks Inc. from Natick, Massachusetts. Their book program provided the license for various components of MATLAB, the paramount current tool for linear algebra computations, which we use in this book.

Last, but not least, I wish to thank my wife, Doina, a source of strength and loving support.

About the Author

Dan Simovici is a Professor of Computer Science and the Director of the Computer Science Graduate Program at the University of Massachusetts Boston. His main research interests are in the applications of algebraic and information-theoretical methods in data mining and machine learning. He has published over 200 research papers and several books whose subjects range from theoretical computer science to applications of mathematical methods in machine learning. Dr. Simovici served as a visiting professor at Tohoku University in Japan and the University of Lille, France, and he is currently the Editor-in-Chief of the *Journal for Multiple-Valued Logic and Soft Computing.*

Contents

Chapter 1

Preliminaries

1.1 Introduction

We include in this chapter a presentation of basic set-theoretical notions: functions, permutations, and algebraic structures that are intended to make this work as self-contained as possible. We assume that the reader is familiar with certain important classes of relations such as equivalences and partial orders, as presented, for example, in [53] or [152]. Also, we assume familiarity with basic notions and results concerning directed and undirected graphs.

We use the following standard notations for several numerical sets:

$\mathbb{C}$	the set of complex numbers	$\mathbb{R}$	the set of real numbers
$\mathbb{R}_{\geqslant 0}$	the set of non-negative real numbers	$\mathbb{R}_{>0}$	the set of positive real numbers
$\mathbb{R}_{\leqslant 0}$	the set of non-positive real numbers	$\mathbb{R}_{<0}$	the set of negative real numbers
$\widehat{\mathbb{C}}$	the set $\mathbb{C} \cup \{\infty\}$		
$\widehat{\mathbb{R}}$	the set $\mathbb{R} \cup \{-\infty, +\infty\}$	$\widehat{\mathbb{R}}_{\geqslant 0}$	the set $\mathbb{R}_{\geqslant 0} \cup \{+\infty\}$
$\widehat{\mathbb{R}}_{\leqslant 0}$	the set $\mathbb{R}_{\leqslant 0} \cup \{-\infty\}$	$\widehat{\mathbb{R}}_{>0}$	the set $\mathbb{R}_{<>0} \cup \{+\infty\}$
$\mathbb{Q}$	the set of rational numbers	$\mathbb{I}$	the set of irrational numbers
$\mathbb{Z}$	the set of integers	$\mathbb{N}$	the set of natural numbers

1.2 Functions

In this section, we review several types of functions and discuss some results that are used in later chapters.

The set of functions defined on S and ranging on T is denoted by $S \longrightarrow T$. If f belongs to this set of functions, we write $f : S \longrightarrow T$.

The *image of a subset* P of S under a function $f : S \longrightarrow T$ is the set $f(P) = \{t \in T \mid f = f(p) \text{ for some } p \in P\}$. It is easy to verify that for any two subsets P, Q of S we have

$$f(P \cup Q) = f(P) \cup f(Q),$$
$$f(P \cap Q) \subseteq f(P) \cap f(Q).$$

The *preimage of a subset* U of T under a function $f : S \longrightarrow T$ is the set $f^{-1}(U) = \{s \in S \mid f(s) \in U\}$. Again, it is easy to verify that for any two subsets U, V of T, we have

$$f^{-1}(U \cup V) = f^{-1}(U) \cup f^{-1}(V),$$
$$f^{-1}(U \cap V) = f^{-1}(U) \cap f^{-1}(V).$$

Definition 1.1. Let S, T be two sets. A function $f : S \longrightarrow T$ is:

(i) an *injection* if $f(s) = f(s')$ implies $s = s'$;

(ii) a *surjection* if for every $t \in T$ there exists $s \in S$ such that $f(s) = t$;

(iii) a *bijection*, if it is both an injection and a surjection.

The set $S \longrightarrow S$ contains the *identity function* 1_S defined by $1_S(s) = s$ for $s \in S$.

Definition 1.2. Let S, T, U be three sets and let $f : S \longrightarrow T$ and $g : T \longrightarrow U$ be two functions. The *composition* or the *product* of f and g is the function $gf : S \longrightarrow U$ defined by $(gf)(s) = g(f(s))$ for $s \in S$.

Note that if $f : S \longrightarrow T$, then $f1_S = f$ and $1_T f = f$.

It is easy to verify that for four sets S, T, U, V and $f : S \longrightarrow T$, $g : T \longrightarrow U$, $h : U \longrightarrow V$, we have $h(gf) = (hg)f$ (associativity of function composition).

Theorem 1.1. *A function $f : S \longrightarrow T$ is a surjection if and only if for any two functions $g_1 : T \longrightarrow U$ and $g_2 : T \longrightarrow U$ the equality $g_1 f = g_2 f$ implies $g_1 = g_2$.*

Proof. Let $f : S \longrightarrow T$ be a surjection such that $g_1 f = g_2 f$, and let $t \in T$. There exists $s \in S$ such that $t = f(s)$. Then

$$g_1(t) = g_1(f(s)) = (g_1 f)(s) = (g_2 f)(s) = g_2(f(s)) = g_2(t)$$

for every $t \in T$, which implies $g_1 = g_2$.

Conversely, suppose that $g_1 f = g_2 f$ implies $g_1 = g_2$ for any g_1, g_2, and assume that f is not a surjection. Then there exists $t \in T$ such that $t \neq f(s)$ for every $s \in S$. Since $t \notin f(S)$, it is possible to define $g_1, g_2 : T \longrightarrow U$ such that $g_1(t) \neq g_2(t)$, yet $g_1(f(s)) = g_2(f(s))$. This contradicts the initial supposition, so f must be a surjection. □

A similar result holds for injections:

Theorem 1.2. *A function $f : S \longrightarrow T$ is an injection if and only if for any two functions $k_1 : X \longrightarrow S$ and $g_2 : X \longrightarrow S$ the equality $fk_1 = fk_2$ implies $k_1 = k_2$.*

Proof. The argument is left to the reader. □

1.3 Sequences

Definition 1.3. Let S be a set. An *S-sequence of length n* is a function $\boldsymbol{s} : \{0, \ldots, n-1\} \longrightarrow S$. The number n is the *length of the sequence* $\boldsymbol{s}$ and is denoted as $|\boldsymbol{s}|$.

The set of S-sequences of length n is denoted as $\mathbf{Seq}_n(S)$; the set of all sequences on S is

$$\mathbf{Seq}(S) = \bigcup_{n \geqslant 0} \mathbf{Seq}_n(S).$$

A S-sequence $\boldsymbol{s}$ of length n is denoted by $(\boldsymbol{s}(0), \ldots, \boldsymbol{s}(n-1))$. The element $\boldsymbol{s}(j)$ is the jth symbol of $\boldsymbol{s}$.

Example 1.1. If $S = \{\uparrow, \downarrow, -\}$, then

$$(\uparrow, \uparrow, -, \downarrow, -)$$

is an S-sequence of length 5.

For a finite set S, with $|S| = m$, there exist m^n S-sequences of length n. In particular, taking $n = 0$ it follows that there exists a unique S-sequence of length 0 denoted as $\lambda = ()$ and referred to as the *null sequence.*

If $\boldsymbol{u} \in \mathbf{Seq}_p(S)$, $\boldsymbol{v} \in \mathbf{Seq}_q(S)$, $\boldsymbol{u} = (u_0, \ldots, u_{p-1})$, and $\boldsymbol{v} = (v_0, \ldots, v_{q-1})$, the *concatenation* of $\boldsymbol{u}$ and $\boldsymbol{v}$ is the sequence $\boldsymbol{uv}$ given by

$$\boldsymbol{uv} = (u_0, \ldots, u_{p-1}, v_0, \ldots, v_{q-1}).$$

This means that $|\boldsymbol{uv}| = |\boldsymbol{u}|\ |\boldsymbol{v}|$.

It is easy to verify that sequence concatenation is associative. This means that if $\boldsymbol{u}, \boldsymbol{v}, \boldsymbol{w} \in \mathbf{Seq}(S)$, we have

$$(\boldsymbol{uv})\boldsymbol{w} = \boldsymbol{u}(\boldsymbol{vw}),$$

as the reader can easily verify. The null sequence plays the role of the unit element, that is,

$$\boldsymbol{u}\lambda = \lambda\boldsymbol{u} = \boldsymbol{u}.$$

If S contains more than one element, then sequence concatenation is not commutative, as the next example shows.

Example 1.2. Let $S = \{0, 1\}$, $\boldsymbol{u} = (0, 1, 1)$, and $\boldsymbol{v} = (1, 0, 0, 1)$. We have

$$\boldsymbol{uv} = (0, 1, 1, 1, 0, 0, 1) \text{ and } \boldsymbol{vu} = (1, 0, 0, 1, 0, 1, 1).$$

It is clear that $\boldsymbol{uv} \neq \boldsymbol{vu}$.

If $\boldsymbol{s} = (j_1, \ldots, j_n) \in \mathbf{Seq}(\mathbb{N})$, we refer to each pair (j_p, j_q) such that $p < q$ and $j_p > j_q$ as an *inversion*. The set of inversions of $\boldsymbol{s}$ is denoted by $INV(\boldsymbol{s})$ and $|INV(\boldsymbol{s})|$ is denoted by $inv(\boldsymbol{s})$.

Definition 1.4. For $S \subseteq \mathbb{N}$, a sequence $\boldsymbol{s} \in \mathbf{Seq}(\mathbb{N})$ is *strict* if $\boldsymbol{s} = (a_0, \ldots, a_{n-1})$ and $a_0 < a_1 < \cdots < a_{n-1}$, where $n \geqslant 1$.

In other words, a sequence $\boldsymbol{s} \in \mathbf{Seq}(n)$ is strict if it consists of distinct elements and $inv(\boldsymbol{s}) = 0$.

The sequence obtained from the sequence $\boldsymbol{s} \in \mathbf{Seq}(\mathbb{N})$ by sorting its elements in increasing order is denoted as $(\boldsymbol{s})$.

Example 1.3. If $\boldsymbol{s} = (4, 1, 8, 2, 1, 6)$, then $(\boldsymbol{s}) = (1, 1, 2, 4, 6, 8)$.

If $\boldsymbol{s}$ and $\boldsymbol{t}$ are two strict sequences in $\mathbf{Seq}(\mathbb{N})$, their concatenation is not strict, in general, even if they have no elements in common.

Example 1.4. Let $\boldsymbol{s} = (1, 5, 8)$ and $\boldsymbol{t} = (2, 7, 9, 11)$. The sequence $\boldsymbol{st} = (1, 5, 8, 2, 7, 9, 11)$ is clearly not strict and we have $inv(\boldsymbol{st}) = 3$. The sequence $(\boldsymbol{st})$ is $(1, 2, 5, 7, 8, 9, 11)$.

If $\boldsymbol{s}$ and $\boldsymbol{t}$ are two strict sequences, the number of inversions in the sequence $\boldsymbol{st}$ is denoted by $inv(\boldsymbol{s}, \boldsymbol{t})$.

Theorem 1.3. *Let* $\boldsymbol{s}, \boldsymbol{t}, \boldsymbol{u} \in \mathbf{Seq}(\mathbb{N})$. *We have*

$$inv((\boldsymbol{st}), (\boldsymbol{u})) = inv(\boldsymbol{s}, \boldsymbol{u}) + inv(\boldsymbol{t}, \boldsymbol{u}),$$
$$inv((\boldsymbol{s}), (\boldsymbol{tu})) = inv(\boldsymbol{s}, \boldsymbol{t}) + inv(\boldsymbol{s}, \boldsymbol{u}).$$

Proof. Since $(\boldsymbol{st})$ and $(\boldsymbol{u})$ are sorted sequences, there are no inversions in $(\boldsymbol{st})$ or in $(\boldsymbol{u})$. Therefore, inversions in $(\boldsymbol{st})\boldsymbol{u}$ may occur only because an inversion occurs between a component of $\boldsymbol{s}$ and a component of $\boldsymbol{u}$, or between a component of $\boldsymbol{t}$ and a component of $(\boldsymbol{u})$. This justifies the first equality. The second equality has a similar argument. □

1.4 Permutations

The notion of permutation that we discuss in this section is essential for the study of determinants and exterior algebras.

Definition 1.5. A *permutation* of a set S is a bijection $\phi : S \longrightarrow S$.

A permutation ϕ of a finite set $S = \{s_1, \ldots, s_n\}$ is completely described by the sequence $(\phi(s_1), \ldots, \phi(s_n))$. No two distinct components of such a sequence may be equal because of the injectivity of ϕ, and all elements of the set S appear in this sequence because ϕ is surjective. Therefore, the number of permutations equals the number of such sequences, so there are $n(n-1)\cdots 2 \cdot 1$ permutations of a finite set S with $|S| = n$.

The number $n(n-1)\cdots 2 \cdot 1$ is usually denoted by $n!$. This notation is extended by defining $0! = 1$, which is consistent with the interpretation of $n!$ as the number of bijections of a set that has n elements.

The *set of permutations* of the set $S_n = \{1, \ldots, n\}$ is denoted by $PERM_n$. If $\phi \in PERM_n$ is such a permutation, we write

$$\phi : \begin{pmatrix} 1 & \cdots & i & \cdots & n \\ a_1 & \cdots & a_i & \cdots & a_n \end{pmatrix},$$

where $a_i = \phi(i)$ for $1 \leqslant i \leqslant n$. To simplify the notation, we shall specify ϕ just by the sequence of distinct numbers $\boldsymbol{s} = (a_1, \ldots, a_i, \ldots, a_n)$.

The permutation $\iota_n \in PERM_n$ is defined as $\iota_n(j) = j$ for $1 \leqslant j \leqslant n$ and is known as the *identity permutation.*

Definition 1.6. Let $\phi, \psi \in PERM_n$ be two permutations. Their *composition* or *product* is the permutation $\psi\phi$ defined by $(\psi\phi)(j) = \psi(\phi(j))$ for $1 \leqslant j \leqslant n$.

It is clear that the composition of two permutations of S_n is a permutation of S_n.

Example 1.5. Let $\psi, \phi \in PERM_4$ be the permutations

$$\phi : \begin{pmatrix} 1 & 2 & 3 & 4 \\ 3 & 1 & 4 & 2 \end{pmatrix} \text{ and } \psi : \begin{pmatrix} 1 & 2 & 3 & 4 \\ 4 & 2 & 1 & 3 \end{pmatrix}.$$

The permutations $\psi\phi$ and $\phi\psi$ are given by

$$\psi\phi : \begin{pmatrix} 1 & 2 & 3 & 4 \\ 1 & 4 & 3 & 2 \end{pmatrix} \text{ and } \phi\psi : \begin{pmatrix} 1 & 2 & 3 & 4 \\ 2 & 1 & 3 & 4 \end{pmatrix}.$$

Thus, the composition of permutations is not commutative.

Note that every permutation $\phi \in PERM_n$ has an inverse ϕ^{-1} because ϕ is a bijection.

Example 1.6. Let $\phi \in PERM_4$ be the permutation

$$\phi : \begin{pmatrix} 1 & 2 & 3 & 4 \\ 3 & 1 & 4 & 2 \end{pmatrix}.$$

Its inverse is

$$\phi^{-1} : \begin{pmatrix} 1 & 2 & 3 & 4 \\ 2 & 4 & 1 & 3 \end{pmatrix}.$$

Theorem 1.4. *Let* $PERM_n = \{\phi_1, \ldots, \phi_{n!}\}$. *If* $\psi \in PERM_n$, *then* $\{\psi\phi_1, \ldots, \psi\phi_{n!}\} = PERM_n$.

Proof. Note that $\psi\phi_p = \psi\phi_q$ implies $\phi_p = \phi_q$ for $1 \leqslant p \leqslant q \leqslant n!$. The statement follows immediately. □

Definition 1.7. Let S be a finite set, ϕ be a permutation of S, and $x \in S$. The *cycle of* x is the set of elements of the form $C_{\phi,x} = \{\phi^i(x) \mid i \in \mathbb{N}\}$. The number $|C_{\phi,x}|$ is the *length of the cycle.*

Cycles of length 1 are said to be *trivial.*

Theorem 1.5. *The cycles of a permutation* ϕ *of a finite set* S *form a partition* π_ϕ *of* S.

Proof. Let S be a finite set. Since $C_{\phi,x} \subseteq S$, it is clear that $C_{\phi,x}$ is a finite set. If $|C_{\phi,x}| = \ell$, then

$$C_{\phi,x} = \{x, \phi(x), \ldots, \phi^{\ell-1}(x)\}.$$

Note that each pair of elements $\phi^i(x)$ and $\phi^j(x)$ is distinct for $0 \leqslant i, j \leqslant \ell - 1$ and $i \neq j$, because otherwise we would have $|C_{\phi,x}| < \ell$. Moreover, $\phi^\ell(x) = x$.

If $z \in C_{\phi,x}$, then $z = \phi^k(x)$ for some k, $0 \leqslant k \leqslant \ell - 1$, where $\ell = |C_{\phi,x}|$. Since $x = \phi^\ell(x)$, it follows that $x = \phi^{\ell-k}(z)$, which shows that $x \in C_{\phi,z}$. Thus, $C_{\phi,x} = C_{\phi,z}$. □

Definition 1.8. A *k-cyclic permutation* or a *cycle* of a finite set S is a permutation ϕ such that π_ϕ consists of a cycle of length k and a number of $|S| - k$ cycles of length 1.

A *transposition* of S is a 2-cyclic permutation.

Note that if ϕ is a transposition of S, then $\phi^2 = 1_S$.

Theorem 1.6. *Let* S *be a finite set,* ϕ *be a permutation, and*

$$\pi_\phi = \{C_{\phi,x_1}, \ldots, C_{\phi,x_m}\}$$

be the cycle partition associated with ϕ. *Define the cyclic permutations* $\psi_1, \ldots, \psi_m$ *of* S *as*

$$\psi_p(t) = \begin{cases} \phi(t) & \text{if } t \in C_{\phi,x_p}, \\ t & \text{otherwise.} \end{cases}$$

Then, $\psi_p\psi_q = \psi_q\psi_p$ *for every* p, q *such that* $1 \leqslant p, q \leqslant m$.

Proof. Observe first that $u \in C_{\phi,x}$ if and only if $\phi(x) \in C_{\phi,x}$ for any cycle $C_{\phi,x}$.

We can assume that $p \neq q$. Then the cycles C_{ϕ,x_p} and C_{ϕ,x_q} are disjoint. If $u \notin C_{\phi,x_p} \cup C_{\phi,x_q}$, then $\psi_p(\psi_q(u)) = \psi_p(u) = u$ and $\psi_q(\psi_p(u)) = \psi_q(u) = u$.

Suppose now that $u \in C_{\phi,x_p} - C_{\phi,x_q}$. We have $\psi_p(\psi_q(u)) = \psi_p(u) = \phi(u)$. On the other hand, $\psi_q(\psi_p(u)) = \psi_q(\phi(u)) = \phi(u)$ because $\phi(u) \notin C_{\phi,x_q}$. Thus, $\psi_p(\psi_q(u)) = \psi_q(\psi_p(u))$. The case where $u \in C_{\phi,x_q} - C_{\phi,x_p}$ is treated similarly. Also, note that $C_{\phi,x_p} \cap C_{\phi,x_q} = \emptyset$, so, in all cases, we have $\psi_p(\psi_q(x)) = \psi_q(\psi_p(u))$. □

The set of cycles $\{\psi_1, \ldots, \psi_m\}$ is the *cyclic decomposition of the permutation* ϕ.

Definition 1.9. An *adjacent transposition* is a transposition that changes the places of two adjacent elements.

Example 1.7. The permutation $\phi \in PERM_5$ given by

$$\phi : \begin{pmatrix} 1 & 2 & 3 & 4 & 5 \\ 1 & 3 & 2 & 4 & 5 \end{pmatrix}$$

is an adjacent transposition of the set $\{1, 2, 3, 4, 5\}$ because it changes the position of the elements 2 and 3.

On the other hand, the permutation

$$\psi : \begin{pmatrix} 1 & 2 & 3 & 4 & 5 \\ 1 & 5 & 3 & 4 & 2 \end{pmatrix}$$

is a transposition but not an adjacent transposition of the same set because the pair of elements involved are not consecutive.

Theorem 1.7. *Any transposition of a finite set is a product of adjacent transpositions.*

Proof. Let $\phi_{k,\ell}$ be a transposition that swaps k and ℓ, where $k < \ell$. We can move k to ℓ one step at a time and then move ℓ to where k was:

$$\phi_{k,\ell} = \phi_{k,k+1}\phi_{k+1,k+2} \cdots \phi_{\ell-1,\ell}\phi_{\ell-2,\ell-1} \cdots \phi_{k,k+1}.$$

□

For a permutation $\phi \in PERM_n$ specified by the sequence $\boldsymbol{s} = (j_1, \ldots, j_n)$, the set $INV(\boldsymbol{s})$ is denoted by $INV(\phi)$ and $|INV(\phi)|$ is denoted by $inv(\phi)$.

Definition 1.10. The sign of a permutation is the function $sign\ (\phi) : PERM_n \longrightarrow \{-1, 1\}$ defined as

$$sign\ (\phi) = (-1)^{inv(\phi)}.$$

If $sign\ (\phi) = 1$, we say that ϕ is an *even permutation*; otherwise, that is, if $sign\ (\phi) = -1$, we refer to ϕ as an *odd permutation.*

Theorem 1.8. *If* $\phi, \psi \in PERM_n$, *then* $sign\ (\phi\psi) = sign\ (\phi) sign\ (\psi)$.

Proof. The theorem can be proven by showing that $inv(\phi) + inv(\psi)$ has the same parity as $inv(\phi\psi)$.

Suppose that

$$\phi = \begin{pmatrix} 1 & 2 & \cdots & n \\ \phi(1) & \phi(2) & \cdots & \phi(n) \end{pmatrix} \text{ and}$$

$$\psi\phi = \begin{pmatrix} \phi(1) & \phi(2) & \cdots & \phi(n) \\ \psi(\phi(1)) & \psi(\phi(2)) & \cdots & \psi(\phi(n)) \end{pmatrix}.$$

and consider the following cases:

(i) if $i < j$, $\phi(i) < \phi(j)$, and $\psi(\phi(i)) < \psi(\phi(j))$, then no inversions are generated in ϕ, ψ, and $\psi\phi$;
(ii) if $i < j$, $\phi(i) < \phi(j)$, and $\psi(\phi(i)) > \psi(\phi(j))$, ϕ has no inversion, ψ has an inversion, and so does $\psi\phi$;
(iii) if $i < j$, $\phi(i) > \phi(j)$, and $\psi(\phi(i)) > \psi(\phi(j))$, ϕ has an inversion, ψ has no inversion, and $\psi\phi$ has an inversion;
(iv) if $i < j$, $\phi(i) > \phi(j)$, and $\psi(\phi(i)) < \psi(\phi(j))$, then ϕ has an inversion, ψ has an inversion, and $\psi\phi$ has no inversions.

In each of these cases, $inv(\phi) + inv(\psi)$ differs from $inv(\psi\phi)$ by an even number. □

A *descent* of a sequence with distinct numbers $\boldsymbol{s} = (j_1, \ldots, j_n)$ is a number k such that $1 \leqslant k \leqslant n-1$ and $j_k > j_{k+1}$. The set of descents of $\boldsymbol{s}$ is denoted by $D(\boldsymbol{s})$ and the set of descents of a permutation ϕ specified by $\boldsymbol{s}$ is denoted by $D(\phi)$.

Example 1.8. Let $\phi \in PERM_6$ be

$$\phi : \begin{pmatrix} 1 & 2 & 3 & 4 & 5 & 6 \\ 4 & 2 & 5 & 1 & 6 & 3 \end{pmatrix}.$$

We have

$$INV(\phi) = \{(4,2),(4,1),(4,3),(2,1),(5,1),(5,3),(6,3)\},$$

and $inv(\phi) = 7$. Furthermore, $D(\phi) = \{1,3,5\}$.

It is easy to see that the following conditions are equivalent for a permutation ϕ of the finite set S:
(i) $\phi = 1_S$;
(ii) $inv(\phi) = 0$;
(iii) $D(\phi) = \emptyset$.

Theorem 1.9. *Every permutation $\phi \in PERM_n$ is a composition of transpositions.*

Proof. If $D(\phi) = \emptyset$, then $\phi = 1_{\{1,\ldots,n\}}$ and the statement is vacuous. Suppose therefore that $D(\phi) \neq \emptyset$, and let $k \in D(\phi)$, which means that (j_k, j_{k+1}) is an inversion ϕ. Let ψ be the adjacent transposition that exchanges j_k and j_{k+1}. It is clear that $inv(\psi\phi) = inv(\phi) - 1$. Thus, if $\psi_1, \ldots, \psi_p$ are the transpositions that correspond to all adjacent inversions of ϕ, where $p = inv(\phi)$, it follows that $\psi_p \cdots \psi_1 \phi$ has 0 inversions and, as observed above, $\psi_p \cdots \psi_1 \phi = 1_S$. Since $\psi^2 = 1_S$ for every transposition ψ, we have $\phi = \psi_p \cdots \psi_1$, which gives the desired conclusion. □

Corollary 1.1. *If a permutation $\phi \in \Phi_n$ can be factored as a product of p transpositions, then ϕ^{-1} can be factored as the same number of transpositions.*

Proof. This follows immediately from Theorem 1.9. □

Theorem 1.10. *If ϕ is a permutation of the finite set S, then $inv(\phi)$ is the least number of adjacent transpositions that constitute a factorization of ϕ, and the number of adjacent transpositions involved in any other factorization of ϕ as a product of adjacent transpositions differs from $inv(\phi)$ by an even number.*

Proof. Let $\phi = \psi_q \cdots \psi_1$ be a factorization of ϕ as a product of adjacent transpositions. Then $\psi_1 \cdots \psi_q \phi = 1_S$, and we can define the sequence of permutations $\phi_\ell = \psi_\ell \cdots \psi_1 \phi$ for $1 \leqslant \ell \leqslant q$. Since each ψ_i is an adjacent transposition, we have $inv(\phi_{\ell+1}) - inv(\phi_\ell) = 1$ or $inv(\phi_{\ell+1}) - inv(\phi_\ell) = -1$. If

$$|\{\ell \mid 1 \leqslant \ell \leqslant q-1 \text{ and } inv(\phi_{\ell+1}) - inv(\phi_\ell) = 1\}| = r,$$

then $|\{\ell \mid 1 \leqslant \ell \leqslant q-1 \text{ and } inv(\phi_{\ell+1}) - inv(\phi_\ell) = -1\}| = q - r$, so $inv(\phi) + r - (q - r) = 0$, which means that $q = inv(\phi) + 2r$. This implies the desired conclusion. □

Theorem 1.11. *A transposition is an odd permutation.*

Proof. Suppose that $\phi \in PERM_n$ is the transposition

$$\phi : \begin{pmatrix} 1 & \cdots & i & \cdots & j \cdots & n \\ 1 & \cdots & j & \cdots & i \cdots & n \end{pmatrix},$$

so $j > i$.

If p and q form an inversion in ϕ, it follows that we must have $i \leqslant p < q \leqslant j$.

For $i < p < q < j$, the pair (p, q) has no contribution to inversions.

If $i < q \leq j$, the pair (i, q) contributes $j - i$ inversions because $\phi(i) = j > \phi(q)$. If $i \leqslant p < j$ and $\phi(q) > i = \phi(j)$, there are $(j - 1) - i$ inversions. Thus, the total number of inversions of ϕ is $(j - i) + (j - 1) - i = 2(j - i) - 1$ and this number is odd. □

Next, we introduce the *Levi-Civita* symbols that are useful in the study of permutations.

Definition 1.11. Let $(i_1 \cdots i_n)$ be a permutation of the set $\{1, 2, \ldots, n\}$; The *Levi-Civita symbols*[1] $\epsilon_{i_1 i_2 \cdots i_n}$ and $\epsilon^{i_1 i_2 \cdots i_n}$ are defined as

---

[1]Tullio Levi-Civita (March 29, 1873–December 29, 1941) was an Italian mathematician, well-known for his work on tensor calculus and its applications to the theory of relativity. His work included foundational papers in both pure and applied mathematics, celestial mechanics, analytic mechanics, and hydrodynamics. He was born in Padua, graduated in 1892 from the University of Padua Faculty of Mathematics where he became a professor in 1898. In 1918, be was appointed at the University of Rome.

(i) $\epsilon_{1\,2\,\cdots\,n} = \epsilon^{1\,2\,\cdots\,n} = 1$;
(ii) $\epsilon_{\cdots\, i_p\, \cdots\, i_q\, \cdots} = -\epsilon_{\cdots\, i_q\, \cdots\, i_p\, \cdots}$ and $\epsilon^{\cdots\, i_p\, \cdots\, i_q\, \cdots} = -\epsilon^{\cdots\, i_q\, \cdots\, i_p\, \cdots}$ (the *antisymmetry property*);
(iii) when all indices $i_1, \ldots, i_n$ are distinct, we have

$$\epsilon_{i_1\cdots i_n} = \epsilon^{i_1\cdots i_n} = (-1)^p,$$

where $p = inv(i_1 \cdots i_n)$.

When two indices i_p and i_q are equal, the antisymmetry of $\epsilon_{i_1 i_2 \cdots i_n}$ and $\epsilon^{i_1 i_2 \cdots i_n}$ implies $\epsilon_{i_1 i_2 \cdots i_n} = \epsilon^{i_1 i_2 \cdots i_n} = 0$.

Example 1.9. For Levi-Civita symbols $\epsilon_{i_1 i_2 i_3}$ with $n = 3$, we have 27 components. Since an equality of any of the indices implies that the number $\epsilon_{i_1 i_2 i_3}$ is 0, it follows that only six of these indices are non-zero. The non-zero values are

$$\begin{aligned}\epsilon_{123} &= \epsilon_{231} = \epsilon_{312} = 1,\\ \epsilon_{132} &= \epsilon_{213} = \epsilon_{321} = -1.\end{aligned}$$

1.5 Combinatorics

Let S be a finite nonempty set, $S = \{s_1, \ldots, s_n\}$. We seek to count the sequences of S having length k without repetitions.

Suppose initially that $k \geqslant 1$. For the first place in a sequence $\boldsymbol{s}$ of length k, we have n choices. Once an element of S has been chosen for the first place, we have $n-1$ choices for the second place because the sequence may not contain repetitions, etc. For the kth component of $\boldsymbol{s}$, there are $n - 1 + k$ choices. Thus, the number of sequences of length k without repetitions is given by $n(n-1)\cdots(n-k+1)$. We shall denote this number by $A(n, k)$.

There exists only one sequence of length 0, namely the empty sequence, so we extend the definition of A by $A(n, 0) = 1$ for every $n \in \mathbb{N}$.

Let S be a finite set with $|S| = n$. A *k-combination* of S is a subset M of S such that $|M| = k$.

Define the equivalence relation $\sim$ on $\mathbf{Seq}(S)$ by $\boldsymbol{s} \sim \boldsymbol{t}$ if there exists a bijection f such that $\boldsymbol{s} = \boldsymbol{t}f$.

If T is a subset of S such that $|T| = k$, there exists a bijection $\boldsymbol{t} : \{0, \ldots, k-1\} \longrightarrow T$; clearly, this is a sequence without repetitions and there exist $A(n,k)$ such sequences. Note that if $\boldsymbol{u}$ is an equivalent sequence (that is, if $\boldsymbol{t} \sim \boldsymbol{u}$), then the range of this sequence is again the set T and there are $k!$ such sequences (due to the existence of the $k!$ permutations f) that correspond to the same set T. Therefore, we may conclude that $\mathcal{P}_k(S)$ contains $\frac{A(n,k)}{k!}$ elements. We denote this number by $\binom{n}{k}$ and we refer to it as the (n,k)-*binomial coefficient*. We can write $\binom{n}{k}$ using factorials as follows:

$$\begin{aligned}\binom{n}{k} &= \frac{A(n,k)}{k!} = \frac{n(n-1)\cdots(n-k+1)}{k!}\\ &= \frac{n(n-1)\cdots(n-k+1)(n-k)\cdots 2\cdot 1}{k!(n-k)!}\\ &= \frac{n!}{k!(n-k)!}.\end{aligned}$$

We mention the following useful identities:

$$k\binom{n}{k} = n\binom{n-1}{k-1}, \tag{1.1}$$

$$\binom{n}{m} = \frac{n}{m}\binom{n-1}{m-1}. \tag{1.2}$$

Equality (1.1) can be extended as

$$k(k-1)\cdots(k-\ell)\binom{n}{k} = n(n-1)\cdots(n-\ell)\binom{n-\ell-1}{k-\ell-1} \tag{1.3}$$

for $0 \leqslant \ell \leqslant k-1$.

The set of polynomials with complex (real) coefficients in the nondeterminate x is denoted by $\mathbb{C}[x]$ (respectively, $\mathbb{R}[x]$).

Consider now the n-degree polynomial $p \in \mathbb{R}[x]$:

$$p(x) = (x+a_0)\cdots(x+a_{n-2})(x+a_{n-1}).$$

The coefficient of x^{n-k} consists of the sum of all monomials of the form $a_{i_0}\cdots a_{i_{k-1}}$, where the subscripts $i_0,\ldots,i_{k-1}$ are distinct. Thus, the coefficient of x^{n-k} contains $\binom{n}{k}$ terms corresponding to the k-element subsets of the set $\{0,\ldots,n-1\}$. Consequently, the coefficient of x^{n-k} in the power $(x+a)^n$ can be obtained from the similar coefficient in $p(x)$ by taking $a_0 = \cdots = a_{n-1} = a$; thus, the coefficient is $\binom{n}{k}a^k$. This allows us to write

$$(x+a)^n = \sum_{k=0}^{n} \binom{n}{k} x^{n-k} a^k. \tag{1.4}$$

This equality is known as *Newton's binomial formula* and has numerous applications.

Example 1.10. If we take $x = a = 1$ in Formula (1.4), we obtain the identity

$$2^n = \sum_{k=0}^{n} \binom{n}{k}. \tag{1.5}$$

Note that this equality can be obtained directly by observing that the right member, enumerates the subsets of a set having n elements by their cardinality k.

A similar interesting equality can be obtained by taking $x = 1$ and $a = -1$ in Formula (1.4). This yields

$$0 = \sum_{k=0}^{n} \binom{n}{k}(-1)^k = \binom{n}{0} + \binom{n}{2} + \binom{n}{4} + \cdots$$
$$-\binom{n}{0} - \binom{n}{2} - \binom{n}{4} - \cdots.$$

This inequality shows that each set contains an equal number of subsets having an even or odd number of elements.

Example 1.11. Consider the equality $(x+a)^n = (x+a)^{n-1}(x+a)$. The coefficient of $x^{n-k}a^k$ in the left member is $\binom{n}{k}$. In the right member, $x^{n-k}a^k$ has the coefficient $(\binom{n-1}{k} + \binom{n-1}{k-1})$, so we obtain the equality

$$\binom{n}{k} = \binom{n-1}{k} + \binom{n-1}{k-1}, \tag{1.6}$$

for $0 \leqslant k \leqslant n-1$.

Multinomial coefficients are generalizations of binomial coefficients that can be introduced as follows. The nth power of the sum $x_1 + \cdots + x_k$ can be written as

$$(x_1 + \cdots + x_k)^n = \sum_{(r_1,\ldots,r_k)} c(n, r_1, \ldots, r_k) x_1^{r_1} \cdots x_k^{r_k},$$

where the sum involves all $(r_1, \ldots, r_k) \in \mathbb{N}^k$ such that $\sum_{i=1}^k r_i = n$. By analogy with the binomial coefficients, we denote $c(n, r_1, \ldots, r_k)$ by $\binom{n}{r_1,\ldots,r_n}$. As we did with binomial coefficients in Example 1.11, starting from the equality $(x_1 + \cdots + x_k)^n = (x_1 + \cdots + x_k)^{n-1}(x_1 + \cdots + x_k)$, the coefficient of the monomial $x_1^{r_1} \cdots x_k^{r_k}$ in the right member is $\binom{n}{r_1,\ldots,r_n}$. In the left member, the same coefficient is

$$\sum_{i=1}^{k} \binom{n-1}{r_1, \ldots, r_i - 1, \ldots, r_n},$$

so we obtain the identity

$$\binom{n}{r_1, \ldots, r_n} = \sum_{i=1}^{k} \binom{n-1}{r_1, \ldots, r_i - 1, \ldots, r_n}, \tag{1.7}$$

a generalization of the identity (1.6).

1.6 Groups, Rings, and Fields

The notion of operation on a set is needed for introducing various algebraic structures on sets.

Definition 1.12. Let $n \in \mathbb{N}$. An *n-ary operation on a set S* is a function $f : S^n \longrightarrow S$. The number n is the *arity* of the operation f.

If $n = 0$, we have the special case of *zero-ary operations*. A zero-ary operation is a function $f : S^0 = \{\emptyset\} \longrightarrow S$, which is essentially a constant element of S, $f()$. Operations of arity 1 are referred to as *unary operations*.

Binary operations (of arity 2) are frequently used. For example, the union, intersection, and difference of subsets of a set S are binary operations on the set $\mathcal{P}(S)$.

If f is a binary operation on a set, we denote the result $f(x, y)$ of the application of f to x, y by xfy rather than $f(x, y)$.

We now introduce certain important types of binary operations.

Definition 1.13. A binary operation f on a set S is

(i) *associative* if $(xfy)fz = xf(yfz)$ for every $x, y, z \in S$,
(ii) *commutative* if $xfy = yfx$ for every $x, y, \in S$, and
(iii) *idempotent* if $xfx = x$ for every $x \in S$.

Example 1.12. Set union and intersection are both associative, commutative, and idempotent operations on every set of the form $\mathcal{P}(S)$.

The addition of real numbers "+" is an associative and commutative operation on $\mathbb{R}$; however, "+" is not idempotent.

The binary operation $g : \mathbb{R}^2 \longrightarrow \mathbb{R}$ given by $g(x, y) = \frac{x+y}{2}$ for $x, y \in \mathbb{R}$ is a commutative and idempotent operation of $\mathbb{R}$ that is not associative. Indeed, we have $(xgy)gz = \frac{x+y+2z}{4}$ and $xg(ygz) = \frac{2x+y+z}{4}$.

Example 1.13. The binary operations $\max\{x, y\}$ and $\min\{x, y\}$ are associative, commutative, and idempotent operations on the set $\mathbb{R}$.

Next, we introduce special elements relative to a binary operation on a set.

Definition 1.14. Let f be a binary operation on a set S.

(i) An element u is a *unit for* f if $xfu = ufx = x$ for every $x \in S$.
(ii) An element z is a *zero for* f if $zfu = ufz = z$ for every $x \in S$.

Note that if an operation f has a unit, then this unit is unique. Indeed, suppose that u and u' were two units of the operation f. According to Definition 1.14, we would have $ufx = xfu = x$ and, in particular, $ufu' = u'fu = u'$. Applying the same definition to u' yields $u'fx = xfu' = x$ and, in particular, $u'fu = ufu' = u$. Thus, $u = u'$.

Similarly, if an operation f has a zero, then this zero is unique. Suppose that z and z' were two zeros for f. Since z is a zero, we have $zfx = xfz = z$ for every $x \in S$; in particular, for $x = z'$, we have $zfz' = z'fz = z$. Since z' is zero, we also have $z'fx = xfz' = z'$ for every $x \in S$; in particular, for $x = z$, we have $z'fz = zfz' = z'$, and this implies $z = z'$.

Definition 1.15. Let f be a binary associative operation on S such that f has the unit u. An element x has an *inverse* relative to f if there exists $y \in S$ such that $xfy = yfx = u$.

An element x of S has at most one inverse relative to f. Indeed, suppose that both y and y' are inverses of x. Then we have

$$y = yfu = yf(xfy') = (yfx)fy' = ufy' = y',$$

which shows that y coincides with y'.

If the operation f is denoted by "+", then we refer to the inverse of x as the *additive inverse* of x, or the *opposite element* of x; similarly, when f is denoted by "·", we refer to the inverse of x as the *multiplicative inverse of* x. The additive inverse of x is usually denoted by $-x$, while the multiplicative inverse of x is denoted by x^{-1}.

Definition 1.16. An element x of a set S equipped with a binary operation "$*$" is *idempotent* if $x * x = x$.

Observe that every unit and every zero of a binary operation is an idempotent element.

Definition 1.17. Let $\mathcal{I} = \{f_i | i \in I\}$ be a set of operations on a set S indexed by a set I. An *algebra type* is a mapping $\theta : I \longrightarrow \mathbb{N}$.

An *algebra of type* θ is a pair $\mathcal{A} = (A, \mathcal{I})$ such that

(i) A is a set, and
(ii) the operation f_i has arity $\theta(i)$ for every $i \in I$.

The algebra $\mathcal{A} = (A, \mathcal{I})$ is *finite* if the set A is finite. The set A is referred to as the *carrier* of the algebra $\mathcal{A}$.

If the indexing set I is finite, we say that the type θ is a *finite type* and refer to $\mathcal{A}$ as an *algebra of finite type*.

If $\theta : I \longrightarrow \mathbb{N}$ is a finite algebra type, we assume, in general, that the indexing set I has the form $(0, 1, \ldots, n-1)$. In this case, we denote θ by the sequence $(\theta(0), \theta(1), \ldots, \theta(n-1))$.

Definition 1.18. A *groupoid* is an algebra of type (2), $\mathcal{A} = (A, \{f\})$. If f is an associative operation, then we refer to this algebra as a *semigroup*.

In other words, a groupoid is a set equipped with a binary operation f.

Example 1.14. The algebra $(\mathbb{R}, \{f\})$ where $f(x, y) = \frac{x+y}{2}$ is a groupoid. However, it is not a semigroup because f is not an associative operation.

Example 1.15. Define the binary operation g on $\mathbb{R}$ by $xgy = \ln(e^x + e^y)$ for $x, y \in \mathbb{R}$. Since

$$\begin{aligned}(xgy)gz &= \ln(e^{xgy+e^z}) = \ln(e^x + e^y + e^z),\\ xg(ygz) &= \ln(x + e^{ygz}) = \ln(e^x + e^y + e^z),\end{aligned}$$

for every $x, y, z \in \mathbb{R}$, it follows that g is an associative operation. Thus, $(\mathbb{R}, g)$ is a semigroup. It is easy to verify that this semigroup has no unit element.

Definition 1.19. A *monoid* is an algebra of type $(0, 2)$, $\mathcal{A} = (A, \{e, f\})$, where e is a zero-ary operation, f is a binary operation, and e is the unit element for f.

Example 1.16. Let $\gcd(m, n)$ be the greatest common divisor of the numbers $m, n \in \mathbb{N}$.

The algebras $(\mathbb{N}, \{1, \cdot\})$ and $(\mathbb{N}, \{0, \gcd\})$ are monoids. In the first case, the binary operation is the multiplication of natural numbers, the unit element is 1, and the algebra is clearly a monoid. In the second case, the unit element is 0.

We claim that gcd is an associative operation. Let $m, n, p \in \mathbb{N}$. We need to verify that $\gcd(m, \gcd(n, p)) = \gcd(\gcd(m, n), p)$.

Let $k = \gcd(m, \gcd(n, p))$. Then $(k, m) \in \delta$ and $(k, \gcd(n, p)) \in \delta$, where δ is the divisibility relation. Since $\gcd(n, p)$ divides evenly both n and p, it follows that $(k, n) \in \delta$ and $(k, p) \in \delta$. Thus, k divides $\gcd(m, n)$, and therefore k divides $h = \gcd(\gcd(m, n), p)$.

Conversely, h being $\gcd(\gcd(m, n), p)$, it divides both $\gcd(m, n)$ and p. Since h divides $\gcd(m, n)$, it follows that it divides both m and p. Consequently, h divides $\gcd(n, p)$ and therefore divides $k = \gcd(m, \gcd(n, p))$. Since k and h are both natural numbers that divide each other evenly, it follows that $k = h$, which allows us to conclude that gcd is an associative operation. Since n divides 0 evenly, for any $n \in \mathbb{N}$, it follows that $\gcd(0, n) = \gcd(n, 0) = n$, which shows that 0 is the unit for gcd.

Definition 1.20. A *group* is an algebra of type $(0, 2, 1)$, $\mathcal{A} = (A, \{e, f, h\})$, where e is a zero-ary operation, f is a binary operation, e is the unit element for f, and h is a unary operation such that $f(h(x), x) = f(x, h(x)) = e$ for every $x \in A$.

Note that if we have $xfy = yfx = e$, then $y = h(x)$. Indeed, we can write

$$h(x) = h(x)fe = h(x)f(xfy) = (h(x)fx)fy = efy = y.$$

We refer to the unique element $h(x)$ as the *inverse* of x. The usual notation for $h(x)$ is x^{-1}.

Definition 1.21. A group $\mathcal{A} = (A, \{e, f, h\})$ is *Abelian* if $xfy = yfx$ for all $x, y \in A$.

Abelian groups are also known as *commutative groups*. Traditionally, the zero-ary operation of Abelian groups is denoted by 0, the binary operation by "+", and the inverse of an element x is denoted by $-x$. Thus, we usually write an Abelian group $\mathcal{A}$ as $(A, \{0, +, -\})$.

Example 1.17. The algebra $(\mathbb{Z}, \{0, +, -\})$ is an Abelian group where "+" is the usual addition of integers, and the additive inverse of an integer n is $-n$.

Example 1.18. The set of permutations of $\{1, \ldots, n\}$ is the *symmetric group* $\mathcal{S}_n = (PERM_n, \iota_n, \cdot,)$, where $\cdot$ stands for the permutation composition.

Example 1.19. Let $\equiv_n \subseteq \mathbb{Z} \times \mathbb{Z}$ be the equivalence relation defined on $\mathbb{Z}$ by

$$\equiv_n = \{(p, q) \in \mathbb{Z} \times \mathbb{Z} \mid n \text{ evenly divides } p - q\}.$$

In each equivalence class $[p]$ there exists a least non-negative element m such that $0 \leqslant m \leqslant n - 1$, due to the properties of the division of integers. This allows us to consider the quotient set $\mathbb{Z}/\equiv_n = \{[0], \ldots, [n-1]\}$. For example, if $n = 2$, the quotient set contains the classes $[0]$ and $[1]$. All even integers, belong to the class $[0]$ and all odd integers, to the class $[1]$.

We denote the quotient set $\mathbb{Z}/\equiv_n$ by $\mathbb{Z}_n$. The sum of two equivalence classes $[p]$ and $[q]$ is the class $[p+q]$. It is easy to verify that this addition is well-defined for if $r \in [p]$ and $s \in [q]$, then n divides evenly $r-p$ and $s-q$ and, therefore, it divides evenly $r+s-(p+q)$.

For example, $\mathbb{Z}_4$ consists of $\{[0],[1],[2],[3]\}$ and the addition is defined by the table

+	[0]	[1]	[2]	[3]
[0]	[0]	[1]	[2]	[3]
[1]	[1]	[2]	[3]	[0]
[2]	[2]	[3]	[0]	[1]
[3]	[3]	[0]	[1]	[2]

It is easy to verify that $(\mathbb{Z}_n, \{[0],+,-\})$ is an Abelian group, where $-[p]$ is $[n-p]$ for $0 \leqslant p \leqslant n-1$ and $[n]=[0]$.

Definition 1.22. A *ring* is an algebra of type $(0,2,1,2)$, $\mathcal{A} = (A, \{e,f,h,g\})$, such that $\mathcal{A} = (A,\{e,f,h\})$ is an Abelian group and g is a binary associative operation such that

$$xg(ufv) = (xgu)f(xgv),$$
$$(ufv)gx = (ugx)f(vgx),$$

for every $x,u,v \in A$. These equalities are known as *left* and *right distributivity laws*, respectively.

The operation f is known as the *ring addition*, while $\cdot$ is known as the *ring multiplication*. Frequently, these operations are denoted by "+" and "·", respectively.

Example 1.20. The algebra $(\mathbb{Z},\{0,+,-,\cdot\})$ is a ring. The distributive laws amount to the well-known distributive properties

$$p\cdot(q+r) = (p\cdot q)+(p\cdot r),$$
$$(q+r)\cdot p = (q\cdot p)+(r\cdot p),$$

for $p,q,r \in \mathbb{Z}$, of integer addition and multiplication.

Example 1.21. A more interesting type of ring is defined on the set of numbers of the form $m+n\sqrt{2}$, where m and n are integers. The ring operations are given by

$$(m+n\sqrt{2})+(p+q\sqrt{2}) = m+p+(n+q)\sqrt{2},$$
$$(m+n\sqrt{2})\cdot(p+q\sqrt{2}) = m\cdot p+2\cdot n\cdot q+(m\cdot q+n\cdot p)\sqrt{2}.$$

If the multiplicative operation of a ring has a unit element 1, then we say that the ring is a *unitary ring.* We consider a unitary ring as an algebra of type $(0, 0, 2, 1, 2)$ by regarding the multiplicative unit as another zero-ary operation.

Observe, for example, that the ring $(\mathbb{Z}, \{0, 1, +, -, \cdot\})$ is a unitary ring. Also, note that the set of even numbers also generates a ring $(\{2k \mid k \in \mathbb{Z}\}, \{0, +, -, \cdot\})$. However, no multiplicative unit exists in this ring.

Example 1.22. Let $(S, \{0, 1, +, -, \cdot\})$ be a commutative ring and let $s = (s_0, s_1, \ldots)$ be a sequence of elements of S. The *support* of the sequence s is the set $\mathsf{supp}(s) = \{i \in \mathbb{N} \mid s_i \neq 0\}$. A *polynomial* over S is a sequence that has a finite support. If $\mathsf{supp}(s) = \emptyset$, then s is the *zero polynomial.* The *degree of a polynomial* p is the number $\mathsf{deg}(p) = \max \mathsf{supp}(p)$. The degree of the zero polynomial is 0.

The *addition of the polynomials* $p = (p_0, p_1, \ldots)$ and $q = (q_0, q_1, \ldots)$ produces the polynomial

$$p + q = (p_0 + q_0, p_1 + q_1, \ldots).$$

The *product of the polynomials* p and q is

$$pq = \left(p_0q_0, p_0q_1 + p_1q_0, \ldots, \sum_{i=0}^{m} p_iq_{m-i}, \ldots\right).$$

The "usual notation" for polynomials involves considering a symbol λ referred to as an *indeterminate.* Then the polynomial $p = (p_0, p_1, \ldots)$ is denoted as

$$p(\lambda) = p_0 + p_1\lambda + \cdots + p_n\lambda^n,$$

where $n = \mathsf{deg}(p)$. The set of polynomials over a ring S in the indeterminate λ is denoted by $S[\lambda]$.

The reader accustomed to the usual notation for polynomials will realize that the addition and multiplication defined above correspond to the usual addition and multiplication of polynomials. If is easy to verify that the set of polynomials in the indeterminate λ over S denoted by $S[\lambda]$ is itself a ring, denoted by $S[\lambda]$, with the addition and multiplication defined above.

Example 1.23. Example 1.22 can be extended to polynomials of several variables.

Let $\mathbb{N}^k = (\boldsymbol{z}_0, \boldsymbol{z}_1, \ldots)$ be the set of k-tuples of natural numbers listed in a fixed order. A *polynomial in k variables* over S is a sequence $p = (p_{\boldsymbol{z}_0}, p_{\boldsymbol{z}_1}, \ldots, p_{\boldsymbol{z}_n}, \ldots)$ such that the support set $\mathsf{supp}(p) = \{\boldsymbol{z} \in \mathbb{N}^k \mid p_{\boldsymbol{z}} \neq 0\}$ is finite. If $\mathsf{supp}(p) = \emptyset$, then p is the *zero polynomial.* The set of polynomials in k variables over S is denoted by $S[\lambda_1, \ldots, \lambda_k]$.

If $|\mathsf{supp}(p)| = 1$, then p is a *monomial.* Let $\lambda_1, \ldots, \lambda_k$ be k indeterminates. A monomial p such that $\mathsf{supp}(p) = \{(a_1, \ldots, a_k)\}$ is written as

$$p(\lambda_1, \ldots, \lambda_k) = \lambda_1^{a_1} \lambda_2^{a_2} \cdots \lambda_k^{a_k}$$

and the number $\mathsf{deg}(p) = a_1 + \cdots + a_k$ is the *degree* of the monomial p.

If p, q are two polynomials in k variables, $p = (p_{\boldsymbol{z}_0}, p_{\boldsymbol{z}_1}, \ldots, p_{\boldsymbol{z}_n}, \ldots)$ and $q = (q_{\boldsymbol{z}_0}, q_{\boldsymbol{z}_1}, \ldots, q_{\boldsymbol{z}_n}, \ldots)$, their sum is the polynomial $p + q$ given by

$$(p+q)(\lambda_1, \ldots, \lambda_k) = (p_{\boldsymbol{z}_0} + q_{\boldsymbol{z}_0}, p_{\boldsymbol{z}_1} + q_{\boldsymbol{z}_1}, \ldots, p_{\boldsymbol{z}_n} + q_{\boldsymbol{z}_n}, \ldots).$$

The product of p and q is the polynomial pq, where

$$(pq)_{\boldsymbol{z}} = \sum_{\boldsymbol{u}+\boldsymbol{v}=\boldsymbol{z}} p_{\boldsymbol{u}} q_{\boldsymbol{v}}.$$

Definition 1.23. A polynomial $p \in S[\lambda_1, \ldots, \lambda_k]$ is *symmetric* if $p(\lambda_1, \ldots, \lambda_k) = p(\lambda_{\phi(1)}, \ldots, \lambda_{\phi(k)})$ for every permutation $\phi \in PERM_k$.

Example 1.24. The *elementary symmetric polynomials* in $\lambda_1, \ldots, \lambda_n$ are defined by

$$\begin{aligned}
s_0(\lambda_1, \ldots, \lambda_n) &= 1, \\
s_1(\lambda_1, \ldots, \lambda_n) &= \sum_{i=1}^{n} \lambda_i, \\
s_2(\lambda_1, \ldots, \lambda_n) &= \sum_{i<j} \lambda_i \lambda_j, \\
&\vdots \\
s_n(\lambda_1, \ldots, \lambda_n) &= \lambda_1 \cdots \lambda_n.
\end{aligned}$$

A well-known result of elementary algebra, known as *Viéte's Theorem*, is that the coefficients $c_1, \ldots, c_n$ in the equation

$$\lambda^n + c_1\lambda^{n-1} + \cdots + c_{n-1}\lambda + c_n = 0$$

are linked to the values of the elementary symmetric polynomials of the roots as follows:

$$\begin{aligned} s_1(\lambda_1, \ldots, \lambda_n) &= -c_1, \\ s_2(\lambda_1, \ldots, \lambda_n) &= c_2, \\ &\vdots \\ s_n(\lambda_1, \ldots, \lambda_n) &= (-1)^n c_n. \end{aligned}$$

Example 1.25. The *complete symmetric polynomial* of degree k, $c_k(\lambda_1, \ldots, \lambda_n)$ is the sum of all monomials of degree k. For example, we have

$$c_3(\lambda_1, \lambda_2) = \lambda_1^3 + \lambda_1^2\lambda_2^1 + \lambda_1^1\lambda_2^2 + \lambda_2^3.$$

The special case $k = 0$ is defined by $c_0(\lambda_1, \ldots, \lambda_n) = 1$.

Rings with commutative multiplicative operations are known as *commutative rings*. All examples of rings considered so far are commutative rings.

Theorem 1.12. *Let $(S, \{0, 1, +, -, \cdot\})$ be a unitary ring. If x is an idempotent element of S, then $1 - x$ is also idempotent.*

Proof. Let x be an idempotent element of S. Since $x^2 = x$, we have $(1-x)^2 = (1-x)(1-x) = 1 - x - x + x^2 = 1 - x$, which shows that $1 - x$ is also idempotent. □

Definition 1.24. A *field* is a pair $\mathcal{F} = (\mathbb{F}, \{0, 1, +, -, \cdot\})$ such that

(i) $(\mathbb{F}, \{0, 1, +, -, \cdot\})$ is a commutative and unitary ring;
(ii) every element x distinct from 0 has an inverse relative to the operation $\cdot$.

Example 1.26. The pair $\mathcal{R} = (\mathbb{R}, \{0, 1, +, -, \cdot\})$ is a field where the sum and product are the usual operations involving real numbers. Indeed, the multiplication "$\cdot$" is a commutative operation and

1 is a multiplicative unit. In addition, each element $x \neq 0$ has the inverse $\frac{1}{x}$.

Another important example is the field of complex numbers $\mathcal{C} = (\mathbb{C}, \{0, 1, +, -, \cdot\})$, where the sum and product are the usual operations involving complex numbers. Namely, if $u, v \in \mathbb{C}$, $u = a+ib$, and $v = c + id$, where $a, b, c, d \in \mathbb{R}$, then

$$\begin{aligned} u + v &= (a + c) + i(b + d), \\ uv &= (ac - bd) + i(bc + ad). \end{aligned}$$

Example 1.27. Let $\mathcal{F}_2 = (\{0, 1\}, \{0, 1, +, -, \cdot\})$ be the two-element field whose binary operations are defined by the following tables:

$+$	0	1
0	0	1
1	1	0

$\cdot$	0	1
0	0	0
1	0	1

In this field, we have $-x = x$ because $x + x = 0$, so each element is its own opposite. The inverse of 1 is 1.

If $(\mathbb{F}, \{0, 1, +, -, \cdot\})$ is a field, then its *characteristic number* is the least number n such that $1n = 0$, where $1n = \underbrace{1 + 1 + \cdots + 1}_{n} = 0$. If no such number n exists, we say that the field has 0 as a characteristic number.

Example 1.28. The field $\mathcal{F}_2$ introduced in Example 1.27 has a characteristic number equal to 2 because $1 + 1 = 0$. The characteristic of the field $\mathcal{R}$ of real numbers is 0.

If x, y belong to a field $\mathbb{F}$ and $x \cdot y = 0$, then at least one of these elements is 0. Indeed, suppose that $x \cdot y = 0$ and we have $x \neq 0$. Then, x^{-1} exists, hence $x^{-1}(xy) = 0$, or $(x^{-1}x)y = 1y = 0$, which implies $y = 0$. Similarly, $y \neq 0$ implies $x = 0$.

If a field $\mathcal{F} = (\mathbb{F}, \{0, 1, +, -, \cdot\})$ has an n as its characteristic number, then n must be a prime. Indeed, suppose that this is not the case, so $n = pq$, and n is the least number with this property. If p were not prime, we could write $n = pq$, where p and q are at least equal to 2 and this would imply $1n = 1(pq) = (1p)(1q) = 0$. By the property of field mentioned above we would have either $1p = 0$ or

$1q = 0$, and this would contradict the fact that n is the least number that satisfies $1n = 0$. Thus, n must be a prime number.

Unless stated otherwise, we assume that all fields have the characteristic 0.

1.7 Closure and Interior Systems

Closure and interior systems are useful concepts in topology, convex set theory, and linear algebra.

Definition 1.25. Let S be a set. A *closure system on* S is a collection $\mathcal{C}$ of subsets of S such that $S \in \mathcal{C}$, and for every collection $\mathcal{D} \subseteq \mathcal{C}$, we have $\bigcap \mathcal{D} \in \mathcal{C}$.

Example 1.29. Let $\mathcal{C}$ be the collection of all intervals $[a, b] = \{x \in \mathbb{R} \mid a \leqslant x \leqslant b\}$ with $a, b \in \mathbb{R}$ and $a \leqslant b$, together with the empty set and the set $\mathbb{R}$. Note that $\bigcup \mathcal{C} = \mathbb{R} \in \mathcal{C}$, so the first condition of Definition 1.25 is satisfied.

Let $\mathcal{D}$ be a nonempty sub-collection of $\mathcal{C}$. If $\emptyset \in \mathcal{D}$, then $\bigcap \mathcal{D} = \emptyset \in \mathcal{C}$. If $\mathcal{D} = \{\mathbb{R}\}$, then $\bigcap \mathcal{D} = \mathbb{R} \in \mathcal{C}$. Therefore, we need to consider only the case when $\mathcal{D} = \{[a_i, b_i] \mid i \in I\}$. Then $\bigcap \mathcal{D} = \emptyset$, unless $a = \sup\{a_i \mid i \in I\}$ and $b = \inf\{b_i \mid i \in I\}$ both exist and $a \leqslant b$, in which case, $\bigcap \mathcal{D} = [a, b]$. Thus, $\mathcal{C}$ is a closure system.

Definition 1.26. A mapping $\boldsymbol{K} : \mathcal{P}(S) \longrightarrow \mathcal{P}(S)$ is a *closure operator* on a set S if it satisfies the following conditions:
(i) $U \subseteq \boldsymbol{K}(U)$ (*expansiveness*);
(ii) $U \subseteq V$ implies $\boldsymbol{K}(U) \subseteq \boldsymbol{K}(V)$ (*monotonicity*);
(iii) $\boldsymbol{K}(\boldsymbol{K}(U)) = \boldsymbol{K}(U)$ (*idempotency*),
for $U, V \in \mathcal{P}(S)$.

Example 1.30. Let $\boldsymbol{K} : \mathcal{P}(\mathbb{R}) \longrightarrow \mathcal{P}(\mathbb{R})$ be defined by

$$\boldsymbol{K}(U) = \begin{cases} \emptyset & \text{if } U = \emptyset \\ [a, b] & \text{if both } a = \inf U \text{ and } b = \sup U \text{ exist} \\ \mathbb{R} & \text{otherwise,} \end{cases}$$

for $U \in \mathcal{P}(\mathbb{R})$. We leave to the reader the verification that $\boldsymbol{K}$ is a closure operator.

Closure operators generate closure systems, as we show next.

Lemma 1.1. *Let $\boldsymbol{K} : \mathcal{P}(S) \longrightarrow \mathcal{P}(S)$ be a closure operator. Define the family of sets $\mathcal{C}_{\boldsymbol{K}} = \{H \in \mathcal{P}(S) \mid H = \boldsymbol{K}(H)\}$. Then $\mathcal{C}_{\boldsymbol{K}}$ is a closure system on S.*

Proof. Since $S \subseteq \boldsymbol{K}(S) \subseteq S$, we have $S \in \mathcal{C}_{\boldsymbol{K}}$, so $\bigcup \mathcal{C}_{\boldsymbol{K}} = S \in \mathcal{C}_{\boldsymbol{K}}$.

Let $\mathcal{D} = \{D_i \mid i \in I\}$ be a collection of subsets of S such that $D_i = \boldsymbol{K}(D_i)$ for $i \in I$. Since $\bigcap \mathcal{D} \subseteq D_i$, we have $\boldsymbol{K}(\bigcap \mathcal{D}) \subseteq \boldsymbol{K}(D_i) = D_i$ for every $i \in I$. Therefore, $\boldsymbol{K}(\bigcap \mathcal{D}) \subseteq \bigcap \mathcal{D}$, which implies $\boldsymbol{K}(\bigcap \mathcal{D}) = \bigcap \mathcal{D}$. This proves our claim. □

Note that $\mathcal{C}_{\boldsymbol{K}}$, as defined in Lemma 1.1, equals the range of $\boldsymbol{K}$. Indeed, if $L \in \mathsf{range}(\boldsymbol{K})$, then $L = \boldsymbol{K}(H)$ for some $H \in \mathcal{P}(S)$, so $\boldsymbol{K}(L) = \boldsymbol{K}(\boldsymbol{K}(H)) = \boldsymbol{K}(H) = L$, which shows that $L \in \mathcal{C}_{\boldsymbol{K}}$. The reverse inclusion is obvious.

We refer to the sets in $\mathcal{C}_{\boldsymbol{K}}$ as the *$\boldsymbol{K}$-closed subsets of S*.

In the reverse direction to Lemma 1.1, we show that every closure system generates a closure operator.

Lemma 1.2. *Let $\mathcal{C}$ be a closure system on the set S. Define the mapping $\boldsymbol{K}_{\mathcal{C}} : \mathcal{P}(S) \longrightarrow \mathcal{P}(S)$ by $\boldsymbol{K}_{\mathcal{C}}(H) = \bigcap\{L \in \mathcal{C} \mid H \subseteq L\}$. Then $\boldsymbol{K}_{\mathcal{C}}$ is a closure operator on the set S.*

Proof. Note that the collection $\{L \in \mathcal{C} \mid H \subseteq L\}$ is not empty since it contains at least S, so $\boldsymbol{K}_{\mathcal{C}}(H)$ is defined and it is clearly the smallest element of $\mathcal{C}$ that contains H. Also, by the definition of $\boldsymbol{K}_{\mathcal{C}}(H)$, it follows immediately that $H \subseteq \boldsymbol{K}_{\mathcal{C}}(H)$ for every $H \in \mathcal{P}(S)$.

Suppose that $H_1, H_2 \in \mathcal{P}(S)$ are such that $H_1 \subseteq H_2$. Since

$$\{L \in \mathcal{C} \mid H_2 \subseteq L\} \subseteq \{L \in \mathcal{C} \mid H_1 \subseteq L\},$$

we have

$$\bigcap\{L \in \mathcal{C} \mid H_1 \subseteq L\} \subseteq \bigcap\{L \in \mathcal{C} \mid H_2 \subseteq L\},$$

so $\boldsymbol{K}_{\mathcal{C}}(H_1) \subseteq \boldsymbol{K}_{\mathcal{C}}(H_2)$.

We have $\boldsymbol{K}_{\mathcal{C}}(H) \in \mathcal{C}$ for every $H \in \mathcal{P}(S)$ because $\mathcal{C}$ is a closure system. Therefore, $\boldsymbol{K}_{\mathcal{C}}(H) \in \{L \in \mathcal{C} \mid \boldsymbol{K}_{\mathcal{C}}(H) \subseteq L\}$, so $\boldsymbol{K}_{\mathcal{C}}(\boldsymbol{K}_{\mathcal{C}}(H)) \subseteq \boldsymbol{K}_{\mathcal{C}}(H)$. Since the reverse inclusion clearly holds, we obtain $\boldsymbol{K}_{\mathcal{C}}(\boldsymbol{K}_{\mathcal{C}}(H)) = \boldsymbol{K}_{\mathcal{C}}(H)$. □

Definition 1.27. Let $\mathcal{C}$ be a closure system on a set S and let T be a subset of S. The $\mathcal{C}$-*set generated by* T is the set $\boldsymbol{K}_{\mathcal{C}}(T)$.

Note that $\boldsymbol{K}_{\mathcal{C}}(T)$ is the least set in $\mathcal{C}$ that includes T.

Theorem 1.13. *Let S be a set. For every closure system $\mathcal{C}$ on S, we have $\mathcal{C} = \mathcal{C}_{\boldsymbol{K}_{\mathcal{C}}}$. For every closure operator $\boldsymbol{K}$ on S, we have $\boldsymbol{K} = \boldsymbol{K}_{\mathcal{C}_{\boldsymbol{K}}}$.*

Proof. Let $\mathcal{C}$ be a closure system on S and let $H \subseteq M$. Then we have the following equivalent statements:

(1) $H \in \mathcal{C}_{\boldsymbol{K}_{\mathcal{C}}}$;
(2) $\boldsymbol{K}_{\mathcal{C}}(H) = H$;
(3) $H \in \mathcal{C}$.

The equivalence between (2) and (3) follows from the fact that $\boldsymbol{K}_{\mathcal{C}}(H)$ is the smallest element of $\mathcal{C}$ that contains H.

Conversely, let $\boldsymbol{K}$ be a closure operator on S. To prove the equality of $\boldsymbol{K}$ and $\boldsymbol{K}_{\mathcal{C}_{\boldsymbol{K}}}$, consider the following list of equal sets, where $H \subseteq S$:

(1) $\boldsymbol{K}_{\mathcal{C}_{\boldsymbol{K}}}(H)$;
(2) $\bigcap\{L \in \mathcal{C}_{\boldsymbol{K}} \mid H \subseteq L\}$;
(3) $\bigcap\{L \in \mathcal{P}(S) \mid H \subseteq L = \boldsymbol{K}(L)\}$;
(4) $\boldsymbol{K}(H)$.

We need to justify only the equality of the last two members of the list. Since $H \subseteq \boldsymbol{K}(H) = \boldsymbol{K}(\boldsymbol{K}(H))$, we have $\boldsymbol{K}(H) \in \{L \in \mathcal{P}(S) \mid H \subseteq L = \boldsymbol{K}(L)\}$. Thus, $\bigcap\{L \in \mathcal{P}(S) \mid H \subseteq L = \boldsymbol{K}(L)\} \subseteq \boldsymbol{K}(H)$. To prove the reverse inclusion, note that for every $L \in \{L \in \mathcal{P}(S) \mid H \subseteq L = \boldsymbol{K}(L)\}$, we have $H \subseteq L$, so $\boldsymbol{K}(H) \subseteq \boldsymbol{K}(L) = L$. Therefore, $\boldsymbol{K}(H) \subseteq \bigcap\{L \in \mathcal{P}(S) \mid H \subseteq L = \boldsymbol{K}(L)\}$. $\square$

Theorem 1.13 shows the existence of a natural bijection between the set of closure operators on a set S and the set of closure systems on S.

Definition 1.28. Let $\mathcal{C}$ be a closure system on a set S and let T be a subset of S. The $\mathcal{C}$-closure of the set T is the set $\boldsymbol{K}_{\mathcal{C}}(T)$.

As we observed before, $\boldsymbol{K}_{\mathcal{C}}(T)$ is the smallest element of $\mathcal{C}$ that contains T.

Example 1.31. Let $\boldsymbol{K}$ be the closure operator given in Example 1.30. Since the closure system $\mathcal{C}_{\boldsymbol{K}}$ equals the range of $\boldsymbol{K}$, it follows that the members of $\mathcal{C}_{\boldsymbol{K}}$, the $\boldsymbol{K}$-closed sets, are $\emptyset$, $\mathbb{R}$, and all closed intervals $[a, b]$ with $a \leqslant b$. Thus, $\mathcal{C}_{\boldsymbol{K}}$ is the closure system $\mathcal{C}$ introduced in Example 1.29. Therefore, $\boldsymbol{K}$ and $\mathcal{C}$ correspond to each other under the bijection of Theorem 1.13.

Definition 1.29. Let S be a set and let F be a set of operations on S. A subset P of S is *closed under* F or F*-closed* if P is closed under f for every $f \in F$, that is, for every operation $f \in F$, if f is n-ary and $p_0, \ldots, p_{n-1} \in P$, then $f(p_0, \ldots, p_{n-1}) \in P$.

Note that S itself is closed under F. Further, if $\mathcal{C}$ is a nonempty collection of F-closed subsets of S, then $\bigcap \mathcal{C}$ is also F-closed.

Example 1.32. Let F be a set of operations on a set S. The collection of all F-closed subsets of a set S is a closure system.

Let S be a set. The *dual* of the poset $(\mathcal{P}(S), \subseteq)$ is the poset $(\mathcal{P}(S), \supseteq)$.

The dual of the notion of closure operator on a set S is a mapping $\boldsymbol{I} : \mathcal{P}(S) \longrightarrow \mathcal{P}(S)$ that satisfies the following conditions:

(1) $U \supseteq \boldsymbol{I}(U)$ (*contraction*);
(2) $U \supseteq V$ implies $\boldsymbol{I}(U) \supseteq \boldsymbol{I}(V)$ (*monotonicity*);
(3) $\boldsymbol{I}(\boldsymbol{I}(U)) = \boldsymbol{I}(U)$ (*idempotency*),

for $U, V \in \mathcal{P}(S)$. Such a mapping is known as *interior operator on the set* S.

The dual notion for the notion of closure system on a set introduced in Definition 1.25 is introduced next.

Definition 1.30. An *interior system on a set* S is a collection $\mathcal{I}$ of subsets of S such that

(i) $\emptyset \in \mathcal{I}$, and
(ii) for every sub-collection $\mathcal{D}$ of $\mathcal{I}$ we have $\bigcup \mathcal{D} \in \mathcal{I}$.

Theorem 1.14. *Let* $\boldsymbol{I} : \mathcal{P}(S) \longrightarrow \mathcal{P}(S)$ *be an interior operator. Define the family of sets* $\mathcal{I}_{\boldsymbol{I}} = \{U \in \mathcal{P}(S) \mid U = \boldsymbol{I}(U)\}$. *Then* $\mathcal{I}_{\boldsymbol{I}}$ *is an interior system on* S.

Conversely, if $\mathcal{I}$ *is an interior system on the set* S. *Define the mapping* $\boldsymbol{I}_{\mathcal{I}} : \mathcal{P}(S) \longrightarrow \mathcal{P}(S)$ *by* $\boldsymbol{I}_{\mathcal{I}}(U) = \bigcup\{V \in \mathcal{I} \mid V \subseteq U\}$. *Then* $\boldsymbol{I}_{\mathcal{I}}$ *is an interior operator on the set* S.

Moreover, for every interior system $\mathcal{I}$ *on* S, *we have* $\mathcal{I} = \mathcal{I}_{\boldsymbol{I}_{\mathcal{I}}}$. *For every interior operator* $\boldsymbol{I}$ *on* S, *we have* $\boldsymbol{I} = \boldsymbol{I}_{\mathcal{I}_{\boldsymbol{I}}}$.

Proof. This statement follows by duality from Lemmas 1.1 and 1.2 and from Theorem 1.13. $\square$

We refer to the sets in $\mathcal{I}_{\boldsymbol{I}}$ as the *$\boldsymbol{I}$-open subsets of S*.

Theorem 1.15. *Let $\boldsymbol{K} : \mathcal{P}(S) \longrightarrow \mathcal{P}(S)$ be a closure operator on the set S. Then the mapping $\boldsymbol{L} : \mathcal{P}(S) \longrightarrow \mathcal{P}(S)$ given by $\boldsymbol{L}(U) = S - \boldsymbol{K}(S - U)$ for $U \in \mathcal{P}(S)$ is an interior operator on S.*

Proof. Since $S - U \subseteq \boldsymbol{K}(S - U)$, it follows that $\boldsymbol{L}(U) \subseteq S - (S - U) = U$, which proves property (i) of Definition 1.30.

Suppose that $U \subseteq V$, where for $U, V \in \mathcal{P}(S)$. Then we have $S - V \subseteq S - U$, so $\boldsymbol{K}(S - V) \subseteq \boldsymbol{K}(S - U)$ by the monotonicity of closure operators. Therefore,

$$\boldsymbol{L}(U) = S - \boldsymbol{K}(S - U) \subseteq S - \boldsymbol{K}(S - V) = \boldsymbol{L}(V),$$

which proves the monotonicity of $\boldsymbol{L}$.

Finally, observe that we have $\boldsymbol{L}(\boldsymbol{L}(U)) \subseteq \boldsymbol{L}(U)$ because of the contraction property already proven for $\boldsymbol{L}$. Thus, we need only show that $\boldsymbol{L}(U) \subseteq \boldsymbol{L}(\boldsymbol{L}(U))$ to prove the idempotency of $\boldsymbol{L}$. This inclusion follows immediately from

$$\boldsymbol{L}(\boldsymbol{L}(U)) = \boldsymbol{L}(S - \boldsymbol{K}(S - U)) \supseteq \boldsymbol{L}(S - (S - U)) = \boldsymbol{L}(U).$$

$\square$

From Theorem 1.15, by duality, we can prove that if $\boldsymbol{L}$ is an interior operator on a set S, then $\boldsymbol{K} : \mathcal{P}(S) \longrightarrow \mathcal{P}(S)$ defined as $\boldsymbol{K}(U) = S - \boldsymbol{L}(S - U)$ for $U \in \mathcal{P}(S)$ is a closure operator on the same set.

Exercises and Supplements

(1) Let $S_n : \mathbb{R} \longrightarrow \mathbb{R}$ be the function given by

$$S_n(t) = \sum_{i=0}^{n} s_i(\lambda_1, \ldots, \lambda_n) t^i,$$

where the sum is extended to all $n + 1$ elementary symmetric polynomials in n variables. Prove that $S_n(t) = \prod_{i=1}^{n}(1 + \lambda_i t)$.

(2) Let $p_r(\lambda_1, \lambda_n) = \lambda_1^r + \cdots + \lambda_n^r$ be Newton's symmetric polynomial of degree r in n variables and let P_n be the function given by

$$P_n(t) = \sum_{r \geqslant 1} p_r(\lambda_1, \ldots, \lambda_n) t^{r-1},$$

be the generating function for Newton's polynomials. Prove that

$$P_n(t) = \sum_{i=1}^{n} \frac{\lambda_i}{1 - \lambda_i t}.$$

(3) Let $C_n(t)$ be the generating function of complete symmetric polynomials in n variables, where

$$C_n(t) = \sum_{k \geqslant 0} c_k(\lambda_1, \ldots, \lambda_n) t^k.$$

Prove that:
(a)

$$C_n(t) = \sum_{a_1, \ldots, a_n \geqslant 0} \lambda_1^{a_1} \cdots \lambda_n^{a_n} t^{a_1 + \cdots + a_n} = \frac{1}{\prod_{i=1}^{n} (1 - \lambda_i t)};$$

(b) $C_n(t) S_n(-t) = 1$;
(c) if $k \geqslant 1$, we have $\sum_{j=0}^{k} (-1)^j s_j(\lambda_1, \ldots, \lambda_n)$ $p_{n-j}(\lambda_1, \ldots, \lambda_n) = 0$, where $s_j(\lambda_1, \ldots, \lambda_n) = 0$ if $j > n$.

Solution: For the first part we compute the coefficient of t^b for $b \in \mathbb{N}$ in both sides of the equality. In the left member this coefficient is $\sum\{\lambda_1^{a_1} \cdots \lambda_n^{a_n} \mid a_1 + \cdots + a_n = b$ because c_k is the sum of all monomials of degree k in $\lambda_1, \ldots, \lambda_n$. In the right member we have a product of n series of the form

$$\frac{1}{1 - \lambda_i t} = 1 + \lambda_i t + \cdots + \lambda_i^p t^p + \cdots .$$

Therefore, the coefficient of t^b in this product equals the value of the coefficient for the left member for every b, which establishes the equality. The second part follows immediately from the first part. The equality $C_n(t) S_n(-t) = 1$ can be written as

$$\left(\sum_{i=0}^{n} s_i(\lambda_1, \ldots, \lambda_n)(-1)^i t^i \right) \cdot \left(\sum_{k \geqslant 0} c_k(\lambda_1, \ldots, \lambda_n) t^k \right) = 1.$$

The coefficient of t^n in the left member is

$$\sum_{j=0}^{k}(-1)^j s_j(\lambda_1, \ldots, \lambda_n)p_{n-j}(\lambda_1, \ldots, \lambda_n)$$

while the coefficient of t^n in the right-hand member is 0.

(4) If $\pi, \sigma \in PERM_n$, prove that $sign\,(\pi\sigma) = sign\,(\pi) sign\,(\sigma)$.

(5) If $\sigma \in PERM_n$, prove that $sign\,(\sigma^{-1}) = sign\,(\sigma)$.

(6) Let $\pi \in PERM_n$ and $\sigma \in PERM_{n+m}$. Define the permutation $\tilde{\pi} \in PERM_{n+m}$ as

$$\tilde{\pi}(k) = \begin{cases} \pi(k) & \text{if } 1 \leqslant k \leqslant n, \\ k & \text{if} n+1 \leqslant k \leqslant n+m. \end{cases}$$

Prove that $sign\,(\tilde{\pi}\sigma) = sign\,(\pi) sign\,(\sigma)$.

(7) Let $\theta \in PERM_{p+q}$ be defined as

$$\theta : \begin{pmatrix} 1 & 2 & \cdots & p & p+1 & \cdots & p+q \\ p+1 & p+2 & \cdots & p+q & 1 & \cdots & q \end{pmatrix}.$$

Prove that $sign\,(\theta) = (-1)^{pq}$.

(8) Prove that if $\theta \in PERM_n$ can be written as products of transpositions, $\theta = \tau_1\tau_2 \cdots \tau_r = \tau'_1\tau'_2 \cdots \tau'_s$, then $r \equiv s(\mod 2)$.

(9) Prove that

$$\epsilon_{i_1 \cdots i_n} = \begin{cases} 1 & \text{if } (i_1, \ldots, i_n) \text{ is an even permutation} \\ -1 & \text{if } (i_1, \ldots, i_n) \text{ is an odd permutation} \\ 0 & \text{otherwise.} \end{cases}$$

(10) Prove that the Levi-Civita symbols satisfy the identity

$$\sum_{k=1}^{3} \epsilon_{ijk}\epsilon_{\ell mk} = \epsilon_{ij1}\epsilon_{\ell m1} + \epsilon_{ij2}\epsilon_{\ell m2} + \epsilon_{ij3}\epsilon_{\ell m3}.$$

(11) Prove that

$$\sum_{i=1}^{3}\sum_{j=1}^{3}\sum_{k=1}^{3} \epsilon_{ijk}\epsilon_{ijk} = 6.$$

Let X be a finite set, $X = \{x_1, \ldots, x_n\}$. A selection of r objects from X where each object can be selected more than once is a *combination of n objects taken r at a time with repetition.* Two combinations taken r at a time with repetition are identical if they have the same elements repeated the same number of times regardless of the order.

For instance, if $X = \{x_1, x_2, x_3, x_4\}$, then there are 20 combinations of the elements of X taken three at a time with repetition:

$$x_1x_1x_1, x_1x_1x_2, x_1x_1x_3, x_1x_1x_4, x_1x_2x_2,$$
$$x_1x_2x_3, x_1x_2x_4, x_1x_3x_3, x_1x_3x_4, x_1x_4x_4,$$
$$x_2x_2x_2, x_2x_2x_3, x_2x_2x_4, x_2x_3x_3, x_2x_3x_4,$$
$$x_2x_4x_4, x_3x_3x_3, x_3x_3x_4, x_3x_4x_4, x_4x_4x_4.$$

(12) Show that the following numbers are equal:
   (a) the number of combinations of n objects taken r at a time;
   (b) the number of non-negative integer solutions of the equation $x_1 + x_2 + \cdots + x_n = r$;
   (c) the number $\binom{n+r-1}{n-1} = \binom{n+r-1}{r}$.

   Solution: We prove only that the number of non-negative integer solutions of the equation $x_1 + x_2 + \cdots + x_n = r$ equals $\binom{n+r-1}{n-1}$. A solution of the equation $x_1 + x_2 + \cdots + x_n = r$ can be represented as a binary string of length $n + r - 1$ as

$$\underbrace{11\cdots1}_{x_1}0\underbrace{11\cdots1}_{x_2}0\cdots\underbrace{11\cdots1}_{x_n}$$

   containing $x_1 + \cdots + x_n = r$ digits equal to 1 and $n-1$ digits equal to 0. Since there are $\binom{n+r-1}{n-1}$ such binary strings, the conclusion follows. Note that in the example that precedes this supplement, the number of combinations of X taken three at a time with repetition is $\binom{4+3-1}{3} = 20$.

Bibliographical Comments

There are many sources for general algebra [3, 20, 108] and combinatorics [154, 155, 163, 166] that contain lots of material that would amply satisfy the needs of the reader.

Chapter 2

Linear Spaces

2.1 Introduction

This chapter is dedicated to the study of linear spaces, which are mathematical structures that play a central role in linear algebra. A linear space is defined in connection with a field and makes use of two operations: an additive operation between its elements, and an external multiplication operation that involves elements of the field and those of the linear space.

2.2 Linear Spaces

Definition 2.1. Let $\mathcal{F} = (\mathbb{F}, \{0, 1, +, -, \cdot\})$ be a field. An $\mathbb{F}$-*linear space* is a pair $(\mathcal{L}, s)$ such that $\mathcal{L} = (V, \{\mathbf{0}_V, +, -\})$ is an Abelian group and $s : \mathbb{F} \times V \longrightarrow V$ is a function referred to as the *scalar multiplication* that satisfies the following conditions:

(1) $s(a + b, \boldsymbol{x}) = s(a, \boldsymbol{x}) + s(b, \boldsymbol{x})$;
(2) $s(a, \boldsymbol{x} + \boldsymbol{y}) = s(a, \boldsymbol{x}) + s(a, \boldsymbol{y})$;
(3) $s(ab, \boldsymbol{x}) = s(a, s(b, \boldsymbol{x}))$;
(4) $s(1, \boldsymbol{x}) = \boldsymbol{x}$

for every $a, b \in \mathbb{F}$ and $\boldsymbol{x}, \boldsymbol{y} \in V$.

The elements of the field $\mathbb{F}$ are referred to as *scalars*, while the elements of V are referred to as *vectors*.

The result of the scalar multiplication $s(a, \boldsymbol{v})$ is denoted simply by $a\boldsymbol{v}$. This allows us to write the previous equalities as

(1) $(a+b)\boldsymbol{v} = a\boldsymbol{v} + b\boldsymbol{v}$;
(2) $a(\boldsymbol{v}+\boldsymbol{w}) = a\boldsymbol{v} + a\boldsymbol{w}$;
(3) $(ab)\boldsymbol{v} = a(b\boldsymbol{v})$;
(4) $1\boldsymbol{v} = \boldsymbol{v}$

for $a, b \in \mathbb{F}$ and $\boldsymbol{v}, \boldsymbol{w} \in L$.

We omit the explicit mention of the scalar multiplication function from the definition of an $\mathbb{F}$-linear space $\mathcal{V} = (V, s)$ and we refer to $\mathcal{V}$ simply as V.

If the field $\mathbb{F}$ is irrelevant, or it is clearly designated from the context, we refer to an $\mathbb{F}$-linear space just as a linear space. On the other hand, if $\mathbb{F}$ is the real field $\mathbb{R}$ or the complex field $\mathbb{C}$, we refer to an $\mathbb{R}$-linear space as a *real linear space* and to a $\mathbb{C}$-linear space as a *complex linear space*.

Example 2.1. If $\mathcal{F} = (\mathbb{F}, \{0, 1, +, -, \cdot\})$ is a field, then the one-element linear space $V = \{\mathbf{0}_V\}$, where $a\mathbf{0}_V = \mathbf{0}_V$ for every $a \in \mathbb{F}$, is the *zero* $\mathbb{F}$*-linear space*, or, for short, the *zero linear space*.

The field $\mathbb{F}$ itself is an $\mathbb{F}$-linear space, where the Abelian group is $(\mathbb{F}, \{0, +, -\})$ and scalar multiplication coincides with the scalar multiplication of $\mathbb{F}$.

Example 2.2. The set of all sequences of real numbers, $\mathbf{Seq}(\mathbb{R})$, is a real linear space, where the sum of two sequences $\boldsymbol{x} = (x_0, x_1, \ldots)$ and $\boldsymbol{y} = (y_0, y_1, \ldots)$ is the sequence $\boldsymbol{x} + \boldsymbol{y}$ defined by $\boldsymbol{x} + \boldsymbol{y} = (x_0 + y_0, x_1 + y_1, \ldots)$ and the multiplication of $\boldsymbol{x}$ by a scalar a is $a\boldsymbol{x} = (ax_0, ax_1, \ldots)$.

A related real linear space is the set $\mathbf{Seq}_n(\mathbb{R})$ of all sequences of real numbers having length n, where the sum and the scalar multiplications are defined in a similar manner. Namely, if $\boldsymbol{x} = (x_0, x_1, \ldots, x_{n-1})$ and $\boldsymbol{y} = (y_0, y_1, \ldots, y_{n-1})$, the sequence $\boldsymbol{x} + \boldsymbol{y}$ is defined by $\boldsymbol{x} + \boldsymbol{y} = (x_0 + y_0, x_1 + y_1, \ldots, x_{n-1} + y_{n-1})$ and the multiplication of $\boldsymbol{x}$ by a scalar a is $a\boldsymbol{x} = (ax_0, ax_1, \ldots, ax_{n-1})$. This linear space is denoted by $\mathbb{R}^n$ and its zero element is denoted by $\mathbf{0}_n$.

Example 2.3. If the real field $\mathbb{R}$ is replaced by the complex field $\mathbb{C}$, we obtain the linear space $\mathbf{Seq}(\mathbb{C})$ of all sequences of complex numbers. Similarly, we have the complex linear space $\mathbb{C}^n$, which consists of all sequences of length n of complex numbers.

Example 2.4. Let V be an $\mathbb{F}$-linear space and let S be a non-empty set. The set V^S that consists of all functions of the form $f : S \longrightarrow V$ is an $\mathbb{F}$-linear space. The addition of functions is defined by

$$(f + g)(s) = f(s) + g(s),$$

while the multiplication by a scalar is given by $(af)(s) = af(s)$, for $s \in S$ and $a \in \mathbb{F}$. We leave to the reader the task of verifying that the definition of a linear space is satisfied.

Example 2.5. Let S be a set and let $\mathcal{F}_2$ be the two element field defined in Example 1.27. Define the scalar multiplication of a subset T of S by an element of the field as

$$0 \cdot T = \emptyset \text{ and } 1 \cdot T = T$$

for every $T \in \mathcal{P}(S)$.

The sum of two subsets U and V is defined as their symmetric difference

$$U + V = (U - V) \cup (V - U).$$

With these definitions the set of subsets of S is an $\mathcal{F}_2$-linear space, as the reader can easily verify.

Informally, a subspace of an $\mathbb{F}$-linear space is a subset of the $\mathbb{F}$-linear space that behaves exactly like an $\mathbb{F}$-linear space.

Definition 2.2. A non-empty subset U of an $\mathbb{F}$-linear space V is a *subspace* of V if

(i) $\boldsymbol{x} + \boldsymbol{y} \in U$ for all $\boldsymbol{x}, \boldsymbol{y} \in U$,
(ii) $a\boldsymbol{x} \in U$ for $a \in \mathbb{F}$ and $\boldsymbol{x} \in U$.

If U is a subspace of an $\mathbb{F}$-linear space V, then U is itself an $\mathbb{F}$-linear space.

Example 2.6. The subset $\{\boldsymbol{0}_V\}$ of any $\mathbb{F}$-linear space V is a subspace of V named the *zero subspace*. This is the smallest subspace of V.

Theorem 2.1. *If $\mathcal{V} = \{V_i \mid i \in I\}$ is a collection of subspaces of an $\mathbb{F}$-linear space V, then $\bigcap \mathcal{V}$ is a subspace of V.*

Proof. Suppose that $\boldsymbol{x}, \boldsymbol{y} \in \bigcap \mathcal{V}$. Then, $\boldsymbol{x}, \boldsymbol{y} \in V_i$, so $\boldsymbol{x} + \boldsymbol{y} \in V_i$ and $a\boldsymbol{x} \in V_i$ for every $i \in I$. Thus, $\boldsymbol{x} + \boldsymbol{y} \in \bigcup \mathcal{V}$ and $a\boldsymbol{x} \in \bigcup \mathcal{V}$, which allows us to conclude that $\bigcap \mathcal{V}$ is a subspace of L. □

Since V itself is a subspace of V, it follows that the collection of subspaces of a linear space is a closure system $\mathcal{V}$. If $\boldsymbol{K}_{\mathsf{sub}}$ is the closure operator induced by $\mathcal{V}$, then for every subset X of V, $\boldsymbol{K}_{\mathsf{sub}}(X)$ is the smallest subspace of V that contains X.

If U is a subspace of a linear space V and $\boldsymbol{x} \in V$, we denote the set $\{\boldsymbol{x} + \boldsymbol{u} \mid \boldsymbol{u} \in U\}$ by $\boldsymbol{x} + U$.

The following statements are immediate for an $\mathbb{F}$-linear space V:

(i) the sets V and $\{\mathbf{0}_V\}$ are subspaces of V;

(ii) each subspace U of V contains $\mathbf{0}_V$.

2.3 Linear Independence

Let $\mathbb{F}$ be a field and let I be a non-empty set. If $\varphi : I \longrightarrow F$ is a function, define the *support* of φ as the subset of I given by

$$\mathsf{supp}(\varphi) = \{i \in I \mid \varphi(i) \neq 0\}.$$

Definition 2.3. Let $\mathbb{F}$ be a field, I be a non-empty set, and let $\varphi : I \longrightarrow F$ be a function that has finite support.

If V is an $\mathbb{F}$-linear space, the *linear combination* determined by φ is an element $\boldsymbol{w}_\varphi$ of V that can be written as

$$\boldsymbol{w} = \sum_{i \in \mathsf{supp}(\varphi)} \varphi(i)\boldsymbol{x}_i,$$

where $\boldsymbol{x}_i \in V$ for $i \in \mathsf{supp}(\varphi)$.

In the special case when $\mathsf{supp}(\varphi) = \emptyset$, we define $\boldsymbol{w}_\varphi = \mathbf{0}_V$.

If $\mathsf{supp}(\varphi) = \{1, \ldots, n\}$ and $\{\boldsymbol{x}_1, \ldots, \boldsymbol{x}_n\} \subseteq X$, then a X-*linear combination* is an element $\boldsymbol{x}$ of V that can be written as

$$\boldsymbol{x} = a^1\boldsymbol{x}_1 + \cdots + a^n\boldsymbol{x}_n,$$

where $a^i = \varphi(i)$ for $1 \leqslant i \leqslant n$ are scalars called the *coefficients of the linear combination* $\boldsymbol{w}$.

The set of all X-linear combinations is denoted by $\langle X \rangle$ and is referred to as the *set spanned by* X.

Theorem 2.2. *Let V be an $\mathbb{F}$-linear space. If $X \subseteq V$, then $\langle X \rangle$ is the smallest subspace of V that contains the set X. In other words, we have*

(i) *$\langle X\rangle$ is a subspace of V;*
(ii) *$X \subseteq \langle X\rangle$;*
(iii) *if $X \subseteq M$, where M is a subspace of V, then $\langle X\rangle \subseteq M$.*

Proof. It is clear that if $\boldsymbol{u}$ and $\boldsymbol{v}$ are two X-linear combinations, then $\boldsymbol{u}+\boldsymbol{v}$ and $a\boldsymbol{u}$ are also X-linear combinations, so $\langle X\rangle$ is a subspace of V. Also, for $\boldsymbol{x} \in X$, we can write $1\boldsymbol{x} = \boldsymbol{x}$, so $X \subseteq \langle X\rangle$.

Finally, suppose that $X \subseteq M$, where M is a subspace of V and $a_1\boldsymbol{x}_1 + \cdots + a_n\boldsymbol{x}_n \in \langle X\rangle$, where $\boldsymbol{x}_1, \ldots, \boldsymbol{x}_n \in X$. Since $X \subseteq M$, we have $\boldsymbol{x}_1, \ldots, \boldsymbol{x}_n \in M$, hence $a_1\boldsymbol{x}_1 + \cdots + a_n\boldsymbol{x}_n \in M$ because M is a subspace. Thus, $\langle X\rangle \subseteq M$. □

Corollary 2.1. *Let V be an $\mathbb{F}$-linear space. If $X \subseteq V$, then $\langle X\rangle$ equals $\boldsymbol{K}_{\text{sub}}(X)$, the subspace of V generated by X.*

Proof. This statement follows from Theorem 2.2. □

From now on, we use the notation $\langle X\rangle$ instead of $\boldsymbol{K}_{\text{sub}}(X)$.

Definition 2.4. Let V be an $\mathbb{F}$-linear space. A finite subset $U = \{\boldsymbol{x}_1, \ldots, \boldsymbol{x}_n\}$ of V is *linearly dependent* if $\mathbf{0}_V = a_1\boldsymbol{x}_1 + \cdots + a_n\boldsymbol{x}_n = \mathbf{0}_V$, where at least one element a_i of $\mathbb{F}$ is not equal to 0.

If this condition is not satisfied, then U is said to be *linearly independent.*

A set U that consists of one vector $\boldsymbol{x} \neq \mathbf{0}_V$ is linearly independent.

The subset $U = \{\boldsymbol{x}_1, \ldots, \boldsymbol{x}_n\}$ of V is linearly independent if $a_1\boldsymbol{x}_1 + \cdots + a_n\boldsymbol{x}_n = \mathbf{0}_V$ implies $a_1 = \cdots = a_n = 0$. Also, note that a set U that is linearly independent does not contain $\mathbf{0}_V$.

Example 2.7. Let V be an $\mathbb{F}$-linear space. If $\boldsymbol{u} \in V$, then the set $V_{\boldsymbol{v}} = \{a\boldsymbol{u} \mid a \in F\}$ is a linear subspace of V. Moreover, if $\boldsymbol{u} \neq \mathbf{0}_V$, then the set $\{\boldsymbol{u}\}$ is linearly independent. Indeed, if $a\boldsymbol{u} = \mathbf{0}_V$ and $a \neq 0$, then multiplying both sides of the above equality by a^{-1} we obtain $(a^{-1}a)\boldsymbol{u} = \boldsymbol{a}^{-1}\mathbf{0}$, or equivalently, $\boldsymbol{u} = \mathbf{0}_V$, which contradicts the initial assumption. Thus, $\{\boldsymbol{u}\}$ is a linearly independent set.

Definition 2.4 is extended to arbitrary subsets of a linear space.

Definition 2.5. Let V be an $\mathbb{F}$-linear space. A subset W of V is *linearly dependent* if it contains a finite subset U that is linearly dependent.

A subset W is *linearly independent* if it is not linearly dependent.

Thus, W is linearly independent if every finite subset of W is linearly independent. Further, any subset of a linearly independent subset is linearly independent and any superset of a linearly dependent set is linearly dependent.

Example 2.8. For every $\mathbb{F}$-linear space V, the set $\{\mathbf{0}_V\}$ is linearly dependent because we have $1\mathbf{0}_V = \mathbf{0}_V$.

Example 2.9. Consider the $\mathbb{F}$-linear space $\mathsf{SF}(I,\mathbb{F})$ that consists of the set of functions that map the non-empty set I into $\mathbb{F}$. For $i \in I$, define the function $e_i : I \longrightarrow \mathbb{F}$ as

$$e_i(j) = \begin{cases} 1 & \text{if } j = i, \\ 0 & \text{otherwise.} \end{cases}$$

The set $E = \{e_i \mid i \in I\}$ is linearly independent in the linear space $\mathsf{SF}(I,\mathbb{F})$. Indeed, let $\{e_{i_k} \mid 1 \leqslant k \leqslant p\}$ be a finite subset of E and assume that

$$a_{i_1}e_{i_1} + \cdots + a_{i_p}e_{i_p} = z,$$

where z is the function defined by $z(i) = 0$ for $i \in I$. Thus, choosing $i = i_k$, we have

$$a_{i_1}e_{i_1}(i_k) + \cdots + a_{i_p}e_{i_p}(i_k) = z(i_k).$$

All terms in the left member equal 0 with the exception of $a_{i_k}e_{i_k}(i_k) = a_{i_k} = 0$.

Theorem 2.3. *Let V be an $\mathbb{F}$-linear space and let W be a linearly independent subset of V. If $\boldsymbol{y} \in \langle W\rangle$, that is*

$$\boldsymbol{y} = a_1\boldsymbol{x}_1 + \cdots + a_n\boldsymbol{x}_n,$$

for some finite subset $\{\boldsymbol{x}_1, \ldots, \boldsymbol{x}_n\}$ of W, then the coefficients $a_1, \ldots, a_n$ are uniquely determined.

Proof. Suppose that $\boldsymbol{y}$ can be alternatively written as

$$\boldsymbol{y} = b_1\boldsymbol{x}_1 + \cdots + b_n\boldsymbol{x}_n,$$

for some $b_1, \ldots, b_n \in \mathbb{F}$. Since W is linearly independent, this implies

$$(a_1 - b_1)\boldsymbol{x}_1 + \cdots + (a_n - b_n)\boldsymbol{x}_n = \mathbf{0}_V,$$

which, in turn, yields $a_1 - b_1 = \cdots = a_n - b_n = 0$. Thus, we have $a_i = b_i$ for $1 \leqslant i \leqslant n$. □

2.4 Linear Mappings

Linear mappings between linear spaces are functions that are compatible with the algebraic operations of linear spaces, as introduced next.

Definition 2.6. Let $\mathbb{F}$ be a field and let V and W be two $\mathbb{F}$-linear spaces. A *linear mapping* is a function $h : V \longrightarrow W$ such that

$$h(a\boldsymbol{x} + b\boldsymbol{y}) = ah(\boldsymbol{x}) + bh(\boldsymbol{y})$$

for each of the scalars $a, b \in \mathbb{F}$ and $\boldsymbol{x}, \boldsymbol{y} \in V$.

An *affine mapping* is a function $f : V \longrightarrow W$ such that there exists a linear mapping $h : V \longrightarrow W$ and $\boldsymbol{b} \in W$ such that $f(\boldsymbol{x}) = h(\boldsymbol{x}) + \boldsymbol{b}$ for $\boldsymbol{x} \in V$.

Linear mappings are also referred to as *linear space homomorphisms*, *linear morphisms*, or *linear operators*.

The set of morphisms between two $\mathbb{F}$-linear spaces V and W is denoted by $\mathsf{Hom}(V, W)$. The set of affine mappings between two $\mathbb{F}$-linear spaces V and W is denoted by $\mathsf{Aff}(V, W)$.

Definition 2.7. Let $h, g \in \mathsf{Hom}(V, W)$ be two linear mappings between the $\mathbb{F}$-linear spaces V and W. The *sum* of h and g is the mapping $h + g$ defined by

$$(h + g)(\boldsymbol{x}) = h(\boldsymbol{x}) + g(\boldsymbol{x})$$

for $\boldsymbol{x} \in V$.

If $a \in \mathbb{F}$, the product af is defined as $(af)(\boldsymbol{x}) = af(\boldsymbol{x})$ for $\boldsymbol{x} \in V$.

If V, W are two $\mathbb{F}$-linear spaces, then the set $\mathsf{Hom}(V, W)$ is never empty because the zero morphism $\boldsymbol{0}_{V,W} : V \longrightarrow W$ defined as $\boldsymbol{0}_{V,W}(\boldsymbol{x}) = \boldsymbol{0}_V$ for $\boldsymbol{x} \in V$ is always an element of $\mathsf{Hom}(V, W)$.

Note that

$$\begin{aligned}
(f + g)(a\boldsymbol{x} + b\boldsymbol{y}) &= f(a\boldsymbol{x} + b\boldsymbol{y}) + g(a\boldsymbol{x} + b\boldsymbol{y}) \\
&= af(\boldsymbol{x}) + bf(\boldsymbol{y}) + ag(\boldsymbol{x}) + bg(\boldsymbol{y}) \\
&= f(a\boldsymbol{x} + b\boldsymbol{y}) + g(a\boldsymbol{x} + b\boldsymbol{y}),
\end{aligned}$$

for all $a, b \in F$ and $\boldsymbol{x}, \boldsymbol{y} \in L$. This shows that the sum of two linear mappings is also a linear mapping.

Theorem 2.4. $\mathsf{Hom}(V, W)$ *equipped with the sum and product defined above is an* $\mathbb{F}$*-linear space.*

Proof. The zero element of $\mathsf{Hom}(V, W)$ is the mapping $\mathbf{0}_{V,W}$. We leave to the reader to verify that $\mathsf{Hom}(V, W)$ satisfies the properties mentioned in Definition 2.1. □

Definition 2.8. Let V be an $\mathbb{F}$-linear space. A *linear form* on V is a morphism in $\mathsf{Hom}(V, \mathbb{F})$, where the field $\mathbb{F}$ is regarded as an $\mathbb{F}$-linear space.

2.5 Bases in Linear Spaces

Definition 2.9. A *basis* of an $\mathbb{F}$-linear space V is a linearly independent subset B such that $\langle B \rangle = V$.

If an $\mathbb{F}$-linear space V has a finite basis, then we say that V is a *linear space of finite type.*

Example 2.10. Let $\mathbb{F}$ be a field and let I be a non-empty set. In Example 2.9, we saw that the set $E = \{e_i \mid i \in I\}$ is a linear independent set in the linear space $\mathsf{SF}(I, \mathbb{F})$.

Let $\varphi : I \longrightarrow \mathbb{F}$ and suppose that $\mathsf{supp}(\varphi) = \{j_1, \ldots, j_p\}$. Then, $\varphi(i) = \sum_{i=1}^{p} \varphi(j_k) e_{j_k}(i)$ for $i \in I$. Therefore, E is a basis in $\mathsf{SF}(I, \mathbb{F})$.

Example 2.11. Let $\boldsymbol{e}_i$ be the vector in $\mathbb{C}^n$ given by

$$\boldsymbol{e}_i = \begin{pmatrix} 0 \\ \vdots \\ 0 \\ 1 \\ 0 \\ \vdots 0 \end{pmatrix},$$

having one non-zero component equal to 1 on its ith position. Note that for $\boldsymbol{x} \in \mathbb{C}^n$, we have

$$\boldsymbol{x} = x_1 \boldsymbol{e}_1 + \cdots + x_n \boldsymbol{e}_n,$$

which means that $\mathbb{C}^n = \langle\{\boldsymbol{e}_1, \ldots, \boldsymbol{e}_n\}\rangle$. If $x_1\boldsymbol{e}_1 + \cdots + x_n\boldsymbol{e}_n = \boldsymbol{0}_n$, it follows immediately that $x_1 = \cdots = x_n = 0$, which implies that $\{\boldsymbol{e}_1, \ldots, \boldsymbol{e}_n\}$ is a basis for $\mathbb{C}^n$.

Theorem 2.5. *Every non-zero $\mathbb{F}$-linear space V has a basis.*

Proof. Let V be a non-zero $\mathbb{F}$-linear space and let U be a set such that $\langle U\rangle = L$. Note that at least one such set exists because $\langle V\rangle = V$. The set U contains at least an element distinct from $\boldsymbol{0}_V$ because $\langle\{\boldsymbol{0}_V\}\rangle = \{\boldsymbol{0}_V\}$.

Let $\mathcal{I}_U$ be the collection of linearly independent subsets of U. Since for every $\boldsymbol{x} \in U$ such that $\boldsymbol{x} \neq \boldsymbol{0}_V$ the set $\{\boldsymbol{x}\}$ is linearly independent, it follows that $\mathcal{I}_U$ is a non-empty collection.

We claim that in the partially ordered set $(\mathcal{I}_U, \subseteq)$ every chain has an upper bound. Indeed, let $\{K_i \mid i \in J\}$ be a chain of independent subsets in $(\mathcal{I}_U, \subseteq)$. We claim that $K = \bigcup\{K_i \mid i \in J\}$ is a linearly independent set. Indeed, suppose that $\{x_k \mid 1 \leqslant k \leqslant n\}$ is a finite subset of K. For every l, $1 \leqslant k \leqslant n$ there exists $i_k \in J$ such that $x_k \in K_{i_k}$. Since $\{K_i \mid i \in J\}$ is a chain, there exists a set K' among the sets $K_{i_1}, \ldots, K_{i_n}$ that includes all others. Thus, $\{x_k \mid 1 \leqslant k \leqslant n\} \subseteq K'$, which implies that the set $\{x_k \mid 1 \leqslant k \leqslant n\}$ is linearly independent. Consequently, the set K is linearly independent and, therefore, it is an upper bound for the chain $\{K_i \mid i \in J\}$. By Zorn's Lemma (see, for example, Section 4.10 of [152]), the partially ordered set $(\mathcal{I}_U, \subseteq)$ contains an element maximal B. Clearly, B is linearly independent.

To prove that B is a basis, we need to show only that B spans the entire linear space V, that is, that $\langle B\rangle = V$. To this end, it suffices to prove that the set U of generators of V is included in $\langle B\rangle$. Let $\boldsymbol{x} \in U - B$ and let $X = B \cup \{\boldsymbol{x}\}$. Then, since B is maximal, it follows that X is linearly dependent. Therefore, there exists a linear combination $a\boldsymbol{x} + \sum_{i=1}^{p} a_i\boldsymbol{x}_i = \boldsymbol{0}_V$ such that $a \neq 0$. Since $\mathbb{F}$ is a field, there exists an inverse a^{-1} of a. Therefore, $\boldsymbol{x} = -a^{-1}\sum_{i=1}^{p} a_i\boldsymbol{x}_i$, which implies $\boldsymbol{x} \in \langle B\rangle$. Thus, we may conclude that $U \subseteq \langle B\rangle$, so $V = \langle U\rangle \subseteq \langle B\rangle = V$. □

Corollary 2.2. *Let U be a subset of an $\mathbb{F}$-linear space V such that $\langle U\rangle = V$. If $B \subseteq U$ is a linearly independent set, then there is a basis Z of V such that $B \subseteq Z \subseteq U$.*

Proof. Let $\mathfrak{I}_U$ be the collection of linearly independent subsets of U. By Zorn's Lemma, for every element B of $\mathfrak{I}_U$ there exists a maximal element Z of $(\mathfrak{I}_U, \subseteq)$ such that $B \subseteq Z$. The desired basis is Z. □

Corollary 2.3 (Independent set extension corollary). *Let V be an $\mathbb{F}$-linear space. If S is a linearly independent set, then there exists a basis B of V such that $S \subseteq B$.*

Proof. Since S is a linearly independent set, if $\langle T \rangle = V$, then $S \cup T$ also generates V. The statement follows by Corollary 2.2. □

If an $\mathbb{F}$-linear space V has a finite basis, then we say that V is a linear space of *finite type.*

Lemma 2.1. *Let V be a finite type $\mathbb{F}$-linear space and let T be a finite subset of V that is not linearly independent. If $k = |T| \geqslant 2$ and $(\boldsymbol{t}_1, \ldots, \boldsymbol{t}_k)$ is a list of the vectors in T, then there exists a number j such that $2 \leqslant j \leqslant m$ and $\boldsymbol{t}_j$ is a linear combination of its predecessors in the sequence. Furthermore, we have $\langle T - \{\boldsymbol{t}_j\} \rangle = \langle T \rangle$.*

Proof. Suppose that T is linearly dependent. Then there exists a linear combination $\sum_{i=1}^{k} a^i \boldsymbol{t}_i = \boldsymbol{0}_V$ such that some of the scalars $a^1, \ldots, a^k$ are different from 0. Let j be the largest number such that $1 \leqslant j \leqslant k$ and $a_j \neq 0$. The definition of j implies that $a^1 \boldsymbol{t}_1 + \cdots + a^j \boldsymbol{t}_j = \boldsymbol{0}_V$, so $\boldsymbol{t}_j = -\sum_{i=1}^{j-1} \frac{a^i}{a^j} \boldsymbol{t}_i$, which shows that $\boldsymbol{t}_j$ is a linear combination of its predecessors in the list. Consequently, the set of linear combinations of the vectors in $T - \{\boldsymbol{t}_j\}$ equals $\langle T \rangle$. □

Theorem 2.6 (The Replacement theorem). *Let V be a finite-type $\mathbb{F}$-linear space such that the set S spans the linear space V and $|S| = n$.*

If U is a linearly independent set in V such that $|U| = m$, then $m \leqslant n$ and there exists a subset S' of S such that S' contains $n - m$ vectors and $U \cup S'$ spans the space V.

Proof. Suppose that $S = \{\boldsymbol{w}_1, \ldots, \boldsymbol{w}_n\}$ and $U = \{\boldsymbol{u}_1, \ldots, \boldsymbol{u}_m\}$. The argument is by induction on m.

The basis case, $m = 0$, is immediate.

Suppose the statement holds for m and let $U = \{\boldsymbol{u}_1, \ldots, \boldsymbol{u}_m, \boldsymbol{u}_{m+1}\}$ be a linearly independent set that contains $m+1$ vectors.

The set $\{\boldsymbol{u}_1, \ldots, \boldsymbol{u}_m\}$ is linearly independent, so by the inductive hypothesis $m \leqslant n$ and there exists a subset S' of S that contains $n-m$ vectors such that $\{\boldsymbol{u}_1, \ldots, \boldsymbol{u}_m\} \cup S'$ spans the space V. Without loss of generality we may assume that $S' = \{\boldsymbol{w}_1, \ldots, \boldsymbol{w}_{n-m}\}$. Thus, $\boldsymbol{u}_{m+1}$ is a linear combination of the vectors of $\{\boldsymbol{u}_1, \ldots, \boldsymbol{u}_m, \boldsymbol{w}_1, \ldots, \boldsymbol{w}_{n-m}\}$, so we have

$$\boldsymbol{u}_{m+1} = a^1\boldsymbol{u}_1 + \cdots + a^m\boldsymbol{u}_m + b^1\boldsymbol{w}_1 + \cdots + b^{n-m}\boldsymbol{w}_{n-m}.$$

We have $m+1 \leqslant n$ because, otherwise, $m+1 = n$ and $\boldsymbol{u}_{m+1}$ would be a linear combination of $\boldsymbol{u}_1, \ldots, \boldsymbol{u}_m$, thereby contradicting the linear independence of the set U.

The set $\{\boldsymbol{u}_1, \ldots, \boldsymbol{u}_m, \boldsymbol{u}_{m+1}, \boldsymbol{w}_1, \ldots, \boldsymbol{w}_{n-m}\}$ is not linearly independent. Let $\boldsymbol{v}$ be the first member of the sequence $(\boldsymbol{u}_1, \ldots, \boldsymbol{u}_m, \boldsymbol{u}_{m+1}, \boldsymbol{w}_1, \ldots, \boldsymbol{w}_{n-m})$ that is a linear combination of its predecessors. Then, $\boldsymbol{v}$ cannot be one of the $\boldsymbol{u}_i$ (with $1 \leqslant i \leqslant m$) because this would contradict the linear independence of the set U. Therefore, there exists k such that $\boldsymbol{w}_k$ is a linear combination of its predecessors and $1 \leqslant k \leqslant n-m$. By Lemma 2.1, we can remove this element from the set $\{\boldsymbol{u}_1, \ldots, \boldsymbol{u}_m, \boldsymbol{u}_{m+1}, \boldsymbol{w}_1, \ldots, \boldsymbol{w}_{n-m}\}$ without affecting the set spanned. □

Corollary 2.4. *Let V be a finite-type $\mathbb{F}$-linear space and let B, C be two bases of L. Then $|B| = |C|$.*

Proof. Since B is a linearly independent set, and $\langle C\rangle = V$, by Theorem 2.6 we have $|B| \leqslant |C|$. The reverse inequality, $|C| \leqslant |B|$, is obtained by asserting that C is linearly independent and $\langle C\rangle = V$. Thus, $|B| = |C|$. □

Corollary 2.4 allows the introduction of the notion of dimension for a linear space.

Definition 2.10. The *dimension* of a finite-type linear space V is the number of elements of any basis of V.

The dimension of V is denoted by $\dim(V)$.

The dimension of the zero $\mathbb{F}$-linear space $\{\mathbf{0}\}$ is 0.

If a linear space V is not of finite type, then we say that $\dim(V)$ is of *infinite type*.

Theorem 2.7. *Let V be an $\mathbb{F}$-linear space of finite type having the basis $B = \{\boldsymbol{x}_1, \ldots, x_n\}$ and let $Y = \{\boldsymbol{y}_1, \ldots, \boldsymbol{y}_n\}$ be a subset of an $\mathbb{F}$-linear space W. If $f : B \longrightarrow Y$ is a function between B and W, then there exists a unique extension of f as a linear mapping $f : V \longrightarrow W$ such that $f(\boldsymbol{x}_i) = \boldsymbol{y}_i$ for $1 \leqslant i \leqslant n$.*

Proof. If $\boldsymbol{x} \in V$, we have $\boldsymbol{x} = a_1\boldsymbol{x}_1 + \cdots + a_n\boldsymbol{x}_n$ because $\{\boldsymbol{x}_1, \ldots, \boldsymbol{x}_n\}$ is a basis of V. Define $f(\boldsymbol{x})$ as $f(\boldsymbol{x}) = \sum_{i=1}^{n} a_i\boldsymbol{y}_i$. The uniqueness of the expression of $\boldsymbol{x}$ as a linear combination of the elements of B makes f well-defined. The linearity of f is immediate. For uniqueness of the extension of f, observe that the value of f is determined by the values of $f(\boldsymbol{x}_i)$. □

Example 2.12. Let S be a non-empty, finite set. The linear space $\mathbb{C}^S$ has dimension $|S|$. Indeed, for each $t \in S$, consider the function $f_t : S \longrightarrow \mathbb{C}$ defined by

$$f_t(s) = \begin{cases} 1 & \text{if } s = t, \\ 0 & \text{otherwise,} \end{cases}$$

for $t \in S$. If $S = \{t_1, \ldots, t_n\}$, then the set of functions $\{f_{t_1}, \ldots, f_{t_n}\}$ is linearly independent, for if $c^1 f_{t_1}(s) + \cdots + c^n f_{t_n}(s) = 0$, then by taking $s = t_k$ we obtain $c^k = 0$ for any k, $1 \leqslant k \leqslant n$. Furthermore, if $f : S \longrightarrow \mathbb{C}$ is a function and $f(t_i) = c^i$, then $f = \sum_{i=1}^{n} c^i f_{t_i}$, so $\{f_{t_1}, \ldots, f_{t_n}\}$ is a basis for $\mathbb{C}^S$.

Example 2.13. Let V, W be two linear spaces of finite type with $\dim(V) = p$ and $\dim(W) = q$. Then, $\dim(\mathsf{Hom}(V, W)) = pq$.

Suppose that $\{\boldsymbol{x}_1, \ldots, \boldsymbol{x}_p\}$ is a basis in V and $\{\boldsymbol{y}_1, \ldots, \boldsymbol{y}_q\}$ is a basis in W. As we have shown in Theorem 2.7, for every i such that $1 \leqslant i \leqslant p$ and j such that $1 \leqslant j \leqslant q$, there exists a unique linear mapping $f_{ij} : \{\boldsymbol{x}_1, \ldots, \boldsymbol{x}_p\} \longrightarrow W$ such that

$$f_{ij}(\boldsymbol{x}_k) = \begin{cases} \boldsymbol{y}_j & \text{if } i = k, \\ \boldsymbol{0}_M & \text{otherwise,} \end{cases}$$

for $1 \leqslant k \leqslant p$.

Note that if $\boldsymbol{x} = \sum_{k=1}^{p} a_k \boldsymbol{x}_k$, the linearity of f_{ij} implies

$$f_{ij}(\boldsymbol{x}) = f_{ij}\left(\sum_{k=1}^{p} a_k \boldsymbol{x}_k\right) = \sum_{k=1}^{p} a_k f_{ij}(\boldsymbol{x}_k) = a_i f_{ij}(\boldsymbol{x}_i).$$

We claim that the set $\{f_{ij} \mid 1 \leqslant i \leqslant p, 1 \leqslant j \leqslant q\}$ is a basis for $\mathsf{Hom}(V, W)$.

Let $f : V \longrightarrow W$ be a linear mapping. If $\boldsymbol{x} \in V$, we can write $\boldsymbol{x} = \sum_{i=1}^{p} a_i \boldsymbol{x}_i$, so $f(\boldsymbol{x}) = \sum_{i=1}^{p} a_i f(\boldsymbol{x}_i)$. In turn, since $\{\boldsymbol{y}_1, \ldots, \boldsymbol{y}_q\}$ is a basis in W, $f(\boldsymbol{x}_i) = \sum_{j=1}^{q} b_{ij} \boldsymbol{y}_j$, for some $b_{ij} \in \mathbb{F}$. This allows us to write

$$f(\boldsymbol{x}) = \sum_{i=1}^{p} a_i \sum_{j=1}^{q} b_{ij} \boldsymbol{y}_j = \sum_{i=1}^{p} \sum_{j=1}^{q} a_i b_{ij} \boldsymbol{y}_j = \sum_{i=1}^{p} \sum_{j=1}^{q} a_i b_{ij} f_{ij}(\boldsymbol{x}),$$

which shows that each linear mapping in $\mathsf{Hom}(V, W)$ is a linear combination of functions f_{ij}.

Furthermore, the set $\{f_{ij} \mid 1 \leqslant i \leqslant p, 1 \leqslant j \leqslant p\}$ is linearly independent in $\mathsf{Hom}(V, W)$. Indeed, suppose that $\sum_{i=1}^{p} \sum_{j=1}^{q} c_{ij} f_{ij}(\boldsymbol{x}) = \boldsymbol{0}_W$. Then, for $\boldsymbol{x} = \boldsymbol{x}_i$ we have $\sum_{j=1}^{q} c_{ij} \boldsymbol{y}_j = \boldsymbol{0}_W$, which implies $c_{ij} = 0$. We may conclude that $\dim(\mathsf{Hom}(V, W)) = \dim(V)\dim(W)$.

Theorem 2.8. *If W is a subspace of a finite-type linear space V, then* $\dim(W) \leqslant \dim(V)$.

Proof. If U is a linearly independent set in the subspace W, then it is clear that U is linearly independent in V. There exists a basis B of V such that $U \subseteq B$ and $|B| = \dim(V)$. Therefore, $\dim(W) \leqslant \dim(V)$. □

Definition 2.11. Let V and W be two $\mathbb{F}$-linear spaces and let $h : V \longrightarrow W$ be a linear mapping. The *kernel* of h is the subset $\mathsf{Ker}(h)$ of V given by

$$\mathsf{Ker}(h) = \{\boldsymbol{v} \in V \mid h(\boldsymbol{v}) = \boldsymbol{0}_W\}.$$

The *image* of h is the subset $\mathsf{Im}(h)$ of W given by

$$\mathsf{Im}(h) = \{\boldsymbol{w} \in W \mid h(\boldsymbol{v}) = \boldsymbol{w} \text{ for some } \boldsymbol{v} \in V\}.$$

The notion of subspace is closely linked to the notion of linear mapping as we show next.

Theorem 2.9. *Let V, W be two $\mathbb{F}$-linear spaces. If $h : V \longrightarrow W$ is a linear mapping, then $\mathsf{Im}(h)$ is a subspace of W and $\mathsf{Ker}(h)$ is a subspace of V.*

Proof. Let $\boldsymbol{w}_1$ and $\boldsymbol{w}_2$ be two elements of $\mathsf{Im}(h)$. There exist $\boldsymbol{v}_1, \boldsymbol{v}_2 \in V$ such that $\boldsymbol{w}_1 = h(\boldsymbol{v}_1)$ and $\boldsymbol{w}_2 = h(\boldsymbol{v}_2)$. Since h is a linear mapping, we have

$$\boldsymbol{w}_1 + \boldsymbol{w}_2 = h(\boldsymbol{v}_1) + h(\boldsymbol{v}_2) = h(\boldsymbol{v}_1 + \boldsymbol{v}_2).$$

Thus, $\boldsymbol{w}_1 + \boldsymbol{w}_2 \in \mathsf{Im}(h)$. Further, if $a \in \mathbb{F}$ and $\boldsymbol{w} \in W$, then $\boldsymbol{w} = h(\boldsymbol{v})$ for some $\boldsymbol{v} \in V$ and we have $a\boldsymbol{w} = ah(\boldsymbol{v}) = h(a\boldsymbol{v})$, so $a\boldsymbol{w} \in \mathsf{Im}(h)$. Thus, $\mathsf{Im}(h)$ is indeed a subspace of W.

Suppose now that $\boldsymbol{s}$ and $\boldsymbol{t}$ belong to $\mathsf{Ker}(h)$, that is $h(\boldsymbol{s}) = h(\boldsymbol{t}) = \mathbf{0}_W$. Then $h(\boldsymbol{s} + \boldsymbol{t}) = h(\boldsymbol{s}) + h(\boldsymbol{t}) = \mathbf{0}_W$, so $\boldsymbol{s} + \boldsymbol{t} \in \mathsf{Ker}(h)$. Also, $h(a\boldsymbol{s}) = ah(\boldsymbol{s}) = a\mathbf{0}_W = \mathbf{0}_W$, which allows us to conclude that $\mathsf{Ker}(h)$ is a subspace of W. □

Theorem 2.10. *Let V and W be two linear spaces, where $\dim(V) = n$, and let $h : V \longrightarrow W$ be a linear mapping. Then, we have*

$$\dim(\mathsf{Ker}(h)) + \dim(\mathsf{Im}(h)) = n.$$

Proof. Suppose that $\{\boldsymbol{e}_1, \ldots, \boldsymbol{e}_m\}$ is a basis for the subspace $\mathsf{Ker}(h)$ of V. By Corollary 2.3, each such basis can be extended to a basis

$$\{\boldsymbol{e}_1, \ldots, \boldsymbol{e}_m, \boldsymbol{e}_{m+1}, \ldots, \boldsymbol{e}_n\}$$

of the linear space V. Any $\boldsymbol{v} \in V$ can be written as

$$\boldsymbol{v} = \sum_{i=1}^{n} a^i \boldsymbol{e}_i.$$

Since $\{\boldsymbol{e}_1, \ldots, \boldsymbol{e}_m\} \subseteq \mathsf{Ker}(h)$, we have $h(\boldsymbol{e}_i) = \mathbf{0}_W$ for $1 \leqslant i \leqslant m$, so

$$h(\boldsymbol{v}) = \sum_{i=m+1}^{n} a^i h(\boldsymbol{e}_i).$$

This means that the set $\{h(\boldsymbol{e}_{m+1}), \ldots, h(\boldsymbol{e}_n)\}$ spans the subspace $\mathsf{Im}(h)$ of W. We show now that this set is linearly independent. Indeed, suppose that $\sum_{i=m+1}^{n} b^i h(\boldsymbol{e}_i) = \mathbf{0}_W$. This implies

$h(\sum_{i=m+1}^{n} b^i \boldsymbol{e}_i) = \boldsymbol{0}_W$, that is, $\sum_{i=m+1}^{n} b^i \boldsymbol{e}_i \in \mathsf{Ker}(h)$. Since $\{\boldsymbol{e}_1, \ldots, \boldsymbol{e}_m\}$ is a basis for $\mathsf{Ker}(h)$, there exist m scalars $c^1, \ldots, c^m$ such that

$$\sum_{i=m+1}^{n} b^i \boldsymbol{e}_i = c^1 \boldsymbol{e}_1 + \cdots + c^m \boldsymbol{e}_m.$$

The fact that $\{\boldsymbol{e}_1, \ldots, \boldsymbol{e}_m, \boldsymbol{e}_{m+1}, \ldots, \boldsymbol{e}_n\}$ is a basis for V implies that $c^1 = \cdots = c^m = b^{m+1} = \cdots = b^n = 0$, so the set $\{h(\boldsymbol{e}_{m+1}), \ldots, h(\boldsymbol{e}_n)\}$ is linearly independent and, therefore, a basis for $\mathsf{Im}(h)$. Thus, $\dim(\mathsf{Im}(h)) = n - m$, which concludes the argument. □

Definition 2.12. Let V and W be two $\mathbb{F}$-linear spaces and let $h \in \mathsf{Hom}(V, W)$. The *rank* of h is $\mathsf{rank}(h) = \dim(\mathsf{Im}(h))$; the *nullity* of h is $\mathsf{nullity}(h) = \dim(\mathsf{Ker}(h))$.

The *spark* of a linear mapping $h : V \longrightarrow W$ is the minimum size of a subset S of $\mathsf{Im}(h)$ such that $\boldsymbol{0}_W \in \langle S \rangle$.

Theorem 2.10 can now be rephrased by saying that if $h : V \longrightarrow W$ is a linear mapping and V is a linear space of finite type, then

$$\dim(V) = \mathsf{rank}(h) + \mathsf{nullity}(h).$$

Theorem 2.11. *Let $h : V \longrightarrow W$ be a linear mapping between the linear spaces V and W. Then $\mathsf{rank}(h) \leqslant \min\{\dim(V), \dim(W)\}$.*

Proof. From Theorem 2.10 it follows that $\mathsf{rank}(h) \leqslant \dim(V)$. On the other hand, $\mathsf{rank}(h) = \dim(\mathsf{Im}(h)) \leqslant \dim(W)$ because $\mathsf{Im}(h)$ is a subspace of W, so the inequality of the theorem follows. □

Example 2.14. Let V, W be two $\mathbb{F}$-linear spaces. For $h \in V^*$ and $\boldsymbol{y} \in W$, define the mapping $\ell_{h,\boldsymbol{y}} : V \longrightarrow W$ as $\ell_{h,\boldsymbol{y}}(\boldsymbol{x}) = h(\boldsymbol{x})\boldsymbol{y}$ for $\boldsymbol{x} \in V$. It is easy to verify that $\ell_{h,\boldsymbol{y}}$ is a linear mapping, that is, $\ell_{h,\boldsymbol{y}} \in \mathsf{Hom}(V, W)$. Furthermore, we have $\mathsf{rank}(\ell_{h,\boldsymbol{y}}) = 1$ because $\mathsf{Im}(\ell_{h,\boldsymbol{y}})$ consists of the multiples of the vector $\boldsymbol{y}$.

Let $f \in \mathsf{Hom}(V, W)$ be a linear mapping of rank r, which means that $\dim(\mathsf{Im}(f)) = r$. There exists a basis $\{\boldsymbol{y}_1, \ldots, \boldsymbol{y}_r\}$ in $\mathsf{Im}(f)$ such that for every $\boldsymbol{x} \in V$, $f(\boldsymbol{x})$ can be uniquely written as

$$f(\boldsymbol{x}) = \sum_{i=1}^{r} a_i \boldsymbol{y}_i.$$

Let $h_i \in V^*$ be the linear form defined as $h_i(\boldsymbol{x}) = a_i$ for $1 \leqslant i \leqslant r$. Then, $f(\boldsymbol{x}) = \sum_{i=1}^{r} h_i(\boldsymbol{x})\boldsymbol{y}_i$, hence f is the sum of r linear forms of rank 1.

2.6 Isomorphisms of Linear Spaces

Definition 2.13. Let V and W be two $\mathbb{F}$-linear spaces. An *isomorphism* between these linear spaces is a linear mapping $h : V \longrightarrow W$, which is a bijection.

If an isomorphism exists between two $\mathbb{F}$-linear spaces V and W, we say that these linear spaces are *isomorphic* and we write $V \cong W$.

Two $\mathbb{F}$-linear spaces that are isomorphic are indiscernible from an algebraic point of view.

Theorem 2.12. *Let V and W be two $\mathbb{F}$-linear spaces and let $h \in \mathsf{Hom}(V, W)$. Then $\mathsf{Im}(h) \cong (V/\mathsf{Ker}(h))$.*

Proof. Define the mapping $g : V/\mathsf{Ker}(h) \longrightarrow \mathsf{Im}(h)$ by $g([\boldsymbol{x}]) = h(\boldsymbol{x})$ for $\boldsymbol{x} \in V$. We show that g is an isomorphism.

The mapping g is well-defined since if $\boldsymbol{u} \in [\boldsymbol{x}]$, then $\boldsymbol{u} \sim_{\mathsf{Ker}(h)} \boldsymbol{x}$, so $\boldsymbol{u} - \boldsymbol{x} \in \mathsf{Ker}(h)$. Therefore, $h(\boldsymbol{u} - \boldsymbol{x}) = \mathbf{0}_W$, hence $h(\boldsymbol{u}) = h(\boldsymbol{x})$.

We leave it to the reader to verify that g is a linear mapping. Further, it is clear that g is surjective. To prove that g is injective, suppose that $g([\boldsymbol{x}]) = g([\boldsymbol{y}])$. This amounts to $h(\boldsymbol{x}) = h(\boldsymbol{y})$, which is equivalent to $h(\boldsymbol{x} - \boldsymbol{y}) = \mathbf{0}_W$. Thus, $\boldsymbol{x} - \boldsymbol{y} \in \mathsf{Ker}(h)$, which implies $[\boldsymbol{x}] = [\boldsymbol{y}]$. In other words, g is injective and, therefore, is a bijection. This shows that the linear spaces $\mathsf{Im}(h)$ and $V/\mathsf{Ker}(h)$ are isomorphic. □

Corollary 2.5. *Let V and W be two $\mathbb{F}$-linear spaces and let $h : V \longrightarrow W$ be a surjective morphism of linear spaces. Then, $W \cong V/\mathsf{Ker}(h)$.*

Proof. This is an immediate consequence of Theorem 2.12. □

Theorem 2.13. *Isomorphism is an equivalence relation between linear spaces.*

Proof. It is clear that every linear space V is isomorphic to itself, which follows from the fact that the identity map 1_V is an isomorphism.

Suppose now that the linear spaces V and W are isomorphic and let $h : V \longrightarrow W$ be an isomorphism. It is easy to verify that the inverse mapping $h^{-1} : W \longrightarrow V$ is a linear morphism and a bijection, so the existence of an isomorphism is symmetric.

Finally, if $h : V \longrightarrow W$ and $g : W \longrightarrow U$ are isomorphisms, then gh is an isomorphism from V to U, so the existence of an isomorphism is transitive. Thus, isomorphism is an equivalence. □

Theorem 2.14. *If V, W are two finite-dimensional $\mathbb{F}$-linear spaces and $V \cong W$, then* $\dim(V) = \dim(W)$.

Proof. Suppose that $\dim(V) = n$ and that $B = \{\boldsymbol{x}_1, \ldots, \boldsymbol{x}_n\}$ is a basis for V. We claim that if $f : V \longrightarrow W$ is an isomorphism, then $B' = \{f(\boldsymbol{x}_1), \ldots, f(\boldsymbol{x}_n)\}$ is a basis for W.

Let $\boldsymbol{y} \in W$. Since f is a surjection, there exists $\boldsymbol{x} \in V$ such that $\boldsymbol{y} = f(\boldsymbol{x})$. Then, $\boldsymbol{x} = a_1\boldsymbol{x}_1 + \cdots + a_n\boldsymbol{x}_n$ for some $a_1, \ldots, a_n$ in $\mathbb{F}$ because B is a basis for V, hence $\boldsymbol{y} = f(\boldsymbol{x}) = f(a_1\boldsymbol{x}_1 + \cdots + a_n\boldsymbol{x}_n) = a_1 f(\boldsymbol{x}_1) + \cdots + a_n f(x_n)$. This shows that B' spans the space W.

To prove that B' is linearly independent, assume that $\mathbf{0}_W = c_1 f(\boldsymbol{x}_1) + \cdots + c_n f(\boldsymbol{x}_n) = f(c_1\boldsymbol{x}_1 + \cdots + c_n\boldsymbol{x}_n)$. Since f is injective, we have $c_1\boldsymbol{x}_1 + \cdots + c_n\boldsymbol{x}_n = \mathbf{0}_V$, which implies $c_1 = \cdots = c_n = 0$. Thus, B' is also linearly independent, hence B' is a basis of W. We conclude that $\dim(V) = |B| = |B'| = \dim(W)$. □

Theorem 2.15. *If V, W are two finite-dimensional $\mathbb{F}$-linear spaces and* $\dim(V) = \dim(W)$, *then $V \cong W$.*

Proof. To prove that $V \cong W$, it suffices to show that any of these two spaces is isomorphic to $\mathbb{F}^n$.

Suppose that $\dim(V) = n$ and that $B = \{\boldsymbol{x}_1, \ldots, \boldsymbol{x}_n\}$ is a basis for V. Define the mapping $h : V \longrightarrow \mathbb{F}^n$ as

$$h(\boldsymbol{x}) = \begin{pmatrix} a_1 \\ \vdots \\ a_n \end{pmatrix},$$

where $\boldsymbol{x} = a_1\boldsymbol{x}_1 + \cdots + a_n\boldsymbol{x}_n$. Since the expression of $\boldsymbol{x}$ in the basis B is unique, this function is a well-defined injective morphism. The function is also a surjection, so it is an isomorphism. □

Corollary 2.6. *If V, W are two finite-dimensional $\mathbb{F}$-linear spaces, then $V \cong W$ if and only if $\dim(V) = \dim(W)$.*

Proof. This follows from Theorems 2.14 and 2.15. □

Theorem 2.16. *Let $h : V \longrightarrow W$ be a linear mapping between the $\mathbb{F}$-linear spaces V and W such that $\dim(V) = \dim(W) = n$. The following statements are equivalent:*

(i) *h is surjective;*
(ii) *h is an isomorphism;*
(iii) *h is injective.*

Proof. (i) implies (ii): Suppose that h is surjective, that is, $\mathsf{Im}(h) = W$. Since $\dim(\mathsf{Ker}(h)) + \dim(\mathsf{Im}(h)) = n$, it follows that $\dim(\mathsf{Ker}(h)) = 0$, so $\mathsf{Ker}(h) = \{\mathbf{0}_V\}$. Therefore, by Theorem 2.23, h is an injection and, therefore, an isomorphism.

(ii) implies (iii): This implication is immediate.

(iii) implies (i): Suppose that h is injective. Then $\dim(\mathsf{Im}(h)) = n$, so $\mathsf{Im}(h) = W$, which means that h is a surjection. □

Theorem 2.17. *Let V be a subspace of the linear space W. Then*

$$\dim(W/V) = \dim(W) - \dim(V).$$

Proof. We apply Theorem 2.10 to the mapping $h_V : W \longrightarrow W/V$. We noted that $\mathsf{Ker}(h_V) = V$ and $\mathsf{Im}(h_V) = W/V$. Therefore, $\dim(W) = \dim(V) + \dim(W/V)$, which yields our equality. □

Let X be a non-empty set, $\mathbb{F}$ be the real or the complex field, and let $\mathsf{FREE}_{\mathbb{F}}(X)$ be a subset of $\mathbb{F}^X$ that consists of the mappings $f : X \longrightarrow \mathbb{F}$ such that the set $\text{supp}(f) = \{x \in X \mid f(x) \neq 0\}$ is finite. Addition and scalar multiplication are defined as usual. If $f, g \in \mathsf{FREE}_{\mathbb{F}}(X)$, then $\text{supp}(f+g) \subseteq \text{supp}(f) \cup \text{supp}(g)$ and

$$\text{supp}(af) = \begin{cases} \text{supp}(f) & \text{if } a \neq 0, \\ \{0\} & \text{if } a = 0. \end{cases}$$

This turns $\mathsf{FREE}_{\mathbb{F}}(X)$ into an $\mathbb{F}$-linear space. The zero element of $\mathsf{FREE}_{\mathbb{F}}(X)$ denoted by $\mathbf{0}_*$ is given by $\mathbf{0}_*(x) = 0$ for $x \in X$.

Definition 2.14. Let X be a non-empty set. The $\mathbb{F}$*-free linear space over* X is the linear space $\mathsf{FREE}_{\mathbb{F}}(X)$ constructed above.

If the field $\mathbb{F}$ is clear from context, the subscript $\mathbb{F}$ is omitted.

Example 2.15. The free $\mathbb{R}$-linear space on the set $\{1, \ldots, n\}$ is isomorphic to $\mathbb{R}^n$.

For $z \in X$, let δ_z be the function defined by

$$\delta_z(x) = \begin{cases} 1 & \text{if } x = z, \\ 0 & \text{otherwise.} \end{cases}$$

Since $\mathsf{supp}(\delta_z) = \{z\}$, it is clear that $\delta_z \in \mathsf{FREE}(X)$.

The notation δ_z is known as *Kronecker delta.*[1]

A related notation is $\delta_{uv} : X \times X \longrightarrow \{0, 1\}$, which denotes a function given by

$$\delta_{uv} = \begin{cases} 1 & \text{if } u = v, \\ 0 & \text{otherwise.} \end{cases}$$

The set $\{\delta_z \mid z \in X\}$ is a basis in $\mathsf{FREE}(X)$. Indeed, if $\mathsf{supp}(f) = \{z_1, \ldots, z_n\}$, then $f(z) = \sum_{i=1}^n f(z_i)\delta_{z_i}(z)$, or $f = \sum_{i=1}^n f(z_i)\delta_{z_i}$.

The set $\{\delta_z \mid z \in X\}$ is linearly independent because, if $\sum_{j=1}^n a_j \delta_{z_j}(x) = 0$, then choosing $x = z_i$ we obtain $a_j = 0$ for $1 \leqslant j \leqslant n$.

Example 2.16. The free $\mathbb{R}$-linear space $\mathsf{FREE}(\mathbb{N})$ consists of all sequences $r_0 n_0 + r_1 n_1 + \cdots$, where the set $\{r_i \in \mathbb{R} \mid i \in \mathbb{N} \text{ and } r_i \neq 0\}$ is finite. The zero element of this space is $0n_0 + 0n_1 + \cdots$.

Theorem 2.18 (Universal property of h_X). Let X be a set and let $h_X : X \longrightarrow \mathsf{FREE}(X)$ be the mapping defined by $h(x) = \delta_x$ for $x \in X$. If L is an $\mathbb{F}$-vector space and $\phi : X \longrightarrow L$ is a mapping, there exists a unique linear mapping $\psi : \mathsf{FREE}(X) \longrightarrow L$ such that the following diagram is commutative:

[1]Leopold Kronecker (December 7, 1823 in Liegnitz–December 29, 1891 in Berlin) was a German mathematician who worked on number theory, algebra, and logic. He studied at the Universities of Bonn, Breslau, Berlin, where he defended his dissertation in algebraic number theory in 1845. He was elected a member of the Berlin Academy in 1861. There are numerous concepts in mathematics named after Kronecker: Kronecker delta, Kronecker product, and many others.

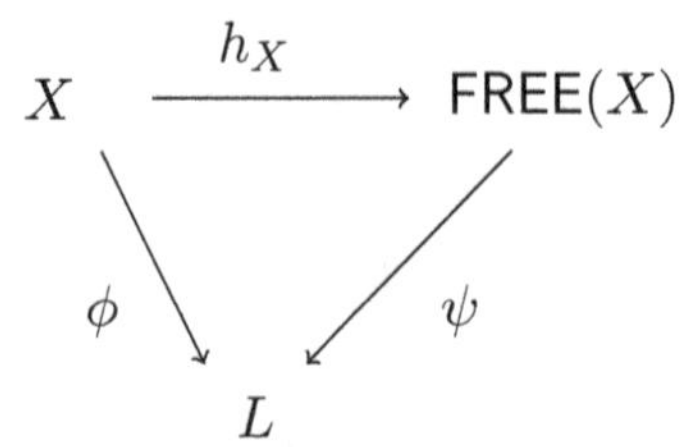

Proof. Define ψ as the function that maps the element δ_z of the basis of $\mathsf{FREE}(X)$ in $\phi(z)$. It is immediate that ψ is a linear mapping and that the diagram is commutative. To prove uniqueness, suppose that ψ_1 is another linear mapping such that $\phi = \psi h_X = \psi_1 h_X$. Then for $x \in X$, we have $\phi(x) = \psi(h_X(x)) = \psi_1(h_X(x))$, which implies $\phi(x) = \psi(\delta_x) = \psi_1(\delta_x)$. Since the set $\{\delta_x \mid x \in X\}$is a basis in $\mathsf{FREE}(X)$, it follows that $\psi = \psi_1$. □

Theorem 2.19. *Let V and W be two $\mathbb{F}$-linear spaces and let $g, h \in \mathsf{Hom}(V, W)$ be two linear mappings. If X is a subset of V such that $g(\boldsymbol{x}) = h(\boldsymbol{x})$ for every $\boldsymbol{x} \in X$, then $g(\boldsymbol{x}) = h(\boldsymbol{x})$ for every $\boldsymbol{x} \in \langle X \rangle$.*

Proof. Consider the set

$$\mathsf{EQ}(g, h) = \{\boldsymbol{u} \in V \mid g(\boldsymbol{u}) = h(\boldsymbol{u})\}.$$

If $\boldsymbol{u}, \boldsymbol{v} \in \mathsf{EQ}(g, h)$, then $g(a\boldsymbol{u}+b\boldsymbol{v}) = ag(\boldsymbol{u})+bg(\boldsymbol{v}) = ah(\boldsymbol{u})+bh(\boldsymbol{u}) = h(a\boldsymbol{u} + b\boldsymbol{v})$, so $a\boldsymbol{u} + b\boldsymbol{v} \in \mathsf{EQ}(g, h)$ for every $a, b \in \mathbb{F}$, which implies that $\mathsf{EQ}(g, h)$ is a subspace of V. Since $X \subseteq \mathsf{EQ}(g, h)$, it follows that $\langle X \rangle \subseteq \mathsf{EQ}(g, h)$, which yields the desired conclusion. □

We refer to $\mathsf{EQ}(g, h)$ as the *equalizer* of g and h.

Theorem 2.20. *Let V and W be two $\mathbb{F}$-linear spaces. A morphism $h \in \mathsf{Hom}(V, W)$ is injective if and only if $h(\boldsymbol{x}) = \boldsymbol{0}_W$ implies $\boldsymbol{x} = \boldsymbol{0}_V$.*

Proof. Let h be a morphism such that $h(\boldsymbol{x}) = \boldsymbol{0}_W$ implies $\boldsymbol{x} = \boldsymbol{0}_V$. If $h(\boldsymbol{x}) = h(\boldsymbol{y})$, by the linearity of h we have $h(\boldsymbol{x} - \boldsymbol{y}) = \boldsymbol{0}_W$, which implies $\boldsymbol{x} - \boldsymbol{y} = \boldsymbol{0}_V$, that is, $\boldsymbol{x} = \boldsymbol{y}$. Thus, h is injective.

Conversely, suppose that h is injective. If $\boldsymbol{x} \neq \boldsymbol{0}_V$, then $h(\boldsymbol{x}) \neq h(\boldsymbol{0}_V) = \boldsymbol{0}_W$. Thus, $h(\boldsymbol{x}) = \boldsymbol{0}_W$ implies $\boldsymbol{x} = \boldsymbol{0}_V$. □

An *endomorphism* of an $\mathbb{F}$-linear space V is a morphism $h : V \longrightarrow V$. The set of endomorphisms of V is denoted by $\mathsf{End}(V)$. Often, we refer to endomorphisms of V as *linear operators* on V.

Let $\mathbb{F}$ be a field and let V be an $\mathbb{F}$-linear space. Define the mapping $h_a : V \longrightarrow V$ by $h_a(\boldsymbol{x}) = a\boldsymbol{x}$ for $\boldsymbol{x} \in V$. It is easy to verify that h_a is a linear operator on L. This mapping is known as a *homothety on* V.

If $a = 1$, then h_1 is given by $h_1(\boldsymbol{x}) = \boldsymbol{x}$ for $\boldsymbol{x} \in V$; this is the identity morphism of V, which is usually denoted by 1_V.

For $a = 0$, we obtain the *zero endomorphism of* V denoted by 0_V and given by $0_V(\boldsymbol{x}) = \boldsymbol{0}_V$ for $\boldsymbol{x} \in V$.

Example 2.17. Let V be an $\mathbb{F}$-linear space and let $\boldsymbol{z} \in V$. The *translation* generated by $\boldsymbol{z} \in V$ is the mapping $t_{\boldsymbol{z}} : V \longrightarrow V$ defined by $t_{\boldsymbol{z}}(\boldsymbol{x}) = \boldsymbol{x} + \boldsymbol{z}$ for $\boldsymbol{x} \in V$. A translation is a bijection but not a morphism unless $\boldsymbol{z} = \boldsymbol{0}_V$. Its inverse is $t_{-\boldsymbol{z}}$.

Definition 2.15. Let V be an $\mathbb{F}$-linear space and let U and Z be two subsets of L. Define the subset $U + Z$ of V as

$$U + Z = \{\boldsymbol{u} + \boldsymbol{z} \mid \boldsymbol{u} \in U \text{ and } \boldsymbol{z} \in Z\}.$$

For $a \in \mathbb{F}$, the set aU is

$$aU = \{a\boldsymbol{u} \mid \boldsymbol{u} \in U\}.$$

Theorem 2.21. *Let V, W, U be three $\mathbb{F}$-linear spaces. The following properties of compositions of linear mappings hold:*

(i) *If $f \in \mathsf{Hom}(V, W)$ and $g \in \mathsf{Hom}(W, U)$, then $gf \in \mathsf{Hom}(V, U)$.*

(iii) *If $f \in \mathsf{Hom}(V, W)$ and $g_0, g_1 \in \mathsf{Hom}(W, U)$, then*

$$f(g_0 + g_1) = fg_0 + fg_1.$$

(iii) *If $f_0, f_1 \in \mathsf{Hom}(V, W)$ and $g \in \mathsf{Hom}(W, U)$, then*

$$(f_0 + f_1)g = f_0 g + f_1 g.$$

Proof. We prove only the second part of the theorem and leave the proofs of the remaining parts to the reader.

Let $\boldsymbol{x} \in V$. Then, $f(g_0 + g_1)(\boldsymbol{x}) = f((g_0 + g_1)(\boldsymbol{x})) = f(g_0(\boldsymbol{x}) + g_1(\boldsymbol{x})) = f(g_0(\boldsymbol{x})) + f(g_1(\boldsymbol{x}))$ for $\boldsymbol{x} \in V$, which yields the desired equality. □

Corollary 2.7. *Let V, W be two $\mathbb{F}$-linear spaces. The algebra $(\mathsf{Hom}(V,W), \{h_0, +, -\})$ is an Abelian group that has the zero morphism $0_{V,W}$ as its zero-ary operation and the addition of linear mappings as its binary operation; the opposite of a linear mapping h is the mapping $-h$.*

Moreover, $(\mathsf{End}(V), \{0_V, 1_V, +, -, \cdot\})$ is a unitary ring, where the multiplication is defined as the composition of linear mappings.

Proof. The first part of the statement is a simple verification that the operations $h_0, +, -$ satisfy the definition of Abelian groups. The second part of the corollary follows immediately from Theorem 2.21. □

An endomorphism h of a field is *idempotent* if $h^2 = h$, that is, if $h(h(\boldsymbol{x})) = h(\boldsymbol{x})$ for every $\boldsymbol{x} \in M$.

Corollary 2.8. *If h is an idempotent endomorphism of the field $(End(V), \{h_0, 1_V, +, -, \cdot\})$, then $1 - h$ is also an idempotent endomorphism.*

Proof. This statement is a direct consequence of Corollary 2.7 and of Theorem 1.12. □

Definition 2.16. Let h be an endomorphism of linear space V. The *mth iteration* of h (for $m \in \mathbb{N}$) is defined as

(i) $h^0 = 1_V$;
(ii) $h^{m+1}(\boldsymbol{x}) = h(h^m(\boldsymbol{x}))$ for $m \in \mathbb{N}$.

For every $m \geqslant 1$, h^m is an endomorphism of V; this can be shown by a straightforward proof by induction on m.

Example 2.18. Let $\mathbb{F}$ be a field and $\mathbb{F}[\lambda]$ be the set of polynomials in the indeterminate λ with coefficients in $\mathbb{F}$. If $p \in \mathbb{F}[\lambda]$, we have $p(\lambda) = p_0 + p_1\lambda + \cdots + p_n\lambda^n$.

For an endomorphism of an $\mathbb{F}$-linear space V, $h : V \longrightarrow V$ define the function $f = p(h)$ as

$$f(\boldsymbol{x}) = p_0 + p_1 h(\boldsymbol{x}) + \cdots + p_n h^n(\boldsymbol{x}),$$

for $\boldsymbol{x} \in V$.

Theorem 2.22. *Let V be an $\mathbb{F}$-linear space and let $p \in \mathbb{F}[\lambda]$ be a polynomial. If $h : V \longrightarrow V$ is an endomorphism of V, then $p(h)$ is an affine mapping on V.*

Proof. Let $p(\lambda) = p_0 + p_1\lambda + \cdots + p_n\lambda^n$. Since h^m is an endomorphism of V for every $m \geqslant 1$ and the sum of endomorphisms is an endomorphism, it follows that $q(h)$ is an endomorphism, where $q(\lambda) = p_1\lambda + \cdots + p_n\lambda^n$. Thus, f is an affine mapping because $f(\boldsymbol{x}) = p_0 + g(\boldsymbol{x})$ for $\boldsymbol{x} \in V$. □

Note that if $p(\mathbf{0}_V) = 0$ and h is an endomorphism, then $p(h)$ is also an endomorphism.

2.7 Constructing Linear Spaces

This section is concerned with methods of constructing new linear spaces.

Definition 2.17. Let S be a subspace of an $\mathbb{F}$-linear space V.

The *equivalence generated by* S is the relation $\sim_S$ on the set V defined by $\boldsymbol{x} \sim_S \boldsymbol{y}$ if $\boldsymbol{x} - \boldsymbol{y} \in S$.

The *quotient set* V/S is the set of equivalence classes of $\sim_S$,

$$V/S = \{[\boldsymbol{x}] \mid \boldsymbol{x} \in V\}.$$

It is easy to verify that $\sim_S$ is indeed an equivalence relation on V.

The quotient set has a natural structure of linear space when we define the addition of classes as

$$[\boldsymbol{x}] + [\boldsymbol{y}] = [\boldsymbol{x} + \boldsymbol{y}]$$

and the scalar multiplication as $a[\boldsymbol{x}] = [a\boldsymbol{x}]$ for $a \in \mathbb{F}$ and $\boldsymbol{x}, \boldsymbol{y} \in V$.

These operations are well-defined. Indeed, if $\boldsymbol{u} \in [\boldsymbol{x}]$ and $\boldsymbol{v} \in [\boldsymbol{y}]$, then $\boldsymbol{x} - \boldsymbol{u} \in S$ and $\boldsymbol{y} - \boldsymbol{v} \in S$. Therefore, $\boldsymbol{x} + \boldsymbol{y} - (\boldsymbol{u} + \boldsymbol{v}) \in S$, so $[\boldsymbol{u} + \boldsymbol{v}] = [\boldsymbol{x} + \boldsymbol{y}]$. Similarly, $a\boldsymbol{x} - a\boldsymbol{u} = a(\boldsymbol{x} - \boldsymbol{u}) \in S$, so $[a\boldsymbol{u}] = [a\boldsymbol{x}]$. We leave it to the reader to check that they satisfy the definition of a linear space.

Definition 2.18. The *quotient space* of an $\mathbb{F}$-linear space V by a subspace S is the $\mathbb{F}$-linear space defined on the set $V/\sim_S$ whose operations are defined as above.

The quotient of V by S is denoted by V/S.

Observe that the surjective mapping $h_S : V \longrightarrow V/S$, also known as the *canonical morphism* (or *canonical surjection*) of S defined by $h_S(\boldsymbol{x}) = [\boldsymbol{x}]$ for $\boldsymbol{x} \in V$, is a linear mapping. We have $\mathsf{Ker}(h_S) = S$ and $\mathsf{Im}(h_S) = V/S$.

We saw that for a linear mapping $h : V \longrightarrow W$, the set $\mathsf{Im}(h)$ is a subspace of W. The quotient linear space $W/\mathsf{Im}(h)$ is referred to as the *co-kernel of* h and is denoted by $\mathsf{Coker}(h)$.

If h is a surjective mapping, $\mathsf{Im}(h) = W$ and, therefore, $\mathsf{Coker}(h) = \{[\mathbf{0}_W]\}$. The co-kernel of a linear mapping h is the zero subspace if and only if h is surjective.

Theorem 2.23. *Let V, W, Z be two $\mathbb{F}$-linear spaces. If $h \in \mathsf{Hom}(V, W)$, then the following three statements are equivalent:*

(i) *h is an injection;*

(ii) *if $f, g : Z \longrightarrow V$ are two linear mappings such that $hf = hg$, then $f = g$;*

(iii) $\mathsf{Ker}(h) = \{\mathbf{0}_V\}$.

Proof. (i) implies (ii): Let h be an injection and suppose that $hf = hg$, where $f, g : Z \longrightarrow V$ are two linear operators. For $\boldsymbol{z} \in Z$, we have $h(f(\boldsymbol{z})) = h(g(\boldsymbol{z}))$, which yields $f(\boldsymbol{z}) = g(\boldsymbol{z})$ for $\boldsymbol{z} \in Z$ because h is injective. Thus, $f = g$.

(ii) implies (iii): Let $i : \mathsf{Ker}(h) \longrightarrow V$ be the linear application defined by $i(\boldsymbol{x}) = \boldsymbol{x}$ for $\boldsymbol{x} \in \mathsf{Ker}(h)$.

Let $k : \mathsf{Ker}(h) \longrightarrow W$ be the linear function defined by $k(\boldsymbol{v}) = \mathbf{0}_W$. Note that $h(i(\boldsymbol{z})) = h(k(\boldsymbol{z})) = \mathbf{0}_W$. Thus, $i(\boldsymbol{z}) = k(\boldsymbol{z}) = \mathbf{0}_W$, which implies $\boldsymbol{z} = \mathbf{0}_V$, so $\mathsf{Ker}(h) = \{\mathbf{0}_V\}$.

(iii) implies (i): Suppose that $\mathsf{Ker}(h) = \{\mathbf{0}_V\}$ and that $h(\boldsymbol{x}) = h(\boldsymbol{y})$. Therefore, $h(\boldsymbol{x} - \boldsymbol{y}) = \mathbf{0}_W$, so $\boldsymbol{x} - \boldsymbol{y} \in \mathsf{Ker}(h) = \{\mathbf{0}_V\}$, which means that $\boldsymbol{x} = \boldsymbol{y}$, that is, h is an injection. □

A similar characterization exists for surjective linear mappings.

Theorem 2.24. *Let V, W be two $\mathbb{F}$-linear spaces. If $h \in Hom(V, W)$, then the following statements are equivalent:*

(i) *h is a surjection;*

(ii) *if U is an $\mathbb{F}$-linear space and $f, g \in \mathsf{Hom}(W, U)$ are such that $fh = gh$, then $f = g$.*

Proof. (i) implies (ii): Suppose that h is a surjective linear mapping. Then, for every $\boldsymbol{y} \in W$ there exists $\boldsymbol{x}$ such that $\boldsymbol{y} = h(\boldsymbol{x})$. Thus, we have $f(\boldsymbol{y}) = f(h(\boldsymbol{x})) = g(h(\boldsymbol{x})) = g(\boldsymbol{y})$ for every $\boldsymbol{y} \in W$, so $f = g$.

(ii) implies (i): If condition (ii) is satisfied, let $U = \mathsf{Coker}(h)$ and define $f, g : W \longrightarrow \mathsf{Coker}(h)$ as $f(\boldsymbol{w}) = [h(w)]$ in $\mathsf{Coker}(h) = W/\mathsf{Im}(h)$ and $g(\boldsymbol{w}) = \mathbf{0}_{\mathsf{Coker}(h)}$. Then $f(h(v)) = g(h(v))$ means that $[h(v)] = \mathbf{0}_{\mathsf{Coker}(h)}$, hence $\mathsf{Coker}(h) = [\mathbf{0}_W]$, hence h is surjective. □

Theorem 2.25. *Let V, W, U be three linear spaces and let $h : V \longrightarrow W$ and $g : V \longrightarrow U$ be two linear mappings. If $\mathsf{Ker}(h) \subseteq \mathsf{Ker}(g)$ and h is surjective, then there exists a unique $k \in \mathsf{Hom}(W, U)$ such that $g = kh$.*

Proof. The statement of the theorem asserts the existence and uniqueness of the morphism $k : W \longrightarrow U$ that makes the diagram

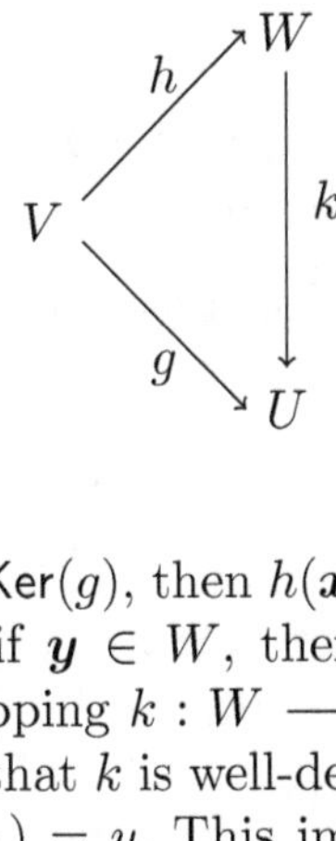

commutative.

Note that if $\mathsf{Ker}(h) \subseteq \mathsf{Ker}(g)$, then $h(\boldsymbol{x}) = \mathbf{0}_W$ implies $g(\boldsymbol{x}) = \mathbf{0}_U$.

Since h is surjective, if $\boldsymbol{y} \in W$, there exists $\boldsymbol{x} \in V$ such that $h(\boldsymbol{x}) = \boldsymbol{y}$. Define the mapping $k : W \longrightarrow U$ as $k(\boldsymbol{y}) = g(\boldsymbol{x})$, where $h(\boldsymbol{x}) = \boldsymbol{y}$. We verify first that k is well-defined. Indeed, suppose that $\boldsymbol{x}_1 \in V$ is such that $h(\boldsymbol{x}_1) = y$. This implies $h(\boldsymbol{x}) = h(\boldsymbol{x}_1)$, which is equivalent to $h(\boldsymbol{x} - \boldsymbol{x}_1) = \mathbf{0}_W$. Therefore, $g(\boldsymbol{x} - \boldsymbol{x}_1) = \mathbf{0}_U$, which means that $g(\boldsymbol{x}_1) = g(\boldsymbol{x})$, so k is well-defined.

The mapping k is linear. Indeed, let $a_1, a_2 \in \mathbb{F}$ and $\boldsymbol{y}_1, \boldsymbol{y}_2 \in W$. Suppose that $\boldsymbol{x}_1, \boldsymbol{x}_2 \in V$ are such that $h(\boldsymbol{x}_i) = \boldsymbol{y}_i$ for $i = 1, 2$. Then $h(a_1\boldsymbol{x}_1 + a_2\boldsymbol{x}_2) = a_1\boldsymbol{y}_1 + a_2\boldsymbol{y}_2$. This means that $k(a_1\boldsymbol{y}_1 + a_2\boldsymbol{y}_2) = g(a_1\boldsymbol{x}_1 + a_2\boldsymbol{x}_2)$, so by the linearity of g, we have

$$k(a_1\boldsymbol{y}_1 + a_2\boldsymbol{y}_2) = a_1 g(\boldsymbol{x}_1) + a_2 g(\boldsymbol{x}_2) = a_1 k(\boldsymbol{y}_1) + a_2 k(\boldsymbol{y}_2),$$

which proves that k is linear.

The definition of h implies that $k(h(\boldsymbol{x})) = g(\boldsymbol{x})$ for every $\boldsymbol{x} \in V$, so $g = kh$.

Suppose that k_1 is a morphism in $\mathsf{Hom}(W, U)$ such that $g = k_1 h$, so $k_1 h = kh$. Since h is a surjection, by Theorem 2.24, we obtain $k_1 = k$. □

Theorem 2.26. *Let $h : V \longrightarrow W$ be a linear mapping between the $\mathbb{F}$-spaces V and W and let S be a subspace of V. There exists a unique linear mapping $\overline{h} : V/S \longrightarrow W$ such that $\overline{h}([\boldsymbol{x}]) = h(\boldsymbol{x})$ (where $[\boldsymbol{x}] = \boldsymbol{x} + S$ for all $\boldsymbol{x} \in S$) if and only if $h(\boldsymbol{s}) = \mathbf{0}_W$ for all $\boldsymbol{s} \in S$.*

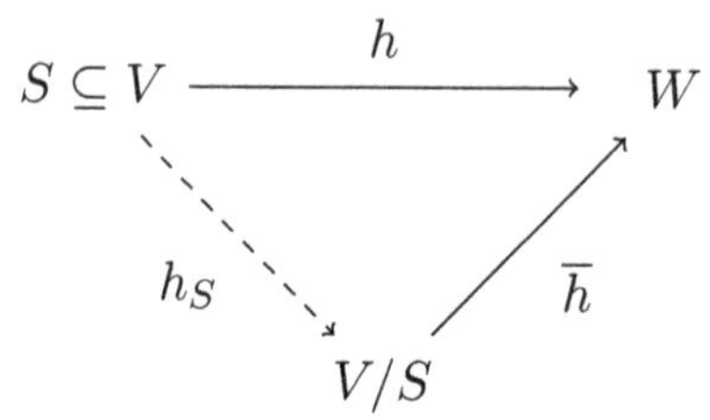

If $\overline{h}$ exists and is unique, then for every linear mapping $\overline{g} : V/S \longrightarrow W$ there exists a unique linear transformation $g : V \longrightarrow W$ that generates $\overline{g}$ in this manner.

Proof. Suppose that $\overline{h}$ exists such that $\overline{h}([\boldsymbol{x}]) = h(\boldsymbol{x})$. Then, if $\boldsymbol{s} \in S$, we have,

$$h(\boldsymbol{s}) = \overline{h}(\boldsymbol{s} + S) = \overline{h}(\mathbf{0}_{V/S}) = \mathbf{0}_W.$$

Conversely, suppose now that $h(\boldsymbol{s}) = \mathbf{0}_W$ for all $\boldsymbol{s} \in S$. Define $\overline{h}$ as $\overline{h}([\boldsymbol{x}]) = h(\boldsymbol{x})$. Note that $\overline{h}$ is well-defined for, if $[\boldsymbol{x}] = [\boldsymbol{y}]$, we have $\boldsymbol{x} - \boldsymbol{y} \in S$, hence $h(\boldsymbol{x} - \boldsymbol{y}) = \mathbf{0}_W$, or $h(\boldsymbol{x}) = h(\boldsymbol{y})$.

The final part of the theorem follows by noting that the equality $\overline{h}(\boldsymbol{x} + S) = h(\boldsymbol{x})$ shows that any of the mappings $h, \overline{h}$ is determined by the other. □

Let $\mathsf{SUBSP}(V)$ be the collection of subspaces of a linear space V. If this set is equipped with the inclusion relation $\subseteq$ (which is a partial order), then for any two subspaces K, L both $\sup\{K, L\}$ and $\inf\{K, L\}$ exist and are given by

$$\sup\{K, L\} = \{\boldsymbol{x} + \boldsymbol{y} \mid \boldsymbol{x} \in K \text{ and } \boldsymbol{y} \in L\}, \tag{2.1}$$

$$\inf\{K, L\} = K \cap L. \tag{2.2}$$

Let $H = \{\boldsymbol{x} + \boldsymbol{y} \mid \boldsymbol{x} \in K \text{ and } \boldsymbol{y} \in L\}$. Observe that we have both $K \subseteq H$ and $L \subseteq H$ because $\mathbf{0}$ belongs to both K and L.

If $\boldsymbol{u}$ and $\boldsymbol{v}$ belong to H, then $\boldsymbol{u} = \boldsymbol{x}_1 + \boldsymbol{y}_1$ and $\boldsymbol{v} = \boldsymbol{x}_2 + \boldsymbol{y}_2$, where $\boldsymbol{x}_1, \boldsymbol{x}_2 \in K$ and $\boldsymbol{y}_1, \boldsymbol{y}_2 \in L$. Since $\boldsymbol{x}_1 - \boldsymbol{x}_2 \in K$ and $\boldsymbol{y}_1 - \boldsymbol{y}_2 \in L$ (because K and L are subspaces), it follows that $\boldsymbol{u} - \boldsymbol{v} = \boldsymbol{x}_1 + \boldsymbol{y}_1 - (\boldsymbol{x}_2 + \boldsymbol{y}_2) = (\boldsymbol{x}_1 - \boldsymbol{x}_2) + (\boldsymbol{y}_1 - \boldsymbol{y}_2) \in H$.

We have $a\boldsymbol{u} = a\boldsymbol{x}_1 + a\boldsymbol{x}_2 \in H$ because $a\boldsymbol{x}_1 \in K$ and $a\boldsymbol{x}_2 \in L$. Thus, H is a subspace of V and is an upper bound of $\{K, L\}$ in the partially ordered set $(\mathsf{SUBSP}(V), \subseteq)$.

If G is a subspace of V that contains both K and L, then $\boldsymbol{x}+\boldsymbol{y} \in G$ for $\boldsymbol{x} \in K$ and $\boldsymbol{y} \in L$, so $H \subseteq G$. Thus, $H = \sup\{K, L\}$.

We denote $H = \sup\{K, L\}$ by $K + L$.

The lattice of subspaces $\mathsf{SUBSP}(V)$ of a linear space V is actually a complete lattice because the collection of subspaces of a linear space is a closure system.

Next, we prove the *modularity* of $\mathsf{SUBSP}(V)$.

Theorem 2.27. *Let V be an $\mathbb{F}$-linear space. For any $P, Q, R \in$ SUBSP(V) such that $Q \subseteq P$, we have $P \cap (Q + R) = Q + (P \cap R)$.*

Proof. Note that $Q \subseteq P \cap (Q+R)$, $P \cap R \subseteq P \cap (Q+R)$. Therefore, we have the inclusion $Q + (P \cap R) \subseteq P \cap (Q + R) =$, which leaves us with the reverse inclusion to prove.

Let $\boldsymbol{z} \in P \cap (Q + R)$. This implies $\boldsymbol{z} \in P$ and $\boldsymbol{z} = \boldsymbol{x} + \boldsymbol{y}$, where $\boldsymbol{x} \in Q \subseteq P$ and $\boldsymbol{y} \in R$. Therefore, $\boldsymbol{y} = \boldsymbol{z} - \boldsymbol{x} \in P$, so $\boldsymbol{y} \in P \cap R$. Consequently, $\boldsymbol{z} \in Q + (P \cap R)$, so $P \cap (Q + R) \subseteq Q + (P \cap R)$. □

Theorem 2.35 can now be reformulated as

Corollary 2.9. *Let K, L be two subspaces of a linear space V. We have*

$$\dim(\sup\{K, L\}) + \dim(\inf\{K, L\}) = \dim(K) + \dim(L).$$

An immediate consequence of Corollary 2.9 is the following inequality valid for two linear subspaces K, L of a linear space V, namely:

$$\dim(K + L) \leqslant \dim(K) + \dim(L). \tag{2.3}$$

Theorem 2.28. *Let V be an n-dimensional linear space and let W be a subspace of V. Then the space V/W is finite-dimensional and* $\dim(V/W) = \dim(V) - \dim(W)$.

Proof. Let $\{\boldsymbol{w}_1, \ldots, \boldsymbol{w}_n\}$ be a basis of W and let $\{\boldsymbol{w}_1, \ldots, \boldsymbol{w}_n, \boldsymbol{v}_1, \ldots, \boldsymbol{v}_k\}$ be its extension to a basis for V, where $\dim(W) = n$ and $\dim(V) = n + k$.

The set $\{\boldsymbol{v}_1 + W, \ldots, \boldsymbol{v}_k + W\}$ is a basis of V/W. Indeed, suppose that

$$a_1(\boldsymbol{v}_1 + W) + \cdots + a_k(\boldsymbol{v}_k + W) = W$$

for $a_1, \ldots, a_k \in \mathbb{F}$, where W is the zero element of V/W. This equality amounts to

$$(a_1\boldsymbol{v}_1 + \cdots + a_k\boldsymbol{v}_k) + W = W,$$

which implies $a_1\boldsymbol{v}_1 + \cdots + a_k\boldsymbol{v}_k \in W$. Thus, we can write

$$a_1\boldsymbol{v}_1 + \cdots + a_k\boldsymbol{v}_k = b_1\boldsymbol{w}_1 + \cdots + b_n\boldsymbol{w}_n$$

for some $b_1, \ldots, b_n \in \mathbb{F}$. Since

$$a_1\boldsymbol{v}_1 + \cdots + a_k\boldsymbol{v}_k - b_1\boldsymbol{w}_1 - \cdots - b_n\boldsymbol{w}_n = \mathbf{0}_V,$$

and $\{\boldsymbol{w}_1, \ldots, \boldsymbol{w}_n, \boldsymbol{v}_1, \ldots, \boldsymbol{v}_k\}$ is a basis, it follows that $a_1 = \cdots = a_k = b_1 = \cdots = b_n = 0$. This shows that $\{\boldsymbol{v}_1 + W, \ldots, \boldsymbol{v}_k + W\}$ is linearly independent.

Let $\boldsymbol{v} + W \in V/W$. Since the set $\{\boldsymbol{w}_1, \ldots, \boldsymbol{w}_n, \boldsymbol{v}_1, \ldots, \boldsymbol{v}_k\}$ spans V, we can write

$$\boldsymbol{v} = a_1\boldsymbol{w}_1 + \cdots + a_n\boldsymbol{w}_n + b_1\boldsymbol{v}_1 + \cdots + b_k\boldsymbol{v}_k$$

for some $a_1, \ldots, a_n, b_1, \ldots, b_k \in \mathbb{F}$. Therefore, we can write

$$\begin{aligned}\boldsymbol{v} + W &= (a_1\boldsymbol{w}_1 + \cdots + a_n\boldsymbol{w}_n + b_1\boldsymbol{v}_1 + \cdots + b_k\boldsymbol{v}_k) + W \\ &= (b_1\boldsymbol{v}_1 + \cdots + b_k\boldsymbol{v}_k) + W \\ &= b_1(\boldsymbol{v}_1 + W) + \cdots + b_k(\boldsymbol{v}_k + W).\end{aligned}$$

This shows that the set $\{\boldsymbol{v}_1 + W, \ldots, \boldsymbol{v}_k + W\}$ spans V/W, and we conclude that this set forms a basis of V/W. Therefore, $\dim(V/W) = k = \dim(V) - \dim(W)$. □

Definition 2.19. Let $\mathcal{V} = \{V_i \mid i \in I\}$ be a family of $\mathbb{F}$-linear spaces indexed by a set I. The *direct product* of $\mathcal{V}$ is the linear space $\prod_{i \in I} V_i$ that consists of all functions $f : I \longrightarrow \bigcup_{i \in I} V_i$ such that $f(i) \in V_i$ for $i \in I$.

The addition and scalar multiplication are defined by

$$(f + g)(i) = f(i) + g(i)$$
$$(af)(i) = af(i)$$

for every $f, g \in I \longrightarrow \bigcup_{i \in I} V_i$ and $a \in \mathbb{F}$.

The *support* of $f \in \prod_{i \in I} V_i$ is the set $\mathsf{supp}(f) = \{i \mid f(i) \neq \mathbf{0}_{V_i}\}$. We have

$$\mathsf{supp}(f + g) \subseteq \mathsf{supp}(f) + \mathsf{supp}(g),$$
$$\mathsf{supp}(af) \subseteq \mathsf{supp}(f)$$

for every $f, g \in I \longrightarrow \bigcup_{i \in I} V_i$ and $a \in \mathbb{F}$.

The notion of *direct sum* of linear spaces has different meanings depending on the nature of the linear spaces involved.

Definition 2.20. Let $\mathcal{V} = \{V_i \mid i \in I\}$ be a family of $\mathbb{F}$-linear spaces.

The *external direct sum* of $\mathcal{V}$ is the $\mathbb{F}$-linear space $\sum_{i \in I} V_i$ that consists of all functions $f : I \longrightarrow \bigcup_{i \in I} V_i$ such that $f(i) \in V_i$ for $i \in I$ *that have a finite support.*

The addition and scalar multiplication are defined exactly as in the case of the members of the direct product.

It is clear that the direct sum of a family of $\mathbb{F}$-linear spaces is a subspace of their direct product. Also, for finite families of linear spaces, the direct product is identical to the direct sum.

If V_1, V_2 are subspaces of an $\mathbb{F}$-linear space V, then their intersection is non-empty because $\mathbf{0}_V \in V_1 \cap V_2$. Moreover, it is easy to see that $V_1 \cap V_2$ is also a subspace of V.

Let V_1, V_2 be two subspaces of a linear space V. Their *internal sum*, or simply their *sum*, is the subset $V_1 + V_2$ of V defined by

$$V_1 + V_2 = \{\boldsymbol{x} + \boldsymbol{y} \mid \boldsymbol{x} \in V_1 \text{ and } \boldsymbol{y} \in V_2\}.$$

It is immediate to verify that $V_1 + V_2$ is a subspace of V and that $\mathbf{0}_V \in V_1 \cap V_2$.

Theorem 2.29. *Let V_1, V_2 be two subspaces of the $\mathbb{F}$-linear space V. If $V_1 \cap V_2 = \{\mathbf{0}_V\}$, then any vector $\boldsymbol{x} \in V_1 + V_2$ can be uniquely written as $\boldsymbol{x} = \boldsymbol{x}_1 + \boldsymbol{x}_2$, where $\boldsymbol{x}_1 \in V_1$ and $\boldsymbol{x}_2 \in V_2$.*

Proof. By the definition of the sum $V_1 + V_2$, it is clear that any vector $\boldsymbol{x} \in V_1 + V_2$ can be written as $\boldsymbol{x} = \boldsymbol{x}_1 + \boldsymbol{x}_2$, where $\boldsymbol{v}_1 \in V_1$ and $\boldsymbol{v}_2 \in V_2$. We need to prove only the uniqueness of $\boldsymbol{x}_1$ and $\boldsymbol{x}_2$.

Suppose that $\boldsymbol{x} = \boldsymbol{x}_1 + \boldsymbol{x}_2 = \boldsymbol{y}_1 + \boldsymbol{y}_2$, where $\boldsymbol{x}_1, \boldsymbol{y}_1 \in V_1$ and $\boldsymbol{x}_2, \boldsymbol{y}_2 \in V_2$. This implies $\boldsymbol{x}_1 - \boldsymbol{y}_1 = \boldsymbol{y}_2 - \boldsymbol{x}_2$ and, since $\boldsymbol{x}_1 - \boldsymbol{y}_1 \in V_1$ and $\boldsymbol{y}_2 - \boldsymbol{x}_2 \in V_2$, it follows that $\boldsymbol{x}_1 - \boldsymbol{y}_1 = \boldsymbol{y}_2 - \boldsymbol{x}_2 = \mathbf{0}_V$ by hypothesis. Therefore, $\boldsymbol{x}_1 = \boldsymbol{y}_1$ and $\boldsymbol{x}_2 = \boldsymbol{y}_2$. □

Theorem 2.30. *Let V_1, V_2 be two subspaces of the $\mathbb{F}$-linear space V. If every vector $\boldsymbol{x} \in V_1 + V_2$ can be uniquely written as $\boldsymbol{x} = \boldsymbol{x}_1 + \boldsymbol{x}_2$, then $V_1 \cap V_2 = \mathbf{0}_V$.*

Proof. Suppose that the uniqueness of the expression of $\boldsymbol{x}$ holds but $\boldsymbol{z} \in V_1 \cap V_2$ and $\boldsymbol{z} \neq \mathbf{0}_V$. If $\boldsymbol{x} = \boldsymbol{x}_1 + \boldsymbol{x}_2$, then we can also write $\boldsymbol{x} = (\boldsymbol{x}_1 + \boldsymbol{z}) + (\boldsymbol{x}_2 - \boldsymbol{z})$, where $\boldsymbol{x}_1 + \boldsymbol{z} \in V_1$ and $\boldsymbol{x}_2 - \boldsymbol{z} \in V_2$, $\boldsymbol{x}_1 + \boldsymbol{z} \neq \boldsymbol{x}_1$ and $\boldsymbol{x}_2 - \boldsymbol{z} \neq \boldsymbol{x}_2$, and this contradicts the uniqueness property. □

Corollary 2.10. *Let V be an $\mathbb{F}$-linear space and let V_1, V_2 be two subspaces of V. The decomposition of a vector $\boldsymbol{x} \in V$ as $\boldsymbol{x} = \boldsymbol{x}_1 + \boldsymbol{x}_2$, where $\boldsymbol{x}_1 \in V_1$ and $\boldsymbol{x}_2 \in V_2$, is unique if and only if $V_1 \cap V_2 = \{\mathbf{0}_V\}$.*

Proof. This statement follows from Theorems 2.29 and 2.30. □

If V_1 and V_2 are subspaces of the $\mathbb{F}$-linear space V and $V_1 \cap V_2 = \{\mathbf{0}_V\}$, we refer to $V_1 + V_2$ as the *direct sum* of the subspaces V_1 and V_2. The direct sum $V_1 + V_2$ is denoted by $V_1 \oplus V_2$.

Theorem 2.31. *Let V be an $\mathbb{F}$-linear space and U_0 be a subspace of V. There exists a subspace U_1 of V such that $V = U_0 \oplus U_1$.*

Proof. Let $\{\boldsymbol{e}_i \mid i \in I\}$ be a basis of the subspace U_0 and let $\{\boldsymbol{e}_i \mid i \in I \cup J\}$ be its completion to a basis of the entire linear space V, where $I \cap J = \emptyset$. If U_1 is the subspace generated by $\{\boldsymbol{e}_i \mid i \in J\}$, then $V = U_0 + U_1$. □

Subspaces of linear spaces are related to idempotent endomorphisms as shown next.

Theorem 2.32. *Let V be an $\mathbb{F}$-linear space. If $h \in \mathsf{End}(V)$ is an idempotent endomorphism of V, then $V = \mathsf{Ker}(h) \oplus \mathsf{Im}(h)$.*

Proof. Observe that both $\mathsf{Ker}(h)$ and $\mathsf{Im}(h)$ are subspaces of V. Furthermore, suppose that $\boldsymbol{x} \in \mathsf{Ker}(h) \cap \mathsf{Im}(h)$. Since $\boldsymbol{x} \in \mathsf{Im}(h)$, we have $\boldsymbol{x} = h(\boldsymbol{y})$ for some $\boldsymbol{y} \in V$. On the other hand, $\boldsymbol{x} \in \mathsf{Ker}(h)$ means that $h(\boldsymbol{x}) = \mathbf{0}_V$. Thus, $h(h(\boldsymbol{y})) = h(\boldsymbol{y})$ implies $h(\boldsymbol{x}) = \boldsymbol{x}$, which yields $\boldsymbol{x} = \mathbf{0}_V$. This allows us to conclude that $\mathsf{Ker}(h) \cap \mathsf{Im}(h) = \{\mathbf{0}_V\}$.

If $\boldsymbol{z} \in V$, we have $\boldsymbol{z} = (\boldsymbol{z} - h(\boldsymbol{z})) + h(\boldsymbol{z})$. Observe that $h(\boldsymbol{z}) \in \mathsf{Im}(h)$ and $h(\boldsymbol{z} - h(\boldsymbol{z})) = h(\boldsymbol{z}) - h(h(\boldsymbol{z})) = \mathbf{0}_V$, so $\boldsymbol{z} - h(\boldsymbol{z}) \in \mathsf{Ker}(h)$ because h is idempotent. We conclude that $V = \mathsf{Ker}(h) \oplus \mathsf{Im}(h)$. □

Theorem 2.33. *Let V be an $\mathbb{F}$-linear space. If U and W are two subspaces of V such that $V = U \oplus W$, then there exists an idempotent endomorphism h of V such that $U = \mathsf{Ker}(h)$ and $W = \mathsf{Im}(h)$.*

Proof. Since V is a direct sum of U and W, each $\boldsymbol{v} \in V$ can be uniquely written as $\boldsymbol{v} = \boldsymbol{u} + \boldsymbol{w}$, where $\boldsymbol{u} \in U$ and $\boldsymbol{w} \in W$.

Let π_1 and π_2 be the canonical projections and let h_1 and h_2 be the canonical injections of the subspaces U and W shown in what follows, where $h_1(\boldsymbol{u}) = \boldsymbol{u}$, $h_2(\boldsymbol{w}) = \boldsymbol{w}$ and $p_1(\boldsymbol{u} + \boldsymbol{w}) = \boldsymbol{u}$, $p_2(\boldsymbol{u} + \boldsymbol{w}) = \boldsymbol{w}$ for $\boldsymbol{u} \in U$ and $\boldsymbol{w} \in W$.

$$U \underset{h_1}{\overset{p_1}{\rightleftarrows}} V \underset{h_2}{\overset{p_2}{\rightleftarrows}} W$$

Define the endomorphism $g_1 \in \mathsf{End}(V)$ as $g_1 = h_1 p_1$. Note that $\boldsymbol{v} \in \mathsf{Ker}(g_1)$, where $\boldsymbol{v} = \boldsymbol{u} + \boldsymbol{w}$ for $\boldsymbol{u} \in U$ and $\boldsymbol{w} \in W$ if and only if $g_1(\boldsymbol{v}) = \mathbf{0}_V$. This, in turn, is equivalent to $h_1(p_1(\boldsymbol{v})) = h_1(\boldsymbol{u}) = \boldsymbol{u} = \mathbf{0}_V$, so $\boldsymbol{v} = \boldsymbol{w} \in W$. This shows that $\mathsf{Ker}(g_1) = W$.

On the other hand, $\boldsymbol{z} \in \mathsf{Im}(g_1)$ means that $\boldsymbol{z} = g_1(\boldsymbol{v}) = h_1 p_1(\boldsymbol{u} + \boldsymbol{v}) = h_1(\boldsymbol{u}) = \boldsymbol{u}$, so $\mathsf{Im}(g_1) = U$. It is immediate to verify that g_1 is idempotent. □

In Theorem 2.31, we saw that if U is a subspace of an $\mathbb{F}$-linear space V, there exists another subspace W of V such that $V = U \oplus W$. By Theorem 2.33, there exists an idempotent endomorphism h of V such that $U = \mathsf{Ker}(h)$ and $W = \mathsf{Im}(h)$.

Definition 2.21. If V is an $\mathbb{F}$-linear space and U, W are two subspaces of V such that $V = U \oplus W$, then we say that U, W are *complementary subspaces*.

Theorem 2.34. *Let V be an $\mathbb{F}$-linear space and let U, W be two subspaces of L. The following statements are equivalent:*

(i) *U, W are complementary spaces;*
(ii) *$V = \{\boldsymbol{u} + \boldsymbol{w} \mid \boldsymbol{u} \in U$ and $\boldsymbol{w} \in W\}$ and $U \cap W = \{\boldsymbol{0}\}$;*
(iii) *If B_U and B_W are two bases of U and W, respectively, then $B_U \cup B_W$ is a basis of V and $B_U \cap B_W = \emptyset$.*

Proof. (i) implies (ii): Suppose that $V = U \oplus W$. Then, every $\boldsymbol{v} \in V$ can be uniquely written as a sum $\boldsymbol{v} = \boldsymbol{u} + \boldsymbol{w}$ with $\boldsymbol{u} \in U$ and $\boldsymbol{w} \in W$, so $V = \{\boldsymbol{u} + \boldsymbol{w} \mid \boldsymbol{u} \in U \text{ and } \boldsymbol{w} \in W\}$. If $\boldsymbol{x} = \boldsymbol{u} + \boldsymbol{w}$, where $\boldsymbol{u} \in U$ and $\boldsymbol{w} \in W$, and $\boldsymbol{t} \in U \cap W$, $\boldsymbol{t} \neq \boldsymbol{0}_L$, then we can also write $\boldsymbol{x} = (\boldsymbol{u} - \boldsymbol{t}) + (\boldsymbol{t} + \boldsymbol{w})$, which contradicts the uniqueness of the representation of $\boldsymbol{x}$. Thus, $\boldsymbol{t} = \boldsymbol{0}_V$.

(ii) implies (iii): Since $U \cap W = \{\boldsymbol{0}_V\}$, we have $B_U \cap B_W = \emptyset$ because $\boldsymbol{0}_V$ does not belong to either B_U or B_W. Every $\boldsymbol{v} \in V$ can be written as $\boldsymbol{v} = \boldsymbol{u} + \boldsymbol{w}$, where $\boldsymbol{u} \in U$ and $\boldsymbol{w} \in W$. Thus, $\boldsymbol{u} = \sum\{a_i \boldsymbol{u}_i \mid a_i \in \mathbb{F} \text{ and } \boldsymbol{u}_i \in U, i \in I\}$, $\boldsymbol{w} = \sum\{b_i \boldsymbol{w}_i \mid b_i \in \mathbb{F} \text{ and } \boldsymbol{w}_i \in U, i \in I\}$, so $\boldsymbol{v} = \boldsymbol{u} + \boldsymbol{w}$ can be written as a linear combination of $B_U \cup B_W$.

(iii) implies (i): Suppose now that B_U and B_W are two bases of subspaces U and W, respectively, such that $B_U \cup B_W$ is a basis of V and $B_U \cap B_W = \emptyset$.

Since $B_U \cup B_W$ is a basis of V, it is clear that every $\boldsymbol{v} \in V$ is a sum of the form $\boldsymbol{u} + \boldsymbol{w}$, where $\boldsymbol{u} \in U$ and $\boldsymbol{w} \in W$. Suppose that $\boldsymbol{t} \in U \cap W$ and $\boldsymbol{t} \neq \boldsymbol{0}_L$. Then, $\boldsymbol{t}$ can be expressed both as a linear combination of B_U and as a linear combination of B_W, $\boldsymbol{u} = \sum_{i \in I} a_i \boldsymbol{u}_i = \sum_{j \in J} b_j \boldsymbol{w}_j$ and not all a_i and not all b_j equal 0. This implies $\sum_{i \in I} a_i \boldsymbol{u}_i - \sum_{j \in J} b_j \boldsymbol{w}_j = \boldsymbol{0}_V$, which contradicts the linear independence of the set $B_U \cup B_W$. □

A generalization of Theorem 2.34 is given next.

Theorem 2.35. *Let V be an $\mathbb{F}$-linear space and let U, W be two subspaces of V. The set $T = \{\boldsymbol{t} \in V \mid \boldsymbol{t} = \boldsymbol{u} + \boldsymbol{w}, \boldsymbol{u} \in U, \boldsymbol{w} \in W\}$ is a subspace of V and*

$$\dim(T) = \dim(U) + \dim(W) - \dim(U \cap W).$$

Proof. It is straightforward to verify that T is indeed a subspace of V. Suppose that $B_Z = \{\boldsymbol{z}_1, \ldots, \boldsymbol{z}_k\}$ is a basis for the subspace $Z = U \cap W$. By the Extension corollary (Corollary 2.3), B can be extended to a basis for U, $B_U = B_Z \cup \{\boldsymbol{u}_1, \ldots, \boldsymbol{u}_p\}$ and to a basis for W, $B_W = B_Z \cup \{\boldsymbol{w}_1, \ldots, \boldsymbol{w}_q\}$.

It is clear that $B = B_U \cup B_W$ generates the subspace T. The set B is linear independent. Indeed, suppose that

$$a_1\boldsymbol{z}_1 + \cdots + a_k\boldsymbol{z}_k + b_1\boldsymbol{u}_1 + \cdots + b_p\boldsymbol{u}_p + c_1\boldsymbol{w}_1 + \cdots + c_q\boldsymbol{w}_q = \boldsymbol{0}_V,$$

for some $a_1, \ldots, a_k, b_1, \ldots, b_p, c_1, \ldots, c_q \in \mathbb{F}$, so

$$c_1\boldsymbol{w}_1 + \cdots + c_q\boldsymbol{w}_q = -\left(a_1\boldsymbol{z}_1 + \cdots + a_k\boldsymbol{z}_k + b_1\boldsymbol{u}_1 + \cdots + b_p\boldsymbol{u}_p\right) \in U$$

and, of course, $c_1\boldsymbol{w}_1 + \cdots + c_q\boldsymbol{w}_q \in W$. Thus, $c_1\boldsymbol{w}_1 + \cdots + c_q\boldsymbol{w}_q \in U \cap W$, so

$$c_1\boldsymbol{w}_1 + \cdots + c_q\boldsymbol{w}_q = d_1\boldsymbol{z}_1 + \cdots + d_k\boldsymbol{z}_k,$$

which implies $c_1 = \cdots = c_q = d_1 = \cdots = d_k = 0$ because B_W is a basis for W. Using this fact, we obtain $a_1\boldsymbol{z}_1 + \cdots + a_k\boldsymbol{z}_k + b_1\boldsymbol{u}_1 + \cdots + b_p\boldsymbol{u}_p = \boldsymbol{0}_V$. Since B_U is a basis, this implies $a_1 = \cdots = a_k = b_1 = \cdots = b_p = 0$. We conclude that $B_U \cup B_W$ is indeed a basis for T and

$$\begin{aligned}\dim(T) = |B| &= |B_U| + |B_V| - |B_U \cap B_V| \\ &= \dim(U) + \dim(W) - \dim(U \cap W).\end{aligned}$$

□

Corollary 2.11. *Let U, W be two subspaces of an $\mathbb{F}$-linear space V with $\dim(V) = n$. We have*

$$\dim(U) + \dim(W) - \dim(U \cap W) \leqslant n.$$

Proof. The inequality follows from Theorem 2.35 by observing that $\dim(T) \leqslant \dim(L) = n$. □

Theorem 2.36. *An $\mathbb{F}$-linear space V is a direct sum of the family of subspaces $\{V_i \mid i \in I\}$ if and only if $V = \bigcup_{i \in I} V_i$ and for each $i \in I$ we have*

$$V_i \cap \bigcup_{j \in I - \{i\}} V_j = \{\boldsymbol{0}_L\}. \tag{2.4}$$

Proof. Suppose that V is the direct sum of the family of its subspaces $\{V_i \mid i \in I\}$. It is clear that $V = \bigcup_{i\in I} V_i$. If $\boldsymbol{z} \in V_i \cap \bigcup_{j\in I-\{i\}} V_j$, then $\boldsymbol{z} = \boldsymbol{x}_i$ for some $\boldsymbol{x}_i \in V_i$ and $\boldsymbol{z} = \boldsymbol{x}_{j_1} + \cdots \boldsymbol{x}_{j_\ell}$, where $j_p \neq i$ for $1 \leqslant p \leqslant \ell$. By the uniqueness of direct sum representation, we have $\boldsymbol{x}_i = \boldsymbol{0}_V$, and $\boldsymbol{z} = \boldsymbol{0}_V$, hence Equality (2.4) holds.

Conversely, suppose that $V = \bigcup_{i\in I} V_i$ and Equality (2.4) holds. Suppose that a vector $\boldsymbol{x} \in V$ can be written as

$$\boldsymbol{x} = \boldsymbol{x}_{j_1} + \cdots + \boldsymbol{x}_{j_n}$$

and as

$$\boldsymbol{x} = \boldsymbol{t}_{k_1} + \cdots + \boldsymbol{t}_{k_m},$$

where $\boldsymbol{x}_{j_p} \in V_{j_p}$ for $1 \leqslant p \leqslant n$ and $\boldsymbol{t}_{k_q} \in V_{k_q}$ for $1 \leqslant q \leqslant m$. Without loss of generality we may assume that $n = m$ (by adding to the expressions of $\boldsymbol{x}$ an appropriate number of zero terms). Thus, we can assume that

$$\boldsymbol{x} = \boldsymbol{x}_{i_1} + \cdots + \boldsymbol{x}_{i_n} = \boldsymbol{t}_{i_1} + \cdots + \boldsymbol{t}_{i_n},$$

where x_{i_ℓ} and $\boldsymbol{t}_{i_\ell}$ both belong to the subspace V_{i_ℓ} for $1 \leqslant \ell \leqslant n$. Therefore, $x_{i_\ell} - \boldsymbol{t}_{i_\ell} \in V_{i_\ell}$ for $1 \leqslant \ell \leqslant n$. Since

$$(\boldsymbol{x}_{i_1} - \boldsymbol{t}_{i_1}) + \cdots + (\boldsymbol{x}_{i_n} - \boldsymbol{t}_{i_n}) = \boldsymbol{0}_V,$$

if follows that each difference $\boldsymbol{x}_{i_r} - \boldsymbol{t}_{i_r}$ is a sum of vectors from other subspaces. This is possible if and only if $\boldsymbol{x}_{i_r} - \boldsymbol{t}_{i_r} = \boldsymbol{0}_V$, hence, $\boldsymbol{x}_{i_r} = \boldsymbol{t}_{i_r}$ for $1 \leqslant r \leqslant n$. □

If $\{\boldsymbol{x}_i \mid i \in I\}$ and $\{\boldsymbol{y}_j \mid j \in J\}$ are bases in the linear spaces V and W, respectively, then the set

$$\{(\boldsymbol{x}_i, \boldsymbol{0}_W) \mid i \in I\} \cup \{(\boldsymbol{0}_V, \boldsymbol{y}_j) \mid j \in J\}$$

is a basis in $V + W$, as the reader can easily verify. If both V and W are of finite type, then so is $V + W$ and $\dim(V + W) = \dim(V) + \dim(W)$.

The direct sum of two linear spaces allows us to introduce four linear mappings: the *injections* $i_1 : V \longrightarrow V+W$, $i_2 : W \longrightarrow V+W$, and the *projections* $p_1 : V+W \longrightarrow V$ and $p_2 : V+W \longrightarrow W$ given by

$$\begin{aligned} i_1(\boldsymbol{x}) &= (\boldsymbol{x}, \mathbf{0}_W), & i_2(\boldsymbol{y}) &= (\mathbf{0}_V, \boldsymbol{y}), \\ p_1(\boldsymbol{x}, \boldsymbol{y}) &= \boldsymbol{x}, & p_2(\boldsymbol{x}, \boldsymbol{y}) &= \boldsymbol{y}, \end{aligned}$$

for $\boldsymbol{x} \in V$ and $\boldsymbol{y} \in W$. It is immediate to verify that i_1, i_2 are injective and p_1, p_2 are surjective mappings.

The definitions of i_1, i_2, p_1, and p_2 imply immediately

$$\begin{aligned} p_1 i_1 &= 1_V & p_2 i_2 &= 1_W, \\ p_1 i_2 &= 0_W & p_2 i_1 &= 0_V. \end{aligned}$$

Additionally, we have

$$\begin{aligned} (i_1 p_1 + i_2 p_2)(\boldsymbol{x}, \boldsymbol{y}) &= i_1 p_1(\boldsymbol{x}, \boldsymbol{y}) + i_2 p_2(\boldsymbol{x}, \boldsymbol{y}) \\ &= (\boldsymbol{x}, \mathbf{0}_W) + (\mathbf{0}_V, \boldsymbol{y}) = (\boldsymbol{x}, \boldsymbol{y}), \end{aligned}$$

for every $(\boldsymbol{x}, \boldsymbol{y}) \in V \times W$.

Theorem 2.37. *Let V, W, U be three linear vector spaces for which there exist the linear mappings $h_1 : V \longrightarrow U$, $h_2 : W \longrightarrow U$ and $g_1 : U \longrightarrow V$, $g_2 : U \longrightarrow W$ such that*

$$\begin{aligned} g_1 h_1 &= 1_V, & g_2 h_2 &= 1_W, \\ g_1 h_2 &= 0_{W,V}, & g_2 h_1 &= 0_{V,W}, \end{aligned}$$

and $h_1 g_1 + h_2 g_2 = 1_U$. Then, there exists an isomorphism $h : V + W \longrightarrow U$ such that

$$\begin{aligned} h_1 &= h i_1 & g_1 &= p_1 h^{-1}, \\ h_2 &= h i_2 & g_2 &= p_2 h^{-1}. \end{aligned}$$

Proof. The linear mappings mentioned above are shown in the following commutative diagram.

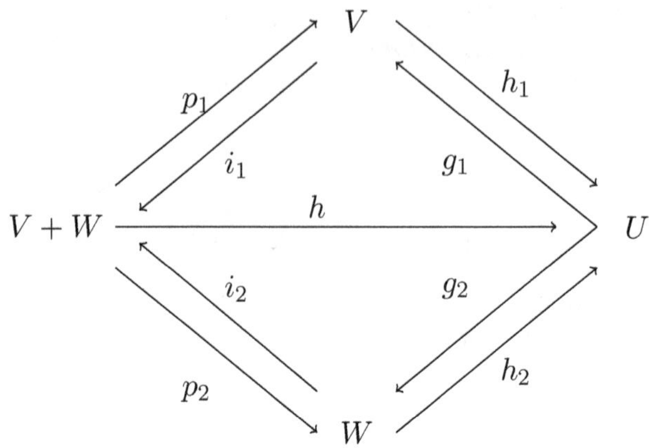

Define the linear mappings $k : U \longrightarrow V + W$ as $k(\boldsymbol{z}) = (g_1(\boldsymbol{z}), g_2(\boldsymbol{z}))$ for $\boldsymbol{z} \in U$ and $\ell : V + W \longrightarrow U$ as $\ell(\boldsymbol{x}, \boldsymbol{y}) = h_1(\boldsymbol{x}) + h_2(\boldsymbol{y})$ for $\boldsymbol{x} \in V$ and $\boldsymbol{y} \in W$.

Note that

$$\begin{aligned}\ell(k(\boldsymbol{z})) &= \ell(g_1(\boldsymbol{z}), g_2(\boldsymbol{z})) = h_1(g_1(\boldsymbol{z})) + h_2(g_2(\boldsymbol{z})) \\ &= (h_1g_1 + h_2g_2)(\boldsymbol{z}) = \boldsymbol{z},\end{aligned}$$

and

$$\begin{aligned}k(\ell(\boldsymbol{x}, \boldsymbol{y})) &= k(h_1(\boldsymbol{x}) + h_2(\boldsymbol{y})) = k(h_1(\boldsymbol{x})) + k(h_2(\boldsymbol{y})) \\ &= (g_1(h_1(\boldsymbol{x})), g_2(h_1(\boldsymbol{x}))) + (g_1(h_2(\boldsymbol{y})), g_2(h_2(\boldsymbol{y}))) \\ &= (\boldsymbol{x}, 0_{L+M}) + (0_{M+L}, \boldsymbol{y}) = (\boldsymbol{x}, \boldsymbol{y}).\end{aligned}$$

This shows that the linear mappings ℓ and k are inverse isomorphisms. □

2.8 Dual Linear Spaces

Let V be an $\mathbb{F}$-linear space. The set of linear forms defined on V is denoted by V^*. This set has the natural structure of an $\mathbb{F}$-linear space known as the *dual of the space* V. The elements of V^* are also referred to as *covariant vectors* or *covectors*.

We will refer to the vectors of the original linear space as *contravariant vectors*. The reason for adopting the designation of "covariant" and "contravariant" terms for the vectors of V^* and V, respectively, will be discussed in Section 3.7.

Definition 2.22. Let $B = \{\boldsymbol{u}_i \in V \mid 1 \leqslant i \leqslant n\}$ be a basis in an n-dimensional $\mathbb{F}$-linear space V, and let $\boldsymbol{f} \in V^*$ be a covector. The numbers $a_i = \boldsymbol{f}(\boldsymbol{u}_i)$, where $1 \leqslant i \leqslant n$ are the *components* of the covector $\boldsymbol{f}$ relative to the basis B of V.

Theorem 2.38. *Let $B = \{\boldsymbol{u}_i \in L \mid 1 \leqslant i \leqslant n\}$ be a basis in an n-dimensional $\mathbb{F}$-linear space V. If $\{a_i \in \mathbb{F} \mid 1 \leqslant i \leqslant n\}$ is a set of scalars, then there is a unique covector $\boldsymbol{f} \in V^*$ such that $\boldsymbol{f}(\boldsymbol{u}_i) = a_i$ for $1 \leqslant i \leqslant n$.*

Proof. Since B is a basis in V, we can write $\boldsymbol{v} = \sum_{i=1}^{n} c_i \boldsymbol{u}_i$ for every $\boldsymbol{v} \in V$. Thus,

$$\boldsymbol{f}(\boldsymbol{v}) = \boldsymbol{f}\left(\sum_{i=1}^{n} c_i \boldsymbol{u}_i\right) = \sum_{i=1}^{n} c_i a_i,$$

which shows that the covector $\boldsymbol{f}$ is uniquely determined by the n-tuple of scalars $\boldsymbol{a} = \begin{pmatrix} a_1 \\ \vdots \\ a_n \end{pmatrix}$. □

Let $B = \{\boldsymbol{u}_i \in V \mid 1 \leqslant i \leqslant n\}$ $\tilde{B} = \{\tilde{\boldsymbol{u}}_i \in V \mid 1 \leqslant i \leqslant n\}$ be two bases in the linear space V, where $\tilde{\boldsymbol{u}}_i = \sum_{j=1}^{n} c_{ij} \boldsymbol{u}_j$ for $1 \leqslant i \leqslant n$. The components of a covector $\boldsymbol{f} \in V^*$ in the basis $\tilde{B}$ can be written as

$$\boldsymbol{f}(\tilde{u}_i) = \boldsymbol{f}\left(\sum_{j=1}^{n} c_{ij} \boldsymbol{u}_j\right) = \sum_{j=1}^{n} c_{ij} \boldsymbol{f}(\boldsymbol{u}_j).$$

This equality shows that the components of a covector transform in the same manner as the basis. This justifies the use of the term *covariant* applied to these components.

Theorem 2.39. *Let V be an n-dimensional $\mathbb{F}$-linear space. Then, its dual V^* is isomorphic to $\mathbb{F}^n$, and, thus,* $\dim(V^*) = \dim(V) = n$.

Proof. The function $h : \mathbb{F}^n \longrightarrow \mathsf{Hom}(V, \mathbb{F})$ that maps the vector

$$\boldsymbol{a} = \begin{pmatrix} a_1 \\ \vdots \\ a_n \end{pmatrix}$$

to the function $\boldsymbol{f}$ defined as in Theorem 2.38 is an isomorphism, as it can be checked easily. □

Theorem 2.38 shows that a linear form $\boldsymbol{f} \in V^*$ is uniquely determined by its values on the basis of the space V. This allows us to prove the following extension theorem.

Theorem 2.40. *Let U be a subspace of a finite-dimensional $\mathbb{F}$-linear space V. A linear function $g : U \longrightarrow \mathbb{F}$ belongs to U^* if and only if there exists a linear form $\boldsymbol{f} \in V^*$ such that g is the restriction of $\boldsymbol{f}$ to U.*

Proof. If g is the restriction of $\boldsymbol{f}$ to U, then it is immediate that $g \in U^*$. Conversely, let $g \in U^*$ and let $B = \{\boldsymbol{u}_1, \ldots, \boldsymbol{u}_p\}$ be a basis of U, where $\dim(U) = p$. Consider an extension of B to a basis of the entire space $B_1 = \{\boldsymbol{u}_1, \ldots, \boldsymbol{u}_p, \boldsymbol{u}_{p+1}, \ldots, \boldsymbol{u}_n\}$, where $n = \dim(L)$ and define the linear form $\boldsymbol{f} : V \longrightarrow \mathbb{F}$ by

$$\boldsymbol{f}(\boldsymbol{u}_i) = \begin{cases} g(\boldsymbol{u}_i) & \text{if } i \leqslant p, \\ 0 & \text{if } p+1 \leqslant i \leqslant n. \end{cases}$$

Since $\boldsymbol{f}$ and g coincide for all members of the basis of U, it follows that g is the restriction of $\boldsymbol{f}$ to U. □

We refer to $\boldsymbol{f}$ as the *zero-extension* of the linear form g defined on the subspace U.

Theorem 2.41. *If $B = \{\boldsymbol{v}_1, \ldots, \boldsymbol{v}_n\}$ is a basis of the $\mathbb{F}$-linear space V, then the set of linear forms $F = \{\boldsymbol{f}^j \mid 1 \leqslant j \leqslant n\}$ defined by*

$$\boldsymbol{f}^j(\boldsymbol{v}_i) = \begin{cases} 1 & \text{if } i = j, \\ 0 & \text{otherwise} \end{cases}$$

is a basis of the dual linear space V^*.

Proof. The set $F = \{\boldsymbol{f}^1, \ldots, \boldsymbol{f}^n\}$ spans the entire dual space V^*. Indeed, let $\boldsymbol{f} \in V^*$ be defined by $\boldsymbol{f}(\boldsymbol{v}_i) = a_i$ for $1 \leqslant i \leqslant n$. If $\boldsymbol{v} = \sum_{i=1}^{n} c^i \boldsymbol{v}_i$, then

$$\boldsymbol{f}(\boldsymbol{v}) = \boldsymbol{f}\left(\sum_{i=1}^{n} c^i \boldsymbol{v}_i\right) = \sum_{i=1}^{n} c^i \boldsymbol{f}(\boldsymbol{v}_i) = \sum_{i=1}^{n} c^i a_i.$$

On the other hand,

$$\sum_{i=1}^{n} a_i \boldsymbol{f}^i(\boldsymbol{v}) = \sum_{i=1}^{n} a_i c^j \boldsymbol{f}^i(\boldsymbol{v}_j) = \sum_{i=1}^{n} a_i c^i,$$

due to the definition of the linear forms $\boldsymbol{f}_1, \ldots, \boldsymbol{f}_n$. Therefore, $\boldsymbol{f} = a_1 \boldsymbol{f}^1 + \cdots + a_n \boldsymbol{f}^n$, which shows that $\langle F \rangle = L^*$.

To prove that the set F is linearly independent in V^*, suppose that $a_1 \boldsymbol{f}^1 + \cdots + a_n \boldsymbol{f}_n = \boldsymbol{0}_{V^*}$. This implies $a_1 \boldsymbol{f}^1(\boldsymbol{v}) + \cdots + a_n \boldsymbol{f}^n(\boldsymbol{v}) = \boldsymbol{0}_V$ for every $\boldsymbol{v} \in V$. Choosing $\boldsymbol{v} = \boldsymbol{v}_j$, we obtain $a_j \boldsymbol{f}^j(\boldsymbol{v}_j) = 0$, hence $a_j = 0$, and this can be shown for $1 \leqslant j \leqslant n$, which implies the linear independence. □

The basis $F = \{\boldsymbol{f}^1, \ldots, \boldsymbol{f}^n\}$ of V^* constructed in Theorem 2.41 is the *dual basis* of the basis $B = \{\boldsymbol{v}_1, \ldots, \boldsymbol{v}_n\}$ of V. We refer to the pair (B, F) as a pair of *dual bases.*

In general, we will index vectors in a linear space V using subscripts and covectors in the dual linear space using superscripts.

Corollary 2.12. *The dual of an n-dimensional $\mathbb{F}$-linear space V is an n-dimensional linear space.*

Proof. This statement follows immediately from Theorem 2.41. □

Let $E = \{\boldsymbol{e}_1, \ldots, \boldsymbol{e}_n\}$ be a basis of V and let $F = \{\boldsymbol{f}^1, \ldots, \boldsymbol{f}^n\}$ be its dual basis in V^*. If $\boldsymbol{x} = a_1 \boldsymbol{e}_1 + \cdots + a_n \boldsymbol{e}_n$, then, we have

$$\boldsymbol{f}^i(\boldsymbol{x}) = \boldsymbol{f}^i(a_1 \boldsymbol{e}_1 + \cdots + a_n \boldsymbol{e}_n) = \sum_{j=1}^{n} a_j \boldsymbol{f}^i(\boldsymbol{e}_j) = a_i \tag{2.5}$$

for $1 \leqslant i \leqslant n$. Thus, we have

$$\boldsymbol{x} = \sum_{i=1}^{n} \boldsymbol{f}^i(\boldsymbol{x}) \boldsymbol{e}_i. \tag{2.6}$$

Similarly, if $\boldsymbol{f} \in V^*$ and $\boldsymbol{f} = b_1\boldsymbol{f}^1 + \cdots + b_n\boldsymbol{f}^n$, then $(\boldsymbol{f}, \boldsymbol{e}_i) = b_i$, and

$$\boldsymbol{f} = \sum_{i=1}^{n} (\boldsymbol{f}, \boldsymbol{e}_i)\boldsymbol{f}^i. \tag{2.7}$$

Example 2.19. Let $P_2[x]$ be the linear space of polynomials of degree 2 in x, that consists of polynomials of the form $p(x) = ax^2 + bx + c$. The set $\{p_0, p_1, p_2\}$ given by $p_0(x) = 1$, $p_1(x) = x$ and $p_2(x) = x^2$ is a basis in $P_2[x]$. Note that, we have

$$\begin{aligned} c &= p(0), \\ b &= \frac{1}{2}(p(1) - p(-1)), \\ a &= \frac{1}{2}(p(1) + p(-1) - 2p(0)). \end{aligned}$$

If $f : P_2[x] \longrightarrow \mathbb{R}$ is a linear form, we have

$$\begin{aligned} f(p) &= af(x^2) + bf(x) + cf(0) \\ &= \frac{1}{2}(p(1) + p(-1) - 2p(0))f(x^2) \\ &\quad + \frac{1}{2}(p(1) - p(-1))f(x) + p(0)f(1). \end{aligned}$$

Therefore, a basis in $P_2[x]^*$ consists of the functions

$$\begin{aligned} f^0(p) &= p(0), \\ f^1(p) &= \frac{1}{2}(p(1) - p(-1)), \\ f^2(p) &= \frac{1}{2}(p(1) + p(-1) - 2p(0)). \end{aligned}$$

Theorem 2.42. *Let V be a finite-dimensional $\mathbb{F}$-linear space and let $\boldsymbol{v} \in V - \{\mathbf{0}_V\}$. There exists a linear form $\boldsymbol{f}_{\boldsymbol{v}}$ in V^* such that $\boldsymbol{f}_{\boldsymbol{v}}(\boldsymbol{v}) = 1$.*

Proof. Since $\boldsymbol{v} \neq \mathbf{0}_V$, there is a basis B in V that includes $\boldsymbol{v}$. Then $\boldsymbol{f}_{\boldsymbol{v}}$ can be defined as the linear form that corresponds to $\boldsymbol{v}$. □

We saw that the dual V^* of an $\mathbb{F}$-linear space V is an $\mathbb{F}$-linear space. The construction of the dual may be repeated, and V^{**}, the dual of the dual $\mathbb{F}$-linear space V^* is an $\mathbb{F}$-linear space. In the case of finite-dimensional linear spaces, we have $\dim(V^{**}) = \dim(V^*) = \dim(V)$, and all these spaces are isomorphic.

Theorem 2.43. *Let V be a finite-dimensional $\mathbb{F}$-linear space. Then, the dual V^{**} of the dual V^* of L is an $\mathbb{F}$-linear space isomorphic to V.*

Proof. The space V^{**} consists of linear functions of the form $\phi : V^* \longrightarrow F$ that map linear forms $\boldsymbol{f} \in V^*$ into scalars in F.

Let $\Psi : V \longrightarrow V^{**}$ be the mapping given by $\Psi(\boldsymbol{v}) = \phi$, where ϕ is the linear form defined by $\phi(\boldsymbol{f}) = \boldsymbol{f}(\boldsymbol{v})$ for all $\boldsymbol{v} \in V$ and $\boldsymbol{f} \in V^*$.

The mapping Ψ is injective. Indeed, suppose that $\Psi(\boldsymbol{u}) = \Psi(\boldsymbol{v}) = \phi$ for $\boldsymbol{u}, \boldsymbol{v} \in V$. This implies $\phi(\boldsymbol{f}) = \boldsymbol{f}(\boldsymbol{u}) = \boldsymbol{f}(\boldsymbol{v})$ for every $\boldsymbol{f} \in V^*$, which, in turn, implies $\boldsymbol{u} = \boldsymbol{v}$.

Suppose that $\Psi(\boldsymbol{v}) = \mathbf{0}_{V^{**}}$. Then, $\boldsymbol{f}(\boldsymbol{v}) = 0$ for every $\boldsymbol{f}$, which implies $\boldsymbol{v} = \mathbf{0}_V$ (by taking $f = 1_V$). Thus, by Theorem 2.16, $\mathsf{Ker}(\Psi) = \{\mathbf{0}_V\}$, which implies that Ψ is an isomorphism. □

Theorem 2.43 allows us to identify V^{**} with V when necessary.

Definition 2.23. Let V be an $\mathbb{F}$-linear space let U be a subset of V. The *annihilator* of U is the subset $\mathsf{Ann}(U)$ of the dual space V^* given by

$$\begin{aligned}\mathsf{Ann}(U) &= \{\boldsymbol{f} \in V^* \mid \boldsymbol{f}(\boldsymbol{x}) = 0 \text{ for every } \boldsymbol{x} \in U\} \\ &= \{\boldsymbol{f} \in V^* \mid U \subseteq \mathsf{Ker}(\boldsymbol{f})\}.\end{aligned}$$

The annihilator of a subset T of the dual space V^* of a linear space V is the set

$$\begin{aligned}\mathsf{Ann}(T) &= \{\boldsymbol{v} \in V \mid \boldsymbol{f}(\boldsymbol{v}) = 0 \text{ for every } \boldsymbol{f} \in T\} \\ &= \bigcap\{\mathsf{Ker}(\boldsymbol{f}) \mid \boldsymbol{f} \in T\}.\end{aligned}$$

Theorem 2.44. *Let V be an $\mathbb{F}$-linear space. For any subsets $U \subseteq V$ and $T \subseteq V^*$, $\mathsf{Ann}(U)$ and $\mathsf{Ann}(T)$ are subspaces of V^* and V, respectively.*

Proof. Let $\boldsymbol{f}, \tilde{\boldsymbol{f}} \in \mathsf{Ann}(U)$ and let $a, b \in \mathbb{F}$. We have $(a\boldsymbol{f}+b\tilde{\boldsymbol{f}})(\boldsymbol{x}) = a\boldsymbol{f}(\boldsymbol{x}) + b\tilde{\boldsymbol{f}}(\boldsymbol{x}) = 0$ for every $\boldsymbol{x} \in U$, so $a\boldsymbol{f} + b\tilde{\boldsymbol{f}} \in \mathsf{Ann}(U)$. The argument for $\mathsf{Ann}(T)$ is similar. □

Theorem 2.45. *If V is a finite-dimensional $\mathbb{F}$-linear space, let $U \subseteq V$ and $T \subseteq V^*$ be subspaces of V and V^*, respectively. The dual space U^* is isomorphic to the quotient space $V^*/\mathsf{Ann}(U)$ and the dual space T^* is isomorphic to the quotient space $V^{**}/\mathsf{Ann}(T)$.*

The quotient linear spaces $(V/U)^*$ and $(V^*/T)^*$ are isomorphic to the subspaces $\mathsf{Ann}(U)$ and $\mathsf{Ann}(T)$, respectively.

Proof. Recall that the zero-extension of a linear form defined on a subspace U of V was introduced in Theorem 2.40. Let $\Phi : V^* \longrightarrow U^*$, where $\Phi(\boldsymbol{f}) = \boldsymbol{g}$ if $\boldsymbol{f}$ is the zero-extension of $\boldsymbol{g}$ for $\boldsymbol{g} \in U^*$. It is easy to verify that Φ is a surjective linear mapping. Furthermore, $\mathsf{Ker}(\Phi)$ consists of those linear forms $\boldsymbol{f}$ in V^* such that $\Phi(\boldsymbol{f})$ is the zero linear form on U, so $\boldsymbol{f}(\boldsymbol{x}) = 0$ for every $\boldsymbol{x} \in U$. Consequently, $\mathsf{Ker}(\Phi) = \mathsf{Ann}(U)$. By Corollary 2.5, the quotient space $V * /\mathsf{Ann}(U)$ is isomorphic to U^*.

Proving that T^* is isomorphic to $V^{**}/\mathsf{Ann}(T)$ is left to the reader.

For the second part of the theorem, define a mapping $\Psi : (V/U)^* \longrightarrow \mathsf{Ann}(U)$ as follows. Consider a linear form $\phi \in \mathsf{Hom}(V/U, F)$ and the canonical mapping $h_U : V \longrightarrow V/U$, and define $\Psi(\phi) = \phi h_U$. Observe that $\phi(h_U(\boldsymbol{x})) = 0$ for $\boldsymbol{x} \in U$, so $\phi h_U \in \mathsf{Ann}(U)$. It is immediate that Ψ is a linear mapping.

If $\boldsymbol{g} \in \mathsf{Ann}(U)$, this means that $\mathsf{Ker}(h_U) = U \subseteq \mathsf{Ker}(\boldsymbol{g})$, so by Theorem 2.25, since h_U is surjective, there exists a unique factorization $\boldsymbol{g} = k h_U$, so $k = \phi$ and $\boldsymbol{g} = \Psi(\phi)$. Thus, Ψ is an isomorphism.

We leave to the reader the proof of the isomorphism between $(V^*/T)^*$ and $\mathsf{Ann}(T)$. □

Definition 2.24. Let V, W be two $\mathbb{F}$-linear spaces and let $h \in \mathsf{Hom}(V, W)$ be a linear mapping. The *dual* of h is the morphism $h^* \in \mathsf{Hom}(W^*, V^*)$ defined by $h^*(\boldsymbol{f}) = \boldsymbol{f}h$, shown in the commutative diagram that follows.

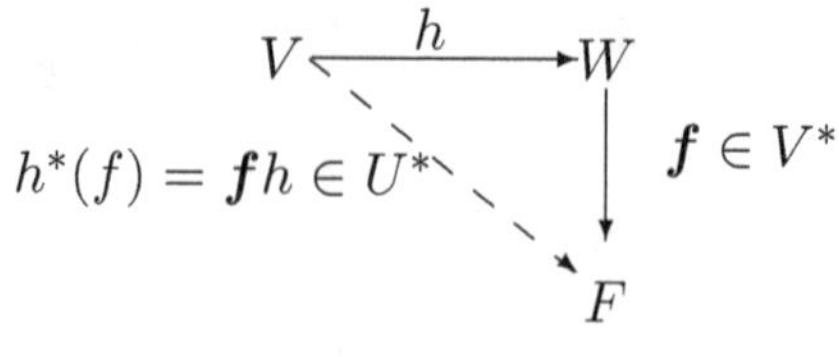

To verify the correctness of Definition 2.24, we need to show that h^* is indeed a morphism between the linear spaces W^* and V^*.

Let $\boldsymbol{f}, \boldsymbol{g} \in W^*$ and let $a, b \in \mathbb{F}$. The linear form $\phi = a\boldsymbol{f} + b\boldsymbol{g}$ is defined by $\phi(\boldsymbol{v}) = a\boldsymbol{f}(\boldsymbol{v}) + b\boldsymbol{g}(\boldsymbol{v})$ for $\boldsymbol{v} \in W$. Then, $h^*(\phi) = h^*(a\boldsymbol{f} + b\boldsymbol{g}) = (a\boldsymbol{f} + b\boldsymbol{g})h$. Since $(a\boldsymbol{f} + b\boldsymbol{g})h(\boldsymbol{u}) = a\boldsymbol{f}(h(\boldsymbol{u})) + b(\boldsymbol{g}(h(\boldsymbol{u}) = a(h^*\boldsymbol{f})(\boldsymbol{u}) + b(h^*\boldsymbol{g})(\boldsymbol{u})$ for $\boldsymbol{u} \in U$, it follows that $h^*(a\boldsymbol{f} + b\boldsymbol{g}) = ah^*\boldsymbol{f} + bh^*\boldsymbol{g}$, which proves that h^* is a morphism.

Theorem 2.46. *If V, W are two $\mathbb{F}$-linear spaces, the mapping $\Phi : \mathsf{Hom}(V, W) \longrightarrow \mathsf{Hom}(W^*, V^*)$ defined by $\Phi(h) = h^*$ for $h \in \mathsf{Hom}(V, W)$ is linear.*

Proof. Let $h_0, h_1 \in \mathsf{Hom}(V, W)$ and let $a, b \in \mathbb{F}$. We need to prove that $(ah_0 + bh_1)^* = ah_0^* + bh_1^*$.

Let $f : W \longrightarrow \mathbb{F}$ be a linear form. By the definition of dual morphisms, we have

$$(ah_0 + bh_1)^*(f) = f(ah_0 + bh_1).$$

Therefore, for every $\boldsymbol{v} \in W$, we have

$$\begin{aligned}(ah_0 + bh_1)^*(f)(\boldsymbol{v}) &= f(ah_0(\boldsymbol{v}) + bh_1(\boldsymbol{v})) = afh_0(\boldsymbol{v}) + bfh_1(\boldsymbol{v}) \\ &= ah_0^*(f)(\boldsymbol{v}) + bh_1^*(\boldsymbol{v}),\end{aligned}$$

which proves that Φ is a linear mapping. □

We leave it to the reader to verify that the dual of 1_V is 1_{V^*} and that for any two linear mappings $h_0 \in \mathsf{Hom}(U, V)$ and $h_1 \in \mathsf{Hom}(V, W)$, we have $(h_1h_0)^* = h_1^*h_0^*$.

Theorem 2.47. *Let V be a finite-dimensional $\mathbb{F}$-linear space and let Z be a subspace of V. Then $\dim(\mathsf{Ann}(Z)) = \dim(V) - \dim(Z)$ and $\mathsf{Ann}(\mathsf{Ann}(Z)) = Z$.*

Proof. If $\dim(V) = n$ and Z is a subspace of V, then $\dim(V/Z) = n - \dim(Z)$ by Theorem 2.17. Since $(V/Z)^* \cong \mathsf{Ann}(Z)$, it follows that $\dim(\mathsf{Ann}(Z)) = \dim(V/Z)^* = n - \dim(Z)$. Thus, $\dim(V) = \dim(Z) + \dim(\mathsf{Ann}(Z))$.

For the second part of the theorem, observe that by Definition 2.23 we have

$$\begin{aligned}\mathsf{Ann}(\mathsf{Ann}(Z)) &= \bigcap\{\mathsf{Ker}(f) \mid f \in \mathsf{Ann}(Z)\}\\ &= \bigcap\{\mathsf{Ker}(f) \mid Z \subseteq \mathsf{Ker}(f)\},\end{aligned}$$

which implies $Z \subseteq \mathsf{Ann}(\mathsf{Ann}(Z))$.

Note that both Z and $\mathsf{Ann}(\mathsf{Ann}(Z))$ are subspaces of V. The first part of the theorem implies $\dim(\mathsf{Ann}(\mathsf{Ann}(Z))) = \dim(V) - \dim(\mathsf{Ann}(Z))$. On the other hand, we saw that $\dim(Z) + \dim(\mathsf{Ann}(Z)) = \dim(V)$, so $\dim(Z) = \dim(\mathsf{Ann}(\mathsf{Ann}(Z))$. This implies $\mathsf{Ann}(\mathsf{Ann}(Z)) = Z$. □

Theorem 2.48. *Let V and W be two finite-dimensional $\mathbb{F}$-linear spaces. If $h \in \mathsf{Hom}(V, W)$, then $\mathsf{rank}(h) = \mathsf{rank}(h^*)$.*

Proof. Note that if $f \in \mathsf{Ker}(h^*)$ is equivalent to $h^*f(\boldsymbol{u}) = 0$, it is equivalent to having $f(h(\boldsymbol{u})) = 0$ for every $\boldsymbol{u} \in V$, that is, with $f(\mathsf{Im}(h)) = 0$. Thus, $\mathsf{Ker}(h^*) = \mathsf{Ann}(\mathsf{Im}(h))$, so $\dim(\mathsf{Ker}(h^*)) = \dim(\mathsf{Ann}(\mathsf{Im}(h)))$.

Since

$$\begin{aligned}\dim(W) &= \dim(\mathsf{Ker}(h^*)) + \dim(\mathsf{Im}(h^*))\\ &= \dim(\mathsf{Ann}(\mathsf{Im}(h)) + \dim(\mathsf{Im}(h^*)),\end{aligned}$$

it follows that $\dim(\mathsf{Im}(h^*)) = \dim(W) - \dim(\mathsf{Ann}(\mathsf{Im}(h)) = \dim(\mathsf{Im}(h))$, by Theorem 2.47. Thus, $\mathsf{rank}(h^*) = \mathsf{rank}(h)$. □

2.9 Topological Linear Spaces

We are examining now the interaction between the algebraic structure of linear spaces and topologies that can be defined on linear spaces that are compatible in a certain sense with the algebraic structure. Compatibility, in this case, is defined as the continuity of addition and scalar multiplication.

Definition 2.25. Let $\mathbb{F}$ be the real field $\mathbb{R}$ or complex field $\mathbb{C}$. An $\mathbb{F}$-*topological linear space* is a topological space $(V, \mathcal{O})$ such that

(i) V is an $\mathbb{F}$-linear space;
(ii) the vector addition is a continuous function between V^2 and V;
(iii) the scalar multiplication is a continuous function between $F \times V$ and V.

Unless stated otherwise, we assume that the field F is either the real or the complex field.

Theorem 2.49. *Let* $(V, \mathcal{O})$ *be an* $\mathbb{F}$-topological linear space *and let* $\boldsymbol{z} \in V$. *The translation mapping* $t_{\boldsymbol{z}} : V \longrightarrow V$ *is a homeomorphism.*

Proof. It is immediate that $t_{\boldsymbol{z}}$ is a bijection whose inverse is $t_{-\boldsymbol{z}}$. The continuity of both $t_{\boldsymbol{z}}$ and $t_{-\boldsymbol{z}}$ follows from the continuity of the vector addition of V. □

Definition 2.26. The mapping $t_{\boldsymbol{z}}$ introduced in Theorem 2.49 is the *translation generated by* $\boldsymbol{z}$.

Example 2.20. If $a \neq 0$, then each homothety h_a of a topological linear space $(V, \mathcal{O})$ is a homeomorphism. Indeed, the inverse of h_a is $h_{a^{-1}}$. The continuity of both h_a and $h_{a^{-1}}$ follows from the continuity of the scalar multiplication of V.

Theorem 2.50. *Let* $(V, \mathcal{O})$ *be a topological linear space. If* W *is a neighborhood of* $\mathbf{0}_V$, *then* $t_{\boldsymbol{x}}(W)$ *is a neighborhood of* $\boldsymbol{x}$. *Moreover, every neighborhood of* x *can be obtained by a translation of a neighborhood of* $\mathbf{0}$.

Proof. Since W is a neighborhood of the origin, there exists an open subset L of V such that $\mathbf{0} \in L \subseteq W$. This implies $\boldsymbol{x} = t_{\boldsymbol{x}}(\mathbf{0}) \in t_{\boldsymbol{x}}(L) \subseteq t_{\boldsymbol{x}}(W)$. It follows that $t_{\boldsymbol{x}}(L)$ is an open set and this, in turn, implies that $t_{\boldsymbol{x}}(W)$ is a neighborhood of $\boldsymbol{x}$.

Conversely, let U be a neighborhood of $\boldsymbol{x}$ and let K be an open set such that $\boldsymbol{x} \in K \subseteq U$. Then, we have $\mathbf{0} = t_{-\boldsymbol{x}}(\boldsymbol{x}) \in t_{-\boldsymbol{x}}(K) \subseteq t_{-\boldsymbol{x}}(U)$. Since $t_{-\boldsymbol{x}}(K)$ is an open set, it follows that $t_{-\boldsymbol{x}}(K)$ is a neighborhood of $\mathbf{0}$ and the desired conclusion follows from the fact that $U = t_{\boldsymbol{x}}(t_{-\boldsymbol{x}}(U))$. □

Theorem 2.50 shows that in a topological linear space $(V, \mathcal{O})$ the neighborhoods of any point are obtained by translating the neighborhoods of $\mathbf{0}_V$.

Corollary 2.13. *If $\mathcal{F}_x$ is a fundamental system of neighborhoods of x in the topological linear space $(V, \mathcal{O})$, then $\mathcal{F}_x$ can be obtained by a translation of a fundamental system of neighborhoods $\mathcal{F}_0$ of $\mathbf{0}_V$.*

Proof. This statement follows immediately from Theorem 2.50. □

The next theorem shows that a linear function between two topological linear spaces is continuous if and only if it is continuous in the zero element of the first space.

Theorem 2.51. *Let $(V_1, \mathcal{O}_1)$ and $(V_2, \mathcal{O}_2)$ be two topological $\mathbb{F}$-linear spaces having $\mathbf{0}_1$ and $\mathbf{0}_2$ as zero elements, respectively. A linear operator $f \in \mathsf{Hom}(V_1, V_2)$ is continuous in $\boldsymbol{x} \in V_1$ if and only if it is continuous in $\mathbf{0}_1 \in V_1$.*

Proof. Let f be a function that is continuous in a point $\boldsymbol{x} \in V_1$. If $U \in neigh_{\mathbf{0}_2}(\mathcal{O}_2)$, then $f(\boldsymbol{x}) + U$ is a neighborhood of $f(\boldsymbol{x})$. Since f is continuous, there exists a neighborhood W of $\boldsymbol{x}$ such that $f(W) \subseteq f(\boldsymbol{x}) + U$.

Observe that the set $-\boldsymbol{x} + W$ is a neighborhood of $\mathbf{0}_1$. Moreover, any neighborhood of $\mathbf{0}_1$ has this form. If $\boldsymbol{t} \in -\boldsymbol{x}+W$, then $\boldsymbol{t}+\boldsymbol{x} \in W$ and, therefore, $f(\boldsymbol{t}) + f(\boldsymbol{x}) = f(\boldsymbol{t} + \boldsymbol{x}) \in f(\boldsymbol{x}) + U$. This shows that $f(\boldsymbol{t}) \in U$, which proves that f is continuous in $\mathbf{0}_1$.

Conversely, suppose that f is continuous in $\mathbf{0}_1$. Let $\boldsymbol{x} \in V_1$ and let $Z \in neigh_{f(\boldsymbol{x})}(\mathcal{O}_2)$. The set $-f(\boldsymbol{x}) + Z$ is a neighborhood of $\mathbf{0}_2$ in V_2. The continuity of f in $\mathbf{0}_1$ implies the existence of a neighborhood T of $\mathbf{0}_1$ such that $f(T) \subseteq -f(\boldsymbol{x})+Z$. Note that $\boldsymbol{x}+T$ is a neighborhood of $\boldsymbol{x}$ in V_1 and every neighborhood of $\boldsymbol{x}$ in V_1 has this form. Since $f(\boldsymbol{x} + T) \subseteq Z$, it follows that f is continuous in $\boldsymbol{x}$. □

Corollary 2.14. *Let $(V_1, \mathcal{O}_1)$ and $(V_2, \mathcal{O}_2)$ be two topological $\mathbb{F}$-linear spaces. A linear operator $f \in \mathsf{Hom}(V_1, V_2)$ is either continuous on V_1 or is discontinuous in every point of V_1.*

Proof. This statement is a direct consequence of Theorem 2.51. □

Theorem 2.52. *Let C and D be two subsets of $\mathbb{R}^n$ such that C is compact and D is closed. Then the set $C + D = \{\boldsymbol{x} + \boldsymbol{y} \mid \boldsymbol{x} \in C, \boldsymbol{y} \in D\}$ is closed.*

Proof. Let $\boldsymbol{x} \in \boldsymbol{K}(C + D)$. There exists a sequence $(\boldsymbol{x}_0, \boldsymbol{x}_1, \ldots)$ such that $\boldsymbol{x}_i \in C + D$ and $\lim_{n\to\infty} \boldsymbol{x}_n = \boldsymbol{x}$. The definition of $C + D$ means that there is a sequence $(\boldsymbol{u}_0, \boldsymbol{u}_1, \ldots) \in \mathbf{Seq}_\infty(C)$ and a sequence $(\boldsymbol{v}_0, \boldsymbol{v}_1, \ldots) \in \mathbf{Seq}_\infty(D)$ such that $\boldsymbol{x}_i = \boldsymbol{u}_i + \boldsymbol{v}_i$ for $i \in \mathbb{N}$.

Since C is compact, the sequence $(\boldsymbol{v}_0, \boldsymbol{v}_1, \ldots)$ contains a convergent subsequence $(\boldsymbol{u}_{i_0}, \boldsymbol{u}_{i_1}, \ldots)$. Let $\boldsymbol{u} = \lim_{m\to\infty} \boldsymbol{u}_{i_m}$. Clearly, $\lim_{m\to\infty} \boldsymbol{x}_{i_m} = \boldsymbol{x}$. Since D is a closed set, $\lim_{m\to\infty} \boldsymbol{v}_{i_m} = \boldsymbol{x} - \boldsymbol{u} \in D$. Therefore, $\boldsymbol{x} = \boldsymbol{u} + \boldsymbol{v} \in C + D$, so $\boldsymbol{K}(C + D) = C + D$, which means that $C + D$ is closed. □

2.10 Isomorphism Theorems

Theorem 2.53 (Noether's first isomorphism theorem). *Let V be an $\mathbb{F}$-linear space, and let K and L be two subspaces of V such that $K \subseteq L$. Then L/K is a linear subspace of M/K and $M/L \cong (M/K)/(L/K)$.*

Proof. Consider the canonical linear mappings $h_K : V \longrightarrow V/K$ and $h_L : V \longrightarrow V/L$ for which we have $\mathsf{Ker}(h_K) = K \subseteq L = \mathsf{Ker}(h_L)$. Since $K \subseteq L$, by Theorem 2.25, there exists a linear mapping $e : V/K \longrightarrow V/L$ such that $h_L = eh_K$, that is $e(h_K(\boldsymbol{x})) = h_L(\boldsymbol{x})$, which means that $e([\boldsymbol{x}]_K) = [\boldsymbol{x}]_L$ for every $\boldsymbol{x} \in V$. We denoted the $\sim_K$-class of $\boldsymbol{x}$ by $[\boldsymbol{x}]_K$ and the $\sim_L$-class of the same element by $[x]_L$. Since $\mathsf{Ker}(e) = L/K$ and $\mathsf{Im}(e) = V/L$, we obtain that V/L is isomorphic to $(V/K)/(L/K)$ by Theorem 2.12. □

Theorem 2.54 (Noether's second isomorphism theorem). *Let V be an $\mathbb{F}$-linear space, and let K and L be two subspaces of V. Then $(L + K)/L \cong H/(L \cap K)$.*

Proof. Consider the linear mapping $h : K \longrightarrow K + L$ defined by $h(\boldsymbol{x}) = \boldsymbol{x}$ for every $\boldsymbol{x} \in K$ and the canonical linear mapping $g : K + L \longrightarrow (K + L)/L$. Define the linear mapping $e = gh$. We have $\mathsf{Ker}(e) = K \cap L$ and $\mathsf{Im}(e) = (K + L)/L$, so by Theorem 2.12, the subspaces $(L + K)/L$ and $H/(L \cap K)$ are isomorphic. □

2.11 Multilinear Functions

The notion of linear mapping can be extended as follows.

Definition 2.27. Let $V_1, \ldots, V_n, U$ be real linear spaces. A *multilinear function* is a mapping $f : V_1 \times \cdots \times V_n \longrightarrow U$ that is linear in *each of its arguments when the other arguments are held fixed.*

In other words, f satisfies the following conditions:

$$f\left(\boldsymbol{x}_1, \ldots, \boldsymbol{x}_{i-1}, \sum_{j=1}^{k} a_j \boldsymbol{x}_i^j, \boldsymbol{x}_{i+1}, \ldots, \boldsymbol{x}_n\right)$$

$$= \sum_{j=1}^{k} a_j f(\boldsymbol{x}_1, \ldots, \boldsymbol{x}_{i-1}, \boldsymbol{x}_i^j, \boldsymbol{x}_{i+1}, \ldots, \boldsymbol{x}_n),$$

for every $\boldsymbol{x}_i, \boldsymbol{x}_i^j \in V_i$ and $a_1, \ldots, a_k \in \mathbb{R}$.

If $V_1 = \cdots = V_n = V$ and $U = \mathbb{R}$, the mapping f is called an *n-linear form* on V or an n-tensor on V. The set of n-tensors on V is denoted as $\mathfrak{T}_n(V)$.

When $n = 2$, $V_1 = V_2 = \mathbb{R}^m$, and $U = \mathbb{R}$, we refer to f as a *bilinear form on* $\mathbb{R}^m$. A bilinear form $f : \mathbb{R}^m \times \mathbb{R}^m \longrightarrow \mathbb{R}$ is *skew-symmetric* if $f(\boldsymbol{x}, \boldsymbol{y}) = -f(\boldsymbol{y}, \boldsymbol{x})$ for $\boldsymbol{x}, \boldsymbol{y} \in \mathbb{R}^m$.

More generally, a *tensor* t *over the linear space* V is a multilinear function

$$\mathsf{t} : \underbrace{V^* \times \cdots \times V^*}_{p} \times \underbrace{V \times \cdots \times V}_{q} \longrightarrow \mathbb{R}.$$

In this case, we say t is covariant of order p and contravariant of order q.

The set of real multilinear functions defined on the linear spaces $V_1, \ldots, V_n$ and ranging in the real linear space V is denoted by $\mathfrak{M}(V_1, \ldots, V_n; V)$. The set of real multilinear forms is $\mathfrak{M}(V_1, \ldots, V_n; \mathbb{R})$.

Multilinear functions are defined on the Cartesian product of the sets $V_1, \ldots, V_n$, *not on the direct product of the linear spaces* $V_1, \ldots, V_n$.

Example 2.21. Let V be a linear space and let V^* be its dual. Consider the following four pair of vectors:

(i) $\boldsymbol{u}$ in V and $\boldsymbol{v} \in V$;
(ii) $\boldsymbol{u}$ in V and $\boldsymbol{g} \in V^*$;
(iii) $\boldsymbol{f}$ in V^* and $\boldsymbol{v} \in V$;
(iv) $\boldsymbol{f}$ in V^* and $\boldsymbol{g} \in V^*$.

Each of these pairs generates a bilinear function as follows:

(i) $\phi_{\boldsymbol{u},\boldsymbol{v}} : V^* \times V^* \longrightarrow \mathbb{R}$ given by

$$\phi_{\boldsymbol{u},\boldsymbol{v}}(\boldsymbol{h}, \boldsymbol{l}) = \boldsymbol{h}(\boldsymbol{u})\boldsymbol{l}(\boldsymbol{v});$$

(ii) $\phi_{\boldsymbol{u},\boldsymbol{g}} : V^* \times V \longrightarrow \mathbb{R}$ given by

$$\phi_{\boldsymbol{u},\boldsymbol{g}}(\boldsymbol{h}, \boldsymbol{t}) = \boldsymbol{h}(\boldsymbol{u})\boldsymbol{g}(\boldsymbol{t});$$

(iii) $\phi_{\boldsymbol{f},\boldsymbol{v}} : V \times V^* \longrightarrow \mathbb{R}$ given by

$$\phi_{\boldsymbol{v},\boldsymbol{f}}(\boldsymbol{x}, \boldsymbol{h}) = \boldsymbol{f}(\boldsymbol{x})\boldsymbol{h}(\boldsymbol{v});$$

(iv) $\phi_{\boldsymbol{f},\boldsymbol{g}} : V \times V \longrightarrow \mathbb{R}$ given by

$$\phi_{\boldsymbol{f},\boldsymbol{g}}(\boldsymbol{u}, \boldsymbol{v}) = \boldsymbol{f}(\boldsymbol{u})\boldsymbol{g}(\boldsymbol{v}).$$

The function $\phi_{\boldsymbol{f},\boldsymbol{v}} : V \times V^* \longrightarrow \mathbb{R}$ defined as $\phi_{\boldsymbol{f},\boldsymbol{v}}(\boldsymbol{x}, \boldsymbol{h}) = \boldsymbol{f}(\boldsymbol{x})\boldsymbol{h}(\boldsymbol{v})$ for $\boldsymbol{x} \in V$ and $\boldsymbol{h} \in V^*$ is linear in $\boldsymbol{x}$ by the linearity of $\boldsymbol{f}$ and is linear in the second variable $\boldsymbol{h}$ by the definition of V^*. Therefore, $h \in \mathcal{M}(V, V^*; \mathbb{R})$.

If $f : U \times W \longrightarrow V$ is a bilinear function, we have

$$f(\mathbf{0}_U, \boldsymbol{w}) = \mathbf{0}_V \text{ and } f(\boldsymbol{u}, \mathbf{0}_W) = \mathbf{0}_V$$

for every $\boldsymbol{w} \in W$ and $\boldsymbol{u} \in U$.

Definition 2.28. Let V, W be two complex linear spaces. A function $f : V \times W \longrightarrow \mathbb{C}$ is said to be *Hermitian* bilinear if it is linear in the first variable and skew-linear in the second, that is, it satisfies the following equalities:

$$f(a_1\boldsymbol{x}_1 + a_2\boldsymbol{x}_2, \boldsymbol{y}) = a_1 f(\boldsymbol{x}_1, \boldsymbol{y}) + a_2 f(\boldsymbol{x}_2, \boldsymbol{y}),$$
$$f(\boldsymbol{x}, b_1\boldsymbol{y}_1 + b_2\boldsymbol{y}_2) = \bar{b}_1 f(\boldsymbol{x}, \boldsymbol{y}_1) + \bar{b}_2 f(\boldsymbol{x}, \boldsymbol{y}_2),$$

for $\boldsymbol{x}_1, \boldsymbol{x}_2, \boldsymbol{x} \in V$, $\boldsymbol{y}, \boldsymbol{y}_1, \boldsymbol{y}_2 \in W$, and $a_1, a_2, b_1, b_2 \in \mathbb{C}$.

Example 2.22. Multilinearity is distinct from the notion of linearity on a product of linear spaces. For instance, the mapping $h : \mathbb{R}^2 \longrightarrow \mathbb{R}$ defined by $h(x, y) = x + y$ is linear but not bilinear. On the other hand, the mapping $g : \mathbb{R}^2 \longrightarrow \mathbb{R}$ given by $h(x, y) = xy$ is bilinear but not linear.

Definition 2.29. Let $V_1, \ldots, V_n, V$ be real linear spaces.

If $f, g \in \mathfrak{M}(V_1, \ldots, V_n; L)$ are two multilinear functions, their *sum* is the function $f + g$ defined by

$$(f + g)(\boldsymbol{x}_1, \ldots, \boldsymbol{x}_n) = f(\boldsymbol{x}_1, \ldots, \boldsymbol{x}_n) + g(\boldsymbol{x}_1, \ldots, \boldsymbol{x}_n),$$

and the product af, where $a \in \mathbb{F}$ is the function af given by

$$(af)(\boldsymbol{x}_1, \ldots, \boldsymbol{x}_n) = af(\boldsymbol{x}_1, \ldots, \boldsymbol{x}_n)$$

for $\boldsymbol{x}_i \in V_i$ and $1 \leqslant i \leqslant n$.

It is immediate to verify that $\mathfrak{M}(V_1, \ldots, V_n; V)$ is an $\mathbb{R}$-linear space relative to these operations.

Let $f : V_1 \times V_2 \longrightarrow V$ be a real bilinear function. Observe that for $\boldsymbol{x} \in V_1$ and $\boldsymbol{y} \in V_2$, we have

$$\begin{aligned} f(\boldsymbol{x}, \mathbf{0}_{V_2}) &= f(\boldsymbol{x}, 0\boldsymbol{y}) = 0f(\boldsymbol{x}, \boldsymbol{y}) = \mathbf{0}_V \text{ and} \\ f(\mathbf{0}_{V_1}, \boldsymbol{y}) &= f(0\boldsymbol{x}, \boldsymbol{y}) = 0f(\boldsymbol{x}, \boldsymbol{y}) = \mathbf{0}_V. \end{aligned} \tag{2.8}$$

Example 2.23. Let V be an $\mathbb{R}$-linear space and let $\langle \cdot, \cdot \rangle : V^* \times V \longrightarrow \mathbb{R}$ be the function given by $\langle h, \boldsymbol{y} \rangle = h(\boldsymbol{y})$ for $h \in V^*$ and $\boldsymbol{y} \in V$. It is immediate that $\langle \cdot, \cdot \rangle$ is a bilinear function because

$$\begin{aligned} \langle ah + bg, \boldsymbol{y} \rangle &= a\langle h, \boldsymbol{y} \rangle + b\langle g, \boldsymbol{y} \rangle, \\ \langle h, a\boldsymbol{y} + b\boldsymbol{z} \rangle &= a\langle h, \boldsymbol{y} \rangle + b\langle h, \boldsymbol{z} \rangle, \end{aligned}$$

for $a, b \in \mathbb{R}$, $h, g \in V^*$, and $\boldsymbol{y}, \boldsymbol{z} \in V$.

Moreover, we have $\langle h, \boldsymbol{y} \rangle = 0$ for every $\boldsymbol{y} \in V$ if and only if $h = \mathbf{0}_{V^*}$ and $\langle h, \boldsymbol{y} \rangle = 0$ for every $h \in V^*$ if and only if $\boldsymbol{y} = \mathbf{0}_V$.

Example 2.24. Let $V_1, \ldots, V_n, V$ be $\mathbb{R}$-linear spaces, $\boldsymbol{a}_i \in V_i$ for $1 \leqslant i \leqslant n$, and let $g_i \in V_i^*$. Define the function $G : V_1 \times \cdots \times V_n \longrightarrow \mathbb{R}$ as

$$G(\boldsymbol{a}_1, \ldots, \boldsymbol{a}_n) = g_1(\boldsymbol{a}_1) \cdots g_n(\boldsymbol{a}_n)$$

for $\boldsymbol{a}_i \in V_i$ and $1 \leqslant i \leqslant n$.

The function G is multilinear. Indeed, if $\boldsymbol{a}_i, \boldsymbol{b}_i \in V_i$ and $a \in \mathbb{R}$, it is immediate to verify that

$$\begin{aligned} &G(\boldsymbol{a}_1, \ldots, \boldsymbol{a}_i + \boldsymbol{b}_i, \ldots, \boldsymbol{a}_n) \\ &\quad = G(\boldsymbol{a}_1, \ldots, \boldsymbol{a}_i, \ldots, \boldsymbol{a}_n) + G(\boldsymbol{a}_1, \ldots, \boldsymbol{b}_i, \ldots, \boldsymbol{a}_n), \end{aligned}$$

and

$$G(\boldsymbol{a}_1, \ldots, a\boldsymbol{a}_i, \ldots, \boldsymbol{a}_n) = aG(\boldsymbol{a}_1, \ldots, \boldsymbol{a}_i, \ldots, \boldsymbol{a}_n).$$

Note, however, that G is not a linear function because

$$G(a\boldsymbol{a}_1, \ldots, a\boldsymbol{a}_n) = a^n G(\boldsymbol{a}_1, \ldots, \boldsymbol{a}_n)$$

for $a \in \mathbb{R}$.

Example 2.25. The function $f : \mathbb{R}^2 \longrightarrow \mathbb{R}$ defined by $f(x_1, x_2) = x_1 x_2$ is bilinear because it is linear in each of its variables, separately, but is not linear in the ensemble of its arguments. Indeed, we have

$$\begin{aligned} f(x_1 + y_1, x_2) &= f(x_1, x_2) + f(y_1, x_2), \\ f(x_1, x_2 + y_2) &= f(x_1, x_2) + f(x_1, y_2), \end{aligned}$$

for every $x_1, x_2, y_1, y_2 \in \mathbb{R}$, which shows the bilinearity of f. However, we have

$$\begin{aligned} f(x_1 + x_2, y_1 + y_2) &= x_1 y_1 + x_1 y_2 + x_2 y_1 + x_2 y_2 \\ &\neq f(x_1, y_1) + f(x_2, y_2), \end{aligned}$$

which means that f is not a linear function.

Let $h : U \times V \longrightarrow W$ be a bilinear mapping. For $\boldsymbol{a} \in U$ and $\boldsymbol{b} \in V$, define $h^{\boldsymbol{a}} : V \longrightarrow W$ and $h_{\boldsymbol{b}} : U \longrightarrow W$ as

$$h^{\boldsymbol{a}}(\boldsymbol{v}) = h(\boldsymbol{a}, \boldsymbol{v}) \text{ for } \boldsymbol{v} \in V,$$

and

$$h_{\boldsymbol{b}}(\boldsymbol{u}) = h(\boldsymbol{u}, \boldsymbol{b}) \text{ for } \boldsymbol{u} \in U.$$

Then, we have $h^{\boldsymbol{a}} \in \mathsf{Hom}(V, W)$ and $h_{\boldsymbol{b}} \in \mathsf{Hom}(U, W)$ because of the bilinearity of the mapping h.

Conversely, if $f : U \longrightarrow \mathsf{Hom}(V, W)$ is a linear mapping, then the function $h : U \times V \longrightarrow W$ defined by $h(\boldsymbol{u}, \boldsymbol{v}) = f(\boldsymbol{u})(\boldsymbol{v})$ is bilinear and $h^{\boldsymbol{u}} = f(\boldsymbol{u})$.

If $W = \mathbb{R}$, then h is a bilinear form, $h^{\boldsymbol{a}} \in V^*$ and $h_{\boldsymbol{b}} \in U^*$. Thus, we have a mapping $\Phi : \mathfrak{M}(U, V; \mathbb{R}) \longrightarrow \mathsf{Hom}(U, V^*)$ given by $\Phi(h)(\boldsymbol{a}) = h^{\boldsymbol{a}}$ for $\boldsymbol{a} \in U$, and a mapping $\Psi : \mathfrak{M}(U, V; \mathbb{R}) \longrightarrow \mathsf{Hom}(V, U^*)$ given by $\Psi(h)(\boldsymbol{b}) = h_{\boldsymbol{b}}$ for $\boldsymbol{b} \in V$.

Theorem 2.55. *Let U, V be two real linear spaces and let $\mathfrak{M}(U, V; \mathbb{R})$ be the linear space of bilinear forms defined on $U \times V$. The linear spaces $\mathfrak{M}(U, V; \mathbb{R})$, $\mathsf{Hom}(U, V^*)$, and $\mathsf{Hom}(V, U^*)$ are isomorphic.*

Proof. It is immediate that Φ is a linear mapping because for $c, d \in \mathbb{R}$ and $h_1, h_2 \in \mathfrak{M}(U, V; \mathbb{R})$, we have

$$\begin{aligned}
\Phi(ch_1 + dh_2)(\boldsymbol{a})(\boldsymbol{v}) &= ((ch_1 + dh_2)^{\boldsymbol{a}})(\boldsymbol{v}) \\
&= (ch_1 + dh_2)(\boldsymbol{a}, \boldsymbol{v}) = ch_1(\boldsymbol{a}, \boldsymbol{v}) + dh_2(\boldsymbol{a}, \boldsymbol{v}) \\
&= ch_1^{\boldsymbol{a}}(\boldsymbol{v}) + dh_2^{\boldsymbol{a}}(\boldsymbol{v}) \\
&= c\Phi(h_1)(\boldsymbol{a})(\boldsymbol{v}) + d\Phi(h_2)(\boldsymbol{a})(\boldsymbol{v}),
\end{aligned}$$

or

$$\Phi(ch_1 + dh_2) = c\Phi(h_1) + d\Phi(h_2).$$

Note that Φ maps $h : U \longrightarrow V$ into the linear form that transforms $\boldsymbol{a}$ into $h^{\boldsymbol{a}}$ for $\boldsymbol{a} \in U$. Thus, if $\Phi(h_1) = \Phi(h_2)$, we have both h_1 and h_2 yield equal values for $\boldsymbol{a} \in U$, so $h_1 = h_2$, which proves the injectivity of Φ.

Let $f \in \mathsf{Hom}(U, V^*)$. For every $\boldsymbol{a} \in U$ there exists a linear form $g : V \longrightarrow \mathbb{R}$ such that $f(\boldsymbol{a}) = g$, or $f(\boldsymbol{a})(\boldsymbol{v}) = g(\boldsymbol{v})$ for every $\boldsymbol{v} \in V$. The mapping $h : U \times V \longrightarrow \mathbb{R}$ defined by $h(\boldsymbol{u}, bfv) = f(\boldsymbol{u})(\boldsymbol{v})$ is bilinear and $\Phi(h)(\boldsymbol{u})(\boldsymbol{v}) = h^{\boldsymbol{u}}(\boldsymbol{v}) = h(\boldsymbol{u}, \boldsymbol{v}) = f(\boldsymbol{u})(\boldsymbol{v})$, which means that $\Phi(h) = f$. Thus, Φ is also surjective and, therefore, it is an isomorphism between the linear spaces $\mathfrak{M}(U, V; \mathbb{R})$ and $\mathsf{Hom}(U, V^*)$.

The existence of an isomorphism targeted to $\mathsf{Hom}(V, U^*)$ has a similar argument. □

Corollary 2.15. *Let V, W be two $\mathbb{R}$-linear spaces. Then,* $\dim(\mathfrak{M}(V, W; \mathbb{R})) = \dim(V) \cdot \dim(W)$.

Proof. Since $\dim(W^*) = \dim(W) = n$, we have $\dim(\mathsf{Hom}(V, W^*)) = mn$. The result follows immediately from Theorem 2.55. □

Let U, V, W be three $\mathbb{R}$-linear spaces of finite dimensions having the bases $\{\boldsymbol{u}_1, \ldots, \boldsymbol{u}_m\}$, $\{\boldsymbol{v}_1, \ldots, \boldsymbol{v}_n\}$, and $\{\boldsymbol{w}_1, \ldots, \boldsymbol{w}_p\}$, respectively, and let $f : U \times V \longrightarrow W$ be a bilinear function. If $\boldsymbol{u} = \sum_{i=1}^{m} a_i \boldsymbol{u}_i \in U$, $\boldsymbol{v} = \sum_{j=1}^{n} b_j \boldsymbol{v}_j$, then

$$f(\boldsymbol{u}, \boldsymbol{v}) = \sum_{i=1}^{m} \sum_{j=1}^{n} a_i b_j f(\boldsymbol{u}_i, \boldsymbol{v}_j).$$

Since $f(\boldsymbol{u}_i, \boldsymbol{v}_j) \in W$, there exist c_{ij}^k such that $f(\boldsymbol{u}_i, \boldsymbol{v}_j) = \sum_{k=1}^{p} c_{ij}^k \boldsymbol{w}_k$, hence

$$f(\boldsymbol{u}, \boldsymbol{v}) = \sum_{i=1}^{m} \sum_{j=1}^{n} \sum_{k=1}^{p} a_i b_j c_{ij}^k w_k.$$

Thus, the set $\{c_{ij}^k \in \mathbb{R} \mid 1 \leqslant i \leqslant m, 1 \leqslant j \leqslant n, 1 \leqslant k \leqslant p\}$ (which contains mnp elements) determines a bilinear function relative to the chosen bases in U, V, and W. The numbers c_{ij}^k are known as the *structural constants* of f.

In the special case of bilinear forms, when $W = \mathbb{R}$, $f : U \times V \longrightarrow \mathbb{R}$ is a bilinear form and can be written using structural constants c_{ij} as in

$$f(\boldsymbol{u}, \boldsymbol{v}) = \sum_{i=1}^{m} \sum_{j=1}^{n} a_i b_j c_{ij}, \tag{2.9}$$

where $\boldsymbol{u} = \sum_{i=1}^{m} a_i \boldsymbol{u}_i \in U$, $\boldsymbol{v} = \sum_{j=1}^{n} b_j \boldsymbol{v}_j$, and c_{ij} are the structural constants. The matrix of f is $C_f = (c_{ij})$ and the form itself can be written as $f(\boldsymbol{u}, \boldsymbol{v}) = \boldsymbol{u}' C_f \boldsymbol{v}$ for $\boldsymbol{u} \in U$ and $\boldsymbol{v} \in V$.

Example 2.26. Unlike the case $n = 1$, the set of values of a multilinear function $f : V_1 \times \cdots \times V_n \longrightarrow W$ is not a subspace of W in general. Indeed, consider a two-dimensional $\mathbb{R}$-linear space U having a basis $\{\boldsymbol{u}_1, \boldsymbol{u}_2\}$, a four-dimensional $\mathbb{R}$-linear space W having the basis $\{\boldsymbol{w}_1, \boldsymbol{w}_2, \boldsymbol{w}_3, \boldsymbol{w}_4\}$, and the bilinear function $f : U \times U \longrightarrow W$ defined as

$$f(\boldsymbol{u}, \boldsymbol{v}) = u_1 v_1 \boldsymbol{w}_1 + u_1 v_2 \boldsymbol{w}_2 + u_2 v_1 \boldsymbol{w}_3 + u_2 v_2 \boldsymbol{w}_4,$$

where $\boldsymbol{u} = u_1 \boldsymbol{u}_1 + u_2 \boldsymbol{u}_2$ and $\boldsymbol{v} = v_1 \boldsymbol{u}_1 + v_2 \boldsymbol{u}_2$.

Let S be the set of all vectors of the form $\boldsymbol{s} = f(\boldsymbol{u}, \boldsymbol{v})$. By the definition of S, there exist $\boldsymbol{u}, \boldsymbol{v} \in U$ such that

$$s_1 = u_1v_1, s_2 = u_1v_2, s_3 = u_2v_1, s_4 = u_2v_2,$$

hence $s_1s_4 = s_2s_3$ for any $\boldsymbol{s} \in S$.

Define the vectors $\boldsymbol{z}, \boldsymbol{t}$ in W as

$$\begin{aligned} \boldsymbol{z} &= 2\boldsymbol{w}_1 + 2\boldsymbol{w}_2 + \boldsymbol{w}_3 + \boldsymbol{w}_4, \\ \boldsymbol{t} &= \boldsymbol{w}_1 + \boldsymbol{w}_3. \end{aligned}$$

Note that we have both $\boldsymbol{z} \in S$ and $\boldsymbol{t} \in S$. However,

$$\boldsymbol{x} = \boldsymbol{z} - \boldsymbol{t} = \boldsymbol{w}_1 + 2\boldsymbol{w}_2 + \boldsymbol{w}_4$$

does not belong to S because $x_1x_4 = 1$ and $x_2x_3 = 0$. This example given in [71] shows that the range of a bilinear function is not necessarily a subspace of the space of values of the function.

Example 2.27. Let $f : \mathbb{R}^2 \times \mathbb{R}^2 \longrightarrow \mathbb{R}$ be a bilinear form. Since the vectors

$$\boldsymbol{e}_1 = \begin{pmatrix} 1 \\ 0 \end{pmatrix} \text{ and } \boldsymbol{e}_1 = \begin{pmatrix} 0 \\ 1 \end{pmatrix}$$

form a basis in $\mathbb{R}^2$, f can be written as

$$\begin{aligned} & f(a\boldsymbol{e}_1 + b\boldsymbol{e}_2, c\boldsymbol{e}_1 + d\boldsymbol{e}_2) \\ &\quad = af(\boldsymbol{e}_1, c\boldsymbol{e}_1 + d\boldsymbol{e}_2) + bf(\boldsymbol{e}_2, c\boldsymbol{e}_1 + d\boldsymbol{e}_2) \\ &\quad = acf(\boldsymbol{e}_1, \boldsymbol{e}_1) + adf(\boldsymbol{e}_1, \boldsymbol{e}_2) + bcf(\boldsymbol{e}_2, \boldsymbol{e}_1) + bdf(\boldsymbol{e}_2, \boldsymbol{e}_2) \\ &\quad = \alpha f(\boldsymbol{e}_1, \boldsymbol{e}_1) + \beta f(\boldsymbol{e}_1, \boldsymbol{e}_2) + \gamma f(\boldsymbol{e}_2, \boldsymbol{e}_1) + \delta f(\boldsymbol{e}_2, \boldsymbol{e}_2), \end{aligned}$$

where

$$\alpha = ac, \beta = ad, \gamma = bc, \delta = bd.$$

Thus, the multilinearity of f implies $\alpha\delta = \beta\gamma$.

Theorem 2.56. *Let V, W, U be three finite-dimensional linear spaces having the bases $\{\boldsymbol{v}_i \mid 1 \leqslant i \leqslant \ell\}$, $\{\boldsymbol{w}_j \mid 1 \leqslant i \leqslant m\}$, and $\{\boldsymbol{u}_k \mid 1 \leqslant k \leqslant p\}$, and the dual bases $\{\boldsymbol{v}_i^* \mid 1 \leqslant i \leqslant \ell\}$, $\{\boldsymbol{w}_j^* \mid 1 \leqslant i \leqslant m\}$, $\{\boldsymbol{u}_k^* \mid 1 \leqslant k \leqslant p\}$, respectively. Then, the functions $\phi_{ij}^k : V \times W \longrightarrow U$ defined as $\phi_{ij}^k(\boldsymbol{v}, \boldsymbol{w}) = \boldsymbol{v}_i^*(\boldsymbol{v})\boldsymbol{w}_j^*(\boldsymbol{w})\boldsymbol{u}_k$ are bilinear and form a basis of $\mathfrak{M}(V \times W, U)$.*

Proof. The bilinearity of the functions ϕ_{ij}^k is immediate. If $f \in \mathfrak{M}(V \times W, U)$, $\boldsymbol{v} \in V$, and $\boldsymbol{w} \in W$, we have

$$
\begin{aligned}
f(\boldsymbol{v}, \boldsymbol{w}) &= f\left(\sum_i \boldsymbol{v}_i^*(\boldsymbol{v}), \sum_j \boldsymbol{w}_j^*(\boldsymbol{w})\boldsymbol{w}_j\right) \\
&= \sum_i \sum_j \boldsymbol{v}_i^*(\boldsymbol{v})\boldsymbol{w}_j^*(\boldsymbol{w})\boldsymbol{w}_j f(\boldsymbol{v}_i, \boldsymbol{w}_j) \\
&= \sum_i \sum_j \sum_k \boldsymbol{v}_i^*(\boldsymbol{v})\boldsymbol{w}_j^*(\boldsymbol{w})\boldsymbol{w}_j \boldsymbol{u}^*(f(\boldsymbol{v}_i, \boldsymbol{w}_j))\boldsymbol{u}_k \\
&= \sum_i \sum_j \sum_k \boldsymbol{u}^*(f(\boldsymbol{v}_i, \boldsymbol{w}_j))\phi_{ij}^k(\boldsymbol{v}, \boldsymbol{w}).
\end{aligned}
$$

Thus, the set of mappings ϕ_{ij}^k spans $\mathfrak{M}(V \times W, U)$ and the set of numbers $\boldsymbol{u}_k^*(f(v_i, w_j))$ uniquely determines the bilinear function f, which shows that the set of functions ϕ_{ij}^k is linearly independent. □

Corollary 2.16. *We have* $\dim(\mathfrak{M}(V \times W, U)) = \dim(V)\dim(W)\dim(U)$.

Proof. The equality is an immediate consequence of Theorem 2.56. □

Exercises and Supplements

(1) Let V be an $\mathbb{R}$-linear space. Prove that the following three statements that concern a non-empty subset P of V are equivalent:
  (a) P is a subspace of V;
  (b) if $\boldsymbol{x}, \boldsymbol{y} \in P$ and $a \in \mathbb{F}$, then $\boldsymbol{x} + \boldsymbol{y} \in P$ and $a\boldsymbol{x} \in P$;
  (c) $\boldsymbol{x}, \boldsymbol{y} \in P$ and $a, b \in \mathbb{F}$ imply $a\boldsymbol{x} + b\boldsymbol{y} \in P$.

(2) Let h be an endomorphism of an $\mathbb{R}$-linear space V. Prove that h^m is an endomorphism of V for every $m \in nn$ and $m \geqslant 1$.

(3) Let V and W be two linear spaces such that $B = \{\boldsymbol{v}_1, \ldots, \boldsymbol{v}_n\}$ is a basis of V and let $f \in \mathsf{Hom}(V, W)$. Prove that f is an injective linear mapping if and only if the set $\{f(\boldsymbol{v}_1), \ldots, f(\boldsymbol{v}_n)\}$ is linearly independent.

Solution: Suppose that f is injective and that $a_1 f(\boldsymbol{v}_1) + \cdots + a_n f(\boldsymbol{v}_n) = \mathbf{0}_W$. Then, since f is linear, we have $f(a_1\boldsymbol{v}_1 + \cdots + a_n\boldsymbol{v}_n) = \mathbf{0}_W$. The injectivity of f implies $a_1\boldsymbol{v}_1 + \cdots + a_n\boldsymbol{v}_n = \mathbf{0}_V$ and this, in turn, implies $a_1 = \cdots = a_n = 0$. Thus, $\{f(\boldsymbol{v}_1), \ldots, f(\boldsymbol{v}_n)\}$ is linearly independent.

Conversely, suppose that the set $\{f(\boldsymbol{v}_1), \ldots, f(\boldsymbol{v}_n)\}$ is linearly independent, and that $f(\boldsymbol{u}) = f(\boldsymbol{v})$ for some $\boldsymbol{u}, \boldsymbol{v} \in V$. Since B is a basis in V, we can write

$$\boldsymbol{u} = b_1\boldsymbol{v}_1 + \cdots + b_n\boldsymbol{v}_n \text{ and } \boldsymbol{v} = c_1\boldsymbol{v}_1 + \cdots + c_n\boldsymbol{v}_n,$$

which, in turn, imply

$$b_1 f(\boldsymbol{v}_1) + \cdots + b_n f(\boldsymbol{v}_n) = c_1 f(\boldsymbol{v}_1) + \cdots + c_n f(\boldsymbol{v}_n),$$

or

$$(b_1 - c_1) f(\boldsymbol{v}_1) + \cdots + (b_n - c_n) f(\boldsymbol{v}_n) = \mathbf{0}_W.$$

The linear independence of $\{f(\boldsymbol{v}_1), \ldots, f(\boldsymbol{v}_n)\}$ means that $b_i = c_i$ for $1 \leqslant i \leqslant n$ and this implies $\boldsymbol{u} = \boldsymbol{v}$. Thus, f is injective.

(4) Let V be a linear space such that $\dim(V) = n$. Prove that no finite set U that contains $n-1$ vectors may span V; prove that no finite set of $n+1$ vectors can be linearly independent.
(5) Prove that every subset of a linearly independent set of a linear space is independent.
(6) Let U and W be two finite, linearly independent subsets of a linear space V such that $|U| \geqslant |W|$. Prove that there exists $\boldsymbol{u} \in U - W$ such that $W \cup \{\boldsymbol{u}\}$ is a linearly independent set.
(7) Let U, V be two linear subspaces of $\mathbb{C}^n$. If $\dim(U) + \dim(V) > n$, prove that the subspace $U \cap V$ contains a vector $\boldsymbol{x} \neq \mathbf{0}_n$.

Solution: We know that $\mathbf{0}_n \in U \cap V$ for any two subspaces U and V.

Suppose that $\dim(U) + \dim(V) > n$ and that $U \cap V = \{\mathbf{0}_n\}$ and let $\boldsymbol{u}_1, \ldots, \boldsymbol{u}_p$ and $\boldsymbol{v}_1, \ldots, \boldsymbol{v}_q$ be two bases of U and V, respectively, where $p = \dim(U)$, $q = \dim(V)$, and $p + q > n$. The set $\{\boldsymbol{u}_1, \ldots, \boldsymbol{u}_p, \boldsymbol{v}_1, \ldots, \boldsymbol{v}_q\}$ is linearly dependent, so there exist $b_1, \ldots, b_p, c_1, \ldots, c_q$ such that not all these numbers are null and

$$b_1\boldsymbol{u}_1 + \cdots + b_p\boldsymbol{u}_p + c_1\boldsymbol{v}_1 + \cdots + c_q\boldsymbol{v}_q = \mathbf{0}_n.$$

Let $\boldsymbol{u} = b_1\boldsymbol{u}_1 + \cdots + b_p\boldsymbol{u}_p \in U$ and $\boldsymbol{v} = c_1\boldsymbol{v}_1 + \cdots + c_q\boldsymbol{v}_q \in V$. The previous equality means that $\boldsymbol{u} + \boldsymbol{v} = \mathbf{0}_n$. Since $\boldsymbol{u} = -\boldsymbol{v}$, both $\boldsymbol{u}$ and $\boldsymbol{v}$ belong to $U \cap V$, so $\boldsymbol{u} = \boldsymbol{v} = \mathbf{0}_n$. This contradicts the linear independence of at least one of the bases, $\{\boldsymbol{u}_1, \ldots, \boldsymbol{u}_p\}$ and $\{\boldsymbol{v}_1, \ldots, \boldsymbol{v}_q\}$.

The notion of linear independence can be extended to subspaces of finite-dimensional linear spaces. Namely, if $V_1, \ldots, V_k$ are subspaces of a finite-dimensional linear space V, then we say that $V_1, \ldots, V_k$ are *linearly independent* if for $\boldsymbol{x}_i \in V_i$, $1 \leqslant i \leqslant k$, the equality $\sum_{i=1}^{k} \boldsymbol{x}_i = \mathbf{0}_V$ implies $\boldsymbol{x}_i = \mathbf{0}_V$ for $1 \leqslant i \leqslant k$.

(8) Prove that the subspaces $V_1, \ldots, V_k$ of a linear space V are linearly independent if and only if any set $\{\boldsymbol{x}_i \mid \boldsymbol{x}_i \in V_i - \{\mathbf{0}_V\}$ for $1 \leqslant i \leqslant k\}$ is linearly independent.

(9) Let U be a subspace of a linear space V. Prove that there exists a non-zero linear form $\ell : V \longrightarrow \mathbb{R}$ such that $\mathsf{Ker}(\ell) = U$.

(10) Let h be an endomorphism of $\mathbb{R}^n$ and let $\boldsymbol{x}$ be a vector in $\mathbb{R}^n - \{\mathbf{0}_n\}$ such that there exists a least integer p such that $h^p(\boldsymbol{x}) = \mathbf{0}$. Prove that the set $\{\boldsymbol{x}, h(\boldsymbol{x}), h^2(\boldsymbol{x}), \ldots, h^{p-1}(\boldsymbol{x})\}$ is linearly independent.

Solution: Suppose that $a_0\boldsymbol{x} + a_1 h(\boldsymbol{x}) + \cdots + a_{p-1}h^{p-1}(\boldsymbol{x}) = \mathbf{0}_n$. Then, we have

$$
\begin{aligned}
& a_0 h(\boldsymbol{x}) + a_1 h^2(\boldsymbol{x}) + \cdots + a_{p-2}h^{p-1}(\boldsymbol{x}) = \mathbf{0}_n \\
& a_0 h^2(\boldsymbol{x}) + a_1 h^3(\boldsymbol{x}) + \cdots + a_{p-3}h^{p-1}(\boldsymbol{x}) = \mathbf{0}_n \\
& \vdots \\
& a_0 h^{p-3}(\boldsymbol{x}) + a_1 h^{p-2}(\boldsymbol{x}) + a_2 h^{p-1}(\boldsymbol{x}) = \mathbf{0}_n \\
& a_0 h^{p-2}(\boldsymbol{x}) + a_1 h^{p-1}(\boldsymbol{x}) = \mathbf{0}_n \\
& a_0 h^{p-1}(\boldsymbol{x}) = \mathbf{0}_n.
\end{aligned}
$$

The last equality implies $a_0 = 0$. Substituting a_0 in the previous equality yields $a_1 = 0$, etc. Eventually, we obtain $a_0 = a_1 = \cdots = a_{p-1} = 0$, which proves that S is linearly independent.

(11) Let $(S, \mathcal{O})$ be a topological space and let U and W be two subsets of S. If U is open and $U \cap W = \emptyset$, then prove that $U \cap \boldsymbol{K}(W) = \emptyset$.

(12) Let $(S, \mathcal{O})$ be a topological space and let $\boldsymbol{I}$ be its interior operator. Prove that the poset of open sets $(\mathcal{O}, \subseteq)$ is a complete lattice, where $\sup \mathcal{L} = \bigcup \mathcal{L}$ and $\inf \mathcal{L} = \boldsymbol{I}\left(\bigcap \mathcal{L}\right)$ for every family of open sets $\mathcal{L}$.

(13) Let $(S, \mathcal{O})$ be a topological space and let $\boldsymbol{I}$ be its interior operator. Prove that the poset of open sets $(\mathcal{O}, \subseteq)$ is a complete lattice, where $\sup \mathcal{L} = \bigcup \mathcal{L}$ and $\inf \mathcal{L} = \boldsymbol{I}\left(\bigcap \mathcal{L}\right)$ for every family of open sets $\mathcal{L}$.

(14) Let $(S, \mathcal{O})$ be a topological space, let $\boldsymbol{K}$ be its interior operator, and let $\mathcal{K}$ be its collection of closed sets. Prove that $(\mathcal{K}, \subseteq)$ is a complete lattice, where $\sup \mathcal{L} = \boldsymbol{K}\left(\bigcup \mathcal{L}\right)$ and $\inf \mathcal{L} = \bigcap \mathcal{L}$ for every family of closed sets.

(15) Let T be a subspace of the topological space $(S, \mathcal{O})$. Let $\boldsymbol{K}_S, \boldsymbol{I}_S$, and ∂_S be the closure, interior, and border operators associated to S and $\boldsymbol{K}_T, \boldsymbol{I}_T$, and ∂_T be the corresponding operators associated to T. Prove that
(a) $\boldsymbol{K}_T(U) = \boldsymbol{K}_S(U) \cap T$,
(b) $\boldsymbol{I}_S(U) \subseteq \boldsymbol{I}_T(U)$, and
(c) $\partial_T U \subseteq \partial_S U$
for every subset U of T.

(16) Let V be an $\mathbb{R}$-linear space, S a subspace of V with $\dim(S) = k$, and suppose that the quotient space V/S is of dimension p. Prove that $\dim(V) = k + p$ and that a basis for V can be obtained by supplementing a basis of S with p representatives of the classes of the quotient space (one per class).

(17) Let V be an $\mathbb{R}$-linear space, and let S and T be complementary subspaces of V. Prove that the restriction of the function $f : V \longrightarrow V/S$ given by $f(x) = x + S$ to T is an isomorphism between T and S/V.

(18) Let V_1, V_2, V be $\mathbb{R}$-linear spaces. Prove that if $f : V_1 \times V_2 \longrightarrow V$ is both a linear and bilinear function, then $f(\boldsymbol{x}, \boldsymbol{y}) = \boldsymbol{0}_V$ for $\boldsymbol{x} \in V_1$ and $\boldsymbol{y} \in V_2$.

Solution: The bilinearity of f implies

$$f(\boldsymbol{x}, \boldsymbol{0}_{V_2}) = f(\boldsymbol{0}_{V_1}, \boldsymbol{y}) = \boldsymbol{0}_V$$

by Equalities (2.8). On the other hand, since f is linear, we have

$$f(\boldsymbol{x}, \boldsymbol{y}) = f(\boldsymbol{0}_{V_1}, \boldsymbol{y}) + f(\boldsymbol{x}, \boldsymbol{0}_{V_2}) = \boldsymbol{0}_V.$$

The next Supplement is a generalization of Corollary 2.15.

(19) Let $V_1, \ldots, V_m, V$ be m linear spaces with $\dim(V_i) = n_i$ for $1 \leqslant i \leqslant m$ and $\dim(V) = n$. Prove that $\dim(\mathfrak{M}(V_1, \ldots, V_n; V)) = n\prod_{i=1}^m n_i$.

Solution: Let $\Gamma(n_1, \ldots, n_m)$ be the set of sequences of integers $(a_1, \ldots, a_m)$ of length m such that $1 \leqslant a_i \leqslant n_i$ for $1 \leqslant i \leqslant m$. Suppose that the linear space V_t has the base $\{\boldsymbol{e}_{t1}, \ldots, \boldsymbol{e}_{tn_t}\}$ and V has the base $\{\boldsymbol{u}_1, \ldots, \boldsymbol{u}_n\}$.
For each sequence $\alpha \in \Gamma(n_1, \ldots, n_m)$, define $n\prod_{t=1}^m n_t$ functions:

$$\phi_{\alpha,j}(\boldsymbol{v}_1, \ldots, \boldsymbol{v}_m) = \left(\prod_{t=1}^m \xi_{t\alpha(t)}\right) u_j,$$

where $\boldsymbol{v}_t = \sum_{s=1}^{n_t} \xi_{ts}\boldsymbol{e}_{ts}$. Each of these functions is in $\mathfrak{M}(V_1, \ldots, V_n; V)$. Indeed, observe that if $\boldsymbol{v}_k$ is replaced by $c\boldsymbol{v}_k + d\tilde{\boldsymbol{v}}_k$, then the product $\prod_{t=1}^m \xi_{t\alpha(t)}$ is replaced by

$$\begin{aligned}
&\xi_{1\alpha(1)} \cdots \xi_{(k-1)\alpha(k-1)}(c\xi_{k\alpha(k)} + d\tilde{\xi}_{k\alpha(k)}) \\
&\quad \times \xi_{(k+1)\alpha(k+1)} \cdots \xi_m\alpha(m) \\
&= c\xi_{1\alpha(1)} \cdots \xi_{(k-1)\alpha(k-1)}\xi_{k\alpha(k)} \cdots \xi_{m\alpha(m)} \\
&\quad + d\xi_{1\alpha(1)} \cdots \xi_{(k-1)\alpha(k-1)}\tilde{\xi}_{k\alpha(k)} \cdots \xi_{m\alpha(m)}.
\end{aligned}$$

Therefore, each of the $n\prod_{t=1}^m n_t$ functions $\phi_{\alpha,j}$ belong to $\mathfrak{M}(V_1, \ldots, V_n; V)$.

Note that for $\phi, \theta \in \mathfrak{M}(V_1, \ldots, V_n; V)$, we have $\phi = \theta$ if and only if $\phi(\boldsymbol{e}_c) = \theta(\boldsymbol{e}_c)$ for every $c \in \Gamma(n_1, \ldots, n_m)$ because of the multilinearity of ϕ and θ.
Let ϕ be an arbitrary multilinear function such that

$$\phi(\boldsymbol{e}_\gamma) = \sum_{j=1}^n c_{\gamma j}\boldsymbol{u}_j$$

for $\gamma \in \Gamma(n_1, \ldots, n_m)$. Define

$$\theta = \sum_{\beta \in \Gamma(n_1,\ldots,n_m)} \sum_{j=1}^n c_{\beta j}\phi_{\beta,j}.$$

We have

$$\begin{aligned}\theta(\boldsymbol{e}_\gamma) &= \sum_{\beta\in\Gamma(n_1,\ldots,n_m)} \sum_{j=1}^{n} c_{\beta j}\phi_{\beta,j}(\boldsymbol{e}_\gamma)\\ &= \sum_{\beta\in\Gamma(n_1,\ldots,n_m)} \sum_{j=1}^{n} c_{\beta j}\delta_{\beta,\gamma}\boldsymbol{u}_j\\ &= \sum_{j=1}^{n} \sum_{\beta\in\Gamma(n_1,\ldots,n_m)} c_{\beta j}\delta_{\beta,\gamma}\boldsymbol{u}_j\\ &= \sum_{j=1}^{n} c_{\gamma\, j}\boldsymbol{u}_j = \phi(\boldsymbol{e}_\gamma).\end{aligned}$$

Thus, an arbitrary ϕ is expressed as a linear combination of $\phi_{\beta j}$. This means that the subspaces generated by the functions $\phi_{\alpha j}$ are $\mathfrak{M}(V_1,\ldots,V_n;V)$.

If $\theta = \sum_{\beta\in\Gamma(n_1,\ldots,n_m)} \sum_{j=1}^{n} d_{\beta j}\phi_{\beta,j} = 0$, then

$$\begin{aligned}\theta(\boldsymbol{e}_\gamma) &= \sum_{\beta\in\Gamma(n_1,\ldots,n_m)} \sum_{j=1}^{n} d_{\beta j}\delta_{\beta\gamma}\boldsymbol{u}_j\\ &= \sum_{j=1}^{n} d_{\gamma j}\boldsymbol{u}_j = 0,\end{aligned}$$

which implies $d_{\gamma j} = 0$ for $1 \leqslant j \leqslant n$.

(20) Let $f \in \mathfrak{M}(\mathbb{R}^2,\mathbb{R}^2;\mathbb{R})$. Prove that

$$f(\boldsymbol{u}_1+\boldsymbol{u}_2,\boldsymbol{v}_1+\boldsymbol{v}_2) = f(\boldsymbol{u}_1,\boldsymbol{v}_1)+f(\boldsymbol{u}_1,\boldsymbol{v}_2)+f(\boldsymbol{u}_2,\boldsymbol{v}_1)+f(\boldsymbol{u}_2,\boldsymbol{v}_2).$$

(21) Let $f \in \mathfrak{M}(\mathbb{R}^2,\mathbb{R}^2;\mathbb{R})$ and let $(\boldsymbol{s},\boldsymbol{u})$ and $(\boldsymbol{t},\boldsymbol{v})$ be two distinct bases in $\mathbb{R}^2$. Suppose that $\boldsymbol{w},\boldsymbol{x},\boldsymbol{y},\boldsymbol{z}$ are vectors in $\mathbb{R}^2$ such that

$$\begin{aligned}&\boldsymbol{w} = a\boldsymbol{s} + b\boldsymbol{u}, \boldsymbol{x} = c\boldsymbol{t} + d\boldsymbol{v},\\ &\boldsymbol{y} = e\boldsymbol{s} + g\boldsymbol{u}, \boldsymbol{z} = h\boldsymbol{t} + k\boldsymbol{v},\end{aligned}$$

for $a,b,c,d,e,g,h,k \in \mathbb{R}$. Prove that if $adgh \neq bcek$, then $f(\boldsymbol{s},\boldsymbol{v})$ and $f(\boldsymbol{u},\boldsymbol{t})$ can be determined from $f(\boldsymbol{s},\boldsymbol{t})$, $f(\boldsymbol{u},\boldsymbol{v})$, $f(\boldsymbol{w},\boldsymbol{x})$, and $f(\boldsymbol{y},\boldsymbol{z})$.

(22) Prove the following equalities involving Levi-Civita and Kronecker symbols:

(a)

$$\sum_{k=1}^{3} \epsilon_{ijk}\epsilon_{\ell mk} = \delta_{i\ell}\delta_{jm} - \delta_{im}\delta_{jl},$$

where all indices vary in the set $\{1, 2, 3\}$;

(b)

$$\sum_{j=1}^{3}\sum_{k=1}^{3} \epsilon_{ijk}\epsilon_{\ell jk} = 2\delta_{i\ell},$$

where all indices vary in the set $\{1, 2, 3\}$.

(23) Let V, W be two finite-dimensional linear spaces having the bases $\{\boldsymbol{v}_i \mid 1 \leqslant i \leqslant m\}$ and $\{\boldsymbol{w}_j \mid 1 \leqslant i \leqslant n\}$, respectively. If $\{\boldsymbol{f}^i \mid 1 \leqslant i \leqslant \ell\}$ is the dual basis in V^* of $\{\boldsymbol{v}_i \mid 1 \leqslant i \leqslant \ell\}$, then the set of morphisms

$$\{h^i_j : V \longrightarrow W \mid h^i_j(\boldsymbol{v}) = \boldsymbol{f}^i(\boldsymbol{v})\boldsymbol{w}_j \text{ for } 1 \leqslant i \leqslant, 1 \leqslant j \leqslant n\}$$

is a basis in $\mathsf{Hom}(V, W)$.

Solution: It is immediate that h^i_j are homomorphisms in $\mathsf{Hom}(V, W)$ for $1 \leqslant i \leqslant m$ and $1 \leqslant j \leqslant n$. Let h be an arbitrary homomorphism in $\mathsf{Hom}(V, W)$. If $\boldsymbol{v} \in V$, then $\boldsymbol{v} = a_1\boldsymbol{v}_1 + \cdots + a_m\boldsymbol{v}_m$, hence $h(\boldsymbol{v}) = \sum_{i=1}^{m} a_i h(\boldsymbol{v}_i)$.
Since $h(\boldsymbol{v}_i) \in W$, we can write: $h(\boldsymbol{v}_i) = \sum_{j=1}^{n} b_{ij}\boldsymbol{w}_j$, for $1 \leqslant i \leqslant m$. Therefore,

$$h(\boldsymbol{v}) = \sum_{i=1}^{m} a_i h(\boldsymbol{v}_i) = \sum_{i=1}^{m} a_i \sum_{j=1}^{n} b_{ij}\boldsymbol{w}_j = \sum_{i=1}^{m}\sum_{j=1}^{n} a_i b_{ij}\boldsymbol{w}_j.$$

Since $\{\boldsymbol{f}^i \mid 1 \leqslant i \leqslant \ell\}$ is the dual basis of $\{\boldsymbol{v}_i \mid 1 \leqslant i \leqslant \ell\}$, we have

$$h^i_j(\boldsymbol{v}) = \boldsymbol{f}^i(\boldsymbol{v})\boldsymbol{w}_j = \boldsymbol{f}^i\left(\sum_{\ell=1}^{m} a_\ell \boldsymbol{v}_\ell\right)\boldsymbol{w}_j = a_i\boldsymbol{w}_j.$$

Thus, $h(v) = \sum_{i=1}^{m}\sum_{j=1}^{n} b_{ij} h^i_j(\boldsymbol{v})$, hence

$$\{h^i_j : V \longrightarrow W \mid h^i_j(\boldsymbol{v}) = \boldsymbol{f}^i(\boldsymbol{v})\boldsymbol{w}_j \text{ for } 1 \leqslant i \leqslant, 1 \leqslant j \leqslant n\}$$

spans the linear space $\mathsf{Hom}(V, W)$.

If $\sum_{i=1}^{m}\sum_{j=1}^{n} c_{ij}h_j^i$ is the zero morphism of $\mathsf{Hom}(V,W)$, then $\sum_{i=1}^{m}\sum_{j=1}^{n} c_{ij}h_j^i(\boldsymbol{v}_i) = \boldsymbol{0}_W$, which amounts to $\sum_{i=1}^{m}\sum_{j=1}^{n} c_{ij}\boldsymbol{f}^i(\boldsymbol{v})\boldsymbol{w}_j = \boldsymbol{0}_W$. Since $\{\boldsymbol{w}_j \mid 1 \leqslant i \leqslant n\}$ is a basis in W, we obtain $\sum_{i=1}^{m} c_{ij}\boldsymbol{f}^i(\boldsymbol{v}) = \boldsymbol{0}_V$, and therefore, $\sum_{i=1}^{m} c_{ij}v_i\boldsymbol{v}_i = \boldsymbol{0}_V$, which implies $c_{ij} = 0$. Thus, $\{h_j^i : V \longrightarrow W \mid h_j^i(\boldsymbol{v}) = \boldsymbol{f}^i(\boldsymbol{v})\boldsymbol{w}_j \text{ for } 1 \leqslant i \leqslant, 1 \leqslant j \leqslant n\}$ is a basis in $\mathsf{Hom}(V,W)$.

(24) Let V, W, U be three linear spaces and let $f : V \times W \longrightarrow U$ be a bilinear mapping.
Define the mapping $\phi : V \times W \longrightarrow U$ as $\phi(\boldsymbol{z}) = f(\pi_1(\boldsymbol{z}), \pi_2(\boldsymbol{z}))$, where π_1 and π_2 are the canonical projections of $V \times W$ on V and W, respectively. Prove that $\phi(\boldsymbol{z}_1 + \boldsymbol{z}_2) + \phi(\boldsymbol{z}_1 - \boldsymbol{z}_2) = 2(\phi(z_1) + \phi(z_2))$ and $\phi(a\boldsymbol{z}) = a^2\phi(\boldsymbol{z})$ for every $a \in \mathbb{R}$.

(25) Let $f : V^3 \longrightarrow \mathbb{R}$ be a real multilinear form. Prove that if f is symmetric in the first two arguments and is skew-symmetric in the last two arguments, then $f(\boldsymbol{x}, \boldsymbol{y}, \boldsymbol{z}) = 0$ for $\boldsymbol{x}, \boldsymbol{y}, \boldsymbol{z} \in V$.

Solution: We have

$$
\begin{aligned}
f(\boldsymbol{x},\boldsymbol{y},\boldsymbol{z}) &= f(\boldsymbol{y},\boldsymbol{x},\boldsymbol{z}) \\
&\qquad \text{(by the symmetry in the first two arguments)} \\
&= -f(\boldsymbol{y},\boldsymbol{z},\boldsymbol{x}) \\
&\qquad \text{(by the skew-symmetry in the last two arguments)} \\
&= -f(\boldsymbol{z},\boldsymbol{y},\boldsymbol{x}) \\
&\qquad \text{(by the symmetry in the first two arguments)} \\
&= f(\boldsymbol{z},\boldsymbol{x},\boldsymbol{y}) \\
&\qquad \text{(by the skew-symmetry in the last two arguments)} \\
&= f(\boldsymbol{x},\boldsymbol{z},\boldsymbol{y}) \\
&\qquad \text{(by the symmetry in the first two arguments)} \\
&= -f(\boldsymbol{x},\boldsymbol{y},\boldsymbol{z}) \\
&\qquad \text{(by the skew-symmetry in the last two arguments)},
\end{aligned}
$$

hence $2f(\boldsymbol{x}, \boldsymbol{y}, \boldsymbol{z}) = 0$.

(26) Prove that every bilinear form f over $\mathbb{R}$ or $\mathbb{C}$ is uniquely expressible as a sum $f = f_1 + f_2$, where f_1 is symmetric and f_2 is skew-symmetric.

(27) Prove that every symmetric bilinear form f over $\mathbb{R}$ or $\mathbb{C}$ is determined by its values of the form $f(\boldsymbol{v}, \boldsymbol{v})$.

Solution: For any $\boldsymbol{v}$ and $\boldsymbol{w}$, we can write

$$\begin{aligned}
&\frac{1}{2}\left(f(\boldsymbol{v}+\boldsymbol{w}, \boldsymbol{v}+\boldsymbol{w})-f(\boldsymbol{v}, \boldsymbol{v})-f(\boldsymbol{w}, \boldsymbol{w})\right)\\
&\quad=\frac{1}{2}(f(\boldsymbol{v}, \boldsymbol{w})+f(\boldsymbol{w}, \boldsymbol{v}))=f(\boldsymbol{v}, \boldsymbol{w}),
\end{aligned}$$

which leads to the desired conclusion.

The *generalized Kronecker symbol* $\delta^{i_1\cdots i_p}_{j_1\cdots j_p}$ is defined as

$$\delta^{i_1\cdots i_p}_{j_1\cdots j_p}=\begin{cases}1 & \text{if } \begin{pmatrix} i_1 & \cdots & i_p\\ j_1 & \cdots & j_p\end{pmatrix} \text{ is an even permutation}\\ -1 & \text{if } \begin{pmatrix} i_1 & \cdots & i_p\\ j_1 & \cdots & j_p\end{pmatrix} \text{ is an odd permutation}\\ 0 & \text{in all other cases.}\end{cases}$$

Thus, if $\{i_1, \ldots, i_p\}$ or $\{j_1, \ldots, j_p\}$ does not consist of distinct integers or $\{i_1, \ldots, i_p\} \neq \{j_1, \ldots, j_p\}$, we have

$$\delta^{i_1\cdots i_p}_{j_1\cdots j_p}=0.$$

(28) Verify that $\epsilon^{i_1 i_2\cdots i_n}=\delta^{i_1 i_2\cdots i_n}_{12\cdots n}$ and $\epsilon_{i_1 i_2\cdots i_n}=\delta^{12\cdots n}_{i_1 i_2\cdots i_n}$.

(29) Prove that

$$\sum_{i_1=1}^{n}\cdots\sum_{i_k=1}^{n}\delta^{i_1\cdots i_k}_{i_1\cdots i_k}=\frac{n!}{(n-k)!}.$$

(30) Prove that

$$\begin{aligned}
\delta^{i_1\cdots i_p}_{j_1\cdots j_p}&=\sum_{\phi\in PERM_p}(-1)^{inv(\phi)}\delta^{i_1\cdots i_p}_{j_{\phi(1)}\cdots j_{\phi(p)}}\\
&=\sum_{\phi\in PERM_p}(-1)^{inv(\phi)}\delta^{i_{\phi(1)}\cdots i_{\phi(p)}}_{j_1\cdots j_p}.
\end{aligned}$$

Bibliographical Comments

The books of MacLane and Birkhoff [108], Artin [3], and van der Waerden [165] contain a vast amount of material from many areas of algebra presented in a lucid and readable manner. Supplement 2.11 appears in [110].

Chapter 3

Matrices

3.1 Introduction

Matrices are rectangular arrays and their elements belong typically to a field. Numeric matrices (that is, matrices whose elements belong to $\mathbb{R}$ or $\mathbb{C}$) serve as representations of linear transformations between linear spaces in the context of certain linear bases in these spaces.

Historically, determinants (discussed in Chapter 5) preceded matrices. The term *matrix* was introduced by Sylvester,[1] who regarded matrices as generators of determinants and therefore adopted the Latin word *matrix* (in Latin, a neutral noun *matrix, matricis*) (which signifies *womb*) to designate arrays of numbers.

We begin with matrices whose entries belong to arbitrary sets and then focus on matrices whose components belong to fields. Then, we present several classes of matrices, discuss matrix partitioning, and the notion of invertible matrix. The fundamental relationship between matrices and linear transformations between finite-dimensional linear spaces is explored and properties of linear mappings are transferred to matrices.

[1]James Joseph Sylvester was born on September 3, 1814 in London and died on March 15, 1897 in the same city. Sylvester made fundamental contributions in algebra, number theory, and Combinatorics. He studied at the University of London and at St. John's College in Cambridge. Sylvester taught at the University College London, the Royal Military Academy, Johns Hopkins University in Baltimore, and Oxford.

The notion of matrix rank is introduced starting from the dimension of the range of linear mappings attached to matrices. We discuss linear systems and the application of matrices in solving such systems. We conclude with a presentation of Kronecker, Hadamard, and Khatri–Rao matrix products.

3.2 Matrices with Arbitrary Elements

We define a class of two-argument finite functions that is ubiquitous in mathematics and is central for linear algebra and its applications.

Definition 3.1. A *matrix on* $\mathbb{C}$ is a function

$$A : \{1, \ldots, m\} \times \{1, \ldots, n\} \longrightarrow \mathbb{C}.$$

The pair (m, n) is the *format* of the matrix A.

If $A(i, j) \in \mathbb{R}$ for $1 \leqslant i \leqslant m$ and $1 \leqslant j \leqslant n$, then we say that A is a matrix on $\mathbb{R}$.

Matrices can be conceptualized as two-dimensional arrays as follows:

$$\begin{pmatrix} A(1,1) & A(1,2) & \ldots & A(1,n) \\ A(2,1) & A(2,2) & \ldots & A(2,n) \\ \vdots & \vdots & \ldots & \vdots \\ A(i,1) & A(i,2) & \ldots & A(i,n) \end{pmatrix}.$$

If $A : \{1, \ldots, m\} \times \{1, \ldots, n\} \longrightarrow \mathbb{C}$, we say that A is an $(m \times n)$-matrix on $\mathbb{C}$. The set of all such matrices are denoted by $\mathbb{C}^{m \times n}$.

The element $A(i, j)$ of the matrix $A \in \mathbb{C}^{m \times n}$ is denoted either by A_{ij} or by a_{ij}, for $1 \leqslant i \leqslant m$ and $1 \leqslant j \leqslant n$.

Definition 3.2. A *row* in a matrix is $\mathbb{C}^{1 \times n}$; a *column* in a matrix is $\mathbb{C}^{n \times 1}$.

Rows or columns are denoted by small bold-faced letters: $\boldsymbol{r}$, $\boldsymbol{s}$, *etc.*

A matrix $A \in \mathbb{C}^{m\times n}$ can be regarded as consisting of m rows, where the ith row is a sequence of the form

$$(a_{i1}, a_{i2}, \ldots, a_{in}),$$

for $1 \leqslant i \leqslant n$, or as a collection of n columns, where the jth column has the form

$$\begin{pmatrix} a_{1j} \\ a_{2j} \\ \vdots \\ a_{mj} \end{pmatrix}$$

for $1 \leqslant j \leqslant m$. The ith row of a matrix is denoted by $A(i,:)$; similarly, the jth column of A is denoted by $A(:,j)$.

The *main diagonal* of a matrix $A \in \mathbb{C}^{m\times n}$ is the set $\{a_{ii} \mid 1 \leqslant i \leq \min\{m,n\}\}$. Note that the set $\{a_{ij} \mid i-j=k\}$ for $1 \leq k \leqslant n-1$ consists of elements located on the kth diagonal above the main diagonal of the matrix A. Similarly, the set $\{a_{ij} \mid j-i=k\}$ for $1 \leqslant k \leqslant n-1$ consists of elements located on the kth diagonal below the main diagonal.

Definition 3.3. A *square matrix* on $\mathbb{C}$ is an $(n \times n)$-matrix on the set $\mathbb{C}$ for some $n \geqslant 1$. The number n is referred to as the *order* of the matrix A.

An $(n \times n)$-square matrix $A = (a_{ij})$ on $\mathbb{C}$ is *symmetric* if $a_{ij} = a_{ji}$ for every i, j such that $1 \leqslant i, j \leqslant n$.

A is *skew-symmetric* if $a_{ij} = -a_{ij}$ for $1 \leqslant i, j \leqslant n$.

Example 3.1. The (3×3)-matrix

$$A = \begin{pmatrix} 1 & 0.5 & 1 \\ 0.5 & 1 & 2 \\ 1 & 2 & 0.3 \end{pmatrix}$$

over the set of reals $\mathbb{R}$ is symmetric.

The matrix

$$B = \begin{pmatrix} 0 & 1 & -2 & & \\ -1 & 0 & 32 & -3 & 0 \end{pmatrix}$$

is skew-symmetric.

Example 3.2. Let $f : \mathbb{R}^m \times \mathbb{R}^m \longrightarrow \mathbb{R}$ be a skew-symmetric form defined by $f(\boldsymbol{x}, \boldsymbol{y}) = \boldsymbol{x}' A \boldsymbol{y}$, where $A \in \mathbb{R}^{m \times m}$. We have

$$\sum_{i,j=1}^{m} a_{ij} x_i y_j = - \sum_{i,j=1}^{m} a_{ij} y_i x_j.$$

Changing the summation indices in the second term of the above equality yields

$$\sum_{i,j=1}^{m} (a_{ij} + a_{ji}) x_i y_j = 0$$

for every $\boldsymbol{x}, \boldsymbol{y} \in \mathbb{R}^m$, which implies $a_{ij} + a_{ji} = 0$, hence $a_{ij} = -a_{ji}$. This shows that the matrix A that defines f is skew-symmetric.

Theorem 3.1. *Let $A \in \mathbb{R}^{n \times n}$. We have $\boldsymbol{x}' A \boldsymbol{x} = 0$ for every $\boldsymbol{x} \in \mathbb{R}^n$ if and only if A is a skew-symmetric matrix.*

Proof. Since $\boldsymbol{x}' A \boldsymbol{x} = \sum_{i=1}^{n} \sum_{j=1}^{n} x_i a_{ij} x_j = \sum_{i<j} (a_{ij} + a_{ji}) x_i x_j$ for every $\boldsymbol{x}$, it follows that $a_{ij} = -a_{ji}$, which shows that A is skew-symmetric. □

Definition 3.4. The *transpose of an* $(m \times n)$*-matrix* A is an $(n \times m)$-matrix A' defined by $A'(i, j) = a_{ji}$ for $1 \leqslant i \leqslant m$ and $1 \leqslant j \leqslant n$.

Thus, the transpose A' of a matrix A has as rows the transposed columns of A; equivalently, the columns of A' are the transposed rows of A.

It is easy to verify that, for any matrix $A \in \mathbb{C}^{m \times n}$, we have

$$(A')' = A.$$

Also, the transpose of a row $\boldsymbol{r} \in \mathbb{C}^{1 \times n}$ is a column $\boldsymbol{r}' \in \mathbb{C}^{n \times 1}$.

Definition 3.5. Let A be an $m \times n$-matrix and let $K = \{i_1, \ldots, i_k\} \subseteq \{1, \ldots, m\}$ and $H = \{j_1, \ldots, j_h\} \subseteq \{1, \ldots, m\}$, where $i_1 < \cdots < i_k$ and $j_1 < \cdots < j_h$. The *submatrix of* A determined by K and H is the $k \times h$ matrix B defined by $B(p, q) = p_{i_p j_q}$ for $1 \leqslant p \leqslant k$ and $1 \leqslant q \leqslant h$. We denote this submatrix by $A \begin{bmatrix} K \\ H \end{bmatrix} = A \begin{bmatrix} i_1 & \cdots & i_k \\ j_1 & \cdots & j_h \end{bmatrix}$.

Example 3.3. Consider the 5×4-matrix

$$A = \begin{pmatrix} 1 & 2 & 3 & 4 & 5 \\ -1 & 2 & 5 & 1 & 2 \\ 5 & 6 & 7 & 8 & 9 \\ 9 & 5 & 3 & 1 & 0 \\ 0 & 2 & 4 & 6 & 8 \end{pmatrix}.$$

The submatrix determined by the sets $\{1, 3, 4\}$ and $\{2, 3\}$ is

$$A \begin{bmatrix} 1, 3, 4 \\ 2, 3 \end{bmatrix} = \begin{pmatrix} 2 & 3 \\ 6 & 7 \\ 5 & 3 \end{pmatrix}.$$

Definition 3.6. If each of the sequences $(i_1, \ldots, i_k)$ and $(j_1, \ldots, j_h)$ consists of consecutive integers, then the submatrix $A \begin{bmatrix} i_1 & \cdots & i_k \\ j_1 & \cdots & j_h \end{bmatrix}$ is referred to as a *contiguous submatrix*.

Let $A \in \mathbb{C}^{m \times n}$ be a matrix. Any submatrix of the form $A \begin{bmatrix} i_1 & \cdots & i_k \\ i_1 & \cdots & i_k \end{bmatrix}$ is called a *principal submatrix* of A.

A *leading submatrix* of A is a submatrix of the form $A \begin{bmatrix} 1 & \cdots & k \\ 1 & \cdots & h \end{bmatrix}$, where $1 \leqslant k \leqslant m$ and $1 \leqslant h \leqslant n$.

A *trailing submatrix of* A has the form $A \begin{bmatrix} m-k & \cdots & m \\ n-h & \cdots & n \end{bmatrix}$, where $0 \leqslant k \leqslant m-1$ and $0 \leqslant h \leqslant n-1$.

Example 3.4. Consider the matrix $A \in \mathbb{C}^{4 \times 5}$:

$$A = \begin{pmatrix} a_{11} & a_{12} & a_{13} & a_{14} & a_{15} \\ a_{21} & a_{22} & a_{23} & a_{24} & a_{25} \\ a_{31} & a_{32} & a_{33} & a_{34} & a_{35} \\ a_{41} & a_{42} & a_{43} & a_{44} & a_{45} \end{pmatrix}.$$

Then, the principal matrix $A \begin{bmatrix} 1\,3\,4 \\ 1\,3\,4 \end{bmatrix}$ is

$$\begin{pmatrix} a_{11} & a_{13} & a_{14} \\ a_{31} & a_{33} & a_{34} \\ a_{41} & a_{43} & a_{44} \end{pmatrix}.$$

A leading submatrix of A is

$$\begin{pmatrix} a_{11} & a_{12} & a_{13} \\ a_{21} & a_{22} & a_{23} \\ a_{31} & a_{32} & a_{33} \end{pmatrix}.$$

3.3 Fields and Matrices

In this section, we focus on matrices whose elements belong to a field $\mathbb{F}$, where $\mathbb{F}$ is the complex field $\mathbb{C}$ or the real field $\mathbb{R}$.

Definition 3.7. The $(n \times n)$-*unit matrix* on the field $\mathbb{F}$ is the square matrix $I_n \in \mathbb{F}^{n \times n}$ given by

$$I_n = \begin{pmatrix} 1 & 0 & 0 & \cdots & 0 \\ 0 & 1 & 0 & \cdots & 0 \\ 0 & 0 & 1 & \cdots & 0 \\ \vdots & \vdots & \vdots & \cdots & \vdots \\ 0 & 0 & 0 & \cdots & 1 \end{pmatrix}.$$

The $(m \times n)$-*zero matrix* is the $(m \times n)$-matrix $O_{m,n} \in \mathbb{F}^{n \times n}$ given by

$$O_{m,n} = \begin{pmatrix} 0 & 0 & 0 & \cdots & 0 \\ 0 & 0 & 0 & \cdots & 0 \\ 0 & 0 & 0 & \cdots & 0 \\ \vdots & \vdots & \vdots & \cdots & \vdots \\ 0 & 0 & 0 & \cdots & 0 \end{pmatrix}.$$

The $(m \times n)$-*complete matrix* is the $(m \times n)$-matrix $J_{m,n}$ given by

$$J_{m,n} = \begin{pmatrix} 1 & 1 & 1 & \cdots & 1 \\ 1 & 1 & 1 & \cdots & 1 \\ 1 & 1 & 1 & \cdots & 1 \\ \vdots & \vdots & \vdots & \cdots & \vdots \\ 1 & 1 & 1 & \cdots & 1 \end{pmatrix}.$$

The one-column matrix $O_{m,1}$ is denoted by $\mathbf{0}_m$. Similarly, the one column matrix having m rows

$$\begin{pmatrix} 1 \\ \vdots \\ 1 \end{pmatrix}$$

is denoted by $\mathbf{1}_m$. The subscripts are omitted whenever there is no ambiguity.

Definition 3.8. A *diagonal matrix* is a matrix $D \in \mathbb{F}^{m \times n}$ such that $i \neq j$ implies $d_{ij} = 0$. If $p \leqslant \min\{m, n\}$, then we denote the diagonal matrix

$$D = \begin{pmatrix} d_1 & 0 & 0 & 0 & \cdots & 0 \\ 0 & d_2 & 0 & 0 & \cdots & 0 \\ \vdots & \vdots & \vdots & \vdots & \cdots & \vdots \\ 0 & 0 & 0 & 0 & \cdots & 0 \\ 0 & 0 & 0 & 0 & \cdots & 0 \end{pmatrix}$$

by $\mathsf{diag}(d_1, \ldots, d_p)$.

Example 3.5. The $\mathbb{R}^{3 \times 4}$ matrix

$$\begin{pmatrix} 2 & 0 & 0 & 0 \\ 0 & 3 & 0 & 0 \\ 0 & 0 & 0 & 0 \end{pmatrix}$$

is denoted by $\mathsf{diag}(2, 3)$.

Definition 3.9. Let $A = (a_{ij})$ and $B = (b_{ij})$ be two matrices in $\mathbb{F}^{m \times n}$.

The *sum* of the matrices A and B is the matrix $A + B$ having the same format and defined by

$$(A + B)(i, j) = a_{ij} + b_{ij}$$

for $1 \leqslant i \leqslant m$ and $1 \leqslant j \leqslant n$.

Example 3.6. Let $A, B \in \mathbb{R}^{2 \times 3}$ be two matrices given by

$$A = \begin{pmatrix} 1 & -2 & 3 \\ 0 & 2 & -1 \end{pmatrix} \text{ and } B = \begin{pmatrix} -1 & 2 & 3 \\ 1 & 4 & 2 \end{pmatrix}.$$

Their sum is the matrix

$$A + B = \begin{pmatrix} 0 & 0 & 6 \\ 1 & 6 & 1 \end{pmatrix}.$$

It is easy to verify that the matrix sum is an associative and commutative operation on $\mathbb{F}^{m\times n}$; that is,

$$A + (B + C) = (A + B) + C,$$
$$A + B = B + A$$

for all $A, B, C \in \mathbb{F}^{m\times n}$.

The zero matrix $O_{m,n}$ acts as an additive unit on the set $\mathbb{F}^{m\times n}$; that is,

$$A + O_{m,n} = O_{m,n} + A$$

for every $A \in \mathbb{F}^{m\times n}$.

The additive inverse, or the *opposite of a matrix* $A = (a_{ij}) \in \mathbb{F}^{m\times n}$, is the matrix $-A$ given by $(-A)(i,j) = -a_{ij}$ for $1 \leqslant i \leqslant m$ and $1 \leqslant j \leqslant n$.

Example 3.7. The opposite of $A \in \mathbb{R}^{2\times 3}$, given by

$$A = \begin{pmatrix} 1 & -2 & 3 \\ 0 & 2 & -1 \end{pmatrix}$$

is the matrix

$$-A = \begin{pmatrix} -1 & 2 & -3 \\ 0 & -2 & 1 \end{pmatrix}.$$

It is immediate that $A + (-A) = O_{2,3}$.

The previous discussion shows that the set of matrices $\mathbb{F}^{m\times n}$ defined on a unit ring $(S, \{0, 1, +, -, \cdot\})$ is an Abelian group $(\mathbb{F}^{m\times n}, \{O_{m,n}, +, -\})$.

Definition 3.10. Let $A \in \mathbb{F}^{m\times n}$ and $B \in \mathbb{F}^{n\times p}$ be two matrices. The *product* of the matrices A, B is the matrix $C \in \mathbb{F}^{m\times p}$ defined by

$$C(i,k) = \sum_{j=1}^{n} a_{ij} b_{jk},$$

where $1 \leqslant i \leqslant m$ and $1 \leqslant k \leqslant p$. The product of the matrices A, B is denoted by AB.

Matrix multiplication of A and B is possible only if number of columns of A is equal to the number of rows of the second matrix B. Any pair of matrices (A, B) that satisfies this condition is said to be *conformant*.

Theorem 3.2. *Matrix multiplication is associative.*

Proof. Let $A = (a_{ij}) \in \mathbb{F}^{m\times n}$, $B = (b_{jk}) \in \mathbb{F}^{n\times p}$, and $C = (c_{ik}) \in \mathbb{F}^{p\times r}$ be three matrices. We need to prove that $(AB)C = A(BC)$.

By applying the definition of the matrix product, we have

$$
\begin{aligned}
((AB)C)(i,\ell) &= \sum_{k=1}^{p} (AB)(i,k)C(k,\ell) \\
&= \sum_{k=1}^{p} \left(\sum_{j=1}^{n} a_{ij} b_{jk} \right) c_{k\ell} \\
&= \sum_{j=1}^{n} a_{ij} \sum_{k=1}^{p} b_{jk} c_{k\ell} \\
&= \sum_{j=1}^{n} a_{ij} (BC)(j,\ell) \\
&= (A(BC))(i,\ell)
\end{aligned}
$$

for $1 \leqslant i \leqslant m$ and $1 \leqslant \ell \leqslant r$, which shows that matrix multiplication is indeed associative. □

Theorem 3.3. *If* $A \in \mathbb{F}^{m\times n}$, *then* $I_m A = A I_n = A$.

Proof. The statement follows immediately from the definition of a matrix product. □

Thus, I_n is a unit relative to matrix multiplication considered as an operation on the set of square matrices $\mathbb{F}^{n\times n}$.

The product of matrices is not commutative. Indeed, consider the matrices $A, B \in \mathbb{Z}^{2\times 2}$ defined by

$$
A = \begin{pmatrix} 0 & 1 \\ 2 & 3 \end{pmatrix} \text{ and } B = \begin{pmatrix} -1 & 1 \\ 1 & 0 \end{pmatrix}.
$$

We have

$$AB = \begin{pmatrix} 1 & 0 \\ 1 & 2 \end{pmatrix} \text{ and } BA = \begin{pmatrix} 2 & 2 \\ 0 & 1 \end{pmatrix},$$

so $AB \neq BA$.

The above discussion shows that for a field F the set $F^{m\times n}$ is an F-linear space.

Definition 3.11. A matrix $A \in \mathbb{F}^{n\times n}$ is *upper triangular* if $j < i$ implies $a_{ij} = 0$ and is *strictly upper triangular* if $j \leqslant i$ implies $a_{ij} = 0$.

A is *lower triangular* (*strictly lower triangular*) if A' is upper triangular (strictly upper triangular).

Example 3.8. The matrix $L \in \mathbb{Z}^{4\times 4}$ given by

$$L = \begin{pmatrix} 1 & 0 & 0 & 0 \\ -1 & 2 & 0 & 0 \\ 3 & 1 & 2 & 0 \\ -7 & 6 & 1 & -6 \end{pmatrix}$$

is a lower triangular matrix. The matrix $U \in \mathbb{Z}^{4\times 4}$

$$U = \begin{pmatrix} 1 & 2 & 3 & 4 \\ 0 & -1 & 2 & 3 \\ 0 & 0 & -1 & -2 \\ 0 & 0 & 0 & 2 \end{pmatrix}$$

is an upper triangular matrix.

Theorem 3.4. *The product of two lower (upper) triangular matrices $L \in \mathbb{F}^{m\times n}$ and $H \in \mathbb{F}^{n\times p}$ is a lower (upper) triangular matrix in $\mathbb{F}^{m\times p}$.*

Proof. Since L is lower triangular, $i < j$ implies $l_{ij} = 0$. Similarly, since H is lower triangular, $j < k$ implies $h_{jk} = 0$.

Suppose now that $i < k$. We have

$$\begin{aligned} (LH)_{ik} &= \textstyle\sum_{j=1}^{n} l_{ij}h_{jk} = \sum_{j=1}^{i} l_{ij}h_{jk} \quad &&(\text{because } l_{ij} = 0 \text{ when } j > i) \\ &= 0 &&(\text{because } j \leqslant i < k \text{ implies } h_{jk} = 0). \end{aligned}$$

Therefore, LH is a lower triangular matrix.

The argument for upper triangular matrices is similar. □

A simple and useful observation is contained in the next theorem.

Theorem 3.5. *Let A and B be two matrices in $\mathbb{C}^{n\times n}$.*

If $A = BR$, where R is an upper triangular matrix, then the first q columns of A are linear combinations of the first q columns of B for $1 \leqslant q \leqslant n$.

If $A = LB$, where L is a lower triangular matrix, then the first q rows of A are linear combinations of the first q rows of B for $1 \leqslant q \leqslant n$.

Proof. Indeed, the equality $A = BR$ can be written as

$$(\boldsymbol{a}_1\ \boldsymbol{a}_2 \cdots \boldsymbol{a}_n) = (\boldsymbol{b}_1\ \boldsymbol{b}_2 \cdots \boldsymbol{b}_n) \begin{pmatrix} r_{11} & r_{12} & \cdots & r_{1n} \\ 0 & r_{22} & \cdots & r_{2n} \\ \vdots & \vdots & \cdots & \vdots \\ 0 & 0 & \cdots & r_{rr} \end{pmatrix},$$

so

$$\begin{aligned} \boldsymbol{a}_1 &= r_{11}\boldsymbol{b}_1, \\ \boldsymbol{a}_2 &= r_{12}\boldsymbol{b}_1 + r_{22}\boldsymbol{b}_2 \\ &\vdots \\ \boldsymbol{a}_n &= r_{1n}\boldsymbol{b}_1 + r_{2n}\boldsymbol{b}_2 + \cdots + r_{nn}\boldsymbol{b}_n. \end{aligned}$$

By applying a transposition it is easy to see that if $A = LB$, where L is a lower triangular matrix, the first q rows of A are linear combinations of the first q rows of B. □

Let $S_{A,q}$ be the *subspace generated by the first q columns of a matrix* $A \in \mathbb{C}^{m\times n}$. This notation is used in the next statement.

3.4 Invertible Matrices

Let $A \in \mathbb{C}^{n\times n}$ be a square matrix. Suppose that there exist two matrices U and V such that $AU = I_n$ and $VA = I_n$. This implies

$$V = VI_n = V(AU) = (VA)U = I_nU = U.$$

Thus, if $AU = VA = I_n$, the two matrices involved, U and V, must be equal.

Definition 3.12. A matrix $A \in \mathbb{C}^{n\times n}$ is invertible if there exists a matrix $B \in \mathbb{C}^{n\times n}$ such that $AB = BA = I_n$.

Suppose that C is another matrix such that $AC = CA = I_n$. By the associativity of the matrix product we have $C = CI_n = C(AB) = (CA)B = I_nB = B$. Therefore, if A is invertible, there is exactly one matrix B such that $AB = BA = I_n$. We denote the matrix B by A^{-1} and we refer to it as the *inverse of the matrix, A*.

Note that $A \in \mathbb{C}^{n\times n}$ is a unitary matrix if and only if $A^{-1} = A^{\mathsf{H}}$.

Theorem 3.6. *If $A, B \in \mathbb{C}^{n\times n}$ are two invertible matrices, then the product AB is invertible and $(AB)^{-1} = B^{-1}A^{-1}$.*

Proof. Applying the definition of the inverse of a matrix, we obtain

$$(AB)(B^{-1}A^{-1}) = A(BB^{-1})A^{-1} = AI_nA^{-1} = AA^{-1} = I_n,$$

which implies $(AB)^{-1} = B^{-1}A^{-1}$. □

Theorem 3.7. *If $A \in \mathbb{C}^{n\times n}$ is invertible, then A^{H} is invertible and $(A^{\mathsf{H}})^{-1} = (A^{-1})^{\mathsf{H}}$.*

Proof. Since $AA^{-1} = I_n$, we have $(A^{-1})^{\mathsf{H}}A^{\mathsf{H}} = I_n$, which shows that $(A^{-1})^{\mathsf{H}}$ is the inverse of A^{H} and $(A^{\mathsf{H}})^{-1} = (A^{-1})^{\mathsf{H}}$. □

Example 3.9. Let

$$A = \begin{pmatrix} a_{11} & a_{12} \\ a_{21} & a_{22} \end{pmatrix}$$

be a matrix in $\mathbb{R}^{2\times 2}$. We seek to determine conditions under which A is invertible. Suppose that

$$X = \begin{pmatrix} x_{11} & x_{12} \\ x_{21} & x_{22} \end{pmatrix}$$

is a matrix in $\mathbb{R}^{2\times 2}$ such that $AX = I_2$. This matrix equality amounts to four scalar equalities:

$$\begin{aligned} a_{11}x_{11} + a_{12}x_{21} &= 1, \\ a_{11}x_{12} + a_{12}x_{22} &= 0, \\ a_{21}x_{11} + a_{22}x_{21} &= 1, \\ a_{21}x_{12} + a_{22}x_{22} &= 0, \end{aligned}$$

which, under certain conditions, can be solved with respect to $x_{11}, x_{12}, x_{21}, x_{22}$.

By multiplying the first equality by a_{22} and the third by $-a_{12}$ and adding the resulting equalities, we obtain

$$(a_{11}a_{22} - a_{12}a_{21})x_{11} = -a_{22}.$$

Thus, if $a_{11}a_{22} - a_{12}a_{21} \neq 0$, we have

$$x_{11} = -\frac{a_{22}}{a_{11}a_{22} - a_{12}a_{21}}.$$

The same condition, $a_{11}a_{22} - a_{12}a_{21} \neq 0$, suffices to allow us to obtain the value of the remaining components of X, as the reader can easily verify. Thus, A is an invertible matrix if and only if $a_{11}a_{22} - a_{12}a_{21} \neq 0$.

Example 3.10. Let

$$\phi : \begin{pmatrix} 1 & \cdots & k & \cdots & n \\ a_1 & \cdots & a_k & \cdots & a_n \end{pmatrix},$$

be a permutation of the set $\{1, \ldots, n\}$, where $a_k = \phi(k)$ for $1 \leqslant k \leqslant n$.

The matrix of this permutation is the square matrix $P_\phi = (p_{ij}) \in \{0,1\}^{n\times n}$, where

$$p_{ij} = \begin{cases} 1 & \text{if } j = \phi(i), \\ 0 & \text{otherwise} \end{cases} \tag{3.1}$$

for $1 \leqslant i, j \leqslant n$.

The set of invertible matrices in $\mathbb{R}^{n\times n}$ is a group relative to matrix multiplication known as the *general linear group* $\mathsf{GL}(n, \mathbb{R})$. Similarly, the set of invertible matrices in $\mathbb{C}^{n\times n}$ forms the group $\mathsf{GL}(n, \mathbb{C})$.

Note that the matrix of the permutation ι_n is I_n.

Also, if ϕ, ψ are two permutations of the set $\{1, \ldots, n\}$, then $P_{\psi\phi} = P_\phi P_\psi$. Indeed, since $(P_\phi P_\psi)_{ij} = \sum_{k=1}^n (P_\phi)_{ik}(P_\psi)_{kj}$, observe that only the term $(P_\phi)_{ik}(P_\psi)_{kj}$ in which $k = \phi(i)$ and $j = \psi(k)$ is different from 0. Thus, $(P_\phi P_\psi)_{ij} \neq 0$ if and only if $j = \psi(\phi(i))$, which means that $P_{\psi\phi} = P_\phi P_\psi$.

Thus, if ϕ and ϕ^{-1} are two inverse permutations in $PERM_n$, we have $P_\phi P_{\phi^{-1}} = I_n$, so P_ϕ is invertible and $P_\phi^{-1} = P_{\phi^{-1}}$.

For instance, if $\phi \in PERM_4$ is the permutation considered in Example 1.6,

$$\phi : \begin{pmatrix} 1 & 2 & 3 & 4 \\ 3 & 1 & 4 & 2 \end{pmatrix},$$

then

$$P_\phi = \begin{pmatrix} 0 & 0 & 1 & 0 \\ 1 & 0 & 0 & 0 \\ 0 & 0 & 0 & 1 \\ 0 & 1 & 0 & 0 \end{pmatrix}.$$

Its inverse is

$$P_\phi^{-1} = P_{\phi^{-1}} = \begin{pmatrix} 0 & 1 & 0 & 0 \\ 0 & 0 & 0 & 1 \\ 1 & 0 & 0 & 0 \\ 0 & 0 & 1 & 0 \end{pmatrix}.$$

It is easy to verify that the inverse of a permutation matrix P_ϕ coincides with its transpose $(P_\phi)'$.

Observe that if $A \in \mathbb{R}^{n\times n}$ having the rows $\boldsymbol{r}_1, \ldots, \boldsymbol{r}_n$ and P_ϕ is a permutation matrix, then $P_\phi A$ is the matrix whose rows are $\boldsymbol{r}_{\phi(1)}, \boldsymbol{r}_{\phi(2)}, \ldots, \boldsymbol{r}_{\phi(n)}$. Similarly, if the columns of A are $\mathbf{c}_1, \ldots, \mathbf{c}_n$, the columns of the matrix AP_ϕ are $\mathbf{c}_{\phi(1)}, \ldots, \mathbf{c}_{\phi(n)}$. In other words, $P_\phi A$ is obtained from A be permuting its rows according to the permutation ϕ and AP_ϕ is obtained from A by permuting the columns according to the same permutation.

Since every column and row of a permutation matrix contains exactly one 1, it follows that each such matrix is also a doubly stochastic matrix.

Theorem 3.8. *Let $A \in \mathbb{R}^{n\times n}$ be a lower (upper) triangular matrix such that $a_{ii} \neq 0$ for $1 \leqslant i \leqslant n$. The matrix A is invertible and its inverse is a lower (upper) triangular matrix having diagonal elements equal to the reciprocal of the diagonal elements of A.*

Proof. Let A be a lower triangular matrix

$$A = \begin{pmatrix} a_{11} & 0 & 0 & \cdots & 0 \\ a_{21} & a_{22} & 0 & \cdots & 0 \\ \vdots & \vdots & \vdots & \cdots & \vdots \\ a_{n1} & a_{n2} & a_{n3} & \cdots & a_{nn} \end{pmatrix},$$

where $a_{ii} \neq 0$ for $1 \leqslant i \leqslant n$. The proof is by induction on $n \geqslant 1$.

The base case, $n = 1$, is immediate, since the inverse of the matrix (a_{11}) is $\left(\frac{1}{a_{11}}\right)$.

Suppose that the statement holds for matrices in $\mathbb{R}^{(n-1)\times(n-1)}$. Then A can be written as

$$A = \left(\begin{array}{c|c} B & \mathbf{0}_{n-1} \\ \hline a_{n1}\ a_{n2}\ \cdots\ a_{n\,n-1} & a_{nn} \end{array}\right),$$

where $B \in \mathbb{R}^{(n-1)\times(n-1)}$ is a lower triangular matrix. By the inductive hypothesis, this matrix is invertible, its inverse B^{-1} is also lower triangular, and the diagonal elements of B^{-1} are the reciprocal elements of the corresponding diagonal elements of B. The matrix

$$\left(\begin{array}{c|c} B^{-1} & \mathbf{0}_{n-1} \\ \hline \boldsymbol{v} & \frac{1}{a_{nn}} \end{array}\right)$$

is the inverse of A, where $\boldsymbol{v} = -\frac{1}{a_{nn}}\boldsymbol{a}'B^{-1}$, and $\boldsymbol{a}' = (a_{n1}, a_{n2}, \ldots, a_{n\,n-1})$, as the reader can easily verify.

A similar argument can be used for upper triangular matrices. □

Theorem 3.9. *Let $A \in \mathbb{R}^{n\times n}$ be an invertible matrix, Then, its transpose A' is invertible and $(A')^{-1} = (A^{-1})'$.*

Proof. Observe that

$$A'(A^{-1})' = (A^{-1}A)' = I_n' = I_n.$$

Therefore, $(A')^{-1} = (A^{-1})'$. □

If $A \in \mathbb{R}^{n\times n}$ is an invertible matrix, we have $AA^{-1} = I_n$, so $\mathsf{trace}(A)\mathsf{trace}(A^{-1}) = \mathsf{trace}(I_n) = n$, by Theorem 3.14. This implies

$$\mathsf{trace}(A^{-1}) = \frac{n}{\mathsf{trace}(A)}. \tag{3.2}$$

The trace of a matrix $A \in \mathbb{C}^{n\times n}$ can be obtained as

$$\mathsf{trace}(A) = \sum_{i=1}^{n} \mathbf{e}_i' A \mathbf{e}_i.$$

Theorem 3.10. *Let A and B be two matrices in $\mathbb{C}^{n\times n}$. If $A = BR$, where R is an invertible upper triangular matrix, then $S_{A,q} = S_{B,q}$ for every q, $1 \leqslant q \leqslant n$.*

Proof. By Theorem 3.5, $A = BR$ implies $S_{A,q} \subseteq S_{B,q}$ for every q, $1 \leqslant q \leqslant n$. Since R is invertible, $B = AR^{-1}$, so $S_{B,q} \subseteq S_{A,q}$ for every q, $1 \leqslant q \leqslant n$ by the same theorem. This implies the desired equality. □

It is interesting to compute two matrix products that can be formed starting from the columns $\boldsymbol{u}$ and $\boldsymbol{v}$ given by

$$\boldsymbol{u} = \begin{pmatrix} u_1 \\ u_2 \\ \vdots \\ u_n \end{pmatrix} \text{ and } \boldsymbol{v} = \begin{pmatrix} v_1 \\ v_2 \\ \vdots \\ v_n \end{pmatrix}.$$

Note that $\boldsymbol{u}'\boldsymbol{v} \in \mathbb{F}^{1\times 1}$, that is,

$$\boldsymbol{u}'\boldsymbol{v} = u_1v_1 + u_2v_2 + \cdots + u_nv_n.$$

This product is known as the *inner product* of $\boldsymbol{u}$ and $\boldsymbol{v}$.

Theorem 3.11. *If $\boldsymbol{u} \in \mathbb{R}^{n\times 1}$ and $\boldsymbol{u}'\boldsymbol{u} = 0$, then $\boldsymbol{u} = \mathbf{0}$.*

Proof. Let

$$\boldsymbol{u} = \begin{pmatrix} u_1 \\ u_2 \\ \vdots \\ u_n \end{pmatrix}.$$

We have $\boldsymbol{u}'\boldsymbol{u} = u_1^2 + u_2^2 + \cdots + u_n^2$, so $\boldsymbol{u}'\boldsymbol{u} = 0$ implies $u_1 = u_2 = \cdots = u_n = 0$, that is, $\boldsymbol{u} = \mathbf{0}$. □

The mth power of a square matrix $A \in \mathbb{F}^{n\times n}$ (where $m \in \mathbb{N}$) can be defined inductively as follows.

Definition 3.13. The 0th power of any matrix $A \in \mathbb{F}^{n\times n}$ is $A^0 = I_n$. The $(m+1)$st power of A is the matrix $A^{m+1} = A^m A$ for $m \geqslant 0$.

Example 3.11. Let

$$A = \begin{pmatrix} a & b \\ -b & a \end{pmatrix}.$$

We have $A^0 = I_2$, $A^1 = A^0 A = I_2 A = A$,

$$A^2 = A^1 A = \begin{pmatrix} a & b \\ -ba & \end{pmatrix} \begin{pmatrix} a & b \\ -ba & \end{pmatrix} = \begin{pmatrix} a^2 - b^2 & 2ab \\ -2ab & a^2 - b^2 \end{pmatrix}.$$

Example 3.12. Let $E_n \in \mathbb{C}^{n\times n}$ be the matrix

$$E_n = \begin{pmatrix} \mathbf{0}_{n-1} & I_{n-1} \\ 0 & \mathbf{0}'_{n-1} \end{pmatrix}.$$

We claim that the pth power of this matrix is given by

$$(E_n)^p = \begin{pmatrix} \mathbf{0}_{n-1} & \cdots & \mathbf{0}_{n-1} & I_{n-p} \\ 0 & \cdots & 0 & \mathbf{0}'_{n-p} \end{pmatrix}.$$

Note that

$$E_n = (\mathbf{0}\ \mathbf{e}_1\ \cdots\ \mathbf{e}_{n-1}) = \begin{pmatrix} \mathbf{e}'_2 \\ \vdots \\ \mathbf{e}'_n \\ \mathbf{0}' \end{pmatrix}.$$

The equality that we need to prove is equivalent to

$$(E_n)^p = \begin{pmatrix} \mathbf{e}'_{p+1} \\ \vdots \\ \mathbf{e}'_n \\ \mathbf{0}' \\ \vdots \\ \mathbf{0}' \end{pmatrix}.$$

The proof is by induction on $p \geqslant 1$. The base case, $p = 1$, is immediate. Suppose that the equality holds for $p - 1$. Since

$$e_i' e_j = \begin{cases} 1 & \text{if } i = j, \\ 0 & \text{otherwise} \end{cases}$$

for $1 \leqslant i, j \leqslant n$, we have

$$(E_n)^p = (E_n)^{p-1} E_n = \begin{pmatrix} e_p' \\ \vdots \\ e_n' \\ \mathbf{0}' \\ \vdots \\ \mathbf{0}' \end{pmatrix} (\mathbf{0} \; e_1 \; \cdots \; e_{n-1})$$

(by the inductive hypothesis)

$$= \begin{pmatrix} e_{p+1}' \\ \vdots \\ e_n' \\ \mathbf{0}' \\ \vdots \\ \mathbf{0}' \end{pmatrix}.$$

The previous equality implies immediately $(E_n)^n = O_{n,n}$.

Theorem 3.12. *Let $T \in \mathbb{C}^{n \times n}$ be an upper (a lower) triangular matrix. Then T^k is an upper (a lower) triangular matrix.*

Proof. By Theorem 3.4, the product of two upper (lower) triangular matrices is an upper (lower) triangular matrix. The current statement follows immediately. □

Definition 3.14. A matrix $A \in \mathbb{C}^{n \times n}$ is *nilpotent* if there is $m \in \mathbb{N}$ such that $A^m = O_{n,n}$. The *nilpotency* of A is the number $\mathsf{nilp}(A) = \min\{m \in \mathbb{N} \mid A^m = O_{n,n}\}$.

If $A \in \mathbb{C}^{n \times n}$ is a nilpotent matrix, we have $\mathsf{nilp}(A) = m$ if and only if $A^m = O_{n,n}$ but $A^{m-1} \neq O_{n,n}$.

Example 3.13. Let a and b be two positive numbers in $\mathbb{R}$. The matrix $A \in \mathbb{R}^{3\times 3}$ given by

$$A = \begin{pmatrix} 0 & a & 0 \\ 0 & 0 & b \\ 0 & 0 & 0 \end{pmatrix}$$

is nilpotent because

$$A^2 = \begin{pmatrix} 0 & 0 & ab \\ 0 & 0 & 0 \\ 0 & 0 & 0 \end{pmatrix}. \text{ and } A^3 = \begin{pmatrix} 0 & 0 & 0 \\ 0 & 0 & 0 \\ 0 & 0 & 0 \end{pmatrix}.$$

Thus, $\mathsf{nilp}(A) = 3$.

Definition 3.15. A matrix $A \in \mathbb{F}^{n\times n}$ is *idempotent* if $A^2 = A$.

Example 3.14. The matrix

$$A = \begin{pmatrix} 0.5 & 1 \\ 0.25 & 0.5 \end{pmatrix}$$

is idempotent, as the reader can easily verify.

Matrix product is distributive with respect to matrix addition.

Theorem 3.13. *Let $A \in \mathbb{F}^{m\times n}$, $B, C \in \mathbb{F}^{n\times p}$, and $D \in \mathbb{F}^{p\times q}$. We have*

$$A(B + C) = AB + AC,$$
$$(B + C)D = BD + CD.$$

Proof. We have

$$\begin{aligned}(A(B + C))_{ik} &= \sum_{j=1}^{n} a_{ij}(b_{jk} + c_{jk}) \\ &= \sum_{j=1}^{n} a_{ij}b_{jk} + \sum_{j=1}^{n} a_{ij}c_{jk} \\ &= (AB)_{ik} + (AC)_{ik},\end{aligned}$$

for $1 \leqslant i \leqslant m$ and $1 \leqslant k \leqslant p$, which proves the first equality. The proof of the second equality is equally straightforward and is left to the reader. □

Definition 3.16. The *$(m \times n)$-vectorization mapping* is the mapping $\mathsf{vec} : \mathbb{C}^{m\times n} \longrightarrow \mathbb{C}^{mn}$ defined by

$$\mathsf{vec}(A) = \begin{pmatrix} a_{11} \\ \vdots \\ a_{m1} \\ \vdots \\ a_{1n} \\ \vdots \\ a_{mn} \end{pmatrix},$$

obtained by reading A column-wise.

The following equality is immediate for a matrix $A \in \mathbb{C}^{m\times n}$:

$$\mathsf{vec}(A) = \begin{pmatrix} Ae_1 \\ Ae_2 \\ \vdots \\ Ae_n \end{pmatrix}. \tag{3.3}$$

The vectorization mapping vec is an isomorphism between the linear space $\mathbb{C}^{m\times n}$ and the linear space $\mathbb{C}^{mn}$, as the reader can easily verify.

Example 3.15. For the matrix I_n, we have

$$\mathsf{vec}(I_n) = \begin{pmatrix} e_1 \\ e_2 \\ \vdots \\ e_n \end{pmatrix}.$$

The MATLAB implementation of vec is discussed in Example 15.32.

Definition 3.17. Let $A = (a_{ij}) \in \mathbb{C}^{n\times n}$ be a square matrix. The *trace* of A is the number $\mathsf{trace}(A)$ given by

$$\mathsf{trace}(A) = a_{11} + a_{22} + \cdots + a_{nn}.$$

Theorem 3.14. *Let A and B be two square matrices in $\mathbb{C}^{n\times n}$. We have*
(i) $\mathsf{trace}(aA) = a\,\mathsf{trace}(A)$,
(ii) $\mathsf{trace}(A+B) = \mathsf{trace}(A) + \mathsf{trace}(B)$, *and*
(iii) $\mathsf{trace}(AB) = \mathsf{trace}(BA)$.

Proof. The first two parts are direct consequences of the definition of the trace. For the last part, we can write

$$\mathsf{trace}(AB) = \sum_{i=1}^{n}(AB)_{ii} = \sum_{i=1}^{n}\sum_{j=1}^{n} a_{ij}b_{ji}.$$

Exchanging the subscripts i and j and, then the order of the summations, we have

$$\sum_{i=1}^{n}\sum_{j=1}^{n} a_{ij}b_{ji} = \sum_{j=1}^{n}\sum_{i=1}^{n} a_{ji}b_{ij} = \sum_{i=1}^{n}\sum_{j=1}^{n} b_{ij}a_{ji} = \sum_{i=1}^{n}(BA)_{ii},$$

which proves the desired equality. □

Note that the elements on the diagonal of a skew-symmetric matrix are 0 and, therefore, its trace equals 0.

Let A, B, C be three matrices in $\mathbb{C}^{n\times n}$. We have

$$\mathsf{trace}(ABC) = \mathsf{trace}((AB)C) = \mathsf{trace}(C(AB)) = \mathsf{trace}(CAB),$$

and

$$\mathsf{trace}(ABC) = \mathsf{trace}(A(BC)) = \mathsf{trace}((BC)A) = \mathsf{trace}(BCA).$$

However, it is important to notice that the third part of Theorem 3.14 *cannot* be extended to arbitrary permutations of a product of matrices. Consider, for example, the matrices

$$A = \begin{pmatrix} 1 & 0 \\ 1 & 1 \end{pmatrix}, B = \begin{pmatrix} 1 & 1 \\ 1 & 0 \end{pmatrix}, \text{ and } C = \begin{pmatrix} 1 & 1 \\ 0 & 1 \end{pmatrix}.$$

We have

$$ABC = \begin{pmatrix} 1 & 2 \\ 2 & 3 \end{pmatrix} \text{ and } ACB = \begin{pmatrix} 2 & 1 \\ 3 & 1 \end{pmatrix},$$

so $\mathsf{trace}(ABC) = 4$ and $\mathsf{trace}(ACB) = 3$.

Definition 3.18. A matrix $A \in \mathbb{R}^{m\times n}$ is *non-negative* if $a_{ij} \geqslant 0$ for $1 \leqslant i \leqslant m$ and $1 \leqslant j \leqslant n$. This is denoted by $A \geqslant O_{m,n}$.

A is *positive* if $a_{ij} > 0$ for $1 \leqslant i \leqslant m$ and $1 \leqslant j \leqslant n$. This is denoted by $A > 0_{m,n}$.

If $B, C \in \mathbb{R}^{m\times n}$, we write $B \geqslant C$ ($B > C$) if $B - C \geqslant O_{m,n}$ ($B - C > O_{m,n}$, respectively).

The sets of non-negative (non-positive, positive, negative) $m \times n$-matrices is denoted by $\mathbb{R}^{m\times n}_{\geqslant 0}$ ($\mathbb{R}^{m\times n}_{\leqslant 0}$, $\mathbb{R}^{m\times n}_{>0}$, $\mathbb{R}^{m\times n}_{<0}$, respectively).

Example 3.16. The diagonal matrix I_n is non-negative but not positive.

3.5 Special Classes of Matrices

Definition 3.19. The matrix $A \in \mathbb{F}^{m\times n}$ has *lower bandwidth* p if $i > j+p$ implies $a_{ij} = 0$. A has *upper bandwidth* q if $j > i+q$ implies $a_{ij} = 0$.

A is a (p,q)-*band matrix* if it has lower bandwidth p and upper bandwidth q.

A *tridiagonal matrix* is a $(1,1)$-band matrix.

A *lower Hessenberg matrix* is an $(m-1,1)$-band matrix, while an *upper Hessenberg matrix* is a $(1,n-1)$-band matrix.

Example 3.17. The matrix $A \in \mathbb{R}^{4\times 6}$ defined by

$$\begin{pmatrix} 1 & 2 & 0 & 0 & 0 & 0 \\ 1 & 2 & 3 & 0 & 0 & 0 \\ 2 & 1 & 3 & 5 & 0 & 0 \\ 0 & 2 & 1 & 4 & 3 & 0 \end{pmatrix}$$

is a $(2,1)$-band matrix. The matrix $B \in \mathbb{R}^{4\times 6}$ given by

$$\begin{pmatrix} 1 & 2 & 0 & 0 & 0 & 0 \\ 1 & 2 & 2 & 0 & 0 & 0 \\ 0 & 1 & 3 & 2 & 0 & 0 \\ 0 & 0 & 1 & 4 & 2 & 0 \end{pmatrix}$$

is a tridiagonal matrix.

The matrix

$$\begin{pmatrix} 1 & 2 & 3 & 4 & 5 & 6 \\ 1 & 2 & 2 & 3 & 4 & 4 \\ 0 & 1 & 3 & 2 & 5 & 0 \\ 0 & 0 & 1 & 4 & 2 & 0 \end{pmatrix}$$

is an upper Hessenberg matrix. On the other hand, the matrix

$$\begin{pmatrix} 1 & 2 & 0 & 0 & 0 & 0 \\ 1 & 2 & 3 & 0 & 0 & 0 \\ 2 & 1 & 3 & 5 & 0 & 0 \\ 3 & 2 & 1 & 4 & 3 & 0 \end{pmatrix}$$

is a lower Hessenberg matrix.

Note that a matrix $L \in \mathbb{F}^{m\times n}$ is lower triangular if it is an $(m-1, 0)$-band matrix. Similarly, U is upper triangular if it is a $(0, n-1)$-band matrix. In other words, L is lower triangular if its upper bandwidth is 0, that is, if $j > i$ implies $l_{ij} = 0$; U is upper triangular if its lower bandwidth is 0, that is, $i > j$ implies $u_{ij} = 0$.

Next, we define several classes of matrices whose components are real numbers.

Definition 3.20. A *Toeplitz matrix* is a matrix $T \in \mathbb{R}^{n\times n}$ such that the elements located in any line parallel to the main diagonal of T (including, of course, the main diagonal) are equal.

Example 3.18. Let the matrix

$$T = \begin{pmatrix} 1 & 2 & 3 & 4 & 5 \\ 6 & 1 & 2 & 3 & 4 \\ 7 & 6 & 1 & 2 & 3 \\ 8 & 7 & 6 & 1 & 2 \\ 9 & 8 & 7 & 6 & 1 \end{pmatrix}$$

be a 5×5 Toeplitz matrix.

The elements of a Toeplitz matrix $T \in \mathbb{R}^{n\times n}$ are completely determined by its first row and its first column (which must have their first components equal). Therefore, such a matrix is fully defined by a set of $2n - 1$ numbers.

Definition 3.21. A *circulant matrix* is a Toeplitz matrix C of the form

$$C = \begin{pmatrix} c_1 & c_2 & c_3 & \cdots & c_n \\ c_n & c_1 & c_2 & \cdots & c_{n-1} \\ c_{n-1} & c_n & c_1 & \cdots & c_{n-2} \\ \vdots & \vdots & \vdots & \cdots & \vdots \\ c_2 & c_3 & c_4 & \cdots & c_1 \end{pmatrix}.$$

Note that if $C = (c_{ij})$, then $c_{ij} = c_{(j-i+1) \bmod n}$.

Definition 3.22. A *Hankel matrix* is a matrix $H \in \mathbb{R}^{n\times n}$ such that the elements located in any line parallel to the skew-diagonal of H (including, of course, the skew-diagonal) are equal.

Example 3.19. Let T be the matrix

$$T = \begin{pmatrix} 1 & 2 & 3 & 4 & 5 \\ 2 & 3 & 4 & 5 & 6 \\ 3 & 4 & 5 & 6 & 7 \\ 4 & 5 & 6 & 7 & 8 \\ 5 & 6 & 7 & 8 & 9 \end{pmatrix}$$

be a 5×5 Hankel matrix.

Next, we introduce a class of matrices that is essential in the study of the theory of Markov chains.

Definition 3.23. A *stochastic matrix* is a matrix $A \in \mathbb{R}^{n\times n}$ such that $a_{ij} \geqslant 0$ for $1 \leqslant i, j \leqslant n$ and $\sum_{j=1}^{n} a_{ij} = 1$ for every i, $1 \leqslant i \leqslant n$.

A *doubly stochastic matrix* is a matrix $A \in \mathbb{R}^{n\times n}$ such that both A and A' are stochastic.

The rows of a stochastic matrix can be regarded as discrete probability distributions.

Example 3.20. The matrix $A \in \mathbb{R}^{3\times 3}$ defined by

$$A = \begin{pmatrix} \frac{1}{2} & 0 & \frac{1}{2} \\ \frac{1}{3} & \frac{1}{2} & \frac{1}{6} \\ 0 & \frac{2}{3} & \frac{1}{3} \end{pmatrix}$$

is a stochastic matrix.

Let $\mathbb{C}$ be the field of complex numbers. A *complex* matrix is a matrix $A \in \mathbb{C}^{m\times n}$.

Definition 3.24. The conjugate of a matrix $A \in \mathbb{C}^{m\times n}$ is the matrix $\bar{A} \in \mathbb{C}^{m\times n}$, where $A(i,j) = \overline{A(i,j)}$ for $1 \leqslant i \leqslant m$ and $1 \leqslant j \leqslant n$.

The notion of symmetry is extended to accommodate complex matrices.

Definition 3.25. The *transpose conjugate* of the matrix $A \in \mathbb{C}^{m\times n}$ or its *Hermitian adjoint* is the matrix $B \in \mathbb{C}^{n\times m}$ given by $B = \bar{A}' = \overline{(A')}$.

The transpose conjugate of A is denoted by A^{H}.

Example 3.21. Let $A \in \mathbb{C}^{3\times 2}$ be the matrix

$$A = \begin{pmatrix} 1+i & 2 \\ 2-i & i \\ 0 & 1-2i \end{pmatrix}.$$

The matrix A^{H} is given by

$$A^{\mathsf{H}} = \begin{pmatrix} 1-i & 2+i & 0 \\ 2 & -i & 1+2i \end{pmatrix}.$$

Using Hermitian conjugates, several important classes of matrices are defined.

Definition 3.26. The matrix $A \in \mathbb{C}^{n\times n}$ is

(i) *Hermitian* if $A = A^{\mathsf{H}}$;
(ii) *skew-Hermitian* if $A^{\mathsf{H}} = -A$;
(iii) *normal* if $AA^{\mathsf{H}} = A^{\mathsf{H}}A$;
(iv) *unitary* if $AA^{\mathsf{H}} = A^{\mathsf{H}}A = I_n$.

The set of unitary matrices in $\mathbb{C}^{n\times n}$ constitutes the unitary group $\mathsf{UG}(n,\mathbb{C})$.

It is immediate that all unitary, Hermitian, and skew-Hermitian matrices are normal. However, there are normal matrices outside these three classes.

Example 3.22. The matrix

$$A = \begin{pmatrix} 1 & -1 \\ 1 & 1 \end{pmatrix}$$

is not Hermitian or skew-Hermitian, and it is not unitary because

$$AA^{\mathsf{H}} = A^{\mathsf{H}}A = \begin{pmatrix} 2 & 0 \\ 0 & 2 \end{pmatrix}.$$

However, A is normal.

Example 3.23. Let $\alpha, \beta, \gamma, \delta$, and θ be five real numbers such that $\alpha - \beta - \gamma + \delta$ is a multiple of 2π. The matrix

$$M_{\alpha,\beta,\gamma,\delta}(\theta) = \begin{pmatrix} e^{i\alpha}\cos\theta & -e^{i\beta}\sin\theta \\ e^{i\gamma}\sin\theta & e^{i\delta}\cos\theta \end{pmatrix}$$

introduced in [57] is unitary because

$$\begin{aligned} &M_{\alpha,\beta,\gamma,\delta}(\theta)^{\mathsf{H}} M_{\alpha,\beta,\gamma,\delta}(\theta) \\ &= \begin{pmatrix} e^{-i\alpha}\cos\theta & e^{-i\gamma}\sin\theta \\ -e^{-i\beta}\sin\theta & e^{-i\delta}\cos\theta \end{pmatrix} \begin{pmatrix} e^{i\alpha}\cos\theta & -e^{i\beta}\sin\theta \\ e^{i\gamma}\sin\theta & e^{i\delta}\cos\theta \end{pmatrix} = \begin{pmatrix} 1 & 0 \\ 0 & 1 \end{pmatrix}. \end{aligned}$$

It is immediate that all unitary matrices are invertible, and their inverse is equal to their Hermitian conjugate. Furthermore, the product of two unitary matrices is a unitary matrix. Indeed, suppose that $A, B \in \mathbb{C}^{n\times n}$ are unitary matrices, that is, $AA^{\mathsf{H}} = BB^{\mathsf{H}} = I_n$. Then $(AB)(AB)^{\mathsf{H}} = ABB^{\mathsf{H}}A^{\mathsf{H}} = AA^{\mathsf{H}} = I_n$, hence AB is a unitary matrix.

If $A \in \mathbb{R}^{n\times n}$ is a unitary real matrix, we refer to A as an *orthogonal matrix* or an *orthonormal matrix* for reasons that we will discuss in Section 6.11.

If $A \in \mathbb{R}^{n\times n}$ is a matrix with real entries, then its Hermitian adjoint coincides with the transposed matrix A'. Thus, a real matrix is Hermitian if and only if it is symmetric.

Observe that if $\boldsymbol{z} \in \mathbb{C}^n$ and

$$\boldsymbol{z} = \begin{pmatrix} z_1 \\ \vdots \\ z_n \end{pmatrix},$$

then $\boldsymbol{z}^{\mathsf{H}}\boldsymbol{z} = \overline{z}_1 z_1 + \cdots + \overline{z}_n z_n = \sum_{i=1}^n |z_i|^2$.

Theorem 3.15. *Let $A \in \mathbb{C}^{n \times n}$. The following statements hold:*

(i) *the matrices $A + A^{\mathsf{H}}$, AA^{H}, and $A^{\mathsf{H}}A$ are Hermitian and $A - A^{\mathsf{H}}$ is skew-Hermitian;*
(ii) *if A is a Hermitian matrix, then so is A^k for $k \in \mathbb{N}$;*
(iii) *if A is Hermitian and invertible, then so is A^{-1};*
(iv) *if A is Hermitian, then a_{ii} are real numbers for $1 \leqslant i \leq n$.*

Proof. All statements follow directly from the definition of Hermitian matrices. □

Theorem 3.16. *If $A \in \mathbb{C}^{n \times n}$ there exists a unique pair of Hermitian matrices (H_1, H_2) such that $A = H_1 + iH_2$.*

Proof. Let

$$H_1 = \frac{1}{2}(A + A^{\mathsf{H}}) \text{ and } H_2 = -\frac{i}{2}(A - A^{\mathsf{H}}).$$

It is immediate that both H_1 and H_2 are Hermitian and that $H_1 + iH_2 = A$.

Suppose that $A = H_3 + iH_4$, where H_3 and H_4 are Hermitian. Then, we have

$$\begin{aligned} 2H_1 = A + A^{\mathsf{H}} &= H_3 + iH_4 + H_3^{\mathsf{H}} - iH_4^{\mathsf{H}} \\ &= 2H_3, \end{aligned}$$

so $H_1 = H_3$. Therefore, $H_2 = H_4$, so the matrices H_1 and H_2 are uniquely determined. □

Theorem 3.17. *If $A \in \mathbb{C}^{n \times n}$, there exists a unique pair of matrices (H, S) such that H is Hermitian, S is skew-Hermitian, and $A = H + S$.*

Proof. By Theorem 3.16, A can be written as $A = H_1 + iH_2$, where H_1 and H_2 are Hermitian matrices. Choose $H = H_1$ and $S = iH_2$. By Exercise 3.17, S is skew-Hermitian. The uniqueness of the pair (H, S) is immediate. □

Next, we discuss a characterization of Hermitian matrices.

Theorem 3.18. *A matrix $A \in \mathbb{C}^{n\times n}$ is Hermitian if and only if $\boldsymbol{x}^{\mathsf{H}} A\boldsymbol{x}$ is a real number for every $\boldsymbol{x} \in \mathbb{C}^n$.*

Proof. Suppose that A is Hermitian. Then

$$\overline{\boldsymbol{x}^{\mathsf{H}} A\boldsymbol{x}} = \overline{\boldsymbol{x}^{\mathsf{H}} A^{\mathsf{H}} \boldsymbol{x}} = \boldsymbol{x}' A' \overline{\boldsymbol{x}} = \boldsymbol{x}' A' (\boldsymbol{x}^{\mathsf{H}})' = \boldsymbol{x}^{\mathsf{H}} A\boldsymbol{x},$$

so $\boldsymbol{x}^{\mathsf{H}} A\boldsymbol{x}$ is a real number because it is equal to its conjugate.

Conversely, suppose that $\boldsymbol{x}^{\mathsf{H}} A\boldsymbol{x}$ is a real number for every $\boldsymbol{x} \in \mathbb{C}^n$. This implies that

$$(\boldsymbol{x} + \boldsymbol{y})^{\mathsf{H}} A(\boldsymbol{x} + \boldsymbol{y}) = \boldsymbol{x}^{\mathsf{H}} A\boldsymbol{x} + \boldsymbol{x}^{\mathsf{H}} A\boldsymbol{y} + \boldsymbol{y}^{\mathsf{H}} A\boldsymbol{x} + \boldsymbol{y}^{\mathsf{H}} A\boldsymbol{y}$$

is a real number, so $\boldsymbol{x}^{\mathsf{H}} A\boldsymbol{y} + \boldsymbol{y}^{\mathsf{H}} A\boldsymbol{x}$ is real for every $\boldsymbol{x}, \boldsymbol{y} \in \mathbb{C}^n$.

Let $\boldsymbol{x} = \boldsymbol{e}_p$ and $\boldsymbol{y} = \boldsymbol{e}_q$. Then $a_{pq} + a_{qp}$ is a real number. If we choose $\boldsymbol{x} = -i\boldsymbol{e}_p$ and $\boldsymbol{y} = \boldsymbol{e}_j$, it follows that $-ia_{pq} + ia_{qp}$ is a real number. Thus, $\Im(a_{pq}) = -\Im(a_{qp})$ and $\Re(a_{pq}) = \Re(a_{qp})$, which leads to $a_{pq} = \overline{a}_{qp}$ for $1 \leqslant p, q \leqslant n$. These equalities are equivalent to $A = A^{\mathsf{H}}$, so A is Hermitian. □

Observe that for any matrix $B \in \mathbb{C}^{m\times n}$, the matrix $B^{\mathsf{H}} B$ is normal since $(B^{\mathsf{H}} B)^{\mathsf{H}} B^{\mathsf{H}} B = B^{\mathsf{H}} BB^{\mathsf{H}} B$ and $B^{\mathsf{H}} B(B^{\mathsf{H}} B)^{\mathsf{H}} = B^{\mathsf{H}} BB^{\mathsf{H}} B$.

Example 3.24. All Hermitian or skew-Hermitian matrices are normal. Indeed, if A is Hermitian, then $A = A^{\mathsf{H}}$ and the normality condition is obviously satisfied. If A is skew-Hermitian, then $A^{\mathsf{H}} = -A$ and the normality follows from $(-A)A = A(-A) = -A^2$.

If A is a real, symmetric matrix, then A is obviously normal.

Theorem 3.19. *A matrix $A \in \mathbb{C}^{n\times n}$ is normal and upper triangular (or lower triangular) if and only if A is a diagonal matrix.*

Proof. Clearly, any diagonal matrix is both normal and upper (and lower) triangular. Therefore, we need to show only that if A is both triangular and normal, then A is diagonal. We make the argument for the case when A is upper triangular.

Since $A^{\mathsf{H}}A = AA^{\mathsf{H}}$, we have the equality $(A^{\mathsf{H}}A)_{pp} = (AA^{\mathsf{H}})_{pp}$ for $1 \leqslant p \leqslant n$.

We prove by induction on p that the non-diagonal elements of A are 0. For the base case, $p = 1$, the conditions of the theorem imply

$$\bar{a}_{11}a_{11} = \sum_{j=1}^{n} a_{1j}\bar{a}_{1j} = a_{11}\bar{a}_{11} + \sum_{j=2}^{n} a_{1j}\bar{a}_{1j}.$$

Since $\bar{a}_{11}a_{11} = a_{11}\bar{a}_{11} = |a_{11}|^2$, it follows that $\sum_{j=2}^{n} a_{1j}\bar{a}_{1j} = \sum_{j=2}^{n} |a_{1j}|^2 = 0$, so $a_{1j} = 0$ for $2 \leqslant j \leqslant n$. Thus, all non diagonal element located on the first line of A (or the first column of A^{H}) are zero.

Suppose now that all non-diagonal elements of the first $p-1$ lines of A are 0. For the pth diagonal element of $(A^{\mathsf{H}}A)_{pp}$, we have

$$\begin{aligned}(A^{\mathsf{H}}A)_{pp} &= \sum_{i=1}^{n} \bar{a}_{ip}a_{ip} \\ &= \sum_{i=p}^{n} \bar{a}_{ip}a_{ip} \\ &\quad \text{(by the inductive hypothesis)} \\ &= \bar{a}_{pp}a_{pp} \\ &\quad \text{(because } A \text{ is an upper diagonal matrix).}\end{aligned}$$

This allows us to write

$$\bar{a}_{pp}a_{pp} = \sum_{j=p}^{n} a_{pj}\bar{a}_{pj} = a_{pp}\bar{a}_{pp} + \sum_{j=p+1}^{n} a_{pj}\bar{a}_{pj},$$

so

$$\sum_{j=p+1}^{n} a_{pj}\bar{a}_{pj} = \sum_{j=p+1}^{n} |a_{pj}|^2 = 0,$$

which implies $a_{p\,p+1} = \cdots = a_{pn} = 0$. Thus, all non-diagonal elements on the line p are 0. □

Theorem 3.20. *If $U \in \mathbb{C}^{n \times n}$ is a unitary matrix, then the matrix $Z \in \mathbb{R}^{n \times n}$ given by $z_{ij} = |u_{ij}|^2$ for $1 \leqslant i, j \leqslant n$ is a doubly stochastic matrix.*

Proof. Since U is a unitary matrix, we have $UU^{\mathsf{H}} = U^{\mathsf{H}}U = I_n$. Thus,

$$\sum_{j=1}^{n} z_{ij} = \sum_{j=1}^{n} |u_{ij}|^2 = \sum_{j=1}^{n} u_{ij}\overline{u_{ij}}$$

$$= \sum_{j=1}^{n} u_{ij}(U^{\mathsf{H}})_{ji} = (UU^{\mathsf{H}})_{ii} = 1.$$

The equality $\sum_{i=1}^{n} z_{ij} = 1$ can be established in a similar manner, so Z is indeed a doubly stochastic matrix. □

Let $A \in \mathbb{C}^{m\times n}$ be a matrix. The *matrix of the absolute values of* A is the matrix $\mathsf{abs}(A) \in \mathbb{R}^{m\times n}$ defined by

$$(\mathsf{abs}(A))_{ij} = |a_{ij}|$$

for $1 \leqslant i \leqslant m$ and $1 \leqslant j \leqslant n$. In particular, if $\boldsymbol{x} \in \mathbb{C}^n$, we have $(\mathsf{abs}(\boldsymbol{x}))_j = |x_j|$.

Theorem 3.21. *Let $A \in \mathbb{C}^{m\times n}$ and $B \in \mathbb{C}^{n\times p}$ be two matrices. We have $\mathsf{abs}(AB) \leqslant \mathsf{abs}(A)\mathsf{abs}(B)$.*

Proof. Since $(AB)_{ik} = \sum_{j=1}^{n} a_{ij}b_{jk}$, it follows that

$$|(AB)_{ik}| = \left|\sum_{j=1}^{n} a_{ij}b_{jk}\right| \leq \sum_{j=1}^{n} |a_{ij}b_{jk}| = \sum_{j=1}^{n} |a_{ij}|\,|b_{jk}|,$$

for $1 \leqslant i \leqslant m$ and $1 \leqslant k \leqslant p$. This amounts to $\mathsf{abs}(AB) \leqslant \mathsf{abs}(A)\mathsf{abs}(B)$. □

Theorem 3.22. *For $A \in \mathbb{C}^{n\times n}$ we have $\mathsf{abs}(A^k) \leqslant (\mathsf{abs}(A))^k$ for every $k \in \mathbb{N}$.*

Proof. The proof is by induction on k. The base case, $k = 0$, is immediate. Suppose that the inequality holds for k. We have

$$\begin{aligned}\mathsf{abs}(A^{k+1}) &= \mathsf{abs}(A^k\ A)\\ &\leq \mathsf{abs}(A^k)\mathsf{abs}(A)\\ &\quad \text{(by Theorem 3.21)}\end{aligned}$$

$$\leq (\mathsf{abs}(A))^k \mathsf{abs}(A)$$
$$\text{(by the inductive hypothesis)}$$
$$= (\mathsf{abs}(A))^{k+1},$$

which completes the induction case.The factor of S'*A*S tends to be sparser than the factor of A. □

3.6 Partitioned Matrices and Matrix Operations

Let $A \in \mathbb{F}^{m\times n}$ be a matrix and suppose that $m = m_1 + \cdots + m_p$ and $n = n_1 + \cdots + n_q$, where $\mathbb{F}$ is the real or the complex field.

A *partitioning of A* is a collection of matrices $A_{hk} \in \mathbb{F}^{m_h \times n_k}$ such that A_{hk} is the contiguous submatrix

$$A\begin{bmatrix} m_1 + \cdots + m_{h-1} + 1, \ldots, m_1 + \cdots + m_{h-1} + m_h \\ n_1 + \cdots + n_{k-1} + 1, \ldots, n_1 + \cdots + n_k \end{bmatrix},$$

for $1 \leqslant h \leqslant p$ and $1 \leqslant k \leqslant q$.

If $\{A_{hk} \mid 1 \leqslant h \leqslant p \text{ and } 1 \leqslant k \leqslant q\}$ is a partitioning of A, A is written as

$$A = \begin{pmatrix} A_{11} & A_{12} & \cdots & A_{1q} \\ A_{21} & A_{22} & \cdots & A_{2q} \\ \vdots & \vdots & \cdots & \vdots \\ A_{p1} & A_{p2} & \cdots & A_{pq} \end{pmatrix}.$$

The matrices A_{hk} are referred to as the *blocks* of the partitioning. All blocks located in a column must have the same number of columns; all blocks located in a row must have the same number of rows.

Example 3.25. The matrix $A \in \mathbb{F}^{5\times 6}$ given by

$$A = \begin{pmatrix} a_{11} & a_{12} & a_{13} & a_{14} & a_{15} & a_{16} \\ a_{21} & a_{22} & a_{23} & a_{24} & a_{25} & a_{26} \\ a_{31} & a_{32} & a_{33} & a_{34} & a_{35} & a_{36} \\ a_{41} & a_{42} & a_{43} & a_{44} & a_{45} & a_{46} \\ a_{51} & a_{52} & a_{53} & a_{54} & a_{55} & a_{56} \end{pmatrix}$$

can be partitioned as

$$\left(\begin{array}{ccc|c|cc} a_{11} & a_{12} & a_{13} & a_{14} & a_{15} & a_{16} \\ a_{21} & a_{22} & a_{23} & a_{24} & a_{25} & a_{26} \\ a_{31} & a_{32} & a_{33} & a_{34} & a_{35} & a_{36} \\ \hline a_{41} & a_{42} & a_{43} & a_{44} & a_{45} & a_{46} \\ a_{51} & a_{52} & a_{53} & a_{54} & a_{55} & a_{56} \end{array}\right).$$

Thus, if we introduce the matrices

$$A_{11} = \begin{pmatrix} a_{11} & a_{12} & a_{13} \\ a_{21} & a_{22} & a_{23} \\ a_{31} & a_{32} & a_{33} \end{pmatrix}, A_{12} = \begin{pmatrix} a_{14} \\ a_{24} \\ a_{34} \end{pmatrix}, A_{13} = \begin{pmatrix} a_{15} & a_{16} \\ a_{25} & a_{26} \\ a_{35} & a_{36} \end{pmatrix},$$

$$A_{21} = \begin{pmatrix} a_{41} & a_{42} & a_{43} \\ a_{51} & a_{52} & a_{53} \end{pmatrix}, A_{22} = \begin{pmatrix} a_{45} \\ a_{55} \end{pmatrix}, A_{23} = \begin{pmatrix} a_{45} & a_{46} \\ a_{55} & a_{56} \end{pmatrix},$$

the matrix A can be written as

$$A = \begin{pmatrix} A_{11} & A_{12} & A_{13} \\ A_{21} & A_{22} & A_{23} \end{pmatrix}.$$

Partitioning matrices is useful because matrix operations can be performed on block submatrices in a manner similar to scalar operations as we show next.

Theorem 3.23. *Let $A \in \mathbb{F}^{m\times n}$ and $B \in \mathbb{F}^{n\times p}$ be two matrices. Suppose that the matrices A, B are partitioned as*

$$A = \begin{pmatrix} A_{11} & \cdots & A_{1k} \\ \vdots & \cdots & \vdots \\ A_{h1} & \cdots & A_{hk} \end{pmatrix} \text{ and } B = \begin{pmatrix} B_{11} & \cdots & B_{1\ell} \\ \vdots & \cdots & \vdots \\ B_{k1} & \cdots & B_{k\ell} \end{pmatrix},$$

where $A_{rs} \in \mathbb{F}^{m_r\times n_s}$, $B_{st} \in \mathbb{F}^{n_s\times p_t}$ for $1 \leqslant r \leqslant h$, $1 \leqslant s \leqslant k$ and $1 \leqslant t \leqslant \ell$. Then, the product $C = AB$ can be partitioned as

$$C = \begin{pmatrix} C_{11} & \cdots & C_{1\ell} \\ \vdots & \cdots & \vdots \\ C_{h1} & \cdots & C_{hl} \end{pmatrix},$$

where $C_{uv} = \sum_{t=1}^{k} A_{ut}B_{tv}$, $1 \leqslant u \leqslant h$, and $1 \leqslant v \leqslant \ell$.

Proof. Note that $m_1 + \cdots + m_h = m$ and $p_1 + \cdots + p_\ell = p$. For a pair (i, j) such that $1 \leqslant i \leqslant m$ and $1 \leqslant j \leqslant n$, let u be the least number such that $i \leqslant m_1 + \cdots + m_u$ and let v be the least number such that $j \leqslant p_1 + \cdots + p_v$. The definition of u and v implies $m_1 + \cdots + m_{u-1} + 1 \leqslant i \leq m_1 + \cdots + m_u$ and $p_1 + \cdots + p_{v-1} + 1 \leqslant j \leqslant p_1 + \cdots + p_v$. This implies that the c_{ij} element of the product is located in the submatrix $C_{uv} = \sum_{t=1}^{k} A_{ut} B_{tv}$ of C. By the definition of the matrix product, we have

$$\begin{aligned} c_{ij} &= \sum_{g=1}^{n} a_{ig} b_{gj} \\ &= \sum_{g=1}^{n_1} a_{ig} b_{gj} + \sum_{g=n_1+1}^{n_1+n_2} a_{ig} b_{gj} + \cdots + \sum_{g=n_1+\cdots+n_{k-1}+1}^{n_1+\cdots+n_s} a_{ig} b_{gj}. \end{aligned}$$

Observe that the vectors $(a_{i1}, \ldots, a_{in_1})$ and $(b_{1j}, \ldots, b_{n_1 j})'$ represent the line number $i - (m_1 + \cdots + m_{u-1} + 1)$ and the column number $j - (p_1 + \cdots + p_{v-1} + 1)$ of the matrix A_{u1} and B_{1v}, *etc.* Similarly,

$$(a_{i,n_1+\cdots+n_{k-1}+1}, \ldots, a_{i,n_1+\cdots+n_s})$$

and

$$(b_{n_1+\cdots+n_{k-1}+1,j}, \ldots, b_{n_1+\cdots+n_s,j})'$$

represent the line number $i - (m_1 + \cdots + m_{u-1} + 1)$ and the column number $j - (p_1 + \cdots + p_{v-1} + 1)$ of the matrix A_{uk} and B_{kv}, which shows that c_{ij} is computed correctly as an element of the block C_{uv}. □

3.7 Change of Bases

Let $B = \{\boldsymbol{e}_1, \ldots, \boldsymbol{e}_n\}$ and $\tilde{B} = \{\tilde{\boldsymbol{e}}_1, \ldots, \tilde{\boldsymbol{e}}_n\}$ be two bases in $\mathbb{C}^n$. These bases define the matrices

$$M_B = (\boldsymbol{e}_1 \ \cdots \ \boldsymbol{e}_n) \text{ and } M_{\tilde{B}} = (\tilde{\boldsymbol{e}}_1 \ \cdots \ \tilde{\boldsymbol{e}}_n)$$

in $\mathbb{C}^{n \times n}$ whose columns are the vectors of the bases B and $\tilde{B}$, respectively.

As vectors of each base can be expressed in terms of the other base, we can write

$$\mathbf{e}_i = \sum_{j=1}^{n} c_{ij}\tilde{\mathbf{e}}_j, \text{ for } 1 \leqslant i \leqslant n,$$

$$\tilde{\mathbf{e}}_h = \sum_{k=1}^{n} d_{hk}\mathbf{e}_k, \text{ for } 1 \leqslant h \leqslant n.$$

In matrix formulation the previous equalities can be written as

$$M_B = CM_{\tilde{B}} \text{ and } M_{\tilde{B}} = DM_B.$$

Theorem 3.24. *The matrices $C = (c_{ij})$ and $D = (d_{hk})$ defined above are inverse to each other.*

Proof. We have

$$\mathbf{e}_i = \sum_{j=1}^{n} c_{ij}\tilde{\mathbf{e}}_j = \sum_{j=1}^{n} c_{ij}\left(\sum_{k=1}^{n} d_{jk}\mathbf{e}_k\right) = \sum_{j=1}^{n}\sum_{k=1}^{n} c_{ij}d_{jk}\mathbf{e}_k$$

for $1 \leqslant i \leqslant n$. Equating the coefficients of vectors of B yields the following equalities:

$$\sum_{j=1}^{n} c_{ij}d_{ji} = 1, \text{ and } \sum_{j=1}^{n} c_{ij}d_{jk} = 0 \text{ if } k \neq i.$$

If $P = CD$, these equalities amount to $p_{ii} = 0$ and $p_{ik} = \sum_{j=1}^{n} c_{ij}d_{jk} = 0$ if $i \neq k$, which means that $P = I_n$. Thus, the matrices C and D are inverse to each other. □

Let $\boldsymbol{x} = \sum_{i=1}^{n} \tilde{x}_i\tilde{\mathbf{e}}_i$ be the expression of the vector $\boldsymbol{x}$ in the basis $\tilde{B}$. Since $\tilde{\mathbf{e}}_i = \sum_{k=1}^{n} d_{ik}\mathbf{e}_k$, it follows that

$$\boldsymbol{x} = \sum_{i=1}^{n} \tilde{x}_i\tilde{\mathbf{e}}_i = \sum_{j=1}^{n} x_j\mathbf{e}_j = \sum_{j=1}^{n} x_j \sum_{i=1}^{n} c_{ji}\tilde{\mathbf{e}}_i = \sum_{i=1}^{n}\sum_{j=1}^{n} x_j c_{ji}\tilde{\mathbf{e}}_i,$$

which implies

$$\tilde{x}_i = \sum_{j=1}^{n} c_{ji}x_j \tag{3.4}$$

for $1 \leqslant i \leqslant n$. The components of $\boldsymbol{x}$ in the new basis $\tilde{B}$, $\tilde{x}_i$ can be expressed via the components of $\boldsymbol{x}$ in the old basis B using the matrix C (that was used to express the old basis B in terms of the new basis $\tilde{B}$ as $M_B = CM_{\tilde{B}}$). This justifies the term *contravariant components* applied to the components x_k because the bases and the components transform in opposite directions.

Similarly, we can write

$$\boldsymbol{x} = \sum_{j=1}^{n} x_j \boldsymbol{e}_j = \sum_{i=1}^{n} \tilde{x}_i \tilde{\boldsymbol{e}}_i = \sum_{i=1}^{n} \tilde{x}_i \sum_{j=1}^{n} d_{ij} \boldsymbol{e}_j = \sum_{j=1}^{n} \sum_{i=1}^{n} \tilde{x}_i d_{ij} \boldsymbol{e}_j,$$

hence

$$x_j = \sum_{i=1}^{n} \tilde{x}_i d_{ij} \boldsymbol{e}_j \tag{3.5}$$

for $1 \leqslant j \leqslant n$.

The effects of a change of basis on covectors have been discussed in Section 2.8.

3.8 Matrices and Bilinear Forms

The link between matrices and bilinear form is discussed next.

Let $C \in \mathbb{R}^{m \times n}$ be a matrix. If $\boldsymbol{x} \in \mathbb{R}^m$ and $\boldsymbol{y} \in \mathbb{R}^n$, the function $f_C : L \times M \longrightarrow \mathbb{R}$ defined by $f_C(\boldsymbol{x}, \boldsymbol{y}) = \boldsymbol{x}' C \boldsymbol{y}$ can be easily seen to be bilinear. The next theorem shows that all bilinear functions between two finite-dimensional spaces can be defined in this manner.

Theorem 3.25. *Let V, W be two finite-dimensional $\mathbb{R}$-linear spaces. If $f : V \times W \longrightarrow \mathbb{R}$ is a bilinear form, then there is a matrix $C_f \in \mathbb{R}^{m \times n}$ such that*

$$f(\boldsymbol{x}, \boldsymbol{y}) = \boldsymbol{x}' C_f \boldsymbol{y}$$

for all $\boldsymbol{x} \in V$ and $\boldsymbol{y} \in W$.

Proof. Suppose that $B = \{\boldsymbol{x}_1, \ldots, \boldsymbol{x}_m\}$ and $B' = \{\boldsymbol{y}_1, \ldots, \boldsymbol{y}_n\}$ are bases in V and W, respectively. Let $\boldsymbol{x} = a_1 \boldsymbol{x}_1 + \cdots + a_m \boldsymbol{x}_m$ be the expression of $\boldsymbol{x} \in V$ in the base B. Similarly, let $\boldsymbol{y} = b_1 \boldsymbol{y}_1 + \cdots + b_n \boldsymbol{y}_n$

be the expression of $\boldsymbol{y} \in W$ in B'. The bilinearity of f implies

$$\begin{aligned} f(\boldsymbol{x}, \boldsymbol{y}) &= f\left(\sum_{i=1}^{m} a_i \boldsymbol{x}_i, \sum_{j=1}^{n} b_j \boldsymbol{y}_j\right) \\ &= \sum_{i=1}^{m}\sum_{j=1}^{n} a_i b_j f(\boldsymbol{x}_i, \boldsymbol{y}_j). \end{aligned}$$

If C_f is the matrix

$$C_f = \begin{pmatrix} f(\boldsymbol{x}_1, \boldsymbol{y}_1) & \cdots & f(\boldsymbol{x}_1, \boldsymbol{y}_n) \\ \vdots & \vdots & \vdots \\ f(\boldsymbol{x}_m, \boldsymbol{y}_1) & \cdots & f(\boldsymbol{x}_m, \boldsymbol{y}_n) \end{pmatrix},$$

then $f(\boldsymbol{x}, \boldsymbol{y}) = \boldsymbol{x}' C_f \boldsymbol{y}$, where

$$\boldsymbol{x} = \begin{pmatrix} a_1 \\ \vdots \\ a_m \end{pmatrix} \text{ and } \boldsymbol{y} = \begin{pmatrix} b_1 \\ \vdots \\ b_n \end{pmatrix}.$$

□

Let $A \in \mathbb{R}^{n\times n}$ be a symmetric matrix. The *quadratic form* associated to the matrix A is the function $f_A : \mathbb{R}^n \longrightarrow \mathbb{R}$ defined as $f_A(\boldsymbol{x}) = \boldsymbol{x}' A \boldsymbol{x}$ for $\boldsymbol{x} \in \mathbb{R}^n$.

The *polar form* of the quadratic form f_A is the bilinear form $\tilde{f}_A$ defined by $\tilde{f}_A(\boldsymbol{x}, \boldsymbol{y}) = \boldsymbol{x}' A \boldsymbol{y}$ for $\boldsymbol{x}, \boldsymbol{y} \in \mathbb{R}^n$.

Since $\boldsymbol{x}' A \boldsymbol{y}$ and $\boldsymbol{y}' A \boldsymbol{x}$ are scalars, they are equal and we have

$$\begin{aligned} f_A(\boldsymbol{x} + \boldsymbol{y}) &= (\boldsymbol{x} + \boldsymbol{y})' A (\boldsymbol{x} + \boldsymbol{y}) \\ &= \boldsymbol{x}' A \boldsymbol{x} + \boldsymbol{y}' A \boldsymbol{y} + \boldsymbol{x}' A \boldsymbol{y} + \boldsymbol{y}' A \boldsymbol{x} \\ &= f_A(\boldsymbol{x}) + f_A(\boldsymbol{y}) + 2\tilde{f}_A(\boldsymbol{x}, \boldsymbol{y}), \end{aligned}$$

which allows us to express the polar form of f_A as

$$\tilde{f}_A(\boldsymbol{x}, \boldsymbol{y}) = \frac{1}{2}(f_A(\boldsymbol{x} + \boldsymbol{y}) - f_A(\boldsymbol{x}) - f_A(\boldsymbol{y})).$$

If $A \in \mathbb{C}^{n\times n}$ is a Hermitian matrix, the *quadratic Hermitian form* associated to A is the function $f_A : \mathbb{C}^n \longrightarrow \mathbb{C}$ defined as

$f_A(\boldsymbol{x}) = \boldsymbol{x}^{\mathsf{H}} A \boldsymbol{x}$ for $\boldsymbol{x} \in \mathbb{C}^n$. Note that

$$\overline{f_A(\boldsymbol{x})} = \boldsymbol{x}^{sH} A^{\mathsf{H}} \boldsymbol{x} = \boldsymbol{x}^{sH} A \boldsymbol{x} = f_A(\boldsymbol{x})$$

for $\boldsymbol{x} \in \mathbb{C}^n$. This allows us to conclude that values of the quadratic Hermitian form associated to a Hermitian matrix are real numbers.

Let us consider the values of a quadratic Hermitian form $f_A(\boldsymbol{x})$ when $\|\boldsymbol{x}\| = 1$. Since $f_A(\boldsymbol{x})$ is a continuous function and the set $\{\boldsymbol{x} \mid \|\boldsymbol{x}\| = 1\}$ is compact, the function f_A attains its supremum. Thus, there exists $\boldsymbol{z}_1 \in \mathbb{C}^n$ with $\|\boldsymbol{z}_1\| = 1$ such that $f_A(\boldsymbol{z}_1)$ attains its maximum M_1.

Consider again the maximization of $f_A(\boldsymbol{x})$ subjected to the restrictions $\|\boldsymbol{x}\| = 1$ and $(\boldsymbol{x}, \boldsymbol{z}_1) = 0$. By the same reasoning as above, there exists a unit vector $\boldsymbol{z}_2$ such that $\|\boldsymbol{z}_2\| = 1$ and $\boldsymbol{z}_2 \perp \boldsymbol{z}_2$ where the maximum M_2 is attained, etc.

We obtain an orthonormal sequence of vectors $\boldsymbol{z}_1, \boldsymbol{z}_2, \ldots, \boldsymbol{z}_n$ known as the *sequence of principal directions* of A. The matrix

$$\begin{pmatrix} \boldsymbol{z}'_1 \\ \vdots \\ \boldsymbol{z}'_n \end{pmatrix}$$

is obviously unitary.

3.9 Generalized Inverses of Matrices

The next theorem involves a generalization of the notion of inverse of a matrix that is applicable to rectangular matrices.

Theorem 3.26. *If $A \in \mathbb{C}^{m\times n}$, there exists at most one matrix $M \in \mathbb{C}^{n\times m}$ such that the matrices MA and AM are Hermitian, $AMA = A$, and $MAM = M$.*

Proof. Suppose that both M and P satisfy the equalities of the theorem. We have

$$\begin{aligned} P &= PAP = P(AP) = P(AP)^{\mathsf{H}} \\ &= PP^{\mathsf{H}}A^{\mathsf{H}} = PP^{\mathsf{H}}(AMA)^{\mathsf{H}} \end{aligned}$$

$$\begin{aligned}
&= PP^{\mathsf{H}}A^{\mathsf{H}}M^{\mathsf{H}}A^{\mathsf{H}} = PP^{\mathsf{H}}A^{\mathsf{H}}(AM)^{\mathsf{H}}\\
&= P(AP)^{\mathsf{H}}(AM)^{\mathsf{H}} = PAPAM\\
&= P(APA)M = PAM = PAMAM\\
&= (PA)^{\mathsf{H}}(MA)^{\mathsf{H}}M = A^{\mathsf{H}}P^{\mathsf{H}}A^{\mathsf{H}}M^{\mathsf{H}}M\\
&= (APA)^{\mathsf{H}}M^{\mathsf{H}}M = A^{\mathsf{H}}M^{\mathsf{H}}M\\
&= (MA)^{\mathsf{H}}M = MAM = M.
\end{aligned}$$

Thus, $P = M$ and the uniqueness of M is proven. □

The matrix M introduced by Theorem 3.26 is known as the *Moore-Penrose pseudoinverse of* A and is denoted by $A^\dagger$.

Theorem 3.27. *Let $A \in \mathbb{C}^{n\times n}$. If A is invertible, then $A^\dagger$ exists and equals A^{-1}.*

Proof. If A is invertible, it is clear that the matrices AA^{-1} and $A^{-1}A$ are Hermitian because both are equal to I_n. Furthermore, $AA^{-1}A = A$ and $A^{-1}AA^{-1} = A^{-1}$, so $A^{-1} = A^\dagger$ by Theorem 3.26. □

The Moore–Penrose pseudoinverse of a matrix may exist even if the inverse does not. For example, we have $O_{m,n}^\dagger = O_{n,m}$, as it can be easily seen. However, $O_{n,n}$ is not invertible.

The symmetry of the equalities defining the Moore Penrose pseudoinverse of a matrix show that $(A^\dagger)^\dagger = A$.

3.10 Matrices and Linear Transformations

Let $h \in \mathsf{Hom}(\mathbb{C}^m, \mathbb{C}^n)$ be a linear transformation between the linear spaces $\mathbb{C}^m$ and $\mathbb{C}^n$. Consider a basis in $\mathbb{C}^m$, $R = \{\boldsymbol{r}_1, \ldots, \boldsymbol{r}_m\}$, and a basis in $\mathbb{C}^n$, $S = \{\boldsymbol{s}_1, \ldots, \boldsymbol{s}_n\}$.

The function h is completely determined by the images of the elements of the basis R, that is, by the set $\{h(\boldsymbol{r}_1), \ldots, h(\boldsymbol{r}_m)\}$. If

$$h(\boldsymbol{r}_j) = a_{1j}\boldsymbol{s}_1 + a_{2j}\boldsymbol{s}_2 + \cdots + a_{nj}\boldsymbol{s}_n = \sum_{i=1}^{n} a_{ij}\boldsymbol{s}_i,$$

then, for $\boldsymbol{x} = x_1\boldsymbol{r}_1 + \cdots + x_m\boldsymbol{r}_m$, we have by linearity

$$\begin{aligned} h(\boldsymbol{x}) &= x_1 h(\boldsymbol{r}_1) + \cdots + x_m h(\boldsymbol{r}_m) \\ &= \sum_{j=1}^{m} x_j h(\boldsymbol{r}_j) \\ &= \sum_{j=1}^{m}\sum_{i=1}^{n} x_j a_{ij} \boldsymbol{s}_i. \end{aligned}$$

In other words, we have $a_{ij} = (h(\boldsymbol{r}_j))_i$ for $1 \leqslant i \leqslant n$ and $1 \leqslant j \leqslant m$.

In a more compact form, we can write

$$h(\boldsymbol{x}) = (\boldsymbol{s}_1 \ \cdots \ \boldsymbol{s}_n) \begin{pmatrix} a_{11} & \cdots & a_{1m} \\ \vdots & \cdots & \vdots \\ a_{n1} & \cdots & a_{nm} \end{pmatrix} \begin{pmatrix} x_1 \\ \vdots \\ x_m \end{pmatrix}.$$

The image of a vector $\boldsymbol{x} = x_1\boldsymbol{r}_1 + \cdots + x_m\boldsymbol{r}_m$ under h equals $A_h\boldsymbol{x}$, where A_h is

$$A_h = \begin{pmatrix} a_{11} & \cdots & a_{1m} \\ \vdots & \cdots & \vdots \\ a_{n1} & \cdots & a_{nm} \end{pmatrix} \in \mathbb{C}^{n\times m}.$$

Clearly, the matrix A_h attached to $h : \mathbb{C}^m \longrightarrow \mathbb{C}^n$ depends on the bases chosen for the linear spaces $\mathbb{C}^m$ and $\mathbb{C}^n$ and $(A_h)_{ij}$ equals $(h(\boldsymbol{r}_j))_i$, the ith component of image $h(\boldsymbol{r}_j)$ of the basis vector $\boldsymbol{r}_j$.

Let now $h : \mathbb{C}^n \longrightarrow \mathbb{C}^n$ be an endomorphism of $\mathbb{C}^n$ and let $R = \{\boldsymbol{r}_1, \ldots, \boldsymbol{r}_n\}$ and $S = \{\boldsymbol{s}_1, \ldots, \boldsymbol{s}_n\}$ be two bases of $\mathbb{C}^n$. The vectors $\boldsymbol{s}_i$ can be expressed as linear combinations of the vectors $\boldsymbol{r}_1, \ldots, \boldsymbol{r}_n$ as follows:

$$\boldsymbol{s}_i = p_{i1}\boldsymbol{r}_1 + \cdots + p_{in}\boldsymbol{r}_n \tag{3.6}$$

for $1 \leqslant i \leqslant n$, which implies

$$h(\boldsymbol{s}_i) = p_{i1}h(\boldsymbol{r}_1) + \cdots + p_{in}h(\boldsymbol{r}_n) \tag{3.7}$$

for $1 \leqslant i \leqslant n$. Therefore, the matrix associated to a linear form $h : \mathbb{C}^m \longrightarrow \mathbb{C}$ is a column vector $\boldsymbol{r}$. In this case we can write $h(\boldsymbol{x}) = \boldsymbol{r}^{\mathsf{H}}\boldsymbol{x}$ for $\boldsymbol{x} \in \mathbb{R}^n$.

Theorem 3.28. *Let $h \in \mathsf{Hom}(\mathbb{C}^m, \mathbb{C}^n)$. The matrix $A_{h^*} \in \mathbb{C}^{m\times n}$ is the transpose of the matrix A_h, that is, we have $A_{h^*} = A_h'$.*

Proof. By the previous discussion, if $\ell_1, \ldots, \ell_n$ is a basis of the space $(\mathbb{C}^n)^*$, then the jth column of the matrix $A_{h^*} \in \mathbb{C}^{m\times n}$ is obtained by expressing the linear form $h^*(\ell_j) = \ell_j h$ in terms of a basis in the dual space $(\mathbb{C}^m)^*$. Therefore, we need to evaluate the linear form $\ell_j h \in (\mathbb{C}^m)^*$.

Let $\{\boldsymbol{p}_1, \ldots, \boldsymbol{p}_m\}$ be a basis in $\mathbb{C}^m$ and let $\{g_1, \ldots, g_m\}$ be its dual in $(\mathbb{C}^m)^*$. Also, let $\{\boldsymbol{q}_1, \ldots, \boldsymbol{q}_n\}$ be a basis in $\mathbb{C}^n$, and let $\{\ell_1, \ldots, \ell_n\}$ be its dual $(\mathbb{C}^n)^*$.

Observe that if $\boldsymbol{v} \in \mathbb{C}^m$ can be expressed as $\boldsymbol{v} = \sum_{j=1}^m v_j \boldsymbol{p}_j$, then

$$g_p(\boldsymbol{v}) = g_p\left(\sum_{j=1}^m v_j \boldsymbol{p}_j\right) = \sum_{j=1}^m v_j g_p(\boldsymbol{p}_j) = v_p,$$

because $\{g_1, \ldots, g_m\}$ is the dual of $\{\boldsymbol{p}_1, \ldots, \boldsymbol{p}_m\}$ in $(\mathbb{C}^m)^*$.

On the other hand, we can write

$$\begin{aligned}
\ell_j(h(\boldsymbol{v})) &= \ell_j\left(\sum_{p=1}^m v_p h(\boldsymbol{p}_p)\right) = \ell_j\left(\sum_{p=1}^m v_p \sum_{i=1}^n a_{ip}\boldsymbol{q}_i\right) \\
&= \ell_j\left(\sum_{p=1}^m \sum_{i=1}^n v_p a_{ip}\boldsymbol{q}_i\right) = \sum_{p=1}^m \sum_{i=1}^n v_p a_{ip}\ell_j(\boldsymbol{q}_i) \\
&= \sum_{p=1}^m v_p a_{jp} = \sum_{p=1}^m a_{jp} g_p(\boldsymbol{v}).
\end{aligned}$$

Thus, $h^*(\ell_j) = \sum_{p=1}^m a_{jp} g_p$ for every j, $1 \leq j \leqslant m$. This means that the jth column of the matrix A_{h^*} is the transposed jth row of the matrix A_h, so $A_{h^*} = (A_h)'$. □

Matrix multiplication corresponds to the composition of linear mappings, as we show next.

Theorem 3.29. *Let $h \in \mathsf{Hom}(\mathbb{C}^m, \mathbb{C}^n)$ and $g \in \mathsf{Hom}(\mathbb{C}^n, \mathbb{C}^p)$. Then,*

$$A_{gh} = A_g A_h.$$

Proof. If $\boldsymbol{p}_1, \ldots, \boldsymbol{p}_m$ is a basis for $\mathbb{C}^m$, then $A_{gh}(\boldsymbol{p}_i) = gh(\boldsymbol{p}_i) = g(h(\boldsymbol{p}_i)) = g(A_h \boldsymbol{p}_i) = A_g(A_h(\boldsymbol{p}_i))$ for every i, where $1 \leqslant i \leqslant n$. This proves that $A_{gh} = A_g A_h$. □

Thus, if h is an idempotent endomorphism of a linear space, the matrix A_h is idempotent.

The inverse direction, from matrices to linear operators, is introduced next.

Definition 3.27. Let $A \in \mathbb{C}^{n\times m}$ be a matrix. The *linear operator associated to* A is the mapping $h_A : \mathbb{C}^m \longrightarrow \mathbb{C}^n$ given by $h_A(\boldsymbol{x}) = A\boldsymbol{x}$ for $\boldsymbol{x} \in \mathbb{C}^m$.

If $\{\boldsymbol{e}_1, \ldots, \boldsymbol{e}_m\}$ is a basis for $\mathbb{C}^m$, then $h_A(\boldsymbol{e}_i)$ is the ith column of the matrix A.

It is immediate that $A_{h_A} = A$ and $h_{A_h} = h$.

Attributes of a matrix A are usually transferred to the linear operator h_A. For example, if A is Hermitian, we say that h_A is Hermitian.

Definition 3.28. Let $A \in \mathbb{C}^{n\times m}$ be a matrix. The *range of* A is the subspace $\mathsf{Im}(h_A)$ of $\mathbb{C}^n$. The *null space of* A is the subspace $\mathsf{Ker}(h_A)$.

The range of A and the null space of A are denoted by $\mathsf{range}(A)$ and $\mathsf{null}(A)$, respectively.

Clearly, $C_{A,n} = \mathsf{range}(A)$. The null space of $A \in \mathbb{C}^{m\times n}$ consists of those $\boldsymbol{x} \in \mathbb{C}^n$ such that $A\boldsymbol{x} = \boldsymbol{0}$.

Let $\{\boldsymbol{p}_1, \ldots, \boldsymbol{p}_m\}$ be a basis of $\mathbb{C}^m$. Since $\mathsf{range}(A) = \mathsf{Im}(h_A)$, it follows that this subspace is generated by the set $\{h_A(\boldsymbol{p}_1), \ldots, h_A(\boldsymbol{p}_m)\}$, that is, by the columns of the matrix A. For this reason, the subspace $\mathsf{range}(A)$ is also known as the *column subspace* of A.

Theorem 3.30. *Let $A, B \in \mathbb{C}^{m\times n}$ be two matrices. Then*

$$\mathsf{range}(A + B) \subseteq \mathsf{range}(A) + \mathsf{range}(B).$$

Proof. Let $u \in \mathsf{range}(A + B)$. There exists $v \in \mathbb{C}^n$ such that $u = (A + B)v = Av + Bv$. If $x = Av$ and $y = Bv$, we have $x \in \mathsf{range}(A)$ and $y \in \mathsf{range}(B)$, so $u = x + y \in \mathsf{range}(A) + \mathsf{range}(B)$. □

Several facts concerning idempotent endomorphisms that were presented in Chapter 2 can now be formulated in terms of matrices. For example, Theorem 2.32 applied to $\mathbb{C}^n$ states that if A is an idempotent matrix, then $\mathbb{C}^n = \mathsf{null}(A) \boxplus \mathsf{range}(A)$. Conversely, by Theorem 2.33, if U and W are two subspaces of $\mathbb{C}^n$ such that $\mathbb{C}^n = U \boxplus W$, then there exists an idempotent matrix $A \in \mathbb{C}^{n\times n}$ such that $U = \mathsf{null}(A)$ and $W = \mathsf{range}(A)$.

Multilinear functions can be represented by generalizations of matrices.

Let V, W, and U be finite-dimensional linear spaces having the bases $\{\boldsymbol{v}_1, \ldots, \boldsymbol{v}_n\}$, $\{\boldsymbol{w}_1, \ldots, \boldsymbol{w}_m\}$, and $\{\boldsymbol{u}_1, \ldots, \boldsymbol{u}_p\}$, respectively, and let $f : V \times W \longrightarrow U$ be a multilinear function. For $\boldsymbol{v} = a^1\boldsymbol{v}_1 + \cdots + a^n\boldsymbol{v}_n$ and $\boldsymbol{w} = b^1\boldsymbol{w}_1 + \cdots + b^m\boldsymbol{w}_m$, by the multilinearity of f we can write

$$\begin{aligned} f(\boldsymbol{v}, \boldsymbol{w}) &= f\left(a^1\boldsymbol{v}_1 + \cdots + a^n\boldsymbol{v}_n, b^1\boldsymbol{w}_1 + \cdots + b^m\boldsymbol{w}_m\right) \\ &= \sum_{i=1}^{n}\sum_{j=1}^{m} a^i b^j f(\boldsymbol{v}_i, \boldsymbol{w}_j). \end{aligned}$$

Since $f(\boldsymbol{v}_i, \boldsymbol{w}_j) \in U$, we can further write

$$f(\boldsymbol{v}_i, \boldsymbol{w}_j) = \sum_{k=1}^{p} c_{ij}^k \boldsymbol{u}^k,$$

for $1 \leqslant i \leqslant n$ and $1 \leqslant j \leqslant m$. Thus,

$$f(\boldsymbol{v}, \boldsymbol{w}) = \sum_{i=1}^{n}\sum_{j=1}^{m}\sum_{k=1}^{p} a^i b^j c_{ij}^k \boldsymbol{u}^k.$$

Thus, f is completely specified by nmp numbers c_{ij}^k. Conversely, every set of nmp numbers specifies a multilinear function.

3.11 The Notion of Rank

Definition 3.29. The *rank* of a matrix A is the same number denoted by $\mathsf{rank}(A)$ given by $\mathsf{rank}(A) = \dim(\mathsf{range}(A)) = \dim(\mathsf{Im}(h_A))$.

Thus, the rank of A is the maximal size of a set of linearly independent columns of A.

Theorem 2.10 applied to the linear mapping $h_A : \mathbb{C}^m \longrightarrow \mathbb{C}^n$ means that for $A \in \mathbb{C}^{n\times m}$, we have

$$\dim(\mathsf{null}(A)) + \mathsf{rank}(A) = m. \tag{3.8}$$

Observe that if $A \in \mathbb{C}^{m\times m}$ is non-singular, then $A\boldsymbol{x} = \mathbf{0}_m$ implies $\boldsymbol{x} = \mathbf{0}_m$. Thus, if $\boldsymbol{x} \in \mathsf{null}(A) \cap \mathsf{range}(A)$, it follows that $A\boldsymbol{x} = \mathbf{0}$, so the subspaces $\mathsf{null}(A)$ and $\mathsf{range}(A)$ are complementary.

Example 3.26. For the matrix

$$A = \begin{pmatrix} 1 & 0 & 2 \\ 1 & -1 & 1 \\ 2 & 1 & 5 \\ 1 & 2 & 4 \end{pmatrix},$$

we have $\mathsf{rank}(A) = 2$. Indeed, if $\boldsymbol{c}_1, \boldsymbol{c}_2, \boldsymbol{c}_3$ are its columns, then it is easy to see that $\{\boldsymbol{c}_1, \boldsymbol{c}_2\}$ is a linearly independent set, and $\boldsymbol{c}_3 = 2\boldsymbol{c}_1 + \boldsymbol{c}_2$. Thus, the maximal size of a set of linearly independent columns of A is 2.

Example 3.27. Let $A \in \mathbb{C}^{n\times m}$ and $B \in \mathbb{C}^{p\times q}$. For the matrix $C \in \mathbb{C}^{(n+p)\times(m+q)}$ defined by

$$C = \begin{pmatrix} A & O_{n,q} \\ O_{p,m} & B \end{pmatrix},$$

we have $\mathsf{rank}(C) = \mathsf{rank}(A) + \mathsf{rank}(B)$.

Suppose that $\mathsf{rank}(C) = \ell$ and let $\boldsymbol{c}_1, \ldots, \boldsymbol{c}_\ell$ be a maximal set of linearly independent columns of C. Without loss of generality we may assume that the first k columns are among the first m columns of A and the remaining $\ell - k$ columns are among the last q columns of C. The first k columns of C correspond to k linearly independent columns of A, while the last $\ell - k$ columns correspond to $\ell - k$ linearly independent columns of B. Thus, $\mathsf{rank}(C) = k \leqslant \mathsf{rank}(A) + \mathsf{rank}(B)$.

Conversely, suppose that $\mathsf{rank}(A) = s$ and $\mathsf{rank}(B) = t$. Let $\boldsymbol{a}_{i_1}, \ldots, \boldsymbol{a}_{i_s}$ be a maximal set of linearly independent columns of A

and let $\boldsymbol{b}_{j_1}, \ldots, \boldsymbol{b}_{j_t}$ be a maximal set of linearly independent columns of B. Then, it is easy to see that the vectors

$$\begin{pmatrix} \boldsymbol{a}_{i_1} \\ \boldsymbol{0}_n \end{pmatrix}, \ldots, \begin{pmatrix} \boldsymbol{a}_{i_s} \\ \boldsymbol{0}_n \end{pmatrix}, \ldots, \begin{pmatrix} \boldsymbol{0}_n \\ \boldsymbol{b}_{j_1} \end{pmatrix}, \ldots, \begin{pmatrix} \boldsymbol{0}_n \\ \boldsymbol{b}_{j_t} \end{pmatrix}$$

constitute a linearly independent set of columns of C, so $\mathsf{rank}(A) + \mathsf{rank}(B) \leqslant \mathsf{rank}(C)$. Thus, $\mathsf{rank}(C) = \mathsf{rank}(A) + \mathsf{rank}(B)$.

Example 3.28. Let $\boldsymbol{x}$ and $\boldsymbol{y}$ be two vectors in $\mathbb{C}^n - \{\boldsymbol{0}\}$. The matrix $\boldsymbol{x}\boldsymbol{y}^{\mathsf{H}}$ has rank 1. Indeed, if $\boldsymbol{y}^{\mathsf{H}} = (y_1, y_2, \ldots, y_n)$, then we can write

$$\boldsymbol{x}\boldsymbol{y}^{\mathsf{H}} = (y_1\boldsymbol{x}\ y_2\boldsymbol{x}\ \cdots\ y_n\boldsymbol{x}),$$

which implies that the maximum number of linearly independent columns of $\boldsymbol{x}\boldsymbol{y}^{\mathsf{H}}$ is 1.

Example 3.29. Let $A, B \in \mathbb{C}^{n\times m}$. We have

$$\mathsf{rank}(A + B) \leqslant \mathsf{rank}(A) + \mathsf{rank}(B).$$

Let $A = (\boldsymbol{a}_1\ \boldsymbol{a}_2\ \cdots\ \boldsymbol{a}_m)$ and $B = (\boldsymbol{b}_1\ \boldsymbol{b}_2\ \cdots\ \boldsymbol{b}_m)$ be two matrices, where $\boldsymbol{a}_1, \ldots, \boldsymbol{a}_m, \boldsymbol{b}_1, \ldots, \boldsymbol{b}_m \in \mathbb{C}^n$. Clearly, we have

$$A + B = (\boldsymbol{a}_1 + \boldsymbol{b}_1\ \boldsymbol{a}_2 + \boldsymbol{b}_2\ \cdots\ \boldsymbol{a}_m + \boldsymbol{b}_m).$$

If $\boldsymbol{x} \in \mathsf{Im}(A + B)$, we can write

$$\boldsymbol{x} = x_1(\boldsymbol{a}_1 + \boldsymbol{b}_1) + x_2(\boldsymbol{a}_2 + \boldsymbol{b}_2) + \cdots + x_m(\boldsymbol{a}_m + \boldsymbol{b}_m) = \boldsymbol{y} + \boldsymbol{z},$$

where

$$\boldsymbol{y} = x_1\boldsymbol{a}_1 + \cdots + x_m\boldsymbol{a}_m \in \mathsf{Im}(A),$$
$$\boldsymbol{z} = x_1\boldsymbol{b}_1 + \cdots + x_m\boldsymbol{b}_m \in \mathsf{Im}(B).$$

Thus, $\mathsf{Im}(A + B) \subseteq \mathsf{Im}(A) + \mathsf{Im}(B)$. Since the dimension of the sum of two subspaces of a linear space is less or equal to the dimension of the sum of these subspaces, the result follows.

For a matrix $A \in \mathbb{C}^{n\times m}$ the range of the matrix $A' \in \mathbb{C}^{m\times n}$ is the subspace of $\mathbb{C}^n$ generated by the rows of the original matrix A and coincides with the subspace $\mathsf{Im}(h_A^*)$ of the dual mapping h_A^*.

By Theorem 2.48, we have $\mathsf{rank}(h_A^*) = \mathsf{rank}(h_A)$, so the rank of the transposed matrix A' is equal to $\mathsf{rank}(A)$. Thus,

$$\dim(\mathsf{null}(A')) + \mathsf{rank}(A) = n, \tag{3.9}$$

and the maximal size of a set of linearly independent rows of A coincides with the rank of A.

The above discussion also shows that if $A \in \mathbb{C}^{n\times m}$, then $\mathsf{rank}(A) \leqslant \min\{m, n\}$.

Theorem 3.31. *Let $A \in \mathbb{C}^{m\times n}$ be a matrix. We have* $\mathsf{rank}(A) = \mathsf{rank}(\overline{A})$.

Proof. Suppose that $A = (\boldsymbol{a}_1, \ldots, \boldsymbol{a}_n)$ and that the set $\{\boldsymbol{a}_{i_1}, \ldots, \boldsymbol{a}_{i_p}\}$ is a set of linearly independent columns of A. Then, the set $\{\overline{\boldsymbol{a}_{i_1}}, \ldots, \overline{\boldsymbol{a}_{i_p}}\}$ is a set of linearly independent columns of $\overline{A}$. This implies $\mathsf{rank}(\overline{A}) = \mathsf{rank}(A)$. □

Corollary 3.1. *We have* $\mathsf{rank}(A) = \mathsf{rank}(A^{\mathsf{H}})$ *for every matrix* $A \in \mathbb{C}^{m\times n}$.

Proof. Since $A^{\mathsf{H}} = \overline{A'}$, the statement follows immediately. □

Definition 3.30. A matrix $A \in \mathbb{C}^{n\times m}$ is a *full-rank* matrix if $\mathsf{rank}(A) = \min\{m, n\}$.

If $A \in \mathbb{C}^{m\times n}$ is a full-rank matrix and $m \geqslant n$, then the n columns of the matrix are linearly independent; similarly, if $n \geqslant m$, the m rows of the matrix are linearly independent.

A matrix that is not a full-rank is said to be *degenerate*. A degenerate square matrix is said to be *singular*. A *non-singular matrix* $A \in \mathbb{C}^{n\times n}$ is a matrix that is not singular and, therefore, has $\mathsf{rank}(A) = n$.

Theorem 3.32. *A matrix $A \in \mathbb{C}^{n\times n}$ is non-singular if and only if it is invertible.*

Proof. Suppose that A is non-singular, that is, $\mathsf{rank}(A) = n$. In other words, the set of columns $\{\boldsymbol{c}_1, \ldots, \boldsymbol{c}_n\}$ of A is linearly independent, and therefore, is a basis of $\mathbb{C}^n$. Then, each of the vectors $\boldsymbol{e}_i$ can be expressed as a unique combination of the columns of A, that is

$$\boldsymbol{e}_i = b_{1i}\boldsymbol{c}_1 + b_{2i}\boldsymbol{c}_2 + \cdots + b_{ni}\boldsymbol{c}_n,$$

for $1 \leqslant i \leqslant n$. These equalities can be written as

$$(\boldsymbol{c}_1 \ \cdots \ \boldsymbol{c}_n) \begin{pmatrix} b_{11} & \cdots & b_{1n} \\ b_{21} & \cdots & b_{2n} \\ \vdots & \cdots & \vdots \\ b_{n1} & \cdots & b_{nn} \end{pmatrix} = I_n.$$

Consequently, the matrix A is invertible and

$$A^{-1} = \begin{pmatrix} b_{11} & \cdots & b_{1n} \\ b_{21} & \cdots & b_{2n} \\ \vdots & \cdots & \vdots \\ b_{n1} & \cdots & b_{nn} \end{pmatrix}.$$

Suppose now that A is invertible and that

$$d_1\boldsymbol{c}_1 + \cdots + d_n\boldsymbol{c}_n = \mathbf{0}.$$

This is equivalent to

$$A \begin{pmatrix} d_1 \\ \vdots \\ d_n \end{pmatrix} = \mathbf{0}.$$

Multiplying both sides by A^{-1} implies

$$\begin{pmatrix} d_1 \\ \vdots \\ d_n \end{pmatrix} = \mathbf{0},$$

so $d_1 = \cdots = d_n = 0$, which means that the set of columns of A is linearly independent, so $\mathsf{rank}(A) = n$. □

Corollary 3.2. *A matrix* $A \in \mathbb{C}^{n\times n}$ *is non-singular if and only if* $A\boldsymbol{x} = \mathbf{0}$ *implies* $\boldsymbol{x} = \mathbf{0}$ *for* $\boldsymbol{x} \in \mathbb{C}^n$.

Proof. If A is non-singular then, by Theorem 3.32, A is invertible. Therefore, $A\boldsymbol{x} = \boldsymbol{0}$ implies $A^{-1}(Ax) = A^{-1}\boldsymbol{0}$, so $\boldsymbol{x} = \boldsymbol{0}$.

Conversely, suppose that $A\boldsymbol{x} = \boldsymbol{0}$ implies $\boldsymbol{x} = \boldsymbol{0}$. If $A = (\boldsymbol{c}_1 \ \cdots \ \boldsymbol{c}_n)$ and $\boldsymbol{x} = (x_1, \ldots, x_n)'$, the previous implication means that $x_1\boldsymbol{c}_1 + \cdots + x_n\boldsymbol{c}_n = \boldsymbol{0}$ implies $x_1 = \cdots = x_n = 0$, so $\{\boldsymbol{c}_1, \ldots, \boldsymbol{c}_n\}$ is linearly independent. Therefore, $\mathsf{rank}(A) = n$, so A is non-singular. □

Let $A \in \mathbb{C}^{n\times m}$ be a matrix. It is easy to see that the square matrix $B = A^{\mathsf{H}}A \in \mathbb{C}^{m\times m}$ is symmetric.

Theorem 3.33. *Let $A \in \mathbb{C}^{n\times m}$ be a matrix and let $B = A^{\mathsf{H}}A$. The matrices A and B have the same rank.*

Proof. We prove that $\mathsf{null}(A) = \mathsf{null}(B)$. If $A\boldsymbol{u} = \boldsymbol{0}$, then $B\boldsymbol{u} = A^{\mathsf{H}}(A\boldsymbol{0}) = \boldsymbol{0}$, so $\mathsf{null}(A) \subseteq \mathsf{null}(B)$. If $\boldsymbol{v} \in \mathsf{null}(B)$, then $A^{\mathsf{H}}A\boldsymbol{v} = 0$, which implies that $\boldsymbol{v}^{\mathsf{H}}A^{\mathsf{H}}A\boldsymbol{v} = 0$. This, in turn can be written as $(A\boldsymbol{v})^{\mathsf{H}}(A\boldsymbol{v}) = 0$, so, by Theorem 3.11, we have $A\boldsymbol{v} = \boldsymbol{0}$, which means that $\boldsymbol{v} \in \mathsf{null}(A)$. We conclude that $\mathsf{null}(A) = \mathsf{null}(A^{\mathsf{H}}A)$. The equalities

$$\dim(\mathsf{null}(A)) + \mathsf{rank}(A) = m,$$
$$\dim(\mathsf{null}(A)) + \mathsf{rank}(A^{\mathsf{H}}A) = m,$$

imply that $\mathsf{rank}(A^{\mathsf{H}}A) = m$. □

Corollary 3.3. *Let $A \in \mathbb{C}^{n\times m}$ be a matrix of full-rank. If $m \geqslant n$, then the matrix $A^{\mathsf{H}}A$ is non-singular; if $n \leqslant m$, then AA^{H} is non-singular.*

Proof. Suppose that $m \geqslant n$. Then, $\mathsf{rank}(A^{\mathsf{H}}A) = \mathsf{rank}(A) = m$ because A is a full-rank matrix. Thus, $A^{\mathsf{H}}A \in \mathbb{C}^{m\times m}$ is non-singular. The argument for the second part of the corollary is similar. □

Example 3.30. Let $A = (\boldsymbol{a}_1 \cdots \boldsymbol{a}_m) \in \mathbb{C}^{n\times m}$. Since $AA^{\mathsf{H}} = \boldsymbol{a}_1\boldsymbol{a}_1^{\mathsf{H}} + \cdots + \boldsymbol{a}_m\boldsymbol{a}_m^{\mathsf{H}}$, it follows that the rank of the matrix $\boldsymbol{a}_1\boldsymbol{a}_1^{\mathsf{H}} + \cdots + \boldsymbol{a}_m\boldsymbol{a}_m^{\mathsf{H}}$ equals the rank of the matrix A and, therefore, it cannot exceed m.

Theorem 3.34 (Sylvester's rank theorem). Let $A \in \mathbb{C}^{m\times n}$ and $B \in \mathbb{C}^{n\times p}$ be two matrices. We have

$$\mathsf{rank}(AB) = \mathsf{rank}(B) - \dim(\mathsf{null}(A) \cap \mathsf{range}(B)).$$

Proof. Both $\mathsf{null}(A)$ and $\mathsf{range}(B)$ are subspaces of $\mathbb{C}^n$. Therefore, $\mathsf{null}(A) \cap \mathsf{range}(B)$ is a subspace of $\mathbb{C}^n$. If $\boldsymbol{u}_1, \ldots, \boldsymbol{u}_k$ is a basis of the subspace $\mathsf{null}(A) \cap \mathsf{range}(B)$, then there exists a basis $\boldsymbol{u}_1, \ldots, \boldsymbol{u}_k, \boldsymbol{u}_{k+1}, \ldots, \boldsymbol{u}_l$ of the subspace $\mathsf{range}(B)$.

The set $\{A\boldsymbol{u}_{k+1}, \ldots, A\boldsymbol{u}_l\}$ is linearly independent. Indeed, suppose that there exists a linear combination

$$a_1 A\boldsymbol{u}_{k+1} + \cdots + a_{l-k} A\boldsymbol{u}_l = \mathbf{0}.$$

Then, $A(a_1\boldsymbol{u}_{k+1} + \cdots + a_{l-k}\boldsymbol{u}_l) = \mathbf{0}$, so $a_1\boldsymbol{u}_{k+1} + \cdots + a_{l-k}\boldsymbol{u}_l \in \mathsf{null}(A)$. Since $\boldsymbol{u}_{k+1}, \ldots, \boldsymbol{u}_l \in \mathsf{range}(B)$, it follows that $a_1\boldsymbol{u}_{k+1} + \cdots + a_{l-k}\boldsymbol{u}_l \in \mathsf{null}(A) \cap \mathsf{range}(B)$. Since $\boldsymbol{u}_1, \ldots, \boldsymbol{u}_k$ is a basis of the subspace $\mathsf{null}(A) \cap \mathsf{range}(B)$, we have

$$a_1\boldsymbol{u}_{k+1} + \cdots + a_{l-k}\boldsymbol{u}_l = d_1\boldsymbol{u}_1 + \cdots + d_k\boldsymbol{u}_k$$

for some $d_1, \ldots, d_k \in \mathbb{C}$, which implies

$$a_1\boldsymbol{u}_{k+1} + \cdots + a_{l-k}\boldsymbol{u}_l - d_1\boldsymbol{u}_1 - \cdots - d_k\boldsymbol{u}_k = \mathbf{0}.$$

Since $\boldsymbol{u}_1, \ldots, \boldsymbol{u}_k, \boldsymbol{u}_{k+1}, \cdots, \boldsymbol{u}_l$ is a basis of $\mathsf{range}(B)$, it follows that $a_1 = \cdots = a_{l-k} = d_1 = \cdots = d_k = 0$, so $A\boldsymbol{u}_{k+1}, \ldots, A\boldsymbol{u}_l$ is indeed linear independent.

Next, we show that $A\boldsymbol{u}_{k+1}, \ldots, A\boldsymbol{u}_l$ spans the subspace $\mathsf{range}(AB)$. Since $\boldsymbol{u}_j \in \mathsf{range}(B)$, it is clear that $A\boldsymbol{u}_j \in \mathsf{range}(AB)$ for $k+1 \leqslant j \leqslant l$. If $\boldsymbol{w} \in \mathsf{range}(AB)$, then $\boldsymbol{w} = AB\boldsymbol{x}$ for some $\boldsymbol{x} \in \mathbb{C}^p$. Since $B\boldsymbol{x} \in \mathsf{range}(B)$, we can write $B\boldsymbol{x} = b_1\boldsymbol{u}_1 + \cdots + b_k\boldsymbol{u}_k + b_{k+1}\boldsymbol{u}_{k+1} + \cdots + b_l\boldsymbol{u}_l$, which implies

$$\boldsymbol{w} = AB\boldsymbol{x} = b_{k+1}A\boldsymbol{u}_{k+1} + \cdots + b_l A\boldsymbol{u}_l,$$

because $A\boldsymbol{u}_1 = \cdots = A\boldsymbol{u}_k = \mathbf{0}$, as $\boldsymbol{u}_1, \ldots, \boldsymbol{u}_k$ belong to $\mathsf{null}(A)$. Thus, $A\boldsymbol{u}_{k+1}, \ldots, A\boldsymbol{u}_l$ spans the subspace $\mathsf{range}(AB)$, which allows us to conclude that this linearly independent set is a basis for this subspace that contains $l-k$ elements. This allows us to conclude that $\mathsf{rank}(AB) = \dim(\mathsf{range}(AB)) = \mathsf{rank}(B) - \dim(\mathsf{null}(A) \cap \mathsf{range}(B))$ □

Corollary 3.4. *Let $A \in \mathbb{C}^{m\times n}$. If $R \in \mathbb{C}^{m\times m}$ and $Q \in \mathbb{C}^{n\times n}$ are invertible matrices, then*

$$\mathsf{rank}(A) = \mathsf{rank}(RA) = \mathsf{rank}(AQ) = \mathsf{rank}(RAQ).$$

Proof. Note that $\mathsf{rank}(R) = m$ and $\mathsf{rank}(Q) = n$. Thus, $\mathsf{null}(R) = \{\mathbf{0}_m\}$ and $\mathsf{null}(Q) = \{\mathbf{0}_n\}$. By Sylvester's rank theorem we have

$$\begin{aligned}\mathsf{rank}(RA) &= \mathsf{rank}(A) - \dim(\mathsf{null}(R) \cap \mathsf{range}(A)) \\ &= \mathsf{rank}(A) - \dim(\{\mathbf{0}\}) = \mathsf{rank}(A).\end{aligned}$$

On the other hand, we have

$$\begin{aligned}\mathsf{rank}(AQ) &= \mathsf{rank}(Q) - \dim(\mathsf{null}(A) \cap \mathsf{range}(Q)) \\ &= n - \dim(\mathsf{null}(A)) = \mathsf{rank}(A),\end{aligned}$$

because $\mathsf{range}(Q) = \mathbb{C}^n$.

The last equality of the theorem follows from the first two. □

Corollary 3.5 (Frobenius[2] inequality). Let $A \in \mathbb{C}^{m\times n}$, $B \in \mathbb{C}^{n\times p}$, and $C \in \mathbb{C}^{p\times q}$ be three matrices. Then,

$$\mathsf{rank}(AB) + \mathsf{rank}(BC) \leqslant \mathsf{rank}(B) + \mathsf{rank}(ABC).$$

Proof. By Sylvester's rank theorem, we have

$$\begin{aligned}\mathsf{rank}(ABC) &= \mathsf{rank}(BC) - \dim(\mathsf{null}(A) \cap \mathsf{range}(BC)), \\ \mathsf{rank}(AB) &= \mathsf{rank}(B) - \dim(\mathsf{null}(A) \cap \mathsf{range}(B)).\end{aligned}$$

These equalities imply

$$\begin{aligned}\mathsf{rank}(ABC) + \mathsf{rank}(B) = \mathsf{rank}(AB) &+ \mathsf{rank}(BC) \\ &- \dim(\mathsf{null}(A) \cap \mathsf{range}(BC)) \\ &+ \dim(\mathsf{null}(A) \cap \mathsf{range}(B)).\end{aligned}$$

Since $\mathsf{null}(A) \cap \mathsf{range}(BC) \subseteq \mathsf{null}(A) \cap \mathsf{range}(B)$, we have

$$\dim(\mathsf{null}(A) \cap \mathsf{range}(BC)) \leqslant \dim(\mathsf{null}(A) \cap \mathsf{range}(B)),$$

which implies the desired inequality. □

[2]Ferdinand Georg Frobenius was born on October 26, 1849 in Charlottenburg-Berlin and died on August 3, 1917 in Berlin. He studied at the University of Göttingen and Berlin and taught at the University of Berlin and ETH in Zürich. Frobenius has contributed to the study of elliptic functions, algebra, and mathematical physics.

Corollary 3.6. *Let $A \in \mathbb{C}^{m\times n}$ and $B \in \mathbb{C}^{n\times p}$ be two matrices. We have*

$$\dim(\mathsf{null}(AB)) = \dim(\mathsf{null}(B)) + \dim(\mathsf{null}(A) \cap \mathsf{range}(B)).$$

Proof. By Equality (3.8), we have

$$\dim(\mathsf{null}(AB)) + \mathsf{rank}(AB) = p,$$
$$\dim(\mathsf{null}(B)) + \mathsf{rank}(B) = p.$$

An application of Sylvester's rank theorem implies

$$\dim(\mathsf{null}(AB)) = \dim(\mathsf{null}(B)) + \dim(\mathsf{null}(A) \cap \mathsf{range}(B)).$$ □

Corollary 3.7. *Let $A \in \mathbb{C}^{m\times n}$ and $B \in \mathbb{C}^{n\times p}$ be two matrices. We have*

$$\mathsf{rank}(A)+\mathsf{rank}(B)-n \leqslant \mathsf{rank}(AB) \leqslant \min\{\mathsf{rank}(A), \mathsf{rank}(B)\}, \quad (3.10)$$

and

$$\max\{\dim(\mathsf{null}(A)), \dim(\mathsf{null}(B))\} \leq \dim(\mathsf{null}(AB)) \leq \dim(\mathsf{null}(A)) + \dim(\mathsf{null}(B)).$$

Proof. Since $\dim(\mathsf{null}(A)\cap\mathsf{rank}(B)) \leqslant \dim(\mathsf{null}(A)) = n - \mathsf{rank}(A)$, it follows that $\mathsf{rank}(AB) \geqslant \mathsf{rank}(B) - (n - \mathsf{rank}(A)) = \mathsf{rank}(A) + \mathsf{rank}(B) - n$.

For the second inequality, observe that Sylvester's rank Theorem, implies immediately $\mathsf{rank}(AB) \leqslant \mathsf{rank}(B)$. Also, $\mathsf{rank}(AB) = \mathsf{rank}((AB)') = \mathsf{rank}(B'A') \leqslant \mathsf{rank}(A') = \mathsf{rank}(A)$, so $\mathsf{rank}(AB) \leq \min\{\mathsf{rank}(A), \mathsf{rank}(B)\}$.

The second part of the Corollary follows from the first part. □

Inequality (3.10) is also known as the *Sylvester rank inequality.*

Corollary 3.8. *If $A \in \mathbb{C}^{m\times n}$ is a full-rank matrix with $m \geqslant n$, then $\mathsf{rank}(AB) = \mathsf{rank}(B)$ for any $B \in \mathbb{C}^{n\times p}$.*

Proof. Since $m \geqslant n$, we have $\mathsf{rank}(A) = n$; therefore, the n columns of A are linearly independent so $\mathsf{null}(A) = \{\mathbf{0}\}$. By Sylvester's Rank theorem, we have $\mathsf{rank}(AB) = \mathsf{rank}(B)$. □

Theorem 3.35 (The full-rank factorization theorem). Let $A \in \mathbb{C}^{m\times n}$ be a matrix with $\mathsf{rank}(A) = r > 0$. There exists $B \in \mathbb{C}^{m\times r}$ and $C \in \mathbb{C}^{r\times n}$ such that $A = BC$.

Furthermore, if $A = DE$, where $D \in \mathbb{C}^{m\times r}$, $E \in \mathbb{C}^{r\times n}$, then both D and E are full-rank matrices, that is, we have $\mathsf{rank}(D) = \mathsf{rank}(E) = r$.

Proof. Let $\{\boldsymbol{b}_1, \ldots, \boldsymbol{b}_r\} \subseteq \mathbb{C}^m$ be a basis for the $\mathsf{range}(A)$. Define $B = (\boldsymbol{b}_1 \ \cdots \ \boldsymbol{b}_r) \in \mathbb{C}^{m\times r}$. The columns of A, $\boldsymbol{a}_1, \ldots, \boldsymbol{a}_n$ can be written as $\boldsymbol{a}_i = c_{1i}\boldsymbol{b}_1 + \cdots c_{ri}\boldsymbol{b}_r$ for $1 \leqslant i \leqslant n$, which amounts to

$$A = (\boldsymbol{a}_1 \ \cdots \ \boldsymbol{a}_n) = (\boldsymbol{b}_1 \ \cdots \ \boldsymbol{b}_r) \begin{pmatrix} c_{11} & \cdots & c_{1r} \\ \vdots & \cdots & \vdots \\ c_{r1} & \cdots & c_r \end{pmatrix}.$$

Thus, $A = BC$, where

$$C = \begin{pmatrix} c_{11} & \cdots & c_{1r} \\ \vdots & \cdots & \vdots \\ c_{r1} & \cdots & c_r \end{pmatrix}.$$

Suppose now that $A = DE$, where $D \in \mathbb{C}^{m\times r}$, $E \in \mathbb{C}^{r\times n}$. It is clear that we have both $\mathsf{rank}(D) \leqslant r$ and $\mathsf{rank}(E) \leqslant r$. On the other hand, by Corollary 3.7, $r = \mathsf{rank}(A) = \mathsf{rank}(DE) \leqslant \min\{\mathsf{rank}(D), \mathsf{rank}(E)\}$ implies $r \leqslant \mathsf{rank}(D)$ and $r \leq \mathsf{rank}(E)$, so $\mathsf{rank}(D) = \mathsf{rank}(E) = r$. □

Corollary 3.9. *Let $A \in \mathbb{C}^{m\times n}$ be a matrix such that $\mathsf{rank}(A) = r > 0$, and let $A = BC$ be a full-rank factorization of A.*

If the columns of B constitute a basis of the column space of A, then C is uniquely determined. Furthermore, if the rows of C constitute a basis of the row space of A, then B is uniquely determined.

Proof. This statement is an immediate consequence of the full-rank factorization theorem. □

Corollary 3.10. *If $A \in \mathbb{C}^{m\times n}$ is a matrix with $\mathsf{rank}(A) = r > 0$, then A can be written as*

$$A = \boldsymbol{b}_1\boldsymbol{c}_1' + \cdots + \boldsymbol{b}_r\boldsymbol{c}_r',$$

where $\{\boldsymbol{b}_1, \ldots, \boldsymbol{b}_r\} \subseteq \mathbb{C}^m$ and $\{\boldsymbol{c}_1, \ldots, \boldsymbol{c}_r\} \subseteq \mathbb{C}^n$ are linearly independent sets.

Proof. The corollary follows from Theorem 3.35 by adopting the set of columns of B as $\{\boldsymbol{b}_1, \ldots, \boldsymbol{b}_r\}$ and the transposed rows of C as $\{\boldsymbol{c}_1, \ldots, \boldsymbol{c}_r\}$. □

Theorem 3.36. *Let $A \in \mathbb{C}^{m\times n}$ be a full-rank matrix. If $m \geqslant n$, then there exists a matrix $D \in \mathbb{C}^{n\times m}$ such that $DA = I_n$. If $n \geqslant m$, then there exists a matrix $E \in \mathbb{C}^{n\times m}$ such that $AE = I_m$.*

Proof. Suppose that $A = (\boldsymbol{a}_1 \ \cdots \ \boldsymbol{a}_n) \in \mathbb{C}^{m\times n}$ is a full-rank matrix and $m \geqslant n$. Then, the n columns of A are linearly independent and, by Corollary 2.3, we can extend the set of columns to a basis of $\mathbb{C}^m$, $\{\boldsymbol{a}_1, \ldots, \boldsymbol{a}_n, \boldsymbol{d}_1, \ldots, \boldsymbol{d}_{m-n}\}$. The matrix $T = (\boldsymbol{a}_1 \ \cdots \ \boldsymbol{a}_n \ \boldsymbol{d}_1 \ \cdots \ \boldsymbol{d}_{m-n})$ is invertible, so there exists

$$T^{-1} = \begin{pmatrix} \boldsymbol{t}_1 \\ \vdots \\ \boldsymbol{t}_n \\ \boldsymbol{t}_{n+1} \vdots \\ \boldsymbol{t}_m \end{pmatrix},$$

such that $T^{-1}T = I_m$. If we define

$$D = \begin{pmatrix} \boldsymbol{t}_1 \\ \vdots \\ \boldsymbol{t}_n \end{pmatrix},$$

it is immediate that $DA = I_n$.

The argument for the second part is similar. □

Definition 3.31. Let $A \in \mathbb{C}^{m\times n}$. A *left inverse* of A is a matrix $D \in \mathbb{C}^{n\times m}$ such that $DA = I_n$. A *right inverse* of A is a matrix $E \in \mathbb{C}^{n\times m}$ such that $AE = I_m$.

Theorem 3.36 can now be restated as follows. Let $A \in \mathbb{C}^{m\times n}$ be a full-rank matrix. If $m \geqslant n$, then A has a left inverse; if $n \geqslant m$, then A has a right inverse.

Corollary 3.11. *Let $A \in \mathbb{C}^{n\times n}$ be a square matrix. The following statements are equivalent.*

(i) *A has a left inverse;*
(ii) *A has a right inverse;*
(iii) *A has an inverse.*

Proof. It is clear that (iii) implies both (i) and (ii). Suppose now that A has a left inverse, so $DA = I_n$. Then, the columns of A, $\mathbf{c}_1, \ldots, \mathbf{c}_n$ are linearly independent, for if $a_1\mathbf{c}_1 + \cdots a_n\mathbf{c}_n = \mathbf{0}$, we have $a_1 D\mathbf{c}_1 + \cdots + a_n D\mathbf{c}_n = a_1\mathbf{e}_1 + \cdots + a_n\mathbf{c}_n = \mathbf{0}$, which implies $a_1 = \cdots = a_n = 0$. Thus, $\mathsf{rank}(A) = n$, so A has an inverse.

In a similar manner (using the rows of A), we can show that (ii) implies (iii). □

Theorem 3.37. *Let $A \in \mathbb{C}^{m\times n}$ be a matrix with $\mathsf{rank}(A) = r > 0$. There exists a non-singular matrix $G \in \mathbb{C}^{m\times m}$ and a non-singular matrix $H \in \mathbb{C}^{n\times n}$ such that*

$$A = G\begin{pmatrix} I_r & O_{r,n-r} \\ O_{m-r,r} & O_{m-r,n-r} \end{pmatrix} H.$$

Proof. By the Full-Rank factorization theorem (Theorem 3.35), there are two full-rank matrices $B \in \mathbb{C}^{m\times r}$ and $C \in \mathbb{C}^{r\times n}$ such that $A = BC$. Let $\{\boldsymbol{b}_1, \ldots, \boldsymbol{b}_r\}$ be the columns of B and let $\boldsymbol{c}'_1, \ldots, \boldsymbol{c}'_r$ be the rows of C. It is clear that both sets of vectors are linearly independent and, therefore, for the first set, there exist $\boldsymbol{b}_{r+1}, \ldots, \boldsymbol{b}_m$ such that $\{\boldsymbol{b}_1, \ldots, \boldsymbol{b}_m\}$ is a basis of $\mathbb{C}^m$; for the second set, we have the vectors $\boldsymbol{c}'_{r+1}, \ldots, \boldsymbol{c}'_n$ such that $\{\boldsymbol{c}'_1, \ldots, \boldsymbol{c}'_n\}$ is a basis for $\mathbb{R}^n$. Define $G = (\boldsymbol{b}_1, \ldots, \boldsymbol{b}_m)$ and

$$H = \begin{pmatrix} \boldsymbol{c}'_1 \\ \vdots \\ \boldsymbol{c}'_n \end{pmatrix}.$$

Clearly, both G and H are non-singular and

$$A = G\begin{pmatrix} I_r & O_{r,n-r} \\ O_{m-r,r} & O_{m-r,n-r} \end{pmatrix} H.$$

□

Next we examine the relationships between full-rank factorization and the Moore–Penrose pseudoinverse.

Theorem 3.38. *Let $A \in \mathbb{C}^{m\times n}$ be a matrix with $\mathsf{rank}(A) = r > 0$ and let $A = BC$ be a full-rank factorization of A, where $B \in \mathbb{C}^{m\times r}$ and $C \in \mathbb{C}^{r\times n}$ are full-rank matrices. The following statements hold:*

(1) the matrices $B^{\mathsf{H}}B \in \mathbb{C}^{r\times r}$ and $CC^{\mathsf{H}} \in \mathbb{C}^{r\times r}$ are non-singular;

(2) the matrix $B^{\mathsf{H}}AC^{\mathsf{H}}$ is non-singular;

(3) the Moore–Penrose pseudoinverse of A is given by

$$A^{\dagger} = C^{\mathsf{H}}(CC^{\mathsf{H}})^{-1}(B^{\mathsf{H}}B)^{-1}B^{\mathsf{H}}.$$

Proof. The first part of the theorem follows from Corollary 3.3.

For Part (ii), note that

$$B^{\mathsf{H}}AC^{\mathsf{H}} = B^{\mathsf{H}}(BC)C^{\mathsf{H}} = (B^{\mathsf{H}}B)(CC^{\mathsf{H}}),$$

so $B^{\mathsf{H}}AC^{\mathsf{H}}$ is non-singular as a product of two non-singular matrices.

For Part (iii), observe that

$$(B^{\mathsf{H}}AC^{\mathsf{H}})^{-1} = ((B^{\mathsf{H}}B)(CC^{\mathsf{H}}))^{-1} = (CC^{\mathsf{H}})^{-1}(B^{\mathsf{H}}B)^{-1}.$$

Therefore,

$$C^{\mathsf{H}}(B^{\mathsf{H}}AC^{\mathsf{H}})^{-1}B^{\mathsf{H}} = C^{\mathsf{H}}(CC^{\mathsf{H}})^{-1}(B^{\mathsf{H}}B)^{-1}B^{\mathsf{H}},$$

and it is easy to verify that the matrix $C^{\mathsf{H}}(CC^{\mathsf{H}})^{-1}(B^{\mathsf{H}}B)^{-1}B^{\mathsf{H}}$ satisfies the conditions of Theorem 3.26. □

Lemma 3.1. *If $A \in \mathbb{C}^{m\times n}$ is a matrix and $\boldsymbol{x} \in \mathbb{C}^m$, $\boldsymbol{y} \in \mathbb{C}^n$ are two vectors such that $\boldsymbol{x}^{\mathsf{H}}A\boldsymbol{y} \neq 0$, then $\mathsf{rank}(A\boldsymbol{y}\boldsymbol{x}^{\mathsf{H}}A) = 1$.*

Proof. By the associative property of matrix product, we have $A\boldsymbol{y}\boldsymbol{x}^{\mathsf{H}}A = A(\boldsymbol{y}\boldsymbol{x}^{\mathsf{H}})A$, so $\mathsf{rank}(A\boldsymbol{y}\boldsymbol{x}^{\mathsf{H}}A) \leqslant \min\{\mathsf{rank}(\boldsymbol{y}\boldsymbol{x}^{\mathsf{H}}, \mathsf{rank}(A)\} = 1$, by Corollary 3.7.

We claim that $A\boldsymbol{y}\boldsymbol{x}^{\mathsf{H}}A \neq O_{m,n}$. Suppose that $A\boldsymbol{y}\boldsymbol{x}^{\mathsf{H}}A = O_{m,n}$. This implies $\boldsymbol{x}^{\mathsf{H}}A\boldsymbol{y}\boldsymbol{x}^{\mathsf{H}}A\boldsymbol{y} = 0$. If $z = \boldsymbol{x}^{\mathsf{H}}A\boldsymbol{y}$, the previous equality amounts to $z^2 = 0$, which yields $z = \boldsymbol{x}^{\mathsf{H}}A\boldsymbol{y} = 0$. This contradicts the hypothesis of the lemma, so $A\boldsymbol{y}\boldsymbol{x}^{\mathsf{H}}A \neq O_{m,n}$, which implies $\mathsf{rank}(A\boldsymbol{y}\boldsymbol{x}^{\mathsf{H}}A) \geqslant 1$. This allows us to conclude that $\mathsf{rank}(A\boldsymbol{y}\boldsymbol{x}^{\mathsf{H}}A) = 1$. □

The rank-1 matrix $A\boldsymbol{y}\boldsymbol{x}^{\mathsf{H}}A$ discussed in Lemma 3.1 plays a central role in the next statement.

Theorem 3.39 (Wedderburn's theorem). Let $A \in \mathbb{C}^{m\times n}$ be a matrix. If $\boldsymbol{x} \in \mathbb{C}^m$ and $\boldsymbol{y} \in \mathbb{C}^n$ are two vectors such that $\boldsymbol{x}^{\mathsf{H}} A\boldsymbol{y} \neq 0$ and B is the matrix

$$B = A - \frac{1}{\boldsymbol{x}^{\mathsf{H}} A\boldsymbol{y}} A\boldsymbol{y}\boldsymbol{x}^{\mathsf{H}} A,$$

then $\mathsf{rank}(B) = \mathsf{rank}(A) - 1$.

Proof. Observe that if $\boldsymbol{z} \in \mathsf{null}(A)$, then $A\boldsymbol{z} = \mathbf{0}$. Therefore, we have

$$B\boldsymbol{z} = -\frac{1}{\boldsymbol{x}^{\mathsf{H}} A\boldsymbol{y}} A\boldsymbol{y}\boldsymbol{x}^{\mathsf{H}} A\boldsymbol{z} = \mathbf{0},$$

so $\mathsf{null}(A) \subseteq \mathsf{null}(B)$. Conversely, if $\boldsymbol{z} \in \mathsf{null}(B)$, we have

$$A\boldsymbol{z} - \frac{1}{\boldsymbol{x}^{\mathsf{H}} A\boldsymbol{y}} A\boldsymbol{y}\boldsymbol{x}^{\mathsf{H}} A\boldsymbol{z} = \mathbf{0},$$

which can be written as

$$\begin{aligned} A\boldsymbol{z} &= \frac{1}{\boldsymbol{x}^{\mathsf{H}} A\boldsymbol{y}} A\boldsymbol{y}(\boldsymbol{x}^{\mathsf{H}} A\boldsymbol{z}) \\ &= \frac{\boldsymbol{x}^{\mathsf{H}} A\boldsymbol{z}}{\boldsymbol{x}^{\mathsf{H}} A\boldsymbol{y}} A\boldsymbol{y}. \end{aligned}$$

Thus, we obtain

$$A(\boldsymbol{z} - k\boldsymbol{y}) = \mathbf{0},$$

where

$$k = \frac{\boldsymbol{x}^{\mathsf{H}} A\boldsymbol{z}}{\boldsymbol{x}^{\mathsf{H}} A\boldsymbol{y}}.$$

Since $A\boldsymbol{y} \neq \mathbf{0}$, this shows that a basis of $\mathsf{null}(B)$ can be obtained by adding $\boldsymbol{y}$ to a basis of $\mathsf{null}(A)$. Therefore, $\dim(\mathsf{null}(B)) = \dim(\mathsf{null}(A)) + 1$, so $\mathsf{rank}(B) = \mathsf{rank}(A) - 1$. $\square$

Theorem 3.40. *A square matrix $A \in \mathbb{C}^{n\times n}$ generates an increasing sequence of null spaces*

$$\{\mathbf{0}\} = \mathsf{null}(A^0) \subseteq \mathsf{null}(A^1) \subseteq \cdots \subseteq \mathsf{null}(A^k) \subseteq \cdots$$

and a decreasing sequence of subspaces

$$\mathbb{C}^n = \mathsf{range}(A^0) \supseteq \mathsf{range}(A^1) \supseteq \cdots \supseteq \mathsf{range}(A^k) \supseteq \cdots$$

Furthermore, there exists a number ℓ such that

$$\mathsf{null}(A^0) \subset \mathsf{null}(A^1) \subset \cdots \subset \mathsf{null}(A^\ell) = \mathsf{null}(A^{\ell+1}) = \cdots$$

and

$$\mathsf{range}(A^0) \supset \mathsf{range}(A^1) \supset \cdots \supset \mathsf{range}(A^\ell) = \mathsf{range}(A^{\ell+1}) = \cdots$$

Proof. The proof of the existence of the increasing sequence of null subspaces and the decreasing sequence of ranges is immediate. Since $\mathsf{null}(A^k) \subseteq \mathbb{C}^n$ for every k, there exists a least number p such that $\mathsf{range}(A^p) = \mathsf{range}(A^{p+1})$. Therefore, $\mathsf{range}(A^{p+i}) = A^i\mathsf{range}(A^p) = A^i\mathsf{range}(A^{p+1}) = \mathsf{range}(A^{p+i+1})$ for every $i \in \mathbb{N}$. Thus, once two consecutive subspaces $\mathsf{range}(A^\ell)$ and $\mathsf{range}(A^{\ell+1})$ are equal, the sequence of range subspaces stops growing.

By Equality (3.8), we have $\dim(\mathsf{range}(A^k)) + \dim(\mathsf{null}(A^k)) = n$, so the sequence of null spaces stabilizes at the same number ℓ. □

Definition 3.32. The *index* of a square matrix $A \in \mathbb{C}^{n\times n}$ is the number ℓ defined in Theorem 3.40.

We denote the index of a matrix $A \in \mathbb{C}^{n\times n}$ by $\mathsf{index}(A)$.

Observe that if $A \in \mathbb{C}^{n\times n}$ is a non-singular matrix, then $\mathsf{index}(A) = 0$ because in this case $\mathbb{C}^n = \mathsf{range}(A^0) = \mathsf{range}(A)$.

Theorem 3.41. *Let $A \in \mathbb{C}^{n\times n}$ be a square matrix. The following statements are equivalent:*

(i) $\mathsf{range}(A^k) \cap \mathsf{null}(A^k) = \{\mathbf{0}\}$;
(ii) $\mathbb{C}^n = \mathsf{range}(A^k) \boxplus \mathsf{null}(A^k)$;
(iii) $k \geqslant \mathsf{index}(A)$.

Proof. We prove this theorem by showing that (i) and (ii) are equivalent, (i) implies (iii), and (iii) implies (ii).

Suppose that the first statement holds. By Theorem 2.35, the set $T = \{\boldsymbol{t} \in V \mid \boldsymbol{t} = \boldsymbol{u} + \boldsymbol{v}, \boldsymbol{u} \in \mathsf{range}(A^k), \boldsymbol{v} \in \mathsf{null}(A^k)\}$ is a subspace of $\mathbb{C}^n$ and $\dim(T) = \dim(\mathsf{range}(A^k)) + \dim(\mathsf{null}(A^k)) = n$. Therefore, $T = \mathbb{C}^n$, so $\mathbb{C}^n = \mathsf{range}(A^k) \boxplus \mathsf{null}(A^k)$. The second statement clearly implies the first.

Suppose now that $\mathbb{C}^n = \mathsf{range}(A^k) \boxplus \mathsf{null}(A^k)$. Then

$$\mathsf{range}(A^k) = A^k\mathbb{C}^n = A\mathsf{range}(A^k) = \mathsf{range}(A^{k+1}),$$

so $k \geqslant \mathsf{index}(A)$.

Conversely, if $k \geqslant \mathsf{index}(A)$ and $\boldsymbol{x} \in \mathsf{range}(A^k) \cap \mathsf{null}(A^k)$, then $\boldsymbol{x} = A^k\boldsymbol{y}$ and $A^k\boldsymbol{x} = \boldsymbol{0}$, so $A^{2k}\boldsymbol{y} = \boldsymbol{0}$. Thus, $\boldsymbol{y} \in \mathsf{null}(A^{2k}) = \mathsf{null}(A^k)$, which means that $\boldsymbol{x} = A^k\boldsymbol{y} = \boldsymbol{0}$. Thus, the first statement holds. □

The notion of spark of linear mappings can be transferred to matrices.

Definition 3.33. Let $A \in \mathbb{C}^{m\times n}$ be a matrix. The *spark* of A is the minimum size of a set of columns that is linearly dependent.

If the set of columns of A is linearly independent, then $\mathsf{spark}(A) = n + 1$.

Note that for $A \in \mathbb{C}^{m\times n}$, we have $1 \leqslant \mathsf{spark}(A) \leqslant n + 1$.

3.12 Matrix Similarity and Congruence

Define the *similarity* relation "$\sim$" on the set of square matrices $\mathbb{C}^{n\times n}$ by $A \sim B$ if there exists an invertible matrix X such that $A = XBX^{-1}$.

If X is a unitary matrix, then we say that A and B are *unitarily similar* and we write $A \sim_u B$, so $\sim_u$ is a subset of $\sim$. In this case, we have $A = XBX^{\mathsf{H}}$.

Theorem 3.42. *The relations "$\sim$" and "$\sim_u$" are equivalence relations.*

Proof. We have $A \sim A$ because $A = I_nA(I_n)^{-1}$, so $\sim$ is a reflexive relation. To prove that $\sim$ is symmetric suppose that $A = XBX^{-1}$. Then, $B = X^{-1}AX$ and, since X^{-1} is invertible, we have $B \sim A$.

Finally, to verify the transitivity, let A, B, C be such that $A = XBX^{-1}$ and $B = YCY^{-1}$, where X and Y are two invertible matrices. This allows us to write

$$A = XBX^{-1} = XYCY^{-1}X^{-1} = (XY)C(XY)^{-1},$$

which proves that $A \sim C$.

We leave to the reader the similar proof concerning $\sim_u$. □

Theorem 3.43. *If $A \sim_u B$, where $A, B \in \mathbb{C}^{n\times n}$, then $A^{\mathsf{H}}A \sim_u B^{\mathsf{H}}B$.*

Proof. Since $A \sim_u B$, there exists a unitary matrix X such that $A = XBX^{-1} = XBX^{\mathsf{H}}$. Then, $A^{\mathsf{H}} = XB^{\mathsf{H}}X^{\mathsf{H}}$, so $A^{\mathsf{H}}A = XB^{\mathsf{H}}X^{\mathsf{H}}XBX^{\mathsf{H}} = XB^{\mathsf{H}}BX^{\mathsf{H}}$. Thus, $A^{\mathsf{H}}A$ is unitarily similar to $B^{\mathsf{H}}B$. □

The similarity relation can be extended to sets of rectangular matrices as follows.

Definition 3.34. Let A, B be two matrices in $\mathbb{C}^{m\times n}$. Then, A and B are similar (written $A \sim B$) if there exist two non-singular matrices $G \in \mathbb{C}^{m\times m}$ and $H \in \mathbb{C}^{n\times n}$ such that $A = GBH$.

It is easy to verify that $\sim$ is an equivalence relation on $\mathbb{C}^{m\times n}$.

Theorem 3.44. *Let A and B be two matrices in $\mathbb{C}^{m\times n}$. We have $A \sim B$ if and only if $\mathsf{rank}(A) = \mathsf{rank}(B)$.*

Proof. By Theorem 3.37, if $A \in \mathbb{C}^{m\times n}$ is a matrix with $\mathsf{rank}(A) = r > 0$, then

$$A \sim \begin{pmatrix} I_r & O_{r,n-r} \\ O_{m-r,r} & O_{m-r,n-r} \end{pmatrix}.$$

Thus, for every two matrices $A, B \in \mathbb{C}^{n\times m}$ of rank r, we have $A \sim B$ because both are similar to

$$\begin{pmatrix} I_r & O_{r,n-r} \\ O_{m-r,r} & O_{m-r,n-r} \end{pmatrix}.$$

Conversely, suppose that $A \sim B$, that is, $A = GBH$, where $G \in \mathbb{C}^{m\times m}$ and $H \in \mathbb{C}^{n\times n}$ are non-singular matrices. By Corollary 3.4, we have $\mathsf{rank}(A) = \mathsf{rank}(B)$. □

Definition 3.35. A matrix $A \in \mathbb{C}^{n\times n}$ is *diagonalizable* if there exists a diagonal matrix D such that $A \sim D$.

Let $\mathcal{M}$ be a class of matrices. A is $\mathcal{M}$-*diagonalizable* if there exists a matrix $M \in \mathcal{M}$ such that $A = MDM^{-1}$.

For example, if A is $\mathcal{M}$-diagonalizable and $\mathcal{M}$ is the class of unitary matrices, we say that A is *unitarily diagonalizable.*

Let $f : \mathbb{C}^n \longrightarrow \mathbb{C}$ be a polynomial given by

$$f(z) = a_0 z^n + a_1 z^{n-1} + \cdots + a_n,$$

where $a_0, a_1, \ldots, a_n \in \mathbb{C}$. If $A \in \mathbb{C}^{m\times m}$, then the matrix $f(A)$ is defined by

$$f(A) = a_0 A^n + a_1 A^{n-1} + \cdots + a_n I_m.$$

Theorem 3.45. *If $T \in \mathbb{C}^{m\times m}$ is an upper (a lower) triangular matrix and f is a polynomial, then $f(T)$ is an upper (a lower) triangular matrix. Furthermore, if the diagonal elements of T are $t_{11}, t_{22}, \ldots, t_{mm}$, then the diagonal elements of $f(T)$ are $f(t_{11}), f(t_{22}), \ldots, f(t_{mm})$, respectively.*

Proof. By Theorem 3.12, any power T^k of T is an upper (a lower) triangular matrix. Since the sum of upper (lower) triangular matrices is upper (lower) triangular, if follows that $f(T)$ is an upper triangular (a lower triangular) matrix.

An easy argument by induction on k (left to the reader) shows that if the diagonal elements of T are $t_{11}, t_{22}, \ldots, t_{mm}$, then the diagonal elements of T^k are $t_{11}^k, t_{22}^k, \ldots, t_{mm}^k$. The second part of the theorem follows immediately. □

Theorem 3.46. *Let $A, B \in \mathbb{C}^{m\times m}$. If $A \sim B$ and f is a polynomial, Then $f(A) \sim f(B)$.*

Proof. Let X be an invertible matrix such that $A = XBX^{-1}$. It is straightforward to verify that $A^k = XB^kX^{-1}$ for $k \in \mathbb{N}$. This implies that $f(A) = Xf(B)X^{-1}$, so $f(A) \sim f(B)$. Then $f(A) \sim f(B)$. □

Definition 3.36. Let A and B be two matrices in $\mathbb{C}^{n\times n}$. The matrices A and B are *congruent* if there exists an invertible matrix $X \in \mathbb{C}^{n\times n}$ such that $B = XAX^{\mathsf{H}}$. This is denoted by $A \sim_{\mathsf{H}} B$.

The relation $\sim_{\mathsf{H}}$ is an equivalence on $\mathbb{C}^{n\times n}$. We have $A \sim_{\mathsf{H}} A$ because $A = I_nAI_n^{\mathsf{H}}$. If $A \sim_{\mathsf{H}} B$, then $B = XAX^{\mathsf{H}}$, so $A = X^{-1}B(X^{\mathsf{H}})^{-1} = X^{-1}B(X^{-1})^{\mathsf{H}}$, which implies $B \sim_{\mathsf{H}} A$. Finally, $\sim_{\mathsf{H}}$ is transitive because if $B = XAX^{\mathsf{H}}$ and $C = YBY^{\mathsf{H}}$, where X and Y are invertible matrices, then $C = (YX)A(YX)^{\mathsf{H}}$ and YX is an invertible matrix.

It is immediate that any two congruent matrices have the same rank.

To recapitulate the definitions of the important similarity relations discussed in this section, consider the following list:

(i) A and B are *similar matrices*, $A \sim B$, if there exists an invertible matrix X such that $A = XBX^{-1}$;

(ii) A and B are *congruent matrices*, $A \sim_{\mathsf{H}} B$, if there exists an invertible matrix $X \in \mathbb{C}^{n\times n}$ such that $B = XAX^{\mathsf{H}}$;

(iii) A and B are *unitarily similar*, $A \sim_u B$, if there exists a unitary matrix U such that $A = UBU^{-1}$.

Since every unitary matrix is invertible and its inverse equals its conjugate Hermitian matrix, it follows that $\sim_u$ is a subset of both $\sim$ and $\sim_{\mathsf{H}}$.

3.13 Linear Systems and LU Decompositions

Consider the following set of linear equalities

$$\begin{aligned}
a_{11}x_1 + \ldots + a_{1n}x_n &= b_1,\\
a_{21}x_1 + \ldots + a_{2n}x_n &= b_2,\\
\vdots \quad & \quad \vdots\\
a_{m1}x_1 + \ldots + a_{mn}x_n &= b_m,
\end{aligned}$$

where a_{ij} and b_i belong to a field F. This set constitutes a *system of linear equations.* Solving this system means finding $x_1, \ldots, x_n$ that satisfy all equalities.

The system can be written succinctly in a matrix form as $A\boldsymbol{x} = \boldsymbol{b}$, where

$$A = \begin{pmatrix} a_{11} & \cdots & a_{1n} \\ a_{21} & \cdots & a_{2n} \\ \vdots & \cdots & \vdots \\ a_{m1} & \cdots & a_{mn} \end{pmatrix}, \boldsymbol{b} = \begin{pmatrix} b_1 \\ b_2 \\ \vdots \\ b_m \end{pmatrix},$$

and

$$\boldsymbol{x} = \begin{pmatrix} x_1 \\ x_2 \\ \vdots \\ x_n \end{pmatrix}.$$

If the set of solutions of a system $A\boldsymbol{x} = \boldsymbol{b}$ is not empty, we say that the system is *consistent.* Note that $A\boldsymbol{x} = \boldsymbol{b}$ is consistent if and only if $\boldsymbol{b} \in \mathsf{range}(A)$.

Let $A\boldsymbol{x} = \boldsymbol{b}$ be a linear system in matrix form, where $A \in \mathbb{C}^{m\times n}$. The matrix $[A \;\; \boldsymbol{b}] \in \mathbb{C}^{m\times(n+1)}$ is the *augmented matrix* of the system $A\boldsymbol{x} = \boldsymbol{b}$.

Theorem 3.47. *Let $A \in \mathbb{C}^{m\times n}$ be a matrix and let $\boldsymbol{b} \in \mathbb{C}^{n\times 1}$. The linear system $A\boldsymbol{x} = \boldsymbol{b}$ is consistent if and only if $\mathsf{rank}(A \; \boldsymbol{b}) = \mathsf{rank}(A)$.*

Proof. If $A\boldsymbol{x} = \boldsymbol{b}$ is consistent and $\boldsymbol{x}' = (x_1, \ldots, x_n)$ is a solution of this system, then $\boldsymbol{b} = x_1\boldsymbol{c}_1 + \cdots + x_n\boldsymbol{c}_n$, where $\boldsymbol{c}_1, \ldots, \boldsymbol{c}_n$ are the columns of A. This implies $\mathsf{rank}([A \;\; \boldsymbol{b}]) = \mathsf{rank}(A)$.

Conversely, if $\mathsf{rank}(A \; \boldsymbol{b}) = \mathsf{rank}(A)$, the vector $\boldsymbol{b}$ is a linear combination of the columns of A, which means that $A\boldsymbol{x} = \boldsymbol{b}$ is a consistent system. □

Definition 3.37. A *homogeneous linear system* is a linear system of the form $A\boldsymbol{x} = \boldsymbol{0}_m$, where $A \in \mathbb{C}^{m\times n}$, $\boldsymbol{x} \in \mathbb{C}^{n,1}$, and $\boldsymbol{0} \in \mathbb{C}^{m\times 1}$.

Clearly, any homogeneous system $A\boldsymbol{x} = \boldsymbol{0}_m$ has the solution $\boldsymbol{x} = \boldsymbol{0}_n$. This solution is referred to as the *trivial solution.* The set of solutions of such a system is $\mathsf{null}(A)$, the null space of the matrix A.

Let $\boldsymbol{u}$ and $\boldsymbol{v}$ be two solutions of the system $A\boldsymbol{x} = \boldsymbol{b}$. Then $A(\boldsymbol{u} - \boldsymbol{v}) = \mathbf{0}_m$, so $\boldsymbol{z} = \boldsymbol{u} - \boldsymbol{v}$ is a solution of the homogeneous system $A\boldsymbol{x} = \mathbf{0}_m$, or $\boldsymbol{z} \in \mathsf{null}(A)$. Thus, the set of solutions of $A\boldsymbol{x} = \boldsymbol{b}$ can be obtained as a "translation" of the null space of A by any particular solution of $A\boldsymbol{x} = \boldsymbol{b}$. In other words, the set of solutions of $A\boldsymbol{x} = \boldsymbol{b}$ is $\{\boldsymbol{x} + \boldsymbol{z} \mid \boldsymbol{z} \in \mathsf{null}(A)\}$.

Thus, for $A \in \mathbb{C}^{m\times n}$, the system $A\boldsymbol{x} = \boldsymbol{b}$ has a unique solution if and only if $\mathsf{null}(A) = \{\mathbf{0}_n\}$, that is, according to Equality (3.8), if $\mathsf{rank}(A) = n$.

Theorem 3.48. *Let $A \in \mathbb{C}^{n\times n}$. Then, A is invertible (which is to say that $\mathsf{rank}(A) = n$) if and only if the system $A\boldsymbol{x} = \boldsymbol{b}$ has a unique solution for every $\boldsymbol{b} \in \mathbb{C}^n$.*

Proof. If A is invertible, then $\boldsymbol{x} = A^{-1}\boldsymbol{b}$, so the system $A\boldsymbol{x} = \boldsymbol{b}$ has a unique solution.

Conversely, if the system $A\boldsymbol{x} = \boldsymbol{b}$ has a unique solution for every $\boldsymbol{b} \in \mathbb{C}^n$, let $\boldsymbol{c}_1, \ldots, \boldsymbol{c}_n$ be the solution of the systems $A\boldsymbol{x} = \boldsymbol{e}_1, \ldots, A\boldsymbol{x} = \boldsymbol{e}_n$, respectively. Then, we have

$$A(\boldsymbol{c}_1 | \cdots | \boldsymbol{c}_n) = I_n,$$

which shows that A is invertible and $A^{-1} = (\boldsymbol{c}_1 | \cdots | \boldsymbol{c}_n)$. □

Corollary 3.12. *A homogeneous linear system $A\boldsymbol{x} = \mathbf{0}$, where $A \in \mathbb{C}^{n\times n}$ has a non-trivial solution if and only if A is a singular matrix.*

Proof. This statement follows from Theorem 3.48. □

Thus, by calculating the inverse of A we can solve any linear system of the form $A\boldsymbol{x} = \boldsymbol{b}$. In Chapter 5, we discuss this type of calculation in detail.

Definition 3.38. A matrix $A \in \mathbb{C}^{n\times n}$ is *diagonally dominant* if $|a_{ii}| > \sum\{|a_{ik}| \mid 1 \leqslant k \leqslant n \text{ and } k \neq i\}$.

Theorem 3.49. *A diagonally dominant matrix is non-singular.*

Proof. Suppose that $A \in \mathbb{C}^{n\times n}$ is a diagonally dominant matrix that is singular. By Corollary 3.12, the homogeneous system $A\boldsymbol{x} = \mathbf{0}$ has a non-trivial solution $\boldsymbol{x} \neq \mathbf{0}$. Let x_k be a component of $\boldsymbol{x}$ that has the largest absolute value. Since $\boldsymbol{x} \neq \mathbf{0}$, we have $|x_k| > 0$. We can write

$$a_{kk}x_k = -\sum\{a_{kj}x_j \mid 1 \leqslant j \leqslant n \text{ and } j \neq k\},$$

which implies

$$\begin{aligned}|a_{kk}|\,|x_k| &= \left|\sum\{a_{kj}x_j \mid 1 \leqslant j \leqslant n \text{ and } j \neq k\}\right| \\ &\leqslant \sum\{|a_{kj}|\,|x_j| \mid 1 \leqslant j \leqslant n \text{ and } j \neq k\}\Big| \\ &\leqslant |x_k|\sum\{|a_{kj}| \mid 1 \leqslant j \leqslant n \text{ and } j \neq k\}.\end{aligned}$$

Thus, we obtain

$$|a_{kk}| \leqslant \sum\{|a_{kj}| \mid 1 \leqslant j \leqslant n \text{ and } j \neq k\},$$

which contradicts the fact that A is diagonally dominant. □

Definition 3.39. The *sparsity* of a vector $\boldsymbol{v} \in \mathbb{C}^n$ is the number $\mathsf{sparse}(\boldsymbol{v})$ of components of $\boldsymbol{v}$ equal to 0.

Let $\nu_0 : \mathbb{R}^n \longrightarrow \mathbb{R}$ be the function defined by

$$\nu_0(\boldsymbol{x}) = \{i \mid 1 \leqslant i \leqslant n, x_i \neq 0\},$$

which gives the number of non-zero components of $\boldsymbol{x}$. It is immediate to verify that $\nu_0(\boldsymbol{x}+\boldsymbol{y}) \leqslant \nu_0(\boldsymbol{x}) + \nu_0(\boldsymbol{y})$ for $\boldsymbol{x}, \boldsymbol{y} \in \mathbb{R}^n$.

Note that $\nu_0(\boldsymbol{v}) = n - \mathsf{sparse}(\boldsymbol{v})$. If $\boldsymbol{x} \in \mathsf{null}(A)$, then $\nu_0(\boldsymbol{x}) \geqslant \mathsf{spark}(A)$ because vectors in this subspace are linear combinations of columns of A that equal $\mathbf{0}_n$ and at least $\mathsf{spark}(A)$ columns are needed to produce the vector $\mathbf{0}_n$.

The following result was obtained in [43]:

Theorem 3.50. *If a linear system $A\boldsymbol{x} = \boldsymbol{b}$ has a solution $\boldsymbol{x}$ such that $\nu_0(\boldsymbol{x}) < \frac{1}{2}\mathsf{spark}(A)$, then $\boldsymbol{x}$ is a sparsest solution of the system.*

Proof. Let $\boldsymbol{y}$ be a solution of the same system. We have $A\boldsymbol{x} - A\boldsymbol{y} = A(\boldsymbol{x}-\boldsymbol{y}) = \mathbf{0}_n$. By the definition of $\mathsf{spark}(A)$ we have $\nu_0(\boldsymbol{x}) + \nu_0(\boldsymbol{y}) \geqslant \nu_0(\boldsymbol{x}-\boldsymbol{y}) \geqslant \mathsf{spark}(A)$. Thus, the number of non-zero components of the vector $\boldsymbol{x}-\boldsymbol{y}$ cannot exceed the sum of the number of non-zero components within each of the vectors $\boldsymbol{x}$ and $\boldsymbol{y}$. Since $\boldsymbol{x}_0$ satisfies $\nu_0(\boldsymbol{x}_0) < \mathsf{spark}(A)/2$, it follows that any other solution has more than $\mathsf{spark}(A)/2$ non-zero components. □

3.14 The Row Echelon Form of Matrices

We begin with a class of linear systems that can easily be solved.

Definition 3.40. A matrix $C \in \mathbb{C}^{m \times n}$ is in row echelon form if the following conditions are satisfied:

(i) rows that contain nonzero elements precede zero rows (that is, rows that contain only zeros);

(ii) if c_{ij} is the first nonzero element of the row i, all elements in the jth column located below c_{ij}, that is, entries of the form c_{kj} with $k > j$ are zero (see Figure 3.1);

(iii) if $i < \ell$, c_{ij_i} is the first non-zero element of the row i, and $c_{\ell j_\ell}$ is the first non-zero element of the row ℓ, then $j_i < j_\ell$ (see Figure 3.2).

The first non-zero element of a row i (if it exists) is called the *pivot of the row i*.

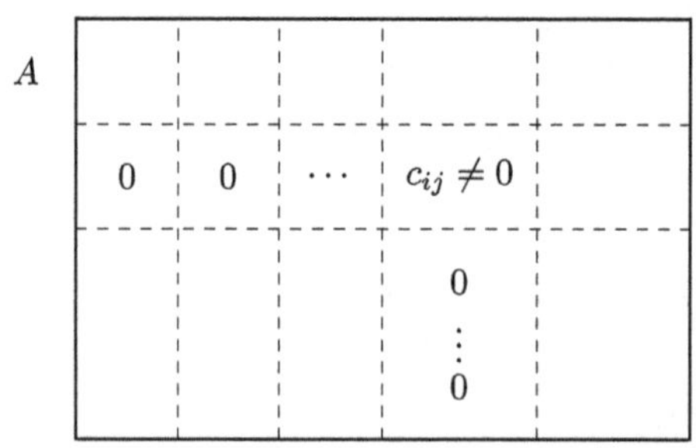

Fig. 3.1 Condition (ii) of Definition 3.40.

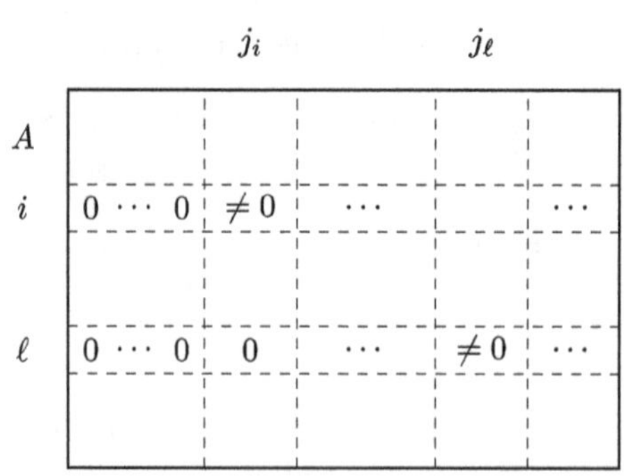

Fig. 3.2 Condition (iii) of Definition 3.40.

Example 3.31. Let $C \in \mathbb{R}^{4\times 5}$ be the matrix

$$C = \begin{pmatrix} 1 & 2 & 0 & 2 & 0 \\ 0 & 2 & 3 & 0 & 1 \\ 0 & 0 & 0 & -1 & 2 \\ 0 & 0 & 0 & 0 & 0 \end{pmatrix}.$$

It is clear that C is in row echelon form; the pivots of the first, second, and third rows are $c_{11} = 1$, $c_{22} = 2$, and $c_{34} = -1$.

Theorem 3.51. *Let $C \in \mathbb{C}^{m\times n}$ be a matrix in row echelon form such that the rows that contain non-zero elements are the first r rows. Then,* $\mathsf{rank}(C) = r$.

Proof. Let $\mathbf{c}_1, \ldots, \mathbf{c}_r$ be the non-zero rows of C. Suppose that the row $\mathbf{c}_i$ has the first non-zero element in the column j_i for $1 \leqslant i \leqslant r$. By the definition of the echelon form, we have $j_1 < j_2 < \cdots < j_r$. Suppose that $a_1\mathbf{c}_1 + \cdots + a_r\mathbf{c}_r = \mathbf{0}$. This equality can be written as

$$\begin{aligned} a_1 c_{1j_1} &= 0, \\ a_1 c_{1j_2} + a_2 c_{2j_2} &= 0, \\ &\vdots \\ a_1 c_{1n} + a_2 c_{2n} + \ldots + a_r c_{rn} &= 0. \end{aligned}$$

Since $c_{1j_1} \neq 0$, we have $a_1 = 0$. Substituting a_1 by 0 in the second equality implies $a_2 = 0$ because $c_{2j_2} \neq 0$, *etc.* Thus, we obtain $a_1 = a_2 = \ldots = a_r = 0$, which proves that the rows $\mathbf{c}_1, \ldots, \mathbf{c}_r$ are linearly independent. Since this is a maximal set of rows of C that is linearly independent, it follows that $\mathsf{rank}(C) = r$. □

Linear systems whose augmented matrices are in row echelon form can be easily solved using a process called *back substitution*. Consider the following augmented matrix in row echelon form of a system with

m equations and n unknowns:

$$\begin{pmatrix} 0 & \cdots & 0 & a_{1j_1} & \cdots & \cdots & \cdots & a_{1n} & b_1 \\ 0 & \cdots & 0 & 0 & \cdots & a_{2j_2} & \cdots & a_{2n} & b_2 \\ \vdots & \cdots & \vdots & \vdots & \cdots & \vdots & \cdots & \vdots & \vdots \\ 0 & \cdots & 0 & 0 & \cdots & \cdots & a_{rj_r} & \cdots & b_r \\ 0 & \cdots & 0 & 0 & \cdots & \cdots & 0 & \cdots & b_{r+1} \\ \vdots & \cdots & \vdots & \vdots & \cdots & \vdots & \cdots & \vdots & \vdots \\ 0 & \cdots & 0 & 0 & \cdots & \cdots & 0 & \cdots & b_m \end{pmatrix}.$$

The system of equations has the following form:

$$\begin{aligned} a_{1j_1}x_{j_1} + \cdots + a_{1n}x_n &= b_1 \\ a_{2j_2}x_{j_2} + \cdots + a_{2n}x_n &= b_2 \\ &\vdots \\ a_{rj_r}x_{j_r} + \cdots + a_{rn}x_n &= b_r \\ 0 &= b_{r+1} \\ \vdots &= \vdots \\ 0 &= b_m. \end{aligned}$$

The variables $x_{j_1}, x_{j_2}, \ldots, x_{j_r}$ that correspond to the columns where the pivot elements occur are referred to as the *basic variables* or *principal variables.* The remaining variables are *non-basic* or *non-principal.*

Note that we have $r \leqslant \min\{m, n\}$. If $r < m$ and there exists $b_\ell \neq 0$ for $r < \ell \leqslant m$, then the system is inconsistent and no solutions exist.

If $r = m$ or $b_\ell = 0$ for $r < \ell \leqslant m \leqslant n$, one can choose the variables that do not correspond to the pivot elements, $\{x_i \mid i \notin \{j_1, j_2, \ldots, j_r\}$, as parameters and express the basic variables as functions of these parameters. The process starts with the last basic variable, x_{j_r} (because every other variable in the equation $a_{rj_r}x_{j_r} + \cdots + a_{rn}x_n = b_r$ is a parameter), and then substitutes this variable in the previous equality. This allows us to express $x_{j_{r-1}}$ as a function of parameters, *etc.* This explains the term *back substitution* previously introduced. If $r = n$, then no parameters exist.

To conclude, if $r < m$, the system has a solution if and only if $b_j = 0$ for $j > r$. If $r = m$, the system has a solution. This solution is unique if $r = n$.

Definition 3.41. A linear system $A\boldsymbol{x} = \boldsymbol{b}$, where $A \in \mathbb{R}^{m\times n}$, $\boldsymbol{b} \in \mathbb{R}^m$, and $m \leqslant n$, is said to be in *explicit form* if A contains the columns of the matrix I_m.

Example 3.32. The linear system $A\boldsymbol{x} = \boldsymbol{b}$ is defined by

$$A = \begin{pmatrix} 1 & 0 & 1 & -1 \\ 0 & 1 & -1 & 1 \end{pmatrix} \text{ and } \boldsymbol{b} = \begin{pmatrix} 2 \\ 1 \end{pmatrix}.$$

Note that A is a full-rank matrix because $\mathsf{rank}(A) = 2$. The system is already in explicit form relative to the variables x_1 and x_2 because the first two columns of the matrix $[A|b]$ are the columns of I_2.

There are six possible explicit forms of this system corresponding to the $\binom{4}{2} = 6$ subsets of the set of variables $\{x_1, x_2, x_3, x_4\}$. Note, however, that if the columns chosen form a submatrix of rank less than 2, the explicit form does not exist. This is the case when we chose the explicit form relative to x_3 and x_4. Therefore, this system has five explicit forms relative to the sets of variables

$$\{x_1, x_2\}, \{x_1, x_3\}, \{x_1, x_4\}, \{x_2, x_3\}, \text{ and } \{x_2, x_4\}.$$

In general, a system with m equations and n variables has maximum $\binom{n}{m}$ explicit forms.

Example 3.33. Consider the system

$$\begin{aligned} x_1 + 2x_2 + \qquad\qquad 2x_4 &= b_1 \\ 2x_2 + 3x_3 + \qquad x_5 &= b_2 \\ -x_4 + x_5 &= b_3 \\ 0 &= b_4 \end{aligned}.$$

The matrix $A \in \mathbb{R}^{4\times 5}$ is not of full rank because $\mathsf{rank}(A) = 3$.

The augmented matrix of this system is

$$\begin{pmatrix} 1 & 2 & 0 & 2 & 0 & b_1 \\ 0 & 2 & 3 & 0 & 1 & b_2 \\ 0 & 0 & 0 & -1 & 2 & b_3 \\ 0 & 0 & 0 & 0 & 0 & b_4 \end{pmatrix}.$$

The basic variables are x_1, x_2, and x_4. If $b_4 = 0$, the system is consistent. Under this assumption we can choose x_3 and x_5 as parameters. Let $x_3 = p$ and $x_4 = q$.

The third equation yields $x_4 = q - b_3$. Similarly, the second equation implies $x_2 = 0.5(b_2 - 3p - q)$. Substituting these values in the first equation allows us to write $x_1 = b_1 - b_2 + 2b_3 - 3p - q$.

Further transformations of this system allow us to construct an equivalent linear system whose matrix contains the columns of the matrix I_3. Subtracting the second row from the first yields

$$\begin{pmatrix} 1 & 0 & -3 & 2 & -1 & b_1 - b_2 \\ 0 & 2 & 3 & 0 & 1 & b_2 \\ 0 & 0 & 0 & -1 & 2 & b_3 \\ 0 & 0 & 0 & 0 & 0 & b_4 \end{pmatrix}.$$

Then, dividing the second row by 2 will produce

$$\begin{pmatrix} 1 & 0 & -3 & 2 & -1 & b_1 - b_2 \\ 0 & 1 & 3/2 & 0 & 1/2 & b_2/2 \\ 0 & 0 & 0 & -1 & 2 & b_3 \\ 0 & 0 & 0 & 0 & 0 & b_4 \end{pmatrix},$$

which creates the second column of I_3. Next, multiply the third row by -1:

$$\begin{pmatrix} 1 & 0 & -3 & 2 & -1 & b_1 - b_2 \\ 0 & 1 & \frac{3}{2} & 0 & \frac{1}{2} & \frac{b_2}{2} \\ 0 & 0 & 0 & 1 & -2 & -b_3 \\ 0 & 0 & 0 & 0 & 0 & b_4 \end{pmatrix}.$$

Finally, multiply the third by -2 and add it to the first row:

$$\begin{pmatrix} 1 & 0 & -3 & 0 & 1 & b_1 - b_2 + 2b_3 \\ 0 & 1 & \frac{3}{2} & 0 & \frac{1}{2} & \frac{b_2}{2} \\ 0 & 0 & 0 & 1 & -2 & -b_3 \\ 0 & 0 & 0 & 0 & 0 & b_4 \end{pmatrix}.$$

Thus, if we choose for the basic variables

$$x_1 = b_1 - b_2 + 2b_3, x_2 = \frac{b_2}{2}, \text{ and } x_4 = -b_3$$

and for the non-basic variables $x_3 = x_5 = 0$, we obtain a solution of the system.

The extended echelon form of a system can be achieved by applying certain transformations on the rows of the augmented matrix of the system (which amount to transformations involving the equations of the system). In preparation, a few special invertible matrices are introduced in the next examples.

Example 3.34. Consider the matrix

$$T^{(i)\leftrightarrow(j)} = \begin{pmatrix} 1 & & \cdots & & \cdots \\ \vdots & \vdots & \vdots & \vdots & \vdots \\ \cdots & 0 & \cdots & 1 & \cdots \\ \vdots & \vdots & \vdots & \vdots & \vdots \\ \cdots & 1 & \cdots & 0 & \cdots \\ \vdots & \vdots & \vdots & \vdots & \vdots \\ & & \cdots & & 1 \end{pmatrix},$$

where line i contains exactly one 1 in position j and line j contains exactly one 1 in position i. If $T^{(i)\leftrightarrow(j)} \in \mathbb{C}^{p\times p}$ and $A \in \mathbb{C}^{p\times q}$, it is easy to see that the matrix $T^{(i)\leftrightarrow(j)}A$ is obtained from the matrix A by permuting the lines i and j.

For instance, consider the matrix $T^{(2)\leftrightarrow(4)} \in \mathbb{C}^{4\times 4}$ defined by

$$T^{(2)\leftrightarrow(4)} = \begin{pmatrix} 1 & 0 & 0 & 0 \\ 0 & 0 & 0 & 1 \\ 0 & 0 & 1 & 0 \\ 0 & 1 & 0 & 0 \end{pmatrix}$$

and the matrix $A \in \mathbb{F}^{4\times 5}$. We have

$$T^{(2)\leftrightarrow(4)} A = \begin{pmatrix} 1 & 0 & 0 & 0 \\ 0 & 0 & 0 & 1 \\ 0 & 0 & 1 & 0 \\ 0 & 1 & 0 & 0 \end{pmatrix} \begin{pmatrix} a_{11} & a_{12} & a_{13} & a_{14} & a_{15} \\ a_{21} & a_{22} & a_{23} & a_{24} & a_{25} \\ a_{31} & a_{32} & a_{33} & a_{34} & a_{35} \\ a_{41} & a_{42} & a_{43} & a_{44} & a_{45} \end{pmatrix}$$

$$= \begin{pmatrix} a_{11} & a_{12} & a_{13} & a_{14} & a_{15} \\ a_{41} & a_{42} & a_{43} & a_{44} & a_{45} \\ a_{31} & a_{32} & a_{33} & a_{34} & a_{35} \\ a_{21} & a_{22} & a_{23} & a_{24} & a_{25} \end{pmatrix}.$$

The inverse of $T^{(i)\leftrightarrow(j)}$ is $T^{(i)\leftrightarrow(j)}$ itself.

Example 3.35. Let $T^{a(i)} \in \mathbb{C}^{p\times p}$ be the matrix

$$T^{a(i)} = \begin{pmatrix} 1 & 0 & \cdots & 0 & \cdots 0 \\ 0 & 1 & \cdots & 0 & \cdots 0 \\ \vdots & \vdots & \cdots & \vdots & \cdots 0 \\ \cdots & \cdots & \cdots & a & \cdots 0 \\ \vdots & \vdots & \cdots & \vdots & \cdots 0 \\ 0 & 0 & \cdots & 0 & \cdots 1 \end{pmatrix}$$

that has $a \in F - \{0\}$ on the ith diagonal element, 1 on the remaining diagonal elements, and 0 everywhere else. The product $T^{a(i)}A$ is obtained from A by multiplying the ith row by a. The inverse of this matrix is $T^{\frac{1}{a}(i)}$.

As an example, consider the matrix $T^{3(2)} \in \mathbb{C}^{4\times 4}$ given by

$$T^{3(2)} = \begin{pmatrix} 1 & 0 & 0 & 0 \\ 0 & 3 & 0 & 0 \\ 0 & 0 & 1 & 0 \\ 0 & 0 & 0 & 1 \end{pmatrix}.$$

If $A \in \mathbb{C}^{4\times 5}$, the matrix $T^{3(2)}A$ is obtained from A by multiplying its second line by 3. We have

$$\begin{pmatrix} 1 & 0 & 0 & 0 \\ 0 & 3 & 0 & 0 \\ 0 & 0 & 1 & 0 \\ 0 & 0 & 0 & 1 \end{pmatrix} \begin{pmatrix} a_{11} & a_{12} & a_{13} & a_{14} & a_{15} \\ a_{21} & a_{22} & a_{23} & a_{24} & a_{25} \\ a_{31} & a_{32} & a_{33} & a_{34} & a_{35} \\ a_{41} & a_{42} & a_{43} & a_{44} & a_{45} \end{pmatrix}$$

$$= \begin{pmatrix} a_{11} & a_{12} & a_{13} & a_{14} & a_{15} \\ 3a_{21} & 3a_{22} & 3a_{23} & 3a_{24} & 3a_{25} \\ a_{31} & a_{32} & a_{33} & a_{34} & a_{35} \\ a_{41} & a_{42} & a_{43} & a_{44} & a_{45} \end{pmatrix}.$$

Example 3.36. Let $T^{(i)+a(j)} \in \mathbb{C}^{p\times p}$ be the matrix whose entries are identical to the matrix I_p with the exception of the element located in row i and column j that equals a:

$$T^{(i)+a(j)} = \begin{pmatrix} 1 & 0 & \cdots & \cdots & \cdots & 0 & \cdots 0 \\ 0 & 1 & \cdots & \cdots & \cdots & 0 & \cdots 0 \\ \vdots & \vdots & \cdots & \vdots & \cdots & \vdots & \cdots 0 \\ 0 & 0 & \cdots & a & \cdots & 1 & \cdots 0 \\ \vdots & \vdots & \cdots & \vdots & \cdots & \vdots & \cdots 1 \end{pmatrix}.$$

The result of the multiplication $T^{(i)+a(j)}A$ is a matrix that can be obtained from A by adding the jth line of A multiplied by a to the ith line of A. The inverse of the matrix $T^{(i)+a(j)}A$ is $T^{(i)-a(j)}A$.

For example, we have

$$T^{(4)+2(2)}A = \begin{pmatrix} 1 & 0 & 0 & 0 \\ 0 & 1 & 0 & 0 \\ 0 & 0 & 1 & 0 \\ 0 & 2 & 0 & 1 \end{pmatrix} \begin{pmatrix} a_{11} & a_{12} & a_{13} & a_{14} & a_{15} \\ a_{21} & a_{22} & a_{23} & a_{24} & a_{25} \\ a_{31} & a_{32} & a_{33} & a_{34} & a_{35} \\ a_{41} & a_{42} & a_{43} & a_{44} & a_{45} \end{pmatrix}$$

$$= \begin{pmatrix} a_{11} & a_{12} & a_{13} & a_{14} & a_{15} \\ a_{21} & a_{22} & a_{23} & a_{24} & a_{25} \\ a_{31} & a_{32} & a_{33} & a_{34} & a_{35} \\ a_{41}+2a_{21} & a_{42}+2a_{22} & a_{43}+2a_{23} & a_{44}+2a_{24} & a_{45}+2a_{25} \end{pmatrix}.$$

It is easy to see that if one multiplies a matrix A at the right by $T^{(i)\leftrightarrow(j)}$, $T^{a(i)}$, and $T^{(i)+a(j)}$, the effect on A consists of exchanging the columns i and j, multiplying the ith column by a, and adding the jth column multiplied by a to the ith column, respectively.

Definition 3.42. Let $\mathbb{F}$ be a field, $A, C \in \mathbb{F}^{m\times n}$ and let $\boldsymbol{b}, \boldsymbol{d} \in \mathbb{F}^{m\times 1}$. Two systems of linear equations $A\boldsymbol{x} = \boldsymbol{b}$ and $C\boldsymbol{x} = \boldsymbol{d}$ are *equivalent* if they have the same set of solutions.

If $A\boldsymbol{x} = \boldsymbol{b}$ is a system of linear equations in matrix form, where $A \in \mathbb{C}^{m\times n}$ and $\boldsymbol{b} \in \mathbb{C}^{m\times 1}$, and $T \in \mathbb{C}^{m\times m}$ is a matrix that has an inverse, then the systems $A\boldsymbol{x} = \boldsymbol{b}$ and $(TA)\boldsymbol{x} = (T\boldsymbol{b})$ are equivalent. Indeed, any solution of $A\boldsymbol{x} = \boldsymbol{b}$ satisfies the system $(TA)\boldsymbol{x} = (T\boldsymbol{b})$. Conversely, if $(TA)\boldsymbol{x} = (T\boldsymbol{b})$, by multiplying this equality by T^{-1} to the left, we get $(T^{-1}T)A\boldsymbol{x} = (T^{-1}T)\boldsymbol{b}$, that is, $A\boldsymbol{x} = \boldsymbol{b}$.

The matrices $T^{(i)\leftrightarrow(j)}$, $T^{a(i)}$, and $T^{(i)+a(j)}$ introduced in Examples 3.34–3.36 play a special role in Algorithm 3.14.1 that transforms a linear system $A\boldsymbol{x} = \boldsymbol{b}$ into an equivalent system in row echelon form. These transformations are known as *elementary transformation matrices*.

Example 3.37. Consider the linear system

$$\begin{aligned} x_1 + 2x_2 + 3x_3 &= 4 \\ x_1 + 2x_2 + x_3 &= 3 \\ x_1 + 3x_2 + x_3 &= 1. \end{aligned}$$

Algorithm 3.14.1: Algorithm for the Row Echelon Form of a Matrix

Data: A matrix $A \in \mathbb{F}^{p \times q}$
Result: A row echelon form of A

```
r = 1;
c = 1;
while r ⩽ p and c ⩽ q do
    while A(*, c) = 0 do
        c = c + 1
    end
    j = r;
    while A(j, c) = 0 do
        j = j+1
    end
    if j ≠ r then
        exchange line r with line j
    end
    multiply line r by 1/A(r,c);
    for each k = r + 1 to p do
        add line r multiplied by −A(k, c) to line k
    end
    r = r + 1;
    c = c + 1;
end
```

The augmented matrix of this system is

$$[A|\boldsymbol{b}] = \begin{pmatrix} 1 & 2 & 3 & 4 \\ 1 & 2 & 1 & 3 \\ 1 & 3 & 1 & 1 \end{pmatrix}.$$

By subtracting the first row from the second and the third, we obtain the matrix

$$T^{(3)-1(1)}T^{(2)-1(1)}[A|\boldsymbol{b}] = \begin{pmatrix} 1 & 2 & 3 & 4 \\ 0 & 0 & -2 & -1 \\ 0 & 1 & -2 & -3 \end{pmatrix}.$$

Next, the second and third row are exchanged yielding the matrix

$$T^{(2)\leftrightarrow(3)}T^{(3)-1(1)}T^{(2)-1(1)}[A|\boldsymbol{b}] = \begin{pmatrix} 1 & 2 & 3 & 4 \\ 0 & 1 & -2 & -3 \\ 0 & 0 & -2 & -1 \end{pmatrix}.$$

To obtain a 1 in the pivot of the third row, we multiply the third row by $-\frac{1}{2}$:

$$T^{-0.5(3)}T^{(2)\leftrightarrow(3)}T^{(3)-1(1)}T^{(2)-1(1)}[A|\boldsymbol{b}] = \begin{pmatrix} 1 & 2 & 3 & 4 \\ 0 & 1 & -2 & -3 \\ 0 & 0 & 1 & 0.5 \end{pmatrix},$$

which is the row echelon form of the matrix $[A|\boldsymbol{b}]$.

To achieve the row echelon form, we needed to multiply the matrix $[A|\boldsymbol{b}]$ by the matrix

$$T = T^{-0.5(3)}T^{(2)\leftrightarrow(3)}T^{(3)-1(1)}T^{(2)-1(1)}.$$

The solutions of the system can now be obtained by back substitution from the linear system

$$\begin{aligned} x_1 + 2x_2 + 3x_3 &= 4, \\ x_2 - 2x_3 &= -3, \\ x_3 &= 0.5. \end{aligned}$$

The last equation yields $x_3 = 0.5$. Substituting x_3 in the second equation implies $x_2 = -2$; finally, from the first equality we have $x_1 = 6.5$.

Theorem 3.52. *Let $T^{a(i)}$, $T^{(p)\leftrightarrow(q)}$, and $T^{(i)+a(j)}$ be the matrices in $\mathbb{R}^{m\times m}$ that correspond to the row transformations applied to matrices in $\mathbb{R}^{m\times n}$, where $i \neq j$ and $p \neq q$. We have*

$$T^{a(i)}T^{(p)\leftrightarrow(q)} = \begin{cases} T^{(p)\leftrightarrow(q)}T^{a(i)} & \text{if } i \notin \{p,q\}, \\ T^{(p)\leftrightarrow(q)}T^{a(q)} & \text{if } i = p, \\ T^{(p)\leftrightarrow(q)}T^{a(p)} & \text{if } i = q, \end{cases}$$

$$T^{(i)+a(j)}T^{(p)\leftrightarrow(q)} = \begin{cases} T^{(p)\leftrightarrow(q)}T^{(i)+a(j)} & \text{if } \{i,j\}\cap\{p,q\}=\emptyset, \\ T^{(q)+a(j)}T^{(p)\leftrightarrow(q)} & \text{if } i=p \text{ and } j\neq q, \\ T^{(i)+a(p)}T^{(p)\leftrightarrow(q)} & \text{if } i\neq p \text{ and } j=q, \\ T^{(q)+a(p)}T^{(p)\leftrightarrow(q)} & \text{if } i=p \text{ and } j=q. \end{cases}$$

Proof. The equalities of the theorem follow immediately from the definitions of the matrices. □

The matrices that describe elementary transformations are of two types: *lower triangular* matrices of the form $T^{a(i)}$ or $T^{(i)+a(j)}$ or *permutation matrices* of the form $T^{(p)\leftrightarrow(q)}$.

If all pivots encountered in the construction of the row echelon form of the matrix A are not zero, then there is no need to use any permutation matrix $T^{(p)\leftrightarrow(q)}$ among the matrices that multiply A at the left. Thus, there is a lower matrix T and an upper triangular matrix U such that $TA = U$. The matrix T is a product of invertible matrices and, therefore, it is invertible. Since the inverse $L = T^{-1}$ of a lower triangular matrix is lower triangular, as we saw in Theorem 3.8, it follows that $A = LU$; in other words, A can be decomposed into a product of a lower triangular and an upper triangular matrix. This factorization of matrices is known as an *LU-decomposition* of A.

Example 3.38. Let $A \in \mathbb{R}^{3\times 3}$ be the matrix

$$A = \begin{pmatrix} 1 & 0 & 1 \\ 2 & 1 & 1 \\ 1 & -1 & 2 \end{pmatrix}.$$

Initially, we add the first row multiplied by -2 to the second row, and the same first row, multiplied by -1, to the third row. This amounts to

$$T^{(3),-(1)}T^{(2,-2(1)}A = \begin{pmatrix} 1 & 0 & 1 \\ 0 & 1 & -1 \\ 0 & -1 & 1 \end{pmatrix}.$$

Next, we add the second row to the third to produce the matrix

$$T^{(3)+(2)}T^{(3),-(1)}T^{(2,-2(1)}A = \begin{pmatrix} 1 & 0 & 1 \\ 0 & 1 & -1 \\ 0 & 0 & 0 \end{pmatrix},$$

which is an upper triangular matrix. By Theorem 3.51, we can conclude that $\mathsf{rank}(A) = 2$. We can write

$$A = (T^{(2,-2(1)})^{-1}(T^{(3),-(1)})^{-1}(T^{(3)+(2)})^{-1}\begin{pmatrix}1 & 0 & 1\\ 0 & 1 & -1\\ 0 & 0 & 0\end{pmatrix}.$$

Thus, the lower triangular matrix we are seeking is

$$L = (T^{(2)-2(1)})^{-1}(T^{(3),-(1)})^{-1}(T^{(3)+(2)})^{-1} = T^{(2)+2(1)}T^{(3)+(1)}T^{(3)-(2)}$$
$$= \begin{pmatrix}1 & 0 & 0\\ 2 & 1 & 0\\ 0 & 0 & 1\end{pmatrix}\begin{pmatrix}1 & 0 & 0\\ 0 & 1 & 0\\ 1 & 0 & 1\end{pmatrix}\begin{pmatrix}1 & 0 & 0\\ 0 & 1 & 0\\ 0 & -1 & 1\end{pmatrix} = \begin{pmatrix}1 & 0 & 0\\ 2 & 1 & 0\\ 1 & -1 & 1\end{pmatrix},$$

which shows that A can be written as

$$A = \begin{pmatrix}1 & 0 & 0\\ 2 & 1 & 0\\ 1 & -1 & 1\end{pmatrix}\begin{pmatrix}1 & 0 & 1\\ 0 & 1 & -1\\ 0 & 0 & 0\end{pmatrix}.$$

Suppose that during the construction of the matrix U some of the elementary transformation matrices are permutation matrices of the form $T^{(p)\leftrightarrow(q)}$. By Theorem 3.52, matrices of the form $T^{(p)\leftrightarrow(q)}$ can be shifted to the right. Therefore, instead of the previous factorization of the matrix A, we have a lower triangular matrix T and a permutation matrix, which results as a product of all permutation matrices of the form $T^{(p)\leftrightarrow(q)}$ used in the algorithm such that $TPA = U$. In this case, we obtain an LU-factorization of PA instead of A.

We give now an alternative characterization of the notion of matrix rank. Note that if T and S are matrices of any of the elementary transformations, then $\mathsf{rank}(TA) = \mathsf{rank}(A)$ and $\mathsf{rank}(AS) = \mathsf{rank}(A)$ by Corollary 3.4.

Theorem 3.53. *Let $A \in \mathbb{R}^{m\times n}$ be a matrix. If $\mathsf{rank}(A) = k$, then the largest non-singular square submatrix B of A is a $k \times k$-matrix.*

Proof. Since $\mathsf{rank}(A) = k$, there is a maximal linearly independent set of k columns of A, $\{\boldsymbol{c}_{i_1}, \ldots, \boldsymbol{c}_{i_k}\}$, and a maximal linearly independent set of k of rows $\{\boldsymbol{r}_{j_1}, \ldots, \boldsymbol{r}_{j_k}\}$.

Let K be the submatrix of A defined by

$$K = A \begin{bmatrix} j_1 & \cdots & j_k \\ i_1 & \cdots & i_k \end{bmatrix}.$$

We claim that K is non-singular, that is, $\mathsf{rank}(K) = k$. Using right multiplications by elementary transformation matrices, we can produce a matrix D that has the columns $\boldsymbol{c}_{i_1}, \ldots, \boldsymbol{c}_{i_k}$ on the first k positions. Since the remaining columns are linear combinations of these first k columns, using the same type of column transformations we can transform each of the remaining columns into $\mathbf{0}_m$. Thus, there exists a non-singular matrix S such that $D = AS = ((\boldsymbol{c}_{i_1} \cdots \boldsymbol{c}_{i_k} | O_{m,n-k})$ and $\mathsf{rank}(D) = \mathsf{rank}(A) = k$. The matrix D has the rows numbered $i_1, \ldots, i_k$ as a maximal set of linearly independent rows. Now, by applying elementary transformations, these rows can be brought in the first k places and the remaining $m - r$ rows can be nullified. Thus, there is an invertible matrix T such that

$$E = TAS = \begin{pmatrix} K & O \\ O & O \end{pmatrix},$$

where K is a non-singular $k \times k$-matrix because $\mathsf{rank}(E) = \mathsf{rank}(A)$. Thus, A contains a submatrix of rank k.

Suppose that G is a non-singular square submatrix of A. By a series of row elementary and column elementary transformations (described by the matrices P and Q, respectively) we have

$$PAQ = \begin{pmatrix} G & H \\ L & M \end{pmatrix}$$

and $\mathsf{rank}(PAQ) = \mathsf{rank}(A) = k$. Consider now the invertible matrices P_1 and Q_1 defined by

$$P_1 = \begin{pmatrix} I & O \\ -LG^{-1} & I \end{pmatrix} \text{ and } Q_1 = \begin{pmatrix} I & -G^{-1}H \\ O & I \end{pmatrix}.$$

Since P_1 is lower triangular and Q_1 is upper triangular having all diagonal elements equal to 1, both P_1 and Q_1 are invertible. Therefore, $\mathsf{rank}(P_1 PAQQ_1) = \mathsf{rank}(A)$. On the other hand,

$$P_1 PAQQ_1 = \begin{pmatrix} I & O \\ -LG^{-1} & I \end{pmatrix} \begin{pmatrix} G & H \\ L & M \end{pmatrix} \begin{pmatrix} I & -G^{-1}H \\ 0 & I \end{pmatrix}$$

$$= \begin{pmatrix} G & O \\ O & M - LG^{-1}H) \end{pmatrix}.$$

Therefore, $\mathsf{rank}(A) = \mathsf{rank}(G) + \mathsf{rank}(M - LG^{-1}H) \geqslant \mathsf{rank}(G)$, so k is the maximal rank of a nonsingular submatrix of A. □

3.15 The Kronecker and Other Matrix Products

Definition 3.43. Let $A \in \mathbb{C}^{m\times n}$ and $B \in \mathbb{C}^{p\times q}$ be two matrices. The *Kronecker product* of these matrices is the matrix $A \otimes B \in \mathbb{C}^{mp\times nq}$ defined by

$$A \otimes B = \begin{pmatrix} a_{11}B & a_{12}B & \cdots & a_{1n}B \\ a_{21}B & a_{22}B & \cdots & a_{2n}B \\ \vdots & \vdots & \ddots & \vdots \\ a_{m1}B & a_{m2}B & \cdots & a_{mn}B \end{pmatrix}.$$

The Kronecker product $A \otimes B$ creates mn copies of the matrix B and multiplies each copy by the corresponding element of A. Note that the Kronecker product can be defined between any two matrices regardless of their format.

If $\boldsymbol{x} \in \mathbb{C}^m$ and $\boldsymbol{y} \in \mathbb{C}^n$, we have

$$\boldsymbol{x} \otimes \boldsymbol{y} = \mathsf{vec}(\boldsymbol{y}\boldsymbol{x}'), \tag{3.11}$$

and

$$\boldsymbol{x} \otimes \boldsymbol{y}' = \boldsymbol{x}\boldsymbol{y}' = \boldsymbol{y}' \otimes \boldsymbol{x}. \tag{3.12}$$

Example 3.39. Consider the matrices

$$A = \begin{pmatrix} a_{11} & a_{12} \\ a_{21} & a_{22} \end{pmatrix} \text{ and } B = \begin{pmatrix} b_{11} & b_{12} & b_{13} \\ b_{21} & b_{22} & b_{23} \\ b_{31} & b_{32} & b_{33} \end{pmatrix}.$$

Their Kronecker product is

$$A \otimes B = \left(\begin{array}{ccc|ccc} a_{11}b_{11} & a_{11}b_{12} & a_{11}b_{13} & a_{12}b_{11} & a_{12}b_{12} & a_{12}b_{13} \\ a_{11}b_{21} & a_{11}b_{22} & a_{11}b_{23} & a_{12}b_{21} & a_{12}b_{22} & a_{12}b_{23} \\ a_{11}b_{31} & a_{11}b_{32} & a_{11}b_{33} & a_{12}b_{31} & a_{12}b_{32} & a_{12}b_{33} \\ \hline a_{21}b_{11} & a_{21}b_{12} & a_{21}b_{13} & a_{22}b_{11} & a_{22}b_{12} & a_{22}b_{13} \\ a_{21}b_{21} & a_{21}b_{22} & a_{21}b_{23} & a_{22}b_{21} & a_{22}b_{22} & a_{22}b_{23} \\ a_{21}b_{31} & a_{21}b_{32} & a_{21}b_{33} & a_{22}b_{31} & a_{22}b_{32} & a_{22}b_{33} \end{array}\right).$$

Let $C \in \mathbb{C}^{mp \times nq}$ be the Kronecker product of the matrices $A \in \mathbb{C}^{m \times n}$ and $B \in \mathbb{C}^{p \times q}$. We seek to express the value of c_{ij}, where $1 \leqslant i \leqslant mp$ and $1 \leqslant j \leq nq$. It is easy to see that

$$c_{ij} = a_{\lceil \frac{i}{p} \rceil, \lceil \frac{j}{q} \rceil} b_{i-p\left(\lceil \frac{i}{p} \rceil - 1\right), j-q\left(\lceil \frac{j}{q} \rceil - 1\right)}. \tag{3.13}$$

Conversely, we have

$$a_{rs}b_{vw} = (A \otimes B)_{p(r-1)+v, q(s-1)+w} \tag{3.14}$$

for $1 \leqslant r \leqslant m$, $1 \leqslant s \leqslant n$ and $1 \leqslant v \leqslant p$, $1 \leqslant w \leqslant q$.

Theorem 3.54. *The Kronecker product is associative. In other words, if $A \in \mathbb{C}^{K \times L}$, $B \in \mathbb{C}^{M \times N}$, and $C \in \mathbb{C}^{R \times S}$, then $(A \otimes B) \otimes C = A \otimes (B \otimes C)$.*

Proof. The product $a_{kl}b_{mn}$ is the entry $((k-1)M + m, (\ell - 1)N + {} + n)$ of $A \otimes B$. Therefore, the product $(a_{kl}b_{mn})c_{rs}$ is the entry $(((k - 1)M + m - 1)R + r, ((\ell - 1)N = n - 1)S + s)$ of $(A \otimes B) \otimes C$.

On the other hand, the product $a_{kl}(b_{mn}c_{rs})$ is the entry of $A \otimes (B \otimes C)$ that occupies the position

$$((k-1)MR + (m-1)R + r, (\ell - 1)NS + (n-1)S + s),$$

which is identical to

$$((k-1)M + (m-1)R + r, ((\ell - 1)N + (n-1)S + s).$$

Thus, the product $a_{kl}b_{mn}c_{op}$ occupies the same position in $(A \otimes B) \otimes C$ and in $A \otimes (B \otimes C)$. Therefore, $(A \otimes B) \otimes C = A \otimes (B \otimes C)$. □

Theorem 3.55. *Let $A_1, \ldots, A_n$ and $B_1, \ldots, B_n$ be matrices. Then, we have*

$$(A_1 \otimes B_1)(A_2 \otimes B_2) \cdots (A_n \otimes B_n) = (A_1 A_2 \cdots A_n) \otimes (B_1 B_2 \cdots B_n).$$

Proof. Note that

$$\begin{aligned}
&(A \otimes B)(C \otimes D)\\
&= \begin{pmatrix} a_{11}B & \cdots & a_{1n}B \\ \vdots & \cdots & \vdots \\ a_{m1}B & \cdots & a_{mn}B \end{pmatrix} \begin{pmatrix} c_{11}D & \cdots & a_{1p}D \\ \vdots & \cdots & \vdots \\ c_{n1}D & \cdots & c_{np}D \end{pmatrix}\\
&= \begin{pmatrix} \sum_j a_{1j}c_{j1}BD & \cdots & \sum_j a_{1j}c_{jp}BD \\ \vdots & \cdots & \vdots \\ \sum_j a_{mj}c_{j1}BD & \cdots & \sum_j a_{mj}c_{jp}BD \end{pmatrix}\\
&= \begin{pmatrix} \sum_j a_{1j}c_{j1} & \cdots & \sum_j a_{1j}c_{jp} \\ \vdots & \cdots & \vdots \\ \sum_j a_{mj}c_{j1} & \cdots & \sum_j a_{mj}c_{jp} \end{pmatrix} \otimes BD = (AC) \otimes (BD).
\end{aligned}$$

Repeated multiplication yields the desired equality. □

The next theorem contains a few other elementary properties of Kronecker's product.

Theorem 3.56. *For any complex matrices A, B, C, D, we have the following:*
(i) $(A \otimes B)' = A' \otimes B'$,
(ii) $(A \otimes B) \otimes C = A \otimes (B \otimes C)$,
(iii) $A \otimes B + A \otimes C = A \otimes (B + C)$,
(iv) $A \otimes D + B \otimes D = (A + B) \otimes D$,
(v) $(A \otimes B)' = A' \otimes B'$,
(vi) $(A \otimes B)^{\mathsf{H}} = A^{\mathsf{H}} \otimes B^{\mathsf{H}}$,
when the usual matrix sum and multiplication are well-defined in each of the above equalities.

Proof. The proof is straightforward and is left to the reader. □

Example 3.40. Let $\boldsymbol{x} \in \mathbb{C}^n$ and $\boldsymbol{y} \in \mathbb{C}^m$. We have

$$\boldsymbol{x} \otimes \boldsymbol{y} = \begin{pmatrix} x_1\boldsymbol{y} \\ \vdots \\ x_n\boldsymbol{y} \end{pmatrix} \text{ and } \boldsymbol{y} \otimes \boldsymbol{x} = \begin{pmatrix} y_1\boldsymbol{x} \\ \vdots \\ y_m\boldsymbol{x} \end{pmatrix}.$$

Note that the Kronecker product is not commutative. For example, we have $\boldsymbol{x} \otimes \boldsymbol{y} \neq \boldsymbol{y} \otimes \boldsymbol{x}$. We have

$$\begin{pmatrix} 1 \\ 2 \\ 3 \end{pmatrix} \otimes \begin{pmatrix} 4 \\ 5 \end{pmatrix} = \begin{pmatrix} 5 \\ 4 \\ 10 \\ 8 \\ 15 \\ 12 \end{pmatrix},$$

but

$$\begin{pmatrix} 4 \\ 5 \end{pmatrix} \begin{pmatrix} 1 \\ 2 \\ 3 \end{pmatrix} = \begin{pmatrix} 4 \\ 8 \\ 12 \\ 5 \\ 10 \\ 15 \end{pmatrix}.$$

Example 3.41. Let $D \in \mathbb{C}^{p\times p}$. The Kronecker product $C = I_m \otimes D$ is given by

$$\begin{aligned} c_{ij} &= I_{\lceil \frac{i}{p} \rceil, \lceil \frac{j}{p} \rceil} d_{i-p\left(\lceil \frac{i}{p} \rceil -1\right), j-p\left(\lceil \frac{j}{p} \rceil -1\right)} \\ &= \begin{cases} d_{i-p(k-1),j-p(k-1)} & \text{if } \lceil \frac{i}{p} \rceil = \lceil \frac{j}{p} \rceil = k \\ 0 & \text{otherwise,} \end{cases} \end{aligned}$$

for $1 \leqslant i, j \leqslant mp$.

Let now $E \in \mathbb{C}^{m\times m}$. The Kronecker product $L = E \otimes I_p$ is given by

$$l_{ij} = \begin{cases} e_{\lceil \frac{i}{p} \rceil, \lceil \frac{j}{p} \rceil} & \text{if } p\lceil \frac{i}{p} \rceil - i = \lceil \frac{j}{p} \rceil - j, \\ 0 & \text{otherwise,} \end{cases}$$

for $1 \leqslant i, j \leqslant mp$.

Theorem 3.57. *If $A \in \mathbb{C}^{n\times n}$ and $B \in \mathbb{C}^{m\times m}$ are two invertible matrices, then $A \otimes B$ is invertible and $(A \otimes B)^{-1} = A^{-1} \otimes B^{-1}$.*

Proof. Since

$$(A \otimes B)(A^{-1} \otimes B^{-1}) = (AA^{-1} \otimes BB^{-1}) = I_n \otimes I_m,$$

the theorem follows by observing that $I_n \otimes I_m = I_{nm}$. □

In a sequence of Kronecker products of column and row vectors, a column vector can be permuted with a row vector without altering the final result. This is formalized in the next statement.

Theorem 3.58. *Let $\boldsymbol{u} \in \mathbb{C}^m$ and let $\boldsymbol{v} \in \mathbb{C}^n$. We have*

$$\boldsymbol{u} \otimes \boldsymbol{v}' = \boldsymbol{v}' \otimes \boldsymbol{u}.$$

Proof. Suppose that

$$\boldsymbol{u} = \begin{pmatrix} u_1 \\ \vdots \\ u_m \end{pmatrix} \text{ and } \boldsymbol{v} = \begin{pmatrix} v_1 \\ \vdots \\ v_n \end{pmatrix}.$$

We have

$$\boldsymbol{u} \otimes \boldsymbol{v}' = \begin{pmatrix} u_1\boldsymbol{v}' \\ \cdots \\ u_m\boldsymbol{v}' \end{pmatrix} = \begin{pmatrix} u_1v_1 & \cdots & u_1v_n \\ \vdots & \vdots & \vdots \\ u_mv_1 & \cdots & u_mv_n \end{pmatrix}.$$

Also,

$$\boldsymbol{v}' \otimes \boldsymbol{u} = (v_1, \cdots, v_n) \begin{pmatrix} u_1 \\ \vdots \\ u_m \end{pmatrix} = \begin{pmatrix} v_1u_1 & \cdots & v_nu_1 \\ \vdots & \vdots & \vdots \\ v_1u_m & \cdots & v_nu_m \end{pmatrix},$$

which establishes our equality. □

Example 3.42. Suppose that $\boldsymbol{u}_1 \in \mathbb{C}^m$, $\boldsymbol{u}_2 \in \mathbb{C}^p$, and $\boldsymbol{v} \in \mathbb{C}^m$. Then, we have

$$\boldsymbol{u}_1 \otimes \boldsymbol{u}_2 \otimes \boldsymbol{v}' = \boldsymbol{u}_1 \otimes \boldsymbol{v}' \otimes \boldsymbol{u}_2 = \boldsymbol{v}' \otimes \boldsymbol{u}_1 \otimes \boldsymbol{u}_2.$$

Note that such transformations can be applied to Kronecker products of column and row vectors *provided we do not change the order of the column and the row vectors.*

Theorem 3.59. *Let $A \in \mathbb{C}^{n\times n}$ and $B \in \mathbb{C}^{m\times m}$ be two normal (unitary) matrices. Their Kronecker product $A \otimes B$ is also a normal (a unitary) matrix.*

Proof. By Theorem 3.56, we can write

$$\begin{aligned}(A\otimes B)'(A\otimes B) &= (A'\otimes B')(A\otimes B)\\ &= (A'A\otimes B'B)\\ &= (AA'\otimes BB')\\ &\quad \text{(because both } A \text{ and } B \text{ are normal)}\\ &= (A\otimes B)(A\otimes B)',\end{aligned}$$

which implies that $A \otimes B$ is normal. □

Definition 3.44. Let $A \in \mathbb{C}^{m\times m}$ and $B \in \mathbb{C}^{n\times n}$ be two square matrices. Their *Kronecker sum* is the matrix $A \oplus B \in \mathbb{C}^{mn\times mn}$ defined by

$$A \oplus B = (A \otimes I_n) + (I_m \otimes B).$$

The *Kronecker difference* is the matrix $A \ominus B \in \mathbb{C}^{mn\times mn}$ defined by

$$A \ominus B = (A \otimes I_n) - (I_m \otimes B).$$

The element $(A \oplus B)_{ij}$ is given by

$$(A \oplus B)_{ij} = (A \otimes I_n)_{ij} + (I_m \otimes B)_{ij}.$$

Thus, $(A \oplus B)$ can be computed by applying the formulas developed in Example 3.41:

$$(A\otimes I_n)_{ij} = \begin{cases} a_{\lceil \frac{i}{n}\rceil,\lceil \frac{j}{n}\rceil} & \text{if } n\lceil \frac{i}{n}\rceil - i = n\lceil \frac{j}{n}\rceil - j,\\ 0 & \text{otherwise,}\end{cases}$$

$$(I_m\otimes B)_{ij} = \begin{cases} b_{i-n(k-1),j-n(k-1)} & \text{if } \lceil \frac{i}{n}\rceil = \lceil \frac{j}{n}\rceil = k,\\ 0 & \text{otherwise,}\end{cases}$$

for $1 \leqslant i, j \leqslant mn$. Similar fact can be proven about Kronecker difference by replacing B by $-B$ in the formula involving the Kronecker sum.

Definition 3.45. Let $A, B \in \mathbb{C}^{m\times n}$. The *Hadamard product* of A and B is the matrix $A \odot B \in \mathbb{C}^{m\times n}$ defined by

$$A \odot B = \begin{pmatrix} a_{11}b_{11} & a_{12}b_{12} & \cdots & a_{1n}b_{1n} \\ a_{21}b_{21} & a_{22}b_{22} & \cdots & a_{2n}b_{2n} \\ \vdots & \vdots & \ddots & \vdots \\ a_{m1}b_{m1} & a_{m2}b_{m2} & \cdots & a_{mn}b_{mn} \end{pmatrix}.$$

The *Hadamard quotient* $A \oslash B$ is defined only if $b_{ij} \neq 0$ for $1 \leqslant i \leqslant m$ and $1 \leqslant j \leqslant n$. In this case,

$$A \oslash B = \begin{pmatrix} \frac{a_{11}}{b_{11}} & \frac{a_{12}}{b_{12}} & \cdots & \frac{a_{1n}}{b_{1n}} \\ \frac{a_{21}}{b_{21}} & \frac{a_{22}}{b_{22}} & \cdots & \frac{a_{2n}}{b_{2n}} \\ \vdots & \vdots & \ddots & \vdots \\ \frac{a_{m1}}{b_{m1}} & \frac{a_{m2}}{b_{m2}} & \cdots & \frac{a_{mn}}{b_{mn}} \end{pmatrix}.$$

Theorem 3.60. *If $A, B, C \in \mathbb{C}^{m\times n}$ and $c \in \mathbb{C}$, we have*

(i) $A \odot B = B \odot A$;
(ii) $A \odot J_{m,n} = J_{m,n} \odot A = A$;
(iii) $A \odot (B + C) = A \odot B + A \odot C$;
(iv) $A \odot (cB) = c(A \odot B)$.

Proof. The proof is straightforward and is left to the reader. □

Note that the Hadamard product of two matrices $A, B \in \mathbb{C}^{m\times n}$ is a submatrix of the Kronecker product $A \otimes B$.

Example 3.43. Let $A, B \in \mathbb{C}^{2\times 3}$ be the matrices

$$A = \begin{pmatrix} a_{11} & a_{12} & a_{13} \\ a_{21} & a_{22} & a_{23} \end{pmatrix} \text{ and } B = \begin{pmatrix} b_{11} & b_{12} & b_{13} \\ b_{21} & b_{22} & b_{23} \end{pmatrix}.$$

The Kronecker product of these matrices is $A \otimes B \in \mathbb{C}^{4\times 9}$ given by

$$A\otimes B = \begin{pmatrix} a_{11}b_{11} & a_{11}b_{12} & a_{11}b_{13} & a_{12}b_{11} & a_{12}b_{12} & a_{12}b_{13} & a_{13}b_{11} & a_{13}b_{12} & a_{13}b_{13} \\ a_{11}b_{21} & a_{11}b_{22} & a_{11}b_{23} & a_{12}b_{21} & a_{12}b_{22} & a_{12}b_{23} & a_{13}b_{21} & a_{13}b_{22} & a_{13}b_{23} \\ a_{21}b_{11} & a_{21}b_{12} & a_{21}b_{13} & a_{22}b_{11} & a_{22}b_{12} & a_{22}b_{13} & a_{23}b_{11} & a_{23}b_{12} & a_{23}b_{13} \\ a_{21}b_{21} & a_{21}b_{22} & a_{21}b_{23} & a_{22}b_{21} & a_{22}b_{22} & a_{22}b_{23} & a_{23}b_{21} & a_{23}b_{22} & a_{23}b_{23} \end{pmatrix}.$$

The Hadamard product of the same matrices is

$$A \odot B = \begin{pmatrix} a_{11}b_{11} & a_{12}b_{12} & a_{13}b_{13} \\ a_{21}b_{21} & a_{22}b_{22} & a_{23}b_{23} \end{pmatrix},$$

and we can regard the Hadamard product as a submatrix of the Kronecker product $A \otimes B$,

$$A \odot B = (A \otimes B) \begin{bmatrix} 1, 5, 9 \\ 4, 4, 4 \end{bmatrix}.$$

Another matrix product involves matrices that have the same number of columns.

Definition 3.46. Let $A \in \mathbb{C}^{m \times n}$ and $B \in \mathbb{C}^{p \times n}$ be two matrices that have the same number n of columns,

$$A = (\boldsymbol{a}_1 \ \cdots \ \boldsymbol{a}_n) \text{ and } B = (\boldsymbol{b}_1 \ \cdots \ \boldsymbol{b}_n).$$

The *Khatri–Rao* product of A and B (or the column-wise Kronecker product) is the matrix

$$A * B = (\boldsymbol{a}_1 \otimes \boldsymbol{b}_1 \ \boldsymbol{a}_2 \otimes \boldsymbol{b}_2 \ \cdots \ \boldsymbol{a}_n \otimes \boldsymbol{b}_n).$$

Example 3.44. The Khatri–Rao product of the matrices

$$A = \begin{pmatrix} 1 & 2 & 3 \\ 4 & 5 & 6 \end{pmatrix} \text{ and } B = \begin{pmatrix} 1 & 0 & 2 \\ 2 & 1 & 3 \\ -1 & 2 & 1 \end{pmatrix}$$

is the matrix $(\boldsymbol{a}_1 \otimes \boldsymbol{b}_1 \ \boldsymbol{a}_2 \otimes \boldsymbol{b}_2 \ \boldsymbol{a}_3 \otimes \boldsymbol{b}_3)$, which equals

$$\begin{pmatrix} 1 & 0 & 6 \\ 2 & 2 & 9 \\ -1 & 4 & 3 \\ 4 & 0 & 12 \\ 8 & 5 & 18 \\ -4 & 10 & 6 \end{pmatrix}.$$

Note that for any vectors $\boldsymbol{a}, \boldsymbol{b}$, we have $\boldsymbol{a} \otimes \boldsymbol{b} = \boldsymbol{a} * \boldsymbol{b}$.

3.16 Outer Products

Definition 3.47. Let $\boldsymbol{u} \in \mathbb{C}^m$ and $\boldsymbol{v} \in \mathbb{C}^n$. The *outer product* of the vectors $\boldsymbol{u}$ and $\boldsymbol{v}$ is the matrix $\boldsymbol{u} \circ \boldsymbol{v} \in \mathbb{C}^{m\times n}$ defined by $\boldsymbol{u} \circ \boldsymbol{v} = \boldsymbol{u}\boldsymbol{v}^{\mathsf{H}}$.

As we saw in Example 3.28, the outer product of two vectors is a matrix of rank 1.

For $\boldsymbol{u} \in \mathbb{C}^m$ and $\boldsymbol{v} \in \mathbb{C}^n$, we have $\boldsymbol{v} \circ \boldsymbol{u} = \boldsymbol{v}\boldsymbol{u}^{\mathsf{H}} = (\boldsymbol{u}\boldsymbol{v}^{\mathsf{H}})^{\mathsf{H}} = (\boldsymbol{u} \circ \boldsymbol{v})^{\mathsf{H}}$. Therefore, the outer product is not commutative because for $\boldsymbol{u} \in \mathbb{C}^m$ and $\boldsymbol{v} \in \mathbb{C}^n$, we have $\boldsymbol{u} \circ \boldsymbol{v} \in \mathbb{C}^{m\times n}$ and $\boldsymbol{v} \circ \boldsymbol{u} \in \mathbb{C}^{n\times m}$.

Note that when $m = n$, we have $\boldsymbol{u}\boldsymbol{v}^{\mathsf{H}} = \mathsf{trace}(\boldsymbol{u} \circ \boldsymbol{v})$.

Example 3.45. Let

$$\boldsymbol{u} = \begin{pmatrix} u_1 \\ u_2 \\ u_3 \end{pmatrix} \text{ and } \boldsymbol{v} = \begin{pmatrix} v_1 \\ v_2 \end{pmatrix}.$$

We have

$$\boldsymbol{u} \circ \boldsymbol{v} = \begin{pmatrix} u_1v_1 & u_1v_2 \\ u_2v_1 & u_2v_2 \\ u_3v_1 & u_3v_2 \end{pmatrix} \text{ and } \boldsymbol{v} \circ \boldsymbol{u} = \begin{pmatrix} v_1u_1 & v_1u_2 & v_1u_3 \\ v_2u_1 & v_2u_2 & v_2u_3 \end{pmatrix}.$$

Contrast this with the Kronecker products:

$$\boldsymbol{u} \otimes \boldsymbol{v} = \begin{pmatrix} u_1v_1 \\ u_1v_2 \\ u_2v_1 \\ u_2v_2 \\ u_3v_1 \\ u_3v_2 \end{pmatrix} \text{ and } \boldsymbol{v} \otimes \boldsymbol{u} = \begin{pmatrix} v_1u_1 \\ v_1u_2 \\ v_1u_3 \\ v_2u_1 \\ v_2u_2 \\ v_2u_3 \end{pmatrix}.$$

Note that the entries of the Kronecker product $\boldsymbol{u} \otimes \boldsymbol{v}$ can be obtained by reading the entries of $\boldsymbol{u} \circ \boldsymbol{v}$ row-wise.

3.17 Associative Algebras

Definition 3.48. An $\mathbb{F}$-*associative algebra* is a pair (V, m), where V is an $\mathbb{F}$-linear space and $m : V \times V \longrightarrow V$ is a bilinear mapping that is associative (which means that $m(\boldsymbol{x}, m(\boldsymbol{y}, \boldsymbol{z})) = m(m(\boldsymbol{x}, \boldsymbol{y}), \boldsymbol{z})$ for all $\boldsymbol{x}, \boldsymbol{y}, \boldsymbol{z} \in V$).

If there exists $\boldsymbol{u} \in V$ such that $m(\boldsymbol{v}, \boldsymbol{u}) = \boldsymbol{v}$ for every $\boldsymbol{v} \in V$, then $\boldsymbol{u}$ is referred to as the unit element and (V, m) is a *unital associative algebra.*

The dimension of an $\mathbb{F}$-associative algebra (V, m) is $\dim(V)$.

We denote $m(\boldsymbol{v}_1, \boldsymbol{v}_2)$ as $\boldsymbol{v}_1\boldsymbol{v}_2$.

In a unital associative algebra there exists a unique unit element.

Example 3.46. The set $\mathbb{R}^{n\times n}$ of square matrices with real elements is a unital associative algebra, where $m(A, B) = AB$. The unit element is the matrix I_n.

Example 3.47. Let $\mathsf{Hom}(V, V)$ be the linear space of homomorphisms of an $\mathbb{F}$-linear space V. For $h, k \in \mathsf{Hom}(V, V)$, define $m(h, k)$ as the composition of mappings hk. It is easy to verify that $(\mathsf{Hom}(V, V), m)$ is a unital associative algebra having 1_V as its unit element.

Definition 3.49. An *associative algebra morphism* between the associative algebras (V, m) and (W, m') is a linear mapping $h : V \longrightarrow W$ such that $h(m(\boldsymbol{u}, \boldsymbol{v})) = m'(h(\boldsymbol{u}), h(\boldsymbol{v}))$ for every $\boldsymbol{u}, \boldsymbol{v} \in V$.

If (V, m) and (W, m') have unit elements u and u', respectively, then $h(u) = u'$.

If h is a bijective morphism, we refer to h as an *isomorphism of associative algebras.*

Let $B = \{\boldsymbol{e}_1, \ldots, \boldsymbol{e}_n\}$ be a basis in V, where (V, m) is an associative algebra. If $\boldsymbol{x}, \boldsymbol{y} \in V$, $\boldsymbol{x} = \sum_{i=1}^{n} a^i \boldsymbol{e}_i$, and $\boldsymbol{y} = \sum_{j=1}^{n} b^j \boldsymbol{e}_j$, then

$$\boldsymbol{x}\boldsymbol{y} = \sum_{i=1}^{n} \sum_{j=1}^{n} a^i b_j \boldsymbol{e}_i \boldsymbol{e}_j.$$

Since $\boldsymbol{e}_i\boldsymbol{e}_j \in V$, it is possible to write

$$\boldsymbol{e}_i\boldsymbol{e}_j = \sum_{k=1}^{n} \gamma_{ij}^k \boldsymbol{e}_k,$$

where γ_{ij}^k are n^3 scalars in $\mathbb{F}$. We refer to these numbers as the *structural coefficients* of the associative algebra (V, m).

Theorem 3.61. *Let γ_{ij}^k be structural coefficients of the associative algebra (V, m) having the basis $\{\mathbf{e}_1, \ldots, \mathbf{e}_n\}$. These coefficients satisfy the following n^4 equalities:*

$$\sum_q \gamma_{jk}^q \gamma_{iq}^r = \sum_p \gamma_{ij}^p \gamma_{pk}^r$$

for i, h, j, k between 1 and n.

Proof. Since $\mathbf{e}_i\mathbf{e}_j = \sum_p \gamma_{ij}^p \mathbf{e}_p$ and $\mathbf{e}_j\mathbf{e}_k = \sum_q \gamma_{jk}^q \mathbf{e}_q$, we can write

$$\begin{aligned} \mathbf{e}_i(\mathbf{e}_j\mathbf{e}_k) &= \sum_q \gamma_{jk}^q \mathbf{e}_i\mathbf{e}_q = \sum_q \gamma_{jk}^q \sum_r \gamma_{iq}^r \mathbf{e}_r \\ &= \sum_q \sum_r \gamma_{jk}^q \gamma_{iq}^r \mathbf{e}_r = \sum_r \sum_q \gamma_{jk}^q \gamma_{iq}^r \mathbf{e}_r \\ (\mathbf{e}_i\mathbf{e}_j)\mathbf{e}_k &= \sum_p \gamma_{ij}^p \mathbf{e}_p\mathbf{e}_k = \sum_p \gamma_{ij}^p \sum_r \gamma_{pk}^r \mathbf{e}_r \\ &= \sum_p \sum_r \gamma_{ij}^p \gamma_{pk}^r \mathbf{e}_r = \sum_r \sum_p \gamma_{ij}^p \gamma_{pk}^r \mathbf{e}_r. \end{aligned}$$

The associativity property implies the following equalities:

$$\sum_q \gamma_{jk}^q \gamma_{iq}^r = \sum_p \gamma_{ij}^p \gamma_{pk}^r$$

for i, h, j, k between 1 and n. □

An associative algebra (V, m) over a finite-dimensional linear space V is completely defined by its multiplicative table. Starting from a basis $\{\mathbf{e}_1, \ldots, \mathbf{e}_n\}$ place at the intersection of the ith row and the jth column the expansion of the product $\mathbf{e}_i\mathbf{e}_j$ in terms of the basis:

	$\mathbf{e}_1$	$\cdots$	$\mathbf{e}_j$	$\cdots$	$\mathbf{e}_n$
$\mathbf{e}_1$	$\sum_k \gamma_{11}^k \mathbf{e}_k$	$\cdots$	$\sum_k \gamma_{1j}^k \mathbf{e}_k$	$\cdots$	$\sum_k \gamma_{1n}^k \mathbf{e}_k$
$\vdots$	$\vdots$	$\vdots$	$\vdots$	$\vdots$	$\vdots$
$\mathbf{e}_i$	$\sum_k \gamma_{i1}^k \mathbf{e}_k$	$\cdots$	$\sum_k \gamma_{ij}^k \mathbf{e}_k$	$\cdots$	$\sum_k \gamma_{in}^k \mathbf{e}_k$
$\vdots$	$\vdots$	$\vdots$	$\vdots$	$\vdots$	$\vdots$
$\mathbf{e}_n$	$\sum_k \gamma_{n1}^k \mathbf{e}_k$	$\cdots$	$\sum_k \gamma_{nj}^k \mathbf{e}_k$	$\cdots$	$\sum_k \gamma_{nn}^k \mathbf{e}_k$

Example 3.48. Let V be a four-dimensional real linear space having the basis $\{\boldsymbol{1}, \boldsymbol{i}, \boldsymbol{j}, \boldsymbol{k}\}$ whose elements have the form

$$\boldsymbol{v} = a\boldsymbol{1} + b\boldsymbol{i} + c\boldsymbol{j} + d\boldsymbol{k},$$

where $a, b, c, d \in \mathbb{R}$. The multiplication table of this algebra is given by

	$\boldsymbol{1}$	$\boldsymbol{i}$	$\boldsymbol{j}$	$\boldsymbol{k}$
$\boldsymbol{1}$	$\boldsymbol{1}$	$\boldsymbol{i}$	$\boldsymbol{j}$	$\boldsymbol{k}$
$\boldsymbol{i}$	$\boldsymbol{i}$	$-\boldsymbol{1}$	$\boldsymbol{k}$	$-\boldsymbol{j}$
$\boldsymbol{j}$	$\boldsymbol{j}$	$-\boldsymbol{k}$	$-\boldsymbol{1}$	$\boldsymbol{i}$
$\boldsymbol{k}$	$\boldsymbol{k}$	$\boldsymbol{j}$	$-\boldsymbol{i}$	$-\boldsymbol{1}$

If $\boldsymbol{q} \in V$, we can write $\boldsymbol{q} = a\boldsymbol{1} + b\boldsymbol{i} + c\boldsymbol{j} + d\boldsymbol{k}$ for some $a, b, c, d \in \mathbb{R}$. Such an element is said to be a *quaternion*. Its *conjugate* is the quaternion $\overline{\boldsymbol{q}} = a\boldsymbol{1} - b\boldsymbol{i} - c\boldsymbol{j} - d\boldsymbol{k}$. It is easy to see that $\boldsymbol{q}\overline{\boldsymbol{q}} = (a^2 + b^2 + c^2 + d^2)\boldsymbol{1}$.

Definition 3.50. A *subalgebra* of an associative algebra (V, m) is a subspace U of V such that $m(\boldsymbol{u}_1, \boldsymbol{u}_2) \in U$ for every $\boldsymbol{u}_1, \boldsymbol{u}_2 \in U$.

If U is a subalgebra of (V, m) and $m' = m \upharpoonright_U$, then (U, m') can be regarded as an associative algebra. The intersection of any family of subalgebras of an associative algebra is a subalgebra.

An *ideal* of an associative algebra (V, m) is a subspace I of V such that $m(\boldsymbol{v}, \boldsymbol{i}) \in I$ and $m(\boldsymbol{i}, \boldsymbol{v}) \in I$ for every $\boldsymbol{v} \in V$ and $\boldsymbol{i} \in I$.

Let $\{I_j \mid j \in J\}$ be a family of ideals of an associative algebra (V, m). It is easy to see that $\bigcap_{j \in J}$ is an ideal of (V, m). Moreover, if $S \subseteq V$, the intersection of all ideals that contain S is again an ideal denoted by I_S.

If I is an ideal of an associative algebra (V, m), we can consider the quotient linear space V/I. Define the multiplication $\overline{m}$ on V/I as $\overline{m}([\boldsymbol{x}], [\boldsymbol{y}]) = [m(\boldsymbol{x}, \boldsymbol{y})]$. Note that $\overline{m}$ is well-defined because if $\tilde{\boldsymbol{x}} \in [\boldsymbol{x}]$ and $\tilde{\boldsymbol{y}} \in [\boldsymbol{y}]$, we have $\tilde{\boldsymbol{x}} - \boldsymbol{x} \in I$ and $\tilde{\boldsymbol{y}} - \boldsymbol{y} \in I$. In other words, there exist $\boldsymbol{u}, \boldsymbol{v} \in I$ such that $\tilde{\boldsymbol{x}} = \boldsymbol{x} + \boldsymbol{u}$ and $\tilde{[\boldsymbol{y}]} = \boldsymbol{y} + \boldsymbol{v}$. This implies

$$\begin{aligned} m(\tilde{\boldsymbol{x}}, \tilde{\boldsymbol{y}}) &= m(\boldsymbol{x} + \boldsymbol{u}, \boldsymbol{y} + \boldsymbol{v}) \\ &= m(\boldsymbol{x}, \boldsymbol{y}) + m(\boldsymbol{x}, \boldsymbol{v}) + m(\boldsymbol{u}, \boldsymbol{y}) + m(\boldsymbol{u}, \boldsymbol{v}) \\ &= m(\boldsymbol{x}, \boldsymbol{y}) + \boldsymbol{z}, \end{aligned}$$

where $\boldsymbol{z} \in I$. Therefore, $[m(\tilde{\boldsymbol{x}}, \tilde{\boldsymbol{y}})] = [m(\boldsymbol{x}, \boldsymbol{y})]$ and

$$\overline{m}([\tilde{\boldsymbol{x}}], [\tilde{\boldsymbol{y}}]) = [m(\tilde{\boldsymbol{x}}, \tilde{\boldsymbol{y}})] = [m(\boldsymbol{x}, \boldsymbol{y})],$$

which shows that $\overline{m}$ is well-defined. The associative algebra $(V/I, \overline{m})$ is the *factor algebra* of V by I.

Definition 3.51. A *graded linear space* is a vector space V that can be written as a direct sum of the form $V = \sum_{n \in \mathbb{N}} V_n$, where each V_n is a vector space.

The linear space V_n is the *set of elements of degree* n.

Example 3.49. Let $\mathbb{R}[x, y]$ be the linear space of polynomials with real coefficients in the indeterminates x and y. If $\mathbb{R}[x, y]_n$ is the set of all homogeneous polynomials of degree n in x and y, then $\mathbb{R}[x, y]$ is a graded linear space.

Generalizing Example 3.49, if V is a direct sum, $V = \sum_{n \in \mathbb{N}} V_n$, we refer to members of the summand V_n as *homogeneous elements of degree* n. If $\boldsymbol{v}$ is $\boldsymbol{v} = \sum_{n \in \mathbb{N}} \boldsymbol{v}_n$ where $\boldsymbol{v}_n \in V_n$, $\boldsymbol{v}_n$ is the *homogeneous component of* $\boldsymbol{v}$ *of degree* n.

A *graded associative algebra* is a graded vector space $V = \sum_{n \in \mathbb{N}} V_n$ such that $m(V_p, V_q) \subseteq V_{p+q}$ for $p, q \in \mathbb{N}$.

A subalgebra W of the graded associative algebra V is a *graded subalgebra* if $W = \sum_{n \in \mathbb{N}} (W \cap V_n)$.

Exercises and Supplements

(1) Let $A \in \mathbb{C}^{m \times n}$ and $B \in \mathbb{C}^{n \times p}$ be two conformant matrices. Prove that:
   (a) computing the matrix $G = AB$ using standard matrix multiplication requires mnp number multiplications;
   (b) if $C \in \mathbb{C}^{p \times q}$ and we compute the matrix $D = (AB)C = A(BC)$ by the standard method, the first modality $D = (AB)C$, requires $mp(n + q)$ multiplications, while the second, $D = A(BC)$, requires $nq(m + p)$ multiplications.

(2) Let $\{i_1, \ldots, i_k\}$ be a subset of $\{1, \ldots, n\}$ and let $\{j_1, \ldots, j_q\} = \{1, \ldots, n\} - \{i_1, \ldots, i_k\}$, where $j_1 < \cdots < j_q$ and $k + q = n$.

Let $A \in \mathbb{C}^{n\times n}$ be a matrix and let $B = A\begin{bmatrix} i_1 \cdots i_k \\ i_1 \cdots i_k \end{bmatrix}$ be a principal submatrix of A. Prove that if $\boldsymbol{y} \in \mathbb{C}^k$, then $\boldsymbol{y}^{\mathsf{H}}B\boldsymbol{y} = \boldsymbol{x}^{\mathsf{H}}A\boldsymbol{x}$, where

$$x_r = \begin{cases} y_i & \text{if } i = j_r, \\ 0 & \text{otherwise} \end{cases}$$

for $1 \leqslant r \leqslant n$.

(3) Let $X = (\boldsymbol{x}_1 \ \cdots \ \boldsymbol{x}_n)$ be a matrix in $\mathbb{C}^{n\times n}$ and let $C = \mathsf{diag}(c_1,\ldots,c_n) \in \mathbb{C}^{n\times n}$. Prove that $XC = c_1\boldsymbol{x}_1\boldsymbol{e}_1' + \cdots + c_1\boldsymbol{x}_n\boldsymbol{e}_n'$.

(4) Let $X = (\boldsymbol{x}_1 \ \cdots \ \boldsymbol{x}_n)$ and $Y = (\boldsymbol{y}_1 \ \cdots \ \boldsymbol{y}_n)$ be two matrices in $\mathbb{C}^{n\times n}$, and let $C = \mathsf{diag}(c_1,\ldots,c_n)$, $D = \mathsf{diag}(d_1,\ldots,d_n)$. Prove that

$$XCYD = \sum_{i=1}^{n}\sum_{j=1}^{n} c_i d_j y_{ji}\boldsymbol{x}_i\boldsymbol{e}_j',$$

and that $\mathsf{trace}(XCYD) = \sum_{i=1}^{n}\sum_{j=1}^{n} c_i d_j y_{ji} x_{ji}$.

(5) Let $a_1,\ldots,a_n$ be n complex numbers. Prove that

$$\prod_{i=1}^{n} \mathsf{diag}(1,\ldots,1,a_i,1,\ldots,1) = \mathsf{diag}(a_1,\ldots,a_n).$$

(6) Let $D = \mathsf{diag}(d_1,\ldots,d_n) \in \mathbb{C}^{n\times n}$ and let $A \in \mathbb{C}^{n\times n}$. Prove that $(DAD)_{ij} = d_i a_{ij} d_j$ for $1 \leqslant i,j \leqslant n$.

(7) Let $S^{n\times n}$ be the set of $n\times n$ symmetric matrices in $\mathbb{R}^{n\times n}$. Prove that $S^{n\times n}$ is a subspace of $\mathbb{R}^{n\times n}$ and $\dim(S^{n\times n}) = \frac{n(n+1)}{2}$.

(8) Let $A \in \mathbb{R}^{n\times n}$ be a symmetric matrix and let $B \in \mathbb{R}^{n\times n}$ be a matrix such that $\boldsymbol{x}'B\boldsymbol{x} = \sum_{i=1}^{n}\sum_{j=1}^{n} a_{ij}(x_i - x_j)^2$ for every $\boldsymbol{x} \in \mathbb{R}^n$. Prove that:
   (a) for $1 \leqslant k \leqslant n$, we have $b_{kk} = 2\sum\{a_{ik} \mid 1 \leqslant i \leqslant n \text{ and } i \neq k\}$;
   (b) $\sum_{i=1}^{n}\sum_{j=1}^{n} b_{ij} = 0$.

(9) Let $\psi \in PERM_n$ and let A be a square matrix, $A \in \mathbb{C}^{n\times n}$. Prove that $(P_\psi A)_{ij} = a_{\psi(i)j}$ and $(AP_\psi)_{ij} = a_{i\psi^{-1}(j)}$ for $1 \leqslant i,j \leqslant n$.

Let $\phi \in PERM_n$ be a permutation and let $A \in \mathbb{C}^{n\times n}$ be a square matrix. A *ϕ-diagonal of A* is a set of n elements $D_\phi(A) = \{a_{1\phi(1)}, a_{2\phi(2)}, \ldots, a_{n\phi(n)}\}$. A *permutation diagonal* of A is a ϕ-diagonal of A for some $\phi \in PERM_n$.

(10) Let $A \in \mathbb{C}^{n\times n}$ be a square matrix and let $\psi \in PERM_n$ be a permutation. Prove that if $D_\phi(A)$ is a permutation diagonal of A, the same set is a permutation diagonal $D_{\psi^{-1}\phi}(B)$ of $B = AP_\psi$.

Solution: Note that $(AP_\psi)_{ij} = \sum_{k=1}^n a_{ik}p_{kj} = a_{i\psi^{-1}(j)}$. Therefore, we have

$$\begin{aligned} D_\phi(AP_\psi) &= \{(AP_\psi)_{1\phi(1)}, \ldots, (AP_\psi)_{n\phi(n)}\} \\ &= \{a_{1\psi^{-1}(\phi(1))}, \ldots, a_{n\psi^{-1}(\phi(n))}\}, \end{aligned}$$

which allows us to conclude that $D_\phi(A)$ coincides with the permutation diagonal $D_{\psi^{-1}\phi}(B)$ of $B = AP_\psi$.

(11) Let $A \in \mathbb{C}^{n\times n}$ be an upper-Hessenberg matrix and let $U \in \mathbb{C}^{n\times n}$ be an upper-triangular matrix. Prove that both AU and UA are upper-Hessenberg matrices.

Solution: Let $B = AU$. For $i > j+1$, we have

$$b_{ij} = \sum_{k=1}^n a_{ik}u_{kj} = \sum_{k=1}^{i-2} a_{ik}u_{kj} + \sum_{k=i-1}^{n} a_{ik}u_{kj}.$$

Since A is upper-Hessenberg, we have $\sum_{k=1}^{i-2} a_{ik}u_{kj} = 0$; since U is upper triangular, $\sum_{k=i-1}^n a_{ik}u_{kj} = 0$ because $j < i-1$. Thus, $b_{ij} = 0$, so B is indeed upper-Hessenberg. The argument for UA is similar.

(12) Let $A, B \in \mathbb{C}^{n\times n}$ be two Hermitian matrices. Prove that AB is a Hermitian matrix if and only if $AB = BA$.

Solution: Suppose that $AB = BA$. Then, $(AB)^{\mathsf{H}} = (BA)^{\mathsf{H}} = A^{\mathsf{H}}B^{\mathsf{H}} = AB$, hence AB is a Hermitian matrix. Conversely, if AB is Hermitian, we have $AB = (AB)^{\mathsf{H}} = B^{\mathsf{H}}A^{\mathsf{H}} = BA$.

(13) Let A, B be two matrices in $\mathbb{C}^{n\times n}$. Suppose that $B = C + D$, where C is a Hermitian matrix and D is a skew-Hermitian matrix. Prove that if A is Hermitian and $AB = BA$, then $AC = CA$ and $AD = DA$.

(14) Prove that if $A = \mathsf{diag}(A_1, \ldots, A_k)$ is a block-diagonal matrix, then A is Hermitian if and only if every block A_i is Hermitian.

(15) Let $A \in \mathbb{C}^{n\times n}$ be a matrix. Prove that each ϕ-diagonal of A contains a 0 if and only if A contains a zero submatrix $B \in \mathbb{C}^{p\times q}$ such that $p + q = n + 1$. This fact is known as the *Frobenius–König Theorem.*

Solution: Suppose that A contains the submatrix

$$B = A\begin{bmatrix} i_1, \ldots, i_p \\ j_1, \ldots, j_q \end{bmatrix} = O_{p,q},$$

where $p + q = n + 1$ and

$$D_\phi = \{a_{1\phi(1)}, a_{2\phi(2)}, \ldots, a_{n\phi(n)}\} = \{a_{\phi^{-1}(1)1}, \ldots, \ldots, a_{\phi^{-1}(n)n}\}$$

is a ϕ-diagonal without zero entries. Then none of the components $a_{\phi^{-1}(j_1)j_1}, \ldots, \ldots, a_{\phi^{-1}(j_q)j_q}$ are 0, so they must be located in the rows $\{1, \ldots, n\} - \{i_1, \ldots, i_p\}$. Therefore, $q + p \leqslant n$ and this contradicts the fact that $p + q = n + 1$.
Conversely, suppose that every diagonal of A contains a 0. We show by induction on $n \geqslant 1$ that A contains a zero submatrix. The base case, $n = 1$ is immediate. Suppose that the implication holds for matrices of size less than n. If $A = O_{n,n}$, the reverse implication is trivial. Therefore, we can assume that A contains a non-zero component. Without loss of generality we may assume that $a_{nn} \neq 0$.
Every diagonal of the submatrix

$$S = A\begin{bmatrix} 1, \cdots, n-1 \\ 1, \ldots, n-1 \end{bmatrix}$$

contains a 0 and, therefore, by inductive hypothesis, S contains a zero submatrix of format $r \times s$, where $r + s = n$. It is obvious that this submatrix of S is also a zero-submatrix of A. There exist two permutations $\sigma, \tau \in PERM_n$ such that

$$P_\sigma A P_\tau = \begin{matrix} C & O_{r,s} \\ D & E \end{matrix},$$

where $C \in \mathbb{C}^{r\times r}$ and $E \in \mathbb{C}^{s\times s}$. Let $\eta \in PERM_n$. An η-diagonal of A has the form

$$D_\eta(A) = \{a_{1\eta(1)}, \ldots, a_{n\eta(n)}\}.$$

Since P_σ permutes the rows of A and P_τ permutes the columns of A, the set D_η corresponds to the set

$$\{a_{\sigma(1)\tau(\eta(1))}, \ldots, a_{\sigma(n)\tau(\eta(n))}\} = \{a_{1\sigma^{-1}(\tau(\eta(1)))}, \ldots, a_{n\sigma^{-1}(\tau(\eta(n)))}\},$$

which is a diagonal set of $P_\sigma AP_\tau$. Thus, the collection of diagonal sets of A and $P_\sigma AP_\tau$ are the same. If all elements of the diagonal set of $P_\sigma AP_\tau$ that correspond to C are non-zero, then the remaining elements of a diagonal are located in E and must contain a 0. Therefore, if all elements of a diagonal of C are non-zero, then each a diagonal of E must contain a 0. Thus, either all diagonals of C contain a 0, or all diagonals of E contain a 0. In the first case, by inductive hypothesis, C contains a $k\times h$ zero submatrix with $k+h = r+1$. This implies that the first r rows of $P_\sigma AP_\tau$ contain a $k \times (h+s)$ zero submatrix and $k+h+s = r+1+s = n+1$. Similarly, in the second case, E contains a $p\times q$ zero-matrix with $p+q = s+1$. Therefore, the last s columns of $P_\sigma AP_\tau$ contain a zero submatrix of format $(r+p)\times q$ and $r+p+q = r+s+1 = n+1$.

(16) Prove that row exchanges for matrices in $\mathbb{R}^{m\times n}$ can be expressed as a sequence of row multiplications and additions of a row multiplied by a constant to another row by showing that

$$T^{(i)\leftrightarrow(j)} = T^{(i)-(j)}T^{-(j)}T^{(j)-(i)}T^{((i)+(j)}$$

for $1 \leqslant i, j \leqslant m$.

(17) Let $A \in \mathbb{C}^{m\times n}$ be a matrix. If $\boldsymbol{v} = \mathsf{vec}(A) \in \mathbb{C}^{mn}$, prove that

$$v_j = a_{j-m\lfloor\frac{j-1}{m}\rfloor,\lceil\frac{j}{m}\rceil}$$

for $1 \leqslant j \leqslant mn$.

(18) Prove that the matrices A, B given by

$$A = \begin{pmatrix} 1 & 0 \\ 0 & 1 \end{pmatrix} \text{ and } B = \begin{pmatrix} 1 & 1 \\ 0 & 1 \end{pmatrix}$$

are not similar.

(19) Let $\mathcal{F}$ be the set of functions $f : \mathbb{C} - \{-\frac{d}{z}\} \longrightarrow \mathbb{C}$ given by

$$f(z) = \frac{az+b}{cz+d}.$$

Denote by M_f the matrix

$$M_f = \begin{pmatrix} a & b \\ c & d \end{pmatrix}.$$

Prove that
(a) if $f, g \in \mathcal{F}$, then $M_{fg} = M_f M_g$;
(b) $f(f(z)) - (a+d)f(z) + ad - bc = 0$ for every $z \in C$.

(20) Recall that $J_{n,n} \in \mathbb{R}^{n \times n}$ is the complete $n \times n$ matrix, that is the matrix having all components equal to 1. Prove that for every number $m \in \mathbb{N}$ and $m \geqslant 1$, we have

$$J_{n,n}^m = n^{m-1} J_{n,n}.$$

(21) Let $A, B \in \mathbb{C}^{n \times n}$ be two matrices. Prove that if $AB = BA$, then we have the following equality known as *Newton's binomial*:

$$(A+B)^n = \sum_{k=0}^{n} \binom{n}{k} A^{n-k} B^k.$$

Give an example of matrices $A, B \in \mathbb{C}^{2 \times 2}$ such that $AB \neq BA$ for which the above formula does not hold.

(22) Let $A \in \mathbb{C}^{n \times n}$ be a circulant matrix whose first row is $(c_1, \ldots, c_n)$ and let $P \in \mathbb{C}^{n \times n}$ be the circulant matrix whose first row is $(0, 1, 0, \ldots, 0)$. Prove that $A = c_1 P^0 + c_1 P^1 + \cdots + c_n P^{n-1}$.

(23) Let $A \in \mathbb{R}^{n \times n}$ and $B \in \mathbb{R}^{n \times n}$ be two matrices. Prove that if AB is invertible, then both A and B are invertible.

Solution: Since AB is invertible, we have $AB(AB)^{-1} = I_n$. Thus, A is invertible and $A^{-1} = B(AB)^{-1}$. Similarly, since $(AB)^{-1}AB = I_n$, it follows that B is invertible and $B^{-1} = (AB)^{-1}A$.

(24) Let $c \in (0, 1]$ and let $A \in \mathbb{R}^{n \times n}$ be the matrix $A = cI_n + (1 - c)J_n$. Prove that A is a non-singular matrix.

Solution: Suppose that $\boldsymbol{x} \in \mathbb{R}^n$ is a vector such that $A\boldsymbol{x} = \boldsymbol{0}_n$. This amounts to

$$c\boldsymbol{x} + (1-c)J_n\boldsymbol{x} = c\boldsymbol{x} + (1-c)\xi\boldsymbol{1}_n,$$

where $\xi = \sum_{i=1}^n x_i$. Thus, $\boldsymbol{x} = -\frac{(1-c)\xi}{c}\boldsymbol{1}_n$, which means that all components of $\boldsymbol{x}$ must equal $-\frac{(1-c)\xi}{c}$. This implies $\xi = -n\frac{(1-c)\xi}{c}$, so $\xi = 0$ which implies $\boldsymbol{x} = \boldsymbol{0}_n$. Thus, A is non-singular.

(25) Let $A, B \in \mathbb{C}^{n\times n}$ be two invertible matrices. Prove that

$$B^{-1} - A^{-1} = B^{-1}(A-B)A^{-1}.$$

Conclude that $\mathsf{rank}(A-B) = \mathsf{rank}(B^{-1} - A^{-1})$.

(26) Let $A \in \mathbb{R}^{n\times n}$ be a non-singular matrix. Prove that the set $\{\boldsymbol{x}_1, \ldots, \boldsymbol{x}_m\} \subseteq \mathbb{R}^n$ is a linearly independent set if and only if $\{A\boldsymbol{x}_1, \ldots, A\boldsymbol{x}_m\}$ is a linearly independent set.

(27) Prove that the matrix $T^{(i)+a(j)} \in \mathbb{R}^{n\times n}$ can be expressed as

$$T^{(i)+a(j)} = I_n + a\boldsymbol{e}_i\boldsymbol{e}_j'.$$

(28) Let

$$U = \begin{pmatrix} O_{k,k} & \cdots \\ O_{n-k,k} & W \end{pmatrix}$$

be a matrix in $\mathbb{C}^{n\times n}$, where W is an upper triangular matrix, and let Y be an upper triangular matrix such that $y_{k+1,k+1} = 0$. Prove that $(UY)_{ij} = O$ for $1 \leqslant i, j \leqslant k+1$.

Solution: Note that Y can be written as

$$Y = \begin{pmatrix} Y_{11} & Y_{12} \\ O_{n-k,k} & Y_{22} \end{pmatrix},$$

where Y_{11} and Y_{22} are upper triangular matrices. Thus,

$$UY = \begin{pmatrix} O_{k,k} & BY_{22} \\ O_{n-k,k} & WY_{22} \end{pmatrix}.$$

Thus, the first k columns of the product UY consist only of Os.

The $k+1$st column of UY is

$$U\begin{pmatrix} y_{1\,k+1} \\ \vdots \\ y_{k\,k+1} \\ 0 \\ 0 \\ \vdots \\ 0 \end{pmatrix}$$

because Y is an upper triangular, matrix. If $p < k$, then $(UY)_{p\,k+1} = 0$ because the first k columns of U contain zeros.

(29) Let m, n be two positive natural numbers such that $m = pq$ and $n = rs$, where $p, q, r, s \in \mathbb{N}$. Prove that there exists a bijection between the set $\mathbb{F}^{m\times n}$ and the set $(\mathbb{F}^{p\times r})^{q\times s}$. Interpret the existence of this bijection in terms of matrices.

(30) Let $A \in \mathbb{C}^{n\times n}$ be a strictly upper triangular matrix. Prove that A is nilpotent.

Solution: Since A is strictly upper triangular, we have $a_{ij} = 0$ for $1 \leqslant j \leqslant i$ and every i, $1 \leqslant i \leqslant n$. We prove by induction on m that for $A^m = (a_{ij}^{(m)})$, we have $a_{ij}^{(m)} = 0$ for $1 \leqslant j \leqslant i+m-1$ and every i, $1 \leqslant i \leqslant n$.
The base case $m = 1$ is immediate. Suppose that the statement holds for m. We show that $a_{ij}^{(m+1)} = 0$ for $1 \leqslant j \leqslant i+m$. We have

$$\begin{aligned} a_{ij}^{(m+1)} &= \sum_{k=1}^{n} a_{ik}^{(m)} a_{kj} \\ &= \sum_{k=i+m}^{n} a_{ik}^{(m)} a_{kj} = 0 \end{aligned}$$

because $a_{kj} = 0$ when $j \leqslant i+m \leqslant k$. Thus, $A^n = 0$ and A is nilpotent.

(31) Let $A \in \mathbb{C}^{n\times n}$ be a nilpotent matrix. Prove that $\mathsf{rank}(A) \leqslant n\frac{\mathsf{nilp}(A)-1}{\mathsf{nilp}(A)}$.

A *Hadamard matrix* is a matrix $H \in \mathbb{R}^{n\times n}$ such that $h_{ij} \in \{-1, 1\}$ for $1 \leqslant i, j \leqslant n$ and $HH' = nI_n$.

(32) Verify that the matrices

$$\begin{pmatrix} 1 & 1 \\ 1 & -1 \end{pmatrix}$$

and

$$\begin{pmatrix} 1 & 1 & 1 & 1 \\ 1 & 1 & -1 & -1 \\ 1 & -1 & -1 & 1 \\ 1 & -1 & 1 & -1 \end{pmatrix}$$

are Hadamard matrices.

(33) Let $H = (\boldsymbol{h}_1 \ \cdots \ \boldsymbol{h}_j \ \cdots \ \boldsymbol{h}_n) \in \mathbb{R}^{n\times n}$ be a Hadamard matrix. Prove that the matrix $(\boldsymbol{h}_1 \ \cdots \ -\boldsymbol{h}_j \ \cdots \ \boldsymbol{h}_n)$ is also a Hadamard matrix.

(34) Let $A = (a_{ij})$ be an $(m \times n)$-matrix of real numbers. Prove that

$$\max_j \min_i a_{ij} \leqslant \min_i \max_j a_{ij}$$

(the *minimax inequality*).

Solution: Observe that $a_{ij_0} \leqslant \max_j a_{ij}$ for every i and j_0, so $\min_i a_{ij_0} \leqslant \min_i \max_j a_{ij}$, again for every j_0. Thus, $\max_j \min_i a_{ij} \leqslant \min_i \max_j a_{ij}$.

(35) Let $A \in \mathbb{R}^{n\times n}$ and let $\Phi_A : \mathbb{R}^{n\times n} \longrightarrow \mathbb{R}^{n\times}$ be the function defined by $\Phi_A(X) = AX - XA$, for $X \in \mathbb{R}^{n\times n}$. Prove that:

(a) Φ_A is linear, that is, for any $a, b \in \mathbb{R}$ and $X, Y \in \mathbb{R}^{n\times n}$, we have $\Phi_A(aX + bY) = a\Phi_A(X) + b\Phi_A(Y)$;

(b) $\Phi_A(XY) = \Phi_A(X)Y + X\Phi_A(Y)$;

(c) if X is an invertible matrix, then $\Phi_A(X) = -X\Phi_A(X^{-1})X$;

(d) $\Phi_A(\Phi_B(X)) - \Phi_B(\Phi_A(X)) = \Phi_{AB}(X) - \Phi_{BA}(X)$;

(e) $\mathsf{trace}(\Phi_A(X)) = 0$

for every $X \in \mathbb{R}^{n\times n}$.

(36) Let A be a matrix in $\mathbb{C}^{m\times n}$ such that $\mathsf{rank}(A) = r$. Prove that A can be factored as $A = PC$, where $P \in \mathbb{C}^{m\times r}$, $C \in \mathbb{C}^{r\times n}$, and $\mathsf{rank}(P) = r$ and as $A = DQ$, where $D \in \mathbb{C}^{m\times r}$, $Q \in \mathbb{C}^{r\times n}$ and $\mathsf{rank}(Q) = r$.

Solution: Let $\boldsymbol{p}_1, \ldots, \boldsymbol{p}_r$ be a basis for $\mathsf{range}(A)$, where $A = (\boldsymbol{a}_1 \ \cdots \ \boldsymbol{a}_n)$. Every column $\boldsymbol{a}_i$ of A can be written as a linear combination $\boldsymbol{a}_i = c_{i1}\boldsymbol{p}_1 + \cdots + c_{ir}\boldsymbol{p}_r$ for $1 \leqslant i \leqslant n$. In matrix form these equalities amount to $A = PC$. The argument for the second part is similar and involves the rows of A.

(37) Let $\{A_i \in \mathbb{R}^{m\times n_i} \mid 1 \leqslant i \leqslant k\}$ be a set of matrices and let $\ell = \sum_{i=1}^{k} n_i$. Prove that the subspaces $\{\mathsf{range}(A_i) \mid 1 \leqslant i \leqslant k\}$ are linearly independent if and only if

$$\mathsf{rank}(A_1|A_2|\cdots|A_k) = \sum_{i=1}^{k} \mathsf{rank}(A_i).$$

Hint: Note that $\mathsf{rank}(A_i) = \dim(\mathsf{range}(A_i))$ and $\mathsf{rank}(A_1|A_2|\cdots|A_k) = \dim(\mathsf{range}(A_1) + \cdots + \mathsf{range}(A_k))$.

(38) Let $\pi = \{B_1, \ldots, B_k\}$ be a partition of a set $S = \{s_1, \ldots, s_n\}$. Define the characteristic matrix of π, $B \in \mathbb{R}^{n\times k}$ as

$$b_{ij} = \begin{cases} 1 & \text{if } s_i \in B_j \\ 0 & \text{otherwise} \end{cases}$$

for $1 \leqslant i \leqslant n$ and $1 \leqslant j \leqslant k$. Prove that $B'B$ is a diagonal matrix, $(B'B)_{jj} = |B_j|$ for $1 \leqslant j \leqslant k$ and that $B'B$ is invertible.

(39) Let A and B be two matrices in $\mathbb{C}^{p\times q}$. Prove that $\mathsf{rank}(A+B) \leqslant \mathsf{rank}(A\ B) \leqslant \mathsf{rank}(A) + \mathsf{rank}(B)$.

Solution: We have $\mathsf{rank}(A\ B) = \mathsf{rank}(A\ A+B) \geqslant \mathsf{rank}(A+B)$ because adding the first q columns of the matrix $(A\ B)$ to the last q columns does not change the rank of a matrix, and the rank of a submatrix is not larger than the rank of the matrix. On the other hand, $\mathsf{rank}(A\ B) = \mathsf{rank}((A\ O_{p,q}) + (O_{p,q}\ B)) \leqslant \mathsf{rank}(A\ O_{p,q}) + \mathsf{rank}(O_{p,q}\ B) = \mathsf{rank}(A) + \mathsf{rank}(B)$.

(40) If $A \in \mathbb{C}^{p\times q}$ and $B \in \mathbb{C}^{q\times r}$, then prove that $\mathsf{rank}(AB) \leqslant \min\{\mathsf{rank}(A), \mathsf{rank}(B)\}$.

(41) Let $A \in \mathbb{R}^{n\times k}$ be a matrix, where $n \geqslant k$. If $A'A = I_k$, prove that $\mathsf{rank}(AA') = \mathsf{rank}(A') = \mathsf{rank}(A) \geqslant k$.

Solution: By Sylvester's rank theorem, we have

$$k = \mathsf{rank}(A'A) = \mathsf{rank}(A) - \dim(\mathsf{null}(A') \cap \mathsf{range}(A)),$$
$$\mathsf{rank}(AA') = \mathsf{rank}(A') - \dim(\mathsf{null}(A) \cap \mathsf{range}(A')).$$

The first equality implies $\mathsf{rank}(A) \geqslant k$.
Let $\boldsymbol{t} \in \mathsf{null}(A) \cap \mathsf{range}(A')$. We have $A\boldsymbol{t} = \mathbf{0}$ and $\boldsymbol{t} = A'\boldsymbol{z}$ for some $\boldsymbol{z} \in \mathbb{R}^n$. Therefore, $\boldsymbol{t}'\boldsymbol{t} = \boldsymbol{t}'A'\boldsymbol{z} = (A\boldsymbol{t})'\boldsymbol{z} = 0$, so $\boldsymbol{t} = \mathbf{0}_k$. Thus, $\dim(\mathsf{null}(A)\cap\mathsf{range}(A')) = 0$, so $\mathsf{rank}(AA') = \mathsf{rank}(A') = \mathsf{rank}(A) \geqslant k$.

(42) Let $A \in \mathbb{R}^{3\times 3}$ be a matrix such that $A^2 = O_{3,3}$. Prove that $\mathsf{rank}(A) \leqslant 1$.

(43) Let $Z \in \mathbb{R}^{n\times n}$, $W \in \mathbb{R}^{m\times m}$ and let $U, V \in \mathbb{R}^{n\times m}$ be four matrices such that each of the matrices W, Z, $Z + UWV'$, and $W^{-1} + UWV'$ has an inverse.
Prove that

$$(Z + UWV')^{-1} = Z^{-1} - Z^{-1}U(W^{-1} + V'Z^{-1}U)^{-1}V'Z^{-1}$$

(the *Woodbury–Sherman–Morrison identity*).

Solution: Consider the following system of matrix equations:

$$ZX + UY = I_n,$$
$$V'X - W^{-1}Y = O_{m,n},$$

where $X \in \mathbb{R}^{n\times n}$ and $Y \in \mathbb{R}^{m\times n}$.
The second equation implies $V'X = W^{-1}Y$, so $UWV'X = UY$. Substituting UY in the first equation yields $ZX + UWV'X = I_n$. Therefore, we have $(Z+UWV')X = I_n$, which implies

$$X = (Z + UWV')^{-1}. \tag{3.15}$$

On the other hand, we have $X = Z^{-1}(I_n - UY)$ from the first equation. Substituting X in the second equation yields

$$V'Z^{-1}(I_n - UY) = W^{-1}Y,$$

which is equivalent to

$$\begin{aligned}V'Z^{-1} &= +W^{-1}Y + V'Z^{-1}UY \\ &= (W^{-1} + V'Z^{-1}U)Y.\end{aligned}$$

Thus, we have

$$Y = (W^{-1} + V'Z^{-1}U)^{-1}V'Z^{-1}.$$

Substituting the values of X and Y in the first equality implies

$$ZX + U(W^{-1} + V'Z^{-1}U)^{-1}V'Z^{-1} = I_n.$$

Therefore,

$$ZX = I_n - U(W^{-1} + V'Z^{-1}U)^{-1}V'Z^{-1}, \tag{3.16}$$

which implies

$$X = Z^{-1} - Z^{-1}U(W^{-1} + V'Z^{-1}U)^{-1}V'Z^{-1}. \tag{3.17}$$

The Woodbury–Sherman–Morrison identity follows immediately from Equalities (3.15) and (3.17).

(44) None of the following matrices A, B, C are invertible:

$$A = \begin{pmatrix} 0 & 0 \\ 0 & 0 \end{pmatrix}, \quad B = \begin{pmatrix} 1 & 0 \\ 1 & 1 \end{pmatrix}, \quad C = \begin{pmatrix} 1 & 1 \\ 1 & 1 \end{pmatrix}.$$

However, their pseudoinverses exist and are given by

$$A^\dagger = \begin{pmatrix} 0 & 0 \\ 0 & 0 \end{pmatrix}, \quad B^\dagger = \begin{pmatrix} \frac{1}{2} & \frac{1}{2} \\ 0 & 0 \end{pmatrix}, \quad C^\dagger = \begin{pmatrix} \frac{1}{4} & \frac{1}{4} \\ \frac{1}{4} & \frac{1}{4} \end{pmatrix},$$

respectively.

(45) Let $A \in \mathbb{R}^{m \times n}$ be a matrix whose columns form an orthonormal set (that is, $A'A = I_n$). Prove that the Moore–Penrose pseudoinverse of A is $A^\dagger = A'$.

Solution: Let $M = A'$. Note that $MA = A'A = I_n$ and $AM = AA'$. Clearly, MA is a symmetric matrix and so is AM. Furthermore, we have $AMA = A(A'A) = A$ and $MAM = A'AA' = A' = M$, so $A^\dagger = A'$.

(46) Let $A \in \mathbb{R}^{n\times n}$ be an invertible matrix, and let $\boldsymbol{c}, \boldsymbol{d} \in \mathbb{R}^n$ be two vectors such that $\boldsymbol{d}'A^{-1}\boldsymbol{c} \neq -1$. Prove that

$$(A + \boldsymbol{c}\boldsymbol{d}')^{-1} = A^{-1} - \frac{1}{1 + \boldsymbol{d}'A^{-1}\boldsymbol{c}} A^{-1}\boldsymbol{c}\boldsymbol{d}'A^{-1}.$$

Hint: Apply the Woodbury–Sherman–Morrison identity.

(47) Let $M \in \mathbb{R}^{n\times n}$ be a partitioned matrix,

$$M = \begin{pmatrix} A & B \\ C & D \end{pmatrix},$$

where $A \in \mathbb{R}^{m\times m}$ and $m < n$. Prove that, if all inverse matrices mentioned in what follows exist and $Q = (A - BD^{-1}C)^{-1}$, then

$$M^{-1} = \begin{pmatrix} Q & -QBD^{-1} \\ -D^{-1}CQ & D^{-1} + D^{-1}CQBD^{-1} \end{pmatrix}.$$

Hint: Multiply M by M^{-1} given above.

(48) Let $\boldsymbol{c}_1, \ldots, \boldsymbol{c}_p$ be p vectors in $\mathbb{R}^n$ and $k_1, \ldots, k_p$ be p numbers. Find sufficient conditions for the existence of a symmetric matrix X, where $X \in \mathbb{R}^{n\times n}$ such that

$$\boldsymbol{c}_i' X \boldsymbol{c}_i = k_i$$

for $1 \leqslant i \leqslant p$.

(49) Let $A \in \mathbb{C}^{n\times n}$ be a Hermitian matrix such that $\mathsf{rank}(A) = r$. Prove that there is a permutation matrix P and a principal submatrix B of A such that $B \in \mathbb{C}^{r\times r}$

$$P^{\mathsf{H}}AP = \begin{pmatrix} I_r \\ F \end{pmatrix} B (I_r \; F^{\mathsf{H}})$$

and $F \in \mathbb{C}^{(n-r)\times r}$.

Solution: Since $\mathsf{rank}(A) = r$, there is a sequence of r linearly independent columns; since A is Hermitian, the rows that have the same numbers are also linearly independent. The elements at the intersection of those rows yield an $r \times r$ submatrix B of

A having rank R. There exists a permutation matrix P such that

$$P^{\mathsf{H}}AP = \begin{pmatrix} B & C^{\mathsf{H}} \\ C & D \end{pmatrix}.$$

Since $P^{\mathsf{H}}AP$ has rank r and so does $(B \;\; C^{\mathsf{H}})$, it follows that $(B \;\; C^{\mathsf{H}})$ generates the rows of $P^{\mathsf{H}}AP$. Consequently, $C = FB$ for some matrix F, so $D = FC^{\mathsf{H}}$, which implies

$$P^{\mathsf{H}}AP = \begin{pmatrix} B & B^{\mathsf{H}}F^{\mathsf{H}} \\ FB & FB^{\mathsf{H}}F^{\mathsf{H}} \end{pmatrix} = \begin{pmatrix} I_r \\ F \end{pmatrix} B(I_r \; F^{\mathsf{H}}).$$

(50) Let $A \in \mathbb{C}^{n\times n}$ be a matrix such that $\mathsf{rank}(A) = r$. Prove that A has a submatrix $B \in \mathbb{C}^{r\times r}$ such that $\mathsf{rank}(B) = r$.

If we choose $A = -I$ in the Equality of Exercise 3.17, we have $(I - \boldsymbol{c}\boldsymbol{d}')^{-1} = I + \frac{1}{1-\boldsymbol{d}'\boldsymbol{c}}\boldsymbol{c}\boldsymbol{d}'$, for $\boldsymbol{c}, \boldsymbol{d} \in \mathbb{R}^n$, such that $\boldsymbol{d}'\boldsymbol{c} \neq 1$. Matrices of the form $I - \boldsymbol{c}\boldsymbol{d}'$, with $\boldsymbol{d}'\boldsymbol{c} \neq 1$, are known as *elementary matrices.*

(51) Prove that
   (a) $T^{(i)\leftrightarrow(j)}$ is an elementary matrix obtained by taking $\boldsymbol{c} = \boldsymbol{d} = \boldsymbol{e}_i - \boldsymbol{e}_j$.
   (b) $T^{a(i)}$, where $a \neq 0$, is an elementary matrix obtained by taking $\boldsymbol{c} = (1-a)\boldsymbol{e}_i$ and $\boldsymbol{d} = \boldsymbol{e}_i$.
   (c) $T^{(i)+a(j)}$ is an elementary matrix obtained by taking $\boldsymbol{c} = -a\boldsymbol{e}_i$ and $\boldsymbol{d} = \boldsymbol{e}_j$.

(52) Prove that if $S, T \in \mathbb{C}^{2\times 2}$ are two Toeplitz matrices, then $TS = ST$. Does this property hold for matrices $T, S \in \mathbb{C}^{n\times n}$, where $n \geqslant 3$?

(53) Prove that every permutation matrix is a unitary matrix.

(54) Prove that if $S \in \mathbb{C}^{n\times n}$ is a skew-Hermitian matrix, then $I_n + S$ is an invertible matrix and the matrix $(I_n - S)(I_n + S)^{-1}$ is a unitary matrix.

(55) Prove that if $A \in \mathbb{C}^{n\times n}$ is a Hermitian matrix, then $z = \boldsymbol{x}^{\mathsf{H}}A\boldsymbol{x}$ is a real number for every vector $\boldsymbol{x} \in \mathbb{C}^n$.

(56) Let $A \in \mathbb{C}^{n\times n}$ be a Hermitian matrix and let $B \in \mathbb{C}^{n\times n}$ be a skew-Hermitian matrix. Prove that iA is skew-Hermitian, and iB is Hermitian.

(57) Prove that if $A \in \mathbb{C}^{n\times n}$, then $\mathsf{trace}(AA^{\mathsf{H}}) = 0$ if and only if $A = O_{n,n}$.

Solution: By Equality (6.12), we have $\|A\|_F^2 = \mathsf{trace}(AA^{\mathsf{H}})$. Therefore, if $\mathsf{trace}(AA^{\mathsf{H}}) = 0$, it follows that $\|A\|_F^2 = 0$, so $A = O_{n,n}$. The reverse implication is immediate.

(58) Let $A \in \mathbb{C}^{n\times n}$. Prove that the following conditions are equivalent:

(a) A is Hermitian;
(b) $\boldsymbol{x}^{\mathsf{H}} A\boldsymbol{x} \in \mathbb{R}$ for any $\boldsymbol{x} \in \mathbb{C}^n$;
(c) $A^2 = A^{\mathsf{H}}A$.

(59) Let $R^{(p,q)} \in \mathbb{R}^{n\times n}$ be the matrix defined by $R = (\mathbf{0} \cdots \mathbf{0}\ \boldsymbol{e}_p\ \mathbf{0} \cdots \mathbf{0})$, where $\boldsymbol{e}_p$ occurs in the qth position.

(a) Prove that if $p \neq q$, we have $(I + aR^{(p,q)})^{-1} = I - aR^{(p,q)}$ for every $a \in \mathbb{R}$.
(b) Prove that if $T \in \mathbb{R}^{n\times n}$ and $p < q$, then

$$R^{(p,q)}T = \begin{pmatrix} 0 & \cdots & 0 & \cdots & 0 & \cdots & 0 \\ \vdots & \cdots & \vdots & \cdots & \vdots & \cdots & \vdots \\ 0 & \cdots & 0 & \cdots & t_{qq} & \cdots & t_{qn} \\ \vdots & \cdots & \vdots & \cdots & \vdots & \cdots & \vdots \\ 0 & \cdots & 0 & \cdots & 0 & \cdots & 0 \end{pmatrix},$$

where the elements $t_{qq}, \ldots, t_{qn}$ occur in line p, and

$$TR^{(p,q)} = \begin{pmatrix} 0 & \cdots & t_{1p} & \cdots & 0 \\ \vdots & \cdots & \vdots & \cdots & \vdots \\ 0 & \cdots & t_{pp} & \cdots & 0 \\ 0 & \cdots & 0 & \cdots & 0 \\ \vdots & \cdots & \vdots & \cdots & \vdots \\ 0 & \cdots & t_{1p} & \cdots & 0 \end{pmatrix},$$

where $t_{1p}, \ldots, t_{pp}$ occur in the qth column.

(c) Prove that if $T \in \mathbb{R}^{n\times n}$ and $p < q$, then $R^{(p,q)}TR^{(p,q)} = O$.
(d) Prove that if $T \in \mathbb{R}^{n\times n}$ is an upper triangular matrix and $p < q$, then $((I + aR^{p,q})^{-1}T(I + aR^{p,q}))_{ij} = t_{ij}$ unless $i = p$ and $j \geqslant q$ or $j = q$ and $i \leqslant p$; moreover, prove that $((I + aR^{p,q})^{-1}T(I + aR^{p,q}))_{pq} = t_{pq} + a(t_{pp} - t_{qq})$.

Solution: We discuss only the last part of this supplement. Observe that

$$\begin{aligned}(I + aR^{(p,q)})^{-1}T(I + aR^{(p,q)}) &= (I - aR^{(p,q)})T(I + aR^{(p,q)}) \\ &= T - aR^{(p,q)}T + aTR^{(p,q)}.\end{aligned}$$

Taking into account the first three parts, the claims made in the last part are immediately justified.

(60) Let $Z^{(k)} \in \mathbb{C}^{n\times n}$ be the matrix defined by

$$(Z^{(k)})_{ij} = \begin{cases} 1 & \text{if } i = j \text{ and } i \neq k \\ 0 & \text{otherwise,} \end{cases}$$

for $1 \leqslant i, j \leqslant n$. If $A \in \mathbb{C}^{n\times n}$, prove that $AZ^{(k)}$ is obtained from A by replacing the kth column by zeros; similarly, $Z^{(k)}A$ is obtained from A by replacing the kth row by zeros.

(61) Let K be a subset of $\{1, \ldots, n\}$ and let $k = |K|$.
Define the matrix $C^{(n,K)}$ as the $n \times k$ matrix obtained from I_n by eliminating all columns whose numbers do not occur in K; the matrix $R^{(n,K)}$ is the $k \times n$ matrix obtained from I_n by eliminating all rows whose numbers do not occur in K.
   (a) If $A \in \mathbb{C}^{n\times n}$, prove that $AC^{(n,K)}$ is obtained from A by eliminating all columns whose numbers do not occur in K; similarly, $R^{(n,K)}A$ is obtained from A by erasing all rows whose numbers do not occur in K.
   (b) Let $B = A\begin{bmatrix} i_1 & \cdots & i_k \\ j_1 & \cdots & j_h \end{bmatrix}$ be a submatrix of $A \in \mathbb{C}^{m\times n}$. Prove that

$$B = R^{(m,\{i_1,\ldots,i_k\}}AC^{(n,\{j_1,\ldots,j_h\})}.$$

(62) Prove that $\mathsf{rank}(A^{\mathsf{H}}A) = \mathsf{rank}(AA^{\mathsf{H}}) = \mathsf{rank}(A)$ for any matrix $A \in \mathbb{C}^{m\times n}$ using Sylvester's Theorem.

(63) Prove that if $A \in \mathbb{C}^{m\times n}$, then $A^{\mathsf{H}}A$ is singular if and only if $m < n$.

(64) Prove or disprove:
   (a) If $\mathbf{0} \in W$, where $W = \{\boldsymbol{w}_1, \ldots, \boldsymbol{w}_n\}$, then W is linearly independent.
   (b) If $W = \{\boldsymbol{w}_1, \ldots, \boldsymbol{w}_n\}$ is linearly independent and $\boldsymbol{w}$ is not a linear combination of the vectors of W, then $W \cup \{\boldsymbol{w}\}$ is linearly independent.

(c) If $W = \{\boldsymbol{w}_1, \ldots, \boldsymbol{w}_n\}$ is linearly dependent, then any of $\boldsymbol{w}_i$ is a linear combination of the others.
(d) If $\boldsymbol{y}$ is not a linear combination of $\{\boldsymbol{w}_1, \ldots, \boldsymbol{w}_n\}$, then $\{\boldsymbol{y}, \boldsymbol{w}_1, \ldots, \boldsymbol{w}_n\}$ is linearly independent.
(e) If any $n-1$ vectors of the set $W = \{\boldsymbol{w}_1, \ldots, \boldsymbol{w}_n\}$ are linearly dependent, then W is linearly independent.

(65) Let $A \in \mathbb{C}^{n\times n}$ be a matrix such that $I + A$ is invertible. The *Cayley transform* of A is the matrix $\mathsf{C}(A) = (I - A)(I + A)^{-1}$.
(a) Prove that if $\mathsf{C}(A)$ exists, then $\mathsf{C}(A)$ is a unitary matrix if and only if A is skew-Hermitian matrix.
(b) Prove that $\mathsf{C}(\mathsf{C}(A)) = A$ if all transforms exist.
(c) Prove that

$$\mathsf{C}\begin{pmatrix} 0 & \tan\alpha \\ -\tan\alpha & 0 \end{pmatrix} = \begin{pmatrix} \cos 2\alpha & -\sin 2\alpha \\ \sin 2\alpha & \cos 2\alpha \end{pmatrix}.$$

(66) Prove that if $A \in \mathbb{C}^{n\times n}$ is an idempotent matrix, then $\mathsf{range}(A) \cap \mathsf{null}(A) = \{\mathbf{0}\}$.

(67) Let $A \in \mathbb{C}^{n\times n}$ be an idempotent matrix. Prove that $I - A$ is an idempotent matrix and that $\mathsf{rank}(A) + \mathsf{rank}(I - A) = n$.

Solution: We leave to the reader the proof of the idempotency of $I - A$. To prove the desired equality, start from Equality (3.8), that is, from $\mathsf{rank}(A) + \dim(\mathsf{null}(A)) = n$. Thus, it suffices to show that $\mathsf{rank}(I - A) = \dim(\mathsf{null}(A))$. In turn, it is sufficient to show that $\mathsf{range}(I - A) = \mathsf{null}(A)$.
Let $\boldsymbol{u} \in \mathsf{range}(I - A)$. We have $\boldsymbol{u} = (I - A)\boldsymbol{x}$ for some $\boldsymbol{x}$, which implies $A\boldsymbol{u} = (A - A^2)\boldsymbol{x} = \mathbf{0}$. Thus, $\boldsymbol{u} \in \mathsf{null}(A)$.
Conversely, if $\boldsymbol{u} \in \mathsf{null}(A)$, then $A\boldsymbol{u} = \mathbf{0}$, and $\boldsymbol{u} = I\boldsymbol{u} - A\boldsymbol{u} = (I - A)\boldsymbol{u}$, so $\boldsymbol{u} \in \mathsf{range}(I - A)$.

(68) Let $\boldsymbol{c} \in \mathbb{C}^n$ be a vector such that $\mathbf{1}_n'\boldsymbol{c} = 1$ and let $Q_{\boldsymbol{c}} \in \mathbb{C}^{n\times n}$ be the matrix defined by $Q_{\boldsymbol{c}} = I_n - \mathbf{1}_n\boldsymbol{c}'$. Prove that $Q_{\boldsymbol{c}}\mathbf{1}_n = \mathbf{0}_n$ and $Q_{\boldsymbol{c}}$ is idempotent.

Solution: It is clear that $\boldsymbol{c}'\mathbf{1}_n = 1$. Therefore,

$$Q_{\boldsymbol{c}}^2 = (I_n - \mathbf{1}_n\boldsymbol{c}')(I_n - \mathbf{1}_n\boldsymbol{c}') = I_n - 2\mathbf{1}_n\boldsymbol{c}' + \mathbf{1}_n\boldsymbol{c}'\mathbf{1}_n\boldsymbol{c}' = I_n - \mathbf{1}_n\boldsymbol{c}' = Q_{\boldsymbol{c}}.$$

(69) Let $\boldsymbol{c}$ be a vector in $\mathbb{C}^n$. Prove that if $\mathbf{1}_n'\boldsymbol{c} = 1$, then $\mathbf{1}_n\boldsymbol{c}'$ is an idempotent matrix.

An *involutive matrix* is a matrix $A \in \mathbb{C}^{n\times n}$ such that $A^2 = I_n$.

(70) Prove that if $B \in \mathbb{C}^{n\times n}$ is an idempotent matrix, then $A = 2B - I_n$ is an involutive matrix.

(71) Let $A \in \mathbb{C}^{n\times n}$ be an involutive matrix and let $S = \{\boldsymbol{x} \in \mathbb{C}^n \mid A\boldsymbol{x} = \boldsymbol{x}\}$ and $T = \{\boldsymbol{x} \in \mathbb{C}^n \mid A\boldsymbol{x} = -\boldsymbol{x}\}$. Prove that both S and T are subspaces of $\mathbb{C}^n$ and $\mathbb{C}^n = S \boxplus T$.

(72) Let $A \in \mathbb{R}^{n\times n}$ be the matrix obtained from I_n by swapping the ith column $\boldsymbol{e}_i$ with the jth column $\boldsymbol{e}_j$. Prove that A is an involutive matrix and XA is obtained from X by swapping the ith column $\boldsymbol{x}_i$ with the jth column $\boldsymbol{x}_j$.

(73) Let $A \in \mathbb{C}^{n\times m}$ be a matrix, and let $\boldsymbol{c} \in \mathbb{C}^n$ and $\boldsymbol{d} \in \mathbb{C}^m$. Prove that

$$\begin{aligned} \mathsf{rank}(A) - 1 \leq \mathsf{rank}(A) - \mathsf{rank}(\boldsymbol{c}\boldsymbol{d}^{\mathsf{H}}) &\leqslant \mathsf{rank}(A - \boldsymbol{c}\boldsymbol{d}^{\mathsf{H}}) \\ &\leq \mathsf{rank}(A) + \mathsf{rank}(\boldsymbol{c}\boldsymbol{d}^{\mathsf{H}}) \leqslant \mathsf{rank}(A) + 1. \end{aligned}$$

(74) Let $A \in \mathbb{C}^{m\times n}$ be a matrix, $\boldsymbol{u} \in \mathbb{C}^m$ and $\boldsymbol{v} \in \mathbb{C}^n$ be two vectors, and let $a \in \mathbb{C}$. Prove that if

$$\mathsf{rank}(A - a\boldsymbol{u}\boldsymbol{v}^{\mathsf{H}}) < \mathsf{rank}(A),$$

then $\boldsymbol{u} \in \mathsf{range}(A)$ and $\boldsymbol{v} \in \mathsf{range}(A^{\mathsf{H}})$.

Solution: Observe that we can write

$$a\boldsymbol{u}\boldsymbol{v}^{\mathsf{H}} = (av_1\boldsymbol{u} \ \cdots \ av_n\boldsymbol{u}) = \begin{pmatrix} au_1\boldsymbol{v}^{\mathsf{H}} \\ \vdots \\ au_m\boldsymbol{v}^{\mathsf{H}} \end{pmatrix}.$$

Thus, if $\boldsymbol{w}_1, \ldots, \boldsymbol{w}_m$ are the rows of the matrix A, we have

$$A - a\boldsymbol{u}\boldsymbol{v}^{\mathsf{H}} = \begin{pmatrix} \boldsymbol{w}_1 - au_1\boldsymbol{v}^{\mathsf{H}} \\ \vdots \\ \boldsymbol{w}_m - au_m\boldsymbol{v}^{\mathsf{H}} \end{pmatrix}.$$

Suppose that $\mathsf{rank}(A) = r$ and let $\boldsymbol{w}_{i_1}, \ldots, \boldsymbol{w}_{i_r}$ be a maximal set of linearly independent rows. Since $\mathsf{rank}(A - a\boldsymbol{u}\boldsymbol{v}^{\mathsf{H}}) < r$, there is a linear combination

$$\alpha_1(\boldsymbol{w}_{i_1} - a u_{i_1} \boldsymbol{v}^{\mathsf{H}}) + \cdots + \alpha_r(\boldsymbol{w}_{i_r} - a u_{i_r} \boldsymbol{v}^{\mathsf{H}}) = \boldsymbol{0},$$

such that not all coefficients α_i are 0. The previous equality can be written as

$$\alpha_1 \boldsymbol{w}_{i_1} + \cdots + \alpha_r \boldsymbol{w}_{i_r} = (a\alpha_1 u_{i_1} + \cdots + a\alpha_r u_{i_r})\boldsymbol{v}^{\mathsf{H}},$$

which shows that $a\alpha_1 u_{i_1} + \cdots + a\alpha_r u_{i_r} \neq 0$ and that $\boldsymbol{v}^{\mathsf{H}}$ is a linear combination of the rows of A. Thus, $\boldsymbol{v} \in \mathsf{range}(A^{\mathsf{H}})$. Proving that $\boldsymbol{u} \in \mathsf{range}(A)$ is entirely similar.

(75) Prove that if $A \in \mathbb{C}^{n\times n}$ is a normal matrix and $U \in \mathbb{C}^{n\times n}$ is a unitary matrix, then $U^{\mathsf{H}}AU$ is a normal matrix.

(76) Let $A \in \mathbb{C}^{n\times n}$ be a normal matrix and let $a, b \in C$. Prove that the matrix $B = aA + bI_n$ is also normal.

(77) Let $A, B \in \mathbb{C}^{n\times n}$ be two matrices such that $AB = pA + qB$, where $p, q \in \mathbb{C} - \{0\}$. Prove that $AB = BA$.

(78) Let $A \in \mathbb{C}^{m\times n}$, $B \in \mathbb{C}^{p\times q}$, $C \in \mathbb{C}^{n\times r}$, and $D \in \mathbb{C}^{q\times s}$. Prove that

$$(A \otimes B)(C \otimes D) = AC \otimes BD.$$

(79) Let $A \in \mathbb{C}^{n\times n}$ and $B \in \mathbb{C}^{m\times n}$. Prove that
   (a) $(A \otimes B)^{\mathsf{H}} = A^{\mathsf{H}} \otimes B^{\mathsf{H}}$;
   (b) if A and B are Hermitian, then so is $A \otimes B$;
   (c) if A and B are invertible, then so is $A \otimes B$ and $(A \otimes B)^{-1} = A^{-1} \otimes B^{-1}$.

(80) Prove that if A and B are two square submatrices in $\mathbb{C}^{n\times n}$, then $A * B$ is a principal submatrix of $A \otimes B$.

(81) Prove that $\mathsf{trace}(A \otimes B) = \mathsf{trace}(A)\mathsf{trace}(B)$, for any matrices A and B.

(82) Prove that $\mathsf{rank}(A \otimes B) = \mathsf{rank}(A)\mathsf{rank}(B)$ for any matrices A and B.

(83) Let $A \in \mathbb{C}^{m\times n}$ and $B \in \mathbb{C}^{p\times q}$. Prove that if A and B are Hermitian, then so is $A \otimes B$.

Solution: By the last part of Theorem 3.56 we have $(A \otimes B)^{\mathsf{H}} = A^{\mathsf{H}} \otimes B^{\mathsf{H}} = A \otimes B$, which gives the desired conclusion.

(84) Let $\psi : \{1, \ldots, m_1\} \times \{1, \ldots, m_k\} \longrightarrow \{1, \ldots, m_1 \cdots m_k\}$ be the function defined by

$$\psi(i_1, \ldots, i_k) = 1 + \sum_{p=1}^{k} (i_p - 1) \prod_{j=1}^{p-1} m_j,$$

where $1 \leqslant i_\ell \leqslant m_\ell$ and $1 \leqslant \ell \leqslant k$. Prove that ψ is a bijection.

Solution: It suffices to show that ψ is injective. Suppose that $\psi(i_1, \ldots, i_k) = \psi(i'_1, \ldots, i'_k)$. By the definition of ψ, we have

$$\begin{aligned}&(i_1 - 1) + (i_2 - 1)m_1 + (i_3 - 1)m_1m_2 + \cdots + (i_k - 1)m_1 \cdots m_k\\&= (i'_1 - 1) + (i'_2 - 1)m_1 + (i'_3 - 1)m_1m_2 + \cdots + (i'_k - 1)m_1 \cdots m_k.\end{aligned}$$

Since $i_1 \leqslant m_1$, by reducing the previous equality modulo m_1 it follows that $i_1 = i'_1$ and

$$\begin{aligned}&(i_2 - 1) + (i_3 - 1)m_2 + \cdots + (i_k - 1) \cdots m_k\\&= (i'_2 - 1) + (i'_3 - 1)m_2 + \cdots + (i'_k - 1)m_2 \cdots m_k.\end{aligned}$$

The same argument involving m_2 implies $i_2 = i'_2$, *etc.*

(85) Let $\boldsymbol{e}_i^{(n)}$ be the vector in $\mathbb{R}^n$ whose unique non-zero component is in position i, where $1 \leqslant i \leqslant n$. Prove that $\boldsymbol{e}_i^{(m)} \otimes \boldsymbol{e}_j^{(n)} = \boldsymbol{e}_{m(i-1)+j}^{(mn)}$.
Extend this result by proving that

$$\boldsymbol{e}_{i_1}^{(m_1)} \otimes \boldsymbol{e}_{i_2}^{(m_2)} \otimes \cdots \otimes \boldsymbol{e}_{i_k}^{(m_k)} = \boldsymbol{e}_{\psi(i_1,\ldots,i_k)}^{\prod_{r=1}^k m_r},$$

where ψ is the function defined in Supplement 3.17.

(86) Prove that:

(a) $\boldsymbol{e}_i^{(m)} (\boldsymbol{e}_j^{(n)})' = \boldsymbol{e}_i^{(m)} \otimes (\boldsymbol{e}_j^{(n)})' = E_{ij}^{(m,n)}$, where $E_{ij}^{(m,n)}$ is the matrix in $\mathbb{R}^{m\times n}$ which has a unique non-zero entry in position (i, j);

(b) if $A = (a_{ij})$ is a matrix in $\mathbb{R}^{m\times n}$, then

$$A = \sum_{i=1}^{m} \sum_{j=1}^{n} a_{ij} \boldsymbol{e}_i^{(m)} (\boldsymbol{e}_j^{(n)})' = \sum_{i=1}^{m} \sum_{j=1}^{n} a_{ij} \boldsymbol{e}_i^{(m)} \otimes (\boldsymbol{e}_j^{(n)})'.$$

(87) Let $A \in \mathbb{R}^{m \times n}$ and $B \in \mathbb{R}^{p \times q}$. Prove that

$$A \otimes B = \sum_{i=1}^{m} \sum_{j=1}^{n} \sum_{h=1}^{p} \sum_{k=1}^{q} a_{ij} b_{hk} \boldsymbol{e}_{m(i-1)+h}^{(mp)} \otimes (\boldsymbol{e}_{(q(k-1)+j}^{(qn)})'.$$

Solution: We can write

$$\begin{aligned}
A \otimes B &= \sum_{i=1}^{m} \sum_{j=1}^{n} a_{ij} \boldsymbol{e}_i^{(m)} \otimes (\boldsymbol{e}_j^{(n)})' \sum_{h=1}^{p} \sum_{k=1}^{q} b_{hk} \boldsymbol{e}_h^{(p)} \otimes (\boldsymbol{e}_k^{(q)})' \\
&= \sum_{i=1}^{m} \sum_{j=1}^{n} \sum_{h=1}^{p} \sum_{k=1}^{q} a_{ij} b_{hk} \boldsymbol{e}_i^{(m)} \otimes (\boldsymbol{e}_j^{(n)})' \otimes \boldsymbol{e}_h^{(p)} \otimes (\boldsymbol{e}_k^{(q)})' \\
&= \sum_{i=1}^{m} \sum_{j=1}^{n} \sum_{h=1}^{p} \sum_{k=1}^{q} a_{ij} b_{hk} \boldsymbol{e}_i^{(m)} \otimes \boldsymbol{e}_h^{(p)} \otimes (\boldsymbol{e}_j^{(n)})' \otimes (\boldsymbol{e}_k^{(q)})' \\
&= \sum_{i=1}^{m} \sum_{j=1}^{n} \sum_{h=1}^{p} \sum_{k=1}^{q} a_{ij} b_{hk} \boldsymbol{e}_{m(i-1)+h}^{(mp)} \otimes (\boldsymbol{e}_{q(k-1)+j}^{(qn)})'.
\end{aligned}$$

(88) Let $A^{(r)} \in \mathbb{R}^{m_r \times p_r}$ for $1 \leqslant r \leqslant N$ be N matrices. Extend Supplement 3.17 for the matrix $A^{(1)} \otimes \cdots \otimes A^{(N)} \in \mathbb{R}^{\prod_{r=1}^{N} m_r \times \prod_{r=1}^{N} p_r}$.

(89) Let A, X, B be three matrices such that $A \in \mathbb{R}^{m \times p}$, $X \in \mathbb{R}^{p \times q}$, and $B \in \mathbb{R}^{q \times n}$. Prove that
(a) we have $\mathsf{vec}(AXB) = (B' \otimes A)\mathsf{vec}(X)$;
(b) if the matrix $B' \otimes A$ is invertible, then the equation $AXB = C$ is solvable in X.

Solution: Suppose that

$$B = \begin{pmatrix} b_{11} & \cdots & b_{1n} \\ \vdots & \vdots & \vdots \\ b_{q1} & \cdots & b_{qn} \end{pmatrix}.$$

By the definition of Kronecker products, we have

$$B' \otimes A = \begin{pmatrix} b_{11}A & \cdots & b_{q1}A \\ \vdots & \vdots & \vdots \\ b_{1n}A & \cdots & b_{qn}A \end{pmatrix}.$$

By writing X columnwise, $X = (\boldsymbol{x}_1 \ \cdots \ \boldsymbol{x}_q)$, we further have

$$\mathsf{vec}(X) = \begin{pmatrix} \boldsymbol{x}_1 \\ \vdots \\ \boldsymbol{x}_q \end{pmatrix}$$

and

$$(B' \otimes A)\mathsf{vec}(X) = \begin{pmatrix} b_{11}A\boldsymbol{x}_1 + \cdots + b_{q1}A\boldsymbol{x}_q \\ \vdots \\ b_{1n}A\boldsymbol{x}_1 + \cdots + b_{qn}A\boldsymbol{x}_q \end{pmatrix}.$$

On the other hand, we have

$$XB = (b_{11}\boldsymbol{x}_1 + \cdots + b_{q1}\boldsymbol{x}_q, \ldots, b_{1n}\boldsymbol{x}_1 + \cdots + b_{qn}\boldsymbol{x}_q),$$

which implies

$$AXB = (b_{11}A\boldsymbol{x}_1 + \cdots + b_{q1}A\boldsymbol{x}_q, \ldots, b_{1n}A\boldsymbol{x}_1 + \cdots + b_{qn}A\boldsymbol{x}_q).$$

This shows that $\mathsf{vec}(AXB) = (B' \otimes A)\mathsf{vec}(X)$. The second part of the supplement is immediate.

(90) Let $U \in \mathbb{C}^{m \times n}$ and $V \in \mathbb{C}^{n \times p}$ be two matrices. Prove that:
(a) $\mathsf{vec}(UV) = (I_p \otimes U)\mathsf{vec}(V)$;
(b) $\mathsf{vec}(UV) = (V' \otimes I_m)\mathsf{vec}(U)$;
(c) $\mathsf{vec}(UV) = \mathsf{diag}(U, \ldots, U)\mathsf{vec}V$.

Solution: This follows immediately from Supplement 3.17.

(91) Prove that $\boldsymbol{x} \in \mathbb{C}^m$ and $\boldsymbol{y} \in \mathbb{C}^n$, then $\boldsymbol{x} \otimes \boldsymbol{y} = \mathsf{vec}(\boldsymbol{y}\boldsymbol{x}')$ and $\boldsymbol{x} \otimes \boldsymbol{y}' = \boldsymbol{x}\boldsymbol{y}' = \boldsymbol{y}' \otimes \boldsymbol{x}$.

(92) Let $A, B \in \mathbb{R}^{n \times n}$. Prove that

$$\mathsf{trace}(AB) = (\mathsf{vec}(A'))'\mathsf{vec}(B).$$

Solution: Since $\mathsf{trace}(AB) = \sum_{i=1}^n \boldsymbol{e}_i' AB\boldsymbol{e}_i$, the result follows from Equality (3.3).

Let $\boldsymbol{d}_i \in \mathbb{C}^m$ be the ith column of the matrix I_m and $\boldsymbol{e}_j$ be the jth column of the matrix I_n. Define the matrix $H_{ij} = \boldsymbol{d}_i \boldsymbol{e}_j' \in \mathbb{C}^{m\times n}$, which has an 1 in its (i,j) position and 0 elsewhere. Then, any matrix $A \in \mathbb{C}^{m\times n}$ can be written as

$$A = \sum_{i=1}^{m} \sum_{j=1}^{n} a_{ij} H_{ij}. \tag{3.18}$$

This implies

$$A' = \sum_{i=1}^{m} \sum_{j=1}^{n} a_{ij} H_{ij}'. \tag{3.19}$$

(93) Let $\boldsymbol{e}_i$ be the ith column of the matrix I_n for $1 \leqslant i \leqslant n$ and let $E_{[ij]} = \boldsymbol{e}_i \boldsymbol{e}_j'$. Prove that:
 (a) the single nonzero entry of $E_{[ij]}$ is $(E_{[ij]})_{ij} = 1$;
 (b) $E_{[ij]}\boldsymbol{e}_j = \boldsymbol{e}_i$;
 (c) $\sum_{i=1}^{n} E_{[ii]} = I_n$;
 (d) $\mathsf{vec}(I_n) = \sum_{i=1}^{n} \boldsymbol{e}_i \otimes \boldsymbol{e}_i$;
 (e) $E_{[ij]} \otimes E_{[kl]} = \mathsf{vec}(E_{[kl]})(\mathsf{vec}(E_{[ij]})'$;
 (f) $\sum_{i=1}^{n} \sum_{j=1}^{n} (E_{[ij]} \otimes E_{[ij]}) = (\mathsf{vec}(I_n))(\mathsf{vec}(I_n))'$.

(94) Prove that if $\boldsymbol{e}_i^m \in \mathbb{C}^m$ and $\boldsymbol{e}_j^n \in \mathbb{C}^n$, then $\boldsymbol{e}_i^m \otimes \boldsymbol{e}_j^n = \boldsymbol{e}_{(i-1)n+j}^{mn}$.

Define the matrix $K_{mn} \in \mathbb{R}^{mn\times mn}$ as

$$K_{mn} = \sum_{i=1}^{m} \sum_{j=1}^{n} (H_{ij} \otimes H_{ij}').$$

K_{mn} is a square mn-dimensional matrix partitioned into mn submatrices in $\mathbb{R}^{m\times n}$ such that the (i,j) submatrix has a 1 in position (j,i) and 0 elsewhere. K_{mn} is referred to as the *mn-commutation matrix* (see [109], which contains a detailed study of these matrices).

For example, K_{23} is a 6×6 matrix partitioned into six submatrices, as follows:

$$K_{23} = \sum_{i=1}^{2} \sum_{j=1}^{3} (H_{ij} \otimes H_{ij}')$$

$$= \left(\begin{array}{cc|cc|cc} 1 & 0 & 0 & 0 & 0 & 0 \\ 0 & 0 & 1 & 0 & 0 & 0 \\ 0 & 0 & 0 & 0 & 1 & 0 \\ \hline 0 & 1 & 0 & 0 & 0 & 0 \\ 0 & 0 & 0 & 1 & 0 & 0 \\ 0 & 0 & 0 & 0 & 0 & 1 \end{array}\right).$$

(95) Prove that if $A \in \mathbb{R}^{m \times n}$, we have $K_{mn}\mathsf{vec}(A) = \mathsf{vec}(A')$.

Solution: We have

$$\begin{aligned}
\mathsf{vec}(A') &= \mathsf{vec}\left(\sum_{ij} \boldsymbol{a}_{ij} H'_{ij}\right) = \mathsf{vec}\left(\sum_{ij} (\boldsymbol{d}'_i A \boldsymbol{e}_j)(\boldsymbol{e}_j \boldsymbol{d}'_i)\right) \\
&= \mathsf{vec}\left(\sum_{ij} \boldsymbol{e}_j \boldsymbol{d}'_i A \boldsymbol{e}_j \boldsymbol{d}'_i\right) = \sum_{ij} \mathsf{vec}(H'_{ij} A H'_{ij}) \\
&= \sum_{ij} H_{ij} H'_{ij} \mathsf{vec}(A) \\
&\quad \text{(by Supplement 3.17)} \\
&= K_{mn}\mathsf{vec}(A).
\end{aligned}$$

Here $\boldsymbol{d}_i \in \mathbb{C}^m$ is the ith column of the matrix I_m and $\boldsymbol{e}_j$ is the jth column of the matrix I_n.

(96) Prove the following properties of the commutation matrix K_{mn} defined above:

(a) $K_{mn} = \sum_{j=1}^{n} (\boldsymbol{e}'_j \otimes I_m \otimes \boldsymbol{e}_j) = \sum_{i=1}^{m} (\boldsymbol{d}_i \otimes I_n \otimes \boldsymbol{d}'_i)$;
(b) $K'_{mn} = K_{nm}$;
(c) $K^{-1}_{mn} = K_{nm}$;
(d) $K_{1n} = K_{n1} = I_n$;
(e) $\mathsf{trace}(K_{mn}) = 1 + \gcd(m-1, n-1)$;
(f) if $A \in \mathbb{C}^{n \times s}$ and $B \in \mathbb{C}^{m \times t}$, then $K_{mn}(A \otimes B)K_{st} = B \otimes A$.

Solution: `Part (a)`: The definition of K_{mn} implies $K_{mn} = \sum_{ij}(H_{ij} \otimes H'_{ij}) = \sum_{ij}(\boldsymbol{d}_i \boldsymbol{e}'_j \otimes \boldsymbol{e}_j \boldsymbol{d}'_i)$. By applying Exercise 91,

we have

$$\begin{aligned}
\sum_{ij}(\boldsymbol{d}_i\boldsymbol{e}_j' \otimes \boldsymbol{e}_j\boldsymbol{d}_i') &= \sum_{ij}(\boldsymbol{e}_j' \otimes \boldsymbol{d}_i \otimes \boldsymbol{d}_i' \otimes \boldsymbol{e}_j)\\
&= \sum_j \left(\boldsymbol{e}_j' \otimes \left(\sum_i \boldsymbol{d}_i \otimes \boldsymbol{d}_i'\right) \otimes \boldsymbol{e}_j\right)\\
&= \sum_j \left(\boldsymbol{e}_j' \otimes \left(\sum_i \boldsymbol{d}_i\boldsymbol{d}_i'\right) \otimes \boldsymbol{e}_j\right)\\
&= \sum_j (\boldsymbol{e}_j' \otimes I_m \otimes \boldsymbol{e}_j).
\end{aligned}$$

Similarly,

$$\begin{aligned}
\sum_{ij}(\boldsymbol{d}_i\boldsymbol{e}_j' \otimes \boldsymbol{e}_j\boldsymbol{d}_i') &= \sum_{ij}(\boldsymbol{d}_i \otimes \boldsymbol{e}_j' \otimes \boldsymbol{e}_j \otimes \boldsymbol{d}_i')\\
&= \sum_i \left(\boldsymbol{d}_i \otimes \left(\sum_j \boldsymbol{e}_j' \otimes \boldsymbol{e}_j\right) \otimes \boldsymbol{d}_i'\right)\\
&= \sum_i \left(\boldsymbol{d}_i \otimes \left(\sum_j \boldsymbol{e}_j\boldsymbol{e}_j'\right) \boldsymbol{d}_i'\right)\\
&= \sum_i (\boldsymbol{d}_i \otimes I_n \otimes \boldsymbol{d}_i').
\end{aligned}$$

`Part (b):` We have

$$K_{mn}' = \sum_{ij}(\boldsymbol{d}_i\boldsymbol{e}_j' \otimes \boldsymbol{e}_j\boldsymbol{d}_i')' = \sum_{ji}(\boldsymbol{e}_j\boldsymbol{d}_i' \otimes \boldsymbol{d}_i\boldsymbol{e}_j') = K_{mn}.$$

`Part (c):` We need to verify that $K_{mn}'K_{mn} = I_{mn}$. This equality follows from

$$\begin{aligned}
K_{mn}'K_{mn} &= \sum_{ji}(\boldsymbol{e}_j\boldsymbol{d}_i' \otimes \boldsymbol{d}_i\boldsymbol{e}_j') \sum_{st}(\boldsymbol{d}_s\boldsymbol{e}_t' \otimes \boldsymbol{e}_t\boldsymbol{d}_s')\\
&= \sum_{jist}(\boldsymbol{e}_j\boldsymbol{d}_i'\boldsymbol{d}_s\boldsymbol{e}_t' \otimes \boldsymbol{d}_i\boldsymbol{e}_j'\boldsymbol{e}_t\boldsymbol{s}_s')\\
&= \sum_{ji}(\boldsymbol{e}_j\boldsymbol{e}_j' \otimes \boldsymbol{d}_i\boldsymbol{d}_i') = \left(\sum_j \boldsymbol{e}_j\boldsymbol{e}_j'\right)\left(\sum_i \boldsymbol{d}_i\boldsymbol{d}_i'\right) = I_{mn}.
\end{aligned}$$

`Part (d):` We have

$$K_{1n} = \sum_j (e'_j \otimes e_j) = I_n = \sum_j (e_j \otimes e'_j) = K_{n1}.$$

`Part (e):` Note that

$$\begin{aligned}\mathsf{trace}(K_{mn}) &= \mathsf{trace}\left(\sum_{ij}(d_i e'_j \otimes e_j d'_i)\right)\\ &= \sum_{ij} \mathsf{trace}((d_i \otimes e_j)(e'_j \otimes d'_i))\\ &= \sum_{ij}(e_j \otimes d_i)'(d_i \otimes e_j).\end{aligned}$$

The mn-vector $e_j \otimes d_i$ has a unique 1 in its *((j-1)m +i)*th position and 0s elsewhere, and the vector $d_i \oplus e_j$ has a unique 1 in its $((i-1)n+j)$th position. Therefore, we have

$$(e_j \otimes d_i)'(d_i \otimes e_j) = \begin{cases} 1 & \text{if}(j-1)+i = (i-1)n+j,\\ 0 & \text{otherwise.}\end{cases}$$

Let $\{\mathcal{E}_1, \ldots, \mathcal{E}_p\}$ be a set of equalities and let $B[\mathcal{E}_i \mid 1 \leqslant i \leqslant p]$ be the number of valid equalities in this set. We have

$$\begin{aligned}\mathsf{trace}(K_{mn}) &= \sum_{ij}(e_j \otimes d_i)'(d_i \otimes e_j)\\ &= B[(j-1)m+i = (i-1)n+j \mid 1 \leqslant i\\ &\leqslant m, 1 \leqslant j \leqslant n]\\ &= B[(j-1)(m-1) = (i-1)(n-1) \mid 1 \leqslant i\\ &\leqslant m, 1 \leqslant j \leqslant n]\\ &= 1 + B[(j-1)(m-1) = (i-1)(n-1) \mid 2 \leqslant i\\ &\leqslant m, 2 \leqslant j \leqslant n]\\ &= 1 + B[j(m-1) = i(n-1) \mid 1 \leqslant i \leqslant m-1,\end{aligned}$$

$$1 \leqslant j \leqslant n-1]$$

$$= 1 + B\left[\frac{m-1}{n-1} = \frac{i}{j} \mid 1 \leqslant i \leqslant m-1,\right.$$

$$\left. 1 \leqslant j \leqslant n-1\right].$$

Let

$$m' = \frac{m}{\gcd(m,n)} \text{ and } n' = \frac{n}{\gcd(m,n)}.$$

Any pair (i,j) such that $\frac{i}{j} = \frac{m}{n}$ where $1 \leqslant i \leqslant m$ and $1 \leqslant j \leqslant n$ must be of the form $i = \alpha m'$ and $j = \alpha n'$, where α is a positive rational number smaller or equal to $\gcd(m,n)$. We can write $\alpha = \frac{p}{q}$, where p, q are positive integers and $\gcd(p,q) = 1$. Then, $i = \frac{pm'}{q}$ and $j = \frac{pn'}{q}$. The numbers i and j are integers if and only if pm' and pn' are both divisible by q. Since $\gcd(p,q) = 1$, the only common divisor of m' and n' is 1, which implies $q = 1$. This, in turn, implies $i = pm'$ and $j = pn'$, where $1 \leqslant p \leqslant \gcd(m,n)$. Thus, there are $\gcd(m,n)$ pairs (i,j) such that $\frac{i}{j} = \frac{m}{n}$, that is,

$$\left| B\left[\frac{m-1}{n-1} = \frac{i}{j} \mid 1 \leqslant i \leqslant m-1, 1 \leqslant j \leqslant n-1\right]\right| = \gcd(m,n).$$

`Part (f)`: Let $X \in \mathbb{R}^{s\times t}$. We have

$$\begin{aligned} K_{mn}(A \otimes B)K_{st}\mathsf{vec}(X) &= K_{mn}(A \otimes B)\mathsf{vec}(X') \\ &= (\text{by Supplement 3.17}) \\ &= K_{mn}\mathsf{vec}(BX'A') = \mathsf{vec}(AXB') \\ &= (B \otimes A)\mathsf{vec}(X), \end{aligned}$$

which gives the desired result.

(97) Prove that for every $A \in \mathbb{C}^{m\times n}$, $\mathsf{rank}(A) = \mathsf{rank}(A^{\mathsf{H}})$.
If $m = n$, prove that $\mathsf{rank}(A - aI_n) = \mathsf{rank}(A^{\mathsf{H}} - \bar{a}I_n)$ for any $a \in \mathbb{C}$.

(98) Let $U \in \mathbb{R}^{n\times n}$ be the upper triangular matrix

$$U = \begin{pmatrix} 1 & 1 & \cdots & 1 & 1 \\ 0 & 1 & \cdots & 1 & 1 \\ \vdots & \vdots & \cdots & \vdots & \vdots \\ 0 & 0 & \cdots & 0 & 1 \end{pmatrix}$$

and let $V \in \mathbb{R}^{n\times n}$ be the matrix

$$V = \begin{pmatrix} 1 & -1 & 0 & \cdots & 0 & 0 \\ 0 & 1 & -1 & \cdots & 1 & 1 \\ \vdots & \vdots & \vdots & \cdots & \vdots & \vdots \\ 0 & 0 & 0 & \cdots & -1 & 1 \\ 0 & 0 & 0 & \cdots & 0 & 1 \end{pmatrix}.$$

(a) Verify that $UV = I_n$.

(b) Let $\boldsymbol{a}, \boldsymbol{b} \in \mathbb{R}^n$ and let $s_k = \sum_{i=1}^k a_i$. Using Part (a), prove *Abel's equality*:

$$\sum_{i=1}^{n} a_i b_i = \sum_{i=1}^{n-1} s_i(b_i - b_{i+1}) + s_n b_n.$$

Solution: Since Part (a) is immediate, we discuss only Part (b). Note that if $\boldsymbol{s} = \begin{pmatrix} s_1 \\ \vdots \\ s_n \end{pmatrix}$, then $\boldsymbol{s} = U'\boldsymbol{a}$. Also, we have

$$V\boldsymbol{b} = \begin{pmatrix} b_1 - b_2 \\ b_2 - b_3 \\ \vdots \\ b_{n-1} - b_n \\ b_n \end{pmatrix}.$$

This allows us to write

$$\sum_{i=1}^{n-1} s_i(b_i - b_{i-1}) + s_n b_n = \boldsymbol{s}'(V\boldsymbol{b}) = \boldsymbol{a}'UV\boldsymbol{b} = \boldsymbol{a}'\boldsymbol{b},$$

which yields Abel's equality.

(99) Let $\boldsymbol{x}, \boldsymbol{y}, \boldsymbol{z} \in \mathbb{R}^n$ such that $x_1 \geqslant x_2 \geqslant \cdots \geqslant x_n \geqslant 0$ and $\sum_{j=1}^i y_j \leqslant \sum_{j=1}^i z_j$ for $1 \leqslant i \leqslant n$. Prove that $\boldsymbol{x}'\boldsymbol{y} \leqslant \boldsymbol{x}'\boldsymbol{z}$.

Solution: Let $\boldsymbol{a} \in \mathbb{R}^n$ be defined by $a_i = z_i - y_i$. To apply Abel's equality to the vectors $\boldsymbol{a}$ and $\boldsymbol{x}$, let $s_i = \sum_{j=1}^i z_j - \sum_{j=1}^i y_j \geqslant 0$ for $1 \leqslant i \leqslant n$. We have

$$\boldsymbol{x}'\boldsymbol{z} - \boldsymbol{x}'\boldsymbol{y} = \sum_{i=1}^n a_i x_i = \sum_{i=1}^{n-1} s_i(x_i - x_{i+1}) + s_n x_n \geqslant 0,$$

which gives the desired inequality.

(100) Let $\boldsymbol{a}, \boldsymbol{b} \in \mathbb{R}^n$ be two vectors such that $a_1 \geqslant a_2 \geqslant \cdots \geqslant a_n \geqslant 0$ and $b_1 \geqslant b_2 \geqslant \cdots \geqslant b_n \geqslant 0$ and let $P \geqslant O_{n,n}$ be a matrix in $\mathbb{R}^{n\times n}$ such that $P\mathbf{1}_n \leqslant \mathbf{1}_n$ and $P'\mathbf{1}_n \leqslant \mathbf{1}_n$. Prove that $\boldsymbol{a}'P\boldsymbol{b} \leqslant \boldsymbol{a}'\boldsymbol{b}$.

Solution: Observe that

$$\sum_{j=1}^n \sum_{i=1}^k p_{ij} = \sum_{i=1}^k \sum_{j=1}^n p_{ij} \leqslant k,$$

because $\sum_{j=1}^n p_{ij} \leqslant 1$ for every i. This allows us to write

$$\sum_{j=1}^k \sum_{i=1}^k p_{ij} + \sum_{j=k+1}^n \sum_{i=1}^k p_{ij} \leqslant k,$$

hence,

$$\begin{aligned}\sum_{j=k+1}^n \sum_{i=1}^k p_{ij} &\leqslant k - \sum_{j=1}^k \sum_{i=1}^k p_{ij}\\ &= \sum_{j=1}^k \left(1 - \sum_{i=1}^k p_{ij}\right).\end{aligned}$$

Since $b_1 \geqslant b_2 \geqslant \cdots \geqslant b_n \geqslant 0$, we get a stronger inequality by introducing the factors $b_1, \ldots, b_n$ as follows:

$$\sum_{j=k+1}^n \sum_{i=1}^k p_{ij} b_j \leq \sum_{j=1}^k b_j \left(1 - \sum_{i=1}^k p_{ij}\right). \tag{3.20}$$

Let $\mathbf{c} = P\mathbf{b}$. We have

$$\begin{aligned}\sum_{i=1}^{k} c_i &= \sum_{i=1}^{k}\sum_{j=1}^{n} p_{ij}b_j = \sum_{j=1}^{n} b_j \sum_{i=1}^{k} p_{ij}\\ &= \sum_{j=1}^{k} b_j \sum_{i=1}^{k} p_{ij} + \sum_{j=k+1}^{n} b_j \sum_{i=1}^{k} p_{ij}.\end{aligned}$$

Combining the last inequality and Inequality (3.20), we have

$$\begin{aligned}\sum_{i=1}^{k} c_i &\leqslant \sum_{j=1}^{k} b_j \sum_{i=1}^{k} p_{ij} + \sum_{j=k+1}^{n} b_j \sum_{i=1}^{k} p_{ij}\\ &\leqslant \sum_{j=1}^{k} b_j \sum_{i=1}^{k} p_{ij} + \sum_{j=1}^{k} b_j \left(1 - \sum_{i=1}^{k} p_{ij}\right) = \sum_{j=1}^{k} b_j\end{aligned}$$

for $1 \leqslant k \leqslant n$. The result follows from Supplement 3.17.

(101) Let $A \in \mathbb{C}^{m\times n}$ be a matrix. If $X \in \mathbb{C}^{n\times k}$ and $Y \in \mathbb{C}^{m\times k}$ are two matrices such that the matrix $R = Y^{\mathsf{H}}AX \in \mathbb{C}^{k\times k}$ is invertible and $B = A - AXR^{-1}Y^{\mathsf{H}}A$, then $\mathsf{rank}(B) = \mathsf{rank}(A) - \mathsf{rank}(AXR^{-1}Y^{\mathsf{H}}A)$.

(102) Let

$$A(\alpha) = \begin{pmatrix} 1 & 1 \\ 1 & \alpha \end{pmatrix},$$

where $\alpha \in \mathbb{R}$. Define the function $f : \mathbb{R} \longrightarrow \mathbb{R}$ as $f(\alpha) = \mathsf{rank}(A_\alpha)$. Prove that f is not continuous in 1.

(103) Let $A \in \mathbb{R}^{n\times n}$ be a matrix such that $A \geqslant O_{n,n}$. Prove that A is a stochastic matrix if and only if $A\mathbf{1}_n = \mathbf{1}$; prove that A is doubly stochastic if and only if $A\mathbf{1}_n = \mathbf{1}$ and $\mathbf{1}'_n A \mathbf{1}'_n$.

(104) Prove that if $A, B \in \mathbb{R}^{n\times n}$ are (doubly) stochastic matrices, then AB is a (doubly) stochastic matrix. Also, prove that for every $c \in [0,1]$, the matrix $cA+(1-c)B$ is a (doubly) stochastic matrix.

(105) Prove that if $A \in \mathbb{R}^{n\times n}$ is a (doubly) stochastic matrix, then A^m is a (doubly) stochastic matrix for $m \in \mathbb{N}$.

(106) Let $A \in \mathbb{R}^{n\times n}$ be a doubly stochastic matrix and let $\boldsymbol{x}, \boldsymbol{y} \in \mathbb{R}^n$ be two vectors such that $x_1 \geqslant x_2 \geqslant \cdots \geqslant x_n \geqslant 0$ and $y_1 \geqslant y_2 \geqslant \cdots \geqslant y_n \geqslant 0$. Prove that $\boldsymbol{x}'A\boldsymbol{y} \leqslant \boldsymbol{x}'\boldsymbol{y}$.

Solution: Note that there exist n nonnegative numbers $a_1, \ldots, a_n$ and n nonnegative numbers $b_1, \ldots, b_n$ such that $x_p = a_r + \cdots + a_n$ and $y_p = b_r + \cdots + b_n$ for $1 \leqslant r \leqslant n$.
We have

$$\boldsymbol{x}'\boldsymbol{y} - \boldsymbol{x}'A\boldsymbol{y} = \boldsymbol{x}(I - A)\boldsymbol{y} = \boldsymbol{a}'H'(I - A)AH\boldsymbol{b},$$

because $\boldsymbol{x} = H\boldsymbol{a}$ and $\boldsymbol{y} = H\boldsymbol{b}$, where

$$H = \begin{pmatrix} 1 & 1 & 1 & \cdots & 1 \\ 0 & 1 & 1 & \cdots & 1 \\ \vdots & \vdots & \vdots & \vdots & \vdots \\ 0 & 0 & 0 & \cdots & 1 \end{pmatrix}.$$

Thus, it suffices to show that $H'(I - A)H \geqslant 0_{n,n}$. We have

$$\begin{aligned}(H'AH)_{ij} &= \sum_{p=1}^{n}\sum_{q=1}^{n} h_{pi}(I - A)_{pq}h_{qj} \\ &= \sum_{p=1}^{i}\sum_{q=1}^{j}(I - A)_{pq}.\end{aligned}$$

If $i \leqslant j$, we have

$$(H'AH)_{ij} = i - \sum_{p=1}^{i}\sum_{q=1}^{j} a_{pq} \geqslant 0,$$

because no sum $\sum_{q=1}^{j} a_{pq}$ can exceed 1 and there are i such sums $(1 \leqslant p \leqslant i)$. The case when $j \leqslant i$ is similar.

(107) Let $A \in \mathbb{R}^{n\times n}$ be a matrix such that $AP = PA$ for every permutation matrix P. Prove that there exist $a, b \in \mathbb{R}$ such that $A = aI_n + bJ_n$.

Solution: Suppose that $P = P_\phi$, where $\phi \in PERM_n$. Taking into account Equality (3.1), $(AP)_{ij} = (PA)_{ij}$ is equivalent to

$a_{i\phi^{-1}(j)} = a_{\phi(i)j}$ for every permutation $\phi \in PERM_n$. If ϕ is the transposition that exchanges i and j, then the last equality amounts to $a_{ii} = a_{jj}$ for $1 \leqslant i, j \leqslant n$, so all elements on the main diagonal are equal.

Let now a_{ik} and $a_{\ell j}$ be two elements outside the main diagonal, so $i \neq k$ and $\ell \neq j$. Let ϕ be a permutation such that $\phi(k) = j$ and $\phi(i) = \ell$. Since $\phi^{-1}(j) = k$, it follows that $a_{ik} = a_{\ell j}$, so any off-diagonal elements of A are also equal. Thus, A has the desired form.

(108) Let $A, B \in \mathbb{R}^{n\times n}$ be two matrices such that $A > 0$ and $B > 0$ and let $D(A, B)$ be the *divergence* defined by

$$D(A,B) = \sum_{i=1}^{n}\sum_{j=1}^{n}\left(a_{ij}\ln\frac{a_{ij}}{b_{ij}} - a_{ij} + b_{ij}\right).$$

Prove that:

(a) $D(A, B) \geqslant 0$ and that $D(A, B) = 0$ implies $A = B$.

(b) If both A and B are stochastic matrices, then $D(A, B) = \sum_{i=1}^{n}\sum_{j=1}^{n} a_{ij}\ln\frac{a_{ij}}{b_{ij}}$. (In this case $D(A, B)$ is the *Kullback–Leibler divergence*, denoted by $\mathsf{KL}(A, B)$.)

Solution: Let $f : \mathbb{R}_{>0} \longrightarrow \mathbb{R}$ be the function defined by $f(x) = x - 1 - \ln x$. We have $f(1) = 0$, $f'(x) = 1 - \frac{1}{x}$, and $f''(x) = \frac{1}{x^2}$, so f has a minimum for $x = 1$. Thus, $f(x) \geqslant f(1) = 0$, which allows us to conclude that $x - 1 - \ln x \geqslant 0$ for $x > 0$; also, $x - 1 - \ln x = 0$ if and only if $x = 1$. Therefore, we have

$$\frac{b_{ij}}{a_{ij}} - 1 - \ln\frac{b_{ij}}{a_{ij}} \geqslant 0,$$

or $b_{ij} - a_{ij} + a_{ij}\ln\frac{a_{ij}}{b_{ij}} \geqslant 0$, where the equality takes place only when $a_{ij} = b_{ij}$. This implies immediately the inequality of the first part. The second part follows immediately from the definition of stochastic matrices.

(109) Give an example of a non-invertible matrix $A \in \mathbb{C}^{n\times n}$ that satisfies the condition $|a_{ii}| \geqslant \sum\{|a_{ik}| \mid 1 \leqslant k \leqslant n \text{ and } k \neq i\}$ for $1 \leqslant i \leqslant n$. Note that this condition is weaker than the condition of diagonal dominance used in Theorem 3.49.

(110) Prove that $\mathsf{rank}(A^{\dagger}) = \mathsf{rank}(A)$ for every $A \in \mathbb{C}^{m\times n}$.

(111) Let $A \in \mathbb{C}^{m\times n}$ and $B \in \mathbb{C}^{n\times p}$. If $\mathsf{rank}(A) = \mathsf{rank}(B) = n$, then prove that $(AB)^\dagger = B^\dagger A^\dagger$.

(112) Let $Q_{r,p}$ be the collection of r-element subsets of the set $\{1,\ldots,p\}$ and let $A \in \mathbb{R}^{m\times n}$ be a matrix such that $\mathsf{rank}(A) = r$. Denote by $\mathcal{I}(A) = \{I \in Q_{r,m} \mid \mathsf{rank}(A(I,:)) = r\}$, $\mathcal{J}(A) = \{J \in Q_{r,n} \mid \mathsf{rank}(A(:,J)) = r\}$, and $\mathcal{N}(A) = \{(I,J) \in Q_{r,m} \times Q_{r,n} \mid \mathsf{rank}\left(A\begin{bmatrix} I \\ J\end{bmatrix}\right) = r$. Prove that:

(a) $\mathcal{I}(A), \mathcal{J}(A)$ and $\mathcal{N}(A)$ denote the maximal sets of linearly independent rows and columns, and of maximal nonsingular matrices, respectively;

(b) $\mathcal{N}(A) = \mathcal{I}(A) \times \mathcal{J}(A)$;

(c) if $0 \leqslant k \leqslant \frac{r}{2}$, then the intersection of any $r-k$ linearly independent rows and $r-k$ linearly independent columns is a matrix of rank at least equal to $r-2k$.

Solution: The inclusion $\mathcal{N}(A) \subseteq \mathcal{I}(A) \times \mathcal{J}(A)$ is immediate. By the Full-rank factorization theorem (Theorem 3.35) there are two full-rank matrices $B \in \mathbb{C}^{m\times r}$ and $C \in \mathbb{C}^{r\times n}$ such that $A = BC$. Then, every matrix $A\begin{bmatrix} I \\ J\end{bmatrix}$ can be factored as

$$A\begin{bmatrix} I \\ J\end{bmatrix} = C(I,:)B(:,J),$$

which implies $\mathcal{I}(A) \times \mathcal{J}(A) \subseteq \mathcal{N}(A)$.
For the third part, note that dropping a row and a column of a matrix decreases the rank by at most 2.

The *Pauli matrices* are the matrices $P_1, P_2, P_3 \in \mathbb{C}^{2\times 2}$ defined by

$$P_1 = \begin{pmatrix} 0 & 1 \\ 1 & 0 \end{pmatrix}, P_2 = \begin{pmatrix} 0 & -i \\ i & 0 \end{pmatrix}, P_3 = \begin{pmatrix} 1 & 0 \\ 0 & -1 \end{pmatrix}.$$

(113) Prove that $P_k^2 = I_2$ and $P_kP_h + P_hP_k = 2\delta_{hk}I_2$ for $k,k \in \{1,2,3\}$.

(114) Prove that the set $\{I_2, P_1, P_2, P_3\}$ is a basis in $\mathbb{C}^{2\times 2}$.

(115) Prove that

$$P_1^2 = P_2^2 = P_3^2 = -iP_1P_2P_3 = I_2,$$

and $\mathsf{trace}(P_k) = 0$ for $1 \leqslant k \leqslant 3$.

(116) Let $\boldsymbol{u} \in \mathbb{C}^m$ and $\boldsymbol{v} \in \mathbb{C}^n$. Prove that $\det(\boldsymbol{u} * \boldsymbol{v}) = 0$.

(117) Prove that if $\boldsymbol{u}, \boldsymbol{v}, \boldsymbol{w} \in \mathbb{C}^n$, then
   (a) $(\boldsymbol{u} * \boldsymbol{v})^{\mathsf{H}} = \boldsymbol{v} * \boldsymbol{u}$;
   (b) $(\boldsymbol{v} + \boldsymbol{w}) * \boldsymbol{u} = \boldsymbol{v} * \boldsymbol{u} + \boldsymbol{w} * \boldsymbol{u}$;
   (c) $\boldsymbol{u} * (\boldsymbol{v} + \boldsymbol{w}) = \boldsymbol{u} * \boldsymbol{v} + \boldsymbol{u} * \boldsymbol{w}$;
   (d) $c(\boldsymbol{u} * \boldsymbol{v}) = (c\boldsymbol{u}) * \boldsymbol{v} = \boldsymbol{u} * (c\boldsymbol{v})$.

Let $\boldsymbol{x}, \boldsymbol{y} \in \mathbb{R}^n$. To compute the product $\boldsymbol{x}'\boldsymbol{y} = \sum_{i=1}^n x_i y_i$, a number of n multiplications and $n-1$ additions of real numbers are required. The main component of this time is given by the number of multiplications, so minimizing this number is important in reducing computing time.

(118) For $\boldsymbol{x}, \boldsymbol{y} \in \mathbb{R}^n$, define ξ and η as

$$\xi = \sum_{j=1}^{\lfloor n/2 \rfloor} x_{2j-1}x_{2j}, \eta = \sum_{j=1}^{\lfloor n/2 \rfloor} y_{2j-1}y_{2j}.$$

Note that the computation of ξ and η requires $2\lfloor n/2 \rfloor$ products.

Prove that $\boldsymbol{x}'\boldsymbol{y}$ is given by

$$\boldsymbol{x}'\boldsymbol{y} = \begin{cases} \sum_{j=1}^{\lfloor n/2 \rfloor} (x_{2j-1} + y_{2j})(x_{2j} + y_{2j-1}) - \xi - \eta & \text{if } n \text{ is even,} \\ \sum_{j=1}^{\lfloor n/2 \rfloor} (x_{2j-1} + y_{2j})(x_{2j} + y_{2j-1}) - \xi - \eta + x_n y_n & \text{if } n \text{ is odd.} \end{cases}$$

Thus, the total number of products needed to compute $\boldsymbol{x}'\boldsymbol{y}$ is upper bounded by $2\lfloor n/2 \rfloor + \lfloor (n+1)/2 \rfloor$.

(119) Let $A \in \mathbb{R}^{m \times n}$ and $B \in \mathbb{R}^{n \times p}$. If we write

$$A = \begin{pmatrix} \boldsymbol{a}_1' \\ \vdots \\ \boldsymbol{a}_m \end{pmatrix} \text{ and } B = (\boldsymbol{b}_1, \ldots, \boldsymbol{b}_p),$$

computing the matrix product AB using the standard computation is equivalent to performing mp inner products of n-dimensional vectors that require mpn multiplications.

Prove that computing these mp inner products using the technique described in Exercise 3.17 requires $(m+p)n + (mp-m-p)\lfloor (n+1)/2 \rfloor$ multiplications. For large values of m, n, p, this number is roughly half of mnp.

(120) Let

$$A = \begin{pmatrix} a_{11} & a_{12} \\ a_{21} & a_{22} \end{pmatrix} \text{ and } B = \begin{pmatrix} b_{11} & b_{12} \\ b_{21} & b_{22} \end{pmatrix}$$

be two matrices in $\mathbb{R}^{2\times 2}$. The standard method for computing their product requires eight multiplications. Define the following seven numbers:

$$\begin{aligned} I &= (a_{11} + a_{22})(b_{11} + b_{22}) \\ II &= (a_{21} + a_{22})b_{11} \\ III &= a_{11}(b_{12} - b_{22}) \\ IV &= a_{22}(-b_{11} + b_{21}) \\ V &= (a_{11} + a_{12})b_{22} \\ VI &= (-a_{11} + a_{21})(b_{11} + b_{12}) \\ VII &= (a_{12} - a_{22})(b_{21} + b_{22}). \end{aligned}$$

Verify that their product $C = AB$ can be written as

$$\begin{aligned} c_{11} &= I + IV - V + VII \\ c_{21} &= II + IV \\ c_{12} &= III + V \\ c_{22} &= I + III - II + VI, \end{aligned}$$

using a total of seven multiplications and 18 additions.

(121) Let $A, B \in \mathbb{R}^{n\times n}$ be two matrices, where $n = 2^k$, and

$$A = \begin{pmatrix} A_{11} & A_{12} \\ A_{21} & A_{22} \end{pmatrix} \text{ and } B = \begin{pmatrix} B_{11} & B_{12} \\ B_{21} & B_{22} \end{pmatrix},$$

and $A_{ij}, B_{ij} \in \mathbb{R}^{2^{k-1}\times 2^{k-1}}$. Using multiplications of block-matrices, the product $C = AB \in \mathbb{R}^{n\times n}$ can be written as

$$C = \begin{pmatrix} C_{11} & C_{12} \\ C_{21} & C_{22,} \end{pmatrix},$$

where $C_{ij} \in \mathbb{R}^{n/2\times n/2}$. Using the result contained in Exercise 3.17, prove that to compute C the number of multiplications required is $O(n^{\log 7})$.

Solution: Denote by $\tau(n)$ the number of operations required to multiply two $n \times n$ matrices, where $n = 2^k$. We have $\tau(n) = 7\tau\left(\frac{n}{2}\right) + 18\left(\frac{n}{2}\right)\Big)^2$ for $n \geqslant 2$. This implies $\tau(n) = O(7^{\log n}) = O(n^{\log 7})$.

Let $A \in \mathbb{C}^{n\times n}$ be a square matrix. Its *directed graph* is the graph $\mathcal{G}_A$ having $\{1, \ldots, n\}$ as its set of vertices. An edge (i, j) exists in $\mathcal{G}_A$ if and only if $a_{ij} \neq 0$.

(122) Draw the graph of the matrix

$$A = \begin{pmatrix} 0 & 1 & 2 & 0 \\ 2 & 1 & 1 & 2 \\ 0 & 2 & 0 & 1 \\ 0 & 0 & 0 & 2 \end{pmatrix}.$$

(123) Let $\mathcal{G} = (V, E)$ be a graph having V as a set of vertices and E as a set of edges. Prove that the relation $\gamma_{\mathcal{G}}$ on V that consists of all pairs of vertices (x, y) such that there is a path that joins x to y is an equivalence on V.

Let $\mathcal{G} = (V, E)$ be a directed graph and let $V_1, \ldots, V_k$ be the set of equivalence classes relative to the equivalence $\gamma_{\mathcal{G}}$.

The *condensed graph of* $\mathcal{G}$ is the digraph $\mathsf{C}(\mathcal{G}) = (\{V_1, \ldots, V_E\}, K)$, having strong components as its vertices such that $(V_i, V_j) \in K$ if and only if there exist $v_i \in V_i$ and $v_j \in V_j$ such that $(v_i, v_j) \in E$.

(124) Let $\mathcal{G} = (V, E)$ be a directed graph. Prove that the condensed graph $\mathsf{C}(\mathcal{G})$ is acyclic.

Solution: Suppose that $(V_{i_1}, \ldots, V_{i_\ell}, V_{i_1})$ is a cycle in the graph $\mathsf{C}(\mathcal{G})$ that consists of ℓ distinct vertices. By the definition of $\mathsf{C}(\mathcal{G})$ we have a sequence of vertices $(v_{i_1}, \ldots, v_{i_\ell}, v_{i_1})$ in $\mathcal{G}$ such that the $(v_{i_p}, v_{i_{p+1}}) \in E$ for $1 \leqslant p \leqslant \ell - 1$ and $(v_{i_\ell}, v_{i_1}) \in E$. Therefore, $(v_{i_1}, \ldots, v_{i_\ell}, v_{i_1})$ is a cycle, and for any two vertices u, v of this cycle, there is a path from u to v and a path from v to u. In other words, for any pair of vertices (u, v) of this cycle we have $(u, v) \in \gamma_{\mathcal{G}}$, so $V_{i_1} = \cdots = V_{i_\ell}$, which contradicts our initial assumption.

(125) Let $\phi \in PERM_n$ and let P_ϕ be its permutation matrix. Prove that if $A \in \mathbb{C}^{n\times n}$ and the vertices of the graph $\mathcal{G}_A$ are renumbered by replacing each number j by $\phi(j)$, the resulting graph corresponds to the matrix $B = P_\phi A P_\phi'$.

A matrix $A \in \mathbb{C}^{n\times n}$ is *reducible* if there exists a permutation matrix P_ϕ such that $A = P_\phi \begin{pmatrix} U & V \\ O_{p,q} & W \end{pmatrix} P_\phi'$, where $U \in \mathbb{C}^{p\times p}$, $V \in \mathbb{C}^{p\times q}$, and $W \in \mathbb{C}^{q\times q}$. Otherwise, A is irreducible.

(126) Prove that a matrix $A \in \mathbb{C}^{n\times n}$ is irreducible if and only if there exists no partition $\{I, J\}$ of the set $\{1, \ldots, n\}$ such that $a_{ij} = 0$ when $i \in I$ and $j \in J$.

Let $A \in \mathbb{C}^{n\times n}$. The *degree of reducibility* of A is k, where $0 \leqslant k \leqslant n-1$ if there exists a partition $\{I_1, \ldots, I_{k+1}\} \in PART_n$ such that any submatrix $A\begin{bmatrix} I_i \\ I_i \end{bmatrix}$ is irreducible for $1 \leqslant i \leqslant k+1$ and $a_{pq} = 0$ whenever $p \in I_i$, $q \in I_j$, and $i \neq j$. The degree of reducibility of A is denoted by $red(A)$.

(127) Let $A \in \mathbb{R}^{n\times n}$ be a matrix. Prove that $\mathcal{G}_A$ is a strongly connected digraph if and only if A is irreducible.

Solution: Let $A \in \mathbb{R}^{n\times n}$ be a reducible matrix. There exists a partition $\{I, J\}$ of the set $\{1, \ldots, n\}$ such that $a_{ij} = 0$ when $i \in I$ and $j \in J$. Therefore, there is no edge from a vertex in J to a vertex in I, which implies that there exists a vertex $i \in I$ and a vertex $j \in J$ such that there is no path leading from j to i. Thus, $\mathcal{G}_A$ is not strongly connected. This shows that if $\mathcal{G}_A$ is strongly connected, then A is irreducible.

Conversely, suppose that $\mathcal{G}_A$ is not strongly connected and let $V_1, \ldots, V_k$ be the strong connected components of $\mathcal{G}_A$, where $k > 1$. Since the condensed digraph $\mathsf{C}(\mathcal{G}_A)$ is acyclic (by Supplement 3.17), we may assume without loss of generality that its vertices $V_1, \ldots, V_k$ are numbered in topological order. In other words, the existence of an edge (V_i, V_j) in $\mathsf{C}(\mathcal{G})$ implies $i < j$.

Assume initially that the vertices of the strong component V_i are

$$v_p, v_{p+1}, \ldots, v_{p+|V_i|-1},$$

where $p = 1+\sum_{j=1}^{i-1} |V_j|$ for $1 \leqslant i \leqslant j \leqslant k$. Under this assumption we have $a_{pq} = 0$ if $v_p \in V_i$, $v_q \in V_\ell$, and $i > \ell$. In other words, the matrix A has the form

$$A = \begin{pmatrix} A_{11} & A_{12} & \cdots & A_{1k} \\ O & A_{22} & \cdots & A_{2k} \\ \vdots & \vdots & \cdots & \vdots \\ O & O & \cdots & A_{kk} \end{pmatrix},$$

where A_{ii} is the incidence matrix of the subgraph induced by V_i. Thus, A is not irreducible.

If the vertices of $\mathcal{G}_A$ are not numbered according to the previous assumptions, let ϕ be a permutation that rearranges the vertices in the needed order. Then $P_\phi A P'_\phi$ has the necessary form and, again, A is not irreducible.

(128) Prove that a matrix $A \in \mathbb{C}^{n\times n}$ is irreducible if and only if its transpose is irreducible.

(129) Let $A \in \mathbb{R}^{n\times n}$ be a non-negative matrix. For $m \geqslant 1$, we have $(A^m)_{ij} > 0$ if and only if there exists a path of length m in the graph $\mathcal{G}_A$ from i to j.

Solution: The argument is by induction on $m \geq 1$. The base case, $m = 1$, is immediate. Suppose that the statement holds for numbers less than m. Then $(A^m)_{ij} = \sum_{k=1}^{n}(A^{m-1})_{ik}A_{kj}$. $(A^m)_{ij} > 0$ if and only if there is a positive term $(A^{m-1})_{ik}A_{kj}$ in the right-hand sum because all terms are non-negative. By the inductive hypothesis, this is the case if and only if there exists a path of length $m-1$ joining i to k and an edge joining k to j, that is, a path of length m joining i to j.

(130) Let $A \in \mathbb{R}^{n\times n}$ be an irreducible matrix such that $A \geqslant O_{n,n}$. If $k_i > 0$ for $1 \leqslant i \leqslant n-1$, then $\sum_{i=0}^{n-1} k_i A^i > O_{n,n}$.

Solution: Since A is an irreducible matrix, the graph $\mathcal{G}_A$ is strongly connected. Thus, there exists a path of length no larger than $n-1$ that joins any two distinct vertices i and

j of the graph $\mathcal{G}_A$. This implies that for some $m \leqslant n-1$ we have $(A^m)_{ij} > 0$. Since

$$\left(\sum_{i=0}^{n-1} k_i A^i\right)_{ij} = \sum_{i=0}^{n-1} k_i (A^m)_{ij}, \tag{3.21}$$

and all numbers that occur in this equality are non-negative, it follows that for $i \neq j$ we have $\left(\sum_{i=0}^{n-1} k_i A^i\right)_{ij} > 0$. If $i = j$, the same inequality follows from the fact that $k_0 I_n > O_{n,n}$.

(131) Let $A \in \mathbb{R}^{n \times n}$ be an irreducible matrix such that $A \geqslant O_{n,n}$. Prove that $(I_n + A)^{n-1} > O_{n,n}$.

Solution: If we choose $k_i = \binom{n}{i}$ for $0 \leqslant i \leqslant n-1$ in Inequality (3.21), the desired inequality follows.

(132) Let $A \in \mathbb{R}^{p \times \ell}$, $B \in \mathbb{R}^{q \times \ell}$, $C \in \mathbb{R}^{r \times \ell}$ be three matrices having the same number of columns. Prove that:

(a) the Khatri–Rao product is associative, that is,

$$(A * B) * C = A * (B * C);$$

(b)

$$(A * B)'(A * B) = (A'A) * (B'B);$$

(c)

$$(A*)^\dagger = ((A'A) * (B'B))^\dagger (A * B)'.$$

(133) Let $A \in \mathbb{R}^{m \times n}$ and $B \in \mathbb{R}^{p \times q}$ be two matrices. Prove that if $\boldsymbol{x} \in \mathbb{R}^{nq}$, $\boldsymbol{y} \in \mathbb{R}^{mp}$, then we have $\boldsymbol{y} = (A \otimes B)\boldsymbol{x}$ if and only if $Y = BXA'$, where $X = (\boldsymbol{x}_1\ \boldsymbol{x}_2\ \cdots\ \boldsymbol{x}_n) \in \mathbb{R}^{q \times n}$ and $Y = (\boldsymbol{y}_1\ \boldsymbol{y}_2\ \cdots\ \boldsymbol{y}_m) \in \mathbb{R}^{p \times m}$.

Solution: We begin by partitioning $\boldsymbol{x}$ and $\boldsymbol{y}$ as

$$\boldsymbol{x} = \begin{pmatrix} \boldsymbol{x}_1 \\ \vdots \\ \boldsymbol{x}_n \end{pmatrix} \text{ and } \boldsymbol{y} = \begin{pmatrix} \boldsymbol{y}_1 \\ \vdots \\ \boldsymbol{y}_m \end{pmatrix},$$

where $\boldsymbol{x}_1, \ldots, \boldsymbol{x}_n \in \mathbb{R}^q$ and $\boldsymbol{y}_1, \ldots, \boldsymbol{y}_m \in \mathbb{R}^p$. Then we have

$$\boldsymbol{y} = (A \otimes B)\boldsymbol{x} = \begin{pmatrix} a_{11}B & a_{12}B & \cdots & a_{1n}B \\ a_{21}B & a_{22}B & \cdots & a_{2n}B \\ \vdots & \vdots & \ddots & \vdots \\ a_{m1}B & a_{m2}B & \cdots & a_{mn}B \end{pmatrix} \begin{pmatrix} \boldsymbol{x}_1 \\ \boldsymbol{x}_2 \\ \vdots \\ \boldsymbol{x}_n \end{pmatrix},$$

which implies

$$\begin{aligned} \boldsymbol{y}_i &= a_{i1}B\boldsymbol{x}_1 + \cdots a_{1n}B\boldsymbol{x}_n \\ &= (B\boldsymbol{x}_1 \ \cdots \ B\boldsymbol{x}_n) \begin{pmatrix} a_{i1} \\ \vdots \\ a_{in} \end{pmatrix} \\ &= B(\boldsymbol{x}_1 \ \cdots \ \boldsymbol{x}_n) \begin{pmatrix} a_{i1} \\ \vdots \\ a_{in} \end{pmatrix} \\ &= BX\boldsymbol{a}_i, \end{aligned}$$

where $\boldsymbol{a}_i' = (a_{i1} \ \cdots \ a_{in})$. By the definition of Y, we have

$$Y = (BX\boldsymbol{a}_1 \ \cdots \ BX\boldsymbol{a}_m) = BX(\boldsymbol{a}_1 \ \cdots \ bfa_m) = BXA'.$$

(134) Let $\boldsymbol{u} \in \mathbb{C}^I$, $\boldsymbol{v} \in \mathbb{C}^J$, and $\boldsymbol{w} \in \mathbb{C}^K$. Prove that $\boldsymbol{x} = \boldsymbol{u} \otimes \boldsymbol{v} \otimes \boldsymbol{w}$ if and only if $u_i v_j w_k = x_{k+(j-1)K+(i-1)JK}$ for $1 \leqslant i \leqslant I$, $1 \leqslant j \leqslant J$, and $1 \leqslant k \leqslant K$.

Bibliographical Comments

Wedderburn's Theorem (Theorem 3.39) was obtained in [169]. In Supplement 23 of Chapter 5, we discuss a converse of this statement, a result of Egerváry [42]. Theorem 3.35 (the Full-Rank factorization theorem) was formulated in [112].

Supplement 3.17 is an extension of Wedderburn's theorem obtained in [29]. Supplement 3.17 is a result of Ky Fan [49]. A comprehensive reference of the Moore–Penrose pseudoinverse of a matrix is the monograph [11].

Fundamental references to linear algebra include the two-volume book by Gantmacher [168]. More advanced topics are treated in [18] and [78, 79]. Concentrated and useful sources are [111] and [177].

Properties of Kronecker products are studied in [109, 118, 133, 134].

The computation of $\boldsymbol{x}'\boldsymbol{y}$ described in Exercise 3.17 was discovered in [172] and initiated a research stream for more economical ways for multiplying matrices.

The reduction of the number of necessary multiplications of two 2×2 matrices from eight to seven shown in Exercise 3.17 was obtained by Strassen in [161].

Chapter 4

MATLAB Environment

4.1 Introduction

MATLAB, which stands for "matrix laboratory" [116], is a formidable tool for anybody interested in linear algebra and its applications.

4.2 The Interactive Environment of MATLAB

MATLAB is an interactive system. Commands can be entered at the prompt >> in the command window shown in Figure 4.1.

It is important to remember that

- MATLAB is case sensitive;
- variables are not typed;
- typing the name of a variable causes the value of the variable to be printed;
- ending a command with a semicolon will suppress the screen display of the results of the command.

In Figure 4.1 we show the command window of MATLAB .

Commas and semicolons separate statements that are placed on the same line as in'

```
x = 1 ; y =2; z =3;
```

As we mentioned above, semicolons suppress the output. To continue a line on the next line, one ends the line with three periods (...).

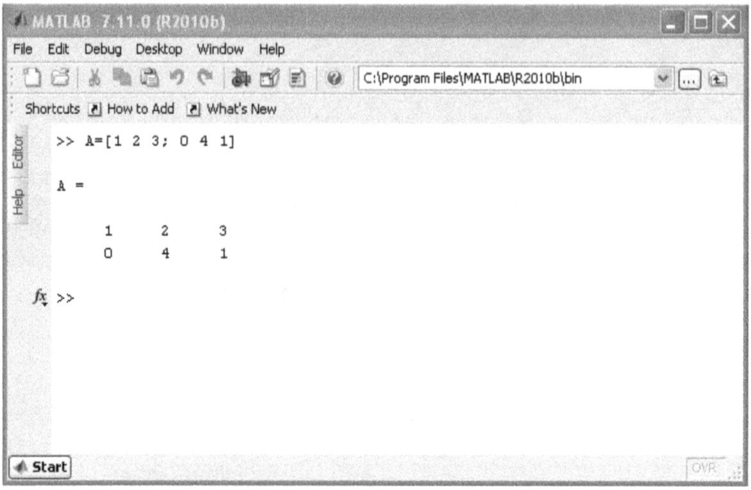

Fig. 4.1 The command window of MATLAB .

MATLAB is equipped with extensive help and documentation facilities. To display a help topic, it suffices to type in the command `help` followed by the name of the topic or to access the MATLAB documentation using the command `doc`.

4.3 Number Representation and Arithmetic Computations

In MATLAB , we deal with two representations of integers: the *unsigned integers* and the *signed integers*. The classes of unsigned integers are denoted by `unitk`, where `k` may assume the values 8, 16, 32, and 64 corresponding to representations over one, two, four, and eight bytes, respectively.

Example 4.1. To create a one-byte unsigned integer having the value 77, we write

```
>> x = uint8(77)
x =
   77
```

To obtain information on `x`, we type `whos('x')`. MATLAB returns the main characteristics of `x`

```
>> whos('x')
  Name        Size              Bytes  Class    Attributes
  x           1x1                   1  uint8
```

Since `x` is a scalar, MATLAB regards `x` as a 1×1-matrix.

The function `size` returns the sizes of each dimension of an array.

Example 4.2. For the matrix A defined by

```
A = [1,2,3;0,5,1]
```

we obtain

```
d =
    2   3
```

The function `unique` applied to an array A produces an array that contains the same data as A with no repetitions and in sorted order.

Example 4.3. The application of `unique` to the array has the following effect:

```
A = [1 2 1 1 0];
>> unique(A)
ans =
   0 1 2
```

The call

```
[m,n] = size(A)
```

returns the size of the matrix `A` in separate variables `m` and `n`. More generally,

```
[d1,d2,...,dn]=size(A)
```

returns the sizes of the first n dimensions of the array `A` in separate variables.

If we need to return the size of the dimension of `A` specified by the scalar `dim` into the variable `m`, we write

```
m = size(A,dim)
```

The function `isa` determines if its first argument belongs to the type specified by its second argument.

Example 4.4. The type of `x` can be verified by typing

```
>> isa(x,'integer')
ans =
     1
>> isa(x,'uint8')
ans =
     1
>> isa(x,'unit32')
ans =
     0
```

The limits of any of the above integer types can be determined by using the functions `intmin` and `intmax`, as follows:

Example 4.5.

```
>> intmin('uint32')
ans =
           0
>> intmax('uint32')
ans =
  4294967295
```

For any of these unsigned types, a number larger than `intmax` is mapped to `intmax` and a number lower than `intmin` is mapped to `intmin`:

```
>> uint8(1000)
ans =
  255
>> uint8(-1000)
ans =
    0
```

Signed integers are represented in MATLAB using the types of the form `intk`, where `k` ranges among 8,16,32, and 64. The leftmost digit is reserved for the sign (1 for a negative integer and 0 for a positive integer). Thus, we have

```
>> intmax('int8')
ans =
  127
>> intmin('int8')
ans =
 -128
```

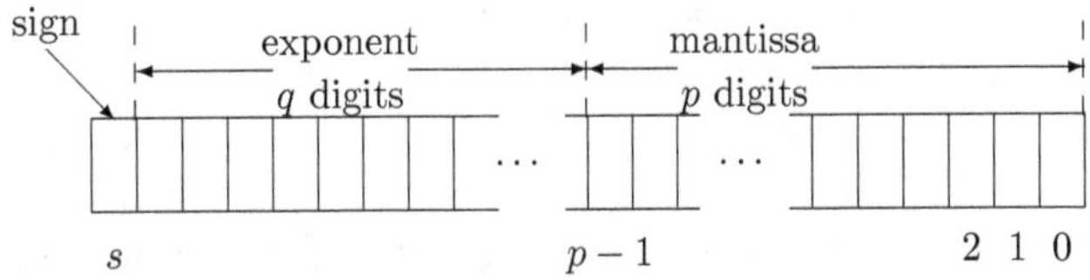

Fig. 4.2 Representation of floating point numbers.

MATLAB makes use of the floating point representation of real numbers. This representation uses a number of $s+1$ binary digits and comprises three parts: the *sign bit*, the sequence of *exponent* bits, and the sequence of *mantissa* (or *significant*) bits, as shown in Figure 4.2. The bits are numbered from right to left starting with 0. We assume that the mantissa uses p bits and the exponent uses q bits, where $p+q=s$. Depending on the desired precision, MATLAB supports a single-precision format or a double-precision format and the latter format (as specified by the IEEE Standard 754) is the default representation. For double-precision numbers, we have $s=63$. The 64th bit is the bit sign b_{63} defined as

$$b_{63}=\begin{cases}0 & \text{if } x\geq 0,\\ 1 & \text{if } x<0,\end{cases}$$

$q=11$ bits are reserved for the exponent, and $p=52$ bits are used by the significant. The number represented by the sequence of bits $(b_{63},b_{62},\ldots,b_0)$ is

$$x=(-1)^{b_{63}}\cdot\left(1+\sum_{i=}^{52}b_{52-i}2^{-i}\right)\cdot 2^{y-1034},$$

where y is the equivalent of $(b_1\cdots b_{11})_2$.

For single-precision numbers, we have $s=32$, the exponent uses $q=8$ bits (biased by 127), and the significant uses the remaining $p=23$ bits. Accordingly, single-precision numbers are given by

$$x=(-1)^{b_{31}}\cdot\left(1+\sum_{i=}^{22}b_{22-i}2^{-i}\right)\cdot 2^{y-127},$$

where y is the equivalent of $(b_1\cdots b_8)_2$. Single-precision numbers require less memory than double-precision numbers, but they are represented to less precision.

Double-precision numbers are created with assignments such as `x = 19.43`, because the default format in MATLAB is double precision. To represent the same number in single precision, we need to write `y = single(19.43)`.

It is important to realize that the set of real numbers that are representable in any of these formats is *finite*. For example, in the case of the double-precision format, we have at our disposal 64 bits, which means that only 2^{64} real numbers have an exact representation as double-precision numbers. For single-precision numbers, the set of real numbers that have exact representations consists of 2^{32} numbers. The remaining real numbers can be represented only with a degree of approximation, which has considerable consequences for numerical computing.

Since the set of reals that can be represented exactly on any floating-point system is finite, it follows that there exists a small gap between each double-precision number and the next larger double-precision number. The gap, which limits the precision of computations, can be determined using the `eps` function. If x is a double-precision number, there are no other double-precision numbers in the interval $(x, x + \texttt{eps}(x))$.

Example 4.6. The distance between 7 and the next double-precision number can be determined as follows:

```
>> eps(7)
ans =
  8.8818e-016
```

As x increases so does `eps(x)`. We have

```
>> eps(70)
ans =
  1.4211e-014
>> eps(700)
ans =
  1.1369e-013
>> eps(7000)
ans =
  9.0949e-013
```

Entering **eps** without arguments is equivalent to **eps(1)**:

```
>> eps
ans =
   2.2204e-16
```

Similar considerations hold for single-precision numbers. Here the gaps between numbers are wider because there are fewer exactly representable numbers.

Example 4.7. If we define y by `y = single(7)`, then `eps(y)` returns `4.7684e-007`, a value larger than `eps(7)` computed above.

Example 4.8. Let $x = 123/1256$. This number is not a sum of powers of 2, so it cannot be represented exactly as a double-precision number. Consider the following MATLAB dialog:

```
>> x =123/1256
x =
    0.0979
>> y = 123 -1256 * x
y =
 -1.4211e-014
```

The approximative representation of x means that y is computed, in turn, approximatively, which explains the fact that MATLAB does not return 0 for y.

Rounding of decimal numbers can cause unexpected results. This phenomenon is known as *roundoff error* and it can be seen in the next example.

Example 4.9. Define x as `x = 0.1`. Using the equality test == we have the following results:

```
>> x + x == 0.2
ans =
     1
>> x + x + x == 0.3
ans =
     0
>> x + x + x + x == 0.4
ans =
     1
```

```
>> x + x + x + x + x + x + x + x == 0.8
ans =
     0
```

Another risk in performing floating point computations is the inadvertent subtraction of two large and close numbers. This may result in a *catastrophic cancellation* as we show next.

Example 4.10. Consider the following MATLAB computation.

```
>> a = 5
a =
     5
>> b = 5e24
b =
  5.0000e+024
>> c = a + b - b
c =
     0
```

MATLAB returns `0` `c` rather than `5` since the numbers $a+b$ and b have the same floating-point representation.

The range of double-precision numbers is determined using the functions `realmin` and `realmax`:

```
>> rangeDouble = 'Double-precision numbers range between
%g and %g';
>> sprintf(rangeDouble,realmin,realmax)
ans =
Double-precision numbers range between 2.22507e-308
and 1.79769e+308
```

When these functions are called with the argument `‘single’`, the corresponding values for the single-precision type are returned:

```
>> rangeSingle = ‘Single-precision numbers range between
%g and %g’;
>> sprintf(rangeSingle,realmin(‘single’), realmax(‘single’))
ans =
Single-precision numbers range between 1.17549e-038
and 3.40282e+038
```

MATLAB operates with an extended set of reals; the values `Inf` and `-Inf` represent real numbers outside the representation ranges, as shown next.

```
realmax('single') + .0001e+038
ans =
   Inf
-realmax('single') - .0001e+038
ans =
  -Inf
```

Values that are not real or complex numbers are represented by the symbol `NaN`, an acronym for "Not a Number." Expressions like `0/0` and `Inf/Inf` yield `NaN`, as do any arithmetic operations involving a NaN:

```
x = 0/0
x =
   NaN
```

The imaginary unit of complex numbers is represented in MATLAB by either of two letters: i or j. A complex number can be created in MATLAB by writing

```
z = 3 + 4i
```

or, equivalently, by using the **`complex`** function:

```
>> z = complex(3,4)
z =
   3.0000 + 4.0000i
```

The real and imaginary parts of a complex number can be obtained using the **`real`** and **`imag`** functions:

```
>> x = real(z)
x =
     3
>> y = imag(z)
y =
     4
```

The functions **`complex`**, **`real`**, and **`imag`** can be applied to arrays, where they act componentwise:

```
>> x= [1 2 3];
>> y = [4 5,6];
>> z=complex(x,y)
z =
   1.0000 + 4.0000i   2.0000 + 5.0000i   3.0000 + 6.0000i
>> zreal = real(z)
```

```
zreal =
     1     2     3
>> zimag = imag(z)
zimag =
     4     5     6
```

Conversions from types are possible using built-in MATLAB functions named after the target type. For example, to convert other numeric data to double precision, we can use the MATLAB function `double`.

Example 4.11. A signed integer created by `y = int64 (-123456789122)` is converted to double-precision floating point by `x = double(y)`.

Arithmetic operators involve the usual arithmetic operators. In MATLAB these operators are overloaded, which means that they can be applied both to numbers and to matrices whose formats accommodate the requirements of these operations. These operators include:

Operator	Significance
+	Addition or unary plus
-	Subtraction or unary minus
*	Numeric or matrix multiplication
^	Numeric or matrix power
\	Backslash or left matrix division
/	Slash or right matrix division
'	Transpose (for real matrices) or Hermitian conjugate (for complex matrices)
.'	Nonconjugated transpose
.*	Array multiplication (element-wise)
.\^	Array power (element-wise)
.\	Left array divide (element-wise)
./	Right array divide (element-wise)

The slash or right matrix division `B/A` is equivalent to BA^{-1}, while the backslash or left matrix division `A\B` is equivalent to $A^{-1}B$.

4.4 Matrices and Multidimensional Arrays

MATLAB accommodates both real and complex matrices. Matrices can be entered row-wise from the MATLAB console making sure that

(i) elements of a row are separated by spaces or commas;
(ii) rows are separated by a semi-colon;
(iii) the matrix is enclosed between square brackets.

For example, the matrix $A \in \mathbb{C}^{2\times 3}$

$$A = \begin{pmatrix} 1+i & 1-i & i \\ 2 & i & 3-2i \end{pmatrix}$$

is entered as

```
>>A = [1+i  1-i  i; 2 i  3-2i]
```

MATLAB prints the content of the matrix as

```
 A  =
    1. + i         1. - i        i
    2.             i             3. - 2.i
```

To inspect the content of the matrix, we type its name at the prompt

```
>>A
```

and, again, MATLAB prints the matrix A, as above.

Lines can be continued by placing ... at the end of the line. For example, we can write

```
>>B = [1  -11  22;...
>>      2  22   33;...
>>      3  44   55]
```

to define the matrix

$$B = \begin{pmatrix} 1 & -11 & 22 \\ 2 & 22 & 33 \\ 3 & 44 & 55 \end{pmatrix}.$$

Scalars can be entered in a rather straightforward manner. To enter $x = 1 + 3i$, one writes

```
>>x = 1+3i
```

The complex unit may be denoted either by i or by j.

MATLAB has a special notation for designating contiguous submatrices known as the *colon notation.* To designate the ith row of a matrix A, we can use the notation `A(i,:)`. Similarly, the jth column is designated by `A(:,j)`.

The submatrix of A that consists of all the rows between the ith row and the kth row is specified by `A(i:k,:)`. Similarly, the submatrix that consists of the columns of A between the jth column and the hth column is `A(:,j:h)`.

Example 4.12. Let $A \in \mathbb{R}^{3\times 4}$ be the matrix

$$A = \begin{pmatrix} 1 & 2 & 3 & 4 \\ 5 & 6 & 7 & 8 \\ 9 & 10 & 11 & 12 \end{pmatrix}.$$

Then `A(:,2:3)` returns

```
ans =
     2     3
     6     7
    10    11
```

The colon notation is also used for specifying sequences. A vector that consists of the members of the arithmetic progression $i, i+r, i+2r, \ldots, i+pr$, where $i+pr <= j$, is designated by the triple $i : s : j$. If s is omitted, then we assume that $s = 1$.

Example 4.13. Consider the following sequences generated using the colon notation:

```
>> 1:2:8
ans =
     1     3     5     7
>> 1:2:9
ans =
     1     3     5     7     9
>> 5:-0.6:1
ans =
    5.0000    4.4000    3.8000    3.2000    2.6000
    2.0000    1.4000
>> 1:5
ans =
     1     2     3     4     5
```

Special matrix can be generated using built-in functions as shown in what follows. For example, the function `eye(m,n)` generates a matrix that contains a unit submatrix I_p, where $p = \min\{m, n\}$, as shown next.

```
>>I = eye(3,3)
 I  =
     1.     0.     0.
     0.     1.     0.
     0.     0.     1.

>>I = eye(4,3)
 I  =
     1.     0.     0.
     0.     1.     0.
     0.     0.     1.
     0.     0.     0.

>>I = eye(3,4)
 I  =
     1.     0.     0.     0.
     0.     1.     0.     0.
     0.     0.     1.     0.
```

Random matrices having elements uniformly distributed in the interval $(0, 1)$ can be generated using the function `rand`. For example,

```
>>A = rand(p,q)
```

creates a random matrix $A \in \mathbb{R}^{p\times q}$. For example,

```
->A = rand(2,3)
```

produces the result

```
 A  =
     0.3616361     0.5664249     0.3321719
     0.2922267     0.4826472     0.5935095
```

Random number producing functions can be used to produce multidimensional arrays whose entries belong to certain intervals.

Example 4.14. To create a $3 \times 4 \times 2$ 3-dimensional array whose entries belong to the $(0, 1)$ interval, we can write

```
-> A = rand(3,4,2)
```

resulting in the array

```
A(:,:,1) =

    0.8147    0.9134    0.2785    0.9649
    0.9058    0.6324    0.5469    0.1576
    0.1270    0.0975    0.9575    0.9706


A(:,:,2) =

    0.9572    0.1419    0.7922    0.0357
    0.4854    0.4218    0.9595    0.8491
    0.8003    0.9157    0.6557    0.9340
```

The number of elements of an array can be determined using the function `numel`. which returns the number of elements of an array A. Its effect is equivalent to that of `\prod(size(A))`.

Experimentation in data mining is greatly helped by the capability of MATLAB for generating pseudorandom numbers having a variety of distributions.

The function `rand` produces uniformly distributed pseudorandom numbers in the interval $(0, 1)$. Depending on the number of arguments, this function can return an $n \times n$ matrix of such numbers (for `rand(n)`), or an $m \times n$ matrix (when using the call `rand(m,n)`). Calls of the form `rand(n,'double')` or `rand(n,'single')` return matrices of numbers that belong to the specified types.

Example 4.15. To generate a 3×3 matrix of pseudorandom numbers in $(0, 1)$, we write `A = rand(3)`. This may yield

```
A =
    0.4505    0.9133    0.5383
    0.0838    0.1524    0.9961
    0.2290    0.8258    0.0782
```

The function `randi` generates a matrix in $\mathbb{R}^{m \times n}$ whose entries are pseudorandom integers from a uniform discrete distribution on an interval $[h, k]$ if called as `randi([h k],[m,n])`.

Example 4.16. The call `A = randi([5 10],[3 3])` returns a 3×3 matrix with integer components in the interval $[5, 10]$:

```
A =
     7     5    10
     5     9     5
    10     9     7
```

If interval `[h,k]` is replaced with a single argument `l`, then the **`randi`** returns a matrix with pseudorandom components in the interval `[1,l]`. For instance, the call `B = randi(10,2,4)` returns a 2×4 matrix of numbers in the interval $[1, 10]$.

The function **`permute`** rearranges the dimensions of an array in the order specified by its second argument that is a vector.

Example 4.17. Let A be a three-dimensional array whose entries belong to the interval $[0, 15]$ produced by the statement `A = randi([0 15],[3 4 2])`:

```
A(:,:,1) =

    10     6    11     0
    12    10     0     1
    11     2     4    13

A(:,:,2) =

    11     0    12     7
     5     7    12     7
    15     6     2    10
```

To permute the first and third dimensions of the three-dimensional array A defined above, we write

```
B=permute(A, [3 2 1])
```

resulting in the array `B` having the dimension vector `[2 4 3]`:

```
B(:,:,1) =

    10     6    11     0
    11     0    12     7


B(:,:,2) =

    12    10     0     1
     5     7    12     7
```

```
B(:,:,3) =

    11     2     4    13
    15     6     2    10
```

The repositioning of the elements of array `A` is shown in Figure 4.3.

To generate pseudorandom numbers following a normal distribution, one can use the function **randn**. The call **randn(m)** returns an $m \times m$-matrix following the normal distribution $N(0,1)$; using a syntax similar to **rand** and **randi**, it is possible to generate matrices of other formats. An additional parameter for the functions of this family located last on their list of parameters allow the generation of values that belong to a more restricted type. For example, a call like **randi(10,100,1,'uint32')** returns an array of 100 4-byte integers, while **randn(10,'double')** returns an array of **double** numbers.

The sequence of pseudorandom numbers produced by any of the generating functions is determined by the internal state of an internal uniform pseudorandom number generator. Resetting this generator to the same state allows computations to be repeated.

The function **normrand** can be used to produce randomly distributed arrays as normally distributed. If `M` and `S` are arrays, then **normrand** returns an array of random numbers chosen from a normal distribution with mean `M` and standard deviation `S` having the same format as `M` and `S`. If either `M` or `S` are scalars, the result is an array having the format of the other parameter. The function **normrnd(M,S,[p,q])** (that we use next) returns a $p \times q$ array.

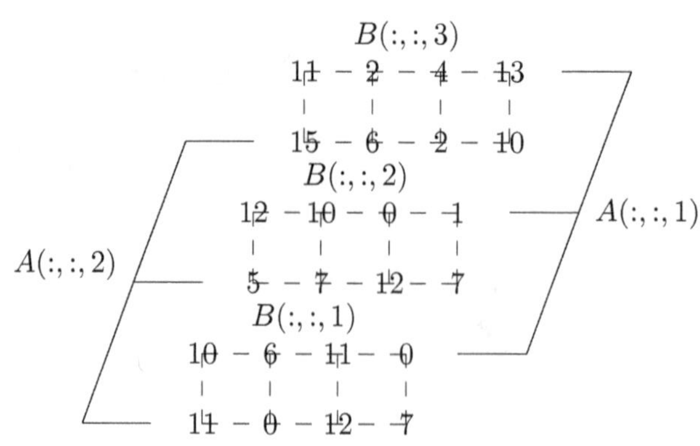

Fig. 4.3 Rearranging of the elements of an array.

Example 4.18. We begin by generating 15 random points in $\mathbb{R}^2$ using

```
U = normrnd(0,1,[15 2])
```

Starting from U we produce another set of 15 points V applying a rotation by 30 degrees, a scaling by 0.5, and a translation by 2 and we add some noise. To this end, we write

```
>> S = [sqrt(3)/2 -1/2; 1/2 sqrt(3)/2];
>> V = normrnd(0.5*U*S+2,0.05,[15 2])
```

Submatrices can be extracted by indicating ranges of indices or by using the placeholder :, which stands for all rows or columns, respectively. To extract the second column of the matrix A, we write

```
>>A(:,2)
 ans  =
    0.5664249
    0.4826472
```

The following is a list of several built-in functions that create or process matrices.

Function	Description
`ones`(m, n)	Creates a matrix in $\mathbb{R}^{m\times n}$ containing 1s
`zeros`(m, n)	Creates a matrix in $\mathbb{R}^{m\times n}$ containing 0s
`eye`(m, n)	Creates a matrix in $\mathbb{R}^{m\times n}$ having 1s on its main diagonal
`toeplitz`$(\boldsymbol{u})$	Creates a Toeplitz matrix whose first row is $\boldsymbol{u}$
`diag`$(\boldsymbol{u})$	Creates a diagonal matrix having $\boldsymbol{u}$ on its main diagonal
`diag`(A)	Yields the diagonal of matrix A
`triu`(A)	Gives the upper part of A
`tril`(A)	Gives the lower part of A
`linspace`(a, b, n)	Creates a vector in $\mathbb{R}^n$ whose components divide $[a, b]$ in $n-1$ equal subintervals
`kron`(A, B)	Computes the Kronecker product of A and B
`rank`(A)	Rank of a matrix A
A'	The Hermitian adjoint A^{H} of A
$A * B$	The product of matrices A and B
$A + B$	The sum of matrices A and B

Operations introduced by a dot apply elementwise. As an example, consider the following operations:

$A.^2$	A matrix that contains the squares of the components of A
$A. * B$	The Hadamard product of A and B
$A./B$	A matrix that contains the ratios of corresponding components of A and B

If A is a real matrix, then A' is the transpose of A; for complex matrices, A' denotes the Hermitian adjoint A^{H}.

Occasionally, we need to make sure that numerical representation of a symmetric matrix $A \in \mathbb{R}^{n\times n}$ does not affect its symmetry. This can be achieved by the statement `A = (A + A')/2`.

The function `diag` mentioned above has several variants. For example, `diag(V,k)`, where V is an n-dimensional vector and k is an integer returns a square matrix of size $n + |k|$ having the components of V on its kth diagonal. The value $k = 0$ corresponds to the main diagonal, positive values correspond to diagonals above the main diagonal, and negative values of k refer to diagonals below the main diagonal. The effect of `diag(V)` is identical to the effect of `diag(V,0)`.

If X is a matrix, `diag(X,k)` is a vector whose components are the elements of the kth diagonal of X.

Example 4.19. The expression

```
diag(ones(4,1)) + diag(ones(3,1),1) + diag(ones(3,1),-1)
```

generates the tridiagonal matrix

```
ans =
     1     1     0     0
     1     1     1     0
     0     1     1     1
     0     0     1     1
```

The function `blkdiag` produces a block-diagonal matrix from its matrix input arguments.

Example 4.20. If we start with the matrices defined by

```
A = [1 2;3 4];
B = [5 6 7; 8 9 10; 11 12 13];
C = [14 15;16 17];
```

the call `blkdiag(A,B,C)` will produce the matrix

```
ans =
     1     2     0     0     0     0     0
     3     4     0     0     0     0     0
     0     0     5     6     7     0     0
     0     0     8     9    10     0     0
     0     0    11    12    13     0     0
     0     0     0     0     0    14    15
     0     0     0     0     0    16    17
```

Block matrices can be formed using vertical or horizontal concatenation of matrices. To concatenate vertically two matrices A and B, the number of columns of A must equal the number of columns of B and the operation is realized as `C = [A ; B]`. Horizontal concatenation requires equality of the numbers of rows and can be obtained as `E = [A D]`.

Example 4.21. Let A, B, and D be the matrices

```
>> A=[1 2 3; 4 5 6]

A =
     1     2     3
     4     5     6

>> B= [7 8 9; 10 11, 12; 13 14 15]

B =
     7     8     9
    10    11    12
    13    14    15

>> D = [17 18; 19 20]

D =
    17    18
    19    20
```

The vertical concatenation of A and B and the horizontal concatenation of A and D are shown as follows:

```
>> C = [A ; B]
```

```
C =

     1     2     3
     4     5     6
     7     8     9
    10    11    12
    13    14    15

>> E = [A D]

E =

     1     2     3    17    18
     4     5     6    19    20
```

The function **repmat** replicates a matrix `A` (referred to as a *tile*) and creates a new matrix `B`, which consists of an $m \times n$ tiling of copies of `A` when we use the call `B = repmat(A,m,n)`. The format of B is $(pm) \times (qn)$, when the format of `A` is $p \times q$. The statement `repmat(A,n)` creates an $n \times n$ tiling.

Example 4.22. Let `A` be the matrix

```
A =

     1     2     3
     4     5     6
```

To create a 3×2 tiling using `A` as a tile, we write `B = repmat(A,3,2)`, which results in

```
B =

     1     2     3     1     2     3
     4     5     6     4     5     6
     1     2     3     1     2     3
     4     5     6     4     5     6
     1     2     3     1     2     3
     4     5     6     4     5     6
```

Real matrices can be sorted in ascending or descending order using the function `sort`. If `v` is a vector, `sort(v)` sorts the elements of `v` in ascending order. If `X` is a matrix, `sort(X)` sorts each column of `X` in ascending order. This function is polymorphic: if `X` is an array of strings, then `sort(X)` sorts the strings in ASCII dictionary order.

Another variant of sort, `sort(X,d,m)`, sorts `X` on the dimension `d`, in either ascending or descending order, as specified by the third parameter `m`, which can assume the values `'ascend'` or `'descend'`. If this function is called with two output parameters as in `[Y,I] = sort(X,d,m)`, then the function returns additionally an index matrix.

Example 4.23. Starting from the unidimensional array

```
>> X=[ 9 2 8 5 11 3 7]
X =
     9     2     8     5    11     3     7
```

the function call `[Y,I] = sort(X)` returns the matrices

```
Y =
     2     3     5     7     8     9    11
I =
     2     6     4     7     3     1     5
```

Clearly, `Y` contains the sorted element of `X`, while `I` gives the position of each of the elements of `Y` in the original matrix `X`.

If `X` is a complex matrix, the elements are sorted in the order of their absolute values and elements that tie for the absolute value are sorted in the order of their angle.

A *sparse matrix* is a matrix whose elements are, to a large extent, equal to 0. Such a matrix can be represented by the position and value of its non-zero elements using the MATLAB function `sparse`.

Example 4.24. Staring from the matrix

```
A =
     1     0     0     2
     3     1     0     0
```

the function call `B = sparse(A)` returns `B`, the representation of `A` as a sparse matrix:

```
B =
   (1,1)        1
   (2,1)        3
   (2,2)        1
   (1,4)        2
```

The format of `B` is the same as the format of `A`.

All matrix operations can be applied to sparse matrices, or to mixtures of sparse and full matrices. Operations on sparse matrices return sparse matrices and operations on full matrices return full matrices. In most cases, operations on mixtures of sparse and full matrices return full matrices. The exceptions include situations where the result of a mixed operation is structurally sparse. For instance, the Hadamard product `A .* S` is at least as sparse as `S`.

Example 4.25. Consider the matrix

```
D =
     5     2     4
     0     1     2
     0     0     1
     1     0     3
```

Its sparse form is

```
E =
   (1,1)        5
   (4,1)        1
   (1,2)        2
   (2,2)        1
   (1,3)        4
   (2,3)        2
   (3,3)        1
   (4,3)        3
```

The sparse form of the matrix `A*D` can be computed as `sparse(A*D)` and is

```
   (1,1)        7
   (2,1)       15
   (1,2)        2
   (2,2)        7
   (1,3)       10
   (2,3)       14
```

The same result can be obtained by multiplying the sparse forms of the matrices `A` and `D`, that is, by writing `B*E`.

The function call `S = sparse(i,j,s,m,n,nzmax)` uses three vectors of equal length `i`, `j`, and `s` to generate an $m \times n$ sparse matrix `S` such that S(i(k),j(k)) = s(k), with space allocated for `nzmax` non-zeros. Any elements of `s` that are zero are ignored, along with the corresponding values of `i` and `j`. Any elements of `s` that have duplicate values of `i` and `j` are added together.

To convert a sparse matrix to a full representation, we can use the function `full`. Its use is illustrated in Example 4.27.

The function call `[B,d] = spdiags(A)` starts with a rectangular matrix A of format $m \times n$ and returns a sparse matrix containing all non-zero diagonals of A. The components of the vector $\boldsymbol{d}$ indicate the position of these diagonals.

Example 4.26. If A is the 3×4 matrix

```
A =
        1     2     3     4
        5     6     7     8
        9    10    11    12
```

that has 6 diagonals, the call `[B,d] = spdiags(A)` produces the results

```
B =
        0     0     1     2     3     4
        0     5     6     7     8     0
        9    10    11    12     0     0


d =
       -2
       -1
        0
        1
        2
        3
```

Note that the first column of B is the -2nd diagonal of A that consists of 9, the second column is the -1st diagonal of A, etc.

Other useful formats of `spdiags` exist. We mention just one, `A = spdiags(B,d,m,n)` that creates an $m \times n$ sparse matrix A from the columns of B and places them along the diagonals specified by the vector $\boldsymbol{d}$.

Example 4.27. Starting with the matrix B and the vector $\boldsymbol{d}$ given by

```
B =

     1     2     3     4
     5     6     7     8
     9    10    11    12
    13    14    15    16

d =

    -2    -1     0     1
```

the call `A = spdiags(B,d,5,4)` produces the sparse matrix

```
A =

   (1,1)          3
   (2,1)          2
   (3,1)          1
   (1,2)          8
   (2,2)          7
   (3,2)          6
   (4,2)          5
   (2,3)         12
   (3,3)         11
   (4,3)         10
   (5,3)          9
   (3,4)         16
   (4,4)         15
   (5,4)         14
```

The normal representation of matrices can be obtained by applying the function `full`. Thus, `C = full(A)` produces

```
C =

     3     8     0     0
     2     7    12     0
     1     6    11    16
     0     5    10    15
     0     0     9    14
```

Note that if the length of the diagonals of C is insufficient to accommodate the full columns of B, these columns are truncated.

Multidimensional arrays generalize matrices. Such arrays can be created starting with matrices and then extending these matrices.

Example 4.28. After creating the matrix A as

```
>> A = [1 0 2 1; -1 1 0 3; 1 2 3 4]

A =

     1     0     2     1
    -1     1     0     3
     1     2     3     4
```

a second matrix with the same format as the initial matrix can be added by writing:

```
>> A(:,:,2) = [5 1 6 7; 8 9 0,2;-1 3 1 0]
```

This results in an multidimensional array having the format $3 \times 4 \times 2$ that is displayed as

```
A(:,:,1) =

     1     0     2     1
    -1     1     0     3
     1     2     3     4


A(:,:,2) =

     5     1     6     7
     8     9     0     2
    -1     3     1     0
```

Multidimensional arrays can be created using the concatenation function `cat`.

Example 4.29. To add a third page to the previously created array `A`, we write

```
A = cat(3,A,[6 7 0 8; -1 -2 -3 0; 4 5 6 3])
```

resulting in

```
A(:,:,1) =

     1     0     2     1
    -1     1     0     3
     1     2     3     4
```

```
A(:,:,2) =

     5     1     6     7
     8     9     0     2
    -1     3     1     0


A(:,:,3) =

     6     7     0     8
    -1    -2    -3     0
     4     5     6     3
```

A fourth page of this array that consists of repeated occurrences of 8 can be added by writing

```
A(:,:,4) = 8
```

The functions `prod` and `sum` return the product of the elements and the sum of the elements of an array, respectively. The sum of the elements of a matrix A can be computed along any of its dimensions by using the function `sum(A,d)`, where d is the dimension. The function `sum(A)` computes the sum of all elements of A. For `prod`, similar conventions apply.

Example 4.30. For the matrix

```
X =
     5     8     1     4
     6     2     1     3
```

we can execute the following computations:

```
>> sum(X)
ans =
    11    10     2     7
>> sum(X,1)
ans =
    11    10     2     7
>> sum(X,2)
ans =
    18
    12
```

For the three-dimensional array created in Example 4.28, `prod(A,[1,2])` returns a $1 \times 1 \times 3$ array whose elements are the products of each page of A; similarly, —sum(A,[1,2])— returns an array whose elements are the sums of each page of A:

```
>> sum(A,[1,2])

ans(:,:,1) =

    17


ans(:,:,2) =

    41


ans(:,:,3) =

    33
```

The `repmat` function can be used to produce multidimensional arrays. Its format in this case is `repmat(A,r)`, where A is an array and `r` is a vector that specifies the repetition scheme.

Example 4.31. Let `A` be the matrix defined by

```
A = [1 2 3; 4 5 6]
```

Copies of the matrix `A` are repeated in a $2 \times 3 \times 2$ multiarray by writing

```
B = repmat(A,[2 3 2])
```

This results in

```
B(:,:,1) =

     1     2     3     1     2     3     1     2     3
     4     5     6     4     5     6     4     5     6
     1     2     3     1     2     3     1     2     3
     4     5     6     4     5     6     4     5     6
```

```
B(:,:,2) =

     1     2     3     1     2     3     1     2     3
     4     5     6     4     5     6     4     5     6
     1     2     3     1     2     3     1     2     3
     4     5     6     4     5     6     4     5     6
```

If, instead, we have written

```
C = repmat(A,[2 3 1])
```

this would have resulted in

```
C =

     1     2     3     1     2     3     1     2     3
     4     5     6     4     5     6     4     5     6
     1     2     3     1     2     3     1     2     3
     4     5     6     4     5     6     4     5     6
```

Obviously, the last dimension (1) of C is useless and can be eliminated using the function `squeeze` which removes dimensions of length 1 of an array:

```
D = squeeze(C)
```

4.5 Cell Arrays

A cell array is an array whose components called *cells* can contain any type of data.

A cell array can be created using the constructor {}.

Example 4.32. To create a 4×2-cell array containing numbers, strings, and arrays, we can write

```
cellAr = {1, 9, 4, 3; 'Febr', randi(2,3), {11; 12; 13},
 randi(2,2)}
```

resulting in

```
cellAr =
  2 x 4 cell array
    {[   1]}    {[         9]}    {[      4]}    {[          3]}
    {'Febr'}    {3 X 3 double}     {3 x 1 cell}
  {2 x 2 double}
```

An empty cell array can be created using `C = {}`; to create an empty cell array of format $2 \times 3 \times 4$, we can write `D = cell(2,3,4)`.

Example 4.33. A cell array of text and data can be created by writing

```
A = {'one','two','three';1,2,3}
```

resulting in

```
A =
  2 x 3 cell array
    {'one'}    {'two'}    {'three'}
    {[  1]}    {[  2]}    {[    3]}
```

Components of a cell array can be accessed individually using curly braces, or as a set of cells using small parentheses.

Example 4.34. Here `A` refers to the cell array introduced in Example 4.33. To access the component (2,3) of `A`, we write `A{2,3}`.

To access the leftmost four cells of the cell array A, we write `A(1:2,1:2)` and obtain

```
  2 x 2 cell array

    {'one'}    {'two'}
    {[  1]}    {[  2]}
```

4.6 Solving Linear Systems

The function `inv` computes the inverse of an invertible square matrix.

Example 4.35. Let `A` be the invertible matrix

```
A =

     1     2
     2     3
```

Its inverse is given by

```
>> inv(A)

ans =

   -3.0000    2.0000
    2.0000   -1.0000
```

On the other hand, if `inv` is applied to a singular matrix

```
A =
        1     2
        2     4
```

an error message is posted:

```
Warning: Matrix is singular to working precision.
```

If A is nonsingular, the function `inv` can be used to solve the system $A\boldsymbol{x} = \boldsymbol{b}$ by writing `x = inv(A)*b`, although a better method is described as follows:

Example 4.36. For

$$A = \begin{pmatrix} 1 & 2 \\ 2 & 3 \end{pmatrix} \text{ and } \boldsymbol{b} = \begin{pmatrix} 13 \\ 23 \end{pmatrix},$$

the solution of the system $A\boldsymbol{x} = \boldsymbol{b}$ is

```
x = inv(A)*b


x =
      7.0000
      3.0000
```

This is not the best way for solving a system of linear equations. In certain circumstances (which we discuss in Section 6.18), this method produces errors and has a poor time performance. A better approach for solving a linear system $A\boldsymbol{x} = \boldsymbol{b}$ is to use the backslash operator `x = A \ b` or `x = mldivide(A,b)`. The term `mldivide` is related to the position of the matrix A at the left of $\boldsymbol{x}$.

Example 4.37. Define A and $\boldsymbol{b}$ as

```
>> A = [5 11 2; 10 6 -4; -2 9 7]
A =
        5    11     2
       10     6    -4
       -2     9     7
>> b=[53;26;48]
b =
      53
      26
      48
```

Then either `x=A\b` or `x=mldivide(A,b)` produces

```
x =
     1.0000
     4.0000
     2.0000
```

The system $\boldsymbol{x}A = \boldsymbol{c}$, where $A \in \mathbb{R}^{n\times m}$ and $\boldsymbol{c}' \in \mathbb{R}^m$ can be solved using either `x = A / b` or `x = mrdivide(A,b)`.

It is easy to see that these operations are related by

$$A\backslash b = (A'/b')'.$$

4.7 Control Structures

Relational expressions allow comparisons between numbers using the following relational operators:

Relational operator	Significance
`<`	Less than
`>`	Greater than
`<=`	Less or equal
`>=`	Greater or equal
`==`	Equal to
`~=`	Not equal

The result of a comparison is a *logical value* defined as one of the numbers 1 or 0, where 1 is equivalent to **true** and 0 is equivalent to **false**. If two scalars are compared, the result is a logical value; if two arrays having the same format are compared, the result is an array of logical values (having the same format as the numerical arrays), which contains logical values that result from the componentwise comparisons of the numerical arrays.

Example 4.38. The result of a comparison between the arrays `x` and `y` is the following array `ans`:

```
>> x = [1 5 2 4 9 6]
x =
     1     5     2     4     9     6
>> y = [7 3 1 3 2 9]
y =
     7     3     1     3     2     9
>> x > y
```

```
ans =
     0     1     1     1     1     0
```

Example 4.39. A scalar `a` can be compared with an array `B`. The result is an array of logical values having the same format as `B`.

```
a =
     5
>> B = [1 5 2; 4 9 6; 6 1 7; 7 3 1]
B =
     1     5     2
     4     9     6
     6     1     7
     7     3     1
>> a <= B
ans =
     0     1     0
     0     1     1
     1     0     1
     1     0     0
```

Logical values result from application of a relational operator or through the application of certain functions and, as we saw, can be grouped in arrays. An array of logical values can be used to extract certain components of a numerical array.

The function `logical` converts a numerical array to a logical array. For example, `M = logical(eye(3))` results in the following array of logical values:

```
M =
     1     0     0
     0     1     0
     0     0     1
```

If `A` is an array and `M` is a logical array having the same format, then `A(M)` returns the values of `A` at the indices where the components of `M` are equal to 1, as in

```
>> A = [1 2 3; 4 5 6; 7 8 9]
A =
     1     2     3
     4     5     6
     7     8     9
>> A(M)
```

```
ans =
     1
     5
     9
```

The `find` function applied to an array produces a list of the positions where the components of the array are non-zero.

For a one-dimensional array `X`, `find(X)` returns a one-dimensional array `I`, as in the following example:

```
X =
     2     0    -1     0     0     3
>> I = find(X)
I =
     1     3     6
```

If `X` is a matrix, `I` contains the places of the non-zero components of `X` where `X` is regarded as an array obtained by concatenating its columns vertically. For example, we have

```
X =
     1     0     3     0     2
     0     0     2     1     1
>> I = find(X)
I =
     1
     5
     6
     8
     9
    10
```

The call `[I,J] = find(X)` returns the row and column indices of the non zero components of `X`, as follows:

```
I =
     1
     1
     2
     2
     1
     2
J =
     1
     3
```

```
        3
        4
        5
        5
```

Example 4.40. Let A and B be two matrices:

```
>> A = [1 2 5; 3 0 2; 6 1 7]
A =
        1      2      5
        3      0      2
        6      1      7
>> B = [5 9 1; 4 3 8; 5 6 7]
B =
        5      9      1
        4      3      8
        5      6      7
```

The answer to the comparison `A < B` is the logical matrix

```
        1      1      0
        1      1      1
        0      1      0
```

If a matrix is regarded as an array obtained by concatenating its columns vertically, then `C = find(A < B)` will return

```
C =
        1
        2
        4
        5
        6
        8
```

MATLAB has four basic control structures: **`if-then-else`**, **`for`**, **`while`** and **`switch`**. Their semantics, discussed next, is somewhat different from similar control structures in common programming languages.

The syntactic diagram of the **`if-then-else`** structure is shown in Figure 4.4.

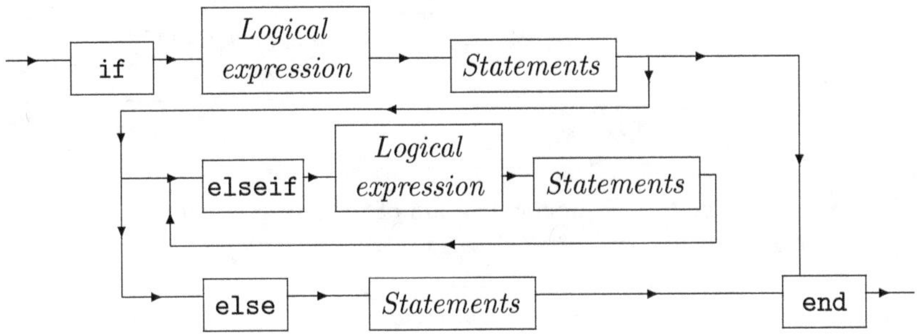

Fig. 4.4 Syntactic diagram of the **if-then-else** structure.

Example 4.41. The following piece of MATLAB code

```
if(sum(sum(A < B))== size(A,1)*size(A,2))
    disp('A is less than B')
elseif(sum(sum(A > B))== size(A,1)*size(A,2))
    disp('A is greater than B')
elseif(sum(sum(A == B))==size(A,1)*size(A,2))
    disp('A and B are equal')
else
    disp('A and B are incomparable')
end
```

applied to two matrices A and B having the same format will return

```
A and B are incomparable
```

when

$$A = \begin{pmatrix} 1 & 2 \\ 3 & 4 \end{pmatrix} \text{ and } B = \begin{pmatrix} 2 & 1 \\ 6 & 6 \end{pmatrix}.$$

The function `sum` when applied to a matrix returns a vector that contains the sum of the columns of the matrix. Therefore, `sum(sum(A))` returns the sum of elements of `A`.

The calls `size(A,1)` and `size(A,2)` return the dimensions of `A`. Note that a condition like `sum(sum(A < B))== size (A,1)*size(A,2)` is satisfied if all entries of the matrix `A < B` are

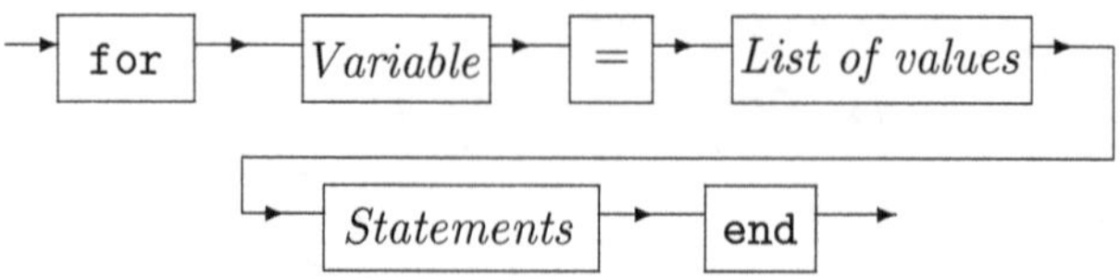

Fig. 4.5 Syntactic diagram of the **for** structure.

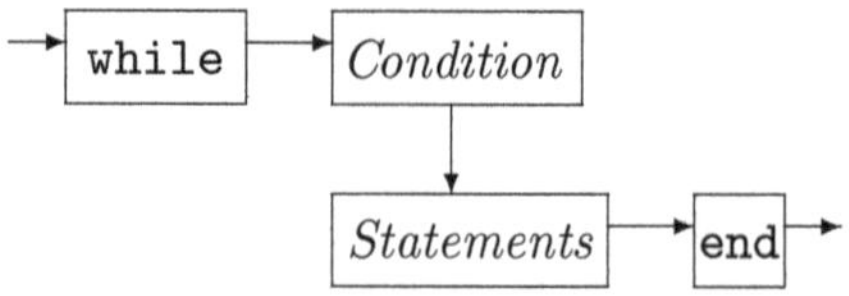

Fig. 4.6 Syntactic diagram of the **while** structure.

equal to 1, that is, if every element of A is less than the corresponding element of B.

The syntax of the `for` structure is shown in Figure 4.5.

The list of values can be entered as a vector or as a colon expression.

Example 4.42. Either

```
>> for i=[5 7 9 11]
     disp(sqrt(i))
   end
```

or

```
>> for i=5:2:12
     disp(sqrt(i))
   end
```

displays the square roots of the odd numbers between 5 and 12:

```
    2.2361
    2.6458
    3
    3.3166
```

The syntax diagram of the `while` structure is shown in Figure 4.6.

Example 4.43. The sum of the terms of the harmonic series that are not smaller than 0.01 is computed by the following code fragment:

```
s = 0;
n = 1;
while(1/n >= 0.01)
    s = s + 1/n;
    n = n + 1;
end;
fprintf('Sum is %g\n',s)
```

which returns

```
Sum is 5.18738
```

The `switch` structure has the following syntactic diagram:

The statements following the first `case` where the switch expression matches the case expression are executed; the second list of statements is executed when the second case expression matches the switch expression. If no case expression is a match for the switch expression, then the statements that follow `otherwise` are executed, if this option exists. Unlike the similar C structure, only one case is executed.

The switch expression can be a scalar or a string.

Example 4.44. The following code fragment will display `cat` (Figure 4.7):

```
>> animal = cougar;
>> switch lower(animal)
          case {'ostrich','emu','turkey'}
               disp('bird');
          case {'tiger','lion','cougar','leopard'}
               disp('cat');
          case {'alligator','crocodile','frog'}
               disp('reptile');
   end
```

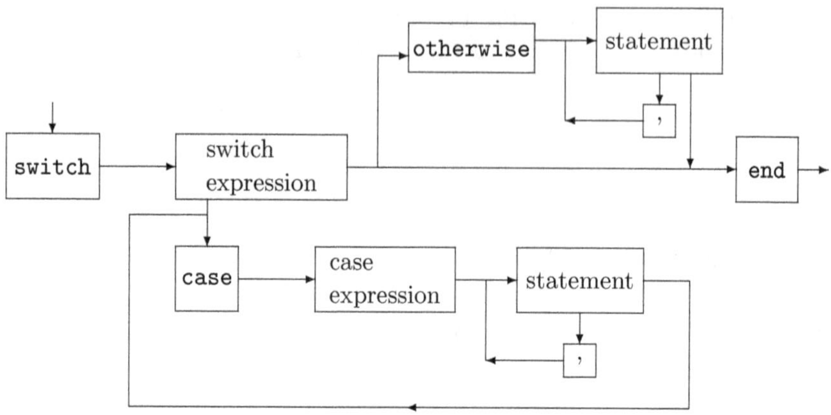

Fig. 4.7 Syntactic diagram of the `switch` structure.

4.8 Indexing

Access to individual components of vectors and matrices can be done by *indexing*. For example, to access the component a_{ij} of matrix A, we write `A(i,j)`.

The index of a vector can be another vector.

Example 4.45. Let `v = [1:3:22]` and let `u = [1 4 5]`. Then, we have

```
>> v=1:3:22
v =
      1     4     7    10    13    16    19    22
>> u=[1 4 5]
u =
      1     4     5
>> v(u)
ans =
      1    10    13
```

This technique can be applied to swapping the two halves of `v` by writing

```
>> v([5:8 1:4])
ans =
     13    16    19    22     1     4     7    10
```

The special operator `end` allows us to access the final position of a vector as follows:

Example 4.46. For the vector `v` defined in Example 4.45, the expression `v(1:2:end)` returns

```
ans =
     1     7    13    19
```

that is, the components of `v` located in odd-numbered positions.

A popular technique in MATLAB is the *logical indexing.* If the indexing expression is of logical type, an indexing expression will extract those elements of the array that make that expression `true`.

Example 4.47. To extract the even number components of the vector `v` defined in Example 4.45, we can write `v(mod(v,2)==0)`. This will result in

```
ans =
     4    10    16    22
```

Example 4.48. Suppose that we have a matrix that has entries that are not defined; such entries can be represented by the special value NaN (not a number) and can be recognized by the logical function `isnan` that returns `true` if its input is NaN. Starting from the matrix

```
>> X = [1 NaN 3 4; 8 -2 NaN NaN; 0 1 NaN 5]
X =
     1   NaN     3     4
     8    -2   NaN   NaN
     0     1   NaN     5
```

the statement `X(isnan(X))=0` results in the substitution of all NaN values by 0:

```
X =
     1     0     3     4
     8    -2     0     0
     0     1     0     5
```

4.9 Functions

Functions can be created in files having the extension `.m` which have the same name as the functions they define. In Example 4.49, we define the function **kronsum**; thus, the file is named **kronsum.m**.

The first line of the file has the form

$$\texttt{function}\ [Y_1,\ldots,Y_q] = f(X_1,\ldots,X_p)$$

where $X_1,\ldots,X_p$ are the input arguments, f is the name of the function, and $Y_1,\ldots,Y_q$ are the values computed by the function.

Example 4.49. Kronecker's sum, $A \oplus B$, is computed by the function **kronsum** as follows:

```
function [S] = kronsum(A,B)
%KRONSUM computes the Kronecker sum of matrices A and B
        if (ndims(A) ~= 2 || ndims(B) ~= 2)
           return;
        end
        [rowsA,colsA] = size(A);
        [rowsB,colsB] = size(B);
        if (rowsA ~= colsA) || (rowsB ~= colsB)
           return;
        end
        S = kron(eye(colsB),A) + kron(B,eye(colsA));
end
```

We ensure that the arguments presented to **kronsum** are matrices by using the function **ndims** that returns the number of dimensions of the arguments. The formats of the arguments are computed using the function **size** and the computation proceeds only if the two arguments are square matrices. Finally, the last line of the function computes effectively the value of the result.

Note the presence of comment lines introduced by %. The first comment line is returned when we use the command `lookfor` or request help.

Example 4.50. The function **datagen** serves as a generator of datasets in $\mathbb{R}^2$ that contain a prescribed number of points that are grouped around given centers.

```
function [T] = datagen(spec)
%DATAGEN produces a set of points in R^2 starting from a
%        k x 4 matrix called spec. Points are grouped in
%        k clusters; the centers of the j-th cluster has
%        coordinates [spec(j,1) spec(j,2)]; the j-th cluster
%        contains spec(j,3) points and the diameter is
%        spec(j,4)

% determine the number of clusters
noc = size(spec,1);

% npgen gives the number of points currently generated
npgen = 0;

for j=1:noc
       centr = [spec(j,1) spec(j,2)];
           T(npgen+1:npgen+spec(j,3),:)=...
             spec(j,4)*randn(spec(j,3),2)+...
             repmat(centr,spec(j,3),1);
             npgen = npgen+spec(j,3);
end
```

Starting from the matrix `spec`

```
spec =
    1.0000    1.0000   50.0000    1.0000
    1.0000    8.0000   40.0000    1.3000
    5.0000    4.0000   50.0000    1.0000
```

and calling the function defined above `T=datagen(spec)`, we obtain the set of objects shown in Figure 4.8.

4.10 Matrix Computations

As a "matrix laboratory," MATLAB offers built-in functions for a wide variety of matrix computations. We offer a few examples that involve commonly used functions

The function `trace` can be applied only to a square matrix A and `trace(A)` returns $\mathsf{trace}(A)$.

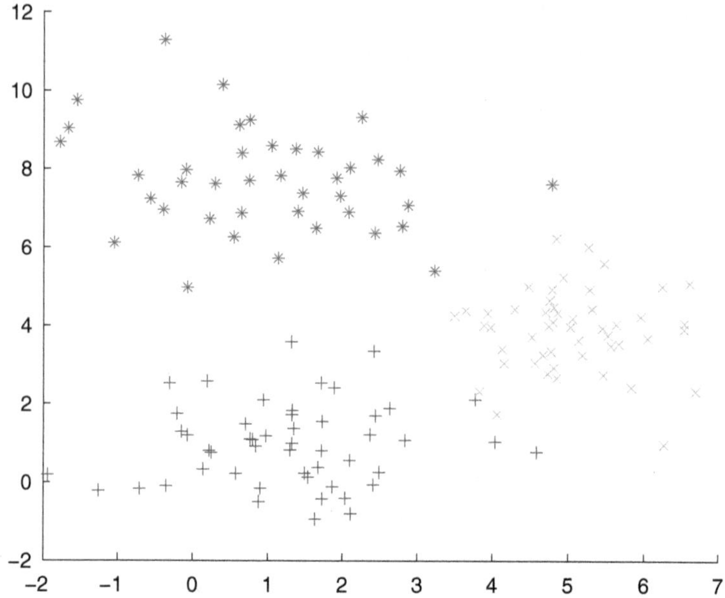

Fig. 4.8 Dataset obtained with the function **datagen**.

The function **abs** applied to a matrix A returns the matrix of absolute values of the elements of A.

Example 4.51. Let A be the matrix

```
A =
   1.0000 + 1.0000i   3.0000 + 4.0000i
   2.0000 - 5.0000i  -7.0000
```

We obtain

```
>> B = abs(A)

B =
    1.4142    5.0000
    5.3852    7.0000
```

The function **rref** produces the reduced row echelon form of a matrix A, when called as `R = rref(A)`. A variant of this function, `[R,r] = rref(A)`, also yields a vector `r` so that `r` indicates the non-zero pivots, `length(r)` is the rank of A, and `A(:,r)` is a basis for the range of A. Roundoff errors may cause this algorithm to produce a rank for A that is different from the actual rank.

A pivot tolerance `tol` used by the algorithm to determine negligible columns can be specified using `rref(A,tol)`.

Example 4.52. Starting from the matrix

```
A =
     1     2     3     4     5     6
     7     8     9    10    11    12
     1     3     5     7     9    11
```

the function call `[R,r]=rref(A)` returns

```
R =
     1     0    -1    -2    -3    -4
     0     1     2     3     4     5
     0     0     0     0     0     0

r =
     1     2
```

showing that the rank of A is 2.

Example 4.53. The echelon form of a matrix A can be used to construct a basis of the null space of the matrix. To this end, consider the following function `nullspbasis`, where `colpiv` and `colnonpiv` designate the pivot and non-pivot columns of A, respectively.

```
function B = nullspbasis(A)
tol = sqrt(eps);
[R, colpiv] = rref(A,tol);
         [m, n] = size(A);
         r = length(colpiv);
         colnonpiv = 1:n;
         colnonpiv(colpiv) = [];
         B = zeros(n, n-r);
         B(colnonpiv,:) = eye(n-r);
         B(colpiv,:) = -R(1:r, colnonpiv);
```

For the matrix A from Example 4.52, we obtain the basis B of $\mathsf{null}(A)$ given by

```
B =
     1     2     3     4
    -2    -3    -4    -5
     1     0     0     0
```

```
     0     1     0     0
     0     0     1     0
     0     0     0     1
```

The LU decomposition of a matrix A is computed by the function call `[L,U,P] = lu(A)`, which returns a lower triangular matrix L, an upper triangular matrix U, and a permutation matrix P such that $PA = LU$.

Example 4.54. Starting from the matrix

```
>> A=[1 0 1; 2 1 1; 1 -1 2]
A =
     1     0     1
     2     1     1
     1    -1     2
```

considered in Example 3.38 and applying the function `lu`, we obtain

```
>> [L,U,P]=lu(A)
L =
    1.0000         0         0
    0.5000    1.0000         0
    0.5000    0.3333    1.0000
U =
    2.0000    1.0000    1.0000
         0   -1.5000    1.5000
         0         0         0
P =
     0     1     0
     0     0     1
     1     0     0
```

When the function `lu` is called with two output arguments, the first argument contains the matrix $P'L$, where P' is the permutation matrix discussed before. Of course, the first output argument is not a lower triangular matrix, as shown next.

```
>> [L,U] = lu(A)
L =
    0.5000    0.3333    1.0000
    1.0000         0         0
    0.5000    1.0000         0
```

```
U =
    2.0000    1.0000    1.0000
         0   -1.5000    1.5000
         0         0         0
```

4.11 Matrices and Images in MATLAB

Binary matrices can be used to encode black and white images in the binary format using the Image Processing toolbox. The function `imshow(A)` displays the binary image of a matrix A in a figure such that pixels with the value 0 (zero) are displayed as black and pixels with value 1, as white.

Example 4.55. Consider the matrix $A \in \{0, 1\}^{8\times 8}$ defined as

```
A = [1 0 1 0 1 0 1 0;
     0 1 0 1 0 1 0 1;
     1 0 1 0 1 0 1 0;
     0 1 0 1 0 1 0 1;
     1 0 1 0 1 0 1 0;
     0 1 0 1 0 1 0 1;
     1 0 1 0 1 0 1 0;
     0 1 0 1 0 1 0 1];
\end{pgmdisplay}
which represents the pattern of squares of a chess board.
The squares themselves are specified using the matrix
\begin{PGMdisplay}
>> B=ones(50,50);
```

and the board is generated as the Kronecker product of the matrices `A` and `B`:

```
>> C=kron(A,B);
```

The matrix C has the format 400×400.

Using the function imshow of the MATLAB image processing toolbox as in

```
>> imshow(C,'Border','tight')
```

the result can be visualized and saved as a `pdf` file shown in Figure 4.9.

Fig. 4.9 Chessboard generated in MATLAB .

Example 4.56. The `pgm` format (an acronym of "Portable Gray Map") is the simplest gray scale graphic image representation. The content of this file is shown in Figure 4.10(b). On its first row, `P2` identifies the file type. The digit image has been discretized in a 10×10 matrix and the format of this matrix is shown in the third line; the fourth line contains the maximum gray value and the rest of the file contains the matrix $A \in \mathbb{R}^{10\times 10}$ (having non-negative integers as entries), which we subject to SVD analysis in Chapter 9 (Figure 4.10).

Exercises and Supplements

(1) Write a MATLAB function that when applied to a matrix A returns bases for both the null space of A and the range of A.
(2) Write a MATLAB function that starts with three real numbers x, y, z and an integer n and returns a tridiagonal $n \times n$-matrix having all elements on the main diagonal equal to a, all elements immediately located under a diagonal equal to b, and all elements immediately above the diagonal equal to c.
(3) Write a MATLAB function that returns a true value if the number of non-zero components of a vector of integers is odd.

(a)

```
P2
# four.pgm
10 10
16
16 16 16 16 16  4 14 16 16 16
16 16 16 14  3  3 12 16 16 16
16 16 13  1  1  3  4 16 16 16
16 14  1  2 15  4  8 16 16 16
16  8  0  2  0  0  5 16 16 16
16  3  0  0  0  0  0  6 12 16
16 16 16 16  3  3 16 16 16 16
16 16 16 16  5  5 16 16 16 16
16 16 16 16  5  9 16 16 16 16
16 16 16 16 16 16 16 16 16 16
```

(b)

Fig. 4.10 Digit 4 representation (a) and the corresponding matrix (b).

To measure the time used by MATLAB operations, one can use a pair of MATLAB functions named `tic` and `toc`. The function `tic` (with no argument) starts a stopwatch; `toc` reads the stopwatch and displays the elapsed time in seconds since the most recent invocation of `toc`.

A similar role can be played by the function `cputime`, which returns the CPU time in seconds that was used by the MATLAB computation since this computation began. For example,

```
t = cputime;
.
.
.
cputime -t
```

returns the cpu time elapsed between these consecutive calls of `cputime`.

(4) Write a MATLAB function that starts from two integer parameters n and k with $k \leq n$ and returns a matrix $S \in \mathbb{R}^{k \times m}$ whose columns contain the distinct subsets of the set $\{1, \ldots, n\}$ with no more than k elements, so $m = 1 + \binom{n}{1} + \cdots + \binom{n}{k}$. For example, if $n = 4$ and $k = 3$, the function should return the matrix

$$\begin{matrix} 0 & 1 & 2 & 3 & 4 & 1 & 1 & 1 & 2 & 2 & 3 & 1 & 1 & 1 & 2 \\ 0 & 0 & 0 & 0 & 0 & 2 & 3 & 4 & 3 & 4 & 4 & 2 & 2 & 3 & 3 \\ 0 & 0 & 0 & 0 & 0 & 0 & 0 & 0 & 0 & 0 & 0 & 3 & 4 & 4 & 4 \end{matrix}$$

Compute and plot the time used by the algorithm for various values of n and k.

(5) Write a MATLAB function that tests if two integer matrices commute; examine the difficulties of writing such a function that deals with real-number matrices.

(6) Write a MATLAB script that picks n points at random on the circle of radius 1 and draws the corresponding polygonal contour.

Hint: Generate uniformly n angles $\alpha_1, \ldots, \alpha_n$ in the interval $[0, 2\pi]$ and then graph the points $(\cos \alpha_i, \sin \alpha_i)$.

The MATLAB function `fzero` can be used to determine the zeros z of a nonlinear single-argument function $f : \mathbb{R} \longrightarrow \mathbb{R}$ located near a number $u \in R$ using two arguments: a string s describing the function to investigate and the number u. For example, to find the zeros of the function f given by $f(x) = x^3 - 6x^2 + 11x - 6$, we write

```
>> z = fzero('x^3 -6*x^2 + 11*x -6',1.3)
```

which results in

```
z =
    1.0000
```

An alternative technique is to use an *anonymous function*; such a function can be written as

```
F = @(x)x^3 -6*x^2 + 11*x -6
```

Then, we could write

```
z = fzero(F,4)
```

which results in

```
z =
    3
```

(7) Experiment with the family of functions $f_a : \mathbb{R} \longrightarrow \mathbb{R}$ defined by $f(x) = x^3 - 10x^2 + 21x + a$ for $a \in \mathbb{R}$ and determine the number of zeros for various values of a.

(8) Consider the complex row vector:

```
>> v = [1+i 2-i 1.5+2*i]
```

Which MATLAB expression will compute v^{H} correctly, `v'` or `v.'` and why?

(9) The least element of a matrix may occur in several positions. Write a MATLAB function that has a matrix A as an argument and replaces its minimal element in each position of the matrix.

(10) Write a MATLAB function that generates a random $m \times n$ matrix with integer entries distributed uniformly between a and b.

(11) Write a MATLAB function that will start with a vector of complex numbers and return a vector having its components in the set $\{1, 2, 3, 4\}$ corresponding to the orthants where the image of each complex number is placed.

(12) Write a MATLAB function that will start with a vector of complex numbers and circularly shift its components one position to the left or to the right, as specified by a parameter of the function.

(13) Write a MATLAB function that accepts four square matrices A_1, A_2, A_3, and A_4 in $\mathbb{R}^{n \times n}$ and outputs the minimum and the maximum trace of a product $A_{i_1}A_{i_2}A_{i_3}A_{i_4}$, where

$$\begin{pmatrix} 1 & 2 & 3 & 4 \\ i_1 & i_2 & i_3 & i_4 \end{pmatrix}$$

is a permutation.

(14) Write a MATLAB function that starts with two vectors that represent the sequences of coefficients of two polynomials and generates the sequence of coefficients of the product of these polynomials.

(15) The *Lagrange interpolating polynomial* is a polynomial of degree no larger than $n - 1$ whose graph passes through the points $(x_1, y_1), \ldots, (x_n, y_n)$ in $\mathbb{R}^2$ and is given by

$$p(x) = \sum_{i=1}^{n} y_i \frac{\prod\{(x - x_j) \mid 1 \leq j \leq n \text{ and } j \neq i\}}{\prod\{(x_i - x_j) \mid 1 \leq j \leq n \text{ and } j \neq i\}}.$$

Given two vectors $\boldsymbol{x}, \boldsymbol{y} \in \mathbb{R}^n$ write a MATLAB function that computes the coefficients of the Lagrange interpolating polynomial.

Bibliographical Comments

MATLAB's popularity in the technical and research communities has generated a substantial literature. We mention, as especially useful, such titles as [64, 75, 116, 167] and [83]. Space limitations did not allow us to present the outstanding visualization capabilities of MATLAB for scientific data. The reader should consult [75].

Chapter 5

Determinants

5.1 Introduction

Determinants are a class of numerical multilinear functions defined on the set of square matrices. They play an important role in theoretical considerations of linear algebra and are useful for symbolic computations. As we shall see, determinants can be used to solve certain small and well-behaved linear systems; however, they are of limited use for large or numerically difficult linear systems.

Historically, determinants appeared long before matrices related to solving linear systems. In modern times, determinants were introduced by Leibniz at the end of the 17th century and Cramer formula appeared in 1750. The term "determinant" was introduced by Gauss in 1801.

5.2 Determinants and Multilinear Forms

Theorem 5.1. *Let $\mathbb{F}$ be a field, M be an $\mathbb{F}$-linear space, and let $f : M^n \longrightarrow \mathbb{F}$ be a skew-symmetric $\mathbb{F}$-multilinear form.*

If two arguments of f are interchanged, then the value of f is multiplied by -1, that is,

$$f(\boldsymbol{x}_1, \ldots, \boldsymbol{x}_i, \ldots, \boldsymbol{x}_j, \ldots, \boldsymbol{x}_n) = -f(\boldsymbol{x}_1, \ldots, \boldsymbol{x}_j, \ldots, \boldsymbol{x}_i, \ldots, \boldsymbol{x}_n)$$

for $\boldsymbol{x}_1, \ldots, \boldsymbol{x}_n \in M$.

Proof. Since f is a multilinear form, we have

$$\begin{aligned}
&f(\boldsymbol{x}_1,\ldots,\boldsymbol{x}_i+\boldsymbol{x}_j,\ldots,\boldsymbol{x}_i+\boldsymbol{x}_j,\ldots,\boldsymbol{x}_n)\\
&= f(\boldsymbol{x}_1,\ldots,\boldsymbol{x}_i,\ldots,\boldsymbol{x}_i,\ldots,\boldsymbol{x}_n)+f(\boldsymbol{x}_1,\ldots,\boldsymbol{x}_i,\ldots,\boldsymbol{x}_j,\ldots,\boldsymbol{x}_n)\\
&\quad +f(\boldsymbol{x}_1,\ldots,\boldsymbol{x}_j,\ldots,\boldsymbol{x}_i,\ldots,\boldsymbol{x}_n)+f(\boldsymbol{x}_1,\ldots,\boldsymbol{x}_j,\ldots,\boldsymbol{x}_j,\ldots,\boldsymbol{x}_n)\\
&= f(\boldsymbol{x}_1,\ldots,\boldsymbol{x}_i,\ldots,\boldsymbol{x}_j,\ldots,\boldsymbol{x}_n)+f(\boldsymbol{x}_1,\ldots,\boldsymbol{x}_j,\ldots,\boldsymbol{x}_j,\ldots,\boldsymbol{x}_n).
\end{aligned}$$

By the defining property of skew-symmetry, we have the equalities

$$\begin{aligned}
f(\boldsymbol{x}_1,\ldots,\boldsymbol{x}_i+\boldsymbol{x}_j,\ldots,\boldsymbol{x}_i+\boldsymbol{x}_j,\ldots,\boldsymbol{x}_n)&=0,\\
f(\boldsymbol{x}_1,\ldots,\boldsymbol{x}_i,\ldots,\boldsymbol{x}_i,\ldots,\boldsymbol{x}_n)&=0,\\
f(\boldsymbol{x}_1,\ldots,\boldsymbol{x}_j,\ldots,\boldsymbol{x}_j,\ldots,\boldsymbol{x}_n)&=0,
\end{aligned}$$

which yield

$$f(\boldsymbol{x}_1,\ldots,\boldsymbol{x}_i,\ldots,\boldsymbol{x}_j,\ldots,\boldsymbol{x}_n)=-f(\boldsymbol{x}_1,\ldots,\boldsymbol{x}_j,\ldots,\boldsymbol{x}_i,\ldots,\boldsymbol{x}_n),$$

for $\boldsymbol{x}_1,\ldots,\boldsymbol{x}_n \in M$. □

Corollary 5.1. *Let V be an $\mathbb{F}$-linear space and let $f : V^n \longrightarrow \mathbb{F}$ be a skew-symmetric multilinear form.*

If $\boldsymbol{x}_i = \boldsymbol{x}_j$ for $i \neq j$, then $f(\boldsymbol{x}_1,\ldots,\boldsymbol{x}_i,\ldots,\boldsymbol{x}_j,\ldots,\boldsymbol{x}_n) = 0$.

Proof. This follows immediately from Theorem 5.1. □

Theorem 5.1 has the following useful extension.

Theorem 5.2. *Let V be an $\mathbb{F}$-linear space and let $f : V^n \longrightarrow \mathbb{F}$ be a skew-symmetric $\mathbb{F}$-multilinear form.*

If $\phi \in PERM_n$ is a permutation given by

$$\phi : \begin{pmatrix} 1 & \cdots & i & \cdots & n \\ j_1 & \cdots & j_i & \cdots & j_n \end{pmatrix},$$

then $f(\boldsymbol{x}_{j_1},\ldots,\boldsymbol{x}_{j_n}) = (-1)^{inv(\phi)} f(\boldsymbol{x}_1,\ldots,\boldsymbol{x}_n)$ for $\boldsymbol{x}_1,\ldots,\boldsymbol{x}_n \in M$.

Proof. The argument is by induction on $p = inv(\phi)$. The basis case, $p = 0$ is immediate because in this case, ϕ is the identity mapping.

Suppose that the argument holds for permutations that have no more than p inversions and let ϕ be a permutation that has $p+1$ inversions. Then, as we saw in the proof of Theorem 1.9, there exists an adjacent transposition ψ such that for the permutation ϕ' defined as $\phi' = \psi\phi$ we have $inv(\phi') = inv(\phi) - 1$. Suppose that ϕ' is the permutation

$$\phi' : \begin{pmatrix} 1 & 2 & \cdots & \ell & \ell+1 & \cdots & n \\ j_1 & j_2 & \cdots & j_\ell & j_{\ell+1} & \cdots & j_n \end{pmatrix}$$

and ψ is the adjacent transposition that exchanges j_ℓ and $j_{\ell+1}$, so

$$\phi : \begin{pmatrix} 1 & 2 & \cdots & \ell & \ell+1 & \cdots & n \\ j_1 & j_2 & \cdots & j_{\ell+1} & j_\ell & \cdots & j_n \end{pmatrix}.$$

By the inductive hypothesis,

$$f(\boldsymbol{x}_{j_1}, \ldots, \boldsymbol{x}_{j_\ell}, \boldsymbol{x}_{j_{\ell+1}}, \ldots, \boldsymbol{x}_{j_n}) = (-1)^{inv(\phi')} f(\boldsymbol{x}_1, \ldots, \boldsymbol{x}_n)$$

and

$$\begin{aligned} & f(\boldsymbol{x}_{j_1}, \ldots, \boldsymbol{x}_{j_{\ell+1}}, \boldsymbol{x}_{j_\ell}, \ldots, \boldsymbol{x}_{j_n}) \\ & = -f(\boldsymbol{x}_{j_1}, \ldots, \boldsymbol{x}_{j_\ell}, \boldsymbol{x}_{j_{\ell+1}}, \ldots, \boldsymbol{x}_{j_n}) \\ & = -(-1)^{inv(\phi')} f(\boldsymbol{x}_1, \ldots, \boldsymbol{x}_n) = (-1)^{inv(\phi)} f(\boldsymbol{x}_1, \ldots, \boldsymbol{x}_n), \end{aligned}$$

which concludes the argument. □

Theorem 5.3. *Let $\mathbb{F}$ be a field, V be an $\mathbb{F}$-linear space, $f : V^n \longrightarrow \mathbb{F}$ be a skew-symmetric $\mathbb{F}$-multilinear form, and let $a \in \mathbb{F}$.*

If $i \neq j$ and $\boldsymbol{x}_1, \ldots, \boldsymbol{x}_n \in V^n$, then

$$f(\boldsymbol{x}_1, \ldots, \boldsymbol{x}_n) = f(\boldsymbol{x}_1, \ldots, \boldsymbol{x}_i + a\boldsymbol{x}_j, \ldots, \boldsymbol{x}_n).$$

Proof. Suppose that $i < j$. Then, by the linearity of f, we have

$$\begin{aligned} & f(\boldsymbol{x}_1, \ldots, \boldsymbol{x}_i + a\boldsymbol{x}_j, \ldots, \boldsymbol{x}_n) \\ & = f(\boldsymbol{x}_1, \ldots, \boldsymbol{x}_i, \ldots, \boldsymbol{x}_n) + af(\boldsymbol{x}_1, \ldots, \boldsymbol{x}_j, \ldots, \boldsymbol{x}_j, \ldots, \boldsymbol{x}_n) \\ & = f(\boldsymbol{x}_1, \ldots, \boldsymbol{x}_i, \ldots, \boldsymbol{x}_n), \end{aligned}$$

by Corollary 5.1. □

Theorem 5.4. *Let V be a linear space and let $f : V^n \longrightarrow \mathbb{R}$ be a skew-symmetric linear form on V. If $\boldsymbol{x}_1, \ldots, \boldsymbol{x}_n$ are linearly dependent, then $f(\boldsymbol{x}_1, \ldots, \boldsymbol{x}_n) = 0$.*

Proof. Suppose that $\boldsymbol{x}_1, \ldots, \boldsymbol{x}_n$ are linearly dependent, that is, one of the vectors can be expressed as a linear combination of the remaining vectors. Suppose that $\boldsymbol{x}_n = a_1\boldsymbol{x}_1 + \cdots + a_{n-1}\boldsymbol{x}_{n-1}$. Then,

$$\begin{aligned} f(\boldsymbol{x}_1, \ldots, \boldsymbol{x}_{n-1}, \boldsymbol{x}_n) &= f(\boldsymbol{x}_1, \ldots, \boldsymbol{x}_{n-1}, a_1\boldsymbol{x}_1 + \cdots + a_{n-1}\boldsymbol{x}_{n-1}) \\ &= \sum_{i=1}^{n-1} a_i f(\boldsymbol{x}_1, \ldots, \boldsymbol{x}_i, \ldots, \boldsymbol{x}_{n-1}, \boldsymbol{x}_i) = 0, \end{aligned}$$

by Corollary 5.1. □

Theorem 5.5. *Let V be an n-dimensional $\mathbb{F}$-linear space and let $\{\mathbf{e}_1, \ldots, \mathbf{e}_n\}$ be a basis in V. There exists a unique, skew-symmetric multilinear form $d_n : V^n \longrightarrow \mathbb{R}$ such that $d_n(\mathbf{e}_1, \ldots, \mathbf{e}_n) = 1$.*

Proof. Let $\boldsymbol{u}_1, \ldots, \boldsymbol{u}_n$ be n vectors such that

$$\boldsymbol{u}_i = a_i^1\mathbf{e}_1 + a_i^2\mathbf{e}_2 + \cdots + a_i^n\mathbf{e}_n$$

for $1 \leqslant i \leqslant n$. If d_n is a skew symmetric multilinear form, $d_n : V^n \longrightarrow \mathbb{R}$, then

$$\begin{aligned} & d_n(\boldsymbol{u}_1, \boldsymbol{u}_2, \ldots, \boldsymbol{u}_n) \\ &= d_n\left(\sum_{j_1=1}^{n} a_1^{j_1}\mathbf{e}_{j_1}, \sum_{j_2=1}^{n} a_2^{j_2}\mathbf{e}_{j_2}, \ldots, \sum_{j_n=1}^{n} a_n^{j_n}\mathbf{e}_{j_n}\right) \\ &= \sum_{j_1=1}^{n}\sum_{j_2=1}^{n}\cdots\sum_{j_n=1}^{n} a_1^{j_1}a_2^{j_2}\cdots a_n^{j_n} d_n(\mathbf{e}_{j_1}, \mathbf{e}_{j_2}, \ldots, \mathbf{e}_{j_n}). \end{aligned}$$

We need to retain only the terms of this sum in which the arguments of $d_n(\mathbf{e}_{j_1}, \mathbf{e}_{j_2}, \ldots, \mathbf{e}_{j_n})$ are pairwise distinct (because the term where $j_p = j_q$ for $p \neq q$ is zero, by Corollary 5.1). In other words, only terms in which the list $(j_1, \ldots, j_n)$ is a permutation of $(1, \ldots, n)$ have a non-zero contribution to the sum. By Theorem 5.2, we can write

$$\begin{aligned} & d_n(\boldsymbol{u}_1, \boldsymbol{u}_2, \ldots, \boldsymbol{u}_n) \\ &= d_n(\mathbf{e}_1, \mathbf{e}_2, \ldots, \mathbf{e}_n) \sum_{j_1, \ldots, j_n} (-1)^{inv(j_1, \ldots, j_n)} a_1^{j_1}a_2^{j_2}\cdots a_n^{j_n}. \end{aligned}$$

where the sum extends to all $n!$ permutations $(j_1, \ldots, j_n)$ of $(1, \ldots, n)$. Since $d_n(\boldsymbol{e}_1, \ldots, \boldsymbol{e}_n) = 1$, it follows that

$$d_n(\boldsymbol{u}_1, \boldsymbol{u}_2, \ldots, \boldsymbol{u}_n) = \sum_{j_1,\ldots,j_n} (-1)^{inv(j_1,\ldots,j_n)} a_1^{j_1} a_2^{j_2} \cdots a_n^{j_n}.$$

□

Note that $d_n(\boldsymbol{u}_1, \boldsymbol{u}_2, \ldots, \boldsymbol{u}_n)$ is expressed using the elements of the matrix A, where

$$A = \begin{pmatrix} a_1^1 & a_1^2 & \cdots & a_1^n \\ a_2^1 & a_2^2 & \cdots & a_2^n \\ \vdots & \vdots & \cdots & \vdots \\ a_n^1 & a_n^2 & \cdots & a_n^n \end{pmatrix}.$$

Definition 5.1. Let $A = (a_i^j) \in \mathbb{C}^{n \times n}$ be a square matrix. The *determinant of* A is the number $\det(A)$ defined as

$$\det(A) = \sum_{j_1,\ldots,j_n} (-1)^{inv(j_1,\ldots,j_n)} a_1^{j_1} a_2^{j_2} \cdots a_n^{j_n}. \tag{5.1}$$

The determinant of A is denoted either by $\det(A)$ or by

$$\begin{vmatrix} a_1^1 & a_1^2 & \cdots & a_1^n \\ a_2^1 & a_2^2 & \cdots & a_2^n \\ \vdots & \vdots & \cdots & \vdots \\ a_n^1 & a_n^2 & \cdots & a_n^n \end{vmatrix}.$$

Equality (5.1) is known as the *Leibniz formula*.

Note that $\det(A)$ can be written using the Levi-Civita symbols introduced in Definition 1.11 as

$$\det(A) = \sum_{j_1,\ldots,j_n} \epsilon_{j_1,\ldots,j_n} a_1^{j_1} a_2^{j_2} \cdots a_n^{j_n}. \tag{5.2}$$

Theorem 5.6. *Let* $A \in \mathbb{C}^{n \times n}$ *be a matrix. We have* $\det(A') = \det(A)$.

Proof. The definition of A' allows us to write

$$\det(A') = \sum_{j_1,\ldots,j_n} (-1)^{inv(j_1,\ldots,j_n)} a^1_{j_1} a^2_{j_2} \cdots a^n_{j_n},$$

where the sum extends to all permutations of $(1,\ldots,n)$. Due to the commutativity of numeric multiplication, we can rearrange the term $a^1_{j_1} a^2_{j_2} \cdots a^n_{j_n}$ as $a^{k_1}_1 a^{k_2}_2 \cdots a^{k_n}_n$, where

$$\phi : \begin{pmatrix} 1 & 2 & \cdots & n \\ j_1 & j_2 & \cdots & j_n \end{pmatrix} \text{ and } \psi : \begin{pmatrix} 1 & 2 & \cdots & n \\ k_1 & k_2 & \cdots & k_n \end{pmatrix}$$

are inverse permutations. Since both ϕ and ψ have the same parity, it follows that

$$(-1)^{inv(j_1,\ldots,j_n)} a^1_{j_1} a^2_{j_2} \cdots a^n_{j_n} = (-1)^{inv(k_1,\ldots,k_n)} a^{k_1}_1 a^{k_2}_2 \cdots a^{j_n}_n,$$

which implies $\det(A') = \det(A)$. $\square$

Corollary 5.2. *If $A \in \mathbb{C}^{n\times n}$, then $\det(A^{\mathsf{H}}) = \overline{\det(A)}$. Furthermore, if A is a Hermitian matrix, $\det(A)$ is a real number.*

Proof. Let $\bar{A}$ be the matrix obtained from A by replacing each a_{ij} by its conjugate. Since conjugation of complex numbers permutes with both the sum and product of complex numbers, it follows that $\det(\bar{A}) = \overline{\det(A)}$. Thus, $\det(A^{\mathsf{H}}) = \det(\bar{A})' = \det(\bar{A}) = \overline{\det(A)}$.

The second part of the corollary follows from the equality $\det(A) = \overline{\det(A)}$. $\square$

Corollary 5.3. *If $A \in \mathbb{C}^{n\times n}$ is a unitary matrix, then $|\det(A)| = 1$.*

Proof. Since A is unitary, we have $A^{\mathsf{H}}A = AA^{\mathsf{H}} = I_n$. By Theorem 5.8, $\det(AA^{\mathsf{H}}) = \det(A)\det(A^{\mathsf{H}}) = \det(A)\overline{\det(A)} = |\det(A)|^2 = 1$. Thus, $|\det(A)| = 1$. $\square$

Example 5.1. Let $A \in \mathbb{R}^{3\times 3}$ be the matrix

$$A = \begin{pmatrix} a^1_1 & a^2_1 & a^3_1 \\ a^1_2 & a^2_2 & a^3_2 \\ a^1_3 & a^2_3 & a^3_3 \end{pmatrix}.$$

The number $\det(A)$ is the sum of six terms corresponding to the six permutations of the set $\{1, 2, 3\}$, as follows:

Permutation ϕ	$inv(\phi)$	Term
$(1,2,3)$	0	$a_1^1 a_2^2 a_3^3$
$(3,1,2)$	2	$a_1^3 a_2^1 a_3^2$
$(2,3,1)$	2	$a_1^2 a_2^3 a_3^1$
$(2,1,3)$	1	$-a_1^2 a_2^1 a_3^3$
$(3,2,1)$	3	$-a_1^3 a_2^2 a_3^1$
$(1,3,2)$	1	$-a_1^1 a_2^3 a_3^2$

Thus, we have

$$\begin{aligned}\det(A) &= a_1^1 a_2^2 a_3^3 + a_1^3 a_2^1 a_3^2 + a_1^2 a_2^3 a_3^1 \\ &= -a_1^2 a_2^1 a_3^3 - a_1^3 a_2^2 a_3^1 - a_1^1 a_2^3 a_3^2.\end{aligned}$$

The number of terms $n!$ grows very fast with the size n of the determinant. For instance, for $n = 10$, we have $10! = 3,682,800$ terms. Thus, direct computations of determinants are very expensive.

The definition of the determinant that makes use of skew-symmetric multilinear forms has the advantage of yielding quite simple proofs for many elementary properties of determinants.

Theorem 5.7. *The following properties of* $\det(A)$ *hold for any* $A \in \mathbb{C}^{n\times n}$:

(i) $\det(A)$ *is a linear function of the rows of* A *(of the columns of* A*);*
(ii) *if* A *has two equal rows (two equal columns), then* $\det(A) = 0$*;*
(iii) *if two rows (columns) are permuted, then* $\det(A)$ *is changing signs;*
(iv) *if a row of a matrix, multiplied by a constant, is added to another row, then* $\det(A)$ *remains unchanged; the same holds if instead of rows we consider columns;*
(v) *if a row (column) equals* $\mathbf{0}$*, then* $\det(A) = 0$.

Proof. We begin with the above statements that involve rows of A. Let $A = (a_i^j)$ and let $\boldsymbol{x}_i = (a_i^1, \ldots, a_i^n) \in \mathbb{C}^n$ be the ith row of the matrix A. We saw that

$$\det(A) = f(\boldsymbol{x}_1, \ldots, \boldsymbol{x}_n),$$

where f is the skew-symmetric multilinear form defined by $f(\boldsymbol{e}_1, \ldots, \boldsymbol{e}_n) = 1$.

The linearity in each argument follows immediately from the linearity of f.

Part (ii) follows from Corollary 5.1. The third part follows from skew-symmetry of f. Theorem 5.3 implies Part (iv). Finally, the last statement is immediate. The corresponding statements concerning columns of A follow from Theorem 5.6 because the columns of A are the transposed rows of A'. □

Theorem 5.8. *Let $A, B \in \mathbb{C}^{n\times n}$ be two matrices. We have $\det(AB) = \det(A)\det(B)$.*

Proof. Let $\boldsymbol{a}_1, \ldots, \boldsymbol{a}_n$ and $\boldsymbol{b}_1, \ldots, \boldsymbol{b}_n$ be the rows of the matrices A and B, respectively. We assume that $\boldsymbol{a}_i = (a_i^1, \ldots, a_i^n)$ for $1 \leqslant i \leqslant n$. Then, the rows $\boldsymbol{c}_1, \ldots, \boldsymbol{c}_n$ of the matrix $C = AB$ are given by $\boldsymbol{c}_i = a_i^j \boldsymbol{b}_j$, where $1 \leqslant j \leqslant n$, as it can be easily seen.

If $d_n : (\mathbb{C}^n)^n \longrightarrow \mathbb{C}$ is the skew-symmetric multilinear function that defines the determinant whose existence and uniqueness were shown in Theorem 5.5, then we have

$$
\begin{aligned}
\det(AB) &= d_n(\boldsymbol{c}_1, \ldots, \boldsymbol{c}_i, \ldots, \boldsymbol{c}_n) \\
&= d_n\left(\sum_{j_1=1}^n a_1^{j_1}\boldsymbol{b}_{j_1}, \ldots, \sum_{j=1}^n a_i^{j_i}\boldsymbol{b}_{j_i}, \ldots, \sum_{j_n=1}^n a_n^{j_n}\boldsymbol{b}_{j_n}\right) \\
&= \sum_{j_1=1}^n \cdots \sum_{j_1=1}^n \cdots \sum_{j_1=1}^n a_1^{j_1} \cdots a_i^{j_i} \cdots a_n^{j_n} d_n(\boldsymbol{b}_{j_1}, \ldots, \\
&\qquad \boldsymbol{b}_{j_i}, \ldots, \boldsymbol{b}_{j_n}),
\end{aligned}
$$

due to the linearity of d_n. Observe now that only the sequences $(j_1, \ldots, j_n)$ that represent permutations of the set $\{1, \ldots, n\}$ contribute to the sum because d_n is skew-symmetric. Furthermore, if $(j_1, \ldots, j_n)$ represents a permutation ϕ, then $d_n(\boldsymbol{b}_{j_1}, \ldots, \boldsymbol{b}_{j_i}, \ldots, \boldsymbol{b}_{j_n}) = (-1)^{inv(\phi)} d_n(\boldsymbol{b}_1, \ldots, \boldsymbol{b}_n)$. Thus, we can write

$$
\begin{aligned}
&\det(AB) \\
&= \sum_{j_1=1}^n \cdots \sum_{j_1=1}^n \cdots \sum_{j_1=1}^n a_1^{j_1} \cdots a_i^{j_i} \cdots a_n^{j_n} d_n(\boldsymbol{b}_{j_1}, \ldots, \boldsymbol{b}_{j_i}, \ldots, \boldsymbol{b}_{j_n})
\end{aligned}
$$

$$= \left(\sum_{j_1=1}^{n} \cdots \sum_{j_1=1}^{n} \cdots \sum_{j_1=1}^{n} (-1)^{inv(j_1,\ldots,j_n)} a_1^{j_1} \cdots a_i^{j_i} \cdots a_n^{j_n} \right)$$

$$\times d_n(\boldsymbol{b}_1, \ldots, \boldsymbol{b}_n)$$

$$= \det(A)\det(B).$$

□

Lemma 5.1. *Let* $B \in \mathbb{R}^{(n+1)\times(n+1)}$ *be the matrix*

$$B = \begin{pmatrix} 1 & 0 & 0 & \cdots & 0 \\ 0 & a_1^1 & a_1^2 & \cdots & a_1^n \\ \vdots & \vdots & \vdots & \cdots & \vdots \\ 0 & a_n^1 & a_n^2 & \cdots & a_n^n \end{pmatrix}.$$

We have $\det(B) = \det(A)$, *where*

$$A = \begin{pmatrix} a_1^1 & a_1^2 & \cdots & a_1^n \\ \vdots & \vdots & \cdots & \vdots \\ a_n^1 & a_n^2 & \cdots & a_n^n \end{pmatrix}.$$

Proof. Note that if $B = (b_i^j)$, then

$$b_1^j = \begin{cases} 1 & \text{if } j = 1 \\ 0 & \text{otherwise,} \end{cases}$$

and

$$b_i^1 = \begin{cases} 1 & \text{if } i = 1 \\ 0 & \text{otherwise.} \end{cases}$$

Also, if $i > 1$ and $j > 1$, then $b_i^j = a_{i-1}^{j-1}$ for $2 \leqslant i, j \leqslant n+1$. By the definition of the determinant, each term of the sum that defines $\det(B)$ must include an element of the first row. However, only the first element of this row is non-zero, so

$$\begin{aligned}\det(B) &= \sum_{(j_1,j_2,\ldots,j_{n+1})} (-1)^{inv(j_1,j_2,\ldots,j_{n+1})} b_1^{j_1} b_2^{j_2} \cdots b_{n+1}^{j_{n+1}},\\ &= \sum_{(j_2,\ldots,j_{n+1})} (-1)^{inv(1,j_2,\ldots,j_{n+1})} a_1^{j_2-1} \cdots a_{n-1}^{j_n-1} a_n^{j_{n+1}-1},\end{aligned}$$

where $(j_2,\ldots,j_{n+1})$ is a permutation of the set $\{2,\ldots,n+1\}$. Since

$$inv(1,j_2,\ldots,j_{n+1}) = inv(j_2,\ldots,j_{n+1}),$$

it follows that

$$\det(B) = \sum_{(j_2,\ldots,j_{n+1})} (-1)^{inv(j_2,\ldots,j_{n+1})} a_1^{j_2-1} a_2^{j_3-1} \cdots a_n^{j_{n+1}-1}.$$

Observe now that if $(j_2,\ldots,j_{n+1})$ is a permutation of the set $\{2,\ldots,n+1\}$, then $(k_1,\ldots,k_n)$, where $k_i = j_{i+1}-1$ for $1 \leqslant i \leqslant n$ is a permutation of $(1,\ldots,n)$ that has the same number of inversions as $(j_2,\ldots,j_{n+1})$. Therefore,

$$\det(B) = \sum_{(k_1,\ldots,k_n)} (-1)^{inv(k_1,\ldots,k_n)} a_1^{k_1} a_2^{k_2} \cdots a_n^{k_n} = \det(A).$$

□

Lemma 5.2. *Let $A \in \mathbb{R}^{n\times n}$ be a matrix partitioned as*

$$A = \left(\begin{array}{ccc|ccc} a_1^1 & \cdots & a_1^q & a_1^{q+1} & \cdots & a_1^n\\ \vdots & \cdots & \vdots & \vdots & \cdots & \vdots\\ a_p^1 & \cdots & a_p^q & a_p^{q+1} & \cdots & a_p^n\\ \hline a_{p+1}^1 & \cdots & a_{p+1}^q & a_{p+1}^{q+1} & \cdots & a_{p+1}^n\\ \vdots & \cdots & \vdots & \vdots & \cdots & \vdots\\ a_n^1 & \cdots & a_n^q & a_n^{q+1} & \cdots & a_n^n \end{array}\right)$$

and let $B \in \mathbb{R}^{(n+1)\times(n+1)}$ be defined by

$$B = \left(\begin{array}{ccc|c|ccc} a_1^1 & \cdots & a_1^q & 0 & a_1^{q+1} & \cdots & a_1^n\\ \vdots & \cdots & \vdots & \vdots & \vdots & \cdots & \vdots\\ a_p^1 & \cdots & a_p^q & 0 & a_p^{q+1} & \cdots & a_p^n\\ \hline 0 & \cdots & 0 & 1 & 0 & \cdots & 0\\ \hline a_{p+1}^1 & \cdots & a_{p+1}^q & 0 & a_{p+1}^{q+1} & \cdots & a_{p+1}^n\\ \vdots & \cdots & \vdots & \vdots & \vdots & \cdots & \vdots\\ a_n^1 & \cdots & a_n^q & 0 & a_n^{q+1} & \cdots & a_n^n \end{array}\right).$$

Then $\det(B) = (-1)^{p+q}\det(A)$.

Proof. By permuting the $(p+1)$st row of B with each of the p rows preceding it in the matrix B and, then, by permuting the $(q+1)$st column with each of the q columns preceding it, we obtain the matrix C given by

$$C = \begin{pmatrix} 1 & 0 & 0 & 0 & 0 & 0 & 0 \\ 0 & a_1^1 & \cdots & a_1^q & a_1^{q+1} & \cdots & a_1^n \\ \vdots & \vdots & \cdots & \vdots & \vdots & \cdots & \vdots \\ 0 & a_p^1 & \cdots & a_p^q & a_p^{q+1} & \cdots & a_p^n \\ 0 & a_{p+1}^1 & \cdots & a_{p+1}^q & a_{p+1}^{q+1} & \cdots & a_{p+1}^n \\ \vdots & \vdots & \cdots & \vdots & \vdots & \cdots & \vdots \\ 0 & a_n^1 & \cdots & a_n^q & a_n^{q+1} & \cdots & a_n^n \end{pmatrix}.$$

By the third part of Theorem 5.7, each of these row or column permutations multiplies $\det(B)$ by -1, so $\det(C) = (-1)^{p+q}\det(B)$. By Lemma 5.1, we have $\det(C) = \det(A)$, so $\det(B) = (-1)^{p+q}$ $\det(A)$. □

Definition 5.2. Let $A \in \mathbb{C}^{m\times n}$. A *minor* of *order* k of A is a determinant of the form

$$det\left(A\begin{bmatrix} i_1 & \cdots & i_k \\ j_1 & \cdots & k_k \end{bmatrix}\right).$$

A *principal minor of order* k of A is a determinant of the form

$$\det\left(A\begin{bmatrix} i_1 & \cdots & i_k \\ i_1 & \cdots & i_k \end{bmatrix}\right).$$

The *leading principal minor* of order k is the determinant

$$\det\left(A\begin{bmatrix} 1 & \cdots & k \\ 1 & \cdots & k \end{bmatrix}\right).$$

For $A \in \mathbb{C}^{n\times n}$, $\det(A)$ is the unique principal minor of order n, and the principal minors of order 1 of A are just the diagonal entries of A: $a_1^1, \ldots, a_n^n$.

Theorem 5.9. *Let $A \in \mathbb{C}^{n\times n}$. Define the matrix $A_i^j \in \mathbb{C}^{(n-1)\times(n-1)}$ as*

$$A_i^j = A\begin{bmatrix} 1 & \cdots & i-1 & i+1 & \cdots & n \\ 1 & \cdots & j-1 & j+1 & \cdots & h \end{bmatrix},$$

that is, the matrix obtained from A by removing the ith row and the jth column. Then, we have

$$\sum_{j=1}^{n}(-1)^{i+j}a_i^j \det(A_\ell^j) = \begin{cases} \det(A) & \text{if } i = \ell, \\ 0 & \text{otherwise,} \end{cases}$$

for every i, ℓ, $1 \leqslant i, \ell \leqslant n$.

Proof. Let $\boldsymbol{x}_i$ be the ith row of A, which can be expressed as

$$\boldsymbol{x}_i = \sum_{j=1}^{n} a_i^j \boldsymbol{e}_j,$$

where $\boldsymbol{e}_1, \ldots, \boldsymbol{e}_n$ is a basis of $\mathbb{R}^n$ such that $d_n(\boldsymbol{e}_1, \ldots, \boldsymbol{e}_n) = 1$.

By the linearity of d_n, we have

$$\begin{aligned} d_n(A) &= d_n(\boldsymbol{x}_1, \ldots, \boldsymbol{x}_n) \\ &= d_n\left(\boldsymbol{x}_1, \ldots, \boldsymbol{x}_{i-1}, \sum_{j=1}^{n} a_i^j \boldsymbol{e}_j, \boldsymbol{x}_{i+1}, \ldots, \boldsymbol{x}_n\right) \\ &= \sum_{j=1}^{n} a_i^j d_n(\boldsymbol{x}_1, \ldots, \boldsymbol{x}_{i-1}, \boldsymbol{e}_j, \boldsymbol{x}_{i+1}, \ldots, \boldsymbol{x}_n). \end{aligned}$$

The determinant $d_n(\boldsymbol{x}_1, \ldots, \boldsymbol{x}_{i-1}, \boldsymbol{e}_j, \boldsymbol{x}_{i+1}, \ldots, \boldsymbol{x}_n)$ corresponds to a matrix $D^{(i,j)}$ obtained from A by replacing the ith row by the sequence

$$(0, \ldots, 0, 1, 0, \ldots, 0),$$

whose unique nonzero component is on the jth position. Next, by multiplying the ith row by $-a_k^j$ and adding the result to the kth row for $1 \leqslant k \leqslant i-1$ and $i+1 \leqslant k \leqslant n$, we obtain a matrix $E^{(i,j)}$ that coincides with the matrix A with the following exceptions:

(i) the elements of row i are 0 with the exception of the jth element of this row that equals 1, and

(ii) the elements of column j are 0 with the exception of the element mentioned above.

Clearly, $\det(D^{(i,j)}) = \det(E^{(i,j)})$. By applying Lemma 5.2, we obtain $\det(E^{(i,j)}) = (-1)^{i+j}\det(A_i^j)$, so

$$d_n(A) = \sum_{j=1}^{n} a_i^j d_n(E^{(i,j)}) = \sum_{j=1}^{n}(-1)^{i+j} a_i^j \det(A_i^j),$$

which is the first case of the desired formula.

Suppose now that $i \neq \ell$. The same determinant could be computed by using an expansion on the ℓth row, as follows:

$$d_n(A) = \sum_{j=1}^{n}(-1)^{i+j} a_\ell^j \det(A_\ell^j).$$

Then $\sum_{j=1}^{n}(-1)^{i+j} a_i^j \det(A_\ell^j)$ is the determinant of a matrix obtained from A by replacing the ℓth row by the ith row and such a determinant is 0 because the new matrix has two identical rows. This proves the second case of the equality of the theorem. □

The equality of the theorem is known as the *Laplace expansion of* $\det(A)$ *by row* i.

Since the determinant of a matrix A equals the determinant of A', $\det(A)$ can be expanded by the jth row as

$$\det(A) = \sum_{i=1}^{n}(-1)^{i+j} a_i^j \det(A_i^j)$$

for every $1 \leqslant j \leqslant n$. Thus, we have

$$\sum_{i=1}^{n}(-1)^{i+j} a_i^j \det(A_\ell^j) = \begin{cases} \det(A) & \text{if } i = \ell, \\ 0 & \text{if } i \neq \ell. \end{cases}$$

This formula is the *Laplace expansion of* $\det(A)$ *by column* j.

The number $\mathsf{cof}(a_i^j) = (-1)^{i+j}\det(A_i^j)$ is the *cofactor* of a_{ij} in either kind of Laplace expansion. Thus, both types of Laplace expansions can be succinctly expressed by the equalities

$$\det(A) = \sum_{j=1}^{n} a_i^j \mathsf{cof}(a_i^j) = \sum_{i=1}^{n} a_i^j \mathsf{cof}(a_i^j) \tag{5.3}$$

for all $i, j \in \{1, \ldots, n\}$.

Cofactors of the form $\mathsf{cof}(a_i^i)$ are known as *principal cofactors* of A.

Example 5.2. Let $\boldsymbol{a} = (a_1, \ldots, a_n)$ be a sequence of n real numbers. The *Vandermonde determinant* $V_{\boldsymbol{a}}$ contains on its ith row the successive powers of a_i, namely $a_i^0 = 1, a_i^1 = a_i, \ldots, a_i^n$:

$$V_{\boldsymbol{a}} = \begin{vmatrix} 1 & a_1 & (a_1)^2 & \cdots & (a_1)^{n-1} \\ 1 & a_2 & (a_2)^2 & \cdots & (a_2)^{n-1} \\ \vdots & \vdots & & \cdots & \vdots \\ 1 & a_n & (a_n)^2 & \cdots & (a_n)^{n-1} \end{vmatrix}.$$

By subtracting the first line from the remaining lines, we have

$$\begin{aligned} V_{\boldsymbol{a}} &= \begin{vmatrix} 1 & a_1 & a_1^2 & \cdots & a_1^{n-1} \\ 0 & a_2 - a_1 & (a_2)^2 - (a_1)^2 & \cdots & (a_2)^{n-1} - (a_1)^{n-1} \\ \vdots & \vdots & & \cdots & \vdots \\ 0 & a_n - a_1 & (a_n)^2 - (a_1)^2 & \cdots & (a_n)^{n-1} - (a_1)^{n-1} \end{vmatrix} \\ &= \begin{vmatrix} a_2 - a_1 & (a_2)^2 - (a_1)^2 & \cdots & (a_2)^{n-1} - (a_1)^{n-1} \\ \vdots & \vdots & \cdots & \vdots \\ a_n - a_1 & (a_n)^2 - (a_1)^2 & \cdots & (a_n)^{n-1} - (a_1)^{n-1} \end{vmatrix}. \end{aligned}$$

Factoring now $a_{i+1} - a_1$ from the ith line of the new determinant for $1 \leqslant i \leqslant n$ yields

$$V_{\boldsymbol{a}} = (a_2 - a_1) \cdots (a_n - a_1) \begin{vmatrix} 1 & a_2 + a_1 & \cdots & \sum_{i=0}^{n-2} a_2^{n-2-i} a_1^i \\ \vdots & \vdots & \cdots & \vdots \\ 1 & a_n + a_1 & \cdots & \sum_{i=0}^{n-2} a_n^{n-2-i} a_1^i \end{vmatrix}.$$

Consider two successive columns of this determinant:

$$\mathbf{c}_k = \begin{pmatrix} \sum_{i=0}^{k-1} a_2^{k-1-i} a_1^i \\ \vdots \\ \sum_{i=0}^{k-1} a_n^{k-1-i} a_1^i \end{pmatrix} \text{ and } \mathbf{c}_{k+1} = \begin{pmatrix} \sum_{i=0}^{k} a_2^{k-i} a_1^i \\ \vdots \\ \sum_{i=0}^{k} a_n^{k-i} a_1^i \end{pmatrix}.$$

Observe that

$$\mathbf{c}_{k+1} = \begin{pmatrix} a_2^k \\ \vdots \\ a_n^k \end{pmatrix} + a_1 \mathbf{c}_k,$$

it follows that by subtracting from each column $\mathbf{c}_{k+1}$ from the previous column multiplied by a_1 (from right to left), we obtain

$$\begin{aligned} V_{\mathbf{a}} &= (a_2 - a_1) \cdots (a_n - a_1) \begin{vmatrix} 1 & a_2 & \cdots & a_2^{n-2} \\ \vdots & \vdots & \cdots & \vdots \\ 1 & a_n & \cdots & a_n^{n-2} \end{vmatrix} \\ &= (a_2 - a_1) \cdots (a_n - a_1) V_{(a_2, \ldots, a_n)}. \end{aligned}$$

By applying repeatedly this formula, we obtain

$$V_{\mathbf{a}} = \prod_{p>q} (a_p - a_q),$$

where $1 \leqslant p, q \leqslant n$.

Theorem 5.8 can be extended to products of rectangular matrices.

Theorem 5.10. *Let $A \in \mathbb{C}^{m \times n}$ and $B \in \mathbb{C}^{n \times m}$ be two matrices, where $m \leqslant n$. We have*

$$\begin{aligned} &\det(AB) \\ &= \sum \left\{ \det \left(A \begin{bmatrix} 1 & \cdots & m \\ k_1 & \cdots & k_m \end{bmatrix} \right) \det \left(B \begin{bmatrix} k_1 & \cdots & k_m \\ 1 & \cdots & m \end{bmatrix} \right) \right. \\ &\qquad \left. \mid 1 \leqslant k_1 < k_2 < \cdots < k_m \leqslant n \right\}. \end{aligned}$$

This equality is known as the Cauchy–Binet formula.

Proof. Let $\boldsymbol{a}_1, \ldots, \boldsymbol{a}_n$ be the rows of the matrix A and let $C = AB$. The first column of the matrix AB equals $\sum_{k_1=1}^{n} \boldsymbol{a}_{k_1} b_{k_1 1}$. Since $\det(C)$ is linear, we can write

$$\det(C) = \sum_{k_1=1}^{n} b_{k_1 1} \begin{vmatrix} a_{k_1}^1 & c_1^2 & \cdots & c_1^n \\ a_{k_1}^2 & c_2^2 & \cdots & c_2^n \\ \vdots & \vdots & \cdots & \vdots \\ a_{k_1}^m & c_m^2 & \cdots & c_m^n \end{vmatrix}.$$

Similarly, the second row of C equals $\sum_{k_2=1}^{n} \boldsymbol{a}_{k_2} b_{k_2 1}$. A further decomposition yields the sum

$$\det(C) = \sum_{k_1=1}^{n} \sum_{k_2=1}^{n} b_{k_1 1} b_{k_2 2} \begin{vmatrix} a_{k_1}^1 & a_{k_2}^1 & \cdots & c_1^n \\ a_{k_1}^2 & a_{k_2}^2 & \cdots & c_2^n \\ \vdots & \vdots & \cdots & \vdots \\ a_{k_1}^m & a_{k_2}^2 & \cdots & c_m^n \end{vmatrix},$$

and so on. Eventually, we can write

$$\det(C) = \sum_{k_1=1}^{n} \cdots \sum_{k_m=1}^{n} b_{k_1}^1 \cdots b_{k_m}^m \begin{vmatrix} a_1^{k_1} & \cdots & a_1^{k_m} \\ \vdots & \vdots & \vdots \\ a_m^{k_1} & \cdots & a_m^{k_m} \end{vmatrix},$$

due to the multilinearity of the determinants. Only terms involving distinct numbers $k_1, \ldots, k_m$ are retained in this sum because any term with $k_p = k_q$ equals 0. Suppose that $\{k_1, \ldots, k_m\} = \{h_1, \ldots, h_m\}$, where $h_1 < \cdots < h_m$ and ϕ is the bijection defined by $k_i = \phi(h_i)$ for $1 \leqslant i \leqslant m$. Then

$$\begin{vmatrix} a_1^{k_1} & \cdots & a_1^{k_m} \\ \vdots & \vdots & \vdots \\ a_m^{k_1} & \cdots & a_m^{k_m} \end{vmatrix} = (-1)^{inv(k_1, \ldots, k_m)} \begin{vmatrix} a_1^{h_1} & \cdots & a_1^{h_m} \\ \vdots & \vdots & \vdots \\ a_m^{h_1} & \cdots & a_m^{h_m} \end{vmatrix},$$

which allows us to write

$$\det(C) = \sum_{h_1 < \cdots < h_m} \begin{vmatrix} a_1^{h_1} & \cdots & a_1^{h_m} \\ \vdots & \vdots & \vdots \\ a_m^{h_1} & \cdots & a_m^{h_m} \end{vmatrix} \sum_{\phi} (-1)^{inv(\phi)} b_{\phi(h_1)}^1 \cdots b_{\phi(h_m)}^m,$$

where ϕ is a permutation of the set $\{h_1, \ldots, h_m\}$. The last equality is equivalent to the Cauchy–Binet formula. □

Example 5.3. Let $A \in \mathbb{C}^{2\times n}$ and $B \in \mathbb{C}^{n\times 2}$ be the matrices

$$A = \begin{pmatrix} a_1 & \cdots & a_n \\ b_1 & \cdots & b_n \end{pmatrix}$$

and

$$B = \begin{pmatrix} c_1 & d_1 \\ \vdots & \vdots \\ c_n & d_n \end{pmatrix}.$$

Note that

$$AB = \begin{pmatrix} \sum_{i=1}^n a_i c_i & \sum_{i=1}^n a_i d_i \\ \sum_{i=1}^n b_i c_i & \sum_{i=1}^n b_i d_i \end{pmatrix}.$$

An application of the Binet–Cauchy formula implies

$$\left(\sum_{i=1}^n a_i c_i\right)\left(\sum_{i=1}^n b_i d_i\right) - \left(\sum_{i=1}^n a_i d_i\right)\left(\sum_{i=1}^n b_i c_i\right)$$
$$= \sum_{1\leqslant i<j\leqslant n} (a_i b_j - a_j b_i)(c_i d_j - d_i c_j).$$

This equality is known as *Lagrange's Identity.*

Let $C \in \mathbb{C}^{n\times n}$ be a square matrix and let $\boldsymbol{b}$ be a vector in $\mathbb{C}^n$. Denote by $(C \overset{i}{\leftarrow} \boldsymbol{b})$ the matrix obtained from C by replacing the ith column by $\boldsymbol{b}$.

Example 5.4. Let $A \in \mathbb{C}^{n\times n}$ be a matrix and let

$$\boldsymbol{e}_p = \begin{pmatrix} 0 \\ \vdots \\ 0 \\ 1 \\ 0 \\ \vdots \\ 0 \end{pmatrix}$$

be the vector whose unique non zero component is the 1 located in the pth position. Then $(A \overset{q}{\leftarrow} \boldsymbol{e}_p)$ is the (p,q)-minor of A and $(-1)^{p+q}\det(A \overset{q}{\leftarrow} \boldsymbol{e}_p)$ is the cofactor of a_{pq}.

Theorem 5.11. *Let* $\{g_i^j : \mathbb{R} \longrightarrow \mathbb{R} \mid 1 \leqslant i, j \leqslant n\}$ *be a collection of* n^2 *differentiable functions and let* $G(x)$ *be the matrix defined by*

$$G(x) = \begin{pmatrix} g_1^1(x) & \cdots & g_1^n(x) \\ \vdots & \cdots & \vdots \\ g_n^1(x) & \cdots & g_n^n(x) \end{pmatrix}$$

for $x \in \mathbb{R}$. *The derivative of the function* $\det(G(x))$ *is given by*

$$(\det(G(x))' = \sum_{i=1}^{n} \det(G(x) \xleftarrow{i} \boldsymbol{g}_i(x)'),$$

where

$$\boldsymbol{g}_i(x)' = \begin{pmatrix} (g_1^i)' \\ \vdots \\ (g_n^i)' \end{pmatrix}$$

is the column of the derivatives of the functions positioned in column i *of the matrix* $G(x)$, *for* $1 \leqslant i \leqslant n$.

Proof. By the definition of determinants, we have

$$\det(G(x)) = \sum_{j_1,\ldots,j_n} (-1)^{inv(j_1,\ldots,j_n)} g_1^{j_1}(x) g_2^{j_2}(x) \cdots g_n^{j_n}(x).$$

Therefore, we can write

$$\begin{aligned}\det(G(x))' &= \sum_{j_1,\ldots,j_n} (-1)^{inv(j_1,\ldots,j_n)} g_1^{j_1}(x)' g_2^{j_2}(x) \cdots g_n^{j_n}(x) \\ &\quad + \sum_{j_1,\ldots,j_n} (-1)^{inv(j_1,\ldots,j_n)} g_1^{j_1}(x) g_2^{j_2}(x)' \cdots g_n^{j_n}(x) \\ &\quad + \sum_{j_1,\ldots,j_n} (-1)^{inv(j_1,\ldots,j_n)} g_1^{j_1}(x) g_2^{j_2}(x) \cdots g_n^{j_n}(x)' \\ &= \sum_{i=1}^{n} \det(G(x) \xleftarrow{i} \boldsymbol{g}_i(x)'),\end{aligned}$$

which concludes the argument. □

Example 5.5. Let $A = (a_{ij}) \in \mathbb{R}^{3\times 3}$ and let $G(x) = \det(A - xI_3)$. We have

$$
\begin{aligned}
&(\det(G(x))' \\
&= \det(G(x) \overset{1}{\leftarrow} (-\boldsymbol{e}_1)) + \det(G(x) \overset{2}{\leftarrow} (-\boldsymbol{e}_2) + \det(G(x) \overset{3}{\leftarrow} (-\boldsymbol{e}_3)) \\
&= \begin{vmatrix} -1 & a_1^2 & a_1^3 \\ 0 & a_2^2 - x & a_2^3 \\ 0 & a_3^2 & a_3^3 - x \end{vmatrix} + \begin{vmatrix} a_1^1 - x & 0 & a_1^3 \\ a_2^1 & -1 & a_2^3 \\ a_3^1 & 0 & a_3^3 - x \end{vmatrix} + \begin{vmatrix} a_1^1 - x & a_1^2 & 0 \\ a_2^1 & a_2^2 - x & 0 \\ a_3^1 & a_3^2 & -1 \end{vmatrix} \\
&= -\begin{vmatrix} a_2^2 - x & a_2^3 \\ a_3^2 & a_3^3 - x \end{vmatrix} - \begin{vmatrix} a_1^1 - x & a_1^3 \\ a_3^1 & a_3^3 - x \end{vmatrix} - \begin{vmatrix} a_1^1 - x & a_1^2 \\ a_2^1 & a_2^2 - x \end{vmatrix}.
\end{aligned}
$$

Applying the same technique we compute the second derivative

$$
\begin{aligned}
(\det(G(x))'' &= -\begin{vmatrix} -1 & a_2^3 \\ 0 & a_3^3 - x \end{vmatrix} - \begin{vmatrix} a_2^2 - x & 0 \\ a_3^2 & -1 \end{vmatrix} \\
&\quad - \begin{vmatrix} -1 & a_1^3 \\ 0 & a_3^3 - x \end{vmatrix} - \begin{vmatrix} a_1^1 - x & 0 \\ a_3^1 & -1 \end{vmatrix} \\
&\quad - \begin{vmatrix} -1 & a_1^2 \\ 0 & a_2^2 - x \end{vmatrix} - \begin{vmatrix} a_1^1 - x & 0 \\ a_2^1 & -1 \end{vmatrix} \\
&= 2(a_1^1 + a_2^2 + a_3^3 - 3x).
\end{aligned}
$$

Note that $G(0) = \det(A)$, $-G'(0)$ equals the sum of order 2 principal minors of A, while $G''(0)$ is twice the sum of order 1 principal minors of A.

This observation can be generalized to square matrices of any size: if $A \in \mathbb{R}^{n\times n}$, then for the kth derivative of the function $G(x) = \det(A - xI_n)$, we have

$$G^{(k)}(0) = (-1)^k k! S_{n-k}(A),$$

where $S_p(A)$ is the sum of all order-p principal minors of A (see Exercise 5.6).

Example 5.6. Let $Q_n(a, b)$ be the determinant of order n:

$$
Q_n(a, b) = \begin{vmatrix} a & 1 & 1 & 1 & \cdots & 1 \\ 1 & b & 1 & 0 & \cdots & 0 \\ 1 & 0 & b & 0 & \cdots & 0 \\ \vdots & \vdots & \vdots & \vdots & \cdots & \vdots \\ 1 & 0 & 0 & 0 & \cdots & b \end{vmatrix}.
$$

We have $Q_2(a,b) = ab - 1$. To compute the value of $Q_n(a,b)$, note that by expanding this determinant by its last column, we have

$$\begin{aligned} Q_n(a,b) &= (-1)^{n+1} \begin{vmatrix} 1 & b & 0 & 0 & \cdots 0 \\ 1 & 0 & b & 0 & \cdots 0 \\ \vdots & \vdots & \vdots & \vdots & \cdots & \vdots \\ 1 & 0 & 0 & 0 & \cdots b \\ 1 & 0 & 0 & 0 & \cdots 0 \end{vmatrix} + bQ_{n-1}(a,b) \\ &= -b^{n-2} + bQ_{n-1}(a,b). \end{aligned}$$

It is easy to verify that $Q_n(a,b) = b^{n-1}a - (n-1)b^{n-2}$ for $n \geqslant 2$.

5.3 Cramer's Formula

In principle, determinants can be used for solving linear systems of equation and we discuss a formula that allows us to do just that. Let $A \in \mathbb{C}^{n\times n}$, $\boldsymbol{b} \in \mathbb{C}^n$, and consider the linear system $A\boldsymbol{x} = \boldsymbol{b}$. The columns of the matrix A are denoted by $\boldsymbol{a}_1, \ldots, \boldsymbol{a}_n$, that is, $A = (\boldsymbol{a}_1 \ \cdots \ \boldsymbol{a}_n)$. Note that

$$\begin{aligned} A(I_n \overset{i}{\leftarrow} \boldsymbol{x}) &= A(\boldsymbol{e}_1\ \boldsymbol{e}_2 \ \cdots \ \boldsymbol{e}_{i-1}\ \boldsymbol{x}\ \boldsymbol{e}_{i+1} \ \cdots \ \boldsymbol{e}_n) \\ &= (\boldsymbol{a}_1\ \boldsymbol{a}_2 \ \cdots \ \boldsymbol{a}_{i-1}\ \boldsymbol{b}\ \boldsymbol{a}_{i+1} \ \cdots \ \boldsymbol{a}_n) \\ &= (A \overset{i}{\leftarrow} \boldsymbol{b}). \end{aligned} \tag{5.4}$$

By expanding the determinant

$$\det(I_n \overset{i}{\leftarrow} \boldsymbol{x}) = \begin{vmatrix} 1 & 0 & \cdots & x_1 & \cdots & 0 \\ 0 & 1 & \cdots & x_2 & \cdots & 0 \\ \vdots & \vdots & \cdots & \vdots & \cdots & \vdots \\ 0 & 0 & \cdots & x_i & \cdots & 0 \\ \vdots & \vdots & \cdots & \vdots & \cdots & \vdots \\ 0 & 0 & \cdots & x_n & \cdots & 1 \end{vmatrix}$$

by its ith row, we obtain $\det(I_n \overset{i}{\leftarrow} \boldsymbol{x}) = x_i$. Thus, computing the determinants on both sides of Equality (5.4), we obtain

$$\det(A)x_i = \det(A \overset{i}{\leftarrow} \boldsymbol{b}),$$

for $1 \leqslant i \leqslant n$. This formula for computing the components of the solution of the system $A\boldsymbol{x} = \boldsymbol{b}$ is known as *Cramer's formula.*[1]

Definition 5.3. Let $A \in \mathbb{R}^{n\times n}$ be a square matrix. The *adjoint matrix* of A is the matrix

$$\mathsf{adj}(A) = \begin{pmatrix} \mathsf{cof}(a_{11}) & \cdots & \mathsf{cof}(a_{n1}) \\ \vdots & \cdots & \vdots \\ \mathsf{cof}(a_{1n}) & \cdots & \mathsf{cof}(a_{nn}) \end{pmatrix},$$

that is, the transposed matrix of the matrix whose entries are the cofactors of the elements of A.

Example 5.7. For the matrix

$$A = \begin{pmatrix} 1 & 0 & 2 \\ 0 & -1 & 1 \\ 1 & 2 & 1 \end{pmatrix}$$

the matrix of cofactors is

$$C = \begin{pmatrix} -3 & 1 & 1 \\ 4 & -1 & -2 \\ 2 & -1 & -1 \end{pmatrix},$$

so the adjoint matrix is

$$\mathsf{adj}(A) = \begin{pmatrix} -3 & 4 & 2 \\ 1 & -1 & -1 \\ 1 & -2 & -1 \end{pmatrix}.$$

[1]Gabriel Cramer (1704–1752) was a Swiss mathematician who studied and taught at the University of Geneva. His method of solving linear systems was published in 1750.

Theorem 5.12. *If $A \in \mathbb{R}^{n\times n}$, then*

$$A\ \mathit{adj}(A) = \det(A) I_n.$$

Proof. The Equalities (5.3) allow us to write

$$(A\ \mathsf{adj}(A))_{ij} = \sum_{k=1}^{n} a_{ik}(\mathsf{adj}(A))_{kj} == \sum_{k=1}^{n} a_{ik}\mathsf{cof}(a_{jk})$$

$$= \begin{cases} \det(A) & \text{if } i = j \\ 0 & \text{otherwise.} \end{cases}$$

Therefore, we have $A\ \mathsf{adj}(A) = \det(A) I_n$. □

This allows us to give an explicit formula for computing the inverse of a non-singular matrix A as

$$A^{-1} = \frac{1}{\det(A)} \mathsf{adj}(A). \tag{5.5}$$

By Theorem 5.8, if A is a non-singular matrix, we have $\det(A)\det(A^{-1}) = 1$.

Example 5.8. For the matrix $A \in \mathbb{R}^{3\times 3}$ introduced in Example 5.7, we have $det(A) = -1$ and

$$A^{-1} = \frac{1}{\det(A)} \mathsf{adj}(A) = \begin{pmatrix} 3 & -4 & -2 \\ -1 & 1 & 1 \\ -1 & 2 & 1 \end{pmatrix}.$$

5.4 Partitioned Matrices and Determinants

Lemma 5.3. *Let $A \in \mathbb{C}^{n\times n}$, $B \in \mathbb{C}^{m\times m}$ be two square matrices and let $D \in \mathbb{C}^{(m+n)\times(m+n)}$ be the matrix*

$$D = \begin{pmatrix} A & O_{m,n} \\ C & B \end{pmatrix},$$

where $C \in \mathbb{C}^{m\times n}$. Then we have $\det(D) = \det(A)\det(B)$.

If the matrix $F \in \mathbb{C}^{(m+n)\times(m+n)}$ is

$$F = \begin{pmatrix} A & E \\ O_{m\times n} & B \end{pmatrix},$$

where $E \in \mathbb{C}^{n\times m}$, then $\det(F) = \det(A)\det(B)$.

Proof. Suppose that $D = (d_{ij})$. The definition of $\det(D)$ implies that

$$\det(D) = \sum_{j_1,\ldots,j_n,j_{n+1},\ldots,j_{n+m}} (-1)^{inv(j_1,\ldots,j_{n+m})} d_{1j_1} \cdots d_{n+m\, j_{n+m}}.$$

Each term of this sum involves factors chosen from each row (specified by the first subscript of d_{ij}). Note that any term

$$d_{1j_1} \cdots d_{nj_n} d_{n+1\, j_{n+1}} \cdots d_{n+m\, j_{n+m}}$$

in which any of the first n subscripts $j_1, \ldots, j_n$ is at least equal to $n+1$ equals 0. Therefore, non-zero terms are those in which $(j_1, \ldots, j_n)$ is a permutation of $(1, \ldots, n)$. In such terms $(j_{n+1}, \ldots, j_{n+m})$ is a permutation of $(n + 1, \ldots, n + m)$. By the definition of the matrix D, the product $d_{1j_1} \cdots d_{nj_n}$ is actually $a_{1j_1} \cdots a_{nj_n}$; the product $d_{n+1\, j_{n+1}} \cdots d_{n+m\, j_{n+m}}$ equals $b_{1k_1} \cdots b_{mk_m}$, where $k_1 = j_{n+1} - n, \ldots, k_m = j_{n+m} - n$ and

$$inv(j_1, \ldots, j_n, j_{n+1}, \ldots, j_{n+m}) = inv(j_1, \ldots, j_n) + inv(k_1, \ldots, k_m).$$

This allows us to write

$$\begin{aligned}\det(D) &= \left(\sum_{j_1 \cdots j_n} a_{1j_1} \cdots a_{nj_n}\right) \cdot \left(\sum_{k_1 \cdots k_m} b_{1k_1} \cdots b_{mk_m}\right)\\ &= \det(A)\det(B).\end{aligned}$$

The proof of the second part of the lemma is similar. □

Theorem 5.13. *Let A be a block upper (or lower) triangular partitioned matrix given by*

$$A = \begin{pmatrix} A_{11} & A_{12} & \cdots & A_{1m} \\ O & A_{22} & \cdots & A_{2m} \\ \vdots & \vdots & \cdots & \vdots \\ O & O & \cdots & A_{mm} \end{pmatrix},$$

where $A_{ii} \in \mathbb{R}^{p_i \times p_i}$ for $1 \leqslant i \leqslant m$. Then

$$\det(A) = \det(A_{11})\det(A_{22}) \cdots \det(A_{mm}).$$

If A is a block lower triangular matrix

$$A = \begin{pmatrix} A_{11} & O & \cdots & O \\ A_{21} & A_{22} & \cdots & O \\ \vdots & \vdots & \cdots & \vdots \\ A_{m1} & A_{m2} & \cdots & A_{mm} \end{pmatrix},$$

the same equality holds.

Proof. The argument is by induction on m, where $m \geqslant 2$. The base case, $m = 2$ was shown in Lemma 5.3. Suppose that the statement holds for partitioned upper diagonal matrices having $m - 1$ diagonal blocks. Note that if

$$A = \begin{pmatrix} A_{11} & A_{12} & \cdots & A_{1m} \\ O & A_{22} & \cdots & A_{2m} \\ \vdots & \vdots & \cdots & \vdots \\ O & O & \cdots & A_{mm} \end{pmatrix},$$

we can also regard A as a partitioned matrix

$$A = \begin{pmatrix} B & C \\ O & A_{mm} \end{pmatrix},$$

where B is a partitioned upper diagonal matrix having $m-1$ diagonal blocks. Then, by the base case and by the inductive hypothesis, we have

$$\det(A) = \det(B)\det(A_{mm}) = \det(A_1)\cdots\det(A_{m-1})\det(A_{mm}).$$

□

Theorem 5.14. *Let $A \in \mathbb{C}^{m\times m}$, $D \in \mathbb{C}^{n\times n}$ be two square matrices and let $E \in \mathbb{C}^{(m+n)\times(m+n)}$ be the matrix*

$$E = \begin{pmatrix} A & B \\ C & D \end{pmatrix},$$

where $B \in \mathbb{C}^{m\times n}$ and $C \in \mathbb{C}^{n\times m}$. If the matrix A is invertible, then $\det(E) = \det(A)\det(D - CA^{-1}B)$.

Proof. If A is invertible, then

$$\begin{pmatrix} I_m & O \\ -CA^{-1} & I_n \end{pmatrix} \begin{pmatrix} A & B \\ C & D \end{pmatrix} = \begin{pmatrix} A & B \\ 0 & D - CA^{-1}B \end{pmatrix},$$

which implies

$$\det(E) = \det \begin{pmatrix} A & B \\ 0 & D - CA^{-1}B \end{pmatrix}.$$

Theorem 5.13 implies the desired equality. □

Theorem 5.15. *Let $A, B, C, D \in \mathbb{C}^{m\times m}$ be four square matrices such that $AC = CA$ and let $E \in \mathbb{C}^{2m\times 2m}$ be the matrix*

$$E = \begin{pmatrix} A & B \\ C & D \end{pmatrix}.$$

We have $\det(E) =$ *det*$(AD - CB)$.

Proof. Suppose initially that A is invertible, so $\det(E) = \det(A^{-1})\det(D - CA^{-1}B)$ by Theorem 5.14. Then,

$$\begin{aligned}\det(E) &= \det(A)\det(D - CA^{-1}B) = \det(AD - ACA^{-1}B) \\ &= \det(AD - CAA^{-1}B) = \det(AD - CB).\end{aligned}$$

If A is not invertible, that is, if A is singular, consider the continuous function $f : \mathbb{R} \longrightarrow \mathbb{R}$ defined by $f(x) = \det(A + xI)$ for $x \in \mathbb{R}$. There exists $\delta > 0$ such that if $x \in (0, \delta)$, then $f(x) \neq 0$, which means that $\det(A + xI) \neq 0$, which implies that $A + xI$ is an invertible matrix. Note that if $AC = CA$, then $(A + xI)C = C(A + xI)$, so the first part of the argument can be applied to the matrix

$$E_x = \begin{pmatrix} A + xI & B \\ C & D \end{pmatrix}.$$

This implies $\det(E_x) = \det((A + xI)D - CB)$ if $x \in (0, \delta)$. If x tends toward 0, by the continuity of $f(x)$ it follows that

$$\det(E) = \lim_{x\to 0} \det((A + xI)D - CB) = \det(AD - BC),$$

which concludes our argument. □

The argument presented in the second part of the proof of Theorem 5.15 is known as a *continuity argument.*

Theorem 5.16. *Let $A \in \mathbb{C}^{n\times n}$ be a square matrix. The rank of A equals the size of the largest non-zero minor of A.*

Proof. This theorem is a direct consequence of Theorem 3.53. □

5.5 Resultants

A *monomial* in the variables $x_1, \ldots, x_n$ is a product $x_1^{a_1} x_2^{a_2} \cdots x_n^{a_n}$, where $a_1, \ldots, a_n$ are non-negative integers. A more condensed notation for the previous monomial is x^a, where $x = (x_1, \ldots, x_n)$ and $a = (a_1, a_2, \ldots, a_n)$. The *total degree* of the monomial is $|a| = a_1 + a_2 + \cdots + a_n$.

The set of polynomials with complex (real) coefficients having $\{x_0, x_1, \ldots, x_n\}$ as set of variables is denoted by $\mathbb{C}[x_0, \ldots, x_n]$ (by $\mathbb{R}[x_0, \ldots, x_n]$, respectively).

Definition 5.4. A polynomial in several variables is *homogeneous* if all monomials that occur as terms in the polynomial have the same total degree; otherwise, the polynomial is said to be *inhomogeneous.*

Example 5.9. The polynomial

$$f(x, y) = x^4 + 3x^3y - 2x^2y^2 + 4y^4$$

is homogeneous. All four monomials that occur in this polynomial have total degree 4.

An inhomogeneous polynomial can be transformed into a homogeneous polynomial by introducing a new variable.

Example 5.10. Let $f(x, y) = x^2 + xy^2 - 5$ be a polynomial in two variables. By substituting x with $\frac{x}{z}$ and y with $\frac{y}{z}$, we have

$$f\left(\frac{x}{z}, \frac{y}{z}\right) = \frac{x^2}{z^2} + \frac{xy^2}{z^3} - 5.$$

The polynomial

$$g(x, y, z) = z^3 f\left(\frac{x}{z}, \frac{y}{z}\right) = x^2z + xy^2 - 5z^3$$

is clearly homogeneous. Then (x_0, y_0) is a solution of $f(x, y) = 0$, if and only if (x_0, y_0, z_0) is a solution for $g(x, y, z) = 0$ for any $z_0 \neq 0$.

The polynomial transformation described above is known as *homogenization.* Polynomials f and g are said to be *associated.*

Theorem 5.17. *A non-null polynomial f is homogeneous of degree n if and only if the equality*

$$f(tx_1, \ldots, tx_r) = t^n f(x_1, \ldots, x_r) \tag{5.6}$$

holds.

Proof. The necessity of the condition is immediate. To prove the sufficiency suppose that f can be written as

$$f = F_{n_1} + F_{n_2} + \cdots + F_{n_k},$$

where F_{n_i} is homogeneous of degree n_i and is not 0. Equality (5.6) becomes

$$t^{n_1}F_{n_1} + t^{n_2}F_{n_2} + \cdots + t^{n_k}F_{n_k} = t^n F_{n_1} + t^n F_{n_2} + \cdots + t^n F_{n_k},$$

hence $t^{n_i} = t^n$ for $1 \leqslant i \leqslant k$. Therefore, $k = 1$ and $n_1 = n$. □

Theorem 5.18. *Any factor of a homogeneous polynomial is homogeneous.*

Proof. Let h be a homogeneous polynomial that can be factored as $h = fg$. Suppose that f is not homogeneous. Then f can be written as $f = F_i + F_{i+1} + \cdots + F_{i+j}$, where F_p is a homogeneous of degree p for $i \leqslant p \leqslant i + j$, $F_i \neq 0$, $F_{i+j} \neq 0$, and $j \geqslant 0$. Similarly, g can be written as $g = G_k + G_{k+1} + \cdots + G_{k+\ell}$, where G_q is a homogeneous of degree q for $1 \leqslant q \leqslant \ell$, $G_K \neq 0$, $G_{k+\ell} \neq 0$, and $\ell \geqslant 0$. Since $F_i G_i \neq 0$ and $F_{i+j}G_{k+\ell} \neq 0$, and $\mathsf{deg}(F_i G_k) = i + k < i + j + k + \ell = \mathsf{deg}(F_{i+j}G(k+\ell)$, this contradicts the homogeneity of h. □

Corollary 5.4. *Let g be the associated homogeneous polynomial of a polynomial $f \in \mathbb{C}[x_1, \ldots, x_n]$. Any factor of g is associated with a factor of f and vice versa. Therefore, f is irreducible if and only if its associate is irreducible.*

Proof. This follows from Theorem 5.18. □

Let

$$f(x) = a_0 x^n + a_1 x^{n-1} + \cdots + a_n \text{ and } g(x) = b_0 x^m + b_1 x^{m-1} + \cdots + b_m \quad (5.7)$$

be two polynomials in $\mathbb{C}[x]$ such that $a_0 \neq 0$ and $b_0 \neq 0$. Their roots are the complex numbers $\alpha_1, \ldots, \alpha_n$, and $\beta_1, \ldots, \beta_m$, respectively.

Polynomials f and g have a common root if and only if the expression $R_{f,g}$ known as the *resultant of* f and g and defined as

$$\begin{aligned}R_{f,g} = a_0^m b_0^n &(\alpha_1 - \beta_1)(\alpha_1 - \beta_2)\cdots(\alpha_1 - \beta_m)\\ &(\alpha_2 - \beta_1)(\alpha_2 - \beta_2)\cdots(\alpha_2 - \beta_m)\\ &\vdots\\ &(\alpha_n - \beta_1)(\alpha_n - \beta_2)\cdots(\alpha_n - \beta_m)\end{aligned}$$

equals 0. Note that $R_{f,g}$ can be written as

$$R_{f,g} = a_0^m g(\alpha_1)g(\alpha_2)\cdots g(\alpha_n) \tag{5.8}$$

$$= (-1)^{mn} b_0^n f(\beta_1)f(\beta_2)\cdots f(\beta_m). \tag{5.9}$$

Also, we have $R_{f,g} = (-1)^{mn} R_{gf}$.

Theorem 5.19. *If f, g_1, and g_2 are three polynomials, then $R_{f,g_1g_2} = R_{f,g_1}R_{f,g_2}$.*

Proof. Suppose that the degrees of g_1 and g_2 are m_1 and m_2, respectively. Then,

$$\begin{aligned}R_{f,g_1g_2} &= a_0^{m_1+m_2}(g_1(\alpha_1)g_2(\alpha_1))\cdots(g_1(\alpha_n)g_2(\alpha_n))\\ &= a_0^{m_1}g_1(\alpha_1)\cdots g_1(\alpha_n)\cdots a_0^{m_2}g_2(\alpha_1)\cdots g_2(\alpha_n) = R_{f,g_1}R_{f,g_2}.\end{aligned}$$

□

Theorem 5.20. *Let $f, g \in \mathbb{C}[x]$ be the following polynomials:*

$$\begin{aligned}f(x) &= a_0x^n + a_1x^{n-1} + \cdots + a_n,\\ g(x) &= b_0x^m + b_1x^{m-1} + \cdots + b_m.\end{aligned}$$

Their resultant equals the determinant of order $m+n$ (known as the Sylvester determinant *of f and g):*

$$R_{f,g} = \begin{vmatrix} a_0 & a_1 & a_2 & \cdots & a_n & 0 & \cdots & 0\\ 0 & a_0 & a_1 & \cdots & a_{n-1} & a_n & \cdots & 0\\ \vdots & \vdots & \vdots & \vdots & \vdots & \vdots & \vdots & \vdots\\ \cdots & \cdots & \cdots & \cdots & \cdots & \cdots & \cdots & a_n\\ b_0 & b_1 & b_2 & \cdots & b_m & 0 & \cdots & 0\\ 0 & b_0 & b_1 & \cdots & b_{m-1} & b_m & \cdots & 0\\ \vdots & \vdots & \vdots & \vdots & \vdots & \vdots & \vdots & \vdots\\ \cdots & \cdots & \cdots & \cdots & \cdots & \cdots & \cdots & b_m \end{vmatrix}. \tag{5.10}$$

Proof. The first m rows of this determinant consist of the sequence of coefficients of f and each following row is shifted to the right by one position, while the last n rows of this determinant consist of the sequence of coefficients of g where each row is shifted to the right by one position.

If x is a root of f, we have the following equalities:

$$\begin{aligned}
a_0x^n + a_1x^{n-1} + a_2x^{n-2} + \cdots + [a_n - f(x)] &= 0,\\
a_0x^{n+1} + a_1x^n + a_2x^{n-1} + \cdots + [a_n - f(x)]x &= 0,\\
&\vdots\\
a_0x^{n+m-1} + a_1x^{n+m-2} + a_2x^{n+m-3} + \cdots + [a_n - f(x)]x^{m-1} &= 0.
\end{aligned}$$

Further, if x equals one of the roots of g, we have

$$\begin{aligned}
b_0x^m + b_1x^{m-1} + \cdots + b_m &= 0\\
b_0x^{m+1} + b_1x^m + \cdots + b_mx &= 0\\
&\vdots\\
b_0x^{m+n-1} + b_1x^{m+n-2} + \cdots + b_mx^m &= 0.
\end{aligned}$$

Consider now the following system of $n+m$ linear equations having as indeterminates $x^{n+m-1}, x^{n+m-2}, \ldots, x, 1$:

$$\begin{array}{llr}
a_0x^{n+m-1} + a_1x^{n+m-2} + a_2x^{n+m-3} + \cdots & & +(a_n - u)x^{m-1} = 0\\
\quad a_0x^{n+m-2} + a_1x^{n+m-3} + \cdots & & +(a_n - u)x^{m-2} = 0\\
\vdots & & \\
\cdots\cdots & & +(a_n - u) = 0\\
b_0x^{n+m-1} + b_1x^{n+m-2} + b_2x^{n+m-3} + \cdots & & +b_mx^{n-1} = 0\\
\quad b_0x^{n+m-2} + \cdots & & +b_nx^{n-1} = 0\\
\vdots & & \\
\cdots\cdots & & +b_{m-1}x + b_m = 0.
\end{array}$$

If the roots of g are $\beta_1, \ldots, \beta_m$, this system is compatible if $u = f(\beta_i)$ for some i, $1 \leqslant i \leqslant m$.

This means that the determinant

$$\begin{vmatrix} a_0 & a_1 & a_2 & \cdots & a_n - u & 0 & \cdots & 0 \\ 0 & a_0 & a_1 & \cdots & a_{n-1} & a_n - u & \cdots & 0 \\ \vdots & \vdots & \vdots & \vdots & \vdots & \vdots & \vdots & \vdots \\ \cdots & \cdots & \cdots & \cdots & \cdots & \cdots & \cdots & a_n \\ b_0 & b_1 & b_2 & \cdots & b_m & 0 & \cdots & 0 \\ 0 & b_0 & b_1 & \cdots & b_{m-1} & b_m & \cdots & 0 \\ \vdots & \vdots & \vdots & \vdots & \vdots & \vdots & \vdots & \vdots \\ \cdots & \cdots & \cdots & \cdots & \cdots & \cdots & \cdots & b_m \end{vmatrix}$$

equals 0 when $u = f(\beta_i)$ for some i, $1 \leqslant i \leqslant m$.

If this determinant is regarded as a polynomial in u of degree m, then its roots will equal $f(\beta_1), \ldots, f(\beta_m)$. Since the coefficient of u^m equals $(-1)^{m(n+1)} b_0^n$, it follows that the constant term of the polynomial (obtained by taking $u = 0$) equals $(-1)^m(-1)^{m(n+1)} b_0^n f(\beta_1) f(\beta_2) \cdots f(\beta_m) = R_{f,g}$. □

Example 5.11. Let f and g be the polynomials

$$\begin{aligned} f(x) &= x^3 - 2x^2 - x + 2, \\ g(x) &= x^2 - 5x + 6. \end{aligned}$$

The resultant of these polynomials is the determinant of order 5

$$R_{f,g} = \begin{vmatrix} 1 & -2 & -1 & 2 & 0 \\ 0 & 1 & -2 & -1 & 2 \\ 1 & -5 & 6 & 0 & 0 \\ 0 & 1 & -5 & 6 & 0 \\ 0 & 0 & 1 & -5 & 6 \end{vmatrix}$$

which is easily seen to equal 0.

Example 5.12. Let $f(x) = a_0x^2 + a_1x + a_2$ and $g(x) = b_0x^2 + b_1x + b_2$ be two polynomials of degree 2. Their resultant is the determinant of order 4:

$$R_{fg} = \begin{vmatrix} a_0 & a_1 & a_2 & 0 \\ 0 & a_0 & a_1 & a_2 \\ b_0 & b_1 & b_2 & 0 \\ 0 & b_1 & b_2 & b_2 \end{vmatrix}.$$

We assumed that a_0 and b_0, the leading coefficients of the polynomials f and g, are non-zero. If this is not the case, all we can say is that f and g have degrees at most equal to n and m, respectively. Sylvester's determinant is 0 when $a_0 \neq 0$ and $b_0 \neq 0$ if f and g have a common root, but this determinant is also 0 if $a_0 = b_0 = 0$.

Theorem 5.21. *The Sylvester determinant for the polynomials $f(x) = a_0x^n + a_1x^{n-1} + \cdots + a_n$ and $g(x) = b_0x^m + b_1x^{m-1} + \cdots + b_m$ is 0 when both polynomials have a common root or both leading coefficients a_0 and b_0 are equal to 0.*

Proof. Suppose that only one of the coefficients a_0 or b_0 equals 0, say $b_0 = 0$; furthermore, suppose that $b_0 = b_1 = \cdots = b_{i-1} = 0$ and $b_i \neq 0$. Expanding Sylvester's determinant after its first i rows, note that the determinant equals a_0^i multiplied with Sylvester's determinant $R_{f,\tilde{g}}$ of f and $\tilde{g} = b_ix^{m-i} + b_{i+1}x^{m-i-1} + \cdots + b_m$. If $R_{f,\tilde{g}} = 0$, then f and $\tilde{g}$ have a common root, hence f and g have a common root. □

Theorem 5.22. *Let $f, g \in \mathbb{C}[x]$ be the polynomials*

$$\begin{aligned} f(x) &= a_0x^n + a_1x^{n-1} + \cdots + a_n, \\ g(x) &= b_0x^m + b_1x^{m-1} + \cdots + b_m. \end{aligned}$$

There exist $\phi, \psi \in \mathbb{C}[x]$ of degrees $m-1$ and $n-1$, respectively, such that

$$R_{f,g} = \phi(x)f(x) + \psi(x)g(x). \tag{5.11}$$

Proof. Starting with Sylvester's determinant of f and g, by multiplying the next to the last column by x, the previous column by x^2, etc., and the first column by x^{n+m-1} and adding these columns to the last column, we obtain

$$R_{f,g} = \begin{vmatrix} a_0 & a_1 & a_2 & \cdots & a_n - u & 0 & \cdots & x^{m-1}f(x) \\ 0 & a_0 & a_1 & \cdots & a_{n-1} & a_n - u & \cdots & x^{m-2}f(x) \\ \vdots & \vdots & \vdots & \vdots & \vdots & \vdots & \vdots & \vdots \\ \cdots & \cdots & \cdots & \cdots & \cdots & \cdots & \cdots & f(x) \\ b_0 & b_1 & b_2 & \cdots & b_m & 0 & \cdots & x^{n-1}g(x) \\ 0 & b_0 & b_1 & \cdots & b_{m-1} & b_m & \cdots & x^{n-2}g(x) \\ \vdots & \vdots & \vdots & \vdots & \vdots & \vdots & \vdots & \vdots \\ \cdots & \cdots & \cdots & \cdots & \cdots & \cdots & \cdots & g(x). \end{vmatrix}.$$

Expanding this determinant by the last column, we obtain $R_{f,g} = \phi(x)f(x) + \psi(x)g(x)$. □

An alternative way to obtain Equality (5.11) is the following.

Lemma 5.4. *Let*

$$f(x) = a_0x^n + a_1x^{n-1} + \cdots + a_n,$$
$$g(x) = b_0x^m + b_1x^{m-1} + \cdots + b_m,$$

be polynomials with complex coefficients such that $a_0 \neq 0$ and $b_0 \neq 0$.

f and g have a common root if and only if there exist two polynomials f_1 and g_1 of degrees $n-1$ and $m-1$, respectively, such that $fg_1(x) - gf_1(x) = 0$.

Proof. If f and g have a common root α, then we can write $f(x) = (x-\alpha)f_1(x)$ and $g(x) = (x-\alpha)g_1(x)$, where the degree of f_1 is $n-1$ and the degree of g_1 is $m-1$. Then,

$$(fg_1)(x) - (gf_1)(x) = (x-\alpha)f_1(x)g_1(x) - (x-\alpha)g_1(x)f_1(x) = 0.$$

Conversely, if $(fg_1)(x) - (gf_1)(x) = 0$, then f and g have a common root because the equality $(fg_1)(x) = (gf_1)(x)$ implies that $(gf_1)(x) = 0$, for the n roots $\alpha_1, \ldots, \alpha_n$ of f. However, since the degree of f_1 is $n-1$, the right member equals 0 for $n-1$ of these values, which means that g will be 0 for at least one of them. □

An extension of the previous result is given next.

Theorem 5.23. *Let*

$$f(x) = a_0x^n + a_1x^{n-1} + \cdots + a_n,$$
$$g(x) = b_0x^m + b_1x^{m-1} + \cdots + b_m,$$

be polynomials with complex coefficients such that $a_0 \neq 0$ and $b_0 \neq 0$. These polynomials have k common roots if and only if there exist two polynomials f_k and g_k of degrees $n-k$ and $m-k$, respectively, such that $fg_k(x) - gf_k(x) = 0$.

Proof. If f and g have k common roots, then we can write $f(x) = h(x)f_k(x)$ and $g(x) = h(x)g_k(x)$, where h is a polynomial of degree k and f_k and g_k are polynomials of degrees $n-k$ and $m-k$, respectively, The equality follows immediately.

Conversely, if we have $fg_k(x) - gf_k(x) = 0$, then $f(x) = \frac{g(x)f_k(x)}{g_k(x)}$. If the polynomials $f_k(x)$ and $g_k(x)$ are relatively prime, then $g(x)$ is divisible by $g_k(x)$, which allows us to write $g(x) = h(x)g_k(x)$, where h is of degree k. Then $f(x) = h(x)f_k(x)$, and the roots of h are roots for f and g.

If $f_k(x)$ and $g_k(x)$ have a greatest common divisor $u(x)$ of degree ℓ, we can write $f_k(x) = \tilde{f}(x)u(x)$ and $g_k(x) = \tilde{g}(x)u(x)$, where $\tilde{f}$ is of degree $n-k-\ell$ and $\tilde{g}$ is of degree $m-k-\ell$. Since $f(x)\tilde{g}(x) - g(x)\tilde{f}(x) = 0$, the previous argument shows that $f(x)$ and $g(x)$ have at least $k+\ell$ roots, which concludes the argument. □

If a polynomial $f \in \mathbb{C}[x]$ of degree n has a root x_0 of multiplicity k, this means that we can write $f(x) = (x - x_0)^k g(x)$, where g is a polynomial of degree $n - k$ and $g(x_0) \neq 0$. Thus, for the derivative f', we have

$$f'(x) = (x - x_0)^{k-1}(kg(x) + (x - x_0)g'(x)] = (x - x_0)^{k-1}h(x).$$

Since $h(x_0) \neq 0$, it follows that x_0 is a root of multiplicity $k - 1$ of $h(x)$.

It is known from elementary algebra that a polynomial $f \in \mathbb{C}[x]$ has a root x_0 of multiplicity k if and only if

$$f(x_0) = f'(x_0) = \cdots = f^{(k-1)}(x_0) = 0 \text{ and } f^{(k)}(x_0) \neq 0.$$

Thus, a polynomial f has a multiple root if and only if $R_{f,f'} = 0$.

Definition 5.5. For $f(x) = a_0x^n + \cdots + a_n$, the *discriminant* of f is $D_f = \frac{1}{a_0}R_{f,f'}$.

It is clear that a polynomial f has multiple roots if and only if $D_f = 0$.

Taking into account Equality (5.9), if $\alpha_1, \ldots, \alpha_n$ are the roots of f, we can write

$$D_f = \frac{1}{a_0}R_{f,f'} = a_0^{n-2}f'(\alpha_1)\cdots f'(\alpha_n),$$

where $\alpha_1, \ldots, \alpha_n$ are the roots of f. Since $f(x) = a_0(x - x_1)\cdots(x - x_n)$, this implies

$$f'(x_k) = a_0(x_k - x_1)\cdots(x_k - x_{k-1})(x_k - x_{k+1})\cdots(x_k - x_n),$$

hence,

$$\begin{aligned}D_f = (-1)^{\frac{n(n-1)}{2}} a_0^{2n-2} (x_2 - x_1)^2 (x_3 - x_1)^2 \cdots (x_n - x_1)^2 \\ (x_3 - x_2)^2 \cdots (x_n - x_2)^2 \\ \cdots \\ (x_n - x_{n-1})^2.\end{aligned}$$

This implies

$$D_f = (-1)^{\frac{n(n-1)}{2}} a_0^{2n-2} V_{x_1,\ldots,x_n}^2.$$

Example 5.13. Let $f(x) = a_0 x^2 + a_1 x + a_2$. Since $f'(x) = 2a_0 x + a_1$, it follows that

$$D_f = \frac{1}{a_0} \begin{vmatrix} a_0 & a_1 & a_2 \\ 2a_0 & a_1 & 0 \\ 0 & 2a_0 & a_1 \end{vmatrix} = a_1^2 - 4a_0 a_2,$$

which is the familiar expression of the discriminant of a trinomial.

The notion of resultant can be formulated for homogeneous polynomials. Instead of the polynomials $f(x) = a_0 x^n + a_1 x^{n-1} + \cdots + a_n$ and $g(x) = b_0 x^m + b_1 x^{m-1} + \cdots + b_m$, let F and G be the associated homogeneous polynomials defined as

$$\begin{aligned}F(x, y) &= a_0 x^n + a_1 x^{n-1} y + \cdots + a_n y^n, \\ G(x, y) &= b_0 x^m + b_1 x^{m-1} y + \cdots + b_m y^m.\end{aligned} \tag{5.12}$$

The resultant of F and G, $R_{F,G}$ is defined as $R_{f,g}$.

Theorem 5.24. *Let F, G be two polynomials with complex coefficients. There is a pair $(x, y) \in \mathbb{C}^2 - \{(0, 0)\}$ such that $F(x, y) = G(x, y) = 0$ if and only if $R_{F,G} = 0$.*

Proof. Note that if $(u, v) \in \mathbb{C}^2$ is a solution for $F(u, v) = G(u, v) = 0$, then so is $(\lambda u, \lambda v)$.

If $a_0 = b_0 = 0$, then $R_{F,G} = 0$ and we have the non-trivial solution $(1, 0)$. Suppose now that $a_0 \neq 0$ and $b_0 \neq 0$. If $R(F, G) = 0$, by taking $y = 1$ the system reverts to the System (5.7). The common factor of f and g implies that f and g must have a common root $x = u$ and then $(x, y) = (u, 1)$ is the desired non-trivial solution.

Conversely, if (u, v) is a non-trivial solution, then $a_0 \neq 0$ and $b_0 \neq 0$ implies $v \neq 0$, which, in turn, implies that $(u/v, 1)$ is also a solution, so u/v is a solution of System (5.7). This implies $R_{F,G} = 0$.

Let us examine the case that occurs when exactly one of a_0, b_0 is 0, say $a_0 \neq 0$, and $b_0 = \cdots = b_{r-1} = 0$ but $b_r \neq 0$. If we set $y = 1$ in the System (5.12), we get polynomials of degree n and $m - r$, respectively. In this case, $R_{F,G} = a_0^r R_{f,g}$. We leave it to the reader to prove that in this case too, $R_{F,G} = 0$ if and only if $F = 0, G = 0$ has a non-trivial solution. □

Consider now a system of n homogeneous polynomial equations

$$\begin{aligned} f_1(x_1, \ldots, x_n) &= 0, \\ &\vdots \\ f_n(x_1, \ldots, x_n) &= 0, \end{aligned} \tag{5.13}$$

where $f_1, \ldots, f_n$ are homogeneous polynomial equations of degrees $r_1, \ldots, r_n$ in n variables.

We seek a homogeneous polynomial $Res_{f_1,\ldots,f_n}$ in the coefficients of the System (5.13) such that it has non-zero solutions if and only these coefficients satisfy the condition $Res_{f_1,\ldots,f_n} = 0$. The resultant is a homogeneous polynomial of degree

$$r_1 \cdots r_n \left(\frac{1}{r_1} + \cdots + \frac{1}{r_n} \right).$$

To find the resultant of System (5.13), consider another system

$$\begin{aligned} x_1^{s_1} \cdots x_n^{s_n} f_1(x_1, \ldots, x_n) &= 0, \\ &\vdots \\ x_1^{s_1} \cdots x_n^{s_n} f_n(x_1, \ldots, x_n) &= 0, \end{aligned} \tag{5.14}$$

obtained from System (5.13) by multiplying by all possible monomials of fixed degree $q = s_1 + \cdots + s_n$.

If we regard each polynomial in System (5.14) as a linear combination of monomials, then in the monomial basis, the left side of the system is written as a rectangular $n\binom{q+n-1}{q} \times \binom{r+q+n-1}{r+q}$-matrix. This is the *generalized Sylvester matrix*.

If we choose q such that $n\binom{q+n-1}{q} = \binom{r+q+n-1}{r+q}$, then the resultant is the determinant of the resulting square Sylvester matrix.

As observed in [117], this is possible in at least two cases, namely, for any n and $r = 1$, which requires $q = 0$, or for any r and $n = 2$.

In the first case, the original system is considered without any additional multiplication by monomials. In the second case, we have to choose $q = r - 1$ to make the Sylvester matrix square.

Example 5.14. If $n = 2, r = 2$, and $q = 1$, the system of equations is

$$\begin{aligned}F(x_1, x_2) &= f_{11}x_1^2 + f_{12}x_1x_2 + f_{22}x_2^2 = 0,\\ G(x_1, x_2) &= g_{11}x_1^2 + g_{12}x_1x_2 + g_{22}x_2^2 = 0.\end{aligned}$$

Multiplying these equations by all monomials of degree $q = 1$, that is, by x_1 and x_2, we obtain the system

$$\begin{aligned}&f_{11}x_1^3 + f_{12}x_1^2x_2 + f_{22}x_1x_2^2 + 0x_2^3 = 0,\\ &0x_1^3 + f_{11}x_1^2x_2 + f_{12}x_1x_2^2 + f_{22}x_2^3 = 0,\\ &g_{11}x_1^3 + g_{12}x_1^2x_2 + g_{22}x_1x_2^2 + 0x_2^3 = 0,\\ &0x_1^3 + g_{11}x_1^2x_2 + g_{12}x_1x_2^2 + g_{22}x_2^3 = 0.\end{aligned}$$

These polynomials are linear combinations of four monomials x_1^3, $x_1^2x_2$, $x_1x_2^2$, and x_2^3 and can be represented by a square matrix

$$\begin{pmatrix} f_{11} & f_{12} & f_{22} & 0\\ 0 & f_{11} & f_{21} & f_{22}\\ g_{11} & g_{12} & g_{22} & 0\\ 0 & g_{11} & g_{21} & g_{22}\end{pmatrix}.$$

Theorem 5.25. *Let*

$$\begin{aligned}F(x, y) &= a_0x^\ell + a_1x^{\ell-1}y + \cdots + a_\ell y^\ell,\\ G(x, y) &= a_0x^m + a_1x^{m-1}y + \cdots + a_my^m,\end{aligned}$$

be homogeneous polynomials in $\mathbb{C}[x, y]$ *of total degrees* ℓ *and* m, *respectively. We have* $Res_{F,G} = 0$ *if and only if* $F = G = 0$ *have a non-trivial solution (that is a solution distinct from* $\mathbf{0}_2$*).*

Proof. If $a_0 = b_0 = 0$, then $Res_{F,G} = 0$ and we have the non-trivial solution $x = 1$ and $y = 0$.

Suppose that $a_0 \neq 0$ and $b_0 \neq 0$. If $R_{F,G} = 0$, let f, g be the polynomials obtained from F and G, respectively, by setting $y = 1$.

We have $R_{f,g} = 0$. Thus, f and g have a common root $x = \alpha$ and $(\alpha, 1)$ is a non-trivial solution of the system. Conversely, if (u, v) is a non-trivial solution of the system, then $a_0 \neq 0$ and $b_0 \neq 0$ implies $v \neq 0$, which in turn implies that $\left(\frac{u}{v}, 1\right)$ is also a solution. Thus, $\frac{u}{v}$ is a common root of f and g, hence $Res_{F,G} = 0$.

Suppose now that $a_0 \neq 0$, $b_0 = \cdots = b_{r-1} = 0$, and $b_r \neq 0$. By taking $y = 1$, we obtain the polynomials f and g of degrees ℓ and $m - r$, respectively. In this case, $Res_{F,G} = a_0^r R_{f,g}$. By a similar argument to the previous case, $Res_{F,G} = 0$ if and only if the system $F = 0, G = 0$ has a non-trivial solution. □

The previous theorem can be extended to a finite set of homogeneous polynomials (see [31, 63]).

Theorem 5.26. *Let $F_1, \ldots, F_n \in \mathbb{C}[x_0, \ldots, x_n]$ be homogeneous polynomials, where F_i has positive degree d_i for $1 \leqslant i \leqslant n$. There exists an irreducible polynomial (referred to as* resultant*) in the coefficients of $F_1, \ldots, F_n$ which vanishes if and only if the polynomial system $F_1(\boldsymbol{x}) = \cdots = F_n(\boldsymbol{x})$ has a nontrivial solution.*

If $F_i(\boldsymbol{x}) = x_i^{d_i}$ for $1 \leqslant i \leqslant n$, then the value of the resultant is 1.

Proof. Proofs can be found in [63, 165]. □

5.6 MATLAB Computations

The determinant of a square matrix $A \in \mathbb{C}^{n \times n}$ can be computed in MATLAB using the function |det|(A).

Example 5.15.

$$A = \begin{pmatrix} 1 & 2 & 3 \\ 4 & 5 & 6 \\ 7 & 8 & 9 \end{pmatrix}.$$

To compute $d = \det(A)$, we write

```
>> A = [1 2 3; 4 5 6; 7 8 9]
A =
     1     2     3
     4     5     6
     7     8     9
```

```
>> d = det(A)
d =
       0


A =
       1      5      2
       4      1      6
       3      1      4
>> d=det(A)
d =
      10
>> B=inv(A)
B =
        -1/5               -9/5               14/5
         1/5               -1/5                1/5
         1/10               7/5              -19/10.
```

The `Symbolic Math Toolbox` of MATLAB provides functions for solving, plotting, and manipulating mathematics. The function **syms** creates symbolic variables and functions.

Example 5.16. To create the variables x and y, one could write

```
syms x y
```

Polynomials can be defined as shown in what follows. For example, to re-create the polynomials defined in Example 5.11, one could write

```
f = x^3 - 2x^2 - x + 2
g = x^2-5x +6
```

The resultant of these polynomials in the variable `x` is obtained with

```
res = resultant(f,g,x)
```

and, as expected, the result is 0.

Suppose now that f and g are two polynomials in `x` and the parameter m as in

```
f = x^3 - (m+1)x^2 - mx + 2
g = x^2-(2m+3) x + 2m+4
```

Clearly, the previous polynomials are obtained by taking $m = 1$.

We begin by computing the resultant as a polynomial in m:

```
>> syms x m
>> f = x^3 -(m+1)* x^2 - m*x + 2

f =
x^3 + (- m - 1)*x^2 - m*x + 2

>> g=x^2 -(2*m +3)*x + 2*m + 4

g =
x^2 + (- 2*m - 3)*x + 2*m + 4

>> res = resultant(f,g,x)

res =
- 8*m^4 - 24*m^3 - 8*m^2 + 24*m + 16
```

To determine the values of m that make the resultant equal to 0, we write

```
>> mRoots = solve(res)
```

which returns:

```
mRoots =
 -2
-1
-1
 1
```

Exercises and Supplements

(1) Let $\phi \in PERM_n$ be

$$\phi : \begin{pmatrix} 1 & \cdots & i & \cdots & n \\ a_1 & \cdots & a_i & \cdots & a_n \end{pmatrix},$$

and let

$$v_p(\phi) = |\{(i_k, i_l) \mid i_l = p, k < l, i_k > i_l\}$$

be the number of inversions of ϕ that have p as their second component, for $1 \leqslant p \leqslant n$. Prove that

(a) $v_p \leqslant n - p$ for $1 \leqslant p \leqslant n$;
(b) for every sequence of numbers $(v_1, \ldots, v_n) \in \mathbb{N}^n$ such that $v_p \leqslant n - p$ for $1 \leqslant p \leqslant n$ there exists a unique permutation ϕ that has $(v_1, \ldots, v_n) \in \mathbb{N}^n$ as its sequence of inversions.

(2) Let p be the polynomial defined as

$$p(x_1, \ldots, x_n) = \prod_{i<j}(x_i - x_j).$$

For a permutation $\phi \in PERM_n$ given by

$$\phi : \begin{pmatrix} 1 & \cdots & i & \cdots & n \\ a_1 & \cdots & a_i & \cdots & a_n \end{pmatrix},$$

let $p_\phi = p(a_1, \ldots, a_n)$.
Prove the following:
(a) $(-1)^{inv}(\phi) = \frac{p_\phi}{p_{\iota_n}}$ for any permutation ϕ;
(b) $(-1)^{inv(\psi\phi)} = (-1)^{inv(\psi)}(-1)^{inv(\phi)}$.

(3) Let $A \in \mathbb{R}^{n \times n}$ be a symmetric matrix. Prove that if $\mathsf{trace}(A) = 0$ and the sum of principal minors of order 2 equals 0, then $A = O_{n,n}$.

Solution: A principal minor of order 2 has the form $m_{ij} = a_{ii}a_{jj} - a_{ij}^2$ for $1 \leqslant i < j \leqslant n$ (because A is a symmetric matrix) and there are $\binom{n}{2}$ such minors. Therefore, the sum of these minors is

$$M = \sum_{i<j} a_{ii}a_{jj} - \sum_{i<j} a_{ij}^2 = 0.$$

Since $\mathsf{trace}(A) = 0$, we have $\mathsf{trace}(A)^2 = \sum_{i=1}^n a_{ii}^2 + 2\sum_{i<j} a_{ii}a_{jj} = 0$. These equalities imply

$$0 = \sum_{i=1}^n a_{ii}^2 + 2\sum_{i<j} a_{ii}a_{jj} = \sum_{i=1}^n a_{ii}^2 + 2\sum_{i<j} a_{ij}^2,$$

which yields

$$a_{11} = \cdots = a_{nn} = a_{12} = \cdots = a_{n-1\,n} = 0.$$

Thus, $A = O_{n,n}$.

(4) Let D_n be the determinant having n rows, defined by

$$D_n = \begin{vmatrix} 1 & 0 & 0 & \cdots & 0 & 0 \\ a & 1 & 0 & \cdots & 0 & 0 \\ \vdots & \vdots & \vdots & \vdots & \vdots & \vdots \\ 0 & 0 & 0 & \cdots & 1 & 0 \\ 0 & 0 & 0 & \cdots & a & 1 \end{vmatrix},$$

where $n \geqslant 2$. Prove that $D_n = 1$ for every $n \geqslant 2$.

(5) Using elementary properties of determinants, prove that

$$\begin{vmatrix} a^2 + x^2 & ab + xy & ac + xz \\ ab + xy & b^2 + y^2 & bc + yz \\ ac + xz & bc + yz & c^2 + z^2 \end{vmatrix} = 0.$$

(6) Let $T_n(a)$ be the tridiagonal determinant

$$T_n(a) = \begin{vmatrix} a & 1 & 0 & 0 & \cdots & 0 & 0 \\ 1 & a & 1 & 0 & \cdots & 0 & 0 \\ 0 & 1 & a & 1 & \cdots & 0 & 0 \\ \vdots & \vdots & \vdots & \vdots & \vdots & 0 & 0 \\ 0 & 0 & 0 & 0 & \cdots & 1 & a \end{vmatrix}$$

having n rows, where $n \geqslant 2$.
Prove that $T_2(a) = a^2 - 1$ and that $T_{n+2}(a) = aT_{n+1}(a) - T_n(a)$ for $n \geqslant 2$.

(7) Let $A \in \mathbb{R}^{n \times n}$ and $c \in \mathbb{R}$. Prove that $\det(cA) = c^n \det(A)$.

(8) Prove that if a matrix $A \in \mathbb{R}^{n \times n}$ is skew-symmetric and n is an odd number, then $\det(A) = 0$.

Solution: Since $A' = -A$ and n is odd, it follows that $\det(A') = (-1)^n \det(A) = -\det(A)$. On the other hand, $\det(A') = \det(A)$, hence $\det(A) = 0$.

(9) Prove that $\det(aI_n + bJ_{n,n}) = a^n + na^{n-1}b$ for $n \geqslant 1$.

(10) Let $A \in \mathbb{R}^{n \times n}$ be a skew-symmetric matrix and let $B = A + bJ_{n,n}$, where $b \in \mathbb{R}$. Prove that if n is even, then $\det(A) = det(B)$.

(11) Let $A \in \mathbb{R}^{n \times n}$ be a non-singular matrix. Prove that $\det(\mathsf{adj}(A)) = \det(A)^{n-1}$.

Solution: The equality proven in Theorem 5.12 implies $\det(A)\det(\mathsf{adj}(A)) = \det(\det(A)I_n) = \det(A)^n$, which implies the desired equality because $\det(A) \neq 0$.

(12) Let $U, V \in \mathbb{C}^{n\times m}$ be two matrices. Prove that $\det(I_n + UV^{\mathsf{H}}) = \det(I_m + V^{\mathsf{H}}U)$ (Sylvester's identity).

Solution: Starting from the matrix equalities

$$\begin{pmatrix} I_n & -U \\ V^{\mathsf{H}} & I_m \end{pmatrix} = \begin{pmatrix} I_n & O_{n,m} \\ V^{\mathsf{H}} & I_m \end{pmatrix} \cdot \begin{pmatrix} I_n & -U \\ O_{m,n} & I_m + V^{\mathsf{H}}U \end{pmatrix}$$
$$= \begin{pmatrix} I_n & -U \\ O_{m,n} & I_m \end{pmatrix} \cdot \begin{pmatrix} I_n + UV^{\mathsf{H}} & O_{n,m} \\ V^{\mathsf{H}} & I_m \end{pmatrix},$$

which are immediate, and taking the determinants of both sides, the Sylvester's identity follows immediately taking into account that

$$\det \begin{pmatrix} I_n & U \\ O_{m,n} & I_m + V^{\mathsf{H}}U \end{pmatrix} = \det(I_m + V^{\mathsf{H}}U)$$

and

$$\det \begin{pmatrix} I_n + UV^{\mathsf{H}} & O_{n,m} \\ V^{\mathsf{H}} & I_m \end{pmatrix} = \det(I_n + UV^{\mathsf{H}}).$$

We used the fact that the determinant of a block upper triangular matrix equals the product of the determinants of matrices situated on the diagonal.

(13) Let $\boldsymbol{x}, \boldsymbol{y} \in \mathbb{C}^n$. Prove the following:
(a) $\det(I_n + \boldsymbol{x}\boldsymbol{y}^{\mathsf{H}}) = 1 + \boldsymbol{y}^{\mathsf{H}}\boldsymbol{x}$ and $\det(I_n - \boldsymbol{x}\boldsymbol{y}^{\mathsf{H}}) = 1 - \boldsymbol{y}^{\mathsf{H}}\boldsymbol{x}$;
(b) if $A \in \mathbb{C}^{n\times n}$ is an invertible matrix, then $\det(A + \boldsymbol{x}\boldsymbol{y}^{\mathsf{H}}) = \det(A)(1 + \boldsymbol{y}^{\mathsf{H}}A^{-1}\boldsymbol{x})$.

Solution: The first part is a direct consequence of Supplement 5.6 by taking $m = 1$.

For the second part, we have $A + \boldsymbol{x}\boldsymbol{y}^{\mathsf{H}} = A(I_n + A^{-1}\boldsymbol{x}\boldsymbol{y}^{\mathsf{H}})$, which yields the desired inequality.

(14) Let $v_1, \ldots, v_n$ be n complex numbers such that $\sum_{k=1}^{n} |v_k|^2 = 2$. Prove that

$$\begin{vmatrix} 1 - v_1\bar{v}_1 & v_1\bar{v}_2 & v_1\bar{v}_3 & \cdots & v_1\bar{v}_n \\ v_2\bar{v}_1 & 1 - v_2\bar{v}_2 & v_2\bar{v}_3 & \cdots & v_2\bar{v}_n \\ \vdots & \vdots & \vdots & \cdots & \vdots \\ v_n\bar{v}_1 & v_n\bar{v}_2 & v_n\bar{v}_3 & \cdots & 1 - v_n\bar{v}_n \end{vmatrix} = -1.$$

Solution: Note that the matrix whose determinant is to be computed equals $I_n - \boldsymbol{v}\boldsymbol{v}^{\mathsf{H}}$. Thus, by Supplement 5.6, we have $\det(I_n - \boldsymbol{v}\boldsymbol{v}^{\mathsf{H}}) = 1 + \boldsymbol{v}^{\mathsf{H}}\boldsymbol{v} = -1$.

(15) Let $D = \mathsf{diag}(a_1, \ldots, a_n)$ and let $A = D + \mathbf{1}_n\mathbf{1}_n'$. Prove that

$$\det(A) = \prod_{i=1}^{n} a_i \left(1 + \sum_{i=1}^{n} \frac{1}{a_i}\right).$$

Solution: By Part (b) of Supplement 5.6, we have

$$\det(A) = \det(D + \mathbf{1}_n\mathbf{1}_n^{\mathsf{H}}) = \det(D)(1 + \mathbf{1}_n^{\mathsf{H}} D^{-1}\mathbf{1}_n),$$

which amounts to the formula we need to prove.

(16) Using the notations introduced in Supplement 43 of Chapter 3 to prove the Woodbury–Sherman–Morrison identity for determinants:

$$\det(Z + UWV') = \det(Z)\det(W)\det(W^{-1} + V'Z^{-1}U).$$

Solution: We can write

$$\begin{aligned} &\det(I_n - U(W^{-1} + V'Z^{-1}U)^{-1}V'Z^{-1}) \\ &= \det(I_n - V'Z^{-1}U(W^{-1} + V'Z^{-1}U)^{-1}) \\ &= \text{(by Sylvester's identity)} \\ &= \det((W^{-1} + V'Z^{-1}U)(W^{-1} + V'Z^{-1}U)^{-1} \\ &\quad -V'Z^{-1}U(W^{-1} + V'Z^{-1}U)^{-1}) \\ &= \det((W^{-1} + V'Z^{-1}U - V'Z^{-1}U)\det(W^{-1} + V'Z^{-1}U)^{-1} \\ &= \det(W^{-1})\det(W^{-1} + V'Z^{-1}U)^{-1}. \end{aligned}$$

Equality (3.16) implies

$$\det(Z)\det(X) = \det(I_n - U(W^{-1} + V'Z^{-1}U)^{-1}V'Z^{-1}).$$

Since $X = (Z + UWV')^{-1}$, it follows that

$$\frac{\det(Z)}{\det(Z + UWV')} = \frac{1}{\det(W^{-1} + V'Z^{-1}U)\det(W)},$$

which is the desired equality.

(17) Let $A \in \mathbb{R}^{n\times n}$. Prove that for the kth derivative of the function $G(x) = \det(A - xI_n)$, we have

$$G^{(k)}(0) = (-1)^k k! S_{n-k}(A),$$

where $S_p(A)$ is the sum of all order-p principal minors of A.

(18) Let $\boldsymbol{x} = (x_1, \ldots, x_n)$ and $\boldsymbol{y} = (y_1, \ldots, y_n)$ be two sequences of real numbers such that $x_i + y_j \neq 0$ for $1 \leqslant i, j \leqslant n$ and $n \geqslant 2$. The *Cauchy matrix* of these sequences is the matrix $C_{\boldsymbol{x},\boldsymbol{y}}$ given by

$$(C_{\boldsymbol{x},\boldsymbol{y}})_{ij} = \frac{1}{x_i + y_j}$$

for $1 \leqslant i, j \leqslant n$. Prove that

$$\det(C_{\boldsymbol{x},\boldsymbol{y}}) = \frac{\prod_{1\leqslant j<i\leqslant n}(x_i - x_j)\prod_{1\leqslant j<i\leqslant n}(y_i - y_j)}{\prod_{1\leqslant i,j\leqslant n}(x_i + y_j)}.$$

Solution: The argument is by induction on $n \geqslant 2$. The base step, $n = 2$, is immediate.

Suppose the formula holds for sequences of length $n - 1$. To prove the formula for sequences of length n, we proceed as above by subtracting the last column from the remaining columns. Then, for the sequences $\boldsymbol{x} = (x_1, \ldots, x_n)$ and $\boldsymbol{y} = (y_1, \ldots, y_n)$, we have

$$\det(C_{\boldsymbol{x},\boldsymbol{y}}) = \begin{vmatrix} \frac{y_n-y_1}{(x_1+y_1)(x_1+y_n)} & \frac{y_n-y_2}{(x_1+y_2)(x_1+y_n)} & \cdots & \frac{1}{x_1+y_n} \\ \frac{y_n-y_1}{(x_2+y_1)(x_2+y_n)} & \frac{y_n-y_2}{(x_2+y_2)(x_2+y_n)} & \cdots & \frac{1}{x_2+y_n} \\ \vdots & \vdots & \vdots & \vdots \\ \frac{y_n-y_1}{(x_n+y_1)(x_n+y_n)} & \frac{y_n-y_2}{(x_n+y_2)(x_n+y_n)} & \cdots & \frac{1}{x_n+y_n} \end{vmatrix}.$$

This equality can be written as

$$\det(C_{\boldsymbol{x},\boldsymbol{y}}) = \prod_{1\leqslant k<n}(y_n - y_k)\prod_{1\leqslant \ell\leqslant n}(x_\ell + y_n)$$
$$\times \begin{vmatrix} \frac{1}{x_1+y_1} & \frac{1}{x_1+y_2} & \cdots & \frac{1}{x_1+y_{n-1}} & 1 \\ \frac{1}{x_2+y_1} & \frac{1}{x_2+y_2} & \cdots & \frac{1}{x_2+y_{n-1}} & 1 \\ \vdots & \vdots & \vdots & \vdots & \\ \frac{1}{x_n+y_1} & \frac{1}{x_n+y_2} & \cdots & \frac{1}{x_n+y_{n-1}} & 1 \end{vmatrix}.$$

By subtracting the last row from the remaining rows and factoring out, we obtain

$$\begin{vmatrix} \frac{1}{x_1+y_1} & \frac{1}{x_1+y_2} & \cdots & \frac{1}{x_1+y_{n-1}} & 1 \\ \frac{1}{x_2+y_1} & \frac{1}{x_2+y_2} & \cdots & \frac{1}{x_2+y_{n-1}} & 1 \\ \vdots & \vdots & \vdots & \vdots & \\ \frac{1}{x_n+y_1} & \frac{1}{x_n+y_2} & \cdots & \frac{1}{x_n+y_{n-1}} & 1 \end{vmatrix}$$
$$= \frac{\prod_{h=1}^{n-2}(x_n - x_h)}{\prod_{k=1}^{n-1}(x_n + y_k)} \begin{vmatrix} \frac{1}{x_1+y_1} & \frac{1}{x_1+y_2} & \cdots & \frac{1}{x_1+y_{n-1}} & 0 \\ \frac{1}{x_2+y_1} & \frac{1}{x_2+y_2} & \cdots & \frac{1}{x_2+y_{n-1}} & 0 \\ \vdots & \vdots & \vdots & \vdots & \\ \frac{1}{x_{n-1}+y_1} & \frac{1}{x_{n-1}+y_2} & \cdots & \frac{1}{x_{n-1}+y_{n-1}} & 0 \\ 1 & 1 & \cdots & 1 & 1 \end{vmatrix}$$
$$= \frac{\prod_{h=1}^{n-1}(x_n - x_h)}{\prod_{k=1}^{n-1}(x_n + y_k)} \det(C_{\boldsymbol{u},\boldsymbol{v}}),$$

where $\boldsymbol{u} = (x_1, \ldots, x_{n-1})$ and $\boldsymbol{v} = (y_1, \ldots, y_{n-1})$. By applying the inductive hypothesis, we obtain the desired equality.

To produce a Cauchy matrix in MATLAB, we can use the command

```
C = gallery('cauchy',x,y),
```

where $\boldsymbol{x}$ and $\boldsymbol{y}$ are row vectors in $\mathbb{R}^n$. For a scalar n, `gallery('cauchy',n)` is the same as `gallery('cauchy',1:n)`.

By writing `C = gallery('cauchy',1:5)`, we obtain

```
C =
    0.5000    0.3333    0.2500    0.2000    0.1667
    0.3333    0.2500    0.2000    0.1667    0.1429
```

```
0.2500    0.2000    0.1667    0.1429    0.1250
0.2000    0.1667    0.1429    0.1250    0.1111
0.1667    0.1429    0.1250    0.1111    0.1000
```

MATLAB offers an extensive collection of matrices that is very useful in assessing its capabilities and in other experiments. This collection is known as the *gallery*. Information about various matrices offered by this collection can be obtained using the help facility. For example, by typing `help private/cauchy`, one obtains information for producing a Cauchy matrix. A list of all members of the `gallery` can be obtained by typing `help gallery`.

(19) Generate in MATLAB a Cauchy matrix C and verify experimentally the value of $\det(C)$.

(20) The *n-Hilbert matrix* H_n is a special Cauchy matrix, where $x_i = y_i = i - \frac{1}{2}$ for $1 \leqslant i \leqslant n$. Let $h_n = 1!2! \cdots (n-1)!$ for $n \in \mathbb{N}$ and $n \geqslant 1$. Prove the following:
   (a) the determinant of the Hilbert matrix H_n is $\det(H_n) = \frac{h_n^4}{h_{2n}}$;
   (b) the number $\frac{1}{\det(H_n)}$ is an integer.

(21) Let $A \in \mathbb{R}^{n \times n}$ be defined by $a_{ij} = i + j$ for $1 \leqslant i, j \leqslant n$. Prove that if $n \geqslant 3$, $\det(A) = 0$.

(22) Let $A \in \mathbb{C}^{m \times n}$ be a matrix, $\boldsymbol{x} \in \mathbb{C}^n$ and $\boldsymbol{y} \in \mathbb{C}^m$. Prove that if $\mathsf{rank}(A - aA\boldsymbol{x}\boldsymbol{y}^{\mathsf{H}}A) = \mathsf{rank}(A) - 1$, then $\boldsymbol{y}^{\mathsf{H}}A\boldsymbol{x} \neq 0$ and $a = \frac{1}{\boldsymbol{y}^{\mathsf{H}}A\boldsymbol{x}}$.

   Solution: Suppose that $\mathsf{rank}(A) = r$. By the full-rank decomposition theorem, A can be factored as $A = DE$, where $D \in \mathbb{C}^{m \times r}$, $E \in \mathbb{C}^{r \times n}$, and both D and E are full-rank matrices, so $\mathsf{rank}(D) = \mathsf{rank}(E) = r$. Then, $A - aA\boldsymbol{x}\boldsymbol{y}^{\mathsf{H}}A = DE - aDE\boldsymbol{x}\boldsymbol{y}^{\mathsf{H}}DE = D(I - aE\boldsymbol{x}\boldsymbol{y}^{\mathsf{H}}D)E$. This implies $\mathsf{rank}(I - aE\boldsymbol{x}\boldsymbol{y}^{\mathsf{H}}D) \geqslant r - 1$. By Sylvester's identity discussed in Supplement 12, we have $\det(I - aE\boldsymbol{x}\boldsymbol{y}^{\mathsf{H}}D) = 1 - a\boldsymbol{y}^{\mathsf{H}}DE\boldsymbol{x} = 1 - a\boldsymbol{y}^{\mathsf{H}}A\boldsymbol{x}$. This shows that $\mathsf{rank}(I - aE\boldsymbol{x}\boldsymbol{y}^{\mathsf{H}}D) = r - 1$ if and only if $\boldsymbol{y}^{\mathsf{H}}A\boldsymbol{x} \neq 0$ and $a = \frac{1}{\boldsymbol{y}^{\mathsf{H}}A\boldsymbol{x}}$.

The next supplement contains a converse of Theorem 3.39 obtained in [42].

(23) Let $A \in \mathbb{C}^{m \times n}$ be a matrix, $\boldsymbol{u} \in \mathbb{C}^m$ and $\boldsymbol{v} \in \mathbb{C}^n$ be two vectors, $a \in \mathbb{C} - \{0\}$, and let $B = A - \frac{1}{a}\boldsymbol{u}\boldsymbol{v}^{\mathsf{H}}$. We have $\mathsf{rank}(B) < \mathsf{rank}(A)$, if and only if there are vectors $\boldsymbol{x} \in \mathbb{C}^m$ and $\boldsymbol{y} \in \mathbb{C}^n$ such that

$\boldsymbol{u} = A\boldsymbol{y}$, $\boldsymbol{v} = A^{\mathsf{H}}\boldsymbol{x}$, and $a = \boldsymbol{x}^{\mathsf{H}}A\boldsymbol{y}$, in which case $\mathsf{rank}(B) = \mathsf{rank}(A) - 1$.

Solution: This statement follows from Supplements 74 of Chapter 3 and from Supplement 22 of this chapter.

(24) Let d_n be the determinant defined by

$$d_n = \begin{vmatrix} \cos\alpha & \cos 2\alpha & \dots & \cos n\alpha \\ \cos(n+1)\alpha & \cos(n+2)\alpha & \dots & \cos 2n\alpha \\ \vdots & \vdots & \dots & \vdots \\ \cos(n^2-n+1)\alpha & \cos(n^2-n+2)\alpha & \dots & \cos n^2\alpha \end{vmatrix}.$$

Prove that for $n \geqslant 3$, we have $d_n = 0$.
Hint: add the third column to the first column.

Let $A \in \mathbb{C}^{m\times n}$ be a partitioned matrix given by

$$A = \begin{pmatrix} A_{11} & A_{12} \\ A_{21} & A_{22} \end{pmatrix},$$

where A_{11} is an invertible square matrix, $A_{11} \in \mathbb{C}^{p\times p}$. Note that $A_{21} \in \mathbb{C}^{(m-p)\times p}$, $A_{12} \in \mathbb{C}^{p\times(n-p}$, and $A_{22} \in \mathbb{C}^{(m-p)\times(n-p)}$. Therefore, the matrix $B = A_{22} - A_{21}A_{11}^{-1}A_{12} \in \mathbb{C}^{(m-p)\times(n-p)}$ is well-defined. We refer to B as *Schur's complement of* A_{11} relative to A and it is denoted by A/A_{11}.

(25) Let $A \in \mathbb{C}^{n\times n}$ be a square partitioned matrix given by

$$A = \begin{pmatrix} A_{11} & A_{12} \\ A_{21} & A_{22} \end{pmatrix},$$

where A_{11} is an invertible matrix. Prove that

(a) we have $\det(A/A_{11}) = \frac{\det(A)}{\det(A_{11})}$;

(b) we have the following equality:

$$A = \begin{pmatrix} I & O \\ A_{21}A_{11}^{-1} & I \end{pmatrix} \begin{pmatrix} A_{11} & O \\ O & A_{22} - A_{21}A_{11}^{-1}A_{12} \end{pmatrix} \times \begin{pmatrix} I & A_{11}^{-1}A_{12} \\ O & I \end{pmatrix}.$$

Solution: Observe that we can write

$$\begin{pmatrix} A_{11} & A_{12} \\ A_{21} & A_{22} \end{pmatrix} \begin{pmatrix} A_{11}^{-1} & -A_{11}^{-1}A_{12} \\ O & I \end{pmatrix}$$
$$= \begin{pmatrix} I & O \\ A_{21}A_{11}^{-1} & A_{22} - A_{21}A_{11}^{-1}A_{12} \end{pmatrix},$$

By Theorem 5.13, we have $\det(A)\det(A_{11}^{-1}) = \frac{\det(A)}{\det(A_{11})} = \det(A_{22} - A_{21}A_{11}^{-1}A_{12})$, which gives the equality of the first part. The decomposition of the second part can be verified directly.

(26) Let

$$A = \begin{pmatrix} A_{11} & \cdots & A_{1m} \\ \vdots & \vdots & \vdots \\ A_{m1} & \cdots & A_{mm} \end{pmatrix}$$

be a partitioned matrix such that each matrix

$$B_p = \begin{pmatrix} A_{11} & \cdots & A_{1p} \\ \vdots & \vdots & \vdots \\ A_{p1} & \cdots & A_{pp} \end{pmatrix}$$

is invertible for $1 \leqslant p \leqslant m-1$.

Prove that A can be uniquely written as a product $A = LDU$ such that the following conditions are satisfied:

(a) L is a lower block triangular matrix whose diagonal blocks are unit matrices;

(b) D is a block diagonal matrix, $D = \mathsf{diag}(D_1, \ldots, D_m)$, whose diagonal blocks are invertible matrices such that $D_1 = A_{11} = B_1$, and $D_j = B_j/B_{j-1}$ for $1 \leqslant p \leqslant m$;

(c) U is an upper block diagonal matrix whose diagonal blocks are unit matrices.

Solution: The proof of the existence of the prescribed decomposition is by induction on m. The base case, $m = 1$, is immediate.

Suppose that the statement holds $m-1$ and let

$$\hat{A} = \begin{pmatrix} A_{11} & \cdots & A_{1\,m-1} \\ \vdots & \vdots & \vdots \\ A_{m-1\,1} & \cdots & A_{m-1\,m-1} \end{pmatrix}.$$

By the inductive hypothesis, there exists a lower block triangular matrix $\hat{L}$ whose diagonal blocks are unit matrices, a block diagonal matrix, $\hat{D} = \mathsf{diag}(D_1, \ldots, D_m - 1)$ whose diagonal blocks are invertible matrices such that $D_1 = A_{11} = B_1$, $D_j = B_j/B_{j-1}$ for $1 \leqslant j \leqslant m-1$, and an upper block diagonal matrix $\hat{U}$ whose diagonal blocks are unit matrices such that $\hat{A} = \hat{L}\hat{D}\hat{U}$.
Since

$$A = \begin{pmatrix} \hat{A} & K \\ H & A_{mm} \end{pmatrix},$$

where $H = (A_{m1} \ \cdots \ A_{m\,m-1})$ and

$$K = \begin{pmatrix} A_{1m} \\ \vdots \\ A_{m-1\,m} \end{pmatrix},$$

we can write

$$\begin{aligned} A &= \begin{pmatrix} \hat{A} & K \\ H & A_{mm} \end{pmatrix} \\ &= \begin{pmatrix} I & O \\ H\hat{A}^{-1} & I \end{pmatrix} \begin{pmatrix} \hat{A} & O \\ O & A/\hat{A} \end{pmatrix} \begin{pmatrix} I & \hat{A}^{-1}K \\ O & I \end{pmatrix} \end{aligned}$$

by the second part of Supplement 5.6. Since $\hat{A} = \hat{L}\hat{D}\hat{U}$, we have

$$\begin{pmatrix} \hat{A} & O \\ O & A/\hat{A} \end{pmatrix} = \begin{pmatrix} \hat{L} & O \\ O & I \end{pmatrix} \begin{pmatrix} \hat{D} & O \\ O & A/\hat{A} \end{pmatrix} \begin{pmatrix} \hat{U} & O \\ O & I \end{pmatrix}.$$

Observe that the matrix

$$\begin{pmatrix} \hat{L} & O \\ O & I \end{pmatrix}$$

is a lower triangular matrix whose diagonal blocks are unit matrices because L has this property; similarly,

$$\begin{pmatrix} \hat{U} & O \\ O & I \end{pmatrix}$$

is an upper triangular matrix whose diagonal blocks are unit matrices. Thus,

$$A = \left(\begin{pmatrix} I & O \\ H\hat{A}^{-1} & I \end{pmatrix} \begin{pmatrix} \hat{L} & O \\ O & I \end{pmatrix} \right) \begin{pmatrix} \hat{D} & O \\ O & A/\hat{A} \end{pmatrix} \times \left(\begin{pmatrix} \hat{U} & O \\ O & I \end{pmatrix} \begin{pmatrix} I & \hat{A}^{-1}K \\ O & I \end{pmatrix} \right),$$

which has the prescribed form. If $A = \tilde{L}\tilde{D}\tilde{U}$ is another decomposition of A with the same properties, then $LDU = \tilde{L}\tilde{D}\tilde{U}$ implies $(\tilde{L})^{-1}LD = \tilde{D}\tilde{U}U^{-1}$ because the matrices $\tilde{L}$ and U are nonsingular. Note that $(\tilde{L})^{-1}LD$ is a lower block triangular matrix and $\tilde{D}\tilde{U}U^{-1}$ is an upper block triangular matrix, so both matrices are block diagonal. Also, $(\tilde{L})^{-1}LD = D$ and $\tilde{D}\tilde{U}U^{-1} = \tilde{D}$, so $D = \tilde{D}$.

Since L has a lower block triangular block structure, we can write

$$(\tilde{L})^{-1}L = \begin{pmatrix} I_1 & O & O & \cdots & O \\ L_{21} & I_2 & O & \cdots & O \\ L_{31} & L_{32} & I_3 & \cdots & O \\ \vdots & \vdots & \vdots & \vdots & O \\ L_{k1} & L_{k2} & L_{k3} & \cdots & I_k \end{pmatrix}$$

and

$$\tilde{U}U^{-1} = \begin{pmatrix} I_1 & U_{12} & U_{13} & \cdots & U_{1k} \\ O & I_2 & U_{23} & \cdots & U_{2k} \\ O & O & I_3 & \cdots & U_{3k} \\ \vdots & \vdots & \vdots & \vdots & \vdots \\ O & O & O & \cdots & I_k \end{pmatrix}.$$

Let $D = \mathsf{diag}(D_1, \cdots, D_k)$. Since $(\tilde{L})^{-1}LD = D$

$$\begin{pmatrix} D_1 & O & O & \cdots & O \\ D_1L_{21} & D_2 & O & \cdots & O \\ D_1L_{31} & D_2L_{32} & D_3 & \cdots & O \\ \vdots & \vdots & \vdots & \vdots & O \\ D_1L_{k1} & D_2L_{k2} & D_3L_{k3} & \cdots & D_k \end{pmatrix}$$

$$= \begin{pmatrix} D_1 & O & O & \cdots & O \\ O & D_2 & O & \cdots & O \\ O & O & O & \cdots & O \\ \vdots & \vdots & \vdots & \vdots & \vdots \\ O & O & O & \cdots & D_k \end{pmatrix},$$

it follows that $D_pL_{qp} = 0$ if $q > p$, so $L_{qp} = O$ because D_p is non-singular for $1 \leqslant p \leqslant k$. Thus, $(\tilde{L})^{-1}L = I$ and therefore $\tilde{L} = L$. Similarly, $\tilde{U} = U$, which proves the uniqueness of the decomposition.

(27) A matrix $A \in \mathbb{C}^{n\times n}$ is *strongly non-singular* if all its principal matrices are non-singular. Prove that if $A \in \mathbb{C}^{n\times n}$ is strongly non-singular, then there exists a unique factorization $A = LDU$ such that the following conditions are satisfied:
   (a) L is a lower block triangular matrix whose diagonal elements are equal to 1;
   (b) D is a diagonal matrix, $D = \mathsf{diag}(d_1, \ldots, d_n)$, such that $d_1 = a_{11}$ and $d_k = \frac{A_k}{A_{k-1}}$, where $A_k = A\begin{bmatrix} 1 \cdots k \\ 1 \cdots k \end{bmatrix}$;
   (c) U is an upper diagonal matrix whose diagonal elements are equal to 1.

   Hint: This follows immediately from Supplement 5.6.

(28) Prove that if $H \in \mathbb{R}^{n\times n}$ is a Hadamard matrix, then $|\det(H)| = n^{\frac{n}{2}}$.

(29) Prove that if $H \in \mathbb{R}^{n\times n}$ is a Hadamard matrix and $n > 2$, then n is a multiple of 4.

(30) Let $a_1, \ldots, a_n$ be n numbers and let $b_{ij} = |a_i - a_j|$ for $1 \leqslant i, j \leqslant n$. Compute the determinant of the matrix $B = (b_{ij})$.

(31) Let $S = \{(x_i, y_i) \in \mathbb{C}^2 \mid 1 \leqslant i \leqslant n\}$ be a set of n pairs of complex numbers such that $i \neq j$ implies $x_i \neq x_j$ and let $p(x) =$

$c_0 + c_1 x + \cdots + c_{n-1} x^{n-1}$ be a polynomial such that $p(x_i) = y_i$ for $1 \leqslant i \leqslant n$. Prove that

(a) the coefficients c_i of p are given by the formula

$$c_i = \frac{V_{\boldsymbol{x}} \stackrel{i}{\leftarrow} \boldsymbol{y}}{V_{\boldsymbol{x}}},$$

where $V_{\boldsymbol{x}} \stackrel{i}{\leftarrow} \boldsymbol{y}$ is the determinant obtained from $V_{\boldsymbol{x}}$ by replacing the ith column by $\boldsymbol{y}$;

(b) the polynomial $p(x)$ can be written as

$$p(x) = \sum_{j=1}^{n} y_j p_j(x),$$

where $p_j(x) = \prod\{\frac{x - x_k}{x_j - x_k} \mid k \in \{1, \ldots, n\} - \{j\}\}$. The polynomial p is known as the *Lagrange interpolation polynomial* for S.

(32) Let $A \in \mathbb{C}^{n \times n}$ be a matrix such that $A\mathbf{1}_n = A'\mathbf{1}_n == \mathbf{0}_n$. Prove that all cofactors $\mathsf{cof}(a_{ij})$ of A are equal.

(33) Let

$$M = \left(\begin{array}{c|c} A & B \\ \hline C & D \end{array}\right)$$

be a partitioned matrix, where $A \in \mathbb{R}^{m \times m}$, $B \in \mathbb{R}^{m \times p}$, $C \in \mathbb{R}^{p \times m}$, and $D \in \mathbb{R}^{p \times p}$. Prove that if $U \in \mathbb{R}^{m \times m}$ and $V \in \mathbb{R}^{p \times p}$ are orthogonal matrices, then

$$\begin{aligned} \det(M) &= \det \begin{pmatrix} UA & UB \\ C & D \end{pmatrix} = \det \begin{pmatrix} A & BV \\ C & DV \end{pmatrix} \\ &= \det \begin{pmatrix} AU & B \\ CU & D \end{pmatrix} = \det \begin{pmatrix} A & B \\ VC & VD \end{pmatrix}. \end{aligned}$$

Solution: Since

$$\begin{pmatrix} U & O_{m,p} \\ O_{p,m} & I_p \end{pmatrix} \begin{pmatrix} A & B \\ C & D \end{pmatrix} = \begin{pmatrix} UA & UB \\ C & D \end{pmatrix},$$

the result follows immediately by observing that

$$\det \begin{pmatrix} U & O_{m,p} \\ O_{p,m} & I_p \end{pmatrix} = 1$$

because U is an orthogonal matrix. The remaining equalities can be shown in a similar manner.

(34) Let $A \in \mathbb{R}^{n\times n}$ be a skew-symmetric matrix. If n is an odd number, prove that $\det(A) = 0$.

Solution: Since $A' = -A$ and n is odd, it follows that $\det(A') = (-1)^n \det(A) = -\det(A)$. Since $\det(A') = \det(A)$, it follows that $\det(A) = 0$.

(35) Let $A \in \mathbb{C}^{m\times n}$ be a matrix of rank r and let $A = BC$ be a full-rank factorization of A, where $B \in \mathbb{C}^{m\times r}$ and $C \in \mathbb{C}^{r\times n}$. Denote by

$$D_r^2(A) = \sum_{(I,J)\in\mathcal{N}(A)} \left(\det\left(A\begin{bmatrix} I \\ J \end{bmatrix}\right)\right)^2.$$

Prove that

$$\begin{aligned} D_r^2(A) &= \sum_{I\in\mathfrak{I}(A)} D_r^2(A_{(I,:)}) = \sum_{J\in\mathfrak{J}(A)} D_r^2(A_{(:,J)}) \\ &= \sum_{I\in\mathfrak{I}(A)} D_r^2(A_{(I,:)}) = D_r^2(B)D_r^2(C). \end{aligned}$$

Solution: The first two equalities are immediate. The last equality follows from the second part of Supplement 112 of Chapter 3 and from Cauchy–Binet formula (Theorem 5.10).

(36) Let $A \in \mathbb{R}^{n\times n}$ be a matrix such that $a_{ii} > 0$ for $1 \leqslant i \leqslant n$, such that $a_{ii} > \sum\{|a_{ik}| \mid 1 \leqslant k \leqslant n \text{ and } k \neq i\}$ and $1 \leqslant i \leqslant n$. Prove that $\det(A) > 0$.

Solution: The argument is by induction on $n \geqslant 2$. The basis step is obviously true. Suppose that it holds for $n-1$. Since $a_{11} > 0$, subtract from the raised ith row of A the first row multiplied by $\frac{a_{i1}}{a_{11}}$. This zeroes the first column with the exception of its first element, so we can write $\det(A) = a_{11}\det(B)$, where $B \in \mathbb{C}^{(n-1)\times(n-1)}$ is given by

$$B = \begin{pmatrix} b_{22} & \cdots & b_{2n} \\ \vdots & \cdots & \vdots \\ b_{n2} & \cdots & b_{nn} \end{pmatrix},$$

and is defined by $b_{pq} = a_{pq} - \frac{a_{p1}a_{1q}}{a_{11}}$ for $2 \leqslant p, q \leqslant n$. Since $\sum_{q=2}^{n} a_{1q} < a_{11}$, we have

$$\sum_{q=2}^{n} \left| \frac{a_{p1}a_{1q}}{a_{11}} \right| \leqslant |a_{p1}|.$$

Therefore,

$$\begin{aligned}
&\frac{a_{p1}a_{1p}}{a_{11}} + \sum_{q=2,q\neq p}^{n} |b_{pq}| \\
&\leqslant \frac{a_{p1}a_{1p}}{a_{11}} + \sum_{q=2,q\neq p}^{n} |a_{pq}| + \sum_{q=2,q\neq p} \left| \frac{a_{p1}a_{1q}}{a_{11}} \right| \\
&\leqslant \sum_{q=2,q\neq p}^{n} |a_{pq}| + \sum_{q=2}^{n} \left| \frac{a_{p1}a_{1q}}{a_{11}} \right| \\
&\leqslant \sum_{q=2,q\neq p} |a_{pq}| + |a_{p1}| = \sum_{q=1,q\neq p} |a_{pq}| < a_{pp}.
\end{aligned}$$

Thus, $b_{pp} > \sum_{q=2,q\neq p} |b_{pq}|$. By the induction hypothesis, $\det(B) > 0$, so $\det(A) > 0$.

(37) Let $A, B \in \mathbb{C}^{n\times n}$ be two matrices and let p be the polynomial $p(x) = \det(A + xB)$. Prove that the degree of f does not exceed the rank of B.

Solution: By Theorem 3.37, if $\mathsf{rank}(B) = r$, there exists the non-singular matrices $G, H \in \mathbb{C}^{n\times n}$ such that $B = GL_rH$ and

$$L_r = \begin{pmatrix} I_r & O_{r,n-r} \\ O_{n-r,r} & O_{n-r,n-r} \end{pmatrix}.$$

Therefore,

$$\begin{aligned}
p(x) &= \det(A + xGL_rH) = \det(G(G^{-1}AH^{-1} + xL_r)H) \\
&= \det(G)\det(H)\det(M + xL_r),
\end{aligned}$$

where $M = G^{-1}AH^{-1}$. The highest degree of a term in $\det(M + xL_r)$ cannot exceed r and the result follows immediately.

(38) Let $f(x_1, \ldots, x_r)$ be a homogeneous polynomial of degree n. Prove that

$$\sum x_i \frac{\partial f}{\partial x_i} = nf.$$

(This result is known as Euler's theorem).

Solution: The homogeneity of F means that $F(tx_1, \ldots, tx_r) = t^n F(x_1, \ldots, x_r)$. Differentiating with respect to t yields

$$\sum x_i \frac{\partial F}{\partial x_i} = nt^{n-1} f(x_1, \ldots, x_r),$$

and the desired result follows taking $t = 1$.

(39) Let $f_1, \ldots, f_k$ be k polynomials in $\mathbb{R}[x]$. Define the polynomials

$$F(x) = a_1 f_1(x) + a_2 f_2(x) + \cdots + a_k f_k(x),$$
$$G(x) = b_1 f_1(x) + b_2 f_2(x) + \cdots + b_k f_k(x).$$

Prove that the polynomials have a common root if and only if $R_{F,G} = 0$ for all $a_1, \ldots, a_k, b_1, \ldots, b_k$.

(40) The *discriminant* of a polynomial $f \in \mathbb{R}[x]$ is $D_f = \frac{1}{a_0} R_{f,f'}$, where a_0 is the leading coefficient of f. Prove that f has roots of multiplicity at least 2 if and only if $D_f = 0$.

(41) Let f_1, f_2 be two polynomials in $\mathbb{R}[x]$ of degrees k_1 and k_2, respectively. Prove that $D_{f_1 f_2} = (-1)^{k_1 k_2} D_{f_1} D_{f_2} R^2_{f_1, f_2}$.

(42) Let g, h be two polynomials of degrees k_1 and k_2, respectively. Prove that

$$D_{gh} = (-1)^{k_1 k_2} D_g D_h R^2_{g,h}.$$

(43) Prove that the polynomial $f \in \mathbb{C}[x]$ defined by $f(x) = ax^3 + bX^2 + cx + d$ has the expression

$$b^2c^2 - 4ac^3 - 4b^3d - 27a^2d^2 + 18abcd$$

as its discriminant.

(44) Let $f \in \mathbb{C}[x]$ be a polynomial. Prove that $D_{f+a} = D_f$ for any constant $a \in \mathbb{C}$. If $g(x) = f(ax)$, prove that $D_g = a^{n(n-1)} D_f$

A *homogeneous linear system* has the form

$$\begin{aligned} a_{11}x_1 + a_{12}x_2 + \cdots a_{1n}x_n &= 0, \\ \vdots &= 0 \\ a_{m1}x_1 + a_{m2}x_2 + \cdots a_{mn}x_n &= 0, \end{aligned}$$

or $A\boldsymbol{x} = \mathbf{0}_m$, where $A = (a_{ij}) \in \mathbb{R}^{m\times n}$ and $\boldsymbol{x} \in \mathbb{R}^n$.

Observe that each such system has the *trivial solution* $x_1 = x_2 = \cdots = x_n = 0$.

(45) Prove that the set of solutions of the homogeneous system $A\boldsymbol{x} = \mathbf{0}_m$ is a subspace of $\mathbb{R}^n$.

(46) Let $A \in \mathbb{R}^{n\times n}$ be a matrix. Prove that the system $A\boldsymbol{x} = \mathbf{0}_n$ has a non-trivial solution if and only if $\det(A) = 0$.

(47) Let $A \in \mathbb{R}^{3\times 4}$ and consider the system $A\boldsymbol{x} = \mathbf{0}_3$, where $\boldsymbol{x} \in \mathbb{R}^4$. Denote by A_k the matrix in $\mathbb{R}^{3\times 3}$ obtained from A by eliminating the kth column, where $1 \leqslant k \leqslant 4$. Prove that if the system $A\boldsymbol{x} = \mathbf{0}_3$ has a non-trivial solution $\boldsymbol{x} \in \mathbb{R}^4$, then

$$\frac{x_1}{\det(A_1)} = \frac{x_2}{-\det(A_2)} = \frac{x_3}{\det(A_3)} = \frac{x_4}{-\det(A_4)}.$$

Solution: The system $A\boldsymbol{u} = \mathbf{0}_3$ has the explicit form

$$\begin{aligned} a_{11}x_1 + a_{12}x_2 + a_{13}x_3 + a_{14}x_4 &= 0 \\ a_{21}x_1 + a_{22}x_2 + a_{23}x_3 + a_{24}x_4 &= 0 \\ a_{31}x_1 + a_{32}x_2 + a_{33}x_3 + a_{34}x_4 &= 0. \end{aligned}$$

Equivalently, we can write

$$\begin{aligned} a_{11}x_1 + a_{12}x_2 + a_{13}x_3 &= -a_{14}x_4 \\ a_{21}x_1 + a_{22}x_2 + a_{23}x_3 &= -a_{24}x_4 \\ a_{31}x_1 + a_{32}x_2 + a_{33}x_3 &= -a_{34}x_4. \end{aligned}$$

Thus, we have

$$x_1 = -x_4 \frac{\begin{vmatrix} a_{14} & a_{12} & a_{13} \\ a_{24} & a_{22} & a_{23} \\ a_{34} & a_{32} & a_{33} \end{vmatrix}}{\det(A_4)} = -x_4 \frac{\det(A_1)}{\det(A_4)}.$$

Similar formulas are

$$x_2 = x_4 \frac{\det(A_2)}{\det(A_4)},$$

and

$$x_3 = -x_4 \frac{\det(A_3)}{\det(A_4)}.$$

Thus,

$$\frac{x_1}{\det(A_1)} = \frac{x_2}{-\det(A_2)} = \frac{x_3}{\det(A_3)} = \frac{x_4}{-\det(A_4)}.$$

Bibliographical Comments

An encyclopedic reference on Schur's complement and its applications in numerical analysis, probabilities, and statistics and other areas can be found in [178].

The result contained in Supplement 22 appears in [42]. Supplement 35 is a result of [10].

Chapter 6

Norms and Inner Products

6.1 Introduction

A norm is a real-valued function defined on a linear space intended to model the "length" of the vector.

Norms generate metrics on linear spaces and these metrics, in turn, generate topologies that are useful in constructing algorithms on these spaces. Elementary notions of set topology used in this chapter can be found in [152].

The other fundamental notion discussed in this chapter is the notion of inner product spaces. Inner products are capable of generating norms and allow the introduction of the concept of orthogonality and of unitary and orthogonal (or orthonormal) matrices. Also, we introduce positive definite matrices and discuss two decomposition results for matrices: the Cholesky decomposition for positive definite matrices and the QR factorizations.

6.2 Basic Inequalities

Lemma 6.1. *Let $p, q \in \mathbb{R} - \{0, 1\}$ such that $\frac{1}{p} + \frac{1}{q} = 1$. Then we have $p > 1$ if and only if $q > 1$. Furthermore, one of the numbers p, q belongs to the interval $(0, 1)$ if and only if the other number is negative.*

Proof. The statement follows immediately from the equality $q = \frac{p}{p-1}$. □

Lemma 6.2. *Let $p, q \in \mathbb{R} - \{0, 1\}$ be two numbers such that $\frac{1}{p} + \frac{1}{q} = 1$ and $p > 1$. Then, for every $a, b \in \mathbb{R}_{\geqslant 0}$, we have*

$$ab \leqslant \frac{a^p}{p} + \frac{b^q}{q},$$

where the equality holds if and only if $a = b^{-\frac{1}{1-p}}$.

Proof. By Lemma 6.1, we have $q > 1$. Consider the function $f(x) = \frac{x^p}{p} + \frac{1}{q} - x$ for $x \geqslant 0$. We have $f'(x) = x^{p-1} - 1$, so the minimum is achieved when $x = 1$ and $f(1) = 0$. Thus,

$$f\left(ab^{-\frac{1}{p-1}}\right) \geqslant f(1) = 0,$$

which amounts to

$$\frac{a^p b^{-\frac{p}{p-1}}}{p} + \frac{1}{q} - ab^{-\frac{1}{p-1}} \geqslant 0.$$

By multiplying both sides of this inequality by $b^{\frac{p}{p-1}}$, we obtain the desired inequality. □

Observe that if $\frac{1}{p} + \frac{1}{q} = 1$ and $p < 1$, then $q < 0$. In this case, we have the reverse inequality

$$ab \geqslant \frac{a^p}{p} + \frac{b^q}{q}, \tag{6.1}$$

which can be shown by observing that the function f has a maximum in $x = 1$. The same inequality holds when $q < 1$ and therefore $p < 0$.

Theorem 6.1 (The Hölder inequality). Let $a_1, \ldots, a_n$ and $b_1, \ldots, b_n$ be $2n$ nonnegative numbers, and let p and q be two numbers such that $\frac{1}{p} + \frac{1}{q} = 1$ and $p > 1$. We have

$$\sum_{i=1}^{n} a_i b_i \leqslant \left(\sum_{i=1}^{n} a_i^p\right)^{\frac{1}{p}} \cdot \left(\sum_{i=1}^{n} b_i^q\right)^{\frac{1}{q}}.$$

Proof. If $a_1 = \cdots = a_n = 0$ or if $b_1 = \cdots = b_n = 0$, then the inequality is clearly satisfied. Therefore, we may assume that at least one of $a_1, \ldots, a_n$ and at least one of $b_1, \ldots, b_n$ is non-zero. Define the numbers

$$x_i = \frac{a_i}{(\sum_{i=1}^n a_i^p)^{\frac{1}{p}}} \text{ and } y_i = \frac{b_i}{(\sum_{i=1}^n b_i^q)^{\frac{1}{q}}}$$

for $1 \leqslant i \leqslant n$. Lemma 6.2 applied to x_i, y_i yields

$$\frac{a_i b_i}{(\sum_{i=1}^n a_i^p)^{\frac{1}{p}} (\sum_{i=1}^n b_i^q)^{\frac{1}{q}}} \leqslant \frac{1}{p} \frac{a_i^p}{\sum_{i=1}^n a_i^p} + \frac{1}{q} \frac{b_i^p}{\sum_{i=1}^n b_i^p}.$$

Adding these inequalities, we obtain

$$\sum_{i=1}^n a_i b_i \leqslant \left(\sum_{i=1}^n a_i^p \right)^{\frac{1}{p}} \left(\sum_{i=1}^n b_i^q \right)^{\frac{1}{q}}$$

because $\frac{1}{p} + \frac{1}{q} = 1$. □

The nonnegativity of the numbers $a_1, \ldots, a_n, b_1, \ldots, b_n$ can be relaxed by using absolute values. Indeed, we can easily prove the following variant of Theorem 6.1.

Theorem 6.2. *Let $a_1, \ldots, a_n$ and $b_1, \ldots, b_n$ be $2n$ numbers and let p and q be two numbers such that $\frac{1}{p} + \frac{1}{q} = 1$ and $p > 1$. We have*

$$\left| \sum_{i=1}^n a_i b_i \right| \leqslant \left(\sum_{i=1}^n |a_i|^p \right)^{\frac{1}{p}} \cdot \left(\sum_{i=1}^n |b_i|^q \right)^{\frac{1}{q}}.$$

Proof. By Theorem 6.1, we have

$$\sum_{i=1}^n |a_i||b_i| \leqslant \left(\sum_{i=1}^n |a_i|^p \right)^{\frac{1}{p}} \cdot \left(\sum_{i=1}^n |b_i|^q \right)^{\frac{1}{q}}.$$

The needed equality follows from the fact that

$$\left| \sum_{i=1}^n a_i b_i \right| \leqslant \sum_{i=1}^n |a_i||b_i|.$$

□

Corollary 6.1 (The Cauchy–Schwarz inequality for $\mathbb{R}^n$). Let $a_1, \ldots, a_n$ and $b_1, \ldots, b_n$ be $2n$ real numbers. We have

$$\left|\sum_{i=1}^{n} a_i b_i\right| \leqslant \sqrt{\sum_{i=1}^{n} a_i^2} \cdot \sqrt{\sum_{i=1}^{n} b_i^2}.$$

Proof. The inequality follows immediately from Theorem 6.2 by taking $p = q = 2$. □

Theorem 6.3 (Minkowski's inequality). Let $a_1, \ldots, a_n$ and $b_1, \ldots, b_n$ be 2n nonnegative real numbers. If $p \geqslant 1$, we have

$$\left(\sum_{i=1}^{n} (a_i + b_i)^p\right)^{\frac{1}{p}} \leqslant \left(\sum_{i=1}^{n} a_i^p\right)^{\frac{1}{p}} + \left(\sum_{i=1}^{n} b_i^p\right)^{\frac{1}{p}}.$$

If $p < 1$, the inequality sign is reversed.

Proof. For $p = 1$, the inequality is immediate. Therefore, we can assume that $p > 1$. Note that

$$\sum_{i=1}^{n} (a_i + b_i)^p = \sum_{i=1}^{n} a_i (a_i + b_i)^{p-1} + \sum_{i=1}^{n} b_i (a_i + b_i)^{p-1}.$$

By Hölder's inequality for p, q such that $p > 1$ and $\frac{1}{p} + \frac{1}{q} = 1$, we have

$$\begin{aligned}\sum_{i=1}^{n} a_i (a_i + b_i)^{p-1} &\leqslant \left(\sum_{i=1}^{n} a_i^p\right)^{\frac{1}{p}} \left(\sum_{i=1}^{n} (a_i + b_i)^{(p-1)q}\right)^{\frac{1}{q}} \\ &= \left(\sum_{i=1}^{n} a_i^p\right)^{\frac{1}{p}} \left(\sum_{i=1}^{n} (a_i + b_i)^p\right)^{\frac{1}{q}}.\end{aligned}$$

Similarly, we can write

$$\sum_{i=1}^{n} b_i (a_i + b_i)^{p-1} \leqslant \left(\sum_{i=1}^{n} b_i^p\right)^{\frac{1}{p}} \left(\sum_{i=1}^{n} (a_i + b_i)^p\right)^{\frac{1}{q}}.$$

Adding the last two inequalities yields

$$\sum_{i=1}^{n} (a_i + b_i)^p \leqslant \left(\left(\sum_{i=1}^{n} a_i^p\right)^{\frac{1}{p}} + \left(\sum_{i=1}^{n} b_i^p\right)^{\frac{1}{p}}\right) \left(\sum_{i=1}^{n} (a_i + b_i)^p\right)^{\frac{1}{q}},$$

which is equivalent to the inequality

$$\left(\sum_{i=1}^{n}(a_i + b_i)^p\right)^{\frac{1}{p}} \leqslant \left(\sum_{i=1}^{n} a_i^p\right)^{\frac{1}{p}} + \left(\sum_{i=1}^{n} b_i^p\right)^{\frac{1}{p}}.$$

□

6.3 Metric Spaces

Definition 6.1. A function $d : S^2 \longrightarrow \mathbb{R}_{\geqslant 0}$ is a *metric* if it has the following properties:

(i) $d(x, y) = 0$ if and only if $x = y$ for $x, y \in S$;
(ii) $d(x, y) = d(y, x)$ for $x, y \in S$;
(iii) $d(x, y) \leqslant d(x, z) + d(z, y)$ for $x, y, z \in S$.

The pair (S, d) will be referred to as a *metric space*.

If property (i) is replaced by the weaker requirement that $d(x, x) = 0$ for $x \in S$, then we refer to d as a *semimetric* on S. Thus, if d is a semimetric, $d(x, y) = 0$ does not necessarily imply $x = y$ and we can have for two distinct elements x, y of S, $d(x, y) = 0$. If d is a semimetric, then we refer to the pair (S, d) as a *semimetric space*.

Example 6.1. Let S be a nonempty set. Define the mapping $d : S^2 \longrightarrow \mathbb{R}_{\geqslant 0}$ by

$$d(u, v) = \begin{cases} 1 & \text{if } u \neq v, \\ 0 & \text{otherwise} \end{cases}$$

for $x, y \in S$. It is easy to see that d satisfies the definiteness property. To prove that d satisfies the triangular inequality, we need to show that

$$d(x, y) \leqslant d(x, z) + d(z, y)$$

for all $x, y, z \in S$. This is clearly the case if $x = y$. Suppose that $x \neq y$, so $d(x, y) = 1$. Then, for every $z \in S$, we have at least one of the inequalities $x \neq z$ or $z \neq y$, so at least one of the numbers $d(x, z)$ or $d(z, y)$ equals 1. Thus, d satisfies the triangular inequality. The metric d introduced here is the *discrete metric* on S.

Example 6.2. Consider the mapping $d : (\mathbf{Seq}_n(S))^2 \longrightarrow \mathbb{R}_{\geqslant 0}$ defined by

$$d(\boldsymbol{p}, \boldsymbol{q}) = |\{i \mid 0 \leqslant i \leqslant n-1 \text{ and } \boldsymbol{p}(i) \neq \boldsymbol{q}(i)\}|$$

for all sequences $\boldsymbol{p}, \boldsymbol{q}$ of length n on the set S.

It is easy to see that d is a metric. We justify here only the triangular inequality. Let $\boldsymbol{p}, \boldsymbol{q}, \boldsymbol{r}$ be three sequences of length n on the set S. If $\boldsymbol{p}(i) \neq \boldsymbol{q}(i)$, then $\boldsymbol{r}(i)$ must be distinct from at least one of $\boldsymbol{p}(i)$ and $\boldsymbol{q}(i)$. Therefore,

$$\begin{aligned} &\{i \mid 0 \leqslant i \leqslant n-1 \text{ and } \boldsymbol{p}(i) \neq \boldsymbol{q}(i)\} \\ &\subseteq \{i \mid 0 \leqslant i \leqslant n-1 \text{ and } \boldsymbol{p}(i) \neq \boldsymbol{r}(i)\} \cup \{i \mid 0 \leqslant i \\ &\leqslant n-1 \text{ and } \boldsymbol{r}(i) \neq \boldsymbol{q}(i)\}, \end{aligned}$$

which implies the triangular inequality.

Example 6.3. For $\boldsymbol{x} \in \mathbb{R}^n$ and $\boldsymbol{y} \in \mathbb{R}^n$, the *Euclidean metric* is the mapping

$$d_2(\boldsymbol{x}, \boldsymbol{y}) = \sqrt{\sum_{i=1}^{n} (x_i - y_i)^2}.$$

The first two conditions of Definition 6.1 are obviously satisfied.

To prove the third inequality, let $\boldsymbol{x}, \boldsymbol{y}, \boldsymbol{z} \in \mathbb{R}^n$. Choosing $a_i = x_i - y_i$ and $b_i = y_i - z_i$ for $1 \leqslant i \leqslant n$ in Minkowski's inequality implies

$$\sqrt{\sum_{i=1}^{n} (x_i - z_i)^2} \leqslant \sqrt{\sum_{i=1}^{n} (x_i - y_i)^2} + \sqrt{\sum_{i=1}^{n} (y_i - z_i)^2},$$

which amounts to $d(\boldsymbol{x}, \boldsymbol{z}) \leqslant d(\boldsymbol{x}, \boldsymbol{y}) + d(\boldsymbol{y}, \boldsymbol{z})$. Thus, we conclude that d is indeed a metric on $\mathbb{R}^n$.

We frequently use the notions of closed sphere and open sphere.

Definition 6.2. Let (S, d) be a metric space. The *closed sphere* centered in $x \in S$ of radius r is the set

$$B_d[x, r] = \{y \in S | d(x, y) \leqslant r\}.$$

The *open sphere* centered in $x \in S$ of radius r is the set

$$B_d(x, r) = \{y \in S | d(x, y) < r\}.$$

Definition 6.3. Let (S, d) be a metric space. The *diameter* of a subset U of S is the number $diam_{S,d}(U) = \sup\{d(x, y) \mid x, y \in U\}$. The set U is *bounded* if $diam_{S,d}(U)$ is finite.

The *diameter* of the metric space (S, d) is the number

$$diam_{S,d} = \sup\{d(x, y) \mid x, y \in S\}.$$

If the metric space is clear from the context, then we denote the diameter of a subset U just by $diam(U)$.

If (S, d) is a finite metric space, then $diam_{S,d} = \max\{d(x, y) \mid x, y \in S\}$.

A mapping $d : S \times S \longrightarrow \hat{\mathbb{R}}_{\geqslant 0}$ can be extended to the set of subsets of S by defining $d(U, V)$ as

$$d(U, V) = \inf\{d(u, v) \mid u \in U \text{ and } v \in V\} \tag{6.2}$$

for $U, V \in \mathcal{P}(S)$.

Observe that, even if d is a metric, then its extension is not, in general, a metric on $\mathcal{P}(S)$ because it does not satisfy the triangular inequality. Instead, we can show that for every U, V, W we have

$$d(U, W) \leqslant d(U, V) + diam(V) + d(V, W).$$

Indeed, by the definition of $d(U, V)$ and $d(V, W)$, for every $\epsilon > 0$, there exist $u \in U$, $v, v' \in V$, and $w \in W$ such that

$$d(U, V) \leqslant d(u, v) \leqslant d(U, V) + \tfrac{\epsilon}{2},$$
$$d(V, W) \leqslant d(v', w) \leqslant d(V, W) + \tfrac{\epsilon}{2}.$$

By the triangular axiom, we have

$$d(u, w) \leqslant d(u, v) + d(v, v') + d(v', w).$$

Hence,

$$d(u, w) \leqslant d(U, V) + diam(V) + d(V, W) + \epsilon,$$

which implies

$$d(U, W) \leqslant d(U, V) + diam(V) + d(V, W) + \epsilon$$

for every $\epsilon > 0$. This yields the needed inequality.

Definition 6.4. Let (S, d) be a metric space. The sets $U, V \in \mathcal{P}(S)$ are *separate* if $d(U, V) > 0$.

We denote the number $d(\{u\}, V) = \inf\{d(u, v) \mid v \in V\}$ by $d(u, V)$. It is clear that $u \in V$ implies $d(u, V) = 0$.

The notion of dissimilarity is a generalization of the notion of metric.

Definition 6.5. A *dissimilarity on a set* S is a function $d : S^2 \longrightarrow \mathbb{R}_{\geqslant 0}$ satisfying the following conditions:

(i) $d(x, x) = 0$ for all $x \in S$;

(ii) $d(x, y) = d(y, x)$ for all $x, y \in S$.

The pair (S, d) is a *dissimilarity space.*

A related concept is the notion of similarity.

Definition 6.6. A *similarity on a set* S is a function $s : S^2 \longrightarrow \mathbb{R}_{\geqslant 0}$ satisfying the following conditions:

(i) $s(x, y) \leqslant s(x, x) = 1$ for all $x, y \in S$;

(ii) $s(x, y) = s(y, x)$ for all $x, y \in S$.

The pair (S, s) is a *similarity space.*

Example 6.4. Let $d : S^2 \longrightarrow \mathbb{R}_{\geqslant 0}$ be a metric on the set S. Then $s : S^2 \longrightarrow \mathbb{R}_{\geqslant 0}$ defined by $s(x, y) = 2^{-d(x,y)}$ for $x, y \in S$ is a dissimilarity, such that $s(x, x) = 1$ for every $x, y \in S$.

6.4 Norms

In this chapter, we study norms on real or complex linear spaces.

Definition 6.7. A *seminorm* on an $\mathbb{F}$-linear space V is a mapping $\nu : V \longrightarrow \mathbb{R}$ that satisfies the following conditions:

(i) $\nu(\boldsymbol{x} + \boldsymbol{y}) \leqslant \nu(\boldsymbol{x}) + \nu(\boldsymbol{y})$ (subadditivity), and

(ii) $\nu(a\boldsymbol{x}) = |a|\nu(\boldsymbol{x})$ (positive homogeneity),

for $\boldsymbol{x}, \boldsymbol{y} \in V$ and $a \in F$.

By taking $a = 0$ in the second condition of the definition, we have $\nu(\mathbf{0}) = 0$ for every seminorm on a real or complex space.

A seminorm can be defined on every linear space. Indeed, if B is a basis of V, $B = \{\boldsymbol{v}_i \mid i \in I\}$, J is a finite subset of I, and

$\boldsymbol{x} = \sum_{i \in I} x_i \boldsymbol{v}_i$, define $\nu_J(\boldsymbol{x})$ as

$$\nu_J(\boldsymbol{x}) = \begin{cases} 0 & \text{if } \boldsymbol{x} = \boldsymbol{0}, \\ \sum_{j \in J} |a_j| & \text{otherwise} \end{cases}$$

for $\boldsymbol{x} \in V$. We leave to the reader the verification of the fact that ν_J is indeed a seminorm.

Theorem 6.4. *If V is a real or complex linear space and $\nu : V \longrightarrow \mathbb{R}$ is a seminorm on V, then*

$$\nu(\boldsymbol{x} - \boldsymbol{y}) \geqslant |\nu(\boldsymbol{x}) - \nu(\boldsymbol{y})|,$$

for $\boldsymbol{x}, \boldsymbol{y} \in V$.

Proof. We have $\nu(\boldsymbol{x}) \leqslant \nu(\boldsymbol{x} - \boldsymbol{y}) + \nu(\boldsymbol{y})$, so

$$\nu(\boldsymbol{x}) - \nu(\boldsymbol{y}) \leqslant \nu(\boldsymbol{x} - \boldsymbol{y}). \tag{6.3}$$

Since $\nu(\boldsymbol{x} - \boldsymbol{y}) = |-1|\nu(\boldsymbol{y} - \boldsymbol{x}) \geqslant \nu(\boldsymbol{y}) - \nu(\boldsymbol{x})$, we have

$$-(\nu(\boldsymbol{x}) - \nu(\boldsymbol{y})) \leqslant \nu(\boldsymbol{x}) - \nu(\boldsymbol{y}). \tag{6.4}$$

Inequalities (6.3) and (6.4) give the desired inequality. □

Corollary 6.2. *If $p : V \longrightarrow \mathbb{R}$ is a seminorm on V, then $p(\boldsymbol{x}) \geqslant 0$ for $\boldsymbol{x} \in V$.*

Proof. By choosing $\boldsymbol{y} = \boldsymbol{0}$ in the inequality of Theorem 6.4, we have $\nu(\boldsymbol{x}) \geqslant |\nu(\boldsymbol{x})| \geqslant 0$. □

Definition 6.8. A *norm* on an $\mathbb{F}$-linear space V is a seminorm $\nu : V \longrightarrow \mathbb{R}$ such that $\nu(\boldsymbol{x}) = 0$ implies $\boldsymbol{x} = \boldsymbol{0}$ for $\boldsymbol{x} \in V$.

The pair (V, ν) is referred to as a *normed linear space*.

Example 6.5. The set of real-valued continuous functions defined on the interval $[-1, 1]$ is a real linear space. The addition of two such functions f, g, is defined by $(f + g)(x) = f(x) + g(x)$ for $x \in [-1, 1]$; the multiplication of f by a scalar $a \in \mathbb{R}$ is $(af)(x) = af(x)$ for $x \in [-1, 1]$.

Define $\nu(f) = \sup\{|f(x)| \mid x \in [-1, 1]\}$. Since $|f(x)| \leqslant \nu(f)$ and $|g(x)| \leqslant \nu(g)$ for $x \in [-1, 1]\}$, it follows that $|(f + g)(x)| \leqslant |f(x)| + |g(x)| \leqslant \nu(f) + \nu(g)$. Thus, $\nu(f + g) \leqslant \nu(f) + \nu(g)$. We leave to the reader the verification of the remaining properties of Definition 6.7.

We denote $\nu(f)$ by $\|f\|$.

Theorem 6.5. *For $p \geqslant 1$, the function $\nu_p : \mathbb{R}^n \longrightarrow \mathbb{R}_{\geqslant 0}$ defined by*

$$\nu_p(\boldsymbol{x}) = \left(\sum_{i=1}^{n} |x_i|^p\right)^{\frac{1}{p}},$$

where $\boldsymbol{x} = \begin{pmatrix} x_1 \\ \vdots \\ x_n \end{pmatrix} \in \mathbb{R}^n$*, is a norm on* $\mathbb{R}^n$*.*

Proof. We must prove that ν_p satisfies the conditions of Definition 6.7 and that $\nu_p(\boldsymbol{x}) = 0$ implies $\boldsymbol{x} = \mathbf{0}$.

Let

$$\boldsymbol{x} = \begin{pmatrix} x_1 \\ \vdots \\ x_n \end{pmatrix} \text{ and } \boldsymbol{y} = \begin{pmatrix} y_1 \\ \vdots \\ y_n \end{pmatrix}.$$

Minkowski's inequality applied to the nonnegative numbers $a_i = |x_i|$ and $b_i = |y_i|$ amounts to

$$\left(\sum_{i=1}^{n} (|x_i| + |y_i|)^p\right)^{\frac{1}{p}} \leqslant \left(\sum_{i=1}^{n} |x_i|^p\right)^{\frac{1}{p}} + \left(\sum_{i=1}^{n} |y_i|^p\right)^{\frac{1}{p}}.$$

Since $|x_i + y_i| \leqslant |x_i| + |y_i|$ for every i, we have

$$\left(\sum_{i=1}^{n} (|x_i + y_i|)^p\right)^{\frac{1}{p}} \leqslant \left(\sum_{i=1}^{n} |x_i|^p\right)^{\frac{1}{p}} + \left(\sum_{i=1}^{n} |y_i|^p\right)^{\frac{1}{p}},$$

that is, $\nu_p(\boldsymbol{x} + \boldsymbol{y}) \leqslant \nu_p(\boldsymbol{x}) + \nu_p(\boldsymbol{y})$. We leave to the reader the verification of the remaining conditions. Thus, ν_p is a norm on $\mathbb{R}^n$. □

Example 6.6. The mapping $\nu_1 : \mathbb{R}^n \longrightarrow \mathbb{R}$ given by

$$\nu_1(\boldsymbol{x}) = |x_1| + |x_2| + \cdots + |x_n|,$$

for $\boldsymbol{x} = \begin{pmatrix} x_1 \\ \vdots \\ x_n \end{pmatrix}$, is a norm on $\mathbb{R}^n$.

Example 6.7. A special norm on $\mathbb{R}^n$ is the function $\nu_\infty : \mathbb{R}^n \longrightarrow \mathbb{R}_{\geqslant 0}$ given by

$$\nu_\infty(\boldsymbol{x}) = \max\{|x_i| \mid 1 \leqslant i \leqslant n\}. \tag{6.5}$$

We verify here that ν_∞ satisfies the first condition of Definition 6.7. We start from the inequality

$$|x_i + y_i| \leqslant |x_i| + |y_i| \leqslant \nu_\infty(\boldsymbol{x}) + \nu_\infty(\boldsymbol{y})$$

for every i, $1 \leqslant i \leqslant n$. This in turn implies

$$\nu_\infty(\boldsymbol{x} + \boldsymbol{y}) = \max\{|x_i + y_i| \mid 1 \leqslant i \leqslant n\} \leqslant \nu_\infty(\boldsymbol{x}) + \nu_\infty(\boldsymbol{y}),$$

which gives the desired inequality.

This norm can be regarded as a limit case of the norms ν_p. Indeed, let $\boldsymbol{x} \in \mathbb{R}^n$ and let $M = \max\{|x_i| \mid 1 \leqslant i \leqslant n\} = |x_{\ell_1}| = \cdots = |x_{\ell_k}|$ for some $\ell_1, \ldots, \ell_k$, where $1 \leqslant \ell_1, \ldots, \ell_k \leqslant n$. Here $x_{\ell_1}, \ldots, x_{\ell_k}$ are the components of $\boldsymbol{x}$ that have the maximal absolute value and $k \geqslant 1$. We can write

$$\lim_{p\to\infty} \nu_p(\boldsymbol{x}) = \lim_{p\to\infty} M \left(\sum_{i=1}^{n} \left(\frac{|x_i|}{M} \right)^p \right)^{\frac{1}{p}} = \lim_{p\to\infty} M(k)^{\frac{1}{p}} = M,$$

which justifies the notation ν_∞.

We use the alternative notation $\|\boldsymbol{x}\|_p$ for $\nu_p(\boldsymbol{x})$. We refer to $\|\boldsymbol{x}\|_2$ as the *Euclidean norm* of $\boldsymbol{x}$ and we denote this norm simply by $\|\boldsymbol{x}\|$ when there is no risk of confusion.

Example 6.8. For $p \geqslant 1$, let ℓ_p be the set that consists of sequences of real numbers $\boldsymbol{x} = (x_0, x_1, \ldots)$ such that the series $\sum_{i=0}^{\infty} |x_i|^p$ is convergent. We can show that ℓ_p is a linear space.

Let $\boldsymbol{x}, \boldsymbol{y} \in \ell_p$ be two sequences in ℓ_p. Using Minkowski's inequality, we have

$$\sum_{i=0}^{n} |x_i + y_i|^p \leqslant \sum_{i=0}^{n} (|x_i| + |y_i|)^p \leqslant \sum_{i=0}^{n} |x_i|^p + \sum_{i=0}^{n} |y_i|^p,$$

which shows that $\boldsymbol{x} + \boldsymbol{y} \in \ell_p$. It is immediate that $\boldsymbol{x} \in \ell_p$ implies $a\boldsymbol{x} \in \ell_p$ for every $a \in \mathbb{R}$ and $\boldsymbol{x} \in \ell_p$.

The following statement shows that any norm defined on a linear space generates a metric on the space.

Theorem 6.6. *Each norm $\nu : V \longrightarrow \mathbb{R}_{\geqslant 0}$ on a real linear space V generates a metric on the set V defined by $d_\nu(\boldsymbol{x}, \boldsymbol{y}) = \nu(\boldsymbol{x} - \boldsymbol{y})$ for $\boldsymbol{x}, \boldsymbol{y} \in V$.*

Proof. Note that if $d_\nu(\boldsymbol{x}, \boldsymbol{y}) = \nu(\boldsymbol{x} - \boldsymbol{y}) = 0$, it follows that $\boldsymbol{x} - \boldsymbol{y} = \mathbf{0}$; that is, $\boldsymbol{x} = \boldsymbol{y}$.

The symmetry of d_ν is obvious and so we need to verify only the triangular axiom. Let $\boldsymbol{x}, \boldsymbol{y}, \boldsymbol{z} \in L$. Applying the subadditivity of norms, we have

$$\nu(\boldsymbol{x} - \boldsymbol{z}) = \nu(\boldsymbol{x} - \boldsymbol{y} + \boldsymbol{y} - \boldsymbol{z}) \leqslant \nu(\boldsymbol{x} - \boldsymbol{y}) + \nu(\boldsymbol{y} - \boldsymbol{z}),$$

or, equivalently, $d_\nu(\boldsymbol{x}, \boldsymbol{z}) \leqslant d_\nu(\boldsymbol{x}, \boldsymbol{y}) + d_\nu(\boldsymbol{y}, \boldsymbol{z})$, for every $\boldsymbol{x}, \boldsymbol{y}, \boldsymbol{z} \in L$, which concludes the argument. □

We refer to d_ν as the *metric induced by the norm ν on the linear space V*.

Observe that the norm ν can be expressed using d_ν as

$$\nu(\boldsymbol{x}) = d_\nu(\boldsymbol{x}, \mathbf{0}) \tag{6.6}$$

for $\boldsymbol{x} \in V$.

For $p \geqslant 1$, then d_p denotes the metric d_{ν_p} induced by the norm ν_p on the linear space $\mathbb{R}^n$ known as the *Minkowski metric* on $\mathbb{R}^n$.

If $p = 2$, we have the *Euclidean metric* on $\mathbb{R}^n$ given by

$$d_2(\boldsymbol{x}, \boldsymbol{y}) = \sqrt{\sum_{i=1}^{n} |x_i - y_i|^2} = \sqrt{\sum_{i=1}^{n} (x_i - y_i)^2}.$$

For $p = 1$, we have

$$d_1(\boldsymbol{x}, \boldsymbol{y}) = \sum_{i=1}^{n} |x_i - y_i|.$$

This metric is known also as the *city-block metric.*

The norm ν_∞ generates the metric d_∞ given by

$$d_\infty(\boldsymbol{x}, \boldsymbol{y}) = \max\{|x_i - y_i| \mid 1 \leqslant i \leqslant n\},$$

also known as the *Chebyshev metric.*

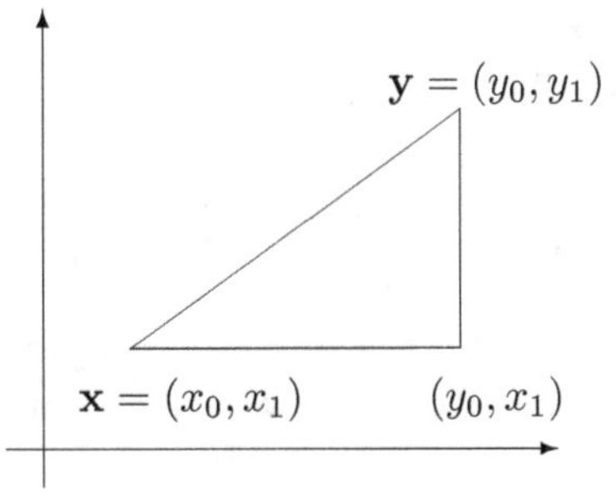

Fig. 6.1 The distances $d_1(\boldsymbol{x}, \boldsymbol{y})$ and $d_2(\boldsymbol{x}, \boldsymbol{y})$.

A representation of these metrics can be seen in Figure 6.1 for the special case of $\mathbb{R}^2$. If $\boldsymbol{x} = (x_0, x_1)$ and $\boldsymbol{y} = (y_0, y_1)$, then $d_2(\boldsymbol{x}, \boldsymbol{y})$ is the length of the hypotenuse of the right triangle and $d_1(\boldsymbol{x}, \boldsymbol{y})$ is the sum of the lengths of the two legs of the triangle.

We can reformulate now the notion of bounded set introduced in Definition 6.3 for general metric spaces in terms of norms, defined on linear spaces.

Theorem 6.7. *Let ν be a norm on a linear space V. A subset U of V is bounded in the metric space (V, d_ν) if and only if there is $b \in \mathbb{R}_{\geqslant 0}$ such that $\nu(\boldsymbol{u}) \leqslant b$ for $\boldsymbol{u} \in U$.*

Proof. Suppose that U is a bounded set in the sense of Definition 6.3, that is, there exists $c \in \mathbb{R}_{\geqslant 0}$ such that $\sup\{d_\nu(x, y) \mid x, y \in U\} = c$. Let $\boldsymbol{x}_0$ be a fixed element of U. Since $\nu(\boldsymbol{x}) = d_\nu(\boldsymbol{x}, \boldsymbol{0}) \leqslant d_\nu(\boldsymbol{x}, \boldsymbol{x}_0) + d(\boldsymbol{x}_0, \boldsymbol{0}) = d_\nu(\boldsymbol{x}, \boldsymbol{x}_0) + \nu(\boldsymbol{x}_0)$, it is immediate that $\nu(\boldsymbol{x}) \leqslant c + \nu(\boldsymbol{x}_0)$, so we can define b as $b = c + \nu(\boldsymbol{x}_0)$.

Conversely, suppose that $\nu(\boldsymbol{u}) \leqslant b$ for $\boldsymbol{u} \in U$. Then $d_\nu(\boldsymbol{x}, \boldsymbol{y}) \leqslant d_\nu(\boldsymbol{x}, \boldsymbol{0}) + d_\nu(\boldsymbol{0}, \boldsymbol{y}) \leqslant 2b$, for every $\boldsymbol{x}, \boldsymbol{y} \in U$. Thus, U is bounded in the metric space (V, d_ν). □

Theorem 6.8 (Projections on closed sets theorem). *Let U be a closed subset of $\mathbb{R}^n$ such that $U \neq \emptyset$ and let $\boldsymbol{x}_0 \in \mathbb{R}^n - U$. Then there exists $\boldsymbol{x}_1 \in U$ such that $\|\boldsymbol{x} - \boldsymbol{x}_0\|_2 \geqslant \|\boldsymbol{x}_1 - \boldsymbol{x}_0\|_2$ for every $\boldsymbol{x} \in U$.*

Proof. Let $d = \inf\{\|\boldsymbol{x} - \boldsymbol{x}_0\|_2 \mid \boldsymbol{x} \in U\}$ and let $U_n = U \cap B\left(\boldsymbol{x}_0, d + \frac{1}{n}\right)$. Note that the sets form a descending sequence of bounded and closed sets $U_1 \supseteq U_2 \supseteq \cdots \supseteq U_n \supseteq \cdots$. Since U_1 is compact, $\bigcap_{n\geqslant 1} U_n \neq \emptyset$. Let $\boldsymbol{x}_1 \in \bigcap_{n\geqslant 1} U_n$. Since $U_n \subseteq U$ for every n, it follows that $\boldsymbol{x}_1 \in U$.

Note that $\|\boldsymbol{x}_1 - \boldsymbol{x}_0\|_2 \leqslant d + \frac{1}{n}$ for every n because $\boldsymbol{x}_1 \in U_n = U \cap B\left(\boldsymbol{x}_0, d + \frac{1}{n}\right)$. This implies $\|\boldsymbol{x}_1 - \boldsymbol{x}_0\|_2 \leqslant d \leqslant \|\boldsymbol{x} - \boldsymbol{x}_0\|_2$ for every $\boldsymbol{x} \in U$. □

Theorem 6.9 to follow allows us to compare the norms ν_p (and the metrics of the form d_p) that were introduced on $\mathbb{R}^n$. We begin with a preliminary result.

Lemma 6.3. *Let $a_1, \ldots, a_n$ be n positive numbers. If p and q are two positive numbers such that $p \leqslant q$, then*

$$(a_1^p + \cdots + a_n^p)^{\frac{1}{p}} \geqslant (a_1^q + \cdots + a_n^q)^{\frac{1}{q}}.$$

Proof. Let $f : \mathbb{R}^{>0} \longrightarrow \mathbb{R}$ be the function defined by

$$f(r) = (a_1^r + \cdots + a_n^r)^{\frac{1}{r}}.$$

Since

$$\ln f(r) = \frac{\ln (a_1^r + \cdots + a_n^r)}{r},$$

it follows that

$$\frac{f'(r)}{f(r)} = -\frac{1}{r^2} \ln (a_1^r + \cdots + a_n^r) + \frac{1}{r} \cdot \frac{a_1^r \ln a_1 + \cdots + a_n^r \ln a_r}{a_1^r + \cdots + a_n^r}.$$

To prove that $f'(r) < 0$, it suffices to show that

$$\frac{a_1^r \ln a_1 + \cdots + a_n^r \ln a_r}{a_1^r + \cdots + a_n^r} \leqslant \frac{\ln (a_1^r + \cdots + a_n^r)}{r}.$$

This last inequality is easily seen to be equivalent to

$$\sum_{i=1}^{n} \frac{a_i^r}{a_1^r + \cdots + a_n^r} \ln \frac{a_i^r}{a_1^r + \cdots + a_n^r} \leqslant 0,$$

which holds because

$$\frac{a_i^r}{a_1^r + \cdots + a_n^r} \leqslant 1$$

for $1 \leqslant i \leqslant n$. □

Theorem 6.9. *Let p and q be two positive numbers such that $p \leqslant q$. For every $\boldsymbol{u} \in \mathbb{R}^n$, we have $\|\boldsymbol{u}\|_p \geqslant \|\boldsymbol{u}\|_q$.*

Proof. This statement follows immediately from Lemma 6.3. □

Corollary 6.3. *Let p, q be two positive numbers such that $p \leqslant q$. For every $\boldsymbol{x}, \boldsymbol{y} \in \mathbb{R}^n$, we have $d_p(\boldsymbol{x}, \boldsymbol{y}) \geqslant d_q(\boldsymbol{x}, \boldsymbol{y})$.*

Proof. This statement follows immediately from Theorem 6.9. □

Example 6.9. For $p = 1$ and $q = 2$, the inequality of Theorem 6.9 becomes

$$\sum_{i=1}^{n} |u_i| \leqslant \sqrt{\sum_{i=1}^{n} |u_i|^2},$$

which is equivalent to

$$\frac{\sum_{i=1}^{n} |u_i|}{n} \leqslant \sqrt{\frac{\sum_{i=1}^{n} |u_i|^2}{n}}. \tag{6.7}$$

Theorem 6.10. *Let $p \geqslant 1$. For every $\boldsymbol{x} \in \mathbb{R}^n$, we have*

$$\|\boldsymbol{x}\|_\infty \leqslant \|\boldsymbol{x}\|_p \leqslant n\|\boldsymbol{x}\|_\infty.$$

Proof. Starting from the definition of ν_p, we have

$$\|\boldsymbol{x}\|_p = \left(\sum_{i=1}^{n} |x_i|^p\right)^{\frac{1}{p}} \leqslant n^{\frac{1}{p}} \max_{1 \leqslant i \leqslant n} |x_i| = n^{\frac{1}{p}} \|\boldsymbol{x}\|_\infty.$$

The first inequality is immediate. □

Corollary 6.4. *Let p and q be two numbers such that $p, q \geqslant 1$. There exist two constants $c, d \in \mathbb{R}_{>0}$ such that*

$$c\|\boldsymbol{x}\|_q \leqslant \|\boldsymbol{x}\|_p \leqslant d\|\boldsymbol{x}\|_q$$

for $\boldsymbol{x} \in \mathbb{R}^n$.

Proof. Since $\|\boldsymbol{x}\|_\infty \leqslant \|\boldsymbol{x}\|_p$ and $\|x\|_q \leqslant n\|\boldsymbol{x}\|_\infty$, it follows that $\|x\|_q \leqslant n\|\boldsymbol{x}\|_p$. Exchanging the roles of p and q, we have $\|x\|_p \leqslant n\|\boldsymbol{x}\|_q$, so

$$\frac{1}{n}\|\boldsymbol{x}\|_q \leqslant \|\boldsymbol{x}\|_p \leqslant n\|\boldsymbol{x}\|_q$$

for every $\boldsymbol{x} \in \mathbb{R}^n$. □

Corollary 6.5. *For every $\boldsymbol{x}, \boldsymbol{y} \in \mathbb{R}^n$ and $p \geqslant 1$, we have $d_\infty(\boldsymbol{x}, \boldsymbol{y}) \leqslant d_p(\boldsymbol{x}, \boldsymbol{y}) \leqslant n d_\infty(\boldsymbol{x}, \boldsymbol{y})$. Further, for $p, q > 1$, there exist $c, d \in \mathbb{R}_{>0}$ such that*

$$c d_q(\boldsymbol{x}, \boldsymbol{y}) \leqslant d_p(\boldsymbol{x}, \boldsymbol{y}) \leqslant c d_q(\boldsymbol{x}, \boldsymbol{y})$$

for $\boldsymbol{x}, \boldsymbol{y} \in \mathbb{R}^n$.

Proof. This statement follows from Theorem 6.10. □

Corollary 6.3 implies that if $p \leqslant q$, then the closed sphere $B_{d_p}(\boldsymbol{x}, r)$ is included in the closed sphere $B_{d_q}(\boldsymbol{x}, r)$. For example, we have

$$B_{d_1}(\mathbf{0}, 1) \subseteq B_{d_2}(\mathbf{0}, 1) \subseteq B_{d_\infty}(\mathbf{0}, 1).$$

In Figures 6.2 (a)–(c), we represent the closed spheres $B_{d_1}(\mathbf{0}, 1)$, $B_{d_2}(\mathbf{0}, 1)$, and $B_{d_\infty}(\mathbf{0}, 1)$.

A useful consequence of Theorem 6.1 is the following statement:

Theorem 6.11. *Let $x_1, \ldots, x_m$ and $y_1, \ldots, y_m$ be $2m$ nonnegative numbers such that $\sum_{i=1}^m x_i = \sum_{i=1}^m y_i = 1$ and let p and q be two positive numbers such that $\frac{1}{p} + \frac{1}{q} = 1$. We have*

$$\sum_{j=1}^m x_j^{\frac{1}{p}} y_j^{\frac{1}{q}} \leqslant 1.$$

Proof. The Hölder inequality applied to $x_1^{\frac{1}{p}}, \ldots, x_m^{\frac{1}{p}}$ and $y_1^{\frac{1}{q}}, \ldots, y_m^{\frac{1}{q}}$ yields the needed inequality

$$\sum_{j=1}^m x_j^{\frac{1}{p}} y_j^{\frac{1}{q}} \leqslant \sum_{j=1}^m x_j \sum_{j=1}^m y_j = 1.$$

□

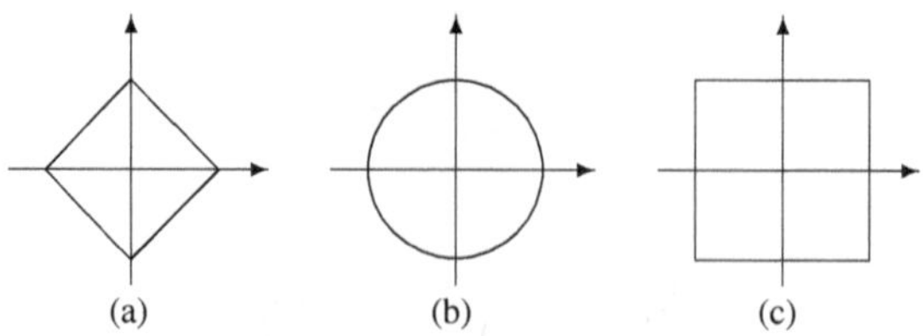

Fig. 6.2 Spheres $B_{d_p}(\mathbf{0}, 1)$ for $p = 1, 2, \infty$.

Theorem 6.11 allows the formulation of a generalization of the Hölder inequality.

Theorem 6.12. *Let A be an $n \times m$ matrix, $A = (a_{ij})$, having positive entries such that $\sum_{j=1}^{m} a_{ij} = 1$ for $1 \leqslant i \leqslant n$. If $\mathbf{p} = (p_1, \ldots, p_n)$ is an n-tuple of positive numbers such that $\sum_{i=1}^{n} p_i = 1$, then*

$$\sum_{j=1}^{m} \prod_{i=1}^{n} a_{ij}^{p_i} \leqslant 1.$$

Proof. The argument is by induction on $n \geqslant 2$. The basis case, $n = 2$, follows immediately from Theorem 6.11 by choosing $p = \frac{1}{p_1}$, $q = \frac{1}{p_2}$, $x_j = a_{1j}$, and $y_j = a_{2j}$ for $1 \leqslant j \leqslant m$.

Suppose that the statement holds for n, let A be an $(n+1) \times m$-matrix having positive entries such that $\sum_{j=1}^{m} a_{ij} = 1$ for $1 \leqslant i \leqslant n+1$, and let $\mathbf{p} = (p_1, \ldots, p_n, p_{n+1})$ be such that $p_1 + \cdots + p_n + p_{n+1} = 1$.

It is easy to see that

$$\sum_{j=1}^{m} \prod_{i=1}^{n+1} a_{ij}^{p_i} \leqslant \sum_{j=1}^{m} a_{1j}^{p_1} a_{n-1\,j}^{p_{n-1}} (a_{nj} + a_{n+1\,j})^{p_n + p_{n+1}}.$$

By applying the inductive hypothesis, we have $\sum_{j=1}^{m} \prod_{i=1}^{n+1} a_{ij}^{p_i} \leqslant 1$. □

A more general form of Theorem 6.12 is given next.

Theorem 6.13. *Let A be an $n \times m$ matrix, $A = (a_{ij})$, having positive entries. If $\mathbf{p} = (p_1, \ldots, p_n)$ is an n-tuple of positive numbers such that $\sum_{i=1}^{n} p_i = 1$, then*

$$\sum_{j=1}^{m} \prod_{i=1}^{n} a_{ij}^{p_i} \leqslant \prod_{i=1}^{n} \left(\sum_{j=1}^{m} a_{ij} \right)^{p_i}.$$

Proof. Let $B = (b_{ij})$ be the matrix defined by

$$b_{ij} = \frac{a_{ij}}{\sum_{j=1}^{m} a_{ij}}$$

for $1 \leqslant i \leqslant n$ and $1 \leqslant j \leqslant m$. Since $\sum_{j=1}^{m} b_{ij} = 1$, we can apply Theorem 6.12 to this matrix. Thus, we can write

$$\begin{aligned}\sum_{j=1}^{m}\prod_{i=1}^{n} b_{ij}^{p_i} &= \sum_{j=1}^{m}\prod_{i=1}^{n}\left(\frac{a_{ij}}{\sum_{j=1}^{m} a_{ij}}\right)^{p_i} \\ &= \sum_{j=1}^{m}\prod_{i=1}^{n}\frac{a_{ij}^{p_i}}{\left(\sum_{j=1}^{m} a_{ij}\right)^{p_i}} \\ &= \frac{\sum_{j=1}^{m}\prod_{i=1}^{n} a_{ij}^{p_i}}{\prod_{i=1}^{n}\left(\sum_{j=1}^{m} a_{ij}\right)^{p_i}} \leqslant 1.\end{aligned}$$

□

We now give a generalization of Minkowski's inequality (Theorem 6.3). First, we need a preliminary result.

Lemma 6.4. *If $a_1, \ldots, a_n$ and $b_1, \ldots, b_n$ are positive numbers and $r < 0$, then*

$$\sum_{i=1}^{n} a_i^r b_i^{1-r} \geqslant \left(\sum_{i=1}^{n} a_i\right)^r \cdot \left(\sum_{i=1}^{n} b_i\right)^{1-r}.$$

Proof. Let $c_1, \ldots, c_n, d_1, \ldots, d_n$ be $2n$ positive numbers such that $\sum_{i=1}^{n} c_i = \sum_{i=1}^{n} d_i = 1$. Inequality (6.1) applied to the numbers $a = c_i^{\frac{1}{p}}$ and $b = d_i^{\frac{1}{q}}$ yields

$$c_i^{\frac{1}{p}} d_i^{\frac{1}{q}} \geqslant \frac{c_i}{p} + \frac{d_i}{q}.$$

Summing these inequalities produces the inequality

$$\sum_{i=1}^{n} c_i^{\frac{1}{p}} d_i^{\frac{1}{q}} \geqslant 1,$$

or

$$\sum_{i=1}^{n} c_i^r d_i^{1-r} \geqslant 1,$$

where $r = \frac{1}{p} < 0$. Choosing $c_i = \frac{a_i}{\sum_{i=1}^{n} a_i}$ and $d_i = \frac{b_i}{\sum_{i=1}^{n} b_i}$, we obtain the desired inequality. □

Theorem 6.14. *Let A be an $n \times m$ matrix, $A = (a_{ij})$, having positive entries, and let p and q be two numbers such that $p > q$ and $p \neq 0$, $q \neq 0$. We have*

$$\left(\sum_{j=1}^{m}\left(\sum_{i=1}^{n} a_{ij}^{p}\right)^{\frac{q}{p}}\right)^{\frac{1}{q}} \geqslant \left(\sum_{i=1}^{n}\left(\sum_{j=1}^{m} a_{ij}^{q}\right)^{\frac{p}{q}}\right)^{\frac{1}{p}}.$$

Proof. Define

$$E = \left(\sum_{j=1}^{m}\left(\sum_{i=1}^{n} a_{ij}^{p}\right)^{\frac{q}{p}}\right)^{\frac{1}{q}},$$

$$F = \left(\sum_{i=1}^{n}\left(\sum_{j=1}^{m} a_{ij}^{q}\right)^{\frac{p}{q}}\right)^{\frac{1}{p}},$$

and $u_i = \sum_{j=1}^{m} a_{ij}^{q}$ for $1 \leqslant i \leqslant n$.

There are three distinct cases to consider related to the position of 0 relative to p and q.

Suppose initially that $p > q > 0$. We have

$$\begin{aligned} F^{p} = \sum_{i=1}^{n} u_i^{\frac{p}{q}} &= \sum_{i=1}^{n} u_i u_i^{\frac{p}{q}-1} \\ = \sum_{i=1}^{n}\sum_{j=1}^{m} a_{ij}^{q} u_i^{\frac{p}{q}-1} &= \sum_{j=1}^{m}\sum_{i=1}^{n} a_{ij}^{q} u_i^{\frac{p}{q}-1}. \end{aligned}$$

By applying the Hölder inequality, we have

$$\begin{aligned} \sum_{i=1}^{n} a_{ij}^{q} u_i^{\frac{p}{q}-1} &\leqslant \left(\sum_{i=1}^{n}(a_{ij}^{q})^{\frac{p}{q}}\right)^{\frac{q}{p}} \cdot \left(\sum_{i=1}^{n}\left(u_i^{\frac{p}{q}-1}\right)^{\frac{p}{p-q}}\right)^{1-\frac{q}{p}} \qquad (6.8) \\ &= \left(\sum_{i=1}^{n} a_{ij}^{p}\right)^{\frac{q}{p}} \cdot \left(\sum_{i=1}^{n} u_i^{\frac{p}{q}}\right)^{1-\frac{q}{p}}, \end{aligned}$$

which implies $F^{p} \leqslant E^{q} F^{p-q}$. This, in turn, gives $F^{q} \leqslant E^{q}$, which implies the generalized Minkowski inequality.

Suppose now that $0 > p > q$, so $0 < -p < -q$. Applying the generalized Minkowski inequality to the positive numbers $b_{ij} = \frac{1}{a_{ij}}$ gives the inequality

$$\left(\sum_{j=1}^{m}\left(\sum_{i=1}^{n} b_{ij}^{-q}\right)^{\frac{p}{q}}\right)^{-\frac{1}{p}} \geqslant \left(\sum_{i=1}^{n}\left(\sum_{j=1}^{m} b_{ij}^{-p}\right)^{\frac{q}{p}}\right)^{-\frac{1}{q}},$$

which is equivalent to

$$\left(\sum_{j=1}^{m}\left(\sum_{i=1}^{n} a_{ij}^{q}\right)^{\frac{p}{q}}\right)^{-\frac{1}{p}} \geqslant \left(\sum_{i=1}^{n}\left(\sum_{j=1}^{m} a_{ij}^{p}\right)^{\frac{q}{p}}\right)^{-\frac{1}{q}}.$$

A last transformation gives

$$\left(\sum_{j=1}^{m}\left(\sum_{i=1}^{n} a_{ij}^{q}\right)^{\frac{p}{q}}\right)^{\frac{1}{p}} \leqslant \left(\sum_{i=1}^{n}\left(\sum_{j=1}^{m} a_{ij}^{p}\right)^{\frac{q}{p}}\right)^{\frac{1}{q}},$$

which is the inequality to be proven.

Finally, suppose that $p > 0 > q$. Since $\frac{q}{p} < 0$, Inequality (6.9) is replaced by the opposite inequality through the application of Lemma 6.4:

$$\sum_{i=1}^{n} a_{ij}^{q} u_i^{\frac{p}{q}-1} \geqslant \left(\sum_{i=1}^{n} a_{ij}^{p}\right)^{\frac{q}{p}} \cdot \left(\sum_{i=1}^{n} u_i^{\frac{p}{q}}\right)^{1-\frac{q}{p}}.$$

This leads to $F^p \geqslant E^q F^{p-q}$ or $F^q \geqslant E^q$. Since $q < 0$, this implies $F \leqslant E$. □

6.5 The Topology of Normed Linear Spaces

Every norm ν defined on a linear space V generates a metric $d : V^2 \longrightarrow \mathbb{R}_{\geqslant 0}$ given by $d(\boldsymbol{x}, \boldsymbol{y}) = \nu(\boldsymbol{x} - \boldsymbol{y})$. Therefore, any normed space can be equipped with the topology of a metric space, using the metric defined by the norm. Since this topology is induced by a metric, any normed space is a Hausdorff space. Further, if $\boldsymbol{v} \in V$, then the collection of subsets $\{B_d(\boldsymbol{v}, r) \mid r > 0\}$ is a fundamental system of neighborhoods for $\boldsymbol{v}$.

By specializing the definition of local continuity of functions between metric spaces, a function $f : V \longrightarrow W$ between two normed spaces (V, ν) and (W, ν') is continuous in $\boldsymbol{x}_0 \in V$ if for every $\epsilon > 0$ there exists $\delta > 0$ such that $\nu(\boldsymbol{x} - \boldsymbol{x}_0) < \delta$ implies $\nu'(f(\boldsymbol{x}) - f(\boldsymbol{x}_0)) < \epsilon$.

A sequence $(\boldsymbol{x}_0, \boldsymbol{x}_1, \ldots)$ of elements of V converges to $\boldsymbol{x}$ if for every $\epsilon > 0$ there exists $n_\epsilon \in \mathbb{N}$ such that $n \geqslant n_\epsilon$ implies $\nu(\boldsymbol{x}_n - \boldsymbol{x}) < \epsilon$.

Theorem 6.15. *In a normed linear space (V, ν), the norm, the multiplication by scalars, and the vector addition are continuous functions.*

Proof. By Theorem 6.4, we have $\nu(\boldsymbol{x} - \boldsymbol{y}) \geqslant |\nu(\boldsymbol{x}) - \nu(\boldsymbol{y})|$ for every $\boldsymbol{x}, \boldsymbol{y} \in V$. Therefore, if $\lim_{n\to\infty} \boldsymbol{x}_n = \boldsymbol{x}$, we have $\nu(\boldsymbol{x}_n - \boldsymbol{x}) \geqslant |\nu(\boldsymbol{x}_n) - \nu(\boldsymbol{x})|$, which implies $\lim_{n\to\infty} \nu(\boldsymbol{x}_n) = \nu(\boldsymbol{x})$. Thus, the norm is continuous.

Suppose now that $\lim_{n\to\infty} a_n = a$ and $\lim_{n\to\infty} \boldsymbol{x}_n = \boldsymbol{x}$, where (a_n) is a sequence of scalars. Since the sequence $(\boldsymbol{x}_n)$ is bounded, we have

$$\begin{aligned}
\nu(a\boldsymbol{x} - a_n\boldsymbol{x}_n) &\leqslant \nu(a\boldsymbol{x} - a_n\boldsymbol{x}) + \nu(a_n\boldsymbol{x} - a_n\boldsymbol{x}_n) \\
&\leqslant |a - a_n|\nu(\boldsymbol{x}) + a_n\nu(\boldsymbol{x} - \boldsymbol{x}_n),
\end{aligned}$$

which implies that $\lim_{n\to\infty} a_n\boldsymbol{x}_n = a\boldsymbol{x}$. This shows that the multiplication by scalars is a continuous function.

To prove that the vector addition is continuous, let $(\boldsymbol{x}_n)$ and $(\boldsymbol{y}_n)$ be two sequences in V such that $\lim_{n\to\infty} \boldsymbol{x}_n = \boldsymbol{x}$ and $\lim_{n\to\infty} \boldsymbol{y}_n = \boldsymbol{y}$. Note that

$$\nu\left((\boldsymbol{x} + \boldsymbol{y}) - (\boldsymbol{x}_n + \boldsymbol{y}_n)\right) \leqslant \nu(\boldsymbol{x} - \boldsymbol{x}_n) + \nu(\boldsymbol{y} - \boldsymbol{y}_n),$$

which implies that $\lim_{n\to\infty}(\boldsymbol{x}_n + \boldsymbol{y}_n) = \boldsymbol{x} + \boldsymbol{y}$. Thus, the vector addition is continuous. □

Definition 6.9. Two norms ν and ν' on a linear space V are *equivalent* if they generate the same topology.

Theorem 6.16. *Let V be a linear space and let $\nu : V \longrightarrow \mathbb{R}_{\geqslant 0}$ and $\nu' : V \longrightarrow \mathbb{R}_{\geqslant 0}$ be two norms on V that generate the topologies $\mathcal{O}$ and $\mathcal{O}'$ on V, respectively.*

The topology $\mathcal{O}'$ is finer than the topology $\mathcal{O}$ (that is, $\mathcal{O} \subseteq \mathcal{O}'$) if and only if there exists $c \in \mathbb{R}_{>0}$ such that $\nu(\boldsymbol{v}) \leqslant c\nu'(\boldsymbol{v})$ for every $\boldsymbol{v} \in V$.

Proof. Suppose that $\mathcal{O} \subseteq \mathcal{O}'$. Then, any open sphere $B_\nu(\mathbf{0}, r_0) = \{\boldsymbol{x} \in V \mid \nu(\boldsymbol{x}) < r_0\}$ (in $\mathcal{O}$) must be an open set in $\mathcal{O}'$. Therefore, there exists an open sphere $B_{\nu'}(\mathbf{0}, r_1)$ such that $B_{\nu'}(\mathbf{0}, r_1) \subseteq B_\nu(\mathbf{0}, r_0)$. This means that for $r_0 \in \mathbb{R}_{\geqslant 0}$ and $\boldsymbol{v} \in V$ there exists $r_1 \in \mathbb{R}_{\geqslant 0}$ such that $\nu'(\boldsymbol{v}) < r_1$ implies $\nu(\boldsymbol{v}) < r_0$ for every $\boldsymbol{u} \in V$. In particular, for $r_0 = 1$, there is $k > 0$ such that $\nu'(\boldsymbol{v}) < k$ implies $\nu(\boldsymbol{v}) < 1$, which is equivalent to

$$c\nu'(\boldsymbol{v}) < 1 \text{ implies } \nu(\boldsymbol{v}) < 1,$$

for every $\boldsymbol{v} \in V$ and $c = \frac{1}{k}$.

For $\boldsymbol{w} = \frac{1}{c+\epsilon} \frac{\boldsymbol{v}}{\nu'(\boldsymbol{v})}$, where $\epsilon > 0$, it follows that

$$c\nu'(\boldsymbol{w}) = c\nu' \left(\frac{1}{c+\epsilon} \frac{\boldsymbol{v}}{\nu'(\boldsymbol{v})} \right) = \frac{c}{c+\epsilon} < 1,$$

so

$$\nu(\boldsymbol{w}) = \nu \left(\frac{1}{c+\epsilon} \frac{\boldsymbol{v}}{\nu'(\boldsymbol{v})} \right) = \frac{1}{c+\epsilon} \frac{\nu(\boldsymbol{v})}{\nu'(\boldsymbol{v})} < 1.$$

Since this inequality holds for every $\epsilon > 0$, it follows that $\nu(\boldsymbol{v}) \leqslant c\nu'(\boldsymbol{v})$.

Conversely, suppose that there exists $c \in \mathbb{R}_{>0}$ such that $\nu(\boldsymbol{v}) \leqslant c\nu'(\boldsymbol{v})$ for every $\boldsymbol{v} \in V$. Since

$$\left\{ \boldsymbol{v} \mid \nu'(\boldsymbol{v}) \leqslant \frac{r}{c} \right\} \subseteq \{\boldsymbol{v} \mid \nu(\boldsymbol{v}) \leqslant r\}$$

for $\boldsymbol{v} \in V$ and $r > 0$, it follows that $\mathcal{O} \subseteq \mathcal{O}'$. □

Corollary 6.6. *Let V be a linear space and let $\nu : V \longrightarrow \mathbb{R}_{\geqslant 0}$ and $\nu' : V \longrightarrow \mathbb{R}_{\geqslant 0}$ be two norms on V. Then ν and ν' are equivalent norms if and only if there exist $a, b \in \mathbb{R}_{>0}$ such that $a\nu(\boldsymbol{v}) \leqslant \nu'(\boldsymbol{v}) \leqslant b\nu(\boldsymbol{v})$ for $\boldsymbol{v} \in V$.*

Proof. This statement follows directly from Theorem 6.16. □

Example 6.10. By Corollary 6.4, any two norms ν_p and ν_q on $\mathbb{R}^n$ (with $p, q \geqslant 1$) are equivalent.

Continuous linear operators between normed spaces have a simple characterization.

Theorem 6.17. *Let (V, ν) and (V', ν') be two normed $\mathbb{F}$-linear spaces where $\mathbb{F}$ is either $\mathbb{R}$ or $\mathbb{C}$. A linear operator $f : V \longrightarrow V'$ is continuous if and only if there exists $M \in \mathbb{R}_{>0}$ such that $\nu'(f(\boldsymbol{x})) \leqslant M\nu(\boldsymbol{x})$ for every $\boldsymbol{x} \in V$.*

Proof. Suppose that $f : V \longrightarrow V'$ satisfies the condition of the theorem. Then

$$f\left(B_\nu\left(\mathbf{0}, \frac{r}{M}\right)\right) \subseteq B_{\nu'}(\mathbf{0}, r)$$

for every $r > 0$, which means that f is continuous in $\mathbf{0}$ and, therefore, it is continuous everywhere (by Theorem 2.51).

Conversely, suppose that f is continuous. Then there exists $\delta > 0$ such that $f(B_\nu(\mathbf{0}, \delta)) \subseteq B_{\nu'}(f(x), 1)$, which is equivalent to $\nu(\boldsymbol{x}) < \delta$, implies $\nu'(f(\boldsymbol{x})) < 1$. Let $\epsilon > 0$ and let $\boldsymbol{z} \in V$ be defined by

$$\boldsymbol{z} = \frac{\delta}{\nu(\boldsymbol{x}) + \epsilon}\boldsymbol{x}.$$

We have $\nu(\boldsymbol{z}) = \frac{\delta\nu(\boldsymbol{x})}{\nu(\boldsymbol{x})+\epsilon} < \delta$. This implies $\nu'(f(\boldsymbol{z})) < 1$, which is equivalent to

$$\frac{\delta}{\nu(\boldsymbol{x}) + \epsilon}\nu'(f(\boldsymbol{x})) < 1$$

because of the linearity of f. This means that

$$\nu'(f(\boldsymbol{x})) < \frac{\nu(\boldsymbol{x}) + \epsilon}{\delta}$$

for every $\epsilon > 0$, so $\nu'(f(\boldsymbol{x})) \leqslant \frac{1}{\delta}\nu(\boldsymbol{x})$. □

Lemma 6.5. *Let (V, ν) and (V', ν') be two normed $\mathbb{F}$-linear spaces where $\mathbb{F}$ is either $\mathbb{R}$ or $\mathbb{C}$. A linear function $f : V \longrightarrow V'$ is not injective if and only if there exists $\boldsymbol{u} \in V - \{\mathbf{0}_V\}$ such that $f(\boldsymbol{u}) = \mathbf{0}_{V'}$.*

Proof. It is clear that the condition of the lemma is sufficient for failing injectivity. Conversely, suppose that f is not injective. There exist $\boldsymbol{t}, \boldsymbol{v} \in V$ such that $\boldsymbol{t} \neq \boldsymbol{v}$ and $f(\boldsymbol{t}) = f(\boldsymbol{v})$. The linearity of f implies $f(\boldsymbol{t} - \boldsymbol{v}) = \mathbf{0}_{V'}$. By defining $\boldsymbol{u} = \boldsymbol{t} - \boldsymbol{v} \neq \mathbf{0}_V$, we have the desired element $\boldsymbol{u}$. □

Theorem 6.18. *Let (V, ν) and (V', ν') be two normed $\mathbb{F}$-linear spaces where $\mathbb{F}$ is either $\mathbb{R}$ or $\mathbb{C}$. A linear function $f : V \longrightarrow V'$ is injective if and only if there exists $m \in \mathbb{R}_{>0}$ such that $\nu'(f(\boldsymbol{x})) \geqslant m\nu(\boldsymbol{x})$ for every $\boldsymbol{x} \in V$.*

Proof. Suppose that f is not injective. By Lemma 6.5, there exists $\boldsymbol{u} \in V - \{\mathbf{0}_V\}$ such that $f(\boldsymbol{u}) = \mathbf{0}_{V'}$, so $\nu'(f(\boldsymbol{u})) < m\nu(\boldsymbol{u})$ for any $m > 0$. Thus, the condition of the theorem is sufficient for injectivity.

Suppose that f is injective, so the inverse function $f^{-1} : V' \longrightarrow V$ is a linear function. By Theorem 6.17, there exists $M > 0$ such that

$$\nu(f^{-1}(\boldsymbol{y})) \leqslant M\nu'(\boldsymbol{y})$$

for every $\boldsymbol{y} \in V'$. Choosing $\boldsymbol{y} = f(\boldsymbol{x})$ yields $\nu(\boldsymbol{x}) \leqslant M\nu'(f(\boldsymbol{x}))$, so $\nu'(f(\boldsymbol{x})) \geqslant m\nu(\boldsymbol{x})$ for $m = \frac{1}{M}$, which concludes the argument. □

Corollary 6.7. *Every linear function $f : \mathbb{C}^m \longrightarrow \mathbb{C}^n$ is continuous.*

Proof. Suppose that both $\mathbb{C}^m$ and $\mathbb{C}^n$ are equipped with the norm ν_1. If $\boldsymbol{x} \in \mathbb{C}^m$, we can write $\boldsymbol{x} = x_1\boldsymbol{e}_1 + \cdots + x_m\boldsymbol{x}_m$ and the linearity of f implies

$$\begin{aligned}\nu_1(f(\boldsymbol{x})) &= \nu_1\left(f\left(\sum_{i=1}^m x_i\boldsymbol{e}_i\right)\right) = \nu_1\left(\sum_{i=1}^m x_i f(\boldsymbol{e}_i)\right)\\ &\leqslant \sum_{i=1}^m |x_i|\nu_1(f(\boldsymbol{e}_i)) \leqslant M\sum_{i=1}^m |x_i| = M\nu_1(\boldsymbol{x}),\end{aligned}$$

where $M = \sum_{i=1}^m \nu_1(f(\boldsymbol{e}_i))$. By Theorem 6.17, the continuity of f follows. □

Next, we introduce a norm on the linear space $\mathsf{Hom}(\mathbb{C}^m, \mathbb{C}^n)$ of linear functions from $\mathbb{C}^m$ to $\mathbb{C}^n$. Recall that if $f : \mathbb{C}^m \longrightarrow \mathbb{C}^n$ is a linear function and ν, ν' are norms on $\mathbb{C}^m$ and $\mathbb{C}^n$, respectively, then there exists a non-negative constant m such that $\nu'(f(\boldsymbol{x})) \leqslant M\nu(\boldsymbol{x})$ for every $\boldsymbol{x} \in \mathbb{C}^m$. Define the *norm of* f, $\mu(f)$, as

$$\mu(f) = \inf\{M \in \mathbb{R}_{\geqslant 0} \mid \nu'(f(\boldsymbol{x})) \leqslant M\nu(\boldsymbol{x}) \text{ for every } \boldsymbol{x} \in \mathbb{C}^m\}. \tag{6.9}$$

Theorem 6.19. *The mapping μ defined by Equality (6.9) is a norm on the linear space of linear functions* $\mathsf{Hom}(\mathbb{C}^m, \mathbb{C}^n)$.

Proof. Let f, g be two functions in $\mathsf{Hom}(\mathbb{C}^m, \mathbb{C}^n)$. There exist M_f and M_g in $\mathbb{R}_{\geqslant 0}$ such that $\nu'(f(\boldsymbol{x})) \leqslant M_f\nu(\boldsymbol{x})$ and $\nu'(g(\boldsymbol{x})) \leqslant M_g\nu(\boldsymbol{x})$ for every $\boldsymbol{x} \in V$. Thus,

$$\nu'((f+g)(\boldsymbol{x})) = \nu'(f(\boldsymbol{x}) + g(\boldsymbol{x})) \leqslant \nu'(f(\boldsymbol{x})) + \nu'(g(\boldsymbol{x})) \leqslant (M_f + M_g)\nu(\boldsymbol{x}),$$

so

$$M_f + M_g \in \{M \in \mathbb{R}_{\geqslant 0} \mid \nu'((f+g)(\boldsymbol{x})) \leqslant M\nu(\boldsymbol{x}) \text{ for every } \boldsymbol{x} \in V\}.$$

Therefore,

$$\mu(f+g) \leqslant \mu(f) + \mu(g).$$

We leave to the reader the verification of the remaining norm properties of μ. □

Since the norm μ defined by Equality (6.9) depends on the norms ν and ν', we denote it by $N(\nu, \nu')$.

Theorem 6.20. *Let $f : \mathbb{C}^m \longrightarrow \mathbb{C}^n$ and $g : \mathbb{C}^n \longrightarrow \mathbb{C}^p$ and let $\mu = N(\nu, \nu')$, $\mu' = N(\nu', \nu'')$ and $\mu'' = N(\mu, \mu'')$, where ν, ν', ν'' are norms on $\mathbb{C}^m$, $\mathbb{C}^n$, and $\mathbb{C}^p$, respectively. We have $\mu''(gf) \leqslant \mu(f)\mu'(g)$.*

Proof. Let $\boldsymbol{x} \in \mathbb{C}^m$. We have

$$\nu'(f(\boldsymbol{x})) \leqslant (\mu(f) + \epsilon')\nu(\boldsymbol{x})$$

for every $\epsilon' > 0$. Similarly, for $\boldsymbol{y} \in \mathbb{C}^n$,

$$\nu''(g(\boldsymbol{y})) \leqslant (\mu'(g) + \epsilon'')\nu'(\boldsymbol{y})$$

for every $\epsilon'' > 0$. These inequalities imply

$$\nu''(g(f(\boldsymbol{x})) \leqslant (\mu'(g) + \epsilon'')\nu'(f(\boldsymbol{x})) \leqslant (\mu'(g) + \epsilon'')\mu(f) + \epsilon')\nu(\boldsymbol{x}).$$

Thus, we have

$$\mu''(gf) \leqslant (\mu'(g) + \epsilon'')\mu(f) + \epsilon')$$

for every ϵ' and ϵ''. This allows us to conclude that $\mu''(fg) \leqslant \mu(f)\mu'(g)$. □

Equivalent definitions of the norm $\mu = N(\nu, \nu')$ are given next.

Theorem 6.21. *Let $f : \mathbb{C}^m \longrightarrow \mathbb{C}^n$ and let ν and ν' be two norms defined on $\mathbb{C}^m$ and $\mathbb{C}^n$, respectively. If $\mu = N(\nu, \nu')$, we have*

(i) $\mu(f) = \inf\{M \in \mathbb{R}_{\geqslant 0} \mid \nu'(f(\boldsymbol{x})) \leqslant M\nu(\boldsymbol{x})$ *for every* $\boldsymbol{x} \in \mathbb{C}^m\}$;
(ii) $\mu(f) = \sup\{\nu'(f(\boldsymbol{x})) \mid \nu(\boldsymbol{x}) \leqslant 1\}$;
(iii) $\mu(f) = \max\{\nu'(f(\boldsymbol{x})) \mid \nu(\boldsymbol{x}) \leqslant 1\}$;
(iv) $\mu(f) = \max\{\nu'(f(\boldsymbol{x})) \mid \nu(\boldsymbol{x}) = 1\}$;
(v) $\mu(f) = \sup\left\{\frac{\nu'(f(\boldsymbol{x}))}{\nu(\boldsymbol{x})} \mid \boldsymbol{x} \in \mathbb{C}^m - \{\boldsymbol{0}_m\}\right\}$.

Proof. The first equality is the definition of $\mu(f)$.

Let ϵ be a positive number. By the definition of the infimum, there exists M such that $\nu'(f(\boldsymbol{x})) \leqslant M\nu(\boldsymbol{x})$ for every $\boldsymbol{x} \in \mathbb{C}^m$ and $M \leqslant \mu(f)+\epsilon$. Thus, for any $\boldsymbol{x}$ such that $\nu(\boldsymbol{x}) \leqslant 1$, we have $\nu'(f(\boldsymbol{x})) \leqslant M \leqslant \mu(f) + \epsilon$. Since this inequality holds for every ϵ, it follows that $\nu'(f(\boldsymbol{x})) \leqslant \mu(f)$ for every $\boldsymbol{x} \in \mathbb{C}^m$ with $\nu(\boldsymbol{x}) \leqslant 1$.

Furthermore, if ϵ' is a positive number, we claim that there exists $\boldsymbol{x}_0 \in \mathbb{C}^m$ such that $\nu(\boldsymbol{x}_0) \leqslant 1$ and $\mu(f) - \epsilon' \leqslant \nu'(f(\boldsymbol{x}_0)) \leqslant \mu(f)$. Suppose that this is not the case. Then, for every $\boldsymbol{z} \in \mathbb{C}^n$ with $\nu(\boldsymbol{z}) \leqslant 1$, we have $\nu'(f(\boldsymbol{z})) \leqslant \mu(f) - \epsilon'$. If $\boldsymbol{x} \in \mathbb{C}^n$, then $\nu\left(\frac{1}{\nu(\boldsymbol{x})}\boldsymbol{x}\right) = 1$, so $\nu'(f(\boldsymbol{x})) \leqslant (\mu(f) - \epsilon')\nu(\boldsymbol{x})$, which contradicts the definition of $\mu(f)$. This allows us to conclude that $\mu(f) = \sup\{\nu'(f(\boldsymbol{x})) \mid \nu(\boldsymbol{x}) \leqslant 1\}$, which proves the second equality.

Observe that the third equality (where we replaced sup by max) holds because the closed sphere $B(\boldsymbol{0}, 1)$ is a compact set in $\mathbb{R}^n$. Thus, we have

$$\mu(A) = \max\{\nu'(f(\boldsymbol{x})) \mid \nu(\boldsymbol{x}) \leqslant 1\}. \tag{6.10}$$

For the fourth equality, since $\{\boldsymbol{x} \mid \nu(\boldsymbol{x}) = 1\} \subseteq \{\boldsymbol{x} \mid \nu(\boldsymbol{x}) \leqslant 1\}$, it follows that

$$\max\{\nu'(f(\boldsymbol{x})) \mid \nu(\boldsymbol{x}) = 1\} \leqslant \max\{\nu'(f(\boldsymbol{x})) \mid \nu(\boldsymbol{x}) \leqslant 1\} = \mu(f).$$

By the third equality, there exists a vector $\boldsymbol{z} \in \mathbb{R}^n - \{\boldsymbol{0}\}$ such that $\nu(\boldsymbol{z}) \leqslant 1$ and $\nu'(f(\boldsymbol{z})) = \mu(f)$. Thus, we have

$$\mu(f) = \nu(\boldsymbol{z})\nu\left(f\left(\frac{\boldsymbol{z}}{\nu(\boldsymbol{z})}\right)\right) \leqslant \nu\left(f\left(\frac{\boldsymbol{z}}{\nu(\boldsymbol{z})}\right)\right).$$

Since $\nu\left(\frac{\boldsymbol{z}}{\nu(\boldsymbol{z})}\right) = 1$, it follows that $\mu(A) \leqslant \max\{\nu'(f(\boldsymbol{x})) \mid \nu(\boldsymbol{x})=1\}$. This yields the desired conclusion.

Finally, to prove the last equality observe that for every $\boldsymbol{x} \in \mathbb{C}^m - \{\mathbf{0}_m\}$, $\frac{1}{\nu(\boldsymbol{x})}\boldsymbol{x}$ is a unit vector. Thus, $\nu'(f(\frac{1}{\nu(\boldsymbol{x})}\boldsymbol{x}) \leqslant \mu(f)$, by the fourth equality. On the other hand, by the third equality, there exists $\boldsymbol{x}_0$ such that $\nu(\boldsymbol{x}_0) = 1$ and $\nu'(f(\boldsymbol{x}_0)) = \mu(f)$. This concludes the argument. □

6.6 Norms for Matrices

In Chapter 3, we saw that the set $\mathbb{C}^{m\times n}$ is a linear space. Therefore, it is natural to consider norms defined on matrices. We discuss two basic methods for defining norms for matrices. The first approach treats matrices as vectors (through the vec mapping). The second regards matrices as representations of linear operators, and defines norms for matrices starting from operator norms.

The vectorization mapping vec was introduced in Definition 3.16. Its use allows us to treat a matrix $A \in \mathbb{C}^{m\times n}$ as a vector from $\mathbb{C}^{mn}$. Using vector norms on $\mathbb{C}^{mn}$, we can define vectorial norms of matrices.

Definition 6.10. Let ν be a vector norm on the space $\mathbb{R}^{mn}$. The *vectorial matrix norm* $\mu^{(m,n)}$ *on* $\mathbb{R}^{m\times n}$ is the mapping $\mu^{(m,n)} : \mathbb{R}^{m\times n} \longrightarrow \mathbb{R}_{\geqslant 0}$ defined by

$$\mu^{(m,n)}(A) = \nu(\mathsf{vec}(A))$$

for $A \in \mathbb{R}^{m\times n}$.

Vectorial norms of matrices are defined without regard for matrix products. The link between linear transformations of finite-dimensional linear spaces and Theorem 6.20 suggests the introduction of an additional condition. Since every matrix $A \in \mathbb{C}^{m\times n}$ corresponds to a linear transformation $h_A : \mathbb{C}^m \longrightarrow \mathbb{C}^n$, if ν and ν' are norms on $\mathbb{C}^m$ and $\mathbb{C}^n$, respectively, it is natural to define a norm on $\mathbb{C}^{m\times n}$ as $\mu(A) = \mu(h_A)$, where $\mu = N(\nu, \nu')$ is a norm on the space of linear transformations between $\mathbb{C}^m$ and $\mathbb{C}^n$.

Suppose that ν, ν', and ν'' are vector norms defined on $\mathbb{C}^m, \mathbb{C}^n$, and $\mathbb{C}_p$, respectively. By Theorem 6.20, $\mu''(gf) \leqslant \mu(f)\mu'(g)$, where $\mu = N(\nu, \nu')$, $\mu' = N(\nu', \nu'')$, and $\mu'' = N(\mu, \mu'')$, so $\mu''(AB) \leqslant \mu(A)\mu'(B)$. This suggests the following definition.

Definition 6.11. A *consistent family of matrix norms* is a family of functions $\mu^{(m,n)} : \mathbb{C}^{m\times n} \longrightarrow \mathbb{R}_{\geqslant 0}$, where $m, n \in \mathbb{P}$, that satisfies the following conditions:

(i) $\mu^{(m,n)}(A) = 0$ if and only if $A = O_{m,n}$;

(ii) $\mu^{(m,n)}(A+B) \leqslant \mu^{(m,n)}(A)+\mu^{(m,n)}(B)$ (the *subadditivity property*);

(iii) $\mu^{(m,n)}(aA) = |a|\mu^{(m,n)}(A)$;

(iv) $\mu^{(m,p)}(AB) \leqslant \mu^{(m,n)}(A)\mu^{(n,p)}(B)$ for every matrix $A \in \mathbb{R}^{m\times n}$ and $B \in \mathbb{R}^{n\times p}$ (the *submultiplicative property*).

If the format of the matrix A is clear from context or is irrelevant, then we shall write $\mu(A)$ instead of $\mu^{(m,n)}(A)$.

Example 6.11. Let $P \in \mathbb{C}^{n\times n}$ be an idempotent matrix. If μ is a matrix norm, then either $\mu(P) = 0$ or $\mu(P) \geqslant 1$.

Indeed, since P is idempotent, we have $\mu(P) = \mu(P^2)$. By the submultiplicative property, $\mu(P^2) \leqslant (\mu(P))^2$, so $\mu(P) \leqslant (\mu(P))^2$. Consequently, if $\mu(P) \neq 0$, then $\mu(P) \geqslant 1$.

Some vectorial matrix norms turn out to be actual matrix norms; others fail to be matrix norms. This point is illustrated by the next two examples.

Example 6.12. Consider the vectorial matrix norm μ_1 induced by the vector norm ν_1. We have $\mu_1(A) = \sum_{i=1}^{n}\sum_{j=1}^{m}|a_{ij}|$ for $A \in \mathbb{R}^{m\times n}$. Actually, this is a matrix norm. To prove this fact, consider the matrices $A \in \mathbb{R}^{m\times p}$ and $B \in \mathbb{R}^{p\times n}$. We have

$$\begin{aligned}
\mu_1(AB) &= \sum_{i=1}^{m}\sum_{j=1}^{n}\left|\sum_{k=1}^{p} a_{ik}b_{kj}\right| \leqslant \sum_{i=1}^{m}\sum_{j=1}^{n}\sum_{k=1}^{p}|a_{ik}b_{kj}| \\
&\leqslant \sum_{i=1}^{m}\sum_{j=1}^{n}\sum_{k'=1}^{p}\sum_{k''=1}^{p}|a_{ik'}||b_{k''j}| \\
&\quad \text{(because we added extra non-negative terms to the sums)} \\
&= \left(\sum_{i=1}^{m}\sum_{k'=1}^{p}|a_{ik'}|\right)\cdot\left(\sum_{j=1}^{n}\sum_{k''=1}^{p}|b_{k''j}|\right) \\
&= \mu_1(A)\mu_1(B).
\end{aligned}$$

We denote this vectorial matrix norm by the same notation as the corresponding vector norm, that is, by $\|A\|_1$.

The vectorial matrix norm μ_2 induced by the vector norm ν_2 is also a matrix norm. Indeed, using the notations as above, we have

$$\begin{aligned}(\mu_2(AB))^2 &= \sum_{i=1}^{m}\sum_{j=1}^{n}\left|\sum_{k=1}^{p} a_{ik}b_{kj}\right|^2 \\ &\leqslant \sum_{i=1}^{m}\sum_{j=1}^{n}\left(\sum_{k=1}^{p}|a_{ik}|^2\right)\left(\sum_{l=1}^{p}|b_{lj}|^2\right) \\ &\quad \text{(by Cauchy–Schwarz inequality)} \\ &\leqslant (\mu_2(A))^2(\mu_2(B))^2.\end{aligned}$$

The vectorial norm of $A \in \mathbb{C}^{m\times n}$,

$$\mu_2(A) = \left(\sum_{i=1}^{n}\sum_{j=1}^{m}|a_{ij}|^2\right)^{\frac{1}{2}},$$

denoted also by $\|A\|_F$, is known as the *Frobenius norm*. For $A \in \mathbb{R}^{m\times n}$, we have

$$\|A\|_F = \sqrt{\sum_{i=1}^{n}\sum_{j=1}^{m}a_{ij}^2}.$$

It is easy to see that for real matrices we have

$$\|A\|_F^2 = \mathsf{trace}(AA') = \mathsf{trace}(A'A). \tag{6.11}$$

For complex matrices, the corresponding equality is

$$\|A\|_F^2 = \mathsf{trace}(AA^{\mathsf{H}}) = \mathsf{trace}(A^{\mathsf{H}}A). \tag{6.12}$$

Note that $\|A^{\mathsf{H}}\|_F^2 = \|A\|_F^2$ for every A.

Example 6.13. The vectorial norm μ_∞ induced by the vector norm ν_∞ is denoted by $\|A\|_\infty$ and is given by

$$\|A\|_\infty = \max_{i,j} |a_{ij}|$$

for $A \in \mathbb{C}^{n\times n}$. This is *not* a matrix norm. Indeed, let a, b be two positive numbers and consider the matrices

$$A = \begin{pmatrix} a & a \\ a & a \end{pmatrix} \text{ and } B = \begin{pmatrix} b & b \\ b & b \end{pmatrix}.$$

We have $\|A\|_\infty = a$ and $\|B\|_\infty = b$. However, since

$$AB = \begin{pmatrix} 2ab & 2ab \\ 2ab & 2ab \end{pmatrix},$$

we have $\|AB\|_\infty = 2ab$ and the submultiplicative property of matrix norms is violated.

A technique that always produces matrix norms starting from vector norms is introduced in the next theorem.

Definition 6.12. Let ν_m be a norm on $\mathbb{C}^m$ and ν_n be a norm on $\mathbb{C}^n$ and let $A \in \mathbb{C}^{n\times m}$ be a matrix. The *operator norm* of A is the number $\mu^{(n,m)}(A) = \mu^{(n,m)}(h_A)$, where $\mu^{(n,m)} = N(\nu_m, \nu_n)$.

Theorem 6.22. *Let $\{\nu_n \mid n \geqslant 1\}$ be a family of vector norms, where ν_n is a vector norm on $\mathbb{C}^n$. The family of norms $\{\mu^{(n,m)} \mid n, m \geqslant 1\}$ is consistent.*

Proof. It is easy to see that the family of norms $\{\mu^{(n,m)} \mid n, m \geqslant 1\}$ satisfies the first three conditions of Definition 6.11 because the corresponding operator norms satisfy similar conditions. For example, if $\mu^{(n,m)}(A) = 0$ for $A \in \mathbb{C}^{n\times m}$, this means that $\mu^{(n,m)}(h_A) = 0$, so $\nu_n(A\boldsymbol{x}) = 0$ for every $\boldsymbol{x} \in \mathbb{C}^m$, such that $\nu_m(\boldsymbol{x}) \leqslant 1$. This implies $A\boldsymbol{x} = \mathbf{0}_n$ for every $\boldsymbol{x} \in \mathbb{R}^m$, which, in turn, implies $A = O_{n,m}$. Since $\mu(O_{m,n}) = 0$, the first condition is satisfied.

For the fourth condition of Definition 6.11 and $A \in \mathbb{C}^{n\times m}$ and $B \in \mathbb{C}^{m\times p}$, we have

$$\begin{aligned}
\mu^{(n,p)}(AB) &= \sup\{\nu_n((AB)\boldsymbol{x}) \mid \nu_p(\boldsymbol{x}) \leqslant 1\}\\
&= \sup\{\nu_n(A(B\boldsymbol{x})) \mid \nu_p(\boldsymbol{x}) \leqslant 1\}\\
&= \sup\left\{\nu_n\left(A\frac{B\boldsymbol{x}}{\nu_m(B\boldsymbol{x})}\right)\nu_m(B\boldsymbol{x}) \middle| \nu_p(\boldsymbol{x}) \leqslant 1\right\}\\
&\leqslant \mu^{(n,m)}(A)\sup\{\nu_m(B\boldsymbol{x}) \mid \nu_p(\boldsymbol{x}) \leqslant 1\}\\
&\qquad \left(\text{because } \nu_m\left(\tfrac{B\boldsymbol{x}}{\nu(B\boldsymbol{x})}\right) = 1\right)\\
&= \mu^{n,m}(A)\mu^{m,p}(B).
\end{aligned}$$

□

Theorem 6.21 implies the following equivalent definitions of $\mu^{(n,m)}(A)$.

Theorem 6.23. *Let ν_n be a norm on $\mathbb{C}^n$ for $n \geqslant 1$. The following equalities hold for $\mu^{(n,m)}(A)$, where $A \in \mathbb{C}^{(n,m)}$:*

$$\begin{aligned}
\mu^{(n,m)}(A) &= \inf\{M \in \mathbb{R}_{\geqslant 0} \mid \nu_n(A\boldsymbol{x}) \leqslant M\nu_m(\boldsymbol{x}) \textit{ for every } \boldsymbol{x} \in \mathbb{C}^m\}\\
&= \sup\{\nu_n(A\boldsymbol{x}) \mid \nu_m(\boldsymbol{x}) \leqslant 1\}\\
&= \max\{\nu_n(A\boldsymbol{x}) \mid \nu_m(\boldsymbol{x}) \leqslant 1\}\\
&= \max\{\nu'(f(\boldsymbol{x})) \mid \nu(\boldsymbol{x}) = 1\}\\
&= \sup\left\{\frac{\nu'(f(\boldsymbol{x}))}{\nu(\boldsymbol{x})} \mid \boldsymbol{x} \in \mathbb{C}^m - \{\mathbf{0}_m\}\right\}.
\end{aligned}$$

Proof. The theorem is simply a reformulation of Theorem 6.21. □

Corollary 6.8. *Let μ be the matrix norm on $\mathbb{C}^{n\times n}$ induced by the vector norm ν. We have $\nu(A\boldsymbol{u}) \leqslant \mu(A)\nu(\boldsymbol{u})$ for every $\boldsymbol{u} \in \mathbb{C}^n$.*

Proof. The inequality is obviously satisfied when $\boldsymbol{u} = \mathbf{0}_n$. Therefore, we may assume that $\boldsymbol{u} \neq \mathbf{0}_n$ and let $\boldsymbol{x} = \frac{1}{\nu(\boldsymbol{u})}\boldsymbol{u}$. Clearly, $\nu(\boldsymbol{x}) = 1$ and Equality (6.10) implies that

$$\nu\left(A\frac{1}{\nu(\boldsymbol{u})}\boldsymbol{u}\right) \leqslant \mu(A)$$

for every $\boldsymbol{u} \in \mathbb{C}^n - \{\mathbf{0}_n\}$. This implies immediately the desired inequality. □

If μ is a matrix norm induced by a vector norm on $\mathbb{R}^n$, then $\mu(I_n) = \sup\{\nu(I_n\boldsymbol{x}) \mid \nu(\boldsymbol{x}) \leqslant 1\} = 1$. This necessary condition can be used for identifying matrix norms that are not induced by vector norms.

The operator matrix norm induced by the vector norm $\|\cdot\|_p$ is denoted by $|\!|\!|\cdot|\!|\!|_p$.

Example 6.14. To compute $|\!|\!|A|\!|\!|_1 = \sup\{\|A\boldsymbol{x}\|_1 \mid \|\boldsymbol{x}\|_1 \leqslant 1\}$, where $A \in \mathbb{R}^{n\times n}$, suppose that the columns of A are the vectors $\boldsymbol{a}_1, \ldots, \boldsymbol{a}_n$, that is

$$\boldsymbol{a}_j = \begin{pmatrix} a_{1j} \\ a_{2j} \\ \vdots \\ a_{nj} \end{pmatrix}.$$

Let $\boldsymbol{x} \in \mathbb{R}^n$ be a vector whose components are $x_1, \ldots, x_n$. Then, $\boldsymbol{Ax} = x_1\boldsymbol{a}_1 + \cdots + x_n\boldsymbol{a}_n$, so

$$\begin{aligned} \|A\boldsymbol{x}\|_1 &= \|x_1\boldsymbol{a}_1 + \cdots + x_n\boldsymbol{a}_n\|_1 \\ &\leqslant \sum_{j=1}^{n} |x_j|\|\boldsymbol{a}_j\|_1 \\ &\leqslant \max_j \|\boldsymbol{a}_j\|_1 \sum_{j=1}^{n} |x_j| \\ &= \max_j \|\boldsymbol{a}_j\|_1 \cdot \|\boldsymbol{x}\|_1. \end{aligned}$$

Thus, $|\!|\!|A|\!|\!|_1 \leqslant \max_j \|\boldsymbol{a}_j\|_1$.

Let $\boldsymbol{e}_j$ be the vector whose components are 0 with the exception of its jth component that is equal to 1. Clearly, we have $\|\boldsymbol{e}_j\|_1 = 1$ and $\boldsymbol{a}_j = A\boldsymbol{e}_j$. This, in turn implies $\|\boldsymbol{a}_j\|_1 = \|A\boldsymbol{e}_j\|_1 \leqslant |\!|\!|A|\!|\!|_1$ for $1 \leqslant j \leqslant n$. Therefore, $\max_j \|\boldsymbol{a}_j\|_1 \leqslant |\!|\!|A|\!|\!|_1$, so

$$|\!|\!|A|\!|\!|_1 = \max_j \|\boldsymbol{a}_j\|_1 = \max_j \sum_{i=1}^{n} |a_{ij}|.$$

In other words, $|\!|\!|A|\!|\!|_1$ equals the maximum column sum of the absolute values.

Example 6.15. Consider now a matrix $A \in \mathbb{R}^{n\times n}$. We have

$$\begin{aligned}\|A\boldsymbol{x}\|_\infty &= \max_{1\leqslant i\leqslant n}\left|\sum_{j=1}^n a_{ij}x_j\right| \\ &\leqslant \max_{1\leqslant i\leqslant n}\sum_{j=1}^n |a_{ij}x_j| \\ &\leqslant \max_{1\leqslant i\leqslant n}\|\boldsymbol{x}\|_\infty\sum_{j=1}^n |a_{ij}|.\end{aligned}$$

Consequently, if $\|\boldsymbol{x}\|_\infty \leqslant 1$, we have $\|A\boldsymbol{x}\|_\infty \leqslant \max_{1\leqslant i\leqslant n}\sum_{j=1}^n |a_{ij}|$. Thus, $|\!|\!|A|\!|\!|_\infty \leqslant \max_{1\leqslant i\leqslant n}\sum_{j=1}^n |a_{ij}|$.

The converse inequality is immediate if $A = O_{n,n}$. Therefore, assume that $A \neq O_{n\times n}$, and let $(a_{p1},\ldots,a_{pn})$ be any row of A that has at least one element distinct from 0. Define the vector $\boldsymbol{z} \in \mathbb{R}^n$ by

$$z_j = \begin{cases}\frac{|a_{pj}|}{a_{pj}} & \text{if } a_{pj} \neq 0,\\ 1 & \text{otherwise}\end{cases}$$

for $1 \leqslant j \leqslant n$. It is clear that $z_j \in \{-1, 1\}$ for every j, $1 \leqslant j \leqslant n$ and, therefore, $\|\boldsymbol{z}\|_\infty = 1$. Moreover, we have $|a_{pj}| = a_{pj}z_j$ for $1 \leqslant j \leqslant n$. Therefore, we can write

$$\begin{aligned}\sum_{j=1}^n |a_{pj}| = \sum_{j=1}^n a_{pj}z_j &\leqslant \left|\sum_{j=1}^n a_{pj}z_j\right| \\ &\leqslant \max_{1\leqslant i\leqslant n}\left|\sum_{j=1}^n a_{ij}z_j\right| \\ &= \|A\boldsymbol{z}\|_\infty \leqslant \max\{\|A\boldsymbol{x}\|_\infty \mid \|\boldsymbol{x}\|_\infty \leqslant 1\} = |\!|\!|A|\!|\!|_\infty.\end{aligned}$$

Since this holds for every row of A, it follows that $\max_{1\leqslant i\leqslant n}$ $\sum_{j=1}^n |a_{ij}| \leqslant |\!|\!|A|\!|\!|_\infty$, which proves that

$$|\!|\!|A|\!|\!|_\infty = \max_{1\leqslant i\leqslant n}\sum_{j=1}^n |a_{ij}|.$$

In other words, $|\!|\!|A|\!|\!|_\infty$ equals the maximum row sum of the absolute values.

Example 6.16. Let $D = \mathsf{diag}(d_1, \ldots, d_n) \in \mathbb{C}^{n\times n}$ be a diagonal matrix. If $\boldsymbol{x} \in \mathbb{C}^n$, we have

$$D\boldsymbol{x} = \begin{pmatrix} d_1 x_1 \\ \vdots \\ d_n x_n \end{pmatrix},$$

so

$$\begin{aligned} \| D \|_2 &= \max\{\|D\boldsymbol{x}\|_2 \mid \|\boldsymbol{x}\|_2 = 1\} \\ &= \max\{\sqrt{(d_1x_1)^2 + \cdots + (d_nx_n)^2} \mid x_1^2 + \cdots + x_n^2 = 1\} \\ &= \max\{|d_i| \mid 1 \leqslant 1 \leqslant n\}. \end{aligned}$$

The next result shows that certain norms are invariant with respect to multiplication by unitary matrices. We refer to these norms as *unitarily invariant norms.*

Theorem 6.24. *Let $U \in \mathbb{C}^{n\times n}$ be a unitary matrix. The following statements hold:*

(i) $\|U\boldsymbol{x}\|_2 = \|\boldsymbol{x}\|_2$ *for every* $\boldsymbol{x} \in \mathbb{C}^n$;
(ii) $\| UA \|_2 = \| A \|_2$ *for every* $A \in \mathbb{C}^{n\times p}$;
(iii) $\|UA\|_F = \|A\|_F$ *for every* $A \in \mathbb{C}^{n\times p}$.

Proof. For the first part of the theorem, note that

$$\|U\boldsymbol{x}\|_2^2 = (U\boldsymbol{x})^{\mathsf{H}}U\boldsymbol{x} = \boldsymbol{x}^{\mathsf{H}}U^{\mathsf{H}}U\boldsymbol{x} = \boldsymbol{x}^{\mathsf{H}}\boldsymbol{x} = \|\boldsymbol{x}\|_2^2,$$

because $U^{\mathsf{H}}A = I_n$.

The second part of the theorem is as follows:

$$\begin{aligned} \| UA \|_2 &= \max\{\|(UA)\boldsymbol{x}\|_2 \mid \|\boldsymbol{x}\|_2 = 1\} \\ &= \max\{\|U(A\boldsymbol{x})\|_2 \mid \|\boldsymbol{x}\|_2 = 1\} \\ &= \max\{\|A\boldsymbol{x}\|_2 \mid \|\boldsymbol{x}\|_2 = 1\} \\ &\quad \text{(by Part (i))} \\ &= \| A \|_2. \end{aligned}$$

For the Frobenius norm, note that

$$\begin{aligned} \|UA\|_F &= \sqrt{\mathsf{trace}((UA)^{\mathsf{H}}UA)} = \sqrt{\mathsf{trace}(A^{\mathsf{H}}U^{\mathsf{H}}UA)} \\ &= \sqrt{\mathsf{trace}(A^{\mathsf{H}}A)} = \|A\|_F, \end{aligned}$$

by Equality (6.11). □

Corollary 6.9. *If $U \in \mathbb{C}^{n\times n}$ is a unitary matrix, then $\|U\|_2 = 1$.*

Proof. Since $\|U\|_2 = \sup\{\|U\boldsymbol{x}\|_2 \mid \|\boldsymbol{x}\|_2 \leqslant 1\}$, by Part (ii) of Theorem 6.24,

$$\|U\|_2 = \sup\{\|\boldsymbol{x}\|_2 \mid \|\boldsymbol{x}\|_2 \leqslant 1\} = 1. \qquad \square$$

Corollary 6.10. *Let $A, U \in \mathbb{C}^{n\times n}$. If U is a unitary matrix, then*

$$\|U^{\mathsf{H}} A U\|_F = \|A\|_F.$$

Proof. Since U is a unitary matrix, so is U^{H}. By Part (iii) of Theorem 6.24,

$$\|U^{\mathsf{H}} A U\|_F = \|AU\|_F = \|U^{\mathsf{H}} A^{\mathsf{H}}\|_F^2 = \|A^{\mathsf{H}}\|_F^2 = \|A\|_F^2,$$

which proves the corollary. $\square$

Example 6.17. Let $S = \{\boldsymbol{x} \in \mathbb{R}^n \mid \|\boldsymbol{x}\|_2 = 1\}$ be the surface of the sphere in $\mathbb{R}^n$. The image of S under the linear transformation h_U that corresponds to the unitary matrix U is S itself. Indeed, by Theorem 6.24, $\|h_U(\boldsymbol{x})\|_2 = \|\boldsymbol{x}\|_2 = 1$, so $h_U(\boldsymbol{x}) \in S$ for every $\boldsymbol{x} \in S$. Also, note that h_U restricted to S is a bijection because $h_{U^{\mathsf{H}}}(h_U(\boldsymbol{x})) = \boldsymbol{x}$ for every $\boldsymbol{x} \in \mathbb{R}^n$. More details on transformations of S are given in Supplement 19 of Chapter 9.

Theorem 6.25. *Let $A \in \mathbb{R}^{n\times n}$. We have*

$$\|A\|_2 \leqslant \|A\|_F.$$

Proof. Let $\boldsymbol{x} \in \mathbb{R}^n$. We have

$$A\boldsymbol{x} = \begin{pmatrix} \boldsymbol{r}_1\boldsymbol{x} \\ \vdots \\ \boldsymbol{r}_n\boldsymbol{x} \end{pmatrix},$$

where $\boldsymbol{r}_1, \ldots, \boldsymbol{r}_n$ are the rows of the matrix A. Thus,

$$\frac{\|A\boldsymbol{x}\|_2}{\|\boldsymbol{x}\|_2} = \frac{\sqrt{\sum_{i=1}^n (\boldsymbol{r}_i\boldsymbol{x})^2}}{\|\boldsymbol{x}\|_2}.$$

By Cauchy–Schwarz inequality, we have $(\boldsymbol{r}_i\boldsymbol{x})^2 \leqslant \|\boldsymbol{r}_i\|_2^2\|\boldsymbol{x}\|_2^2$, so

$$\frac{\|A\boldsymbol{x}\|_2}{\|\boldsymbol{x}\|_2} \leqslant \sqrt{\sum_{i=1}^n \|\boldsymbol{r}_i\|_2^2} = \|A\|_F.$$

This implies $\|A\|_2 \leqslant \|A\|_F$. $\square$

We shall prove in Chapter 9 (in Corollary 9.4) that for every $A \in \mathbb{R}^{n \times n}$, we have

$$\|A\|_F \leqslant \sqrt{n}|||A|||_2. \tag{6.13}$$

6.7 Matrix Sequences and Matrix Series

In this section, we make use of the notion of series of complex numbers and we extend this concept to series of matrices. Recall (see [7], for example) that a series $\sum_{i=1}^{\infty} a_i$ converges absolutely if the series $\sum_{i=1}^{\infty} |a_i|$ converges. Absolute convergence implies convergence. The terms of an absolutely convergent series can be rearranged by altering its sum.

The set of matrices $\mathbb{C}^{m \times p}$ is a $\mathbb{C}$-linear space, and the set of matrices $\mathbb{R}^{m \times p}$ is an $\mathbb{R}$-linear space. Using matrix norms, these spaces can be equipped with a topological structure, as we indicated above.

We focus now on the normed linear space $(\mathbb{R}^{p \times p}, |||\cdot|||)$, where $|||\cdot|||)$ is a matrix norm.

Let $A \in \mathbb{R}^{p \times p}$. We can show by induction on n that

$$|||A^n||| \leqslant (|||A|||)^n. \tag{6.14}$$

The base step, $n = 0$, is immediate. Suppose that the inequality holds for n. We have

$$\begin{aligned}
|||A^{n+1}||| &= |||A^n A||| \\
&\leqslant |||A^n||| |||A||| \\
&\quad \text{(because } |||\cdot||| \text{ is a matrix norm)} \\
&\leqslant (|||A|||)^n |||A||| \\
&\quad \text{(by the inductive hypothesis)} \\
&= (|||A|||)^{n+1},
\end{aligned}$$

which concludes our argument.

If $|||A||| < 1$, the sequence of matrices $(A, A^2, \ldots, A^n, \ldots)$ converges toward the zero matrix $O_{p,p}$. Indeed, $\lim_{n\to\infty} |||A^n - O_{p,p}||| = \lim_{n\to\infty} |||A^n||| \leqslant \lim_{n\to\infty}(|||A|||)^n = 0$, which shows that $\lim_{n\to\infty} A^n = O_{p,p}$.

Definition 6.13. Let $\boldsymbol{A} = (A_0, A_1, \ldots, A_n, \ldots)$ be a sequence of matrices in $\mathbb{R}^{p\times p}$. A *matrix series* having $\boldsymbol{A}$ as its sequence of terms is the sequence of matrices $(S_0, S_1, \ldots, S_n, \ldots)$, where $S_i = \sum_{k=0}^{i} A_k$.

The series $(S_0, S_1, \ldots, S_n, \ldots)$ is denoted also by $A_0 + A_1 + \cdots + A_n + \cdots$.

We say that the series $A_0 + A_1 + \cdots + A_n + \cdots$ *converges to a matrix* S if $\lim_{n\to\infty} S_n = S$. This is also denoted by $A_0 + A_1 + \cdots + A_n + \cdots = S$.

The series $A_0 + A_1 + \cdots + A_n + \cdots$ *converges absolutely* if each of the series $(A_0)_{ij} + (A_1)_{ij} + \cdots + (A_n)_{ij} + \cdots$ converges absolutely for $1 \leqslant i, j \leqslant n$.

The subadditivity property of the norm can be generalized to a series of matrices. Namely, if the series $A_0 + A_1 + \cdots + A_n + \cdots$ converges to S, then

$$\|S\| \leqslant \sum_{i=0}^{\infty} \|A_i\|.$$

Indeed, by the usual subadditivity property,

$$\|A_0 + A_1 + \cdots + A_n\| \geqslant \sum_{i=0}^{n} \|A_i\|.$$

This implies

$$\|A_0 + A_1 + \cdots + A_n\| \geqslant \sum_{i=0}^{\infty} \|A_i\|,$$

for every $n \in \mathbb{N}$, so $\|S\| \geqslant \sum_{i=0}^{\infty} \|A_i\|$, due to the continuity of the matrix norm.

Example 6.18. Let $A \in \mathbb{R}^{p\times p}$ be a matrix such that $\|A\| < 1$. We claim that the matrix $I - A$ is invertible and $A^0 + A^1 + \cdots + A^n + \cdots = (I - A)^{-1}$.

Suppose that $I - A$ is not invertible. Then, the system $(I - A)\boldsymbol{x} = \boldsymbol{0}$ has a non-trivial solution. This implies $\boldsymbol{x} = A\boldsymbol{x}$, so $\|A\| \geqslant 1$, which contradicts the hypothesis. Thus, $I - A$ is an invertible matrix.

Observe that

$$(A^0 + A^1 + \cdots + A^n)(I - A)^{-1} = I - A^{n+1}.$$

Therefore,

$$\left(\lim_{n\to\infty}(A^0 + A^1 + \cdots + A^n)\right)(I - A)^{-1} = \lim_{n\to\infty}(I - A^{n+1}) = I,$$

since $\lim_{n\to\infty} A^{n+1} = O$. This shows that the series $A^0 + A^1 + \cdots + A^n + \cdots$ converges to the inverse of the matrix $I - A$, so

$$(I - A)^{-1} = \sum_{i=0}^{\infty} A^i.$$

Moreover, we have

$$\begin{aligned}\|(I - A)^{-1}\| &= \left\|\sum_{i=0}^{\infty} A^i\right\| \leqslant \sum_{i=0}^{\infty} \|A^i\| \\ &= \sum_{i=0}^{\infty} (\|A\|)^i = \frac{1}{1 - \|A\|},\end{aligned}$$

because $\|A\| \leqslant 1$.

Example 6.19. Let $(\boldsymbol{x}_0, \boldsymbol{x}_1, \ldots, \boldsymbol{x}_n, \ldots)$ be a sequence of vectors defined inductively by

$$\boldsymbol{x}_{n+1} = A\boldsymbol{x}_n + \boldsymbol{b} \tag{6.15}$$

for $n \in \mathbb{N}$, where $A \in \mathbb{C}^{n\times n}$ and $\boldsymbol{b} \in \mathbb{C}^n$. It is easy to verify that $\boldsymbol{x}_n = A^n\boldsymbol{x}_0 + (A^{n-1} + \cdots + A + I)\boldsymbol{b}$. If $\|A\| < 1$, $\lim_{n\to\infty} A^n = O$, $\lim_{n\to\infty}(A^{n-1} + \cdots + A + I) = (I - A)^{-1}$, so $\lim_{n\to\infty} \boldsymbol{x}_n$ exists. If $\lim_{n\to\infty} \boldsymbol{x}_n = \boldsymbol{x}$, it follows from Equality (6.15) that $\boldsymbol{x} = A\boldsymbol{x} + \boldsymbol{b}$.

Let $\boldsymbol{e}_n = \boldsymbol{x}_n - \boldsymbol{x}$ be the *error sequence*. Clearly, if $\|A\| < 1$, $\lim_{n\to\infty} \boldsymbol{e}_n = \boldsymbol{0}$.

Lemma 6.6. *Let $A_0, \ldots, A_m, \ldots$ be matrices in $\mathbb{C}^{n\times n}$. If $\sum_{m=0}^{\infty} A_m$ is convergent, then $\sum_{m=0}^{\infty} UA_mV$ is convergent for every U, V and $\sum_{m=0}^{\infty} UA_mV = U\left(\sum_{m=0}^{\infty} A_m\right)V$.*

Proof. Suppose that $\sum_{m=0}^{\infty} A_m$ is convergent and let $S = \sum_{m=0}^{\infty} A_m$ be its sum. We can write

$$\left\| \sum_{m=0}^{\infty} U A_m V - USV \right\|_{\infty} = \left\| U \left(\sum_{m=0}^{\infty} A_m - S \right) V \right\|_{\infty}$$
$$\leqslant n^2 \|U\|_{\infty} \left\| \sum_{m=0}^{\infty} A_m - S \right\|_{\infty} \|V\|_{\infty}.$$

Therefore, the convergence of $\sum_{m=0}^{\infty} A_m$ implies the convergence of $\sum_{m=0}^{\infty} U A_m V$ and $\sum_{m=0}^{\infty} U A_m V = U \left(\sum_{m=0}^{\infty} A_m\right) V$. □

Lemma 6.7. *Let μ be a vectorial norm on $\mathbb{C}^{n\times n}$ and let $A_0, \dots, A_m, \dots$ be matrices in $\mathbb{C}^{n\times n}$. The series $\sum_{m=0}^{\infty} A_m$ is absolutely convergent if and only if the series $\sum_{m=0}^{\infty} \mu(A_m)$ is convergent.*

Proof. If the series $\sum_{m=0}^{\infty} \mu(A_m)$ is convergent, then for every i, j such that $1 \leqslant i, j \leqslant n$ we have $|(A_m)_{ij}| \leqslant k\mu(A_m)$ for some positive constant k (see Supplement 6.20) which implies the absolute convergence of $\sum_{m=0}^{\infty} A_m$.

Conversely, suppose that $\sum_{m=0}^{\infty} A_m$ is absolutely convergent. Then, there exists a number c such that $\sum_{m=0}^{p} |(A_m)_{ij}| \leqslant c$ for every $p \in \mathbb{N}$ and $1 \leqslant i, j \leqslant n$. Therefore, we can write

$$\sum_{m=0}^{p} \mu(A_m) \leqslant \sum_{m=0}^{p} \sum_{i,j=1}^{n} |(A_m)_{ij}| \leqslant n^2 c,$$

which allows us to conclude that $\sum_{m=0}^{\infty} \mu(A_m)$ is convergent. □

Theorem 6.26. *Let $A_0, \dots, A_m, \dots$ be matrices in $\mathbb{C}^{n\times n}$. If $\sum_{m=0}^{\infty} A_m$ is absolutely convergent, then $\sum_{m=0}^{\infty} U A_m V$ is absolutely convergent for every U, V.*

Proof. The second part of Supplement 6.20 implies

$$\|U A_m V\|_{\infty} \leqslant n^2 \|P\|_{\infty} \|A_m\|_{\infty} \|V\|_{\infty} \leqslant c \|A_m\|_{\infty},$$

where c does not depend on m. By Lemma 6.7, $\sum_{m=0}^{\infty} \|A_m\|_{\infty}$ is convergent, so $\sum_{m=0}^{\infty} \|U A_m V\|_{\infty}$ is convergent for every U, V, which implies the absolute convergence of $\sum_{m=0}^{\infty} U A_m V$. □

6.8 Conjugate Norms

Let $\nu : \mathbb{C}^n \longrightarrow \mathbb{R}_{\geqslant 0}$. Consider the function $\nu^* : \mathbb{C}^n \longrightarrow \mathbb{R}_{\geqslant 0}$ defined by

$$\nu^*(\boldsymbol{y}) = \max\{|\boldsymbol{y}^{\mathsf{H}}\boldsymbol{x}| \mid \nu(\boldsymbol{x}) = 1\}$$

for $\boldsymbol{y} \in \mathbb{C}^n$.

Theorem 6.27. *The mapping ν^* is a norm on $\mathbb{C}^n$.*

Proof. Let $\boldsymbol{y}, \boldsymbol{z} \in \mathbb{C}^n$. For $\nu(\boldsymbol{x}) = 1$, we have

$$|(\boldsymbol{y} + \boldsymbol{z})^{\mathsf{H}}\boldsymbol{x}| \leqslant |\boldsymbol{y}^{\mathsf{H}}\boldsymbol{x}| + |\boldsymbol{z}^{\mathsf{H}}\boldsymbol{x}| \leqslant \nu^*(\boldsymbol{y}) + \nu^*(\boldsymbol{z}),$$

which implies $\nu^*(\boldsymbol{y} + \boldsymbol{z}) \leqslant \nu^*(\boldsymbol{y}) + \nu^*(\boldsymbol{z})$. Thus, ν^* is subadditive.

The positive homogeneity is immediate.

Suppose that $\nu^*(\boldsymbol{y}) = 0$ but $\boldsymbol{y} \neq \boldsymbol{0}$. Since

$$\nu\left(\frac{\boldsymbol{y}}{\nu(\boldsymbol{y})}\right) = 1,$$

this implies

$$\nu^*(\boldsymbol{y}) \geqslant |\boldsymbol{y}^{\mathsf{H}}\frac{\boldsymbol{y}}{\nu(\boldsymbol{y})}| = \frac{\|\boldsymbol{y}\|_2^2}{\nu(\boldsymbol{y})}.$$

Therefore, if $\boldsymbol{y} \neq \boldsymbol{0}$, then $\nu^*(\boldsymbol{y}) > 0$. Consequently, $\nu^*(\boldsymbol{y}) \neq 0$ implies $\boldsymbol{y} \neq \boldsymbol{0}$, which allows us to conclude that ν^* is a norm. □

Definition 6.14. The norm ν^* is the *conjugate norm* of the norm ν.

Example 6.20. Let ν_p be the norm introduced in Theorem 6.5,

$$\nu_p(\boldsymbol{x}) = \left(\sum_{i=1}^{n} |x_i|^p\right)^{\frac{1}{p}}$$

for $\boldsymbol{x} \in \mathbb{R}^n$.

To compute its dual ν_p^*, we need to compute

$$\begin{aligned}\nu_p^*(\boldsymbol{y}) &= \max\{|\boldsymbol{y}^{\mathsf{H}}\boldsymbol{x} \mid \nu_p(\boldsymbol{x}) = 1\}\\ &= \max\{|y_1x_1 + \cdots + y_nx_n| \sum_{i=1}^{n} |x_i|^p = 1\}\end{aligned}$$

for $\boldsymbol{y} \in \mathbb{C}^n$.

Without loss of generality, we can assume that $y_1, \ldots, y_n, x_1, \ldots, x_n$ belong to $\mathbb{R}_{\geqslant 0}$. By introducing a Lagrange multiplier λ, we consider the function

$$\Phi(y_1, \ldots, y_n, x_1, \ldots, x_n, \lambda) = y_1 x_1 + \cdots + y_n x_n - \lambda \left(\sum_{i=1}^{n} x_i^p - 1 \right).$$

The necessary extremum conditions are

$$\frac{\partial \Phi}{\partial x_i} = y_i - \lambda p x_i^{p-1} = 0$$

for $1 \leqslant i \leqslant n$. Thus, we must have

$$\frac{y_1}{x_1^{p-1}} = \cdots = \frac{y_n}{x_n^{p-1}} = p\lambda.$$

Equivalently, we have

$$\frac{y_1^{\frac{p}{p-1}}}{x_1^p} = \cdots = \frac{y_n^{\frac{p}{p-1}}}{x_n^p} = (p\lambda)^{\frac{p}{p-1}}.$$

For $q = \frac{p}{p-1}$, the last equality can be written as

$$\frac{y_1^q}{x_1^p} = \cdots = \frac{y_n^q}{x_n^p} = (p\lambda)^q = \frac{\nu_q(\boldsymbol{y})^q}{\nu_p(\boldsymbol{x})^p} = \nu_q(\boldsymbol{y})^q,$$

since $\nu_p(\boldsymbol{x}) = 1$. Thus, $p\lambda = \nu_q(\boldsymbol{y})$ and, therefore, $y_i x_i = x_i^p \nu_q(\boldsymbol{y})$ for $1 \leqslant i \leqslant n$.

The maximum of $y_1 x_1 + \cdots + y_n x_n$ for $\nu_p(\boldsymbol{x}) = 1$ is therefore

$$\sum_{i=1}^{n} y_i x_i = \sum_{i=1}^{n} x_i^p \nu_q(\boldsymbol{y}) = \nu_q(\boldsymbol{y}).$$

This shows that the conjugate of the norm ν_p is the norm ν_q, where

$$\frac{1}{p} + \frac{1}{q} = 1.$$

Observe that the conjugate of the Euclidean norm ν_2 is ν_2 itself.

Let U be a subspace of $\mathbb{C}^n$ and let $h : U \longrightarrow \mathbb{C}$ be a linear functional defined on U. We observed that there exists $\boldsymbol{y} \in \mathbb{C}^n$ such that h can be expressed as $h(\boldsymbol{x}) = \boldsymbol{y}^{\mathsf{H}}\boldsymbol{x}$ for $\boldsymbol{x} \in U$. Note that the vector $\boldsymbol{y}$ is not unique because $h(\boldsymbol{x}) = (\boldsymbol{y}+\boldsymbol{z})^{\mathsf{H}}\boldsymbol{x}$ for $\boldsymbol{x} \in U$ for every vector $\boldsymbol{z} \in U^{\perp}$.

Theorem 6.28 (Hahn–Banach theorem). Let h be a linear functional defined on a subspace U of $\mathbb{C}^n$ and let ν be a norm on $\mathbb{C}$ such that

$$\max\{|h(\boldsymbol{x})| \mid \boldsymbol{x} \in U \text{ and } \nu(\boldsymbol{x}) = 1\} = M.$$

There exists an extension $\hat{h}$ of h to $\mathbb{C}^n$ such that

$$\max\{|\hat{h}(\boldsymbol{x})| \mid \boldsymbol{x} \in \mathbb{C}^n \text{ and } \nu(\boldsymbol{x}) = 1\} = M.$$

Proof. If the subspace U coincides with $\mathbb{C}^n$, then there is nothing to prove. Suppose that $U \subset \mathbb{C}^n$ and let $\boldsymbol{w}$ be a vector in $\mathbb{C}^n - U$. Clearly, we have $\boldsymbol{w} \neq \boldsymbol{0}$.

If $\boldsymbol{u} \in U - \{\boldsymbol{0}\}$, we have $|h(\boldsymbol{u})| \leqslant M\nu(\boldsymbol{u})$. The linearity of h implies

$$h(\boldsymbol{u}_1)-h(\boldsymbol{u}_2) = h(\boldsymbol{u}_1-\boldsymbol{u}_2) \leqslant M\nu(\boldsymbol{u}_1-\boldsymbol{u}_2) \leqslant M(\nu(\boldsymbol{u}_1+\boldsymbol{w})+\nu(\boldsymbol{u}_2+\boldsymbol{w}))$$

which yields

$$h(\boldsymbol{u}_1) - M\nu(\boldsymbol{u}_1 + \boldsymbol{w}) \leqslant h(\boldsymbol{u}_2) + M\nu(\boldsymbol{u}_2 + \boldsymbol{w})$$

for any $\boldsymbol{u}_1, \boldsymbol{u}_2 \in U$. Therefore, every number in the set $A = \{h(\boldsymbol{u}_1) - M\nu(\boldsymbol{u}_1 + \boldsymbol{w}) \mid \boldsymbol{u}_1 \in U\}$ is less than or equal to any number of the set $B = \{h(\boldsymbol{u}_2) + M\nu(\boldsymbol{u}_2 + \boldsymbol{w}) \mid \boldsymbol{u}_2 \in U\}$.

We aim to linearly extend h to a linear functional $h_{\boldsymbol{w}}$ defined on the subspace $W = \langle U \cup \{\boldsymbol{w}\}\rangle$. To this end, define $h_{\boldsymbol{w}}(\boldsymbol{w}) = -a$, where a is a number located between the sets A and B. Since $h_{\boldsymbol{w}}(\boldsymbol{u}) = h(\boldsymbol{u})$ for every $\boldsymbol{u} \in U$, we have

$$h_{\boldsymbol{w}}(\boldsymbol{u}) - M\nu(\boldsymbol{u} + \boldsymbol{w}) \leqslant -h_{\boldsymbol{w}}(\boldsymbol{w}) \leqslant h_{\boldsymbol{w}}(\boldsymbol{u}) + M\nu(\boldsymbol{u} + \boldsymbol{w}),$$

which implies

$$h_{\boldsymbol{w}}(\boldsymbol{u} + \boldsymbol{w}) - M\nu(\boldsymbol{u} + \boldsymbol{w}) \leqslant 0 \leqslant h_{\boldsymbol{w}}(\boldsymbol{u} + \boldsymbol{w}) + M\nu(\boldsymbol{u} + \boldsymbol{w}).$$

This is equivalent to $|h_{\boldsymbol{w}}(\boldsymbol{u}+\boldsymbol{w})| \leqslant M\nu(\boldsymbol{u}+\boldsymbol{w})$. For $\alpha \in \mathbb{C} - \{0\}$, define $h_{\boldsymbol{w}}(\boldsymbol{u}+\alpha\boldsymbol{w}) = h(\boldsymbol{u}) + \alpha h_{\boldsymbol{w}}(\boldsymbol{w})$. We have

$$\begin{aligned}|h_{\boldsymbol{w}}(\boldsymbol{u}+\alpha\boldsymbol{w})| &= |\alpha|\left|h_{\boldsymbol{w}}\left(\frac{1}{|\alpha|}\boldsymbol{u}+\boldsymbol{w}\right)\right| \leqslant |\alpha| M\nu\left(\frac{1}{|\alpha|}\boldsymbol{u}+\boldsymbol{w}\right)\\ &= M\nu(\boldsymbol{u}+\alpha\boldsymbol{w}).\end{aligned}$$

If $W \neq \mathbb{C}^n$, this extension can be repeated a finite number of times because $\mathbb{C}^n$ is of finite dimension. Eventually, we obtain a linear functional defined on $\mathbb{C}^n$ with the preservation of the boundedness condition. □

Corollary 6.11. *The conjugate ν^{**} of the conjugate ν^* of a norm ν on $\mathbb{C}^n$ equals ν.*

Proof. Since $\nu^*(\boldsymbol{y}) = \max\{|\boldsymbol{y}^{\mathsf{H}}\boldsymbol{x}| \mid \nu(\boldsymbol{x}) = 1\}$, it follows that

$$|\boldsymbol{y}^{\mathsf{H}}\boldsymbol{x}| \leqslant \nu^*(\boldsymbol{y})\nu(\boldsymbol{x}) \tag{6.16}$$

for $\boldsymbol{x}, \boldsymbol{y} \in \mathbb{C}^n$, an inequality that is a generalization of the Cauchy–Schwarz inequality. Therefore, we have

$$\nu^{**}(\boldsymbol{x}) \leqslant \max\{|\boldsymbol{x}^{\mathsf{H}}\boldsymbol{y} \mid \nu^*(\boldsymbol{y}) = 1\} \leqslant \nu(\boldsymbol{x}).$$

To prove the converse inequality, we need to use the Hahn–Banach Theorem. Consider the linear functional g defined on the subspace $\langle\boldsymbol{x}\rangle$ by $g(\boldsymbol{u}) = a\nu(\boldsymbol{x})$ for $\boldsymbol{u} = a\boldsymbol{x}$ and $a \in \mathbb{C}$. If $\nu(\boldsymbol{u}) = 1$, we have $|a| = \frac{1}{\nu(\boldsymbol{x})}$, so $|g(\boldsymbol{u})| = |a|\nu(\boldsymbol{x}) = 1$. Thus, by the Hahn–Banach Theorem, g can be extended to $\hat{g} : \mathbb{C}^* \longrightarrow \mathbb{C}$ such that $|\hat{(g)}(\boldsymbol{v})| \leqslant 1$ if $\nu(\boldsymbol{v}) = 1$ for $\boldsymbol{v} \in \mathbb{C}$.

If $\hat{g}(\boldsymbol{v}) = \boldsymbol{z}^{\mathsf{H}}\boldsymbol{v}$, then

$$\nu^*(\boldsymbol{z}) = \max\{|\boldsymbol{z}^{\mathsf{H}}\boldsymbol{v}| \mid \nu(\boldsymbol{v}) = 1\} = 1$$

and $|\boldsymbol{z}^{\mathsf{H}}\boldsymbol{v}| = \nu(\boldsymbol{v})$. Therefore,

$$\nu^{**}(\boldsymbol{x}) = \max\{|\boldsymbol{x}^{\mathsf{H}}\boldsymbol{z}| \mid \nu^*(\boldsymbol{z}) = 1\} \geqslant \nu(\boldsymbol{x}),$$

so $\nu^{**}(\boldsymbol{x}) = \nu(\boldsymbol{x})$. □

6.9 Inner Products

Definition 6.15. Let V be a $\mathbb{C}$-linear space. An *inner product* on V is a function $f : V \times V \longrightarrow \mathbb{C}$ that has the following properties:

(i) $f(a\boldsymbol{x} + b\boldsymbol{y}, \boldsymbol{z}) = af(\boldsymbol{x}, \boldsymbol{z}) + bf(\boldsymbol{y}, \boldsymbol{z})$ (linearity in the first argument);

(ii) $f(\boldsymbol{x}, \boldsymbol{y}) = \overline{f(\boldsymbol{y}, \boldsymbol{x})}$ for $\boldsymbol{y}, \boldsymbol{x} \in V$ (conjugate symmetry);

(iii) if $\boldsymbol{x} \neq \boldsymbol{0}_V$, then $f(\boldsymbol{x}, \boldsymbol{x})$ is a positive real number (positivity);

(iv) $f(\boldsymbol{x}, \boldsymbol{x}) = 0$ if and only if $\boldsymbol{x} = \boldsymbol{0}_V$ (definiteness);

for every $\boldsymbol{x}, \boldsymbol{y}, \boldsymbol{z} \in V$ and $a, b \in \mathbb{C}$.

The pair (V, f) is called an *inner product space.*

An alternative terminology [106] for real inner product spaces is *Euclidean spaces*, and *Hermitian spaces* for complex inner product spaces.

For the second argument of an inner product on a $\mathbb{C}$-linear space, we have the property of *conjugate linearity*, that is,

$$f(\boldsymbol{z}, a\boldsymbol{x} + b\boldsymbol{y}) = \bar{a}f(\boldsymbol{z}, \boldsymbol{x}) + \bar{b}f(\boldsymbol{z}, \boldsymbol{y})$$

for every $\boldsymbol{x}, \boldsymbol{y}, \boldsymbol{z} \in V$ and $a, b \in \mathbb{C}$. Indeed, by the conjugate symmetry property, we can write

$$\begin{aligned} f(\boldsymbol{z}, a\boldsymbol{x} + b\boldsymbol{y}) &= \overline{f(a\boldsymbol{x} + b\boldsymbol{y}, \boldsymbol{z})} \\ &= \overline{af(\boldsymbol{x}, \boldsymbol{z}) + bf(\boldsymbol{y}, \boldsymbol{z})} \\ &= \bar{a}\overline{f(\boldsymbol{x}, \boldsymbol{z})} + \bar{b}\overline{f(\boldsymbol{y}, \boldsymbol{z})} \\ &= \bar{a}f(\boldsymbol{z}, \boldsymbol{x}) + \bar{b}f(\boldsymbol{z}, \boldsymbol{y}). \end{aligned}$$

To simplify notations, if there is no risk of confusion, we denote the inner product $f(\boldsymbol{u}, \boldsymbol{v})$ as $(\boldsymbol{u}, \boldsymbol{v})$.

Observe that the conjugate symmetry property on inner products implies that for $\boldsymbol{x} \in V$, $(\boldsymbol{x}, \boldsymbol{x})$ is a real number because $(\boldsymbol{x}, \boldsymbol{x}) = \overline{(\boldsymbol{x}, \boldsymbol{x})}$.

When V is a real linear space, the definition of the inner product becomes simpler because the conjugate of a real number a is a itself. Namely, for real linear spaces, the conjugate symmetry is replaced by the plain symmetry property,

$$(\boldsymbol{x}, \boldsymbol{y}) = (\boldsymbol{y}, \boldsymbol{x})$$

for $\boldsymbol{x}, \boldsymbol{y} \in V$. Thus, in case of real linear spaces, an inner product is linear in both arguments.

Let $W = \{\boldsymbol{w}_1, \ldots, \boldsymbol{w}_n\}$ be a basis in the complex n-dimensional inner product space V. If $\boldsymbol{x} = \sum_{i=1}^n x^i \boldsymbol{w}_i$ and $\boldsymbol{y} = \sum_{j=1}^n y^j \boldsymbol{w}_j$, then

$$(\boldsymbol{x}, \boldsymbol{y}) = \sum_{i=1}^n \sum_{j=1}^n x^i \overline{y^j} (\boldsymbol{w}_i, \boldsymbol{w}_j),$$

due to the bilinearity of the inner product.

If we denote $(\boldsymbol{w}_i, \boldsymbol{w}_j)$ by g_{ij}, then $(\boldsymbol{x}, \boldsymbol{y})$ can be written as

$$(\boldsymbol{x}, \boldsymbol{y}) = \sum_{i=1}^n \sum_{j=1}^n x^i \overline{y^j} g_{ij} \tag{6.17}$$

for $\boldsymbol{x}, \boldsymbol{y} \in V$, when V is a complex inner product linear space.

If V is a real inner product space, then

$$(\boldsymbol{x}, \boldsymbol{y}) = \sum_{i=1}^n \sum_{j=1}^n x^i y^j g_{ij}.$$

Note that in this case (g_{ij}) is a symmetric matrix. The matrix $G = (g_{ij})$ is actually the Gram matrix of the basis W, which will be referred to as the *fundamental matrix* of the basis B.

Definition 6.16. Two vectors $\boldsymbol{u}, \boldsymbol{v} \in \mathbb{C}^n$ are said to be *orthogonal* with respect to an inner product if $(\boldsymbol{u}, \boldsymbol{v}) = 0$. This is denoted by $x \perp y$.

An *orthogonal set of vectors* in an inner product space V equipped with an inner product is a subset W of V such that for every $\boldsymbol{u}, \boldsymbol{v} \in W$, we have $\boldsymbol{u} \perp \boldsymbol{v}$.

Theorem 6.29. *Any inner product on a linear space V generates a norm on that space defined by $\|\boldsymbol{x}\| = \sqrt{(\boldsymbol{x}, \boldsymbol{x})}$ for $\boldsymbol{x} \in V$.*

Proof. Let V be a $\mathbb{C}$-linear space. We need to verify that the norm satisfies the conditions of Definition 6.7. Applying the properties of the inner product, we have

$$\begin{aligned}
\|\boldsymbol{x} + \boldsymbol{y}\|^2 &= (\boldsymbol{x} + \boldsymbol{y}, \boldsymbol{x} + \boldsymbol{y}) \\
&= (\boldsymbol{x}, \boldsymbol{x}) + 2(\boldsymbol{x}, \boldsymbol{y}) + (\boldsymbol{y}, \boldsymbol{y}) \\
&= \|\boldsymbol{x}\|^2 + 2(\boldsymbol{x}, \boldsymbol{y}) + \|\boldsymbol{y}\|^2
\end{aligned}$$

$$\leqslant \|\boldsymbol{x}\|^2 + 2\|\boldsymbol{x}\|\|\boldsymbol{y}\| + \|\boldsymbol{y}\|^2$$
$$= (\|\boldsymbol{x}\| + \|\boldsymbol{y}\|)^2.$$

Because $\|\boldsymbol{x}\| \geqslant 0$, it follows that $\|\boldsymbol{x}+\boldsymbol{y}\| \leqslant \|\boldsymbol{x}\| + \|\boldsymbol{y}\|$, which is the subadditivity property.

If $a \in \mathbb{C}$, then $\|a\boldsymbol{x}\| = \sqrt{(a\boldsymbol{x}, a\boldsymbol{x})} = \sqrt{a\bar{a}(\boldsymbol{x},\boldsymbol{x})} = \sqrt{|a|^2(\boldsymbol{x},\boldsymbol{x})} = |a|\sqrt{(\boldsymbol{x},\boldsymbol{x})} = |a|\|\boldsymbol{x}\|$.

Finally, from the definiteness property of the inner product, it follows that $\|\boldsymbol{x}\| = 0$ if and only if $\boldsymbol{x} = \boldsymbol{0}_V$, which allows us to conclude that $\|\cdot\|$ is indeed a norm. □

The norm induced by the inner product $f(\boldsymbol{x},\boldsymbol{y}) = x^i\overline{y^j}g_{ij}$ introduced in Equality (6.17) is

$$\|\boldsymbol{x}\|^2 = f(\boldsymbol{x},\boldsymbol{x}) = x^i\overline{x^j}g_{ij}.$$

Theorem 6.30. *If W is a set of orthogonal vectors in an n-dimensional $\mathbb{C}$-linear space V and $\boldsymbol{0}_V \notin W$, then W is linearly independent.*

Proof. Let $\boldsymbol{c} = a^1\boldsymbol{w}_1 + \cdots + a^n\boldsymbol{w}_n$ be a linear combination in V such that $a^1\boldsymbol{w}_1 + \cdots + a^n\boldsymbol{w}_n = \boldsymbol{0}_V$. Since $(\boldsymbol{c},\boldsymbol{w}_i) = a_i\|\boldsymbol{w}_i\|^2 = 0$, we have $a_i = 0$ because $\|\boldsymbol{w}_i\|^2 \neq 0$, and this holds for every i, where $1 \leqslant i \leqslant n$. Thus, W is linearly independent. □

Definition 6.17. An *orthonormal set of vectors* in an inner product space V equipped with an inner product is an orthogonal subset W of V such that for every $\boldsymbol{u}$ we have $\|\boldsymbol{u}\| = 1$, where the norm is induced by the inner product.

Corollary 6.12. *If W is an orthonormal set of vectors in an n-dimensional $\mathbb{C}$-linear space V and $|W| = n$, then W is a basis in L.*

Proof. This statement follows immediately from Theorem 6.30. □

If $W = \{\boldsymbol{w}_1,\ldots,\boldsymbol{w}_n\}$ is an orthonormal basis in $\mathbb{C}^n$, we have

$$g_{ij} = (\boldsymbol{w}_i,\boldsymbol{w}_j) = \begin{cases} 0 & \text{if } i \neq j, \\ 1 & \text{if } i = j, \end{cases}$$

which means that the inner product of the vectors $\boldsymbol{x} = x^i\boldsymbol{w}_i$ and $\boldsymbol{y} = y^j\boldsymbol{w}_j$ is given by

$$(\boldsymbol{x},\boldsymbol{y}) = x^i\overline{y^j}(\boldsymbol{w}_i,\boldsymbol{w}_j) = x^i\overline{y^i}. \tag{6.18}$$

Consequently, $\|\boldsymbol{x}\|^2 = \sum_{i=1}^n |x^i|^2$.

The inner product of $\boldsymbol{x}, \boldsymbol{y} \in \mathbb{R}^n$ is

$$(\boldsymbol{x}, \boldsymbol{y}) = x^i y^j (\boldsymbol{w}_i, \boldsymbol{w}_j) = x^i y^i. \tag{6.19}$$

Next, we present a direct proof for the Cauchy–Schwarz inequality.

Theorem 6.31 (Cauchy–Schwarz inequality). Let V be a $\mathbb{C}$-linear space. Then, for every $\boldsymbol{x}, \boldsymbol{y} \in V$, we have

$$|(\boldsymbol{x}, \boldsymbol{y})| \leqslant \|\boldsymbol{x}\|\|\boldsymbol{y}\|.$$

Moreover, the equality takes place if and only if $\{\boldsymbol{x}, \boldsymbol{y}\}$ is a linear dependent set of vectors.

Proof. If $\boldsymbol{y} = \boldsymbol{0}_V$, the inequality obviously holds. So, suppose that $\boldsymbol{y} \neq \boldsymbol{0}_V$. For every $t \in \mathbb{R}$, we have $(\boldsymbol{x} + t\boldsymbol{y}, \boldsymbol{x} + t\boldsymbol{y}) \geqslant 0$. In particular, taking $t = \frac{(\boldsymbol{x},\boldsymbol{y})}{\|\boldsymbol{y}\|^2}$, we have

$$\begin{aligned}
0 \leqslant \|\boldsymbol{x} - t\boldsymbol{y}\|^2 &= (\boldsymbol{x}, \boldsymbol{x}) - \bar{t}(\boldsymbol{x}, \boldsymbol{y}) - t(\boldsymbol{y}, \boldsymbol{x}) + t\bar{t}(\boldsymbol{y}, \boldsymbol{y}) \\
&= \|x\|^2 - \bar{t}(\boldsymbol{x}, \boldsymbol{y}) - t\overline{(\boldsymbol{x}, \boldsymbol{y})} + t\bar{t}\|\boldsymbol{y}\|^2 \\
&= \|x\|^2 - 2\frac{|(\boldsymbol{x}, \boldsymbol{y})|^2}{\|\boldsymbol{y}\|^2} + \frac{|(\boldsymbol{x}, \boldsymbol{y})|^2}{\|\boldsymbol{y}\|^2} \\
&= \|\boldsymbol{x}\|^2 - \frac{|(\boldsymbol{x}, \boldsymbol{y})|^2}{\|\boldsymbol{y}\|^2},
\end{aligned}$$

which implies $|(\boldsymbol{x}, \boldsymbol{y})| \leqslant \|\boldsymbol{x}\|\|\boldsymbol{y}\|$.

If the equality $|(\boldsymbol{x}, \boldsymbol{y})| = \|\boldsymbol{x}\|\|\boldsymbol{y}\|$ takes place, then $\|\boldsymbol{x} - t\boldsymbol{y}\| = 0$, so $\boldsymbol{x} - t\boldsymbol{y} = 0$, which means that $\{\boldsymbol{x}, \boldsymbol{y}\}$ is linearly dependent. □

Not every norm can be induced by an inner product. A characterization of this type of norms in linear spaces, obtained in [85], is presented next.

This equality shown in the next theorem is known as the *parallelogram equality*.

Theorem 6.32. *Let V be a real linear space. A norm $\|\cdot\|$ is induced by an inner product if and only if*

$$\|\boldsymbol{x} + \boldsymbol{y}\|^2 + \|\boldsymbol{x} - \boldsymbol{y}\|^2 = 2(\|\boldsymbol{x}\|^2 + \|\boldsymbol{y}\|^2)$$

for every $\boldsymbol{x}, \boldsymbol{y} \in V$.

Proof. Suppose that the norm is induced by an inner product. In this case, we can write the following for every $\boldsymbol{x}$ and $\boldsymbol{y}$:

$$(\boldsymbol{x}+\boldsymbol{y},\boldsymbol{x}+\boldsymbol{y}) = (\boldsymbol{x},\boldsymbol{x}) + 2(\boldsymbol{x},\boldsymbol{y}) + (\boldsymbol{y},\boldsymbol{y}),$$
$$(\boldsymbol{x}-\boldsymbol{y},\boldsymbol{x}-\boldsymbol{y}) = (\boldsymbol{x},\boldsymbol{x}) - 2(\boldsymbol{x},\boldsymbol{y}) + (\boldsymbol{y},\boldsymbol{y}).$$

Thus,

$$(\boldsymbol{x}+\boldsymbol{y},\boldsymbol{x}+\boldsymbol{y}) + (\boldsymbol{x}-\boldsymbol{y},\boldsymbol{x}-\boldsymbol{y}) = 2(\boldsymbol{x},\boldsymbol{x}) + 2(\boldsymbol{y},\boldsymbol{y}),$$

which can be written in terms of the norm generated as the inner product as

$$\|\boldsymbol{x}+\boldsymbol{y}\|^2 + \|\boldsymbol{x}-\boldsymbol{y}\|^2 = 2(\|\boldsymbol{x}\|^2 + \|\boldsymbol{y}\|^2).$$

Conversely, suppose that the condition of the theorem is satisfied by the norm $\|\cdot\|$. Consider the function $f : V \times V \longrightarrow \mathbb{R}$ defined by

$$f(\boldsymbol{x},\boldsymbol{y}) = \frac{1}{4}\left(\|\boldsymbol{x}+\boldsymbol{y}\|^2 - \|\boldsymbol{x}-\boldsymbol{y}\|^2\right) \tag{6.20}$$

for $\boldsymbol{x},\boldsymbol{y} \in V$. The symmetry of f is immediate, that is, $f(\boldsymbol{x},\boldsymbol{y}) = f(\boldsymbol{y},\boldsymbol{x})$ for $\boldsymbol{x},\boldsymbol{y} \in V$.

The definition of f implies

$$f(\boldsymbol{0},\boldsymbol{y}) = \frac{1}{4}\left(\|\boldsymbol{y}\|^2 - \|-\boldsymbol{y}\|^2\right) = 0. \tag{6.21}$$

We prove that f is a bilinear form that satisfies the conditions of Definition 6.15.

Starting from the parallelogram equality, we can write

$$\|\boldsymbol{u}+\boldsymbol{v}+\boldsymbol{y}\|^2 + \|\boldsymbol{u}+\boldsymbol{v}-\boldsymbol{y}\|^2 = 2(\|\boldsymbol{u}+\boldsymbol{v}\|^2 + \|\boldsymbol{y}\|^2),$$
$$\|\boldsymbol{u}-\boldsymbol{v}+\boldsymbol{y}\|^2 + \|\boldsymbol{u}-\boldsymbol{v}-\boldsymbol{y}\|^2 = 2(\|\boldsymbol{u}-\boldsymbol{v}\|^2 + \|\boldsymbol{y}\|^2).$$

Subtracting these equalities yields

$$\|\boldsymbol{u}+\boldsymbol{v}+\boldsymbol{y}\|^2 + \|\boldsymbol{u}+\boldsymbol{v}-\boldsymbol{y}\|^2 - \|\boldsymbol{u}-\boldsymbol{v}+\boldsymbol{y}\|^2 - \|\boldsymbol{u}-\boldsymbol{v}-\boldsymbol{y}\|^2$$
$$= 2(\|\boldsymbol{u}+\boldsymbol{v}\|^2 - \|\boldsymbol{u}-\boldsymbol{v}\|^2).$$

This equality can be written as

$$f(\boldsymbol{u}+\boldsymbol{y},\boldsymbol{v}) + f(\boldsymbol{u}-\boldsymbol{y},\boldsymbol{v}) = 2f(\boldsymbol{u},\boldsymbol{v}).$$

Choosing $\boldsymbol{y} = \boldsymbol{u}$ implies

$$f(2\boldsymbol{u},\boldsymbol{v}) = 2f(\boldsymbol{u},\boldsymbol{v}), \tag{6.22}$$

due to Equality (6.21).

Let $\boldsymbol{t} = \boldsymbol{u}+\boldsymbol{y}$ and $\boldsymbol{s} = \boldsymbol{u}-\boldsymbol{y}$. Since $\boldsymbol{u} = \frac{1}{2}(\boldsymbol{t}+\boldsymbol{s})$ and $\boldsymbol{y} = \frac{1}{2}(\boldsymbol{t}-\boldsymbol{s})$, we have

$$f(\boldsymbol{t}, \boldsymbol{v}) + f(\boldsymbol{s}, \boldsymbol{v}) = 2f\left(\frac{1}{2}(\boldsymbol{t}+\boldsymbol{s}), \boldsymbol{v}\right) = f(\boldsymbol{t}+\boldsymbol{s}, \boldsymbol{v}),$$

by Equality (6.22).

Next, we show that $f(a\boldsymbol{x}, \boldsymbol{y}) = af(\boldsymbol{x}, \boldsymbol{y})$ for $a \in \mathbb{R}$ and $\boldsymbol{x}, \boldsymbol{y} \in V$. Consider the function $\phi : \mathbb{R} \longrightarrow \mathbb{R}$ defined by $\phi(a) = f(a\boldsymbol{x} + \boldsymbol{y})$.

The basic properties of norms imply that

$$\Big| \|a\boldsymbol{x} + \boldsymbol{y}\| - \|b\boldsymbol{x} + \boldsymbol{y}\| \Big| \leqslant \|(a-b)\boldsymbol{x}\|$$

for every $a, b \in \mathbb{R}$ and $\boldsymbol{x}, \boldsymbol{y} \in V$. Therefore, the function $\phi, \psi : \mathbb{R} \longrightarrow \mathbb{R}$ given by $\phi(a) = \|a\boldsymbol{x} + \boldsymbol{y}\|$ and $\psi(a) = \|a\boldsymbol{x} - \boldsymbol{y}\|$ for $a \in \mathbb{R}$ are continuous. The continuity of these functions implies that the function f defined by Equality (6.20) is continuous relative to a.

Define the set

$$S = \{a \in \mathbb{R} \mid f(a\boldsymbol{x}, \boldsymbol{y}) = af(\boldsymbol{x}, \boldsymbol{y})\}.$$

Clearly, we have $1 \in S$. Further, if $a, b \in S$, then $a + b \in S$ and $a - b \in S$, which implies $\mathbb{Z} \subseteq S$.

If $b \neq 0$ and $b \in S$, then, by substituting $\boldsymbol{x}$ by $\frac{1}{b}\boldsymbol{x}$ in the equality $f(b\boldsymbol{x}, \boldsymbol{y}) = bf(\boldsymbol{x}, \boldsymbol{y})$, we have $f(\boldsymbol{x}, \boldsymbol{y}) = bf(\frac{1}{b}\boldsymbol{x}, \boldsymbol{y})$, so $\frac{1}{b}f(\boldsymbol{x}, \boldsymbol{y}) = f(\frac{1}{b}\boldsymbol{x}, \boldsymbol{y})$. Thus, if $a, b \in S$ and $b \neq 0$, we have $f(\frac{a}{b}\boldsymbol{x}, \boldsymbol{y}) = \frac{a}{b}f(\boldsymbol{x}, \boldsymbol{y})$, so $\mathbb{Q} \subseteq S$. Consequently, $S = \mathbb{R}$. This allows us to conclude that f is linear in its first argument. The symmetry of f implies the linearity in its second argument, so f is bilinear.

Observe that $f(\boldsymbol{x}, \boldsymbol{x}) = \|\boldsymbol{x}\|^2$. The definition of norms implies that $f(\boldsymbol{x}, \boldsymbol{x}) = 0$ if and only if $\boldsymbol{x} = \boldsymbol{0}$, and if $\boldsymbol{x} \neq \boldsymbol{0}$, then $f(\boldsymbol{x}, \boldsymbol{x}) > 0$. Thus, f is indeed an inner product and $\|\boldsymbol{x}\| = \sqrt{f(\boldsymbol{x}, \boldsymbol{x})}$. □

Theorem 6.33. *Let $\boldsymbol{x}, \boldsymbol{y} \in \mathbb{R}^n$ be two vectors such that $x^1 \geqslant x^2 \geqslant \cdots \geqslant x^n$, $y^1 \geqslant y^2 \geqslant \cdots \geqslant y^n$. For every permutation matrix P, we have $\boldsymbol{x}'\boldsymbol{y} \geqslant \boldsymbol{x}'(P\boldsymbol{y})$. For every permutation matrix P, we have $\boldsymbol{x}'\boldsymbol{y} \leqslant \boldsymbol{x}'(P\boldsymbol{y})$.*

Proof. Let ϕ be the permutation that corresponds to the permutation matrix P and suppose that $\phi = \psi_p \ldots \psi_1$, where $p = inv(\phi)$

and $\psi_1, \ldots, \psi_p$ are standard transpositions that correspond to all standard inversions of ϕ (see Theorems 1.9 and 1.10).

Let ψ be a standard transposition of $\{1, \ldots, n\}$,

$$\psi : \begin{pmatrix} 1 & \cdots & i & i+1 & \cdots & n \\ 1 & \cdots & i+1 & i & \cdots & n \end{pmatrix}.$$

We have

$$\boldsymbol{x}'(P\boldsymbol{y}) = x_1y_1 + \cdots + x_{i-1}y_{i-1} + x_iy_{i+1} + x_{i+1}y_i + \cdots + x_ny_n,$$

so the inequality $\boldsymbol{x}'\boldsymbol{y} \geqslant \boldsymbol{x}'(P\boldsymbol{y})$ is equivalent to

$$x_iy_i + x_{i+1}y_{i+1} \geqslant x_iy_{i+1} + x_{i+1}y_i.$$

This, in turn is equivalent to $(x_{i+1} - x_i)(y_{i+1} - y_i) \geqslant 0$, which obviously holds in view of the hypothesis.

As we observed previously, $P_\phi = P_{\psi_1} \cdots P_{\psi_p}$, so

$$\boldsymbol{x}'\boldsymbol{y} \geqslant \boldsymbol{x}'(P_{\psi_p}\boldsymbol{y}) \geqslant \boldsymbol{x}'(P_{\psi_{p-1}}P_{\psi_p}\boldsymbol{y}) \geqslant \cdots \geqslant \boldsymbol{x}'(P_{\psi_1} \cdots P_{\psi_p}\boldsymbol{y}) = \boldsymbol{x}(P\boldsymbol{y}),$$

which concludes the proof of the first part of the theorem.

To prove the second part of the theorem, apply the first part to the vectors $\boldsymbol{x}$ and $-\boldsymbol{y}$. □

Corollary 6.13. *Let $\boldsymbol{x}, \boldsymbol{y} \in \mathbb{R}^n$ be two vectors such that $x_1 \geqslant x_1 \geqslant \cdots \geqslant x_n$, $y_1 \geqslant y_2 \geqslant \cdots \geqslant y_n$. For every permutation matrix P, we have*

$$\|\boldsymbol{x} - \boldsymbol{y}\|_F \leqslant \|\boldsymbol{x} - P\boldsymbol{y}\|_F.$$

If $x_1 \geqslant x_1 \geqslant \cdots \geqslant x_n$ and $y_1 \leqslant y_2 \leqslant \cdots \leqslant y_n$, then for every permutation matrix P, we have

$$\|\boldsymbol{x} - \boldsymbol{y}\|_F \geqslant \|\boldsymbol{x} - P\boldsymbol{y}\|_F.$$

Proof. Note that

$$\begin{aligned} \|\boldsymbol{x} - \boldsymbol{y}\|_F^2 &= \|\boldsymbol{x}\|_F^2 + \|\boldsymbol{y}\|_F^2 - 2\boldsymbol{x}'\boldsymbol{y}, \\ \|\boldsymbol{x} - P\boldsymbol{y}\|_F^2 &= \|\boldsymbol{x}\|_F^2 + \|P\boldsymbol{y}\|_F^2 - 2\boldsymbol{x}'(P\boldsymbol{y}) \\ &= \|\boldsymbol{x}\|_F^2 + \|\boldsymbol{y}\|_F^2 - 2\boldsymbol{x}'(P\boldsymbol{y}) \end{aligned}$$

because $\|P\boldsymbol{y}\|_F^2 = \|\boldsymbol{y}\|_F^2$. Then, by Theorem 6.33, $\|\boldsymbol{x} - \boldsymbol{y}\|_F \leqslant \|\boldsymbol{x} - P\boldsymbol{y}\|_F$.

The argument for the second part of the corollary is similar. □

Corollary 6.14. *Let V be an n-dimensional linear space. If W is an orthogonal (orthonormal) set and $|W| = n$, then W is an orthogonal (orthonormal) basis of L.*

Proof. This statement is an immediate consequence of Theorem 6.30. □

For every $A \in \mathbb{C}^{n\times n}$ and $\boldsymbol{x}, \boldsymbol{y} \in \mathbb{C}^n$, we have

$$(A\boldsymbol{x}, \boldsymbol{y}) = (\boldsymbol{x}, A^{\mathsf{H}}\boldsymbol{y}), \tag{6.23}$$

which follows from the equalities

$$\begin{aligned}
(A\boldsymbol{x}, \boldsymbol{y}) &= \sum_{i=1}^{n} (A\boldsymbol{x})_i \bar{y}_i \\
&= \sum_{i=1}^{n} \sum_{j=1}^{n} a_{ij} x_j \bar{y}_i \\
&= \sum_{j=1}^{n} x_j \sum_{i=1}^{n} a_{ij} \bar{y}_i \\
&= \sum_{j=1}^{n} x_j \sum_{i=1}^{n} \overline{\bar{a}_{ij} y_i} \\
&= (\boldsymbol{x}, A^{\mathsf{H}}\boldsymbol{y}).
\end{aligned}$$

More generally, we have the following definition:

Definition 6.18. A matrix $B \in \mathbb{C}^{n\times n}$ is the *adjoint* of a matrix $A \in \mathbb{C}^{n\times n}$ relative to the inner product $(\cdot,\cdot)$ if $(A\boldsymbol{x}, \boldsymbol{y}) = (\boldsymbol{x}, B\boldsymbol{y})$ for every $\boldsymbol{x}, \boldsymbol{y} \in \mathbb{C}^n$.

A matrix is *self-adjoint* if it equals its own adjoint, that is if $(A\boldsymbol{x}, \boldsymbol{y}) = (\boldsymbol{x}, A\boldsymbol{y})$ for every $\boldsymbol{x}, \boldsymbol{y} \in \mathbb{C}^n$. Thus, a Hermitian matrix is self-adjoint relative to the inner product $(\boldsymbol{x}, \boldsymbol{y}) = \boldsymbol{x}^{\mathsf{H}}\boldsymbol{y}$ for $\boldsymbol{x}, \boldsymbol{y} \in \mathbb{C}^n$. If we use the Euclidean inner product, we omit the reference to this product and refer to the adjoint of A relative to this product simply as the adjoint of A.

Example 6.21. An inner product on $\mathbb{C}^{n\times n}$, the linear space of matrices of format $n \times n$, can be defined as $(X, Y) = \mathsf{trace}(XY^{\mathsf{H}})$ for $X, Y \in \mathbb{C}^{n\times n}$. Note that $\|X\|_F^2 = (X, X^{\mathsf{H}})$ for every $X \in \mathbb{C}^{n\times n}$.

By Theorem 2.38, a linear form f defined on an $\mathbb{R}^n$ can be uniquely written as

$$f(\boldsymbol{v}) = \sum_{i=1}^{n} c_i a_i,$$

where $\{\boldsymbol{e}_1, \ldots, \boldsymbol{e}_n\}$ is a basis of $\mathbb{R}^n$, $\boldsymbol{v} = \sum\{c_i \boldsymbol{e}_i \mid 1 \leqslant i \leqslant n\}$, and $a_i = f(\boldsymbol{e}_i)$ for $1 \leqslant i \leqslant n$. Thus, using the Euclidean inner product, we can write $f(\boldsymbol{v}) = (\boldsymbol{v}, \boldsymbol{a})$ for $\boldsymbol{v} \in \mathbb{R}^n$, where

$$\boldsymbol{a} = \begin{pmatrix} a_1 \\ \vdots \\ a_n \end{pmatrix}.$$

The Cauchy–Schwarz inequality implies that

$$|(\boldsymbol{x}, \boldsymbol{y})| \leqslant \|\boldsymbol{x}\|_2 \|\boldsymbol{y}\|_2.$$

Equivalently, this means that

$$-1 \leqslant \frac{(\boldsymbol{x}, \boldsymbol{y})}{\|\boldsymbol{x}\|_2 \|\boldsymbol{y}\|_2} \leqslant 1.$$

This double inequality allows us to introduce the notion of angle between two vectors $\boldsymbol{x}, \boldsymbol{y}$ of a real linear space V.

Definition 6.19. The *angle* between the vectors $\boldsymbol{x}$ and $\boldsymbol{y}$ is the number $\alpha \in [0, \pi]$ defined by

$$\cos \alpha = \frac{(\boldsymbol{x}, \boldsymbol{y})}{\|\boldsymbol{x}\|_2 \|\boldsymbol{y}\|_2}.$$

This angle will be denoted by $\angle(\boldsymbol{x}, \boldsymbol{y})$.

Example 6.22. Let $\boldsymbol{u} = (u_1, u_2) \in \mathbb{R}^2$ be a unit vector. Since $u_1^2 + u_2^2 = 1$, there exists $\alpha \in [0, 2\pi]$ such that $u_1 = \cos \alpha$ and $u_2 = \sin \alpha$. Thus, for any two unit vectors in $\mathbb{R}^2$, $\boldsymbol{u} = (\cos \alpha, \sin \alpha)$ and $\boldsymbol{v} = (\cos \beta, \sin \beta)$ we have $(\boldsymbol{u}, \boldsymbol{v}) = \cos \alpha \cos \beta + \sin \alpha \sin \beta = \cos(\alpha - \beta)$, where $\alpha, \beta \in [0, 2\pi]$. Consequently, $\angle(\boldsymbol{u}, \boldsymbol{v})$ is the angle in the interval $[0, \pi]$ that has the same cosine as $\alpha - \beta$.

Theorem 6.34 (The Cosine theorem). Let $\boldsymbol{x}$ and $\boldsymbol{y}$ be two vectors in $\mathbb{R}^n$ equipped with the Euclidean inner product. We have

$$\|\boldsymbol{x} - \boldsymbol{y}\|^2 = \|\boldsymbol{x}\|^2 + \|\boldsymbol{y}\|^2 - 2\|\boldsymbol{x}\|\|\boldsymbol{y}\| \cos\alpha,$$

where $\alpha = \angle(\boldsymbol{x}, \boldsymbol{y})$.

Proof. Since the norm is induced by the inner product, we have

$$\begin{aligned}\|\boldsymbol{x} - \boldsymbol{y}\|^2 &= (\boldsymbol{x} - \boldsymbol{y}, \boldsymbol{x} - \boldsymbol{y}) \\ &= (\boldsymbol{x}, \boldsymbol{x}) - 2(\boldsymbol{x}, \boldsymbol{y}) + (\boldsymbol{y}, \boldsymbol{y}) \\ &= \|\boldsymbol{x}\|^2 - 2\|\boldsymbol{x}\|\|\boldsymbol{y}\| \cos\alpha + \|\boldsymbol{y}\|^2,\end{aligned}$$

which is the desired equality. □

If $T \subseteq V$, then the set $T^\perp$ is defined by

$$T^\perp = \{\boldsymbol{v} \in V \mid \boldsymbol{v} \perp \boldsymbol{t} \text{ for every } \boldsymbol{t} \in T\}.$$

Note that $T \subseteq U$ implies $U^\perp \subseteq T^\perp$.

If S, T are two subspaces of an inner product space, then S and T are *orthogonal* if $\boldsymbol{s} \perp \boldsymbol{t}$ for every $\boldsymbol{s} \in S$ and every $\boldsymbol{t} \in T$. This is denoted as $S \perp T$.

Theorem 6.35. *Let V be an inner product space and let $T \subseteq V$. The set $T^\perp$ is a subspace of V. Furthermore, $\langle T \rangle^\perp = T^\perp$.*

Proof. Let $\boldsymbol{x}$ and $\boldsymbol{y}$ be two members of $T^\perp$. We have $(\boldsymbol{x}, \boldsymbol{t}) = (\boldsymbol{y}, \boldsymbol{t}) = 0$ for every $\boldsymbol{t} \in T$. Therefore, for every $a, b \in F$, by the linearity of the inner product, we have

$$(a\boldsymbol{x} + b\boldsymbol{y}, \boldsymbol{t}) = a(\boldsymbol{x}, \boldsymbol{t}) + b(\boldsymbol{y}, \boldsymbol{t}) = 0$$

for $\boldsymbol{t} \in T$, so $a\boldsymbol{x} + b\boldsymbol{t} \in T^\perp$. Thus, $T^\perp$ is a subspace of V.

By a previous observation, since $T \subseteq \langle T \rangle$, we have $\langle T \rangle^\perp \subseteq T^\perp$. To prove the converse inclusion, let $\boldsymbol{z} \in T^\perp$.

If $\boldsymbol{y} \in \langle T \rangle$, $\boldsymbol{y}$ is a linear combination of vectors of T, $\boldsymbol{y} = a_1\boldsymbol{t}_1 + \cdots + a_m\boldsymbol{t}_m$, so $(\boldsymbol{y}, \boldsymbol{z}) = a_1(\boldsymbol{t}_1, \boldsymbol{z}) + \cdots + a_m(\boldsymbol{t}_m, \boldsymbol{z}) = 0$. Therefore, $\boldsymbol{z} \perp \boldsymbol{y}$, which implies $\boldsymbol{z} \in \langle T \rangle^\perp$. This allows us to conclude that $\langle T \rangle^\perp = T^\perp$. □

We refer to $T^\perp$ as the *orthogonal complement* of T.

Note that $T \cap T^\perp \subseteq \{\boldsymbol{0}\}$. If T is a subspace, then this inclusion becomes an equality, that is, $T \cap T^\perp = \{\boldsymbol{0}\}$.

If $\boldsymbol{x}$ and $\boldsymbol{y}$ are orthogonal, by Theorem 6.34, we have

$$\|\boldsymbol{x} - \boldsymbol{y}\|^2 = \|\boldsymbol{x}\|^2 + \|\boldsymbol{y}\|^2,$$

which is the well-known *Pythagorean Theorem.*

Theorem 6.36. *Let T be a subspace of $\mathbb{C}^n$. We have $(T^\perp)^\perp = T$.*

Proof. Observe that $T \subseteq (T^\perp)^\perp$. Indeed, if $\boldsymbol{t} \in T$, then $(\boldsymbol{t}, \boldsymbol{z}) = 0$ for every $\boldsymbol{z} \in T^\perp$, so $\boldsymbol{t} \in (T^\perp)^\perp$.

To prove the reverse inclusion, let $\boldsymbol{x} \in (T^\perp)^\perp$. Theorem 6.37 implies that we can write $\boldsymbol{x} = \boldsymbol{u} + \boldsymbol{v}$, where $\boldsymbol{u} \in T$ and $\boldsymbol{v} \in T^\perp$, so $\boldsymbol{x} - \boldsymbol{u} = \boldsymbol{v} \in T^\perp$.

Since $T \subseteq (T^\perp)^\perp$, we have $\boldsymbol{u} \in (T^\perp)^\perp$, so $\boldsymbol{x} - \boldsymbol{u} \in (T^\perp)^\perp$. Consequently, $\boldsymbol{x} - \boldsymbol{u} \in T^\perp \cap (T^\perp)^\perp = \{\boldsymbol{0}\}$, so $\boldsymbol{x} = \boldsymbol{u} \in T$. Thus, $(T^\perp)^\perp \subseteq T$, which concludes the argument. □

Corollary 6.15. *Let Z be a subset of $\mathbb{C}^n$. We have $(Z^\perp)^\perp = \langle Z \rangle$.*

Proof. Let Z be a subset of $\mathbb{C}^n$. Since $Z \subseteq \langle Z \rangle$, it follows that $\langle Z \rangle^\perp \subseteq Z^\perp$. Let now $\boldsymbol{y} \in Z^\perp$ and let $\boldsymbol{z} = a_1\boldsymbol{z}_1 + \cdots + a_p\boldsymbol{z}_p \in \langle Z \rangle$, where $\boldsymbol{z}_1, \ldots, \boldsymbol{z}_p \in Z$. Since

$$(\boldsymbol{y}, \boldsymbol{z}) = a_1(\boldsymbol{y}, \boldsymbol{z}_1) + \cdots + a_p(\boldsymbol{y}, \boldsymbol{z}_p) = 0,$$

it follows that $\boldsymbol{y} \in \langle Z \rangle^\perp$. Thus, we have $Z^\perp = \langle Z \rangle^\perp$.

This allows us to write $(Z^\perp)^\perp = (\langle Z \rangle^\perp)^\perp$. Since $\langle Z \rangle$ is a subspace of $\mathbb{C}^n$, by Theorem 6.36, we have $(\langle Z \rangle^\perp)^\perp = \langle Z \rangle$, so $(Z^\perp)^\perp = \langle Z \rangle$□

Theorem 6.37. *Let U be a subspace of $\mathbb{C}^n$. Then, $\mathbb{C}^n = U \oplus U^\perp$.*

Proof. If $U = \{\boldsymbol{0}\}$, then $U^\perp = \mathbb{C}^n$ and the statement is immediate. Therefore, we can assume that $U \neq \{\boldsymbol{0}\}$.

In Theorem 6.35 we saw that $U^\perp$ is a subspace of $\mathbb{C}^n$. Thus, we need to show that $\mathbb{C}^n$ is the direct sum of the subspaces U and $U^\perp$. By Corollary 2.10, we need to verify only that every $\boldsymbol{x} \in \mathbb{C}^n$ can be uniquely written as a sum $\boldsymbol{x} = \boldsymbol{u} + \boldsymbol{v}$, where $\boldsymbol{u} \in U$ and $\boldsymbol{v} \in U^\perp$.

Let $\boldsymbol{u}_1, \ldots, \boldsymbol{u}_m$ be an orthonormal basis of U, that is, a basis such that

$$(\boldsymbol{u}_i, \boldsymbol{u}_j) = \begin{cases} 1 & \text{if } i = j, \\ 0 & \text{otherwise} \end{cases}$$

for $1 \leqslant i, j \geqslant m$. Define the vector $\boldsymbol{u}$ as

$$\boldsymbol{u} = (\boldsymbol{x}, \boldsymbol{u}_1)\boldsymbol{u}_1 + \cdots + (\boldsymbol{x}, \boldsymbol{u}_m)\boldsymbol{u}_m$$

and the vector $\boldsymbol{v}$ as $\boldsymbol{v} = \boldsymbol{x} - \boldsymbol{u}$.

The vector $\boldsymbol{v}$ is orthogonal to every vector $\boldsymbol{u}_i$ because

$$(\boldsymbol{v}, \boldsymbol{u}_i) = (\boldsymbol{x} - \boldsymbol{u}, \boldsymbol{u}_i) = (\boldsymbol{x}, \boldsymbol{u}_i) - (\boldsymbol{u}, \boldsymbol{u}_i) = 0.$$

Therefore $\boldsymbol{v} \in U^{\perp}$ and $\boldsymbol{x}$ has the necessary decomposition. To prove that the decomposition is unique, suppose that $\boldsymbol{x} = \boldsymbol{s} + \boldsymbol{t}$, where $\boldsymbol{s} \in U$ and $\boldsymbol{t} \in U_{\perp}$. Since $\boldsymbol{s} + \boldsymbol{t} = \boldsymbol{u} + \boldsymbol{v}$, we have $\boldsymbol{s} - \boldsymbol{u} = \boldsymbol{v} - \boldsymbol{t} \in U \cap U^{\perp} = \{\boldsymbol{0}\}$, which implies $\boldsymbol{s} = \boldsymbol{u}$ and $\boldsymbol{t} = \boldsymbol{v}$. □

Theorem 6.38. *Let S be a subspace of $\mathbb{C}^n$ such that* $\dim(S) = k$. *There exists a matrix $A \in \mathbb{C}^{n\times k}$ having orthonormal columns such that $S = \mathsf{range}(A)$.*

Proof. Let $\boldsymbol{v}_1, \ldots, \boldsymbol{v}_k$ be an orthonormal basis of S. Define the matrix A as $A = (\boldsymbol{v}_1, \ldots, \boldsymbol{v}_k)$. We have $\boldsymbol{x} \in S$, if and only if $\boldsymbol{x} = a_1\boldsymbol{v}_1 + \cdots + a_k\boldsymbol{v}_k$, which is equivalent to $\boldsymbol{x} = A\boldsymbol{a}$, where $\boldsymbol{a} = \begin{pmatrix} a_1 \\ \vdots \\ a_k \end{pmatrix}$. This amounts to $\boldsymbol{x} \in \mathsf{range}(A)$. Thus, we have $S = \mathsf{range}(A)$. □

Clearly, for an orthonormal basis in an n-dimensional space, the associated matrix is the diagonal matrix I_n. In this case, we have

$$(\boldsymbol{x}, \boldsymbol{y}) = x_1y_1 + x_2y_2 + \cdots + x_ny_n$$

for $\boldsymbol{x}, \boldsymbol{y} \in V$.

Observe that if $W = \{\boldsymbol{w}_1, \ldots, \boldsymbol{w}_n\}$ is an orthonormal set and $\boldsymbol{x} \in \langle W \rangle$, which means that $\boldsymbol{x} = a_1\boldsymbol{w}_1 + \cdots + a_n\boldsymbol{w}_n$, then $a_i = (\boldsymbol{x}, \boldsymbol{w}_i)$ for $1 \leqslant i \leqslant n$.

Definition 6.20. Let $W = \{\boldsymbol{w}_1, \ldots, \boldsymbol{w}_n\}$ be an orthonormal set and let $\boldsymbol{x} \in \langle W \rangle$.

The equality

$$\boldsymbol{x} = (\boldsymbol{x}, \boldsymbol{w}_1)\boldsymbol{w}_1 + \cdots + (\boldsymbol{x}, \boldsymbol{w}_n)\boldsymbol{w}_n \tag{6.24}$$

is the *Fourier expansion of* $\boldsymbol{x}$ with respect to the orthonormal set W.

Furthermore, we have *Parseval's equality*:

$$\|\boldsymbol{x}\|^2 = (\boldsymbol{x}, \boldsymbol{x}) = \sum_{i=1}^{n} (\boldsymbol{x}, \boldsymbol{w}_i)^2. \tag{6.25}$$

Thus, if $1 \leqslant q \leqslant n$, we have

$$\sum_{i=1}^{q} (\boldsymbol{x}, \boldsymbol{w}_i)^2 \geqslant \|\boldsymbol{x}\|^2. \tag{6.26}$$

It is easy to see that a square matrix $C \in \mathbb{C}^{n\times n}$ is unitary if and only if its set of columns is an orthonormal set in $\mathbb{C}^n$.

Definition 6.21. Let V be an inner product $\mathbb{C}$-linear n-dimensional space and let $B = \{\boldsymbol{e}_1, \ldots, \boldsymbol{e}_n\}$ be a basis of V.

A set of vectors $\tilde{B} = \{\tilde{\mathbf{e}}_1, \ldots, \tilde{\mathbf{e}}_n\}$ is a *reciprocal set* of B if $(\boldsymbol{e}_i, \tilde{\boldsymbol{e}}_j) = \delta_{ij}$ for $1 \leqslant i, j \leqslant n$.

Theorem 6.39. *Let V be an inner product $\mathbb{C}$-linear n-dimensional space. The reciprocal set $\tilde{B}$ of a basis $B = \{\boldsymbol{e}_1, \ldots, \boldsymbol{e}_n\}$ exists and is unique.*

Proof. Let $U = \langle \boldsymbol{e}_2, \ldots, \boldsymbol{e}_n\rangle$ and let $U^\perp$ be the orthogonal complement of U. Then, $\dim(U^\perp) = 1$ and there exists a non-zero vector $\boldsymbol{w} \in U^\perp$. Note that $(\boldsymbol{e}_1, \boldsymbol{w}) \neq 0$, which allows us to define $\tilde{\mathbf{e}}_1 = \frac{1}{(\boldsymbol{e}_1, \boldsymbol{w})}\boldsymbol{w}$. Thus, $(\tilde{\mathbf{e}}_1, \boldsymbol{e}_1) = 1$ and $(\tilde{\mathbf{e}}_1, \boldsymbol{e}_j) = 0$ for $j \neq 1$. In a similar manner, we can prove the existence of the remaining vectors $\tilde{\mathbf{e}}_2, \ldots, \tilde{\mathbf{e}}_n$ of $\tilde{B}$.

To prove the uniqueness of $\tilde{B}$, suppose that there exist two reciprocal sets of B, $\tilde{B} = \{\tilde{\mathbf{e}}_1, \ldots, \tilde{\mathbf{e}}_n\}$ and $\tilde{D} = \{\tilde{\boldsymbol{d}}_1, \ldots, \tilde{\boldsymbol{d}}_n\}$. Then, $(\tilde{\mathbf{e}}_k - \tilde{\boldsymbol{d}}_k, \boldsymbol{e}_k) = 0$ for $1 \leqslant k \leqslant n$, which means that $\tilde{\mathbf{e}}_k - \tilde{\boldsymbol{d}}_k$ is orthogonal to B. This is equivalent to $(\tilde{\mathbf{e}}_k - \tilde{\boldsymbol{d}}_k, \boldsymbol{v}) = 0$ for all $\boldsymbol{v} \in V$, which implies $\tilde{\mathbf{e}}_k = \tilde{\boldsymbol{d}}_k$. This proves that the reciprocal set of B is unique. □

Theorem 6.40. *The reciprocal set of a basis in an inner product space V is a basis of V.*

Proof. Let $\tilde{B} = \{\tilde{\mathbf{e}}_1, \ldots, \tilde{\mathbf{e}}_n\}$ be the reciprocal set of a basis $B = \{\boldsymbol{e}_1, \ldots, \boldsymbol{e}_n\}$ in V. Suppose that $a_1\tilde{\mathbf{e}}_1 + \cdots + a_n\tilde{\mathbf{e}}_n = \mathbf{0}_V$. Then,

$$(a_1\tilde{\mathbf{e}}_1 + \cdots + a_n\tilde{\mathbf{e}}_n, \boldsymbol{e}_k) = a_k = 0$$

for $1 \leqslant k \leqslant n$, which means that the reciprocal set is linearly independent. Since $|\tilde{B}| = n$, it follows that $\tilde{B}$ is a basis in V. □

6.10 Hyperplanes in $\mathbb{R}^n$

Definition 6.22. Let $\boldsymbol{w} \in \mathbb{R}^n - \{\boldsymbol{0}\}$ and let $a \in \mathbb{R}$. The *hyperplane* determined by $\boldsymbol{w}$ and a is the set

$$H_{\boldsymbol{w},a} = \{\boldsymbol{x} \in \mathbb{R}^n \mid \boldsymbol{w}'\boldsymbol{x} = a\}.$$

If $\boldsymbol{x}_0 \in H_{\boldsymbol{w},a}$, then $\boldsymbol{w}'\boldsymbol{x}_0 = a$, so $H_{\boldsymbol{w},a}$ is also described by the equality

$$H_{\boldsymbol{w},a} = \{\boldsymbol{x} \in \mathbb{R}^n \mid \boldsymbol{w}'(\boldsymbol{x} - \boldsymbol{x}_0) = 0\}.$$

Any hyperplane $H_{\boldsymbol{w},a}$ partitions $\mathbb{R}^n$ into three sets:

$$\begin{aligned} H^{>}_{\boldsymbol{w},a} &= \{\boldsymbol{x} \in \mathbb{R}^n \mid \boldsymbol{w}'\boldsymbol{x} > a\}, \\ H^{0}_{\boldsymbol{w},a} &= H_{\boldsymbol{w},a}, \\ H^{<}_{\boldsymbol{w},a} &= \{\boldsymbol{x} \in \mathbb{R}^n \mid \boldsymbol{w}'\boldsymbol{x} < a\}. \end{aligned}$$

The sets $H^{>}_{\boldsymbol{w},a}$ and $H^{<}_{\boldsymbol{w},a}$ are the *positive* and *negative open* half-spaces determined by $H_{\boldsymbol{w},a}$, respectively. The sets

$$\begin{aligned} H^{\geqslant}_{\boldsymbol{w},a} &= \{\boldsymbol{x} \in \mathbb{R}^n \mid \boldsymbol{w}'\boldsymbol{x} \geqslant a\}, \\ H^{\leqslant}_{\boldsymbol{w},a} &= \{\boldsymbol{x} \in \mathbb{R}^n \mid \boldsymbol{w}'\boldsymbol{x} \leqslant a\} \end{aligned}$$

are the *positive* and *negative closed* half-spaces determined by $H_{\boldsymbol{w},a}$, respectively.

If $\boldsymbol{x}_1, \boldsymbol{x}_2 \in H_{\boldsymbol{w},a}$, we say that the vector $\boldsymbol{x}_1 - \boldsymbol{x}_2$ is located in the hyperplane $H_{\boldsymbol{w},a}$. In this case, $\boldsymbol{w} \perp \boldsymbol{x}_1 - \boldsymbol{x}_2$. This justifies referring to $\boldsymbol{w}$ as the *normal to the hyperplane* $H_{\boldsymbol{w},a}$. Observe that a hyperplane is fully determined by a vector $\boldsymbol{x}_0 \in H_{\boldsymbol{w},a}$ and by $\boldsymbol{w}$.

Let $\boldsymbol{x}_0 \in \mathbb{R}^n$ and let $H_{\boldsymbol{w},a}$ be a hyperplane. We seek $\boldsymbol{x} \in H_{\boldsymbol{w},a}$ such that $\|\boldsymbol{x} - \boldsymbol{x}_0\|_2$ is minimal. Finding $\boldsymbol{x}$ amounts to minimizing the function $f(\boldsymbol{x}) = \|\boldsymbol{x} - \boldsymbol{x}_0\|_2^2 = \sum_{i=1}^n (x_i - x_{0i})^2$ subjected to the constraint $\boldsymbol{w}_1\boldsymbol{x}_1 + \cdots + w_n\boldsymbol{x}_n - a = 0$. Using the Lagrangian $\Lambda(\boldsymbol{x}) = f(\boldsymbol{x}) + \lambda(\boldsymbol{w}'\boldsymbol{x} - a)$ and the multiplier λ, we impose the conditions

$$\frac{\partial \Lambda}{\partial x_i} = 0 \text{ for } 1 \leqslant i \leqslant n,$$

which amount to

$$\frac{\partial f}{\partial x_i} + \lambda w_i = 0$$

for $1 \leqslant i \leqslant n$. These equalities yield $2(x_i - x_{0i}) + \lambda \boldsymbol{w}_i = 0$, so we have $x_i = x_{0i} - \frac{1}{2}\lambda \boldsymbol{w}_i$. Consequently, we have $\boldsymbol{x} = \boldsymbol{x}_0 - \frac{1}{2}\lambda \boldsymbol{w}$.

Since $\boldsymbol{x} \in H_{\boldsymbol{w},a}$, this implies

$$\boldsymbol{w}'\boldsymbol{x} = \boldsymbol{w}'\boldsymbol{x}_0 - \frac{1}{2}\lambda\boldsymbol{w}'\boldsymbol{w} = a.$$

Thus,

$$\lambda = 2\frac{\boldsymbol{w}'\boldsymbol{x}_0 - a}{\boldsymbol{w}'\boldsymbol{w}} = 2\frac{\boldsymbol{w}'\boldsymbol{x}_0 - a}{\|\boldsymbol{w}\|_2^2}.$$

We conclude that the closest point in $H_{\boldsymbol{w},a}$ to $\boldsymbol{x}_0$ is

$$\boldsymbol{x} = \boldsymbol{x}_0 - \frac{\boldsymbol{w}'\boldsymbol{x}_0 - a}{\|\boldsymbol{w}\|_2^2}\boldsymbol{w}.$$

The smallest distance between $\boldsymbol{x}_0$ and a point in the hyperplane $H_{\boldsymbol{w},a}$ is given by

$$\|\boldsymbol{x}_0 - \boldsymbol{x}\| = \frac{\boldsymbol{w}'\boldsymbol{x}_0 - a}{\|\boldsymbol{w}\|_2}.$$

If we define the distance $d(H_{\boldsymbol{w},a}, \boldsymbol{x}_0)$ between $\boldsymbol{x}_0$ and $H_{\boldsymbol{w},a}$ as this smallest distance, we have

$$d(H_{\boldsymbol{w},a}, \boldsymbol{x}_0) = \frac{\boldsymbol{w}'\boldsymbol{x}_0 - a}{\|\boldsymbol{w}\|_2}. \tag{6.27}$$

6.11 Unitary and Orthogonal Matrices

Lemma 6.8. *Let $A \in \mathbb{C}^{n\times n}$. If $\boldsymbol{x}^{\mathsf{H}}A\boldsymbol{x} = 0$ for every $\boldsymbol{x} \in \mathbb{C}^n$, then $A = O_{n,n}$.*

Proof. If $\boldsymbol{x} = \boldsymbol{e}_k$, then $\boldsymbol{x}^{\mathsf{H}}A\boldsymbol{x} = a_{kk}$ for every k, $1 \leqslant k \leqslant n$, so all diagonal entries of A equal 0. Choose now $\boldsymbol{x} = \boldsymbol{e}_k + \boldsymbol{e}_j$. Then

$$\begin{aligned}
&(\boldsymbol{e}_k + \boldsymbol{e}_j)^{\mathsf{H}}A(\boldsymbol{e}_k + \boldsymbol{e}_j)\\
&= \boldsymbol{e}_k^{\mathsf{H}}A\boldsymbol{e}_k + \boldsymbol{e}_k^{\mathsf{H}}A\boldsymbol{e}_j + \boldsymbol{e}_j^{\mathsf{H}}A\boldsymbol{e}_k + \boldsymbol{e}_j^{\mathsf{H}}A\boldsymbol{e}_j\\
&= \boldsymbol{e}_k^{\mathsf{H}}A\boldsymbol{e}_j + \boldsymbol{e}_j^{\mathsf{H}}A\boldsymbol{e}_k\\
&= a_{kj} + a_{jk} = 0.
\end{aligned}$$

Similarly, if we choose $\boldsymbol{x} = \boldsymbol{e}_k + i\boldsymbol{e}_j$, we obtain

$$\begin{aligned}
&(\boldsymbol{e}_k + i\boldsymbol{e}_j)^{\mathsf{H}} A(\boldsymbol{e}_k + i\boldsymbol{e}_j) \\
&= (\boldsymbol{e}_k^{\mathsf{H}} - i\boldsymbol{e}_j^{\mathsf{H}}) A(\boldsymbol{e}_k + i\boldsymbol{e}_j) \\
&= \boldsymbol{e}_k^{\mathsf{H}} A\boldsymbol{e}_k - i\boldsymbol{e}_j^{\mathsf{H}} A\boldsymbol{e}_k + i\boldsymbol{e}_k^{\mathsf{H}} A\boldsymbol{e}_j + \boldsymbol{e}_j^{\mathsf{H}} A\boldsymbol{e}_j \\
&= -ia_{jk} + ia_{kj} = 0.
\end{aligned}$$

The equalities $a_{kj} + a_{jk} = 0$ and $-a_{jk} + a_{kj} = 0$ imply $a_{kj} = a_{jk} = 0$. Thus, all off-diagonal elements of A are also 0, hence $A = O_{n,n}$. □

Theorem 6.41. *A matrix $U \in \mathbb{C}^{n\times n}$ is unitary if $\|U\boldsymbol{x}\|_2 = \|\boldsymbol{x}\|_2$ for every $\boldsymbol{x} \in \mathbb{C}^n$.*

Proof. If U is unitary, we have

$$\|U\boldsymbol{x}\|_2^2 = (U\boldsymbol{x})^{\mathsf{H}} U\boldsymbol{x} = \boldsymbol{x}^{\mathsf{H}} U^{\mathsf{H}} U\boldsymbol{x} = \|\boldsymbol{x}\|_2^2$$

because $U^{\mathsf{H}} U = I_n$. Thus, $\|U\boldsymbol{x}\|_2 = \|\boldsymbol{x}\|_2$.

Conversely, let U be a matrix such that $\|U\boldsymbol{x}\|_2 = \|\boldsymbol{x}\|_2$ for every $\boldsymbol{x} \in \mathbb{C}^n$. This implies $\boldsymbol{x}^{\mathsf{H}} U^{\mathsf{H}} U\boldsymbol{x} = \boldsymbol{x}^{\mathsf{H}}\boldsymbol{x}$, hence $\boldsymbol{x}^{\mathsf{H}}(U^{\mathsf{H}} U - I_n)\boldsymbol{x} = 0$ for $\boldsymbol{x} \in \mathbb{C}^n$. By Lemma 6.8, this implies $U^{\mathsf{H}} U = I_n$, so U is a unitary matrix. □

Corollary 6.16. *The following statements that concern a matrix $U \in \mathbb{C}^{n\times n}$ are equivalent:*

(i) *U is unitary;*
(ii) *$\|U\boldsymbol{x} - U\boldsymbol{y}\|_2 = \|\boldsymbol{x} - \boldsymbol{y}\|_2$ for $\boldsymbol{x}, \boldsymbol{y} \in \mathbb{C}^n$;*
(iii) *$(U\boldsymbol{x}, U\boldsymbol{y}) = (\boldsymbol{x}, \boldsymbol{y})$ for $\boldsymbol{x}, \boldsymbol{y} \in \mathbb{C}^n$.*

Proof. This statement is a direct consequence of Theorem 6.41. □

The counterpart of unitary matrices in the set of real matrices are introduced next.

Definition 6.23. A matrix $A \in \mathbb{R}^{n\times n}$ is *orthogonal* or *orthonormal* if it is unitary.

In other words, a real matrix $A \in \mathbb{R}^{n\times n}$ is orthogonal if and only if $A'A = AA' = I_n$. Clearly, A is orthogonal if and only if A' is orthogonal.

Theorem 6.42. *If* $A \in \mathbb{R}^{n\times n}$ *is an orthogonal matrix, then* $\det(A) \in \{-1, 1\}$.

Proof. By Corollary 5.3, $|\det(A)| = 1$. Since $\det(A)$ is a real number, it follows that $\det(A) \in \{-1, 1\}$. □

Corollary 6.17. *Let A be a matrix in $\mathbb{R}^{n\times n}$. The following statements are equivalent:*

(i) *A is orthogonal;*
(ii) *A is invertible and $A^{-1} = A'$;*
(iii) *A' is invertible and $(A')^{-1} = A$;*
(iv) *A' is orthogonal.*

Proof. The equivalence between these statements is an immediate consequence of definitions. □

Corollary 6.17 implies that the columns of a square matrix form an orthonormal set of vectors if and only if the set of rows of the matrix is an orthonormal set.

Theorem 6.41 specialized to orthogonal matrices shows that a matrix A is orthogonal if and only if it preserves the length of vectors.

Theorem 6.43. *Let S be an r-dimensional subspace of $\mathbb{R}^n$ and let $\{\boldsymbol{u}_1, \ldots, \boldsymbol{u}_r\}$, $\{\boldsymbol{v}_1, \ldots, \boldsymbol{v}_r\}$ be two orthonormal bases of the space S. The orthogonal matrices $B = (\boldsymbol{u}_1 \ \cdots \ \boldsymbol{u}_r) \in \mathbb{R}^{n\times r}$ and $C = (\boldsymbol{v}_1 \ \cdots \ \boldsymbol{v}_r) \in \mathbb{R}^{n\times r}$ of any two such bases are related by the equality $B = CT$, where $T = C'B \in \mathbb{C}^{r\times r}$ is an orthogonal matrix.*

Proof. Since the columns of B form a basis for S, each vector $\boldsymbol{v}_i$ can be written as

$$\boldsymbol{v}_i = \boldsymbol{v}_1 t_{1i} + \cdots + \boldsymbol{v}_r t_{ri}$$

for $1 \leqslant i \leqslant r$. Thus, $B = CT$. Since B and C are orthogonal, we have

$$B^{\mathsf{H}}B = T^{\mathsf{H}}C^{\mathsf{H}}CT = T^{\mathsf{H}}T = I_r,$$

so T is an orthogonal matrix and because it is a square matrix, it is also a unitary matrix. Furthermore, we have $C^{\mathsf{H}}B = C^{\mathsf{H}}CT = T$, which concludes the argument. □

Definition 6.24. A *rotation matrix* is an orthogonal matrix $R \in \mathbb{R}^{n\times n}$ such that $\det(R) = 1$. A *reflection matrix* is an orthogonal matrix $R \in \mathbb{R}^{n\times n}$ such that $\det(R) = -1$.

Example 6.23. In the two-dimensional case, $n = 2$, a rotation is a matrix $R \in \mathbb{R}^{2\times 2}$,

$$R = \begin{pmatrix} r_{11} & r_{12} \\ r_{21} & r_{22} \end{pmatrix},$$

such that

$$\begin{aligned} r_{11}^2 + r_{21}^2 &= 1, \\ r_{12}^2 + r_{22}^2 &= 1, \\ r_{11}r_{12} + r_{21}r_{22} &= 0 \end{aligned}$$

and

$$r_{11}r_{22} - r_{12}r_{21} = 1.$$

This implies

$$r_{22}(r_{11}r_{12} + r_{21}r_{22}) - r_{12}(r_{11}r_{22} - r_{12}r_{21}) = -r_{12},$$

or

$$r_{21}(r_{22}^2 + r_{12}^2) = -r_{12},$$

so $r_{21} = -r_{12}$.

If $r_{21} = -r_{21} = 0$, the above equalities imply that either $r_{11} = r_{22} = 1$ or $r_{11} = r_{22} = -1$. Otherwise, the equality $r_{11}r_{12}+r_{21}r_{22} = 0$ implies $r_{11} = r_{22}$.

Since $r_{11}^2 \leqslant 1$, it follows that there exists θ such that $r_{11} = \cos\theta$. This implies that R has the form

$$R = \begin{pmatrix} \cos\theta & -\sin\theta \\ \sin\theta & \cos\theta \end{pmatrix}.$$

Its effect on a vector

$$\boldsymbol{x} = \begin{pmatrix} x_1 \\ x_2 \end{pmatrix} \in \mathbb{R}^2$$

is to produce the vector $\boldsymbol{y} = R\boldsymbol{x}$, where

$$\boldsymbol{y} = \begin{pmatrix} x_1 \cos\theta - x_2 \sin\theta \\ x_1 \sin\theta + x_2 \cos\theta \end{pmatrix},$$

which is obtained from $\boldsymbol{x}$ by a counterclockwise rotation by angle θ. It is easy to see that $\det(R) = 1$, so the term "rotation matrix" is clearly justified for R. To mark the dependency of R on θ, we will use the notation

$$R(\theta) = \begin{pmatrix} \cos\theta & -\sin\theta \\ \sin\theta & \cos\theta \end{pmatrix}.$$

The same conclusion can be reached by examining Figure 6.3. If the angle of the vector $\boldsymbol{x} = \begin{pmatrix} x_1 \\ x_2 \end{pmatrix}$ with the x_1 axis is α and $\boldsymbol{x}$ is rotated counterclockwise by θ to yield the vector $\boldsymbol{y} = y_1\boldsymbol{e}_1 + y_2\boldsymbol{e}_2$, then $x_1 = r\cos\alpha$, $x_2 = r\sin\alpha$, and

$$\begin{aligned} y_1 &= r\cos(\alpha+\theta) = r\cos\alpha\cos\theta - r\sin\alpha\sin\theta = x_1\cos\theta - x_2\sin\theta, \\ y_2 &= r\sin(\alpha+\theta) = r\sin\alpha\cos\theta + r\cos\alpha\sin\theta = x_1\sin\theta + x_2\cos\theta, \end{aligned}$$

which are the formulas that describe the transformation of $\boldsymbol{x}$ into $\boldsymbol{y}$.

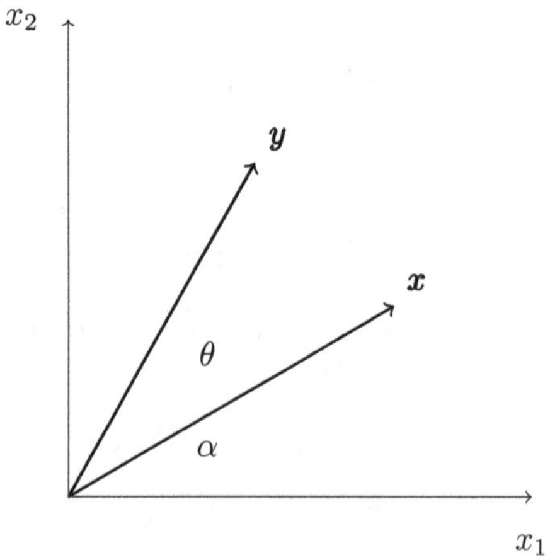

Fig. 6.3 $\boldsymbol{y}$ is obtained by a counterclockwise rotation by an angle θ from $\boldsymbol{x}$.

A extension of Example 6.23 is the *Givens matrix* $G(p,q,\theta) \in \mathbb{R}^{n\times n}$, defined as

$$\begin{pmatrix} 1 & \cdots & \cdots & \cdots & \cdots & \cdots & 0 \\ \vdots & \cdots & \vdots & \cdots & \vdots & \cdots & \vdots \\ 0 & \cdots & \cos\theta & \cdots & \sin\theta & \cdots & 0 \\ \vdots & \cdots & \vdots & \cdots & \vdots & \cdots & \vdots \\ 0 & \cdots & -\sin\theta & \cdots & \cos\theta & \cdots & 0 \\ \vdots & \cdots & \vdots & \cdots & \vdots & \cdots & \vdots \\ 0 & \cdots & \cdots & \cdots & \cdots & \cdots & 1 \end{pmatrix} \begin{matrix} \\ \\ p \\ \\ q \\ \\ \\ \end{matrix}.$$
$$\qquad\qquad p \qquad\qquad q$$

We can write

$$G(p,q,\theta) = (\boldsymbol{e}_1 \ \cdots \ \cos\theta\boldsymbol{e}_p - \sin\theta\boldsymbol{e}_q \ \cdots \ \sin\theta\boldsymbol{e}_p + \cos\theta\ \boldsymbol{e}_q \ \cdots \ \boldsymbol{e}_n).$$

It is easy to verify that $G(p,q,\theta)$ is a rotation matrix since it is orthogonal and $\det(G(p,q,\theta)) = 1$.

Since

$$G(p,q,\theta)\begin{pmatrix} v_1 \\ \vdots \\ v_p \\ \vdots \\ v_q \\ \vdots \\ v_n \end{pmatrix} = \begin{pmatrix} v_1 \\ \vdots \\ \cos\theta v_p + \sin\theta v_q \\ \vdots \\ -\sin\theta v_p + \cos\theta v_q \\ \vdots \\ v_n \end{pmatrix},$$

the multiplication of a vector $\boldsymbol{v}$ be a Givens matrix amounts to a clockwise rotation by θ in the plane of the coordinates (v_p, v_q).

If $v_p \neq 0$, then the rotation described by the Givens matrix can be used to zero the qth component of the resulting vector by taking θ such that $\tan\theta = \frac{v_q}{v_p}$.

It is easy to see that $R(\theta)^{-1} = R(-\theta)$ and that $R(\theta_2)R(\theta_1) = R(\theta_1+\theta_2)$.

Example 6.24. Let $\boldsymbol{v} \in \mathbb{C}^n - \{\mathbf{0}_n\}$ be a unit vector. The *Householder matrix* $H_{\boldsymbol{v}} \in \mathbb{C}^{n\times n}$ is defined by

$$H_{\boldsymbol{v}} = I_n - 2\boldsymbol{v}\boldsymbol{v}^{\mathsf{H}}.$$

The matrix $H_{\boldsymbol{v}}$ is clearly Hermitian. Moreover, we have

$$\begin{aligned} HH^{\mathsf{H}} &= HH = (I_n - 2\boldsymbol{v}\boldsymbol{v}^{\mathsf{H}})^2 \\ &= I_n - 4\boldsymbol{v}\boldsymbol{v}^{\mathsf{H}} + 4\boldsymbol{v}(\boldsymbol{v}^{\mathsf{H}}\boldsymbol{v})\boldsymbol{v}^{\mathsf{H}} \\ &= I_n, \end{aligned}$$

so $H_{\boldsymbol{v}}$ is unitary and involutive. Supplement 14 of Chapter 5 implies that $\det(H_{\boldsymbol{v}}) = -1$, so $H_{\boldsymbol{v}}$ is a reflection.

For a unit vector $\boldsymbol{v} \in \mathbb{R}^n$, $H_{\boldsymbol{v}}$ is an orthogonal and involutive matrix. In this case the vector $H_{\boldsymbol{v}}\boldsymbol{x}$ is a reflection of $\boldsymbol{x}$ relative to the hyperplane $H_{\boldsymbol{v},0}$ defined by $\boldsymbol{v}'\boldsymbol{x} = 0$, which follows from the fact that vector $\boldsymbol{t} = \boldsymbol{x} - H_{\boldsymbol{v}}\boldsymbol{x} = (I_n - H_{\boldsymbol{v}})\boldsymbol{x} = 2\boldsymbol{v}(\boldsymbol{v}'\boldsymbol{x})$ is orthogonal to the hyperplane $\boldsymbol{v}'\boldsymbol{x} = 0$.

6.12 Projection on Subspaces

We have shown in Theorem 2.33 that if U, W are two complementary subspaces of $\mathbb{C}^n$, then there exist idempotent endomorphisms g and h of $\mathbb{C}^n$ such that $W = \mathsf{Ker}(g)$, $U = \mathsf{Im}(g)$, and $U = \mathsf{Ker}(h)$ and $W = \mathsf{Im}(h)$. The vector $g(\boldsymbol{x})$ is the *oblique projection of* $\boldsymbol{x}$ *on* U *along the subspace* W and $h(\boldsymbol{x})$ is the *oblique projection* of $\boldsymbol{x}$ on W along the subspace U.

If g and h are represented by the matrices B_U and B_W, respectively, it follows that these matrices are idempotent, $g(\boldsymbol{x}) = B_U\boldsymbol{x} \in U$, and $h(\boldsymbol{x}) = B_W\boldsymbol{x} \in W$ for $\boldsymbol{x} \in \mathbb{C}^n$. Also, $B_UB_W = B_WB_U = O_{n,n}$.

Let U and W be two complementary subspaces of $\mathbb{C}^n$, $\{\boldsymbol{u}_1, \ldots, \boldsymbol{u}_p\}$ be a basis for U, and let $\{\boldsymbol{w}_1, \ldots, \boldsymbol{w}_q\}$ be a basis for W, where $p + q = n$. Clearly, $\{\boldsymbol{u}_1, \ldots, \boldsymbol{u}_p, \boldsymbol{w}_1, \ldots, \boldsymbol{w}_q\}$ is a basis for $\mathbb{C}^n$ and every $\boldsymbol{x} \in \mathbb{C}^n$ can be written as

$$\boldsymbol{x} = x_1\boldsymbol{u}_1 + \cdots + x_p\boldsymbol{u}_p + x_{p+1}\boldsymbol{w}_1 + \cdots + x_{p+q}\boldsymbol{w}_q.$$

Let $B \in \mathbb{C}^{n\times n}$ be the matrix

$$B = (\boldsymbol{u}_1 \ \cdots \ \boldsymbol{u}_p \ \boldsymbol{w}_1 \ \cdots \ \boldsymbol{w}_q),$$

which is clearly invertible. Note that

$$\begin{aligned} B_U \boldsymbol{u}_i &= \boldsymbol{u}_i \;\; B_U \boldsymbol{w}_j = \mathbf{0}_n, \\ B_W \boldsymbol{u}_i &= \mathbf{0}_n \; B_W \boldsymbol{w}_j = \boldsymbol{w}_j \end{aligned}$$

for $1 \leqslant i \leqslant p$ and $1 \leqslant j \leqslant q$. Therefore, we have

$$B_U B = (\boldsymbol{u}_1 \cdots \boldsymbol{u}_p \, \mathbf{0}_n \cdots \mathbf{0}_n) = B \begin{pmatrix} I_p & O_{p,n-p} \\ O_{n-p,p} & O_{n-p,n-p} \end{pmatrix},$$

so

$$B_U = B \begin{pmatrix} I_p & O_{p,n-p} \\ O_{n-p,p} & O_{n-p,n-p} \end{pmatrix} B^{-1}.$$

Similarly, we can show that

$$B_W = B \begin{pmatrix} O_{n-q,n-q} & O_{n-q,q} \\ O_{q,n-q} & I_q \end{pmatrix} B^{-1}.$$

Note that $B_U + B_W = B I_n B^{-1} = I_n$. Thus, the oblique projection on U along W is given by

$$g(\boldsymbol{x}) = B_U \boldsymbol{x} = B \begin{pmatrix} I_p & O_{p,n-p} \\ O_{n-p,p} & O_{n-p,n-p} \end{pmatrix} B^{-1} \boldsymbol{x}.$$

The similar oblique projection on W along U is

$$h(\boldsymbol{x}) = B_U \boldsymbol{x} = B \begin{pmatrix} O_{n-q,n-q} & O_{n-q,q} \\ O_{q,n-q} & I_q \end{pmatrix} B^{-1} \boldsymbol{x}$$

for $\boldsymbol{x} \in \mathbb{C}^n$. Observe that $g(\boldsymbol{x}) + h(\boldsymbol{x}) = \boldsymbol{x}$, so the projection on W along U is $h(\boldsymbol{x}) = \boldsymbol{x} - g(\boldsymbol{x}) = (I_n - B_U)\boldsymbol{x}$.

A special important type of projections involves pairs of orthogonal subspaces. Let U be a subspace of $\mathbb{C}^n$ with $\dim U = p$ and let $B_U = \{\boldsymbol{u}_1, \ldots, \boldsymbol{u}_p\}$ be an orthonormal basis of U. Taking into account that $(U^\perp)^\perp = U$, by Theorem 2.33 there exists an idempotent endomorphism g of $\mathbb{C}^n$ such that $U = \mathsf{Im}(g)$ and $U^\perp = \mathsf{Ker}(g)$. The proof of Theorem 6.37 shows that this endomorphism is defined by $g(\boldsymbol{x}) = (\boldsymbol{x}, \boldsymbol{u}_1)\boldsymbol{u}_1 + \cdots + (\boldsymbol{x}, \boldsymbol{u}_m)\boldsymbol{u}_m$.

Definition 6.25. Let U be an m-dimensional subspace of $\mathbb{C}^n$ and let $\{\boldsymbol{u}_1, \ldots, \boldsymbol{u}_m\}$ be an orthonormal basis of this subspace. The *orthogonal projection of the vector* $\boldsymbol{x} \in \mathbb{C}^n$ *on the subspace* U is the vector $\mathsf{proj}_U(\boldsymbol{x}) = (\boldsymbol{x}, \boldsymbol{u}_1)\boldsymbol{u}_1 + \cdots + (\boldsymbol{x}, \boldsymbol{u}_m)\boldsymbol{u}_m$.

Theorem 6.44. *Let U be an m-dimensional subspace of $\mathbb{R}^n$ and let $\boldsymbol{x} \in \mathbb{R}^n$. The vector $\boldsymbol{y} = \boldsymbol{x} - \mathsf{proj}_U(\boldsymbol{x})$ belongs to the subspace $U^\perp$.*

Proof. Let $B_U = \{\boldsymbol{u}_1, \ldots, \boldsymbol{u}_m\}$ be an orthonormal basis of U. Note that

$$
\begin{aligned}
(\boldsymbol{y}, \boldsymbol{u}_j) &= (\boldsymbol{x}, \boldsymbol{u}_j) - \left(\sum_{i=1}^{m} (\boldsymbol{x}, \boldsymbol{u}_i)\boldsymbol{u}_i, \boldsymbol{u}_j \right) \\
&= (\boldsymbol{x}, \boldsymbol{u}_j) - \sum_{i=1}^{m} (\boldsymbol{x}, \boldsymbol{u}_i)(\boldsymbol{u}_i, \boldsymbol{u}_j) = 0,
\end{aligned}
$$

due to the orthogonality of the basis B_U. Therefore, $\boldsymbol{y}$ is orthogonal on every linear combination of B_U, that is on the subspace U. □

Theorem 6.45. *Let U be an m-dimensional subspace of $\mathbb{C}^n$ having the orthonormal basis $\{\boldsymbol{u}_1, \ldots, \boldsymbol{u}_m\}$. The orthogonal projection proj_U is given by $\mathsf{proj}_U(\boldsymbol{x}) = B_U B_U^{\mathsf{H}} \boldsymbol{x}$ for $\boldsymbol{x} \in \mathbb{C}^n$, where $B_U \in \mathbb{R}^{n \times m}$ is the matrix $B_U = (\boldsymbol{u}_1 \ \cdots \ \boldsymbol{u}_m) \in \mathbb{C}^{n \times m}$.*

Proof. We can write

$$
\mathsf{proj}_U(\boldsymbol{x}) = \sum_{i=1}^{m} \boldsymbol{u}_i(\boldsymbol{u}_i^{\mathsf{H}} \boldsymbol{x}) = (\boldsymbol{u}_1 \ \cdots \ \boldsymbol{u}_m) \begin{pmatrix} \boldsymbol{u}_1^{\mathsf{H}} \\ \vdots \\ \boldsymbol{u}_m^{\mathsf{H}} \end{pmatrix} \boldsymbol{x} = B_U B_U^{\mathsf{H}} \boldsymbol{x}.
$$

□

Since the basis $\{\boldsymbol{u}_1, \ldots, \boldsymbol{u}_m\}$ is orthonormal, we have $B_U^{\mathsf{H}} B_U = I_m$. Observe that the matrix $B_U B_U^{\mathsf{H}} \in \mathbb{C}^{n \times n}$ is symmetric and idempotent because

$$
(B_U B_U^{\mathsf{H}})(B_U B_U^{\mathsf{H}}) = B_U (B_U^{\mathsf{H}} B_U) B_U^{\mathsf{H}} = B_U B_U^{\mathsf{H}}.
$$

For an m-dimensional subspace U of $\mathbb{C}^n$, we denote by $P_U = B_U B_U^{\mathsf{H}} \in \mathbb{C}^{n \times n}$, where B_U is a matrix of an orthonormal basis of U as defined before. P_U is the *projection matrix* of the subspace U.

Corollary 6.18. *For every non-zero subspace U, the matrix P_U is a Hermitian matrix, and therefore, a self-adjoint matrix.*

Proof. Since $P_U = B_U B_U^{\mathsf{H}}$ where B_U is a matrix of an orthonormal basis of the subspace S, it is immediate that $P_U^{\mathsf{H}} = P_U$. □

The self-adjointness of P_U means that $(\boldsymbol{x}, P_U\boldsymbol{y}) = (P_U\boldsymbol{x}, \boldsymbol{y})$ for every $\boldsymbol{x}, \boldsymbol{y} \in \mathbb{C}^n$.

Corollary 6.19. *Let U be an m-dimensional subspace of $\mathbb{C}^n$ having the orthonormal basis $\{\boldsymbol{u}_1, \ldots, \boldsymbol{u}_m\}$. If $B_U = (\boldsymbol{u}_1 \cdots \boldsymbol{u}_m) \in \mathbb{C}^{n\times m}$, then for every $\boldsymbol{x} \in \mathbb{C}$ we have the decomposition $\boldsymbol{x} = P_U\boldsymbol{x} + Q_U\boldsymbol{x}$, where $P_U = B_U B_U^{\mathsf{H}}$ and $Q_U = I_n - P_U$, $P_U\boldsymbol{x} \in U$ and $Q_U\boldsymbol{x} \in U^{\perp}$.*

Proof. This statement follows immediately from Theorem 6.45. □

Observe that

$$
\begin{aligned}
Q_U^2 &= (I_n - P_U P_U^{\mathsf{H}})(I_n - P_U P_U^{\mathsf{H}}) \\
&= I_n - P_U P_U^{\mathsf{H}} - P_U P_U^{\mathsf{H}} + P_U P_U^{\mathsf{H}} P_U P_U^{\mathsf{H}} = Q_U,
\end{aligned}
$$

so Q_U is an idempotent matrix. The matrix Q_U is the projection matrix on the subspace $U^{\perp}$. Clearly, we have

$$
P_{U^{\perp}} = Q_U = I_n - P_U. \tag{6.28}
$$

It is possible to give a direct argument for the independence of the projection matrix P_U relative to the choice of orthonormal basis in U.

Theorem 6.46. *Let U be an m-dimensional subspace of $\mathbb{C}^n$ having the orthonormal bases $\{\boldsymbol{u}_1, \ldots, \boldsymbol{u}_m\}$ and $\{\boldsymbol{v}_1, \ldots, \boldsymbol{v}_m\}$ and let $B_U = (\boldsymbol{u}_1 \cdots \boldsymbol{u}_m) \in \mathbb{C}^{n\times m}$ and $\tilde{B}_U = (\boldsymbol{v}_1 \cdots \boldsymbol{v}_m) \in \mathbb{C}^{n\times m}$. The matrix $B_U^{\mathsf{H}}\tilde{B}_U \in \mathbb{C}^{m\times m}$ is unitary and $\tilde{B}_U\tilde{B}_U^{\mathsf{H}} = B_U B_U^{\mathsf{H}}$.*

Proof. Since both sets of columns of B_U and $\tilde{B}_U$ are bases for U, there exists a unique square matrix $Q \in \mathbb{C}^{m\times m}$ such that $B_U = \tilde{B}_U Q$. The orthonormality of B_U and $\tilde{B}_U$ implies $B_U^{\mathsf{H}} B_U = \tilde{B}_U^{\mathsf{H}}\tilde{B}_U = I_m$. Thus, we can write

$$
I_m = B_U^{\mathsf{H}} B_U = Q^{\mathsf{H}}\tilde{B}_U^{\mathsf{H}}\tilde{B}_U Q = Q^{\mathsf{H}} Q,
$$

which shows that Q is unitary. Furthermore, $B_U^{\mathsf{H}}\tilde{B}_U = Q^{\mathsf{H}}\tilde{B}_U^{\mathsf{H}}\tilde{B}_U = Q^{\mathsf{H}}$ is unitary and

$$
B_U B_U^{\mathsf{H}} = \tilde{B}_U Q Q^{\mathsf{H}}\tilde{B}_U^{\mathsf{H}} = \tilde{B}_U\tilde{B}_U^{\mathsf{H}}.
$$

□

In Example 6.11, we have shown that if P is an idempotent matrix, then for every matrix norm μ we have $\mu(P) = 0$ or $\mu(P) \geqslant 1$. For orthogonal projection matrices of the form P_U, where U is a non-zero subspace, we have $|||P_U|||_2 = 1$. Indeed, we can write $\boldsymbol{x} = (\boldsymbol{x} - \mathsf{proj}_U(\boldsymbol{x})) + \mathsf{proj}_U(\boldsymbol{x})$, so

$$\|\boldsymbol{x}\|_2^2 = \|\boldsymbol{x} - \mathsf{proj}_U(\boldsymbol{x})\|_2^2 + \|\mathsf{proj}_U(\boldsymbol{x})\|_2^2 \geqslant \|\mathsf{proj}_U(\boldsymbol{x})\|_2^2.$$

Thus, $\|\boldsymbol{x}\|_2 \geqslant \|P_U(\boldsymbol{x})\|$ for any $\boldsymbol{x} \in \mathbb{C}^n$, which implies

$$|||P_U|||_2 = \sup\left\{\frac{\|P_U\boldsymbol{x}\|_2}{\|\boldsymbol{x}\|_2}\middle| \boldsymbol{x} \in \mathbb{C}^n - \{\mathbf{0}\}\right\} \leqslant 1.$$

This implies $|||P_U|||_2 = 1$.

The next theorem shows that the best approximation of a vector $\boldsymbol{x}$ by a vector in a subspace U (in the sense of Euclidean distance) is the orthogonal projection of $\boldsymbol{x}$ on U.

Theorem 6.47. *Let U be an m-dimensional subspace of $\mathbb{C}^n$ and let $\boldsymbol{x} \in \mathbb{C}^n$. The minimal value of $d_2(\boldsymbol{x}, \boldsymbol{u})$, the Euclidean distance between $\boldsymbol{x}$ and an element $\boldsymbol{u}$ of the subspace U, is achieved when $\boldsymbol{u} = \mathsf{proj}_U(\boldsymbol{x})$.*

Proof. We saw that $\boldsymbol{x}$ can be uniquely written as $\boldsymbol{x} = \boldsymbol{y} + \mathsf{proj}_U(\boldsymbol{x})$, where $\boldsymbol{y} \in U^\perp$. Let now $\boldsymbol{u}$ be an arbitrary member of U. We have

$$\begin{aligned} d_2(\boldsymbol{x}, \boldsymbol{u})^2 &= \|\boldsymbol{x} - \boldsymbol{u}\|_2^2 \\ &= \|(\boldsymbol{x} - \mathsf{proj}_U(\boldsymbol{x})) + (\mathsf{proj}_U(\boldsymbol{x}) - \boldsymbol{u})\|_2^2. \end{aligned}$$

Since $\boldsymbol{x} - \mathsf{proj}_U(\boldsymbol{x}) \in U^\perp$ and $\mathsf{proj}_U(\boldsymbol{x}) - \boldsymbol{u} \in U$, it follows that these vectors are orthogonal. Thus, we can write

$$d_2(\boldsymbol{x}, \boldsymbol{u})^2 = \|(\boldsymbol{x} - \mathsf{proj}_U(\boldsymbol{x}))\|_2^2 + \|(\mathsf{proj}_U(\boldsymbol{x}) - \boldsymbol{u})\|_2^2,$$

which implies that $d_2(\boldsymbol{x}, \boldsymbol{u}) \geqslant d_2(\boldsymbol{x} - \mathsf{proj}_U(\boldsymbol{x}))$. □

The orthogonal projections associated with subspaces allow us to define a metric on the collection of subspaces of $\mathbb{C}^n$. Indeed, if S and T are two subspaces of $\mathbb{C}^n$, we define $d_F(S, T) = \|P_S - P_T\|_F$. When using the vector norm $\|\cdot\|_2$ and the metric induced by this norm on $\mathbb{C}^n$, we denote the corresponding metric on subspaces by d_2.

Example 6.25. Let $\boldsymbol{u}, \boldsymbol{w}$ be two distinct unit vectors in the linear space V. The orthogonal projection matrices of $\langle \boldsymbol{u} \rangle$ and $\langle \boldsymbol{w} \rangle$ are $\boldsymbol{u}\boldsymbol{u}'$ and $\boldsymbol{w}\boldsymbol{w}'$, respectively. Thus,

$$d_F(\langle \boldsymbol{u} \rangle, \langle \boldsymbol{w} \rangle) = \|\boldsymbol{u}\boldsymbol{u}' - \boldsymbol{v}\boldsymbol{v}'\|_F.$$

Suppose now that $V = \mathbb{R}^2$. Since $\boldsymbol{u}$ and $\boldsymbol{w}$ are unit vectors in $\mathbb{R}^2$, there exist $\alpha, \beta \in [0, 2\pi]$ such that

$$\boldsymbol{u} = \begin{pmatrix} \cos\alpha \\ \sin\alpha \end{pmatrix} \text{ and } \boldsymbol{w} = \begin{pmatrix} \cos\beta \\ \sin\beta \end{pmatrix}.$$

Thus, we can write

$$\boldsymbol{u}\boldsymbol{u}' - \boldsymbol{v}\boldsymbol{v}' = \begin{pmatrix} \cos^2\alpha - \cos^2\beta & \cos\alpha\sin\alpha - \cos\beta\sin\beta \\ \cos\alpha\sin\alpha - \cos\beta\sin\beta & \sin^2\alpha - \sin^2\beta \end{pmatrix}$$

and

$$d_F(\langle \boldsymbol{u} \rangle, \langle \boldsymbol{w} \rangle) = \sqrt{2}|\sin(\alpha - \beta)|.$$

We could use any matrix norm in the definition of the distance between subspaces. For example, we could replace the Frobenius norm by $|\!|\!| \cdot |\!|\!|_1$ or by $|\!|\!| \cdot |\!|\!|_2$.

Let S be a subspace of $\mathbb{C}^n$ and let $\boldsymbol{x} \in \mathbb{C}^n$. The *distance between* $\boldsymbol{x}$ *and* S *defined by the norm* $\|\cdot\|$ is

$$d(\boldsymbol{x}, S) = \|\boldsymbol{x} - \mathsf{proj}_S(\boldsymbol{x})\| = \|\boldsymbol{x} - P_S\boldsymbol{x}\| = \|(I - P_S)\boldsymbol{x}\|.$$

Theorem 6.48. *Let* S *and* T *be two non-zero subspaces of* $\mathbb{C}^n$ *and let*

$$\begin{aligned} \delta_S &= \max\{d_2(\boldsymbol{x}, T) \mid \boldsymbol{x} \in S, \|\boldsymbol{x}\|_2 = 1\}, \\ \delta_T &= \max\{d_2(\boldsymbol{x}, S) \mid \boldsymbol{x} \in T, \|\boldsymbol{x}\|_2 = 1\}. \end{aligned}$$

We have

$$d_2(S, T) = \max\{\delta_S, \delta_T\}.$$

Proof. If $\boldsymbol{x} \in S$ and $\|\boldsymbol{x}\|_2 = 1$, we have

$$\begin{aligned} d_2(\boldsymbol{x}, T) &= \|\boldsymbol{x} - P_T\boldsymbol{x}\|_2 = \|P_S\boldsymbol{x} - P_T\boldsymbol{x}\|_2 \\ &= \|(P_S - P_T)\boldsymbol{x}\|_2 \leqslant |\!|\!|P_S - P_T|\!|\!|_2. \end{aligned}$$

Therefore, $\delta_S \leqslant |\!|\!|P_S - P_T|\!|\!|_2$. Similarly, $\delta_T \leqslant |\!|\!|P_S - P_T|\!|\!|_2$, so $\max\{\delta_S, \delta_T\} \leqslant d_2(S, T)$.

Note that

$$\begin{aligned}\delta_S &= \max\{\|(I - P_T)\boldsymbol{x}\|_2 \mid \boldsymbol{x} \in S, \|\boldsymbol{x}\|_2 = 1\},\\ \delta_T &= \max\{\|(I - P_S)\boldsymbol{x}\|_2 \mid \boldsymbol{x} \in T, \|\boldsymbol{x}\|_2 = 1\},\end{aligned}$$

so, taking into account that $P_S\boldsymbol{x} \in S$ and $P_T\boldsymbol{x} \in T$ for every $\boldsymbol{x} \in \mathbb{C}^n$, we have

$$\|(I - P_S)P_T\boldsymbol{x}\|_2 \leqslant \delta_S\|P_T\boldsymbol{x}\|_2, \|(I - P_T)P_S\boldsymbol{x}\|_2 \leqslant \delta_T\|P_S\boldsymbol{x}\|_2.$$

We have

$$\begin{aligned}\|P_T(I - P_S)\boldsymbol{x}\|_2^2 &= (P_T(I - P_S)\boldsymbol{x}, P_T(I - P_S)\boldsymbol{x})\\ &= ((P_T)^2(I - P_S)\boldsymbol{x}, (I - P_S)\boldsymbol{x})\\ &= (P_T(I - P_S)\boldsymbol{x}, (I - P_S)\boldsymbol{x})\\ &= (P_T(I - P_S)\boldsymbol{x}, (I - P_S)^2\boldsymbol{x})\\ &= ((I - P_S)P_T(I - P_S)\boldsymbol{x}, (I - P_S)\boldsymbol{x})\end{aligned}$$

because both P_S and $I - P_S$ are idempotent and self-adjoint. Therefore,

$$\begin{aligned}\|P_T(I - P_S)\boldsymbol{x}\|_2^2 &\leqslant \|(I - P_S)P_T(I - P_S)\boldsymbol{x}\|_2\|(I - P_S)\boldsymbol{x}\|_2\\ &\leqslant \delta_T\|P_T(I - P_S)\boldsymbol{x}\|_2\|(I - P_S)\boldsymbol{x}\|_2.\end{aligned}$$

This allows us to infer that

$$\|P_T(I - P_S)\boldsymbol{x}\|_2 \leqslant \delta_T\|(I - P_S)\boldsymbol{x}\|_2.$$

□

We discuss now four fundamental subspaces associated to a matrix $A \in \mathbb{C}^{m\times n}$. The range and the null space of A, $\mathsf{range}(A) \subseteq \mathbb{C}^m$ and $\mathsf{null}(A) \subseteq \mathbb{C}^n$, have been already discussed in Chapter 3. We add now two new subspaces: $\mathsf{range}(A^{\mathsf{H}}) \subseteq \mathbb{C}^n$ and $\mathsf{null}(A^{\mathsf{H}}) \subseteq \mathbb{C}^m$.

Theorem 6.49. *For every matrix* $A \in \mathbb{C}^{m\times n}$, *we have* $(\mathsf{range}(A))^\perp = \mathsf{null}(A^{\mathsf{H}})$.

Proof. The statement follows from the equivalence of the following statements:

(i) $\boldsymbol{x} \in (\mathsf{range}(A))^\perp$;

(ii) $(\boldsymbol{x}, A\boldsymbol{y}) = 0$ for all $y \in \mathbb{C}^n$;
(iii) $\boldsymbol{x}^{\mathsf{H}} A\boldsymbol{y} = 0$ for all $y \in \mathbb{R}^n$;
(iv) $\boldsymbol{y}^{\mathsf{H}} A^{\mathsf{H}}\boldsymbol{x} = 0$ for all $y \in \mathbb{R}^n$;
(v) $A^{\mathsf{H}}\boldsymbol{x} = \mathbf{0}$;
(vi) $\boldsymbol{x} \in \mathsf{null}(A^{\mathsf{H}})$. □

Corollary 6.20. *For every matrix $A \in \mathbb{C}^{m\times n}$, we have* $(\mathsf{range}(A^{H}))^{\perp} = \mathsf{null}(A)$.

Proof. This statement follows from Theorem 6.49 by replacing A by A^{H}. □

Corollary 6.21. *For every matrix $A \in \mathbb{C}^{m\times n}$, we have*

$$\mathbb{C}^m = \mathsf{range}(A) \oplus \mathsf{null}(A^{H}),$$
$$\mathbb{C}^n = \mathsf{null}(A) \oplus \mathsf{range}(A^{H}).$$

Proof. By Theorem 6.37, we have $\mathbb{C}^m = \mathsf{range}(A) \oplus \mathsf{range}(A)^{\perp}$ and $\mathbb{C}^n = \mathsf{null}(A) \oplus \mathsf{null}(A)^{\perp}$. Taking into account Theorem 6.49 and Corollary 6.21, we obtain the desired equalities. □

6.13 Positive Definite and Positive Semidefinite Matrices

Definition 6.26. A matrix $A \in \mathbb{C}^{n\times n}$ is *positive definite* if $\boldsymbol{x}^{\mathsf{H}} A\boldsymbol{x}$ is a real positive number for every $\boldsymbol{x} \in \mathbb{C}^n - \{\mathbf{0}\}$.

Theorem 6.50. *If $A \in \mathbb{C}^{n\times n}$ is positive definite, then A is Hermitian.*

Proof. Let $A \in \mathbb{C}^{n\times n}$ be a matrix. Since $\boldsymbol{x}^{\mathsf{H}} A\boldsymbol{x}$ is a real number, it follows that it equals its conjugate, so $\boldsymbol{x}^{\mathsf{H}} A\boldsymbol{x} = \boldsymbol{x}^{\mathsf{H}} A^{\mathsf{H}}\boldsymbol{x}$ for every $\boldsymbol{x} \in \mathbb{C}^n$. By Theorem 3.16, there exists a unique pair of Hermitian matrices H_1 and H_2 such that $A = H_1 + iH_2$, which implies $A^{\mathsf{H}} = H_1^{\mathsf{H}} - iH_2^{\mathsf{H}}$. Thus, we have

$$\boldsymbol{x}^{\mathsf{H}}(H_1 + iH_2)\boldsymbol{x} = \boldsymbol{x}^{\mathsf{H}}(H_1^{\mathsf{H}} - iH_2^{\mathsf{H}})\boldsymbol{x} = \boldsymbol{x}^{\mathsf{H}}(H_1 - iH_2)\boldsymbol{x},$$

because H_1 and H_2 are Hermitian. This implies $\boldsymbol{x}^{\mathsf{H}} H_2\boldsymbol{x} = 0$ for every $\boldsymbol{x} \in \mathbb{C}^n$, which, in turn, implies $H_2 = O_{n,n}$. Consequently, $A = H_1$, so A is indeed Hermitian. □

Definition 6.27. A matrix $A \in \mathbb{C}^{n\times n}$ is *positive semidefinite* if $\boldsymbol{x}^{\mathsf{H}} A\boldsymbol{x}$ is a non-negative real number for every $\boldsymbol{x} \in \mathbb{C}^n - \{\boldsymbol{0}\}$.

If $\boldsymbol{x}^{\mathsf{H}} A\boldsymbol{x}$ is a non-positive real number for every $\boldsymbol{x} \in \mathbb{C}^n - \{\boldsymbol{0}\}$, then we say that A is a positive definite matrix.

Positive definiteness (positive semidefiniteness) is denoted by $A \succ 0$ ($A \succeq 0$, respectively).

The definition of positive definite (semidefinite) matrix can be specialized for real matrices as follows.

Definition 6.28. A symmetric matrix $A \in \mathbb{R}^{n\times n}$ is *positive definite* if $\boldsymbol{x}'A\boldsymbol{x} > 0$ for every $\boldsymbol{x} \in \mathbb{R}^n - \{\boldsymbol{0}\}$.

If A satisfies the weaker inequality $\boldsymbol{x}'A\boldsymbol{x} \geqslant 0$ for every $\boldsymbol{x} \in \mathbb{R}^n - \{\boldsymbol{0}\}$, then we say that A is *positive semidefinite.*

$A \succ 0$ denotes that A is positive definite and $A \succeq 0$ means that A is positive semidefinite.

Note that in the case of real-valued matrices, we need to require explicitly the symmetry of the matrix because, unlike the complex case, the inequality $\boldsymbol{x}'A\boldsymbol{x} > 0$ for $\boldsymbol{x} \in \mathbb{R}^n - \{\boldsymbol{0}_n\}$ does *not* imply the symmetry of A. For example, consider the matrix

$$A = \begin{pmatrix} a & b \\ -b & a \end{pmatrix},$$

where $a, b \in \mathbb{R}$ and $a > 0$. We have

$$\boldsymbol{x}'A\boldsymbol{x} = (x_1\ x_2)\begin{pmatrix} a & b \\ -b & a \end{pmatrix}\begin{pmatrix} x_1 \\ x_2 \end{pmatrix} = a(x_1^2 + x_2^2) > 0$$

if $\boldsymbol{x} \neq \boldsymbol{0}_2$.

Example 6.26. The symmetric real matrix

$$A = \begin{pmatrix} a & b \\ b & c \end{pmatrix}$$

is positive definite if and only if $a > 0$ and $b^2 - ac < 0$. Indeed, we have $\boldsymbol{x}'A\boldsymbol{x} > 0$ for every $\boldsymbol{x} \in \mathbb{R}^2 - \{\boldsymbol{0}\}$ if and only if $ax_1^2 + 2bx_1x_2 + cx_2^2 > 0$, where $\boldsymbol{x}' = (x_1\ \ x_2)$; elementary algebra considerations lead to $a > 0$ and $b^2 - ac < 0$.

A positive definite matrix is non-singular. Indeed, if $A\boldsymbol{x} = \boldsymbol{0}$, where $A \in \mathbb{R}^{n\times n}$ is positive definite, then $\boldsymbol{x}^{\mathsf{H}}A\boldsymbol{x} = 0$, so $\boldsymbol{x} = \boldsymbol{0}$. By Corollary 3.2, A is non-singular.

Example 6.27. If $A \in \mathbb{C}^{m\times n}$, then the matrices $A^{\mathsf{H}}A \in \mathbb{C}^{n\times n}$ and $AA^{\mathsf{H}} \in \mathbb{C}^{m\times m}$ are positive semidefinite. For $\boldsymbol{x} \in \mathbb{C}^n$, we have

$$\boldsymbol{x}^{\mathsf{H}}(A^{\mathsf{H}}A)\boldsymbol{x} = (\boldsymbol{x}^{\mathsf{H}}A^{\mathsf{H}})(A\boldsymbol{x}) = (A\boldsymbol{x})^{\mathsf{H}}(A\boldsymbol{x}) = \|A\boldsymbol{x}\|_2^2 \geqslant 0.$$

The argument for AA^{H} is similar.

If $\mathsf{rank}(A) = n$, then the matrix $A^{\mathsf{H}}A$ is positive definite because $\boldsymbol{x}^{\mathsf{H}}(A^{\mathsf{H}}A)\boldsymbol{x} = 0$ implies $A\boldsymbol{x} = \boldsymbol{0}$, which, in turn, implies $\boldsymbol{x} = \boldsymbol{0}$.

Theorem 6.51. *If $A \in \mathbb{C}^{n\times n}$ is a positive definite matrix, then any principal submatrix $B = A\begin{bmatrix} i_1 & \cdots & i_k \\ i_1 & \cdots & i_k \end{bmatrix}$ is a positive definite matrix.*

Proof. Let $\boldsymbol{x} \in \mathbb{C}^n - \{\boldsymbol{0}\}$ be a vector such that all components located on positions other than $i_1, \ldots, i_k$ equal 0 and let $\boldsymbol{y} = \boldsymbol{x}\begin{bmatrix} i_1 & \cdots & i_k \\ & 1 & \end{bmatrix} \in \mathbb{C}^k$ be the vector obtained from $\boldsymbol{x}$ by retaining only the components located on positions $i_1, \ldots, i_k$. Since $\boldsymbol{y}^{\mathsf{H}}B\boldsymbol{y} = \boldsymbol{x}^{\mathsf{H}}A\boldsymbol{x} > 0$, it follows that $B \succ 0$. □

Corollary 6.22. *If $A \in \mathbb{C}^{n\times n}$ is a positive definite matrix, then any diagonal element a_{ii} is a real positive number for $1 \leqslant i \leqslant n$.*

Proof. This statement follows immediately from Theorem 6.51 by observing that every diagonal element of A is a 1×1 principal submatrix of A. □

Theorem 6.52. *If $A, B \in \mathbb{C}^{n\times n}$ are two positive semidefinite matrices and a, b are two non-negative numbers, then $aA + bB \succeq 0$.*

Proof. The statement holds because $\boldsymbol{x}^{\mathsf{H}}(aA + bB)\boldsymbol{x} = a\boldsymbol{x}^{\mathsf{H}}A\boldsymbol{x} + b\boldsymbol{x}^{\mathsf{H}}B\boldsymbol{x} \geqslant 0$, due to the fact that A and B are positive semidefinite. □

Theorem 6.53. *Let $A \in \mathbb{C}^{n\times n}$ be a positive definite matrix and let $S \in \mathbb{C}^{n\times m}$. The matrix $S^{\mathsf{H}}AS$ is positive semidefinite and has the same rank as S. Moreover, if $\mathsf{rank}(S) = m$, then $S^{\mathsf{H}}AS$ is positive definite.*

Proof. Since A is positive definite, it is Hermitian and $(S^{\mathsf{H}}AS)^{\mathsf{H}} = S^{\mathsf{H}}AS$ implies that $S^{\mathsf{H}}AS$ is a Hermitian matrix.

Let $\boldsymbol{x} \in \mathbb{C}^m$. We have $\boldsymbol{x}^{\mathsf{H}}S^{\mathsf{H}}AS\boldsymbol{x} = (S\boldsymbol{x})^{\mathsf{H}}A(S\boldsymbol{x}) \geqslant 0$ because A is positive definite. Thus, the matrix $S^{\mathsf{H}}AS$ is positive semidefinite.

If $S\boldsymbol{x} = \boldsymbol{0}$, then $S^{\mathsf{H}}AS\boldsymbol{x} = \boldsymbol{0}$; conversely, if $S^{\mathsf{H}}AS\boldsymbol{x} = \boldsymbol{0}$, then $\boldsymbol{x}^{\mathsf{H}}S^{\mathsf{H}}AS\boldsymbol{x} = 0$, so $S\boldsymbol{x} = \boldsymbol{0}$. This allows us to conclude that $\mathsf{null}(S) = \mathsf{null}(S^{\mathsf{H}}AS)$. Therefore, by Equality (3.8), we have $\mathsf{rank}(S) = \mathsf{rank}(S^{\mathsf{H}}AS)$.

Suppose now that $\mathsf{rank}(S) = m$ and that $\boldsymbol{x}^{\mathsf{H}}S^{\mathsf{H}}AS\boldsymbol{x} = 0$. Since A is positive definite, we have $S\boldsymbol{x} = \boldsymbol{0}$ and this implies $\boldsymbol{x} = \boldsymbol{0}$, because of the assumption made relative to $\mathsf{rank}(S)$. Consequently, $S^{\mathsf{H}}AS$ is positive definite. □

Corollary 6.23. *Let $A \in \mathbb{C}^{n\times n}$ be a positive definite matrix and let $S \in \mathbb{C}^{n\times n}$. If S is non-singular, then so is $S^{\mathsf{H}}AS$.*

Proof. This is an immediate consequence of Theorem 6.53. □

Theorem 6.54. *A Hermitian matrix $B \in \mathbb{C}^{n\times n}$ is positive definite if and only if the mapping $f : \mathbb{C}^n \times \mathbb{C}^n \longrightarrow \mathbb{C}$ given by $f(\boldsymbol{x}, \boldsymbol{y}) = \boldsymbol{x}^{\mathsf{H}}B\boldsymbol{y}$ for $\boldsymbol{x}, \boldsymbol{y} \in \mathbb{C}^n$ defines an inner product on $\mathbb{C}^n$.*

Proof. Suppose that B defines an inner product on $\mathbb{C}^n$. Then, by Property (iii) of Definition 6.15 we have $f(\boldsymbol{x}, \boldsymbol{x}) > 0$ for $\boldsymbol{x} \neq \boldsymbol{0}$, which amounts to the positive definiteness of B.

Conversely, if B is positive definite, then f satisfies the condition from Definition 6.15. We show here only that f has the conjugate symmetry property.

We can write

$$\overline{f(\boldsymbol{y}, \boldsymbol{x})} = \overline{\boldsymbol{y}^{\mathsf{H}}B\boldsymbol{x}} = \boldsymbol{y}'\overline{B}\overline{\boldsymbol{x}}$$

for $\boldsymbol{x}, \boldsymbol{y} \in \mathbb{C}^n$. Since B is Hermitian, $\overline{B} = \overline{B^{\mathsf{H}}} = B'$, so $\overline{f(\boldsymbol{y}, \boldsymbol{x})} = \boldsymbol{y}'B'\overline{\boldsymbol{x}}$. Observe that $\boldsymbol{y}'B'\overline{\boldsymbol{x}}$ is a number (that is, a 1×1 matrix), so

$$(\boldsymbol{y}'B'\overline{\boldsymbol{x}})' = \boldsymbol{x}^{\mathsf{H}}B\boldsymbol{y} = f(\boldsymbol{x}, \boldsymbol{y}).$$

□

Corollary 6.24. *A symmetric matrix $B \in \mathbb{R}^{n\times n}$ is positive definite if and only if the mapping $f : \mathbb{R}^n \times \mathbb{R}^n \longrightarrow \mathbb{R}$ given by $f(\boldsymbol{x}, \boldsymbol{y}) = \boldsymbol{x}'B\boldsymbol{y}$ for $\boldsymbol{x}, \boldsymbol{y} \in \mathbb{R}^n$ defines an inner product on $\mathbb{R}^n$.*

Proof. This follows immediately from Theorem 6.54. □

Definition 6.29. Let $S = (\boldsymbol{v}_1, \ldots, \boldsymbol{v}_m)$ be a sequence of vectors in $\mathbb{R}^n$. The *Gram matrix of* S is the matrix $G_S = (g_{ij}) \in \mathbb{R}^{m\times m}$ defined by $g_{ij} = \boldsymbol{v}_i'\boldsymbol{v}_j$ for $1 \leqslant i,j \leqslant m$.

Note that if $A_S = (\boldsymbol{v}_1 \ \cdots \ \boldsymbol{v}_m) \in \mathbb{R}^{n\times m}$, then $G_S = A_S'A_S$. Also, note that G_S is a symmetric matrix.

Example 6.28. Let

$$\boldsymbol{v}_1 = \begin{pmatrix} 1 \\ 0 \\ -1 \end{pmatrix}, \boldsymbol{v}_2 = \begin{pmatrix} 1 \\ 2 \\ 2 \end{pmatrix}, \boldsymbol{v}_3 = \begin{pmatrix} 2 \\ 1 \\ 0 \end{pmatrix}.$$

The Gram matrix of the set $S = \{\boldsymbol{v}_1, \boldsymbol{v}_2, \boldsymbol{v}_3\}$ is

$$G_S = \begin{pmatrix} 2 & -1 & 2 \\ -1 & 9 & 4 \\ 2 & 4 & 5 \end{pmatrix}.$$

Note that $\det(G_S) = 1$.

The Gram matrix of an arbitrary sequence of vectors is positive semidefinite, as the reader can easily verify.

Theorem 6.55. *Let $S = (\boldsymbol{v}_1, \ldots, \boldsymbol{v}_m)$ be a sequence of m vectors in $\mathbb{R}^n$, where $m \leqslant n$. If S is linearly independent, then the Gram matrix G_S is positive definite.*

Proof. Let S be a linearly independent sequence of vectors, $S = (\boldsymbol{v}_1, \ldots, \boldsymbol{v}_m)$, and let $\boldsymbol{x} \in \mathbb{R}^m$. We have $\boldsymbol{x}'G_S\boldsymbol{x} = \boldsymbol{x}'A_S'A_S\boldsymbol{x} = (A_S\boldsymbol{x})'A_S\boldsymbol{x} = \|A_S\boldsymbol{x}\|_2^2$. Therefore, if $\boldsymbol{x}'G_S\boldsymbol{x} = 0$, we have $A_S\boldsymbol{x} = \mathbf{0}$, which is equivalent to $x_1\boldsymbol{v}_1 + \cdots + x_n\boldsymbol{v}_n = 0$. Since $\{\boldsymbol{v}_1, \ldots, \boldsymbol{v}_m\}$ is linearly independent, it follows that $x_1 = \cdots = x_m = 0$, so $\boldsymbol{x} = \mathbf{0}_n$. Thus, G_S is indeed positive definite. □

Definition 6.30. Let $S = (\boldsymbol{v}_1, \ldots, \boldsymbol{v}_m)$ be a sequence of m vectors in $\mathbb{R}^n$, where $m \leqslant n$. The *Gramian* of S is the number $\det(G_S)$.

Theorem 6.56. *If $S = (\boldsymbol{v}_1, \ldots, \boldsymbol{v}_m)$ is a sequence of m vectors in $\mathbb{R}^n$, then S is linearly independent if and only if* $\det(G_S) \neq 0$.

Proof. Suppose that $\det(G_S) \neq 0$ and that S is not linearly independent. In other words, the numbers $a_1, \ldots, a_m$ exist such that at least one of them is not 0 and $a_1\boldsymbol{v}_1 + \cdots + a_m\boldsymbol{v}_m = \boldsymbol{0}_{\mathbb{R}^n}$. This implies the equalities

$$a_1(\boldsymbol{v}_1, \boldsymbol{v}_j) + \cdots + a_m(\boldsymbol{v}_m, \boldsymbol{v}_j) = 0$$

for $1 \leqslant j \leqslant m$, so the system $G_S\boldsymbol{a} = \boldsymbol{0}_n$ has a non-trivial solution in $a_1, \ldots, a_m$. This implies $\det(G_S) = 0$, which contradicts the initial assumption.

Conversely, suppose that S is linearly independent and $\det(G_S) = 0$. Then the linear system

$$a_1(\boldsymbol{v}_1, \boldsymbol{v}_j) + \cdots + a_m(\boldsymbol{v}_m, \boldsymbol{v}_j) = 0$$

for $1 \leqslant j \leqslant m$, has a non-trivial solution in $a_1, \ldots, a_m$.

Let $\boldsymbol{w} = a_1\boldsymbol{v}_1 + \cdots a_m\boldsymbol{v}_m$. We have $(\boldsymbol{w}, \boldsymbol{v}_i) = 0$ for $1 \leqslant i \leqslant n$. This, in turn, implies $(\boldsymbol{w}, \boldsymbol{w}) = \|\boldsymbol{w}\|_2^2 = 0$, so $\boldsymbol{w} = 0$, which contradicts the linear independence of S. □

Theorem 6.57 (Cholesky's decomposition theorem). Let $A \in \mathbb{C}^{n\times n}$ be a Hermitian positive definite matrix. There exists a unique upper triangular matrix R with real, positive diagonal elements such that $A = R^{\mathsf{H}}R$.

Proof. The argument is by induction on $n \geqslant 1$. The base step, $n = 1$, is immediate.

Suppose that a decomposition exists for all Hermitian positive matrices of order n, and let $A \in \mathbb{C}^{(n+1)\times(n+1)}$ be a symmetric and positive definite matrix. We can write

$$A = \begin{pmatrix} a_{11} & \boldsymbol{a}^{\mathsf{H}} \\ \boldsymbol{a} & B \end{pmatrix},$$

where $B \in \mathbb{C}^{n\times n}$. By Theorem 6.51, $a_{11} > 0$ and $B \in \mathbb{C}^{n\times n}$ is a Hermitian positive definite matrix. It is easy to verify the following identity:

$$A = \begin{pmatrix} \sqrt{a_{11}} & \mathbf{0} \\ \frac{1}{\sqrt{a_{11}}}\boldsymbol{a} & I_n \end{pmatrix} \begin{pmatrix} 1 & \mathbf{0}' \\ \mathbf{0} & B - \frac{1}{a_{11}}\boldsymbol{a}\boldsymbol{a}^{\mathsf{H}} \end{pmatrix} \begin{pmatrix} \sqrt{a_{11}} & \frac{1}{\sqrt{a_{11}}}\boldsymbol{a}^{\mathsf{H}} \\ 0 & I_n \end{pmatrix}. \tag{6.29}$$

Let $R_1 \in \mathbb{C}^{n\times n}$ be the upper triangular non-singular matrix

$$R_1 = \begin{pmatrix} \sqrt{a_{11}} & \frac{1}{\sqrt{a_{11}}}\boldsymbol{a}^{\mathsf{H}} \\ 0 & I_n \end{pmatrix}.$$

This allows us to write

$$A = R_1^{\mathsf{H}} \begin{pmatrix} 1 & \mathbf{0}' \\ \mathbf{0} & A_1 \end{pmatrix} R_1,$$

where $A_1 = B - \frac{1}{a_{11}}\boldsymbol{a}\boldsymbol{a}^{\mathsf{H}}$. Since

$$\begin{pmatrix} 1 & \mathbf{0}' \\ \mathbf{0} & A_1 \end{pmatrix} = (R_1^{-1})^{\mathsf{H}} A R_1^{-1},$$

by Theorem 6.53, the matrix

$$\begin{pmatrix} 1 & \mathbf{0}' \\ \mathbf{0} & A_1 \end{pmatrix}$$

is positive definite, which allows us to conclude that the matrix $A_1 = B - \frac{1}{a_{11}}\boldsymbol{a}\boldsymbol{a}^{\mathsf{H}} \in \mathbb{C}^{n\times n}$ is a Hermitian positive definite matrix.

By the inductive hypothesis, A_1 can be factored as

$$A_1 = P^{\mathsf{H}} P,$$

where P is an upper triangular matrix. This allows us to write

$$\begin{pmatrix} 1 & \mathbf{0}' \\ \mathbf{0} & A_1 \end{pmatrix} = \begin{pmatrix} 1 & \mathbf{0}' \\ \mathbf{0} & P^{\mathsf{H}} \end{pmatrix} \begin{pmatrix} 1 & \mathbf{0}' \\ \mathbf{0} & P \end{pmatrix}.$$

Thus,

$$A = R_1^{\mathsf{H}} \begin{pmatrix} 1 & \mathbf{0}' \\ \mathbf{0} & P^{\mathsf{H}} \end{pmatrix} \begin{pmatrix} 1 & \mathbf{0}' \\ \mathbf{0} & P \end{pmatrix} R_1.$$

If R is defined as

$$R = \begin{pmatrix} 1 & \mathbf{0}' \\ \mathbf{0} & P \end{pmatrix} R_1 = \begin{pmatrix} 1 & \mathbf{0}' \\ \mathbf{0} & P \end{pmatrix} \begin{pmatrix} \sqrt{a_{11}} & \frac{1}{\sqrt{a_{11}}}\boldsymbol{a}^{\mathsf{H}} \\ 0 & I_n \end{pmatrix} = \begin{pmatrix} \sqrt{a_{11}} & \frac{1}{\sqrt{a_{11}}}\boldsymbol{a}^{\mathsf{H}} \\ 0 & P \end{pmatrix},$$

then $A = R^{\mathsf{H}} R$ and R is clearly an upper triangular matrix. □

We refer to the matrix R as the *Cholesky factor* of A.

Corollary 6.25. *If* $A \in \mathbb{C}^{n\times n}$ *is a Hermitian positive definite matrix, then* $\det(A) > 0$.

Proof. By Corollary 5.2, $\det(A)$ is a real number. By Theorem 6.57, $A = R^{\mathsf{H}}R$, where R is an upper triangular matrix with real, positive diagonal elements, so $\det(A) = \det(R^{\mathsf{H}})\det(R) = (\det(R))^2$. Since $\det(R)$ is the product of its diagonal elements, $\det(R)$ is a real, positive number, which implies $\det(A) > 0$. □

Example 6.29. Let A be the symmetric matrix

$$A = \begin{pmatrix} 3 & 0 & 2 \\ 0 & 2 & 1 \\ 2 & 1 & 2 \end{pmatrix}.$$

We leave it to the reader to verify that this matrix is indeed positive definite starting from Definition 6.26.

By Equality (6.29), the matrix A can be written as

$$A = \begin{pmatrix} \sqrt{3} & 0 & 0 \\ 0 & 1 & 0 \\ \frac{2}{\sqrt{3}} & 0 & 1 \end{pmatrix} \begin{pmatrix} 1 & 0 & 0 \\ 0 & 2 & 1 \\ 0 & 1 & \frac{2}{3} \end{pmatrix} \begin{pmatrix} \sqrt{3} & 0 & \frac{2}{\sqrt{3}} \\ 0 & 1 & 0 \\ 0 & 0 & 1 \end{pmatrix},$$

because

$$A_1 = \begin{pmatrix} 2 & 1 \\ 1 & 2 \end{pmatrix} - \frac{1}{3}\begin{pmatrix} 0 \\ 2 \end{pmatrix}\begin{pmatrix} 0 & 2 \end{pmatrix} = \begin{pmatrix} 2 & 1 \\ 1 & \frac{2}{3} \end{pmatrix}.$$

Applying the same equality to A_1, we have

$$A_1 = \begin{pmatrix} \sqrt{2} & 0 \\ \frac{1}{\sqrt{2}} & 1 \end{pmatrix} \begin{pmatrix} 1 & 0 \\ 0 & \frac{1}{6} \end{pmatrix} \begin{pmatrix} \sqrt{2} & \frac{1}{\sqrt{2}} \\ 0 & 1 \end{pmatrix}.$$

Since the matrix $(\frac{1}{6})$ can be factored directly, we have

$$\begin{aligned} A_1 &= \begin{pmatrix} \sqrt{2} & 0 \\ \frac{1}{\sqrt{2}} & 1 \end{pmatrix} \begin{pmatrix} 1 & 0 \\ 0 & \frac{1}{\sqrt{6}} \end{pmatrix} \begin{pmatrix} 1 & 0 \\ 0 & \frac{1}{\sqrt{6}} \end{pmatrix} \begin{pmatrix} \sqrt{2} & \frac{1}{\sqrt{2}} \\ 0 & 1 \end{pmatrix} \\ &= \begin{pmatrix} \sqrt{2} & 0 \\ \frac{1}{\sqrt{2}} & \frac{1}{\sqrt{6}} \end{pmatrix} \begin{pmatrix} \sqrt{2} & \frac{1}{\sqrt{2}} \\ 0 & \frac{1}{\sqrt{6}} \end{pmatrix}. \end{aligned}$$

In turn, this implies

$$\begin{pmatrix} 1 & 0 & 0 \\ 0 & 2 & 1 \\ 0 & 1 & \frac{2}{3} \end{pmatrix} = \begin{pmatrix} 1 & 0 & 0 \\ 0 & \sqrt{2} & 0 \\ 0 & \frac{1}{\sqrt{2}} & \frac{1}{\sqrt{6}} \end{pmatrix} \begin{pmatrix} 1 & 0 & 0 \\ 0 & \sqrt{2} & \frac{1}{\sqrt{2}} \\ 0 & 0 & \frac{1}{\sqrt{6}} \end{pmatrix},$$

which produces the following Cholesky final decomposition of A:

$$\begin{aligned} A &= \begin{pmatrix} \sqrt{3} & 0 & 0 \\ 0 & 1 & 0 \\ \frac{2}{\sqrt{3}} & 0 & 1 \end{pmatrix} \begin{pmatrix} 1 & 0 & 0 \\ 0 & \sqrt{2} & 0 \\ 0 & \frac{1}{\sqrt{2}} & \frac{1}{\sqrt{6}} \end{pmatrix} \begin{pmatrix} 1 & 0 & 0 \\ 0 & \sqrt{2} & \frac{1}{\sqrt{2}} \\ 0 & 0 & \frac{1}{\sqrt{6}} \end{pmatrix} \begin{pmatrix} \sqrt{3} & 0 & \frac{2}{\sqrt{3}} \\ 0 & 1 & 0 \\ 0 & 0 & 1 \end{pmatrix} \\ &= \begin{pmatrix} \sqrt{3} & 0 & 0 \\ 0 & \sqrt{2} & 0 \\ \frac{2}{\sqrt{3}} & \frac{1}{\sqrt{2}} & \frac{1}{\sqrt{6}} \end{pmatrix} \begin{pmatrix} \sqrt{3} & 0 & \frac{2}{\sqrt{3}} \\ 0 & \sqrt{2} & \frac{1}{\sqrt{2}} \\ 0 & 0 & \frac{1}{\sqrt{6}} \end{pmatrix}. \end{aligned}$$

Cholesky's decomposition theorem can be extended to positive semi-definite matrices.

Theorem 6.58 (Cholesky's decomposition theorem for positive semidefinite matrices). *Let $A \in \mathbb{C}^{n\times n}$ be a Hermitian positive semidefinite matrix. There exists an upper triangular matrix R with real, non-negative diagonal elements such that $A = R^{\mathsf{H}}R$.*

Proof. The argument is similar to the one used for Theorem 6.57 and is omitted. □

Observe that for positive semidefinite matrices, the diagonal elements of R are non-negative numbers and the uniqueness of R does not longer holds.

Example 6.30. Let $A = \begin{pmatrix} 1 & -1 \\ -1 & 1 \end{pmatrix}$. Since $\boldsymbol{x}'A\boldsymbol{x} = (x_1 - x_2)$, it is clear that A is a positive semidefinite but not a positive definite matrix. Let R be a matrix of the form

$$R = \begin{pmatrix} r_1 & r \\ 0 & r_2 \end{pmatrix}$$

such that $A = R'R$. It is easy to see that the last equality is equivalent to $r_1^2 = r_2^2 = 1$ and $rr_1 = -1$. Thus, we have for distinct Cholesky factors, the following matrices:

$$\begin{pmatrix} 1 & -1 \\ 0 & 1 \end{pmatrix}, \begin{pmatrix} 1 & -1 \\ 0 & -1 \end{pmatrix}, \begin{pmatrix} -1 & 1 \\ 0 & 1 \end{pmatrix}, \begin{pmatrix} -1 & 1 \\ 0 & -1 \end{pmatrix}.$$

Theorem 6.59. *A Hermitian matrix $A \in \mathbb{C}^{n\times n}$ is positive definite if and only if all its leading principal minors are positive.*

Proof. By Theorem 6.51, if A is positive definite, then every principal submatrix is positive definite, so by Corollary 6.25, each principal minor of A is positive.

Conversely, suppose that $A \in \mathbb{C}^{n\times n}$ is a Hermitian matrix having positive leading principal minors. We prove by induction on n that A is positive definite.

The base case, $n = 1$, is immediate. Suppose that the statement holds for matrices in $\mathbb{C}^{(n-1)\times(n-1)}$. Note that A can be written as

$$A = \begin{pmatrix} B & \mathbf{b} \\ \mathbf{b}^{\mathsf{H}} & a \end{pmatrix},$$

where $B \in \mathbb{C}^{(n-1)\times(n-1)}$ is a Hermitian matrix. Since the leading minors of B are the first $n-1$ leading minors of A, it follows, by the inductive hypothesis, that B is positive definite. Thus, there exists a Cholesky decomposition $B = R^{\mathsf{H}}R$, where R is an upper triangular matrix with real, positive diagonal elements. Since R is invertible, let $\boldsymbol{w} = (R^{\mathsf{H}})^{-1}b$.

The matrix B is invertible. Therefore, by Theorem 5.14, we have $\det(A) = \det(B)(a - \mathbf{b}^{\mathsf{H}}B^{-1}\mathbf{b}) > 0$. Since $\det(B) > 0$, it follows that $a \geqslant \mathbf{b}^{\mathsf{H}}B^{-1}\mathbf{b}$. We observed that if B is positive definite, then so is B^{-1}. Therefore, $a \geqslant 0$ is and we can write $a = c^2$ for some positive c. This allows us to write

$$A = \begin{pmatrix} R^{\mathsf{H}} & \mathbf{0} \\ \boldsymbol{w}^{sH} & c \end{pmatrix} \begin{pmatrix} R & \boldsymbol{w} \\ \mathbf{0}^{\mathsf{H}} & c \end{pmatrix} = C^{\mathsf{H}}C,$$

where C is the upper triangular matrix with positive e

$$C = \begin{pmatrix} R & \boldsymbol{w} \\ \mathbf{0}^{\mathsf{H}} & c \end{pmatrix}.$$

This implies immediately the positive definiteness of A. □

Let $A, B \in \mathbb{C}^{n\times n}$. We write $A \succ B$ if $A - B \succ 0$, that is, if $A - B$ is a positive definite matrix. Similarly, we write $A \succeq B$ if $A - B \succeq O$, that is, if $A - B$ is positive semidefinite.

Theorem 6.60. *Let $A_0, A_1, \ldots, A_m$ be $m+1$ matrices in $\mathbb{C}^{n\times n}$ such that A_0 is positive definite and all matrices are Hermitian. There* *exists $a > 0$ such that for any $t \in [-a, a]$ the matrix $B_m(t) = A_0 + A_1 t + \cdots + A_m t^m$ is positive definite.*

Proof. Since all matrices $A_0, \ldots, A_m$ are Hermitian, note that $\boldsymbol{x}^{\mathsf{H}} A_i \boldsymbol{x}$ are real numbers for $0 \leqslant i \leqslant m$. Therefore, $p_m(t) = \boldsymbol{x}^{\mathsf{H}} B_m(t) \boldsymbol{x}$ is a polynomial in t with real coefficients and $p_m(0) = \boldsymbol{x}^{\mathsf{H}} A_0 \boldsymbol{x}$ is a positive number if $\boldsymbol{x} \neq \mathbf{0}$. Since p_m is a continuous function, there exists an interval $[-a, a]$ such that $t \in [-a, a]$ implies $p_m(t) > 0$ if $\boldsymbol{x} \neq \mathbf{0}$. This shows that $B_m(t)$ is positive definite. □

6.14 The Gram–Schmidt Orthogonalization Algorithm

The Gram–Schmidt algorithm (Algorithm 6.14.1) constructs an orthonormal basis for a subspace U of $\mathbb{C}^n$, starting from an arbitrary basis of $\{\boldsymbol{u}_1, \ldots, \boldsymbol{u}_m\}$ of U. The orthonormal basis is constructed sequentially such that $\langle \boldsymbol{w}_1, \ldots, \boldsymbol{w}_k \rangle = \langle \boldsymbol{u}_1, \ldots, \boldsymbol{u}_k \rangle$ for $1 \leqslant k \leqslant m$.

Algorithm 6.14.1: Gram–Schmidt Orthogonalization Algorithm

Data: A basis $\{\boldsymbol{u}_1, \ldots, \boldsymbol{u}_m\}$ for a subspace U of $\mathbb{C}^n$
Result: An orthonormal basis $\{\boldsymbol{w}_1, \ldots, \boldsymbol{w}_m\}$ for U

$W = O_{n,m}$;
$W(:,1) = W(:,1) + \frac{1}{\|U(:,1)\|_2} U(:,1)$;
for $k = 2$ *to* m **do**
  $P = I_n - W(:, 1:(k-1)) W(:, 1:(k-1))^{\mathsf{H}}$;
  $W(:,k) = W(:,k) + \frac{1}{\|PU(:,k)\|_2} PU(:,k)$;
end
return $W = (\boldsymbol{w}_1 \ \cdots \ \boldsymbol{w}_m)$;

Next, we prove the correctness of the Gram–Schmidt algorithm.

Theorem 6.61. *Let $(\boldsymbol{w}_1, \ldots, \boldsymbol{w}_m)$ be the sequence of vectors constructed by the Gram–Schmidt algorithm starting from the basis $\{\boldsymbol{u}_1, \ldots, \boldsymbol{u}_m\}$ of an m-dimensional subspace U of $\mathbb{C}^n$. The set $\{\boldsymbol{w}_1, \ldots, \boldsymbol{w}_m\}$ is an orthogonal basis of U and $\langle \boldsymbol{w}_1, \ldots, \boldsymbol{w}_k \rangle = \langle \boldsymbol{u}_1, \ldots \boldsymbol{u}_k \rangle$ for $1 \leqslant k \leqslant m$.*

Proof. In the algorithm the matrix W is initialized as $O_{n,m}$. Its columns will contain eventually the vectors of the orthonormal basis $\boldsymbol{w}_1, \ldots, \boldsymbol{w}_m$. The argument is by induction on $k \geqslant 1$.

The base case, $k = 1$, is immediate. Suppose that the statement of the theorem holds for k, that is, the set $\{\boldsymbol{w}_1, \ldots, \boldsymbol{w}_k\}$ is an orthonormal basis for $U_k = \langle \boldsymbol{u}_1, \ldots, \boldsymbol{u}_k \rangle$ and constitutes the set of the initial k columns of the matrix W, that is, $W_k = W(:, 1 : k)$. Then, $P_k = I_n - W_k W_k^{\mathsf{H}}$ is the projection matrix on the subspace $U_k^{\perp}$, so $P_k \boldsymbol{u}_k$ is orthogonal on every $\boldsymbol{w}_i$, where $1 \leqslant i \leqslant k$. Therefore, $\boldsymbol{w}_{k+1} = W(:, (k+1))$ is a unit vector orthogonal on all its predecessors $\boldsymbol{w}_1, \ldots, \boldsymbol{w}_k$, so $\{\boldsymbol{w}_1, \ldots, \boldsymbol{w}_m\}$ is an orthonormal set.

The equality $\langle \boldsymbol{u}_1, \ldots, \boldsymbol{u}_k \rangle = \langle \boldsymbol{w}_1, \ldots, \boldsymbol{w}_k \rangle$ clearly holds for $k = 1$. Suppose that it holds for k. Then, we have

$$\begin{aligned} \boldsymbol{w}_{k+1} &= \frac{1}{\|P_k \boldsymbol{u}_{k+1}\|_2} (\boldsymbol{u}_{k+1} - W_k W_k^{\mathsf{H}} \boldsymbol{u}_{k+1}) \\ &= \frac{1}{\|P_k \boldsymbol{u}_{k+1}\|_2} (\boldsymbol{u}_{k+1} - (\boldsymbol{w}_1 \ \cdots \ \boldsymbol{w}_k) W_k^{\mathsf{H}} \boldsymbol{u}_{k+1}). \end{aligned}$$

Since $\boldsymbol{w}_1, \ldots, \boldsymbol{w}_k$ belong to the subspace $\langle \boldsymbol{u}_1, \ldots, \boldsymbol{u}_k \rangle$ (by inductive hypothesis), it follows that $\boldsymbol{w}_{k+1} \in \langle \boldsymbol{u}_1, \ldots, \boldsymbol{u}_k, \boldsymbol{u}_{k+1} \rangle$, so $\langle \boldsymbol{w}_1, \ldots, \boldsymbol{w}_{k+1} \rangle \subseteq \langle \boldsymbol{u}_1, \ldots, \boldsymbol{u}_k \rangle$. For the converse inclusion, since

$$\boldsymbol{u}_{k+1} = \|P_k \boldsymbol{u}_{k+1}\|_2 \boldsymbol{w}_{k+1} + (\boldsymbol{w}_1 \ \cdots \ \boldsymbol{w}_k) W_k^{\mathsf{H}} \boldsymbol{u}_{k+1},$$

it follows that $\boldsymbol{u}_{k+1} \in \langle \boldsymbol{w}_1, \ldots, \boldsymbol{w}_k, \boldsymbol{w}_{k+1} \rangle$. Thus, $\langle \boldsymbol{u}_1, \ldots, \boldsymbol{u}_k, \boldsymbol{u}_{k+1} \rangle \subseteq \langle \boldsymbol{w}_1, \ldots, \boldsymbol{w}_k, \boldsymbol{w}_{k+1} \rangle$. □

Example 6.31. Let $A \in \mathbb{R}^{3 \times 2}$ be the matrix

$$A = \begin{pmatrix} 1 & 1 \\ 0 & 0 \\ 1 & 3 \end{pmatrix}.$$

It is easy to see that $\mathsf{rank}(A) = 2$. We have $\{\boldsymbol{u}_1, \boldsymbol{u}_2\} \subseteq \mathbb{R}^3$ and we construct an orthogonal basis for the subspace generated by these columns. The matrix W is initialized to $O_{3,2}$

By Algorithm 6.14.1, we begin by defining

$$\boldsymbol{w}_1 = \frac{1}{\|\boldsymbol{u}_1\|_2}\boldsymbol{u}_1 = \begin{pmatrix} \frac{\sqrt{2}}{2} \\ 0 \\ \frac{\sqrt{2}}{2} \end{pmatrix},$$

so

$$W = \begin{pmatrix} \frac{\sqrt{2}}{2} & 0 \\ 0 & 0 \\ \frac{\sqrt{2}}{2} & 0 \end{pmatrix}.$$

The projection matrix is

$$P = I_3 - W(:,1)W(:,1)' = I_3 - \boldsymbol{w}_1\boldsymbol{w}_1' = \begin{pmatrix} \frac{1}{2} & 0 & -\frac{1}{2} \\ 0 & 1 & 0 \\ -\frac{1}{2} & 0 & \frac{1}{2} \end{pmatrix}.$$

The projection of $\boldsymbol{u}_2$ is

$$P\boldsymbol{u}_2 = \begin{pmatrix} -1 \\ 0 \\ 1 \end{pmatrix}$$

and the second column of W becomes

$$\boldsymbol{w}_k = W(:,2) = \frac{\|P\boldsymbol{u}_2\|_2}{P}\boldsymbol{u}_2 = \begin{pmatrix} -\frac{\sqrt{2}}{2} \\ 0 \\ \frac{\sqrt{2}}{2} \end{pmatrix}.$$

Thus, the orthonormal basis we are seeking consists of the vectors

$$\begin{pmatrix} \frac{\sqrt{2}}{2} \\ 0 \\ \frac{\sqrt{2}}{2} \end{pmatrix} \text{ and } \begin{pmatrix} -\frac{\sqrt{2}}{2} \\ 0 \\ \frac{\sqrt{2}}{2} \end{pmatrix}.$$

To decrease the sensitivity of the Gram–Schmidt algorithm to rounding errors, a goal often referred to as *numerical stability*, a modified Gram–Schmidt algorithm has been proposed. In the modified variant, lines 4 and 5 of Algorithm 6.14.1 (in which projection on the subspace $\langle \boldsymbol{w}_1, \ldots, \boldsymbol{w}_k \rangle$ is computed) are replaced by a computation of this projection using the equality

$$I_n - W_k W_k^{\mathsf{H}} = (I_n - \boldsymbol{w}_k \boldsymbol{w}_k^{\mathsf{H}}) \cdots (I_n - \boldsymbol{w}_1 \boldsymbol{w}_1^{\mathsf{H}})$$

of Supplement 6.20. The modified Gram–Schmid algorithm is Algorithm 6.14.2.

Algorithm 6.14.2: The Modified Gram–Schmidt Algorithm

Data: A basis $\{\boldsymbol{u}_1, \ldots, \boldsymbol{u}_m\}$ for a subspace U of $\mathbb{C}^n$
Result: An orthonormal basis $\{\boldsymbol{w}_1, \ldots, \boldsymbol{w}_m\}$ for U

$W = O_{n,m}$;
$W(:,1) = W(:,1) + \frac{1}{\|U(:,1)\|_2} U(:,1)$;
for $k = 2$ *to* m **do**
  $\boldsymbol{t} = U(:,k)$;
  for $j = 1$ *to* k **do**
    $\boldsymbol{t} = (I_n - W(:,j)W(:,j)^{\mathsf{H}})\boldsymbol{t}$;
  end
  $W(:,k) = W(:,k) + \frac{1}{\|\boldsymbol{t}\|_2}\boldsymbol{t}$;
end
return $W = (\boldsymbol{w}_1 \ \cdots \ \boldsymbol{w}_m)$;

Theorem 6.62. *If $L = (\boldsymbol{v}_1, \ldots, \boldsymbol{v}_m)$ is a sequence of m vectors in $\mathbb{R}^n$, we have*

$$\det(G_L) \leqslant \prod_{j=1}^{m} \|\boldsymbol{v}_j\|_2^2.$$

The equality takes place only if the vectors of L are pairwise orthogonal.

Proof. Suppose that L is linearly independent and construct the orthonormal set $\{\boldsymbol{y}_1, \ldots, \boldsymbol{y}_m\}$ as

$$\boldsymbol{y}_j = b_{j1}\boldsymbol{v}_1 + \cdots + b_{jj}\boldsymbol{v}_j$$

for $1 \leqslant j \leqslant m$, using Gram–Schmidt algorithm. Since $b_{jj} \neq 0$, it follows that we can write

$$\boldsymbol{v}_j = c_{j1}\boldsymbol{y}_1 + \cdots + c_{jj}\boldsymbol{y}_j$$

for $1 \leqslant j \leqslant m$ so that $(\boldsymbol{v}_j, \boldsymbol{y}_p) = 0$ if $j < p$ and $(\boldsymbol{v}_j, \boldsymbol{y}_p) = c_{jp}$ if $p \leqslant j$. Thus, we have

$$(\boldsymbol{v}_1, \ldots, \boldsymbol{v}_m) = (\boldsymbol{y}_1, \ldots, \boldsymbol{y}_m) \begin{pmatrix} (\boldsymbol{v}_1, \boldsymbol{y}_1) & (\boldsymbol{v}_2, \boldsymbol{y}_1) & \cdots & (\boldsymbol{v}_m, \boldsymbol{y}_1) \\ 0 & (\boldsymbol{v}_2, \boldsymbol{y}_2) & \cdots & (\boldsymbol{v}_m, \boldsymbol{y}_2) \\ 0 & 0 & \cdots & (\boldsymbol{v}_m, \boldsymbol{y}_3) \\ \vdots & \vdots & \vdots & \vdots \\ 0 & 0 & \cdots & (\boldsymbol{v}_m, \boldsymbol{y}_m) \end{pmatrix}.$$

This implies

$$\begin{pmatrix} (\boldsymbol{v}_1, \boldsymbol{v}_1) & \cdots & (\boldsymbol{v}_1, \boldsymbol{v}_m) \\ \vdots & \cdots & \vdots \\ (\boldsymbol{v}_m, \boldsymbol{v}_1) & \cdots & (\boldsymbol{v}_m, \boldsymbol{v}_m) \end{pmatrix} = \begin{pmatrix} \boldsymbol{v}_1' \\ \vdots \\ \boldsymbol{v}_m' \end{pmatrix} (\boldsymbol{v}_1, \ldots, \boldsymbol{v}_m)$$

$$= \begin{pmatrix} (\boldsymbol{v}_1, \boldsymbol{y}_1) & 0 & 0 \\ (\boldsymbol{v}_2, \boldsymbol{y}_1) & (\boldsymbol{v}_2, \boldsymbol{y}_2) & 0 \\ \vdots & \vdots & \vdots \\ (\boldsymbol{v}_m, \boldsymbol{y}_1) & (\boldsymbol{v}_m, \boldsymbol{y}_2) & (\boldsymbol{v}_m, \boldsymbol{y}_m) \end{pmatrix}$$

$$\times \begin{pmatrix} (\boldsymbol{v}_1, \boldsymbol{y}_1) & (\boldsymbol{v}_2, \boldsymbol{y}_1) & \cdots & (\boldsymbol{v}_m, \boldsymbol{y}_1) \\ 0 & (\boldsymbol{v}_2, \boldsymbol{y}_2) & \cdots & (\boldsymbol{v}_m, \boldsymbol{y}_2) \\ \vdots & \vdots & \vdots & \vdots \\ 0 & 0 & \cdots & (\boldsymbol{v}_m, \boldsymbol{y}_m) \end{pmatrix}.$$

Therefore, we have

$$\det(G_L) = \prod_{i=1}^{m} (\boldsymbol{v}_i, \boldsymbol{y}_i)^2 \leqslant \prod_{i=1}^{m} (\boldsymbol{v}_i, \boldsymbol{v}_i)^2,$$

because $(\boldsymbol{v}_i, \boldsymbol{y}_i)^2 \leqslant (\boldsymbol{v}_i, \boldsymbol{v}_i)^2 (\boldsymbol{y}_i, \boldsymbol{y}_i)^2$ and $(\boldsymbol{y}_i, \boldsymbol{y}_i) = 1$ for $1 \leqslant i \leqslant m$.

To have $\det(G_L) = \prod_{i=1}^{m} (\boldsymbol{v}_i, \boldsymbol{v}_i)^2$, we must have $\boldsymbol{v}_i = k_i \boldsymbol{y}_i$, that is, the vectors $\boldsymbol{v}_i$ must be pairwise orthogonal. □

Corollary 6.26. *Let $A \in \mathbb{R}^{n \times n}$ be a matrix such that $|a_{ij}| \leqslant 1$ for $1 \leqslant i, j \leqslant n$. Then $|\det(A)| \leqslant n^{\frac{n}{2}}$ and the equality holds only if A is a Hadamard matrix.*

Proof. Let $\boldsymbol{a}_i = (a_{i1}, \ldots, a_{in})$ be the ith row of A. We have $\|\boldsymbol{a}_i\|_2 \leqslant \sqrt{n}$, so $|(\boldsymbol{a}_i, \boldsymbol{a}_j)| \leqslant n$ by the Cauchy–Schwartz inequality, for $1 \leqslant i, j \leqslant n$.

Note that $G_L = A'A$, where L is the set of rows of A, so $\det(A)^2 = \det(G_L) \leqslant \prod_{j=1}^{m} \|\boldsymbol{v}_j\|_2^2$, so

$$|\det(A)| \leqslant \prod_{j=1}^{m} \|\boldsymbol{v}_j\|_2 \leqslant n^{\frac{n}{2}}.$$

The equality takes place only when the vectors of A are mutually orthogonal, so when A is a Hadamard matrix. □

We saw that orthonormal sets of vectors that do not contain $\mathbf{0}$ are linearly independent. The Extension Corollary (Corollary 2.3) can now be specialized for orthonormal sets.

Theorem 6.63. *Let V be a finite-dimensional linear space. If U is an orthonormal set of vectors, then there exists a basis T of V that consists of orthonormal vectors such that $U \subseteq T$.*

Proof. Let $U = \{\boldsymbol{u}_1, \ldots, \boldsymbol{u}_m\}$ be an orthonormal set of vectors in V. There is an extension of U, $Z = \{\boldsymbol{u}_1, \ldots, \boldsymbol{u}_m, \boldsymbol{u}_{m+1}, \ldots, \boldsymbol{u}_n\}$ to a basis of V, where $n = \dim(V)$, by the Extension Corollary. Now, apply the Gram–Schmidt algorithm to the set U to produce an orthonormal basis $W = \{\boldsymbol{w}_1, \ldots, \boldsymbol{w}_n\}$ for the entire space V. It is easy to see that $\boldsymbol{w}_i = \boldsymbol{u}_i$ for $1 \leqslant i \leqslant m$, so $U \subseteq W$ and W is the orthonormal basis of V that extends the set U. □

Corollary 6.27. *If $A \in \mathbb{C}^{m\times n}$ is a matrix with $m \geqslant n$ having orthonormal set of columns, then there exists a matrix $B \in \mathbb{C}^{m\times(m-n)}$ such that $(A\ B)$ is an orthogonal (unitary) square matrix.*

Proof. This follows directly from Theorem 6.63. □

Corollary 6.28. *Let U be a subspace of an n-dimensional linear space V such that $\dim(U) = m$, where $m < n$. Then $\dim(U^{\perp}) = n - m$.*

Proof. Let $\boldsymbol{u}_1, \ldots, \boldsymbol{u}_m$ be an orthonormal basis of U, and let

$$\boldsymbol{u}_1, \ldots, \boldsymbol{u}_m, \boldsymbol{u}_{m+1}, \ldots, \boldsymbol{u}_n$$

be its completion to an orthonormal basis for V, which exists by Theorem 6.63. Then $\boldsymbol{u}_{m+1}, \ldots, \boldsymbol{u}_n$ is a basis of the orthogonal complement $U^{\perp}$, so $\dim(U^{\perp}) = n - m$. □

Theorem 6.64. *A subspace U of $\mathbb{R}^n$ is m-dimensional if and only if it is the set of solutions of a homogeneous linear system $A\boldsymbol{x} = \mathbf{0}$, where $A \in \mathbb{R}^{(n-m)\times n}$ is a full-rank matrix.*

Proof. Suppose that U is an m-dimensional subspace of $\mathbb{R}^n$. If $\boldsymbol{v}_1, \ldots, \boldsymbol{v}_{n-m}$ is a basis of the orthogonal complement of U, then

$\boldsymbol{v}_i'\boldsymbol{x} = 0$ for every $\boldsymbol{x} \in U$ and $1 \leqslant i \leqslant n - m$. These conditions are equivalent to the equality

$$(\boldsymbol{v}_1' \; \boldsymbol{v}_2' \; \cdots \; \boldsymbol{v}_{n-m}')\boldsymbol{x} = \mathbf{0},$$

which shows that U is the set of solutions of a homogeneous linear $A\boldsymbol{x} = \mathbf{0}$, where $A = (\boldsymbol{v}_1' \; \boldsymbol{v}_2' \; \cdots \; \boldsymbol{v}_{n-m}')$.

Conversely, if $A \in \mathbb{R}^{(n-m)\times n}$ is a full-rank matrix, then the set of solutions of the homogeneous system $A\boldsymbol{x} = \boldsymbol{b}$ is the null subspace of A and, therefore, it is an m-dimensional subspace. □

6.15 Change of Bases Revisited

Next, we examine the behavior of vector components relative to change of bases in real linear spaces equipped with inner products. In Section 3.7, we saw that if $B = \{\boldsymbol{e}_1, \ldots, \boldsymbol{e}_n\}$ and $\widetilde{B} = \{\tilde{\boldsymbol{e}}_1, \ldots, \tilde{\boldsymbol{e}}_n\}$ are two bases of $\mathbb{C}^n$, then the contravariant components x_k of a vector $\boldsymbol{x}$ relative to the basis B are linked to the contravariant components $\tilde{x}_k$ by equalities of the form $x_k = \sum_{i=1}^n \tilde{x}_i d_{ik}$, where D is the matrix defined by $\tilde{\boldsymbol{e}}_h = \sum_{k=1}^n d_{hk}\boldsymbol{e}_k$.

If the inner product of two vectors of the basis B is $g_{ij} = (\boldsymbol{e}_i, \boldsymbol{e}_j)$ for $1 \leqslant i, j \leqslant n$, the inner product of the vectors $\boldsymbol{u} = \sum_{i=1}^n u_i\boldsymbol{e}_i$ and $\boldsymbol{v} = \sum_{j=1}^n v_j\boldsymbol{e}_j$ can be written as

$$(\boldsymbol{u}, \boldsymbol{v}) = \sum_{i=1}^n \sum_{j=1}^n u_i v_j (\boldsymbol{e}_i, \boldsymbol{e}_j) = \sum_{i=1}^n \sum_{j=1}^n u_i v_j g_{ij},$$

where $g_{ij} = (\boldsymbol{e}_i, \boldsymbol{e}_j)$ for $1 \leqslant i, j \leqslant n$. The matrix $G = (g_{ij})$ is the Gram matrix of vectors of the basis B and, therefore, is a positive definite matrix.

Next, we define the covariant and contravariant components of a vector $\boldsymbol{x}$ in a real n-dimensional linear space L equipped with an inner product.

Definition 6.31. Let L be an n-dimensional inner product real linear space and let $B = \{\boldsymbol{e}_1, \ldots, \boldsymbol{e}_n\}$ be a basis of L. If $\boldsymbol{x} \in L$, the *covariant components* of $\boldsymbol{x}$ are the numbers $x_i = (\boldsymbol{x}, \boldsymbol{e}_i)$ for $1 \leqslant i \leqslant n$.

The contravariant components of $\boldsymbol{x}$ are the numbers x^i defined by $\boldsymbol{x} = x^1\boldsymbol{e}_1 + \cdots + x^n\boldsymbol{e}_n$.

The application of Definition 6.31 yields

$$x_i = (\boldsymbol{x}, \mathbf{e}_i) = \left(\sum_{j=1}^{n} x^j \mathbf{e}_j, \mathbf{e}_i \right)$$
$$= \sum_{j=1}^{n} x^j (\mathbf{e}_j, \mathbf{e}_i) = \sum_{j=1}^{n} g_{ij} x^j$$

for $1 \leqslant i \leqslant n$.

Conversely, the contravariant components of $\boldsymbol{x}$ can be computed from the covariant components x_i by solving the system

$$\begin{pmatrix} x_1 \\ \vdots \\ x_n \end{pmatrix} = G \begin{pmatrix} x^1 \\ \vdots \\ x^n \end{pmatrix},$$

using Cramer's formula

$$x^j = \frac{\det\left(G \overset{j}{\leftarrow} \begin{pmatrix} x_1 \\ \vdots \\ x_n \end{pmatrix} \right)}{\det(G)}.$$

Since

$$\det\left(G \overset{j}{\leftarrow} \begin{pmatrix} x_1 \\ \vdots \\ x_n \end{pmatrix} \right) = \sum_{i=1}^{n} g_{ji} x_i, \tag{6.30}$$

it follows that $x^j = \frac{\sum_{i=1}^{n} g_{ji} x_i}{g}$, where $g = \det(G)$ and g_{ji} is defined by Equality (6.30).

When the basis $\{\mathbf{e}_i \mid 1 \leqslant i \leqslant n\}$ is orthonormal, we have

$$g_{ij} = \begin{cases} 1 & \text{if } i = j, \\ 0 & \text{if } i \neq j, \end{cases}$$

where $1 \leqslant i, j \leqslant n$. In this case, $x_i = x^i$ for $1 \leqslant i \leqslant n$, which means that the contravariant components of $\boldsymbol{x}$ coincide with the covariant components of $\boldsymbol{x}$.

6.16 The QR Factorization of Matrices

We describe two variants of a factorization algorithm for rectangular matrices. The first algorithm allows us to express a matrix as a product of a rectangular matrix with orthogonal columns and an upper triangular invertible matrix (the *thin QR factorization*).

The second algorithm uses Householder matrices and factorizes a matrix as a product of a square orthogonal matrix and an upper triangular matrix with non-negative diagonal entries (the *full QR factorization*).

Theorem 6.65 (The Thin QR factorization theorem). *Let $A \in \mathbb{C}^{m\times n}$ be a full-rank matrix such that $m \geqslant n$. Then A can be factored as $A = QR$, where $Q \in \mathbb{C}^{m\times n}$, $R \in \mathbb{C}^{n\times n}$ such that*

(i) *the columns of Q constitute an orthonormal basis for $\mathsf{range}(A)$, and*

(ii) *$R = (r_{ij})$ is an upper triangular invertible matrix such that its diagonal elements are real non-negative numbers, that is, $r_{ii} \geqslant 0$ for $1 \leqslant i \leqslant n$.*

Proof. Let $\boldsymbol{u}_1, \ldots, \boldsymbol{u}_n$ be the columns of A. Since $\mathsf{rank}(A) = n$, these columns constitute a basis for $\mathsf{range}(A)$. Starting from this set of columns construct an orthonormal basis $\boldsymbol{w}_1, \ldots, \boldsymbol{w}_n$ for the subspace $\mathsf{range}(A)$ using the Gram–Schmidt algorithm. Define Q as the orthogonal matrix

$$Q = (\boldsymbol{w}_1 \ \cdots \ \boldsymbol{w}_n).$$

By the properties of the Gram–Schmidt algorithm, we have $\langle \boldsymbol{u}_1, \ldots, \boldsymbol{u}_k\rangle = \langle \boldsymbol{w}_1, \ldots, \boldsymbol{w}_k\rangle$ for $1 \leqslant k \leqslant n$, so it is possible to write

$$\begin{aligned}\boldsymbol{u}_k &= r_{1k}\boldsymbol{w}_1 + \cdots + r_{kk}\boldsymbol{w}_k \\ &= (\boldsymbol{w}_1 \ \cdots \ \boldsymbol{w}_n)\begin{pmatrix} r_{1k} \\ \vdots \\ r_{kk} \\ 0 \\ \vdots \\ 0 \end{pmatrix} = Q\begin{pmatrix} r_{1k} \\ \vdots \\ r_{kk} \\ 0 \\ \vdots \\ 0 \end{pmatrix}.\end{aligned}$$

We can assume that $r_{kk} \geqslant 0$; otherwise, that is, if $r_{kk} < 0$, replace $\boldsymbol{w}_k$ by $-\boldsymbol{w}_k$. Clearly, this does not affect the orthonormality of the set $\{\boldsymbol{w}_1, \ldots, \boldsymbol{w}_n\}$.

It is clear that $\mathsf{rank}(Q) = n$. Therefore, by Corollary 3.7, since $\mathsf{rank}(A) \leqslant \min\{\mathsf{rank}(Q), \mathsf{rank}(R)\}$, it follows that $\mathsf{rank}(R) = n$, so R is an invertible matrix. Therefore, we have $r_{kk} > 0$ for $1 \leqslant k \leqslant n$. □

Example 6.32. Let us determine a QR factorization for the matrix introduced in Example 6.31. We constructed an orthonormal basis for $\mathsf{range}(A)$ that consists of the vectors

$$\boldsymbol{w}_1 = \begin{pmatrix} \frac{1}{\sqrt{2}} \\ 0 \\ \frac{1}{\sqrt{2}} \end{pmatrix} \text{ and } \boldsymbol{w}_2 = \begin{pmatrix} -\frac{1}{\sqrt{2}} \\ 0 \\ \frac{1}{\sqrt{2}} \end{pmatrix}.$$

Thus, the orthogonal matrix Q is

$$Q = \begin{pmatrix} \frac{1}{\sqrt{2}} & -\frac{1}{\sqrt{2}} \\ 0 & 0 \\ \frac{1}{\sqrt{2}} & \frac{1}{\sqrt{2}} \end{pmatrix}.$$

To compute R, we need to express $\boldsymbol{u}_1$ and $\boldsymbol{u}_2$ as linear combinations of $\boldsymbol{w}_1$ and $\boldsymbol{w}_2$. Since

$$\begin{aligned} \boldsymbol{u}_1 &= \sqrt{2}\boldsymbol{w}_1, \\ \boldsymbol{u}_2 &= 2\sqrt{2}\boldsymbol{w}_1 + \sqrt{2}\boldsymbol{w}_2, \end{aligned}$$

the matrix R is

$$R = \begin{pmatrix} \sqrt{2} & 2\sqrt{2} \\ 0 & \sqrt{2} \end{pmatrix}.$$

Note that the matrix Q obtained by the thin QR decomposition is not a square matrix in general. However, its columns form an orthonormal set. Another variant of the QR decomposition algorithm that produces a first factor Q that is an orthogonal matrix can be obtained using Householder matrices introduced in Example 6.24. We need the following preliminary result.

Theorem 6.66. *Let $\boldsymbol{x}, \boldsymbol{y} \in \mathbb{C}^p$ be two vectors such that $\|\boldsymbol{x}\| = \|\boldsymbol{y}\|$. There exists a Householder matrix $H_{\boldsymbol{v}} \in \mathbb{C}^{p \times p}$ such that $H_{\boldsymbol{v}}\boldsymbol{x} = \boldsymbol{y}$.*

Proof. Let $H_{\boldsymbol{v}} = I_p - 2\boldsymbol{v}\boldsymbol{v}'$, where $\boldsymbol{v} \in \mathbb{C}^p$ is a unit vector. The equality $(I_p - 2\boldsymbol{v}\boldsymbol{v}')\boldsymbol{x} = \boldsymbol{y}$ is equivalent to $\boldsymbol{x} - \boldsymbol{y} = 2\boldsymbol{v}\boldsymbol{v}'\boldsymbol{x} = 2a\boldsymbol{v}$, where $a = \boldsymbol{v}'\boldsymbol{x}$. Then, $\boldsymbol{v} = \frac{1}{2a}(\boldsymbol{x} - \boldsymbol{y})$. Since $\|\boldsymbol{v}\| = 1$, this implies $a = \frac{1}{2}\|\boldsymbol{x} - \boldsymbol{y}\|$, so

$$\boldsymbol{v} = \frac{1}{\|\boldsymbol{x} - \boldsymbol{y}\|}(\boldsymbol{x} - \boldsymbol{y}).$$

□

Corollary 6.29. *If $\boldsymbol{u}, \boldsymbol{w} \in \mathbb{C}^{1\times n}$ are two row vectors such that $\|\boldsymbol{u}\|_2 = \|\boldsymbol{w}\|_2 = 1$, there exists a Householder matrix $H_{\boldsymbol{v}}$ such that $\boldsymbol{u} = \boldsymbol{w}H_{\boldsymbol{v}}$.*

Proof. Clearly, we have $\|\boldsymbol{u}^{\mathsf{H}}\|_2 = \|\boldsymbol{w}^{\mathsf{H}}\|$. Thus, by Theorem 6.66, there exists a matrix $H_{\boldsymbol{v}}$ such that $\boldsymbol{u}^{\mathsf{H}} = H_{\boldsymbol{v}}\boldsymbol{w}^{\mathsf{H}}$, which implies $\boldsymbol{u} = \boldsymbol{w}H_{\boldsymbol{v}}^{\mathsf{H}} = \boldsymbol{w}H_{\boldsymbol{v}}$. □

Theorem 6.66 is used for describing an iterative process that leads to a QR decomposition of a matrix $A \in \mathbb{C}^{m\times n}$, where $m \geqslant n$ and $\mathsf{rank}(A) = n$.

Theorem 6.67 (The Full QR factorization theorem). *Let $A \in \mathbb{C}^{m\times n}$ be a matrix such that $m \geqslant n$ and $\mathsf{rank}(A) = n$. Then A can be factored as*

$$A = Q\begin{pmatrix} R \\ O_{m-n,n} \end{pmatrix},$$

where $Q \in \mathbb{C}^{m\times m}$ and $R \in \mathbb{C}^{n\times n}$ such that

(i) *Q is a unitary matrix, and*

(ii) *$R = (r_{ij})$ is an upper triangular matrix having non-negative diagonal entries.*

Proof. A vector $\boldsymbol{x}$ has the same norm as the vector $\boldsymbol{y} = \|\boldsymbol{x}\|\boldsymbol{e}_1$ (which has only one non-zero component which is non-negative). Therefore, by Theorem 6.66, there exists a Householder matrix $H_{\boldsymbol{v}}$ such that $\boldsymbol{y} = H_{\boldsymbol{v}}\boldsymbol{x}$.

Let $A = (\boldsymbol{c}_1\ \boldsymbol{c}_2\ \cdots\ \boldsymbol{c}_n) \in \mathbb{C}^{m\times n}$. If we multiply A by the Householder matrix $H_{\boldsymbol{v}_1} \in \mathbb{C}^{m\times m}$ that zeroes all components of the first column $\boldsymbol{c}_1$ located below the first element, then we obtain a matrix

$B_1 \in \mathbb{C}^{m\times n}$ having the form

$$B_1 = H_{\boldsymbol{v}_1} A = \begin{pmatrix} * & * & \cdots \\ 0 & * & \cdots \\ \vdots & \vdots & \vdots \\ 0 & * & \cdots \end{pmatrix},$$

where the asterisks represents components that are not necessarily 0. Next, let

$$A_1 = B_1 \begin{bmatrix} 2 & \cdots & m \\ 2 & \cdots & n \end{bmatrix} \in \mathbb{C}^{(m-1)\times(n-1)}$$

and let $H_{\boldsymbol{v}_2} \in \mathbb{C}^{(m-1)\times(m-1)}$ be a Householder matrix that zeroes the elements of the first column of A_1, that is, the subdiagonal elements of the second column of A:

$$H_{\boldsymbol{v}_2} A_1 = \begin{pmatrix} * & * & \cdots \\ 0 & * & \cdots \\ \vdots & \vdots & \vdots \\ 0 & * & \cdots \end{pmatrix} \in \mathbb{C}^{(m-1)\times(n-1)}.$$

Observe that by multiplying B_1 by the matrix

$$\begin{pmatrix} 1 & \mathbf{0}' \\ \mathbf{0} & H_{\boldsymbol{v}_2} \end{pmatrix} \in \mathbb{C}^{m\times m},$$

the first line and the first row of B_1 are left intact and the submatrix

$$\left(\begin{pmatrix} 1 & \mathbf{0}' \\ \mathbf{0} & H_{\boldsymbol{v}_2} \end{pmatrix} B_1\right) \begin{bmatrix} 2 & \cdots & m \\ 2 & \cdots & n \end{bmatrix}$$

coincides with $H_{\boldsymbol{v}_2} A_1$. Define

$$B_2 = \begin{pmatrix} 1 & \mathbf{0}' \\ \mathbf{0} & H_{\boldsymbol{v}_2} \end{pmatrix} B_1 = \begin{pmatrix} 1 & \mathbf{0}' \\ \mathbf{0} & H_{\boldsymbol{v}_2} \end{pmatrix} R_{v_1} A = \begin{pmatrix} * & * & \cdots & * \\ 0 & * & \cdots & * \\ 0 & 0 & \cdots & * \\ \vdots & \vdots & \cdots & * \\ 0 & 0 & \cdots & * \end{pmatrix} \in \mathbb{C}^{m\times n}.$$

Continuing this construction, we obtain a sequence of matrices $B_1, B_2, \ldots, B_{n-1}$ in $\mathbb{C}^{m\times n}$ such that

$$B_{k+1} = \begin{pmatrix} I_k & O \\ O & H_{\boldsymbol{v}_{k+1}} \end{pmatrix} B_k,$$

where $H_{\boldsymbol{v}_{k+1}}$ zeroes the subdiagonal elements of the $(k+1)$st column of the matrix B_k, where $1 \leqslant k \leqslant n-1$.

Thus, we have $B_n = QA$, where B_n is an upper triangular matrix of the form

$$\begin{pmatrix} * & * & \cdots & * \\ 0 & * & \cdots & * \\ 0 & 0 & \cdots & * \\ \vdots & \vdots & \vdots & 0 \\ \vdots & \vdots & \cdots & 0 \\ 0 & 0 & \cdots & 0 \end{pmatrix},$$

and $Q \in \mathbb{C}^{m\times m}$ is a unitary matrix. This allows us to write $A = Q'B_n$, which is the desired decomposition of the matrix A. Note that $B_n \in \mathbb{C}^{m\times n}$ can be written as

$$B_n = \begin{pmatrix} R \\ O_{m-n,n} \end{pmatrix},$$

where R is an upper triangular matrix in $\mathbb{C}^{n\times n}$ having non-negative diagonal entries. □

Observe that by writing $Q' = (Q_0\ Q_1)$, where $Q_0 \in \mathbb{C}^{m\times n}$ and $Q_1 \in \mathbb{C}^{m\times(m-n)}$, we have $A = Q_0 R$, which is the thin QR decomposition discussed in Theorem 6.65.

Corollary 6.30. *Let $A \in \mathbb{C}^{m\times n}$ be a matrix such that $n \geqslant m$ and $\mathsf{rank}(A) = m$. Then, A can be factored as*

$$A = (L\ \ O_{m,n-m})Q$$

where $Q \in \mathbb{C}^{n\times n}$ and $L \in \mathbb{C}^{m\times m}$ such that

(i) *$L = (l_{ij})$ is a lower triangular matrix having non-negative diagonal entries, and*

(ii) *Q is a unitary matrix.*

Proof. The statement is obtained by applying Theorem 6.67 to the matrix $A^{\mathsf{H}} \in \mathbb{C}^{n\times m}$. □

By choosing carefully the sequence of Householder matrices, one can prove similar results.

Theorem 6.68. *Let $A \in \mathbb{C}^{m\times n}$ be a matrix such that $m \geqslant n$ and $\mathsf{rank}(A) = n$. Then, A can be factored as*

$$A = \begin{pmatrix} R \\ O_{m-n,n} \end{pmatrix} S,$$

where $S \in \mathbb{C}^{n\times n}$ and $R \in \mathbb{C}^{n\times n}$ such that

(i) *S is a unitary matrix, and*

(ii) *R is an upper triangular matrix having real non-negative elements on its diagonal ending in the bottom right corner.*

Also, if $\mathsf{rank}(A) = m$, then A can be factored as

$$A = T(L \;\; O_{m,n-m})$$

where $T \in \mathbb{C}^{m\times m}$ and $L \in \mathbb{C}^{m\times m}$ such that

(i) *L is a lower triangular matrix with real non-negative elements on its diagonal ending in the bottom right corner, and*

(ii) *T is a unitary matrix.*

Proof. We prove only the first part of the theorem, which is similar to the proof of the full QR Decomposition Theorem.

Let $\boldsymbol{r}_1, \ldots, \boldsymbol{r}_m$ be the rows of the matrix A. Clearly, the row vector $\|\boldsymbol{r}_m\|\boldsymbol{e}_n^{\mathsf{H}}$ has the same norm as $\boldsymbol{r}_m$, so, by Corollary 6.29, there exists a unit vector $\boldsymbol{v}_m \in \mathbb{R}^n$ such that $\boldsymbol{r}_m H_{\boldsymbol{v}_m} = \|\boldsymbol{r}_m\|\boldsymbol{e}_n^{\mathsf{H}}$. Therefore, we have

$$B_1 = AH_{\boldsymbol{v}_m} = \begin{pmatrix} * & * & \cdots & * & * \\ * & * & \cdots & * & * \\ \vdots & \vdots & \vdots & \vdots & \vdots \\ 0 & 0 & \cdots & 0 & * \end{pmatrix}.$$

Let

$$A_1 = B_1 \begin{bmatrix} 1 & \cdots & (m-1) \\ 1 & \cdots & (n-1) \end{bmatrix} \in \mathbb{C}^{(m-1)\times(n-1)}$$

and let $H_{\boldsymbol{v}_{m-1}} \in \mathbb{C}^{(n-1)\times(n-1)}$ be a Householder matrix that zeroes the first $n-2$ elements of the last row of A_1:

$$A_1 H_{\boldsymbol{v}_2} = \begin{pmatrix} * & * & \cdots \\ * & * & \cdots \\ \vdots & \vdots & \vdots \\ 0 & * & \cdots \end{pmatrix} \in \mathbb{C}^{(m-1)\times(n-1)}.$$

By multiplying $B_1 = AH_{\boldsymbol{v}_m}$ by

$$\begin{pmatrix} H_{\boldsymbol{v}_{m-1}} & \mathbf{0}_{n-1} \\ \mathbf{0}'_{n-1} & 1 \end{pmatrix},$$

the last row and the last column of B_1 are left intact and the submatrix

$$\left(\begin{pmatrix} B_1 H_{\boldsymbol{v}_{m-1}} & \mathbf{0}_{n-1} \\ \mathbf{0}'_{n-1} & 1 \end{pmatrix}\right) \begin{bmatrix} 1 & \cdots & m-1 \\ 1 & \cdots & n-1 \end{bmatrix}$$

coincides with $A_1 H_{\boldsymbol{v}_{m-1}}$.

Continuing this construction, we obtain a sequence of matrices $B_1, B_2, \ldots, B_m$ in $\mathbb{C}^{m\times n}$ such that

$$B_{k+1} = B_k \begin{pmatrix} H_{\boldsymbol{v}_{m-(k+1)}} & \mathbf{0}_{n-1} \\ \mathbf{0}'_{n-1} & 1 \end{pmatrix},$$

where $H_{\boldsymbol{v}_{m-(k+1)}}$ zeroes the elements of the $(k+1)$st row of the matrix B_k placed at the left of the main diagonal, where $1 \leqslant k \leqslant m-1$.

Thus, for the last matrix B_m we have $B_m = AS$, where B_m is an upper triangular matrix and S is a unitary matrix (as a product of unitary matrices).

The second part follows from the first part. □

The subspaces associated with symmetric and idempotent matrices enjoy a special relationship that we discuss next.

Theorem 6.69. *Let $A \in \mathbb{C}^{n\times n}$ be an idempotent matrix. Then,* $\mathsf{null}(A) \perp \mathsf{range}(A)$ *if and only if the matrix A is Hermitian.*

Proof. Let A be an idempotent matrix such that $\mathsf{null}(A) \perp \mathsf{range}(A)$. We have $A\boldsymbol{u} \in \mathsf{range}(A)$. Also, we have

$$A(I_n - A)\boldsymbol{v} = (A - A^2)\boldsymbol{v} = \boldsymbol{0},$$

so $(I_n - A)\boldsymbol{v} \in \mathsf{null}(A)$. This implies $A\boldsymbol{u} \perp (I - A)\boldsymbol{v}$ for every $\boldsymbol{u}, \boldsymbol{v} \in \mathbb{C}^n$. Thus,

$$(A\boldsymbol{u})^{\mathsf{H}}(I_n - A)\boldsymbol{v} = \boldsymbol{u}^{\mathsf{H}}A^{\mathsf{H}}(I_n - A)\boldsymbol{v} = 0$$

for every $\boldsymbol{u}, \boldsymbol{v} \in \mathbb{R}^n$, so $A^{\mathsf{H}}(I_n - A) = O$. Equivalently, $A_{\mathsf{H}} = A^{\mathsf{H}}A$. Therefore, $A = A^{\mathsf{H}}$.

Conversely, if A is both idempotent and Hermitian, then

$$(A\boldsymbol{u})^{\mathsf{H}}(I_n - A)\boldsymbol{v} = \boldsymbol{u}^{\mathsf{H}}A^{\mathsf{H}}(I_n - A)\boldsymbol{v} = \boldsymbol{u}^{\mathsf{H}}A(I_n - A)\boldsymbol{v} = \boldsymbol{u}^{\mathsf{H}}(A - A^2)\boldsymbol{v} = 0$$

for every $\boldsymbol{u}, \boldsymbol{v} \in \mathbb{C}^n$, so $\mathsf{null}(A) \perp \mathsf{range}(A)$. □

Note that if A is Hermitian and idempotent, then the mapping that transforms a vector $\boldsymbol{x}$ into $A\boldsymbol{x}$ projects $\mathbb{C}^n$ onto the subspace $\mathsf{range}(A)$.

Next we examine the possibility of constructing a unitary matrix P for a matrix $A \in \mathbb{C}^{n\times n}$ such that the matrix $P^{\mathsf{H}}AP$ is an upper Hessenberg matrix. To this end, we define a sequence of Householder matrices as follows. Let $A_0 = A$ and write

$$A_0 = \begin{pmatrix} a_{11} & \boldsymbol{b}_0' \\ \boldsymbol{a}_0 & B_0 \end{pmatrix},$$

where $\boldsymbol{a}_0, \boldsymbol{b}_0 \in \mathbb{C}^{n-1}$ and $B_0 \in \mathbb{C}^{(n-1)\times(n-1)}$. Consider the Householder matrix $H_{\boldsymbol{v}_0}$ (where $\boldsymbol{v}_0 \in \mathbb{C}^{n-1}$) such that

$$H_{\boldsymbol{v}_0}\boldsymbol{a}_0 = \begin{pmatrix} * \\ 0 \\ \vdots \\ 0 \end{pmatrix}$$

and define

$$P_1 = \begin{pmatrix} 1 & \boldsymbol{0}' \\ \boldsymbol{0} & H_{\boldsymbol{v}_0} \end{pmatrix}.$$

Then

$$P_1A_0P_1 = \begin{pmatrix} 1 & \mathbf{0}' \\ \mathbf{0} & H_{\boldsymbol{v}_0} \end{pmatrix} \begin{pmatrix} a_{11} & \boldsymbol{b}_0' \\ \boldsymbol{a}_0 & B_0 \end{pmatrix} \begin{pmatrix} 1 & \mathbf{0}' \\ \mathbf{0} & H_{\boldsymbol{v}_0} \end{pmatrix}$$

$$= \begin{pmatrix} a_{11} & \boldsymbol{a}_0' H_{\boldsymbol{v}_0} \\ H_{\boldsymbol{v}_0}\boldsymbol{a}_0 & H_{\boldsymbol{v}_0} B_0 H_{\boldsymbol{v}_0} \end{pmatrix} = \begin{pmatrix} a_{11} & \boldsymbol{a}_0' H_{\boldsymbol{v}_0} \\ * & \\ 0 & \\ \vdots & H_{\boldsymbol{v}_0} B_0 H_{\boldsymbol{v}_0} \\ 0 & \end{pmatrix}.$$

Define $A_1 = H_{\boldsymbol{v}_0} B_0 H_{\boldsymbol{v}_0}$ and repeat the process on this matrix to obtain a unitary matrix $H_{\boldsymbol{v}_1}$ such that the elements of the first column located under the first subdiagonal element in $H_{\boldsymbol{v}_1} A_1 H_{\boldsymbol{v}_1}$ are 0s. The matrix P_2 is given by

$$P_2 = \begin{pmatrix} I_2 & O \\ O & H_{\boldsymbol{v}_2} \end{pmatrix}$$

and it is clearly a unitary matrix. We have

$$P_2^{\mathsf{H}} P_1^{\mathsf{H}} A P_1 P_2 = \begin{pmatrix} * & * & * & \cdots & * \\ * & * & * & \cdots & * \\ 0 & * & * & \cdots & * \\ 0 & 0 & * & \cdots & * \\ \vdots & \vdots & * & \cdots & * \\ 0 & 0 & * & \cdots & * \end{pmatrix}.$$

Continuing in the same manner, we build the matrices $P_1, P_2, \ldots, P_{n-2}$ such that $C = P_{n-2}^{\mathsf{H}} \cdots P_1^{\mathsf{H}} A P_1 \cdots P_{n-2}$ is an upper Hessenberg matrix. If A is a symmetric matrix, then so is C. Therefore, C is actually a tridiagonal matrix, that is, a matrix of the form

$$\begin{pmatrix} * & * & 0 & 0 & \cdots & 0 & 0 & 0 \\ * & * & * & 0 & \cdots & 0 & 0 & 0 \\ * & * & * & 0 & \cdots & 0 & 0 & 0 \\ \vdots & \vdots & \vdots & \vdots & \cdots & \vdots & \vdots & \vdots \\ 0 & 0 & 0 & 0 & \cdots & * & * & * \\ 0 & 0 & 0 & 0 & \cdots & 0 & * & * \end{pmatrix}.$$

6.17 Matrix Groups

One can define on the set of invertible square matrices $\mathbb{C}^{n\times n}$ a group structure, where the group operation is the matrix multiplication, the unit of the group is I_n, and the inverse of a matrix A is A^{-1}. This is the *general linear group* $\mathsf{GL}_n(\mathbb{C})$ and we shall examine several of its subgroups.

Similar linear group of invertible square matrices in $\mathbb{R}^{n\times n}$ will be denoted by $\mathsf{GL}_n(\mathbb{R})$. Since every matrix with real entries also belongs to $\mathbb{C}^n$, it is clear that $\mathsf{GL}_n(\mathbb{R})$ is a subgroup of $\mathsf{GL}_n(\mathbb{C})$.

Subgroups of $\mathsf{GL}_n(\mathbb{C})$ are known as *linear groups.*

Definition 6.32. An *isometry* of $\mathbb{R}^n$ is a mapping $f : \mathbb{R}^n \longrightarrow \mathbb{R}^n$ such that

$$\|\boldsymbol{x} - \boldsymbol{y}\|_2 = \|f(\boldsymbol{x}) - f(\boldsymbol{y})\|_2$$

for $\boldsymbol{x}, \boldsymbol{y} \in \mathbb{C}^n$ (for $\boldsymbol{x}, \boldsymbol{y} \in \mathbb{R}^n$).

Example 6.33. Let $\boldsymbol{z} \in \mathbb{R}^n$ and let $t_{\boldsymbol{z}}$ be the translation defined by $t_{\boldsymbol{z}}(\boldsymbol{x}) = \boldsymbol{x} + \boldsymbol{z}$ for $\boldsymbol{x} \in \mathbb{R}^n$. Since $t_{\boldsymbol{z}}(\boldsymbol{x}) - t_{\boldsymbol{z}}(\boldsymbol{y}) = \boldsymbol{x} - \boldsymbol{y}$, it is clear that any translation is an isometry.

It is straightforward to verify that every isometry is a bijection having as inverse an isometry and, therefore, the set of isometries of $\mathbb{R}^n$ is a group with respect to function composition. We denote this group by $\mathsf{ISO}(\mathbb{R}^n)$.

Theorem 6.70. *Let $f : \mathbb{R}^n \longrightarrow \mathbb{R}^n$. The following statements are equivalent:*

(i) *f is an isometry such that $f(\mathbf{0}_n) = \mathbf{0}_n$;*
(ii) *f preserves the Euclidean inner product, that is $f(\boldsymbol{x}'\boldsymbol{y}) = f(\boldsymbol{x})'f(\boldsymbol{y})$ for every $\boldsymbol{x}, \boldsymbol{y} \in \mathbb{R}^n$;*
(iii) *$f(\boldsymbol{x}) = A\boldsymbol{x}$, where A is an orthogonal matrix.*

Proof. `(i) implies (ii)`: Let f be an isometry such that $f(\mathbf{0}_n) = \mathbf{0}_n$. Since f preserves distances, we have

$$(\boldsymbol{x} - \boldsymbol{y})'(\boldsymbol{x} - \boldsymbol{y}) = (f(\boldsymbol{x}) - f(\boldsymbol{y})')(f(\boldsymbol{x}) - f(\boldsymbol{y})). \tag{6.31}$$

Taking $\boldsymbol{y} = \mathbf{0}_n$ in the above equality, we obtain

$$\boldsymbol{x}'\boldsymbol{x} = f(\boldsymbol{x})'f(\boldsymbol{x}) \tag{6.32}$$

for $\boldsymbol{x} \in \mathbb{R}^n$. Equality (6.31) is equivalent to

$$\boldsymbol{x}'\boldsymbol{x} - 2\boldsymbol{y}'\boldsymbol{x} + \boldsymbol{y}'\boldsymbol{y} = f(\boldsymbol{x})'f(\boldsymbol{x}) - 2f(\boldsymbol{y})'f(\boldsymbol{x}) + f(\boldsymbol{y})'f(\boldsymbol{y})$$

and taking into account Equality (6.32), we have $\boldsymbol{y}'\boldsymbol{x} = f(\boldsymbol{y})'f(\boldsymbol{x})$, which means that f preserves the Euclidean inner product.

`(ii) implies (iii)`: Suppose that $f : \mathbb{R}^n \longrightarrow \mathbb{R}^n$ is a mapping that preserves the inner product. Then, $1 = \boldsymbol{e}_i'\boldsymbol{e}_i = f(\boldsymbol{e}_i)'f(\boldsymbol{e}_i)$ and $0 = \boldsymbol{e}_i'\boldsymbol{e}_j = f(\boldsymbol{e}_i)'f(\boldsymbol{e}_j)$ for $i \neq j$, $1 \leqslant i, j \leqslant n$. Thus, $\{f(\boldsymbol{e}_1), \ldots, f(\boldsymbol{e}_n)\}$ is an orthonormal set and the matrix

$$A = (f(\boldsymbol{e}_1) \ \cdots \ f(\boldsymbol{e}_n))$$

is orthogonal. Since the set of orthogonal matrices is a subgroup of $\mathsf{GL}_n(\mathbb{R})$, the matrix A^{-1} is also orthogonal and, by Corollary 6.16, it preserves the inner product. Thus, the mapping $g : \mathbb{R}^n \longrightarrow \mathbb{R}^n$ given by $g(\boldsymbol{x}) = A^{-1}f(\boldsymbol{x})$ also preserves the inner product and, in addition, $g(\boldsymbol{e}_i) = A^{-1}f(\boldsymbol{e}_i) = \boldsymbol{e}_i$ for $1 \leqslant i \leqslant n$.

Since g preserves the inner product, we have

$$x_i = \boldsymbol{x}'\boldsymbol{e}_i = g(\boldsymbol{x})'g(\boldsymbol{e}_i) = g(\boldsymbol{x})'\boldsymbol{e}_i = (g(\boldsymbol{x}))_i$$

for every $\boldsymbol{x} \in \mathbb{R}^n$ and $1 \leqslant i \leqslant n$, which means that g is the identity mapping on $\mathbb{R}^n$. Consequently, $f(\boldsymbol{x}) = A\boldsymbol{x}$.

`(iii) implies (i)`: This implication follows immediately from Corollary 6.16. □

Theorem 6.71. *An isometry h of $\mathbb{R}^n$ has the form $h(\boldsymbol{x}) = A\boldsymbol{x} + \boldsymbol{b}$ for $\boldsymbol{x} \in \mathbb{R}^n$.*

Proof. Let $\boldsymbol{b} = h(\boldsymbol{0}_n)$. The mapping $f : \mathbb{R}^n \longrightarrow \mathbb{R}^n$ given by $f(\boldsymbol{x}) = h(\boldsymbol{x}) - h(\boldsymbol{0}_n)$ has the property $f(\boldsymbol{0}_n) = \boldsymbol{0}_n$. Moreover, f is an isometry because $f(\boldsymbol{x}) - f(\boldsymbol{y}) = h(\boldsymbol{x}) - h(\boldsymbol{y})$. Therefore, by Theorem 6.70, there exists an orthogonal matrix $A \in \mathbb{R}^{n \times n}$ such that $f(\boldsymbol{x}) = A\boldsymbol{x}$ for $\boldsymbol{x} \in \mathbb{R}^n$. This implies that $h(\boldsymbol{x}) = A\boldsymbol{x} + \boldsymbol{b}$ for $\boldsymbol{x} \in \mathbb{R}^n$, where $\boldsymbol{b} = h(\boldsymbol{0}_n)$. □

Theorem 6.71 means that every isometry is the composition of a multiplication by an orthogonal matrix followed by a translation.

An isometry $f(\boldsymbol{x}) = A\boldsymbol{x} + \boldsymbol{b}$ for $\boldsymbol{x} \in \mathbb{R}^n$ is *orientation-preserving* if the orthogonal matrix A is a rotation matrix, and is *orientation-reversing* if A is a reflection matrix.

Definition 6.33. A *dilation* is a mapping $h_a : \mathbb{R}^n \longrightarrow \mathbb{R}^n$ such that $h_a(\boldsymbol{x}) = a\boldsymbol{x}$ for $\boldsymbol{x} \in \mathbb{R}^n$, where $a \in R$ is the *ratio* of the dilation.

A dilation does not preserve distances (so it is not an isometry), but it preserves angles. Indeed, for $\boldsymbol{u}, \boldsymbol{v} \in \mathbb{R}^n$, we have

$$\begin{aligned}\cos \angle(h_a(\boldsymbol{u}), h_a(\boldsymbol{v})) &= \frac{h_a(\boldsymbol{u})' h_a(\boldsymbol{v}))}{\|h_a(\boldsymbol{u})\|_2 \|h_a(\boldsymbol{v})\|_2} = \frac{\boldsymbol{u}'\boldsymbol{v}}{\|\boldsymbol{u}\|_2 \|\boldsymbol{v}\|_2} \\ &= \cos \angle(\boldsymbol{u}, \boldsymbol{v}).\end{aligned}$$

The set of dilations having a non-zero ratio is easily seen to be a group, where $(h_a)^{-1} = h_{\frac{1}{a}}$.

6.18 Condition Numbers for Matrices

Let $A\boldsymbol{u} = \boldsymbol{b}$ be a linear system, where $A \in \mathbb{C}^{n \times n}$ is a non-singular matrix and $\boldsymbol{b} \in \mathbb{R}^n$. We examine the sensitivity of the solution of this system to small variations of $\boldsymbol{b}$. So, together with the original system, we work with a system of the form $A\boldsymbol{v} = \boldsymbol{b} + \boldsymbol{h}$, where $\boldsymbol{h} \in \mathbb{R}^n$ is the *perturbation* of $\boldsymbol{b}$. Note that $A(\boldsymbol{v} - \boldsymbol{u}) = \boldsymbol{h}$, so $\boldsymbol{v} - \boldsymbol{u} = A^{-1}\boldsymbol{b}$. Using a vector norm $\|\cdot\|$ and its corresponding matrix norm $|||\cdot|||$, we have $\|\boldsymbol{v} - \boldsymbol{u}\| = \|A^{-1}\boldsymbol{h}\| \leqslant |||A^{-1}||| \|\boldsymbol{h}\|$.

Since $\|\boldsymbol{b}\| = \|A\boldsymbol{u}\| \leqslant |||A||| \|\boldsymbol{u}\|$, it follows that

$$\begin{aligned}\frac{\|\boldsymbol{v} - \boldsymbol{u}\|}{\|\boldsymbol{u}\|} &\leqslant \frac{|||A^{-1}||| \|\boldsymbol{h}\|}{\frac{\|\boldsymbol{b}\|}{|||A|||}} \\ &= \frac{|||A||| |||A^{-1}||| \|\boldsymbol{h}\|}{\|\boldsymbol{b}\|}. \qquad (6.33)\end{aligned}$$

Thus, the relative variation of the solution, $\frac{\|\boldsymbol{v}-\boldsymbol{u}\|}{\|\boldsymbol{u}\|}$, is upper bounded by the number $\frac{|||A||| |||A^{-1}||| \|\boldsymbol{h}\|}{\|\boldsymbol{b}\|}$. These considerations justify the following definition.

Definition 6.34. Let $A \in \mathbb{C}^{n \times n}$ be a non-singular matrix. The *condition number of A relative to the matrix norm $|||\cdot|||$* is the number $\mathsf{cond}(A) = |||A||| |||A^{-1}|||$.

Equality (6.33) implies that if the condition number is large, then small variations in $\boldsymbol{b}$ may generate large variations in the solution of the system $A\boldsymbol{u} = \boldsymbol{b}$, especially when $\boldsymbol{b}$ is close to $\boldsymbol{0}$. When this is the case, we say that the system $A\boldsymbol{u} = \boldsymbol{b}$ is *ill-conditioned.* Otherwise, the system $A\boldsymbol{u} = \boldsymbol{b}$ is *well-conditioned.*

Theorem 6.72. *Let $A \in \mathbb{C}^{n\times n}$ be a non-singular matrix. The following statements hold for every matrix norm induced by a vector norm:*

(i) $\mathsf{cond}(A) = \mathsf{cond}(A^{-1})$;
(ii) $\mathsf{cond}(cA) = |c|\mathsf{cond}(A)$;
(iii) $\mathsf{cond}(A) \geqslant 1$.

Proof. We prove here only Part (iii). Since $AA^{-1} = I$, by the properties of a matrix norm induced by a vector norm, we have $\mathsf{cond}(A) = \|A\|\|A^{-1}\| \geqslant \|AA^{-1}\| = \|I_n\| = 1$. □

Let A, B be two non-singular matrices in $C^{n\times n}$ such that $B = aA$, where $a \in \mathbb{C}$. We have $B^{-1} = aA^{-1}$, $\|B\| = |a|\|B\|$, and $\|B^{-1}\| = |a|\|A^{-1}\|$ so $\mathsf{cond}(B) = |a|^2\mathsf{cond}(A)$. On the other hand, $\det(B) = a^n\det(A)$. Thus, if n is large enough and $a < 1$, then $\det(B)$ can be quite close to 0, while the condition number of B may be quite large. This shows that the determinant and the condition number are relatively independent.

Example 6.34. Let $A \in \mathbb{C}^{2\times 2}$ be the matrix

$$A = \begin{pmatrix} a & a+\alpha \\ a+\alpha & a+2\alpha \end{pmatrix},$$

where $a > 0$ and $\alpha < 0$. We have

$$A^{-1} = \begin{pmatrix} -\frac{a+2\alpha}{\alpha^2} & \frac{a+\alpha}{\alpha^2} \\ \frac{a+\alpha}{\alpha^2} & -\frac{a}{\alpha^2} \end{pmatrix},$$

so $\|A\|_1 = a$ and $\|A^{-1}\| = \frac{a}{\alpha^2}$. Thus, $\mathsf{cond}(A) = \left(\frac{a}{\alpha}\right)^2$ and, if $|\alpha|$ is small, a system of the for $A\boldsymbol{u} = \boldsymbol{b}$ may be ill-conditioned.

Ill-conditioned linear systems $A\boldsymbol{u} = \boldsymbol{b}$ may occur when large differences in scale exist among the columns of A, or among the rows of A.

Theorem 6.73. *Let $A = (\boldsymbol{a}_1 \ \cdots \ \boldsymbol{a}_n)$ be an invertible matrix in $\mathbb{C}^{n\times n}$, where $\boldsymbol{a}_1, \ldots, \boldsymbol{a}_n$ are the columns of A. Then*

$$\mathsf{cond}(A) \geqslant \max\left\{\frac{\|\boldsymbol{a}_i\|}{\|\boldsymbol{a}_j\|} \mid 1 \leqslant i, j \leqslant n\right\}.$$

Proof. Since $\mathsf{cond}(A) = \|A\| \|A^{-1}\|$, we have

$$\mathsf{cond}(A) = \frac{\max\{\|A\boldsymbol{x}\| \mid \|\boldsymbol{x}\| = 1\}}{\min\{\|A\boldsymbol{x}\| \mid \|\boldsymbol{x}\| = 1\}},$$

by Supplement 6.20.

Note that $A\boldsymbol{e}_k = \boldsymbol{a}_k$, where $\boldsymbol{a}_k$ is the kth column of A and that $\|\boldsymbol{e}_k\| = 1$. Therefore,

$$\max\{\|A\boldsymbol{x}\| \mid \|\boldsymbol{x}\| = 1\} \geqslant \|\boldsymbol{a}_i\|,$$
$$\min\{\|A\boldsymbol{x}\| \mid \|\boldsymbol{x}\| = 1\} \leqslant \|\boldsymbol{a}_j\|,$$

which implies

$$\mathsf{cond}(A) \geqslant \frac{\|\boldsymbol{a}_i\|}{\|\boldsymbol{a}_j\|}$$

for all $1 \leqslant i, j \leqslant n$. This yields the inequality of the theorem. □

Example 6.35. Let

$$A = \begin{pmatrix} 1 & 0 \\ 1 & \alpha \end{pmatrix},$$

where $\alpha \in \mathbb{R}$ and $\alpha > 0$. The matrix A is invertible and

$$A^{-1} = \begin{pmatrix} 1 & 0 \\ -\frac{1}{\alpha} & \frac{1}{\alpha} \end{pmatrix}.$$

It is easy to see that the condition number of A relative to the Frobenius norm is

$$\mathsf{cond}(A) = \frac{2 + \alpha^2}{\alpha}.$$

Thus, if α is sufficiently close to 0, the condition number can reach arbitrarily large values.

In general, we use as matrix norms, norms of the form $\| \cdot \|_p$. The corresponding condition number of a matrix A is denoted by $\mathsf{cond}_p(A)$.

Example 6.36. Let $A = \mathsf{diag}(a_1, \ldots, a_n)$ be a diagonal matrix. Then, $\|A\|_2 = \max_{1 \leqslant i \leqslant n} |a_i|$. Since $A^{-1} = \mathsf{diag}\left(\frac{1}{a_1}, \ldots, \frac{1}{a_n}\right)$, it follows that $\|A^{-1}\|_2 = \frac{1}{\min_{1 \leqslant i \leqslant n} |a_i|}$, so $\mathsf{cond}_2(A) = \frac{\max_{1 \leqslant i \leqslant n} |a_i|}{\min_{1 \leqslant i \leqslant n} |a_i|}$.

6.19 Linear Space Orientation

Definition 6.35. Let V be an n-dimensional linear-space. An *ordered basis* for V is a sequence $\boldsymbol{B} = (\boldsymbol{u}_1, \ldots, \boldsymbol{u}_n)$ such that $B = \{\boldsymbol{u}_1, \ldots, \boldsymbol{u}_n\}$ is a basis for n.

Since B is a basis, note that its Gram matrix G_B is positive definite by Theorem 6.55; also, $\det(G_B) \neq 0$.

Definition 6.36. The ordered basis $\boldsymbol{B}$ has a *positive orientation* if $\det(G_{\boldsymbol{B}}) > 0$, and a *negative orientation* if $\det(G_{\boldsymbol{B}}) < 0$.

Example 6.37. Let $\{\boldsymbol{e}_1, \boldsymbol{e}_2, \boldsymbol{e}_3\}$ be the standard basis in $\mathbb{R}^3$. The ordered basis $\boldsymbol{B} = (\boldsymbol{e}_1, \boldsymbol{e}_2, \boldsymbol{e}_3)$ has a positive orientation because

$$\det(G_{\boldsymbol{B}}) = \det \begin{pmatrix} 1 & 0 & 0 \\ 0 & 1 & 0 \\ 0 & 0 & 1 \end{pmatrix} = 1 > 0.$$

On the other hand, $\tilde{B} = (\boldsymbol{e}_1, \boldsymbol{e}_3, \boldsymbol{e}_2)$ has a negative orientation because

$$\det(G_{\boldsymbol{B}}) = \det \begin{pmatrix} 1 & 0 & 0 \\ 0 & 0 & 1 \\ 0 & 1 & 0 \end{pmatrix} = -1 < 0.$$

We extend this definition to an arbitrary sequence of vectors $\boldsymbol{T} = (\boldsymbol{t}_1, \ldots, \boldsymbol{t}_n)$ in $\mathbb{R}^n$ by saying that T has a positive orientation if $\det(G_{\boldsymbol{T}}) > 0$ and a negative orientation if $\det(G_{\boldsymbol{T}}) < 0$.

Definition 6.37. Let $\boldsymbol{u}, \boldsymbol{v}$ be two vectors in $\mathbb{R}^3$. The *cross-product* of $\boldsymbol{u}$ and $\boldsymbol{v}$ is the vector $\boldsymbol{u} \times \boldsymbol{v}$ defined as

$$\boldsymbol{u} \times \boldsymbol{v} = (u_2v_3 - u_3v_2)\boldsymbol{e}_1 + (u_3v_1 - u_1v_3)\boldsymbol{e}_2 + (u_1v_2 - u_2v_1)\boldsymbol{e}_3.$$

Note that $\boldsymbol{u} \times \boldsymbol{v}$ can be written as a determinant, that is,

$$\boldsymbol{u} \times \boldsymbol{v} = \begin{vmatrix} \boldsymbol{e}_1 & \boldsymbol{e}_2 & \boldsymbol{e}_3 \\ u_1 & u_2 & u_3 \\ v_1 & v_2 & v_3 \end{vmatrix}.$$

The vector $\boldsymbol{w} = \boldsymbol{u} \times \boldsymbol{v}$ is orthogonal on both $\boldsymbol{u}$ and $\boldsymbol{v}$. Indeed, it is easy to see that

$$(\boldsymbol{w}, \boldsymbol{u}) = (u_2v_3 - u_3v_2)u_1 + (u_3v_1 - u_1v_3)u_2 + (u_1v_2 - u_2v_1)u_3 = 0,$$

and, similarly, $(\boldsymbol{w}, \boldsymbol{v}) = 0$.

Note that the triple $\boldsymbol{T} = (\boldsymbol{u}, \boldsymbol{v}, \boldsymbol{u} \times \boldsymbol{v})$ has a positive orientation when $\boldsymbol{u} \times \boldsymbol{v} \neq \boldsymbol{0}_3$ and when $\boldsymbol{u}$ and $\boldsymbol{v}$ are not collinear. Indeed, since

$$G_{\boldsymbol{T}} = \begin{pmatrix} \|\boldsymbol{u}\|^2 & (\boldsymbol{u}, \boldsymbol{v}) & 0 \\ (\boldsymbol{v}, \boldsymbol{u}) & \|\boldsymbol{v}\|^2 & 0 \\ 0 & 0 & \|\boldsymbol{u} \times \boldsymbol{v}\|^2 \end{pmatrix},$$

we have $\det(G_{\boldsymbol{T}}) = (\|\boldsymbol{u}\|^2\|\boldsymbol{v}\|^2 - (\boldsymbol{u}, \boldsymbol{v})^2)\|\boldsymbol{u} \times \boldsymbol{v}\|^2 > 0$.

Finally,

$$\begin{aligned} \|\boldsymbol{w}\|^2 &= (u_2v_3 - u_3v_2)^2 + (u_3v_1 - u_1v_3)^2 + (u_1v_2 - u_2v_1)^2 \\ &= (u_1^2 + u_2^2 + u_3^2)(v_1^2 + v_2^2 + v_3^2) \\ &\quad -(u_1v_1 + u_2v_2 + u_3v_3)^2 \\ &= \|\boldsymbol{u}\|^2\|\boldsymbol{v}\|^2(1 - \cos^2(\angle(\boldsymbol{u}, \boldsymbol{v}))). \end{aligned}$$

Thus, $\|\boldsymbol{w}\| = \|\boldsymbol{u}\|\|\boldsymbol{v}\| \sin \alpha$, where $\alpha = \angle(\boldsymbol{u}, \boldsymbol{v})$. We note that $\|w\|$ equals the area of the parallelogram formed by the vectors $\boldsymbol{u}$ and $\boldsymbol{v}$.

The following properties are immediate:

(i) $\boldsymbol{u} \times \boldsymbol{v} = -(\boldsymbol{v} \times \boldsymbol{u})$,

(ii) $(a\boldsymbol{u}) \times \boldsymbol{v} = a(\boldsymbol{u} \times \boldsymbol{v})$,

(iii) $(\boldsymbol{u} + \boldsymbol{v}) \times \boldsymbol{w} = \boldsymbol{u} \times \boldsymbol{w} + \boldsymbol{v} \times \boldsymbol{w}$

for $\boldsymbol{u}, \boldsymbol{v}, \boldsymbol{w} \in \mathbb{R}^3$, and $a \in \mathbb{R}$.

Definition 6.38. Let $\boldsymbol{u}, \boldsymbol{v}, \boldsymbol{w} \in \mathbb{R}^3$. The *scalar triple product* of these vectors is the real number $((\boldsymbol{u} \times \boldsymbol{v}), \boldsymbol{w})$ denoted as $(\boldsymbol{u}, \boldsymbol{v}, \boldsymbol{w})$.

It is easy to verify the following properties:

(i) $(\boldsymbol{u}, \boldsymbol{v}, \boldsymbol{w}) = -(\boldsymbol{v}, \boldsymbol{u}, \boldsymbol{w})$,

(ii) $(\boldsymbol{u}, \boldsymbol{v}, \boldsymbol{w}) = (\boldsymbol{v}, \boldsymbol{w}, \boldsymbol{u}) = (\boldsymbol{w}, \boldsymbol{u}, \boldsymbol{v})$,

(iii) $(a\boldsymbol{u}, \boldsymbol{v}, \boldsymbol{w}) = a(\boldsymbol{u}, \boldsymbol{v}, \boldsymbol{w})$,

(iv) $(\boldsymbol{u} + \boldsymbol{t}, \boldsymbol{v}, \boldsymbol{w}) = (\boldsymbol{u}, \boldsymbol{v}, \boldsymbol{w}) + (\boldsymbol{t}, \boldsymbol{v}, \boldsymbol{w})$.

We claim that $(\boldsymbol{u}, \boldsymbol{v}, \boldsymbol{w})$ equals the volume of the parallelepiped constructed on the sequence of vectors $(\boldsymbol{u}, \boldsymbol{v}, \boldsymbol{w})$. Indeed, since $\|\boldsymbol{u} \times \boldsymbol{v}\|$ is the area of the parallelogram formed by the vectors $\boldsymbol{u}$ and $\boldsymbol{v}$, $((\boldsymbol{u} \times \boldsymbol{v}), \boldsymbol{w})$ is the product of this area with the projection of $\boldsymbol{w}$ on a vector perpendicular to the parallelogram determined by $\boldsymbol{u}$ and $\boldsymbol{v}$.

Note that

$$
\begin{aligned}
(\boldsymbol{u}, \boldsymbol{v}, \boldsymbol{w}) &= \left(\begin{vmatrix} \boldsymbol{e}_1 & \boldsymbol{e}_2 & \boldsymbol{e}_3 \\ u_1 & u_2 & u_3 \\ v_1 & v_2 & v_3 \end{vmatrix}, w_1\boldsymbol{e}_1 + w_2\boldsymbol{e}_2 + w_3\boldsymbol{e}_3 \right) \\
&= \begin{vmatrix} w_1 & w_2 & w_3 \\ u_1 & u_2 & u_3 \\ v_1 & v_2 & v_3 \end{vmatrix} = \begin{vmatrix} u_1 & u_2 & u_3 \\ v_1 & v_2 & v_3 \\ w_1 & w_2 & w_3 \end{vmatrix}.
\end{aligned}
$$

Definition 6.39. Let $\boldsymbol{u}, \boldsymbol{v}, \boldsymbol{w} \in \mathbb{R}^3$. The *vector triple product* of these vectors is the vector $\boldsymbol{u} \times (\boldsymbol{v} \times \boldsymbol{w})$.

Note that

$$
\begin{aligned}
(\boldsymbol{u} \times (\boldsymbol{v} \times \boldsymbol{w}))_1 &= u_2(v_1w_2 - v_2w_1) - u_3(v_3w_1 - v_1w_3) \\
&= v_1(u_2w_2 + u_3w_3) - w_1(u_2v_2 + u_3v_3) \\
&= v_1(u_1w_1 + u_2w_2 + u_3w_3) - w_1(u_1v_1 + u_2v_2 + u_3v_3) \\
&= v_1(\boldsymbol{u}, \boldsymbol{w}) - w_1(\boldsymbol{u}, \boldsymbol{v}).
\end{aligned}
$$

Similarly, we have

$$
\begin{aligned}
(\boldsymbol{u} \times (\boldsymbol{v} \times \boldsymbol{w}))_2 &= v_2(\boldsymbol{u}, \boldsymbol{w}) - w_2(\boldsymbol{u}, \boldsymbol{v}), \\
(\boldsymbol{u} \times (\boldsymbol{v} \times \boldsymbol{w}))_3 &= v_3(\boldsymbol{u}, \boldsymbol{w}) - w_3(\boldsymbol{u}, \boldsymbol{v}),
\end{aligned}
$$

which allows us to write

$$
\boldsymbol{u} \times (\boldsymbol{v} \times \boldsymbol{w}) = \boldsymbol{v}(\boldsymbol{u}, \boldsymbol{w}) - \boldsymbol{w}(\boldsymbol{u}, \boldsymbol{v}).
$$

6.20 MATLAB Computations

Vector norms can be computed using the function `norm` which comes in two signatures: `norm(v)` and `norm(v,p)`. The first variant computes $\|\boldsymbol{v}\|_2$; the second computes $\|\boldsymbol{v}\|_p$ for any p, $1 \leqslant p \leqslant \infty$. In addition, `norm(v,inf)` computes $\|\boldsymbol{v}\|_\infty = \max\{|v_i| \mid 1 \leqslant i \leqslant n\}$, where $\boldsymbol{v} \in \mathbb{R}^n$. If one uses $-\infty$ as the second parameter, then `norm(v,-inf)` returns $\min\{|v_i| \mid 1 \leqslant i \leqslant n\}$.

Example 6.38. For the vector

```
v = [2 -3 5 -4]
```

the computation

```
norms = [norm(v,1),norm(v,2),norm(v,2.5),norm(v,inf),
         norm(v,-inf)]
```

returns

```
norms =

   14.0000    7.3485    6.5344    5.0000    2.0000
```

If the first argument of `norm` is a matrix A, `norm(A)` returns $\|A\|_2$. For the two-parameter format, the second parameter is restricted to the values $1, 2$, `inf`, and `'fro'`. Then, `norm(A,2)` is the same as `norm(A)`, `norm(A,1)` is $\|A\|_1$, `(A,inf)` is $\|A\|_\infty$, and `norm(A,'fro')` yields the Frobenius norm of A.

Example 6.39. For the matrix

```
A = [1 -1 2; 3 2 -1; 5 4 2]
```

the following computation is performed:

```
norms = [norm(A,1),norm(A,2),norm(A,inf),norm(A,'fro')]

norms =

    9.0000    7.4783   11.0000    8.0623
```

For matrices whose norm is expensive to compute, an approximative estimation of $\|A\|_2$ can be performed using the function `normest(A)`, or `normest(A,r)`, where r is the relative error; the default for r is 10^{-6}.

The following function implements the Gram–Schmidt algorithm:

```
function [W] = gram(U)
%GRAM implements the classical Gram-Schmidt algorithm
    [n,m] = size(U);
    W = zeros(n,m);
    W(:,1)= (1/norm(U(:,1)))*U(:,1);
    for k = 2:1:m
          P = eye(n) - W*W';
          W(:,k) = W(:,k) + (1/norm(P*U(:,k)))* P*U(:,k);
    end
end
```

An implementation of the modified Gram–Schmidt algorithm is given next.

```
function [W] = modgram(U)
%MODGRAM implements the modified Gram--Schmidt algorithm
    [n,m] = size(U);
    W = zeros(n,m);
    W(:,1)= (1/norm(U(:,1)))*U(:,1);
    for k = 2:1:m
              t = U(:,k);
              for j = 1:1:k
                  t = (eye(n) - W(:,j)*W(:,j)')*t;
              end
          W(:,k) = W(:,k) + (1/norm(t))*t;
    end
end
```

The Cholesky decomposition of a Hermitian positive definite matrix is computed in MATLAB using the function `chol`. The function call `R = chol(A)` returns an upper triangular matrix R, satisfying the equation $R^{\mathsf{H}}R = A$. If A is not positive definite, an error message is generated. The matrix R is computed using the diagonal and the upper triangle of A and the computation makes sense only if A is Hermitian.

Example 6.40. Let A be the symmetric positive definite matrix considered in Example 6.29,

$$A = \begin{pmatrix} 3 & 0 & 2 \\ 0 & 2 & 1 \\ 2 & 1 & 2 \end{pmatrix}.$$

Then `R = chol(A)` yields

```
R =
     1.7321         0    1.1547
          0    1.4142    0.7071
          0         0    0.4082
```

The call `L = chol(A,'lower')` returns a lower triangular matrix `L` from the diagonal and lower triangle of matrix A, satisfying the equation $LL^{\mathsf{H}} = A$. When A is sparse, this syntax of `chol` is faster.

Example 6.41. For the same matrix A as in Example 6.40, `L = chol(A,'lower')` returns

```
L =
      1.7321         0         0
           0    1.4142         0
      1.1547    0.7071    0.4082
```

For added flexibility, `[R,p] = chol(A)` and `[L,p] = chol (A,'lower')` set `p` to 0 if A is positive definite and to a positive number, otherwise, without returning an error message.

The thin QR decomposition of a matrix $A \in \mathbb{C}^{m\times n}$ is obtained using the function `qr` as in

```
[Q R] = qr(A)
```

To obtain the full decomposition, we write

```
[Q R] = qr(A,0)
```

The Hessenberg form of a matrix is computed using the function `hess`. To produce a Hessenberg matrix H and a unitary matrix P such that $P^{\mathsf{H}}AP = C$ (or $A = PCP^{\mathsf{H}}$), one can use

```
[P,H] = hess(A)
```

For example, if

```
A =
      1     2     3     4
      5     6     2     5
     -2     2     4     1
      2    -4     1     2
```

then `[P,H] = hess(A)` will return

```
P =
      1.0000         0         0         0
           0   -0.8704   -0.4694    0.1486
           0    0.3482   -0.3733    0.8599
           0   -0.3482    0.8002    0.4883
H =
      1.0000   -2.0889    1.1421    4.8303
     -5.7446    4.1212   -2.0302   -3.3597
           0    5.7089    2.8879   -2.9554
           0         0    0.1780    4.9909
```

If only the Hessenberg form is desired, one could use the function call `H = hess(A)`.

Note that the matrix

```
B =
     1     2     3     4
     2     2     5     6
     3     5     3     7
     4     6     7     4
```

is symmetric, so its Hessenberg form obtained with `[P1,H1] = hess(B)`,

```
P1 =
    0.6931   -0.6010   -0.3980         0
   -0.6931   -0.4039   -0.5970         0
    0.1980    0.6897   -0.6965         0
         0         0         0    1.0000

H1 =
   -0.9118   -0.8868         0         0
   -0.8868   -2.1872    0.1000         0
         0    0.1000    9.0990  -10.0499
         0         0  -10.0499    4.0000
```

is a tridiagonal matrix.

The condition number of a matrix A is computed using the function `cond(A,p)` which returns the p-norm condition of matrix A. When used with a single parameter, as in `cond(A)`, the 2-norm condition number of A is returned.

Example 6.42. Let A be the matrix

```
>> A=[10.1 6.2; 5.1 3.1]
A =
   10.1000    6.2000
    5.1000    3.1000
```

The condition number `cond(A)` is 567.966, which is quite large indicating significant sensitivity to inverse calculations. The inverse of A is

```
>> inv(A)
ans =
  -10.0000   20.0000
   16.4516  -32.5806
```

If we make a small change in A yielding the matrix

```
>> B=[10.2 6.3;5.1 3.1]
B =
   10.2000    6.3000
    5.1000    3.1000
```

the inverse of B changes completely:

```
>> inv(B)
ans =
   -6.0784   12.3529
   10.0000  -20.0000
```

Values of the condition number close to 1 indicate a well-conditioned matrix, and the opposite is true for large values of the condition number.

Example 6.43. Consider the linear systems:

$$\begin{aligned} 10.1x_1 + 6.2x_2 &= 12 \\ 5.1x_1 + 3.1x_2 &= 6 \end{aligned} \quad \text{and} \quad \begin{aligned} 10.2x_1 + 6.3x_2 &= 12 \\ 5.1x_1 + 3.1x_2 &= 6 \end{aligned}$$

that correspond to $A\boldsymbol{x} = \boldsymbol{b}$ and $B\boldsymbol{x} = \boldsymbol{b}$, where $\boldsymbol{b} = \begin{pmatrix} 12 \\ 6 \end{pmatrix}$. In view of the resemblance of A and B, one would expect their solutions to be close. However, this is not the case. The solution of $A\boldsymbol{x} = \boldsymbol{b}$ is

```
>> x=inv(A)*b
x =
         0
    1.9355
```

while the solution of $B\boldsymbol{x} = \boldsymbol{b}$ is

```
>> x=inv(B)*b
x =
    1.1765
         0
```

Exercises and Supplements

(1) Let ν be a norm on $\mathbb{C}^n$. Prove that there exists a number $k \in \mathbb{R}$ such that for any vector $\boldsymbol{x} \in \mathbb{C}^n$, we have $\nu(\boldsymbol{x}) \leqslant k \sum_{i=1}^{n} |x_i|$.

Solution: Starting from the equality $\boldsymbol{x} = \sum_{i=1}^{n} x_i \boldsymbol{e}_i$, we have $\nu(\boldsymbol{x}) \leqslant \sum_{i=1}^{n} \nu(x_i \boldsymbol{e}_i) = \sum_{i=1}^{n} |x_i| \nu(\boldsymbol{e}_i) \leqslant k \sum_{i=1}^{n} |x_i|$, where $k = \max\{\nu(\boldsymbol{e}_i) \mid 1 \leqslant i \leqslant n\}$.

(2) Prove that the mapping $s : \mathbb{R}^n \times \mathbb{R}^n \longrightarrow \mathbb{R}$ defined by $s(\boldsymbol{x}, \boldsymbol{y}) = d(\boldsymbol{x}, \boldsymbol{y})_2^2$ is not a metric itself. Show that $s(\boldsymbol{x}, \boldsymbol{y}) \leqslant 2(s(\boldsymbol{x}, \boldsymbol{z}) + s(\boldsymbol{z}, \boldsymbol{y}))$ for $\boldsymbol{x}, \boldsymbol{y}, \boldsymbol{z} \in \mathbb{R}^n$.

(3) Prove that if $\boldsymbol{x}, \boldsymbol{y}, \boldsymbol{z}$ are three vectors in $\mathbb{R}^n$ and ν is a norm on V, then $\nu(\boldsymbol{x} - \boldsymbol{y}) \leqslant \nu(\boldsymbol{x} - \boldsymbol{z}) + \nu(\boldsymbol{z} - \boldsymbol{y})$.

(4) Prove that for any vector norm ν on $\mathbb{R}^n$, we have

$$\nu(\boldsymbol{x} + \boldsymbol{y})^2 + \nu(\boldsymbol{x} - \boldsymbol{y})^2 \leqslant 4(\nu(\boldsymbol{x})^2 + \nu(\boldsymbol{y})^2)$$

for every $\boldsymbol{x}, \boldsymbol{y} \in \mathbb{R}^n$.

Solution: Note that $\nu(\boldsymbol{x} + \boldsymbol{y}) \leqslant \nu(\boldsymbol{x}) + \nu(\boldsymbol{y})$; similarly, $\nu(\boldsymbol{x} - \boldsymbol{y}) \leqslant \nu(\boldsymbol{x}) + \nu(\boldsymbol{y})$. Thus,

$$\begin{aligned}\nu(\boldsymbol{x} + \boldsymbol{y})^2 &\leqslant \nu(\boldsymbol{x})^2 + \nu(\boldsymbol{y})^2 + 2\nu(\boldsymbol{x})\nu(\boldsymbol{y}),\\ \nu(\boldsymbol{x} - \boldsymbol{y})^2 &\leqslant \nu(\boldsymbol{x})^2 + \nu(\boldsymbol{y})^2 + 2\nu(\boldsymbol{x})\nu(\boldsymbol{y}),\end{aligned}$$

hence,

$$\nu(\boldsymbol{x} + \boldsymbol{y})^2 + \nu(\boldsymbol{x} - \boldsymbol{y})^2 \leqslant 2\nu(\boldsymbol{x})^2 + 2\nu(\boldsymbol{y})^2 + 4\nu(\boldsymbol{x})\nu(\boldsymbol{y}).$$

The desired inequality follows from

$$2\nu(\boldsymbol{x})\nu(\boldsymbol{y}) \leqslant \nu(\boldsymbol{x})^2 + \nu(\boldsymbol{y})^2.$$

(5) Let $\boldsymbol{x} \in \mathbb{R}^n$. Prove that for every $\epsilon > 0$ there exists $\boldsymbol{y} \in \mathbb{R}^n$ such that the components of the vector $\boldsymbol{x} + \boldsymbol{y}$ are distinct and $\|\boldsymbol{y}\|_2 < \epsilon$.

Solution: Partition the set $\{1, \ldots, n\}$ into the blocks $B_1, \ldots, B_k$ such that all components of $\boldsymbol{x}$ that have an index in B_j have a common value c_j. Suppose that $|B_j| = p_j$. Then $\sum_{j=1}^{k} p_j = n$ and the numbers $\{c_1, c_2, \ldots, c_k\}$ are pairwise distinct. Let $d = \min_{i,j} |c_i - c_j|$. The vector $\boldsymbol{y}$ can be defined as follows. If $B_j = \{i_1, \ldots, i_{p_j}\}$, then

$$y_{i_1} = \eta \cdot 2^{-1}, y_{i_2} = \eta \cdot 2^{-2}, \ldots, y_{i_{p_j}} = \eta \cdot 2^{-p},$$

where $\eta > 0$, which makes the numbers $c_j + y_{i_1}, c_j + y_{i_2}, \ldots, c_j + y_{i_{p_j}}$ pairwise distinct. It suffices to take $\eta < d$ to ensure that the components of $\boldsymbol{x}+\boldsymbol{y}$ are pairwise distinct. Also, note that $\|\boldsymbol{y}\|_2^2 \leqslant \sum_{j=1}^k p_j \frac{\eta^2}{4} = \frac{n\eta^2}{4}$. It suffices to choose η such that $\eta < \min\{d, \frac{2\epsilon}{n}\}$ to ensure that $\|\boldsymbol{y}\|_2 < \epsilon$.

(6) Let $\mu^{(m,n)} : \mathbb{C}^{m\times n} \longrightarrow \mathbb{R}_{\geqslant 0}$ be a vectorial matrix norm. Prove that for every $A \in \mathbb{C}^{m\times n}$ there exists a constant $k \in \mathbb{R}$ such that $\mu^{(m,n)}(A) \leqslant k \sum_{i=1}^m \sum_{j=1}^n |a_{ij}|$.

Hint: Apply Supplement 6.20.

(7) Prove that a matrix $A \in \mathbb{C}^{n\times n}$ is normal if and only if $\|A\boldsymbol{x}\| = \|A^{\mathsf{H}}\boldsymbol{x}\|$ for all $\boldsymbol{x} \in \mathbb{C}^n$.

(8) Let $\{\mu^{(m,n)} \mid m,n \in \mathbb{N}_{>0}\}$ be a consistent family of matrix norms and let $\boldsymbol{a} \in \mathbb{C}^n$ be a vector. Prove that for each $m \in \mathbb{N}$, the function $\nu_m : \mathbb{R}^{m}`\mathbb{R}_{\geqslant 0}$ defined by $\nu_m(\boldsymbol{x}) = \mu^{(m,n)}(\boldsymbol{x}\boldsymbol{a}^{\mathsf{H}})$ is a vector norm that is consistent with the family of matrix norms.

(9) Prove that a matrix $A \in \mathbb{C}^{n\times n}$ is normal if and only if $\|A\boldsymbol{x}\|_2 = \|A^{\mathsf{H}}\boldsymbol{x}\|_2$ for every $\boldsymbol{x} \in \mathbb{C}^n$.

Solution: Suppose that A is normal. Then

$$\begin{aligned}\|A\boldsymbol{x}\|_2^2 &= (A\boldsymbol{x}, A\boldsymbol{x}) = (\boldsymbol{x}, A^{\mathsf{H}}A\boldsymbol{x}) = (\boldsymbol{x}, AA^{\mathsf{H}}\boldsymbol{x})\\ &= (A^{\mathsf{H}}\boldsymbol{x}, A^{\mathsf{H}}\boldsymbol{x}) = \|A^{\mathsf{H}}\boldsymbol{x}\|_2^2.\end{aligned}$$

Conversely, suppose that $\|A\boldsymbol{x}\|_2 = \|A^{\mathsf{H}}\boldsymbol{x}\|_2$ for every $\boldsymbol{x} \in \mathbb{C}^n$. Let $\lambda \in \mathbb{C}$ be such that $|\lambda| = 1$. We have

$$\begin{aligned}\|A(\lambda\boldsymbol{x}+\boldsymbol{y})\|_2^2 &= \|\lambda A\boldsymbol{x} + A\boldsymbol{y}\|_2^2\\ &= (\overline{\lambda}\boldsymbol{x}^{\mathsf{H}}A^{\mathsf{H}} + \boldsymbol{y}^{\mathsf{H}}A^{\mathsf{H}}, \lambda A\boldsymbol{x} + A\boldsymbol{y})\\ &= (\overline{\lambda}\boldsymbol{x}^{\mathsf{H}}A^{\mathsf{H}}, \lambda A\boldsymbol{x}) + (\overline{\lambda}\boldsymbol{x}^{\mathsf{H}}A^{\mathsf{H}}, A\boldsymbol{y})\\ &\quad +(\boldsymbol{y}^{\mathsf{H}}A^{\mathsf{H}}, \lambda A\boldsymbol{x}) + (\boldsymbol{y}^{\mathsf{H}}A^{\mathsf{H}}, A\boldsymbol{y})\\ &= (\boldsymbol{x}^{\mathsf{H}}A^{\mathsf{H}}, A\boldsymbol{x}) + (\boldsymbol{y}^{\mathsf{H}}A^{\mathsf{H}}, A\boldsymbol{y})\\ &\quad +(\overline{\lambda}\boldsymbol{x}^{\mathsf{H}}A^{\mathsf{H}}, A\boldsymbol{y}) + (\boldsymbol{y}^{\mathsf{H}}A^{\mathsf{H}}, \lambda A\boldsymbol{x})\\ &= \|A\boldsymbol{x}\|_2^2 + \|A\boldsymbol{y}\|_2^2 + 2\Re(\lambda(A\boldsymbol{x}, A\boldsymbol{y})),\end{aligned}$$

because $(\overline{\lambda}\boldsymbol{x}^{\mathsf{H}}A^{\mathsf{H}}, A\boldsymbol{y}) + (\boldsymbol{y}^{\mathsf{H}}A^{\mathsf{H}}, \lambda A\boldsymbol{x}) = 2\Re(\lambda(A\boldsymbol{x}, A\boldsymbol{y}))$. Similarly,

$$\|A^{\mathsf{H}}(\lambda\boldsymbol{x}+\boldsymbol{y})\|_2^2 = \|A^{\mathsf{H}}\boldsymbol{x}\|_2^2 + \|A^{\mathsf{H}}\boldsymbol{y}\|_2^2 + 2\Re(\lambda(A^{\mathsf{H}}\boldsymbol{x}, A^{\mathsf{H}}\boldsymbol{y})),$$

hence,

$$\|A\boldsymbol{x}\|_2^2 + \|A\boldsymbol{y}\|_2^2 + 2\Re(\lambda(A\boldsymbol{x}, A\boldsymbol{y})) = \|A^{\mathsf{H}}\boldsymbol{x}\|_2^2 \\ +\|A^{\mathsf{H}}\boldsymbol{y}\|_2^2 + 2\Re(\lambda(A^{\mathsf{H}}\boldsymbol{x}, A^{\mathsf{H}}\boldsymbol{y})).$$

Since $\|A\boldsymbol{x}\|_2^2 = \|A^{\mathsf{H}}\boldsymbol{x}\|_2^2$ and $\|A\boldsymbol{y}\|_2^2 = \|A^{\mathsf{H}}\boldsymbol{y}\|_2^2$, we obtain

$$\begin{aligned}\Re(\lambda(A\boldsymbol{x}, A\boldsymbol{y}) - \lambda(A^{\mathsf{H}}\boldsymbol{x}, A^{\mathsf{H}}\boldsymbol{y})) &= \Re(\lambda(\boldsymbol{x}, A^{\mathsf{H}}A\boldsymbol{y}) - \lambda(\boldsymbol{x}, AA^{\mathsf{H}}\boldsymbol{y})) \\ &= \Re(\lambda(\boldsymbol{x}, (A^{\mathsf{H}}A - AA^{\mathsf{H}})\boldsymbol{y})) = 0,\end{aligned}$$

hence, $|\boldsymbol{x}, (A^{\mathsf{H}}A - AA^{\mathsf{H}})\boldsymbol{y})| = 0$ for every $\boldsymbol{x} \in \mathbb{C}^n$. Therefore, $(A^{\mathsf{H}}A - AA^{\mathsf{H}})\boldsymbol{y} = 0$ for every $\boldsymbol{y} \in \mathbb{C}^n$. This implies $A^{\mathsf{H}}A - AA^{\mathsf{H}} = O_{n,n}$, hence, $A^{\mathsf{H}}A = AA^{\mathsf{H}}$.

(10) Prove that for every matrix $A \in \mathbb{C}^{n\times n}$, we have $\|A\|_2 = \|A^{\mathsf{H}}\|_2$.

(11) Let $U \in \mathbb{C}^{n\times n}$ be a matrix whose set of columns is orthonormal and let $V \in \mathbb{C}^{n\times n}$ be a matrix whose set of rows is orthonormal. Prove that $\|U\|_2 = \|V\|_2 = 1$.

Solution: Since $U^{\mathsf{H}}U = I_n$, we have $\|U\|_2 = \max\{\|U\boldsymbol{x}\|_2 \mid \boldsymbol{x} \in \mathbb{C}^n and \|\boldsymbol{x}\|_2 = 1\}$. For $\boldsymbol{x} \in \mathbb{C}^n$ such that $\|\boldsymbol{x}\|_2 = 1$, we have

$$\|U\boldsymbol{x}\|_2^2 = \boldsymbol{x}^{\mathsf{H}}U^{\mathsf{H}}U\boldsymbol{x} = \boldsymbol{x}^{\mathsf{H}}\boldsymbol{x} = \|\boldsymbol{x}\|_2^2 = 1,$$

which implies $\|U\|_2 = 1$. The similar argument for V follows from Exercise 6.20.

(12) Let $U \in \mathbb{C}^{n\times n}$ be a matrix whose set of columns is orthonormal and let $V \in \mathbb{C}^{n\times n}$ be a matrix whose set of rows is orthonormal. Prove that $\|UA\|_2 = \|AV\|_2 = \|A\|_2$ and, therefore, $\|UAV\|_2 = \|A\|_2$.

Solution: By hypothesis, we have $U^{\mathsf{H}}U = I_n$ and $VV^{\mathsf{H}} = 1$. Therefore, $\|UA\|_2^2 = \max\{\|UA\boldsymbol{x}\|_2^2 \mid \|\boldsymbol{x}\|_2 = 1\} = \max\{\boldsymbol{x}^{\mathsf{H}}A^{\mathsf{H}}U^{\mathsf{H}}UA\boldsymbol{x} \mid \|\boldsymbol{x}\|_2 = 1\} = \max\{\boldsymbol{x}^{\mathsf{H}}A^{\mathsf{H}}A\boldsymbol{x} \mid \|\boldsymbol{x}\|_2 = 1\} = \max\{\|A\boldsymbol{x}\|_2^2 \mid \|\boldsymbol{x}\|_2 = 1\} = \|A\|_2^2$. This allows us to conclude that $\|UA\|_2 = \|A\|_2$.

The second equality follows immediately from the first.

(13) Let $A \in \mathbb{R}^{m\times n}$. Prove that there exists i, $1 \leqslant i \leqslant n$ such that $\|A\boldsymbol{e}_i\|_2^2 \geqslant \frac{1}{n}\|A\|_F^2$.

(14) Let $A, B \in \mathbb{C}^{n\times n}$ be two matrices such that $AB = BA$. Prove that if A is a normal matrix, then $A^{\mathsf{H}}B = BA^{\mathsf{H}}$.

Solution: Let $C = A^{\mathsf{H}}B - BA^{\mathsf{H}}$. We have $C^{\mathsf{H}} = B^{\mathsf{H}}A - AB^{\mathsf{H}}$ and, therefore,

$$\begin{aligned}\mathsf{trace}(CC^{\mathsf{H}}) &= \mathsf{trace}((A^{\mathsf{H}}B - BA^{\mathsf{H}})(B^{\mathsf{H}}A - AB^{\mathsf{H}}))\\ &= \mathsf{trace}(A^{\mathsf{H}}BB^{\mathsf{H}}A) - \mathsf{trace}(A^{\mathsf{H}}BAB^{\mathsf{H}})\\ &\quad -\mathsf{trace}(BA^{\mathsf{H}}B^{\mathsf{H}}A) + \mathsf{trace}(BA^{\mathsf{H}}AB^{\mathsf{H}}).\end{aligned}$$

Since

$$\begin{aligned}\mathsf{trace}(A^{\mathsf{H}}BB^{\mathsf{H}}A) &= \mathsf{trace}(AA^{\mathsf{H}}BB^{\mathsf{H}}) = \mathsf{trace}(A^{\mathsf{H}}ABB^{\mathsf{H}})\\ &= \mathsf{trace}(A^{\mathsf{H}}BAB^{\mathsf{H}}),\\ \mathsf{trace}(BA^{\mathsf{H}}AB^{\mathsf{H}}) &= \mathsf{trace}(ABA^{\mathsf{H}}B^{\mathsf{H}}) = \mathsf{trace}(BAA^{\mathsf{H}}B^{\mathsf{H}})\\ &= \mathsf{trace}(BA^{\mathsf{H}}AB^{\mathsf{H}}),\end{aligned}$$

it follows that $\mathsf{trace}(CC^{\mathsf{H}}) = 0$, so $C = O_{n,n}$. Thus, $A^{\mathsf{H}}B = BA^{\mathsf{H}}$ by Supplement 57 of Chapter 3. This result is known as *Fuglede's theorem* [60].

(15) Let $A, B \in \mathbb{C}^{n\times n}$ be two normal matrices such that $AB = BA$. Prove that AB is a normal matrix.

Solution: Since both A and B are normal, we have

$$\begin{aligned}(AB)(AB)^{\mathsf{H}} &= ABB^{\mathsf{H}}A^{\mathsf{H}} = AB^{\mathsf{H}}BA^{\mathsf{H}} = B^{\mathsf{H}}AA^{\mathsf{H}}B\\ &= B^{\mathsf{H}}A^{\mathsf{H}}AB = (AB)^{\mathsf{H}}(AB),\end{aligned}$$

which proves that AB is normal.

(16) Let $U \in \mathbb{C}^{n\times n}$ be a matrix whose set of columns is orthonormal and let $V \in \mathbb{C}^{n\times n}$ be a matrix whose set of rows is orthonormal. Prove that $\|UAV\|_F = \|A\|_F$.

Solution: Since $\|U\|_F^2 = \mathsf{trace}(U^{\mathsf{H}}U) = n$, it follows that $\|U\|_F = \sqrt{n}$. Similarly, $\|V\|_F = \sqrt{n}$. Consequently, $\|UAV\|_F^2 = \mathsf{trace}(V^{\mathsf{H}}A^{\mathsf{H}}U^{\mathsf{H}}UAV) = \mathsf{trace}(V^{\mathsf{H}}A^{\mathsf{H}}AV) = \mathsf{trace}(VV^{\mathsf{H}}A^{\mathsf{H}}A) = \mathsf{trace}(A^{\mathsf{H}}A) = \|A\|_F^2$, which yields the needed equality.

(17) Let $H \in \mathbb{C}^{n\times n}$ be a non-singular matrix. Prove that the function $f : \mathbb{C}^{n\times n} \longrightarrow \mathbb{R}_{\geqslant 0}$ defined by $f(X) = \|HXH^{-1}\|_2$ for $X \in \mathbb{C}^{n\times n}$ is a matrix norm.

(18) Let $A \in \mathbb{C}^{m\times n}$ and $B \in \mathbb{C}^{p\times q}$ be two matrices. Prove that $\|A \otimes B\|_F^2 = \mathsf{trace}(A'A \otimes B'B)$.

(19) Let $\boldsymbol{x}_0, \boldsymbol{x}$, and $\boldsymbol{y}$ be three members of $\mathbb{R}^n$. Prove that if $t \in [0,1]$ and $\boldsymbol{u} = t\boldsymbol{x} + (1-t)\boldsymbol{y}$, then

$$\|\boldsymbol{x}_0 - \boldsymbol{u}\|_2 \leqslant \max\{\|\boldsymbol{x}_0 - \boldsymbol{x}\|_2, \|\boldsymbol{x}_0 - \boldsymbol{y}\|_2\}.$$

Solution: Since

$$\boldsymbol{x}_0 - \boldsymbol{u} = \boldsymbol{x}_0 - \boldsymbol{y} - t(\boldsymbol{x} - \boldsymbol{y}),$$

we have

$$\|\boldsymbol{x} - \boldsymbol{u}\|_2^2 = \|\boldsymbol{x}_0 - \boldsymbol{y}\|_2^2 - 2t(\boldsymbol{x}_0 - \boldsymbol{y}, \boldsymbol{x} - \boldsymbol{y}) + t^2\|\boldsymbol{x} - \boldsymbol{y}\|_2^2.$$

The graph of the function $f : [0,1] \longrightarrow \mathbb{R}_{\geqslant 0}$ given by $f(t) = \|\boldsymbol{x} - \boldsymbol{u}\|_2^2$ is a segment of a convex parabola. Therefore, $\max_{t\in[0,1]} f(t) = \max\{f(0), f(1)\} = \max\{\|\boldsymbol{x}_0 - \boldsymbol{y}\|_2^2, \|\boldsymbol{x}_0 - \boldsymbol{x}\|_2^2\}$, which leads to the desired conclusion.

(20) Let $\boldsymbol{u}_1, \ldots, \boldsymbol{u}_m$ be m unit vectors in $\mathbb{R}^2$, such that $\|\boldsymbol{u}_i - \boldsymbol{u}_j\| = 1$. Prove that $m \leqslant 6$.

(21) Prove that if $A \in \mathbb{C}^{n\times n}$ is an invertible matrix, then $\mu(A) \geqslant \frac{1}{\mu(A^{-1})}$ for any matrix norm μ.

(22) Let $A \in \mathbb{R}^{m\times n}$ and $B \in \mathbb{R}^{n\times p}$ be two rectangular matrices that have orthonormal sets of columns. Prove that the matrix $AB \in \mathbb{R}^{m\times p}$ also has an orthonormal set of columns.

Solution: By hypothesis, we have $A'A = I_n$ and $B'B = I_p$. Therefore, $(AB)'(AB) = B'A'AB = B'I_nB = B'B = I_p$, which shows that AB has an orthonormal set of columns.

(23) Let $Y \in \mathbb{C}^{n\times p}$ be a matrix that has an orthonormal set of columns, that is, $Y^{\mathsf{H}}Y = I_p$. Prove the following:

(a) $\|Y\|_F = p$;

(b) for every matrix $R \in \mathbb{C}^{p\times q}$ we have $\|YR\|_F = \|R\|_F$.

Solution: We have $Y^{\mathsf{H}}Y = I_p$, so $\|Y\|_F = \sqrt{\mathsf{trace}(Y^{\mathsf{H}}Y)} = \sqrt{\mathsf{trace}(I_p)} = \sqrt{p}$.

For the second part, we can write

$$\begin{aligned}\|YR\|_F^2 &= \mathsf{trace}((YR)^{\mathsf{H}}YR) = \mathsf{trace}(R^{\mathsf{H}}Y^{\mathsf{H}}YR)\\ &= \mathsf{trace}(R^{\mathsf{H}}R) = \|R\|_F^2,\end{aligned}$$

which gives the desired equality.

(24) Let $\mu : \mathbb{C}^{n\times n} \longrightarrow \mathbb{R}_{\geqslant 0}$ be a matrix norm. Prove that there exists a vector norm $\nu : \mathbb{C}^n \longrightarrow \mathbb{R}_{\geqslant 0}$ such that $\nu(A\boldsymbol{x}) \leqslant \mu(A)\nu(\boldsymbol{x})$ for $A \in \mathbb{C}^{n\times n}$ and $\boldsymbol{x} \in \mathbb{C}^n$.

Solution: Let $\boldsymbol{b} \in \mathbb{C}^n - \{\mathbf{0}_n\}$. It is easy to see that the mapping $\nu : \mathbb{C}^n \longrightarrow \mathbb{R}_{\geqslant 0}$ defined by $\nu(\boldsymbol{x}) = \mu(\boldsymbol{x}\boldsymbol{b}')$ for $\boldsymbol{x} \in \mathbb{C}^n$ is a vector norm. Furthermore, we have $\nu(A\boldsymbol{x}) = \mu(A\boldsymbol{x}\boldsymbol{b}') \leqslant \mu(A)\mu(\boldsymbol{x}\boldsymbol{b}') = \mu(A)\nu(\boldsymbol{x})$.

(25) Let $\Phi_H : \mathbb{C}^{m\times n} \longrightarrow \mathbb{C}^{m\times n}$ be the function defined by $\Phi(X) = XHX$, where $H \in \mathbb{C}^{n\times m}$. Prove that

$$\|\Phi_H(X) - \Phi_H(Y)\|_F \leqslant 2\|H\|_F \max\{\|X\|_F, \|Y\|_F\} \|X - Y\|_F$$

Solution: We have

$$\begin{aligned}
&\|\Phi_H(X) - \Phi_H(Y)\|_F \\
&= \|XHX - XHY + XHY - YHY\|_F \\
&\leqslant \|XHX - XHY\|_F + \|XHY - YHY\|_F \\
&\leqslant \|X\|_F\|HX - HY\|_F + \|XH - YH\|_F\|Y\|_F \\
&\leqslant \max\{\|X\|_F, \|Y\|_F\}(\|HX - HY\|_F + \|XH - YH\|_F) \\
&\leqslant 2\|H\|_F \max\{\|X\|_F, \|Y\|_F\}\|X - Y\|_F.
\end{aligned}$$

(26) Let $\boldsymbol{x}, \boldsymbol{y} \in \mathbb{C}^n - \{\mathbf{0}\}$. Prove the following:

(a) $\|\boldsymbol{x}\boldsymbol{y}^{\mathsf{H}}\|_F = \|\boldsymbol{x}\|_2\|\boldsymbol{y}\|_2$;

(b) $|||\boldsymbol{x}\boldsymbol{y}^{\mathsf{H}}|||_1 = \|\boldsymbol{x}\|_1\|\boldsymbol{y}\|_\infty$;

(c) $|||\boldsymbol{x}\boldsymbol{y}^{\mathsf{H}}|||_\infty = \|\boldsymbol{x}\|_\infty\|\boldsymbol{y}\|_1$.

(27) Let $A, B, \hat{A}, \hat{B}$ be four matrices in $\mathbb{C}^{m\times n}$ such that none of the matrices A, B, or $A + B$ equals $O_{m,n}$. Define $\Delta_A = \hat{A} - A$ and $\Delta_B = \hat{B} - B$. Prove that for any matrix norm μ, we have

$$\frac{\mu(\hat{A} + \hat{B} - (A + B))}{\mu(A + B)} \leqslant \frac{\mu(A) + \mu(B)}{\mu(A + B)} \max\left\{\frac{\mu(\Delta_A)}{\mu(A)}, \frac{\mu(\Delta_B)}{\mu(B)}\right\}.$$

Solution: By the triangular property of norms, we have

$$\begin{aligned}
\mu(\Delta_A + \Delta_B) &\leqslant \mu(\Delta_A) + \mu(\Delta_B) \\
&\leqslant \mu(A) \cdot \frac{\mu(\Delta_A)}{\mu(A)} + \mu(B) \cdot \frac{\mu(\Delta_B)}{\mu(B)} \\
&\leqslant (\mu(A) + \mu(B)) \max\left\{\frac{\mu(\Delta_A)}{\mu(A)}, \frac{\mu(\Delta_B)}{\mu(B)}\right\}.
\end{aligned}$$

(28) Let $A, \hat{A} \in \mathbb{C}^{m\times n}$ and $B, \hat{B} \in \mathbb{C}^{n\times p}$ such that none of the matrices A, B or AB is a zero matrix. Define, as above, $\Delta_A = \hat{A} - A$ and $\Delta_B = \hat{B} - B$.

$$\frac{\mu(\hat{A}\hat{B} - AB)}{\mu(AB)} \leqslant \frac{\mu(A)\mu(B)}{\mu(AB)} \left(\frac{\mu(\Delta_A)}{\mu(A)} + \frac{\mu(\Delta_B)}{\mu(B)} + \frac{\mu(\Delta_A)}{\mu(A)} \frac{\mu(\Delta_B)}{\mu(B)} \right).$$

Solution: We have

$$\begin{aligned} \hat{A}\hat{B} - AB &= (A + \Delta_A)(B + \Delta_B) - AB \\ &= A\Delta_B + B\Delta_A + \Delta_A\Delta_B. \end{aligned}$$

By the triangle inequality,

$$\begin{aligned} \mu(\hat{A}\hat{B} - AB) &= \mu(A\Delta_B + B\Delta_A + \Delta_A\Delta_B) \\ &\leqslant \mu(A\Delta_B) + \mu(B\Delta_A) + \mu(\Delta_A\Delta_B) \\ &\leqslant \mu(A)\mu(\Delta_B) + \mu(B)\mu(\Delta_A) + \mu(\Delta_A)\mu(\Delta_B) \\ &\quad \text{(since } \mu \text{ is a matrix norm).} \end{aligned}$$

(29) Prove that for every matrix norm μ induced by a vector norm, we have $\mu(I) = 1$.

(30) Let $A \in \mathbb{C}^{n\times n}$ be a matrix and let μ be a matrix norm induced by a vector norm ν. Prove that if $\mu(A) < 1$, then the matrix $I_n + A$ is non-singular and

$$\frac{1}{1 + \mu(A)} \leqslant \mu((I_n + A)^{-1}) \leqslant \frac{1}{1 - \mu(A)}.$$

Solution: The matrix $I_n + A$ is non-singular because, otherwise, the system $(I_n + A)\boldsymbol{x} = \boldsymbol{0}$ would have a non-zero solution $\boldsymbol{u}$. This would imply $\boldsymbol{u} = -A\boldsymbol{u}$, so $\nu(\boldsymbol{u}) = \nu(A\boldsymbol{u}) \leqslant \mu(A)\nu(\boldsymbol{u})$, which would imply $\mu(A) \geqslant 1$, contradicting the hypothesis of the statement.
Since $I_n = (I_n + A)(I_n + A)^{-1}$, we have $1 = \mu(I_n) \leqslant \mu(I_n + A)\mu((I_n+A)^{-1}) \leqslant (1+\mu(A))\mu((I_n+A)^{-1})$, so $\frac{1}{1+\mu(A)} \leqslant \mu((I_n + A)^{-1})$.

The equality $I_n = (I_n + A)(I_n + A)^{-1}$ can be written as $I_n = (I_n + A)^{-1} + A(I_n + A)^{-1}$. This implies

$$1 \geqslant \mu((I_n+A)^{-1})-\mu(A(I_n+A)^{-1}) \geqslant (1-\mu(A))\mu((I_n+A)^{-1}),$$

so $\mu((I_n + A)^{-1}) \leqslant \frac{1}{1-\mu(A)}$.

(31) Let $A \in \mathbb{R}^{n\times n}$ (or in $\mathbb{C}^{n\times n}$) be a non-singular matrix. Prove that if ν is a norm on $\mathbb{R}^n$ (on $\mathbb{C}^n$, respectively), then ν_A defined by $\nu_A(\boldsymbol{x}) = \nu(A\boldsymbol{x})$ is a norm on $\mathbb{R}^n$ (on $\mathbb{C}^n$).

(32) Prove that the function $\nu_0 : \mathbb{R}^n \longrightarrow \mathbb{R}_{\geqslant 0}$, where $\nu_0(\boldsymbol{x})$ is the number of non-zero components of $\boldsymbol{x}$, is not a norm, although it satisfies the inequality $\nu(\boldsymbol{x}+\boldsymbol{y}) \leqslant \nu(\boldsymbol{x})+\nu(\boldsymbol{y})$ for $\boldsymbol{x}, \boldsymbol{y} \in \mathbb{R}^n$.

(33) Let $A \in \mathbb{R}^{m\times n}$ and let ν be a vector norm on $\mathbb{R}^n$. Prove that if $A \in \mathbb{R}^{m\times n}$, then we have the following equalities:

$$\begin{aligned}\mu(A) &= \sup\{\nu(A\boldsymbol{x}) \mid \nu(\boldsymbol{x}) = 1\}\\ &= \sup\left\{\frac{\nu(A\boldsymbol{x})}{\nu(\boldsymbol{x})}\middle| \boldsymbol{x} \in \mathbb{R}^n - \{\boldsymbol{0}\}\right\}\\ &= \inf\{k \mid \nu(A\boldsymbol{x}) \leqslant k\nu(\boldsymbol{x}), \text{ for every } \boldsymbol{x} \in \mathbb{R}^n\}.\end{aligned}$$

Solution: To prove the first equality, note that

$$\{\nu(A\boldsymbol{x}) \mid \nu(\boldsymbol{x}) = 1\} \subseteq \{\nu(A\boldsymbol{x}) \mid \nu(\boldsymbol{x}) \leqslant 1\}.$$

This implies

$$\sup\{\nu(A\boldsymbol{x}) \mid \nu(\boldsymbol{x}) = 1\} \leqslant \sup\{\nu(A\boldsymbol{x}) \mid \nu(\boldsymbol{x}) \leqslant 1\} = \mu(A).$$

On the other hand, let $\boldsymbol{x}$ be a vector such that $\nu(\boldsymbol{x}) \leqslant 1$. We have $\boldsymbol{x} = \boldsymbol{0}$ if and only if $\nu(\boldsymbol{x}) = 0$ because ν is a norm. Otherwise, $\boldsymbol{x} \neq 0$, and for $\boldsymbol{y} = \frac{1}{\nu(\boldsymbol{x})}\boldsymbol{x}$ we have $\nu(\boldsymbol{y}) = 1$ and $\nu(A\boldsymbol{x}) = \nu(A(\nu(\boldsymbol{x})\boldsymbol{y}) = \nu(\boldsymbol{x})\nu(A\boldsymbol{y}) \leqslant \nu(A\boldsymbol{y})$. Therefore, in either case we have

$$\nu(A\boldsymbol{x}) \leqslant \sup\{\nu(A\boldsymbol{y}) \mid \nu(\boldsymbol{y}) = 1\}$$

for $\nu(\boldsymbol{x}) \leqslant 1$. Thus, we have the reverse inequality,

$$\sup\{\nu(A\boldsymbol{x}) \mid \nu(\boldsymbol{x}) \leqslant 1\} \leqslant \sup\{\nu(A\boldsymbol{y}) \mid \nu(\boldsymbol{y}) = 1\},$$

so

$$\mu(A) = \sup\{\nu(A\boldsymbol{x}) \mid \nu(\boldsymbol{x}) = 1\}.$$

To prove the second equality, observe that

$$\begin{aligned}\sup\left\{\frac{\nu(A\boldsymbol{x})}{\nu(\boldsymbol{x})}\middle|\boldsymbol{x} \neq \mathbf{0}\right\} &= \sup\left\{\nu\left(A\left(\frac{\boldsymbol{x}}{\nu(\boldsymbol{x})}\right)\right|\boldsymbol{x} \neq \mathbf{0}\right\}\\ &= \sup\{\nu(A\boldsymbol{y}) \mid \nu(\boldsymbol{y}) = 1\} = \mu(A),\end{aligned}$$

because $\nu\left(\frac{\boldsymbol{x}}{\nu(\boldsymbol{x})}\right) = 1$ and every vector $\boldsymbol{y}$ with $\nu(\boldsymbol{y}) = 1$ can be written as $\boldsymbol{y} = \frac{1}{\nu(\boldsymbol{x})}\boldsymbol{x}$ for some $\boldsymbol{x} \neq \mathbf{0}$.

We leave the third equality to the reader.

(34) Let $\boldsymbol{x}$ and $\boldsymbol{y}$ be two vectors in $\mathbb{R}^n$. Prove that if $\|a\boldsymbol{x} + b\boldsymbol{y}\| = \|c\boldsymbol{x} + d\boldsymbol{y}\|$, then

$$(a^2 - c^2)\|\boldsymbol{x}\|^2 + (b^2 - d^2)\|\boldsymbol{y}\|^2 + (ab - cd)(\boldsymbol{y}'\boldsymbol{x} + \boldsymbol{x}'\boldsymbol{y}) = 0.$$

(35) Let $U \in \mathbb{C}^{n\times n}$ be a unitary matrix. Prove that $\mathsf{cond}_2(U) = 1$.

(36) Let A and B be two matrices in $\mathbb{C}^{n\times n}$ such that $A \sim B$. If $A = XBX^{-1}$, prove that

$$\frac{1}{\mathsf{cond}_2(X)}\|B\|_2 \leqslant \|A\|_2 \leqslant \mathsf{cond}_2(X)\|B\|_2.$$

(37) Prove that $\|C^{(n,K)}\|_p \leqslant 1$ and $\|R^{(n,K)}\|_p \leqslant 1$ for any $p \geqslant 1$, where $C^{(n,K)}$ and $R^{(n,K)}$ are the matrices defined in Exercise 61.

(38) Let $A \in \mathbb{C}^{n\times n}$ be a matrix and let $B = A\begin{bmatrix} i_1 & \cdots & i_k \\ j_1 & \cdots & j_h \end{bmatrix}$ be a submatrix of A. Prove that $\|B\|_p \leqslant \|A\|_p$ for any $p \geqslant 1$.

Hint: Apply Part (b) of Exercise 61 of Chapter 3.

(39) Prove that if $D = \mathsf{diag}(d_1, \ldots, d_n)$ is a diagonal matrix, then $|||D|||_p = \max\{|d_i| \mid 1 \leqslant i \leqslant n\}$ for every $p \geqslant 1$.

(40) Let $A \in \mathbb{C}^{n\times n}$. We have seen that if A is a unitary matrix, then $\|A\boldsymbol{x}\|_2 = \|\boldsymbol{x}\|_2$ for every vector $\boldsymbol{x} \in \mathbb{C}^n$ (see Theorem 6.24). Prove the inverse statement, that is, if $\|A\boldsymbol{x}\|_2 = \|\boldsymbol{x}\|_2$ for every vector $\boldsymbol{x} \in \mathbb{C}^n$, then A is a unitary matrix.

Solution: Observe that the condition satisfied by A implies $\|A(\boldsymbol{x}+\boldsymbol{y})\|_2 = \|\boldsymbol{x}+\boldsymbol{y}\|_2$ for every $\boldsymbol{x}, \boldsymbol{y} \in \mathbb{C}^n$. This, in turn, implies

$$(A\boldsymbol{x} + A\boldsymbol{y}, A\boldsymbol{x} + A\boldsymbol{y}) = (\boldsymbol{x}+\boldsymbol{y}, \boldsymbol{x}+\boldsymbol{y}),$$

which is equivalent to $(A\boldsymbol{x}, A\boldsymbol{y}) = (\boldsymbol{x}, \boldsymbol{y})$, or $(A\boldsymbol{x})^{\mathsf{H}} A\boldsymbol{y} = \boldsymbol{x}^{\mathsf{H}}\boldsymbol{y}$. Choosing $\boldsymbol{x} = \boldsymbol{e}_i$ and $\boldsymbol{y} = \boldsymbol{e}_j$, the last condition amounts to

$$(A^{\mathsf{H}}A)_{ij} = \begin{cases} 1 & \text{if } i = j \\ 0 & \text{otherwise} \end{cases}$$

for $1 \leqslant i, j \leqslant n$. Thus, $A^{\mathsf{H}}A = I_n$.

(41) Let $\|\cdot\|$ be a unitarily invariant norm. Prove that for every Hermitian matrix $A \in \mathbb{C}^{n\times n}$ and every unitary matrix U, we have

$$\|A - I_n\| \leqslant \|A - U\| \leqslant \|A + I_n\|.$$

(42) Let $\|\cdot\|$ be a unitarily invariant norm. Prove that

$$\left\|\begin{pmatrix} A & O \\ O & D \end{pmatrix}\right\| \left\|\begin{pmatrix} A & B \\ C & D \end{pmatrix}\right\|$$

for all conforming matrices A, B, C, D.

(43) Let $A \in \mathbb{C}^{n\times n}$ be an invertible matrix and let $\|\cdot\|$ be a norm on $\mathbb{C}^n$. Prove that

$$|||A^{-1}||| = \frac{1}{\min\{\|A\boldsymbol{x}\| \mid \|\boldsymbol{x}\| = 1\}},$$

where $|||\cdot|||$ is the matrix norm generated by $\|\cdot\|$.

Solution: We claim that

$$\{\|A^{-1}\boldsymbol{t}\| \mid \|\boldsymbol{t}\| = 1\} = \left\{\frac{1}{\|A\boldsymbol{x}\|} \mid \|\boldsymbol{x}\| = 1\right\}.$$

Let $a = \|A^{-1}\boldsymbol{t}\|$ for some $\boldsymbol{t} \in \mathbb{C}^n$ such that $\|\boldsymbol{t}\| = 1$. Define $\boldsymbol{x}$ as

$$\boldsymbol{x} = \frac{1}{\|A^{-1}\boldsymbol{t}\|}A^{-1}\boldsymbol{t}.$$

Clearly, we have $\|\boldsymbol{x}\| = 1$. In addition,

$$\|A\boldsymbol{x}\| = \frac{\|\boldsymbol{t}\|}{\|A^{-1}\boldsymbol{t}\|} = \frac{1}{\|A^{-1}\boldsymbol{t}\|} = \frac{1}{a},$$

so $a \in \left\{\frac{1}{\|A\boldsymbol{x}\|} \mid \|\boldsymbol{x}\| = 1\right\}$. Thus,

$$\{\|A^{-1}\boldsymbol{t}\| \mid \|\boldsymbol{t}\| = 1\} \subseteq \left\{\frac{1}{\|A\boldsymbol{x}\|} \mid \|\boldsymbol{x}\| = 1\right\}.$$

The reverse inclusion can be shown in a similar way. Therefore,

$$|\!|\!|A^{-1}|\!|\!| = \max\{\|A^{-1}\boldsymbol{t}\| \mid \|\boldsymbol{t}\| = 1\} = \frac{1}{\min\{\|A\boldsymbol{x}\| \mid \|\boldsymbol{x}\| = 1\}}.$$

(44) Prove that if $\mu_1, \ldots, \mu_k$ are matrix norms on $\mathbb{R}^{n\times n}$, then $\mu : \mathbb{R}^{n\times n} \leqslant \mathbb{R}_{\geqslant 0}$ defined by $\mu(A) = \max\{\mu_1(A), \ldots, \mu_k(A)\}$ for $A \in \mathbb{R}^{n\times n}$ is a matrix norm.

Solution: Let A, B be two matrices in $\mathbb{R}^{n\times n}$. We have

$$\begin{aligned}\mu_i(AB) \leqslant \mu_i(A)\mu_i(B) \leqslant \max\{\mu_1(A), \ldots, \mu_k(A)\} \\ \times \max\{\mu_1(B), \ldots, \mu_k(B)\}\end{aligned}$$

for every i, $1 \leqslant i \leqslant k$. Therefore,

$$\begin{aligned}\max \mu_1(AB), \ldots, \mu_k(AB) \leqslant \max\{\mu_1(A), \ldots, \mu_k(A)\} \\ \times \max\{\mu_1(B), \ldots, \mu_k(B)\},\end{aligned}$$

so $\mu(AB) \leqslant \mu(A)\mu(B)$. We leave to the reader the verification of the remaining properties of matrix norms.

(45) Let $S \in \mathbb{C}^{n\times n}$ be a matrix such that

$$s_{ij} = \begin{cases} 0 & \text{if } j \leqslant i, \\ t_{ij}\delta^{j-i} & \text{if } j > i, \end{cases}$$

where $t_{ij} \in \mathbb{C}$ for $1 \leqslant i,j \leqslant n$ and $i < j$ and $\delta < 1$. Prove that there exists a positive number c such that $\|S\|_2 \leqslant c\delta$.

Solution: Let $t = \max\{|t_{ij}| \mid 1 \leqslant i,j \leqslant n, i \neq j\}$ and let $\boldsymbol{x} \in \mathbb{C}^n$ be a vector such that $\|\boldsymbol{x}\|_2 = 1$. We have

$$|(S\boldsymbol{x})_i| = \left|\sum_{j=1}^{n} s_{ij}x_j\right| \leqslant \sum_{j=i+1}^{n} |t_{ij}|\delta^{i-j}|x_j|$$

$$\leqslant t(n-i)\delta \sum_{j+1}^{n} |x_j| \leqslant tn\delta\|\boldsymbol{x}\|_2.$$

This implies $\|S\|_2 \leqslant tn\delta$, so we obtain the desired inequality with $c = tn$.

(46) Let $A \in \mathbb{C}^{n\times n}$, $a \in \mathbb{C}$, $\boldsymbol{b} \in \mathbb{C}^{n-1}$, and $C \in \mathbb{C}^{(n-1)\times(n-1)}$ be such that

$$A = \begin{pmatrix} a & \boldsymbol{b}^{\mathsf{H}} \\ \boldsymbol{b} & C \end{pmatrix}.$$

Prove that $\|\boldsymbol{b}\|_2 \leqslant \|A\|_2$.

Solution: Since $\|A\|_2 = \sup\{\|A\boldsymbol{x}\|_2 \mid \|\boldsymbol{x}\| = 1\}$, by substituting $\begin{pmatrix} 1 \\ \boldsymbol{0}_{n-1} \end{pmatrix}$, the desired inequality follows immediately.

(47) In Example 6.13 we saw that $\|A\|_\infty$ fails to be a matrix norm. However, $\|A\|_\infty$ can be useful due to its simplicity.
Let $A \in \mathbb{C}^{n\times n}$. Prove the following:

(a) if $B \in \mathbb{R}^{n\times n}$ is a matrix such that $\mathsf{abs}(A) \leqslant B$, then $\|A\|_\infty(A) \leqslant \|A\|_\infty$;

(b) if $A_1,\ldots,A_k \in \mathbb{C}^{n\times n}$, then $\|A\|_\infty(A_1\cdots A_k) \leqslant n^{k-1}\prod_{i=1}^{k}\|A\|_\infty(A_i)$.

Solution: If $\mathsf{abs}(A) \leqslant B$, we have $|a_{ij}| \leqslant b_{ij}$ for $1 \leqslant i,j \leqslant n$. Therefore, the elements of B are non-negative and we have $\|A\|_\infty = \max_{i,j}|a_{ij}| \leqslant \max_{i,j} b_{ij} = \|B\|_\infty$.

For the second part, note that

$$\begin{aligned}&(A_1 \cdots A_k)_{ij}\\&= \sum \{(A_1)_{ii_1}(A_2)_{i_1 i_2} \cdots (A_k)_{i_{k-1}j}\\&\qquad \mid (i_1, \ldots, i_{k-1}) \in \{1, \ldots, n\}^{k-1}\}.\end{aligned}$$

Thus,

$$\begin{aligned}&|(A_1 \cdots A_k)_{ij}|\\&\leqslant \sum_{(i_1,\ldots,i_{k-1})} \{|(A_1)_{ii_1}|\,|(A_2)_{i_1 i_2}| \cdots (A_k)_{i_{k-1}j}.\end{aligned}$$

Since the last sum contains n^{k-1} terms and each term is less or equal to $\prod_{i=1}^{k} \|A\|_\infty(A_i)$, the desired inequality follows immediately.

(48) Prove that if $A, B \in \mathbb{R}^{n\times n}$ and $\mathsf{abs}(A) \leqslant abs(B)$, then $\|A\|_F \leqslant \|B\|F$.

(49) Let A and B be two matrices in $\mathbb{C}^{n\times n}$ and let $\|\cdot\|$ be a matrix norm on $\mathbb{C}^{n\times n}$. Prove that if $\|AB - I_n\| \leqslant 1$, then both A and B are invertible.

For $\boldsymbol{x}, \boldsymbol{y} \in \mathbb{C}^n$, we write $\mathsf{abs}(x) \leqslant \mathsf{abs}(\boldsymbol{y})$ if $|x_i| \leqslant |y_i|$ for $1 \leqslant i \leqslant n$. A norm $\nu : \mathbb{C}^n \longrightarrow \mathbb{R}_{\geqslant 0}$ is *monotone* if $\mathsf{abs}(\boldsymbol{x}) \leqslant \mathsf{abs}(\boldsymbol{y})$ implies $\nu(\boldsymbol{x}) \leqslant \nu(\boldsymbol{y})$ for $\boldsymbol{x}, \boldsymbol{y} \in \mathbb{C}^n$; ν is said to be *absolute* if $\nu(\boldsymbol{x}) = \nu(\mathsf{abs}(\boldsymbol{x}))$ for $\boldsymbol{x} \in \mathbb{C}^n$.

(50) Prove that a norm ν on $\mathbb{C}^n$ is monotone if and only if it is absolute.

Solution: If ν is monotone, then it is clearly absolute. Conversely, let ν be an absolute norm and let $\boldsymbol{x}, \boldsymbol{y} \in \mathbb{C}^n$ such that $\boldsymbol{x} = \mathsf{diag}(1, \ldots, 1, a, 1, \ldots, 1)\boldsymbol{y}$, where $a \in [0, 1]$. Note that

$$\boldsymbol{x} = \frac{1+a}{2}\begin{pmatrix} y_1 \\ \vdots \\ y_k \\ \vdots \\ y_n \end{pmatrix} + \frac{1-a}{2}\begin{pmatrix} y_1 \\ \vdots \\ -y_k \\ \vdots \\ y_n \end{pmatrix}.$$

The definition of norms implies that $\nu(\boldsymbol{x}) \leqslant \frac{1+a}{2}\nu(\boldsymbol{y}) + \frac{1-a}{2}\nu(\tilde{\boldsymbol{y}})$, where

$$\tilde{\boldsymbol{y}} = \begin{pmatrix} y_1 \\ \vdots \\ -y_k \\ \vdots \\ y_n \end{pmatrix}.$$

Since ν is absolute, this implies $\nu(\boldsymbol{x}) \leqslant \nu(\boldsymbol{y})$. If $a_1, \ldots, a_n \in [0,1]$ and $\boldsymbol{x} = \mathsf{diag}(a_1, \ldots, a_n)\boldsymbol{y}$, then by Exercise 5 of Chapter 3 and the previous argument, we have $\nu(\boldsymbol{x}) \leqslant \nu(\boldsymbol{y})$.
Let $\boldsymbol{x}, \boldsymbol{y} \in \mathbb{C}^n$ such that $\mathsf{abs}(\boldsymbol{x}) \leqslant \mathsf{abs}(\boldsymbol{y})$, which is equivalent to $|x_i| \leqslant |y_i|$. This allows finding $a_1, \ldots, a_n \in [0,1]$ such that $|x_i| = a_i|y_i|$, which implies $\nu(\mathsf{abs}(\boldsymbol{x})) \leqslant \nu(\mathsf{abs}(\boldsymbol{y}))$. Since ν is absolute, this implies $\nu(\boldsymbol{x}) \leqslant \nu(\boldsymbol{y})$.

A function $g : \mathbb{R}^n \longrightarrow \mathbb{R}_{\geqslant 0}$ is a *symmetric gauge function* if it satisfies the following conditions:

(i) g is a norm on $\mathbb{R}^n$;
(ii) $g(P_\phi \boldsymbol{u}) = g(\boldsymbol{u})$ for every permutation $\phi \in PERM_n$ and every $\boldsymbol{u} \in \mathbb{R}^n$;
(iii) $g(\mathsf{diag}(b_1, \ldots, b_n)\boldsymbol{u}) = g(\boldsymbol{u})$ for every $(b_1, \ldots, b_n) \in \{-1, 1\}^n$ and $\boldsymbol{u} \in \mathbb{R}^n$.

(51) Prove that for any $p \geqslant 1$ the function $\nu_p : \mathbb{R}^n \longrightarrow \mathbb{R}_{\geqslant 0}$ is a symmetric gauge function on $\mathbb{R}^n$.
(52) Let $\|\cdot\|$ be a norm defined on $\mathbb{R}^n$. If $f : \mathbb{R}^n \longrightarrow \mathbb{R}$ is a homogeneous function, that is, a function such that $f(a\boldsymbol{x}) = af(\boldsymbol{x})$ for $a \geqslant 0$, then $\max\{f(\boldsymbol{x}) \mid \|\boldsymbol{x}\| = 1\} = \max\{f(\boldsymbol{x}) \mid \|\boldsymbol{x}\| \leqslant 1\}$.

Solution: It is clear that

$$\max\{f(\boldsymbol{x}) \mid \|\boldsymbol{x}\| = 1\} \leqslant \max\{f(\boldsymbol{x}) \mid \|\boldsymbol{x}\| \leqslant 1\},$$

because the set from the left member is a subset of the set that occurs in the right member. To prove the reverse inequality, let $\boldsymbol{x}$ be a vector such that $\|\boldsymbol{x}\| \leqslant 1$. Then

$$\left\| \frac{\boldsymbol{x}}{\|\boldsymbol{x}\|} \right\| = 1.$$

By the defining property of f, we also have

$$f\left(\frac{\boldsymbol{x}}{\|\boldsymbol{x}\|}\right) = \frac{1}{\|\boldsymbol{x}\|} f(\boldsymbol{x}) \geqslant f(\boldsymbol{x}).$$

Thus,

$$\max\{f(\boldsymbol{x}) \mid \|\boldsymbol{x}\| \leqslant 1\} \leqslant \max\{f(\boldsymbol{x}) \mid \|\boldsymbol{x}\| = 1\}.$$

(53) Let $A \in \mathbb{R}^{m\times p}$. Prove that $\|A\|_F = \|\mathsf{vec}(A)\|_2$.

(54) Let $U \in \mathbb{C}^{m\times n}$, $V \in \mathbb{C}^{n\times p}$, and let $A \in \mathbb{C}^{m\times p}$. Prove that

$$\begin{aligned}\|A - UV\|_F &= \|\mathsf{vec}(A) - (V' \otimes I_m)\mathsf{vec}(U)\|_2 \\ &= \|\mathsf{vec}(A) - (I_n \otimes U)\mathsf{vec}(V)\|_2.\end{aligned}$$

Solution: We have

$$\begin{aligned}\|A - UV\|_F &= \|\mathsf{vec}(A - UV)\|_2 \\ &\quad \text{(by Exercise 6.20)} \\ &= \|\mathsf{vec}(A) - \mathsf{vec}(UV)\|_2 \\ &= \|\mathsf{vec}(A) - (V' \otimes I_m)\mathsf{vec}(U)\|_2 \\ &= \|\mathsf{vec}(A) - (I_n \otimes U)\mathsf{vec}(V)\|_2,\end{aligned}$$

by Supplement 90 of Chapter 3.

(55) Prove that if $A \in \mathbb{C}^{n\times n}$, then

$$|||A|||_2 = \max\{|\boldsymbol{y}^{\mathsf{H}} A\boldsymbol{x}| \mid \|\boldsymbol{x}\|_2 = \|\boldsymbol{y}\|_2 = 1\}.$$

Solution: By the Cauchy–Schwarz inequality, we have

$$|\boldsymbol{y}^{\mathsf{H}} A\boldsymbol{x}| \leqslant \|\boldsymbol{y}\|_2 \|A\boldsymbol{x}\|_2,$$

so

$$\begin{aligned}&\max\{|\boldsymbol{y}^{\mathsf{H}} A\boldsymbol{x}| \mid \|\boldsymbol{x}\|_2 = \|\boldsymbol{y}\|_2 = 1\} \\ &\quad \leqslant \max\{\|A\boldsymbol{x}\|_2 \mid \|\boldsymbol{x}\|_2\} = |||A|||_2.\end{aligned} \tag{6.34}$$

Let $\tilde{\boldsymbol{x}}$ be a unit vector such that $\|A\tilde{\boldsymbol{x}}\|_2 = |||A|||_2$ and let $\tilde{\boldsymbol{y}}$ be the unit vector $\tilde{\boldsymbol{y}} = \frac{1}{\|A\tilde{\boldsymbol{x}}\|_2} A\tilde{\boldsymbol{x}}$. We have

$$\tilde{\boldsymbol{y}}^{\mathsf{H}} A\tilde{\boldsymbol{x}} = \frac{1}{\|A\tilde{\boldsymbol{x}}\|_2} \tilde{\boldsymbol{x}}^{\mathsf{H}} A^{\mathsf{H}} A\boldsymbol{x} = \frac{\|A\tilde{\boldsymbol{x}}\|_2^2}{\|A\tilde{\boldsymbol{x}}\|_2} = \|A\tilde{\boldsymbol{x}}\|_2 = |||A|||_2.$$

Thus, the Inequality (6.34) can be replaced by an equality.

(56) Prove that every symmetric gauge function $g : \mathbb{R}^n \longrightarrow \mathbb{R}_{\geqslant 0}$ is an absolute norm on $\mathbb{R}^n$.

Solution: Let $\boldsymbol{u} \in \mathbb{R}^n$. Define $b_i = 1$ if $u_i \geqslant 0$ and $b_i = -1$ if $u_i < 0$. Then $|u_i| = b_i u_i$ for $1 \leqslant i \leqslant n$, so $\mathsf{abs}(\boldsymbol{u}) = \mathsf{diag}(b_1, \ldots, b_n)\boldsymbol{u}$. By the definition of gauge functions, $g(\boldsymbol{u}) = g(\mathsf{diag}(b_1, \ldots, b_n)\boldsymbol{u}) = g(\mathsf{abs}(u))$, which implies that g is an absolute norm.

(57) For $\boldsymbol{x} \in \mathbb{R}^n$, let $g_k(\boldsymbol{x})$ be the sum of the largest k absolute values of components of $\boldsymbol{x}$. Prove that

(a) g_k is a gauge symmetric function for $1 \leqslant k \leqslant n$;

(b) if $g : \mathbb{R}^n \longrightarrow \mathbb{R}_{\geqslant 0}$ is a gauge symmetric function and $g_k(\boldsymbol{x}) \leqslant g_k(\boldsymbol{y})$ for $1 \leqslant k \leqslant n$, then $g(\boldsymbol{x}) \leqslant g(\boldsymbol{y})$.

(58) Let $A \in \mathbb{R}^{m\times n}$ and let $I \subseteq \{1, \ldots, m\}$ and $J \subseteq \{1, \ldots\}$. Define $A(I, J) = \sum\{a_{ij} \mid i \in I, j \in J\}$, and $\|A\|_C = \max_{I,J} |A(I, J)|$. Prove that $\|\cdot\|_C$ is a norm on $C^{m\times n}$. This norm will be referred to as the *cut norm.*

(59) A (b, I, J)-*cut matrix* is a matrix $B \in \mathbb{R}^{m\times n}$ such that there exist a number $b \in \mathbb{R}$ and two sets I, J, such that $I \subseteq \{1, \ldots, m\}$ and $J \subseteq \{1, \ldots\}$ and

$$b_{ij} = \begin{cases} b & \text{if } i \in I \text{ and } j \in J, \\ 0 & \text{otherwise.} \end{cases}$$

Prove that every cut matrix has rank 1 and that there are 2^{m+n} distinct cut matrices having b as their non-zero entry.

(60) Let $A \in \mathbb{R}^{m\times n}$ be such that there exists a pair of sets (I, J) such that $A(I, J) \geqslant \epsilon|I||J|$ for some $\epsilon > 0$ and let $b = \frac{A(I,J)}{|I||J|}$. If B is the (b, I, J)-cut matrix, prove that $\|A-B\|_F^2 \leqslant \|A\|_F^2 - \epsilon^2|I||J|$.

Solution: Note that in $A - B$, one subtracts from every element of the submatrix $A\begin{bmatrix} S \\ T \end{bmatrix}$ their average. Thus,

$$\begin{aligned} \|A - B\|_F^2 &= \sum\{a_{ij}^2 \mid i \notin I \text{ or } j \notin J\} \\ &\quad + \sum\left\{\left(a_{ij} - \frac{A(I, J)}{|I||J|}\right)^2 \,\middle|\, i \in I, j \in J\right\} \\ &= \|A\|_F^2 - \frac{A(I, J)^2}{|I||J|} \leqslant \|A\|_F^2 - \epsilon^2|I||J|. \end{aligned}$$

(61) Let $A \in \mathbb{R}^{m\times n}$ be such that $|a_{ij}| \leqslant 1$. Prove that for every $\epsilon > 0$ it is possible to construct a sequence of cut matrices $B_1, \ldots, B_p$ such that $\|A - B\|_C \leqslant \epsilon mn$ and $p \leqslant \frac{1}{\epsilon^2}$, where $B = \sum_{k=1}^{p} B_k$.

Solution: If $\|A\|_C \leqslant \epsilon mn$, we take $p = 1$ and $B_1 = O_{m,n}$. Otherwise, $\|A\|_C > \epsilon mn$ and we can apply repeatedly the process described in Supplement 6.20. In this manner, we construct two sequences of matrices $A_1, \ldots, A_p$ and $B_1, \ldots, B_p$ such that

$$A_1 = A - B_1, A_2 = A_1 - B_2, \ldots, A_p = A_{p-1} - B_p.$$

By Supplement 6.20, we have $\|A_{i+1}\|_F^2 \leqslant \|A_i\|_F^2 - \epsilon^2 mn$. Since $\|\cdot\|_F^2$ is non-negative, it is clear that this process can be repeated no more than $\frac{1}{\epsilon^2}$ times until we obtain $\|A - B\|_C \leqslant \epsilon mn$, where $B = B_1 + \cdots + B_p$.

(62) Prove the following extension of the result of Supplement 6.20: if $A \in \mathbb{R}^{m\times n}$, then for every $\epsilon > 0$ there exists a sequence of cut matrices $B_1, \ldots, B_p$ such that $A = B_1 + \cdots + B_p + R$, where $\|R\|_C \leqslant \epsilon mn$. The number p is the *width* of this decomposition, while $\|R\|_C$ is the *error*.

Solution: Let A be a matrix in $\mathbb{R}^{m\times n}$ and let $\tilde{A}$ be the matrix $\tilde{A} = \frac{1}{\|A\|_\infty} A$. We have $|\tilde{a}_{ij}| \leqslant 1$. Choose now $\epsilon_1 = \frac{\epsilon}{\|A\|_\infty}$. By Supplement 6.20, there is a sequence of cut matrices $E_1, \ldots, E_p$ such that $\tilde{A} = E_1 + \cdots + E_p + Q$ and $\|Q\|_C \leqslant \epsilon_1 mn$. Therefore, $A = \|A\|_\infty \tilde{A} = \|A\|_\infty E_1 + \cdots + \|A\|_\infty E_p + \|A\|_\infty Q$ and we define B_i as the cut matrix $B_i = \|A\|_\infty E_i$ for $1 \leqslant i \leqslant p$ and $R = \|A\|_\infty Q$. Note that $\|R\|_C = \|A\|_\infty \|Q\|_C \leqslant \|A\|_\infty \epsilon_1 mn \leqslant \epsilon mn$.

(63) Let $D \in \mathbb{R}^{n\times n}$ be a diagonal matrix such that $d_{ii} \geqslant 0$ for $1 \leqslant i \leqslant n$. Prove that if X is an orthogonal matrix, then $\mathsf{trace}(XD) \leqslant \mathsf{trace}(D)$.

Solution: Since D is a diagonal matrix, we have $\mathsf{trace}(XD) = \sum_{i=1}^{n} x_{ii} d_{ii}$. The orthogonality of X implies $x_{ii} \leqslant 1$, so $x_{ii} d_{ii} \leqslant d_{ii}$ because $d_{ii} \geqslant 0$. Thus,

$$\mathsf{trace}(XD) = \sum_{i=1}^{n} x_{ii} d_{ii} \leqslant \sum_{i=1}^{n} d_{ii} = trace(D).$$

(64) Let $\|\cdot\|$ be a norm on $\mathbb{R}^n$, $D_n = \mathbb{R}^n \times \mathbb{R}^n - \{(\mathbf{0}, \mathbf{0})\} \longrightarrow \mathbb{R}$ and let $f : D_n \longrightarrow \mathbb{R}$ be defined by

$$f(\boldsymbol{x}, \boldsymbol{y}) = \frac{\|\boldsymbol{x}+\boldsymbol{y}\|^2 + \|\boldsymbol{x}-\boldsymbol{y}\|^2}{2(\|\boldsymbol{x}\|^2 + \|\boldsymbol{y}\|^2)}.$$

Prove that

(a) $f(\boldsymbol{x}+\boldsymbol{y}, \boldsymbol{x}-\boldsymbol{y}) = \frac{1}{f(\boldsymbol{x},\boldsymbol{y})}$ for $(\boldsymbol{x}, \boldsymbol{y}) \in D_n$;

(b) if $a = \inf\{f(\boldsymbol{x}, \boldsymbol{y}) \mid (\boldsymbol{x}, \boldsymbol{y}) \in D_n\}$ and $b = \sup\{f(\boldsymbol{x}, \boldsymbol{y}) \mid (\boldsymbol{x}, \boldsymbol{y}) \in D_n\}$, then

$$\frac{1}{2} \leqslant a \leqslant 1 \leqslant b \leqslant 2,$$

and $ab = 1$;

(c) the norm $\|\cdot\|$ is generated by an inner product if and only if $a = b = 1$.

Solution: The first part is immediate. It is clear that $a \leqslant b$. The definition of the norm implies $f(x, y) \leqslant 2$, so $b \leqslant 2$. By the first part we have $a = \frac{1}{b}$, which implies $a \geqslant \frac{1}{2}$, and $a \leqslant b$ implies $a \leqslant 1 \leqslant b$. The last part follows from Theorem 6.32.

(65) Prove that if $\boldsymbol{x}, \boldsymbol{y} \in \mathbb{C}^n$ are such that $\|\boldsymbol{x}\|_2 = \|\boldsymbol{y}\|_2$, then $\boldsymbol{x}+\boldsymbol{y} \perp \boldsymbol{x}-\boldsymbol{y}$.

(66) Let $f : V \times V \longrightarrow \mathbb{R}$ be a bilinear form on a real inner product space such that $\boldsymbol{x} \perp \boldsymbol{y}$ implies $f(\boldsymbol{x}, \boldsymbol{y}) = 0$. Prove that $f(\boldsymbol{x}, \boldsymbol{y}) = c(\boldsymbol{x}, \boldsymbol{y})$ for some $c \in \mathbb{R}$.

(67) Let $f : V \times V\mathbb{R}$ be a bilinear form on the $\mathbb{R}$-linear space V. Define $\boldsymbol{x} \perp_f \boldsymbol{y}$ to mean that $f(\boldsymbol{x}, \boldsymbol{y}) = 0$. Prove the following:

(a) $\boldsymbol{x} \perp_f \boldsymbol{y}$ and $\boldsymbol{x} \perp_f \boldsymbol{z}$ imply $\boldsymbol{x} \perp_f (a\boldsymbol{y} + b\boldsymbol{z})$ for $a, b \in \mathbb{R}$;

(b) $\boldsymbol{x}_1 \perp_f \boldsymbol{y}$ and $\boldsymbol{x}_2 \perp_f \boldsymbol{y}$ imply $(a\boldsymbol{x}_1 + b\boldsymbol{x}_2) \perp_f \boldsymbol{y}$ for $a, b \in \mathbb{R}$;

(c) Let $f : \mathbb{R}^2 \times \mathbb{R}^2 \longrightarrow \mathbb{R}$ be the bilinear form defined by

$$f\left(\begin{pmatrix} x \\ y \end{pmatrix}, \begin{pmatrix} x' \\ y' \end{pmatrix}\right) = xx' + xy' - x'y - yy'.$$

Prove that there exist $\boldsymbol{x}, \boldsymbol{y} \in \mathbb{R}^2$ such that $\boldsymbol{x} \perp_f \boldsymbol{y}$ but $\boldsymbol{y} \not\perp_f \boldsymbol{x}$.

(68) If S is a subspace of $\mathbb{C}^n$, prove that $(S^\perp)^\perp = s$.

(69) Prove that every permutation matrix P_ϕ is orthogonal.

A matrix $A \in \mathbb{C}^{m\times n}$ is *subunitary* if A is a submatrix of a unitary matrix U. If $A \in \mathbb{R}^{m\times n}$, and A is subunitary, then A is a *suborthogonal matrix* (see [14], where suborthogonal matrices were introduced and referred to as suborthonormal matrices).

A matrix $A \in \mathbb{C}^{m\times n}$ is *semiunitary* if it is rowwise or columnwise unitary; if $A \in \mathbb{R}^{m\times n}$, then A is said to be *semiorthogonal*. Clearly, every unitary (orthogonal) matrix is a semiunitary (semiorthogonal) matrix and every semiunitary (semiorthogonal) matrix is a subunitary (suborthogonal) matrix.

(70) Prove that every submatrix of a subunitary (suborthogonal) matrix is subunitary (suborthogonal).

(71) Prove that every suborthogonal matrix can be augmented to a semiorthogonal matrix by adding only rows and by adding only columns to it.

(72) Let $A \in \mathbb{C}^{n\times n}$ be an upper Hessenberg matrix and let $A = QR$ be its QR decomposition. If R is a non-singular matrix, prove that both matrices Q and RQ are upper Hessenberg matrices.

Solution: By Theorem 3.8, R^{-1} is an upper triangular matrix. Since $Q = AR^{-1}$, it follows from Supplement 11 of Chapter 3 that Q is an upper Hessenberg matrix. From the same supplement, it follows that RQ is an upper Hessenberg matrix.

(73) Let $A \in \mathbb{R}^{n\times n}$ be a symmetric matrix. Prove that

$$(\boldsymbol{x}, A\boldsymbol{x}) - (\boldsymbol{y}, A\boldsymbol{y}) = (A(\boldsymbol{x} - \boldsymbol{y}), \boldsymbol{x} + \boldsymbol{y})$$

for every $\boldsymbol{x}, \boldsymbol{y} \in \mathbb{R}^n$.

(74) Let $\boldsymbol{x}, \boldsymbol{y} \in \mathbb{R}^n$ be two unit vectors. Prove that

$$|\sin \angle(\boldsymbol{x}, \boldsymbol{y})| = \frac{\|\boldsymbol{x} + \boldsymbol{y}\|_2 \|\boldsymbol{x} - \boldsymbol{y}\|_2}{2}.$$

(75) Let $\boldsymbol{u}$ and $\boldsymbol{v}$ be two unit vectors in $\mathbb{R}^n$. Prove that if $\alpha = \angle(\boldsymbol{u}, \boldsymbol{v})$, then

$$\|\boldsymbol{u} - \boldsymbol{v}\cos\alpha\|_2 = \sin\alpha.$$

Further, prove that $\boldsymbol{v}\cos\alpha$ is the closest vector in $\langle \boldsymbol{v} \rangle$ to $\boldsymbol{u}$.

Solution: By the definition of the Euclidean norm, we have

$$\begin{aligned}&\|\boldsymbol{u} - \boldsymbol{v}\cos\alpha\|_2^2\\&= (\boldsymbol{u} - \boldsymbol{v}\cos\alpha)'(\boldsymbol{u} - \boldsymbol{v}\cos\alpha) = 1 - 2\boldsymbol{u}'\boldsymbol{v}\cos\alpha + \cos^2\alpha\\&= 1 - \cos^2\alpha = \sin^2\alpha,\end{aligned}$$

which justifies the equality that we need to prove. For the second part, let $\boldsymbol{w} = a\boldsymbol{v}$ be a vector in $\langle \boldsymbol{v} \rangle$. We have

$$\begin{aligned}&\|\boldsymbol{u} - a\boldsymbol{v}\|_2^2\\&= (\boldsymbol{u} - a\boldsymbol{v})'(\boldsymbol{u} - a\boldsymbol{v}) = 1 - 2a\boldsymbol{u}'\boldsymbol{v} + a^2\\&= 1 - 2a\cos\alpha + a^2.\end{aligned}$$

The least value of the function in a is achieved when $a = \cos\alpha$.

(76) Let $\{\boldsymbol{v}_1, \ldots, \boldsymbol{v}_p\} \subseteq \mathbb{R}^n$ be a collection of p unit vectors such that $\angle(\boldsymbol{v}_i, \boldsymbol{v}_j) = \theta$, where $0 < \theta \leqslant \frac{\pi}{2}$ for every pair $(\boldsymbol{v}_i, \boldsymbol{v}_j)$ such that $1 \leqslant i, j \leqslant p$ and $i \neq j$. Prove that $p \leqslant \frac{n(n+1)}{2}$.

Solution: We shall prove that under the assumptions made above, the set $\{A_1, \ldots, A_p\}$ of symmetric matrices $A_i = \boldsymbol{v}_i\boldsymbol{v}_i' \in \mathbb{R}^{n\times n}$ is linearly independent.

Suppose that $a_1A_1 + \cdots + a_pA_p = O_{n\times n}$. This is equivalent to

$$a_1\boldsymbol{a}_1\boldsymbol{a}_1' + \cdots + a_p\boldsymbol{a}_p\boldsymbol{a}_p' = O_{n\times n}.$$

Therefore, by multiplying the last equality by $\boldsymbol{a}_i'$ to the left and by $\boldsymbol{a}_i$ to the right, we obtain

$$a_1\boldsymbol{a}_i'\boldsymbol{a}_1\boldsymbol{a}_1'\boldsymbol{a}_i + \cdots + a_p\boldsymbol{a}_i'\boldsymbol{a}_p\boldsymbol{a}_p'\boldsymbol{a}_i = 0,$$

which amounts to $a_1\cos^2\theta + \cdots + a_{p-1}\cos^2\theta = a_p + a_{p+1}\cos^2\theta + \cdots + a_p\cos^2\theta = 0$. This equality and the similar $p-1$ equalities can be expressed in matrix form as

$$(I_p(1 - \cos^2\theta) + J_p\cos^2\theta)\boldsymbol{a} = O_{p,p},$$

where

$$\boldsymbol{a} = \begin{pmatrix} a_1 \\ \vdots \\ a_p \end{pmatrix}.$$

By Supplement 24 of Chapter 3, $\boldsymbol{a} = \boldsymbol{0}_p$, so $A_1, \ldots, A_p$ are linearly independent. By Supplement 7 of the same chapter, $p \leqslant \frac{n(n+1)}{2}$.

(77) Let $A \in \mathbb{C}^{n\times n}$ be a unitary matrix. If $A = (\boldsymbol{u}_1, \ldots, \boldsymbol{u}_n)$, prove that the set $\{\boldsymbol{u}_1, \ldots, \boldsymbol{u}_n\}$ is orthonormal.

(78) Let $\{\boldsymbol{w}_1, \ldots, \boldsymbol{w}_k\} \subseteq \mathbb{C}^n$ be a set of unit vectors such that $\boldsymbol{w}_i \perp \boldsymbol{w}_j$ for $i \neq j$ and $1 \leqslant i, j \leqslant k$. If $W_k = (\boldsymbol{w}_1 \;\cdots\; \boldsymbol{w}_k) \in \mathbb{C}^{n\times k}$, then prove that

$$I_n - W_k W_k^{\mathsf{H}} = (I_n - \boldsymbol{w}_k \boldsymbol{w}_k^{\mathsf{H}}) \cdots (I_n - \boldsymbol{w}_1 \boldsymbol{w}_1^{\mathsf{H}}).$$

Solution: The straightforward proof is by induction on k.

(79) This exercise refines the result presented in Supplement 36 of Chapter 3.
Let A be a matrix in $\mathbb{C}^{m\times n}$ such that $\mathsf{rank}(A) = r$. Prove that A can be factored as $A = PC$, where $P \in \mathbb{C}^{m\times r}$ is a matrix having an orthonormal set of columns (that is, $P^{\mathsf{H}}P = I_r$), $C \in \mathbb{C}^{r\times n}$, and $\mathsf{rank}(P) = r$. Also, prove that A can be factored as $A = DQ$, where $D \in \mathbb{C}^{m\times r}$, $Q \in \mathbb{C}^{r\times n}$ has an orthonormal set of rows (that is, $QQ^{\mathsf{H}} = I_r$, and $\mathsf{rank}(Q) = r$.

(80) Let $\mathbb{C}^{n\times n}$ be the linear space of complex matrices. Prove the following:
   (a) the set of Hermitian matrices $\mathcal{H}$ and the set of skew-Hermitian matrices $\mathcal{K}$ in $\mathbb{C}^{n\times n}$ are subspaces of $\mathbb{C}^{n\times n}$;
   (b) if $\mathbb{C}^{n\times n}$ is equipped with the inner product defined in Example 6.21, then $\mathcal{K} = \mathcal{H}^{\perp}$.

(81) Give an example of a matrix that has positive elements but is not positive definite.

(82) Prove that if $A \in \mathbb{R}^{n\times n}$ is a positive definite matrix, then A is invertible and A^{-1} is also positive definite.

(83) Let $A \in \mathbb{C}^{n\times n}$ be a positive definite Hermitian matrix. If $A = B + iC$, where $B, C \in \mathbb{R}^{n\times n}$, prove that the real matrix

$$D = \begin{pmatrix} B & -C \\ C & B \end{pmatrix}$$

is positive definite.

(84) Let $A \in \mathbb{R}^{2\times 2}$ be a matrix such that $\boldsymbol{x}'A\boldsymbol{x} > 0$ for every $\boldsymbol{x} \in \mathbb{R}^2 - \{\mathbf{0}\}$. Does it follow that $\boldsymbol{u}^{\mathsf{H}}A\boldsymbol{u} > \mathbf{0}$ for every $\boldsymbol{x} \in \mathbb{C}^{2\times 2} - \{\mathbf{0}\}$?

   Hint: Consider $A = I_2$ and $\boldsymbol{x} = \begin{pmatrix} i \\ 0 \end{pmatrix}$.

(85) Let $\boldsymbol{x}, \boldsymbol{y} \in \mathbb{R}^n$ be two vectors such that $\boldsymbol{x} > \mathbf{0}$. Prove that there exists a number $\epsilon > 0$ such that $\|\boldsymbol{y}\|_2 \leqslant \epsilon$ implies $\boldsymbol{x} + \boldsymbol{y} > \mathbf{0}$.

(86) Let $U \in \mathbb{R}^{n \times k}$ be a matrix such that $U'U = I_k$, where $k \leqslant n$. Prove that for every $\boldsymbol{x} \in \mathbb{R}^n$, we have

$$\|\boldsymbol{x}\|_2 \geqslant \|U'\boldsymbol{x}\|_2.$$

Solution: Let $\boldsymbol{u}_1, \ldots, \boldsymbol{u}_k$ be the columns of the matrix U. We have

$$U'U = \begin{pmatrix} \boldsymbol{u}_1' \\ \vdots \\ \boldsymbol{u}_k' \end{pmatrix} (\boldsymbol{u}_1 \ \cdots \ \boldsymbol{u}_k) = I_k,$$

which shows that $\{\boldsymbol{u}_1, \ldots, \boldsymbol{u}_k\}$ is an orthonormal set of vectors in $\mathbb{R}^n$. Let $\{\boldsymbol{u}_1, \ldots, \boldsymbol{u}_k, \boldsymbol{u}_{k+1}, \ldots, \boldsymbol{u}_n\}$ be the completion of this set to an orthonormal set of $\mathbb{R}^n$. If $\boldsymbol{x} = \sum_{i=1}^n a_i \boldsymbol{u}_i$, then $\|\boldsymbol{x}\|_2^2 = \sum_{i=1}^n a_i^2$.

On the other hand,

$$U'\boldsymbol{x} = \begin{pmatrix} \boldsymbol{u}_1' \\ \vdots \\ \boldsymbol{u}_k' \end{pmatrix} \boldsymbol{x} = \begin{pmatrix} \boldsymbol{u}_1'\boldsymbol{x} \\ \vdots \\ \boldsymbol{u}_k'\boldsymbol{x} \end{pmatrix} = \begin{pmatrix} a_1 \\ \vdots \\ a_k \end{pmatrix},$$

because of the orthonormality of the set $\{\boldsymbol{u}_1, \ldots, \boldsymbol{u}_k\}$, so

$$\|Y'\boldsymbol{x}\|_2^2 = \sum_{i=1}^k a_i^2 \leqslant \sum_{i=1}^n a_i^2 = \|\boldsymbol{x}\|_2^2.$$

(87) Let V be a real linear space and let $\|\cdot\|$ be a norm generated by an inner product defined on V. V is said to be *symmetric relative to the norm* if $\|a\boldsymbol{x} - \boldsymbol{y}\| = \|\boldsymbol{x} - a\boldsymbol{y}\|$ for $a \in \mathbb{R}$ and $\boldsymbol{x}, \boldsymbol{y} \in V$ such that $\|\boldsymbol{x}\| = \|\boldsymbol{y}\| = 1$.

(a) Prove that if a norm on a linear vector space V is induced by an inner product, then V is symmetric relative to that norm.

(b) Prove that V satisfies the *Ptolemy inequality* $\|\boldsymbol{x} - \boldsymbol{y}\|\|\boldsymbol{z}\| \leqslant \|\boldsymbol{y} - \boldsymbol{z}\|\|\boldsymbol{x}\| + \|\boldsymbol{z} - \boldsymbol{x}\|\|\boldsymbol{y}\|$, for $\boldsymbol{x}, \boldsymbol{y}, \boldsymbol{z} \in V$ if and only if V is symmetric.

Solution: Suppose that $\|\boldsymbol{x}\|^2 = (\boldsymbol{x}, \boldsymbol{x})$ for $\boldsymbol{x} \in V$. Then, if $\|\boldsymbol{x}\| = \|\boldsymbol{y}\| = 1$, we have $\|a\boldsymbol{x} - \boldsymbol{y}\|^2 = (a\boldsymbol{x} - \boldsymbol{y}, a\boldsymbol{x} - \boldsymbol{y}) = a^2(\boldsymbol{x}, \boldsymbol{x}) - 2a(\boldsymbol{x}, \boldsymbol{y}) + (\boldsymbol{y}, \boldsymbol{y}) = a^2 + 1 + 2a(\boldsymbol{x}, \boldsymbol{y})$. It is easy to see that we also have $\|\boldsymbol{x} - a\boldsymbol{y}\| = a^2 + 1 + 2a(\boldsymbol{x}, \boldsymbol{y})$, which implies that V is symmetric relative to the norm.
For the second part, suppose that V is symmetric relative to the norm. Observe that Ptolemy inequality is immediate if any of $\boldsymbol{x}, \boldsymbol{y}$, or $\boldsymbol{z}$ is $\boldsymbol{0}$. Therefore, let $\hat{\boldsymbol{x}}, \hat{\boldsymbol{y}}, \hat{\boldsymbol{z}}$ be three non-zero vectors defined by

$$\hat{\boldsymbol{x}} = \frac{1}{\|\boldsymbol{x}\|^2}\boldsymbol{x}, \hat{\boldsymbol{y}} = \frac{1}{\|\boldsymbol{y}\|^2}\boldsymbol{y}, \hat{\boldsymbol{z}} = \frac{1}{\|\boldsymbol{z}\|^2}\boldsymbol{z}.$$

We have

$$\|\hat{\boldsymbol{x}} - \hat{\boldsymbol{y}}\|^2 = \frac{1}{\|\boldsymbol{x}\|^2} - \frac{2(\boldsymbol{x}, \boldsymbol{y})}{\|\boldsymbol{x}\|^2\|\boldsymbol{y}\|^2} + \frac{1}{\|\boldsymbol{y}\|^2} = \frac{\|\boldsymbol{x} - \boldsymbol{y}\|^2}{\|\boldsymbol{x}\|^2\|\boldsymbol{y}\|^2}.$$

Since $\|\hat{\boldsymbol{x}} - \hat{\boldsymbol{y}}\| \leqslant \|\hat{\boldsymbol{x}} - \hat{\boldsymbol{z}}\| + \|\hat{\boldsymbol{z}} - \hat{\boldsymbol{y}}\|$, the Ptolemy inequality follows immediately.

(88) Let $H_{\boldsymbol{u}}$ be the Householder matrix corresponding to the unit vector $\boldsymbol{u} \in \mathbb{R}^n$. If $\boldsymbol{x} \in \mathbb{R}^n$ is written as $\boldsymbol{x} = \boldsymbol{y} + \boldsymbol{z}$, where $\boldsymbol{y} = a\boldsymbol{u}$ and $\boldsymbol{z} \perp \boldsymbol{u}$, then $H_{\boldsymbol{u}}\boldsymbol{x}$ is obtained by a reflection of $\boldsymbol{x}$ relative to the hyperplane that is perpendicular to $\boldsymbol{u}$, that is, $H_{\boldsymbol{u}}\boldsymbol{x} = -\boldsymbol{u} + \boldsymbol{v}$.

Solution: We have

$$H_{\boldsymbol{u}}\boldsymbol{x} = H_{\boldsymbol{u}}(\boldsymbol{y} + \boldsymbol{z}) = H_{\boldsymbol{u}}\boldsymbol{y} + H_{\boldsymbol{u}}\boldsymbol{z}.$$

Then, $H_{\boldsymbol{u}}\boldsymbol{y} = (I_n - 2\boldsymbol{u}\boldsymbol{u}')(a\boldsymbol{u}) = a\boldsymbol{u} - 2a\boldsymbol{u}\boldsymbol{u}'\boldsymbol{u} = a\boldsymbol{u} - 2a\boldsymbol{u} = -a\boldsymbol{u} = -\boldsymbol{y}$. Also, $H_{\boldsymbol{u}}\boldsymbol{z} = (I_n - 2\boldsymbol{u}\boldsymbol{u}')\boldsymbol{z} = \boldsymbol{z} - 2\boldsymbol{u}\boldsymbol{u}'\boldsymbol{z} = \boldsymbol{z}$ because $\boldsymbol{u} \perp \boldsymbol{z}$.

(89) Let $\boldsymbol{x} \in \mathbb{R}^n$ be a unit vector such that $\boldsymbol{x} \notin \{\boldsymbol{e}_1, -\boldsymbol{e}_1\}$, and let

$$\boldsymbol{v} = \frac{1}{\sqrt{2(1 + x_1)}}(\boldsymbol{x} + \boldsymbol{e}_1) \text{ and } \boldsymbol{w} = \frac{1}{\sqrt{2(1 - x_1)}}(\boldsymbol{x} - \boldsymbol{e}_1).$$

Prove that

(a) $\boldsymbol{v}$ and $\boldsymbol{w}$ are unit vectors;
(b) we have $H_{\boldsymbol{v}}\boldsymbol{x} = -\boldsymbol{e}_1$ and $H_{\boldsymbol{w}}\boldsymbol{x} = \boldsymbol{e}_1$.

Solution: Observe that

$$\begin{aligned}\|\boldsymbol{v}\|^2 &= \frac{1}{2(1+x_1)}(\boldsymbol{x}+\boldsymbol{e}_1)(\boldsymbol{x}+\boldsymbol{e}_1)\\ &= \frac{1}{2(1+x_1)}(\|\boldsymbol{x}\|^2 + 2x_1 + \|\boldsymbol{e}_1\|^2) = 1.\end{aligned}$$

The computation for $\|\boldsymbol{w}\|$ is similar. Solving the second part is straightforward and is omitted.

(90) Let

$$A = \begin{pmatrix} a & b \\ b & a \end{pmatrix} \in \mathbb{R}^{2\times 2}.$$

Prove that if A is positive semidefinite but not positive definite, then $|a| = |b|$.

(91) Let $S = \{s_1, \ldots, s_n\}$ be a finite set. For a subset U of S, define $\boldsymbol{c}_U \in \mathbb{R}^n$, the characteristic vector of U, by

$$(c_U)_i = \begin{cases} 1 & \text{if } s_i \in U, \\ 0 & \text{otherwise.} \end{cases}$$

(a) Prove that $(\boldsymbol{c}_U, \boldsymbol{c}_W) = |U \cap V|$ for any two subsets U, W of S.

(b) Let $\mathcal{U} = (U_1, \ldots, U_m)$ be a sequence of subsets of S. The *incidence matrix of* $\mathcal{U}$ is the matrix $C_{\mathcal{U}} = (\boldsymbol{c}_{U_1}, \ldots, \boldsymbol{c}_{U_n})$. If $A_{\mathcal{U}} \in \mathbb{R}^{m\times m}$ is defined by $(A_{\mathcal{U}})_{ij} = |U_i \cap U_j|$ for $1 \leqslant i, j \leqslant m$, prove that $A_{\mathcal{U}} = C'_{\mathcal{U}} C_{\mathcal{U}}$, and therefore, $A_{\mathcal{U}}$ is a positive semidefinite matrix.

(92) Let $Y \in \mathbb{R}^{n\times k}$ be a matrix such that $Y'Y = I_k$, where $k \leqslant n$. Prove that the matrix $I_n - YY'$ is positive semidefinite.

Solution: Let $\boldsymbol{x} \in \mathbb{R}^n$. We have

$$\boldsymbol{x}'(I_n - YY')\boldsymbol{x} = \boldsymbol{x}'\boldsymbol{x} - (\boldsymbol{Y}'\boldsymbol{x})'(\boldsymbol{Y}'\boldsymbol{x}) = \|\boldsymbol{x}\|_2^2 - \|Y'\boldsymbol{x}\|_2^2.$$

The desired inequality follows immediately from Supplement 6.20.

(93) Let A and B be two matrices in $\mathbb{C}^{n\times n}$, where A is Hermitian and B is positive semidefinite. Prove that $\boldsymbol{x}^{\mathsf{H}} A\boldsymbol{x} < 0$ for all $\boldsymbol{x} \in \mathbb{C}^n$ such that $B\boldsymbol{x} = \boldsymbol{0}$ and $\boldsymbol{x} \neq \boldsymbol{0}$ if and only if there exists $a \in \mathbb{R}$ such that $a > 0$ and $A - aB \prec 0$.

Solution: Suppose that there exists $a \in \mathbb{R}$ such that $a > 0$ and $A - aB \prec 0$. This means that $\boldsymbol{x}^{\mathsf{H}}(A - aB)\boldsymbol{x} < 0$. Then, if $B\boldsymbol{x} = \mathbf{0}$, it is clear that $a\boldsymbol{x}^{\mathsf{H}}B\boldsymbol{x} = 0$, so $\boldsymbol{x}^{\mathsf{H}}A\boldsymbol{x} < 0$.
Conversely, let A be Hermitian and let B be positive semidefinite such that $\boldsymbol{x}^{\mathsf{H}}A\boldsymbol{x} < 0$ for all $\boldsymbol{x} \in \mathbb{C}^n$ such that $B\boldsymbol{x} = \mathbf{0}$ and $\boldsymbol{x} \neq \mathbf{0}$. Suppose that for every $a > 0$ there exists $\boldsymbol{x} \neq \mathbf{0}$ such that $\boldsymbol{x}^{\mathsf{H}}A\boldsymbol{x} \geqslant a\boldsymbol{x}^{\mathsf{H}}B\boldsymbol{x}$. In this case, $B\boldsymbol{x} = 0$ implies $\boldsymbol{x}^{\mathsf{H}}A\boldsymbol{x} \geqslant 0$. This contradiction yields the desired implication.

(94) Let $\{\boldsymbol{w}_1, \ldots, \boldsymbol{w}_k\} \subseteq \mathbb{C}^n$ be an orthonormal set of vectors in $\mathbb{C}^n$ and let $\boldsymbol{x} \in \mathbb{C}^n$. If $a_i = (\boldsymbol{x}, \boldsymbol{w}_i)$ for $1 \leqslant i \leqslant k$, prove that

$$\left(\boldsymbol{x} - \sum_{i=1}^{k} a_i\boldsymbol{w}_i, \boldsymbol{x} - \sum_{i=1}^{k} a_i\boldsymbol{w}_i\right) \leqslant (\boldsymbol{x}, \boldsymbol{x}) - \sum_{i=1}^{k} a_i\overline{a_i} \leqslant (\boldsymbol{x}, \boldsymbol{x}).$$

Prove that $(\boldsymbol{x}, \boldsymbol{x}) = \sum_{i=1}^{k} a_i\overline{a_i}$ if and only if $\boldsymbol{x} = \sum_{i=1}^{k} a_i\boldsymbol{w}_i$.

(95) Let $A_0, \ldots, A_n$ be $n+1$ Hermitian matrices and let $B(t) = \sum_{k=0}^{n} A_k t^{n-k}$. Prove that if A_n is positive definite, there exists a positive number ϵ such that $B(t)$ is positive definite for every $t \in [-\epsilon, \epsilon]$.

Solution: The definition of $B(t)$ allows us to write

$$\boldsymbol{x}'B(t)\boldsymbol{x} = \sum_{k=0}^{n} \boldsymbol{x}'A_k\boldsymbol{x}t^{n-k}.$$

Since the function $g(t) = \sum_{k=0}^{n} \boldsymbol{x}'A_k\boldsymbol{x}t^{n-k}$ is a polynomial in t and $g(0) = \boldsymbol{x}'A_n\boldsymbol{x} > 0$, the continuity of the polynomial $g(t)$ implies that there exists $\epsilon > 0$ such that $t \in [\epsilon, \epsilon]$ implies $g(t) > 0$. This produces the desired conclusion.

(96) Let $A \in \mathbb{R}^{n\times n}$ and let $b \in \mathbb{R}^n$. Define the matrix $C \in \mathbb{R}^{n\times n}$ by $c_{ij} = b_i a_{ij} b_j$ for $1 \leqslant i, j \leqslant n$. Prove that A is positive semidefinite if and only if C is positive semidefinite.

Solution: Let $\boldsymbol{x}$ be a vector in $\mathbb{R}^n$. We have

$$\boldsymbol{x}'C\boldsymbol{x} = \sum_{i=1}^{n}\sum_{j=1}^{n} x_i c_{ij} x_j = \sum_{i=1}^{n}\sum_{j=1}^{n} x_i b_i a_{ij} b_j x_j,$$

which is equivalent to $\boldsymbol{x}'C\boldsymbol{x} = \boldsymbol{z}'A\boldsymbol{z}$, where

$$\boldsymbol{z} = \begin{pmatrix} x_1 b_1 \\ \vdots \\ x_n b_n \end{pmatrix}.$$

We leave the remainder of the argument to the reader.

(97) Prove that if $A, B \in \mathbb{R}^{2\times 2}$ are two rotation matrices, then $AB = BA$.

Let $\{\boldsymbol{e}_1, \boldsymbol{e}_2, \boldsymbol{e}_3\}$ and $\{\tilde{\boldsymbol{e}}_1, \tilde{\boldsymbol{e}}_2, \tilde{\boldsymbol{e}_3}\}$ be two orthonormal bases in $\mathbb{R}^3$. We have $(\boldsymbol{e}_i, \boldsymbol{e}_j) = \delta_{ij}$ and $(\tilde{\boldsymbol{e}}_i, \tilde{\boldsymbol{e}}_j) = \delta_{ij}$.

Suppose that $\tilde{\boldsymbol{e}}_j = \sum_{k=1}^{3} r_{jk}\boldsymbol{e}_k$ and $\boldsymbol{e}_j = \sum_{k=1}^{3} \hat{r}_{jk}\tilde{\boldsymbol{e}}_k$. Since both bases are orthonormal, we obtain $r_{j\ell} = (\tilde{\boldsymbol{e}_j}, \boldsymbol{e}_\ell)$ and $\hat{r}_{j\ell} = (\boldsymbol{e}_j, \tilde{\boldsymbol{e}}_\ell)$. Clearly, we have: $r_{j\ell} = \hat{r}_{\ell j}$.

(98) Prove that $r_{j\ell} = \hat{r}_{\ell j}$, $\tilde{\boldsymbol{e}_j} = \sum_{k,\ell} r_{jk}r_{\ell k}\tilde{\boldsymbol{e}}_\ell$, and $\boldsymbol{e}_j = \sum_{k,\ell} r_{kj}r_{k\ell}\boldsymbol{e}_\ell$. This shows that the matrix $R = (r_{j\ell})$ is orthogonal.

(99) A vector $\boldsymbol{v} \in \mathbb{R}^3$ is *isotropic* if its components are the same relative to any basis. Prove that the single isotropic vector in $\mathbb{R}^3$ is $\boldsymbol{0}_3$.

Solution: Suppose that $\boldsymbol{v} \in \mathbb{R}^3$ is isotropic. Then, we have

$$\boldsymbol{v} = \sum_j v_j\boldsymbol{e}_j = \sum_{j,k} v_j r_{kj}\tilde{\boldsymbol{e}}_j = \sum_k \tilde{v}_k\tilde{\boldsymbol{e}}_k,$$

and

$$\boldsymbol{v} = \sum_j \tilde{v}_j\tilde{\boldsymbol{e}}_j = \sum_{j,k} \tilde{v}_j r_{jk}\boldsymbol{e}_k = \sum_k v_k\boldsymbol{e}_k =$$

If $\boldsymbol{v}$ is isotropic, then $\tilde{v}_j = v_j = \sum_k r_{jk}v_k$ for any rotation matrix R. Choosing R as

$$R = \begin{matrix} 0 & 1 & 0 \\ -1 & 0 & 0 \\ 0 & 0 & 1 \end{matrix}$$

we obtain $v_1 = v_2 = 0$. Similarly, choosing

$$R = \begin{matrix} 1 & 0 & 0 \\ 0 & 0 & 1 \\ 0 & -1 & 0 \end{matrix}$$

we obtain $v_2 = v_3 = 0$, hence, $\boldsymbol{v} = \boldsymbol{0}_3$.

(100) Let $\boldsymbol{u} \in \mathbb{R}^n$ be a unit vector. A *rotation with axis* $\boldsymbol{u}$ is an orthogonal matrix A such that $A\boldsymbol{u} = \boldsymbol{u}$. Prove that if $\boldsymbol{v} \perp \boldsymbol{u}$, then $A\boldsymbol{v} \perp \boldsymbol{u}$ and $A'\boldsymbol{v} \perp \boldsymbol{u}$.

(101) Let $\boldsymbol{u}, \boldsymbol{v}$, and $\boldsymbol{w}$ be three unit vectors in $\mathbb{R}^2 - \{\boldsymbol{0}\}$. Prove that $\angle(\boldsymbol{u}, \boldsymbol{v}) \leqslant \angle(\boldsymbol{u}, \boldsymbol{w}) + \angle(\boldsymbol{w}, \boldsymbol{v})$.

Solution: The hypothesis implies the existence of $\alpha, \beta, \gamma \in (0, 2\pi)$ such that

$$\boldsymbol{u} = \begin{pmatrix} \cos\alpha \\ \sin\alpha \end{pmatrix}, \boldsymbol{v} = \begin{pmatrix} \cos\beta \\ \sin\beta \end{pmatrix}, \text{ and } \boldsymbol{w} = \begin{pmatrix} \cos\gamma \\ \sin\gamma \end{pmatrix}.$$

Thus,

$$\begin{aligned} \angle(\boldsymbol{u}, \boldsymbol{v}) &= \arccos(\cos(\alpha - \beta)), \\ \angle(\boldsymbol{u}, \boldsymbol{w}) &= \arccos(\cos(\alpha - \gamma)), \\ \angle(\boldsymbol{w}, \boldsymbol{v}) &= \arccos(\cos(\gamma - \beta)). \end{aligned}$$

Without loss of generality, we may assume that $\alpha \geqslant \gamma \geqslant \beta$. Thus,

$$\arccos(\cos(\alpha - \beta)) = \begin{cases} \alpha - \beta & \text{if } \alpha - \beta \leqslant \pi, \\ 2\pi - \alpha + \beta & \text{if } \alpha - \beta > \pi \end{cases}$$

and similar equalities can be written for $\arccos(\cos(\alpha - \gamma))$ and $\arccos(\cos(\gamma - \beta))$.
If $\alpha - \beta \leqslant \pi$, then $\alpha - \gamma \leqslant \pi$ and $\gamma - \beta \leqslant \pi$, so $\angle(\boldsymbol{u}, \boldsymbol{v}) \leqslant \angle(\boldsymbol{u}, \boldsymbol{w}) + \angle(\boldsymbol{w}, \boldsymbol{v})$.
Otherwise, $\alpha - \beta > \pi$ and several cases may occur. Since $\alpha - \beta = (\alpha - \gamma) + (\gamma - \beta)$, at most one of $\alpha - \gamma$ and $\gamma - \beta$ can be greater than π.
If we have $\alpha - \gamma \leqslant \pi$ and $\gamma - \beta \leqslant \pi$, the inequality to be shown amounts to

$$2\pi - \alpha + \beta \leqslant \alpha - \gamma + \gamma - \beta,$$

which clearly holds.
If $\alpha - \gamma \leqslant \pi$ and $\gamma - \beta \geqslant \pi$, then we have

$$2\pi - \alpha + \beta \leqslant \alpha - \gamma + 2\pi - \gamma + \beta,$$

which amounts to the inequality $\alpha \geqslant \gamma$, which holds according to the initial assumption.

(102) Let T be a subspace of $\mathbb{R}^n$ and let $\boldsymbol{u}$ and $\boldsymbol{v}$ be two unit vectors in $\mathbb{R}^n$ such that $\boldsymbol{v} \in T$. If $\boldsymbol{t} = \mathsf{proj}_T(\boldsymbol{u})$, prove that $\angle(\boldsymbol{u}, \boldsymbol{v}) \geqslant \angle(\boldsymbol{t}, \boldsymbol{v})$.

Solution: Suppose that T is an m-dimensional space, where $m \leqslant n$. Let $\{\boldsymbol{v}, \boldsymbol{v}_1, \ldots, v_{m-1}\}$ be the extension of the set $\{\boldsymbol{v}\}$ to an orthonormal basis of T. Then $\boldsymbol{t} = \mathsf{proj}_U(\boldsymbol{u})$, we have $\boldsymbol{t} = (\boldsymbol{u}, \boldsymbol{v})\boldsymbol{v} + (\boldsymbol{u}, \boldsymbol{v}_1)\boldsymbol{v}_1 + \cdots + (\boldsymbol{u}, \boldsymbol{v}_{m-1})\boldsymbol{v}_{m-1}$, so $(\boldsymbol{t}, \boldsymbol{v}) = (\boldsymbol{u}, \boldsymbol{v})$. Thus, $\|\boldsymbol{t}\|_2 \cos\angle(\boldsymbol{t}, \boldsymbol{v}) = \cos\angle(\boldsymbol{u}, \boldsymbol{v})$, which implies $\cos\angle(\boldsymbol{u}, \boldsymbol{v}) \leqslant \cos\angle(\boldsymbol{t}, \boldsymbol{v})$.

(103) We now extend the result of Supplement 6.20 to $\mathbb{R}^n$ as follows. Let $\boldsymbol{u}, \boldsymbol{v}$, and $\boldsymbol{w}$ be three unit vectors in $\mathbb{R}^n - \{\boldsymbol{0}\}$. Prove that $\angle(\boldsymbol{u}, \boldsymbol{v}) \leqslant \angle(\boldsymbol{u}, \boldsymbol{w}) + \angle(\boldsymbol{w}, \boldsymbol{v})$.

(104) This supplement formulates a reciprocal to Theorem 6.36. Prove that if $A \in \mathbb{C}^{n \times n}$ is a Hermitian and idempotent matrix, then there exists a subspace S of $\mathbb{C}^n$ such that A is the projection matrix P_S.

Solution: Let $\boldsymbol{x} \in \mathbb{R}^n$ be a vector, and let $S = \mathsf{range}(A)$. Then, $\boldsymbol{u} = A\boldsymbol{x} \in S$. If $\boldsymbol{z} = \boldsymbol{x} - \boldsymbol{u}$, we have $(\boldsymbol{z}, \boldsymbol{u}) = \boldsymbol{z}^{\mathsf{H}}\boldsymbol{u} = (\boldsymbol{x}^{\mathsf{H}} - \boldsymbol{u}^{\mathsf{H}})\boldsymbol{u} = (\boldsymbol{x}^{\mathsf{H}} - \boldsymbol{x}^{\mathsf{H}}A^{\mathsf{H}})A\boldsymbol{x} = \boldsymbol{x}^{\mathsf{H}}A^{\mathsf{H}})A\boldsymbol{x} - \boldsymbol{x}^{\mathsf{H}}A^{\mathsf{H}}A\boldsymbol{x} = 0$, because A is Hermitian and idempotent. Thus, $\boldsymbol{z} \perp \boldsymbol{u}$ and $\boldsymbol{z} \in S^{\perp}$. By Theorem 6.37, the decomposition of $\boldsymbol{x} = \boldsymbol{u} + \boldsymbol{z}$ is unique, so $\boldsymbol{u} = A\boldsymbol{x} = \mathsf{proj}_S\boldsymbol{x}$.

Let V be an n-dimensional linear space equipped with an inner product $(\cdot, \cdot)$. The subsets $B = \{\boldsymbol{b}_1, \ldots, \boldsymbol{b}_n\}$ and $C = \{\boldsymbol{c}_1, \ldots, \boldsymbol{c}_n\}$ of V are *reciprocal* if $(\boldsymbol{b}_i, \boldsymbol{c}_j) = 1$ if $i = j$ and $(\boldsymbol{b}_i, \boldsymbol{c}_j) = 0$ if $i \neq j$, for $1 \leqslant i, j \leqslant n$.

(105) Let V be an n-dimensional linear space equipped with an inner product $(\cdot, \cdot)$. If $B = \{\boldsymbol{b}_1, \ldots, \boldsymbol{b}_n\}$ is a basis of V, then there exists a unique reciprocal set of B.

Solution: Let U_i be the subspace of V generated by the set $B - \{\boldsymbol{b}_i\}$ and let $U_i^{\perp}$ be its orthogonal complement. By Corollary 6.28, $\dim(U_i^{\perp}) = 1$ because $\dim(U_i) = n - 1$. Thus, there exists a vector $\boldsymbol{t} \neq \boldsymbol{0}$ in $U_i^{\perp}$. Note that $(\boldsymbol{t}, \boldsymbol{b}_i) \neq 0$ because $\boldsymbol{b}_i \notin U_i$. Define

$$\boldsymbol{c}_i = \frac{1}{(\boldsymbol{t}, \boldsymbol{b}_i)}\boldsymbol{b}_i.$$

Then $(\boldsymbol{b}_i, \boldsymbol{c}_i) = 1$ and $(\boldsymbol{b}_i, \boldsymbol{c}_j) = 0$ if $j \neq i$. This construction can be applied to all i, where $1 \leqslant i \leqslant n$ and this yields a set $C = \{\boldsymbol{c}_1, \ldots, \boldsymbol{c}_n\}$, which is reciprocal to B.

To prove the uniqueness of the set C, assume that $D = \{\boldsymbol{d}_1, \ldots, \boldsymbol{d}_n\}$ is another reciprocal set of the basis B. Then since $(\boldsymbol{b}_i, \boldsymbol{c}_j) = (\boldsymbol{b}_i, \boldsymbol{d}_j)$, it follows that

$$(\boldsymbol{b}_i, \boldsymbol{c}_j - \boldsymbol{d}_j) = 0$$

for every i, j. Since $\boldsymbol{c}_j - \boldsymbol{d}_j$ is orthogonal on all vectors of B, it follows that $\boldsymbol{c}_j - \boldsymbol{d}_j = \boldsymbol{0}$, so $\boldsymbol{c}_j = \boldsymbol{d}_j$. Thus, $D = C$.

(106) Let V be an n-dimensional linear space equipped with an inner product $(\cdot, \cdot)$. If $B = \{\boldsymbol{b}_1, \ldots, \boldsymbol{b}_n\}$ is a basis of V, then the reciprocal set C of B is also a basis of V.

(107) Let $L = (\boldsymbol{v}_1, \ldots, \boldsymbol{v}_n)$ be a sequence of vectors, where $n \geqslant 2$. Prove that the volume $\mathcal{V}_n$ of the parallelepiped constructed on these vectors equals the square root of the Gramian of the sequence $(\boldsymbol{v}_1, \ldots, \boldsymbol{v}_n)$.

Solution: For the base case $n = 2$, the area A of the parallelogram is given by $A = \|\boldsymbol{u}\|_2 \|\boldsymbol{v}\|_2 \sin \alpha$, where $\alpha = \angle(\boldsymbol{u}, \boldsymbol{v})$. In other words,

$$\begin{aligned}
\det(G_{\boldsymbol{u},\boldsymbol{v}}) &= \det \begin{pmatrix} (\boldsymbol{u}, \boldsymbol{u}) & (\boldsymbol{u}, \boldsymbol{v}) \\ (\boldsymbol{u}, \boldsymbol{v}) & (\boldsymbol{v}, \boldsymbol{v}) \end{pmatrix} \\
&= \|\boldsymbol{u}\|_2^2 \|\boldsymbol{v}\|_2^2 - \|\boldsymbol{u}\|_2^2 \|\boldsymbol{v}\|_2^2 \cos^2 \alpha \\
&= \|\boldsymbol{u}\|_2^2 \|\boldsymbol{v}\|_2^2 \sin^2 \alpha = \mathcal{V}_2^2.
\end{aligned}$$

Suppose that the statement holds for sequences of n vectors and let $L = (\boldsymbol{v}_1, \ldots, \boldsymbol{v}_n, \boldsymbol{v}_{n+1})$ be a sequence of $n+1$ vectors. Let $\boldsymbol{v}_{n+1} = \boldsymbol{x} + \boldsymbol{y}$ be the orthogonal decomposition of $\boldsymbol{v}_{n+1}$ on the subspace $U_n = \langle \boldsymbol{v}_1, \ldots, \boldsymbol{v}_n \rangle$, where $\boldsymbol{x} \in U_n$ and $\boldsymbol{y} \perp U_n$. Since $\boldsymbol{x} \in U_n$, there exist $a_1, \ldots, a_n \in \mathbb{R}$ such that $\boldsymbol{x} = a_1 \boldsymbol{v}_1 + \cdots + a_n \boldsymbol{v}_n$. Let

$$\det(G_L) = \begin{vmatrix}
(\boldsymbol{v}_1, \boldsymbol{v}_1) & (\boldsymbol{v}_1, \boldsymbol{v}_2) & \cdots & (\boldsymbol{v}_1, \boldsymbol{v}_n) & (\boldsymbol{v}_1, \boldsymbol{v}_{n+1}) \\
(\boldsymbol{v}_2, \boldsymbol{v}_1) & (\boldsymbol{v}_2, \boldsymbol{v}_2) & \cdots & (\boldsymbol{v}_2, \boldsymbol{v}_n) & (\boldsymbol{v}_2, \boldsymbol{v}_{n+1}) \\
\vdots & \vdots & \vdots & \vdots & \vdots \\
(\boldsymbol{v}_n, \boldsymbol{v}_1) & (\boldsymbol{v}_n, \boldsymbol{v}_2) & \cdots & (\boldsymbol{v}_n, \boldsymbol{v}_n) & (\boldsymbol{v}_n, \boldsymbol{v}_{n+1}) \\
(\boldsymbol{v}_{n+1}, \boldsymbol{v}_1) & (\boldsymbol{v}_{n+1}, \boldsymbol{v}_2) & \cdots & (\boldsymbol{v}_{n+1}, \boldsymbol{v}_n) & (\boldsymbol{v}_{n+1}, \boldsymbol{v}_{n+1})
\end{vmatrix}.$$

By subtracting from the last row the first row multiplied by a_1, the second row multiplied by a_2, etc., the value of the determinant remains the same, and we obtain

$$\det(G_L) = \begin{vmatrix} (\boldsymbol{v}_1, \boldsymbol{v}_1) & (\boldsymbol{v}_1, \boldsymbol{v}_2) & \cdots & (\boldsymbol{v}_1, \boldsymbol{v}_n) & (\boldsymbol{v}_1, \boldsymbol{v}_{n+1}) \\ (\boldsymbol{v}_2, \boldsymbol{v}_1) & (\boldsymbol{v}_2, \boldsymbol{v}_2) & \cdots & (\boldsymbol{v}_2, \boldsymbol{v}_n) & (\boldsymbol{v}_2, \boldsymbol{v}_{n+1}) \\ \vdots & \vdots & \vdots & \vdots & \vdots \\ (\boldsymbol{v}_n, \boldsymbol{v}_1) & (\boldsymbol{v}_n, \boldsymbol{v}_2) & \cdots & (\boldsymbol{v}_n, \boldsymbol{v}_n) & (\boldsymbol{v}_n, \boldsymbol{v}_{n+1}) \\ (\boldsymbol{y}, \boldsymbol{v}_1) & (\boldsymbol{y}, \boldsymbol{v}_2) & \cdots & (\boldsymbol{y}, \boldsymbol{v}_n) & (\boldsymbol{y}, \boldsymbol{v}_{n+1}) \end{vmatrix}.$$

Note that $(\boldsymbol{y}, \boldsymbol{v}_1) = (\boldsymbol{y}, \boldsymbol{v}_2) = \cdots = (\boldsymbol{y}, \boldsymbol{v}_n) = 0$ because $\boldsymbol{y} \perp U_n$ and $(\boldsymbol{y}, \boldsymbol{v}_{n+1}) = (\boldsymbol{y}, \boldsymbol{x} + \boldsymbol{y}) = \|\boldsymbol{y}\|_2^2$, which allows us to further write

$$\det(G_L) = \begin{vmatrix} (\boldsymbol{v}_1, \boldsymbol{v}_1) & (\boldsymbol{v}_1, \boldsymbol{v}_2) & \cdots & (\boldsymbol{v}_1, \boldsymbol{v}_n) & (\boldsymbol{v}_1, \boldsymbol{v}_{n+1}) \\ (\boldsymbol{v}_2, \boldsymbol{v}_1) & (\boldsymbol{v}_2, \boldsymbol{v}_2) & \cdots & (\boldsymbol{v}_2, \boldsymbol{v}_n) & (\boldsymbol{v}_2, \boldsymbol{v}_{n+1}) \\ \vdots & \vdots & \vdots & \vdots & \vdots \\ (\boldsymbol{v}_n, \boldsymbol{v}_1) & (\boldsymbol{v}_n, \boldsymbol{v}_2) & \cdots & (\boldsymbol{v}_n, \boldsymbol{v}_n) & (\boldsymbol{v}_n, \boldsymbol{v}_{n+1}) \\ 0 & 0 & \cdots & 0 & \|\boldsymbol{y}\|_2^2 \end{vmatrix}.$$

$$= \mathcal{V}_n^2 \|\boldsymbol{y}\|_2^2 = \mathcal{V}_{n+1}^2.$$

(108) Let $A \in \mathbb{R}^{m \times n}$ be a matrix such that $\mathsf{rank}(A) = n$. Prove that the R-factor of the QR decomposition of $A = QR$ has positive diagonal elements, it equals the Cholesky factor of $A'A$, and therefore is uniquely determined.

Solution: If $\mathsf{rank}(A) = n$, then, by Theorem 6.57, there exists a unique Cholesky factor of the matrix $A'A$. Suppose that A has the full QR decomposition

$$A = Q \begin{pmatrix} R \\ O_{m-n,n} \end{pmatrix},$$

where $Q \in \mathbb{R}^{m \times m}$ and $R \in \mathbb{R}^{n \times n}$. Then

$$A'A = (R' \; O_{n,m-n}) Q'Q \begin{pmatrix} R \\ O_{m-n,n} \end{pmatrix} = R'R.$$

(109) Let $A \in \mathbb{R}^{n \times n}$ be a symmetric positive definite matrix and let $d_A : \mathbb{R}^n \times \mathbb{R}^n \longrightarrow \mathbb{R}$ be the function defined by $d_A(\boldsymbol{x}, \boldsymbol{y}) =$

$\sqrt{(\boldsymbol{x}-\boldsymbol{y})'A(\boldsymbol{x}-\boldsymbol{y})}$ for $\boldsymbol{x},\boldsymbol{y} \in \mathbb{R}^n$. Prove that d_A is a metric on $\mathbb{R}^n$.

Solution: By Cholesky's Decomposition Theorem, we can factor A as $A = R'R$, where R is an upper triangular matrix R with positive diagonal elements. Therefore, $d_A(\boldsymbol{x},\boldsymbol{y}) = \sqrt{(R(\boldsymbol{x}-\boldsymbol{y}))'(R(\boldsymbol{x}-\boldsymbol{y}))} = \|R(\boldsymbol{x}-\boldsymbol{y})\|_2$ and the desired conclusion follows immediately.

(110) Let $A \in \mathbb{C}^{n\times m}$ be a full-rank matrix such that $m \geqslant n$. Prove that A can be factored as $A = LQ$, where $L \in \mathbb{C}^{n\times n}$ and $Q \in \mathbb{C}^{n\times m}$, such that
  (a) the columns of Q constitute an orthonormal basis for $\mathsf{range}(A^{\mathsf{H}})$, and
  (b) $L = (\ell_{ij})$ is a lower triangular invertible matrix such that its diagonal elements are real non-negative numbers, that is, $\ell_{ii} \geqslant 0$ for $1 \leqslant i \leqslant n$.

(111) Let $f_n : \mathbb{R}^n \longrightarrow \mathbb{R}^n \longrightarrow \mathbb{R}$ be the function defined by $f_n(\boldsymbol{x},\boldsymbol{y}) = \prod_{j=1}^{n}(1 + x_j y_j)$ for $\boldsymbol{x},\boldsymbol{y} \in \mathbb{R}^n$. Prove that if $\boldsymbol{x},\boldsymbol{y} \in C(\mathbf{0}_n, 1)$, then $|f_n(\boldsymbol{x},\boldsymbol{y})| \leqslant e$.

Solution: It is easy to verify the elementary inequality $\ln(1 + |t|) \leqslant |t|$ for $t \in \mathbb{R}$. This allows us to write

$$|f_n(\boldsymbol{x},\boldsymbol{y})| \leqslant e^{\sum_{j=1}^{n}\ln(1+|x_j y_j|)} \leqslant e^{\sum_{i=1}^{n}|x_j y_j|} \leqslant e^{\|\boldsymbol{x}\|_2\|\boldsymbol{y}\|_2}.$$

because $\|\boldsymbol{x}\|_2 \leqslant 1$ and $\|\boldsymbol{y}\|_2 \leqslant 1$.

(112) Prove that if a matrix $A \in \mathbb{C}^{n\times n}$ is normal, then $\mathsf{range}(A) \perp \mathsf{null}(A)$.

(113) We saw that for every matrix A, the matrix $A^{\mathsf{H}}A$ is positive semidefinite. Prove that if $A = QR$ is the full QR decomposition of A, then the Cholesky decomposition of $A^{\mathsf{H}}A$ is $R^{\mathsf{H}}R$.

(114) Let $U \in \mathbb{C}^{n\times n}$ be a unitary matrix. If $U = (X\ Y)$, where $X \in \mathbb{C}^{n\times p}$ and $Y \in \mathbb{C}^{n\times(n-p)}$, prove that $\mathsf{range}(X) = \mathsf{range}(Y)^{\perp}$.

(115) Let $A \in \mathbb{R}^{n\times n}$ be a skew-symmetric matrix. Prove that A^2 is a symmetric negative semi definite matrix.

Solution: We have $(A^2)' = (A')^2 = (-A)^2 = A^2$, so A^2 is symmetric. Furthermore, $\boldsymbol{x}'A^2\boldsymbol{x} = -\boldsymbol{x}'A'A\boldsymbol{x} = -(A\boldsymbol{x})'A\boldsymbol{x} \leqslant 0$ for $\boldsymbol{x} \in \mathbb{R}^n$.

(116) Let $B = \{\boldsymbol{b}_1, \ldots, \boldsymbol{b}_n\}$ and $C = \{\boldsymbol{c}_1, \ldots, \boldsymbol{c}_n\}$ be two orthonormal bases of $\mathbb{R}^n$. The *coherence* of B and C is the number $\mathsf{coh}(B, C) = \max_{1 \leqslant i,j \leqslant n} \boldsymbol{b}_i' \boldsymbol{c}$. Prove that

$$\frac{1}{\sqrt{n}} \leqslant \mathsf{coh}(B, C) \leqslant 1.$$

Solution: Note that the matrix $D = (\boldsymbol{b}_i' \boldsymbol{c}_j) \in \mathbb{R}^{n \times n}$ is an orthonormal matrix and $\|\boldsymbol{d}_k\|_2 = 1$ for each of the columns $\boldsymbol{d}_k$ of D. Thus, not all entries of D can be less than $\frac{1}{\sqrt{n}}$, so $\max_{1 \leqslant i,j \leqslant n} \boldsymbol{b}_i' \boldsymbol{c} \geqslant \frac{1}{\sqrt{n}}$.

(117) Let $A \in \mathbb{R}^{m \times m}$ and $B \in \mathbb{R}^{n \times n}$ be two orthogonal matrices. Prove that their Kronecker product $A \otimes B \in \mathbb{R}^{mn \times mn}$ is also an orthogonal matrix.

Solution: We have

$$\begin{aligned}(A \otimes B)(A \otimes B)' &= (A \otimes B)(A' \otimes B') \\ &\quad \text{(by Theorem 3.56)} \\ &= (AA') \otimes (BB') \\ &\quad \text{(by Theorem 3.55)} \\ &= I_m \otimes I_n = I_{mn}.\end{aligned}$$

(118) Let $U, V \in \mathbb{R}^{n \times n}$ be two orthonormal matrices, where $U = (\boldsymbol{u}_1, \ldots, \boldsymbol{u}_n)$ and $V = (\boldsymbol{v}_1, \ldots, \boldsymbol{v}_n)$, and let $w \in \mathbb{R}^n$.
Since the columns of U and V are bases for $\mathbb{R}^n$, for every $\boldsymbol{x} \in \mathbb{R}^n$ with $\|x\|_2 = 1$ there exist $\boldsymbol{a}, \boldsymbol{b} \in \mathbb{R}^n$ such that $\boldsymbol{x} = U\boldsymbol{a} = V\boldsymbol{b}$. Prove the inequality

$$\|\boldsymbol{a}\|_1 + \|\boldsymbol{b}\|_1 \geqslant \frac{2}{\sqrt{\mathsf{coh}(U, V)}}$$

known an the *uncertainty principle* (see [43]).

Solution: Since both U and V are orthonormal matrices, we have $\|\boldsymbol{x}\|_2 = \|\boldsymbol{a}\|_2 = \|\boldsymbol{b}\|_2$. Therefore,

$$\begin{aligned}1 = \|\boldsymbol{x}\|^2 = \boldsymbol{x}'\boldsymbol{x} = \boldsymbol{a}'U'V\boldsymbol{b} &= \sum_{i=1}^{n} \sum_{j=1}^{n} |a_i|\ |b_j| \boldsymbol{u}_i' \boldsymbol{v}_j \\ &\leqslant \mathsf{coh}(U, V) \|\boldsymbol{a}\|_1 \|\boldsymbol{b}\|_1.\end{aligned}$$

This implies $\|\boldsymbol{a}\|_1\|\boldsymbol{b}\|_1 \geqslant \frac{1}{\mathsf{coh}(U,V)}$, which in turn yields $\|\boldsymbol{a}\|_1 + \|\boldsymbol{b}\|_1 \geqslant \frac{2}{\sqrt{\mathsf{coh}(U,V)}}$.

(119) Let $\boldsymbol{u}, \boldsymbol{v}, \boldsymbol{w}, \boldsymbol{z} \in \mathbb{R}^3$. Prove that

$$((\boldsymbol{u} \times \boldsymbol{v}), (\boldsymbol{w} \times \boldsymbol{z})) = \begin{vmatrix} (\boldsymbol{u}, \boldsymbol{w}) & (\boldsymbol{u}, \boldsymbol{z}) \\ (\boldsymbol{v}, \boldsymbol{w}) & (\boldsymbol{v}, \boldsymbol{z}) \end{vmatrix}.$$

This is *Lagrange's identity.*

(120) Let $\boldsymbol{a}, \boldsymbol{b}, \boldsymbol{c}, \boldsymbol{d} \in \mathbb{R}^3$ be such that $\boldsymbol{a} \perp \boldsymbol{b}$ and $\boldsymbol{c} \perp \boldsymbol{d}$. Prove that the vectors $\boldsymbol{x}$ and $\boldsymbol{y}$ are orthogonal, where $\boldsymbol{x} = (\boldsymbol{b} \times \boldsymbol{c}) \times (\boldsymbol{a} \times \boldsymbol{d})$, and $\boldsymbol{y} = (\boldsymbol{a} \times \boldsymbol{c}) \times (\boldsymbol{b} \times \boldsymbol{d})$.

(121) Let $\boldsymbol{u}, \boldsymbol{v}, \boldsymbol{w} \in \mathbb{R}^3$. Prove that

$$\boldsymbol{u} \times (\boldsymbol{v} \times \boldsymbol{w}) + \boldsymbol{v} \times (\boldsymbol{w} \times \boldsymbol{u}) + \boldsymbol{w} \times (\boldsymbol{u} \times \boldsymbol{v}) = \boldsymbol{0}_{\mathbb{R}^3}.$$

This is *Jacobi's identity.*

(122) Prove that if $\boldsymbol{u}, \boldsymbol{v} \in \mathbb{R}^3$, then $(\boldsymbol{u} \times \boldsymbol{v})_i = \sum_{j,k=1}^{3} \epsilon_{ijk} u_j v_k$ for $1 \leqslant i \leqslant 3$.

(123) Let $\boldsymbol{a}, \boldsymbol{b}, \boldsymbol{c} \in \mathbb{R}^3$ be three vectors that do not belong to the same plane. Prove that the vectors $\boldsymbol{a} + \alpha\boldsymbol{b}$, $\boldsymbol{b} + \beta\boldsymbol{c}$, and $\boldsymbol{c} + \gamma\boldsymbol{a}$ are coplanar if and only if $\alpha\beta\gamma = -1$.

Solution: The scalar triple product of $\boldsymbol{a} + \alpha\boldsymbol{b}$, $\boldsymbol{b} + \beta\boldsymbol{c}$, and $\boldsymbol{c} + \gamma\boldsymbol{a}$ can be rewritten as

$$\begin{aligned}
&(\boldsymbol{a} + \alpha\boldsymbol{b}, \boldsymbol{b} + \beta\boldsymbol{c}, \boldsymbol{c} + \gamma\boldsymbol{a}) \\
&= (\boldsymbol{a}, \boldsymbol{b} + \beta\boldsymbol{c}, \boldsymbol{c} + \gamma\boldsymbol{a}) + \alpha(\boldsymbol{b}, \boldsymbol{b} + \beta\boldsymbol{c}, \boldsymbol{c} + \gamma\boldsymbol{a}) \\
&= (\boldsymbol{a}, \boldsymbol{b} + \beta\boldsymbol{c}, \boldsymbol{c} + \gamma\boldsymbol{a}) + \alpha\beta(\boldsymbol{b}, \boldsymbol{c}, \boldsymbol{c} + \gamma\boldsymbol{a}) \\
&= (\boldsymbol{a}, \boldsymbol{b} + \beta\boldsymbol{c}, \boldsymbol{c} + \gamma\boldsymbol{a}) + \alpha\beta\gamma(\boldsymbol{b}, \boldsymbol{c}, \boldsymbol{a}).
\end{aligned}$$

Similarly,

$$\begin{aligned}
&(\boldsymbol{a}, \boldsymbol{b} + \beta\boldsymbol{c}, \boldsymbol{c} + \gamma\boldsymbol{a}) \\
&= (\boldsymbol{a}, \boldsymbol{b} + \beta\boldsymbol{c}, \boldsymbol{c}) + (\boldsymbol{a}, \boldsymbol{b} + \beta\boldsymbol{c}, \gamma\boldsymbol{a}) \\
&= (\boldsymbol{a}, \boldsymbol{b}, \boldsymbol{c}).
\end{aligned}$$

Thus,

$$\begin{aligned}
&(\boldsymbol{a} + \alpha\boldsymbol{b}, \boldsymbol{b} + \beta\boldsymbol{c}, \boldsymbol{c} + \gamma\boldsymbol{a}) \\
&= (\boldsymbol{a}, \boldsymbol{b}, \boldsymbol{c})(1 + \alpha\beta\gamma).
\end{aligned}$$

Consequently, the vectors $\boldsymbol{a} + \alpha\boldsymbol{b}$, $\boldsymbol{b} + \beta\boldsymbol{c}$, and $\boldsymbol{c} + \gamma\boldsymbol{a}$ are coplanar if and only if their scalar triple product is 0, that is, if and only if $\alpha\beta\gamma = -1$, because the fact that $\boldsymbol{a}, \boldsymbol{b}, \boldsymbol{c}$ are not coplanar implies $(\boldsymbol{a}, \boldsymbol{b}, \boldsymbol{c}) \neq 0$.

Bibliographical Comments

Supplements 58–61 contain concepts and results developed in [59, 87].

Supplement 52 is stated in [113].

Chapter 7

Eigenvalues

7.1 Introduction

The existence of directions that are preserved by linear transformations (which are referred to as eigenvectors) was discovered by Euler in his study of movements of rigid bodies. This work was continued by Lagrange, Cauchy, Fourier, and Hermite. The theme of eigenvectors and eigenvalues acquired increasing significance through its applications in heat propagation and stability theory. Later, Hilbert initiated the study of eigenvalues in functional analysis (in the theory of integral operators). He introduced the terms "eigenvalue" and "eigenvector".[1]

7.2 Eigenvalues and Eigenvectors

Definition 7.1. Let $A \in \mathbb{C}^{n\times n}$ be a square matrix. An *eigenvector* of A is a vector $\boldsymbol{v} \in \mathbb{C}^n - \{\mathbf{0}_n\}$ such that $A\boldsymbol{v} = \lambda\boldsymbol{v}$ for some $\lambda \in \mathbb{C}$.

The complex number λ that satisfies the previous equality is known as an *eigenvalue* of the matrix A and the pair $(\lambda, \boldsymbol{v})$ is known as an *eigenpair* of A.

[1]The term *eigenvalue* is a German–English hybrid formed from the German word *eigen*, which means "own", and the English word "value." Its use is common in the literature and we adopt it here.

The set of eigenvalues of a matrix A will be referred to as the *spectrum* of A and will be denoted by $\mathsf{spec}(A)$.

Example 7.1. Let $A \in \mathbb{C}^{2\times 2}$ be the matrix

$$A = \begin{pmatrix} a & b \\ c & d \end{pmatrix}.$$

The vector $\boldsymbol{v} = \begin{pmatrix} v_1 & v_2 \end{pmatrix} \neq \mathbf{0}_2$ is an eigenvector of A if $A\boldsymbol{v} = \lambda\boldsymbol{v}$, a system equivalent to

$$\begin{aligned} av_1 + bv_2 &= \lambda v_1, \\ cv_1 + dv_2 &= \lambda v_2. \end{aligned}$$

This, in turn, is equivalent to the homogeneous system

$$\begin{aligned} (a - \lambda)v_1 + bv_2 &= 0, \\ cv_1 + (d - v_2) &= 0. \end{aligned}$$

A non-trivial solution exists if and only if $\det(A) = 0$, which is equivalent to

$$(a - \lambda)(d - \lambda) - bc = 0$$

or

$$\lambda^2 - (a + d)\lambda + ad - bc = 0.$$

The roots of this equation are the eigenvalues of A:

$$\lambda_{1,2} = \frac{a + d \pm \sqrt{(a - d)^2 + 4bc}}{2}.$$

Theorem 7.1. *Let $A \in \mathbb{C}^{n\times n}$ be a matrix. If A has n distinct eigenvalues and $\boldsymbol{v}_1, \ldots, \boldsymbol{v}_n$ are corresponding eigenvectors, then $\{\boldsymbol{v}_1, \ldots, \boldsymbol{v}_n\}$ is a linearly independent set.*

Proof. Suppose that $\lambda_1, \ldots, \lambda_n$ are distinct eigenvalues of A and $v_1, \ldots, v_n$ are eigenvectors that correspond to these values.

If $\{\boldsymbol{v}_1, \ldots, \boldsymbol{v}_n\}$ were a linearly dependent set, we would have a linear combination of these vectors

$$a_1\boldsymbol{v}_{p_1} + \cdots + a_k\boldsymbol{v}_{p_k} = \mathbf{0}_n, \tag{7.1}$$

containing a minimal number k of vectors such that not every number $a_1, \ldots, a_k$ is 0. This would yield $a_1A\boldsymbol{v}_{p_1} + \cdots + a_kA\boldsymbol{v}_{p_k} = a_1\lambda_{p_1}\boldsymbol{v}_{p_1} + \cdots + a_k\lambda_{p_k}\boldsymbol{v}_{p_k} = \mathbf{0}_n$. Taking into account Equality (7.1) multiplied by λ_{p_k}, we would have

$$a_1(\lambda_{p_1} - \lambda_{p_k})\boldsymbol{v}_{p_1} + \cdots + a_{k-1}(\lambda_{p_{k-1}}\boldsymbol{v}_{p_{k-1}} = \mathbf{0}_n,$$

which would contradict the minimality of the number of terms in Equality (7.1). Thus, $\{\boldsymbol{v}_1, \ldots, \boldsymbol{v}_n\}$ is a linearly independent set. □

Corollary 7.1. *A matrix $A \in \mathbb{C}^{n\times n}$ has at most n distinct eigenvalues.*

Proof. Since the maximum size of a linearly independent set in $\mathbb{C}^n$ is n, it follows from Theorem 7.1 that A cannot have more than n distinct eigenvalues. □

The set of eigenvectors $S_{A,\lambda}$ that correspond to an eigenvalue λ of A is a subspace of $\mathbb{C}^n$ because $\boldsymbol{u}, \boldsymbol{v} \in S_{A,\lambda}$ implies

$$A(a\boldsymbol{u} + b\boldsymbol{v}) = aA\boldsymbol{u} + bA\boldsymbol{v} = a\lambda\boldsymbol{u} + b\lambda\boldsymbol{v} = \lambda(a\boldsymbol{u} + b\boldsymbol{v}).$$

The subspace $S_{A,\lambda}$ is known as the *invariant subspace of A for the eigenvalue λ*.

Clearly, the invariant subspace of $A \in \mathbb{C}^{n\times n}$ for λ coincides with the null space of the matrix $\lambda I_n - A$.

Not every real matrix $A \in \mathbb{R}^{n\times n}$ has a non-zero invariant subspace in $\mathbb{R}^n$.

Example 7.2. Let $A = \begin{pmatrix} 1 & 0 \\ 0 & -1 \end{pmatrix} \in \mathbb{R}^{2\times 2}$. If $A\boldsymbol{x} = \lambda\boldsymbol{x}$ for $\boldsymbol{x} \in \mathbb{R}^2$, then we have $x_2 = \lambda x_1$ and $-x_1 = \lambda x_2$, which implies $x_1^2 + x_2^2 = 0$. This is equivalent to $\boldsymbol{x} = \mathbf{0}_2$. Thus, A has no non-zero invariant subspace.

The situation is different if we regard A as a matrix in $\mathbb{C}^{2\times 2}$. Under this assumption, the equalities $x_2 = \lambda x_1$ and $-x_1 = \lambda x_2$

imply $\lambda^2 + 1 = 0$ if $\boldsymbol{x} \neq \mathbf{0}_2$. Thus, we have $\lambda_1 = i$ and $\lambda_2 = -i$. In the first case, we have the invariant subspace

$$\left\{ \begin{pmatrix} x_1 \\ x_2 \end{pmatrix} \in \mathbb{C}^2 \mid x_2 = ix_1 \right\},$$

while in the second case the invariant subspace is

$$\left\{ \begin{pmatrix} x_1 \\ x_2 \end{pmatrix} \in \mathbb{C}^2 \mid x_2 = -ix_1 \right\}.$$

Observe that if λ is an eigenvalue for $A \in \mathbb{C}^{n\times n}$, then $a\lambda$ is an eigenvalue of the matrix aA for every $a \in \mathbb{C}$. Thus, $a\mathsf{spec}(A) = \mathsf{spec}(aA)$.

If $(\lambda, \boldsymbol{x})$ is an eigenpair of A, then $\boldsymbol{x}^{\mathsf{H}} A\boldsymbol{x} = \lambda \boldsymbol{x}^{\boldsymbol{H}}\boldsymbol{x}$, so

$$\lambda = \frac{\boldsymbol{x}^{\mathsf{H}} A\boldsymbol{x}}{\boldsymbol{x}^{\mathsf{H}}\boldsymbol{x}}. \tag{7.2}$$

Equality (7.2) can be specialized to the real case by replacing $\boldsymbol{x}^{\mathsf{H}}$ by $\boldsymbol{x}'$. Namely, if $A \in \mathbb{R}^{n\times n}$, λ is an eigenvalue and $\boldsymbol{x}$ is an eigenvector that corresponds to λ, then

$$\lambda = \frac{\boldsymbol{x}' A\boldsymbol{x}}{\boldsymbol{x}'\boldsymbol{x}}. \tag{7.3}$$

The *geometric multiplicity of an eigenvalue* λ of a matrix $A \in \mathbb{R}^{n\times n}$ is denoted by $\mathsf{geomm}(A, \lambda)$ and is equal to $\dim(S_{A,\lambda})$. Equivalently, the geometric multiplicity of λ is

$$\mathsf{geomm}(A, \lambda) = \dim(\mathsf{null}(A - \lambda I_n)) = n - \mathsf{rank}(A - \lambda I_n) \tag{7.4}$$

by Equality (3.8).

Theorem 7.2. *Let* $A \in \mathbb{R}^{n\times n}$. *We have* $0 \in \mathsf{spec}(A)$ *if and only if* A *is a singular matrix. Moreover, in this case,* $\mathsf{geomm}(A, 0) = n - \mathsf{rank}(A) = \dim(\mathsf{null}(A))$.

Proof. The statement is an immediate consequence of Equality (7.4). □

Corollary 7.2. *Let* $A \in \mathbb{R}^{n\times n}$. *If* $0 \in \mathsf{spec}(A)$ *and* $\mathsf{algm}(A, 0) = 1$, *then* $\mathsf{rank}(A) = n - 1$.

Proof. Clearly, we have $\mathsf{geomm}(A, 0) = 1$, so $\mathsf{rank}(A) = n - 1$. $\square$

Theorem 7.3. *Let $A \in \mathbb{C}^{n\times n}$ and let $S \subseteq \mathbb{C}^n$ be an invariant subspace of A. If the columns of a matrix $X \in \mathbb{C}^{n\times p}$ constitute a basis of S, then there exists a unique matrix $L \in \mathbb{C}^{p\times p}$ such that $AX = XL$.*

Proof. Let $X = (\boldsymbol{x}_1 \ \cdots \ \boldsymbol{x}_p)$. Since $A\boldsymbol{x}_1 \in S$, it follows that $A\boldsymbol{x}_1$ can be uniquely expressed as a linear combination of the columns of X, that is,

$$A\boldsymbol{x}_j = \boldsymbol{x}_1 \ell_{1j} + \cdots + \boldsymbol{x}_p \ell_{pj}$$

for $1 \leqslant i \leqslant p$. Thus,

$$A\boldsymbol{x}_j = X \begin{pmatrix} \ell_{1j} \\ \vdots \\ \ell_{pj} \end{pmatrix}.$$

The matrix L is defined by $L = (\ell_{ij})$. $\square$

Corollary 7.3. *Using the notations of Theorem 7.3, the pair $(\lambda, \boldsymbol{v})$ is an eigenpair of the matrix L if and only if $(\lambda, X\boldsymbol{v})$ is an eigenpair of A.*

Proof. The statement is an immediate consequence of Theorem 7.3. $\square$

The matrix L introduced in Theorem 7.3 will be referred to as a *representation* of A on the invariant subspace S. Clearly, L depends on the basis chosen for S, so this representation is not unique. Furthermore, we have $\mathsf{spec}(L) \subseteq \mathsf{spec}(A)$.

Theorem 7.4. *Let $A \in \mathbb{C}^{m\times n}$ be a matrix with $\mathsf{rank}(A) = n$ and let $B \in \mathbb{C}^{p\times q}$ be a matrix such that $\mathsf{range}(B) = \mathsf{range}(A)^{\perp}$. Then $\mathsf{range}(A)$ is an invariant subspace of a matrix $X \in \mathbb{C}^{p\times m}$ if and only if $B^{\mathsf{H}}XA = O_{q,n}$.*

Proof. The following statements are easily seen to be equivalent:
(i) the subspace $\mathsf{range}(A)$ is an invariant subspace of X;
(ii) $X\mathsf{range}(A) \subseteq \mathsf{range}(A)$;
(iii) $X\mathsf{range}(A) \perp \mathsf{range}(A)^{\perp}$;
(iv) $X\mathsf{range}(A) \perp \mathsf{range}(B)$.
The last statement is equivalent to $B^{\mathsf{H}}XA = O_{q,n}$. $\square$

Let $A \in \mathbb{C}^{n\times n}$ be a matrix having the eigenvalues $\lambda_1, \ldots, \lambda_n$. If $\boldsymbol{x}_1, \ldots, \boldsymbol{x}_n$ are n eigenvectors corresponding to these values, then we have

$$A\boldsymbol{x}_1 = \lambda_1\boldsymbol{x}_1, \ldots, A\boldsymbol{x}_n = \lambda_n\boldsymbol{x}_n.$$

By introducing the matrix $X = (\boldsymbol{x}_1 \ \cdots \ \boldsymbol{x}_n) \in \mathbb{C}^{n\times n}$, these equalities can be written in a concentrated form as

$$AX = X\mathsf{diag}(\lambda_1, \ldots, \lambda_n). \tag{7.5}$$

Obviously, since the eigenvalues can be listed in several ways, this equality is not unique.

Suppose now that $\boldsymbol{x}_1, \ldots, \boldsymbol{x}_n$ are unit vectors and that the eigenvalues $\lambda_1, \ldots, \lambda_n$ are distinct. Then X is a unitary matrix, $X^{-1} = X^{\mathsf{H}}$, and we obtain the following equality:

$$A = X\mathsf{diag}(\lambda_1, \ldots, \lambda_n)X^{\mathsf{H}} = \lambda_1\boldsymbol{x}_1\boldsymbol{x}_1^{\mathsf{H}} + \cdots + \lambda_n\boldsymbol{x}_n\boldsymbol{x}_n^{\mathsf{H}}, \tag{7.6}$$

known as the *spectral decomposition* of the matrix A. We will discuss later (in Chapter 9) a far-reaching extension of the spectral decomposition.

7.3 The Characteristic Polynomial of a Matrix

If λ is an eigenvalue of the matrix $A \in \mathbb{C}^{n\times n}$, there exists a non-zero eigenvector $\boldsymbol{x} \in \mathbb{C}^n$ such that $A\boldsymbol{x} = \lambda\boldsymbol{x}$. Therefore, the homogeneous linear system

$$(\lambda I_n - A)\boldsymbol{x} = \boldsymbol{0}_n \tag{7.7}$$

has a non-trivial solution. This is possible if and only if $\det(\lambda I_n - A) = 0$, so eigenvalues are the solutions of the equation

$$\det(\lambda I_n - A) = 0.$$

Note that $\det(\lambda I_n - A)$ is a polynomial of degree n in λ, known as the *characteristic polynomial* of the matrix A. We denote this polynomial by p_A.

Example 7.3. Let

$$A = \begin{pmatrix} a_{11} & a_{12} & a_{13} \\ a_{21} & a_{22} & a_{23} \\ a_{31} & a_{32} & a_{33} \end{pmatrix}$$

be a matrix in $\mathbb{C}^{3\times 3}$. Its characteristic polynomial is

$$\begin{aligned} p_A(\lambda) = \begin{vmatrix} \lambda - a_{11} & -a_{12} & -a_{13} \\ -a_{21} & \lambda - a_{22} & -a_{23} \\ -a_{31} & -a_{32} & \lambda - a_{33} \end{vmatrix} &= \lambda^3 - (a_{11} + a_{22} + a_{33})\lambda^2 \\ &+ (a_{11}a_{22} + a_{22}a_{33} + a_{33}a_{11} - a_{12}a_{21} - a_{23}a_{32} - a_{13}a_{31})\lambda \\ &- (a_{11}a_{22}a_{33} + a_{12}a_{23}a_{31} + a_{13}a_{32}a_{21} \\ &- a_{12}a_{21}a_{33} - a_{23}a_{32}a_{11} - a_{13}a_{31}a_{22}). \end{aligned}$$

Theorem 7.5. *Let $A \in \mathbb{C}^{n\times n}$. Then $\mathsf{spec}(A) = \mathsf{spec}(A')$ and $\mathsf{spec}(A^{\mathsf{H}}) = \{\overline{\lambda} \mid \lambda \in \mathsf{spec}(A)\}$.*

Proof. We have

$$p_{A'}(\lambda) = \det(\lambda I_n - A') = \det((\lambda I_n - A)') = \det(\lambda I_n - A) = p_A(\lambda).$$

Thus, since A and A' have the same characteristic polynomials, their spectra are the same.

For A^{H}, we can write

$$p_{A^{\mathsf{H}}}(\overline{\lambda}) = \det(\overline{\lambda} I_n - A^{\mathsf{H}}) = \det((\lambda I_n - A)^{\mathsf{H}}) = (p_A(\lambda))^{\mathsf{H}},$$

which implies the second part of the theorem. □

Definition 7.2. Let λ be an eigenvalue of a matrix $A \in \mathbb{C}^{n\times n}$. A *left eigenvector* of the matrix A is a vector $\boldsymbol{v} \in \mathbb{C}^n - \{\mathbf{0}\}$ such that $\boldsymbol{v}^{\mathsf{H}} A = \lambda \boldsymbol{v}^{\mathsf{H}}$.

We could have used the obvious term *right eigenvectors* for the eigenvectors of a matrix A. Since in the vast majority of cases we deal with right eigenvectors, we prefer to use the simpler term *eigenvectors* for the right eigenvectors.

Note that if λ is an eigenvalue of $A \in \mathbb{C}^{n\times n}$, the set of solutions of the homogeneous system $(\lambda I_n - A)\boldsymbol{v} = \mathbf{0}$ is non-empty and consists of the non-zero invariant space corresponding to the eigenvalue λ. This is also equivalent to saying that $\mathsf{rank}(\lambda I_n - A) < n$. Since $\mathsf{rank}(\lambda I_n - A) = \mathsf{rank}(\lambda I_n - A^{\mathsf{H}})$, it follows that the linear system $(\lambda I_n - A^{\mathsf{H}})\boldsymbol{v} = \mathbf{0}_n$ has a non-zero solution, so $A^{\mathsf{H}}\boldsymbol{v} = \lambda\boldsymbol{v}$, which is equivalent to $\boldsymbol{v}^{\mathsf{H}}A = \lambda\boldsymbol{v}^{\mathsf{H}}$. Thus, every eigenvalue has both an eigenvector and a left eigenvector.

Equality of spectra of A and A' does not imply that the eigenvectors or the invariant subspaces of the corresponding eigenvalues are identical, as it can be seen from the following example.

Example 7.4. Consider the matrix $A \in \mathbb{C}^{2\times 2}$ defined by

$$A = \begin{pmatrix} a & 0 \\ c & b \end{pmatrix},$$

where $a \neq b$ and $c \neq 0$. It is immediate that $\mathsf{spec}(A) = \mathsf{spec}(A') = \{a, b\}$.

For $\lambda_1 = a$, we have the distinct invariant subspaces:

$$S_{A,a} = \left\{k \begin{pmatrix} a-b \\ c \end{pmatrix} \middle| k \in \mathbb{C}\right\},$$

$$S_{A',a} = \left\{k \begin{pmatrix} 1 \\ 0 \end{pmatrix} \middle| k \in \mathbb{C}\right\},$$

as the reader can easily verify.

If λ_1, λ_2 are two distinct eigenvalues of A, $\boldsymbol{u}$ is a right eigenvector that corresponds to λ_1, and $\boldsymbol{v}$ is a left eigenvector that corresponds to λ_2, then $\boldsymbol{u} \perp \boldsymbol{v}$. Indeed, we have

$$\lambda_1 \boldsymbol{v}^{\mathsf{H}}\boldsymbol{u} = \boldsymbol{v}^{\mathsf{H}}A\boldsymbol{u} = \lambda_2\boldsymbol{v}^{\mathsf{H}}\boldsymbol{u},$$

so $\boldsymbol{v}^{\mathsf{H}}\boldsymbol{u} = 0$.

The leading term of the characteristic polynomial of A is generated by $(\lambda - a_{11})(\lambda - a_{22})\cdots(\lambda - a_{nn})$ and equals λ^n.

The fundamental theorem of algebra implies that p_A has n complex roots, not necessarily distinct. Observe also that, *if A is a matrix with real entries, the roots are paired as conjugate complex numbers.*

Definition 7.3. The *algebraic multiplicity of an eigenvalue* λ of a matrix $A \in \mathbb{C}^{n\times n}$, $\mathsf{algm}(A, \lambda)$, equals k if λ is a root of order k of the equation $p_A(\lambda) = 0$.

If $\mathsf{algm}(A, \lambda) = 1$, we refer to λ as a *simple eigenvalue.*

Example 7.5. Let $A \in \mathbb{R}^{3\times 3}$ be the matrix

$$A = \begin{pmatrix} 1 & 1 & 1 \\ 0 & 1 & 2 \\ 2 & 1 & 0 \end{pmatrix}.$$

The characteristic polynomial of A is

$$p_A(\lambda) = \begin{vmatrix} \lambda - 1 & -1 & -1 \\ 0 & \lambda - 1 & -2 \\ -2 & -1 & \lambda \end{vmatrix} = \lambda^3 - 2\lambda^2 - 3\lambda.$$

Therefore, the eigenvalues of A are $3, 0$, and -1.

The eigenvalues of I_3 are obtained from the equation

$$det(\lambda I_3 - I_3) = \begin{vmatrix} \lambda - 1 & 0 & 0 \\ 0 & \lambda - 1 & 0 \\ 0 & 0 & \lambda - 1 \end{vmatrix} = (\lambda - 1)^3 = 0.$$

Thus, I_3 has one eigenvalue, 1, and $\mathsf{algm}(I_3, 1) = 3$.

Example 7.6. Let $P(a) \in \mathbb{C}^{n\times n}$ be the matrix

$$P(a) = \begin{pmatrix} a & 1 & \cdots & 1 \\ 1 & a & \cdots & 1 \\ \vdots & \vdots & \cdots & \vdots \\ 1 & 1 & \cdots & a \end{pmatrix}.$$

To find the eigenvalues of $P(a)$, we need to solve the equation

$$\begin{vmatrix} \lambda - a & -1 & \cdots & -1 \\ -1 & \lambda - a & \cdots & -1 \\ \vdots & \vdots & \cdots & \vdots \\ -1 & -1 & \cdots & \lambda - a \end{vmatrix} = 0.$$

By adding the first $n-1$ columns to the last and factoring out $\lambda - (a+n-1)$, we obtain the equivalent equation

$$(\lambda - (a+n-1)) \begin{vmatrix} \lambda - a & -1 & \cdots & 1 \\ -1 & \lambda - a & \cdots & 1 \\ \vdots & \vdots & \cdots & \vdots \\ -1 & -1 & \cdots & 1 \end{vmatrix} = 0.$$

Adding the last column from the first $n-1$ columns and expanding the determinant yields the equation

$$(\lambda - (a+n-1))(\lambda - a + 1)^{n-1} = 0,$$

which allows us to conclude that $P(a)$ has the eigenvalue $a+n-1$ with $\mathsf{algm}(P(a), a+n-1) = 1$ and the eigenvalue $a-1$ with $\mathsf{algm}(P(a), a-1) = n-1$.

In the special case when $a = 1$, we have $P(1) = J_{n,n}$. Thus, $J_{n,n}$ has the eigenvalue $\lambda_1 = n$ with algebraic multiplicity 1 and the eigenvalue 0 with algebraic multiplicity $n-1$.

Definition 7.4. A matrix $A \in \mathbb{C}^{n\times n}$ is *simple* if there exists a linearly independent set of n eigenvectors.

If $A \in \mathbb{C}^{n\times n}$ has n distinct eigenvalues, then, by Theorem 7.1, A is a simple matrix. The reverse of this statement is false because there exist simple matrices for which not all eigenvalues are distinct. For example, $\mathsf{spec}(I_n) = \{1\}$, but $\{\boldsymbol{e}_1, \ldots, \boldsymbol{e}_n\}$ is a linearly independent set of distinct eigenvectors.

Theorem 7.6. *Let $A \in \mathbb{R}^{n\times n}$ be a matrix and let $\lambda \in \mathsf{spec}(A)$. Then, for any $k \in \mathbb{P}$, $\lambda^k \in \mathsf{spec}(A^k)$.*

Proof. The proof is by induction on $k \geqslant 1$. The base step, $k = 1$, is immediate. Suppose that $\lambda^k \in \mathsf{spec}(A^k)$, that is $A^k\boldsymbol{x} = \lambda^k\boldsymbol{x}$ for some $\boldsymbol{x} \in V - \{\boldsymbol{0}\}$. Then $A^{k+1}\boldsymbol{x} = A(A^k\boldsymbol{x}) = A(\lambda^k\boldsymbol{x}) = \lambda^k A\boldsymbol{x} = \lambda^{k+1}\boldsymbol{x}$, so $\lambda^{k+1} \in \mathsf{spec}(A^{k+1})$. □

Theorem 7.7. *Let $A \in \mathbb{R}^{n\times n}$ be a non-singular matrix and let $\lambda \in \mathsf{spec}(A)$. We have $\frac{1}{\lambda} \in \mathsf{spec}(A^{-1})$ and the sets of eigenvectors of A and A^{-1} are equal.*

Proof. Since $\lambda \in \mathsf{spec}(A)$ and A is non-singular, we have $\lambda \neq 0$ and $A\boldsymbol{x} = \lambda\boldsymbol{x}$ for some $\boldsymbol{x} \in V - \{\boldsymbol{0}\}$. Therefore, we have $A^{-1}(A\boldsymbol{x}) = \lambda A^{-1}\boldsymbol{x}$, which is equivalent to $\lambda^{-1}\boldsymbol{x} = A^{-1}\boldsymbol{x}$, which implies $\frac{1}{\lambda} \in \mathsf{spec}(A^{-1})$. In addition, this implies that the set of eigenvectors of A and A^{-1} are identical. □

Theorem 7.8. *Let $p_A(\lambda) = \lambda^n + c_1\lambda^{n-1} + \cdots + c_{n-1}\lambda + c_n$ be the characteristic polynomial of the matrix A. Then we have $c_i = (-1)^i S_i(A)$ for $1 \leqslant i \leqslant n$, where $S_i(A)$ is the sum of all principal minors of order i of A.*

Proof. Since $p_A(\lambda) = \lambda^n + c_1\lambda^{n-1} + \cdots + c_{n-1}\lambda + c_n$, it is easy to see that the derivatives of $p_A(\lambda)$ are given by

$$\begin{aligned}
p_A^{(1)}(\lambda) &= n\lambda^{n-1} + (n-1)c_1\lambda^{n-2} + \cdots + c_{n-1},\\
p_A^{(2)}(\lambda) &= n(n-1)\lambda^{n-2} + (n-1)(n-2)c_1\lambda^{n-3} + \cdots + 2c_{n-2},\\
&\vdots\\
p_A^{(k)}(\lambda) &= n(n-1)\cdots(n-k+1)\lambda^{n-k} + \cdots + k!c_{n-k},\\
&\vdots\\
p_A^{(n)}(\lambda) &= n!c_0.
\end{aligned}$$

This implies

$$c_{n-k} = k!p_A^{(k)}(0)$$

for $0 \leqslant k \leqslant n$. On the other hand, the derivatives of $p_A(\lambda)$ can be computed using Theorem 5.11, and taking into account the formula given in Exercise 17 of Chapter 5, we have

$$c_{n-k} = \frac{1}{k!}(-1)^k k! S_{n-k}(A) = (-1)^{n-k} S_{n-k}(A),$$

which implies the statement of the theorem. □

By Viéte's Theorem, taking into account Theorem 7.8, we have

$$\lambda_1 + \cdots + \lambda_n = a_{11} + a_{22} + \cdots + a_{nn} = \mathsf{trace}(A) = -c_1. \tag{7.8}$$

Another interesting fact that follows immediately from Theorem 7.8 is

$$\lambda_1 \cdots \lambda_n = \det(A). \tag{7.9}$$

Theorem 7.9. *Let $p(\lambda) = \lambda^n + a_1\lambda^{n-1} + \cdots + a_{n-1}\lambda + a_n$ be a polynomial in $\mathbb{C}[x]$, where $n \geqslant 2$. Then the matrix $A_p \in \mathbb{C}^{n\times n}$ defined by*

$$A_p = \begin{pmatrix} 0 & 1 & 0 & \cdots & 0\\ 0 & 0 & 1 & \cdots & 0\\ \vdots & \vdots & \vdots & \cdots & \vdots\\ -a_n & -a_{n-1} & -a_{n-2} & \cdots & -a_1 \end{pmatrix}$$

has p as its characteristic polynomial.

Proof. The proof is by induction on $n \geqslant 2$. For the base case, $n = 2$, we have

$$A = \begin{pmatrix} 0 & -1 \\ -a_2 & -a_1 \end{pmatrix},$$

the characteristic polynomial of A is

$$p_A(\lambda) = \begin{vmatrix} \lambda & 1 \\ a_2 & \lambda + a_1 \end{vmatrix} = \lambda^2 + a_1\lambda + a_2,$$

as claimed.

Suppose now that the statement holds for polynomials of degree less than n and let p be a polynomial of degree n. Define q as

$$q(\lambda) = \lambda^{n-1} + a_1\lambda^{n-2} + \cdots + a_{n-1}.$$

By the inductive hypothesis, the characteristic polynomial of the matrix

$$A_q = \begin{pmatrix} 0 & 1 & 0 & \cdots & 0 \\ 0 & 0 & 1 & \cdots & 0 \\ \vdots & \vdots & \vdots & \cdots & \vdots \\ -a_{n-1} & -a_{n-2} & -a_{n-3} & \cdots & -a_1 \end{pmatrix}$$

is $q(\lambda)$. Observe that

$$\det(\lambda I_n - A_p) = \lambda \det(A_q) + a_n,$$

as follows by expanding $\det(\lambda I_n - A_p)$ by its first column. Thus, the characteristic polynomial of A_p is

$$p_{A_p} = \lambda p_{A_q}(\lambda) + a_n = \lambda q(\lambda) + a_n = p(\lambda),$$

which concludes the argument. □

The matrix A_p will be referred to as the *companion matrix* of the polynomial p.

Theorem 7.10. *Let $A \in \mathbb{C}^{m\times n}$ and $B \in \mathbb{C}^{n\times m}$ be two matrices. Then the set of non-zero eigenvalues of the matrices $AB \in \mathbb{C}^{m\times m}$ and $BA \in \mathbb{C}^{n\times n}$ are the same and $\mathsf{algm}(AB, \lambda) = \mathsf{algm}(BA, \lambda)$ for each such eigenvalue.*

Proof. Consider the following straightforward equalities:

$$\begin{pmatrix} I_m & -A \\ O_{n,m} & \lambda I_n \end{pmatrix} \begin{pmatrix} \lambda I_m & A \\ B & I_n \end{pmatrix} = \begin{pmatrix} \lambda I_m - AB & O_{m,n} \\ -\lambda B & \lambda I_n \end{pmatrix},$$

$$\begin{pmatrix} -I_m & O_{m,n} \\ -B & \lambda I_n \end{pmatrix} \begin{pmatrix} \lambda I_m & A \\ B & I_n \end{pmatrix} = \begin{pmatrix} -\lambda I_m & -A \\ O_{n,m} & \lambda I_n - BA \end{pmatrix}.$$

Observe that

$$\det\left(\begin{pmatrix} I_m & -A \\ O_{n,m} & \lambda I_n \end{pmatrix} \begin{pmatrix} \lambda I_m & A \\ B & I_n \end{pmatrix}\right) = \det\left(\begin{pmatrix} -I_m & O_{m,n} \\ -B & \lambda I_n \end{pmatrix} \begin{pmatrix} \lambda I_m & A \\ B & I_n \end{pmatrix}\right),$$

and therefore,

$$\det \begin{pmatrix} \lambda I_m - AB & O_{m,n} \\ -\lambda B & \lambda I_n \end{pmatrix} = \det \begin{pmatrix} -\lambda I_m & -A \\ O_{n,m} & \lambda I_n - BA \end{pmatrix}.$$

The last equality amounts to

$$\lambda^n p_{AB}(\lambda) = \lambda^m p_{BA}(\lambda).$$

Thus, for $\lambda \neq 0$, we have $p_{AB}(\lambda) = p_{BA}(\lambda)$, which gives the desired conclusion. □

Corollary 7.4. *Let*

$$\boldsymbol{a} = \begin{pmatrix} a_1 \\ \vdots \\ a_n \end{pmatrix}$$

be a vector in $\mathbb{C}^n - \{\mathbf{0}\}$. *Then, the matrix* $\boldsymbol{a}\boldsymbol{a}^{\mathsf{H}} \in \mathbb{C}^{n \times n}$ *has one eigenvalue distinct from* 0, *and this eigenvalue is equal to* $\|\boldsymbol{a}\|^2$.

Proof. By Theorem 7.10, the matrix $\boldsymbol{a}\boldsymbol{a}^{\mathsf{H}}$ has the same non-zero eigenvalues as the matrix $\boldsymbol{a}^{\mathsf{H}}\boldsymbol{a} \in \mathbb{C}^{1\times 1}$ and the single eigenvalue of $\boldsymbol{a}^{\mathsf{H}}\boldsymbol{a}$ is $\boldsymbol{a}^{\mathsf{H}}\boldsymbol{a} = \|a\|^2$. □

Theorem 7.11. *Let* $A \in \mathbb{C}^{(m+n)\times(m+n)}$ *be a matrix partitioned as*

$$A = \begin{pmatrix} B & C \\ O_{n,m} & D \end{pmatrix},$$

where $B \in \mathbb{C}^{m\times m}$, $C \in \mathbb{C}^{m\times n}$, *and* $D \in \mathbb{C}^{n\times n}$. *Then* $\mathsf{spec}(A) = \mathsf{spec}(B) \cup \mathsf{spec}(D)$.

Proof. Let $\lambda \in \mathsf{spec}(A)$ and let $\boldsymbol{x} \in \mathbb{C}^{m+n}$ be an eigenvector that corresponds to λ. If

$$\boldsymbol{x} = \begin{pmatrix} \boldsymbol{u} \\ \boldsymbol{v} \end{pmatrix},$$

where $\boldsymbol{u} \in \mathbb{C}^m$ and $\boldsymbol{v} \in \mathbb{C}^n$, then we have

$$A\boldsymbol{x} = \begin{pmatrix} B & C \\ O_{n,m} & D \end{pmatrix} \begin{pmatrix} \boldsymbol{u} \\ \boldsymbol{v} \end{pmatrix} = \begin{pmatrix} B\boldsymbol{u} + C\boldsymbol{v} \\ D\boldsymbol{v} \end{pmatrix} = \lambda \begin{pmatrix} \boldsymbol{u} \\ \boldsymbol{v} \end{pmatrix}.$$

This implies $B\boldsymbol{u} + C\boldsymbol{v} = \lambda\boldsymbol{u}$ and $D\boldsymbol{v} = \lambda\boldsymbol{v}$. If $\boldsymbol{v} \neq \boldsymbol{0}$, then $\lambda \in \mathsf{spec}(D)$; otherwise, $B\boldsymbol{u} = \lambda\boldsymbol{u}$, which yields $\lambda \in \mathsf{spec}(B)$, so $\lambda \in \mathsf{spec}(B) \cup \mathsf{spec}(D)$. Thus, $\mathsf{spec}(A) \subseteq \mathsf{spec}(B) \cup \mathsf{spec}(D)$.

To prove the converse inclusion, note that if $\lambda \in \mathsf{spec}(B)$ and $\boldsymbol{u}$ is an eigenvector of λ, then $B\boldsymbol{u} = \lambda\boldsymbol{u}$, which means that

$$A \begin{pmatrix} \boldsymbol{u} \\ \boldsymbol{0} \end{pmatrix} = \lambda \begin{pmatrix} \boldsymbol{u} \\ \boldsymbol{0} \end{pmatrix},$$

so $\mathsf{spec}(B) \subseteq \mathsf{spec}(A)$. Similarly, $\mathsf{spec}(D) \subseteq \mathsf{spec}(A)$, which implies the equality of the theorem. □

7.4 Spectra of Hermitian Matrices

Theorem 7.12. *All eigenvalues of a Hermitian matrix $A \in \mathbb{C}^{n\times n}$ are real numbers.*

All eigenvalues of a skew-Hermitian matrix are purely imaginary numbers.

Proof. Theorem 3.18 implies that $\boldsymbol{x}^{\mathsf{H}}\boldsymbol{x}$ is a real number for every $\boldsymbol{x} \in \mathbb{C}^n$. Then, by Equality (7.2), λ is a real number.

Suppose now that B is a skew-Hermitian matrix. Then, as above, $\overline{\boldsymbol{x}^{\mathsf{H}}A\boldsymbol{x}} = -\boldsymbol{x}^{\mathsf{H}}A\boldsymbol{x}$, which implies that the real part of $\boldsymbol{x}^{\mathsf{H}}A\boldsymbol{x}$ is 0. Thus, $\boldsymbol{x}^{\mathsf{H}}A\boldsymbol{x}$ is a purely imaginary number and, by the same Equality (7.3), λ is a purely imaginary number. □

Corollary 7.5. *If $A \in \mathbb{R}^{n \times n}$ and A is a symmetric matrix, then all its eigenvalues are real numbers.*

Proof. This statement follows from Theorem 7.12 by observing that the Hermitian adjoint A^{H} of a matrix $A \in \mathbb{R}^{n \times n}$ coincides with its transposed matrix A'. □

Example 7.7. Let $A \in \mathbb{R}^{n \times n}$ be a symmetric and orthogonal matrix. By Corollary 7.5, all its eigenvalues are real numbers. Let $\lambda \in \mathsf{spec}(A)$ and let $\boldsymbol{x}$ be an eigenvector that corresponds to λ. Since $A\boldsymbol{x} = \lambda\boldsymbol{x}$, we have $\boldsymbol{x}'A' = \lambda\boldsymbol{x}'$, hence

$$\lambda^2\boldsymbol{x}'\boldsymbol{x} = (A\boldsymbol{x}')'A\boldsymbol{x}\boldsymbol{x}'A'A\boldsymbol{x} = \boldsymbol{x}'\boldsymbol{x}.$$

Therefore, $\lambda^2 = 1$, hence $\lambda \in \{-1, 1\}$.

Corollary 7.6. *Let $A \in \mathbb{C}^{m \times n}$ be a matrix. The non-zero eigenvalues of the matrices AA^{H} and $A^{\mathsf{H}}A$ are positive numbers and they have the same algebraic multiplicities for the matrices AA^{H} and $A^{\mathsf{H}}A$.*

Proof. By Theorem 7.10, we need to verify only that if λ is a non-zero eigenvalue of $A^{\mathsf{H}}A$, then λ is a positive number. Since $A^{\mathsf{H}}A$ is a Hermitian matrix, by Theorem 7.12, λ is a real number. The equality $A^{\mathsf{H}}A\boldsymbol{x} = \lambda\boldsymbol{x}$ for some eigenvector $\boldsymbol{x} \neq \mathbf{0}$ implies

$$\lambda\|\boldsymbol{x}\|_2^2 = \lambda\boldsymbol{x}^{\mathsf{H}}\boldsymbol{x} = (A\boldsymbol{x})^{\mathsf{H}}A\boldsymbol{x} = \|A\boldsymbol{x}\|_2^2,$$

so $\lambda > 0$. □

Corollary 7.7. *Let $A \in \mathbb{C}^{m \times n}$ be a matrix. The eigenvalues of the matrix $B = A^{\mathsf{H}}A \in \mathbb{C}^{n \times n}$ are real non-negative numbers.*

Proof. The matrix B defined above is clearly Hermitian and, therefore, its eigenvalues are real numbers by Theorem 7.12. Next, if λ is an eigenvalue of B, then by Equality (7.2), we have

$$\lambda = \frac{\boldsymbol{x}^{\mathsf{H}}A^{\mathsf{H}}A\boldsymbol{x}}{\boldsymbol{x}^{\mathsf{H}}\boldsymbol{x}} = \frac{(A\boldsymbol{x})^{\mathsf{H}}A\boldsymbol{x}}{\boldsymbol{x}^{\mathsf{H}}\boldsymbol{x}} = \frac{\|A\boldsymbol{x}\|}{\|\boldsymbol{x}\|} \geqslant 0,$$

where $\boldsymbol{x}$ is an eigenvector that corresponds to λ. □

Note that if A is a Hermitian matrix, then $A^{\mathsf{H}}A = A^2$, hence the spectrum of $A^{\mathsf{H}}A$ is $\{\lambda^2 \mid \lambda \in \mathsf{spec}(A)\}$.

Theorem 7.13. *If $A \in \mathbb{C}^{n\times n}$ is a Hermitian matrix and $\boldsymbol{u}, \boldsymbol{v}$ are two eigenvectors that correspond to two distinct eigenvalues λ_1 and λ_2, then $\boldsymbol{u} \perp \boldsymbol{v}$.*

Proof. We have $A\boldsymbol{u} = \lambda_1\boldsymbol{u}$ and $A\boldsymbol{v} = \lambda_2\boldsymbol{v}$. This allows us to write $\boldsymbol{v}^{\mathsf{H}}A\boldsymbol{u} = \lambda_1\boldsymbol{v}^{\mathsf{H}}\boldsymbol{u}$. Since A is Hermitian, we have

$$\lambda_1\boldsymbol{v}^{\mathsf{H}}\boldsymbol{u} = \boldsymbol{v}^{\mathsf{H}}A\boldsymbol{u} = \boldsymbol{v}^{\mathsf{H}}A^{\mathsf{H}}\boldsymbol{u} = (A\boldsymbol{v})^{\mathsf{H}}\boldsymbol{u} = \lambda_2\boldsymbol{v}^{\mathsf{H}}\boldsymbol{u},$$

which implies $\boldsymbol{v}^{\mathsf{H}}\boldsymbol{u} = 0$, that is, $\boldsymbol{u} \perp \boldsymbol{v}$. □

Theorem 7.14 (Ky Fan's Theorem). Let $A \in \mathbb{C}^{n\times n}$ be a Hermitian matrix such that $\mathsf{spec}(A) = \{\lambda_1, \ldots, \lambda_n\}$, where $\lambda_1 \geqslant \cdots \geqslant \lambda_n$ and let $V = (\boldsymbol{v}^1 \ \cdots \ \boldsymbol{v}^n)$ be the matrix whose columns consist of the corresponding unit eigenvectors of A.

Let $\{\boldsymbol{x}^1, \ldots, \boldsymbol{x}^n\}$ be an orthonormal set of vectors in $\mathbb{C}^n$. For any positive integer $q \leqslant n$, the sums $\sum_{i=1}^{q} \lambda_i$ and $\sum_{i=1}^{q} \lambda_{n+1-i}$ are, respectively, the maximum and minimum of $\sum_{j=1}^{q} \boldsymbol{x}^{j\prime} A\boldsymbol{x}^j$. Namely, the maximum is obtained by choosing the vectors $\boldsymbol{x}^1, \ldots, \boldsymbol{x}^q$ as the first q columns of V; the minimum is obtained by assigning to $\boldsymbol{x}^1, \ldots, \boldsymbol{x}^q$ the last q columns of V.

Proof. Let $\boldsymbol{x}^i = \sum_{p=1}^{n} b_p^i \boldsymbol{v}^p$ be the expressions of $\boldsymbol{x}^i$ relative to the basis V for $1 \leqslant i \leqslant m$. In matrix form, these equalities can be written as

$$(\boldsymbol{x}^1 \ \ldots \ \boldsymbol{x}^n) = (\boldsymbol{v}^1 \ \ldots \ \boldsymbol{v}^n)\begin{pmatrix} b_1^1 & \cdots & b_1^n \\ \vdots & \cdots & \vdots \\ b_n^1 & \cdots & b_n^n \end{pmatrix},$$

where $b_i^p = (\boldsymbol{v}^p)'\boldsymbol{x}^i = (\boldsymbol{x}^i)'\boldsymbol{v}_p$. Thus, for $X = (\boldsymbol{x}^1 \ldots \boldsymbol{x}^n)$ and $V = (\boldsymbol{v}^1 \ \ldots \ \boldsymbol{v}^n)$, we have $X = VB$, where B is the orthonormal matrix

$$B = \begin{pmatrix} b_1^1 & \cdots & b_1^n \\ \vdots & \cdots & \vdots \\ b_n^1 & \cdots & b_n^n \end{pmatrix}.$$

We have

$$\begin{aligned} \boldsymbol{x}^{j\prime} A \boldsymbol{x}^j &= \boldsymbol{x}^{j\prime} A b_p^j \boldsymbol{v}^p = b_p^j \boldsymbol{x}^{j\prime} A \boldsymbol{v}^p \\ &= b_p^j (\boldsymbol{x}^j)' \lambda_p \boldsymbol{v}^p = (b_p^j)^2 \lambda_p \\ y &= \lambda_q \sum_{p=1}^{n} (b_p^j)^2 + \sum_{p=1}^{q} (\lambda_p - \lambda_q)(b_p^j)^2 + \sum_{j=q+1}^{n} (\lambda_j - \lambda_q)(b_p^j)^2. \end{aligned}$$

By Inequality (6.26), this implies

$$(\boldsymbol{x}^j)' A \boldsymbol{x}^j \leqslant \lambda_q + \sum_{p=1}^{q} (\lambda_p - \lambda_q)(b_p^j)^2.$$

Therefore,

$$\sum_{i=1}^{q} \lambda_i - \sum_{j=1}^{q} (\boldsymbol{x}^j)' A \boldsymbol{x}^j \geqslant \sum_{i=1}^{q} (\lambda_i - \lambda_q) \left(1 - \sum_{j=1}^{q} (b_i^j)^2 \right). \tag{7.10}$$

Again, by Inequality (6.26), we have $\sum_{j=1}^{q} (b_i^j)^2 \leqslant \|\boldsymbol{x}_i\|^2 = 1$, so $\sum_{i=1}^{q} (\lambda_i - \lambda_q) \left(1 - \sum_{j=1}^{q} (b_i^j)^2\right) \geqslant 0$. The left member of Inequality (7.10) becomes 0, when $\boldsymbol{x}^i = \boldsymbol{v}^i$, so $\sum_{j=1}^{q} (\boldsymbol{x}^j)' A \boldsymbol{x}^j \leqslant \sum_{i=1}^{q} \lambda_i$. The maximum of $\sum_{j=1}^{q} (\boldsymbol{x}^j)' A \boldsymbol{x}^j$ is obtained when $\boldsymbol{x}^j = \boldsymbol{v}^j$ for $1 \leqslant j \leqslant q$, that is, when X consists of the first q columns of V that correspond to eigenvectors of the top k largest eigenvalues.

The argument for the minimum is similar. □

An equivalent form of Ky Fan's Theorem can be obtained by observing that the orthonormality condition of the set $\{\boldsymbol{x}_1, \ldots, \boldsymbol{x}_q\}$ can be expressed as $X'X = I_q$, where $X \in \mathbb{C}^{n \times q}$ is the matrix $X = (\boldsymbol{x}_1 \ \cdots \ \boldsymbol{x}_q)$. Also, the sum $\sum_{j=1}^{q} \boldsymbol{x}_j' A \boldsymbol{x}_j$ equals $\mathsf{trace}(X'AX)$. Thus, Ky Fan's Theorem is equivalent to the fact that the sums $\sum_{i=1}^{n} \lambda_i$ and $\sum_{i=1}^{q} \lambda_{n+1-i}$ are, respectively, the maximum and minimum of $\mathsf{trace}(X'AX)$, where $X'X = I_q$. In this form, Ky Fan's Theorem is useful for the discussion of principal component analysis in Chapter 13.

7.5 Spectra of Special Matrices

In this section, we examine the spectra of special classes of matrices. We begin with spectra of block upper triangular and block lower triangular matrices.

Theorem 7.15. *Let A be a block upper triangular partitioned matrix given by*

$$A = \begin{pmatrix} A_{11} & A_{12} & \cdots & A_{1m} \\ O & A_{22} & \cdots & A_{2m} \\ \vdots & \vdots & \cdots & \vdots \\ O & O & \cdots & A_{mm} \end{pmatrix},$$

where $A_{ii} \in \mathbb{R}^{p_i \times p_i}$ for $1 \leqslant i \leqslant m$. Then, $\mathsf{spec}(A) = \bigcup_{i=1}^{n} \mathsf{spec}(A_{ii})$.

If A is a block lower triangular matrix

$$A = \begin{pmatrix} A_{11} & O & \cdots & O \\ A_{21} & A_{22} & \cdots & O \\ \vdots & \vdots & \cdots & \vdots \\ A_{m1} & A_{m2} & \cdots & A_{mm} \end{pmatrix},$$

the same equality holds.

Proof. Let A be a block upper triangular matrix. Its characteristic equation is $\det(\lambda I_n - A) = 0$. Observe that the matrix $\lambda I_n - A$ is also a block upper triangular matrix:

$$\lambda I_n - A = \begin{pmatrix} \lambda I_{p_1} - A_{11} & O & \cdots & O \\ -A_{21} & \lambda I_{p_2} - A_{22} & \cdots & O \\ \vdots & \vdots & \cdots & \vdots \\ -A_{m1} & -A_{m2} & \cdots & \lambda I_{p_m} - A_{mm} \end{pmatrix}.$$

By Theorem 5.13, the characteristic polynomial of A can be written as

$$p_A(\lambda) = \prod_{i=1}^{m} \det(\lambda I_{p_i} - A_{ii}) = \prod_{i=1}^{m} p_{A_{ii}}(\lambda).$$

Therefore, $\mathsf{spec}(A) = \bigcup_{i=1}^{n} \mathsf{spec}(A_{ii})$.

The argument for block lower triangular matrices is similar. □

Corollary 7.8. *Let $A \in \mathbb{R}^{n\times n}$ be a block diagonal matrix given by*

$$A = \begin{pmatrix} A_{11} & O & \cdots & O \\ O & A_{22} & \cdots & O \\ \vdots & \vdots & \cdots & \vdots \\ O & O & \cdots & A_{mm} \end{pmatrix},$$

where $A_{ii} \in \mathbb{R}^{n_i\times n_i}$ for $1 \leqslant i \leqslant m$. We have $\mathsf{spec}(A) = \bigcup_{i=1}^n \mathsf{spec}(A_{ii})$ and $\mathsf{algm}(A,\lambda) = \sum_{i=1}^m \mathsf{algm}(A_i,\lambda)$. Moreover, $\boldsymbol{v} \neq \mathbf{0}_n$ is an eigenvector of A if and only if we can write

$$\boldsymbol{v} = \begin{pmatrix} \boldsymbol{v}_1 \\ \vdots \\ \boldsymbol{v}_m \end{pmatrix},$$

where each vector $\boldsymbol{v}_i$ is either an eigenvector of A_i or $\mathbf{0}_{n_i}$ for $1 \leqslant i \leqslant m$, and there exists i such that $\boldsymbol{v}_i \neq \mathbf{0}_{n_i}$.

Proof. This statement follows immediately from Theorem 7.15. □

Theorem 7.16. *Let $A = (a_i^j) \in \mathbb{C}^{n\times n}$ be an upper (lower) triangular matrix. Then, $\mathsf{spec}(A) = \{a_i^i \mid 1 \leqslant i \leqslant n\}$.*

Proof. It is easy to see that the characteristic polynomial of A is $p_A(\lambda) = (\lambda - a_1^1)\cdots(\lambda - a_n^n)$, which implies immediately the theorem. □

Corollary 7.9. *If $A \in \mathbb{C}^{n\times n}$ is an upper triangular matrix and λ is an eigenvalue such that the diagonal entries of A that equal λ occur in $a_{i_1}^{i_1}, \ldots, a_{i_p}^{i_p}$, then $S_{A,\lambda}$ is a p-dimensional subspace of $\mathbb{C}^n$ generated by $\boldsymbol{e}_{i_1}, \ldots, \boldsymbol{e}_{i_p}$.*

Proof. This statement is immediate. □

Corollary 7.10. *We have*

$$\mathsf{spec}(\mathsf{diag}(d_1, \ldots, d_n)) = \{d_1, \ldots, d_n\}.$$

Proof. This statement is a direct consequence of Theorem 7.16. □

Theorem 7.17. *If $A \in \mathbb{C}^{n\times n}$ is a nilpotent matrix, then $\mathsf{spec}(A) = \{0\}$.*

Proof. Let $A \in \mathbb{C}^{n\times n}$ be a nilpotent matrix such that $\mathsf{nilp}(A) = k$. By Theorem 7.6, if $\lambda \in \mathsf{spec}(A)$, then $\lambda^k \in \mathsf{spec}(A^k) = \mathsf{spec}(O) = \{0\}$. Thus, $\lambda = 0$. □

Theorem 7.18. *If $A \in \mathbb{C}^{n\times n}$ is an idempotent matrix, then* $\mathsf{spec}(A) \subseteq \{0,1\}$.

Proof. Let $A \in \mathbb{C}^{n\times n}$ be an idempotent matrix, λ be an eigenvalue of A, and let $\boldsymbol{x}$ be an eigenvector of λ. We have $P^2\boldsymbol{x} = P\boldsymbol{x} = \lambda\boldsymbol{x}$; on the other hand, $P^2\boldsymbol{x} = P(P\boldsymbol{x}) = P(\lambda\boldsymbol{x}) = \lambda P(\boldsymbol{x}) = \lambda^2\boldsymbol{x}$, so $\lambda^2 = \lambda$, which means that $\lambda \in \{0,1\}$. □

Definition 7.5. Let $A \in \mathbb{C}^{p\times p}$. The *spectral radius* of A is the number

$$\rho(A) = \max\{|\lambda| \mid \lambda \in \mathsf{spec}(A)\}.$$

The next statement shows that for a stochastic matrix, the spectral radius is equal to 1.

Theorem 7.19. *At least one eigenvalue of a stochastic matrix is equal to* 1 *and all eigenvalues lie on or inside the unit circle.*

Proof. Let $A \in \mathbb{R}^{n\times n}$ be a stochastic matrix. Then $1 \in \mathsf{spec}(A)$ and $\mathbf{1}$ is an eigenvector that corresponds to the eigenvalue 1 as the reader can easily verify.

If λ is an eigenvalue of A and $A\boldsymbol{x} = \lambda\boldsymbol{x}$, then $\lambda x_i = a_i^j x_j$ for $1 \leqslant n \leqslant n$, which implies

$$|\lambda||x_i| \leqslant a_i^j |x_j|.$$

Since $\boldsymbol{x} \neq \mathbf{0}_n$, let x_p be a component of $\boldsymbol{x}$ such that $|x_p| = \max\{|x_i| \mid 1 \leqslant i \leqslant n\}$. Choosing $i = p$, we have

$$|\lambda| \leqslant \sum_{i=1}^{n} a_i^j \frac{|x_j|}{|x_p} \leqslant \sum_{i=1}^{n} a_i^j = 1,$$

which shows that all eigenvalues of A lie on or inside the unit circle. □

Theorem 7.20. *All eigenvalues of a unitary matrix are located on the unit circle.*

Proof. Let $A \in \mathbb{R}^{n\times n}$ be a unitary matrix and let λ be an eigenvalue of A. By Theorem 6.24, if $\boldsymbol{x}$ is an eigenvector that corresponds to λ, we have

$$\|\boldsymbol{x}\| = \|A\boldsymbol{x}\| = \|\lambda\boldsymbol{x}\| = |\lambda|\|\boldsymbol{x}\|,$$

which implies $|\lambda| = 1$. □

7.6 Geometry of Eigenvalues

Any matrix norm of a matrix $A \in \mathbb{C}^{n\times n}$ provides an upper bound for the absolute value of any eigenvalue. Indeed, if μ is a matrix norm, and $(\lambda, \boldsymbol{x})$ is an eigenpair, then $|\lambda|\mu(\boldsymbol{x}) = (\mu(A\boldsymbol{x}) \leqslant \mu(A)\mu(\boldsymbol{x})$, so $|\lambda| \leqslant \mu(A)$.

The next result allows finding a more precise location of the eigenvalues of a square matrix A in the complex plane.

Theorem 7.21 (Gershgorin's theorem). Let $A \in \mathbb{R}^{n\times n}$ be a square matrix and let $r_i = \sum\{|a_{ij}| \mid 1 \leqslant j \leqslant n \text{ and } j \neq i\}$, for $1 \leqslant i \leqslant n$. Then we have

$$\mathsf{spec}(A) \subseteq \bigcup_{i=1}^{n} \{z \in \mathbb{C} \mid |z - a_{ii}| \leqslant r_i\}.$$

Proof. Let $\lambda \in \mathsf{spec}(A)$ and let us suppose that $A\boldsymbol{x} = \lambda\boldsymbol{x}$, where $\boldsymbol{x} \neq \boldsymbol{0}$.

Let p be such that $|x_p| = \max\{|x_i| \mid 1 \leqslant i \leqslant n\}$. Then $\sum_{j=1}^{n} a_{pj}x_j = \lambda x_p$, which is equivalent to $\sum_{j=1,j\neq p}^{n} a_{pj}x_j = (\lambda - a_{pp})x_p$. This, in turn, implies

$$\begin{aligned} |x_p||\lambda - a_{pp}| &= \left|\sum_{j=1,j\neq p}^{n} a_{pj}x_j\right| \leqslant \sum_{j=1,j\neq p}^{n} |a_{pj}||x_j| \\ &\leqslant |x_p| \sum_{j=1,j\neq p}^{n} |a_{pj}| = |x_p| r_p. \end{aligned}$$

Therefore, $|\lambda - a_{pp}| \leqslant r_p$ for some p. This yields the desired conclusion. □

Definition 7.6. Let $A \in \mathbb{R}^{n \times n}$ be a square matrix and let $r_i = \sum\{a_{ij} \mid 1 \leqslant j \leqslant n \text{ and } j \neq i\}$, for $1 \leqslant i \leqslant n$. A disk of the form

$$D_i(A) = \{z \in \mathbb{C} \mid |z - a_{ii}| \leqslant r_i\}$$

is called a *Gershgorin disk.*

By Theorem 3.17, if $A \in \mathbb{C}^{n \times n}$, then there exists a Hermitian matrix H_A and a skew-Hermitian matrix S_A such that $A = H_A + S_A$. Let $\mu_A = \|A\|_\infty = \max\{|a_{ij} \mid 1 \leqslant i, j \leqslant n\}$ be the generalized ∞-norm of A and let μ_{H_A} and μ_{S_A} be the generalized ∞-norms of H_A and S_A, respectively.

Let $\lambda_1, \ldots, \lambda_n$ be the eigenvalues of A, where $|\lambda_1| \geqslant \cdots \geqslant |\lambda_n|$. The real eigenvalues of H_A and S_A are denoted as $\eta_1, \ldots, \eta_n$ and $\sigma_1, \ldots, \sigma_n$, where $\eta_1 \geqslant \cdots \geqslant \eta_n$ and $\sigma_1 \geqslant \cdots \geqslant \sigma_n$. These notations are needed for the next two theorems.

Theorem 7.22 (Hirsch's first theorem). If $A \in \mathbb{C}^{n \times n}$, then we have $\lambda_k \leqslant n\mu_A$, $\Re(\lambda_k) \leqslant n\mu_{H_A}$, and $\Im(\lambda_k) \leqslant n\mu_{S_A}$ for $1 \leqslant k \leqslant n$.

Proof. Let $\boldsymbol{x}$ be a unit eigenvector that corresponds to an eigenvalue λ_k. We have $A\boldsymbol{x} = \lambda_k \boldsymbol{x}$, which implies $(A\boldsymbol{x}, \boldsymbol{x}) = \lambda_k$ and $(A^{\mathsf{H}}\boldsymbol{x}, \boldsymbol{x}) = (\boldsymbol{x}, A\boldsymbol{x}) = (\boldsymbol{x}, \lambda_k \boldsymbol{x}) = \bar{\lambda}_k$. Thus, we obtain

$$\begin{aligned}
\Re(\lambda_k) &= \frac{\lambda_k + \bar{\lambda}_k}{2} \\
&= \frac{(A\boldsymbol{x}, \boldsymbol{x}) + (A^{\mathsf{H}}\boldsymbol{x}, \boldsymbol{x})}{2} \\
&= \left(\frac{A + A^{\mathsf{H}}}{2}\boldsymbol{x}, \boldsymbol{x}\right) \\
&= H_A \boldsymbol{x},
\end{aligned}$$

and

$$\begin{aligned}
\Im(\lambda_k) &= \frac{\lambda_k - \bar{\lambda}_k}{2i} \\
&= \frac{(A\boldsymbol{x}, \boldsymbol{x}) - (A^{\mathsf{H}}\boldsymbol{x}, \boldsymbol{x})}{2i} \\
&= \left(\frac{A - A^{\mathsf{H}}}{2i}\boldsymbol{x}, \boldsymbol{x}\right) \\
&= S_A \boldsymbol{x}.
\end{aligned}$$

Since $\lambda_k = (A\boldsymbol{x}, \boldsymbol{x})$, we have

$$|\lambda_k| = \left| \sum_{i=1}^{n} \sum_{j=1}^{n} a_{ij} x_i \bar{x}_j \right| \leqslant \sum_{i=1}^{n} \sum_{j=1}^{n} |a_{ij}||x_i||\bar{x}_j|$$

$$\leqslant \mu_A \sum_{i=1}^{n} \sum_{j=1}^{n} |x_i||\bar{x}_j| = \mu_A \left(\sum_{i=1}^{n} |x_i| \right)^2 .$$

By Inequality (6.7), we have

$$|\lambda_k| \leqslant n\mu_A.$$

□

A related theorem is next.

Theorem 7.23 (Hirsch's second theorem). If $A \in \mathbb{C}^{n\times n}$ and H_A is a real matrix, then

$$|\Im(\lambda_k)| \leqslant \mu_{S_A} \sqrt{\frac{n(n-1)}{2}}.$$

Proof. If $H_A = \frac{A+A^{\mathsf{H}}}{2}$ is real, then $\Im(a_{ij}+a_{ji}) = 0$ for $1 \leqslant i, j \leqslant n$. Since $\Im(z) = \frac{z-\bar{z}}{2i}$, it follows that

$$a_{ij} + a_{ji} - \bar{a}_{ij} - \bar{a}_{ji} = 0,$$

so $a_{ij} - \bar{a}_{ji} = -(a_{ji} - \bar{a}_{ij})$ for $1 \leqslant i, j \leqslant n$. Therefore, by the proof of Theorem 7.22,

$$\begin{aligned}
|\Im(\lambda_k)| &= \left| \frac{\lambda_k - \bar{\lambda}_k}{2i} \right| \\
&= \left| \frac{(A\boldsymbol{x}, \boldsymbol{x}) - (A^{\mathsf{H}}\boldsymbol{x}, \boldsymbol{x})}{2i} \right| \\
&= \left| \left(\frac{A - A^{\mathsf{H}}}{2i} \boldsymbol{x}, \boldsymbol{x} \right) \right| \\
&= \left| \sum_{i=1}^{n} \left(\sum_{j=1}^{n} \frac{a_{ij} - \bar{a}_{ji}}{2i} x_j \right) \bar{x}_i \right| \\
&\leqslant \left| \sum_{i<j} \frac{a_{ij} - \bar{a}_{ji}}{2} \right| \cdot \left| \frac{x_j \bar{x}_i - x_i \bar{x}_j}{i} \right| \\
&\leqslant \mu_{S_A} \sum i < j \left| \frac{x_j \bar{x}_i - x_i \bar{x}_j}{i} \right|.
\end{aligned}$$

Note that

$$\overline{\frac{x_j\bar{x}_i - x_i\bar{x}_j}{i}} = \frac{\bar{x}_j x_i - \bar{x}_i x_j}{-i} = \frac{x_j\bar{x}_i - x_i\bar{x}_j}{i},$$

so each of the $\frac{n(n-1)}{2}$ numbers $\frac{x_j\bar{x}_i - x_i\bar{x}_j}{i}$ is real. By Inequality (6.7), we have

$$|\Im(\lambda_k)| \leqslant \mu_{S_A}\sqrt{-\frac{n(n-1)}{2}\sum_{i<j}(x_j\bar{x}_i - x_i\bar{x}_j)^2}.$$

By applying Lagrange's identity (see Example 5.3), we have

$$\begin{aligned}|\Im(\lambda_k)| &\leqslant \mu_{S_A}\sqrt{\frac{n(n-1)}{2}}\sqrt{-\left(\sum_{i=1}^n x_i^2\sum_{i=1}^n \bar{x}_i^2 - \left(\sum_{i=1}^n x_i\bar{x}_i\right)^2\right)}\\ &\leqslant \mu_{S_A}\sqrt{\frac{n(n-1)}{2}}.\end{aligned}$$

□

7.7 Spectra of Kronecker Products and Sums

Theorem 7.24. *Let $A \in \mathbb{C}^{m\times m}$ and $B \in \mathbb{C}^{n\times n}$ be two matrices that have the eigenvalues $\lambda_1, \ldots, \lambda_m$ and $\mu_1, \ldots, \mu_n$, respectively. The Kronecker product $A \otimes B$ has the following eigenvalues:*

$$\lambda_1\mu_1, \ldots, \lambda_1\mu_n, \ldots, \lambda_m\mu_1, \ldots, \lambda_m\mu_n.$$

Proof. Suppose that $A\boldsymbol{v}_i = \lambda_i\boldsymbol{v}_i$ and $B\boldsymbol{u}_j = \mu_j\boldsymbol{u}_j$. Then we have

$$(A \otimes B)(\boldsymbol{v}_i \otimes \boldsymbol{u}_j) = (A\boldsymbol{v}_i) \otimes (B\boldsymbol{u}_j) = \lambda_i\mu_j(\boldsymbol{v}_i \times \boldsymbol{u}_j),$$

which shows that $A \otimes B$ has the eigenvalues $\lambda_i\mu_j$. □

Corollary 7.11. *Let $A \in \mathbb{C}^{n\times n}$ and $B \in \mathbb{C}^{m\times m}$. Then $\mathsf{trace}(A \otimes B) = \mathsf{trace}(A)\mathsf{trace}(B) = \mathsf{trace}(B \otimes A)$ and $\det(A \otimes B) = (\det(A))^m(\det(B))^n = \det(B \otimes A)$.*

Proof. The corollary is an immediate consequence of Theorem 7.24. □

Theorem 7.25. *Let $A \in \mathbb{C}^{n\times n}$ and $B \in \mathbb{C}^{m\times m}$ be two matrices. If* $\mathsf{spec}(A) = \{\lambda_1, \ldots, \lambda_n\}$ *and* $\mathsf{spec}(B) = \{\mu_1, \ldots, \mu_m\}$, *then* $\mathsf{spec}(A \oplus B) = \{\lambda_i + \mu_j \mid 1 \leqslant i \leqslant n, 1 \leqslant j \leqslant m\}$ *and* $\mathsf{spec}(A \ominus B) = \{\lambda_i - \mu_j \mid 1 \leqslant i \leqslant n, 1 \leqslant j \leqslant m\}$.

Proof. Let $\boldsymbol{x}$ and $\boldsymbol{y}$ be two eigenvectors of A and B that correspond to the eigenvalues λ and μ, respectively. Since $A \oplus B = (A \otimes I_m) + (I_n \otimes B)$, we have

$$\begin{aligned}(A \oplus B)(\boldsymbol{x} \otimes \boldsymbol{y}) &= (A \otimes I_m)(\boldsymbol{x} \otimes \boldsymbol{y}) + (I_n \otimes B)(\boldsymbol{x} \otimes \boldsymbol{y}) \\ &= (A\boldsymbol{x} \otimes \boldsymbol{y}) + (\boldsymbol{x} \otimes B\boldsymbol{y}) \\ &= \lambda(\boldsymbol{x} \otimes \boldsymbol{y}) + \mu(\boldsymbol{x} \otimes \boldsymbol{y}) \\ &= (\lambda + \mu)(\boldsymbol{x} \otimes \boldsymbol{y}).\end{aligned}$$

By replacing B by $-B$, we obtain the spectrum of $A \ominus B$. □

7.8 The Power Method for Eigenvalues

Let $A \in \mathbb{R}^{n\times n}$ be a matrix that has eigenvalue $\lambda_1, \ldots, \lambda_n$, such that

$$|\lambda_1| > |\lambda_2| \geqslant \cdots \geqslant |\lambda_n|.$$

We refer to λ_1 as the *dominant eigenvalue* of A. Observe that a dominant eigenvalue is always real for, otherwise, A would have a pair of eigenvalues λ_1 and $\bar{\lambda}_1$ having equal absolute values.

To compute the dominant eigenvalue of a matrix and an associated eigenvector, we can use a technique known as the *power method.*

Let $\boldsymbol{v}_1, \ldots, \boldsymbol{v}_n$ be a sequence of unit eigenvectors corresponding to the eigenvalues $\lambda_1, \ldots, \lambda_n$ of the matrix A, respectively, where λ_1 is the dominant eigenvalue of A and $|\lambda_1| > |\lambda_2| \geqslant \cdots \geqslant |\lambda_n|$.

Define a sequence of vectors $\boldsymbol{x}_0, \boldsymbol{x}_1, \ldots$ as follows. The initial vector $\boldsymbol{x}_0 \in \mathbb{C}^n$ is any unit vector that is not orthogonal on any eigenvector corresponding to λ_1. Then $\boldsymbol{x}_{k+1}$ is the unit vector given by

$$\boldsymbol{x}_{k+1} = \frac{A\boldsymbol{x}_k}{\|A\boldsymbol{x}_k\|}$$

for $k \in \mathbb{N}$. The unit vector $\boldsymbol{x}_k$ can be written as

$$\boldsymbol{x}_k = \frac{A^k\boldsymbol{x}_0}{\|A^k\boldsymbol{x}_0\|}$$

for $k \geqslant 1$. Indeed, in the base case ($k = 1$), this equality clearly holds. Suppose that it holds for k. We have

$$\boldsymbol{x}_{k+1} = \frac{A\boldsymbol{x}_k}{\|A\boldsymbol{x}_k\|} = \frac{A\frac{A^k\boldsymbol{x}_0}{\|A^k\boldsymbol{x}_0\|}}{\|A\frac{A^k\boldsymbol{x}_0}{\|A^k\boldsymbol{x}_0\|}\|} = \frac{A^{k+1}\boldsymbol{x}_0}{\|A^{k+1}\boldsymbol{x}_0\|}.$$

We claim that $\lim_{k\to\infty} \boldsymbol{x}_k = \boldsymbol{v}_1$.

By Theorem 7.1, the set $\{\boldsymbol{v}_1, \ldots, \boldsymbol{v}_n\}$ is linearly independent and this allows us to write

$$\boldsymbol{x}_0 = a_1\boldsymbol{v}_1 + \cdots + a_n\boldsymbol{v}_n.$$

By the assumption made concerning $\boldsymbol{x}_0$ (as not being orthogonal on any eigenvector corresponding to λ_1), we have $a_1 \neq 0$. Thus, $A\boldsymbol{x}_0 = a_1 A\boldsymbol{v}_1 + \cdots + a_n A\boldsymbol{v}_n = a_1\lambda_1\boldsymbol{v}_1 + \cdots + a_n\lambda_n\boldsymbol{v}_n$. A straightforward induction argument on $k \geqslant 1$ shows that

$$\begin{aligned} A^k\boldsymbol{x}_0 &= a_1\lambda_1^k\boldsymbol{v}_1 + \cdots + a_n\lambda_n^k\boldsymbol{v}_n \\ &= a_1\lambda_1^k\left(\boldsymbol{v}_1 + \frac{a_2}{a_1}\left(\frac{\lambda_2}{\lambda_1}\right)^k\boldsymbol{v}_2 + \cdots + \frac{a_n}{a_1}\left(\frac{\lambda_n}{\lambda_1}\right)^k\boldsymbol{v}_n\right). \end{aligned}$$

Since $\frac{|\lambda_j|}{|\lambda_1|} < 1$ for $2 \leqslant j \leqslant n$, it follows that

$$\lim_{k\to\infty}\frac{A^k\boldsymbol{x}_0}{a_1\lambda_1^k} = \lim_{k\to\infty}\left(\boldsymbol{v}_1 + \frac{a_2}{a_1}\left(\frac{\lambda_2}{\lambda_1}\right)^k\boldsymbol{v}_2 + \cdots + \frac{a_n}{a_1}\left(\frac{\lambda_n}{\lambda_1}\right)^k\boldsymbol{v}_n\right) = \boldsymbol{v}_1,$$

and the speed of convergence of $\frac{A^k\boldsymbol{x}_0}{a_1\lambda_1^k}$ to $\boldsymbol{v}_1$ is determined by $\frac{\lambda_2}{\lambda_1}$. This implies

$$\lim_{k\to\infty}\frac{\|A^k\boldsymbol{x}_0\|}{|a_1\lambda_1^k|} = \|\boldsymbol{v}_1\| = 1.$$

Therefore,

$$\lim_{k\to\infty}\boldsymbol{x}_k = \lim_{k\to\infty}\frac{A^k\boldsymbol{x}_0}{\|A^k\boldsymbol{x}_0\|} = \lim_{k\to\infty}\frac{A^k\boldsymbol{x}_0}{|a_1\lambda_1^k|}\cdot\frac{|a_1\lambda_1^k|}{\|A^k\boldsymbol{x}_0\|} = \lim_{k\to\infty}\frac{A^k\boldsymbol{x}_0}{|a_1\lambda_1^k|} = \boldsymbol{v}_1.$$

Since the limit of the sequence $\boldsymbol{x}_0, \ldots, \boldsymbol{x}_k, \ldots$ is $\boldsymbol{v}_1$, we have a method to approximatively determine a unit vector corresponding to the dominating eigenvalue λ_1 of A.

Example 7.8. Let us apply the iteration method to the symmetric matrix

$$A = \begin{pmatrix} 3 & 2 & 1 \\ 2 & 4 & 2 \\ 1 & 2 & 6 \end{pmatrix},$$

beginning with the vector $\boldsymbol{x}_0 = \mathbf{1}_3$. The following MATLAB code computes ten vectors of the sequence $(\boldsymbol{x}_1, \boldsymbol{x}_2, \ldots)$:

```
x = ones(3,1);
sequence(:,1) = x;
for i=1:10
    x = A*x/norm(A*x);
    sequence(:,i+1)=x;
end
sequence
```

and prints the sequence as

```
1.0000 0.4015 0.3844 0.3772 0.3740 0.3726 0.3719 0.3716 0.3715 0.3714 0.3714
1.0000 0.5789 0.5678 0.5621 0.5594 0.5581 0.5576 0.5573 0.5572 0.5571 0.5571
1.0000 0.7097 0.7279 0.7361 0.7398 0.7414 0.7422 0.7425 0.7427 0.7427 0.7428
```

The last vector in the sequence is very close to an eigenvector that corresponds to the eigenvalue $\lambda = 8$.

A variant of the power method considered in [113] replaces the scaling factor $\|A\boldsymbol{x}_k\|$ by $\mathsf{m}(A\boldsymbol{x}_k)$, where $\mathsf{m}(\boldsymbol{v})$ is the first component of $\boldsymbol{v}$ that has the largest absolute value. For example, if $v = (-2, -4, 3, -4)$, then $\mathsf{m}(\boldsymbol{v}) = v_2 = -4$. We have $\mathsf{m}(a\boldsymbol{v}) = a\,\mathsf{m}(\boldsymbol{v})$ for every $a \in \mathbb{R}$.

The power method is limited to the dominant eigenvalue. However, if a close approximative value is known for an eigenvalue λ_i, the power method can be applied to compute an eigenvector associated to this eigenvalue. Observe that if $|\lambda_i - a| < |\lambda_j - a|$ for every $\lambda_j \in \mathsf{spec}(A) - \{\lambda_i\}$, then $\frac{1}{\lambda_i - a}$ is the dominant eigenvalue of the matrix $B = (A - aI_n)^{-1}$.

The justification of the algorithm follows from the next statement.

Theorem 7.26. *Let $A \in \mathbb{C}^{n\times n}$ and let a be a number such that $a \notin \mathsf{spec}(A)$. Then $\boldsymbol{v}$ is an eigenvector of A that corresponds to the eigenvalue λ if and only if $\boldsymbol{v}$ is an eigenvector of the matrix $B = (A - aI)^{-1}$ that corresponds to the eigenvalue $\frac{1}{\lambda - a} \in \mathsf{spec}(B)$.*

Proof. If $\boldsymbol{v}$ is an eigenvector for A that corresponds to the eigenvalue λ, then $A\boldsymbol{v} = \lambda\boldsymbol{v}$, so $(A - aI)\boldsymbol{v} = (\lambda - a)\boldsymbol{v}$. Since $a \notin \mathsf{spec}(A)$, the matrix $A - aI$ is invertible, so $\frac{1}{\lambda - a}\boldsymbol{v} = (A - aI)^{-1}\boldsymbol{v}$, which proves that $\boldsymbol{v}$ is an eigenvector of B that corresponds to the eigenvalue $\frac{1}{\lambda - a}$. The reverse implication is immediate. □

By Theorem 7.26, if a is a good approximation of λ, the power method applied to B will yield an eigenvector $\boldsymbol{v}$ of A that corresponds to λ. This technique is known as the *inverse power method.*

7.9 The QR Iterative Algorithm

Another technique for computing eigenvalues is the *QR iterative algorithm.* The approach involves decomposing a matrix $A \in \mathbb{R}^{n\times n}$ into a product $A = QR$, where Q is an orthogonal matrix and R is an upper triangular matrix. If $A \in \mathbb{C}^{n\times n}$ is a complex matrix, then we seek Q as a unitary matrix.

A sequence of matrices $(A_0, A_1, \ldots)$ is defined, where $A_0 = A \in \mathbb{R}^{n\times n}$. Begin by computing a QR factorization of A_0 as $A_0 = Q_0R_0$. If A_i is factored as $A_i = Q_iR_i$, define A_{i+1} as $A_{i+1} = R_iQ_i$ for $i \in \mathbb{N}$.

Let $P_i = Q_0 \cdots Q_i$ and $S_i = R_i \cdots R_0$ for $i \in \mathbb{N}$. Clearly, P_i is an orthonormal matrix and S_i is an upper triangular matrix for every $i \in \mathbb{N}$. We claim that

$$A_{i+1} = P_i'AP_i. \tag{7.11}$$

Indeed, for $i = 0$, we have $P_0'AP_0 = Q_0'Q_0R_0Q_0 = R_0Q_0 = Q_1$. Suppose that the equality holds for i. Then

$$\begin{aligned} P_{i+1}'AP_{i+1} &= Q_{i+1}'P_i'AP_iQ_{i+1} = Q_{i+1}'A_{i+1}Q_{i+1} \\ &= Q_{i+1}'Q_{i+1}R_{i+1}Q_{i+1} = A_{i+2}, \end{aligned}$$

which concludes the proof of the claim. Therefore, each of the matrices A_i is orthonormally similar to A and $\mathsf{spec}(A_i) = \mathsf{spec}(A)$ for $i \in \mathbb{N}$.

Equality (7.11) can also be written as

$$P_iA_{i+1} = AP_i \tag{7.12}$$

for $i \in \mathbb{N}$.

Note that

$$P_i R_i = P_{i-1} Q_i R_i = P_{i-1} A_i = A P_{i-1},$$

taking into account Equality (7.12). This implies that the subspace $C_{P_i,q}$ generated by the first q columns of P_i equals $C_{AP_{i-1},q}$.

If $\lim_{n\to\infty} P_n$ exists and it equals P, since $P_{n-1}Q_n = P_n$, it follows that $\lim_{n\to\infty} Q_n = I$. Therefore, since $A_n = Q_n R_n$, we have $\lim_{n\to\infty} A_n = \lim_{n\to\infty} R_n = R$. The matrix R is an upper triangular matrix, as the limit of a sequence of upper triangular matrices, has non-negative entries on its main diagonal, which equal the eigenvalues of the matrix A. Therefore, a necessary condition for the algorithm to work is that all eigenvalues of A be real, non-negative numbers. Thus, the value of this initial version of the QR algorithm is limited.

In practical terms, it is desirable to replace the initial matrix A with its upper Hessenberg equivalent H_1 (which is a tridiagonal matrix when A is symmetric). As we saw in Supplement 72 of Chapter 6, if we factor $H_1 = Q_1 R_1$ such that R_1 is non-singular, Q_1 and $R_1 Q_1$ are both upper Hessenberg matrices, so the next matrix $H_2 = R_1 Q_1$ is again an upper Hessenberg matrix, and the process continues in the same manner as in the initial QR algorithm.

Other improvements of the QR involve the double shifting of the matrices H_n by a multiple of I_n. Thus, instead of factoring H_k, one factors $H - aI_n$, where a is an approximation of a real eigenvalue, $H_k - aI_n = Q_k R_k$. Then, the next matrix is $H_{k+1} = R_k Q_k + aI_n$. Therefore,

$$H_{k+1} = R_k Q_k + aI_n = Q_k'(H_k - aI_n)Q_k + aI_n = Q_k' H_k Q_k$$

for $k \geqslant 1$. Thus, $\mathsf{spec}(H_{k+1}) = \mathsf{spec}(H_k)$.

7.10 MATLAB Computations

To compute the eigenvalues of a matrix $A \in \mathbb{C}^{n\times n}$, we can use the command `eig(A)`, which returns an n-dimensional vector containing the eigenvalues of A. The variant `[V,D] = eig(A)` produces a diagonal matrix D of eigenvalues and a full matrix $V \in \mathbb{C}^{n\times n}$ whose columns are the corresponding eigenvectors so that $AV = VD$.

A faster computation of eigenvalues is obtained using the **eigs**. If A is a large, sparse, and square matrix, **eigs(A)** returns a vector that consists of the six largest magnitude eigenvalues.

The function call `[V,D] = eigs(A)` returns a diagonal matrix D that contains six eigenvalues of A having the largest absolute values and a matrix V whose columns are the corresponding eigenvectors. If a parameter `flag` is added, `[V,D,flag] = eigs(A)`, and `flag` is 0, then all the eigenvalues converge; otherwise not all converge. The call `eigs(A,k)` computes the k largest magnitude eigenvalues. Various options can be set by specifying additional parameters. For instance, if A is a symmetric matrix, then the call

```
[V,D] = eigs(A,k,'SA')
```

will compute the k smallest eigenvalues of A, and return a diagonal matrix D that contains the eigenvalues and a matrix V that contains the corresponding eigenvectors.

Exercises and Supplements

(1) Prove that a matrix $A \in \mathbb{C}^{n\times n}$ is non-singular if and only if $0 \notin \mathsf{spec}(A)$.

(2) Prove that $\mathsf{spec}(J_{n,n}) = \{0, n\}$ and that $\mathsf{algm}(A, 0) = n - 1$.

(3) Determine the spectrum of the matrix $A = aI_n + bJ_{n,n}$.

(4) Let $X, Y \in \mathbb{C}^{n\times n}$ be two non-singular matrices. Prove that λ is an eigenvalue of one of the matrices XY^{-1}, YX^{-1} if and only if λ^{-1} is an eigenvalue of the other.

(5) Let $A \in \mathbb{C}^{n\times n}$ be a matrix and let $a, b \in \mathbb{C}$ such that $a \neq 0$. Prove that $\mathsf{spec}(aA + bI_n) = \{a\lambda + b \mid \lambda \in A\}$.

Solution: Let $B = aA + bI_n$. For the characteristic polynomial p_B, we can write

$$p_B(\lambda) = \det(\lambda I_n - B) = \det(\lambda I_n - (aA + bI_n))$$
$$= \det((\lambda - b)I_n - aA) = a^n \det\left(\frac{\lambda - b}{a} I_n - A\right),$$

which shows that $\mathsf{spec}(B)$ has the desired form.

(6) Prove that the non-zero eigenvalues of a real skew-symmetric matrix are purely imaginary numbers.

Solution: Let λ be a non-zero eigenvalue of a real skew-symmetric matrix $A \in \mathbb{R}^{n\times n}$. Since $A\boldsymbol{x} = \lambda\boldsymbol{x}$ for some $\boldsymbol{x} \in \mathbb{R}^n$, it follows that $A^2\boldsymbol{x} = \lambda^2\boldsymbol{x}$. By Supplement 115 of Chapter 6, we have $\lambda^2 \leqslant 0$.

(7) Let S be a k-dimensional invariant subspace for a matrix $A \in \mathbb{C}^{n\times n}$ and let the columns of the matrix $X \in \mathbb{C}^{n\times k}$ form a basis for S. Prove that there is a unique matrix $C \in \mathbb{C}^{k\times k}$ such that $AX = XC$ and that $\mathsf{spec}(C) \subseteq \mathsf{spec}(A)$.

Solution: Suppose that $X = (\boldsymbol{x}_1 \ \cdots \ \boldsymbol{x}_k)$. $A\boldsymbol{x}_i$ is a unique linear combination of the columns of X because $A\boldsymbol{x}_i \in S$, so there exists a unique vector $\boldsymbol{c}_i$ such that $A\boldsymbol{x}_i = X\boldsymbol{c}_i$ for $1 \leqslant i \leqslant k$. The matrix $C = (\boldsymbol{c}_1 \ \cdots \ \boldsymbol{c}_k)$ is the matrix we are seeking. If $\lambda \in \mathsf{spec}(C)$, then $C\boldsymbol{u} = \lambda\boldsymbol{u}$ for some vector $\boldsymbol{u}$, which implies $A(X\boldsymbol{u}) = \lambda(X\boldsymbol{u})$. Thus, $\mathsf{spec}(C) \subseteq \mathsf{spec}(A)$.

(8) Let

$$A = \begin{pmatrix} \sin\alpha & \cos\alpha \\ -\cos\alpha & \sin\alpha \end{pmatrix} \text{ and } B = \begin{pmatrix} \sinh t & \cosh t \\ \cosh t & \sinh t \end{pmatrix}.$$

Prove that $\mathsf{spec}(A) = \{\sin\alpha\}$ and $\mathsf{spec}(B) = \{\sinh t + \cosh t, \sinh t - \cosh t\}$.

(9) Let

$$A = \begin{pmatrix} a_{11} & a_{12} \\ a_{21} & a_{22} \end{pmatrix},$$

be a matrix (that is, $a_{12} = a_{21}$) in $\mathbb{R}^{2\times 2}$. Prove that there exists a rotation matrix

$$R(\theta) = \begin{pmatrix} \cos\theta & \sin\theta \\ -\sin\theta & \cos\theta \end{pmatrix},$$

such that $R(-\theta)AR(\theta)$ is a diagonal matrix having the eigenvalue λ_1 and λ_2 as its diagonal elements.

(10) Let $A \in \mathbb{R}^{2\times 2}$ be a matrix and let

$$\boldsymbol{x} = \begin{pmatrix} \cos\alpha \\ \sin\alpha \end{pmatrix} \in \mathbb{R}^2.$$

As α varies between 0 and 2π, $\boldsymbol{x}$ rotates around the origin. Prove that $A\boldsymbol{x}$ moves on an ellipse centered in the origin. Compute α such that the vectors $\boldsymbol{x}$ and $A\boldsymbol{x}$ are collinear.

(11) Let $A \in \mathbb{C}^{n\times n}$ be a matrix such that $\mathsf{spec}(A) = \{\lambda_1, \ldots, \lambda_n\}$. Prove that $\mathsf{spec}(A + cI_n) = \{\lambda_1 + c, \ldots, \lambda_n + c\}$.
(12) Let $A \in \mathbb{R}^{n\times n}$ be a matrix. Prove that if A has a dominating eigenvalue λ_1, then λ_1 is a real number.
(13) Prove that if $A \in \mathbb{C}^{n\times n}$ has the spectrum $\mathsf{spec}(A) = \{\lambda_1, \ldots, \lambda_n\}$, then $\mathsf{spec}(A - aI_n) = \{\lambda_1 - a, \ldots, \lambda_n - a\}$.
(14) Let $\mathsf{C}(A)$ be the Cayley transform of the matrix $A \in \mathbb{C}^{n\times n}$ introduced in Exercise 65 of Chapter 3.
   (a) Prove that if $\lambda \in \mathsf{spec}(\mathsf{C}(A))$, then $\frac{1-\lambda}{1+\lambda} \in \mathsf{spec}(A)$.
   (b) Prove that if $\mathsf{C}(A)$ exists, then $-1 \notin \mathsf{spec}(\mathsf{C}(A))$.
(15) Let $A \in \mathbb{C}^{m\times m}$ and $B \in \mathbb{C}^{n\times n}$. Prove that $\mathsf{trace}(A \otimes B) = \mathsf{trace}(A)\mathsf{trace}(B)$ and $\det(A \otimes B) = \det(A)^n \det(B)^m$.
(16) Let $A \in \mathbb{C}^{n\times n}$ be a Hermitian matrix such that $\mathsf{spec}(A) = \{\lambda_1, \ldots, \lambda_n\}$ consists of n distinct eigenvalues and let $\boldsymbol{v}_1, \ldots, \boldsymbol{v}_n$ be n unit eigenvectors that correspond to the eigenvalues $\lambda_1, \ldots, \lambda_n$, respectively.
   (a) If $\boldsymbol{w} \in \mathbb{C}^n$ is a unit vector and $c_i = \boldsymbol{w}^{\mathsf{H}}\boldsymbol{u}_i$ for $1 \leqslant i \leqslant n$, prove that $\sum_{i=1}^{n} c_i^2 = 1$.
   (b) Prove that $|\lambda_i - \boldsymbol{w}^{\mathsf{H}}A\boldsymbol{w}| \leqslant \max_j\{|\lambda_i - \lambda_j|\|\boldsymbol{w} - \boldsymbol{u}_i\|_2^2$.
(17) Prove that the matrix $A \in \mathbb{C}^{n\times n}$ has a set of pairwise orthonormal eigenvectors $\{\boldsymbol{u}_1, \ldots, \boldsymbol{u}_n\}$ if and only if A is a normal matrix.
(18) Prove that the set of simple matrices is dense in the metric space $(\mathbb{C}^{n\times n}, d)$, where $d(A, B) = \|A - B\|_2$ for $A, B \in \mathbb{C}^{n\times n}$.
(19) Let $A \in \mathbb{C}^{m\times n}$ and $B \in \mathbb{C}^{n\times m}$. Prove the following:
   (a) the matrices

$$\begin{pmatrix} AB & O_{m,m} \\ B & 0_{n,m} \end{pmatrix} \text{ and } \begin{pmatrix} O_{n,m} & O_{n,n} \\ B & BA \end{pmatrix}$$

   are similar;
   (b) if $m \leqslant n$, then $BA \in \mathbb{C}^{n\times n}$ has the same eigenvalues as $AB \in \mathbb{C}^{m\times m}$ together with $n - m$ zero eigenvalues;
   (c) if $m = n$ and at least one of the matrices A or B is nonsingular, then AB and BA are similar.

Solution: For the first part, note that the matrix

$$M = \begin{pmatrix} I_m & A \\ O_{n,m} & I_n \end{pmatrix}$$

has all its eigenvalues equal to 1 and, therefore, is non-singular. It is easy to see that

$$M^{-1} = \begin{pmatrix} I_m & -A \\ O_{n,m} & I_n \end{pmatrix}$$

and that

$$M^{-1} \begin{pmatrix} AB & O_{m,m} \\ B & 0_{n,m} \end{pmatrix} M = \begin{pmatrix} O_{n,m} & O_{n,n} \\ B & BA \end{pmatrix},$$

which proves the desired similarity.

The last two parts are left to the reader.

A *commuting family of matrices* is a collection of matrices $\mathcal{M} \subseteq \mathbb{C}^{n\times n}$ such that for each $A, B \in \mathcal{M}$, we have $AB = BA$.

An invariant subspace for a commuting family of matrices $\mathcal{M} \subseteq \mathbb{C}^{n\times n}$ is a subspace W of $\mathbb{C}^n$ that is invariant for each matrix in $\mathcal{M}$.

(20) Let $A \in \mathbb{C}^{n\times n}$ and let S be an invariant subspace of A with $\dim(S) \geqslant 1$. Prove that S contains an eigenvector of A.

Solution: Suppose that $\dim(S) = m \leqslant 1$ and let $\boldsymbol{v}_1, \ldots, \boldsymbol{v}_m \in \mathbb{C}^n$ be a set of vectors that form a basis of S. Let $V = (\boldsymbol{v}_1\ \boldsymbol{v}_2\ \cdots\ \boldsymbol{v}_m) \in \mathbb{C}^{n\times m}$ be the matrix whose columns are these vectors. Note that $\mathsf{rank}(V) = m$ so the nullspace of V consists of $\mathbf{0}_m$. Then we have $A\boldsymbol{v}_i \in S$ for every i, $1 \leqslant i \leqslant m$, because S is an invariant subspace of A. Thus, we have $AV = VU$ for some $U \in \mathbb{C}^{m\times m}$. If $\boldsymbol{x} \neq \mathbf{0}_m$ is an eigenvector of U, we have $U\boldsymbol{x} = \lambda\boldsymbol{x}$, hence $VU\boldsymbol{x} = \lambda V\boldsymbol{x}$, or $AV\boldsymbol{x} = \lambda V\boldsymbol{x}$. Thus, $V\boldsymbol{x}$ is an eigenvector of A that is contained in S.

(21) Prove that if $\mathcal{M}$ is a commuting family of matrices in $\mathbb{C}^{n\times n}$, then there exists $\boldsymbol{x} \in \mathbb{C}^n$ such that $\boldsymbol{x}$ is an eigenvector of every $A \in \mathcal{M}$.

Solution: Note that $\mathbb{C}^n$ is an invariant subspace for each matrix in $\mathcal{M}$. Thus, we can assume that there exists an invariant subspace W for $\mathcal{F}$ such that $\dim(W)$ is minimal and positive. We claim that every $\boldsymbol{w} \in W - \{\mathbf{0}_n\}$ is an eigenvector of $\mathcal{F}$. Suppose that this is not the case, so not every non-zero vector in W is an eigenvector of A. Since W is $\mathcal{M}$-invariant, it is also A-invariant and, by Supplement 7.10, there exists an eigenvector of A in W such that $\boldsymbol{x} \neq \mathbf{0}_n$ and $A\boldsymbol{x} = \lambda\boldsymbol{x}$.

Let W_0 be the subspace $W_0 = \{\boldsymbol{y} \in W \mid A\boldsymbol{y} = \lambda\boldsymbol{y}\}$. Clearly, $\boldsymbol{x} \in W_0$. Since we assumed that not every non-zero vector in W is an eigenvector of A, $W_0 \neq W$, so $\dim(W_0) < \dim(W)$.
If $B \in \mathcal{M}$, $\boldsymbol{x} \in W_0$ implies $B\boldsymbol{x} \in W$ because $W_0 \subseteq W$ and W is $\mathcal{M}$-invariant. Since $\mathcal{M}$ is a commuting family, we have

$$A(B\boldsymbol{x}) = (AB)\boldsymbol{x} = (BA)\boldsymbol{x} = B(A\boldsymbol{x}) = B(\lambda\boldsymbol{x}) = \lambda(B\boldsymbol{x}),$$

hence $B\boldsymbol{x} \in W_0$. Therefore, W_0 is $\mathcal{M}$-invariant, which results in a contradiction, since we assumed that $\dim(W)$ is minimal.

Certain spectral properties of Hermitian matrices can be extended to matrices that are self-adjoint with respect to an inner product $f : \mathbb{C}^n \times \mathbb{C}^n \longrightarrow \mathbb{R}$. These extensions are discussed next.

(22) Prove that a matrix $A \in \mathbb{C}^{n\times n}$ that is self-adjoint relative to an inner product $f : \mathbb{C}^n \times \mathbb{C}^n \longrightarrow \mathbb{R}$ has real eigenvalues. Furthermore, if $\lambda, \mu \in \mathsf{spec}(A)$ are two distinct eigenvalues of A and $\boldsymbol{u}, \boldsymbol{v}$ are eigenvectors of A that correspond to λ and μ, respectively, then $\boldsymbol{u}$ and $\boldsymbol{v}$ are orthogonal relative to f, that is, $f(\boldsymbol{u}, \boldsymbol{v}) = 0$.

(23) Let $A \in \mathbb{C}^{n\times n}$ be a matrix such that $\sum_{i=1}^{n} a_{ij} = 1$ for every j, $1 \leqslant j \leqslant n$ and let $\boldsymbol{x} \in \mathbb{C}^n$ be a eigenvector such that $\boldsymbol{1}'\boldsymbol{x} \neq 0$. Prove that the eigenvalue that corresponds to $\boldsymbol{x}$ is 1.

(24) Let $\boldsymbol{u}, \boldsymbol{v} \in \mathbb{C}^n$. Prove that the matrix $\boldsymbol{u}\boldsymbol{v}^{\mathsf{H}} + \boldsymbol{v}\boldsymbol{u}^{\mathsf{H}}$ has at most two non-zero eigenvalues.

Solution: Let $A = (\boldsymbol{v}\ \boldsymbol{u}) \in \mathbb{C}^{n\times 2}$ and $B = \begin{pmatrix} \boldsymbol{u}^{\mathsf{H}} \\ \boldsymbol{v}^{\mathsf{H}} \end{pmatrix} \in \mathbb{C}^{2\times n}$. By Theorem 7.10, the set of non-zero eigenvalues of the matrices $AB = \boldsymbol{v}\boldsymbol{u}^{\mathsf{H}} + \boldsymbol{u}\boldsymbol{v}^{\mathsf{H}} \in \mathbb{C}^{n\times n}$ and $BA \in \mathbb{C}^{2\times 2}$ are the same and $\mathsf{algm}(AB, \lambda) = \mathsf{algm}(BA, \lambda)$ for each such eigenvalue.

(25) Let $A \in \mathbb{C}^{n\times n}$ be a matrix such that $\sum_{i=1}^{n} |a_{ij}| \leqslant 1$ for every j, $1 \leqslant j \leqslant n$. Prove that for every eigenvalue λ of A we have $|\lambda| \leqslant 1$.

(26) For any matrix $A \in \mathbb{C}^{n\times m}$, prove that there exists $c \in \mathbb{C}$ such that the matrices $A + cI_n$ and $A - cI_n$ are invertible. Conclude that every matrix is the sum of two invertible matrices.

(27) Let A, B, and E be three matrices in $\mathbb{C}^{n\times n}$ such that $B = A + E$ and let $\mu \in \mathsf{spec}(B) - \mathsf{spec}(A)$. Prove that if $Q \in \mathbb{C}^{n\times n}$ is an

invertible matrix, then

$$\|Q^{-1}(A - \mu I_n)Q\| \leqslant \|Q^{-1}EQ\|.$$

Solution: Since $\mu \in \mathsf{spec}(B) - \mathsf{spec}(A)$, the matrix $B - \mu I_n$ is singular, while $A - \mu I_n$ is nonsingular. We have

$$\begin{aligned} Q^{-1}(B - \mu I_n)Q &= Q^{-1}(A - \mu I_n + E)Q \\ &= Q^{-1}(A - \mu I_n)Q[I_n + Q^{-1}(A - \mu I_n)^{-1}Q] \\ &\quad \times (Q^{-1}EQ). \end{aligned}$$

Since $B - \mu I_n$ is singular, the matrix

$$[I_n + Q^{-1}(A - \mu I_n)^{-1}Q](Q^{-1}EQ)$$

must be singular, which implies $\|[Q^{-1}(A - \mu I_n)^{-1}Q](Q^{-1}EQ)\| \geqslant 1$, so $\|Q^{-1}(A - \mu I_n)^{-1}Q\|\|Q^{-1}EQ\| \geqslant 1$. This implies the desired inequality.

(28) Let $\boldsymbol{v}$ be an eigenvector of the matrix $A \in \mathbb{C}^{n\times n}$ and let

$$B = \begin{pmatrix} 0 & \boldsymbol{a}^{\mathsf{H}} \\ \boldsymbol{a} & A \end{pmatrix},$$

where $a \in \mathbb{C}^n$. Prove that there exists an eigenvector $\boldsymbol{u}$ of B such that $\boldsymbol{u} = \begin{pmatrix} x \\ \boldsymbol{v} \end{pmatrix}$ for some $x \in \mathbb{C}$ if and only if $\boldsymbol{a}^{\mathsf{H}}\boldsymbol{v} = 0$.

(29) Let $B \in \mathbb{R}^{n\times n}$ be a matrix and let $A = \mathsf{diag}(b_{11}, \ldots, b_{nn})$. If μ is an eigenvalue of B, prove that there exists a number b_{ii} such that $|\mu - b_{ii}| \leqslant \|E\|$.

(30) Let A, B be two matrices in $\mathbb{C}^{n\times n}$. Prove that for every $\epsilon > 0$, there exists $\delta > 0$ such that if $|a_{ij} - b_{ij}| < \delta$ and $\lambda \in \mathsf{spec}(A)$, then there exists $\theta \in \mathsf{spec}(B)$ such that $|\lambda - \theta| < \epsilon$.

(31) Let $A \in \mathbb{C}^{m\times m}$ and $B \in \mathbb{C}^{n\times n}$ be two matrices. Prove that $\mathsf{trace}(A \otimes B) = \mathsf{trace}(A)\mathsf{trace}(B)$ and $\det(A \otimes B) = \det(A)^n \det(B)^m$.

(32) Let $A \in \mathbb{C}^{n\times n}$ be a matrix whose eigenvalues are $\lambda_1, \ldots, \lambda_n$. Define the matrix $A^{[k]} = A \otimes A \otimes \cdots \otimes A \in \mathbb{C}^{n^k\times n^k}$ which is the kth power of A in the sense of the Kronecker product. Prove that the eigenvalues of $A^{[k]}$ have the form $\lambda_{i_1}\lambda_{i_2}\cdots\lambda_{i_k}$, where $1 \leqslant i_1, \ldots, i_k \leqslant n$.

(33) Let $A \in \mathbb{C}^{n\times n}$ be a matrix such that $\mathsf{spec}(A) = \{\lambda_1, \ldots, \lambda_n\}$. Prove that

$$\det(I_n - A) = \prod_{i=1}^{n}(1-\lambda_i) \text{ and } \det(I_n + A) = \prod_{i=1}^{n}(1+\lambda_i).$$

(34) Let $P \in \mathbb{C}^{n\times n}$ be a projection matrix. Prove that
(a) if $\mathsf{rank}(P) = r$, then $\mathsf{algm}(P, 1) = r$ and $\mathsf{algm}(P, 0) = n - r$;
(b) we have $\mathsf{rank}(P) = \mathsf{trace}(P)$.

(35) Let $A \in \mathbb{R}^{n\times n}$ be a matrix such that $\lambda \geqslant 0$ for every $\lambda \in \mathsf{spec}(A)$. Prove that

$$\det(A) \leqslant \left(\frac{\mathsf{trace}(A)}{n}\right)^n.$$

(36) Let $A \in \mathbb{C}^{m\times n}$ and $B = \mathbb{C}^{n\times m}$. Prove that $\lambda^n p_{AB}(\lambda) = \lambda^m p_{BA}(\lambda)$.

(37) Let $\phi : \mathbb{R}^n \longrightarrow \mathbb{R}$ be a quadratic form defined by $\phi(\boldsymbol{x}) = \boldsymbol{x}' A \boldsymbol{x}$ for $\boldsymbol{x} \in \mathbb{R}^n$, where $A \in \mathbb{R}^{n\times n}$ is a symmetric matrix. Prove that if the unit vector $\boldsymbol{x}$ is an extreme for ϕ, then $\boldsymbol{x}$ is an eigenvector for A.

Solution: Since $\boldsymbol{x}$ is a unit vector, we have $\boldsymbol{x}'\boldsymbol{x} - 1 = 0$, so the Lagrangian of this problem is

$$\Lambda(\boldsymbol{x}, \lambda) = \boldsymbol{x}' A \boldsymbol{x} + \lambda(\boldsymbol{x}'\boldsymbol{x} - 1).$$

We have

$$(\nabla\Lambda)(\boldsymbol{x}) = 2A\boldsymbol{x} + 2\lambda\boldsymbol{x} = \mathbf{0}_m,$$

which shows that $\boldsymbol{x}$ must be a unit eigenvector of A.

(38) Let P_ϕ be a permutation matrix, where $\phi \in PERM_n$ is an n-cyclic permutation. If z is a root of order n of 1 (that is, $z^n = 1$), prove that $(1, z, z^2, \ldots, z^{n-1})'$ is an eigenvector of P_ϕ. Further, prove that the eigenvalues of the matrix $P_\phi + P_{\phi^{-1}}$ have the form $2\cos\frac{2\pi k}{n}$ for $1 \leqslant k \leqslant n$.

(39) Let $A \in \mathbb{R}^{n\times n}$ be a matrix. A vector $\boldsymbol{x}$ is said to be *subharmonic* for A if $\boldsymbol{x} \geqslant \mathbf{0}_n$ and $A\boldsymbol{x} \geqslant a\boldsymbol{x}$ for some $a \in \mathbb{R}$. Prove that if $\boldsymbol{x}$ is an eigenvector of A, then $\mathsf{abs}(\boldsymbol{x})$ is a subharmonic vector for $\mathsf{abs}(A)$.

(40) Let $A \in \mathbb{R}^{n\times n}$ be a matrix that can be written as $A = \sum_{i=1}^{r} \boldsymbol{v}_i\boldsymbol{v}_i'$, where $\boldsymbol{v}_i \in \mathbb{R}^n - \{\mathbf{0}_n\}$ and the vectors $\boldsymbol{v}_1, \ldots, \boldsymbol{v}_r$ are pairwise orthogonal. Prove that
(a) A is a symmetric matrix of rank r;
(b) the eigenpairs of A that involve non-zero eigenvalues are $(\sqrt{\|\boldsymbol{v}_i\|}, \boldsymbol{v}_i)$ for $1 \leqslant i \leqslant r$.

(41) Let $A \in \mathbb{C}^{n\times n}$ be a Hermitian matrix and let $r = \max\{\sum_{i=1}^{n} |a_{ij}| \mid 1 \leqslant i \leqslant n\}$. Prove that $\mathsf{spec}(A) \subseteq [-r, r]$.

(42) Let $K_{nn} \in \mathbb{R}^{n^2\times n^2}$ be the commutation matrix introduced in Section 3.17. Prove that $\mathsf{spec}(K_{nn}) = \{-1, 1\}$, $\mathsf{algm}(K_{nn}, 1) = \frac{n(n+1)}{2}$, and $\mathsf{algm}(K_{nn}, -1) = \frac{n(n-1)}{2}$. Also, show that $\det(K_{nn}) = (-1)^{\frac{n(n-1)}{2}}$.

Solution: Since the real matrix K_{nn} is orthogonal and symmetric, $\mathsf{spec}(K_{nn}) = \{-1, 1\}$ by Example 7.7. Therefore, $\mathsf{algm}(K_{nn}, 1) + \mathsf{algm}(K_{nn}, -1) = n^2$ and $\det(K_{nn}) = (-1)^{\mathsf{algm}(K_{nn}, -1)}$.
On the other hand, by Part (e) of Supplement 96, $\mathsf{trace}(K_{nn}) = n = \mathsf{algm}(K_{nn}, 1) - \mathsf{algm}(K_{nn}, -1) = n^2 - 2\mathsf{algm}(A, -1)$. Therefore, $\mathsf{algm}(A, -1) = \frac{n(n-1)}{2}$ and $\mathsf{algm}(K_{nn}, 1) = \frac{n(n+1)}{2}$.
Since $\det(K_{nn})$ equals the product of eigenvalues, the last equality follows immediately.

Bibliographical Comments

Lanczos decomposition presented in Supplement 40 was obtained from [98]; the elementary solution given was developed in [148].

The result discussed in Supplement 7.10 was obtained from [8].

Chapter 8

Similarity and Spectra

8.1 Introduction

This chapter presents the links between spectral properties of matrices and conditions that ensure the existence of diagonal matrices which are equivalent to matrices that enjoy these properties. Further spectral properties are discussed such as the relationships between the geometric and algebraic multiplicities of eigenvalues. The standard Jordan form of matrices is given in the context of λ-matrices and the link between matrix norms and eigenvalues is also examined.

8.2 Diagonalizable Matrices

Diagonalizable matrices were defined in Section 3.12 as square matrices that are similar to diagonal matrices. In the context of spectral theory, we have the following characterization of diagonalizable matrices.

Theorem 8.1. *A matrix $A \in \mathbb{C}^{n\times n}$ is diagonalizable if and only if there exists a linearly independent set $\{\boldsymbol{v}_1, \ldots, \boldsymbol{v}_n\}$ of n eigenvectors of A.*

Proof. Let $A \in \mathbb{C}^{n\times n}$ be such that there exists a set $\{\boldsymbol{v}_1, \ldots, \boldsymbol{v}_n\}$ of n eigenvectors of A that is linearly independent and let P be the matrix $(\boldsymbol{v}_1\ \boldsymbol{v}_2\ \cdots\ \boldsymbol{v}_n)$ that is clearly invertible. We have

$$P^{-1}AP = P^{-1}(A\boldsymbol{v}_1\ A\boldsymbol{v}_2\ \cdots\ A\boldsymbol{v}_n) = P^{-1}(\lambda_1\boldsymbol{v}_1\ \lambda_2\boldsymbol{v}_2\ \cdots\ \lambda_n\boldsymbol{v}_n)$$

$$= P^{-1}P\begin{pmatrix} \lambda_1 & 0 & \cdots & 0 \\ 0 & \lambda_2 & \cdots & 0 \\ \vdots & \vdots & \cdots & \vdots \\ 0 & 0 & \cdots & \lambda_n \end{pmatrix} = \begin{pmatrix} \lambda_1 & 0 & \cdots & 0 \\ 0 & \lambda_2 & \cdots & 0 \\ \vdots & \vdots & \cdots & \vdots \\ 0 & 0 & \cdots & \lambda_n \end{pmatrix}.$$

Therefore, we have $A = PDP^{-1}$, where

$$D = \begin{pmatrix} \lambda_1 & 0 & \cdots & 0 \\ 0 & \lambda_2 & \cdots & 0 \\ \vdots & \vdots & \cdots & \vdots \\ 0 & 0 & \cdots & \lambda_n \end{pmatrix},$$

so $A \sim D$.

Conversely, suppose that A is diagonalizable, so $AP = PD$, where D is a diagonal matrix and P is an invertible matrix, and let $\boldsymbol{v}_1, \ldots, \boldsymbol{v}_n$ be the columns of the matrix P. We have $A\boldsymbol{v}_i = d_{ii}\boldsymbol{v}_i$ for $1 \leqslant i \leqslant n$, so each $\boldsymbol{v}_i$ is an eigenvector of A. Since P is invertible, its columns are linear independent. □

Theorem 8.1 can be restated by saying that a matrix is diagonalizable if and only if it is non-defective.

Corollary 8.1. *If $A \in \mathbb{C}^{n\times n}$ is diagonalizable, then the columns of any matrix P such that $D = P^{-1}AP$ is a diagonal matrix are eigenvectors of A. Furthermore, the diagonal entries of D are the eigenvalues that correspond to the columns of P.*

Proof. This statement follows from the proof of Theorem 8.1. □

Corollary 8.2. *If $A \in \mathbb{C}^{n\times n}$ is a matrix such that $\sum\{\mathsf{geomm}(A, \lambda) \mid \lambda \in \mathsf{spec}(A)\} = n$, then A is diagonalizable.*

Proof. Suppose that $\mathsf{spec}(A) = \{\lambda_1, \ldots, \lambda_k\}$ and let $B_k = \{\boldsymbol{v}^{k1}, \ldots, \boldsymbol{v}^{kp_k}\}$ be a basis of the invariant spaces S_{A,λ_k}, where $\sum_{k=1}^{p} p_k = n$. Then $B = \bigcup_{k=1}^{p} B_k$ is a linearly independent set of eigenvectors, so A is diagonalizable. □

Corollary 8.3. *If the eigenvalues of the matrix $A \in \mathbb{C}^{n\times n}$ are distinct, then A is diagonalizable.*

Proof. By Theorem 7.1, the set of eigenvectors of A is linearly independent. The statement follows immediately from Theorem 8.1. □

Theorem 8.2. *Let $A \in \mathbb{C}^{n\times n}$ be a block diagonal matrix,*

$$A = \begin{pmatrix} A_{11} & O & \cdots & O \\ O & A_{22} & \cdots & O \\ \vdots & \vdots & \cdots & \vdots \\ O & O & \cdots & A_{mm} \end{pmatrix}.$$

A is diagonalizable if and only if every matrix A_{ii} is diagonalizable for $1 \leqslant i \leqslant m$.

Proof. Suppose that A is a block diagonal matrix which is diagonalizable. Furthermore, suppose that $A_{ii} \in \mathbb{C}^{n_i\times n_i}$ and $\sum_{i=1}^{m} n_i = n$. There exists an invertible matrix $P \in \mathbb{C}^{n\times n}$ such that $P^{-1}AP$ is a diagonal matrix $D = \mathsf{diag}(\lambda_1, \ldots, \lambda_n)$. Let $\boldsymbol{p}_1, \ldots, \boldsymbol{p}_n$ be the columns of P, which are eigenvectors of A. Each vector $\boldsymbol{p}_i$ is divided into m blocks $\boldsymbol{p}_i^j$ with $1 \leqslant j \leqslant m$, where $\boldsymbol{p}_i^j \in \mathbb{C}^{n_j}$. Thus, P can be written as

$$P = \begin{pmatrix} \boldsymbol{p}_1^1 & \boldsymbol{p}_2^1 & \cdots & \boldsymbol{p}_n^1 \\ \boldsymbol{p}_1^2 & \boldsymbol{p}_2^2 & \cdots & \boldsymbol{p}_n^2 \\ \vdots & \vdots & \cdots & \vdots \\ \boldsymbol{p}_1^m & \boldsymbol{p}_2^m & \cdots & \boldsymbol{p}_n^m \end{pmatrix}.$$

The equality $A\boldsymbol{p}_i = \lambda_i \boldsymbol{p}_i$ can be expressed as

$$\begin{pmatrix} A_{11} & O & \cdots & O \\ O & A_{22} & \cdots & O \\ \vdots & \vdots & \cdots & \vdots \\ O & O & \cdots & A_{mm} \end{pmatrix} \begin{pmatrix} \boldsymbol{p}_i^1 \\ \boldsymbol{p}_i^2 \\ \vdots \\ \boldsymbol{p}_i^m \end{pmatrix} = \lambda_i \begin{pmatrix} \boldsymbol{p}_i^1 \\ \boldsymbol{p}_i^2 \\ \vdots \\ \boldsymbol{p}_i^m \end{pmatrix},$$

which shows that $A_{jj}\boldsymbol{p}_i^j = \lambda_i \boldsymbol{p}_i^j$ for $1 \leqslant j \leqslant m$. Let $M^j = (\boldsymbol{p}_1^j \ \boldsymbol{p}_2^j \ \cdots \ \boldsymbol{p}_n^j) \in \mathbb{C}^{n_j\times n}$. We claim that $\mathsf{rank}(M^j) = n_j$. Indeed, if $\mathsf{rank}(M^j)$ were less than n_j, we would have fewer that n independent rows M^j for $1 \leqslant j \leqslant m$. This, however, would imply that the rank of P is less than n, which contradicts the invertibility of P. Since there are n_j linearly independent eigenvectors of A_{jj}, it follows that each block A_{jj} is diagonalizable.

Conversely, suppose that each A_{jj} is diagonalizable, that is, there exists an invertible matrix Q_j such that $Q_j^{-1}A_{jj}Q_j$ is a diagonal matrix. Then, it is immediate to verify that the block diagonal matrix

$$Q = \begin{pmatrix} Q_1 & O & \cdots & O \\ O & Q_2 & \cdots & O \\ \vdots & \vdots & \cdots & \vdots \\ O & O & \cdots & Q_m \end{pmatrix}$$

is invertible and $Q^{-1}AQ$ is a diagonal matrix. □

Definition 8.1. Two matrices $A, B \in \mathbb{C}^{n\times n}$ are *simultaneously diagonalizable* if there exists a matrix $T \in \mathbb{C}^{n\times n}$ such that $A = TDT^{-1}$ and $B = TET^{-1}$, where D and E are diagonal matrices in $\mathbb{C}^{n\times n}$.

Lemma 8.1. *If $C \in \mathbb{C}^{n\times n}$ is a diagonalizable matrix and $T \in \mathbb{C}^{n\times n}$ is an invertible matrix, then TCT^{-1} is also diagonalizable.*

Proof. Since C is diagonalizable, there exists a diagonal matrix D and an invertible matrix S such that $SCS^{-1} = D$, so $C = S^{-1}DS$. Then,

$$TCT^{-1} = TS^{-1}DST^{-1} = (TS^{-1})D(TS^{-1})^{-1}.$$

Therefore,

$$(TS^{-1})^{-1}(TCT^{-1})(TS^{-1}) = D,$$

which shows that TCT^{-1} is diagonalizable. □

Theorem 8.3. *Let $A, B \in \mathbb{C}^{n\times n}$ be two diagonalizable matrices. Then A and B are simultaneously diagonalizable if and only if $AB = BA$.*

Proof. Suppose that A and B are simultaneously diagonalizable, so $A = TDT^{-1}$ and $B = TET^{-1}$, where D and E are diagonal matrices. Then $AB = TDT^{-1}TET^{-1} = TDET^{-1}$ and $BA = TET^{-1}TDT^{-1} = TEDT^{-1}$. Since any two diagonal matrices commute, we have $AB = BA$.

Conversely, suppose that $AB = BA$. Since A is diagonalizable, there exists a matrix T such that

$$TAT^{-1} = \begin{pmatrix} \lambda_1 I_{k_1} & O_{k_1,k_2} & \cdots & O_{k_1,k_m} \\ O_{k_2,k_1} & \lambda_2 I_{k_2} & \cdots & O_{k_2,k_m} \\ \vdots & \vdots & \cdots & \vdots \\ O_{k_m,k_1} & O_{k_m,k_2} & \cdots & \lambda_m I_{k_m} \end{pmatrix},$$

where k_i is the multiplicity of λ_i for $1 \leqslant i \leqslant m$. It is easy to see that if A and B commute, then TAT^{-1} and TBT^{-1} also commute. If we write TBT^{-1} as a block matrix $TBT^{-1} = (B_{pq})$, where $B_{pq} \in \mathbb{C}^{k_p \times k_q}$ for $1 \leqslant p, q \leqslant m$, then the fact that the matrices TAT^{-1} and TBT^{-1} commute translates into the equality

$$\begin{pmatrix} \lambda_1 B_{11} & \lambda_1 B_{12} & \cdots & \lambda_1 B_{1m} \\ \lambda_2 B_{21} & \lambda_2 B_{22} & \cdots & \lambda_2 B_{2m} \\ \vdots & \vdots & \cdots & \vdots \\ \lambda_m B_{m1} & \lambda_m B_{m1} & \cdots & \lambda_m B_{mm} \end{pmatrix} = \begin{pmatrix} \lambda_1 B_{11} & \lambda_2 B_{12} & \cdots & \lambda_m B_{1m} \\ \lambda_1 B_{21} & \lambda_2 B_{22} & \cdots & \lambda_m B_{2m} \\ \vdots & \vdots & \cdots & \vdots \\ \lambda_1 B_{m1} & \lambda_2 B_{m1} & \cdots & \lambda_m B_{mm} \end{pmatrix}.$$

Thus, if $\lambda_i \neq \lambda_j$ we have $B_{ij} = O_{k_i,k_j}$, which shows that the matrix TBT^{-1} is a block diagonal matrix

$$TBT^{-1} = \begin{pmatrix} B_{11} & O_{k_1,k_2} & \cdots & O_{k_1,k_m} \\ O_{k_2,k_1} & B_{22} & \cdots & O_{k_2,k_m} \\ \vdots & \vdots & \cdots & \vdots \\ O_{k_m,k_1} & O_{k_m,k_2} & \cdots & B_{mm} \end{pmatrix}.$$

Since B is diagonalizable, it follows that TBT^{-1} is diagonalizable (by Lemma 8.1), so each matrix B_{jj} is diagonalizable. Let W_i be a matrix such that $W_i^{-1} B_{ii} W_i$ is diagonal and let W be the block diagonal matrix

$$W = \begin{pmatrix} W_1 & O & \cdots & O \\ O & W_2 & \cdots & O \\ \vdots & \vdots & \cdots & \vdots \\ O & O & \cdots & W_m \end{pmatrix}.$$

Then, both $W^{-1}TAT^{-1}W$ and $W^{-1}TBT^{-1}W$ are diagonal matrices, and this implies that A and B are simultaneously diagonalizable. □

It is interesting to observe that if $A \in \mathbb{C}^{n\times n}$ is a diagonalizable matrix, then there exists a matrix $B \in \mathbb{C}^{n\times n}$ such that $B^2 = A$. Indeed, suppose that $A \sim D$, where D is a diagonal matrix and, therefore, $A = PDP^{-1}$, where P is an invertible matrix and $D = \mathsf{diag}(d_1, \ldots, d_n)$. Consider the matrix $E \in \mathbb{C}^{n\times n}$ given by $E = \mathsf{diag}(\sqrt{d_1}, \ldots, \sqrt{d_n})$ for which we have $E^2 = D$, and define $B = PEP^{-1}$. Then, we have

$$B^2 = PEP^{-1}PEP^{-1} = PE^2P^{-1} = PDP^{-1} = A.$$

We refer to B as a *square root of* A. Note that for the existence of the square root of a diagonalizable matrix it is essential that we deal with matrices with complex entries. Under certain conditions, square roots exist also for matrices with real entries, as we shall see in Section 8.12.

8.3 Matrix Similarity and Spectra

Theorem 8.4. *If $A, B \in \mathbb{C}^{n\times n}$ and $A \sim B$, then the two matrices have the same characteristic polynomials and, therefore,* $\mathsf{spec}(A) = \mathsf{spec}(B)$.

Proof. Since $A \sim B$, there exists an invertible matrix X such that $A = XBX^{-1}$. Then, the characteristic polynomial $\det(A - \lambda I_n)$ can be rewritten as

$$\begin{aligned}\det(A - \lambda I_n) &= \det(XBX^{-1} - \lambda X I_n X^{-1})\\ &= \det(X(B - \lambda I_n)X^{-1})\\ &= \det(X)\det(B - \lambda I_n)\det(X^{-1})\\ &= \det(B - \lambda I_n),\end{aligned}$$

which implies $\mathsf{spec}(A) = \mathsf{spec}(B)$. □

Theorem 8.5. *If $A, B \in \mathbb{C}^{n\times n}$ and $A \sim B$, then* $\mathsf{trace}(A) = \mathsf{trace}(B)$.

Proof. Since the two matrices are similar, they have the same characteristic polynomials, so both $\mathsf{trace}(A)$ and $\mathsf{trace}(B)$ equal $-c_1$, where c_1 is the coefficient of λ^{n-1} in both $p_A(\lambda)$ and $p_B(\lambda)$. □

Theorem 8.6. *If $A \sim_u B$, where $A, B \in \mathbb{C}^{n\times n}$, then the Frobenius norm of these matrices are equal, that is, $\|A\|_F = \|B\|_F$.*

Proof. We have shown in Theorem 3.43 that $A^{\mathsf{H}}A \sim_u B^{\mathsf{H}}B$. Therefore, these matrices have the same characteristic polynomials which allows us to infer that $\mathsf{trace}(A^{\mathsf{H}}A) \sim_u \mathsf{trace}(B^{\mathsf{H}}B)$. By the analogue of the Equality (6.11) for complex matrices, we obtain the desired conclusion. □

Theorem 8.7. *Let $A \in \mathbb{C}^{n\times n}$ and $B \in \mathbb{C}^{k\times k}$ be two matrices. If there exists a matrix $U \in \mathbb{C}^{n\times k}$ having an orthonormal set of columns such that $AU = UB$, then there exists $V \in \mathbb{C}^{n\times(n-k)}$ such that $(U\,V) \in \mathbb{C}^{n\times n}$ is a unitary matrix and*

$$(U\,V)^{\mathsf{H}}A(U\,V) = \begin{pmatrix} B & (U^{\mathsf{H}}AV) \\ O & (V^{\mathsf{H}}AV) \end{pmatrix}.$$

Proof. Since U has an orthonormal set of columns, by Corollary 6.27, there exists $V \in \mathbb{C}^{n\times(n-k)}$ such that $(U\,V)$ is a unitary matrix. We have

$$U^{\mathsf{H}}AU = U^{\mathsf{H}}UB = I_kB = B,$$
$$V^{\mathsf{H}}AU = V^{\mathsf{H}}UB = OB = O,$$

which allows us to write

$$\begin{aligned}(U\,V)^{\mathsf{H}}A(U\,V) &= (U\,V)^{\mathsf{H}}(AU\ AV) = \begin{pmatrix} U^{\mathsf{H}} \\ V^{\mathsf{H}} \end{pmatrix}(AU\ AV) \\ &= \begin{pmatrix} U^{\mathsf{H}}AU & U^{\mathsf{H}}AV \\ V^{\mathsf{H}}AU & V^{\mathsf{H}}AV \end{pmatrix} = \begin{pmatrix} B & U^{\mathsf{H}}AV \\ O & V^{\mathsf{H}}AV \end{pmatrix}.\end{aligned}$$

□

Corollary 8.4. *Let $A \in \mathbb{C}^{n\times n}$ be a Hermitian matrix and $B \in \mathbb{C}^{k\times k}$ be a matrix.*

If there exists a matrix $U \in \mathbb{C}^{n\times k}$ having an orthonormal set of columns such that $AU = UB$, then there exists $V \in \mathbb{C}^{n\times(n-k)}$ such that $(U\,V)$ is a unitary matrix and

$$(U\,V)^{\mathsf{H}}A(U\,V) = \begin{pmatrix} B & O \\ O & V^{\mathsf{H}}AV \end{pmatrix}.$$

Proof. Since A is Hermitian, we have $U^{\mathsf{H}}AV = U^{\mathsf{H}}A^{\mathsf{H}}V = (V^{\mathsf{H}}AU)^{\mathsf{H}} = O$, which, by Theorem 8.7 produces the desired result. □

Corollary 8.5. *Let $A \in \mathbb{C}^{n\times n}$, λ be an eigenvalue of A, and let $\boldsymbol{u}$ be an eigenvector of A with $\|\boldsymbol{u}\| = 1$ that corresponds to λ. There exists $V \in \mathbb{C}^{n\times(n-1)}$ such that $(\boldsymbol{u}\ V) \in \mathbb{C}^{n\times n}$ is a unitary matrix and*

$$(\boldsymbol{u}\ V)^{\mathsf{H}}A(\boldsymbol{u}\ V) = \begin{pmatrix} \lambda & (\boldsymbol{u}^{\mathsf{H}}AV) \\ O & (V^{\mathsf{H}}AV) \end{pmatrix}.$$

If A is a Hermitian matrix, then

$$(\boldsymbol{u}\ V)^{\mathsf{H}}A(\boldsymbol{u}\ V) = \begin{pmatrix} \lambda & O \\ O & (V^{\mathsf{H}}AV) \end{pmatrix}.$$

Proof. This statement follows from Theorem 8.7 by taking $k = 1$. □

Theorem 8.8 (Schur's triangularization theorem). *Let $A \in \mathbb{C}^{n\times n}$ be a square matrix. There exists a unitary matrix $U \in \mathbb{C}^{n\times n}$ and an upper-triangular matrix $T \in \mathbb{C}^{n\times n}$ such that $A = U^{\mathsf{H}}TU$ and the diagonal elements of T are the eigenvalues of A. Moreover, each eigenvalue λ occurs in the sequence of diagonal values a number of* $\mathsf{algm}(A, \lambda)$ *times.*

Proof. The argument is by induction on $n \geqslant 1$. The base case, $n = 1$, is trivial. So, suppose that the statement is true for matrices in $\mathbb{C}^{(n-1)\times(n-1)}$.

Let $\lambda_1 \in \mathbb{C}$ be an eigenvalue of A, and let $\boldsymbol{u}$ be an eigenvector that corresponds to this eigenvalue. By Corollary 8.5, we have

$$U^{\mathsf{H}}AU = \begin{pmatrix} \lambda_1 & \boldsymbol{u}^{\mathsf{H}}AV \\ O & V^{\mathsf{H}}AV \end{pmatrix},$$

where $U = (\boldsymbol{u}|V)$ is an unitary matrix.

By the inductive hypothesis, since $V^{\mathsf{H}}AV \in \mathbb{C}^{(n-1)\times(n-1)}$, there exists a unitary matrix $S \in \mathbb{C}^{(n-1)\times(n-1)}$ such that $V^{\mathsf{H}}AV = S^{H}WS$, where W is an upper-triangular matrix. Then we have

$$U^{\mathsf{H}}AU = \begin{pmatrix} \lambda_1 & \boldsymbol{u}^{\mathsf{H}}VS^{\mathsf{H}}WS \\ \mathbf{0} & S^{\mathsf{H}}WS \end{pmatrix} = \begin{pmatrix} \lambda_1 & O \\ \mathbf{0} & W \end{pmatrix},$$

which shows that an upper triangular matrix T that is unitarily similar to A can be defined as

$$T = \begin{pmatrix} \lambda_1 & O \\ \mathbf{0} & W \end{pmatrix}.$$

Since $T \sim_u A$, it follows that the two matrices have the same characteristic polynomials and, therefore, the same spectra and algebraic multiplicities for each eigenvalue. □

Corollary 8.6. *If $A \in \mathbb{R}^{n\times n}$ is a matrix such that $\mathsf{spec}(A) = \{0\}$, then A is nilpotent.*

Proof. By Schur's Triangularization Theorem, A is unitarily similar to a strictly upper triangular matrix, $A = U^{\mathsf{H}}TU$, so $A^n = U^{\mathsf{H}}T^nU$. By Supplement 30 of Chapter 3, we have $T^n = O$, so $A^n = O$. □

Example 8.1. Let $A \in \mathbb{R}^{3\times 3}$ be the symmetric matrix

$$A = \begin{pmatrix} 14 & -10 & -2 \\ -10 & -5 & 5 \\ -2 & 5 & 11 \end{pmatrix}$$

whose characteristic polynomial is

$$p_A(\lambda) = \lambda^3 - 20\lambda^2 - 100\lambda + 2000.$$

The eigenvalues of A are $\lambda_1 = 20, \lambda_2 = 10$, and $\lambda_3 = -10$.

It is easy to see that

$$\boldsymbol{v}_1 = \begin{pmatrix} -2 \\ 1 \\ 1 \end{pmatrix}, \boldsymbol{v}_2 = \begin{pmatrix} 1 \\ 0 \\ 2 \end{pmatrix}, \boldsymbol{v}_3 = \begin{pmatrix} -2 \\ -5 \\ 1 \end{pmatrix}$$

are eigenvectors that correspond to the eigenvalues λ_1, λ_2, and λ_3, respectively. The corresponding unit vectors are

$$\boldsymbol{u}_1 = \begin{pmatrix} -\frac{2}{\sqrt{6}} \\ \frac{1}{\sqrt{6}} \\ \frac{1}{\sqrt{6}} \end{pmatrix}, \boldsymbol{u}_2 = \begin{pmatrix} \frac{1}{\sqrt{5}} \\ 0 \\ \frac{2}{\sqrt{5}} \end{pmatrix}, \boldsymbol{u}_3 = \begin{pmatrix} -\frac{2}{\sqrt{30}} \\ -\frac{5}{\sqrt{30}} \\ \frac{1}{\sqrt{30}} \end{pmatrix}.$$

For $U = (\boldsymbol{u}_1\ \boldsymbol{u}_2\ \boldsymbol{u}_3)$, we have

$$U'AU = U'(20\boldsymbol{u}_1\ 10\boldsymbol{u}_2\ \ -10\boldsymbol{u}_3) = \mathsf{diag}(20, 10, -10).$$

Corollary 8.7. *Let $A \in \mathbb{C}^{n\times n}$ and let f be a polynomial. If* $\mathsf{spec}(A) = \{\lambda_1, \ldots, \lambda_n\}$ *(including multiplicities), then* $\mathsf{spec}(f(A)) = \{f(\lambda_1), \ldots, f(\lambda_n)\}$.

Proof. By Schur's Triangularization Theorem, there exists a unitary matrix $U \in \mathbb{C}^{n\times n}$ and an upper triangular matrix $T \in \mathbb{C}^{n\times n}$ such that $A = U^{\mathsf{H}}TU$ and the diagonal elements of T are the eigenvalues of A, $\lambda_1, \ldots, \lambda_n$. Therefore, $Uf(A)U^{-1} = f(T)$, and by Theorem 3.45, the diagonal elements of $f(T)$ are $f(\lambda_1), \ldots, f(\lambda_m)$. Since $f(A) \sim f(T)$, we obtain the desired conclusion because two similar matrices have the same eigenvalues with the same algebraic multiplicities. □

Let $r, s \in \mathbb{N}$ be two numbers such that $1 \leqslant r, s \leqslant n$, $r \neq s$, and let $E_{rs} \in \mathbb{C}^{n\times n}$ be the matrix whose unique non-zero entry is $e_{rs} = 1$. In other words, E_{rs} is defined by

$$(E_{rs})_{ij} = \begin{cases} 1 & \text{if } i = r \text{ and } j = s, \\ 0 & \text{otherwise,} \end{cases}$$

for $1 \leqslant i, j \leqslant n$. Note that

$$E_{rs}\boldsymbol{e}_p = \begin{cases} \boldsymbol{e}_r & \text{if } p = s, \\ \boldsymbol{0} & \text{otherwise,} \end{cases}$$

for $1 \leqslant p \leqslant n$, and

$$\boldsymbol{e}_q' E_{rs} = \begin{cases} \boldsymbol{e}_s' & \text{if } q = r, \\ \boldsymbol{0} & \text{otherwise,} \end{cases}$$

for $1 \leqslant q \leqslant n$.

Lemma 8.2. *The matrix $I_n + cE_{rs}$ is nonsingular, and $(I_n + cE_{rs})^{-1} = I_n - cE_{rs}$. Furthermore, for every $A \in \mathbb{C}^{n\times n}$, if $B = (I_n - cE_{rs})A(I_n + cE_{rs})$, then $B = A - cE_{rs}A + cAE_{rs} - c^2 a_{sr}E_{rs}$.*

Proof. Since $r \neq s$, we have $(E_{rs})^2 = O$. Therefore, $(I_n + cE_{rs})(I_n - cE_{rs}) = I_n$, so $(I_n + cE_{rs})^{-1} = I_n - cE_{rs}$. Also, $E_{rs}AE_{rs}$ is the matrix that has a unique non-zero element, that is,

$$E_{rs}AE_{rs} = a_{sr}E_{rs}.$$

Since $B = (I_n - cE_{rs})A(I_n + cE_{rs})$, we have

$$\begin{aligned} B &= A - cE_{rs}A + cAE_{rs} - c^2 E_{rs}AE_{rs} \\ &= A - cE_{rs}A + cAE_{rs} - c^2 a_{sr}E_{rs}. \end{aligned}$$

□

Lemma 8.3. *If $A \in \mathbb{C}^{n\times n}$ is an upper triangular matrix, $1 \leqslant r < s \leqslant n$, and $B = (I_n - cE_{rs})A(I_n + cE_{rs})$, then $b_{ij} \neq a_{ij}$ only if $1 \leqslant i \leqslant r$ and $j = s$, or if $i = r$ and $s \leqslant j \leqslant n$ (see Figure* 8.1*, where the elements of B that differ from the corresponding elements of A are represented by thick lines) and $b_{rs} = a_{rs} + c(a_{rr} - a_{ss})$.*

Proof. Since A is an upper triangular matrix and $r < s$, we have $a_{sr} = 0$, so by Lemma 8.2,

$$B = A - cE_{rs}A + cAE_{rs}.$$

Since A is an upper triangular matrix, we have

$$E_{rs}A = \begin{pmatrix} 0 & 0 & \cdots\cdots & 0 & \cdots & 0 \\ \vdots & \vdots & \cdots & \vdots & \cdots & \vdots \\ 0 & 0 & \cdots\cdots & a_{ss} & \cdots & a_{sn} \\ \vdots & \vdots & \cdots & \vdots & \cdots & \vdots \\ 0 & 0 & \cdots\cdots & 0 & \cdots & 0 \end{pmatrix},$$

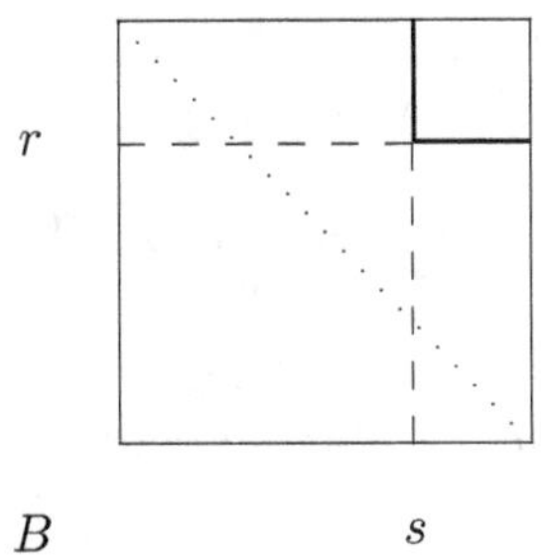

Fig. 8.1 Elements of B that differ from the corresponding elements of A.

where $a_{ss}, \ldots, a_{sn}$ occur in the rth row of $E_{rs}A$, and

$$AE_{rs} = \begin{pmatrix} 0 & 0 & \cdots & a_{1r} & \cdots & 0 \\ \vdots & \vdots & \cdots & \vdots & \cdots & \vdots \\ 0 & 0 & \cdots & a_{rr} & \cdots & 0 \\ 0 & 0 & \cdots & 0 & \cdots & 0 \\ \vdots & \vdots & \cdots & \vdots & \cdots & \vdots \\ 0 & 0 & \cdots & 0 & \cdots & 0 \end{pmatrix},$$

where $a_{1r}, \ldots, a_{rr}$ occur in the sth column. Thus, the elements of B that differ from those of A are located on the rth row at the right of the sth column and on the sth row above the rth row. Also, $b_{rs} = a_{rs} + c(a_{rr} - a_{ss})$. □

Lemma 8.2 implies that if $a_{rr} \neq a_{ss}$, by taking $c = a_{rr}/(a_{ss} - a_{rr})$ we have $b_{rs} = 0$.

Theorem 8.9. *Let $A \in \mathbb{C}^{n\times n}$ be a square matrix having k distinct eigenvalues $\lambda_1, \ldots, \lambda_k$ such that $\mathsf{algm}(\lambda_i) = n_i$ for $1 \leqslant i \leqslant k$ and $\sum_{i=1}^{k} n_i = n$.*

There exists a block-triangular matrix $T \in \mathbb{C}^{n\times n}$ such that $A \sim_u T$,

$$T = \begin{pmatrix} T_1 & O & \cdots & O \\ O & T_2 & \cdots & O \\ \vdots & \vdots & \cdots & \vdots \\ O & O & \cdots & T_k \end{pmatrix},$$

where each $T_i \in \mathbb{C}^{n_i\times n_i}$ is an upper triangular matrix having all its diagonal elements equal to λ_i for $1 \leqslant i \leqslant k$.

Proof. By Schur's Triangularization Theorem, there exists an upper-triangular matrix $V \in \mathbb{C}^{n\times n}$ such that $A \sim V$ and the diagonal elements of V are the eigenvalues of A such that each eigenvalue λ_i occurs in the sequence of diagonal values a number of $\mathsf{algm}(A, \lambda_i)$ times.

By Lemmas 8.2 and 8.3, we can construct a matrix T, similar to V and, therefore, similar to A such that T is a block triangular matrix.

To build the matrix T, we construct a sequence of matrices starting from V by zeroing elements situated above the main diagonal, outside the triangular blocks that correspond to each eigenvalue. This process must be conducted such that the creation of a new zero component will not disturb the already zeroed elements. This can be easily done by numbering the positions to be zeroes in such a manner that all numbers located at the right and above the position numbered k are greater than k, as we show in Example 8.2. □

Example 8.2. Let $A \in \mathbb{C}^{8\times 8}$ be a matrix whose distinct eigenvalues are λ_1, λ_2, and λ_3 such that $\mathsf{algm}(A, \lambda_1) = \mathsf{algm}(A, \lambda_2) = 3$ and $\mathsf{algm}(A, \lambda_3) = 2$. In Figure 8.2, we show the order of zeroing the desired elements of the upper triangular matrix in order to construct a block triangular matrix.

Theorem 8.8 shows that a Schur factorization $A = U^{\mathsf{H}}TU$ exists for every complex matrix A. The next statement presents a property of real matrices that admit real Schur factorizations.

Theorem 8.10. *Let $A \in \mathbb{R}^{n\times n}$ be a real square matrix. If there exists a orthogonal matrix $U \in \mathbb{R}^{n\times n}$ and an upper triangular matrix $T \in \mathbb{R}^{n\times n}$ such that $A = U^{-1}TU$, that is, a real Schur factorization, then the eigenvalues of A are real numbers.*

Proof. If the above factorization exists, we have $T = UAU^{-1}$. Thus, the eigenvalues of A are the diagonal components of T and, therefore, they are real numbers. □

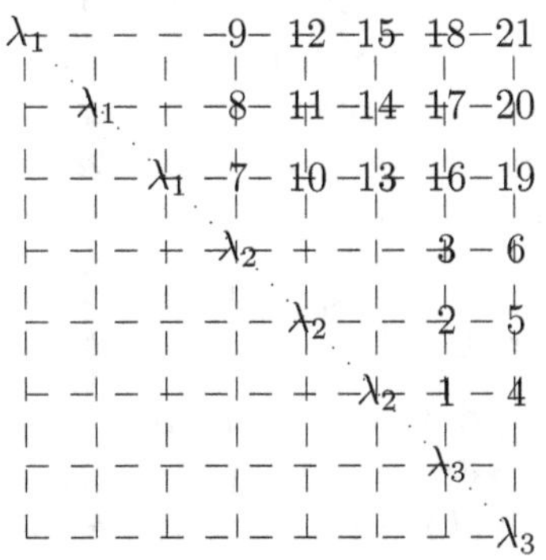

Fig. 8.2 Order of nullification of elements of a block triangular matrix.

Theorem 8.11. *Let $A \in \mathbb{R}^{n \times n}$ be a matrix having n real eigenvalues. Then A has a real Schur factorization, that is, $A = UTU^{-1}$, where $U \in \mathbb{R}^{n \times n}$ is an orthogonal matrix and $T \in \mathbb{R}^{n \times n}$ is an upper triangular matrix that has the eigenvalues of A on its diagonal.*

Proof. The argument is a paraphrase of the argument of Theorem 8.4, taking into account that the eigenvalues are real numbers. □

Corollary 8.8. *If $A \in \mathbb{R}^{n \times n}$ and A is a symmetric matrix, then A is orthogonally diagonalizable.*

Proof. By Corollary 7.5, A has real eigenvalues. Theorem 8.11 implies the existence of a real Schur factorization of A, $A = UTU^{-1}$, where U is an orthogonal matrix and T is both upper triangular and symmetric and, therefore, a diagonal matrix. □

Corollary 8.9. *If $A \in \mathbb{R}^{n \times n}$ is a symmetric matrix, then $\|A\|_F = \sum_{i=1}^{n} \lambda_i^2$, where $\mathsf{spec}(A) = \{\lambda_1, \ldots, \lambda_n\}$.*

Proof. By Corollary 8.8, $A \sim_u D$, where $D = \mathsf{diag}(\lambda_1, \ldots, \lambda_n)$. Therefore, by Theorem 8.6 we have $\|A\|_F = \|D\|_F = \sum_{i=1}^{n} \lambda_i^2$. □

Theorem 8.12. *If $A \in \mathbb{R}^{n \times n}$ is a symmetric matrix, then $\mathsf{geomm}(A, \lambda) = \mathsf{algm}(A, \lambda)$ for every eigenvalue $\lambda \in \mathsf{spec}(A)$.*

Proof. By Corollary 8.8, A is orthonormally diagonalizable, that is, there exists an orthogonal matrix P such that $A = PDP'$, where D is a diagonal matrix. Therefore, $\det(\lambda I_n - A) = \det(\lambda I_n - D)$. In other words, the matrices $\lambda I_n - A$ and $\lambda I_n - D$ have the same characteristic polynomial. Thus, if $\lambda \in \mathsf{spec}(A)$ and $\mathsf{algm}(A, \lambda) = k$, λ occurs in k diagonal positions of D, $p_1, \ldots, p_k$. Thus, $\boldsymbol{e}_{p_1}, \ldots, \boldsymbol{e}_{p_k}$ are k linearly independent eigenvectors in the invariant subspace $S_{D,\lambda}$, which implies that the vectors $P\boldsymbol{e}_{p_1}, \ldots, P\boldsymbol{e}_{p_k}$ are k linearly independent eigenvectors in $S(A, \lambda)$. □

Next, we show that unitary diagonalizability is a characteristic property of normal matrices.

Theorem 8.13 (Spectral theorem for normal matrices). *A matrix $A \in \mathbb{C}^{n \times n}$ is normal if and only if there exists a unitary matrix U and a diagonal matrix D such that $A = U^{\mathsf{H}} DU$, the columns of U^{H} are unit eigenvectors, and the diagonal elements of D are the eigenvalues of A that correspond to these eigenvectors.*

Proof. Suppose that A is a normal matrix. By Schur's triangularization theorem, there exists a unitary matrix $U \in \mathbb{C}^{n\times n}$ and an upper triangular matrix $T \in \mathbb{C}^{n\times n}$ such that $A = U^{-1}TU$. Thus, $T = UAU^{-1} = UAU^{\mathsf{H}}$ and $T^{\mathsf{H}} = UA^{\mathsf{H}}U^{\mathsf{H}}$. Therefore,

$$\begin{aligned} T^{\mathsf{H}}T &= UA^{\mathsf{H}}U^{\mathsf{H}}UAU^{\mathsf{H}} \\ &= UA^{\mathsf{H}}AU^{\mathsf{H}} \\ &\quad \text{(because } U \text{ is unitary)} \\ &= UAA^{\mathsf{H}}U^{\mathsf{H}} \\ &\quad \text{(because } A \text{ is normal)} \\ &= UAU^{\mathsf{H}}UA^{\mathsf{H}}U^{\mathsf{H}} \\ &= TT^{\mathsf{H}}, \end{aligned}$$

so T is a normal matrix. By Theorem 3.19, T is a diagonal matrix, so D's role is played by T.

We leave the proof of the converse implication to the reader. □

Corollary 8.10. *Let $A \in \mathbb{C}^{n\times n}$ be a normal matrix and let $A = U^{\mathsf{H}}DU$ be the factorization whose existence was established in Theorem 8.13.*

If $U^{\mathsf{H}} = (\boldsymbol{u}_1, \ldots, \boldsymbol{u}_n)$ and $D = \mathsf{diag}(\lambda_1, \ldots, \lambda_n)$, where $\lambda_1, \ldots, \lambda_n$ are the eigenvalues of A, then each column vector $\boldsymbol{u}_i$ is an eigenvector of A that corresponds to the eigenvalue λ_i for $1 \leqslant i \leqslant n$.

Proof. We have

$$\begin{aligned} A = U^{\mathsf{H}}DU &= (\boldsymbol{u}_1 \ \cdots \ \boldsymbol{u}_n) \begin{pmatrix} \lambda_1 & 0 & \cdots & 0 \\ 0 & \lambda_2 & \cdots & 0 \\ \vdots & \vdots & \cdots & \vdots \\ 0 & 0 & \cdots & \lambda_n \end{pmatrix} \begin{pmatrix} \boldsymbol{u}_1^{\mathsf{H}} \\ \vdots \\ \boldsymbol{u}_n^{\mathsf{H}} \end{pmatrix} \\ &= \lambda_1 \boldsymbol{u}_1 \boldsymbol{u}_1^{\mathsf{H}} + \cdots + \lambda_n \boldsymbol{u}_n \boldsymbol{u}_n^{\mathsf{H}}. \end{aligned} \tag{8.1}$$

Therefore, $A\boldsymbol{u}_i = \lambda_i \boldsymbol{u}_i$ for $1 \leqslant i \leqslant n$, which proves the statement. □

Equality (8.1) is referred to as *spectral decomposition* of the normal matrix A.

Theorem 8.14 (Spectral theorem for hermitian matrices). *If the matrix $A \in \mathbb{C}^{n\times n}$ is Hermitian or skew-Hermitian, A can be written as $A = U^{\mathsf{H}}DU$, where U is a unitary matrix and D is a diagonal matrix having the eigenvalues of A as its diagonal elements.*

Proof. This statement follows from Theorem 8.13 because any Hermitian or skew-Hermitian matrix is normal. □

Corollary 8.11. *The rank of a Hermitian matrix is equal to the number of non-zero eigenvalues.*

Proof. The statement of the corollary obviously holds for any diagonal matrix. If A is a Hermitian matrix, by Theorems 3.44 and 8.14, we have $\mathsf{rank}(A) = \mathsf{rank}(D)$, where D is a diagonal matrix having the eigenvalues of A as its diagonal elements. This implies the statement of the corollary. □

Let $A \in \mathbb{C}^{n\times n}$ be a Hermitian matrix of rank p. By Theorem 8.14, A can be written as $A = U^{\mathsf{H}}DU$, where U is a unitary matrix and $D = \mathsf{diag}(\lambda_1, \ldots, \lambda_p, 0, \ldots, 0)$, where $\lambda_1, \ldots, \lambda_p$ are the non-zero eigenvalues of A and $\lambda_1 \geqslant \cdots \geqslant \lambda_p > 0$. Thus, if $W \in \mathbb{C}^{p\times n}$ is a matrix that consists of the first p rows of U, we can write

$$
\begin{aligned}
A &= W' \begin{pmatrix} \lambda_1 & 0 & \cdots & 0 \\ 0 & \lambda_2 & \cdots & 0 \\ \vdots & \vdots & \vdots & \vdots \\ 0 & 0 & \cdots & \lambda_p \end{pmatrix} W \\
&= (\boldsymbol{u}'_1 \ \ldots \ \boldsymbol{u}'_p) \begin{pmatrix} \lambda_1 & 0 & \cdots & 0 \\ 0 & \lambda_2 & \cdots & 0 \\ \vdots & \vdots & \vdots & \vdots \\ 0 & 0 & \cdots & \lambda_p \end{pmatrix} \begin{pmatrix} \boldsymbol{u}_1 \\ \vdots \\ \boldsymbol{u}_p \end{pmatrix} \\
&= \lambda_1 \boldsymbol{u}'_1 \boldsymbol{u}_1 + \cdots + \lambda_p \boldsymbol{u}'_p \boldsymbol{u}_p.
\end{aligned}
$$

Note that if A is not Hermitian, $\mathsf{rank}(A)$ may differ from the number of non-zero eigenvalues. For example, the matrix

$$
A = \begin{pmatrix} 0 & 1 \\ 0 & 0 \end{pmatrix}
$$

has no non-zero eigenvalues. However, its rank is 1.

The spectral decomposition (8.1) of Hermitian matrices,

$$A = \lambda_1 \boldsymbol{u}_1 \boldsymbol{u}_1^{\mathsf{H}} + \cdots + \lambda_n \boldsymbol{u}_n \boldsymbol{u}_n^{\mathsf{H}},$$

allows us to extend functions of the form $f : \mathbb{R} \longrightarrow \mathbb{R}$ to Hermitian matrices. Since the eigenvalues of a Hermitian matrix are real numbers, it makes sense to define $f(A)$ as

$$f(A) = f(\lambda_1) \boldsymbol{u}_1 \boldsymbol{u}_1^{\mathsf{H}} + \cdots + f(\lambda_n) \boldsymbol{u}_n \boldsymbol{u}_n^{\mathsf{H}}.$$

In particular, if A is positive semi-definite, we have $\lambda_i \geqslant 0$ for $1 \leqslant i \leqslant n$ and we can define

$$\sqrt{A} = \sqrt{\lambda_1} \boldsymbol{u}_1 \boldsymbol{u}_1^{\mathsf{H}} + \cdots + \sqrt{\lambda_n} \boldsymbol{u}_n \boldsymbol{u}_n^{\mathsf{H}}.$$

Theorem 8.15. *Let $A, B \in \mathbb{C}^{n \times n}$ be two Hermitian matrices, where A is positive definite. There exists a matrix P such that $P^{\mathsf{H}} A P = I_n$ and $P^{\mathsf{H}} B P$ is a diagonal matrix.*

Proof. By Cholesky's Decomposition Theorem (Theorem 6.57), A can be factored as $A = R^{\mathsf{H}} R$, where R is an invertible matrix, so $(R^{\mathsf{H}})^{-1} A R^{-1} = I_n$. The matrix $C = (R^{\mathsf{H}})^{-1} B R^{-1}$ is Hermitian and, therefore, it is unitarily diagonalizable, that is, there exists a unitary matrix U such that $U^{\mathsf{H}} C U$ is diagonal. Since $U^{\mathsf{H}} I_n U = I_n$, we have $P = R^{-1} U$. □

Corollary 8.12. *Let $A, B \in \mathbb{C}^{n \times n}$. If $A \succ B \succ O$, then $A^{-1} \prec B^{-1}$.*

Proof. By Theorem 8.15, there exists a matrix P such that $A = P^{\mathsf{H}} P$ and $B = P^{\mathsf{H}} D P$, where D is a diagonal matrix, $D = \mathsf{diag}(d_1, \ldots, d_n)$. Thus, $\boldsymbol{x}^{\mathsf{H}} A \boldsymbol{x} > \boldsymbol{x}^{\mathsf{H}} B \boldsymbol{x}$ is equivalent to $\boldsymbol{y}^{\mathsf{H}} \boldsymbol{y} > \boldsymbol{y}^{\mathsf{H}} D \boldsymbol{y}$, where $\boldsymbol{y} = P \boldsymbol{x}$, so $A > B$ if and only if $d_i < 1$ for $1 \leqslant i \leqslant n$. Therefore, $A^{-1} = Q^{\mathsf{H}} Q$ and $B^{-1} = Q^{\mathsf{H}} E Q$, where $E = \mathsf{diag}(d_1^{-1}, \ldots, d_n^{-1})$ and $d_i^{-1} > 1$ for $1 \leqslant i \leqslant n$, which implies $A^{-1} < B^{-1}$. □

Definition 8.2. Let $A \in \mathbb{C}^{n \times n}$ be a Hermitian matrix. The triple $\mathfrak{I}(A) = (n_+(A), n_-(A), n_0(A))$, where $n_+(A)$ is the number of positive eigenvalues, $n_-(A)$ is the number of negative eigenvalues, and $n_0(A)$ is the number of zero eigenvalues, is the *inertia* of the matrix A.

The number $\mathsf{sig}(A) = n_+(A) - n_-(A)$ is the *signature* of A.

Example 8.3. If $A = \mathsf{diag}(4, -1, 0, 0, 1)$, then $\mathcal{I}(A) = (2, 1, 2)$ and $\mathsf{sig}(A) = 1$.

Let $A \in \mathbb{C}^{n\times n}$ be a Hermitian matrix. By Theorem 8.14, A can be written as $A = U^{\mathsf{H}}DU$, where U is a unitary matrix and $D = \mathsf{diag}(\lambda_1, \ldots, \lambda_n)$ is a diagonal matrix having the eigenvalues of A (which are real numbers) as its diagonal elements. Without loss of generality, we may assume that the positive eigenvalues of A are $\lambda_1, \ldots, \lambda_{n_+}$, followed by the negative values $\lambda_{n_++1}, \ldots, \lambda_{n_++n_-}$, and the zero eigenvalues $\lambda_{n_++n_-+1}, \ldots, \lambda_n$.

Let θ_j be the numbers defined by

$$\theta_j = \begin{cases} \sqrt{\lambda_j} & \text{if } 1 \leqslant j \leqslant n_+, \\ \sqrt{-\lambda_j} & \text{if } n_+ + 1 \leqslant j \leqslant n_+ + n_-, \\ 1 & \text{if } n_+ + n_- + 1 \leqslant j \leqslant n \end{cases}$$

for $1 \leqslant j \leqslant n$. If $T = \mathsf{diag}(\theta_1, \ldots, \theta_n)$, then we can write $D = T^{\mathsf{H}}GT$, where G is a diagonal matrix, $G = (g_1, \ldots, g_n)$ defined by

$$g_j = \begin{cases} 1 & \text{if } \lambda_j > 0, \\ -1 & \text{if } \lambda_j < 0, \\ 0 & \text{if } \lambda_j = 0, \end{cases}$$

for $1 \leqslant j \leqslant n$. This allows us to write $A = U^{\mathsf{H}}DU = U^{\mathsf{H}}T^{\mathsf{H}}GTU = (TU)^{\mathsf{H}}G(TU)$. The matrix TU is clearly nonsingular, so $A \sim_{\mathsf{H}} G$. The matrix G defined above is the *inertia matrix* of A and these definitions show that any Hermitian matrix is congruent to its inertia matrix.

For a Hermitian matrix $A \in \mathbb{C}^{n\times n}$, let $\mathsf{S}_+(A)$ be the subspace of $\mathbb{C}^n$ generated by $n_+(A)$ orthonormal eigenvectors that correspond to the positive eigenvalues of A. Clearly, we have $\dim(\mathsf{S}_+(A)) = n_+(A)$. This notation is used in the proof of the next theorem.

Theorem 8.16 (Sylvester's inertia theorem). *Let A, B be two Hermitian matrices, $A, B \in \mathbb{C}^{n\times n}$. The matrices A and B are congruent if and only if $\mathcal{I}(A) = \mathcal{I}(B)$.*

Proof. If $\mathcal{I}(A) = \mathcal{I}(B)$, then we have $A = S^{\mathsf{H}}GS$ and $B = T^{\mathsf{H}}GT$, where both S and T are nonsingular matrices. Since $A \sim_{\mathsf{H}} G$ and $B \sim_{\mathsf{H}} G$, we have $A \sim_{\mathsf{H}} B$.

Conversely, suppose that $A \sim_{\mathsf{H}} B$, that is, $A = S^{\mathsf{H}}BS$, where S is a nonsingular matrix. We have $\mathsf{rank}(A) = \mathsf{rank}(B)$, so $n_0(A) = n_0(B)$. To prove that $\mathfrak{I}(A) = \mathfrak{I}(B)$, it suffices to show that $n_+(A) = n_+(B)$.

Let $m = n_+(A)$ and let $\boldsymbol{v}_1, \ldots, \boldsymbol{v}_m$ be m orthonormal eigenvectors of A that correspond to the m positive eigenvalues of this matrix, and let $\mathsf{S}_+(A)$ be the subspace generated by these vectors. If $\boldsymbol{v} \in \mathsf{S}_+(A) - \{\mathbf{0}\}$, then we have $\boldsymbol{v} = a_1\boldsymbol{v}_1 + \cdots + a_m\boldsymbol{v}_m$, so

$$\boldsymbol{v}^{\mathsf{H}}A\boldsymbol{v} = \left(\sum_{j=1}^{m} a_j\boldsymbol{v}_j\right)^{\mathsf{H}} A \left(\sum_{j=1}^{m} a_j\boldsymbol{v}_j\right) = \sum_{j=1}^{m} |a_j|^2 > 0.$$

Therefore, $\boldsymbol{x}^{\mathsf{H}}S^{\mathsf{H}}BS\boldsymbol{x} > 0$, so if $\boldsymbol{y} = S\boldsymbol{x}$, then $\boldsymbol{y}^{\mathsf{H}}B\boldsymbol{y} > 0$, which means that $\boldsymbol{y} \in \mathsf{S}_+(B)$. This shows that $\mathsf{S}_+(A)$ is isomorphic to a subspace of $\mathsf{S}_+(B)$, so $n_+(A) \leqslant n_+(B)$. The reverse inequality can be shown in the same manner, so $n_+(A) = n_+(B)$. □

We can add an interesting detail to the full-rank decomposition of a matrix.

Corollary 8.13. *If $A \in \mathbb{C}^{m\times n}$ and $A = CR$ is the full-rank decomposition of A with $\mathsf{rank}(A) = k$, $C \in \mathbb{C}^{m\times k}$, and $R \in \mathbb{C}^{k\times n}$, then C may be chosen to have orthogonal columns and R to have orthogonal rows.*

Proof. Since the matrix $A^{\mathsf{H}}A \in \mathbb{C}^{n\times n}$ is Hermitian, by Theorem 8.14, there exists a unitary matrix $U \in \mathbb{C}^{n\times k}$ such that $A^{\mathsf{H}}A = U^{\mathsf{H}}DU$, where $D \in \mathbb{C}^{k\times k}$ is a non-negative diagonal matrix. Let $C = AU^{\mathsf{H}} \in C^{n\times k}$ and $R = U$. Clearly, $CR = A$, and R has orthogonal rows because U is unitary. Let $\boldsymbol{c}_p, \boldsymbol{c}_q$ be two columns of C, where $1 \leqslant p, q \leqslant k$ and $p \neq q$. Since $\boldsymbol{c}_p = A\boldsymbol{u}_p$ and $\boldsymbol{c}_q = A\boldsymbol{u}_q$, where $\boldsymbol{u}_p, \boldsymbol{u}_q$ are the corresponding columns of U, we have

$$\boldsymbol{c}_p^{\mathsf{H}}\boldsymbol{c}_q = \boldsymbol{u}_p^{\mathsf{H}}A^{\mathsf{H}}A\boldsymbol{u}_q = \boldsymbol{u}_p^{\mathsf{H}}U^{\mathsf{H}}DU\boldsymbol{u}_q = \boldsymbol{e}_p^{\mathsf{H}}D\boldsymbol{e}_q = 0,$$

because $p \neq q$. □

Theorem 8.17 (Rayleigh–Ritz theorem). *Let $A \in \mathbb{C}^{n\times n}$ be a Hermitian matrix and let $\lambda_1, \lambda_2, \ldots, \lambda_n$ be its eigenvalues, where $\lambda_1 \geqslant \lambda_2 \geqslant \cdots \geqslant \lambda_n$. Define the Rayleigh–Ritz function*

$\mathsf{ral}_A : \mathbb{R}^n - \{\mathbf{0}\} \longrightarrow \mathbb{R}$ *as*

$$\mathsf{ral}_A(\boldsymbol{x}) = \frac{\boldsymbol{x}^{\mathsf{H}} A \boldsymbol{x}}{\boldsymbol{x}^{\mathsf{H}} \boldsymbol{x}}$$

Then

$$\lambda_1 \geqslant \mathsf{ral}_A(\boldsymbol{x}) \geqslant \lambda_n$$

for $\boldsymbol{x} \in \mathbb{C}^n - \{\mathbf{0}_n\}$.

Proof. By Theorem 8.13, since A is Hermitian, there exists a unitary matrix P and a diagonal matrix T such that $A = P^{\mathsf{H}} T P$ and the diagonal elements of T are the eigenvalues of A, that is, $T = \mathsf{diag}(\lambda_1, \lambda_2, \ldots, \lambda_n)$. This allows us to write

$$\boldsymbol{x}^{\mathsf{H}} A \boldsymbol{x} = \boldsymbol{x}^{\mathsf{H}} P^{\mathsf{H}} T P \boldsymbol{x} = (P\boldsymbol{x})^{\mathsf{H}} T P \boldsymbol{x} = \sum_{j=1}^{n} \lambda_i |(P\boldsymbol{x})_i|^2,$$

which implies

$$\lambda_1 \|P\boldsymbol{x}\|^2 \geqslant \boldsymbol{x}^{\mathsf{H}} A \boldsymbol{x} \geqslant \lambda_n \|P\boldsymbol{x}\|^2.$$

Since P is unitary, we also have

$$\|P\boldsymbol{x}\|^2 = \boldsymbol{x}^{\mathsf{H}} P^{\mathsf{H}} P \boldsymbol{x} = \boldsymbol{x}^{\mathsf{H}} \boldsymbol{x},$$

which implies

$$\lambda_1 \boldsymbol{x}^{\mathsf{H}} \boldsymbol{x} \geqslant \boldsymbol{x}^{\mathsf{H}} A \boldsymbol{x} \geqslant \lambda_n \boldsymbol{x}^{\mathsf{H}} \boldsymbol{x},$$

for $\boldsymbol{x} \in \mathbb{C}^n$. □

Corollary 8.14. *Let* $A \in \mathbb{C}^{n \times n}$ *be a Hermitian matrix and let* $\lambda_1, \lambda_2, \ldots, \lambda_n$ *be its eigenvalues, where* $\lambda_1 \geqslant \lambda_2 \geqslant \cdots \geqslant \lambda_n$. *We have*

$$\lambda_1 = \max\{\boldsymbol{x}^{\mathsf{H}} A \boldsymbol{x} \mid \boldsymbol{x}^{\mathsf{H}} \boldsymbol{x} = 1\},$$
$$\lambda_n = \min\{\boldsymbol{x}^{\mathsf{H}} A \boldsymbol{x} \mid \boldsymbol{x}^{\mathsf{H}} \boldsymbol{x} = 1\}.$$

Proof. Note that if $\boldsymbol{x}$ is an eigenvector that corresponds to λ_1, then $A\boldsymbol{x} = \lambda_1 \boldsymbol{x}$, so $\boldsymbol{x}^{\mathsf{H}} A \boldsymbol{x} = \lambda_1 \boldsymbol{x}^{\mathsf{H}} \boldsymbol{x}$; in particular, if $\boldsymbol{x}^{\mathsf{H}} \boldsymbol{x} = 1$, we have $\lambda_1 = \boldsymbol{x}^{\mathsf{H}} A \boldsymbol{x}$, so

$$\lambda_1 = \max\{\boldsymbol{x}^{\mathsf{H}} A \boldsymbol{x} \mid \boldsymbol{x}^{\mathsf{H}} \boldsymbol{x} = 1\}.$$

The equality for λ_n can be shown in a similar manner. □

We discuss next an important result that is a generalization of the Rayleigh–Ritz theorem (Theorem 8.17). Denote by $\mathcal{S}_k(L)$ the collection of all subspaces of dimension k of an $\mathbb{F}$-linear space L.

Theorem 8.18 (Courant-Fischer theorem). *Let $A \in \mathbb{C}^{n\times n}$ be a Hermitian matrix having the eigenvalues $\lambda_1 \geqslant \cdots \geqslant \lambda_n$. We have*

$$\lambda_k = \min_{S\in\mathcal{S}_{n-k+1}(\mathbb{C}^n)} \max_{\boldsymbol{x}\in S}\{\boldsymbol{x}^{\mathsf{H}}A\boldsymbol{x} \mid \|\boldsymbol{x}\|_2 = 1\},$$

and

$$\lambda_k = \max_{S\in\mathcal{S}_k(\mathbb{C}^n)} \min_{\boldsymbol{x}\in S}\{\boldsymbol{x}^{\mathsf{H}}A\boldsymbol{x} \mid \|\boldsymbol{x}\|_2 = 1\}.$$

Proof. By Theorem 8.13, there exists a unitary matrix U and a diagonal matrix D such that $A = U^{\mathsf{H}}DU$ and the diagonal elements of D are the eigenvalues of A, that is, $D = \mathsf{diag}(\lambda_1, \lambda_2, \ldots, \lambda_n)$.

We prove only that

$$\lambda_k = \min_{S\in\mathcal{S}_{n-k+1}(\mathbb{C}^n)} \max_{\boldsymbol{x}\in S}\{\boldsymbol{x}^{\mathsf{H}}D\boldsymbol{x} \mid \|\boldsymbol{x}\|_2 = 1\}.$$

The proof of the second part of the theorem is entirely similar.

Let S be a subspace of $\mathbb{C}^n$ with $\dim(S) = n-k+1$. We have $S \cap \langle \boldsymbol{e}_1, \ldots, \boldsymbol{e}_k\rangle \neq \{\boldsymbol{0}_n\}$ because otherwise the equality $S \cap \langle \boldsymbol{e}_1, \ldots, \boldsymbol{e}_k\rangle = \{\boldsymbol{0}_n\}$ would imply $\dim(S) \leqslant n-k$.

Define

$$\tilde{S} = \{\boldsymbol{y} \in S \mid \|\boldsymbol{y}\| = 1\},$$
$$\hat{S} = \{\boldsymbol{y} \in \tilde{S} \mid \boldsymbol{y} \in \langle \boldsymbol{e}_1, \ldots, \boldsymbol{e}_k\rangle\}.$$

Therefore, $\hat{S}$ consists of vectors of $\tilde{S}$ having the form

$$\boldsymbol{y} = \begin{pmatrix} y_1 \\ \vdots \\ y_k \\ 0 \\ \vdots \\ 0 \end{pmatrix}$$

such that $\sum_{i=1}^{k} y_i^2 = 1$. Thus, for all $\boldsymbol{y} \in \hat{S}$, we have

$$\boldsymbol{y}^{\mathsf{H}} D\boldsymbol{y} = \sum_{i=1}^{k} \lambda_i |y_i|^2 \geqslant \lambda_k \sum_{i=1}^{k} |y_i|^2 = \lambda_k.$$

Since $\hat{S} \subseteq \tilde{S}$, it follows that $\max_{\boldsymbol{y}\in\tilde{S}} \boldsymbol{y}^{\mathsf{H}} D\boldsymbol{y} \geqslant \max_{\boldsymbol{y}\in\hat{S}} \boldsymbol{y}^{\mathsf{H}} D\boldsymbol{y} \geqslant \lambda_k$, so

$$\min_{\dim(S)=n-k+1} \max_{\boldsymbol{x}} \{\boldsymbol{x}^{\mathsf{H}} D\boldsymbol{x} \mid \boldsymbol{x} \in S \text{ and } \|\boldsymbol{x}\|_2 = 1\} \geqslant \lambda_k.$$

Let now S be $S = \langle \boldsymbol{e}_1, \ldots, \boldsymbol{e}_{k-1}\rangle^{\perp}$. Clearly, $\dim(S) = n - k + 1$. A vector $\boldsymbol{y} \in S$ has the form

$$\boldsymbol{y} = \begin{pmatrix} 0 \\ \vdots \\ 0 \\ y_k \\ \vdots \\ y_n \end{pmatrix}.$$

Therefore,

$$\boldsymbol{y}^{\mathsf{H}} D\boldsymbol{y} = \sum_{i=k}^{n} \lambda_i |y_i|^2 \leqslant \lambda_i \sum_{i=k}^{n} |y_i|^2 = \lambda_i$$

for all $\boldsymbol{y} \in \{\boldsymbol{y} \in S \mid \|\boldsymbol{y}\|_2 = 1\}$. This implies

$$\min_{\dim(S)=n-k+1} \max_{\boldsymbol{x}} \{\boldsymbol{x}^{\mathsf{H}} D\boldsymbol{x} \mid \boldsymbol{x} \in S \text{ and } \|\boldsymbol{x}\|_2 = 1\} \leqslant \lambda_k,$$

which yields the desired equality.

The matrices A and D have the same eigenvalues. Also $\boldsymbol{x}^{\mathsf{H}} A\boldsymbol{x} = \boldsymbol{x}^{\mathsf{H}} A\boldsymbol{x} = \boldsymbol{x}^{\mathsf{H}} U^{\mathsf{H}} DU\boldsymbol{x} = (U\boldsymbol{x})^{\mathsf{H}} D(U\boldsymbol{x})$ and $\|U\boldsymbol{x}\|_2 = \|\boldsymbol{x}\|_2$, because U is a unitary matrix. This yields the first equality of the theorem. □

Another form of the Courant–Fisher theorem can be obtained by observing that every p-dimensional subspace S of $\mathbb{C}^n$ is the orthogonal space of an $(n-p)$-dimensional subspace. Therefore, for each p-dimensional subspace S there is a sequence of $n-p$ vectors $\boldsymbol{w}_1, \ldots, \boldsymbol{w}_{n-p}$ (which is a basis of $S^{\perp}$) such that $S = \{\boldsymbol{x} \in \mathbb{C}^n \mid \boldsymbol{x} \perp \boldsymbol{w}_1, \ldots, \boldsymbol{x} \perp \boldsymbol{w}_{n-p}\}$.

Theorem 8.19. *Let $A \in \mathbb{C}^{n\times n}$ be a Hermitian matrix having the eigenvalues $\lambda_1 \geqslant \cdots \geqslant \lambda_n$. We have*

$$\begin{aligned}\lambda_k &= \min_{\boldsymbol{w}_1,\ldots,\boldsymbol{w}_{k-1}} \max_{\boldsymbol{x}} \{\boldsymbol{x}^{\mathsf{H}} A \boldsymbol{x} \mid \boldsymbol{x} \perp \boldsymbol{w}_1, \ldots, \boldsymbol{x} \perp \boldsymbol{w}_{k-1} \text{ and } \|\boldsymbol{x}\|_2 = 1\},\\ &= \max_{\boldsymbol{w}_1,\ldots,\boldsymbol{w}_{n-k}} \min_{\boldsymbol{x}} \{\boldsymbol{x}^{\mathsf{H}} A \boldsymbol{x} \mid \boldsymbol{x} \perp \boldsymbol{w}_1, \ldots, \boldsymbol{x} \perp \boldsymbol{w}_{n-k}, \text{ and } \|\boldsymbol{x}\|_2 = 1\}.\end{aligned}$$

Proof. This statement is equivalent to theorem 8.18, as we observed before. □

Theorem 8.20 (Interlacing theorem). *Let $A \in \mathbb{C}^{n\times n}$ be a Hermitian matrix and let $B = A\begin{bmatrix} i_1 \cdots i_k \\ i_1 \cdots i_k \end{bmatrix}$ be a principal submatrix of A, $B \in \mathbb{C}^{k\times k}$. If $\mathsf{spec}(A) = \{\lambda_1, \ldots, \lambda_n\}$ and $\mathsf{spec}(B) = \{\mu_1, \ldots, \mu_k\}$, where $\lambda_1 \geqslant \cdots \geqslant \lambda_n$ and $\mu_1 \geqslant \cdots \geqslant \mu_k$, then $\lambda_j \geqslant \mu_j \geqslant \lambda_{n-k+j}$ for $1 \leqslant j \leqslant k$.*

Proof. Let $\{j_1, \ldots, j_q\} = \{1, \ldots, n\} - \{i_1, \ldots, i_k\}$, where $j_1 < \cdots < j_q$ and $k + q = n$. By the Courant–Fisher theorem, we have

$$\lambda_j = \min_W \max_{\boldsymbol{x}} \{\boldsymbol{x}^{\mathsf{H}} A \boldsymbol{x} \mid \|\boldsymbol{x}\|_2 = 1 \text{ and } \boldsymbol{x} \in \langle W \rangle^{\perp}\},$$

where W ranges over sets of non-zero vectors in $\mathbb{C}^n$ containing $j-1$ vectors. Therefore,

$$\begin{aligned}\lambda_j &\geqslant \min_W \max_{\boldsymbol{x}} \{\boldsymbol{x}^{\mathsf{H}} A \boldsymbol{x} \mid \|\boldsymbol{x}\|_2 = 1 \text{ and } \boldsymbol{x} \in \langle W \rangle^{\perp}\\ &\qquad \text{and } x \in \langle \boldsymbol{e}_{j_1}, \ldots, \boldsymbol{e}_{j_q} \rangle^{\perp}\}\\ &= \min_U \max_{\boldsymbol{y}} \{\boldsymbol{y}^{\mathsf{H}} B \boldsymbol{y} \mid \|\boldsymbol{y}\|_2 = 1 \text{and} \boldsymbol{y} \in \langle U \rangle^{\perp} = \mu_j\\ &\qquad \text{(by Exercise 2 of Chapter 3)},\end{aligned}$$

where U ranges over sets of non-zero vectors in $\mathbb{C}^k$ containing $j-1$ vectors.

Again, by the Courant–Fisher theorem,

$$\lambda_{n-k+j} = \max_{Z} \min_{\boldsymbol{x}} \{\boldsymbol{x}^{\mathsf{H}} A \boldsymbol{x} \mid \|\boldsymbol{x}\|_2 = 1 \text{ and } \boldsymbol{x} \in \langle Z \rangle^{\perp}\},$$

where Z ranges over sets containing $k-j$ non-zero vectors in $\mathbb{C}^n$. Consequently,

$$\begin{aligned}\lambda_{n-k+j} &\leqslant \max_{Z} \min_{\boldsymbol{x}} \{\boldsymbol{x}^{\mathsf{H}} A \boldsymbol{x} \mid \|\boldsymbol{x}\|_2 = 1 \text{ and } \boldsymbol{x} \in \langle Z \rangle^{\perp} \\ &\qquad \text{and } x \in \langle \boldsymbol{e}_{j_1}, \ldots, \boldsymbol{e}_{j_q} \rangle^{\perp}\} \\ &= \max_{S} \min_{\boldsymbol{y}} \{\boldsymbol{y}^{\mathsf{H}} B \boldsymbol{y} \mid \|\boldsymbol{y}\|_2 = 1 \text{ and } \boldsymbol{y} \in \langle S \rangle^{\perp}\} = \mu_j,\end{aligned}$$

where S ranges over the sets of non-zero vectors in $\mathbb{C}^k$ containing $n-j$ vectors. □

Corollary 8.15. *Let $A \in \mathbb{C}^{n \times n}$ be a Hermitian matrix and let $B = A\begin{bmatrix} i_1 \cdots i_k \\ i_1 \cdots i_k \end{bmatrix}$ be a principal submatrix of A, $B \in \mathbb{C}^{k \times k}$. The set $\mathsf{spec}(B)$ contains no more positive eigenvalues than the number of positive eigenvalues of A and no more negative eigenvalues than the number of negative eigenvalues of A.*

Proof. This observation is a direct consequence of the Interlacing Theorem. □

Theorem 8.21. *Let $A \in \mathbb{C}^{n \times n}$ be a Hermitian matrix having the eigenvalues $\lambda_1 \geqslant \cdots \geqslant \lambda_n$. If $\boldsymbol{u}_1, \ldots, \boldsymbol{u}_n$ are eigenvectors that correspond to $\lambda_1, \ldots, \lambda_n$, respectively, $W = \{\boldsymbol{u}_1, \ldots, \boldsymbol{u}_k\}$ and $Z = \{\boldsymbol{u}_{k+2}, \ldots, \boldsymbol{u}_n\}$, then we have*

$$\begin{aligned}\lambda_{k+1} &= \max_{\boldsymbol{x}} \{\boldsymbol{x}^{\mathsf{H}} A \boldsymbol{x} \mid \|\boldsymbol{x}\|_2 = 1 \text{ and } \boldsymbol{x} \in \langle W \rangle^{\perp}\} \\ &= \min_{\boldsymbol{x}} \{\boldsymbol{x}^{\mathsf{H}} A \boldsymbol{x} \mid \|\boldsymbol{x}\|_2 = 1 \text{ and } \boldsymbol{x} \in \langle Z \rangle^{\perp}\}.\end{aligned}$$

Proof. If $A = U^{\mathsf{H}} D U$, where U is a unitary matrix and D is a diagonal matrix, then $\boldsymbol{u}_i$, the ith column of U^{H}, can be written as $\boldsymbol{u}_i = U^{\mathsf{H}} \boldsymbol{e}_i$. Therefore, by the second part of the proof of the Courant–Fisher theorem, we have $\boldsymbol{x} A \boldsymbol{x} \leqslant \lambda_{k+1}$ if $\boldsymbol{x}$ belongs to the subspace orthogonal to the subspace generated by the first k eigenvectors of A. Consequently, the Courant–Fisher theorem implies the first equality of this theorem. The second equality can be obtained in a similar manner. □

Corollary 8.16. *Let $A \in \mathbb{C}^{n\times n}$ be a Hermitian matrix having the eigenvalues $\lambda_1 \geqslant \cdots \geqslant \lambda_n$. If $\boldsymbol{u}_1, \ldots, \boldsymbol{u}_k$ are eigenvectors that correspond to $\lambda_1, \ldots, \lambda_k$, respectively, then a unit vector $\boldsymbol{x}$ that maximizes $\boldsymbol{x}^{\mathsf{H}} A\boldsymbol{x}$ and belongs to the subspace orthogonal to the subspace generated by the first k eigenvectors of A is an eigenvector that corresponds to λ_{k+1}.*

Proof. Let $\boldsymbol{u}_1, \ldots, \boldsymbol{u}_n$ be the eigenvectors of A and let $\boldsymbol{x} \in \langle \boldsymbol{u}_1, \ldots, \boldsymbol{u}_k\rangle^{\perp}$ be a unit vector. We have $\boldsymbol{x} = \sum_{j=k+1}^{n} a_j \boldsymbol{u}_j$, and $\sum_{j=k+1}^{n} a_j^2 = 1$, which implies

$$\boldsymbol{x}^{\mathsf{H}} A\boldsymbol{x} = \sum_{j=k+1}^{n} \lambda_j a_j^2 = \lambda_{k+1}.$$

This, in turn, implies $a_{k+1} = 1$ and $a_{k+2} = \cdots = a_n = 0$, so $\boldsymbol{x} = \boldsymbol{u}_{k+1}$. □

Theorem 8.22. *Let $A, B \in \mathbb{C}^{n\times}$ be two Hermitian matrices and let $E = B - A$. Suppose that the eigenvalues of A, B, E are $\alpha_1 \geqslant \cdots \geqslant \alpha_n$, $\beta_1 \geqslant \cdots \geqslant \beta_n$, and $\epsilon_1 \geqslant \cdots \geqslant \epsilon_n$, respectively. Then we have $\epsilon_n \leqslant \beta_i - \alpha_i \leqslant \epsilon_1$.*

Proof. Note that E is also Hermitian, so all matrices involved have real eigenvalues. By the Courant–Fisher theorem,

$$\beta_k = \min_{W} \max_{\boldsymbol{x}} \{\boldsymbol{x}^{\mathsf{H}} B\boldsymbol{x} \mid \|\boldsymbol{x}\|_2 = 1 \text{ and } \boldsymbol{w}_i^{\mathsf{H}}\boldsymbol{x} = 0 \text{ for } 1 \leqslant i \leqslant k-1\},$$

where $W = \{\boldsymbol{w}_1, \ldots, \boldsymbol{w}_{k-1}\}$. Thus,

$$\beta_k \leqslant \max_{\boldsymbol{x}} \boldsymbol{x}^{\mathsf{H}} B\boldsymbol{x} = \max_{\boldsymbol{x}} (\boldsymbol{x}^{\mathsf{H}} A\boldsymbol{x} + \boldsymbol{x}^{\mathsf{H}} E\boldsymbol{x}). \tag{8.2}$$

Let U be a unitary matrix such that $U^{\mathsf{H}} AU = \mathsf{diag}(\alpha_1, \ldots, \alpha_n)$. Choose $\boldsymbol{w}_i = U\boldsymbol{e}_i$ for $1 \leqslant i \leqslant k-1$. We have $\boldsymbol{w}_i^{\mathsf{H}}\boldsymbol{x} = \boldsymbol{e}_i^{\mathsf{H}} U^{\mathsf{H}}\boldsymbol{x} = 0$ for $1 \leqslant i \leqslant k-1$.

Define $\boldsymbol{y} = U^{\mathsf{H}}\boldsymbol{x}$. Since U is a unitary matrix, $\|\boldsymbol{y}\|_2 = \|\boldsymbol{x}\|_2 = 1$. Observe that $\boldsymbol{e}_i^{\mathsf{H}}\boldsymbol{y} = y_i = 0$ for $1 \leqslant i \leqslant k$. Therefore, $\sum_{i=k}^{n} y_i^2 = 1$. This, in turn, implies $\boldsymbol{x}^{\mathsf{H}} A\boldsymbol{x} = \boldsymbol{y}^{\mathsf{H}} U^{\mathsf{H}} AU\boldsymbol{y} = \sum_{i=k}^{n} \alpha_i y_i^2 \leqslant \alpha_k$.

From the Inequality (8.2), it follows that

$$\beta_k \leqslant \alpha_k + \max_{\boldsymbol{x}} \boldsymbol{x}^{\mathsf{H}} E\boldsymbol{x} \leqslant \alpha_k + \epsilon_n.$$

Since $A = B - E$, by inverting the roles of A and B we have $\alpha_k \leqslant \beta_k - \epsilon_1$, or $\epsilon_1 \leqslant \beta_k - \alpha_k$, which completes the argument. □

Theorem 8.23 (Hoffman–Wielandt theorem). *Let* $A, B \in \mathbb{C}^{n\times n}$ *be two normal matrices having the eigenvalues* $\alpha_1, \ldots, \alpha_n$ *and* $\beta_1, \ldots, \beta_n$*, respectively. Then there exist permutations* ϕ *and* ψ *in* $PERM_n$ *such that*

$$\sum_{i=1}^{n} |\alpha_i - \beta_{\psi(i)}|^2 \geqslant \|A - B\|_F^2 \geqslant \sum_{i=1}^{n} |\alpha_i - \beta_{\phi(i)}|^2.$$

Proof. Since A and B are normal matrices, they can be diagonalized as $A = U^{\mathsf{H}} D_A U$ and $B = W^{\mathsf{H}} D_B W$, where U and W are unitary matrices and C, D are diagonal matrices, $C = \mathsf{diag}(\alpha_1, \ldots, \alpha_n)$ and $D = \mathsf{diag}(\beta_1, \ldots, \beta_n)$. Then we can write

$$\|A - B\|_F^2 = \|U^{\mathsf{H}} C U - W^{\mathsf{H}} D W\|_F^2 = \mathsf{trace}(E^{\mathsf{H}} E),$$

where $E = U^{\mathsf{H}} C U - W^{\mathsf{H}} D W$. Note that

$$\begin{aligned} E^{\mathsf{H}} E &= (U^{\mathsf{H}} C^{\mathsf{H}} U - W^{\mathsf{H}} D^{\mathsf{H}} W)(U^{\mathsf{H}} C U - W^{\mathsf{H}} D W) \\ &= U^{\mathsf{H}} C^{\mathsf{H}} C U + W^{\mathsf{H}} D^{\mathsf{H}} D W - W^{\mathsf{H}} D^{\mathsf{H}} W U^{\mathsf{H}} C U - U^{\mathsf{H}} C^{\mathsf{H}} U W^{\mathsf{H}} D W \\ &= U^{\mathsf{H}} C^{\mathsf{H}} C U + W^{\mathsf{H}} D^{\mathsf{H}} D W - U^{\mathsf{H}} C^{\mathsf{H}} U W^{\mathsf{H}} D W - (U^{\mathsf{H}} C^{\mathsf{H}} U W^{\mathsf{H}} D W)^{\mathsf{H}} \\ &= U^{\mathsf{H}} C^{\mathsf{H}} C U + W^{\mathsf{H}} D^{\mathsf{H}} D W - 2\Re(U^{\mathsf{H}} C^{\mathsf{H}} U W^{\mathsf{H}} D W). \end{aligned}$$

Observe that

$$\begin{aligned} \mathsf{trace}(\Re(U^{\mathsf{H}} C^{\mathsf{H}} U W^{\mathsf{H}} D W)) &= \Re(\mathsf{trace}(U^{\mathsf{H}} C^{\mathsf{H}} U W^{\mathsf{H}} D W)) \\ &= \Re(\mathsf{trace}(C^{\mathsf{H}} U W^{\mathsf{H}} D W U^{\mathsf{H}})). \end{aligned}$$

Thus, if Z is the unitary matrix $Z = W U^{\mathsf{H}}$, we have

$$\mathsf{trace}(\Re(U^{\mathsf{H}} C^{\mathsf{H}} U W^{\mathsf{H}} D W)) = \Re(\mathsf{trace}(C^{\mathsf{H}} Z^{\mathsf{H}} D Z)).$$

Since $\|C\|_F^2 = \sum_{i=1}^n \alpha_i^2$ and $\|D\|_F^2 = \sum_{i=1}^n \beta_i^2$, we have

$$\begin{aligned} \mathsf{trace}(E^{\mathsf{H}} E) &= \|C\|_F^2 + \|D\|_F^2 - 2\Re(\mathsf{trace}(C^{\mathsf{H}} Z^{\mathsf{H}} D Z)) \\ &= \sum_{i=1}^{n} \alpha_i^2 + \sum_{i=1}^{n} \beta_i^2 - 2\Re\left(\sum_{i=1}^{n}\sum_{j=1}^{n} \bar{a}_i |z_{ij}|^2 \beta_j\right). \end{aligned}$$

Theorem 3.20 implies that the matrix S that has the elements $|z_{ij}|^2$ is doubly stochastic because Z is a unitary matrix. This allows us to write

$$\begin{aligned}\|A-B\|_F^2 &= \mathsf{trace}(E^{\mathsf{H}}E)\\ &\geqslant \sum_{i=1}^n \alpha_i^2 + \sum_{i=1}^n \beta_i^2 - \max_S \Re\left(\sum_{i=1}^n\sum_{j=1}^n \bar{a}_i s_{ij}\beta_j\right),\end{aligned}$$

and

$$\begin{aligned}\|A-B\|_F^2 &= \mathsf{trace}(E^{\mathsf{H}}E)\\ &\leqslant \sum_{i=1}^n \alpha_i^2 + \sum_{i=1}^n \beta_i^2 - \min_S \Re\left(\sum_{i=1}^n\sum_{j=1}^n \bar{a}_i s_{ij}\beta_j\right),\end{aligned}$$

where the maximum and the minimum are taken over the set of all doubly stochastic matrices.

The Birkhoff–von Neumann Theorem states that the polyhedron of doubly–stochastic matrices has the permutation matrices as its vertices. Therefore, the extremes of the linear function

$$f(S) = \Re\left(\sum_{i=1}^n\sum_{j=1}^n \bar{\alpha}_i s_{ij}\beta_j\right)$$

are achieved when S is a permutation matrix. Let P_ϕ be the permutation matrix that gives the maximum of f and let P_ψ be the permutation matrix that gives the minimum.

If $S = P_\phi$, then $\sum_{j=1}^n s_{ij}\beta_j = \beta_{\phi(i)}$, so

$$\begin{aligned}\|A-B\|_F^2 &\geqslant \sum_{i=1}^n \alpha_i^2 + \sum_{i=1}^n \beta_i^2 - \Re\left(\sum_{i=1}^n\sum_{j=1}^n \bar{\alpha}_i\beta_{\phi(j)}\right)\\ &= \sum_{i=1}^n |\alpha_i - \beta_{\phi(i)}|^2.\end{aligned}$$

In the last equality, we used the elementary calculation

$$\begin{aligned}|a-b|^2 &= (a-b)\overline{(a-b)} = (a-b)(\bar{a}-\bar{b})\\ &= a\bar{a} + b\bar{b} - \bar{a}b - \overline{\bar{a}b}\\ &= |a|^2 + |b|^2 - 2\Re(\bar{a}b),\end{aligned}$$

for $a, b \in \mathbb{C}$. Similarly, if $S = P_\psi$, we obtain the other inequality. □

Corollary 8.17. *Let $A, B \in \mathbb{C}^{n \times n}$ be two Hermitian matrices having the eigenvalues $\alpha_1, \ldots, \alpha_n$ and $\beta_1, \ldots, \beta_n$, respectively, where $\alpha_1 \geqslant \cdots \geqslant \alpha_n$ and $\beta_1 \geqslant \cdots \geqslant \beta_n$. Then,*

$$\sum_{i=1}^{n} |\alpha_i - \beta_i|^2 \leqslant \|A - B\|_F^2.$$

If $\alpha_1 \geqslant \cdots \geqslant \alpha_n$ and $\beta_1 \leqslant \cdots \leqslant \beta_n$, then

$$\sum_{i=1}^{n} |\alpha_i - \beta_i|^2 \geqslant \|A - B\|_F^2.$$

Proof. Since A and B are Hermitian, their eigenvalues are real numbers. By the Hoffman–Wielandt Theorem, there exist two permutations $\phi, \psi \in PERM_n$ such that

$$\sum_{i=1}^{n} |\alpha_i - \beta_{\psi(i)}|^2 \geqslant \|A - B\|_F^2 \geqslant \sum_{i=1}^{n} |\alpha_i - \beta_{\phi(i)}|^2.$$

We have

$$\sum_{i=1}^{n} |\alpha_i - \beta_{\phi(i)}|^2 = \|\boldsymbol{a} - P_\phi \boldsymbol{b}\|_F^2$$

and

$$\sum_{i=1}^{n} |\alpha_i - \beta_{\psi(i)}|^2 = \|\boldsymbol{a} - P_\psi \boldsymbol{b}\|_F^2,$$

where

$$\boldsymbol{a} = \begin{pmatrix} \alpha_1 \\ \vdots \\ \alpha_n \end{pmatrix} \text{ and } \boldsymbol{b} = \begin{pmatrix} \beta_1 \\ \vdots \\ \beta_n \end{pmatrix},$$

so

$$\|\boldsymbol{a} - P_\psi \boldsymbol{b}\|_F^2 \geqslant \|A - B\|_F^2 \geqslant \|\boldsymbol{a} - P_\phi \boldsymbol{b}\|_F^2.$$

By Corollary 6.13, since the components of $\boldsymbol{a}$ and $\boldsymbol{b}$ are placed in decreasing order, we have

$$\|\boldsymbol{a} - P_\phi \boldsymbol{b}\|_F \geqslant \|\boldsymbol{a} - \boldsymbol{b}\|_F,$$

so

$$\|\boldsymbol{a} - \boldsymbol{b}\|_F^2 = \sum_{i=1}^{n} |\alpha_i - \beta_i|^2 \leqslant \|\boldsymbol{a} - P_\phi \boldsymbol{b}\|_F \leqslant \|A - B\|_F^2,$$

which proves the first inequality of the corollary.

For the second part, by Corollary 6.13, we have

$$\|A - B\|_F \leqslant \|\boldsymbol{a} - P_\psi \boldsymbol{b}\|_F \leqslant \|\boldsymbol{a} - \boldsymbol{b}\|_F,$$

□

By Theorem 7.16, the characteristic polynomial of an upper triangular matrix T is

$$p_T(\lambda) = (\lambda - \lambda_1) \cdots (\lambda - \lambda_n),$$

where $\lambda_1, \ldots, \lambda_n$ are the eigenvalues of T and, at the same time, the diagonal elements of T.

Lemma 8.4. *Let $T \in \mathbb{C}^{n \times n}$ be an upper triangular matrix and let $p_T(\lambda) = \lambda^n + a_1\lambda^{n-1} + \cdots + a_{n-1}\lambda + a_n$ be its characteristic polynomial. Then*

$$p_T(T) = T^n + a_1 T^{n-1} + \cdots + a_{n-1}T + a_n I_n = O_{n,n}.$$

Proof. We have

$$p_T(T) = (T - \lambda_1 I_n) \cdots (T - \lambda_n I_n).$$

Observe that for any matrix $A \in \mathbb{C}^{n \times n}$, $\lambda_j, \lambda_k \in \mathsf{spec}(A)$, and every eigenvector $\boldsymbol{v}$ of A in S_{A,λ_k}, we have

$$(\lambda_j I_n - A)\boldsymbol{v} = (\lambda_j - \lambda_k)\boldsymbol{v}.$$

Therefore, for $\boldsymbol{v} \in S_{T,\lambda_k}$, we have

$$p_T(T)\boldsymbol{v} = (\lambda_1 I_n - T) \cdots (\lambda_n I_n - T)\boldsymbol{v} = \boldsymbol{0},$$

because $(\lambda_k I - T)\boldsymbol{v} = \boldsymbol{0}$.

By Corollary 7.9, $p_T(T)\boldsymbol{e}_i = \boldsymbol{0}$ for $1 \leqslant i \leqslant n$, so $p_T(T) = O_{n,n}$, so $p_T(T) = O_{n,n}$. □

Theorem 8.24 (Cayley–Hamilton theorem). *If $A \in \mathbb{C}^{n \times n}$ is a matrix, then $p_A(A) = O_{n,n}$.*

Proof. By Schur's Triangularization Theorem, there exists a unitary matrix $U \in \mathbb{C}^{n\times n}$ and an upper–triangular matrix $T \in \mathbb{C}^{n\times n}$ such that $A = U^{\mathsf{H}}TU$ and the diagonal elements of T are the eigenvalues of A. Taking into account that $U^{\mathsf{H}}U = UU^{\mathsf{H}} = I_n$, we can write

$$\begin{aligned}
p_A(A) &= (\lambda_1 I_n - A)(\lambda_2 I_n - A)\cdots(\lambda_n I_n - A)\\
&= (\lambda_1 U^{\mathsf{H}}U - U^{\mathsf{H}}TU)\cdots(\lambda_n U^{\mathsf{H}}U - U^{\mathsf{H}}TU)\\
&= U^{\mathsf{H}}(\lambda_1 I_n - T)UU^{\mathsf{H}}(\lambda_2 I_n - T)U\cdots U^{\mathsf{H}}(\lambda_n I_n - T)U\\
&= U^{\mathsf{H}}(\lambda_1 I_n - T)(\lambda_2 I_n - T)\cdots(\lambda_n I_n - T)U\\
&= U^{\mathsf{H}}p_T(T)U = O_{n,n},
\end{aligned}$$

by Lemma 8.4. □

8.4 The Sylvester Operator

Schur's Theorem allows us to examine the solvability of the matrix equation

$$AX - XB = C,$$

where $A \in \mathbb{C}^{m\times m}$, $B \in \mathbb{C}^{n\times n}$, and $C \in \mathbb{C}^{m\times n}$.

Theorem 8.25. *Let $A \in \mathbb{C}^{m\times m}$ and $B \in \mathbb{C}^{n\times n}$ be two matrices. The* Sylvester operator defined by A and B *is the mapping* $\boldsymbol{S}_{A,B} : \mathbb{C}^{m\times n} \longrightarrow \mathbb{C}^{m\times n}$ *given by*

$$\boldsymbol{S}_{A,B}(X) = AX - XB.$$

The linear mapping $S_{A,B}$ has an inverse if and only if $\mathsf{spec}(A) \cap \mathsf{spec}(B) = \emptyset$.

Proof. The linearity of $S_{A,B}$ is immediate. Suppose that $\lambda \in \mathsf{spec}(A) \cap \mathsf{spec}(B)$, so $A\boldsymbol{u} = \lambda\boldsymbol{u}$ and $\boldsymbol{v}^{\mathsf{H}}B = \lambda\boldsymbol{v}^{\mathsf{H}}$ for some $\boldsymbol{u} \in \mathbb{C}^{m\times 1}$ and $\boldsymbol{v} \in \mathbb{C}^{n\times 1}$. Thus, $\boldsymbol{u}$ is an eigenvector for A and $\boldsymbol{v}$ is a left eigenvector for B that corresponds to the same eigenvalue λ. Define the matrix $X = \boldsymbol{u}\boldsymbol{v}^{\mathsf{H}} \in \mathbb{C}^{m\times n}$. We have

$$AX - XB = A\boldsymbol{u}\boldsymbol{v}^{\mathsf{H}} - \boldsymbol{u}\boldsymbol{v}^{\mathsf{H}}B = (A\boldsymbol{u})\boldsymbol{v}^{\mathsf{H}} - \boldsymbol{u}(\boldsymbol{v}^{\mathsf{H}}B) = \mathbf{0},$$

which means that $\boldsymbol{S}_{A,B}$ is not invertible. Thus, if $S_{A,B}$ is invertible, the spectra of A and B are disjoint.

Conversely, suppose that $\boldsymbol{S}_{A,B}$ is invertible, so the equation $AX - XB = C$ is solvable for any $C \in \mathbb{C}^{m\times n}$. If $B = U^{\mathsf{H}}TU$ is a Schur decomposition of B, where $U \in \mathbb{C}^{n\times n}$ is a unitary matrix and $T \in \mathbb{C}^{n\times n}$

$$T = \begin{pmatrix} t_{11} & t_{12} & \cdots & t_{1n} \\ 0 & t_{22} & \cdots & t_{2n} \\ \vdots & \vdots & \cdots & \vdots \\ 0 & 0 & \cdots & t_{nn} \end{pmatrix}$$

is an upper triangular matrix having as diagonal elements the eigenvalues $t_{11}, t_{22}, \ldots, t_{nn}$ of B, then $AX - XU^{\mathsf{H}}TU = C$, which implies $AXU^{\mathsf{H}} - XU^{\mathsf{H}}T = CU^{\mathsf{H}}$. If $Z = XU^{\mathsf{H}} \in \mathbb{C}^{m\times n}$ and $D = CU^{\mathsf{H}}$, we have $AZ - ZT = D$ and this system has the form

$$A(\boldsymbol{z}_1 \ \cdots \ \boldsymbol{z}_n) - (\boldsymbol{z}_1 \ \cdots \ \boldsymbol{z}_n)T = (\boldsymbol{d}_1 \ \cdots \ \boldsymbol{d}_n).$$

Equivalently, we have

$$\begin{aligned} A\boldsymbol{z}_1 - t_{11}\boldsymbol{z}_1 &= \boldsymbol{d}_1 \\ A\boldsymbol{z}_2 - t_{12}\boldsymbol{z}_1 - t_{22}\boldsymbol{z}_2 &= \boldsymbol{d}_2 \\ &\vdots \\ A\boldsymbol{z}_n - t_{1n}\boldsymbol{z}_1 - t_{2n}\boldsymbol{z}_2 - \cdots - t_{nn}\boldsymbol{z}_n &= \boldsymbol{d}_n. \end{aligned}$$

Note that the first equation of this system $(A - t_{11}I_n)\boldsymbol{z}_1 = \boldsymbol{d}_1$ can be solved for $\boldsymbol{z}_1$ because t_{11}, as an eigenvalue of B, does not belong to $\mathsf{spec}(A)$. Thus, the second equation $A\boldsymbol{z}_2 - t_{22}\boldsymbol{z}_2 = \boldsymbol{d}_2 + t_{12}\boldsymbol{z}_1$ can be resolved with respect to $\boldsymbol{z}_2$, because $t_{22} \notin \mathsf{spec}(A)$, etc. Thus, the matrix Z can be found, so $X = ZU$. □

As observed in [76], special cases of this equation occur in many important problems in linear algebra. For instance, if $n = 1$ and $B = (0)$, then solving $S_{A,0}(\boldsymbol{x}) = \boldsymbol{c}$ amounts to solving the linear system $A\boldsymbol{x} = \boldsymbol{c}$. Solving the equation $S_{A,A}(X) = O_{n,n}$ amounts to finding the matrices X that commute with A, etc.

Theorem 8.26. *Let $A \in \mathbb{C}^{m\times m}$ and $B \in \mathbb{C}^{n\times n}$ be two matrices such that $\mathsf{spec}(B) \subset \{z \in \mathbb{C} \mid |z| < r\}$ and $\mathsf{spec}(A) \subset \{z \in \mathbb{C} \mid |z| > r\}$*

for some $r > 0$. *Then, the solution of the equation* $AX - XB = C$ *is given by the series*

$$X = \sum_{n\in\mathbb{N}} (A^{-1})^{n+1} C B^n.$$

Proof. Since $\mathsf{spec}(A)$ and $\mathsf{spec}(B)$ are finite sets, there exist r_1 and r_2 such that $0 < r_1 < r < r_2$, $\mathsf{spec}(B) \subset \{z \in \mathbb{C} \mid |z| < r_1\}$, and $\mathsf{spec}(A) \subset \{z \in \mathbb{C} \mid |z| > r_2\}$. Then $\mathsf{spec}(A^{-1}) \subset \{z \in \mathbb{C} \mid |z| < 1/r_2\}$.

By Theorem 8.48, there exists a positive integer n_0 such that if $n \geqslant n_0$, $\|B^n\| \leqslant r_1^n$ and $\|A^{-1}\| \leqslant r_2^n$. Thus, for $n \geqslant n_0$, we have $\|A^{-n-1}CB^n\| \leqslant \left(\frac{r_1}{r_2}\right)^n \|A^{-1}C\|$, which shows that the series $\sum_{n\in\mathbb{N}} (A^{-1})^{n+1} C B^n$ is convergent. It is immediate that for this choice of X we have $AX - XB = C$. □

Corollary 8.18. *Let* $A \in \mathbb{C}^{m\times m}$ *and* $B \in \mathbb{C}^{n\times n}$ *be two normal matrices such that* $\mathsf{spec}(B) \subset \{z \in \mathbb{C} \mid |z| < r\}$ *and* $\mathsf{spec}(A) \subset \{z \in \mathbb{C} \mid |z| > r + a\}$ *for some* $r > 0$ *and* $a > 0$. *If* X *is the solution of the Sylvester equation* $AX - XB = C$, *and* $\|\cdot\|$ *is a unitarily invariant matrix norm, then* $\|X\| \leqslant \frac{1}{a}\|C\|$.

Proof. From Theorem 8.26, we have

$$\begin{aligned}\|X\| &\leqslant \sum_{n=0}^{\infty} \|A^{-1}\|^{n+1} \|C\| \|B\|^n \\ &\leqslant \|C\| \sum_{n=0}^{\infty} (r+a)^{-n-1} a^n = \frac{1}{a}\|C\|.\end{aligned}$$

□

The Sylvester operator can be seen as a linear transformation of the linear space $\mathbb{C}^{m\times n}$ (which is isomorphic to $\mathbb{C}^{mn}$). The *separation* of two matrices A and B, denoted by $\mathsf{sep}(A, B)$, was introduced in [157]. This quantity is very useful in studying relationships between invariant spaces of matrices and is defined as

$$\mathsf{sep}(A, B) = \min\{\|\boldsymbol{S}_{A,B}(X)\| \mid \|X\| = 1\}. \tag{8.3}$$

Obviously, the number is dependent on the norm used in its definition. If we wish to specify this norm, we will add a suggestive subscript; for example, if we use the Frobenius norm, we will denote the corresponding separation by $\mathsf{sep}_F(A, B)$; $\mathsf{sep}_2(A, B)$ will be used when we deal with $\|\cdot\|_2$.

Theorem 8.27. *If $A \in \mathbb{C}^{m\times m}$ and $B \in \mathbb{C}^{n\times n}$ are two matrices, then the spectrum of the Sylvester operator $\boldsymbol{S}_{A,B}$ is*

$$\mathsf{spec}(\boldsymbol{S}_{A,B}) = \{\lambda - \mu \mid \lambda \in \mathsf{spec}(A) \textit{ and } \mu \in \mathsf{spec}(B)\}.$$

Proof. Suppose that $\theta \in \mathsf{spec}(\boldsymbol{S}_{A,B})$. There exists a matrix $X \in \mathbb{C}^{m\times n} - \{O_{m,n}\}$ such that $AX - XB = \theta X$, which amounts to $(A - \theta I_m)X - XB = O$. In other words, $\boldsymbol{S}_{A-\theta I_m,B}$ is singular, which implies that $\mathsf{spec}(A-\theta I_m)\cap\mathsf{spec}(B) \neq \emptyset$. This means that there exist $\lambda \in \mathsf{spec}(A)$ and $\mu \in \mathsf{spec}(B)$ such that $\lambda - \theta = \mu$, which implies that $\theta = \lambda - \mu$. Thus, $\mathsf{spec}(\boldsymbol{S}_{A,B}) \subseteq \{\lambda - \mu \mid \lambda \in \mathsf{spec}(A)$ and $\mu \in \mathsf{spec}(B)\}$. The reverse inclusion can be shown by reversing the above implications. □

Corollary 8.19. *If $A \in \mathbb{C}^{m\times m}$ and $B \in \mathbb{C}^{n\times n}$ are two diagonalizable matrices such that $\mathsf{spec}(A) = \{\lambda_1, \ldots, \lambda_m\}$, and $\mathsf{spec}(B) = \{\mu_1, \ldots, \mu_m\}$, and if $\boldsymbol{u}_i$ is an eigenvector of A associated to λ_i and $\boldsymbol{v}_j$ is an eigenvector of B^{H} associated with μ_j, then the matrix $X_{ij} = \boldsymbol{u}_i\boldsymbol{v}_j^{\mathsf{H}}$ is an eigenvector of $\boldsymbol{S}_{A,B}$ associated to $\lambda_i - \mu_j$.*

Proof. We have

$$\begin{aligned}
\boldsymbol{S}_{A,B}(X_{ij}) &= A\boldsymbol{u}_i\boldsymbol{v}_j^{\mathsf{H}} - \boldsymbol{u}_i\boldsymbol{v}_j^{\mathsf{H}}B \\
&= \lambda_i\boldsymbol{u}_i\boldsymbol{v}_j - \boldsymbol{u}_i(B^{\mathsf{H}}\boldsymbol{v}_j)^{\mathsf{H}} \\
&= \lambda_i\boldsymbol{u}_i\boldsymbol{v}_j - \mu_j\boldsymbol{u}_i\boldsymbol{v}_j^{\mathsf{H}} = (\lambda_i - \mu_j)\boldsymbol{u}_i\boldsymbol{v}_j^{\mathsf{H}} = (\lambda_i - \mu_j)X_{ij}.
\end{aligned}$$ □

Corollary 8.20. *Let $A \in \mathbb{C}^{m\times m}$ and $B \in \mathbb{C}^{n\times n}$ be two matrices. We have $\mathsf{sep}(A, B) \leqslant \min\{|\lambda - \mu| \mid \lambda \in \mathsf{spec}(A), \mu \in \mathsf{spec}(B)\}$.*

Proof. The inequality holds if $\mathsf{sep}(A, B) = 0$. Therefore, suppose that $\mathsf{sep}(A, B) > 0$. This implies that $\boldsymbol{S}_{A,B}$ is nonsingular and by Supplement 43 of Chapter 6 we have $\|\boldsymbol{S}_{A,B}^{-1}\|_2 = \frac{1}{\min\{\|\boldsymbol{S}_{A,B}(X)\|_2 \mid \|X\|_2=1\}} = \frac{1}{\mathsf{sep}(A,B)}$. Since the spectral radius of $\boldsymbol{S}_{A,B}^{-1}$ is less than $\|\boldsymbol{S}_{A,B}^{-1}\|$, it follows that $\frac{1}{\mathsf{sep}(A,B)}$ is larger or equal to any eigenvalues of $\boldsymbol{S}_{A,B}^{-1}$, so $\mathsf{sep}(A, B) \leqslant |\lambda - \mu|$ for every $\lambda \in \mathsf{spec}(A)$ and $\mu \in \mathsf{spec}(B)$. □

8.5 Geometric versus Algebraic Multiplicity

Theorem 8.28. *Let* $A \in \mathbb{C}^{n\times n}$ *be a square matrix and let* $\lambda \in \mathsf{spec}(A)$*. The geometric multiplicity* $\mathsf{geomm}(A,\lambda)$ *is less than or equal to the algebraic multiplicity* $\mathsf{algm}(A,\lambda)$*.*

Proof. By Equality (7.4),

$$\mathsf{geomm}(A,\lambda) = \dim(\mathsf{null}(A-\lambda I_n)) = n - \mathsf{rank}(A-\lambda I_n).$$

Let $m = \mathsf{geomm}(A,\lambda)$. Starting from an orthonormal basis $\boldsymbol{u}_1,\ldots,\boldsymbol{u}_m$ of the subspace $\mathsf{null}(A-\lambda I_n)$, define the matrix $U = (\boldsymbol{u}_1 \ \cdots \ \boldsymbol{u}_m)$. We have $(A-\lambda I_n)U = O$, so $AU = \lambda U = U(\lambda I_m)$. Thus, by Theorem 8.7, we have

$$A \sim \begin{pmatrix} \lambda I_m & U^{\mathsf{H}}AV \\ O & V^{\mathsf{H}}AV \end{pmatrix},$$

where $U \in \mathbb{C}^{n\times m}$ and $V \in \mathbb{C}^{n\times(n-m)}$. Therefore, A has the same characteristic polynomial as

$$B = \begin{pmatrix} \lambda I_m & U^{\mathsf{H}}AV \\ O & V^{\mathsf{H}}AV \end{pmatrix},$$

which implies $\mathsf{algm}(A,\lambda) = \mathsf{algm}(B,\lambda)$. Since the algebraic multiplicity of λ in B is at least equal to m, it follows that

$$\mathsf{algm}(A,\lambda) \geqslant m = \mathsf{geomm}(A,\lambda).$$

□

Definition 8.3. If λ is an eigenvalue of A and $\mathsf{geomm}(A,\lambda) = \mathsf{algm}(A,\lambda)$, then we refer to λ as a *semi-simple eigenvalue.*

The matrix A is *defective* if there exist at least one eigenvalue that is not semi-simple. Otherwise, A is said to be *non-defective.*

A is a *non-derogatory* matrix if $\mathsf{geomm}(A,\lambda) = 1$ for every eigenvalue λ.

If λ is a simple eigenvalue of A, we have $\mathsf{algm}(A,\lambda) = 1$, and, since $\mathsf{algm}(A,\lambda) \geqslant \mathsf{geomm}(A,\lambda) \geqslant 1$, it follows that $\mathsf{algm}(A,\lambda) = \mathsf{geomm}(A,\lambda) = 1$, so λ is semi-simple. Thus, the notion of semi-simplicity of an eigenvalue generalizes the notion of simplicity.

Example 8.4. Let A be the matrix

$$A = \begin{pmatrix} 2 & 1 & 0 \\ 0 & 2 & 1 \\ 0 & 0 & 2 \end{pmatrix}.$$

Its characteristic polynomial is

$$p_A(\lambda) = \det(A - \lambda I_3) = \begin{vmatrix} 2-\lambda & 1 & 0 \\ 0 & 2-\lambda & 1 \\ 0 & 0 & 2-\lambda \end{vmatrix} = (2-\lambda)^3,$$

which means that A has the unique value 2 with algebraic multiplicity 3.

If $\boldsymbol{v} = \begin{pmatrix} 1 \\ 0 \\ 0 \end{pmatrix}$ is an eigenvector, then $A\boldsymbol{v} = 2\boldsymbol{v}$, which amounts to $v_1 = 1$, $v_2 = 0$, and $v_3 = 0$. Thus, the geometric multiplicity of 2 is equal to 1, hence the eigenvalue 2 is not semi-simple.

Note that if λ is a simple eigenvalue of A, then $\mathsf{geomm}(A, \lambda) = \mathsf{algm}(A, \lambda) = 1$, so λ is semi-simple.

Example 8.5. Let $A \in \mathbb{R}^{2\times 2}$ be the matrix

$$A = \begin{pmatrix} a & b \\ b & a \end{pmatrix},$$

where $a, b \in \mathbb{R}$ are such that $a \neq 0$ and $b \neq 0$. The characteristic polynomial of A is $p_A(\lambda) = (a-\lambda)^2 - b^2$, so $\mathsf{spec}(A) = \{a+b, a-b\}$ and $\mathsf{algm}(A, a+b) = \mathsf{algm}(A, a-b) = 1$. Thus, both $a+b$ and $a-b$ are simple eigenvalues of A, and, therefore, they are also semi-simple.

Theorem 8.29. *If $A \in \mathbb{R}^{n\times n}$ is a symmetric matrix, each of its eigenvalues is semi-simple.*

Proof. We saw that each symmetric matrix has real eigenvalues and is orthonormally diagonalizable (by Corollary 8.8). Starting from the real Schur factorization $A = UTU^{-1}$, where U is an orthogonal matrix and $T = \mathsf{diag}(t_{11}, \ldots, t_{nn})$ is a diagonal matrix, we can write

$AU = UT$. If we denote the columns of U by $\boldsymbol{u}_1, \ldots, \boldsymbol{u}_n$, then we can write

$$(A\boldsymbol{u}_1, \ldots, A\boldsymbol{u}_n) = (t_{11}\boldsymbol{u}_1, \ldots, t_{nn}\boldsymbol{u}_n),$$

so $A\boldsymbol{u}_i = t_{ii}\boldsymbol{u}_i$ for $1 \leqslant i \leqslant n$. Thus, the diagonal elements of T are the eigenvalues of A and the columns of U are corresponding eigenvectors. Since these eigenvectors are pairwise orthogonal, the dimension of the invariant subspace that corresponds to an eigenvalue equals the algebraic multiplicity of the eigenvalue, so each eigenvalue is semi-simple. □

8.6 λ-Matrices

The set of polynomials over the complex field depending on a non-determinate λ is denoted by $\mathbb{C}[\lambda]$. If $f(\lambda) \in \mathbb{C}[\lambda]$, we denote its degree by $\mathsf{deg}(f)$.

Definition 8.4. A λ-*matrix* is a polynomial of the form

$$G(\lambda) = A_0\lambda^m + A_1\lambda^{m-1} + \cdots + A_m,$$

where $A_0, A_1, \ldots, A_m$ are matrices in $\mathbb{C}^{p\times q}$.

If $A_0 \neq O_{p,q}$, then we say that m is the *degree* of the λ-matrix G and we write $\mathsf{deg}(G) = m$.

A square λ-matrix, that is, a λ-matrix of type $n \times n$, is *regular* if its leading coefficient A_0 is a nonsingular matrix.

The set of λ-matrices of format $p \times q$ over a field F is denoted by $F[\lambda]^{p\times q}$.

In the special case $p = q = 1$, $G(\lambda)$ is a usual polynomial in λ.

Note that a λ-matrix of type $p \times q$ can also be regarded as a matrix whose entries are polynomials in λ, that is, as a member of $\mathbb{C}[\lambda]^{p\times q}$; the converse is also true, that is, every matrix whose entries are polynomials in λ is a λ-matrix.

Example 8.6. Consider the λ-matrix

$$G(\lambda) = A_0\lambda^3 + A_1\lambda^2 + A_2\lambda + A_3,$$

where

$$A_0 = \begin{pmatrix} 0 & 0 & 1 \\ 1 & -1 & 0 \\ 2 & 0 & 0 \end{pmatrix}, A_1 = \begin{pmatrix} 1 & 1 & 1 \\ 2 & 0 & 0 \\ 0 & -2 & 0 \end{pmatrix},$$

and

$$A_2 = \begin{pmatrix} 2 & 3 & 4 \\ 1 & 9 & 4 \\ 3 & 0 & 0 \end{pmatrix}, A_3 = \begin{pmatrix} 0 & 0 & 0 \\ 1 & 0 & 0 \\ 0 & 0 & 1 \end{pmatrix}.$$

$G(\lambda)$ is the matrix

$$G(\lambda) = \begin{pmatrix} \lambda^2 + 2\lambda & \lambda^2 + 3\lambda & \lambda^3 + \lambda^2 + 4\lambda \\ \lambda^3 + 2\lambda^2 + \lambda + 1 & -\lambda^3 + 9\lambda & 4\lambda \\ 2\lambda^3 + 3\lambda & -2\lambda^2 & 1 \end{pmatrix}.$$

Theorem 8.30. *A λ-matrix $G(\lambda)$ is invertible if and only if* $\det(G(\lambda))$ *is a non-zero constant.*

Proof. Suppose that $G(\lambda)$ is invertible, that is, there exists a λ-matrix $H(\lambda)$ such that $G(\lambda)H(\lambda) = I_n$. Then $\det(G(\lambda))\det(H(\lambda)) = 1$, so $\det(G(\lambda))$ is a constant that cannot be 0.

Conversely, suppose that $\det(G(\lambda))$ is a non-zero constant. If we construct $G^{-1}(\lambda)$ using the standard approach, it is immediate that $G^{-1}(\lambda)$ is a λ-matrix, so $G(\lambda)$ is invertible. □

The sum and product of λ-matrices are component-wise defined exactly as the general sum and products of matrices. Note that if $G(\lambda)$ and $H(\lambda)$ are two λ-matrices, the degree of their product is not necessarily equal to the sum of the degrees of the factors. In general, the degree of $G(\lambda)H(\lambda)$ is not larger than the sum of the degrees of the factors.

Example 8.7. Let $G(\lambda), H(\lambda)$ be the λ-matrices defined by

$$G(\lambda) = \begin{pmatrix} 0 & 1 \\ 0 & 0 \end{pmatrix} \lambda + \begin{pmatrix} 1 & 2 \\ -2 & 1 \end{pmatrix}, \text{ and } H(\lambda) = \begin{pmatrix} 1 & 1 \\ 0 & 0 \end{pmatrix} \lambda + \begin{pmatrix} 1 & 3 \\ 3 & 2 \end{pmatrix}.$$

We have

$$GH = \begin{pmatrix} 4 & 3 \\ -2 & -2 \end{pmatrix} \lambda + \begin{pmatrix} 7 & 7 \\ 1 & -4 \end{pmatrix}.$$

Let $G(\lambda) = A_0\lambda^m + A_1\lambda^{m-1} + \cdots + A_m$ be a λ-matrix. The same matrix can also be written as $\tilde{G}(\lambda) = \lambda^m A_0 + \lambda^{m-1}A_1 + \cdots + A_m$. If the matrix H is substituted for λ in $G(\lambda)$ and in $\tilde{G}(\lambda)$, the results

$$\begin{aligned} G(B) &= A_0B^m + A_1B^{m-1} + \cdots + A_m, \\ \tilde{G}(B) &= B^m A_0 + B^{m-1}A_1 + \cdots + A_m \end{aligned}$$

are distinct, in general, because matrices of the form B^j do not commute with $A_0, A_1, \ldots, A_m$. Thus, $G(A)$ is referred to as the *right value* of G in A and $\tilde{G}(A)$ is referred to as the *left value* of G in A.

Theorem 8.31. *Let $G(\lambda)$ and $H(\lambda)$ be two $n \times n$ λ-matrices such that $H(\lambda)$ is a regular matrix. There are two unique $n\times n$ λ-matrices $Q(\lambda)$ and $R(\lambda)$ such that $G(\lambda) = H(\lambda)Q(\lambda)+R(\lambda)$ such that $R(\lambda) = O_{n,n}$ or $\mathsf{deg}(R(\lambda)) < \mathsf{deg}(H(\lambda))$.*

Proof. Suppose that

$$\begin{aligned} G(\lambda) &= A_0\lambda^m + A_1\lambda^{m-1} + \cdots + A_m, \\ F(\lambda) &= B_0\lambda^q + B_1\lambda^{q-1} + \cdots + B_q, \end{aligned}$$

where $A_0 \neq O_{n,n}$ and $\det(B_0) \neq 0$. We assume that $m \geqslant q$; otherwise, we can take $Q(\lambda) = O_{n,n}$ and $R(\lambda) = G(\lambda)$.

We define a sequence of λ-matrices $G^{(1)}(\lambda), G^{(2)}(\lambda)\ldots$ as having a non-increasing sequence of degrees $m_1 \geqslant m_2 \geqslant \cdots$. The coefficients of $G^{(p)}(\lambda)$ are denoted by $A_j^{(p)}$. We define $G^{(1)}(\lambda) = G(\lambda) - H(\lambda)B_0^{-1}A_0\lambda^{m-q}$ and

$$G^{(k)}(\lambda) = G^{(k-1)}(\lambda) - H(\lambda)B_0^{-1}A_0^{(k-1)}\lambda^{m_{k-1}-q},$$

as long as $m_k \geqslant q$. If $m_{k+1} < q$, the computation halts and

$$G^{(k+1)}(\lambda) = G^{(k)}(\lambda) - H(\lambda)B_0^{-1}A_0^{(k)}.$$

This allows us to write

$$\begin{aligned} G(\lambda) = H(\lambda)[&B_0^{-1}A_0\lambda^{m-q} + B_0^{-1}A_0^{(1)}\lambda^{m_1-q} + \cdots + B_0^{-1}A_0^{(k)}] \\ &+ G^{(k+1)}(\lambda). \end{aligned}$$

Therefore, we can take

$$Q(\lambda) = B_0^{-1}A_0\lambda^{m-q} + B_0^{-1}A_0^{(1)}\lambda^{m_1-q} + \cdots + B_0^{-1}A_0^{(k)},$$

and $R(\lambda) = G^{(k+1)}(\lambda)$; clearly, we have $\mathsf{deg}(R(\lambda)) < \mathsf{deg}(H(\lambda)$.

Suppose now that we have both $G(\lambda) = H(\lambda)Q(\lambda) + R(\lambda)$ and $G(\lambda) = H(\lambda)Q_1(\lambda) + R_1(\lambda)$, where $\mathsf{deg}(R(\lambda)) < \mathsf{deg}(H(\lambda))$ and $\mathsf{deg}(R_1(\lambda)) < \mathsf{deg}(H(\lambda))$. These equalities imply

$$H(\lambda)(Q(\lambda) - Q_1(\lambda)) = R_1(\lambda) - R(\lambda).$$

If $Q(\lambda) - Q_1(\lambda) \neq O_{n,n}$, we have $\mathsf{deg}(H(\lambda)(Q(\lambda) - Q_1(\lambda))) \geqslant \mathsf{deg}(H(\lambda))$ because $H(\lambda)$ is regular. However, $\mathsf{deg}(R_1(\lambda) - R(\lambda)) < \mathsf{deg}(H(\lambda))$, which leads to a contradiction. Therefore, we have $Q(\lambda) = Q_1(\lambda)$, which implies $R_1(\lambda) = R(\lambda)$. □

We refer to the λ-matrices $Q(\lambda)$ and $R(\lambda)$ defined by the equality $G(\lambda) = H(\lambda)Q(\lambda) + R(\lambda)$ of Theorem 8.31 as the *left quotient* and *left remainder* of the division of $G(\lambda)$ by $H(\lambda)$.

It is possible to prove in an entirely similar manner, under the same assumptions as the ones of Theorem 8.31, that the matrices $P(\lambda)$ and $S(\lambda)$, such that $G(\lambda) = B(\lambda)H(\lambda) + S(\lambda)$ and $S(\lambda) = O_{n,n}$ or $\mathsf{deg}(S(\lambda)) < \mathsf{deg}(H(\lambda))$, are uniquely determined. In this case we refer to the matrices $P(\lambda)$ and $S(\lambda)$ as the *right quotient* and the *right remainder* of the division of $G(\lambda)$ by $H(\lambda)$.

Corollary 8.21. *The left remainder of the division of the λ-matrix $G(\lambda) = A_0\lambda^m + A_1\lambda^{m-1} + \cdots + A_m$ by $H(\lambda) = \lambda I - C$ is the matrix $G(C)$, where $G(C) = C^m A_0 + C^{m-1}A_1 + \cdots + A_m$. The right remainder of the division of $G(\lambda)$ by $H(\lambda)$ is the matrix $\tilde{G}(C) = A_0C^m + A_1C^{m-1} + \cdots + A_m$.*

Proof. Starting from the equality

$$A_0\lambda^m + A_1\lambda^{m-1} + \cdots + A_m = (\lambda I - C)Q(\lambda) + R,$$

where $Q(\lambda) = Q_0\lambda_{m-1} + \cdots + Q_{m-1}$ and identifying the coefficients of the same powers of λ, we obtain the equalities

$$\begin{aligned}
&A_0 = Q_0, A_1 = Q_1 - CQ_0, A_2 = Q_2 - CQ_1, \ldots,\\
&A_{m-1} = Q_{m-1} - CQ_{m-2}, A_m = R - CQ_{m-1}.
\end{aligned}$$

These equalities imply $R = C^m A_0 + C^{m-1}A_1 + \cdots + CA_{m-1} + A_m = A(C)$. The argument for the right remainder is similar. □

Polynomials with complex coefficients, that is, polynomials in $\mathbb{C}[\lambda]$, are 1×1 λ-matrices.

Definition 8.5. Let $G, H \in \mathbb{C}[\lambda]$ be two polynomials and let $A \in \mathbb{C}^{n\times n}$ be a matrix such that $\det(H(A)) \neq 0$. The matrix $Q(A) = G(A)(H(A))^{-1}$ is the *quotient* of $G(A)$ by $H(A)$.

A *rational matrix function* of A is a quotient of two polynomials $H(A)/G(A)$.

Since $G(A)H(A) = H(A)G(A)$ for every $A \in \mathbb{C}^{n\times n}$, if $H(A)$ is invertible, then $(H(A))^{-1}G(A) = G(A)(H(A))^{-1}$.

Theorem 8.32. *Let $A \in \mathbb{C}^{n\times n}$. There exists a unique polynomial m_A of minimal degree whose leading coefficient is 1 such that $m_A(A) = O_{n\times n}$.*

Proof. Theorem 8.24 involving the characteristic polynomial p_A of A shows that the set of polynomials whose leading coefficient is 1 and that has A as a root is non-empty. Thus, we need to show the uniqueness of a polynomial of minimal degree.

Suppose that f and g are two distinct polynomials with leading coefficient 1, of minimal degree k, such that $f(A) = g(A) = O_{n,n}$. Then $(f - g)(A) = O_{n,n}$ and the degree of the polynomial $f - g$ is less than k. This contradicts the minimality of the degrees of f and g. Thus, $f = g$. □

Definition 8.6. Let $A \in \mathbb{C}^{n\times n}$. The polynomial m_A of minimal degree whose leading coefficient is 1 such that $m_A(A) = O_{n\times n}$ is referred to as the *minimal polynomial* of A.

Theorem 8.33. *$A \in \mathbb{C}^{n\times n}$ is an invertible matrix if and only if $m_A(0) \neq 0$.*

Proof. Let $m_A(\lambda) = \lambda^k + a_1\lambda^{k-1} + \cdots + a_{k-1}\lambda + a_k$ be the minimal polynomial of A.

Suppose that A is an invertible matrix. If $a_k = 0$, then $m_A(\lambda) = \lambda(\lambda^{k-1} + a_1\lambda^{k-2} + \cdots + a_{k-1})$, so $A(A^{k-1} + a_1A^{k-2} + \cdots + a_{k-1}I_n) = O_{n,n}$. By multiplying the last equality by A^{-1} to the left, we obtain $A^{k-1} + a_1A^{k-2} + \cdots + a_{k-1}I_n = O_{n,n}$, which contradicts the minimality of the degree of m_A. Thus, $a_k \neq 0$.

Conversely, suppose that $a_k \neq 0$. Since

$$A^k + a_1 A^{k-1} + \cdots + a_{k-1} A + a_k I_n = O_{n,n},$$

it follows that

$$A(A^{k-1} + a_1 A^{k-2} + \cdots + a_{k-1} I_n) = -a_k I_n$$

Thus, A is invertible and its inverse matrix is

$$A^{-1} = -\frac{1}{a_k}\left(A^{k-1} + a_1 A^{k-2} + \cdots + a_{k-1} I_n\right).$$

□

Definition 8.7. An *annihilating polynomial* of $A \in \mathbb{C}^{n\times n}$ is a polynomial f such that $f(A) = O_{n,n}$.

Theorem 8.34. *If f is an annihilating polynomial for $A \in \mathbb{C}^{n\times n}$, than m_A divides evenly f.*

Proof. Suppose that under the hypothesis of the theorem, m_A does not evenly divide f. Then we can write $f(\lambda) = m_A(\lambda)q(\lambda) + r(\lambda)$, where r is a polynomial of degree smaller than the degree of m_A. Note, however, that $r(A) = f(A) - m_A(A)q(A) = O_{n,n}$, which contradicts the minimality of the degree of m_A. □

Let $A \in \mathbb{C}^{n\times n}$ be a matrix and let $B(\lambda)$ be the matrix whose transpose consists of the cofactors of the elements of the matrix $\lambda I_n - A$. Then,

$$B(\lambda)(\lambda I_n - A) = p_A(\lambda) I_n. \tag{8.4}$$

By substituting $\lambda = A$, we obtain an alternative proof of the Cayley–Hamilton equality $p_A(A) = O_{n,n}$.

The matrix $B(\lambda)$ introduced above allows us to give an explicit form of the minimal polynomial of a matrix $A \in \mathbb{C}^{n\times n}$.

Theorem 8.35. *Let $A \in \mathbb{C}^{n\times n}$ be a matrix. Its minimal polynomial is given by*

$$m_A(\lambda) = \frac{p_A(\lambda)}{d(\lambda)},$$

where $d(\lambda)$ is the greatest common divisor of the elements of the matrix $B(\lambda)$.

Proof. The definition of $d(\lambda)$ means that we can write $B(\lambda) = d(\lambda)C(\lambda)$, where the entries of $C(\lambda)$ are pairwise relatively prime polynomials. Thus, Equality (8.4) can be written as

$$C(\lambda)(\lambda I_n - A) = \frac{p_A(\lambda)}{d(\lambda)} I_n,$$

which shows that $\frac{p_A(\lambda)}{d(\lambda)}$ is an annihilating polynomial for A. Therefore, $\frac{p_A(\lambda)}{d(\lambda)}$ is divisible by the minimal polynomial $m_A(\lambda)$, which allows us to write

$$\frac{p_A(\lambda)}{d(\lambda)} = m_A(\lambda)r(\lambda).$$

Since $C(\lambda)(\lambda I_n - A) = m_A(\lambda)r(\lambda)I_n$, taking into account that $m_A(\lambda)I_n$ is an annihilator for A and, therefore, divisible by $\lambda I_n - A$, it follows that $m_A(\lambda)I_n = M(\lambda)(\lambda I_n - A)$. This implies

$$C(\lambda)(\lambda I_n - A) = \frac{p_A(\lambda)}{d(\lambda)} I_n = m_A(\lambda)r(\lambda)I_n = r(\lambda)M(\lambda)(\lambda I_n - A).$$

Therefore, $C(\lambda) = r(\lambda)M(\lambda)$, so $r(\lambda)$ is a common divisor of the elements of $C(\lambda)$. Since entries of $C(\lambda)$ are pairwise relatively prime polynomials, it follows that $r(\lambda)$ is a constant r_0. By the definition of the polynomials p_A and m_A, it follows that $r_0 = 1$, which implies $m_A(\lambda) = \frac{p_A(\lambda)}{d(\lambda)}$. □

Theorem 8.36. *If A and B are similar matrices in $\mathbb{C}^{n\times n}$, then their minimal polynomials are equal.*

Proof. Suppose that $A \sim B$, that is $A = P^{-1}BP$, where P is an invertible matrix and let m_A and m_B be the two minimal polynomials of A and B, respectively. We have

$$m_B(A) = m_B(P^{-1}BP) = P^{-1}m_B(B)P = O_{n,n},$$

so m_A divides m_B. In a similar manner, we can prove that m_B divides m_A, so $m_A = m_B$ because both polynomials have 1 as leading coefficient. □

Let $C(\lambda) \in \mathbb{C}[\lambda]^{n\times n}$ be a λ-matrix such that $\mathsf{rank}(C) = r$. Thus, $C(\lambda)$ has at least one non-null minor of order r and all minors of order greater than r are null. Denote by $\delta_k(\lambda)$ the polynomial having 1 as leading coefficient that is the greatest common divisor of all minors of order k of $C(\lambda)$ for $1 \leqslant k \leqslant r$. The polynomial $\delta_0(\lambda)$ is defined as equal to 1.

Since every minor of order k is a linear combination of minors of order $k-1$, it follows that $\delta_{k-1}(\lambda)$ divides $\delta_k(\lambda)$ for $1 \leqslant k \leqslant r$.

Definition 8.8. The *invariant factors* of the matrix $C(\lambda)$ with $\mathsf{rank}(C) = r$ are the polynomials $t_0(\lambda), \ldots, t_{r-1}(\lambda)$, where

$$t_{r-k}(\lambda) = \frac{\delta_k(\lambda)}{\delta_{k-1}(\lambda)}$$

for $1 \leqslant k \leqslant r$.

Next, we introduce the notion of *elementary transformation matrices* for λ-matrices.

Definition 8.9. The *row elementary transformation matrices* of λ-matrices are the matrices $T^{a(i)}$, $T^{(p)\leftrightarrow(q)}$, and $T^{(i)+p(\lambda)(j)}$, where $T^{a(i)}$ and $T^{(p)\leftrightarrow(q)}$ were introduced in Examples 3.34 and 3.35, and $T^{(i)+p(\lambda)(j)}$ is defined as

$$T^{(i)+p(\lambda)(j)} = \begin{pmatrix} 1 & 0 & \cdots & \cdots & \cdots & 0 & \cdots 0 \\ 0 & 1 & \cdots & \cdots & \cdots & 0 & \cdots 0 \\ \vdots & \vdots & \cdots & \vdots & \cdots & \vdots & \cdots 0 \\ 0 & 0 & \cdots & p(\lambda) & \cdots & 1 & \cdots 0 \\ \vdots & \vdots & \cdots & \vdots & \cdots & \vdots & \cdots 1 \end{pmatrix},$$

where the polynomial $p(\lambda)$ occurs in row i and column j.

Note that all elementary transformation matrices are invertible.

The effect of a left multiplication of a matrix $G(\lambda)$ by any of these matrices is clearly identical to the effect of left multiplication of a matrix by the usual elementary transformation matrices that affect the rows of a matrix. Similarly, if one multiplies a matrix $G(\lambda)$ at the right by $T^{(i)\leftrightarrow(j)}$, $T^{a(i)}$, and $T^{(i)+p(\lambda)(j)}$, the effect is to exchange columns i and j, multiply the ith column by a, and add the jth column multiplied by a to the ith column, respectively.

Definition 8.10. To λ-matrices $G(\lambda), H(\lambda)$ are *equivalent* if one of them can be obtained from the other my multiplications by elementary transformation matrices.

It is easy to verify that the relation introduced in Definition 8.10 is indeed an equivalence relation. We denote the equivalence of two λ-matrices $G(\lambda)$ and $H(\lambda)$ by $A \sim_\lambda B$.

Theorem 8.37. *Let $G(\lambda) \in \mathbb{C}[\lambda]^{n\times n}$ be a λ-matrix having rank r. There exists an equivalent diagonal λ-matrix*

$$D(\lambda) = \mathsf{diag}(t_{r-1}(\lambda), t_{r-2}(\lambda), \ldots, t_0(\lambda)),$$

where $t_0(\lambda), \ldots, t_{r-2}(\lambda), t_{r-1}(\lambda)$ are the invariant factors of $G(\lambda)$. Furthermore, each invariant factor t_{r-j} is a divisor of the invariant factor t_{r-j+1} for $1 \leqslant j \leqslant r$.

Proof. Let $B^{(0)}(\lambda) \in \mathbb{C}[\lambda]^{n\times n}$ be a λ-matrix that is equivalent to $G(\lambda)$ such that the polynomial $b_{11}(\lambda)$ has minimal degree among all polynomials $b_{ij}(\lambda)$.

We claim that $(B^{(0)}(\lambda))$ can be chosen such that $(B^{(0)}(\lambda))_{11}$ divides all polynomials located on the first row. Indeed, suppose that this is not the case and $(B^{(0)}(\lambda))_{11}$ does not divide $(B^{(0)}(\lambda))_{1j}$. Since $\mathsf{deg}((B^{(0)}(\lambda))_{1j}) \geqslant \mathsf{deg}((B^{(0)}(\lambda))_{11})$, we can write

$$\mathsf{deg}((B^{(0)}(\lambda))_{1j}) = \mathsf{deg}((B^{(0)}(\lambda))_{1j})Q(\lambda) + R(\lambda),$$

where $\mathsf{deg}(R(\lambda)) < \mathsf{deg}((B^{(0)}(\lambda))_{1j})$. Then by subtracting from the jth column the first column multiplied by $Q(\lambda)$, we obtain a matrix equivalent to $B^{(0)}(\lambda)$ (and, therefore, with $G(\lambda)$) that has an element in the position $(1, j)$ of degree smaller than the element in position $(1, 1)$. This contradicts the definition of $B^{(0)}(\lambda)$. A similar argument shows that $(B^{(0)}(\lambda))_{11}$ divides all polynomials located on the first column.

Since the entries in the first row and the first column are divisible by $(B^{(0)}(\lambda))_{11}$, starting from the matrix $B^{(0)}(\lambda)$ we construct, by applying elementary transformations, the matrix $B^{(1)}(\lambda)$ that has the form

$$\begin{pmatrix} b_{11}^{(1)}(\lambda) & 0 & \cdots & 0 \\ 0 & b_{22}^{(1)}(\lambda) & \cdots & b_{2n}^{(1)}(\lambda) \\ \vdots & \vdots & \cdots & \vdots \\ 0 & b_{n2}^{(1)}(\lambda) & \cdots & b_{nn}^{(1)}(\lambda) \end{pmatrix}$$

and we have $B^{(0)}(\lambda) \sim_\lambda B^{(1)(\lambda)}$. Note that the degrees of the components of $B^{(1)(\lambda)}$ outside the first row and the first columns are at least equal to the degree of $(B^{(1)}(\lambda))_{11}$.

Repeating this argument for the matrix $B^{(1)}(\lambda)$ and involving the second row and the second column of this matrix, we obtain the matrix

$$B^{(2)}(\lambda) = \begin{pmatrix} b_{11}^{(1)}(\lambda) & 0 & 0 & 0 \\ 0 & b_{22}^{(2)}(\lambda) & 0 & 0 \\ \vdots & \vdots & \vdots & \vdots \\ 0 & 0 & \cdots & b_{nn}^{(1)}(\lambda) \end{pmatrix},$$

which is equivalent to $B^{(1)}(\lambda)$ and therefore with $G(\lambda)$. We can assume that the degrees of all polynomials of $B^{(2)}$ located below the first row and at the right of the first columns are at least equal to $\mathsf{deg}(b_{11}^{(1)}(\lambda))$.

Without loss of generality we may assume that $b_{11}^{(1)}(\lambda)$ divides the polynomial $b_{22}^{(2)}(\lambda)$. Indeed, if this is not the case, we can write

$$b_{22}^{(2)}(\lambda) = b_{11}^{(1)}(\lambda)u(\lambda) + v(\lambda),$$

where $\mathsf{deg}(v(\lambda)) < \mathsf{deg}(b_{11}^{(1)}(\lambda))$. By multiplying the first column of $B^{(2)}(\lambda)$ by $-u(\lambda)$ and adding it to the second column, we have the equivalent matrix

$$\begin{pmatrix} b_{11}^{(1)}(\lambda) & -b11^{(1)}(\lambda)u(\lambda) & 0 & 0 \\ 0 & b_{22}^{(2)}(\lambda) & 0 & 0 \\ \vdots & \vdots & \vdots & \vdots \\ 0 & 0 & \cdots & b_{nn}^{(1)}(\lambda) \end{pmatrix}.$$

By adding the first row to the second, we have another equivalent matrix

$$\begin{pmatrix} b_{11}^{(1)}(\lambda) & -b11^{(1)}(\lambda)u(\lambda) & 0 & 0 \\ 0 & v(\lambda) & 0 & 0 \\ \vdots & \vdots & \vdots & \vdots \\ 0 & 0 & \cdots & b_{nn}^{(1)}(\lambda) \end{pmatrix}.$$

Finally, by adding the first column multiplied by $u(\lambda)$ to the second column, we obtain yet another equivalent matrix

$$\begin{pmatrix} b_{11}^{(1)}(\lambda) & 0 & 0 & 0 \\ 0 & v(\lambda) & 0 & 0 \\ \vdots & \vdots & \vdots & \vdots \\ 0 & 0 & \cdots & b_{nn}^{(1)}(\lambda) \end{pmatrix}.$$

The existence of this matrix contradicts the assumption made about the matrix $B^{(2)}(\lambda)$ because $\mathsf{deg}(v(\lambda)) < \mathsf{deg}(b_{11}^{(1)}(\lambda)$. Thus, $b_{11}^{(1)}(\lambda)$ divides the polynomial $b_{22}^{(2)}(\lambda)$.

Eventually, we are left with the matrix

$$D(\lambda) = B^{(r)}(\lambda) = \begin{pmatrix} b_{11}^{(1)}(\lambda) & 0 & 0 & \cdots & 0 \\ 0 & b_{22}^{(2)}(\lambda) & 0 & \cdots & 0 \\ \vdots & \vdots & \vdots & \cdots & 0 \\ 0 & 0 & b_{rr}^{(r)}(\lambda) & \cdots & 0 \\ \vdots & \vdots & \vdots & \cdots & 0 \\ 0 & 0 & 0 & \cdots & 0 \end{pmatrix}$$

which is equivalent to $G(\lambda)$. The degrees of the polynomials that occur on the diagonal of this matrix are increasing and each polynomial $b_{jj}^{(j)}(\lambda)$ divides its successor $b_{j+1\,j+1}^{(j+1)}(\lambda)$ for $1 \leqslant j \leqslant r - 1$.

Note that the divisibility properties obtained above imply that

$$\begin{aligned}
\delta_r(\lambda) &= b_{11}^{(1)}(\lambda) b_{22}^{(2)}(\lambda) \cdots b_{rr}^{(r)}(\lambda) \\
\delta_{r-1}(\lambda) &= b_{11}^{(2)}(\lambda) \cdots b_{r-1\,r-1}^{(r-1)}(\lambda) \\
&\vdots \\
\delta_1(\lambda) &= b_{11}^{(1)}(\lambda).
\end{aligned}$$

Thus, the invariant factors of this matrix are

$$\begin{aligned}
t_0(\lambda) &= \frac{\delta_r(\lambda)}{\delta_{r-1}(\lambda)} = b_{rr}^{(r)}(\lambda), \\
&\vdots \\
t_j(\lambda) &= \frac{\delta_{r-j}(\lambda)}{\delta_{r-j-1}(\lambda)} = b_{r-j\,r-j}^{(r-j)}(\lambda) \\
&\vdots \\
t_{r-1}(\lambda) &= \frac{\delta_1(\lambda)}{\delta_0(\lambda)} = b_{11}^{(1)}(\lambda).
\end{aligned}$$

□

Theorem 8.38. *Elementary transformations applied to λ-matrices preserve the invariant factors.*

Proof. It is clear that by multiplying a row (or a column) by a constant or by permuting two rows (or columns), the value of the polynomials $\delta_k(\lambda)$ is not affected.

Suppose that we add the jth row multiplied by the polynomial p to the ith row. Minors of order k can be classified in the following categories:

(i) minors that contain both the ith row and the jth row;
(ii) minors that contain neither the ith row nor the jth row;
(iii) minors that contain the jth row but do not contain the ith row, and
(iv) minors that contain the ith row but do not contain the jth row.

It is clear that minors in the first three categories are not affected by this elementary transformation. If M is a minor of order k of the matrix $G(\lambda)$ that contains the ith row but does not contain the jth row and we add the elements of the jth row, then M is replaced by the sum of two minors of order k. Thus, $\delta_k(\lambda)$ is not affected by this elementary transformation. □

Theorem 8.39. *If two λ-matrices have the same invariant factors, then they are equivalent.*

Proof. Let $G(\lambda)$ and $H(\lambda)$ be two matrices that have the same invariant factors $t_0(\lambda), \ldots, t_{r-1}(\lambda)$. Both matrices are equivalent to the matrix

$$D(\lambda) = \begin{pmatrix} t_0(\lambda) & 0 & 0 & \cdots & 0 \\ 0 & t_1(\lambda) & 0 & \cdots & 0 \\ \vdots & \vdots & \vdots & \cdots & 0 \\ 0 & 0 & t_{r-1}(\lambda) & \cdots & 0 \\ \vdots & \vdots & \vdots & \cdots & 0 \\ 0 & 0 & 0 & \cdots & 0 \end{pmatrix}$$

that has their common invariant factors on its main diagonal and, therefore, are equivalent. □

We saw that a λ-matrix $G(\lambda)$ is invertible if and only if $\det(G(\lambda))$ is a non-zero constant.

Lemma 8.5. *Every invertible λ-matrix $G(\lambda) \in \mathbb{C}[\lambda]^{n\times n}$ is a product of elementary transformation matrices.*

Proof. Since $G(\lambda)$ is an invertible λ-matrix, $\det(A)$ is a non-zero constant, so $\delta_n(\lambda)$ is a non-zero constant c. Therefore, all polynomials $\delta_k(\lambda)$ are equal to c, so all invariant factors of $G(\lambda)$ equal 1. Thus, $G(\lambda)$ is equivalent to matrix I_n, which amounts to the having $G(\lambda)$ equal to a product of elementary transformation matrices. □

Theorem 8.40. *The λ-matrices $G(\lambda)$ and $H(\lambda)$ are equivalent if and only if there exist two invertible λ-matrices $P(\lambda)$ and $Q(\lambda)$ such that $G(\lambda) = P(\lambda)H(\lambda)Q(\lambda)$.*

Proof. If $G(\lambda)$ and $H(\lambda)$ are equivalent, then $H(\lambda)$ can be obtained from $G(\lambda)$ by applying elementary transformations (multiplying by elementary λ-matrices for row transformations and multiplying by elementary λ-matrices for column transformations). Each of these matrices is invertible and this leads to the desired equality.

Conversely, suppose that the equality $G(\lambda) = P(\lambda)H(\lambda)Q(\lambda)$ holds for two invertible λ-matrices $P(\lambda)$ and $Q(\lambda)$. By Lemma 8.5, both $P(\lambda)$ and $Q(\lambda)$ are products of elementary transformation matrices, so $G(\lambda) \sim_\lambda H(\lambda)$. □

Example 8.8. Let

$$A(\lambda) = \begin{pmatrix} \lambda & 2\lambda & 2\lambda \\ \lambda^2+\lambda & 3\lambda^2+\lambda & 2\lambda^2+2\lambda \\ \lambda & 2\lambda & \lambda^3+\lambda \end{pmatrix}$$

be a λ-matrix. This matrix is equivalent to the matrix

$$\begin{pmatrix} \lambda & 0 & 0 \\ 0 & \lambda^2-\lambda & 0 \\ 0 & 0 & \lambda^3-\lambda \end{pmatrix}.$$

The invariant factors of A are $t_0(\lambda) = \lambda$, $t_1(\lambda) = \lambda^2 - \lambda$, and $t_2(\lambda) = \lambda^3 - \lambda$,

Let $t_{r-1}(\lambda)$ be the invariant factor of A having the highest degree and let

$$t_{r-1}(\lambda) = p_1(\lambda) \cdots p_k(\lambda)$$

be the factorization of t_{r-1} as a product of irreducible polynomials over the field F that contains the elements of A.

Definition 8.11. The *elementary divisors* of the λ-matrix $G(\lambda)$ are the irreducible polynomials $p_1(\lambda), \ldots, p_k(\lambda)$ that occur in the factorization of the invariant factor of the highest degree of $G(\lambda)$.

Since $i \leqslant j$ implies that $t_i(\lambda)$ divides $t_j(\lambda)$ for $1 \leqslant i, j \leqslant r-1$, it follows that the factorization of any of the polynomials $t_i(\lambda)$ may contain only these elementary divisors.

Example 8.9. The elementary divisors of the matrix $A(\lambda)$ considered in Example 8.8 are $\lambda, \lambda+1$, and $\lambda - 1$.

8.7 The Jordan Canonical Form

We begin by introducing a type of matrix which is as defective as possible.

Definition 8.12. Let λ be a complex number. An r-*Jordan block associated with* λ is the matrix

$$B_r(\lambda) = \begin{pmatrix} \lambda & 1 & 0 & \cdots & 0 \\ 0 & \lambda & 1 & \cdots & 0 \\ \vdots & \vdots & \ddots & \ddots & \vdots \\ 0 & 0 & 0 & \cdots & 1 \\ 0 & 0 & 0 & \cdots & \lambda \end{pmatrix} \in \mathbb{C}^{r \times r}.$$

The matrix $B_r(\lambda)$ is diagonal if and only if $r = 1$; otherwise, that is, if $r > 1$, the matrix $B_r(\lambda)$ is not even diagonalizable. Also, a block $B_r(\lambda)$ can be written as

$$B_r(\lambda) = \lambda I_r + E_r,$$

where

$$E_r = \begin{pmatrix} \mathbf{0}_{r-1} & I_{r-1} \\ 0 & \mathbf{0}'_{r-1} \end{pmatrix}.$$

The unique eigenvalue of the Jordan block $B_r(\lambda)$ is $\lambda_1 = \lambda$ and $\mathsf{algm}(B_r, \lambda) = n$. On the other hand, the invariant subspace corresponding to λ is the subspace generated by $\boldsymbol{e}_1$, so $\mathsf{geomm}(B_r, \lambda) = 1$. Thus, every r-Jordan block (for $r > 1$) is a defective matrix.

Definition 8.13. A *Jordan segment* associated with the number λ is a block-diagonal matrix $J_{r_1,\ldots,r_k}(\lambda)$ given by

$$J_{r_1,\ldots,r_k}(\lambda) = \begin{pmatrix} B_{r_1}(\lambda) & O & \cdots & O \\ O & B_{r_2}(\lambda) & \cdots & O \\ \vdots & \vdots & \cdots & \vdots \\ O & O & \cdots & B_{r_k}(\lambda) \end{pmatrix},$$

where $B_{r_1}(\lambda), \ldots, B_{r_k}(\lambda)$ are Jordan blocks and $r_1 \geqslant r_2 \geqslant \cdots \geqslant r_k$. The sequence $(r_1, r_2, \ldots, r_k)$ is the *Segrè sequence of* the segment $J_{r_1,\ldots,r_k}(a)$.

Given a Segrè sequence $(r_1, r_2, \ldots, r_k)$ and a number λ, the Jordan segment $J_{r_1,\ldots,r_k}(\lambda)$ is completely determined.

A Jordan segment $J_{r_1,\ldots,r_k}(\lambda)$ is a diagonal matrix if and only if each of its Jordan blocks is unidimensional. Otherwise, a Jordan segment is not even diagonalizable. Indeed, if $J_{r_1,\ldots,r_k}(\lambda) = XDX^{-1}$, where D is a diagonal matrix and X is an invertible matrix, then $D = \mathsf{diag}(\lambda, \ldots, \lambda) = \lambda I$, so $J_{r_1,\ldots,r_k}(\lambda) = \lambda I$, which contradicts the fact that $J_{r_1,\ldots,r_k}(\lambda)$ contains a Jordan block of size larger than 1.

By Theorem 7.15, the spectrum of a Jordan segment $J_{r_1,\ldots,r_k}(\lambda) \in \mathbb{C}^{n\times n}$ consists of a single eigenvalue λ of algebraic multiplicity n. The geometric multiplicity of λ is k and the eigenvectors that generate the invariant subspace of λ are $\boldsymbol{e}_1, \boldsymbol{e}_{r_1+1}, \ldots, \boldsymbol{e}_{r_1+\cdots+r_{k-1}+1}$.

Definition 8.14. A *Jordan matrix* $R \in \mathbb{C}^{n\times n}$ is a block diagonal matrix, whose blocks are Jordan segments,

$$R = (J_{r_{1,1},\ldots,r_{1,k_1}}(\lambda_1), \ldots, J_{r_{p,1},\ldots,r_{p,k_p}}(\lambda_p)).$$

We shall prove that every matrix $A \in \mathbb{C}^{n\times n}$ is similar to a Jordan matrix R such that:

(i) for each eigenvalue λ_ℓ of A we have a Jordan segment $J_{r_{\ell,1},\ldots,r_{\ell,k_\ell}}(\lambda_\ell)$ in R;

(ii) for the Jordan segment $J_{r_{\ell,1},\ldots,r_{\ell,k_\ell}}(\lambda_\ell)$ that corresponds to the eigenvalue λ_ℓ, the number k_ℓ of Jordan blocks equals $\mathsf{geomm}(A, \lambda_\ell)$;

(iii) the algebraic multiplicity of λ_ℓ equals the size of the Jordan segment that corresponds to this eigenvalue, that is, we have

$$\mathsf{algm}(A, \lambda_\ell) = \sum_{i=1}^{k_\ell} r_{\ell,i}.$$

Next we give an algorithmic proof of the fact that for every square matrix $A \in \mathbb{C}^{n\times n}$ there exists a similar Jordan matrix. This proof was obtained in [54].

Theorem 8.41. *Let $T \in \mathbb{C}^{n\times n}$ be an upper triangular matrix. There exists a nonsingular matrix $X \in \mathbb{C}^{n\times n}$ such that*

$$X^{-1}TX = \mathsf{diag}(K_1, \ldots, K_m),$$

where $K_i = \lambda_i I_{p_i} + L_i$, L_i is strictly upper triangular and each λ_i is distinct, for $1 \leqslant i \leqslant m$.

Proof. The argument is by induction on n. The base case, $n = 1$, is immediate.

Suppose that the theorem holds for matrices of size less than n and let $T \in \mathbb{C}^{n\times n}$ be an upper triangular matrix. Without loss of generality we may assume that

$$T = \begin{pmatrix} T_1 & S \\ O & T_2 \end{pmatrix},$$

where T_1 and T_2 have no eigenvalues in common and $T_1 = \lambda_1 I + E$, where E is strictly upper triangular. Since

$$\begin{pmatrix} I & Y \\ O & I \end{pmatrix}\begin{pmatrix} T_1 & S \\ O & T_2 \end{pmatrix}\begin{pmatrix} I & -Y \\ O & I \end{pmatrix} = \begin{pmatrix} T_1 & -T_1Y + S + YT_2 \\ O & T_2 \end{pmatrix},$$

there exists a matrix Y such that

$$\begin{pmatrix} I & Y \\ O & I \end{pmatrix} T \begin{pmatrix} I & -Y \\ O & I \end{pmatrix} = \begin{pmatrix} T_1 & O \\ O & T_2 \end{pmatrix}$$

if and only if $S = T_1Y - YT_2$. By Theorem 8.25, this equation has a solution Y if and only if $\mathsf{spec}(T_1) \cap \mathsf{spec}(T_2) = \emptyset$, which is the case by the assumption we made about T_1 and T_2. By the induction hypothesis, T_2 may be reduced to block diagonal form, that is, there is a matrix Z such that $Z^{-1}T_2Z = \mathsf{diag}(H_1, \ldots, H_p)$. Therefore, we have

$$\begin{pmatrix} I & O \\ O & Z^{-1} \end{pmatrix}\begin{pmatrix} T_1 & O \\ O & T_2 \end{pmatrix}\begin{pmatrix} I & O \\ O & Z \end{pmatrix} = \begin{pmatrix} T_1 & O \\ O & Z^{-1}T_2Z \end{pmatrix}.$$

The last matrix is clearly in the block diagonal form, which also shows that the matrix X is given by

$$X = \begin{pmatrix} I & -Y \\ O & I \end{pmatrix}\begin{pmatrix} I & O \\ O & Z \end{pmatrix} = \begin{pmatrix} I & -YZ \\ O & Z \end{pmatrix}.$$

□

Lemma 8.6. *Let $E \in \mathbb{C}^{k\times k}$ be a matrix of the form*

$$E = \begin{pmatrix} \mathbf{0}_{k-1} & I_{k-1} \\ 0 & \mathbf{0}'_{k-1} \end{pmatrix}.$$

We have

$$E'E = \begin{pmatrix} 0 & \mathbf{0}'_{k-1} \\ \mathbf{0}_{k-1} & I_{k-1} \end{pmatrix},$$

$E\boldsymbol{e}_{i+1} = \boldsymbol{e}_i$ *and* $(I_k - E'E)\boldsymbol{x} = (\boldsymbol{e}_1'\boldsymbol{x})\boldsymbol{e}_1$ *for* $\boldsymbol{x} \in \mathbb{C}^k$.

Proof. The proofs of the equalities of the lemma are straightforward. □

Theorem 8.42. *Let $W \in \mathbb{C}^{n\times n}$ be a strictly upper triangular matrix. Then there is a nonsingular matrix X such that $X^{-1}WX = G$, where $G = \mathsf{diag}(E_1, \ldots, E_m)$ with each E_j given by*

$$E_j = \begin{pmatrix} \mathbf{0} & I_{k_j} \\ 0 & \mathbf{0}' \end{pmatrix}$$

such that $k_{j+1} \leqslant k_j$ for $1 \leqslant j \leqslant m-1$.

Proof. The proof is by induction on $n \geqslant 1$. The statement clearly holds in the base case, $n = 1$.

Assume that the result holds for strictly upper triangular matrices of format $(n-1)\times(n-1)$ and let $W \in \mathbb{C}^{n\times n}$ be an upper triangular matrix. We can write

$$W = \begin{pmatrix} 0 & \boldsymbol{u}' \\ \mathbf{0} & V \end{pmatrix},$$

where $V \in \mathbb{C}^{(n-1)\times(n-1)}$ is an upper triangular matrix. By the inductive hypothesis, there exists a nonsingular matrix Y such that

$$Y^{-1}VY = \begin{pmatrix} E_1 & O \\ O & H \end{pmatrix},$$

where $H = \mathsf{diag}(E_2, \ldots, E_m)$ and the order of E_1 is at least equal to the size of any of the matrices E_i, where $2 \leqslant i \leqslant m$. Then

$$\begin{pmatrix} 1 & 0 \\ 0 & Y^{-1} \end{pmatrix} W \begin{pmatrix} 1 & 0 \\ 0 & Y \end{pmatrix} = \begin{pmatrix} 0 & \boldsymbol{u}'Y \\ 0 & Y^{-1}VY \end{pmatrix}.$$

The matrix

$$\begin{pmatrix} 0 & \boldsymbol{u}'Y \\ 0 & Y^{-1}VY \end{pmatrix}$$

can be written as

$$\begin{pmatrix} 0 & \boldsymbol{u}'Y \\ 0 & Y^{-1}VY \end{pmatrix} = \begin{pmatrix} 0 & \boldsymbol{u}_1' & \boldsymbol{u}_2' \\ \mathbf{0} & E_1 & O \\ \mathbf{0} & O & H \end{pmatrix}$$

by partitioning the vector $\boldsymbol{u}'Y$ as $\boldsymbol{u}'Y = (\boldsymbol{u}_1' \ \boldsymbol{u}_2')$. This allows us to write

$$\begin{pmatrix} 1 & -\boldsymbol{u}_1'E_1' & 0 \\ \mathbf{0} & I & 0 \\ \mathbf{0} & \mathbf{0}' & I \end{pmatrix} \begin{pmatrix} 0 & \boldsymbol{u}_1' & \boldsymbol{u}_2' \\ \mathbf{0} & E_1 & O \\ \mathbf{0} & O & H \end{pmatrix} \begin{pmatrix} 1 & \boldsymbol{u}_1'E_1' & 0 \\ \mathbf{0} & I & \mathbf{0} \\ \mathbf{0} & \mathbf{0} & I \end{pmatrix}$$

$$= \begin{pmatrix} 0 & \boldsymbol{u}_1'(I - E_1'E_1) & u_2' \\ 0 & E_1 & O \\ 0 & O & H \end{pmatrix}.$$

By Lemma 8.6, we have

$$\boldsymbol{u}_1'(I - E_1'E_1) = ((I - E_1'E_1)\boldsymbol{u}_1)' = (\boldsymbol{e}_1'\boldsymbol{u}_1)\boldsymbol{e}_1',$$

so

$$\begin{pmatrix} 0 & \boldsymbol{u}_1'(I - E_1'E_1) & u_2' \\ 0 & E_1 & O \\ 0 & O & H \end{pmatrix} = \begin{pmatrix} 0 & (\boldsymbol{e}_1'\boldsymbol{u}_1)\boldsymbol{e}_1' & u_2' \\ 0 & E_1 & O \\ 0 & O & H \end{pmatrix}.$$

We need to consider two cases depending on whether the number $\boldsymbol{e}_1'\boldsymbol{u}_1$ is 0 or not.

Case 1: Suppose that $\boldsymbol{e}_1'\boldsymbol{u}_1 \neq 0$. We have

$$\begin{pmatrix} 0 & (\boldsymbol{e}_1'\boldsymbol{u}_1)\boldsymbol{e}_1' & u_2' \\ 0 & E_1 & O \\ 0 & O & H \end{pmatrix} \sim \begin{pmatrix} E & \mathbf{e}_1\boldsymbol{u}_2' \\ O & H \end{pmatrix},$$

where

$$E = \begin{pmatrix} 0 & \boldsymbol{e}_1' \\ \mathbf{0} & E_1 \end{pmatrix}.$$

This follows from the equality

$$\begin{pmatrix} \frac{1}{\boldsymbol{e}_1'\boldsymbol{u}_1} & 0 & 0 \\ 0 & I & 0 \\ 0 & 0 & \frac{1}{\boldsymbol{e}_1'\boldsymbol{u}_1}I \end{pmatrix} \begin{pmatrix} 0 & (\boldsymbol{e}_1'\boldsymbol{u}_1)\boldsymbol{e}_1' & u_2' \\ 0 & E_1 & O \\ 0 & O & H \end{pmatrix} \begin{pmatrix} \boldsymbol{e}_1'\boldsymbol{u}_1 & 0 & 0 \\ 0 & I & 0 \\ 0 & 0 & (\boldsymbol{e}_1'\boldsymbol{u}_1)I \end{pmatrix}$$

$$= \begin{pmatrix} 0 & \boldsymbol{e}_1' & \boldsymbol{u}_2' \\ 0 & E_1 & 0 \\ 0 & 0 & H \end{pmatrix}$$

because

$$\mathbf{e}_1\boldsymbol{u}_2' = \begin{pmatrix} \boldsymbol{u}_2' \\ O \end{pmatrix}.$$

Note that the order k of E is strictly greater than the order of and diagonal block of H, so $H^{k-1} = O$.

Define $\boldsymbol{s}_i' = \boldsymbol{u}_2 H^{i-1}$ for $1 \leqslant i \leqslant k$. Then

$$\begin{pmatrix} I & \mathbf{e}_{i+1}\boldsymbol{s}_i' \\ 0 & I \end{pmatrix} \begin{pmatrix} E & \mathbf{e}_i\boldsymbol{s}_i' \\ O & H \end{pmatrix} \begin{pmatrix} I & -\mathbf{e}_{i+1}\boldsymbol{s}_i' \\ 0 & I \end{pmatrix} = \begin{pmatrix} E & \mathbf{e}_{i+1}\boldsymbol{s}_{i+1}' \\ 0 & H \end{pmatrix}$$

for $1 \leqslant i \leqslant k-1$. We have $\boldsymbol{s}_k = \mathbf{0}$ because $H^{k-1} = O$, and it follows that W is similar to the matrix

$$\begin{pmatrix} E & O \\ O & H \end{pmatrix}.$$

Case 2: If $\mathbf{e}_1'\boldsymbol{u}_1 = 0$, by permuting the rows and columns, the matrix W is similar to the matrix

$$\begin{pmatrix} E_1 & 0 & 0 \\ 0 & 0 & \boldsymbol{u}_2' \\ 0 & 0 & H \end{pmatrix}.$$

Then, by the inductive hypothesis, there is a nonsingular matrix Z such that

$$Z^{-1} \begin{pmatrix} 0 & \boldsymbol{u}_2' \\ 0 & H \end{pmatrix} Z = L,$$

where L has the desired block diagonal form. Thus, W is similar to the matrix

$$\begin{pmatrix} E_1 & 0 \\ 0 & L \end{pmatrix}.$$

By applying a permutation of the blocks, we obtain a matrix in the proper form, which completes the proof. □

Theorem 8.43. *Let $A \in \mathbb{C}^{n\times n}$ be a matrix such that* $\mathsf{spec}(A) = \{\lambda_1, \ldots, \lambda_p\}$. *The matrix A is similar to a Jordan matrix*

$$R = (J_{r_{1,1},\ldots,r_{1,k_1}}(\lambda_1), \ldots, J_{r_{p,1},\ldots,r_{p,k_p}}(\lambda_p)),$$

where $\mathsf{algm}(A, \lambda_i) = \sum_{h=1}^{k_i} r_{i,h}$ *and* $\mathsf{geomm}(A, \lambda_i) = k_i$ *for* $1 \leqslant i \leqslant p$.

Proof. By Schur's Triangularization Theorem (Theorem 8.8), there exists a unitary matrix $U \in \mathbb{C}^{n\times n}$ and an upper triangular matrix $T \in \mathbb{C}^{n\times n}$ such that $A = U^{\mathsf{H}}TU$ and the diagonal elements of T are the eigenvalues of A; furthermore, each eigenvalue λ occurs in the sequence of diagonal values a number of $\mathsf{algm}(A, \lambda)$ times.

By Theorem 8.41, the upper triangular matrix T is similar to a block upper triangular matrix $\mathsf{diag}(K_1, \ldots, K_m)$, where $K_i = \lambda_i I_{p_i} + L_i$, L_i is strictly upper triangular, and each λ_i is distinct for $1 \leqslant i \leqslant m$.

Theorem 8.42 implies that each triangular block is similar to a matrix of the required form because for each diagonal block K_i there is a nonsingular X_i such that $X_i^{-1}(\lambda_i I +)X_i = \lambda_i I + \mathsf{diag}(E_1, \ldots, E_{m_i})$. □

Theorem 8.44. *For an eigenvalue λ of a square matrix $A \in \mathbb{C}^{n\times n}$, we have* $\mathsf{algm}(A, \lambda) = 1$ *if and only if*

$$\mathsf{geomm}(A, \lambda) = \mathsf{geomm}(A^{\mathsf{H}}, \lambda) = 1,$$

and for any eigenvector $\boldsymbol{u}$ that corresponds to λ in A and any eigenvector $\boldsymbol{v}$ that corresponds to the same eigenvalue in A^{H}, we have $\boldsymbol{v}^{\mathsf{H}}\boldsymbol{u} \neq 0$.

Proof. Suppose that λ is an eigenvalue of A such that $\mathsf{geomm}(A, \lambda) = \mathsf{geomm}(A^{\mathsf{H}}, \lambda) = 1$ and for any eigenvector $\boldsymbol{u}$ that corresponds to λ in A and any eigenvector $\boldsymbol{v}$ that corresponds to the same eigenvalue in A^{H} we have $\boldsymbol{v}^{\mathsf{H}}\boldsymbol{u} \neq 0$.

Let $B \in \mathbb{C}^{n\times n}$ be a matrix that is similar to A. There exists an invertible matrix X such that $B = XAX^{-1}$. Then, $A - \lambda I \sim B - \lambda I$, so $\dim(S_{A,\lambda}) = \dim(S_{B,\lambda})$, which allows us to conclude that λ has geometric multiplicity 1 in both A and B. Note that

$BX\boldsymbol{u} = XA\boldsymbol{u} = \lambda X\boldsymbol{u}$, so $X\boldsymbol{u}$ is an eigenvector of B. Similarly,

$$B^{\mathsf{H}}(X^{\mathsf{H}})^{-1}\boldsymbol{v} = (X^{\mathsf{H}})^{-1}A^{\mathsf{H}}X^{\mathsf{H}}(X^{\mathsf{H}})^{-1}\boldsymbol{v} = (X^{\mathsf{H}})^{-1}A^{\mathsf{H}}\boldsymbol{v} = \lambda(X^{\mathsf{H}})^{-1}\boldsymbol{v},$$

so $(X^{\mathsf{H}})^{-1}\boldsymbol{v}$ is an eigenvector for B^{H} that corresponds to the same eigenvalue λ. Furthermore, we have

$$((X^{\mathsf{H}})^{-1}\boldsymbol{v})^{\mathsf{H}}X\boldsymbol{u} = \boldsymbol{v}^{\mathsf{H}}X^{-1}X\boldsymbol{u} = \boldsymbol{v}^{\mathsf{H}}\boldsymbol{u} \neq 0.$$

Thus, if A is a matrix that satisfies the conditions of the theorem, then any matrix similar to A satisfies the same conditions.

If λ is a simple eigenvalue of A, then λ is also a simple eigenvalue of the Jordan normal form of A. Therefore, a Jordan segment that corresponds to an eigenvalue λ of geometric multiplicity 1 consists of a single Jordan block $B_1(\lambda)$ of order 1 that has (1) as an eigenvector. The transposed segment has also (1) as an eigenvector. The eigenvector $\boldsymbol{u}$ of A that corresponds to λ can be obtained from (1) by adding zeros corresponding to the remaining components and the eigenvector $\boldsymbol{v}$ can be obtained from (1) in a similar manner. Since $(1)^{\mathsf{H}}(1) \neq 0$, we have $\boldsymbol{v}^{\mathsf{H}}\boldsymbol{u} \neq 0$.

Conversely, if A and λ satisfy the conditions of the theorem, then the same conditions are satisfied by the Jordan normal form C of A. Therefore, the Jordan segment that corresponds to λ in C consists of a single block $B_m(\lambda)$. Suppose that $m > 1$. Then $\boldsymbol{e}_1 \in \mathbb{C}^m$ is an eigenvector of $B_m(\lambda)$, $\boldsymbol{e}_m \in \mathbb{C}^m$ is an eigenvector of $(B_m(\lambda))^{\mathsf{H}}$, and $\boldsymbol{e}_m^{\mathsf{H}}\boldsymbol{e}_1 = 0$, which would imply $\boldsymbol{v}^{\mathsf{H}}\boldsymbol{u} = 0$. Therefore, $m = 1$ and λ is a simple eigenvalue of A. □

8.8 Matrix Norms and Eigenvalues

There exists a simple relationship between the spectral radius and any matrix norm $\|\!|\cdot|\!\|$. Namely, if $\boldsymbol{x}$ is an eigenvector that corresponds to $\lambda \in \mathsf{spec}(A)$, then

$$\|\!|A\boldsymbol{x}|\!\| = |\lambda| \|\!|\boldsymbol{x}|\!\| \leqslant \|\!|A|\!\| \|\!|\boldsymbol{x}|\!\|,$$

which implies $\lambda \leqslant \|\!|A|\!\|$. Therefore,

$$\rho(A) \leqslant \|\!|A|\!\|. \tag{8.5}$$

It is easy to see that if $A \in \mathbb{C}^{n\times n}$ and $a \in \mathbb{R}_{>0}$, then $\rho(aA) = a\rho(A)$.

In Section 6.7, we have seen that for $A \in \mathbb{C}^{p\times p}$, $|||A||| < 1$ implies $\lim_{n\to\infty} A^n = O_{p,p}$. Using the spectral radius we can prove a stronger result:

Theorem 8.45 (Oldenburger's theorem). *If $A \in \mathbb{C}^{p\times p}$, then $\lim_{n\to\infty} A^n = O_{p,p}$ if and only if $\rho(A) < 1$.*

Proof. Let B be a matrix in Jordan normal form that is similar to A. There exists a non-singular matrix U such that $UAU^{-1} = B$, and, therefore, $UA^nU^{-1} = B^n$. Therefore, if $\lim_{n\to\infty} A^n = O_{p,p}$ we have $\lim_{n\to\infty} B^n = O_{p,p}$, so $\lim_{n\to\infty} \lambda^n = O$ for any $\lambda \in \mathsf{spec}(A) = \mathsf{spec}(B)$. This is possible only if $|\lambda| < 1$ for $\lambda \in \mathsf{spec}(A)$, so $\rho(A) < 1$.

Conversely, if $\rho(A) < 1$, then $|\lambda| < 1$ for $\lambda \in \mathsf{spec}(A)$, so for any block $B_r(\lambda)$ of B we have $\lim_{n\to\infty}(B_r(\lambda))^n = O_{r,r}$ by Exercise 8.13. Therefore, $\lim_{n\to\infty} B^n = O_{p,p}$, so $\lim_{n\to\infty} A^n = O_{p,p}$. □

Theorem 8.46. *If $A, B \in \mathbb{R}^{n\times n}$ and $\mathsf{abs}(A) \leqslant B$, then $\rho(A) \leqslant \rho(B)$.*

Proof. Suppose that $\mathsf{abs}(A) \leqslant B$ and $\rho(B) < \rho(A)$. There exists $\alpha \in \mathbb{R}$ such that $0 \leqslant \rho(B) < \alpha < \rho(A)$. If $C = \frac{1}{\alpha}A$ and $D = \frac{1}{\alpha}B$, then $\rho(C) = \frac{1}{\alpha}\rho(A) > 1$ and $\rho(D) = \frac{1}{\alpha}\rho(B) < 1$. Thus, by Theorem 8.45, $\lim_{k\to\infty} D^k = O$.

Since $\mathsf{abs}(A) \leqslant B$, we have $\mathsf{abs}(C) = \frac{1}{\alpha}\mathsf{abs}(A) \leqslant \frac{1}{\alpha}B = D$. Therefore, $\mathsf{abs}(C^k) \leqslant (\mathsf{abs}(C))^k < D^k$, which implies $\lim_{k\to\infty} C^k = 0$. This contradicts the fact that $\rho(C) > 1$. □

Corollary 8.22. *If $A \in \mathbb{R}^{n\times n}$, then $\rho(A) \leqslant \rho(\mathsf{abs}(A))$.*

Proof. This follows immediately from Theorem 8.46 by taking $B = \mathsf{abs}(A)$. □

Theorem 8.47. *Let $A \in \mathbb{R}^{n\times n}$ be a matrix and $\boldsymbol{x} \in \mathbb{R}^n$ be a vector such that $A \geqslant O_{n,n}$ and $\boldsymbol{x} \geqslant \boldsymbol{0}_n$. If $A\boldsymbol{x} > a\boldsymbol{x}$, it follows that $a < \rho(A)$.*

Proof. Since $\rho(A) \geqslant 0$ we assume that $a > 0$. Also, since $A\boldsymbol{x} > a\boldsymbol{x}$, we have $\boldsymbol{x} \neq 0$. The strict inequality $A\boldsymbol{x} > a\boldsymbol{x}$ implies that $A\boldsymbol{x} \geqslant (a+\epsilon)\boldsymbol{x}$ for some $\epsilon > 0$. Therefore, if $B = \frac{1}{a+\epsilon}A$, we have $B\boldsymbol{x} \geqslant \boldsymbol{x}$, so

$$\boldsymbol{x} \leqslant B\boldsymbol{x} \leqslant \cdots \leqslant B^k\boldsymbol{x}.$$

Thus, $\rho(B) \geqslant 1$, so $a < a + \epsilon < \rho(A)$. □

Inequality (8.5) applied to matrix A^k implies $\rho(A^k) \leqslant \|A^k\|$. By Theorem 7.6, we have $(\rho(A))^k \leqslant \|A^k\|$ for every $k \in \mathbb{N}$. Actually, the following statement holds.

Theorem 8.48. *For every $A \in \mathbb{C}^{p\times p}$, we have*

$$\lim_{k\to\infty} \left(\|A^k\|_2\right)^{\frac{1}{k}} = \rho(A).$$

Proof. We need to prove only that $\lim_{k\to\infty} \left(\|A^k\|_2\right)^{\frac{1}{k}} \leqslant \rho(A)$, because the reverse inequality was already proven.

Let T be an upper triangular matrix and let $D = \mathsf{diag}(d_1, \ldots, d_p)$ be a diagonal matrix, where $d_i \neq 0$ for $1 \leqslant i \leqslant p$. Note that

$$D^{-1} = \mathsf{diag}\left(\frac{1}{d_1}, \ldots, \frac{1}{d_p}\right),$$

hence $(D^{-1}TD)_{ij} = t_{ij}\frac{d_j}{d_i}$ for $1 \leqslant i, j \leqslant p$. If $d_i = \delta^i$ for $1 \leqslant i \leqslant p$, where $\delta > 0$, denote the upper triangular matrix $D^{-1}TD$ by T_δ. It follows that

$$(T_\delta)_{ij} = \begin{cases} t_{ii} & \text{if } i = j, \\ t_{ij}\delta^{j-i} & \text{if } j > i, \\ 0 & \text{if } j < i. \end{cases}$$

The matrix T_δ can now be written as $T_\delta = E + S$, where E is a diagonal matrix such that $E = \mathsf{diag}(t_{11}, \ldots, t_{nn})$ and

$$s_{ij} = \begin{cases} 0 & \text{if } j \leqslant i, \\ t_{ij}\delta^{j-i} & \text{if } j > i. \end{cases}$$

If $\delta < 1$, there exists a positive number c such that $\|S\|_2 \leqslant c\delta$. Therefore,

$$\begin{aligned} \|T_\delta\|_2 &\leqslant \|E\|_2 + \|S\|_2 \\ &= \max\{|t_{ii}| \mid 1 \leqslant i \leqslant n\} + \|S\|_2 \\ &\leqslant \max\{|t_{ii}| \mid 1 \leqslant i \leqslant n\} + c\delta \\ &\quad \text{(by Supplement 45).} \end{aligned}$$

By Schur's Triangularization Theorem (Theorem 8.8), A can be factored as $A = U^{\mathsf{H}}TU$, where U is a unitary matrix and T is an upper triangular matrix $T \in \mathbb{C}^{p\times p}$ such that elements of T are the eigenvalues of A and each such eigenvalue occurs in the sequence of diagonal values a number of $\mathsf{algm}(A, \lambda)$ times.

If we construct the matrix T_δ as above, starting from the upper triangular matrix T that results from Schur's decomposition, we have $\|T_\delta\|_2 \leqslant \rho(A) + c\delta$. Since $T = DT_\delta D^{-1}$, it follows that $A = U^{\mathsf{H}}DT_\delta D^{-1}U$, so $A = WT_\delta W^{-1}$, where $W = U^{\mathsf{H}}D$. This implies $A^k = WT_\delta^k(W^{-1})^k$, so $\|A^k\|_2 \leqslant \|T_\delta^k\|_2\|W\|_2\|W^{-1}\|_2 \leqslant (\|T_\delta\|_2)^k\|W\|_2\|W^{-1}\|_2$. Consequently,

$$\left(\|A^k\|_2\right)^{\frac{1}{k}} \leqslant \|T_\delta\|_2(\|W\|_2\|W^{-1}\|_2)^{\frac{1}{k}} \leqslant (\rho(A)+c\delta)(\|W\|_2\|W^{-1}\|_2)^{\frac{1}{k}}.$$

Thus, $\lim_{k\to\infty}\left(\|A^k\|_2\right)^{\frac{1}{k}} \leqslant \rho(A) + c\delta$, and since this equality holds for any $\delta > 0$, it follows that $\lim_{k\to\infty}\left(\|A^k\|_2\right)^{\frac{1}{k}} \leqslant \rho(A)$. □

Another interesting connection exists between the matrix norm $\|A\|_2$ and the spectral radius of the matrix $A'A$.

Theorem 8.49. *Let $A \in \mathbb{C}^{m\times n}$. We have*

$$\|A\|_2 = \max\{\sqrt{\lambda} \mid \lambda \in \mathsf{spec}(A^{\mathsf{H}}A)\}.$$

Proof. Observe that if $\lambda \in \mathsf{spec}(A^{\mathsf{H}}A)$, then $A^{\mathsf{H}}A\boldsymbol{x} = \lambda\boldsymbol{x}$ for $\boldsymbol{x} \neq \boldsymbol{0}$, and therefore, $\boldsymbol{x}^{\mathsf{H}}A^{\mathsf{H}}A\boldsymbol{x} = \lambda\boldsymbol{x}^{\mathsf{H}}\boldsymbol{x}$, or $(A\boldsymbol{x})^{\mathsf{H}}(A\boldsymbol{x}) = \lambda\boldsymbol{x}^{\mathsf{H}}\boldsymbol{x}$. The last equality amounts to $\|A\boldsymbol{x}\|_2^2 = \lambda\|\boldsymbol{x}\|_2^2$, which allows us to conclude that all eigenvalues of $A^{\mathsf{H}}A$ are real and non-negative.

By the definition of $\|A\|_2$, we have

$$\|A\|_2 = \max\{\|A\boldsymbol{x}\|_2 \mid \|\boldsymbol{x}\|_2 = 1\}.$$

This implies that $\|A\|_2^2$ is the maximum of $\|A\boldsymbol{x}\|_2^2$ under the restriction $\|\boldsymbol{x}\|_2^2 = 1$. We apply Lagrange's multiplier method to determine conditions that are necessary for the existence of the maximum. Let $B = (b_{ij}) = A^{\mathsf{H}}A$. Then, consider the function

$$g(x_1, \ldots, x_n) = \sum_{i=1}^{n}\sum_{j=1}^{n} x_i b_{ij} x_j - \ell\left(\sum_{i=1}^{n} x_i^2 - 1\right).$$

For the maximum, we need to have $\frac{\partial g}{\partial x_i} = 0$, which implies

$$\sum_{j=1}^{n} b_{ij}x_j - \ell x_i = 0,$$

for $1 \leqslant i \leqslant n$. This is equivalent to $B\boldsymbol{x} = \ell\boldsymbol{x}$, which means that ℓ is an eigenvalue of $B = A^{\mathsf{H}}A$. The value of $\|A\boldsymbol{x}\|_2^2$ is $\boldsymbol{x}^{\mathsf{H}}A^{\mathsf{H}}A\boldsymbol{x} = \ell\boldsymbol{x}^{\mathsf{H}}\boldsymbol{x} = \ell\|\boldsymbol{x}\|_2^2 = \ell$. Thus, the maximum of $\|A\boldsymbol{x}\|_2^2$ for $\|\boldsymbol{x}\|_2 = 1$ is the largest eigenvalue of $A^{\mathsf{H}}A$, which completes the argument. □

Theorem 8.49 states that

$$\|A\|_2 = \sqrt{\rho(A^{\mathsf{H}}A)}, \tag{8.6}$$

which explains why $\|A\|_2$ is also known as the *spectral norm* of the matrix A. It is interesting to recall that we also have

$$\|A\|_F = \sqrt{\mathsf{trace}(A^{\mathsf{H}}A)},$$

as we saw in Equality (6.12).

Corollary 8.23. *For $A \in \mathbb{R}^{m\times n}$, we have $\|A\|_2^2 \leqslant \|A\|_1\|A\|_\infty$.*

Proof. In Theorem 8.49, we have shown that $\|A\|_2^2$ is an eigenvalue of $A^{\mathsf{H}}A$. Therefore, there exists $\boldsymbol{x} \neq \boldsymbol{0}$ such that $A^{\mathsf{H}}A\boldsymbol{x} = \|A\|_2^2\boldsymbol{x}$, which implies $\|A^{\mathsf{H}}A\boldsymbol{x}\|_1 = \|A\|_2^2\|\boldsymbol{x}\|_1$. Note that $\|A^{\mathsf{H}}A\boldsymbol{x}\|_1 \leqslant \|A^{\mathsf{H}}\|_1\|A\|_1\|\boldsymbol{x}\|_1$, which implies

$$\|A\|_2^2\|\boldsymbol{x}\|_1 \leqslant \|A^{\mathsf{H}}\|_1\|A\|_1\|\boldsymbol{x}\|_1.$$

Thus, we have $\|A\|_2^2 \leqslant \|A^{\mathsf{H}}\|_1\|A\|_1$. The desired inequality follows by observing that $\|A^{\mathsf{H}}\|_1 = \|A\|_\infty$. □

Theorem 8.50 (Weyr's theorem). *Let $f(z) = \sum_{m=0}^{\infty} c_m z^m$ be a power series having the convergence radius r. If $A \in \mathbb{C}^{n\times n}$ is such that $|\lambda| < r$ for every $\lambda \in \mathsf{spec}(A)$, then the power series $\sum_{m=0}^{\infty} c_m A^m$ converges absolutely. If there exists $\lambda \in \mathsf{spec}(A)$ such that $|\lambda| > r$, then the series $\sum_{m=0}^{\infty} c_m A^m$ is divergent.*

Proof. Suppose that $\mathsf{spec}(A) = \{\lambda_1, \ldots, \lambda_n\}$. By Schur's Triangularization Theorem (Theorem 8.8), there exists a unitary matrix

$U \in \mathbb{C}^{n\times n}$ and an upper triangular matrix $T \in \mathbb{C}^{n\times n}$ such that $A = U^{\mathsf{H}}TU$ and the diagonal elements of T are the eigenvalues of A.

The diagonal elements of the matrix $\sum_{m=0}^{p} c_m T^m$ are $\sum_{m=0}^{p} c_m \lambda_i^m$. If there exists an eigenvalue λ_i such that $|\lambda_i| > r$, then the series $\sum_{m=0}^{\infty} c_m T^m$ diverges, which implies that the series $\sum_{m=0}^{\infty} c_m A^m$ diverges.

Suppose now that $|\lambda_i| < r$ for $1 \leqslant i \leqslant n$.

Let $\rho_1, \ldots, \rho_n, \rho$ be $n+1$ distinct numbers such that $|\lambda_i| \leqslant \rho_i \leqslant \rho < r$ and let S an upper triangular matrix such that $\mathsf{abs}(T) \leqslant S$ such that S has $\rho_1, \ldots, \rho_n$ as its diagonal elements. Let U be a matrix such that $U^{-1}SU = \mathsf{diag}(\rho_1, \ldots, \rho_n)$. Then $\mathsf{abs}(T^m) \leqslant U\mathsf{diag}(\rho_1, \ldots, \rho_n)^m U^{-1}$, so $\mathsf{abs}(c_m T^m) \leqslant |c_m| U\mathsf{diag}(\rho_1, \ldots, \rho_n)^m U^{-1}$ for $m \geqslant 0$. Consequently,

$$\begin{aligned}
\|c_m T^m\|_\infty &\leqslant \| |c_m| U\mathsf{diag}(\rho_1, \ldots, \rho_n)^m U^{-1}\|_\infty \\
&\leqslant |c_m| n^2 \|U\|_\infty \|\mathsf{diag}(\rho_1, \ldots, \rho_n)^m\|_\infty \|U^{-1}\|_\infty \\
&\leqslant C|c_m|\rho^m,
\end{aligned}$$

where C is a constant that does not depend on m. Therefore, the series $\sum_{m=0}^{\infty} \|c_m T^m\|$ converges, so the series $\sum_{m=0}^{\infty} c_m T^m$ converges absolutely. This implies that the series $\sum_{m=0}^{\infty} c_m A^m$ converges absolutely. □

Theorem 8.51. *Let $f(x)$ be a rational function, $f(z) = p(z)/q(z)$, where p and q belong to $\mathbb{C}[z]$, $p(z) = p_0 + p_1 z + \cdots + p_k z^k$, $q(z) = q_0 + q_1 z + \cdots + q_h z^h$, and $f(z) = \sum_{j=0}^{\infty} c_j z^j$.*

If the series $\sum_{k=0}^{\infty} c_j z^j$ converges to z, such that $|z| < r$ and $A \in \mathbb{C}^{n\times n}$ is a matrix such that $\mathsf{spec}(A) \subseteq \{z \in \mathbb{C} \mid |z| < r\}$, then $f(A) = (q(A))^{-1} p(A)$ equals $\sum_{j=0}^{\infty} c_j A^j$.

Proof. By hypothesis, if $|z| < r$, then $q(z) \neq 0$. If $\mathsf{spec}(A) = \{\lambda_1, \ldots, \lambda_n\}$, then $\det(q(A)) = q(\lambda_1)\cdots q(\lambda_n) \neq 0$, which means that $q(A)$ is an invertible matrix. This shows that $f(A) = (q(A))^{-1}p(A)$, or $q(A)f(A) = p(A)$. The definition of f implies

$$(q_0 + q_1 z + \cdots + q_h z^h) \sum_{k=0}^{\infty} c_j z^j = p_0 + p_1 z + \cdots + p_k z^k.$$

If $\ell \leqslant k$, we have $q_0c_\ell + q_1c_{\ell-1} + q_hc_{\ell-h} = p_\ell$; otherwise, if $k < \ell$, we have $q_0c_\ell + q_1c_{\ell-1} + q_hc_{\ell-h} = 0$. This implies

$$
\begin{aligned}
& q(A)\sum_{k=0}^{\infty} c_j A^j \\
&= (q_0 I_n + q_1 A + \cdots + q_h A^h) f(A) \\
&= \sum_{j=0}^{\infty} q_0 c_j A^j + \sum_{j=0}^{\infty} q_1 c_j A^{j+1} + \sum_{j=0}^{\infty} q_h c_j A^{j+h} \\
&= \sum_{j=0}^{\infty} q_0 c_j A^j + \sum_{j=0}^{\infty} q_1 c_{j-1} A^j + \sum_{k=0}^{\infty} q_h c_{j-h} A^j, \\
&= \sum_{j=0}^{\infty} (q_0 c_j + q_1 c_{j-1} + \cdots + q_h c_{j-h}) A^j \\
&= \sum_{j=0}^{k} p_j A^j = f(A),
\end{aligned}
$$

where $c_l = 0$ if $l < 0$. Consequently, $\sum_{j=0}^{\infty} c_j A^j = q(A)^{-1} p(A) = f(A)$. □

Definition 8.15. Let $A \in \mathbb{C}^{n\times n}$ be a matrix such that $|\lambda| < r$ for every $\lambda \in \mathsf{spec}(A)$, where r is the convergence radius of the series $f(z) = \sum_{j=0}^{\infty} c_j z^j$. The matrix $f(A)$ is defined as $f(A) = \sum_{j=0}^{\infty} c_j A^j$.

Theorem 8.51 shows that if $f : \mathbb{C} \longrightarrow \mathbb{C}$ is a rational function, $f(z) = \frac{p(z)}{q(z)}$, the definition of the matrix $f(A) \in \mathbb{C}^{n\times n}$ can be given either as $q(A)^{-1}p(A)$ or as the sum $\sum_{j=0}^{\infty} c_j A^j$, where $f(A) = \sum_{j=0}^{\infty} c_j A^j$.

Example 8.10. The series $\sum_{j=0}^{\infty} \frac{z^j}{j!}$ is convergent in $\mathbb{C}$ and its sum is e^z. Thus, following Definition 8.15, e^A is defined as

$$e^A = \sum_{j=0}^{\infty} \frac{1}{j!} A^j.$$

Similarly, considering the series

$$\sin z = z - \frac{1}{3!}z^3 + \frac{1}{5!}z^5 - \cdots, \quad \cos z = 1 - \frac{1}{2!}z^2 + \frac{1}{4!}z^4 - \cdots,$$
$$\sinh z = z + \frac{1}{3!}z^3 + \frac{1}{5!}z^5 + \cdots, \cosh z = 1 + \frac{1}{2!}z^2 + \frac{1}{4!}z^4 + \cdots,$$

which are convergent everywhere, we can define

$$\sin A = A - \frac{1}{3!}A^3 + \frac{1}{5!}A^5 - \cdots, \quad \cos A = 1 - \frac{1}{2!}A^2 + \frac{1}{4!}A^4 - \cdots,$$
$$\sinh A = A + \frac{1}{3!}A^3 + \frac{1}{5!}A^5 + \cdots, \cosh A = 1 + \frac{1}{2!}A^2 + \frac{1}{4!}A^4 + \cdots.$$

Therefore, we obtain the equalities

$$e^{iA} = \cos A + i \sin A \text{ and } e^A = \cosh A + \sinh A,$$

and

$$\sin A = \frac{1}{2}(e^{iA} - e^{-iA}), \cos A = \frac{1}{2}(e^{iA} + e^{-iA}),$$
$$\sinh A = \frac{1}{2}(e^{A} - e^{-A}), \quad \cosh A = \frac{1}{2}(e^{A} + e^{-A}),$$

for every $A \in \mathbb{C}^{n\times n}$.

Example 8.11. Let $A \in \mathbb{C}^{2\times 2}$ be a matrix such that $\mathsf{spec}(A) = \{-\frac{\pi}{4}, \frac{\pi}{4}\}$. Its characteristic polynomial is $p_A(\lambda) = \lambda^2 - \frac{1}{16}$. By the Cayley–Hamilton Theorem, we have $A^2 - \frac{1}{16}I_2 = O_{3,3}$. Thus,

$$\begin{aligned}\sin A &= \sum_{n=0}^{\infty}(-1)^n \frac{1}{(2n+1)!}A^{2n+1} \\ &= A\sum_{n=0}^{\infty}(-1)^n \frac{1}{(2n+1)!}\left(\frac{\pi^2}{2^4}\right)^n \\ &= A\sum_{n=0}^{\infty}(-1)^n \frac{\pi^{2n}}{2^{4n}(2n+1)!}.\end{aligned}$$

Theorem 8.52. *Let*

$$f(z) = \sum_{k=0}^{\infty} a_k z^k, g(z) = \sum_{k=0}^{\infty} b_k z^k, h(z) = \sum_{k=0}^{\infty} c_k z^k$$

be the functions defined by the given series that are convergent for $|z| < r$ *and let* $A \in \mathbb{C}^{n\times n}$ *be a matrix such that* $\mathsf{spec}(A) \subseteq \{z \in C \mid$

$|z| < r\}$. If $h(z) = f(z)g(z)$ for $|z| < r$, then $h(A) = f(A)g(A)$; if $h(z) = f(z) + g(z)$ for $|z| < r$, then $h(A) = f(A) + g(A)$.

Proof. We discuss only the case when $h(z) = f(z)g(z)$. The equality $h(z) = f(z)g(z)$ implies $c_k = a_0 b_k + a_1 b_{k-1} + \cdots + a_{k-1} b_1 + a_k b_0$ for $k \in \mathbb{N}$. The series with real non-negative coefficients, $\sum_{k=0}^{\infty} d_k z^k$, where $d_k = \sum_{j=0}^{k} |a_j|\ |b_{k-j}|$ for $k \in \mathbb{N}$ is convergent for $|z| < r$, which implies that the series $\sum_{k=0}^{\infty} d_k A^k$ is absolutely convergent, by Weyr's Theorem (Theorem 8.50). In turn, by Lemma 6.7, this is equivalent to the convergence of $\sum_{k=0}^{\infty} |d_k| \| A^k \|$. Thus, the series $\sum_{k=0}^{\infty} \sum_{j=0}^{k} a_j b_{k-j} A^k$ is absolutely convergent and its terms can be permuted. Therefore,

$$
\begin{aligned}
h(A) &= \sum_{k=0}^{\infty} c_k A^k = \sum_{k=0}^{\infty} \left(\sum_{j=0}^{k} a_j b_{k-j} \right) A^k \\
&= \sum_{j=0}^{\infty} a_j \sum_{k=j}^{\infty} b_{k-j} A^k = \sum_{j=0}^{\infty} a_j \sum_{\ell=0}^{\infty} b_\ell A^{j+\ell} \\
&= \sum_{j=0}^{\infty} a_j A^j \sum_{\ell=0}^{\infty} b_\ell A^\ell = f(A)g(A).
\end{aligned}
$$

□

Some properties of scalar functions are transmitted to the corresponding matrix functions; others fail to do so, as we show in the following.

Example 8.12. Since $e^z e^{-z} = 1$, we have $e^A e^{-A} = I_n$ for $A \in \mathbb{C}^{n \times n}$, by Theorem 8.52. However, despite the fact that $e^z e^w = e^{z+w}$, the corresponding equality does not hold. For example, let

$$
A = \begin{pmatrix} 1 & 1 \\ 0 & 1 \end{pmatrix} \text{ and } B = \begin{pmatrix} 0 & 1 \\ 1 & 0 \end{pmatrix},
$$

where $a \neq 0$. By the Cayley–Hamiltion Theorem, we have $A^2 = 2A - I_2$ and $B^2 = I_2$. It is easy to see that $A^n = nA - (n-1)I_2$ and $B^n = I_2$ if n is even and $B^n = B$ if n is odd. Thus, the exponentials of these matrices are

$$
e^A = \sum_{n=0}^{\infty} \frac{1}{n!} (nA - (n-1)I_2)
$$

$$= \sum_{n=0}^{\infty} \frac{1}{n!} nA - \sum_{n=0}^{\infty} \frac{1}{n!} nI_2 + \sum_{n=0}^{\infty} \frac{1}{n!} I_2$$
$$= (1+e)A - (1+e)I_2 + I_2 = (1+e)A - eI_2 = \begin{pmatrix} 1 & 1+e \\ 0 & 1 \end{pmatrix},$$

and

$$e^B = 1 + \frac{1}{1!}B + \frac{1}{2!}I_2 + \frac{1}{3!}B + \cdots$$
$$= I_2 \cosh 1 + B \sinh 1 = \frac{1}{2e^2} \begin{pmatrix} e^2+1 & e^2-1 \\ e^2-1 & e^2+1 \end{pmatrix}.$$

Thus,

$$e^A e^B = ((1+e)A - eI_2)(I_2 \cosh 1 + B \sinh 1)$$
$$= \frac{1}{2e} \begin{pmatrix} e^2+2e^2-e & e^3+2e^2+e \\ e^2-1 & e^2+1 \end{pmatrix}.$$

The sum of the matrices is

$$C = A + B = \begin{pmatrix} 1 & 2 \\ 1 & 1 \end{pmatrix},$$

and we have $C^2 = 2C + I_2$. We leave it to the reader to verify that $e^C \neq e^A e^B$.

8.9 Matrix Pencils and Generalized Eigenvalues

Definition 8.16. Let $A, B \in \mathbb{C}^{n \times n}$. The *matrix pencil* determined by A and B is the one-parameter set of matrices

$$\mathsf{Pen}(A, B) = \{A - tB \mid t \in \mathbb{C}\}.$$

The *set of generalized eigenvalues of* $\mathsf{Pen}(A, B)$ is

$$\mathsf{spec}(A, B) = \{\lambda \in \mathbb{C} \mid \det(A - \lambda B) = 0\}.$$

If $\lambda \in \mathsf{spec}(A, B)$ and $A\boldsymbol{x} = \lambda B\boldsymbol{x}$, then $\boldsymbol{x}$ is an *eigenvector* of $\mathsf{Pen}(A, B)$. Note that in this case $\boldsymbol{x} \in \mathsf{null}(A - \lambda B)$. When $B = I_n$, then an (A, I_n)-generalized eigenvalue is an eigenvalue of A and any eigenvector of the pair (A, I_n) is just an eigenvector of A.

Theorem 8.53. *Let A and B be two matrices in $\mathbb{C}^{n\times n}$. If $\mathsf{rank}(B) = n$, then $\mathsf{spec}(A, B)$ consists of n eigenvalues (taking into account their multiplicities).*

Proof. Suppose that $\mathsf{rank}(B) = n$, that is, B is an invertible matrix. Then, $\lambda \in \mathsf{spec}(A, B)$ is equivalent with the existence of an eigenvector $\boldsymbol{x} \neq \boldsymbol{0}_n$ such that $A\boldsymbol{x} = \lambda B\boldsymbol{x}$, so $B^{-1}A\boldsymbol{x} = \lambda\boldsymbol{x}$. Since $\mathsf{rank}(B) = n$, by Corollary 3.4, $\mathsf{rank}(B^{-1}A) = n$ and $\mathsf{spec}(A, B)$ consists of n eigenvalues. □

Corollary 8.24. *If B is a nonsingular matrix, then an (A, B)-generalized eigenvalue is simply an eigenvalue of the matrix $B^{-1}A$.*

Proof. The corollary follows immediately from the proof of Theorem 8.53. □

When $\mathsf{null}(A)\cap\mathsf{null}(B) \neq \{\boldsymbol{0}_n\}$, the existence of (A, B)-generalized eigenvalues becomes trivial. If $\boldsymbol{z} \in (\mathsf{null}(A) \cap \mathsf{null}(B)) - \{\boldsymbol{0}\}$, then we have both $A\boldsymbol{z} = \boldsymbol{0}$ and $B\boldsymbol{z} = \boldsymbol{0}$. Thus, $(A-tB)\boldsymbol{z} = \boldsymbol{0}_n$ for every $t \in \mathbb{C}$, so any complex number is an (A, B)-generalized eigenvalue and any non-zero vector $\boldsymbol{z} \in (\mathsf{null}(A) \cap \mathsf{null}(B)) - \{\boldsymbol{0}\}$ is an eigenvector of the pair (A, B).

On the other hand, if

$$A = \begin{pmatrix} a & 0 \\ 0 & b \end{pmatrix} \text{ and } B = \begin{pmatrix} 0 & c \\ 0 & 0 \end{pmatrix},$$

where $ab \neq 0$, the pencil $\mathsf{Pen}(A, B)$ has no generalized eigenvalues because $\det(A - \lambda B) = ab \neq 0$.

Definition 8.17. Let $A, B \in \mathbb{C}^n$ be two matrices. We refer to $\mathsf{Pen}(A, B)$ as a *regular pencil* when $\det(A - \lambda B)$ is not identically zero.

A regular pencil (A, B) can have only a finite number of eigenvalues because the polynomial $\det(A - \lambda B)$ is of degree not larger than n.

We saw in Theorem 6.54 that if B is a Hermitian and positive definite matrix, then the mapping $f_B : \mathbb{C}^n \times \mathbb{C}^n \longrightarrow \mathbb{R}$ given by $f_B(\boldsymbol{x}, \boldsymbol{y}) = \boldsymbol{x}^{\mathsf{H}} B \boldsymbol{y}$ for $\boldsymbol{x}, \boldsymbol{y} \in \mathbb{C}^n$ defines an inner product on $\mathbb{C}^n$. If $\mathsf{Pen}(A, B)$ is a regular matrix pencil and B is Hermitian, the matrix $D = B^{-1}A$ is self-adjoint with respect to this inner product. Indeed, we have

$$\begin{aligned}(B^{-1}A\boldsymbol{x}, \boldsymbol{y}) &= (B^{-1}A\boldsymbol{x})^{\mathsf{H}} B\boldsymbol{y} = \boldsymbol{x}^{\mathsf{H}} A^{\mathsf{H}} (B^{\mathsf{H}})^{-1} B\boldsymbol{y} = \boldsymbol{x}^{\mathsf{H}} A\boldsymbol{x},\\ (\boldsymbol{x}, (B^{-1}A)\boldsymbol{y}) &= \boldsymbol{x}^{\mathsf{H}} BB^{-1}A\boldsymbol{y} = \boldsymbol{x}^{\mathsf{H}} A\boldsymbol{x},\end{aligned}$$

which justifies our claim.

The self-adjoint matrix $D = B^{-1}A$ has real eigenvalues $\{\lambda_1, \ldots, \lambda_n\}$ and a linearly independent orthonormal family of eigenvectors $\{\boldsymbol{z}_1, \ldots, \boldsymbol{z}_n\}$ (by the inner product f_B). Thus, we have $B^{-1}A\boldsymbol{z}_i = \lambda_i \boldsymbol{z}_i$ and

$$f_B(\boldsymbol{z}_i, \boldsymbol{z}_j) = (\boldsymbol{z}_i)^{\mathsf{H}} B\boldsymbol{z}_j = \begin{cases} 1 & \text{if } i = j,\\ 0 & \text{if } i \neq j,\end{cases}$$

for $1 \leqslant i \leqslant j$ (see Exercise 22 of Chapter 7).

Definition 8.18. Let $\mathsf{Pen}(A, B)$ be a regular matrix pencil, where $A, B \in \mathbb{C}^{n\times n}$. The eigenvalues of this pencil are the eigenvalues of the matrix $D = B^{-1}A$. The vectors $\boldsymbol{z}_i$ are referred to as the *eigenvectors of the pencil.*

For the eigenvectors of a regular pencil $\mathsf{Pen}(A, B)$, we have $A\boldsymbol{z}_i = \lambda_i B\boldsymbol{z}_i$ for $1 \leqslant i \leqslant n$. The matrix $Z = (\boldsymbol{z}_1 \ \cdots \ \boldsymbol{z}_n) \in \mathbb{C}^{n\times n}$ is nonsingular because the set of vectors is linearly independent; we refer to Z as a *principal matrix* of $\mathsf{Pen}(A, B)$. Using this matrix, we have $Z^{\mathsf{H}} BZ = I_n$ and

$$Z^{\mathsf{H}} AZ = Z^{\mathsf{H}} \mathsf{diag}(\lambda_1, \ldots, \lambda_n) BZ = \mathsf{diag}(\lambda_1, \ldots, \lambda_n).$$

For $\boldsymbol{w} \in \mathbb{C}^n$ defined by $\boldsymbol{w} = Z^{-1}\boldsymbol{x}$, we have

$$\boldsymbol{x}^{\mathsf{H}} A\boldsymbol{x} = \boldsymbol{w}^{\mathsf{H}} Z^{\mathsf{H}} AZ\boldsymbol{w} == \boldsymbol{w}^{\mathsf{H}} \mathsf{diag}(\lambda_1, \ldots, \lambda_n)\boldsymbol{w} = \sum_{i=1}^{n} \lambda_i |w_i|^2 \qquad (8.7)$$

and

$$\boldsymbol{x}^{\mathsf{H}} B\boldsymbol{x} = \boldsymbol{w}^{\mathsf{H}} Z^{\mathsf{H}} AZ\boldsymbol{w} = \boldsymbol{w}^{\mathsf{H}} \boldsymbol{w} = \sum_{i=1}^{n} |w_i|^2. \qquad (8.8)$$

Let $\lambda_1 \leqslant \lambda_2 \leqslant \cdots \leqslant \lambda_n$ be the eigenvalues of a regular pencil $\mathsf{Pen}(A, B)$, where $A, B \in \mathbb{R}^{n\times n}$. Consider the *generalized Rayleigh–Ritz quotient* $\mathsf{ral}_{A,B} : \mathbb{R}^n \longrightarrow \mathbb{R}$ defined by

$$\mathsf{ral}_{A,B}(\boldsymbol{x}) = \frac{\boldsymbol{x}'A\boldsymbol{x}}{\boldsymbol{x}'B\boldsymbol{x}}$$

for $\boldsymbol{x} \in \mathbb{R}^n$ such that $\boldsymbol{x}'B\boldsymbol{x} \neq 0$. If Z is a principal matrix of $\mathsf{Pen}(A, B)$, by Equalities (8.7) and (8.7) we can write

$$\mathsf{ral}_{A,B}(\boldsymbol{x}) = \frac{\sum_{i=1}^n \lambda_i w_i^2}{\sum_{i=1}^n w_i^2}.$$

It is clear that $\lambda_1 \leqslant \mathsf{ral}_{A,B}(\boldsymbol{x}) \leqslant \lambda_n$. It is easy to see that we have $\mathsf{ral}_{A,B}(\boldsymbol{x}) = \lambda_1$ if and only if $w_2 = \cdots = w_n = 0$, which means that $\boldsymbol{x} = Z\boldsymbol{e}_1 = \boldsymbol{z}_1$. Similarly, $\mathsf{ral}_{A,B}(\boldsymbol{x}) = \lambda_n$ if and only if $\boldsymbol{x} = \boldsymbol{z}_n$. In other words, the minimum of $\mathsf{ral}_{A,B}(\boldsymbol{x})$ is achieved when $\boldsymbol{x}$ is an eigenvector that corresponds to the least eigenvalue of the pencil, while the maximum is achieved when $\boldsymbol{x}$ is an eigenvector that corresponds to the largest eigenvalue.

Similar results hold for the second smallest or the second largest eigenvalues of a regular pencil. Note that $\boldsymbol{x} = \sum_{i=1}^n w_i \boldsymbol{z}_i$. Thus, if $\boldsymbol{x}$ is orthogonal on $\boldsymbol{z}_1$ (in the sense of the inner product f_B), we have $\boldsymbol{x}'B\boldsymbol{z}_1 = w_1 = 0$, so

$$\mathsf{ral}_{A,B}(\boldsymbol{x}) = \frac{\sum_{i=2}^n \lambda_i w_i^2}{\sum_{i=2}^n w_i^2}$$

and

$$\min\{\mathsf{ral}_{A,B}(\boldsymbol{x}) \mid \boldsymbol{x}B\boldsymbol{z}_1 = 0\} = \lambda_2, \max\{\mathsf{ral}_{A,B}(\boldsymbol{x}) \mid \boldsymbol{x}B\boldsymbol{z}_n = 0\} = \lambda_{n-1}.$$

It is not difficult to show that, in general,

$$\lambda_p = \min\left\{\frac{\boldsymbol{x}'A\boldsymbol{x}}{\boldsymbol{x}'B\boldsymbol{x}}\middle|\boldsymbol{x}'B\boldsymbol{z}_i = 0 \text{ for } 1 \leqslant i \leqslant p-1\right\}$$

and

$$\lambda_{n-p} = \max\left\{\frac{\boldsymbol{x}'A\boldsymbol{x}}{\boldsymbol{x}'B\boldsymbol{x}}\middle|\boldsymbol{x}'B\boldsymbol{z}_i = 0 \text{ for } n-p+1 \leqslant i \leqslant n\right\}.$$

Next, we present an intrinsic characterization of the eigenvalues of a regular pencil, that is, a characterization that avoids the use of eigenvectors.

Let $C \in \mathbb{R}^{k\times n}$ be a matrix. Let us examine the variation of the generalized Rayleigh–Ritz quotient $\mathsf{ral}_{A,B} : \mathbb{R}^n \longrightarrow \mathbb{R}$ when $\boldsymbol{x}'B\boldsymbol{x} \neq 0$ and $C\boldsymbol{x} = \mathbf{0}_k$. In other words, we apply k linear constraints of the form $\boldsymbol{r}_i\boldsymbol{x} = 0$ to $\boldsymbol{x}$, where $\boldsymbol{r}_1, \ldots, \boldsymbol{r}_k$ are the rows of the matrix C.

Choose $k = n-1$ and let $C_{(p)}$ be

$$C_{(p)} = \begin{pmatrix} \boldsymbol{r}_1 \\ \vdots \\ \boldsymbol{r}_{p-1} \\ \boldsymbol{z}'_{p+1}B \\ \vdots \\ \boldsymbol{z}'_n B \end{pmatrix}.$$

It is clear that there exists $\boldsymbol{x}$ such that $C_{(p)}\boldsymbol{x} = \mathbf{0}_{n-1}$ because the system $C_{(p)}\boldsymbol{x} = \mathbf{0}_{n-1}$ has $n-1$ equations. Note that any solution $\boldsymbol{x}$ satisfies $\boldsymbol{z}'_k B\boldsymbol{x} = 0$, which shows that $\boldsymbol{x}$ is orthogonal on the last $n-p$ eigenvectors of $\mathsf{Pen}(A, B)$. Therefore, for the vectors $\boldsymbol{x}$ that satisfy these restrictions, we have

$$\min \frac{\boldsymbol{x}'A\boldsymbol{x}}{\boldsymbol{x}'B\boldsymbol{x}} \leqslant \lambda_p.$$

Consequently, λ_p is the maximum of the minimum of the ratio $\frac{\boldsymbol{x}'A\boldsymbol{x}}{\boldsymbol{x}'B\boldsymbol{x}}$ when $\boldsymbol{x}$ satisfies $p-1$ arbitrary linear restrictions.

8.10 Quadratic Forms and Quadrics

Let $g : V \times W \longrightarrow \mathbb{R}$ be a bilinear form, where V and W are finite-dimensional real linear spaces, $\dim(V) = n$, and $\dim(W) = m$. If $\boldsymbol{e}_1, \ldots, \boldsymbol{e}_n$ and $\boldsymbol{f}_1, \ldots, \boldsymbol{f}_m$ are two bases in V and W, respectively, then for $\boldsymbol{x} \in V$ and $\boldsymbol{y} \in W$ we can write $\boldsymbol{x} = x_1\boldsymbol{e}_1 + \cdots + x_n\boldsymbol{e}_n$ and $\boldsymbol{y} = y_1\boldsymbol{f}_1 + \cdots + y_m\boldsymbol{f}_m$. By the definition of bilinear forms, we have

$$g(\boldsymbol{x}, \boldsymbol{y}) = \sum_{i=1}^{n}\sum_{j=1}^{m} x_i y_j g(\boldsymbol{e}_i, \boldsymbol{f}_j) = \boldsymbol{x}'A\boldsymbol{y}, \tag{8.9}$$

where $A \in \mathbb{R}^{n\times m}$ is the matrix $A = (g(\boldsymbol{e}_i, \boldsymbol{f}_j))$. This matrix will be referred to as the *matrix of the bilinear form.*

The bilinear form f is completely defined by the matrix A as introduced above.

Let V be a real vector space and let $f : V \times V \longrightarrow \mathbb{R}$ be a bilinear form. If $f(\boldsymbol{x}, \boldsymbol{y}) = f(\boldsymbol{y}, \boldsymbol{x})$ for every $\boldsymbol{x}, \boldsymbol{y} \in V$, we say that f is a *symmetric bilinear form.*

Theorem 8.54. *A bilinear form $f : \mathbb{R}^n \times \mathbb{R}^n \longrightarrow \mathbb{R}$ is symmetric if and only if its matrix is symmetric.*

Proof. Let A be the matrix of f. Since f is symmetric, we have $\boldsymbol{x}'A\boldsymbol{y} = \boldsymbol{y}'A\boldsymbol{x}$ for every $\boldsymbol{x}, \boldsymbol{y} \in \mathbb{R}^n$. For $\boldsymbol{x} = \boldsymbol{e}_i$ and $\boldsymbol{y} = \boldsymbol{e}_j$, this implies $a_{ij} = a_{ji}$ for every $1 \leqslant i, j \leqslant n$. Thus, $A' = A$.

Conversely, if $A' = A$, we have $(\boldsymbol{x}'A\boldsymbol{y})' = (\boldsymbol{x}'A'\boldsymbol{y})' = \boldsymbol{y}'A\boldsymbol{x}$, so $(f(\boldsymbol{x}, \boldsymbol{y}))' = f(\boldsymbol{y}, \boldsymbol{x})$. Since $f(, \boldsymbol{y}) \in \mathbb{R}$, the symmetry of f follows. □

Let $f : V \times V$ be a symmetric bilinear form. Define the function $\phi : V \longrightarrow \mathbb{R}$ by $\phi(\boldsymbol{x}) = f(\boldsymbol{x}, \boldsymbol{x})$ for $\boldsymbol{x} \in V$. We have

$$\phi(\boldsymbol{u} + \boldsymbol{v}) = f(\boldsymbol{u} + \boldsymbol{v}, \boldsymbol{u} + \boldsymbol{v}) = f(\boldsymbol{u}, \boldsymbol{u}) + 2f(\boldsymbol{u}, \boldsymbol{v}) + f(\boldsymbol{v}, \boldsymbol{v}), \quad (8.10)$$
$$\phi(\boldsymbol{u} - \boldsymbol{v}) = f(\boldsymbol{u} - \boldsymbol{v}, \boldsymbol{u} - \boldsymbol{v}) = f(\boldsymbol{u}, \boldsymbol{u}) - 2f(\boldsymbol{u}, \boldsymbol{v}) + f(\boldsymbol{v}, \boldsymbol{v}), \quad (8.11)$$

because f is bilinear and symmetric. Therefore, we obtain the equality

$$\phi(\boldsymbol{u} + \boldsymbol{v}) + \phi(\boldsymbol{u} - \boldsymbol{v}) = 2(\phi(\boldsymbol{u}) + \phi(\boldsymbol{v})), \quad (8.12)$$

which is a generalization of the parallelogram equality from Theorem 6.32.

Definition 8.19. A *quadratic form* defined on a real vector space V is a function $\phi : V \longrightarrow \mathbb{R}$ for which there exists a bilinear form $f : V \times V \longrightarrow \mathbb{R}$ such that $\phi(\boldsymbol{x}) = f(\boldsymbol{x}, \boldsymbol{x})$ for $\boldsymbol{x} \in V$.

Equality (8.10) implies $f(\boldsymbol{u}, \boldsymbol{v}) = \frac{1}{2}(\phi(\boldsymbol{u} + \boldsymbol{v}) - \phi(\boldsymbol{u}) - \phi(\boldsymbol{v}))$, which shows that the bilinear form f is completely determined by its corresponding quadratic form.

Unlike bilinear forms, a quadratic form may be defined by several matrices.

Theorem 8.55. *Let ϕ be a quadratic form. We have $\phi(\boldsymbol{x}) = \boldsymbol{x}'A\boldsymbol{x} = \boldsymbol{x}'B\boldsymbol{x}$ for every $\boldsymbol{x} \in \mathbb{R}^n$ if and only if $A + A' = B + B'$.*

Proof. Suppose that $\phi(x) = \boldsymbol{x}'A\boldsymbol{x} = \boldsymbol{x}'B\boldsymbol{x}$ for every $\boldsymbol{x} \in \mathbb{R}^n$. If $\boldsymbol{x} = \boldsymbol{e}_i$ this amounts to $a_{ii} = b_{ii}$ for $1 \leqslant i \leqslant n$.

If $\boldsymbol{x} = \boldsymbol{e}_i + \boldsymbol{e}_j$, this yields $a_{ii} + a_{ij} + a_{ji} + a_{jj} = b_{ii} + b_{ij} + b_{ji} + b_{jj}$, which means that $a_{ij} + a_{ji} = b_{ij} + b_{ji}$ for $1 \leqslant i, j \leqslant n$. Thus, $A + A' = B + B'$.

Conversely, suppose that $A + A' = B + B'$. Note that $\phi(\boldsymbol{x}) = \frac{1}{2}(\phi(x) + \phi(x)')$, because $\phi(\boldsymbol{x}) \in \mathbb{R}$. Thus,

$$\begin{aligned}\phi(\boldsymbol{x}) &= \frac{1}{2}(\boldsymbol{x}'A\boldsymbol{x} + \boldsymbol{x}'A\boldsymbol{x}) \\ &= \boldsymbol{x}'\left(\frac{A + A'}{2}\right)\boldsymbol{x}.\end{aligned}$$

Thus, it is clear that either A or B defines the quadratic form. □

Among the set of matrices that define a quadratic form, there is only one symmetric matrix. This is stated formally in the next theorem.

Corollary 8.25. *If $\phi : \mathbb{R}^n \longrightarrow \mathbb{R}$ is a quadratic form, then there exists a unique symmetric matrix $C \in \mathbb{R}^{n \times n}$ such that $\phi(\boldsymbol{x}) = \boldsymbol{x}'C\boldsymbol{x}$ for $\boldsymbol{x} \in \mathbb{R}^n$.*

Proof. Suppose that $\phi(\boldsymbol{x}) = \boldsymbol{x}'A\boldsymbol{x}$ for $\boldsymbol{x} \in \mathbb{R}^n$. We saw that $\phi(\boldsymbol{x}) = \boldsymbol{x}'\frac{1}{2}\left(A + A'\right)\boldsymbol{x}$ and the matrix $\frac{1}{2}\left(A + A'\right)$ is clearly symmetric. On the another hand, if $A + A' = B + B'$ and B is symmetric, then $B = \frac{1}{2}(A + A')$. □

Since a quadratic form ϕ can be uniquely written as $\phi(\boldsymbol{x}) = \boldsymbol{x}'C\boldsymbol{x}$ for $\boldsymbol{x} \in \mathbb{R}^n$, properties of C can be transferred to ϕ. For example, if C is positive definite (semidefinite), then we say that ϕ is positive definite (semidefinite).

Example 8.13. Let $\boldsymbol{x} \in \mathbb{R}^n$ and let $\phi(\boldsymbol{x}) = \sum_{i=1}^{n}(x_{i+1} - x_i)^2$. It is easy to verify that

$$\phi(x) = \boldsymbol{x}'A\boldsymbol{x},$$

where A is the matrix

$$A = \begin{pmatrix} 1 & -1 & 0 & 0 & \cdots & & & 0 \\ 0 & -1 & 2 & -1 & \cdots & & & 0 \\ \vdots & \vdots & \vdots & \vdots & \cdots & \vdots & \vdots & \vdots \\ 0 & 0 & 0 & -0 & \cdots & 2 & -1 & 0 \\ 0 & 0 & 0 & -0 & \cdots & -1 & 2 & -1 \\ 0 & 0 & 0 & -0 & \cdots & 0 & -1 & 1 \end{pmatrix}$$

and $\boldsymbol{x} \in \mathbb{R}^n$. The definition of ϕ shows that $\boldsymbol{x} \neq \mathbf{0}_n$ implies $\phi(\boldsymbol{x}) > 0$. Thus, both A and ϕ are positive definite.

Observe that the bilinear symmetric function f can be recovered from the associated quadratic form ϕ because

$$f(\boldsymbol{u}, \boldsymbol{v}) = \frac{1}{2}\left(\phi(\boldsymbol{u} + \boldsymbol{v}) - \phi(\boldsymbol{u}) - \phi(\boldsymbol{v})\right),$$

for $\boldsymbol{u}, \boldsymbol{v} \in V$.

Example 8.14. Let $\phi : \mathbb{R}^3 \longrightarrow \mathbb{R}$ be the quadratic form defined by

$$\phi(\boldsymbol{x}) = 4x_1^2 + 8x_2^2 - x_3^2 - 6x_1x_2 + 4x_1x_3 + x_2x_3,$$

for

$$\boldsymbol{x} = \begin{pmatrix} x_1 \\ x_2 \\ x_3 \end{pmatrix} \in \mathbb{R}^3.$$

The symmetric matrix C that defines ϕ is

$$C = \begin{pmatrix} 4 & -3 & 2 \\ -3 & 8 & 0.5 \\ 2 & 0.5 & -1 \end{pmatrix}.$$

The off-diagonal terms of C are the halves of the coefficients of the corresponding linear form coefficients; for instance, since the coefficient of the cross-product term x_1x_2 is -6, we have $c_{12} = c_{21} = -3$.

Theorem 8.56. *A continuous function $\phi : V \longrightarrow \mathbb{R}$ is a quadratic form if and only if it satisfies the Parallelogram Equality (8.12) for every $\boldsymbol{u}, \boldsymbol{v} \in V$.*

Proof. The necessity of the condition was shown already. Therefore, we need to show only that if a continuous ϕ satisfies Equality (8.12), then it is a quadratic form.

Let ϕ be a function such that $\phi(\boldsymbol{u}+\boldsymbol{v})+\phi(\boldsymbol{u}-\boldsymbol{v}) = 2(\phi(\boldsymbol{u})+\phi(\boldsymbol{v}))$ for every $\boldsymbol{u}, \boldsymbol{v} \in V$.

If $\boldsymbol{u} = \boldsymbol{v} = \boldsymbol{0}$, it follows that $\phi(\boldsymbol{0}) = 0$. Choosing $\boldsymbol{u} = \boldsymbol{0}$ in Equality (8.12) implies $\phi(\boldsymbol{v}) + \phi(-\boldsymbol{v}) = 2\phi(\boldsymbol{v})$, so $\phi(-\boldsymbol{v}) = \phi(\boldsymbol{v})$ for $\boldsymbol{v} \in V$, so ϕ is an even function.

Define the symmetric function $f : V \times V \longrightarrow \mathbb{R}$ by

$$f(\boldsymbol{u}, \boldsymbol{v}) = \frac{1}{2}(\phi(\boldsymbol{u}+\boldsymbol{v}) - \phi(\boldsymbol{u}) - \phi(\boldsymbol{v})),$$

for $\boldsymbol{u}, \boldsymbol{v} \in V$. We need to show that f is bilinear. In view of the symmetry of f, it suffices to prove that f is linear in its first argument.

The definition of f allows us to write

$$\begin{aligned}
2f(\boldsymbol{u}_1+\boldsymbol{u}_2, \boldsymbol{v}) &= \phi(\boldsymbol{u}_1+\boldsymbol{u}_2+\boldsymbol{v}) - \phi(\boldsymbol{u}_1+\boldsymbol{u}_2) - \phi(\boldsymbol{v}),\\
2f(\boldsymbol{u}_1, \boldsymbol{v}) &= \phi(\boldsymbol{u}_1+\boldsymbol{v}) - \phi(\boldsymbol{u}_1) - \phi(\boldsymbol{v}),\\
2f(\boldsymbol{u}_2, \boldsymbol{v}) &= \phi(\boldsymbol{u}_2+\boldsymbol{v}) - \phi(\boldsymbol{u}_2) - \phi(\boldsymbol{v}),
\end{aligned}$$

which implies

$$\begin{aligned}
&2(f(\boldsymbol{u}_1+\boldsymbol{u}_2, \boldsymbol{v}) - f(\boldsymbol{u}_1, \boldsymbol{v}) - f(\boldsymbol{u}_2, \boldsymbol{v}))\\
&= (\phi(\boldsymbol{u}_1+\boldsymbol{u}_2+\boldsymbol{v}) + \phi(\boldsymbol{v})) - (\phi(\boldsymbol{u}_1+\boldsymbol{v}) + \phi(\boldsymbol{u}_2+\boldsymbol{v}))\\
&\quad -(\phi(\boldsymbol{u}_1+\boldsymbol{u}_2) - \phi(\boldsymbol{u}_1) - \phi(\boldsymbol{u}_2)).
\end{aligned} \tag{8.13}$$

Since ϕ satisfies the parallelogram equality, we have

$$\phi(\boldsymbol{u}_1+\boldsymbol{u}_2+\boldsymbol{v}) + \phi(\boldsymbol{v}) = \frac{1}{2}(\phi(\boldsymbol{u}_1\boldsymbol{u}_2+2\boldsymbol{v}) + \phi(\boldsymbol{u}_1+\boldsymbol{u}_2)), \tag{8.14}$$

$$\phi(\boldsymbol{u}_1+\boldsymbol{v}) + \phi(\boldsymbol{u}_2+\boldsymbol{v}) = \frac{1}{2}(\phi(\boldsymbol{u}_1\boldsymbol{u}_2+2\boldsymbol{v}) + \phi(\boldsymbol{u}_1-\boldsymbol{u}_2)). \tag{8.15}$$

The last two equalities imply

$$\begin{aligned}&\phi(\boldsymbol{u}_1+\boldsymbol{u}_2+\boldsymbol{v})+\phi(\boldsymbol{v})-\phi(\boldsymbol{u}_1+\boldsymbol{v})-\phi(\boldsymbol{u}_2+\boldsymbol{v})\\&=\frac{1}{2}(\phi(\boldsymbol{u}_1+\boldsymbol{u}_2)-\phi(\boldsymbol{u}_1-\boldsymbol{u}_2))\\&=-\phi(\boldsymbol{u}_1)-\phi(\boldsymbol{u}_2)+\phi(\boldsymbol{u}_1+\boldsymbol{u}_2),\end{aligned}\tag{8.16}$$

using again the parallelogram's equality.

The last equality combined with Equality (8.13) implies

$$f(\boldsymbol{u}_1+\boldsymbol{u}_2,\boldsymbol{v})=f(\boldsymbol{u}_1,\boldsymbol{v})+f(\boldsymbol{u}_2,\boldsymbol{v}).\tag{8.17}$$

We prove now that

$$f(a\boldsymbol{u},\boldsymbol{v})=af(\boldsymbol{u},\boldsymbol{v})\tag{8.18}$$

for $a\in\mathbb{R}$ and $\boldsymbol{u},\boldsymbol{v}\in V$.

Choosing $\boldsymbol{u}_1=\boldsymbol{u}$ and $\boldsymbol{u}_2=-\boldsymbol{u}$, we have $f(\boldsymbol{u},\boldsymbol{v})=-f(-\boldsymbol{u},\boldsymbol{v})$, so Equality (8.18) holds for $a=-1$.

It is easy to verify by induction on $k\in\mathbb{N}$ that Equality (8.17) implies that $f(k\boldsymbol{u},\boldsymbol{v})=kf(\boldsymbol{u},\boldsymbol{v})$ for every $k\in\mathbb{N}$.

Let k,l be two integers, $l\neq 0$. We have $lf(\frac{k}{l}\boldsymbol{u},\boldsymbol{v})=f(k\boldsymbol{u},\boldsymbol{v})=kf(\boldsymbol{u},\boldsymbol{v})$, which implies $f(\frac{k}{l}\boldsymbol{u},\boldsymbol{v})=\frac{k}{l}f(\boldsymbol{u},\boldsymbol{v})$. Thus, $f(r\boldsymbol{u},\boldsymbol{v})=rf(\boldsymbol{u},\boldsymbol{v})$ for any rational number r. The continuity of f allows us to conclude that $f(a\boldsymbol{u},\boldsymbol{v})=af(\boldsymbol{u},\boldsymbol{v})$ for any real number a. Thus, f is a bilinear form.

Since $\phi(\boldsymbol{x})=f(\boldsymbol{x},\boldsymbol{x})$, it follows that ϕ is indeed a quadratic form.

□

Let $\phi:V\longrightarrow\mathbb{R}$ be a quadratic form and let $\boldsymbol{v}_1,\ldots,\boldsymbol{v}_n$ be a basis of V, where V is an n-dimensional real linear space. Since $\boldsymbol{x}=\sum_{i=1}^n x_i\boldsymbol{v}_i$, the quadratic form can be written as

$$\phi(\boldsymbol{x})=\boldsymbol{x}'C\boldsymbol{x}=\sum_{i=1}^n\sum_{j=1}^n x_ix_j\boldsymbol{v}_i'C\boldsymbol{v}_j,$$

where C is the matrix of ϕ. If $i\neq j$ implies $\boldsymbol{v}_i'C\boldsymbol{v}_j=0$, then the basis $\boldsymbol{v}_1,\ldots,\boldsymbol{v}_n$ is referred to as a *canonical basis* of ϕ. Relative to this basis we can write

$$\phi(\boldsymbol{x})=\sum_{i=1}^n c_ix_i^2,$$

where $c_i=\boldsymbol{v}_i'C\boldsymbol{v}_i$.

Definition 8.20. A *quadric* in $\mathbb{R}^n$ is a set of the form

$$\mathfrak{Q} = \{\boldsymbol{x} \in \mathbb{R}^n \mid \boldsymbol{x}'A\boldsymbol{x} + 2\boldsymbol{b}'\boldsymbol{x} + c = 0\},$$

where $A \in \mathbb{R}^{n\times n}$ is a symmetric matrix, $\boldsymbol{b} \in \mathbb{R}^n$, and $c \in \mathbb{R}$.

We will refer both to a quadric $\mathfrak{Q}$ and to the equation that describes it as a quadric.

The *extended matrix of the quadric* $\mathfrak{Q}$ is the symmetric matrix $Q \in \mathbb{R}^{(n+1)\times(n+1)}$ given by

$$Q = \begin{pmatrix} a_{11} & \cdots & a_{1n} & b_1 \\ \vdots & \cdots & \vdots & \vdots \\ a_{n1} & \cdots & a_{nn} & b_n \\ b_1 & \cdots & b_n & c \end{pmatrix} = \begin{pmatrix} A & \boldsymbol{b} \\ \boldsymbol{b}' & c \end{pmatrix}.$$

Consider the quadric

$$\boldsymbol{x}'A\boldsymbol{x} + 2\boldsymbol{b}'\boldsymbol{x} + c = 0, \tag{8.19}$$

where $A \in \mathbb{R}^{n\times n}$ is a symmetric matrix, $\boldsymbol{b} \in \mathbb{R}^n$, and $c \in \mathbb{R}$. By the Spectral Theorem for Hermitian Matrices (applied, in this case, to real, symmetric matrices) there exists an orthogonal matrix U such that $A = U'DU$ and D is a diagonal matrix having the eigenvalues of A as its diagonal elements. Consider an isometry defined by $\boldsymbol{x} = U\boldsymbol{y} + \boldsymbol{r}$. Equation (8.19) can be written as

$$(\boldsymbol{y}'U' + \boldsymbol{r}')A(U\boldsymbol{y} + \boldsymbol{r}) + 2\boldsymbol{b}'(U\boldsymbol{y} + \boldsymbol{r}) + c = 0.$$

An equivalent form of this equality is

$$\begin{aligned} &\boldsymbol{y}'U'AU\boldsymbol{y} + \boldsymbol{r}'AU\boldsymbol{y} + \boldsymbol{y}'U'A\boldsymbol{r} + \boldsymbol{r}'A\boldsymbol{r} + 2\boldsymbol{b}'U\boldsymbol{y} + 2\boldsymbol{b}'\boldsymbol{r} + c \\ &= \boldsymbol{y}'U'AU\boldsymbol{y} + 2(\boldsymbol{r}'A + \boldsymbol{b}')U\boldsymbol{y} + \boldsymbol{r}'A\boldsymbol{r} + 2\boldsymbol{b}'\boldsymbol{r} + c = 0. \end{aligned}$$

We used here the fact that both $\boldsymbol{r}'AU\boldsymbol{y}$ and $\boldsymbol{y}'U'A\boldsymbol{r}$ are scalars and, therefore, they coincide with their transposed form.

Definition 8.21. Let $\boldsymbol{x}'A\boldsymbol{x} + 2\boldsymbol{b}'\boldsymbol{x} + c = 0$ be a quadric, $\boldsymbol{x} = U\boldsymbol{y} + \boldsymbol{r}$ be an isometry (where U is an orthogonal matrix), and let $\boldsymbol{y}'\tilde{A}\boldsymbol{y} + 2\tilde{b}'\boldsymbol{y} + \tilde{c} = 0$ be the transformed equation under the isometry. A function $\Phi : \mathbb{R}^{n\times n} \times \mathbb{R}^n \times \mathbb{R} \longrightarrow \mathbb{R}$ is an *isometric invariant* if for any quadric we have

$$\Phi(A, \boldsymbol{b}, c) = \Phi(\tilde{A}, \tilde{\boldsymbol{b}}, \tilde{c}).$$

Observe that

$$\begin{aligned}\tilde{A} &= U'AU,\\ \tilde{\boldsymbol{b}} &= (\boldsymbol{b} + A\boldsymbol{r})U,\\ \tilde{c} &= \boldsymbol{r}'A\boldsymbol{r} + 2\boldsymbol{b}'\boldsymbol{r} + c.\end{aligned}$$

Since A and $\tilde{A}$ are similar matrices, they have the same characteristic polynomial, which implies that any function of the form $\Phi(A, \boldsymbol{b}, c) = a_i$, where a_i is the ith coefficient of the characteristic polynomial of A, is an isometric invariant.

Theorem 8.57. *Let*

$$Q = \begin{pmatrix} A & \boldsymbol{b} \\ \boldsymbol{b}' & c \end{pmatrix}$$

be the extended matrix of the quadric $\boldsymbol{x}'A\boldsymbol{x} + 2\boldsymbol{b}'\boldsymbol{x} + c = 0$. *Then* $\det(Q)$ *is an isometric invariant.*

Proof. We have to show the equality

$$\begin{aligned}\det(Q) = \det\begin{pmatrix} A & \boldsymbol{b} \\ \boldsymbol{b}' & c \end{pmatrix} &= \det\begin{pmatrix} \tilde{A} & \tilde{\boldsymbol{b}} \\ \tilde{\boldsymbol{b}}' & \tilde{c} \end{pmatrix}\\ &= \det\begin{pmatrix} U'AU & (\boldsymbol{b} + A\boldsymbol{r})U \\ U'(\boldsymbol{b}' + \boldsymbol{r}'A) & \boldsymbol{r}'A\boldsymbol{r} + 2\boldsymbol{b}'\boldsymbol{r} + c \end{pmatrix}.\end{aligned}$$

Taking into account the result shown in Supplement 33 of Chapter 5, it follows that

$$\begin{aligned}&\det\begin{pmatrix} U'AU & (\boldsymbol{b} + A\boldsymbol{r})U \\ U'(\boldsymbol{b}' + \boldsymbol{r}'A) & \boldsymbol{r}'A\boldsymbol{r} + 2\boldsymbol{b}'\boldsymbol{r} + c \end{pmatrix}\\ &= \det\begin{pmatrix} AU & (\boldsymbol{b} + A\boldsymbol{r})U \\ (\boldsymbol{b}' + \boldsymbol{r}'A) & \boldsymbol{r}'A\boldsymbol{r} + 2\boldsymbol{b}'\boldsymbol{r} + c \end{pmatrix}\\ &= \det\begin{pmatrix} A & (\boldsymbol{b} + A\boldsymbol{r}) \\ (\boldsymbol{b}' + \boldsymbol{r}'A) & \boldsymbol{r}'A\boldsymbol{r} + 2\boldsymbol{b}'\boldsymbol{r} + c \end{pmatrix},\end{aligned}$$

because U is an orthogonal matrix. Further elementary transformation yields the following equalities:

$$\det\begin{pmatrix} A & (\boldsymbol{b}+A\boldsymbol{r}) \\ (\boldsymbol{b}'+\boldsymbol{r}'A) & \boldsymbol{r}'A\boldsymbol{r}+2\boldsymbol{b}'\boldsymbol{r}+c \end{pmatrix} = \det\begin{pmatrix} A & \boldsymbol{b}+A\boldsymbol{r} \\ \boldsymbol{b}' & \boldsymbol{b}'\boldsymbol{r}+c \end{pmatrix} = \det\begin{pmatrix} A & \boldsymbol{b} \\ \boldsymbol{b}' & c \end{pmatrix},$$

which show that $\det(\tilde{Q}) = \det(Q)$. □

If the equation $\boldsymbol{r}'A + \boldsymbol{b}' = \boldsymbol{0}'$ has a solution in $\boldsymbol{r}$, we refer to $\boldsymbol{r}$ as a *center of the quadric.*

A quadric may have a unique center, a set of centers, or no center, depending on the number of solutions of the equation $\boldsymbol{r}'A + \boldsymbol{b}' = \boldsymbol{0}'$, which is equivalent to $A\boldsymbol{r} + \boldsymbol{b} = \boldsymbol{0}$, since A is a symmetric matrix. If A is an invertible matrix, then $\mathfrak{Q}$ has a unique center $\boldsymbol{r}_0 = -A^{-1}\boldsymbol{b}$. Choosing $\boldsymbol{r} = \boldsymbol{r}_0$, the quadric is defined by the equation

$$\boldsymbol{y}'D\boldsymbol{y} - \boldsymbol{r}_0'A\boldsymbol{r}_0 + c = 0, \tag{8.20}$$

because $\boldsymbol{r}_0'A\boldsymbol{r}_0 + 2\boldsymbol{b}'\boldsymbol{r}_0 = -\boldsymbol{r}_0'A\boldsymbol{r}_0$. Note that if a center $\boldsymbol{r}_0$ exists, then $\boldsymbol{y} \in \mathfrak{Q}$ if and only if $-\boldsymbol{y} \in \mathfrak{Q}$. Equivalently, $\boldsymbol{r}_0 + U\boldsymbol{x} \in \mathfrak{Q}$ if and only if $\boldsymbol{r}_0 - U\boldsymbol{x} \in \mathfrak{Q}$ and the middle of the line segment determined by $\boldsymbol{r}_0 + U\boldsymbol{x}$ and $\boldsymbol{r}_0 - U\boldsymbol{x}$ is $\boldsymbol{r}_0$. Thus, $\boldsymbol{r}_0$ is a center of symmetry for the quadric.

The set of orthonormal columns of the matrix U for which $A = UDU'$ (or $D = U'AU$, where D is a diagonal matrix) are called the *principal axes* of the quadric $\mathfrak{Q}$.

Quadrics are classified based on the non-nullity and the sign of the eigenvalues of the symmetric matrix A. Thus, we can use for the classification either the diagonal matrix D whose entries are the eigenvalues of A or, equivalently, the inertia of the matrix A.

Let $\boldsymbol{x}'A\boldsymbol{x} + 2\boldsymbol{b}'\boldsymbol{x} + c = 0$ be a quadric having a unique center. Since A is a nonsingular matrix, there are no null eigenvalues of A, so $\mathfrak{I}(A) = (n_+(A), n_-(A), 0)$. There are $n + 1$ possible classes since we can have $0 \leqslant n_+ \leqslant n$ and $n_+ + n_- = n$.

When $n_+ = n$, the quadric is an *n-dimensional ellipsoid.* The equality

$$\boldsymbol{y}'D\boldsymbol{y} - \boldsymbol{r}_0'A\boldsymbol{r}_0 + c = 0$$

can be written as

$$\sum_{i=1}^{n} \lambda_i y_i^2 - \boldsymbol{r}_0' A \boldsymbol{r}_0 + c = 0.$$

If $\boldsymbol{r}_0' A \boldsymbol{r}_0 + c > 0$, the last equality can be written as

$$\sum_{i=1}^{n} \frac{y_i^2}{a_i^2} = 1,$$

where $a_i^2 = \frac{\boldsymbol{r}' A \boldsymbol{r} - c}{\lambda_i}$ for $1 \leqslant i \leqslant n$ and the quadric is a *real ellipsoid* having $a_1, \ldots, a_n$ as its semiaxes.

At the other extreme, suppose that $n_- = n$, which means that $\mathsf{spec}(A)$ consists of negative numbers. If $k = \boldsymbol{r}' A \boldsymbol{r} - c < 0$, then $\mathfrak{Q} = \emptyset$. If $k > 0$, by multiplying by -1 we revert back to the previous case.

Example 8.15. In this example, we consider the case of quadrics $\boldsymbol{x}' A \boldsymbol{x} + 2\boldsymbol{b}' \boldsymbol{x} + c = 0$ in $\mathbb{R}^3$ that have a unique center, that is, for which $\det(A) \neq 0$.

If $n_+ = 3$, we have an ellipsoid, as we saw before. For $n_+ < 3$, the classification of the quadric depends on $\det(Q)$. If $\det(Q) \neq 0$, $n_+ = 2$, and $n_- = 1$, we have a hyperboloid of one sheet, while in the case when $\det(Q) \neq 0$, $n_+ = 1$, and $n_- = 2$, we have a hyperboloid of two sheets. If $\det(Q) = 0$, we have a cone.

After transforming the equation of the quadric $\boldsymbol{x}' A \boldsymbol{x} + 2\boldsymbol{b}' \boldsymbol{x} + c = 0$ to the canonical form given by Equality (8.20):

$$\boldsymbol{y}' D \boldsymbol{y} - k = 0,$$

where $k = \boldsymbol{r}_0' A \boldsymbol{r}_0 - c$. Note that

$$\det(Q) = \det \begin{pmatrix} \lambda_1 & 0 & 0 & 0 \\ 0 & \lambda_2 & 0 & 0 \\ 0 & 0 & \lambda_3 & 0 \\ 0 & 0 & 0 & k \end{pmatrix} = \lambda_1 \lambda_2 \lambda_3 k.$$

Now, the class of the quadric can be easily established, based on the following table:

n_+	n_-	$\det(Q)$	Class
3	0	$\neq 0$	Ellipsoid
3	0	$= 0$	One-point ellipsoid
3	0	< 0	Imaginary ellipsoid
2	1	$\neq 0$	Hyperboloid of one sheet
1	2	$\neq 0$	Hyperboloid of two sheets
2	1	$= 0$	Cone
1	2	$= 0$	Cone

If $\det(A) = 0$, then $\mathsf{rank}(A) \leqslant 2$ and several cases are possible.

If $\det(A) = 0$ and the sum of minors of order 2 of A is non-zero, the rank of A equals 2 (see Supplement 8.13). The set of solutions of system $A\boldsymbol{r} + \boldsymbol{b} = \boldsymbol{0}$ depends on the value of $\det(Q)$. If $\det(Q) = 0$, then the rank of the extended matrix $(A\ \boldsymbol{b})$ equals the rank of A, and the system $A\boldsymbol{r}+\boldsymbol{b} = \boldsymbol{0}$ is consistent. The set of centers of the quadric is a line of centers. Otherwise, that is, if $\det(Q) \neq 0$, the quadric has no center.

If $\det(A) = 0$, the sum of minors of order 2 of A is 0, and $\mathsf{trace}(A) \neq 0$, then $\mathsf{rank}(A) = 1$.

Example 8.16. Let $\mathfrak{Q}$ be the quadric $14x_1^2 - 5x_2^2 + 11x_3^2 - 20x_1x_2 - 4x_1x_3 + 10x_2x_3 = 0$. Its matrix was examined in Example 8.1:

$$A = \begin{pmatrix} 14 & -10 & -2 \\ -10 & -5 & 5 \\ -2 & 5 & 11 \end{pmatrix}.$$

The matrix U is

$$U = \begin{pmatrix} -\frac{2}{\sqrt{6}} & \frac{1}{\sqrt{5}} & -\frac{2}{\sqrt{30}} \\ \frac{1}{\sqrt{6}} & 0 & -\frac{5}{\sqrt{30}} \\ \frac{1}{\sqrt{6}} & \frac{2}{\sqrt{5}} & \frac{1}{\sqrt{30}} \end{pmatrix},$$

so the principal axes of this quadric form correspond to the eigenvectors $\boldsymbol{u}_1, \boldsymbol{u}_2$, and $\boldsymbol{u}_3$ studied in the same example.

8.11 Spectra of Positive Matrices

Let $A \in \mathbb{R}^{n\times n}$ be a matrix with positive entries, $A > O$. Observe that its spectral radius $\rho(A)$ is positive because $\rho(A) = 0$ implies that $\mathsf{spec}(A) = \{0\}$, which implies that A is nilpotent by Corollary 8.6. Since $A > O$, this is impossible, so $\rho(A) > 0$.

Theorem 8.58 (Perron–Frobenius theorem). *Let $A \in \mathbb{R}^{n\times n}$ be a symmetric matrix with positive elements and let λ be its largest eigenvalue. The following statements hold:*

(i) *λ is a positive number;*
(ii) *there exists an eigenvector $\boldsymbol{x}$ that corresponds to λ such that $\boldsymbol{x} > \mathbf{0}_n$;*
(iii) $\mathsf{geomm}(A, \lambda) = 1$;
(iv) *if θ is any other eigenvalue of A, then $|\theta| < \lambda$.*

Proof. Since the eigenvalues of A are real and their sum equals $\mathsf{trace}(A) > 0$, it follows that its largest value λ is positive, which proves Part (i).

Let $\boldsymbol{u}$ be a real unit eigenvector that belongs to λ, so $\sum_{j=1}^{n} a_{ij}u_j = \lambda u_i$. We have $A\boldsymbol{u} = \lambda\boldsymbol{u}$, so $\boldsymbol{u}'A\boldsymbol{u} = \lambda\boldsymbol{u}'\boldsymbol{u} = \lambda\|\boldsymbol{u}\|_2^2 = \lambda$. This allows us to write

$$\lambda = \sum_{i=1}^{n}\sum_{i=1}^{n} a_{ij}u_i u_j,$$

so

$$\lambda = \sum_{i=1}^{n}\sum_{i=1}^{n} a_{ij}u_i u_j = \left|\sum_{i=1}^{n}\sum_{i=1}^{n} a_{ij}u_i u_j\right|,$$

because $\lambda > 0$.

We claim that the vector $\boldsymbol{x} = \mathsf{abs}(\boldsymbol{u})$ is an eigenvector that corresponds to λ and has positive components. Note that

$$\lambda = \left|\sum_{i=1}^{n}\sum_{i=1}^{n} a_{ij}u_i u_j\right| \leqslant \sum_{i=1}^{n}\sum_{i=1}^{n} a_{ij}x_i x_j.$$

By Corollary 8.14, since λ is the largest eigenvalue of A, we have $\sum_{i=1}^{n}\sum_{i=1}^{n} a_{ij}x_i x_j \leqslant \lambda$; the equality takes place only if $\boldsymbol{x}$ is an eigenvector of λ.

Since $\boldsymbol{x} \geqslant \mathbf{0}$, if $x_i = 0$ for some i, then the equality $\lambda x_i = \sum_{j=1}^{n} a_{ij}x_j$ implies that $x_j = 0$ for $1 \leqslant j \leqslant n$, because all numbers a_{ij} are positive. This means that $\boldsymbol{x} = \mathbf{0}$, which is impossible because $\boldsymbol{x}$ is an eigenvector. Thus, we conclude that all components of $\boldsymbol{x}$ are positive. This completes the proof of Part (ii).

For Part (iii), suppose that the geometric multiplicity of λ is greater than 1 and let $\boldsymbol{u}$ and $\boldsymbol{v}$ be two real unit vectors of the invariant subspace $S_{A,\lambda}$ such that $\boldsymbol{u} \perp \boldsymbol{v}$. Note that the vectors $\mathsf{abs}(\boldsymbol{u})$ and $\mathsf{abs}(\boldsymbol{v})$ are also eigenvectors corresponding to λ.

Suppose that $u_i < 0$ for some i. Since $\lambda u_i = \sum_{j=1}^{n} a_{ij}u_j$ and $\lambda|u_i| = \sum_{j=1}^{n} a_{ij}|u_j|$, we have $\lambda(u_i + |u_i|) = 0 = \sum_{j=1}^{n} a_{ij}(u_j + |u_j|)$, and this implies $u_j + |u_j| = 0$ for $1 \leqslant j \leqslant n$. In other words, we have either $u_j = |u_j| > 0$ for every j, or $u_j = -|u_j| < 0$ for every j. The same applies to $\boldsymbol{v}$, so $\boldsymbol{v}'\boldsymbol{u} = \sum_{i=1}^{n} v_i u_i \neq 0$, which contradicts the orthogonality of $\boldsymbol{u}$ and $\boldsymbol{v}$. Thus, $\mathsf{geomm}(A, \lambda) = 1$.

For Part (iv), let $\boldsymbol{w}$ be a unit eigenvector that corresponds to the eigenvalue θ, so $\sum_{i=1}^{n} a_{ij}w_j = \theta w_i$ for $1 \leqslant i \leqslant n$. Again, by Corollary 8.14,

$$\lambda \geqslant \sum_{i=1}^{n}\sum_{j=1}^{n} a_{ij}|w_i||w_j| \geqslant \left|\sum_{i=1}^{n}\sum_{j=1}^{n} a_{ij}w_i w_j\right| = |\theta|.$$

If $\theta = -\lambda$, these inequalities show that $|w_j| = x_j$ for all j and, therefore, there exists i such that $w_i = x_i$. Adding the equalities $\lambda x_i = \sum_{j=0}^{n} a_{ij}x_j$ and $-\lambda w_i = \sum_{j=0}^{n} a_{ij}w_j$ yields $0 = \sum_{j=1}^{n} a_{ij}(x_+ w_j) \geqslant a_{ii}(x_i + w_i)$, which contradicts the fact that $a_{ii} > 0$ and $w_i = x_i > 0$. Thus, $\theta \neq -\lambda$. □

Definition 8.22. Let $A \in \mathbb{R}^{n\times n}$ be a symmetric matrix with positive elements. The number $\rho(A)$ is the *Perron number* of A; the positive vector $\boldsymbol{x}$ with $\|\boldsymbol{x}\|_2 = 1$ that corresponds to the eigenvalue $\rho(A)$ is the *Perron vector* of A.

The Perron–Frobenius Theorem states that the spectral radius $\rho(A)$ of a positive matrix A is always an eigenvalue of A. This property holds also for non-negative matrices.

Lemma 8.7. *Let $A \in \mathbb{R}^{n\times n}$ be a non-negative matrix that is irreducible, where $n > 1$. Its spectral radius $\rho(A)$ is an eigenvalue and there exists a positive eigenvector that corresponds to* *this eigenvalue. No non-negative eigenvector corresponds to any other eigenvalue of A.*

Proof. Since A is an irreducible matrix, the matrix $(I_n + A)^{n-1}$ is positive by Supplement 131 of Chapter 3. This implies

$$((I_n + A)^{n-1})' = (I_n + A')^{n-1} > O_{n,n},$$

so by the Perron–Frobenius theorem, there exists a positive vector $\boldsymbol{y}$ such that $(I + A')^{n-1}\boldsymbol{y} = \rho((I + A')^{n-1})\boldsymbol{y}$. Equivalently, we have

$$\boldsymbol{y}'(I + A)^{n-1} = \rho((I + A)^{n-1})\boldsymbol{y}'. \tag{8.21}$$

Let $\lambda \in \mathsf{spec}(A)$ be an eigenvalue of A such that $|\lambda| = \rho(A)$ and let $\boldsymbol{x}$ be an eigenvector associated to λ. Since $\lambda\boldsymbol{x} = A\boldsymbol{x}$, it follows that $|\lambda|\mathsf{abs}(\boldsymbol{x}) = \mathsf{abs}(A\boldsymbol{x}) \leqslant \mathsf{abs}(A)\mathsf{abs}(\boldsymbol{x})$, so $\rho(A)\mathsf{abs}(\boldsymbol{x}) \leqslant A\mathsf{abs}(\boldsymbol{x})$ because A is a non-negative matrix. It is immediate that $\rho(A)^p\mathsf{abs}(\boldsymbol{x}) \leqslant A^p\mathsf{abs}(\boldsymbol{x})$ for $p \in \mathbb{N}$. In turn, this implies

$$(1 + \rho(A))^{n-1}\mathsf{abs}(\boldsymbol{x}) \leqslant (I_n + A)^{n-1}\mathsf{abs}(\boldsymbol{x}),$$

so, by Equality (8.21), we have

$$(1+\rho(A))^{n-1}(\boldsymbol{y}'\mathsf{abs}(\boldsymbol{x})) \leqslant \boldsymbol{y}'(I_n+A)^{n-1}\mathsf{abs}(\boldsymbol{x}) = \rho((I+A)^{n-1})\boldsymbol{y}'\mathsf{abs}(\boldsymbol{x}).$$

Since $\boldsymbol{y} > \boldsymbol{0}$, we have $\boldsymbol{y}'\mathsf{abs}(\boldsymbol{x}) > 0$ and, therefore,

$$(1 + \rho(A))^{n-1} \leqslant \rho((I + A)^{n-1}). \tag{8.22}$$

By Corollary 8.7, the eigenvalues of $(I + A)^{n-1}$ have the form $(1 + \lambda)^{n-1}$, where $\lambda \in \mathsf{spec}(A)$. Therefore, there exists an eigenvalue μ of A such that

$$|(1 + \mu)^{n-1}| = \rho((I + A)^{n-1}).$$

Since $|\mu| \leqslant \rho(A)$, Equality (8.22) implies

$$(1 + |\mu|)^{n-1} \leqslant (1 + \rho(A))^{n-1} \leqslant \rho((I + A)^{n-1}) = |(1 + \mu)^{n-1}|,$$

so $1 + |\mu| \leqslant |1 + \mu|$. This implies that μ is a positive real number and, therefore, $\mu = \rho(A)$. Consequently,

$$(\rho(A))^k\mathsf{abs}(\boldsymbol{x}) \leqslant A^k\mathsf{abs}(\boldsymbol{x})$$

for $k \geqslant 1$. In particular, for $k = 1$, we have $\rho(A)\mathsf{abs}(\boldsymbol{x}) \leqslant A\mathsf{abs}(\boldsymbol{x}) = \mu\mathsf{abs}(\boldsymbol{x})$.

Since

$$(I + A)^{n-1}\mathsf{abs}(\boldsymbol{x}) = (1 + \mu)^{n-1}\mathsf{abs}(\boldsymbol{x}) = \rho((I + A)^{n-1})\mathsf{abs}(\boldsymbol{x}),$$

and $\mathsf{abs}(\boldsymbol{x}) > \mathbf{0}$ (by Theorem 8.58), it follows that there is only one linearly independent vector associated with μ. Indeed, suppose that we would have two linearly independent vectors, $\boldsymbol{u}$ and $\boldsymbol{v}$, associated with the eigenvalue μ. Since $\boldsymbol{v} \neq 0$, there exists $v_i \neq 0$, so the vector $\boldsymbol{w} = \boldsymbol{u} - \frac{w_i}{v_i}\boldsymbol{v}$ is also an eigenvector of the eigenvalue μ because $\boldsymbol{w} \neq \mathbf{0}$. But $\boldsymbol{w}_i = 0$, which contradicts the fact that $\mathsf{abs}(\boldsymbol{x}) > 0$ for any eigenvector of μ.

Since $A \neq O_{n,n}$, we have $\rho(A) > 0$. Moreover, $\rho(A)$ is a simple eigenvalue of A. Indeed, there is only one linearly independent eigenvector $\boldsymbol{u}$ of A associated with $\rho(A)$ such that $\boldsymbol{u} > 0$. Since A' is also irreducible (by Exercise 128 of Chapter 3), there exists only one linearly independent eigenvector $\boldsymbol{v}$ of A' associated to $\rho(A')$ and $\boldsymbol{v} > 0$. Since $\boldsymbol{v}'\boldsymbol{u} > 0$, it follows that $\rho(A)$ is a simple eigenvalue of A by Theorem 8.44.

Suppose that $\boldsymbol{z}$ is an eigenvector of an eigenvalue ζ of A and $\boldsymbol{z} > 0$, where $\zeta \neq \rho(A)$. We have shown that A' has an eigenvector $\boldsymbol{w} > \mathbf{0}$ such that $A'\boldsymbol{w} = \rho(A)\boldsymbol{w}$. This implies $\boldsymbol{w}'A\boldsymbol{z} = \zeta\boldsymbol{w}'\boldsymbol{z}$. Since $\boldsymbol{w}'A\boldsymbol{z} = (A'\boldsymbol{w})'\boldsymbol{z} = \rho(A)\boldsymbol{w}'\boldsymbol{z}$, we have $\zeta\boldsymbol{w}'\boldsymbol{z} = \rho(A)\boldsymbol{w}'\boldsymbol{z}$. Since $\boldsymbol{w}'\boldsymbol{z} > 0$, it follows that $\zeta = \rho(A)$, which is a contradiction. Thus, no other eigenvalue except $\rho(A)$ has a positive eigenvector. □

Now we prove the existence of a non-negative eigenvector for $\rho(A)$ without the assumption of irreducibility of A made in Lemma 8.7.

Theorem 8.59. *Let $A \in \mathbb{R}^{n\times n}$ be a non-negative matrix such that $n > 1$. Its spectral radius $\rho(A)$ is an eigenvalue of A and there exists a non-negative eigenvector that corresponds to this eigenvalue.*

Proof. Suppose that $red(A) = k$. As we saw in the proof of Supplement 127 of Chapter 3, there exists a permutation matrix P such that

$$PAP' = \begin{pmatrix} B_{11} & B_{12} & \cdots & B_{1k} \\ O & B_{22} & \cdots & B_{2k} \\ \vdots & \vdots & \cdots & \vdots \\ O & O & \cdots & B_{kk} \end{pmatrix},$$

where the matrices $B_{11}, \ldots, B_{kk}$ are irreducible. By Theorem 7.15, $\mathsf{spec}(A) = \bigcup_{i=1}^{k} \mathsf{spec}(B_{ii})$. We have

$$\rho(A) = \rho(PAP') = \max\{\rho(B_{ii} \mid 1 \leqslant\leqslant k\}.$$

Let j be the smallest number such that $\rho(A) = \rho(B_{jj})$. The matrix PAP' can now be written as

$$A = \begin{pmatrix} C & E & F \\ O & B_{jj} & G \\ O & O & D \end{pmatrix},$$

where $C \in \mathbb{C}^{p\times p}$ and $D \in \mathbb{C}^{q\times q}$ are upper triangular matrices or are missing, when B_{jj} is the first diagonal block or the last diagonal block. Since B_{jj} is irreducible, Lemma 8.7 implies that there exists $\boldsymbol{y}_j$ such that $B_{jj}\boldsymbol{y}_j = \rho(A)\boldsymbol{y}_j$

If $j = 1$ and B_{jj} is the first diagonal block, then

$$\begin{pmatrix} \boldsymbol{y}_1 \\ \mathbf{0} \end{pmatrix}$$

is a non-negative eigenvector of PAP'.

If $j > 1$, there exists $\boldsymbol{z} > 0$ such that

$$\begin{pmatrix} \boldsymbol{z} \\ \boldsymbol{y}_1 \\ \mathbf{0} \end{pmatrix}$$

is an eigenvector of PAP'. Since

$$PAP' \begin{pmatrix} \boldsymbol{z} \\ \boldsymbol{y}_j \\ \mathbf{0} \end{pmatrix} = \begin{pmatrix} C\boldsymbol{z} + E\boldsymbol{y}_1 \\ B_{jj}\boldsymbol{y}_j \\ \mathbf{0} \end{pmatrix},$$

it suffices to find $\boldsymbol{z} \geqslant \mathbf{0}$ such that $C\boldsymbol{z} + E\boldsymbol{y}_j = \rho(A)\boldsymbol{z}$. We have $\rho(C) < \rho(A)$, which means that for the matrix $Z = \frac{1}{\rho(A)}C$ we have $\rho(Z) < 1$. Thus, $I+Z+Z^2+\cdots$ converges to $(I-Z)^{-1}$. Since $Z \geqslant O$, the terms of the series $I + Z + Z^2 + \cdots$ are nonnegative matrices, so $(I - Z)^{-1}$ is a non-negative matrix. The desired vector $\boldsymbol{z}$ is

$$\boldsymbol{z} = \frac{1}{\rho(A)}(1 - Z)^{-1}E\boldsymbol{y}_j$$

and $\boldsymbol{z} \geqslant \mathbf{0}$. □

8.12 Spectra of Positive Semidefinite Matrices

In Theorem 7.12, we saw that eigenvalues of Hermitian matrices are real numbers. The next theorem links these eigenvalues to positive definiteness.

Theorem 8.60. *Let $A \in \mathbb{C}^{n\times n}$ be a Hermitian matrix. If A is positive semidefinite, then all its eigenvalues are non-negative; if A is positive definite, then its eigenvalues are positive.*

Proof. Since A is Hermitian, all its eigenvalues are real numbers. Suppose that A is positive semidefinite, that is, $\boldsymbol{x}^{\mathsf{H}} A\boldsymbol{x} \geqslant 0$ for $\boldsymbol{x} \in \mathbb{C}^n$. If $\lambda \in \mathsf{spec}(A)$, then $A\boldsymbol{v} = \lambda\boldsymbol{v}$ for some eigenvector $\boldsymbol{v} \neq \boldsymbol{0}$. The positive semidefiniteness of A implies $\boldsymbol{v}^{\mathsf{H}} A\boldsymbol{v} = \lambda\boldsymbol{v}^{\mathsf{H}}\boldsymbol{v} = \lambda\|\boldsymbol{v}\|_2^2 \geqslant 0$, which implies $\lambda \geqslant 0$. It is easy to see that if A is positive definite, then $\lambda > 0$. □

Theorem 8.61. *Let $A \in \mathbb{C}^{n\times n}$ be a Hermitian matrix. If A is positive semidefinite, then all its principal minors are non-negative real numbers. If A is positive definite, then all its principal minors are positive real numbers.*

Proof. Since A is positive semidefinite, every sub-matrix $A\begin{bmatrix} i_1 & \cdots & i_k \\ i_1 & \cdots & i_k \end{bmatrix}$ is a Hermitian positive semidefinite matrix by Theorem 6.51, so every principal minor is a non-negative real number. The second part of the theorem is proven similarly. □

Corollary 8.26. *Let $A \in \mathbb{C}^{n\times n}$ be a Hermitian matrix. The following statements are equivalent.*

(i) *A is positive semidefinite;*

(ii) *all eigenvalues of A are non-negative numbers;*

(iii) *there exists a Hermitian matrix $C \in \mathbb{C}^{n\times n}$ such that $C^2 = A$;*

(iv) *A is the Gram matrix of a sequence of vectors, that is, $A = B^{\mathsf{H}}B$ for some $B \in \mathbb{C}^{n\times n}$.*

Proof. (i) implies (ii): This was shown in Theorem 8.60.

(ii) implies (iii): Suppose that A is a matrix such that all its eigenvalues are the non-negative numbers $\lambda_1, \ldots, \lambda_n$. By Theorem 8.14,

A can be written as $A = U^{\mathsf{H}} D U$, where U is a unitary matrix and

$$D = \begin{pmatrix} \lambda_1 & 0 & \cdots & 0 \\ 0 & \lambda_2 & \cdots & 0 \\ \vdots & \vdots & \cdots & \vdots \\ 0 & 0 & \cdots & \lambda_n \end{pmatrix}.$$

Define the matrix $\sqrt{D}$ as

$$\sqrt{D} = \begin{pmatrix} \sqrt{\lambda_1} & 0 & \cdots & 0 \\ 0 & \sqrt{\lambda_2} & \cdots & 0 \\ \vdots & \vdots & \cdots & \vdots \\ 0 & 0 & \cdots & \sqrt{\lambda_n} \end{pmatrix}.$$

Clearly, we gave $(\sqrt{D})^2 = D$. Now we can write $A = U\sqrt{D}U^{\mathsf{H}}U\sqrt{D}U^{\mathsf{H}}$, which allows us to define the desired matrix C as $C = U\sqrt{D}U^{\mathsf{H}}$.

(iii) implies (iv): Since C is itself a Hermitian matrix, this implication is obvious.

(iv) implies (i): Suppose that $A = B^{\mathsf{H}}B$ for some matrix $B \in \mathbb{C}^{n\times k}$. Then, for $\boldsymbol{x} \in \mathbb{C}^n$, we have $\boldsymbol{x}^{\mathsf{H}}A\boldsymbol{x} = \boldsymbol{x}^{\mathsf{H}}B^{\mathsf{H}}B\boldsymbol{x} = (B\boldsymbol{x})^{\mathsf{H}}(B\boldsymbol{x}) = \|B\boldsymbol{x}\|_2^2 \geqslant 0$, so A is positive semidefinite. □

8.13 MATLAB Computations

The function **eigalgmult** computes the eigenvalues of a matrix and their algebraic multiplicities:

```
function [eigvalues, repeats] = eigalgmult(A)
tol = sqrt(eps);
v = sort(eig(A));
v = round(v/tol) * tol;
eigvalues = flipud(unique(v));
n = length(v);
d = length(eigvalues);
B = v * ones(1, d);
C = ones(n,1) * eigvalues';
m = abs(C-B) <= tol;
repeats = sum(m);
```

Example 8.17. Let

$$A = \begin{pmatrix} 3 & 1 & 1 \\ 1 & 3 & 1 \\ 1 & 1 & 3 \end{pmatrix}.$$

which is a special case of the matrix discussed in Example 7.6. An application of the function **eigalgmult** returns the eigenvalues 5 and 2 having 1 and 2 as their algebraic multiplicities, respectively:

```
>> A=[3 1 1;1 3 1; 1 1 3]
A =
     3     1     1
     1     3     1
     1     1     3
>> [e,mult]=eigalgmult(A)
e =
     5
     2
mult =
     1     2
```

The function **eiggeommult(A)** computes eigenvectors of A and their geometric multiplicity. Its outputs are two matrices V and G. The diagonal matrix G has on its diagonal the eigenvalues of A. Each value is repeated a number of times equal to the dimensionality of its invariant space, which equals its geometric multiplicity. The matrix V contains eigenvectors corresponding to these eigenvalues.

```
function [V, G] = eiggeommult(A)

[m,n] = size(A);
[eigvalues, r] = eigalgmult(A);

V = []; d = [];
for k = 1 : length(eigvalues);
  v = nullspbasis(A - eigvalues(k)*eye(n));
  [ms, ns] = size(v);
  V = [V v];
  temp = ones(ns, 1) * eigvalues(k);
  d = [d; temp];
end
G = diag(d);
```

Example 8.18. The matrix

$$A = \begin{pmatrix} 2 & 1 & 0 \\ 0 & 2 & 1 \\ 0 & 0 & 2 \end{pmatrix}$$

has one eigenvalue 2 with algebraic multiplicity 3 and geometric multiplicity 1 as follows from the following MATLAB computation:

```
>> A=[2 1 0;0 2 1; 0 0 2]
A =
     2     1     0
     0     2     1
     0     0     2
>> [V,G]=eiggeommult(A)
V =
     1
     0
     0
G =
     2
```

On the other hand, the next matrix has two non-defective eigenvalues, 5 and 2 as follows from the next computation:

```
>> A=[3 1 1;1 3 1; 1 1 3]
A =

     3     1     1
     1     3     1
     1     1     3
>> [V,G]=eiggeommult(A)
V =
     1    -1    -1
     1     1     0
     1     0     1
G =
     5     0     0
     0     2     0
     0     0     2

>>
```

Let A and B be two square matrices, $A, B \in \mathbb{C}^{n\times n}$. To compute the generalized eigenvalues of A and B, we can use `eig(A,B)`, which returns a vector containing these values. Also, `[V,D] = eig(A,B)` yields a diagonal matrix D of generalized eigenvalues and a matrix $V \in \mathbb{C}^{n\times n}$ whose columns are the corresponding eigenvectors so that $AV = BVD$.

The exponential of a matrix can be computed using the function `expm`.

Example 8.19. Let A and B be the matrices considered in Example 8.12,

$$A = \begin{pmatrix} 1 & 1 \\ 0 & 1 \end{pmatrix}, B = \begin{pmatrix} 0 & 1 \\ 1 & 0 \end{pmatrix}, \text{ and } C = A + B = \begin{pmatrix} 1 & 2 \\ 1 & 1 \end{pmatrix}.$$

The exponentials of these matrices obtained with the function `expm` are

```
>> expm(B)
ans =
    2.7183    2.7183
         0    2.7183

>> expm(B)
ans =
    1.5431    1.1752
    1.1752    1.5431

>> expm(C)
ans =
    5.9209    7.4388
    3.7194    5.9209
```

Furthermore, the product $e^A * e^B$ computed as

```
>> expm(A)*expm(B)
ans =
    7.3891    7.3891
    3.1945    4.1945
```

is clearly distinct from e^C.

Similar functions in MATLAB exist for computing the principal logarithm $\log A$ of a matrix A, namely `logm`, and for computing the principal square root of a matrix, `sqrtm(A)`, namely, the unique square root for which every eigenvalue has a non-negative real part.

Exercises and Supplements

(1) Let $A, B \in \mathbb{C}^{n\times n}$ be two matrices. Prove that if B is nonsingular, then $AB^{-1} \sim B^{-1}A$.

(2) Prove that if $A \in \mathbb{C}^{m\times n}$ and $B \in \mathbb{C}^{n\times m}$, then the matrices

$$C = \begin{pmatrix} AB & O_{m,n} \\ B & O_{n,n} \end{pmatrix} \text{ and } D = \begin{pmatrix} 0_{m,m} & 0_{m,n} \\ B & BA \end{pmatrix}$$

are similar.

(3) Let $A \in \mathbb{C}^{m\times m}$ and $B \in \mathbb{C}^{n\times n}$ be two matrices. Prove that $\boldsymbol{S}_{A,B}(X) = (\boldsymbol{S}_{-B^{\mathsf{H}},-A^{\mathsf{H}}}(X^{\mathsf{H}}))^{\mathsf{H}}$ for $X \in \mathbb{C}^{m\times n}$.

(4) The *Lie bracket* of $\mathbb{C}^{n\times n}$ is the mapping $[\cdot,\cdot] : \mathbb{C}^{n\times n} \times \mathbb{C}^{n\times n} \longrightarrow \mathbb{C}^{n\times n}$ given by

$$[X, Y] = \boldsymbol{S}_{X,X}(Y).$$

Prove that $[X, Y] = -[Y, X]$, $[X, X] = O_{n,n}$ and $[X, [Y, Z]] + [Y, [Z, X]] + [Z, [X, Y]] = O_{n,n}$ for any matrices $X, Y, Z \in \mathbb{C}^{n,n}$.

(5) Let $A \in \mathbb{R}^{n\times n}$ be a symmetric matrix having the eigenvalues $\lambda_1 \geqslant \cdots \geqslant \lambda_n$. Prove that

$$\|A - \frac{\lambda_1 + \lambda_n}{2} I_n\|_2 = \frac{\lambda_1 - \lambda_n}{2}.$$

(6) Let $A \in \mathbb{R}^{n\times n}$ be a matrix and let $c \in \mathbb{R}$. Prove that for $\boldsymbol{x}, \boldsymbol{y} \in \mathbb{R}^n - \{\boldsymbol{0}_n\}$, we have $\mathsf{ral}_A(\boldsymbol{x}) - \mathsf{ral}_A(\boldsymbol{y}) = \mathsf{ral}_B(\boldsymbol{x}) - \mathsf{ral}_B(\boldsymbol{y})$, where $B = A + cI_n$.

(7) Let $A \in \mathbb{R}^{n\times n}$ be a symmetric matrix having the eigenvalues $\lambda_1 \geqslant \cdots \geqslant \lambda_n$ and let $\boldsymbol{x}$ and $\boldsymbol{y}$ be two vectors in $\mathbb{R}^n - \{\boldsymbol{0}_n\}$.

Prove that

$$|\mathsf{ral}_A(\boldsymbol{x}) - \mathsf{ral}_A(\boldsymbol{y})| \leqslant (\lambda_1 - \lambda_n) \sin \angle(\boldsymbol{x}, \boldsymbol{y}).$$

Solution: Assume that $\|\boldsymbol{x}\| = \|\boldsymbol{y}\| = 1$ and let $B = A - \frac{\lambda_1+\lambda_n}{2} I_n$. By Exercise 8.13, we have

$$\begin{aligned}|\mathsf{ral}_A(\boldsymbol{x}) - \mathsf{ral}_A(\boldsymbol{y})| &= |\mathsf{ral}_B(\boldsymbol{x}) - \mathsf{ral}_B(\boldsymbol{y})| = |\boldsymbol{x}'B\boldsymbol{x} - \boldsymbol{y}'B\boldsymbol{y}| \\ &= |B(\boldsymbol{x}-\boldsymbol{y})'(\boldsymbol{x}+\boldsymbol{y})|.\end{aligned}$$

By the Cauchy–Schwarz Inequality, we have

$$\begin{aligned}|B(\boldsymbol{x}-\boldsymbol{y})'(\boldsymbol{x}+\boldsymbol{y})| &\leqslant 2\|B\|\frac{\|\boldsymbol{x}-\boldsymbol{y}\|\ \|\boldsymbol{x}+\boldsymbol{y}\|}{2} \\ &= (\lambda_1 - \lambda_n) \sin \angle(\boldsymbol{x}, \boldsymbol{y}).\end{aligned}$$

(8) Prove that if A is a unitary matrix and $1 \notin \mathsf{spec}(A)$, then there exists a skew-Hermitian S such that $A = (I_n - S)(I_n + S)^{-1}$.

(9) Let $f : \mathbb{R}^{n\times n} \longrightarrow \mathbb{R}$ be a function such that $f(AB) = f(BA)$ for $A, B \in \mathbb{R}^{n\times n}$. Prove that if $A \sim B$, then $f(A) = f(B)$.

(10) Let $B_r(\lambda, a) \in \mathbb{C}^{r\times r}$ be the matrix defined by

$$B_r(\lambda, a) = \begin{pmatrix} \lambda & a & 0 & \cdots & 0 \\ 0 & \lambda & a & \cdots & 0 \\ \vdots & \vdots & \ddots & \ddots & \vdots \\ 0 & 0 & 0 & \cdots & a \\ 0 & 0 & 0 & \cdots & \lambda \end{pmatrix} \in \mathbb{C}^{r\times r}.$$

Note that $B_r(\lambda, 1)$ is a Jordan block. Let $a, b \in \mathbb{C} - \{0\}$ be two nonzero numbers. Prove that $B_n(\lambda, a) \sim B_n(\lambda, b)$.

(11) Let λ be a complex number. Prove that the kth power of $B_r(\lambda)$ is given by

$$(B_r(\lambda)^k = \begin{pmatrix} \lambda^k & \binom{k}{1}\lambda^{k-1} & \binom{k}{2}\lambda^{k-2} & \cdots & \binom{k}{r-1}\lambda^{k-r+1} \\ 0 & \lambda^k & \binom{k}{1}\lambda^{k-1} & \cdots & \binom{k}{r-2}\lambda^{k-r+2} \\ \vdots & \vdots & \vdots & \cdots & \vdots \\ 0 & 0 & 0 & \cdots & \lambda^k \end{pmatrix}.$$

(12) Prove that if $|\lambda| < 1$, then $\lim_{k\to\infty} B_r(\lambda)^k = O$.
(13) Let $A \in \mathbb{R}^{n\times n}$ be a symmetric real matrix. Prove that

$$\nabla \mathsf{ral}_A(\boldsymbol{x}) = \frac{2}{\boldsymbol{x}'\boldsymbol{x}}(A\boldsymbol{x} - \mathsf{ral}_A(\boldsymbol{x})\boldsymbol{x}).$$

Also, show that the eigenvectors of A are the stationary points of the function $\mathsf{ral}_A(\boldsymbol{x})$.
(14) Let $A, B \in \mathbb{C}^{n\times n}$ be two Hermitian matrices. Prove that AB is a Hermitian matrix if and only if $AB = BA$.
(15) Let $A \in \mathbb{R}^{3\times 3}$ be a symmetric matrix. Prove that if $\mathsf{trace}(A) \neq 0$, the sum of principal minors of order 2 equals 0, and $\det(A) = 0$, then $\mathsf{rank}(A) = 1$.

Solution: The characteristic polynomial of A is $p_A(\lambda) = \lambda^3 - \mathsf{trace}(A)\lambda^2 = 0$. Thus, $\mathsf{spec}(A) = \{\mathsf{trace}(A), 0\}$, where $\mathsf{algm}(A, 0) = 2$, so $\mathsf{rank}(A) = 1$.
(16) Let $A \in \mathbb{R}^{3\times 3}$ be a symmetric matrix. Prove that if the sum of principal minors of order 2 does not equal 0 but $\det(A) = 0$, then $\mathsf{rank}(A) = 2$.
(17) Let $A \in \mathbb{C}^{n\times n}$ be a Hermitian matrix, $\boldsymbol{u} \in \mathbb{C}^n$ be a vector, and let a be a complex number. Define the Hermitian matrix B as

$$B = \begin{pmatrix} A & \boldsymbol{u} \\ \boldsymbol{u}^{\mathsf{H}} & a \end{pmatrix}.$$

Let $\alpha_1 \leqslant \cdots \leqslant \alpha_n$ be the eigenvalues of A and let $\beta_1 \leqslant \cdots \leqslant \beta_n \leqslant \beta_{n+1}$ be the eigenvalues of B. Prove that

$$\beta_1 \leqslant \alpha_1 \leqslant \beta_2 \leqslant \cdots \leqslant \beta_n \leqslant \alpha_n \leqslant \beta_{n+1}.$$

Solution: Since $B \in \mathbb{C}^{(n+1)\times(n+1)}$, by the Courant–Fisher theorem we have

$$\begin{aligned}\beta_{k+1} &= \min_W \max_{\boldsymbol{x}} \{\boldsymbol{x}^{\mathsf{H}} B\boldsymbol{x} \mid \|\boldsymbol{x}\|_2 = 1 \text{ and } \boldsymbol{x} \in \langle W\rangle^{\perp}\} \\ &= \max_Z \min_{\boldsymbol{x}} \{\boldsymbol{x}^{\mathsf{H}} B\boldsymbol{x} \mid \|\boldsymbol{x}\|_2 = 1 \text{ and } \boldsymbol{x} \in \langle Z\rangle^{\perp}\},\end{aligned}$$

where W denotes ranges of sets of k non-zero arbitrary vectors, and let Z be a subset of $\mathbb{C}^n$ that consists of $n - k$ non-zero arbitrary vectors in $\mathbb{C}^{n+1}$.

Let U be a set of k non-zero vectors in $\mathbb{C}^n$ and let Y be a set of $n-k-1$ vectors in $\mathbb{C}^n$. Define the subsets W_U and Z_Y of $\mathbb{C}^{n+1}$ as

$$W_U = \left\{ \begin{pmatrix} \boldsymbol{u} \\ 0 \end{pmatrix} \middle| \boldsymbol{u} \in U \right\}$$

and

$$Z_Y = \left\{ \begin{pmatrix} \boldsymbol{y} \\ 0 \end{pmatrix} \middle| \boldsymbol{y} \in Y \right\} \cup \{\boldsymbol{e}_{n+1}\}.$$

By restricting the sets W and Z to sets of the form W_U and Z_Y, we obtain the double inequality

$$\begin{aligned}&\max_{Z_Y} \min_{\boldsymbol{x}} \{\boldsymbol{x}^{\mathsf{H}} B\boldsymbol{x} \mid \|\boldsymbol{x}\|_2 = 1 \text{ and } \boldsymbol{x} \in \langle Z_Y \rangle^{\perp}\} \\ &\leqslant \beta_{k+1} \leqslant \min_{W_U} \max_{\boldsymbol{x}} \{\boldsymbol{x}^{\mathsf{H}} B\boldsymbol{x} \mid \|\boldsymbol{x}\|_2 = 1 \text{ and } \boldsymbol{x} \in \langle W_U \rangle^{\perp}\}.\end{aligned}$$

Note that, if $\boldsymbol{x} \in \langle Z_Y \rangle^{\perp}$, then we have $\boldsymbol{x} \perp \boldsymbol{e}_{n+1}$, so $x_{n+1} = 0$. Therefore,

$$\boldsymbol{x}^{\mathsf{H}} B\boldsymbol{x} = (\boldsymbol{y}^{\mathsf{H}} 0) \begin{pmatrix} A & \boldsymbol{u} \\ \boldsymbol{u}^{\mathsf{H}} & a \end{pmatrix} \begin{pmatrix} \boldsymbol{y} \\ 0 \end{pmatrix} = \boldsymbol{y}^{\mathsf{H}} A\boldsymbol{y}.$$

Consequently,

$$\begin{aligned}&\max_{Z_Y} \min_{\boldsymbol{x}} \{\boldsymbol{x}^{\mathsf{H}} B\boldsymbol{x} \mid \|\boldsymbol{x}\|_2 = 1, \boldsymbol{x} \in \langle Z_Y \rangle^{\perp}\} \\ &= \max_{Y} \min_{\boldsymbol{y}} \{\boldsymbol{y}^{\mathsf{H}} A\boldsymbol{y} \mid \|\boldsymbol{y}\|_2 = 1, \boldsymbol{y} \in \langle Y \rangle^{\perp}\} = \alpha_k.\end{aligned}$$

This allows us to conclude that $\alpha_k \leqslant \beta_{k+1}$ for $1 \leqslant k \leqslant n$. On the another hand, if $\boldsymbol{x} \in \langle W_U \rangle^{\perp}$ and

$$\boldsymbol{x} = \begin{pmatrix} \boldsymbol{u} \\ 0 \end{pmatrix},$$

then $\boldsymbol{x}^{\mathsf{H}} B\boldsymbol{x} = \boldsymbol{u}^{\mathsf{H}} A\boldsymbol{u}$ and $\|\boldsymbol{x}\|_2 = \|\boldsymbol{u}\|_2$. Now we can write

$$\begin{aligned}&\min_{W_U} \max_{\boldsymbol{x}} \{\boldsymbol{x}^{\mathsf{H}} B\boldsymbol{x} \mid \|\boldsymbol{x}\|_2 = 1, \boldsymbol{x} \in \langle W_U \rangle^{\perp}\} \\ &= \min_{U} \max_{\boldsymbol{u}} \{\boldsymbol{u}^{\mathsf{H}} A\boldsymbol{u} \mid \|\boldsymbol{u}\|_2 = 1, \text{ and } \boldsymbol{u} \in \langle U \rangle^{\perp}\} = \alpha_{k+1},\end{aligned}$$

so $\beta_{k+1} \leqslant \alpha_{k+1}$ for $1 \leqslant k \leqslant n-1$.

(18) Let $A, B \in \mathbb{C}^{n\times n}$ be two matrices such that $AB = BA$. Prove that A and B have a common eigenvector.

Solution: Let $\lambda \in \mathsf{spec}(A)$ and let $\{\boldsymbol{x}_1, \ldots, \boldsymbol{x}_k\}$ be a basis for $\mathsf{null}(A-\lambda I_n)$. Observe that the matrices $A-\lambda I_n$ and B commute because

$$(A - \lambda I_n)B = AB - \lambda B \text{ and } B(A - \lambda I_n) = BA - B.$$

Therefore, we have

$$(A - \lambda I_n)B\boldsymbol{x}_i = B(A - \lambda I_n)\boldsymbol{x}_i = \mathbf{0},$$

so $(A - \lambda I_n)BX = O_{n,n}$, where $X = (\boldsymbol{x}_1, \ldots, \boldsymbol{x}_k)$. Consequently, $ABX = \lambda BX$. Let $\boldsymbol{y}_1, \ldots, \boldsymbol{y}_m$ be the columns of the matrix BX. The last equality implies that $A\boldsymbol{y}_i = \lambda\boldsymbol{y}_i$, so $\boldsymbol{y}_i \in \mathsf{null}(A - \lambda I_n)$. Since X is a basis of $\mathsf{null}(A - \lambda I_n)$, it follows that each $\boldsymbol{y}_i$ is a linear combination of the columns of X so there exists a matrix P such that $(\boldsymbol{y}_1 \cdots \boldsymbol{y}_m) = (\boldsymbol{x}_1 \cdots \boldsymbol{x}_k)P$, which is equivalent to $BX = XP$. Let $\boldsymbol{w}$ be an eigenvector of P. We have $P\boldsymbol{w} = \mu\boldsymbol{w}$. Consequently, $BX\boldsymbol{w} = XP\boldsymbol{w} = \mu X\boldsymbol{w}$, which proves that $X\boldsymbol{w}$ is an eigenvector of B. Also, $A(X\boldsymbol{w}) = A(BX\boldsymbol{w}) = (\lambda BX)\boldsymbol{w} = \lambda\mu X\boldsymbol{w}$, so $X\boldsymbol{w}$ is also an eigenvector of A.

(19) Let $A \in \mathbb{C}^{n\times n}$ be a matrix and let $\mathsf{spec}(A) = \{\lambda_1, \ldots, \lambda_n\}$. Prove that

(a) $\sum_{p=1}^{n} |\lambda_p|^2 \leqslant \|A\|_F^2$;

(b) the equality $\sum_{p=1}^{n} |\lambda_p|^2 = \|A\|_F^2$ holds if and only if A is normal.

Solution: By Schur's Triangularization Theorem (Theorem 8.8), there exists a unitary matrix $U \in \mathbb{C}^{n\times n}$ and an upper triangular matrix $T \in \mathbb{C}^{n\times n}$ such that $A = U^{\mathsf{H}}TU$ and the diagonal elements of T are the eigenvalues of A. Thus,

$$\|A\|_F^2 = \|T\|_F^2 = \sum_{p=1}^{n} |\lambda_p|^2 + \sum_{i<j} |t_{ij}|^2,$$

which implies the desired inequality.

The equality of the second part follows from the Spectral Theorem for Normal Matrices. The converse implication can be obtained noting that by the first part, the equality of the second part implies $t_{ij} = 0$ for $i < j$, which means that T is actually a diagonal matrix.

(20) Let A and B be two normal matrices in $\mathbb{C}^{n\times n}$. Prove that if AB is a normal matrix, then so is BA.

Solution: By Theorem 7.10, the matrices AB and BA have the same non-zero eigenvalues. Since A and B are normal we have $A^{\mathsf{H}}A = AA^{\mathsf{H}}$ and $B^{\mathsf{H}}B = BB^{\mathsf{H}}$. Thus, we can write

$$\begin{aligned}
\mathsf{trace}((AB)^{\mathsf{H}}AB) &= \mathsf{trace}(B^{\mathsf{H}}A^{\mathsf{H}}AB)\\
&= \mathsf{trace}(B^{\mathsf{H}}AA^{\mathsf{H}}B)\\
&\quad \text{(because } A \text{ is a normal matrix)}\\
&= \mathsf{trace}((B^{\mathsf{H}}A)(A^{\mathsf{H}}B))\\
&\quad \text{(by the associativity of matrix product)}\\
&= \mathsf{trace}((A^{\mathsf{H}}B)(B^{\mathsf{H}}A))\\
&\quad \text{(by the third part of Theorem 3.14)}\\
&= \mathsf{trace}(A^{\mathsf{H}}(BB^{\mathsf{H}})A))\\
&\quad \text{(by the associativity of matrix product)}\\
&= \mathsf{trace}(A^{\mathsf{H}}(B^{\mathsf{H}}B)A))\\
&\quad \text{(because } B \text{ is a normal matrix)}\\
&= \mathsf{trace}((BA)^{\mathsf{H}}BA).
\end{aligned}$$

Since AB is a normal matrix, we have $\sum_{p=1}^{n} |\lambda_p|^2 = \|AB\|_F^2 = \mathsf{trace}((AB)^{\mathsf{H}}AB)$, where $\mathsf{spec}(AB) = \{\lambda_1, \ldots, \lambda_p\}$, by the second part of Supplement 8.13. Taking into account the equalities shown above, it follows that BA is a normal matrix.

(21) Let $A \in \mathbb{R}^{n\times n}$ be a symmetric matrix such that $a_{ij} \in \{0, 1\}$ for $1 \leqslant i, j \leqslant n$. If $d = \frac{|\{a_{ij} \mid a_{ij}=1\}|}{n^2}$, prove that $n\sqrt{d} \geqslant \lambda_1 \geqslant nd$, where λ_1 is the largest eigenvalue of A.

Solution: By the Rayleigh–Ritz theorem (Theorem 8.17), we have $\lambda_1 \mathbf{1}_n'\mathbf{1}_n \geqslant \mathbf{1}_n' A\mathbf{1}_n$. Since $\mathbf{1}_n'\mathbf{1}_n = n$ and $\mathbf{1}_n' A\mathbf{1}_n = n^2 d$, it follows that $\lambda_1 \geqslant nd$. On the other hand, by Corollary 8.9, we have $\sum_{i=1}^{n} \lambda_i^2 = \|A\|_F = n^2 d$, so $\lambda_1 \leqslant n\sqrt{d}$.

(22) Let A, B, and E be three matrices in $\mathbb{C}^{n\times n}$ such that $B = A+E$ and A is a diagonalizable matrix, $A = Q^{-1}\mathsf{diag}(\lambda_1, \ldots, \lambda_n)Q$. Prove that if $\mu \in \mathsf{spec}(B)$, then for every $\lambda_k \in \mathsf{spec}(A)$ we have $|\mu - \lambda_k| \leqslant \|Q\| \; \|Q^{-1}\| \; \|E\|$, where $\|\cdot\|$ of the matrix norms $|\!|\!|\cdot|\!|\!|_p$.

Solution: By Supplement 27, we have $\|Q^{-1}(A - \mu I_n)Q\| \leqslant \|Q^{-1}EQ\|$ for any eigenvalue μ of B. This inequality can be written as

$$\begin{aligned}&\|(\|\mathsf{diag}(\lambda_1, \ldots, \lambda_n) - \mathsf{diag}(\mu, \ldots, \mu))\\ &= \|\mathsf{diag}(\lambda_1 - \mu, \ldots, \lambda_n - \mu)\|\\ &= \max |\lambda_k - \mu| \leqslant \|Q^{-1}EQ\|.\end{aligned}$$

This implies immediately the inequality to be shown.

(23) Let $A \in \mathbb{C}^{m\times m}$ and $B \in \mathbb{C}^{n\times n}$ be two diagonalizable matrices. Prove that the matrix $M \in \mathbb{C}^{mn\times mn}$ that corresponds to the Sylvester operator $\boldsymbol{S}_{A,B} = AX - XB$ is diagonalizable.

Solution: By Theorem 8.1, it suffices to show that the set of eigenvectors $\{X_{ij} \mid 1 \leqslant i \leqslant m, 1 \leqslant j \leqslant n\}$ is linearly independent.
Using the notations of Theorem 8.27, we need to show that the equality $\sum_{i=1}^{m}\sum_{j=1}^{n} a_{ij}\boldsymbol{u}_i\boldsymbol{v}_j^{\mathsf{H}} = \boldsymbol{0}$ implies $a_{ij} = 0$.
If $\tilde{\boldsymbol{u}}_p$ is a left eigenvector of A that corresponds to λ_p and $\tilde{\boldsymbol{v}}_q$ is a right eigenvector that corresponds to λ_q, then $\tilde{\boldsymbol{u}}_p^{\mathsf{H}}\sum_{i=1}^{m}\sum_{j=1}^{n} a_{ij}\boldsymbol{u}_i\boldsymbol{v}_j^{\mathsf{H}}\tilde{\boldsymbol{v}}_q = a_{pq}\tilde{\boldsymbol{u}}_p^{\mathsf{H}}\boldsymbol{u}_p\boldsymbol{v}_q^{\mathsf{H}}\tilde{\boldsymbol{v}}_q = 0$, which yields $a_{pq} = 0$.

(24) Let $A \in \mathbb{C}^{m\times m}$ and $B \in \mathbb{C}^{n\times n}$ be two matrices. Prove that for every $X \in \mathbb{C}^{m\times n}$, we have

$$\mathsf{vec}(\boldsymbol{S}_{A,B}(X)) = (I_n \otimes A - B \otimes I_m)\mathsf{vec}X.$$

(25) Let $U = (\boldsymbol{u}_1, \ldots, \boldsymbol{u}_r) \in \mathbb{C}^{n\times r}$ be a matrix having an orthonormal set of columns. For $A \in \mathbb{C}^{n\times n}$, define the matrix $A_U \in \mathbb{C}^{r\times r}$ by $A_U = U^{\mathsf{H}}AU$.
   (a) Prove that if A is a Hermitian matrix, then A_U is also a Hermitian matrix.
   (b) If A is a Hermitian matrix having the eigenvalues $\lambda_1 \leqslant \cdots \leqslant \lambda_n$ and A_U has the eigenvalues $\mu_1 \leqslant \cdots \leqslant \mu_r$, prove that

$$\lambda_k \leqslant \mu_k \leqslant \lambda_{k+n-r}.$$

(c) Prove that

$$\sum_{i=1}^{r} \lambda_i = \min\{\mathsf{trace}(A_U) \mid U^{\mathsf{H}}U = I_r\},$$

$$\sum_{i=n-r+1}^{n} \lambda_i = \max\{\mathsf{trace}(A_U) \mid U^{\mathsf{H}}U = I_r\}.$$

Solution: Observe that $(A_U)^{\mathsf{H}} = U^{\mathsf{H}}A^{\mathsf{H}}U = U^{\mathsf{H}}AU = A_U$, so A_U is indeed Hermitian and its eigenvalues $\mu_1, \ldots, \mu_r$ are real numbers.
Extend the set of columns $\boldsymbol{u}_1, \ldots, \boldsymbol{u}_r$ of U to an orthonormal basis $\{\boldsymbol{u}_1, \ldots, \boldsymbol{u}_r, \boldsymbol{u}_{r+1}, \ldots, \boldsymbol{u}_n\}$ and let $W \in \mathbb{C}^{n\times n}$ be the matrix whose columns are $\boldsymbol{u}_1, \ldots, \boldsymbol{u}_n$. Since W is a unitary matrix, $\mathsf{spec}(W^{\mathsf{H}}AW) = \mathsf{spec}(A)$ and A_U is a principal submatrix of $\mathsf{spec}(W^{\mathsf{H}}AW)$. The second part follows from Theorem 8.20.
The first equality of the third part follows from the fact that the second part implies

$$\sum_{i=1}^{r} \lambda_i \leqslant \sum_{i=1}^{r} \mu_i = \mathsf{trace}(A_U),$$

where $A_U \in \mathbb{C}^{r\times r}$. If the columns $\boldsymbol{u}_1, \ldots, \boldsymbol{u}_r$ of U are chosen as orthonormal eigenvectors, the above inequality becomes an equality. In this case, we have $U^{\mathsf{H}}U = I_r$ and the first equality of the third part follows. The argument for the second equality is similar.

(26) Let $A, B \in \mathbb{C}^{n\times n}$ be two Hermitian matrices. Suppose that $\mathsf{spec}(A) = \{\xi_1, \ldots, \xi_n\}$, $\mathsf{spec}(B) = \{\zeta_1, \ldots, \zeta_n\}$, and $\mathsf{spec}(A + B) = \{\lambda_1, \ldots, \lambda_n\}$ where $\xi_1 \leqslant \cdots \leqslant \xi_n$, $\zeta_1 \leqslant \cdots \leqslant \zeta_n$, and $\lambda_1 \leqslant \cdots \leqslant \lambda_n$. Prove that for $r \leqslant n$, we have

$$\sum_{i=1}^{r} \lambda_i \leqslant \sum_{i=1}^{r} \xi_i + \sum_{i=1}^{r} \zeta_i$$

and

$$\sum_{i=n-r+1}^{n} \lambda_i \geqslant \sum_{i=n-r+1}^{n} \xi_i + \sum_{i=n-r+1}^{n} \zeta_i.$$

Solution: Supplement 8.13 implies that

$$\begin{aligned}\sum_{i=1}^{r} \lambda_i &= \min\{\mathsf{trace}((A+B)_U) \mid U^{\mathsf{H}}U = I_r\}\\ &\geqslant \min\{\mathsf{trace}(A_U) \mid U^{\mathsf{H}}U = I_r\}\\ &\quad + \min\{\mathsf{trace}(B_U) \mid U^{\mathsf{H}}U = I_r\}\\ &= \sum_{i=1}^{r} \xi_i + \sum_{i=1}^{r} \zeta_i.\end{aligned}$$

For the second part, we can write

$$\begin{aligned}\sum_{i=n-r+1}^{n} \lambda_i &= \max\{\mathsf{trace}((A+B)_U) \mid U^{\mathsf{H}}U = I_r\}\\ &\leqslant \max\{\mathsf{trace}(A_U) \mid U^{\mathsf{H}}U = I_r\}\\ &\quad + \max\{\mathsf{trace}(B_U) \mid U^{\mathsf{H}}U = I_r\}\\ &= \sum_{i=n-r+1}^{n} \xi_i + \sum_{i=n-r+1}^{n} \zeta_i.\end{aligned}$$

(27) Let $A \in \mathbb{R}^{2\times 2}$ be the matrix

$$A = \begin{pmatrix} a & b \\ c & d \end{pmatrix}.$$

Prove that A is diagonalizable if and only if $(a-d)^2 + 4bc \neq 0$.

(28) Let $A \in \mathbb{C}^{n\times n}$ be a matrix. Prove that the following statements are equivalent:
- (a) A is a rank 1 matrix;
- (b) A has exactly one non-zero eigenvalue λ with $\mathsf{algm}(A, \lambda) = 1$;
- (c) There exist $\boldsymbol{x}, \boldsymbol{y} \in \mathbb{C}^n - \{\boldsymbol{0}\}$ such that $A = \boldsymbol{x}\boldsymbol{y}^{\mathsf{H}}$ and $\boldsymbol{x}^{\mathsf{H}}\boldsymbol{y}$ is an eigenvalue of A.

(29) In this exercise we give a generalization of Theorem 8.4. Let $A \in \mathbb{C}^{n\times n}$, $B \in \mathbb{C}^{k\times k}$, and $X \in \mathbb{C}^{n\times k}$ be three matrices such that $AX = XB$. Prove that
- (a) $\mathsf{range}(X)$ is an invariant subspace of A;

(b) if $\boldsymbol{v}$ is an eigenvector of B, then $X\boldsymbol{v}$ is an eigenvector of A;
(c) if $\mathsf{rank}(X) = k$, then $\mathsf{spec}(B) \subseteq \mathsf{spec}(A)$.

The next supplements present a result known as *Weyl's theorem* (Supplement 8.13) and several of its important consequences. For a Hermitian matrix $A \in \mathbb{C}^{n\times n}$, we denote its eigenvalues arranged in increasing order as $\lambda_1(A) \leqslant \cdots \leqslant \lambda_n(A)$.

(30) Let A and B be two Hermitian matrices in $\mathbb{C}^{n\times n}$. Prove that

$$\lambda_1(B) \leqslant \lambda_k(A+B) - \lambda_k(A) \leqslant \lambda_n(B)$$

for $1 \leqslant k \leqslant n$.

Solution: By Corollary 8.14 of the Rayleigh–Ritz theorem, we have

$$\lambda_1(B) \leqslant \boldsymbol{x}^{\mathsf{H}} B\boldsymbol{x} \leqslant \lambda_n(B),$$

for $\boldsymbol{x} \neq \mathbf{0}$ and $\|\boldsymbol{x}\|_2 = 1$. Then, by the Courant–Fisher theorem,

$$\lambda_k(A+B) = \min_W \max_{\boldsymbol{x}} \{\boldsymbol{x}^{\mathsf{H}}(A+B)\boldsymbol{x} \mid \|\boldsymbol{x}\|_2 = 1 \text{ and } \boldsymbol{x} \in \langle W\rangle^{\perp}\},$$

where the minimum is taken over by sets W that contain $k-1$ vectors. Since $\boldsymbol{x}^{\mathsf{H}}(A+B)\boldsymbol{x} = \boldsymbol{x}^{\mathsf{H}}A\boldsymbol{x} + \boldsymbol{x}^{\mathsf{H}}B\boldsymbol{x}$, it follows that

$$\lambda_k(A+B) \geqslant \min_W \max_{\boldsymbol{x}} \{\boldsymbol{x}^{\mathsf{H}}A\boldsymbol{x} + \lambda_n(B) \mid \|\boldsymbol{x}\|_2 = 1 \text{ and}$$
$$\times\, bfx \in \langle W\rangle^{\perp}\} = \lambda_k(A) + \lambda_n(B).$$

Similarly, we have

$$\lambda_k(A+B) \leqslant \min_W \max_{\boldsymbol{x}} \{\boldsymbol{x}^{\mathsf{H}}A\boldsymbol{x} + \lambda_1(B) \mid \|\boldsymbol{x}\|_2 = 1 \text{ and}$$
$$\times\, \boldsymbol{x} \in \langle W\rangle^{\perp}\} = \lambda_k(A) + \lambda_1(B).$$

(31) Let A and E be Hermitian matrices in $\mathbb{C}^{n\times n}$. Prove that $|\lambda_p(A+E) - \lambda_p(A)| \leqslant \rho(E) = \|E\|_2$ for $1 \leqslant p \leqslant n$.

Solution: By Weyl's inequalities (Supplement 8.13), we have

$$\lambda_1(E) \leqslant \lambda_p(A+E) - \lambda_p(A) \leqslant \lambda_n(E)$$

By Definition 7.5, this implies $|\lambda_p(A+E) - \lambda_p(A)| \leqslant \rho(A) = \|A\|_2$ because A is Hermitian.

(32) Let $A \in \mathbb{C}^{n \times n}$ be a Hermitian matrix and let $\boldsymbol{w} \in \mathbb{C}^n$. Then, $\lambda_k(A + \boldsymbol{w}\boldsymbol{w}^{\mathsf{H}}) \leqslant \lambda_{k+1}(A) \leqslant \lambda_{k+2}(A + \boldsymbol{w}\boldsymbol{w}^{\mathsf{H}})$ and $\lambda_k(A) \leqslant \lambda_{k+1}(A + \boldsymbol{w}\boldsymbol{w}^{\mathsf{H}}) \leqslant \lambda_{k+2}(A)$ for $1 \leqslant k \leqslant n-2$.

Solution: Let W be a range over the subsets of $\mathbb{C}^n$ that consist of $n-k-2$ vectors. By the Courant–Fisher theorem (Theorem 8.18),

$$\begin{aligned}
&\lambda_{k+2}(A + \boldsymbol{w}\boldsymbol{w}^{\mathsf{H}}) \\
&\quad = \min_W \max_{\boldsymbol{x}} \{\boldsymbol{x}^{\mathsf{H}}(A + \boldsymbol{w}\boldsymbol{w}^{\mathsf{H}})\boldsymbol{x} \mid \|\boldsymbol{x}\|_2 = 1 \text{ and } \boldsymbol{x} \in \langle W \rangle^{\perp}\} \\
&\quad \geqslant \min_W \max_{\boldsymbol{x}} \{\boldsymbol{x}^{\mathsf{H}}(A + \boldsymbol{w}\boldsymbol{w}^{\mathsf{H}})\boldsymbol{x} \mid \|\boldsymbol{x}\|_2 = 1, \boldsymbol{x} \in \langle W \rangle^{\perp} \text{ and} \\
&\qquad \times \boldsymbol{x} \perp \boldsymbol{w}\} \\
&\quad = \min_W \max_{\boldsymbol{x}} \{\boldsymbol{x}^{\mathsf{H}}A\boldsymbol{x} \mid \|\boldsymbol{x}\|_2 = 1, \boldsymbol{x} \in \langle W \rangle^{\perp} \text{ and } \boldsymbol{x} \perp \boldsymbol{w}\} \\
&\qquad \times (\text{because } \boldsymbol{x}^{\mathsf{H}}\boldsymbol{w} = \boldsymbol{w}^{\mathsf{H}}\boldsymbol{x} = 0) \\
&\quad \geqslant \min_{W_1} \max_{\boldsymbol{x}} \{\boldsymbol{x}^{\mathsf{H}}A\boldsymbol{x} \mid \|\boldsymbol{x}\|_2 = 1 \text{ and } \boldsymbol{x} \in \langle W_1 \rangle^{\perp}\} = \lambda_{k+1}(A),
\end{aligned}$$

where W_1 ranges over the sets that contain $n-k-1$ vectors. For $2 \leqslant k \leqslant n-2$, the same Courant–Fisher theorem yields

$$\begin{aligned}
\lambda_k(A + \boldsymbol{w}\boldsymbol{w}^{\mathsf{H}}) = \max_Z \min_{\boldsymbol{x}} \{&\boldsymbol{x}^{\mathsf{H}}(A + \boldsymbol{w}\boldsymbol{w}^{\mathsf{H}})\boldsymbol{x} \mid \|\boldsymbol{x}\|_2 = 1 \text{ and} \\
&\times \boldsymbol{x} \in \langle Z \rangle^{\perp}\},
\end{aligned}$$

where Z is a set that contains $k-1$ vectors. This implies

$$\begin{aligned}
&\lambda_k(A + \boldsymbol{w}\boldsymbol{w}^{\mathsf{H}}) \\
&\quad \leqslant \max_Z \min_{\boldsymbol{x}} \{\boldsymbol{x}^{\mathsf{H}}(A + \boldsymbol{w}\boldsymbol{w}^{\mathsf{H}})\boldsymbol{x} \mid \|\boldsymbol{x}\|_2 = 1 \text{ and } \boldsymbol{x} \in \langle Z \rangle^{\perp} \\
&\qquad \text{and } \boldsymbol{x} \perp \boldsymbol{w}\} \\
&\quad = \max_Z \min_{\boldsymbol{x}} \{\boldsymbol{x}^{\mathsf{H}}A\boldsymbol{x} \mid \|\boldsymbol{x}\|_2 = 1 \text{ and } \boldsymbol{x} \in \langle Z \rangle^{\perp} \text{ and } \boldsymbol{x} \perp \boldsymbol{w}\} \\
&\quad \leqslant \max_{Z_1} \min_{\boldsymbol{x}} \{\boldsymbol{x}^{\mathsf{H}}A\boldsymbol{x} \mid \|\boldsymbol{x}\|_2 = 1 \text{ and } \boldsymbol{x} \in \langle Z_1 \rangle^{\perp}\} \\
&\quad = \lambda(k+1, A),
\end{aligned}$$

where Z_1 ranges over sets that contain k vectors.

(33) Let A and B be two Hermitian matrices in $\mathbb{C}^{n\times n}$. If $\mathsf{rank}(B) \leqslant r$, prove that $\lambda_k(A+B) \leqslant \lambda_{k+r}(A) \leqslant \lambda_{k+2r}(A+B)$ for $1 \leqslant k \leqslant n-2r$ and $\lambda_k(A) \leqslant \lambda_{k+r}(A+B) \leqslant \lambda_{k+2r}(A)$.

Solution: If B is a Hermitian matrix of rank no larger than r, then by the Spectral Theorem for Hermitian matrices, we have $B = U\mathsf{diag}(\beta_1,\ldots,\beta_r,0,\ldots,0)U^{\mathsf{H}}$, where $U = (\boldsymbol{u}_1 \ \cdots \ \boldsymbol{u}_n)$ is a unitary matrix. This amounts to $B = \beta_1\boldsymbol{u}_1\boldsymbol{u}_1' + \cdots + \beta_r\boldsymbol{u}_r\boldsymbol{u}_r'$. Conversely, every Hermitian matrix of rank no larger than r can be written in this form.
Let W range over the subsets of $\mathbb{R}^n$ that consist of $n-k-2r$ vectors. We have

$$\begin{aligned}
&\lambda_{k+2r}(A+B)\\
&= \min_W \max_{\boldsymbol{x}}\{\boldsymbol{x}^{\mathsf{H}}(A+B)\boldsymbol{x} \mid \|\boldsymbol{x}\|_2 = 1 \text{ and } \boldsymbol{x} \in \langle W\rangle^{\perp}\}\\
&\geqslant \min_W \max_{\boldsymbol{x}}\{\boldsymbol{x}^{\mathsf{H}}(A+B)\boldsymbol{x} \mid \|\boldsymbol{x}\|_2 = 1,\\
&\quad \times \boldsymbol{x} \in \langle W \cup \{\boldsymbol{u}_1,\ldots,\boldsymbol{u}_r\}\rangle^{\perp}\}\\
&= \min_W \max_{\boldsymbol{x}}\{\boldsymbol{x}^{\mathsf{H}}A\boldsymbol{x} \mid \|\boldsymbol{x}\|_2 = 1, \boldsymbol{x} \in \langle W \cup \{\boldsymbol{u}_1,\ldots,\boldsymbol{u}_r\}\rangle^{\perp}\}\\
&\quad (\text{because } \boldsymbol{x}^{\mathsf{H}}\boldsymbol{u}_i = \boldsymbol{u}_i^{\mathsf{H}}\boldsymbol{x} = 0)\\
&\geqslant \min_{W_1} \max_{\boldsymbol{x}}\{\boldsymbol{x}^{\mathsf{H}}A\boldsymbol{x} \mid \|\boldsymbol{x}\|_2 = 1 \text{ and } \boldsymbol{x} \in \langle W_1\rangle^{\perp}\} = \lambda_{k+r}(A),
\end{aligned}$$

where W_1 ranges over the subsets of $\mathbb{R}^n$ that contain $n-k-r$ vectors.
The proof of the remaining inequalities follows the same pattern as above and generalizes the results and proofs of Supplement 8.13.

(34) Let A be a Hermitian matrix in $\mathbb{C}^{n\times n}$ such that $A = UDU^{\mathsf{H}}$, where $U = (\boldsymbol{u}_1 \ \cdots \ \boldsymbol{u}_n)$ is a unitary matrix and $D = (\lambda_1(A),\ldots,\lambda_n(A))$. If $A_j = \sum_{i=j+1}^{n}\lambda_i(A)\boldsymbol{u}_i\boldsymbol{u}_i'$ for $0 \leqslant j \leqslant n-1$, prove that the largest eigenvalue of the matrix $A - A_{n-k}$ is $\lambda_{n-k}(A)$.

Solution: Since $A - A_{n-k} = \sum_{i=1}^{n-k}\lambda_i(A)\boldsymbol{u}_i\boldsymbol{u}_i'$, the statement follows immediately.

(35) Let A and B be two Hermitian matrices in $\mathbb{C}^{n\times n}$. Using the same notations as in Supplement 8.13, prove that for any i, j

such that $1 \leqslant i, j \leqslant n$ and $i + j \geqslant n + 1$, we have

$$\lambda_{j+k-n}(A + B) \leqslant \lambda_j(A) + \lambda_k(B).$$

Also, if $i + j \leqslant n + 1$, then

$$\lambda_j(A) + \lambda_k(B) \leqslant \lambda_{j+k-1}(A + B).$$

(36) Let $A, B \in \mathbb{R}^{n\times n}$ be two symmetric matrices. Suppose that $\mathsf{spec}(A) = \{\alpha_1, \dots, \alpha_n\}$ and $\mathsf{spec}(B) = \{\beta_1, \dots, \beta_n\}$.
(a) Prove that if $\alpha_1 \geqslant \cdots \geqslant \alpha_n$ and $\beta_1 \geqslant \cdots \geqslant \beta_n$, then

$$\mathsf{trace}(AB) \leqslant \sum_{i=1}^{n} \alpha_i \beta_i.$$

(b) Prove that if $\alpha_1 \geqslant \cdots \geqslant \alpha_n$ and $\beta_1 \leqslant \cdots \leqslant \beta_n$, then

$$\mathsf{trace}(AB) \geqslant \sum_{i=1}^{n} \alpha_i \beta_i.$$

Solution: Observe that $\|A - B\|_F^2 = \mathsf{trace}((A - B)'(A - B)) = \mathsf{trace}((A - B)^2)$ due to the symmetry of the matrices A and B. This implies

$$\|A - B\|_F^2 = \mathsf{trace}(A^2 - 2AB + B^2) = \|A\|_F^2 + \|A\|_F^2 - 2\,\mathsf{trace}(AB),$$

so $2\,\mathsf{trace}(AB) = \|A\|_F^2 + \|B\|_F^2 - \|A - B\|_F^2$.
By Corollary 8.17 of the Hoffman–Wieland Theorem, we have $\|A - B\|_F^2 \geqslant \sum_{i=1}^{n} (\alpha_i - \beta_i)^2$, so

$$\begin{aligned}
2\,\mathsf{trace}(AB) &\leqslant \|A\|_F^2 + \|B\|_F^2 - \sum_{i=1}^{n} (\alpha_i - \beta_i)^2 \\
&= \sum_{i=1}^{n} \alpha_i^2 + \sum_{i=1}^{n} \beta_i^2 - \sum_{i=1}^{n} (\alpha_i - \beta_i)^2 \\
&= 2 \sum_{i=1}^{n} \alpha_i \beta_i,
\end{aligned}$$

which yields the first inequality.

If $\alpha_1 \geqslant \cdots \geqslant \alpha_n$ and $\beta_1 \leqslant \cdots \leqslant \beta_n$, then by the second part of Corollary 8.17, we have $\|A - B\|_F^2 \leqslant \sum_{i=1}^n (\alpha_i - \beta_i)^2$, and this implies $\mathsf{trace}(AB) \leqslant \sum_{i=1}^n (\alpha_i - \beta_i)^2$.

(37) Consider the sets of matrices

$$\mathcal{M}_1 = \{YY' \mid Y \in \mathbb{R}^{n\times k} \text{ and } Y'Y = I_k\},$$
$$\mathcal{M}_2 = \{W \in \mathbb{R}^{n\times n} \mid W = W', \mathsf{trace}(W) = k \text{ and } W \text{ and } I_n - W \text{ are positive semidefinite}\}.$$

Prove that
(a) we have $\mathcal{M}_2 = \boldsymbol{K}_{\mathrm{conv}}(\mathcal{M}_1)$;
(b) $\mathcal{M}_1$ is the set of extreme points of the polytope $\mathcal{M}_2$.

Solution: Every convex combination of elements of $\mathcal{M}_1$ lies in $\mathcal{M}_2$. Indeed, let $Z = a_1 Y_1 Y_1' + \cdots + a_p Y_p Y_p'$ be a convex combination of $\mathcal{M}_1$. It is immediate that Z is a symmetric matrix. Furthermore, by Theorem 3.14, we have

$$\mathsf{trace}(Z) = \sum_{h=1}^p a_h \mathsf{trace}(Y_h Y_h') = \sum_{h=1}^p a_h \mathsf{trace}(Y_h' Y_h)$$
$$= \sum_{h=1}^p a_h \mathsf{trace}(I_k) = k.$$

because $\sum_{h=1}^p a_h = 1$. The positive semidefiniteness of YY' follows from Example 6.27, while the positive semidefiniteness of $I_n - YY'$ follows from Supplement 92. Thus, $\boldsymbol{K}_{\mathrm{conv}}(\mathcal{M}_1) \subseteq \mathcal{M}_2$.
Conversely, let $W \in \mathcal{M}_2$. By the spectral theorem for Hermitian matrices (Theorem 8.14) applied to real symmetric matrices, W can be written as $W = U'DU$, where U is a unitary matrix. Clearly, all eigenvalues of W belong to the interval $[0, 1]$.
If the columns of the matrix U' are $\boldsymbol{u}_1, \ldots, \boldsymbol{u}_n$ and the eigenvalues of W are $\lambda_1, \ldots, \lambda_n$, then

$$W = (\boldsymbol{u}_1 \ \cdots \ \boldsymbol{u}_n) \begin{pmatrix} \lambda_1 & 0 & \cdots & 0 \\ 0 & \lambda_2 & \cdots & 0 \\ \vdots & \vdots & \cdots & \vdots \end{pmatrix} \begin{pmatrix} \boldsymbol{u}_1' \\ \vdots \\ \boldsymbol{u}_n' \end{pmatrix},$$

which allows us to write $W = \lambda_1 \boldsymbol{u}_1 \boldsymbol{u}_1' + \cdots + \lambda_r \boldsymbol{u}_r \boldsymbol{u}_r'$, where W has rank r, $\lambda_1, \ldots, \lambda_r$ are the non zero eigenvalues of W, and $\sum_{i=1}^r \lambda_i = \mathsf{trace}(W) = k$. Note that the rank of W is at least k because its eigenvalues reside in the interval $[0, 1]$ and their sum is k.

If the rank of W equals k, then $W = \boldsymbol{u}_1 \boldsymbol{u}_1' + \cdots + \boldsymbol{u}_k \boldsymbol{u}_k'$ because all eigenvalues equal 1. This allows us to write $W = ZZ'$, where $Z \in \mathbb{R}^{n \times k}$ is the matrix $Z = (\boldsymbol{u}_1 \ \cdots \ \boldsymbol{u}_k)$. Since $Z'Z = I_k$, it follows that in this case $W \in \mathcal{M}_1$. In other words, $\mathcal{M}_1$ is exactly the subset of $\mathcal{M}_2$ that consists of rank k matrices.

If $\mathsf{rank}(W) = r > k$, we have $W = \lambda_1 \boldsymbol{u}_1 \boldsymbol{u}_1' + \cdots + \lambda_r \boldsymbol{u}_r \boldsymbol{u}_r'$. Starting from the r matrices $\boldsymbol{u}_i \boldsymbol{u}_i'$, we can form $\binom{r}{k}$ matrices of rank k of the form $\sum_{i \in I} \boldsymbol{u}_i \boldsymbol{u}_i'$ by considering all subsets I of $\{1, \ldots, r\}$ that contain k elements. We have

$$\begin{aligned} W &= \sum_{j=1}^r \lambda_j \boldsymbol{u}_j \boldsymbol{u}_j' \\ &= \sum_{I, |I|=k} \alpha_I \sum_{i \in I} \boldsymbol{u}_i \boldsymbol{u}_i'. \end{aligned}$$

If we match the coefficients of $\boldsymbol{u}_i \boldsymbol{u}_i'$, we have $\lambda_i = \sum_{I, i \in I, |I|=k} \alpha_I$. If we add these equalities, we obtain

$$k = \sum_{i=1}^r \sum_{I, i \in I, |I|=k} \alpha_I.$$

We choose α_I to depend on the cardinality of I and take into account that each α_I occurs k times in the previous sum. This implies $\sum_{I, i \in I, |I|=k} \alpha_I = 1$, so each W is a convex combination of matrices of rank k, so $\boldsymbol{K}_{\text{conv}}(\mathcal{M}_1) = \mathcal{M}_2$. No matrix of rank greater than k can be an extreme point. Since every convex and compact set has extreme elements, only matrices of rank k can play this role. Since the definition of $\mathcal{M}_2$ makes no distinction between the k-rank matrices, it follows that the set of extreme elements coincides with $\mathcal{M}_1$.

(38) Prove that Ky Fan's Theorem can be derived from Supplement 8.13.

(39) Prove that the Jordan block $B_r(a)$ can be written as

$$B_r(a) = a \begin{pmatrix} e_1' \\ e_2' \\ \vdots \\ e_{r-1}' \\ e_r' \end{pmatrix} .+ \begin{pmatrix} e_2' \\ e_3' \\ \vdots \\ e_r' \\ \mathbf{0} \end{pmatrix}.$$

(40) Let $A \in \mathbb{R}^{m \times n}$ be a matrix such that $\mathsf{rank}(A) = r$. Prove that A can be factored as $A = UDV'$, where $U \in \mathbb{R}^{m \times r}$ and $V \in \mathbb{R}^{n \times r}$ are matrices with orthonormal columns and $D \in \mathbb{R}^{r \times r}$ is a diagonal matrix. This factorization is known as the *Lanczos decomposition* of A.

Solution: By Theorem 3.33, the symmetric and square matrix $A'A \in \mathbb{R}^{n \times n}$ has the same rank r as the matrix A. Therefore, by the Spectral Theorem for Hermitian Matrices (Theorem 8.14), there exists an orthogonal matrix $V \in \mathbb{R}^{n \times n}$ such that $A'A = VSV'$, where S is a diagonal matrix $D = (\sigma_1, \ldots, \sigma_r, 0, \ldots, 0)$ and $\sigma_1 \geqslant \sigma_2 \geqslant \cdots \geqslant \sigma_r > 0$. This is equivalent to $V'A'AV = (AV)'AV = S$.
The matrix V can be written as $V = (V_1 \;\; V_0)$, where $V_1 = (\boldsymbol{v}_1 \;\; \cdots \;\; \boldsymbol{v}_r) \in \mathbb{R}^{n \times r}$ consists of the first r columns of the matrix V, that is, from eigenvectors of $A'A$ that correspond to $\sigma_1, \ldots, \sigma_r$. The last $n - r$ columns of V (the columns of V_0) are eigenvectors that correspond to the eigenvalue 0. Since $(AV)'(AV) = S$, it follows that

$$(A\boldsymbol{v}_1 \;\cdots\; A\boldsymbol{v}_r \; AV_0)'(A\boldsymbol{v}_1 \cdots A\boldsymbol{v}_r AV_0) = S,$$

which means that for the vectors $\boldsymbol{z}_i = A\boldsymbol{v}_i$ $(1 \leqslant i \leqslant r)$, we have

$$\boldsymbol{z}_i'\boldsymbol{z}_j = \begin{cases} \sigma_i & \text{if } i = j, \\ 0 & \text{otherwise.} \end{cases}$$

Let $\lambda_i = \sqrt{\sigma_i}$ and let $\boldsymbol{u}_i = \frac{1}{\lambda_i}\boldsymbol{z}_i$ for $1 \leqslant i \leqslant r$. The set of vectors $\{\boldsymbol{u}_1, \ldots, \boldsymbol{u}_r\}$ is orthonormal and $AV = (\lambda_1 \boldsymbol{u}_1, \ldots, \lambda_r \boldsymbol{u}_r) = UD$, where $U = (\boldsymbol{u}_1 \;\; \cdots \;\; \boldsymbol{u}_r)$ and $D = \mathsf{diag}(\lambda_1, \ldots, \lambda_r)$.

Note that $V \in \mathbb{R}^{n\times r}$ and that $A'A\boldsymbol{v} = \mathbf{0}$ for any column $\boldsymbol{v}$ of $\boldsymbol{V}_0$. This implies $AV_0 = O$, so $AV = A(V_1\ V_0) = (AV_1\ O) = (UD\ O)$ and

$$A = (UA\ O)V' = (UA\ O)\begin{pmatrix} V_0' \\ V_1' \end{pmatrix} = UDV'.$$

The *field of values of a matrix* $A \in \mathbb{C}^{n\times n}$ is the set of numbers

$$F(A) = \{\boldsymbol{x}A\boldsymbol{x}^{\mathsf{H}} \mid \boldsymbol{x} \in \mathbb{C}^n \text{ and } \|\boldsymbol{x}\|_2 = 1\}.$$

(41) Prove that $F(A)$ is a convex set.

(42) Prove that $\mathsf{spec}(A) \subseteq F(A)$ for any $A \in \mathbb{C}^{n\times n}$.

(43) If $U \in \mathbb{C}^{n\times n}$ is a unitary matrix and $A \in \mathbb{C}^{n\times n}$, prove that $F(UAU^{\mathsf{H}}) = F(A)$.

(44) Prove that for a normal matrix $A \in \mathbb{C}^{n\times n}$, $F(A)$ equals the convex hull of $\mathsf{spec}(A)$. Infer that A is Hermitian if and only if $F(A)$ is an interval of $\mathbb{R}$.

Solution: Since A is normal, by the Spectral Theorem for Normal Matrices (Theorem 8.13), there exists a unitary matrix U and a diagonal matrix D such that $A = U^{\mathsf{H}}DU$ and the diagonal elements of D are the eigenvalues of A. Then, by Exercise 8.13, $F(A) = F(D)$. Therefore, $z \in F(A)$ if $z = \boldsymbol{x}D\boldsymbol{x}^{\mathsf{H}}$ for some $\boldsymbol{x} \in \mathbb{C}^n$ such that $\|\boldsymbol{x}\|_2 = 1$, so $z = \sum_{k=1}^n |x_k|^2\lambda_k$, where $\mathsf{spec}(A) = \mathsf{spec}(D) = \{\lambda_1, \ldots, \lambda_n\}$, which proves that $F(A)$ is included in the convex closure of $\mathsf{spec}(A)$. The reverse inclusion is immediate.

(45) Let

$$M = \begin{pmatrix} A & B \\ O & C \end{pmatrix}$$

be a block matrix, where A and C are square matrices. Prove that if $\mathsf{spec}(A) \cap \mathsf{spec}(C) = \emptyset$, then there exists a matrix X such that

$$\begin{pmatrix} I & X \\ O & I \end{pmatrix} M \begin{pmatrix} I & X \\ O & I \end{pmatrix}^{-1}$$

is a block diagonal matrix.

Solution: The matrix

$$\begin{pmatrix} I & X \\ O & I \end{pmatrix}$$

is invertible and we have

$$\begin{pmatrix} I & X \\ O & I \end{pmatrix}^{-1} = \begin{pmatrix} I & -X \\ O & I \end{pmatrix}.$$

Thus, we need to find a matrix X such that

$$\begin{pmatrix} I & X \\ O & I \end{pmatrix}\begin{pmatrix} A & B \\ O & C \end{pmatrix}\begin{pmatrix} I & -X \\ O & I \end{pmatrix} = \begin{pmatrix} A & -AX + B + XC \\ O & C \end{pmatrix}.$$

By Theorem 8.25, if $\mathsf{spec}(A) \cap \mathsf{spec}(C) = \emptyset$, there exists X such that $AX - XC = B$.

(46) Prove that $A \sim B$ implies $f(A) \sim f(B)$ for every $A, B \in \mathbb{C}^{n\times n}$ and every polynomial f.

(47) Let $A \in \mathbb{C}^{n\times n}$ be a matrix such that $\mathsf{spec}(A) = \{\lambda_1, \ldots, \lambda_n\}$. Prove that

(a)

$$\sum_{i=1}^{n} |\lambda_i|^2 \leqslant \sum_{i=1}^{n}\sum_{j=1}^{n} |a_{ij}|^2;$$

(b) the matrix A is normal if and only if

$$\sum_{i=1}^{n} |\lambda_i|^2 = \sum_{i=1}^{n}\sum_{j=1}^{n} |a_{ij}|^2.$$

(48) Let $A \in \mathbb{C}^{n\times n}$ such that $A \geqslant O_{n,n}$. Prove that if $\mathbf{1}_n$ is an eigenvector of A, then $\rho(A) = \|A\|_\infty$ and if $\mathbf{1}_n$ is an eigenvector of A', then $\rho(A) = \|A\|_1$.

(49) Let A be a non-negative matrix in $\mathbb{C}^{n\times n}$ and let $\boldsymbol{u} = A\mathbf{1}_n$ and $\boldsymbol{v} = A'\mathbf{1}_n$. Prove that

$$\max\{\min u_i, \min v_j\} \leqslant \rho(A) \leqslant \min\{\max u_i, \max v_j\}.$$

(50) Let $A \in \mathbb{C}^{n\times n}$ be a matrix such that $A \geqslant O_{n,n}$. Prove that if there exists $k \in \mathbb{N}$ such that $A^k > O_{n,n}$, then $\rho(A) > 0$.

(51) Let $A \in \mathbb{C}^{n\times n}$ be a matrix such that $A \geqslant O_{n,n}$. If $A \neq O_{n,n}$ and there exists an eigenvector $\boldsymbol{x}$ of A such that $\boldsymbol{x} > \mathbf{0}_n$, prove that $\rho(A) > 0$.

(52) Prove that $A \in \mathbb{C}^{n\times n}$ is positive semidefinite if and only if there is a set $U = \{\boldsymbol{v}_1, \ldots, \boldsymbol{v}_n\} \subseteq \mathbb{C}^n$ such that $A = \sum_{i=1}^{n} \boldsymbol{v}_i \boldsymbol{v}_i^{\mathsf{H}}$. Furthermore, prove that A is positive definite if and only if there exists a linearly independent set U as above.

(53) Prove that A is positive definite if and only if A^{-1} is positive definite.

(54) Prove that if $A \in \mathbb{C}^{n\times n}$ is a positive semidefinite matrix, then A^k is positive semidefinite for every $k \geqslant 1$.

(55) Let $A \in \mathbb{C}^{n\times n}$ be a Hermitian matrix and let $p_A(\lambda) = \lambda^n + c_1\lambda^{n-1} + \cdots + c_m\lambda^{n-m}$ be its characteristic polynomial, where $c_m \neq 0$. Then, A is positive semidefinite if and only if $c_i \neq 0$ for $0 \leqslant k \leqslant m$ (where $c_0 = 1$) and $c_j c_{j+1} < 0$ for $0 \leqslant j \leqslant m-1$.

(56) Corollary 8.26 can be extended as follows. Let $A \in \mathbb{C}^{n\times n}$ be a positive semidefinite matrix. Prove that for every $k \geqslant 1$ there exists a positive semidefinite matrix B having the same rank as A such that
   (a) $B^k = A$;
   (b) $AB = BA$;
   (c) B can be expressed as a polynomial in A.

Solution:
Since A is Hermitian, its eigenvalues are real nonnegative numbers and, by the Spectral Theorem for Hermitian matrices, there exists a unitary matrix $U \in \mathbb{C}^{n\times n}$ such that $A = U^{\mathsf{H}}\mathsf{diag}(\lambda_1, \ldots, \lambda_n)U$. Let $B = U^{\mathsf{H}}\mathsf{diag}(\lambda_1^{\frac{1}{k}}, \ldots, \lambda_n^{\frac{1}{k}})U$, where $\lambda_i^{\frac{1}{k}}$ is a non-negative root of order k of λ_i. Thus, $B^k = A$, B is clearly positive semidefinite, $\mathsf{rank}(B) = \mathsf{rank}(A)$, and $AB = BA$.
Let

$$p(x) = \sum_{j=1}^{n} \lambda_j^{\frac{1}{k}} \prod_{k=1,k\neq j}^{n} \frac{x - \lambda_k}{\lambda_j - \lambda_k}$$

be a Lagrange interpolation polynomial such that $p(\lambda_j) = \lambda_j^{\frac{1}{k}}$ (see Exercise 31). Then,

$$p(\mathsf{diag}(\lambda_1, \ldots, \lambda_n)) = \mathsf{diag}(\lambda_1^{\frac{1}{k}}, \ldots, \lambda_n^{\frac{1}{k}}),$$

so

$$\begin{aligned}p(A) &= p(U^{\mathsf{H}}\mathsf{diag}(\lambda_1, \ldots, \lambda_n)U) = U^{\mathsf{H}}p(\mathsf{diag}(\lambda_1, \ldots, \lambda_n))U \\ &= U^{\mathsf{H}}\mathsf{diag}(\lambda_1^{\frac{1}{k}}, \ldots, \lambda_n^{\frac{1}{k}})U = B.\end{aligned}$$

(57) Let $A \in \mathbb{R}^{n\times n}$ be a symmetric matrix. Prove that there exists $b \in \mathbb{R}$ such that $A + b(\mathbf{1}\mathbf{1}' - I_n)$ is positive semidefinite, where $\mathbf{1} \in \mathbb{R}^n$.

Solution: We need to find b such that for every $\boldsymbol{x} \in \mathbb{R}^n$ we will have

$$\boldsymbol{x}'(A + b(\mathbf{1}\mathbf{1}' - I_n))\boldsymbol{x} \geqslant 0.$$

We have

$$\boldsymbol{x}'(A + b(\mathbf{1}\mathbf{1}' - I_n))\boldsymbol{x} = \boldsymbol{x}'A\boldsymbol{x} + b\boldsymbol{x}'\mathbf{1}\mathbf{1}'\boldsymbol{x} - b\boldsymbol{x}'\boldsymbol{x} \geqslant 0,$$

which amounts to

$$\boldsymbol{x}'A\boldsymbol{x} + b\left(\left(\sum_{i=1}^n x_i\right)^2 - \|\boldsymbol{x}\|_2^2\right) \geqslant 0.$$

Since A is symmetric, by the Rayleigh–Ritz theorem, we have $\boldsymbol{x}'A\boldsymbol{x} \geqslant \lambda_1\|\boldsymbol{x}\|_2^2$, where λ_1 is the least eigenvalue of A. Therefore, it suffices to take $b \leqslant \lambda_1$ to satisfy the equality for every $\boldsymbol{x}$.

(58) If $A \in \mathbb{R}^{n\times n}$ is a positive definite matrix, prove that there exist $c, d > 0$ such that

$$c\|\boldsymbol{x}\|_2^2 \leqslant \boldsymbol{x}'A\boldsymbol{x} \leqslant d\|\boldsymbol{x}\|_2^2,$$

for every $\boldsymbol{x} \in \mathbb{R}^n$.

(59) Let $A = \mathsf{diag}(A_1, \ldots, A_p)$ and $B = (B_1, \ldots, B_q)$ be two block diagonal matrices. Prove that

$$\mathsf{sep}_F(A, B) = \min\{\mathsf{sep}_F(A_i, B_j) \mid 1 \leqslant i \leqslant p \text{ and } 1 \leqslant j \leqslant q\}.$$

(60) Let $A \in \mathbb{C}^{n\times n}$ be a Hermitian matrix. Prove that if for any $\lambda \in \mathsf{spec}(A)$ we have $\lambda > -a$, then the matrix $A + aI$ is positive-semidefinite.

(61) Let $A \in \mathbb{C}^{m\times m}$ and $B \in \mathbb{C}^{n\times n}$ be two matrices that have the eigenvalues $\lambda_1, \ldots, \lambda_m$ and $\mu_1, \ldots, \mu_n$, respectively. Prove that:
   (a) if A and B are positive definite, then so is $A \otimes B$;
   (b) if $m = n$ and A, B are symmetric positive definite, the Hadamard product $A \odot B$ is positive definite.

Solution: The first part follows immediately from Theorem 7.24.
For the second part, recall that the Hadamard product $A \odot B$ of two square matrices of the same format is a principal submatrix of $A \otimes B$. Then, apply Theorem 8.20.

(62) Let $A \in \mathbb{R}^{n\times n}$ be a real matrix that is symmetric and positive semidefinite) such that $A\mathbf{1}_n = \mathbf{0}_n$. Prove that

$$2 \max_{1\leqslant i\leqslant n} \sqrt{a_{ii}} \leqslant \sum_{j=1}^{n} \sqrt{a_{jj}}.$$

Solution: By Corollary 8.26, A is the Gram matrix of a sequence of vectors $B = (\boldsymbol{b}_1, \ldots, \boldsymbol{b}_n)$, so $A = B'B$. Since $A\mathbf{1}_n = \mathbf{0}_n$, it follows that $(B\mathbf{1}_n)'(B\mathbf{1}_n) = 0$, so $B\mathbf{1}_n = \mathbf{0}_n$. Thus, $\sum_{i=1}^{n} \boldsymbol{b}_i = \mathbf{0}_n$. Then we have

$$\|\boldsymbol{u}_i\|_2 = \| - \sum_{j\neq i} \boldsymbol{u}_j\|_2 \leqslant \sum_{j\neq i} \|\boldsymbol{u}_j\|_2,$$

which implies

$$2 \max_{1\leqslant i\leqslant n} \|\boldsymbol{u}_i\|_2 \leqslant \sum_{j=1}^{n} \|\boldsymbol{u}_j\|_2.$$

This is immediately equivalent to the inequality to be shown.

(63) Let $A, B \in \mathbb{R}^n$. Prove that the function $\psi : \mathbb{R}^n \longrightarrow \mathbb{R}$ defined by $\psi(\boldsymbol{x}) = \|A\boldsymbol{x}\|_2^2 - \|B\boldsymbol{x}\|_2^2$ is a quadratic form and find the symmetric matrix that represents ψ.

Bibliographical Comments

The reader should consult [162]. The solution of Supplement 11 is given in [78]. The proof of the Perron–Frobenius Theorem that we present was obtained in [119]. The proof of Theorem 8.22 is given in [171]. The Hoffman–Wielandt theorem was shown in [77].

The proof of Theorem 8.48 is given in [104]. Ky Fan's Theorem appeared in [48]. The proof of Supplement 8.13 was obtained from Overton and Womersley [124], where the reader can find the solution of Exercise 8.13.

Theorem 8.45 was obtained in [120]; the result of Theorem 8.50 appeared in [170]. The treatment of spectral resolution of matrices follows [160].

Supplement 8.13 is a result of Juhász, which appears in [86]; Supplement 8.13 originated in [90]. Lemma 9.4 is a specialization of a result proved in [160] for Banach spaces.

Chapter 9

Singular Values

9.1 Introduction

The singular value decomposition has been described as the "Swiss Army knife of matrix decompositions" [121] due to its many applications in the study of matrices; from our point of view, singular value decomposition is relevant for dimensionality reduction techniques in data mining.

9.2 Singular Values and Singular Vectors

The notion of singular value introduced in this section allows us to formulate the singular value decomposition (SVD) theorem, which extends a certain property of unitarily diagonalizable matrices.

Let $A \in \mathbb{C}^{n\times n}$ be a square matrix which is unitarily diagonalizable. There exists a diagonal matrix $D = \mathsf{diag}(d_1, \ldots, d_n) \in \mathbb{C}^{n\times n}$ and a unitary matrix $X \in \mathbb{C}^{n\times n}$ such that $A = XDX^{\mathsf{H}}$; equivalently, we have $AX = XD$. If we denote the columns of X by $\boldsymbol{x}_1, \ldots, \boldsymbol{x}_n$, then $A\boldsymbol{x}_i = d_i\boldsymbol{x}_i$, which shows that $\boldsymbol{x}_i$ is a unit eigenvector that corresponds to the eigenvalue d_i for $1 \leqslant i \leqslant n$. Also, we have

$$A = (\boldsymbol{x}_1 \ \cdots \ \boldsymbol{x}_n)\mathsf{diag}(d_1, \ldots, d_n)\begin{pmatrix}\boldsymbol{x}_1^{\mathsf{H}}\\ \vdots \\ \boldsymbol{x}_n^{\mathsf{H}}\end{pmatrix}$$

$$= d_1\boldsymbol{x}_1\boldsymbol{x}_1^{\mathsf{H}} + \cdots + d_n\boldsymbol{x}_n\boldsymbol{x}_n^{\mathsf{H}}.$$

This is the *spectral decomposition of* A, which we already discussed in Chapter 7. Note that $\mathsf{rank}(\boldsymbol{x}_i\boldsymbol{x}_i^{\mathsf{H}}) = 1$ for $1 \leqslant i \leqslant n$.

The SVD theorem extends this decomposition to rectangular matrices.

Theorem 9.1 (SVD Theorem). If $A \in \mathbb{C}^{m\times n}$ is a matrix and $\mathsf{rank}(A) = r$, then A can be factored as $A = UDV^{\mathsf{H}}$, where $U \in \mathbb{C}^{m\times m}$ and $V \in \mathbb{C}^{n\times n}$ are unitary matrices,

$$D = \begin{pmatrix} \sigma_1 & 0 & 0 & \cdots & 0 & 0 \\ 0 & \sigma_2 & 0 & & 0 & 0 \\ \vdots & \vdots & \vdots & \cdots & \vdots & \vdots \\ 0 & 0 & \cdots & \sigma_r & \cdots & 0 \\ 0 & 0 & \cdots & 0 & \cdots & 0 \\ \vdots & \vdots & \vdots & \cdots & \vdots & \vdots \\ 0 & 0 & \cdots & 0 & \cdots & 0 \end{pmatrix} \in \mathbb{C}^{m\times n},$$

and $\sigma_1 \geqslant \ldots \geqslant \sigma_r$ are real positive numbers.

Proof. By Theorem 3.33 and Example 6.27, the square matrix $A^{\mathsf{H}}A \in \mathbb{C}^{n\times n}$ has the same rank as the matrix A and is positive semidefinite. Therefore, there are r positive eigenvalues of this matrix, denoted by $\sigma_1^2, \ldots, \sigma_r^2$, where $\sigma_1 \geqslant \sigma_2 \geqslant \cdots \geqslant \sigma_r > 0$.

Let $\boldsymbol{v}_1, \ldots, \boldsymbol{v}_r$ be the corresponding pairwise orthogonal unit eigenvectors in $\mathbb{C}^n$. We have $A^{\mathsf{H}}A\boldsymbol{v}_i = \sigma_i^2\boldsymbol{v}_i$ for $1 \leqslant i \leqslant r$.

Let V be the matrix $V = (\boldsymbol{v}_1 \;\cdots\; \boldsymbol{v}_r \;\boldsymbol{v}_{r+1} \;\cdots\; \boldsymbol{v}_n)$ obtained by completing the set $\{\boldsymbol{v}_1, \ldots, \boldsymbol{v}_r\}$ to an orthogonal basis for $\mathbb{C}^n$. If $V_1 = (\boldsymbol{v}_1 \;\cdots\; \boldsymbol{v}_r)$ and $V_2 = (\boldsymbol{v}_{r+1} \;\cdots\; \boldsymbol{v}_n)$, we can write $V = (V_1\; V_2)$.

The equalities involving the eigenvectors can now be written as $A^{\mathsf{H}}AV_1 = V_1E^2$, where $E = \mathsf{diag}(\sigma_1, \ldots, \sigma_r)$.

Define $U_1 = AV_1E^{-1} \in \mathbb{C}^{m\times r}$. We have $U_1^{\mathsf{H}} = S^{-1}V_1^{\mathsf{H}}A^{\mathsf{H}}$, so

$$U_1^{\mathsf{H}}U_1 = S^{-1}V_1^{\mathsf{H}}A^{\mathsf{H}}AV_1E^{-1} = E^{-1}V_1^{\mathsf{H}}V_1E^2E^{-1} = I_r,$$

which shows that the columns of U_1 are pairwise orthogonal unit vectors. Consequently, $U_1^{\mathsf{H}}AV_1E^{-1} = I_r$, so $U_1^{\mathsf{H}}AV_1 = E$.

If $U_1 = (\boldsymbol{u}_1, \ldots, \boldsymbol{u}_r)$, let $U_2 = (\boldsymbol{u}_{r+1}, \ldots, \boldsymbol{u}_m)$ be the matrix whose columns constitute the extension of the set $\{\boldsymbol{u}_1, \ldots, \boldsymbol{u}_r\}$ to an orthogonal basis of $\mathbb{C}^m$. Define $U \in \mathbb{C}^{m\times m}$ as $U = (U_1\ U_2)$. Note that

$$\begin{aligned} U^{\mathsf{H}}AV &= \begin{pmatrix} U_1^{\mathsf{H}} \\ U_2^{\mathsf{H}} \end{pmatrix} A(V_1\ V_2) \\ &= \begin{pmatrix} U_1^{\mathsf{H}}AV_1 & U_1^{\mathsf{H}}AV_2 \\ U_2^{\mathsf{H}}AV_1 & U_2^{\mathsf{H}}AV_2 \end{pmatrix} = \begin{pmatrix} U_1^{\mathsf{H}}AV_1 & U_1^{\mathsf{H}}AV_2 \\ U_2^{\mathsf{H}}AV_1 & U_2^{\mathsf{H}}AV_2 \end{pmatrix} \\ &= \begin{pmatrix} U_1^{\mathsf{H}}AV_1 & O \\ O & O \end{pmatrix} = \begin{pmatrix} E & O \\ O & O \end{pmatrix}, \end{aligned}$$

which is the desired decomposition. □

Observe that in the SVD described by Theorem 9.1 known as the *full SVD* of A, the diagonal matrix D has the same format as A, while both U and V are square unitary matrices.

Definition 9.1. Let $A \in \mathbb{C}^{m\times n}$ be a matrix. A number $\sigma \in \mathbb{R}_{>0}$ is a *singular value* of A if there exists a pair of vectors $(\boldsymbol{u}, \boldsymbol{v}) \in \mathbb{C}^n \times \mathbb{C}^m$ such that

$$A\boldsymbol{v} = \sigma\boldsymbol{u} \text{ and } A^{\mathsf{H}}\boldsymbol{u} = \sigma\boldsymbol{v}. \tag{9.1}$$

The vector $\boldsymbol{u}$ is the *left singular vector* and $\boldsymbol{v}$ is the *right singular vector* associated to the singular value σ.

Note that if $(\boldsymbol{u}, \boldsymbol{v})$ is a pair of vectors associated to σ, then $(a\boldsymbol{u}, a\boldsymbol{v})$ is also a pair of vectors associated with σ for every $a \in \mathbb{C}$.

Let $A \in \mathbb{C}^{m\times n}$ and let $A = UDV^{\mathsf{H}}$, where $U \in \mathbb{C}^{m\times m}$, $D = \mathsf{diag}(\sigma_1, \ldots, \sigma_r, 0, \ldots, 0) \in \mathbb{C}^{m\times n}$, and $V \in \mathbb{C}^{n\times n}$. Note that

$$\begin{aligned} A\boldsymbol{v}_j &= UDV^{\mathsf{H}}\boldsymbol{v}_j = UD\boldsymbol{e}_j \\ &\quad \text{(because } V \text{ is a unitary matrix)} \\ &= \sigma_j U\boldsymbol{e}_j = \sigma_j \boldsymbol{u}_j \end{aligned}$$

and

$$\begin{aligned} A^{\mathsf{H}}\boldsymbol{u}_j &= VD^{\mathsf{H}}U^{\mathsf{H}}\boldsymbol{u}_j = VDU^{\mathsf{H}}\boldsymbol{u}_j \\ &\quad VD\boldsymbol{e}_j \\ &\quad \text{(because } U \text{ is a unitary matrix)} \\ &= \sigma_j V\boldsymbol{e}_j = \sigma_j \boldsymbol{v}_j. \end{aligned}$$

Thus, the jth column of the matrix U, $\boldsymbol{u}_j$ and the jth column of the matrix V, $\boldsymbol{v}_j$ are left and right singular vectors, respectively, associated to the singular value σ_j.

Corollary 9.1. *Let $A \in \mathbb{C}^{m\times n}$ be a matrix and let $A = UDV^{\mathsf{H}}$ be the singular value decomposition of A. If $\|\cdot\|$ is a unitarily invariant norm, then*

$$\|A\| = \|D\| = \|\mathsf{diag}(\sigma_1, \ldots, \sigma_r, 0, \ldots, 0)\|.$$

Proof. This statement is a direct consequence of Theorem 9.1 because the matrices $U \in \mathbb{C}^{m\times m}$ and $V \in \mathbb{C}^{n\times n}$ are unitary. □

In other words, the value of a unitarily invariant norm of a matrix depends only on its singular values. As we saw in Theorem 6.24, $|\!|\!|\cdot|\!|\!|_2$ and $\|\cdot\|_F$ are unitarily invariant. Therefore, the Frobenius norm can be written as

$$\|A\|_F = \sqrt{\sum_{i=1}^{r} \sigma_r^2}.$$

Definition 9.2. Two matrices $A, B \in \mathbb{C}^{m\times n}$ are *unitarily equivalent* (denoted by $A \equiv_u B$) if there exist two unitary matrices W_1 and W_2 such that $A = W_1^{\mathsf{H}} B W_2$.

Clearly, if $A \sim_u B$, then $A \equiv_u B$.

Theorem 9.2. *Let A and B be two matrices in $\mathbb{C}^{m\times n}$. If A and B are unitarily equivalent, then they have the same singular values.*

Proof. Suppose that $A \equiv_u B$, that is, $A = W_1^{\mathsf{H}} B W_2$ for some unitary matrices W_1 and W_2. If A has the SVD $A = U^{\mathsf{H}}\mathsf{diag}(\sigma_1, \ldots, \sigma_r, 0, \ldots, 0)V$, then

$$B = W_1 A W_2^{\mathsf{H}} = (W_1 U^{\mathsf{H}})\mathsf{diag}(\sigma_1, \ldots, \sigma_r, 0, \ldots, 0)(V W_2^{\mathsf{H}}).$$

Since $W_1 U^{\mathsf{H}}$ and $V W_2^{\mathsf{H}}$ are both unitary matrices, it follows that they have the same singular values as B. □

Let $\boldsymbol{v} \in \mathbb{C}^n$ be an eigenvector of the matrix $A^{\mathsf{H}}A$ that corresponds to a non-zero, positive eigenvalue σ^2, that is, $A^{\mathsf{H}}A\boldsymbol{v} = \sigma^2\boldsymbol{v}$.

Define $\boldsymbol{u} = \frac{1}{\sigma}A\boldsymbol{v}$. We have $A\boldsymbol{v} = \sigma\boldsymbol{u}$. Also,

$$A^{\mathsf{H}}\boldsymbol{u} = A^{\mathsf{H}}\left(\frac{1}{\sigma}A\boldsymbol{v}\right) = \sigma\boldsymbol{v}.$$

This implies $AA^{\mathsf{H}}\boldsymbol{u} = \sigma^2\boldsymbol{u}$, so $\boldsymbol{u}$ is an eigenvector of AA^{H} that corresponds to the same eigenvalue σ^2.

Conversely, if $\boldsymbol{u} \in \mathbb{C}^m$ is an eigenvector of the matrix AA^{H} that corresponds to a non-zero, positive eigenvalue σ^2, we have $AA^{\mathsf{H}}\boldsymbol{u} = \sigma^2\boldsymbol{u}$. Thus, if $\boldsymbol{v} = \frac{1}{\sigma}A\boldsymbol{u}$, we have $A\boldsymbol{v} = \sigma\boldsymbol{u}$ and $\boldsymbol{v}$ is an eigenvector of $A^{\mathsf{H}}A$ for the eigenvalue σ^2.

The Courant–Fisher theorem (Theorem 8.18) allows the formulation of a similar result for singular values.

Theorem 9.3. *Let A be a matrix, $A \in \mathbb{C}^{m\times n}$. If $\sigma_1 \geqslant \sigma_2 \geqslant \cdots \geqslant \sigma_k \geqslant \cdots$ is the non-increasing sequence of singular values of A, then*

$$\sigma_k = \min_{\dim(S)=n-k+1} \max\{\|A\boldsymbol{x}\|_2 \mid \boldsymbol{x} \in S \text{ and } \|\boldsymbol{x}\|_2 = 1\}$$

$$\sigma_k = \max_{\dim(T)=k} \min\{\|A\boldsymbol{x}\|_2 \mid \boldsymbol{x} \in T \text{ and } \|\boldsymbol{x}\|_2 = 1\},$$

where S and T range over subspaces of $\mathbb{C}^n$.

Proof. We give the argument only for the second equality of the theorem; the first can be shown in a similar manner.

We saw that σ_k equals the kth largest absolute value of the eigenvalue $|\lambda_k|$ of the matrix $A^{\mathsf{H}}A$. By the Courant–Fisher theorem, we have

$$\begin{aligned}\lambda_k &= \max_{dim(T)=k} \min_{\boldsymbol{x}}\{\boldsymbol{x}^{\mathsf{H}}A^{\mathsf{H}}A\boldsymbol{x} \mid \boldsymbol{x} \in T \text{ and } \|\boldsymbol{x}\|_2 = 1\} \\ &= \max_{dim(T)=k} \min_{\boldsymbol{x}}\{\|A\boldsymbol{x}\|_2^2 \mid \boldsymbol{x} \in T \text{ and } \|\boldsymbol{x}\|_2 = 1\},\end{aligned}$$

which implies the second equality of the theorem. □

Theorem 9.3 can be restated as follows;

Theorem 9.4. *Let A be a matrix, $A \in \mathbb{C}^{m\times n}$. If $\sigma_1 \geqslant \sigma_2 \geqslant \cdots \geqslant \sigma_k \geqslant \cdots$ is the non-increasing sequence of singular values of A, then*

$$\begin{aligned}\sigma_k &= \min_{\boldsymbol{w}_1,\ldots,\boldsymbol{w}_{k-1}} \max\{\|A\boldsymbol{x}\|_2 \mid \boldsymbol{x} \perp \boldsymbol{w}_1,\ldots,\boldsymbol{x} \perp \boldsymbol{w}_{k-1} \text{ and } \|\boldsymbol{x}\|_2 = 1\} \\ &= \max_{\boldsymbol{w}_1,\ldots,\boldsymbol{w}_{n-k}} \min\{\|A\boldsymbol{x}\|_2 \mid \boldsymbol{x} \perp \boldsymbol{w}_1,\ldots,\boldsymbol{x} \perp \boldsymbol{w}_{n-k} \text{ and } \|\boldsymbol{x}\|_2 = 1\}.\end{aligned}$$

Proof. The argument is similar to the one used in Theorem 8.19. □

Corollary 9.2. *The smallest singular value of a matrix* $A \in \mathbb{C}^{m\times n}$ *equals*

$$\min\{\|A\boldsymbol{x}\|_2 \mid \boldsymbol{x} \in \mathbb{C}^n \text{ and } \|\boldsymbol{x}\|_2 = 1\}.$$

The largest singular value of a matrix $A \in \mathbb{C}^{m\times n}$ *equals*

$$\max\{\|A\boldsymbol{x}\|_2 \mid \boldsymbol{x} \in \mathbb{C}^n \text{ and } \|\boldsymbol{x}\|_2 = 1\}.$$

Proof. The corollary is a direct consequence of Theorem 9.3. □

The SVD theorem can also be proven by induction on $q = \min\{m, n\}$. In the base case, $q = 1$, we have $A \in \mathbb{C}^{1\times 1}$, or $A \in \mathbb{C}^{m\times 1}$, or $A \in \mathbb{C}^{1\times n}$.

Suppose, for example, that $A = \boldsymbol{a} \in \mathbb{C}^{m\times 1}$, where

$$\boldsymbol{a} = \begin{pmatrix} a_1 \\ \vdots \\ a_m \end{pmatrix}$$

and let $a = \|\boldsymbol{a}\|_2$. We seek $U \in \mathbb{C}^{m\times m}$, $V = (v) \in \mathbb{C}^{1\times 1}$ such that

$$\boldsymbol{a} = U\mathsf{diag}(a)v,$$

where

$$\mathsf{diag}(a) = \begin{pmatrix} a \\ 0 \\ \vdots \\ 0 \end{pmatrix} \in \mathbb{C}^{m\times 1}.$$

The role of the matrix U is played by any unitary matrix which has the first column equal to

$$\begin{pmatrix} \frac{a_1}{a} \\ \frac{a_2}{a} \\ \vdots \\ \frac{a_n}{a} \end{pmatrix},$$

and we can adopt $v = 1$. The remaining base subcases can be treated in a similar manner.

Suppose now that the statement holds when at least one of the numbers m and n is less than q and let us prove the assertion when at least one of m and n is less than $q+1$.

Let $\boldsymbol{u}_1$ be a unit eigenvector of AA^{H} that corresponds to the eigenvalue σ_1^2 and let $\boldsymbol{v}_1 = \frac{1}{\sigma_1}A^{\mathsf{H}}\boldsymbol{u}_1$. We have $\|\boldsymbol{v}_1\|_2 = 1$ and

$$A\boldsymbol{v}_1 = \frac{1}{\sigma_1}AA^{\mathsf{H}}\boldsymbol{u}_1 = \sigma_1\boldsymbol{u}_1,$$

which shows that $(\boldsymbol{v}_1, \boldsymbol{u}_1)$ is a pair of singular vectors corresponding to the singular value σ_1. We have also

$$\boldsymbol{u}_1^{\mathsf{H}}A^{\mathsf{H}}\boldsymbol{v}_1 = \frac{1}{\sigma_1}\boldsymbol{u}_1^{\mathsf{H}}AA^{\mathsf{H}}\boldsymbol{u}_1 = \sigma_1.$$

Define $U = (\boldsymbol{u}_1\ U_1)$ and $V = (\boldsymbol{v}_1\ V_1)$ as unitary matrices having $\boldsymbol{u}_1$ and $\boldsymbol{v}_1$ as their first columns, respectively. Then,

$$\begin{aligned} U^{\mathsf{H}}A^{\mathsf{H}}V &= \begin{pmatrix} \boldsymbol{u}_1^{\mathsf{H}} \\ U_1^{\mathsf{H}} \end{pmatrix} A^{\mathsf{H}} \begin{pmatrix} \boldsymbol{v}_1 & V_1 \end{pmatrix} \\ &= \begin{pmatrix} \boldsymbol{u}_1^{\mathsf{H}}A^{\mathsf{H}} \\ U_1^{\mathsf{H}}A^{\mathsf{H}} \end{pmatrix} \begin{pmatrix} \boldsymbol{v}_1 & V_1 \end{pmatrix} \\ &= \begin{pmatrix} \boldsymbol{u}_1^{\mathsf{H}}A^{\mathsf{H}}\boldsymbol{v}_1 & \boldsymbol{u}_1^{\mathsf{H}}AV_1 \\ U_1^{\mathsf{H}}A\boldsymbol{v}_1 & U_1^{\mathsf{H}}AV_1 \end{pmatrix}. \end{aligned}$$

Since U is a unitary matrix, every column of U_1 is orthogonal to $\boldsymbol{u}_1$. Therefore,

$$U_1^{\mathsf{H}}A\boldsymbol{v}_1 = \frac{1}{\sigma_1}U_1^{\mathsf{H}}AA^{\mathsf{H}}\boldsymbol{u}_1 = \sigma_1 U_1^{\mathsf{H}}\boldsymbol{u}_1 = \mathbf{0},$$

and, similarly,

$$\boldsymbol{u}_1^{\mathsf{H}}A^{\mathsf{H}}V_1 = \sigma_1\boldsymbol{v}_1^{\mathsf{H}}V_1 = \mathbf{0}',$$

because $\boldsymbol{v}_1$ is orthogonal on all columns of V_1. Thus,

$$U^{\mathsf{H}}AV = \begin{pmatrix} \sigma_1 & \mathbf{0}' \\ \mathbf{0} & U_1^{\mathsf{H}}AV_1 \end{pmatrix}.$$

The matrix $U_1^{\mathsf{H}} A V_1$ has fewer rows and columns than $U^{\mathsf{H}} A V$, so we can apply the inductive hypothesis to $B = U_1^{\mathsf{H}} A V_1$. Therefore, by the inductive hypothesis, B can be written as $B = XDY^{\mathsf{H}}$, where X and Y are unitary matrices and D is a diagonal matrix. This allows us to write

$$U^{\mathsf{H}} A V = \begin{pmatrix} \sigma_1 & \mathbf{0}' \\ \mathbf{0} & XDY^{\mathsf{H}} \end{pmatrix} = \begin{pmatrix} 1 & \mathbf{0}' \\ \mathbf{0} & X \end{pmatrix} \begin{pmatrix} \sigma_1 & \mathbf{0}' \\ \mathbf{0} & D \end{pmatrix} \begin{pmatrix} 1 & \mathbf{0} \\ \mathbf{0} & Y^{\mathsf{H}} \end{pmatrix}.$$

Since the matrices

$$\begin{pmatrix} 1 & \mathbf{0}' \\ \mathbf{0} & X \end{pmatrix} \text{ and } \begin{pmatrix} 1 & \mathbf{0} \\ \mathbf{0} & Y^{\mathsf{H}} \end{pmatrix}$$

are unitary, we obtain the desired conclusion.

If $A \in \mathbb{C}^{n \times n}$ is an invertible matrix and σ is a singular value of A, then $\frac{1}{\sigma}$ is a singular value of the matrix A^{-1}.

Example 9.1. Let

$$\boldsymbol{a} = \begin{pmatrix} a_1 \\ \vdots \\ a_n \end{pmatrix}$$

be a non-zero vector in $\mathbb{C}^n$, which can also be regarded as a matrix in $\mathbb{C}^{n \times 1}$. The square of a singular value of A is an eigenvalue of the matrix

$$A^{\mathsf{H}} A = \begin{pmatrix} \bar{a}_1 a_1 & \cdots & \bar{a}_n a_1 \\ \bar{a}_1 a_2 & \cdots & \bar{a}_n a_2 \\ \vdots & \cdots & \vdots \\ \bar{a}_1 a_n & \cdots & \bar{a}_n a_n \end{pmatrix}$$

and we have seen (in Corollary 7.4) that the unique non-zero eigenvalue of this matrix is $\|a\|_2^2$. Thus, the unique singular value of $\boldsymbol{a}$ is $\|a\|_2$.

Example 9.2. Let $A \in \mathbb{R}^{3\times 2}$ be the matrix

$$A = \begin{pmatrix} 0 & 1 \\ 1 & 1 \\ 1 & 0 \end{pmatrix}.$$

The matrices $A^{\mathsf{H}}A$ and $A^{\mathsf{H}}A$ are given by

$$AA^{\mathsf{H}} = \begin{pmatrix} 1 & 1 & 0 \\ 1 & 2 & 1 \\ 0 & 1 & 1 \end{pmatrix} \text{ and } A^{\mathsf{H}}A = \begin{pmatrix} 2 & 1 \\ 1 & 2 \end{pmatrix}.$$

The eigenvalues of $A^{\mathsf{H}}A$ are the roots of the polynomial $\lambda^2 - 4\lambda + 3$, and therefore, they are $\lambda_1 = 3$ and $\lambda_2 = 1$. The eigenvalues of AA^{H} are $3, 1$, and 0.

Unit eigenvectors of $A^{\mathsf{H}}A$ that correspond to 3 and 1 are

$$\boldsymbol{v}_1 = \alpha_1 \begin{pmatrix} \frac{\sqrt{2}}{2} \\ \frac{\sqrt{2}}{2} \end{pmatrix} \text{ and } \boldsymbol{v}_2 = \alpha_2 \begin{pmatrix} \frac{\sqrt{2}}{2} \\ -\frac{\sqrt{2}}{2} \end{pmatrix},$$

respectively, where $\alpha_i \in \{-1, 1\}$ for $i = 1, 2$.

Unit eigenvectors of $A^{\mathsf{H}}A$ that correspond to $3, 1$, and 0 are

$$\boldsymbol{u}_1 = \beta_1 \begin{pmatrix} \frac{\sqrt{6}}{6} \\ \frac{\sqrt{6}}{3} \\ \frac{\sqrt{6}}{6} \end{pmatrix}, \boldsymbol{u}_2 = \beta_2 \begin{pmatrix} \frac{\sqrt{2}}{2} \\ 0 \\ -\frac{\sqrt{2}}{2} \end{pmatrix}, \boldsymbol{u}_3 = \beta_3 \begin{pmatrix} \frac{\sqrt{3}}{3} \\ -\frac{\sqrt{3}}{3} \\ \frac{\sqrt{3}}{3} \end{pmatrix},$$

respectively, where $\beta_i \in \{-1, 1\}$ for $i = 1, 2, 3$.

The choice of the columns of the matrices U and V must be done such that for a pair of eigenvectors (u, v) that correspond to a singular value σ, we have $\boldsymbol{v} = \frac{1}{\sigma}A^{\mathsf{H}}\boldsymbol{u}$ or, equivalently, $\boldsymbol{u} = \frac{1}{\sigma}A\boldsymbol{v}$, as we saw in the proof of Theorem 9.1. For instance, if we choose $\alpha_1 = \alpha_2 = 1$, then

$$\boldsymbol{v}_1 = \begin{pmatrix} \frac{\sqrt{2}}{2} \\ \frac{\sqrt{2}}{2} \end{pmatrix}, \boldsymbol{v}_2 = \begin{pmatrix} \frac{\sqrt{2}}{2} \\ -\frac{\sqrt{2}}{2} \end{pmatrix},$$

and $\boldsymbol{u}_1 = \frac{1}{\sqrt{3}} A\boldsymbol{v}_1$ and $\boldsymbol{u}_2 = A\boldsymbol{v}_2$, that is

$$\boldsymbol{u}_1 = \begin{pmatrix} \frac{\sqrt{6}}{6} \\ \frac{\sqrt{6}}{3} \\ \frac{\sqrt{6}}{6} \end{pmatrix}, \boldsymbol{u}_2 = \begin{pmatrix} -\frac{\sqrt{2}}{2} \\ 0 \\ \frac{\sqrt{2}}{2} \end{pmatrix},$$

which means that $\beta_1 = 1$ and $\beta_2 = -1$; the value of β_3 that corresponds to the eigenvalue of 0 can be chosen arbitrarily. Thus, an SVD of A is

$$A = \begin{pmatrix} \frac{\sqrt{6}}{6} & -\frac{\sqrt{2}}{2} & \frac{\sqrt{3}}{3} \\ \frac{\sqrt{6}}{3} & 0 & -\frac{\sqrt{3}}{3} \\ \frac{\sqrt{6}}{6} & \frac{\sqrt{2}}{2} & \frac{\sqrt{3}}{3} \end{pmatrix} \begin{pmatrix} \sqrt{3} & 0 \\ 0 & 1 \\ 0 & 0 \end{pmatrix} \begin{pmatrix} \frac{\sqrt{2}}{2} & \frac{\sqrt{2}}{2} \\ \frac{\sqrt{2}}{2} & -\frac{\sqrt{2}}{2} \end{pmatrix}.$$

The singular values of a matrix $A \in \mathbb{C}^{m\times n}$ are uniquely determined. However, the matrices U and V of the SVD of A are not unique, as we saw in Example 9.2. Once we choose a column of the matrix V for a singular value σ, the corresponding column of U is determined by $\boldsymbol{u} = \frac{1}{\sigma} A\boldsymbol{v}$.

A variant of the SVD Decomposition Theorem is given next.

Corollary 9.3 (The Thin SVD Decomposition Corollary). Let $A \in \mathbb{C}^{m\times n}$ be a matrix having non-zero singular values $\sigma_1, \sigma_2, \ldots, \sigma_r$, where $\sigma_1 \geqslant \sigma_2 \geqslant \cdots \geqslant \sigma_r > 0$ and $r \leqslant \min\{m, n\}$. Then, A can be factored as $A = UDV^{\mathsf{H}}$, where $U \in \mathbb{C}^{m\times r}$ and $V \in \mathbb{C}^{n\times r}$ are matrices having orthonormal sets of columns and D is the diagonal matrix

$$D = \begin{pmatrix} \sigma_1 & 0 & \cdots & 0 \\ 0 & \sigma_2 & \cdots & 0 \\ \vdots & \vdots & \cdots & \vdots \\ 0 & 0 & \cdots & \sigma_r \end{pmatrix}.$$

Proof. The statement is an immediate consequence of Theorem 9.1. □

The decomposition described in Corollary 9.3 is known as a *thin SVD decomposition* of the matrix A.

Example 9.3. The thin SVD decomposition of the matrix A introduced in Example 9.2,

$$A = \begin{pmatrix} 0 & 1 \\ 1 & 1 \\ 1 & 0 \end{pmatrix}.$$

is

$$A = \begin{pmatrix} \frac{\sqrt{6}}{6} & -\frac{\sqrt{2}}{2} \\ \frac{\sqrt{6}}{3} & 0 \\ \frac{\sqrt{6}}{6} & \frac{\sqrt{2}}{2} \end{pmatrix} \begin{pmatrix} \sqrt{3} & 0 \\ 0 & 1 \end{pmatrix} \begin{pmatrix} \frac{\sqrt{2}}{2} & \frac{\sqrt{2}}{2} \\ \frac{\sqrt{2}}{2} & -\frac{\sqrt{2}}{2} \end{pmatrix}.$$

Since U and V in the thin SVD have orthonormal columns, it is easy to see that

$$U^{\mathsf{H}}U = V^{\mathsf{H}}V = I_p. \tag{9.2}$$

Lemma 9.1. *Let $D \in \mathbb{R}^{n \times n}$ be a diagonal matrix, where $D = \mathsf{diag}(\sigma_1, \ldots, \sigma_r)$ and $\sigma_1 \geqslant \cdots \geqslant \sigma_r$. Then, we have $\|D\|_2 = \sigma_1$, and $\|D\|_F = \sqrt{\sum_{i=1}^r \sigma_i^2}$.*

Proof. By the definition of $\|D\|_2$, we have

$$\begin{aligned} \|D\|_2 &= \max\{\|D\boldsymbol{x}\|_2 \mid \|\boldsymbol{x}\| = 1\} \\ &= \max\left\{ \sqrt{\sum_{i=1}^r \sigma_i^2 |x_i|^2} \,\middle|\, \sum_{i=1}^n |x_i|^2 = 1 \right\}. \end{aligned}$$

Since

$$\sum_{i=1}^r \sigma_i^2 |x_i|^2 \leqslant \sigma_1^2 \left(\sum_{i=1}^r |x_i|^2 \right) \leqslant \sigma_1^2,$$

because $\sum_{i=1}^n |x_i|^2 = 1$, it follows that

$$\max\left\{ \sqrt{\sum_{i=1}^r \sigma_i^2 |x_i|^2} \,\middle|\, \sum_{i=1}^n |x_i|^2 = 1 \right\} = \sigma_1.$$

The second part is immediate. □

Theorem 9.5. *Let $A \in \mathbb{C}^{m\times n}$ be a matrix whose singular values are $\sigma_1 \geqslant \cdots \geqslant \sigma_r$. Then $\|\!|A|\!\|_2 = \sigma_1$, and $\|A\|_F = \sqrt{\sum_{i=1}^r \sigma_i^2}$.*

Proof. Suppose that the SVD of A is $A = UDV^{\mathsf{H}}$, where U and V are unitary matrices. Then, by Theorem 6.24 and Lemma 9.1, we have

$$\|\!|A|\!\|_2 = \|\!|UDV^{\mathsf{H}}|\!\|_2 = \|\!|D|\!\|_2 = \sigma_1,$$

$$\|A\|_F = \|UDV^{\mathsf{H}}\|_F = \|D\|_F = \sqrt{\sum_{i=1}^r \sigma_i^2}.$$

□

Corollary 9.4. *If $A \in \mathbb{C}^{m\times n}$ is a matrix, then $\|\!|A|\!\|_2 \leqslant \|A\|_F \leqslant \sqrt{n}\|\!|A|\!\|_2$.*

Proof. Suppose that $\sigma_1(A)$ is the largest of the singular values of A. Then, since $\|A\|_F = \sqrt{\sum_{i=1}^r \sigma_i^2}$, we have

$$\sigma_1(A) \leqslant \|A\|_F \leqslant \sqrt{n \max_i \sigma_j(A)^2} = \sigma_1(A)\sqrt{n},$$

which is the desired double inequality. □

Theorem 9.6. *Let $A \in \mathbb{C}^{n\times n}$ be an invertible matrix. If the singular values of A are $\sigma_1 \geqslant \cdots \geqslant \sigma_n > 0$, then*

$$\mathsf{cond}(A) = \frac{\sigma_1}{\sigma_n}.$$

Proof. We have shown in Theorem 9.5 that $\|\!|A|\!\|_2 = \sigma_1$. Since the singular values of A^{-1} are

$$\frac{1}{\sigma_n} \geqslant \cdots \geqslant \frac{1}{\sigma_1},$$

it follows that $\|\!|A^{-1}|\!\|_2 = \frac{1}{\sigma_n}$. The desired equality follows immediately. □

Corollary 9.5. *Let $A \in \mathbb{C}^{n\times n}$ be an invertible matrix. We have $\mathsf{cond}(A^{\mathsf{H}}A) = (\mathsf{cond}(A))^2$.*

Proof. Let σ be a singular value of A and let $\boldsymbol{u}, \boldsymbol{v}$ be two left and right singular vectors corresponding to σ, respectively. We have

$$A\boldsymbol{v} = \sigma\boldsymbol{u} \text{ and } A^{\mathsf{H}}\boldsymbol{u} = \sigma\boldsymbol{v}.$$

This implies $A^{\mathsf{H}}A\boldsymbol{v} = \sigma A^{\mathsf{H}}\boldsymbol{u} = \sigma^2\boldsymbol{v}$, which shows that the singular values of the matrix $A^{\mathsf{H}}A$ are the squares of the singular values of A, which produces the desired conclusion. □

Let $A = UDV^{\mathsf{H}}$ be an SVD of A. If we write U and V using their columns as

$$U = (\boldsymbol{u}_1 \ \cdots \ \boldsymbol{u}_m), V = (\boldsymbol{v}_1 \ \cdots \ \boldsymbol{v}_n),$$

then A can be written as

$$\begin{aligned}
A &= UDV^{\mathsf{H}} \\
&= (\boldsymbol{u}_1 \ \cdots \ \boldsymbol{u}_n) \begin{pmatrix} \sigma_1 & 0 & \cdots & \cdots & 0 \\ 0 & \sigma_2 & \cdots & \cdots & 0 \\ \vdots & \vdots & \cdots & \cdots & \vdots \\ 0 & 0 & \cdots & \sigma_r & 0 \\ \vdots & \vdots & \cdots & \cdots & \vdots \\ 0 & 0 & 0 & 0 & 0 \end{pmatrix} \begin{pmatrix} \boldsymbol{v}_1^{\mathsf{H}} \\ \vdots \\ \boldsymbol{v}_m^{\mathsf{H}} \end{pmatrix} \\
&= (\boldsymbol{u}_1 \ \cdots \ \boldsymbol{u}_m) \begin{pmatrix} \sigma_1\boldsymbol{v}_1^{\mathsf{H}} \\ \vdots \\ \sigma_r\boldsymbol{v}_p^{\mathsf{H}} \end{pmatrix} \\
&= \sigma_1\boldsymbol{u}_1\boldsymbol{v}_1^{\mathsf{H}} + \cdots + \sigma_r\boldsymbol{u}_r\boldsymbol{v}_p^{\mathsf{H}}. \qquad (9.3)
\end{aligned}$$

Since $\boldsymbol{u}_i \in \mathbb{C}^m$ and $\boldsymbol{v}_i \in \mathbb{C}^n$, each of the matrices $\boldsymbol{u}_i\boldsymbol{v}_i^{\mathsf{H}}$ is an $m \times n$ matrix of rank 1. Thus, the SVD yields an expression of A as a sum of r matrices of rank 1, where r is the number of non-zero singular values of A.

Theorem 9.7. *The rank-1 matrices of the form $\boldsymbol{u}_i\boldsymbol{v}_i^{\mathsf{H}}$, where $1 \leqslant i \leqslant r$ that occur in Equality (9.3), are pairwise orthogonal. Moreover, $\|\boldsymbol{u}_i\boldsymbol{v}_i^{\mathsf{H}}\|_F = 1$ for $1 \leqslant i \leqslant r$.*

Proof. For $i \neq j$ and $1 \leqslant i, j \leqslant r$, we have

$$\mathsf{trace}\left(\boldsymbol{u}_i \boldsymbol{v}_i^{\mathsf{H}} (\boldsymbol{u}_j \boldsymbol{v}_j^{\mathsf{H}})^{\mathsf{H}}\right) = \mathsf{trace}\left(\boldsymbol{u}_i \boldsymbol{v}_i^{\mathsf{H}} \boldsymbol{v}_j \boldsymbol{u}_j\right) = 0,$$

because the vectors $\boldsymbol{v}_i$ and $\boldsymbol{v}_j$ are orthogonal. Thus, $(\boldsymbol{u}_i \boldsymbol{v}_i^{\mathsf{H}}, \boldsymbol{u}_j \boldsymbol{v}_j^{\mathsf{H}}) = 0$.

By Equality (6.12), we have

$$\begin{aligned}\|\boldsymbol{u}_i \boldsymbol{v}_i^{\mathsf{H}}\|_F^2 &= \mathsf{trace}((\boldsymbol{u}_i \boldsymbol{v}_i^{\mathsf{H}})^{\mathsf{H}} \boldsymbol{u}_i \boldsymbol{v}_i^{\mathsf{H}}) \\ &= \mathsf{trace}(\boldsymbol{v}_i \boldsymbol{u}_i^{\mathsf{H}} \boldsymbol{u}_i \boldsymbol{v}_i^{\mathsf{H}}) = 1,\end{aligned}$$

because the matrices U and V are unitary. □

Theorem 9.7 shows that Equality (9.3) is similar to a Fourier expansion of A. Namely, the matrix is extended in terms of the orthonormal set $\{\boldsymbol{u}_i \boldsymbol{v}_i^{\mathsf{H}} \mid 1 \leqslant i \leqslant r\}$ of the linear space of all matrices of rank no larger than r.

Theorem 9.8. *Let $A \in \mathbb{C}^{m \times n}$ be a matrix that has the singular value decomposition $A = UDV^{\mathsf{H}}$. If* $\mathsf{rank}(A) = r$, *then the first r columns of U form an orthonormal basis for* $\mathsf{range}(A)$, *and the last $n-r$ columns of V constitute an orthonormal basis for* $\mathsf{null}(A)$.

Proof. Since both U and V are unitary matrices, it is clear that $\{\boldsymbol{u}_1, \ldots, \boldsymbol{u}_r\}$, the set of the first r columns of U, and $\{\boldsymbol{v}_{r+1}, \ldots, \boldsymbol{v}_n\}$, the set of the last $n - r$ columns of V, are linearly independent sets. Thus, we only need to show that $\langle \boldsymbol{u}_1, \ldots, \boldsymbol{u}_r \rangle = \mathsf{range}(A)$ and $\langle \boldsymbol{v}_{r+1}, \ldots, \boldsymbol{v}_n \rangle = \mathsf{null}(A)$.

By Equality (9.3), we have

$$A = \sigma_1 \boldsymbol{u}_1 \boldsymbol{v}_1^{\mathsf{H}} + \cdots + \sigma_r \boldsymbol{u}_r \boldsymbol{v}_r^{\mathsf{H}}.$$

If $\boldsymbol{t} \in \mathsf{range}(A)$, then $\boldsymbol{t} = A\boldsymbol{s}$ for some $\boldsymbol{s} \in \mathbb{C}^n$. Therefore, $\boldsymbol{t} = \sigma_1 \boldsymbol{u}_1 (\boldsymbol{v}_1^{\mathsf{H}} \boldsymbol{s}) + \cdots + \sigma_r \boldsymbol{u}_r (\boldsymbol{v}_r^{\mathsf{H}} \boldsymbol{s})$, and, since every product $\boldsymbol{v}_j^{\mathsf{H}} \boldsymbol{s}$ is a scalar for $1 \leqslant j \leqslant r$, it follows that $\boldsymbol{t} \in \langle \boldsymbol{u}_1, \ldots, \boldsymbol{u}_r \rangle$, so $\mathsf{range}(A) \subseteq \langle \boldsymbol{u}_1, \ldots, \boldsymbol{u}_r \rangle$.

To prove the reverse inclusion, note that

$$A\left(\frac{1}{\sigma_i} \boldsymbol{v}_i\right) = \boldsymbol{u}_i,$$

for $1 \leqslant i \leqslant r$, due to the orthogonality of the columns of V. Thus, $\langle \boldsymbol{u}_1, \ldots, \boldsymbol{u}_r \rangle = \mathsf{range}(A)$.

Note that Equality (9.3) implies that $A\boldsymbol{v}_j = 0$ for $r+1 \leqslant j \leqslant n$, so $\langle \boldsymbol{v}_{r+1}, \ldots, \boldsymbol{v}_n \rangle \subseteq \mathsf{null}(A)$. Conversely, suppose that $A\boldsymbol{r} = \boldsymbol{0}$. Since the columns of V form a basis of $\mathbb{C}^n$, we have $\boldsymbol{r} = a_1\boldsymbol{v}_1 + \cdots + a_n\boldsymbol{v}_n$, so $A\boldsymbol{r} = a_1 A\boldsymbol{v}_1 + \cdots + a_r\boldsymbol{v}_r = \boldsymbol{0}$. The linear independence of $\{\boldsymbol{v}_1, \ldots, \boldsymbol{v}_r\}$ implies $a_1 = \cdots = a_r = 0$, so $\boldsymbol{r} = a_{r+1}\boldsymbol{v}_{r+1} + \cdots + a_n\boldsymbol{v}_n$, which shows that $\mathsf{null}(A) \subseteq \langle \boldsymbol{v}_{r+1}, \ldots, \boldsymbol{v}_n \rangle$. Thus, $\mathsf{null}(A) = \langle \boldsymbol{v}_{r+1}, \ldots, \boldsymbol{v}_n \rangle$. □

Corollary 9.6. *Let $A \in \mathbb{C}^{m\times n}$ be a matrix that has the singular value decomposition $A = UDV^{\mathsf{H}}$. If $\mathsf{rank}(A) = r$, then the first r transposed columns of V form an orthonormal basis for the subspace of $\mathbb{R}^n$ generated by the rows of A.*

Proof. This statement follows immediately from Theorem 9.8 applied to A^{H}. □

9.3 Numerical Rank of Matrices

The SVD allows us to find the best approximation of a matrix by a matrix of limited rank. The central result of this section is Theorem 9.9.

Lemma 9.2. *Let $A = \sigma_1\boldsymbol{u}_1\boldsymbol{v}_1^{\mathsf{H}} + \cdots + \sigma_r\boldsymbol{u}_r\boldsymbol{v}_r^{\mathsf{H}}$ be the SVD of a matrix $A \in \mathbb{R}^{m\times n}$, where $\sigma_1 \geqslant \cdots \geqslant \sigma_r > 0$. For every k, $1 \leqslant k \leqslant r$, the matrix $B(k) = \sum_{i=1}^{k} \sigma_i\boldsymbol{u}_i\boldsymbol{v}_i^{\mathsf{H}}$ has rank k.*

Proof. The null space of the matrix $B(k)$ consists of those vectors $\boldsymbol{x}$ such that $\sum_{i=1}^{k} \sigma_i\boldsymbol{u}_i\boldsymbol{v}_i^{\mathsf{H}}\boldsymbol{x} = \boldsymbol{0}$. The linear independence of the vectors $\boldsymbol{u}_i$, and the fact that $\sigma_i > 0$ for $1 \leqslant i \leqslant r$, implies the equalities $\boldsymbol{v}_i^{\mathsf{H}}\boldsymbol{x} = \boldsymbol{0}$ for $1 \leqslant i \leqslant r$. Thus,

$$\mathsf{null}(B(k)) = \mathsf{null}\left((\boldsymbol{v}_1 \ \cdots \ \boldsymbol{v}_k)\right).$$

Since $\boldsymbol{v}_1, \ldots, \boldsymbol{v}_k$ are linearly independent, it follows that dim $(\mathsf{null}(B(k)) = n - k$, which implies $\mathsf{rank}(B(k)) = k$ for $1 \leqslant k \leqslant r$. □

Theorem 9.9 (Eckhart–Young Theorem). Let $A \in \mathbb{C}^{m\times n}$ be a matrix whose sequence of nonzero singular values is $(\sigma_1, \ldots, \sigma_r)$. Assume that $\sigma_1 \geqslant \cdots \geqslant \sigma_r > 0$ and that A can be written as

$$A = \sigma_1\boldsymbol{u}_1\boldsymbol{v}_1^{\mathsf{H}} + \cdots + \sigma_r\boldsymbol{u}_r\boldsymbol{v}_r^{\mathsf{H}}.$$

Let $B(k) \in \mathbb{C}^{m\times n}$ be the matrix defined by

$$B(k) = \sum_{i=1}^{k} \sigma_i \boldsymbol{u}_i \boldsymbol{v}_i^{\mathsf{H}}.$$

If $r_k = \inf\{\|A - X\|_2 \mid X \in \mathbb{C}^{m\times n} \text{ and } \mathsf{rank}(X) \leqslant k\}$, then

$$\|A - B(k)\|_2 = r_k = \sigma_{k+1},$$

for $1 \leqslant k \leqslant r$, where $\sigma_{r+1} = 0$ and $B(k)$ is the best approximation of A among the matrices of rank no larger than k in the sense of the norm $\|\cdot\|_2$.

Proof. Observe that

$$A - B(k) = \sum_{i=k+1}^{r} \sigma_i \boldsymbol{u}_i \boldsymbol{v}_i^{\mathsf{H}},$$

and the largest singular value of the matrix $\sum_{i=k+1}^{r} \sigma_i \boldsymbol{u}_i \boldsymbol{v}_i^{\mathsf{H}}$ is σ_{k+1}. Therefore, by Theorem 9.5,

$$\|A - B(k)\|_2 = \sigma_{k+1}.$$

for $1 \leqslant k \leqslant r$.

We prove now that for every matrix $X \in \mathbb{C}^{m\times n}$ such that $\mathsf{rank}(X) \leqslant k$, we have $\|A - X\|_2 \geqslant \sigma_{k+1}$. Since $\dim(\mathsf{null}(X)) = n - \mathsf{rank}(X)$, it follows that $\dim(\mathsf{null}(X)) \geqslant n - k$. If T is the subspace of $\mathbb{R}^n$ spanned by $\boldsymbol{v}_1, \ldots, \boldsymbol{v}_{k+1}$, we have $\dim(T) = k+1$. Since $\dim(\mathsf{null}(X)) + \dim(T) > n$, the intersection of these subspaces contains a non-zero vector and, without loss of generality, we can assume that this vector is a unit vector $\boldsymbol{x}$.

We have $\boldsymbol{x} = a_1\boldsymbol{v}_1 + \cdots a_k\boldsymbol{v}_k + a_{k+1}\boldsymbol{v}_{k+1}$ because $\boldsymbol{x} \in T$. The orthogonality of $\boldsymbol{v}_1, \ldots, \boldsymbol{v}_k, \boldsymbol{v}_{k+1}$ implies $\|\boldsymbol{x}\|_2^2 = \sum_{i=1}^{k+1} |a_i|^2 = 1$.

Since $\boldsymbol{x} \in \mathsf{null}(X)$, we have $X\boldsymbol{x} = \boldsymbol{0}$, so

$$(A - X)\boldsymbol{x} = A\boldsymbol{x} = \sum_{i=1}^{k+1} a_i A\boldsymbol{v}_i = \sum_{i=1}^{k+1} a_i \sigma_i \boldsymbol{u}_i.$$

Thus, we have

$$\|(A - X)\boldsymbol{x}\|_2^2 = \sum_{i=1}^{k+1} |\sigma_i a_i|^2 \geqslant \sigma_{k+1}^2 \sum_{i=1}^{k+1} |a_i|^2 = \sigma_{k+1}^2,$$

because $\boldsymbol{u}_1, \ldots, \boldsymbol{u}_k$ are also orthonormal. This implies $\|A - X\|_2 \geqslant \sigma_{k+1} = \|A - B(k)\|_2$. □

It is interesting to observe that the matrix $B(k)$ provides an optimal approximation of A not only with respect to $\|\!|\cdot\|\!|_2$ but also relative to the Frobenius norm.

Theorem 9.10. *Using the notations introduced in Theorem 9.9, $B(k)$ is the best approximation of A among matrices of rank no larger than k in the sense of the Frobenius norm.*

Proof. Note that $\|A - B(k)\|_F^2 = \|A\|_F^2 - \sum_{i=1}^{k} \sigma_i^2$. Let X be a matrix of rank k, which can be written as $X = \sum_{i=1}^{k} \boldsymbol{x}_i \boldsymbol{y}_i^{\mathsf{H}}$. Without loss of generality we may assume that the vectors $\boldsymbol{x}_1, \ldots, \boldsymbol{x}_k$ are orthonormal. If this is not the case, we can use the Gram–Schmidt algorithm to express them as linear combinations of orthonormal vectors, replace these expressions in $\sum_{i=1}^{k} \boldsymbol{x}_i \boldsymbol{y}_i^{\mathsf{H}}$, and rearrange the terms. Now, the Frobenius norm of $A - X$ can be written as

$$
\begin{aligned}
&\|A - X\|_F^2 \\
&\quad = \mathsf{trace}\left(\left(A - \sum_{i=1}^{k} \boldsymbol{x}_i \boldsymbol{y}^{\mathsf{H}}\right)^{\mathsf{H}} \left(A - \sum_{i=1}^{k} \boldsymbol{x}_i \boldsymbol{y}^{\mathsf{H}}\right)\right) \\
&\quad = \mathsf{trace}\left(A^{\mathsf{H}} A + \sum_{i=1}^{k} (\boldsymbol{y}_i - A^{\mathsf{H}} \boldsymbol{x}_i)(\boldsymbol{y}_i - A^{\mathsf{H}} \boldsymbol{x}_i)^{\mathsf{H}} - \sum_{i=1}^{k} A^{\mathsf{H}} \boldsymbol{x}_i \boldsymbol{x}_i^{\mathsf{H}} A\right).
\end{aligned}
$$

Taking into account that $\sum_{i=1}^{k} (\boldsymbol{y}_i - A^{\mathsf{H}} \boldsymbol{x}_i)(\boldsymbol{y}_i - A^{\mathsf{H}} \boldsymbol{x}_i)^{\mathsf{H}}$ is a real non-negative number and that $\sum_{i=1}^{k} A^{\mathsf{H}} \boldsymbol{x}_i \boldsymbol{x}_i^{\mathsf{H}} A = \|A\boldsymbol{x}_i\|_F^2$, we have

$$
\begin{aligned}
\|A - X\|_F^2 &\geqslant \mathsf{trace}\left(A^{\mathsf{H}} A - \sum_{i=1}^{k} A^{\mathsf{H}} \boldsymbol{x}_i \boldsymbol{x}_i^{\mathsf{H}} A\right) \\
&= \|A\|_F^2 - \mathsf{trace}\left(\sum_{i=1}^{k} A^{\mathsf{H}} \boldsymbol{x}_i \boldsymbol{x}_i^{\mathsf{H}} A\right).
\end{aligned}
$$

Let $A = U \mathsf{diag}(\sigma_1, \ldots, \sigma_n) V^{\mathsf{H}}$ be the singular value decomposition of A. If $V = (V_1 \; V_2)$, where V_1 has k columns $\boldsymbol{v}_1, \ldots, \boldsymbol{v}_k$, $D_1 = \mathsf{diag}(\sigma_1, \ldots, \sigma_k)$, and $D_2 = \mathsf{diag}(\sigma_{k+1}, \ldots, \sigma_n)$, then we can write

$$A^{\mathsf{H}}A = VD^{\mathsf{H}}U^{\mathsf{H}}UDV^{\mathsf{H}} = (V_1\ V_2)\begin{pmatrix} D_1^2 & O \\ O & D_2^2 \end{pmatrix}\begin{pmatrix} V_1^{\mathsf{H}} \\ V_2^{\mathsf{H}} \end{pmatrix}$$
$$= V_1 D_1^2 V_1^{\mathsf{H}} + V_2 D_2^2 V_2^{\mathsf{H}}.$$

and $A^{\mathsf{H}}A = VD^2V^{\mathsf{H}}$. These equalities allow us to write

$$\begin{aligned}\|A\boldsymbol{x}_i\|_F^2 &= \mathsf{trace}(\boldsymbol{x}_i^{\mathsf{H}}A^{\mathsf{H}}A\boldsymbol{x}_i) \\ &= \mathsf{trace}\left(\boldsymbol{x}_i^{\mathsf{H}}V_1D_1^2V_1^{\mathsf{H}}\boldsymbol{x}_i + \boldsymbol{x}_i^{\mathsf{H}}V_2D_2^2V_2^{\mathsf{H}}\boldsymbol{x}_i\right) \\ &= \|D_1V_1^{\mathsf{H}}\boldsymbol{x}_i\|_F^2 + \|D_2V_2^{\mathsf{H}}\boldsymbol{x}_i\|_F^2 \\ &= \sigma_k^2 + \left(\|D_1V_1^{\mathsf{H}}\boldsymbol{x}_i\|_F^2 - \sigma_k^2\|V_1^{\mathsf{H}}\boldsymbol{x}_i\|_F^2\right) \\ &\quad - \left(\sigma_k^2\|V_2^{\mathsf{H}}\boldsymbol{x}_i\|_F^2 - \|D_2V_2^{\mathsf{H}}\boldsymbol{x}_i\|_F^2)\right) - \sigma_k^2(1 - \|V^{\mathsf{H}}\boldsymbol{x}_i\|).\end{aligned}$$

Since $\|V^{\mathsf{H}}\boldsymbol{x}_i\|_F^1 = 1$ (because $\boldsymbol{x}_i$ is a unit vector and V is a unitary matrix) and $\sigma_k^2\|V_2^{\mathsf{H}}\boldsymbol{x}_i\|_F^2 - \|D_2V_2^{\mathsf{H}}\boldsymbol{x}_i\|_F^2 \geqslant 0$, it follows that

$$\|A\boldsymbol{x}_i\|_F^2 \leqslant \sigma_k^2 + \left(\|D_1V_1^{\mathsf{H}}\boldsymbol{x}_i\|_F^2 - \sigma_k^2\|V_1^{\mathsf{H}}\boldsymbol{x}_i\|_F^2\right).$$

Consequently,

$$\begin{aligned}\sum_{i=1}^k \|A\boldsymbol{x}_i\|_F^2 &\leqslant k\sigma_k^2 + \sum_{i=1}^k \left(\|D_1V_1^{\mathsf{H}}\boldsymbol{x}_i\|_F^2 - \sigma_k^2\|V_1^{\mathsf{H}}\boldsymbol{x}_i\|_F^2\right) \\ &= k\sigma_k^2 + \sum_{i=1}^k\sum_{j=1}^k(\sigma_j^2 - \sigma_k^2)|\boldsymbol{v}_j^{\mathsf{H}}\boldsymbol{x}_i|^2 \\ &= \sum_{j=1}^k\left(\sigma_k^2 + (\sigma_j^2 - \sigma_k^2)\sum_{i=1}^k|\boldsymbol{v}_j\boldsymbol{x}_i|^2\right) \\ &\leqslant \sum_{j=1}^k(\sigma_k^2 + (\sigma_j^2 - \sigma_k^2)) = \sum_{j=1}^k\sigma_j^2,\end{aligned}$$

which concludes the argument. □

Definition 9.3. Let $A \in \mathbb{C}^{m\times n}$. The *numerical rank* of A is the function $\mathsf{numrank}_A : [0,\infty) \longrightarrow \mathbb{N}$ given by

$$\mathsf{numrank}_A(d) = \min\{\mathsf{rank}(B) \mid \|A - B\|_2 \leqslant d\}$$

for $d \geqslant 0$.

Theorem 9.11. *Let $A \in \mathbb{C}^{m\times n}$ be a matrix having the sequence of non-zero singular values $\sigma_1 \geqslant \sigma_2 \geqslant \cdots \geqslant \sigma_r$. Then $\mathsf{numrank}_A(d) = k < r$ if and only if $\sigma_k > d \geqslant \sigma_{k+1}$.*

Proof. Let d be a number such that $\sigma_k > d \geqslant \sigma_{k+1}$. Equivalently, by the Eckhart–Young Theorem, we have

$$\|A - B(k-1)\|_2 > d \geqslant \|A - B(k)\|_2,$$

Since $\|A - B(k-1)\|_2 = \min\{\|A - X\|_2 \mid \mathsf{rank}(X) = k-1\} > d$, it follows that $\min\{\mathsf{rank}(B) \mid \|A-B\|_2 \leqslant d\} = k$, so $\mathsf{numrank}_A(d) = k$.

Conversely, suppose that $\mathsf{numrank}_A(d) = k$. This means that the minimal rank of a matrix B such that $\|A - B\|_2 \leqslant d$ is k. Therefore, $\|A - B(k-1)\|_2 > d$. On the other hand, $d \geqslant \|A - B(k)\|_2$ because there exists a matrix C of rank k such that $d \geqslant \|A - C\|_2$, so $d \geqslant \|A - B(k)\|_2 = \sigma_{k+1}$. Thus, $\sigma_k > d \geqslant \sigma_{k+1}$. □

Recapitulating facts that we have previously established, we have the following equivalent statements concerning a matrix $A \in \mathbb{R}^{m\times n}$ and its rank $r = \mathsf{rank}(A)$:

(i) $r = \dim(\mathsf{rank}(A))$;
(ii) $r = \dim(\mathsf{rank}(A'))$;
(iii) r is the maximal number of linearly independent rows of A;
(iv) r is the maximal number of linearly independent columns of A;
(v) r is minimal with the property $A = \sum_{i=1}^{r} \boldsymbol{x}_i \boldsymbol{y}_i'$, where $\boldsymbol{x}_i \in \mathbb{R}^m$ and $\boldsymbol{y}_i \in \mathbb{R}^n$;
(vi) r is maximal with the property that there exists a non-singular $r \times r$ submatrix of A;
(vii) r is the number of singular positive values of A.

9.4 Updating SVDs

Efficient algorithms for updating the SVD of matrices are particularly useful when we deal with large matrices that cannot be entirely accommodated in the main memory of computers.

Let $A \in \mathbb{C}^{m\times n}$ be a matrix having the thin SVD decomposition $A = UDV^{\mathsf{H}}$, where $\mathsf{rank}(A) = r$, $U \in \mathbb{C}^{m\times r}$, $D = \mathsf{diag}(\sigma_1, \ldots, \sigma_r)$, and $V \in \mathbb{C}^{n\times r}$. As we observed in Equality (9.2), $U^{\mathsf{H}}U = V^{\mathsf{H}}V = I_r$.

Suppose that A is modified by adding to A a matrix of the form XY^{H}, where $X \in \mathbb{C}^{m\times c}$ and $Y \in \mathbb{C}^{n\times c}$. The matrix

$$B = A + XY^{\mathsf{H}} = (U\ X)\begin{pmatrix} D & O_{r,c} \\ O_{c,r} & I_c \end{pmatrix}(V\ Y)^{\mathsf{H}} \tag{9.4}$$

has a new SVD, $B = U_1 D_1 V_1^{\mathsf{H}}$. We discuss a technique (due to Brand [22]) for producing the factors U_1, D_1, and V_1 of the new SVD starting from the factors of the initial SVD rather than using the matrix A.

Let $P_X \in \mathbb{C}^{m\times b}$ be an orthonormal basis for $\mathsf{range}((I_m - UU^{\mathsf{H}})X)$ and let $R_X = P_X^{\mathsf{H}}(I_m - UU^{\mathsf{H}})X \in \mathbb{C}^{b\times c}$. We have

$$(U\ \ X) = (U\ \ P_X)\begin{pmatrix} I_r & U^{\mathsf{H}}X \\ O_{b,r} & R_X \end{pmatrix}.$$

Similarly, let $P_Y \in \mathbb{C}^{n\times d}$ be an orthonormal basis for $\mathsf{range}((I_n - VV^{\mathsf{H}})Y)$, and let $R_Y = P_Y^{\mathsf{H}}(I_n - VV^{\mathsf{H}})Y \in \mathbb{C}^{d\times c}$. Again, we have

$$(V\ \ Y) = (V\ \ P_Y)\begin{pmatrix} I_r & V^{\mathsf{H}}Y \\ O_{d,r} & R_Y \end{pmatrix}.$$

Thus, the Equality (9.4) can be written as

$$\begin{aligned} B &= (U\ X)\begin{pmatrix} D & O_{r,c} \\ O_{c,r} & I_c \end{pmatrix}(V\ Y)^{\mathsf{H}} \\ &= (U\ \ P_X)\begin{pmatrix} I_r & U^{\mathsf{H}}X \\ O_{b,r} & R_X \end{pmatrix}\begin{pmatrix} D & O_{r,c} \\ O_{c,r} & I_c \end{pmatrix}\begin{pmatrix} I_r & O_{r,d} \\ Y^{\mathsf{H}}V & R_Y^{\mathsf{H}} \end{pmatrix}\begin{pmatrix} Y^{\mathsf{H}} \\ P_Y^{\mathsf{H}} \end{pmatrix}. \end{aligned}$$

Furthermore, we have

$$\begin{aligned} &\begin{pmatrix} I_r & U^{\mathsf{H}}X \\ O_{b,r} & R_X \end{pmatrix}\begin{pmatrix} D & O_{r,c} \\ O_{c,r} & I_c \end{pmatrix}\begin{pmatrix} I_r & O_{r,d} \\ Y^{\mathsf{H}}V & R_Y^{\mathsf{H}} \end{pmatrix} \\ &= \begin{pmatrix} D + U^{\mathsf{H}}XY^{\mathsf{H}}V & U^{\mathsf{H}}XR_Y^{\mathsf{H}} \\ R_XY^{\mathsf{H}}V & R_XR_Y^{\mathsf{H}} \end{pmatrix} \\ &= \begin{pmatrix} D & O_{r,c} \\ O_{c,r} & O_{c,c} \end{pmatrix} + \begin{pmatrix} U^{\mathsf{H}}X \\ R_X \end{pmatrix}\begin{pmatrix} V^{\mathsf{H}}Y \\ R_Y \end{pmatrix}^{\mathsf{H}}. \end{aligned}$$

The matrix

$$K = \begin{pmatrix} D & O_{r,c} \\ O_{c,r} & O_{c,c} \end{pmatrix} + \begin{pmatrix} U^{\mathsf{H}}X \\ R_X \end{pmatrix} \begin{pmatrix} V^{\mathsf{H}}Y \\ R_Y \end{pmatrix}^{\mathsf{H}}$$

is sparse and rather small. By diagonalizing K as $K = WEZ^{\mathsf{H}}$, we obtain

$$\begin{aligned} A + XY^{\mathsf{H}} &= (U \;\; P_X)WEZ^{\mathsf{H}} \begin{pmatrix} Y^{\mathsf{H}} \\ P_Y^{\mathsf{H}} \end{pmatrix} \\ &= (UW \;\; P_X W)E \begin{pmatrix} Z^{\mathsf{H}}V^{\mathsf{H}}Y \\ Z^{\mathsf{H}}P_Y^{\mathsf{H}} \end{pmatrix}. \end{aligned}$$

An important special case occurs when $c = 1$, that is, when X and Y are just vectors in $\mathbb{C}^m$ and $\mathbb{C}^n$, respectively. Thus, the matrix A is modified by adding a rank-1 matrix $\boldsymbol{x}\boldsymbol{y}^{\mathsf{H}}$ as follows:

$$B = A + \boldsymbol{x}\boldsymbol{y}^{\mathsf{H}} = (U \; \boldsymbol{x}) \begin{pmatrix} D & \mathbf{0}_r \\ \mathbf{0}_r' & 1 \end{pmatrix} (V \; \boldsymbol{y})^{\mathsf{H}}.$$

The previous matrices P_X and P_Y are now replaced by the vectors

$$\begin{aligned} \boldsymbol{p_x} &= \frac{1}{\|(I - UU^{\mathsf{H}})\boldsymbol{x}\|}(I - UU^{\mathsf{H}})\boldsymbol{x}, \\ \boldsymbol{p_y} &= \frac{1}{\|(I - VV^{\mathsf{H}})\boldsymbol{y}\|}(I - VV^{\mathsf{H}})\boldsymbol{y}, \end{aligned}$$

and R_X, R_Y are now reduced to two vector numbers:

$$\boldsymbol{r_x} = \boldsymbol{p_x}^{\mathsf{H}}(I - UU^{\mathsf{H}})\boldsymbol{x} \in \mathbb{C}^b \text{ and } \boldsymbol{r_y} = \boldsymbol{p_y}^{\mathsf{H}}(I - VV^{\mathsf{H}})\boldsymbol{y} \in \mathbb{C}^d.$$

This allows us to write

$$(U \;\; \boldsymbol{x}) = (U \;\; \boldsymbol{p_x}) \begin{pmatrix} I_r & U^{\mathsf{H}}\boldsymbol{x} \\ O_{b,r} & \boldsymbol{r_x} \end{pmatrix}$$

and

$$(V \;\; \boldsymbol{y}) = (Y \;\; \boldsymbol{p_y}) \begin{pmatrix} I & V^{\mathsf{H}}\boldsymbol{y} \\ O_{d,r} & \boldsymbol{r_y} \end{pmatrix}.$$

In this special case, Equality (9.4) amounts to

$$\begin{aligned}B &= (U\ \boldsymbol{x})\begin{pmatrix} D & \mathbf{0}_r \\ \mathbf{0}'_r & 1 \end{pmatrix}(V\ \boldsymbol{y})^{\mathsf{H}} \\ &= (U\ \boldsymbol{p}_{\boldsymbol{x}})\begin{pmatrix} I_r & U^{\mathsf{H}}\boldsymbol{x} \\ O_{b,r} & \boldsymbol{r}_{\boldsymbol{x}} \end{pmatrix}\begin{pmatrix} D & \mathbf{0}_r \\ \mathbf{0}'_r & 1 \end{pmatrix}\begin{pmatrix} I_r & O_{r,d} \\ Y^{\mathsf{H}}V & \boldsymbol{r}_{\boldsymbol{y}}^{\mathsf{H}} \end{pmatrix}\begin{pmatrix} Y^{\mathsf{H}} \\ \boldsymbol{p}_{\boldsymbol{y}}^{\mathsf{H}} \end{pmatrix}.\end{aligned}$$

Furthermore, we have

$$\begin{aligned}\begin{pmatrix} I_r & U^{\mathsf{H}}\boldsymbol{x} \\ O_{b,r} & \boldsymbol{r}_{\boldsymbol{x}} \end{pmatrix}\begin{pmatrix} D & \mathbf{0}_r \\ \mathbf{0}'_r & 1 \end{pmatrix}\begin{pmatrix} I_r & O_{r,d} \\ Y^{\mathsf{H}}V & \boldsymbol{r}_{\boldsymbol{y}}^{\mathsf{H}} \end{pmatrix} &= \begin{pmatrix} D + U^{\mathsf{H}}\boldsymbol{x}\boldsymbol{y}^{\mathsf{H}}V & U^{\mathsf{H}}\boldsymbol{x}\boldsymbol{r}_{\boldsymbol{y}}^{\mathsf{H}} \\ \boldsymbol{r}_{\boldsymbol{x}}\boldsymbol{y}^{\mathsf{H}}V & \boldsymbol{r}_{\boldsymbol{x}}\boldsymbol{r}_{\boldsymbol{y}}^{\mathsf{H}} \end{pmatrix} \\ &= \begin{pmatrix} D & \mathbf{0}_r \\ \mathbf{0}'_r & 0 \end{pmatrix} + \begin{pmatrix} U^{\mathsf{H}}\boldsymbol{x} \\ \boldsymbol{r}_{\boldsymbol{x}} \end{pmatrix}\begin{pmatrix} V^{\mathsf{H}}\boldsymbol{y} \\ \boldsymbol{r}_{\boldsymbol{y}} \end{pmatrix}^{\mathsf{H}}.\end{aligned}$$

Finally, by diagonalizing $K = WEZ^{\mathsf{H}}$, we have

$$\begin{aligned}A + \boldsymbol{x}\boldsymbol{y}^{\mathsf{H}} &= (U\ \boldsymbol{p}_{\boldsymbol{x}})WEZ^{\mathsf{H}}\begin{pmatrix} \boldsymbol{y}^{\mathsf{H}} \\ \boldsymbol{p}_{\boldsymbol{y}}^{\mathsf{H}} \end{pmatrix} \\ &= (UW\ \boldsymbol{p}_{\boldsymbol{x}}W)E\begin{pmatrix} Z^{\mathsf{H}}\boldsymbol{y}^{\mathsf{H}} \\ Z^{\mathsf{H}}\boldsymbol{p}_{\boldsymbol{y}}^{\mathsf{H}} \end{pmatrix}.\end{aligned}$$

9.5 Polar Form of Matrices

The SVD allows us to define the *polar form* of a matrix, which as an extension of the polar representation of a complex number $z = |z|e^{i\theta}$, where $|z|$ is the modulus of z and θ, is the argument of z.

Theorem 9.12 (Polar Form Theorem). Let $A \in \mathbb{C}^{m\times n}$ be a matrix with singular values $\sigma_1, \sigma_2, \ldots, \sigma_r$, where $\sigma_1 \geqslant \sigma_2 \geqslant \cdots \geqslant \sigma_r > 0$ and $r \leqslant \min\{m, n\}$.

If $m \leqslant n$, then A can be factored as $A = PY$, where $P \in \mathbb{C}^{m\times m}$ is a positive semidefinite matrix such that $P^2 = AA^{\mathsf{H}}$ and $Y \in \mathbb{C}^{m\times n}$ is a matrix having a set of orthonormal rows (that is, it satisfies the equality $YY^{\mathsf{H}} = I_m$).

If $m \geqslant n$, then A can be factored as $A = WQ$, where $W \in \mathbb{C}^{m\times n}$ has orthonormal columns (that is, $W^{\mathsf{H}}W = I_n$ and $Q \in \mathbb{C}^{n\times n}$ is a positive semidefinite matrix.

Proof. By the Thin SVD Decomposition Corollary (Corollary 9.3), A can be factored as $A = UDV^{\mathsf{H}}$, where $U \in \mathbb{C}^{m\times r}$ and $V \in \mathbb{C}^{n\times r}$ are matrices having orthonormal columns and D is the diagonal matrix

$$D = \begin{pmatrix} \sigma_1 & 0 & \cdots & 0 \\ 0 & \sigma_2 & \cdots & 0 \\ \vdots & \vdots & \cdots & \vdots \\ 0 & 0 & \cdots & \sigma_r \end{pmatrix}.$$

Since U has orthonormal columns, we have $U^{\mathsf{H}}U = I_r$, which allows us to write $A = (UDU^{\mathsf{H}})(UV^{\mathsf{H}})$.

Define $P = UDU^{\mathsf{H}} \in \mathbb{C}^{m\times m}$. It is easy to see that P is a positive semidefinite matrix. Also,

$$P^2 = UDU^{\mathsf{H}}UDU^{\mathsf{H}} = UDDU^{\mathsf{H}} = UDV^{\mathsf{H}}VDU^{\mathsf{H}} = AA^{\mathsf{H}}.$$

On the other hand, if we define $Y = UV^{\mathsf{H}}$, we have $YY^{\mathsf{H}} = UV^{\mathsf{H}}VU^{\mathsf{H}} = I_m$, which means that P and Y satisfy the conditions of the theorem.

To prove the second part of the theorem, we apply the first part to the matrix $A^{\mathsf{H}} \in \mathbb{C}^{n\times m}$. Since $A^{\mathsf{H}} = VD^{\mathsf{H}}U^{\mathsf{H}} = VD'U^{\mathsf{H}}$, the polar decomposition of A^{H} is $A^{\mathsf{H}} = P_1Y_1$, where $P_1 = VD'V^{\mathsf{H}}$ is positive semidefinite and $Y_1 = VU^{\mathsf{H}}$ is a matrix such that $Y_1Y_1^{\mathsf{H}} = I_n$. Thus, $A = Y_1^{\mathsf{H}}P_1^{\mathsf{H}}$, which allows us to define $W = Y_1^{\mathsf{H}}$ and $Q = P_1^{\mathsf{H}}$. Note that $W^{\mathsf{H}}W = Y_1Y_1^{\mathsf{H}} = I_n$ and that Q is positive semidefinite. □

Corollary 9.7. *If $A \in \mathbb{C}^{n\times n}$, then A can be factored as $A = PY = WQ$, where $P, Q \in \mathbb{C}^{n\times n}$ are positive semidefinite matrices, and $Y, W \in \mathbb{C}^{n\times n}$ are unitary matrices.*

Proof. This statement follows immediately from Theorem 9.12. □

Note that the positive semidefinite matrix P in the Polar Form Theorem is uniquely determined by the matrix A. Also, the spectrum of P equals the set of singular values of A.

9.6 CS Decomposition

The cosine–sine decomposition of unitary matrices was introduced in Stewart's paper [158] although it was implicit in an article by Davis and Kahan [33]. As we show in this section and in the next, this factorization of unitary matrices referred to as the *CS decomposition* plays an essential role in the study of the geometry of subspaces.

Theorem 9.13 (Paige–Wei CS Decomposition Theorem). Let $A \in \mathbb{C}^{n\times n}$ be a unitary matrix. For any partitioning

$$A = \begin{pmatrix} A_{11} & A_{12} \\ A_{21} & A_{22} \end{pmatrix},$$

where $A_{11} \in \mathbb{C}^{r_1\times c_1}$, $A_{12} \in \mathbb{C}^{r_1\times c_2}$, $A_{21} \in \mathbb{C}^{r_2\times c_1}$, $A_{22} \in \mathbb{C}^{r_2\times c_2}$, and $r_1+r_2 = c_1+c_2 = n$, there exist unitary matrices U_1, U_2, V_1, V_2 such that

(i) $U = \mathsf{diag}(U_1, U_2)$ and $V = \mathsf{diag}(V_1, V_2)$;

(ii) we have

$$U^{\mathsf{H}}AV = D = \begin{pmatrix} D_{11} & D_{12} \\ D_{21} & D_{22} \end{pmatrix}, \tag{9.5}$$

where $D_{ij} \in \mathbb{R}^{r_i\times c_j}$ are real matrices for $1 \leqslant i, j \leqslant 2$, and

(iii) the matrices D_{ij} have the form

$$D_{11} = \mathsf{diag}(I, C, \tilde{O}), \quad D_{12} = \mathsf{diag}(\hat{O}^{\mathsf{H}}, S, I),$$
$$D_{21} = \mathsf{diag}(\hat{O}, S, I), \quad D_{22} = \mathsf{diag}(I, -C, \tilde{O}^{\mathsf{H}}),$$

where $C = \mathsf{diag}(\gamma_1, \ldots, \gamma_s)$, $S = \mathsf{diag}(\sigma_1, \ldots, \sigma_s)$, $C^2 + S^2 = I$, $1 > \gamma_1 \geqslant \cdots \geqslant \gamma_s > 0$, and $1 > \sigma_s \geqslant \cdots \geqslant \sigma_1 > 0$.

The $\hat{O}$ and $\tilde{O}$ matrices are zero matrices and, depending on A and its partitioning, may have no rows or no columns. Some of the unit matrices could be nonexistent and no two need to be equal. The four C and S submatrices are square matrices with the same format and could be nonexistent.

Proof. Let $A_{11} = U_1 D_{11} V_1^{\mathsf{H}}$ be the SVD of A_{11}. Since D is a unitary matrix, no singular value of D_{11} can exceed 1 and $D_{11} = U_1^{\mathsf{H}} A_{11} V_1$ has the prescribed form $D_{11} = \mathsf{diag}(I, C, \tilde{O})$. We have

$$\begin{aligned} \mathsf{diag}(U_1^{\mathsf{H}}, I) A \mathsf{diag}(V_1, I) &= \begin{pmatrix} U_1^{\mathsf{H}} & O \\ O & I \end{pmatrix} \begin{pmatrix} A_{11} & A_{12} \\ A_{21} & A_{22} \end{pmatrix} \begin{pmatrix} V_1 & O \\ O & I \end{pmatrix} \\ &= \begin{pmatrix} U_1^{\mathsf{H}} A_{11} V_1 & U_1^{\mathsf{H}} A_{12} \\ A_{21} V_1 & A_{22} \end{pmatrix} = B, \end{aligned}$$

where B is the unitary matrix

$$B = \begin{pmatrix} D_{11} & B_{12} \\ B_{21} & B_{22} \end{pmatrix}.$$

By Theorem 6.68, there exist unitary matrices U_2 and V_2 such that the matrix $(U_1^{\mathsf{H}} A_{12})V_2 = B_{12}V_2$ is upper triangular and $U_2^{\mathsf{H}}(A_{21}V_1) = U_2^{\mathsf{H}}B_{21}$ is lower triangular, both having real non-negative elements on their diagonal ending in the bottom right corner. This implies

$$\begin{pmatrix} I & O \\ O & U_2^{\mathsf{H}} \end{pmatrix} \begin{pmatrix} D_{11} & B_{12} \\ B_{21} & B_{22} \end{pmatrix} \begin{pmatrix} I & O \\ O & V_2 \end{pmatrix} = C,$$

where C is the matrix

$$C = \begin{pmatrix} D_{11} & B_{12}V_2 \\ U_2^{\mathsf{H}}B_{21} & U_2^{\mathsf{H}}B_{22}V_2 \end{pmatrix}.$$

By the choice of U_2 and V_2, C has the form

$$C = \left(\begin{array}{ccc|c} I & O & O & \\ O & C & O & C_{12} \\ O & O & \tilde{O} & \\ \hline & C_{21} & & C_{22} \end{array}\right) = \left(\begin{array}{ccc|ccc} I & O & O & O & E_{12} & E_{13} \\ O & C & O & O & S & E_{23} \\ O & O & \tilde{O} & O & O & I \\ \hline \hat{O} & O & O & G_{11} & G_{12} & G_{13} \\ F_{21} & S & O & G_{21} & G_{22} & G_{23} \\ F_{31} & F_{32} & I & G_{31} & G_{32} & G_{33} \end{array}\right).$$

The orthonormality of the first three block columns implies $F_{21}^{\mathsf{H}}S + F_{31}^{\mathsf{H}}F_{32} = O$, $F_{31}^{\mathsf{H}} = O$, and $F_{32} = O$, so F_{21}, F_{31}, and F_{32} are zero matrices. Similarly, the orthonormality of the first three block rows implies that E_{12}, E_{12}, E_{23} are zero matrices. Thus, the matrix becomes

$$C = \left(\begin{array}{ccc|ccc} I & O & O & O & O & O \\ O & C & O & O & S & O \\ O & O & \tilde{O} & O & O & I \\ \hline \hat{O} & O & O & G_{11} & G_{12} & G_{13} \\ O & S & O & G_{21} & G_{22} & G_{23} \\ O & O & I & G_{31} & G_{32} & G_{33} \end{array}\right).$$

Orthogonality of the third column block on the fourth, fifth, and sixth column blocks implies that G_{31}, G_{32}, and G_{33} are zero matrices;

orthogonality of the third row block on the fourth and fifth row blocks implies that G_{13} and G_{23} are zero matrices. Thus, C is actually

$$C = \left(\begin{array}{ccc|ccc} I & O & O & O & O & O \\ O & C & O & O & S & O \\ O & O & \tilde{O} & O & O & I \\ \hline \hat{O} & O & O & G_{11} & G_{12} & O \\ O & S & O & G_{21} & G_{22} & O \\ O & O & I & O & O & O \end{array}\right).$$

Orthogonality of the second and fourth blocks of columns implies $SG_{21} = O$. Since S is non-singular, G_{21} is a zero matrix. Similarly, the orthogonality of the second and fourth blocks of rows shows that G_{12} is a zero matrix.

Orthogonality of the fifth and the second blocks of rows yields $CS + SG_{22} = O$, so $G_{22} = -C$ (because C and S are diagonal matrices). Finally, we obtain $G_{11}G_{11}^{\mathsf{H}} = G_{11}^{\mathsf{H}}G_{11} = I$, so G_{11} is unitary. The matrix G_{11} can be transformed to I without affecting the rest of the matrix by replacing U_2^{H} by $\mathsf{diag}(G_{11}^{\mathsf{H}}, I, I)U_2^{\mathsf{H}}$.

The unitary character of the matrix implies $C^2 + S^2 = I$. □

Corollary 9.8 (The Thin CS Decomposition Theorem). Let $A \in \mathbb{C}^{r \times c}$ be a matrix having orthonormal columns. For any partitioning

$$A = \begin{pmatrix} A_1 \\ A_2 \end{pmatrix},$$

where $A_1 \in \mathbb{C}^{r_1 \times c}$, $A_2 \in \mathbb{C}^{r_2 \times c}$, $r_1 \geqslant c$, $r_2 \geqslant c$, there exist unitary matrices $U_1 \in \mathbb{C}^{r_1 \times r_1}$, $U_2 \in \mathbb{C}^{r_2 \times r_2}$, and $V \in \mathbb{C}^{c \times c}$ such that

(i) $U = \mathsf{diag}(U_1, U_2)$;

(ii) we have

$$U^{\mathsf{H}}AV = D = \begin{pmatrix} C \\ S \end{pmatrix},$$

where $C \in \mathbb{C}^{r_i \times n}$ and $S \in \mathbb{C}^{r_2 \times n}$ are real matrices for $1 \leqslant i, j \leqslant 2$, and

(iii) the matrices C and S have the form

$$C = \mathsf{diag}(\cos\theta_1, \ldots, \cos\theta_n) \text{ and } S = \mathsf{diag}(\sin\theta_1, \ldots, \sin\theta_n),$$

where $0 \leqslant \theta_1 \leqslant \theta_2 \leqslant \cdots \leqslant \theta_n \leqslant \frac{\pi}{2}$.

Proof. This statement follows from the Theorem 9.13 for $c_2 = 0$□

Starting from Equality (9.5),

$$U^{\mathsf{H}}AV = D = \begin{pmatrix} D_{11} & D_{12} \\ D_{21} & D_{22} \end{pmatrix},$$

it is possible to obtain other variants of the CS decomposition by multiplying both sides of the equality by appropriate unitary matrices that transform D to the desired aspect. Recall that all permutation matrices are unitary matrices.

Permuting the first r_1 rows or the last r_2 rows in D or changing the sign of any row or column preserves the almost diagonality of the blocks D_{ij} of D.

Theorem 9.14 (Stewart–Sun CS Decomposition). Let $A \in \mathbb{C}^{n\times n}$ be a unitary matrix. For any partitioning

$$A = \begin{pmatrix} A_{11} & A_{12} \\ A_{21} & A_{22} \end{pmatrix},$$

where $2\ell \leqslant n$, $A_{11} \in \mathbb{C}^{\ell\times\ell}$, $A_{12} \in \mathbb{C}^{\ell\times(n-\ell)}$, $A_{21} \in \mathbb{C}^{(n-\ell)\times\ell}$, $A_{22} \in \mathbb{C}^{(n-\ell)\times(n-\ell)}$ there exist unitary matrices $U, V \in \mathbb{C}^{n\times n}$ such that

(i) $U = \mathsf{diag}(U_1, U_2)$ and $V = \mathsf{diag}(V_1, V_2)$, where $U_1, V_1 \in \mathbb{C}^{\ell\times\ell}$;

(ii) we have

$$U^{\mathsf{H}}AV = \begin{pmatrix} \tilde{C} & -\hat{S} & O \\ \hat{S} & \tilde{C} & O \\ O & O & I_{n-2\ell} \end{pmatrix}, \tag{9.6}$$

where $\tilde{C} = \mathsf{diag}(\gamma_1, \ldots, \gamma_\ell)$, $\hat{S} = \mathsf{diag}(\sigma_1, \ldots, \sigma_\ell)$ are two diagonal matrices having non-negative elements and $\tilde{C}^2 + \hat{S}^2 = I_\ell$.

Proof. Since $2\ell \leqslant n$, we have $\ell = r_1 =\leqslant c_1 \leqslant r_2 = n - \ell$ in Equality (9.5). We can write $\tilde{O} = (\tilde{O}^{\square}, \tilde{\tilde{O}})$ and $\hat{O}^{\mathsf{H}} = (\hat{\hat{O}}, \hat{O}^{\square})$, where

$\tilde{O}^{\square}$ and $\hat{O}^{\square}$ are square matrices. The matrix D can now be written as

$$D = \left(\begin{array}{ccc|ccc} I & O & O & \hat{O}^{\mathsf{H}} & O & O \\ O & C & O & O & S & O \\ O & O & \tilde{O} & O & O & O \\ \hline \hat{O} & O & O & I & O & O \\ O & S & O & O & -C & O \\ O & O & I & O & O & \tilde{O}^{\mathsf{H}} \end{array}\right) = \left(\begin{array}{cccc|cccc} I & O & O & O & \hat{\hat{O}} & \hat{O}^{\square} & O & O \\ O & C & O & O & O & O & S & O \\ O & O & \tilde{O}^{\square} & \tilde{\tilde{O}} & O & O & O & O \\ \hline \hat{\hat{O}}^{\mathsf{H}} & O & O & O & I & O & O & O \\ \hat{O}^{\square\mathsf{H}} & O & O & O & O & I & O & O \\ O & S & O & O & O & O & -C & O \\ O & O & I & O & O & O & O & \tilde{O}^{\square\mathsf{H}} \\ O & O & O & I & O & O & O & \tilde{\tilde{O}}^{\mathsf{H}} \end{array}\right).$$

By multiplying the last two columns of the last matrix by -1 yields the matrix

$$\left(\begin{array}{cccc|cccc} I & O & O & O & \hat{\hat{O}} & \hat{O}^{\square} & O & O \\ O & C & O & O & O & O & -S & O \\ O & O & \tilde{O}^{\square} & \tilde{\tilde{O}} & O & O & O & O \\ \hline \hat{\hat{O}}^{\mathsf{H}} & O & O & O & I & O & O & O \\ \hat{O}^{\square\mathsf{H}} & O & O & O & O & I & O & O \\ O & S & O & O & O & O & C & O \\ O & O & I & O & O & O & O & \tilde{O}^{\square\mathsf{H}} \\ O & O & O & I & O & O & O & \tilde{\tilde{O}}^{\mathsf{H}} \end{array}\right)$$

Let now $\tilde{C}$ and $\hat{S}$ be the square matrices

$$\tilde{C} = \mathsf{diag}(I, C, \tilde{O}^{\square}) \text{ and } \hat{S} = \mathsf{diag}(\hat{O}^{\square}, S, I).$$

Permuting rows and columns within the main block, we obtain another variant of the CS decomposition:

$$U^{\mathsf{H}}AV = \left(\begin{array}{c|ccc} \tilde{C} & O & -\hat{S} & O \\ \hline O & I & O & O \\ \hat{S} & O & \tilde{C} & O \\ O & O & O & I \end{array}\right), \tag{9.7}$$

as shown in [129]. If $r_1 = c_1$, one obtains the matrix

$$U^{\mathsf{H}}AV = \left(\begin{array}{cc|c} \tilde{C} & -\hat{S} & O \\ \hline \hat{S} & \tilde{C} & O \\ O & O & I_{n-2\ell} \end{array}\right). \tag{9.8}$$

□

9.7 Geometry of Subspaces

Many of the results discussed in this section and in the next appear in [160].

Let S and T be two subspaces of $\mathbb{C}^n$ having the same dimension, m, and let $\{\boldsymbol{u}_1, \ldots, \boldsymbol{u}_m\}$ and $\{\boldsymbol{v}_1, \ldots, \boldsymbol{v}_m\}$ be two orthonormal bases for S and T, respectively. Consider the matrices $B = (\boldsymbol{u}_1 \ \cdots \ \boldsymbol{u}_m) \in \mathbb{C}^{n\times m}$ and $C = (\boldsymbol{v}_1 \ \cdots \ \boldsymbol{v}_m) \in \mathbb{C}^{n\times m}$. The projection matrices of these subspaces are $P_S = BB^{\mathsf{H}}$ and $P_T = CC^{\mathsf{H}}$, respectively, and we saw that these Hermitian matrices do not depend on the particular choices of the orthonormal bases of the subspaces.

Theorem 9.15. *Let S and T be two subspaces of $\mathbb{C}^n$ such that $\dim(S) = \dim(T) = m$, and let $\{\boldsymbol{u}_1, \ldots, \boldsymbol{u}_m\}$ and $\{\boldsymbol{v}_1, \ldots, \boldsymbol{v}_m\}$ be two orthonormal bases for S and T, respectively. Consider the matrices $B = (\boldsymbol{u}_1 \ \cdots \ \boldsymbol{u}_m) \in \mathbb{C}^{n\times m}$ and $C = (\boldsymbol{v}_1 \ \cdots \ \boldsymbol{v}_m) \in \mathbb{C}^{n\times m}$.*

The singular values $\gamma_1, \ldots, \gamma_m$ of the matrix $B^{\mathsf{H}}C$ do not depend on the choices of the orthonormal bases in the subspaces S and T.

Proof. Let $\tilde{B}$ and $\tilde{C}$ be matrices that correspond to a different choice of bases in the two spaces. By Theorem 9.2, to prove the independence of the singular values on the choice of bases, it suffices to show that the matrices $B^{\mathsf{H}}C$ and $\tilde{B}^{\mathsf{H}}\tilde{C}$ are unitarily equivalent.

By Theorem 6.46, there exists a unitary matrix Q and a unitary matrix P such that $B^{\mathsf{H}} = Q^{\mathsf{H}}\tilde{B}$ and $C = \tilde{C}P$. This allows us to write $B^{\mathsf{H}}C = Q^{\mathsf{H}}\tilde{B}^{\mathsf{H}}\tilde{C}P$, which shows that $B^{\mathsf{H}}C \equiv_u \tilde{B}^{\mathsf{H}}\tilde{C}$. □

Since B and C have orthonormal columns, we have $B^{\mathsf{H}}B = I_m$ and $C^{\mathsf{H}}C = I_m$, so $\|B\|_2 \leqslant 1$ and $\|\tilde{C}\| \leqslant 1$. This implies $\|B^{\mathsf{H}}C\|_2 \leqslant \|B^{\mathsf{H}}\|_2\|C\|_2 \leqslant 1$, so $\gamma_i \in [0, 1]$.

Definition 9.4. Let S and T be two subspaces of $\mathbb{C}^n$ such that $\dim(S) = \dim(T) = m$ and denote the singular values of the matrix

$B^{\mathsf{H}}C$ by $\gamma_1, \ldots, \gamma_m$, where B and C are matrices that define bases of S and T, respectively. The *angles* between S and T are $\theta_1, \ldots, \theta_m \in \left[0, \frac{\pi}{2}\right]$, where $\cos\theta_i = \gamma_i$ for $1 \leqslant i \leqslant k$.

The squares of the singular values of $B^{\mathsf{H}}C$, $\cos^2\theta_i$ for $1 \leqslant i \leqslant m$, are the eigenvalues of the matrices $C^{\mathsf{H}}BB^{\mathsf{H}}C = C^{\mathsf{H}}P_BC$ or $B^{\mathsf{H}}CC^{\mathsf{H}}B = B^{\mathsf{H}}P_CB$.

If we denote a matrix of a basis of the subspace that is orthogonal to $\mathsf{range}(B)$ by $B_\perp$, then the squares of the singular values of $(B_\perp)^{\mathsf{H}}C$ are the eigenvalues of the matrices $C^{\mathsf{H}}B_\perp B_\perp^{\mathsf{H}}C = C^{\mathsf{H}}(I_n - P_B)C = I_n - C^{\mathsf{H}}P_BC$. These squares correspond to $\sin^2\theta_1, \ldots, \sin^2\theta_n$. Thus, for $S = \mathsf{range}(B)$ and $T = \mathsf{range}(C)$, the vector $\mathsf{SIN}(S,T) = \begin{pmatrix} \sin\theta_1 \\ \vdots \\ \sin\theta_p \end{pmatrix}$ of sinuses of angles between the subspaces S and T consists of the singular values of the matrix $B_\perp^{\mathsf{H}}C$.

The minimal angle between the subspaces S and T is given by $\cos\theta_1 = \sigma_1$.

Theorem 9.16. *Let S and T be two subspaces of $\mathbb{R}^n$. The least angle between S and T is determined by the equality $\cos\theta_1 = \|P_S'P_T\|_2$, where P_S and P_T are the projection matrices of S and T, respectively.*

Proof. We have

$$\cos\theta_1 = \min\{\boldsymbol{u}'\boldsymbol{v} \mid \boldsymbol{u} \in S, \boldsymbol{v} \in T, \|\boldsymbol{u}\|_2 = \|\boldsymbol{v}\|_2 \leqslant 1\}.$$

For $\boldsymbol{u} \in S$ and $\boldsymbol{v} \in T$, we have $P_S\boldsymbol{u} = \boldsymbol{u}$ and $P_T\boldsymbol{v} = \boldsymbol{v}$, which implies

$$\begin{aligned} \cos\theta_1 &= \min\{\boldsymbol{u}'P_S'P_T\boldsymbol{v} \mid \boldsymbol{u} \in S, \boldsymbol{v} \in T, \|\boldsymbol{u}\|_2 = \|\boldsymbol{v}\|_2 \leqslant 1\} \\ &= \|P_S'P_T\|_2. \end{aligned}$$

□

Theorem 9.17. *Let $A \in \mathbb{C}^{n\times\ell}$ and $B \in \mathbb{C}^{n\times\ell}$ be two matrices both having orthonormal columns. If $2\ell \leqslant n$, there are unitary matrices $Q \in \mathbb{C}^{n\times n}$, $U_1 \in \mathbb{C}^{\ell\times\ell}$, and $V_1 \in \mathbb{C}^{\ell\times\ell}$ such that*

$$QAU_1 = \begin{pmatrix} I_\ell \\ O_{\ell,\ell} \\ O_{n-2\ell,\ell} \end{pmatrix} \in \mathbb{C}^{n\times\ell} \text{ and } QBV_1 = \begin{pmatrix} C \\ S \\ O_{n-2\ell,\ell} \end{pmatrix} \in \mathbb{C}^{n\times\ell},$$

where $C = \mathsf{diag}(\gamma_1, \ldots, \gamma_\ell)$, $S = \mathsf{diag}(\sigma_1, \ldots, \sigma_\ell)$, $0 \leqslant \gamma_1 \leqslant \cdots \gamma_\ell$, $1 \geqslant \sigma_1 \geqslant \cdots \geqslant \sigma_\ell \geqslant 0$, and $\gamma_i^2 + \sigma_i^2 = 1$ for $1 \leqslant i \leqslant \ell$.

Proof. Expand the matrices A and B to two unitary matrices $\tilde{A} = (A\ A_1) \in \mathbb{C}^{n\times n}$ and $\tilde{B} = (B\ B_1) \in \mathbb{C}^{n\times n}$, respectively, and define the matrix $D \in \mathbb{C}^{n\times n}$ as

$$D = \tilde{A}^{\mathsf{H}}\tilde{B} = \begin{pmatrix} A^{\mathsf{H}}B & A^{\mathsf{H}}B_1 \\ A_1^{\mathsf{H}}B & A_1^{\mathsf{H}}B_1 \end{pmatrix}.$$

We have $A_1 \in \mathbb{C}^{n\times(n-\ell)}$ and $B_1 \in \mathbb{C}^{n\times(n-\ell)}$. The matrix D is unitary as a product of two unitary matrices, $\tilde{A}^{\mathsf{H}}$ and $\tilde{B}$. By applying the CS decomposition, we obtain the existence of the unitary block diagonal matrices $U = \mathsf{diag}(U_1, U_2) \in \mathbb{C}^{n\times n}$ and $V = \mathsf{diag}(V_1, V_2) \in \mathbb{C}^{n\times n}$ such that $U_1, V_1 \in \mathbb{C}^{\ell\times\ell}$, $U_2, V_2 \in \mathbb{C}^{(n-\ell)\times(n-\ell)}$, and

$$U^{\mathsf{H}}DV = \begin{pmatrix} C & -S & O \\ S & C & O \\ O & O & I_{n-2\ell} \end{pmatrix},$$

where $C, S \in \mathbb{C}^{\ell\times\ell}$. In turn, this yields

$$\begin{aligned} U_1^{\mathsf{H}}A^{\mathsf{H}}BV_1 &= C \in \mathbb{C}^{\ell\times\ell}, \\ U_1^{\mathsf{H}}A^{\mathsf{H}}B_1V_2 &= (-S\ O) \in \mathbb{C}^{\ell\times(n-\ell)}, \\ U_2^{\mathsf{H}}A_1^{\mathsf{H}}BV_1 &= \begin{pmatrix} S \\ O \end{pmatrix} \in \mathbb{C}^{(n-\ell)\times\ell}, \\ U_2^{\mathsf{H}}A_1^{\mathsf{H}}B_1V_2 &= \begin{pmatrix} C & O \\ O & I_{n-2\ell} \end{pmatrix} \in \mathbb{C}^{(n-\ell)\times(n-\ell)}. \end{aligned} \tag{9.9}$$

Let

$$Q = \begin{pmatrix} U_1^{\mathsf{H}}A^{\mathsf{H}} \\ U_2^{\mathsf{H}}A_1^{\mathsf{H}} \end{pmatrix}.$$

We have

$$\begin{aligned} QAU_1 &= \begin{pmatrix} U_1^{\mathsf{H}}A^{\mathsf{H}} \\ U_2^{\mathsf{H}}A_1^{\mathsf{H}} \end{pmatrix} AU_1 = \begin{pmatrix} U_1^{\mathsf{H}}A^{\mathsf{H}}AU_1 \\ U_2^{\mathsf{H}}A_1^{\mathsf{H}}AU_1 \end{pmatrix} \\ &= \begin{pmatrix} I_\ell \\ O_{n-\ell,\ell} \end{pmatrix} = \begin{pmatrix} I_\ell \\ O_{\ell,\ell} \\ O_{n-2\ell,\ell} \end{pmatrix} \in \mathbb{C}^{n\times\ell}, \end{aligned}$$

because A has orthonormal columns and U_1 is unitary, and

$$QBV_1 = \begin{pmatrix} U_1^{\mathsf{H}} A^{\mathsf{H}} \\ U_2^{\mathsf{H}} A_1^{\mathsf{H}} x \\ y \end{pmatrix} BV_1 = \begin{pmatrix} U_1^{\mathsf{H}} A^{\mathsf{H}} BV_1 \\ U_2^{\mathsf{H}} A_1^{\mathsf{H}} BV_1 \end{pmatrix} = \begin{pmatrix} C \\ S \\ O \end{pmatrix} \in \mathbb{C}^{n \times \ell}.$$

□

Further results can be derived from Theorem 9.17 and its proof. Observe that the Equalities (9.9) imply

$$(U_2^{\mathsf{H}} A_1^{\mathsf{H}} B_1 V_2 \; U_2^{\mathsf{H}} A_1^{\mathsf{H}} BV_1) = \begin{pmatrix} C & O & S \\ O & I & O \end{pmatrix}$$

so

$$U_2^{\mathsf{H}} A_1^{\mathsf{H}} (B_1 V_2 \; BV_1) = \begin{pmatrix} C & O & S \\ O & I & O \end{pmatrix} \in \mathbb{C}^{n \times \ell},$$

which is equivalent to

$$\begin{pmatrix} V_2^{\mathsf{H}} B_1^{\mathsf{H}} \\ V_1^{\mathsf{H}} B^{\mathsf{H}} \end{pmatrix} A_1 U_2 = \begin{pmatrix} C' & O \\ O & I \\ S' & O \end{pmatrix} \in \mathbb{C}^{\ell \times n}.$$

Also,

$$\begin{pmatrix} V_2^{\mathsf{H}} B_1^{\mathsf{H}} \\ V_1^{\mathsf{H}} B^{\mathsf{H}} \end{pmatrix} B_1 V_2 = \begin{pmatrix} I_{n-\ell} \\ O_{n,n-\ell} \end{pmatrix} \in \mathbb{C}^{n \times (n-\ell)}$$

because B_1 has orthonormal columns and V_2 is unitary. If $\tilde{Q}$ is the matrix

$$\tilde{Q} = \begin{pmatrix} V_2^{\mathsf{H}} B_1^{\mathsf{H}} \\ V_1^{\mathsf{H}} B^{\mathsf{H}} \end{pmatrix},$$

then we have

$$\tilde{Q} A_1 U_2 == \begin{pmatrix} C' & O \\ O & I \\ S' & O \end{pmatrix} \text{ and } \tilde{Q} B_1 V_2 = \begin{pmatrix} I \\ O \end{pmatrix}.$$

A twin of Theorem 9.17 follows.

Theorem 9.18. *Let $A \in \mathbb{C}^{n\times \ell}$ and $B \in \mathbb{C}^{n\times \ell}$ be two matrices both having orthonormal columns. If $\ell \leqslant n < 2\ell$, there are unitary matrices $Q \in \mathbb{C}^{n\times n}$, $U_1 \in \mathbb{C}^{\ell\times \ell}$, and $V_1 \in \mathbb{C}^{\ell\times \ell}$ such that*

$$QAU_1 = \begin{pmatrix} I_{n-\ell} & O_{n-\ell,2\ell-n} \\ O_{2\ell-n,n-\ell} & I_{2\ell-n} \\ O_{n-\ell,n-\ell} & O_{n-\ell,2\ell-n} \end{pmatrix} \in \mathbb{C}^{n\times \ell}$$

and

$$QBV_1 = \begin{pmatrix} C_1 & O_{n-\ell,2\ell-n} \\ O_{2\ell-n,n-\ell} & I_{2\ell-n} \\ S_1 & O_{n-\ell,2\ell-n} \end{pmatrix} \in \mathbb{C}^{n\times \ell},$$

where $C_1 = \mathsf{diag}(\gamma_1, \ldots, \gamma_{n-\ell})$, $S_1 = \mathsf{diag}(\sigma_1, \ldots, \sigma_{n-\ell})$, $0 \leqslant \gamma_1 \leqslant \cdots \leqslant \gamma_{n-\ell}$, $1 \geqslant \sigma_1 \geqslant \cdots \geqslant \sigma_{n-\ell} \geqslant 0$, and $\gamma_i^2 + \sigma_i^2 = 1$ for $1 \leqslant i \leqslant n-\ell$.

Proof. We are using the same notations as in Theorem 9.17 and the discussion that precedes the current theorem. Since $A \in \mathbb{C}^{n\times \ell}$, $B \in \mathbb{C}^{n\times \ell}$, and $\ell \leqslant n < 2\ell$, there exist two matrices $A_0 \in \mathbb{C}^{n\times n-\ell}$ and $B_0 \in \mathbb{C}^{n-\ell}$ such that the matrices $\tilde{A} = (A_0, A)$ and $\tilde{B} = (B_0, B)$ are both unitary. For $\ell' = n - \ell$ we have $0 \leqslant \ell' < n$. Thus, we can apply Theorem 9.17 to the matrices A_0 and B_0; the roles played by the matrices A_1 and B_1 in the previous proof are played now by A and B, and the roles played by U_1 and V_1 are played by U_2 and V_2, respectively. □

Theorem 9.19. *Let S and T be two subspaces of $\mathbb{R}^n$ such that $\dim(S) = \dim(T) = \ell$ where $2\ell \leqslant n$ and let $\sigma_1, \ldots, \sigma_\ell$ be the sinuses of the angles between S and T. If P_S and P_T are the projections on S and T, then the singular values of the matrix $P_S(I_n - P_T)$ are $\sigma_1, \ldots, \sigma_\ell, 0, \ldots, 0$ and the singular values of the matrix $P_S - P_T$ are $\sigma_1, \sigma_1, \ldots, \sigma_\ell, \sigma_\ell, 0, \ldots, 0$.*

Proof. Let $B = (\boldsymbol{u}_1 \ \cdots \ \boldsymbol{u}_\ell) \in \mathbb{C}^{n\times \ell}$ and $C = (\boldsymbol{v}_1 \ \cdots \ \boldsymbol{v}_\ell) \in \mathbb{C}^{n\times \ell}$ be matrices whose columns are orthonormal bases of S and T, respectively. The projection matrices of these subspaces are the Hermitian matrices $P_S = BB^{\mathsf{H}}$ and $P_T = CC^{\mathsf{H}}$. By Theorem 9.17, there are unitary matrices $Q \in \mathbb{C}^{n\times n}$, $U_1 \in \mathbb{C}^{\ell\times\ell}$, and $V_1 \in \mathbb{C}^{\ell\times\ell}$ such that

$$QBU_1 = \begin{pmatrix} I_\ell \\ O_{\ell,\ell} \\ O_{n-2\ell,\ell} \end{pmatrix} \in \mathbb{C}^{n\times\ell} \text{ and } QCV_1 = \begin{pmatrix} C \\ S \\ O_{n-2\ell,\ell} \end{pmatrix} \in \mathbb{C}^{n\times\ell},$$

where $C = \mathsf{diag}(\gamma_1, \ldots, \gamma_\ell)$, $S = \mathsf{diag}(\sigma_1, \ldots, \sigma_\ell)$, $0 \leqslant \gamma_1 \leqslant \cdots \gamma_\ell$, $1 \geqslant \sigma_1 \geqslant \cdots \geqslant \sigma_\ell \geqslant 0$, and $\gamma_i^2 + \sigma_i^2 = 1$ for $1 \leqslant i \leqslant \ell$. Therefore, we have

$$\begin{aligned} &QP_S(I_n - P_T)Q^{\mathsf{H}} \\ &\quad = QBB^{\mathsf{H}}(I_n - CC^{\mathsf{H}})Q^{\mathsf{H}} \\ &\quad = QBB^{\mathsf{H}}Q^{\mathsf{H}} - QBB^{\mathsf{H}}CC^{\mathsf{H}}Q^{\mathsf{H}} \\ &\quad = (QBU_1)(U_1^{\mathsf{H}}B^{\mathsf{H}}Q^{\mathsf{H}}) - QBU_1U_1^{\mathsf{H}}B^{\mathsf{H}}Q^{\mathsf{H}}(QCV_1)(V_1^{\mathsf{H}}C^{\mathsf{H}}Q^{\mathsf{H}}) \\ &\quad = (QBU_1)(QBU_1)^{\mathsf{H}} - (QBU_1)(QBU_1)^{\mathsf{H}}(QCV_1)(QCV_1)^{\mathsf{H}} \\ &\quad = (QBU_1)(QBU_1)^{\mathsf{H}}(I_n - (QCV_1)(QCV_1)^{\mathsf{H}}). \end{aligned}$$

Since

$$\begin{aligned} (QBU_1)(QBU_1)^{\mathsf{H}} &= \begin{pmatrix} I_\ell \\ O_{\ell,\ell} \\ O_{n-2\ell,\ell} \end{pmatrix} (I_\ell \; O_{\ell,\ell} \; O_{\ell,n-2\ell}) \\ &= \begin{pmatrix} I_\ell & O_{\ell,\ell} & O_{\ell,n-2\ell} \\ O_{\ell,\ell} & O_{\ell,\ell} & O_{\ell,n-2\ell} \\ O_{n-2\ell,\ell} & O_{n-2\ell,\ell} & O_{n-2\ell,n-2\ell} \end{pmatrix} \end{aligned}$$

and

$$\begin{aligned} (QCV_1)(QCV_1)^{\mathsf{H}} &= \begin{pmatrix} C \\ S \\ O_{n-2\ell,\ell} \end{pmatrix} (C \; S \; O_{\ell,n-2\ell}) \\ &= \begin{pmatrix} C^2 & CS & O_{\ell,n-2\ell} \\ SC & S^2 & O_{\ell,n-2\ell} \\ O_{n-2\ell,\ell} & O_{n-2\ell,\ell} & O_{n-2\ell,n-2\ell} \end{pmatrix}, \end{aligned}$$

it follows that

$$QP_S(I_n - P_T)Q^{\mathsf{H}}$$

$$= \begin{pmatrix} I_\ell & O_{\ell,\ell} & O_{\ell,n-2\ell} \\ O_{\ell,\ell} & O_{\ell,\ell} & O_{\ell,n-2\ell} \\ O_{n-2\ell,\ell} & O_{n-2\ell,\ell} & O_{n-2\ell,n-2\ell} \end{pmatrix} \begin{pmatrix} I_\ell - C^2 & -CS & O_{\ell,n-2\ell} \\ -SC & I_\ell - S^2 & O_{\ell,n-2\ell} \\ O_{n-2\ell,\ell} & O_{n-2\ell,\ell} & I_{n-2\ell} \end{pmatrix}$$

$$= \begin{pmatrix} I_\ell - C^2 & -CS & O_{\ell,n-2\ell} \\ O_{\ell,\ell} & O_{\ell,\ell} & O_{\ell,n-2\ell} \\ O_{n-2\ell,\ell} & O_{n-2\ell,\ell} & O_{n-2\ell,n-2\ell} \end{pmatrix} = \begin{pmatrix} S \\ O_{\ell,\ell} \\ O_{n-2\ell,\ell} \end{pmatrix} (S \ \ -C \ O_{\ell,n-2\ell})$$

because $I_\ell - C^2 = S^2$. Since the rows of the matrix $(S \ \ -C \ O_{\ell,n-2\ell})$ are orthonormal, it follows that the singular values of $P_S(I_n - P_T)$ are $\sigma_1, \ldots, \sigma_\ell, 0, \ldots, 0$.

For the matrix $P_S - P_T$, we can write

$$\begin{aligned} Q(P_S - P_T)Q^{\mathsf{H}} &= QBB^{\mathsf{H}}Q^{\mathsf{H}} - QCC^{\mathsf{H}})Q^{\mathsf{H}} \\ &= QBU_1U_1^{\mathsf{H}}B^{\mathsf{H}}Q^{\mathsf{H}} - QCV_1V_1^{\mathsf{H}}C^{\mathsf{H}})Q^{\mathsf{H}} \\ &= (QBU_1)(QBU_1)^{\mathsf{H}} - (QCV_1)(QCV_1)^{\mathsf{H}}. \end{aligned}$$

Substituting QBU_1 and QCV_1 yields the equality

$$\begin{aligned} Q(P_S - P_T)Q^{\mathsf{H}} &= \begin{pmatrix} I_\ell & O_{\ell,\ell} & O_{\ell,n-2\ell} \\ O_{\ell,\ell} & O_{\ell,\ell} & O_{\ell,n-2\ell} \\ O_{n-2\ell,\ell} & O_{n-2\ell,\ell} & O_{n-2\ell,n-2\ell} \end{pmatrix} \\ &\quad - \begin{pmatrix} C^2 & CS & O_{\ell,n-2\ell} \\ SC & S^2 & O_{\ell,n-2\ell} \\ O_{n-2\ell,\ell} & O_{n-2\ell,\ell} & O_{n-2\ell,n-2\ell} \end{pmatrix} \\ &= \begin{pmatrix} I_\ell - C^2 & -CS & O_{\ell,n-2\ell} \\ -SC & -S^2 & O_{\ell,n-2\ell} \\ O_{n-2\ell,\ell} & O_{n-2\ell,\ell} & O_{n-2\ell,n-2\ell} \end{pmatrix} \\ &= \begin{pmatrix} S^2 & -CS & O_{\ell,n-2\ell} \\ -SC & -S^2 & O_{\ell,n-2\ell} \\ O_{n-2\ell,\ell} & O_{n-2\ell,\ell} & O_{n-2\ell,n-2\ell} \end{pmatrix}. \end{aligned}$$

The desired conclusion follows from Supplement 9.9. □

Theorem 9.20. *Let S and T be two subspaces of $\mathbb{R}^n$ such that $\dim(S) = \dim(T) = \ell$ where $\ell \leqslant n < 2\ell$ and let $\sigma_1, \ldots, \sigma_{n-\ell}$ be the sinuses of the angles between S and T. If P_S and P_T are the projections on S and T, then the singular values of the matrix $P_S(I_n - P_T)$ are $\sigma_1, \ldots, \sigma_{n-\ell}, 0, \ldots, 0$ and the singular values of the matrix $P_S - P_T$ are $\sigma_1, \sigma_1, \ldots, \sigma_{n-\ell}, \sigma_{n-\ell}, 0, \ldots, 0$.*

Proof. This statement can be obtained from Theorem 9.18 using an argument similar to the one used in the proof of Theorem 9.19□

9.8 Spectral Resolution of a Matrix

In Theorem 7.3, we have shown that if S is an invariant subspace of a matrix $A \in \mathbb{C}^{n\times n}$ such that $\dim(S) = p$, and $X \in \mathbb{C}^{n\times p}$ is a matrix whose set of columns is a basis of S, then there exists a unique matrix $L \in \mathbb{C}^{p\times p}$ that depends only on the basis S such that $AX = XL$.

Definition 9.5. The *representation of a matrix* $A \in \mathbb{C}^{n\times n}$ relative to a basis $B = \{\boldsymbol{x}_1, \ldots, \boldsymbol{x}_p\}$ of an invariant subspace S with $\dim(S) = p$ is the matrix $L \in \mathbb{C}^{p\times p}$ that satisfies the equality $AX = XL$, where $X = (\boldsymbol{x}_1 \ \cdots \ \boldsymbol{x}_p)$.

Note that if $p = 1$, S is a unidimensional subspace and $L = (\lambda)$ has as its unique entry the eigenvalue of that corresponding to this space.

Lemma 9.3. *Let $A \in \mathbb{C}^{n\times n}$ and let $X \in \mathbb{C}^{n\times p}$ be a full-rank matrix such that $\mathsf{rank}(X) = p \leqslant n$. If $Y \in \mathbb{C}^{n\times(n-p)}$ is a matrix such that $\mathsf{range}(Y) = \mathsf{range}(X)^{\perp}$, then $\mathsf{range}(X)$ is an invariant subspace of A if and only if $Y^{\mathsf{H}}AX = O_{n-p,p}$. In this case, $\mathsf{range}(Y)$ is an invariant subspace of A^{H}.*

Proof. Let X and Y be matrices that satisfy the conditions formulated above. The subspace $\mathsf{range}(X)$ is an invariant subspace of A if and only if $A\boldsymbol{x} \in \mathsf{range}(X)$, which is equivalent to $A\boldsymbol{x} \perp \mathsf{range}(Y)$ for every column $\boldsymbol{x}$ of X. This, in turn, is equivalent to $Y^{\mathsf{H}}A\boldsymbol{x} = \mathbf{0}_{n-p}$ for every column $\boldsymbol{x}$ of X and, therefore, to $Y^{\mathsf{H}}AX = O_{n-p,p}$. The last equality can be written as $X^{\mathsf{H}}A^{\mathsf{H}}Y = O_{p,n-p}$, which implies that $\mathsf{range}(Y)$ is an invariant subspace of A^{H}. □

Theorem 9.21. *Let S be an invariant subspace for a matrix $A \in \mathbb{C}^{n\times n}$ such that* $\dim(S) = p$ *and let $X \in \mathbb{C}^{n\times p}$ be a matrix whose columns constitute an orthonormal basis for S. Let $U = (X\ Y)$ be a unitary matrix, where $Y \in \mathbb{C}^{n\times(n-p)}$. We have*

$$U^{\mathsf{H}}AU = \begin{pmatrix} K & L \\ O_{n-p,p} & M \end{pmatrix}, \tag{9.10}$$

where $K \in \mathbb{C}^{p\times p}$, $L \in \mathbb{C}^{p\times(n-p)}$, and $M \in \mathbb{C}^{(n-p)\times(n-p)}$.

Proof. We have

$$(X\ Y)^{\mathsf{H}}A(X\ Y) = \begin{pmatrix} X^{\mathsf{H}}AX & X^{\mathsf{H}}AY \\ Y^{\mathsf{H}}AX & Y^{\mathsf{H}}AY \end{pmatrix}.$$

Since $(X\ \ Y)$ is a unitary matrix, it is clear that $\mathsf{range}(Y) = \mathsf{range}(X)^{\perp}$. By Lemma 9.3, it follows that $Y^{\mathsf{H}}AX = O_{n-p,p}$. Thus, we can define $K = X^{\mathsf{H}}AX$, $M = Y^{\mathsf{H}}AY$, and $L = X^{\mathsf{H}}AY$. □

The Equality (9.10) is known as the *reduced form of S with respect to the unitary matrix $U = (X\ Y)$.*

By Theorem 7.15, $\mathsf{spec}(A) = \mathsf{spec}(K) \cup \mathsf{spec}(M)$. If $\mathsf{spec}(K) \cap \mathsf{spec}(M) = \emptyset$, we say that $S = \mathsf{range}(X)$ is a *simple invariant space.*

Theorem 9.22. *Let S be a simple invariant subspace for a matrix $A \in \mathbb{C}^{n\times n}$ of rank p that has the reduced form*

$$U^{\mathsf{H}}AU = \begin{pmatrix} K & L \\ O_{n-p,p} & M \end{pmatrix}, \tag{9.11}$$

with respect to the unitary matrix $U = (X\ Y)$, where $X \in \mathbb{C}^{n\times p}$ and $Y \in \mathbb{C}^{n\times(n-p)}$.

There are matrices $W \in \mathbb{C}^{n\times(n-p)}$ and $Z \in \mathbb{C}^{n\times p}$ such that $(X\ W)^{-1} = (Z\ Y)^{\mathsf{H}}$ and

$$A = XKZ^{\mathsf{H}} + WMY^{\mathsf{H}}, \tag{9.12}$$

where $K = Z^{\mathsf{H}}AX \in \mathbb{C}^{p\times p}$ and $M = Y^{\mathsf{H}}AW \in \mathbb{C}^{(n-p)\times(n-p)}$.

Proof. We claim that there exists a matrix $Q \in \mathbb{C}^{p\times(n-p)}$ such that

$$\begin{pmatrix} I_p & -Q \\ O_{n-p,p} & I_{n-p} \end{pmatrix} \begin{pmatrix} K & L \\ O_{n-p,p} & M \end{pmatrix} \begin{pmatrix} I_p & Q \\ O_{n-p,p} & I_{n-p} \end{pmatrix} = \begin{pmatrix} K & O_{p,n-p} \\ O_{n-p,p} & M \end{pmatrix}.$$

Indeed, the above equality is equivalent to

$$\begin{pmatrix} K & KQ - QM + L \\ O_{n-p,p} & M \end{pmatrix} = \begin{pmatrix} K & O_{p,n-p} \\ O_{n-p,p} & M \end{pmatrix}$$

and the matrix Q that satisfies the equality $KQ - QM = -L$ exists by Theorem 8.25, because S is a simple invariant space (so $\mathsf{spec}(K)\cap \mathsf{spec}(M) = \emptyset$). This allows us to write

$$\begin{aligned} \begin{pmatrix} K & O_{p,n-p} \\ O_{n-p,p} & M \end{pmatrix} &= \begin{pmatrix} I_p & -Q \\ O_{n-p,p} & I_{n-p} \end{pmatrix} \begin{pmatrix} X^{\mathsf{H}} \\ Y^{\mathsf{H}} \end{pmatrix} A(X\ Y) \begin{pmatrix} I_p & Q \\ O_{n-p,p} & I_{n-p} \end{pmatrix} \\ &= \begin{pmatrix} X^{\mathsf{H}} - QY^{\mathsf{H}} \\ Y^{\mathsf{H}} \end{pmatrix} A(X\ XQ + Y). \end{aligned}$$

Observe that

$$(X\ XQ+Y)^{-1} = \begin{pmatrix} X^{\mathsf{H}} - QY^{\mathsf{H}} \\ Y^{\mathsf{H}} \end{pmatrix},$$

because

$$(X\ XQ+Y) \begin{pmatrix} X^{\mathsf{H}} - QY^{\mathsf{H}} \\ Y^{\mathsf{H}} \end{pmatrix} = XX^{\mathsf{H}} + YY^{\mathsf{H}} = I_n.$$

This follows from the fact that $(X\ Y)$ is a unitary matrix.

Define $W = XQ + Y$ and $Z = X - YQ^{\mathsf{H}}$. We have $(X\ W)^{-1} = (Z\ Y)^{\mathsf{H}}$; also, $Z^{\mathsf{H}}W = O_{p,n-p}$. This allows us to write

$$A = (X\ XQ+Y) \begin{pmatrix} K & O_{p,n-p} \\ O_{n-p,p} & M \end{pmatrix} \begin{pmatrix} X^{\mathsf{H}} - QY^{\mathsf{H}} \\ Y^{\mathsf{H}} \end{pmatrix}, \tag{9.13}$$

which is equivalent to $A = XKZ^{\mathsf{H}} + WMY^{\mathsf{H}}$. □

Note that we have

$$\begin{pmatrix} Z^{\mathsf{H}} \\ Y^{\mathsf{H}} \end{pmatrix} (X\ W) = \begin{pmatrix} I & O \\ O & I \end{pmatrix},$$

so

$$Z^{\mathsf{H}}X = I_p, Z^{\mathsf{H}}W = O_{p,n-p}, Y^{\mathsf{H}}X = O_{n-p,p}, \text{ and } Y^{\mathsf{H}}W = I_{n-p}. \tag{9.14}$$

Equality (9.12) implies

$$AX = XKZ^{\mathsf{H}}X + WMY^{\mathsf{H}}X = X(KZ^{\mathsf{H}}X),$$

because $Y^{\mathsf{H}}X = O_{n,n}$ and

$$AW = XKZ^{\mathsf{H}}W + WMY^{\mathsf{H}}W = W(MY^{\mathsf{H}}W)$$

because $Z^{\mathsf{H}}W = O_{p,n-p}$. The last equality implies that $\mathsf{range}(W)$ is an invariant space of A. Since $(X\ W)$ is non-singular, $\mathsf{range}(X\ W) = \mathbb{C}^n$.

The Equality (9.12) and its equivalent (9.13) are known as *the spectral resolution equalities of A relative to the subspaces $S = \mathsf{range}(X)$ and $T = \mathsf{range}(W)$*. Note that $\mathsf{range}(W)$ is a complementary invariant subspace of the subspace $\mathsf{range}(X)$.

Let $P_1 = XZ^{\mathsf{H}} \in \mathbb{C}^{n\times n}$ and $P_2 = WY^{\mathsf{H}} \in \mathbb{C}^{n\times n}$. The Equalities (9.14) imply

$$P_1^2 = XZ^{\mathsf{H}}XZ^{\mathsf{H}} = P_1, P_2^2 = WY^{\mathsf{H}}WY^{\mathsf{H}} = P_2$$

and

$$P_1P_2 = XZ^{\mathsf{H}}WY^{\mathsf{H}} = O_{n,n}, P_2P_1 = WY^{\mathsf{H}}XZ^{\mathsf{H}} = O_{n,n}.$$

Also, we have $A = P_1AP_1 + P_2AP_2$.

It is clear that $P_1\boldsymbol{s} \in S$ for $\boldsymbol{s} \in S$ and $P_2\boldsymbol{t} \in T$ for $\boldsymbol{t} \in T$, where $S = \mathsf{range}(X)$ and $T = \mathsf{range}(W)$. Also $P_1\boldsymbol{t} = \mathbf{0}_n$ and $P_2\boldsymbol{s} = \mathbf{0}_n$ for $\boldsymbol{s} \in S$ and $\boldsymbol{t} \in T$. A vector $\boldsymbol{x} \in \mathbb{C}^n$ can be written as $\boldsymbol{x} = \boldsymbol{s}+\boldsymbol{t}$, where $\boldsymbol{s} = P_1\boldsymbol{x}$ and $\boldsymbol{t} = P_2\boldsymbol{x}$.

This justifies referring to P_1 as the *projection of S along T*. Recall that the separation of two matrices was introduced in Equality 8.3.

Lemma 9.4. *Let $K \in \mathbb{C}^{p\times p}$, $H \in \mathbb{C}^{p\times(n-p)}$, $G \in \mathbb{C}^{(n-p)\times p}$, and $M \in \mathbb{C}^{(n-p)\times(n-p)}$ be four matrices such that $\mathsf{spec}(K)\cap\mathsf{spec}(M) = \emptyset$. Define γ, η, and δ as*

$$\gamma = \|G\|_F, \eta = \|H\|_F, \text{ and } \delta = \mathsf{sep}(A, B).$$

If $\gamma\eta < \frac{\delta^2}{4}$, then there exists a unique $X \in \mathbb{C}^{(n-p)\times p}$ such that $S_{M,K}(X) = MX - XK = XHX - G$ and

$$\|X\|_F \leqslant \frac{2\gamma}{\delta + \sqrt{\delta^2 - 4\gamma\eta}} < 2\frac{\gamma}{\delta}.$$

Proof. Since $\mathsf{spec}(K) \cap \mathsf{spec}(M) = \emptyset$, $S_{M,K}$ is invertible. Thus, $\delta = \frac{1}{\|S_{M,K}^{-1}\|}$. Let $\Phi : \mathbb{C}^{(n-p)\times p} \longrightarrow \mathbb{C}^{(n-p)\times p}$ be the function defined by $\Phi(X) = XHX$. Note that $\|\Phi(X)\|_F \leqslant \eta\|X\|_F^2$ and that

$$\|\Phi(X) - \Phi(Y)\|_F \leqslant 2\eta \max\{\|X\|_F, \|Y\|_F\}\|X - Y\|_F,$$

by Supplement 25 of Chapter 6.

Let (X_k) be the sequence of matrices in $\mathbb{C}^{(n-p)\times p}$ defined by $X_0 = O_{n-p,p}$ and

$$X_{k+1} = \boldsymbol{S}_{M,K}^{-1}(X_k H X_k - G). \tag{9.15}$$

Note that $\|X_k H X_k\|_F \leqslant \eta\|X_k\|_F$, so $\delta\|X_{k+1}\|_F \leqslant \gamma + \eta\|X_k\|_F^2$.

Consider the sequence of non-negative numbers $\xi_k = \|X_k\|_F$ for $k \in \mathbb{N}$. We have $\xi_{k+1} \leqslant \frac{\gamma}{\delta} + \frac{\eta}{\delta}\xi_k^2$ for $k \in \mathbb{N}$. Let $f : \mathbb{R} \longrightarrow \mathbb{R}$ be the function defined by $f(x) = \frac{\gamma}{\delta} + \frac{\eta}{\delta}x^2$. This function has a minimum $\frac{\gamma}{\delta}$ for $x = 0$ and $\xi_{k+1} = f(\xi_k)$ for $k \in \mathbb{N}$.

The least fixed point of f (that is, the least root of the equation $f(x) = x$) is

$$x_{min} = \frac{\delta - \sqrt{\delta^2 - 4\gamma\eta}}{2\eta} = \frac{2\gamma}{\delta + \sqrt{\delta^2 - 4\gamma\eta}}.$$

It is clear that $x_{\min} \leqslant \frac{2\gamma}{\delta}$.

We claim that (ξ_k) is an increasing sequence and that x_{min} is an upper bound of this sequence. Note that $\xi_0 = 0 \leqslant x_{min}$. Suppose that $\xi_k \leqslant x_{min}$. Then $x_{k+1} = f(x_k) \leqslant f(x_{min}) = x_{min}$, so (x_k) is an increasing and upper bounded sequence. Therefore, (ξ_k) is convergent. Since $\xi_{k+1} = f(\xi_k)$, it follows that $\lim \xi_k = x_{min}$. Therefore, $\|X_k\|_F \leqslant \frac{2\gamma}{\delta}$ for $k \in \mathbb{N}$.

Let $\rho = \frac{4\gamma\eta}{\delta^2}$. Under our assumptions, we have $\rho < 1$.

The sequence of matrices (X_k) is convergent. Indeed, we have

$$\begin{aligned}\|X_{k+1} - X_k\|_F &\leqslant \frac{1}{\delta} 2\eta \max\{\|X_k\|_F, \|X_{k-1}\|_F\} \|X_k - X_{k-1}\|_F \\ &\leqslant \rho \|X_k - X_{k-1}\|_F.\end{aligned}$$

A repeated application of this inequality leads to

$$\|X_{k+1} - X_k\|_F \leqslant \rho^k \|X_1 - X_0\|_F \leqslant \rho^k \frac{\gamma}{\delta},$$

which implies that (X_k) is a Cauchy sequence. Let $X = \lim_{k\to\infty} X_k$. We have

$$\|X - X_k\|_F \leqslant \sum_{i=k}^{\infty} \|X_{i+1} - X_i\| \leqslant \sum_{i=k}^{\infty} \rho^i \frac{\gamma}{\delta} = \frac{\rho^k}{1-\rho} \frac{\gamma}{\delta}.$$

By taking the limit in both sides of Equality (9.15), we have

$$\boldsymbol{S}_{M,K}(X) = XHX - G.$$

□

Lemma 9.5. *Let $U = (U_1\ U_2) \in \mathbb{C}^{n\times n}$ be a unitary matrix such that $U_1 \in \mathbb{C}^{n\times p}$ and $U_2 \in \mathbb{C}^{n\times(n-p)}$ and let $P \in \mathbb{C}^{(n-p)\times p}$ be a matrix. The matrix $\hat{U} = (\hat{U}_1\ \hat{U}_2)$, defined by*

$$\begin{aligned}\hat{U}_1 &= (U_1 + U_2 P)(I_p + P^{\mathsf{H}} P)^{-\frac{1}{2}} \in \mathbb{C}^{n\times p}, \\ \hat{U}_2 &= (U_2 - U_1 P^{\mathsf{H}})(I_{n-p} + P P^{\mathsf{H}})^{\frac{1}{2}} \in \mathbb{C}^{n\times(n-p)},\end{aligned}$$

is a unitary matrix.

Proof. Since U is a unitary matrix, we have $U^{\mathsf{H}} U = U U^{\mathsf{H}} = I_n$, which amount to $U_1^{\mathsf{H}} U_1 = I_p$, $U_1^{\mathsf{H}} U_2 = O_{p,n-p}$, $U_2^{\mathsf{H}} U_1 = O_{n-p,p}$, and

$U_2^{\mathsf{H}}U_2 = I_{n-p}$. Observe that

$$\begin{aligned}&(U_1^{\mathsf{H}} + P^{\mathsf{H}}U_2^{\mathsf{H}})(U_1 + U_2P)\\ &\quad = U_1^{\mathsf{H}}U_1 + P^{\mathsf{H}}U_2^{\mathsf{H}}U_1 + U_1^{\mathsf{H}}U_2P + P^{\mathsf{H}}U_2^{\mathsf{H}}U_2P = I_p + P^{\mathsf{H}}P.\end{aligned}$$

and

$$\begin{aligned}&(U_1^{\mathsf{H}} + P^{\mathsf{H}}U_2^{\mathsf{H}})(U_2 - U_1P^{\mathsf{H}})\\ &\quad = U_1^{\mathsf{H}}U_2 + P^{\mathsf{H}}U_2^{\mathsf{H}}U_2 - U_1^{\mathsf{H}}U_1P^{\mathsf{H}} - P^{\mathsf{H}}U_2^{\mathsf{H}}U_1P^{\mathsf{H}} = O_{p,n-p}.\end{aligned}$$

Thus,

$$\begin{aligned}\hat{U}_1^{\mathsf{H}}\hat{U}_1 &= (I_p + P^{\mathsf{H}}P)^{-\frac{1}{2}}(U_1^{\mathsf{H}} + P^{\mathsf{H}}U_2^{\mathsf{H}})(U_1 + U_2P)(I_p + P^{\mathsf{H}}P)^{-\frac{1}{2}}\\ &= (I_p + P^{\mathsf{H}}P)^{-\frac{1}{2}}(I_p + P^{\mathsf{H}}P)(I + P^{\mathsf{H}}P)^{-\frac{1}{2}} = I_p.\end{aligned}$$

Similarly,

$$\begin{aligned}\hat{U}_1^{\mathsf{H}}\hat{U}_2 = (I_p + P^{\mathsf{H}}P)^{-\frac{1}{2}}&(U_1^{\mathsf{H}} + P^{\mathsf{H}}U_2^{\mathsf{H}})\cdot\\ &(U_2 - U_1P^{\mathsf{H}})(I_{n-p} + PP^{\mathsf{H}})^{\frac{1}{2}} = O_{p,n-p}.\end{aligned}$$

In the same manner, we obtain $\hat{U}_2^{\mathsf{H}}\hat{U}_1 = O_{n-p,p}$ and $\hat{U}_2^{\mathsf{H}}\hat{U}_2 = I_{n-p}$, which proves that $\hat{U}$ is a unitary matrix. □

Lemma 9.6. *Let $A \in \mathbb{C}^{n\times n}$ and let $U \in \mathbb{C}^{n\times n}$ be a unitary matrix such that $U = (U_1\ U_2)$, $U_1 \in \mathbb{C}^{n\times p}$, $U_2 \in \mathbb{C}^{n\times(n-p)}$,*

$$U^{\mathsf{H}}AU = \begin{pmatrix} K & H \\ G & M \end{pmatrix},$$

where $K \in \mathbb{C}^{p\times p}$, $H \in \mathbb{C}^{p\times(n-p)}$, $G \in \mathbb{C}^{(n-p)\times p}$, $M \in \mathbb{C}^{(n-p)\times(n-p)}$, and $\mathsf{spec}(K) \cap \mathsf{spec}(M) = \emptyset$.

If $P \in \mathbb{C}^{(n-p)\times p}$ is a matrix such that $PK - MP = G - PHP$, then $\mathsf{range}(\hat{U}_1)$ and $\mathsf{range}(\hat{U}_2)$ are invariant subspaces of A and A^{H}, respectively, where $\hat{U}_1$ and $\hat{U}_2$ are

$$\begin{aligned}\hat{U}_1 &= (U_1 + U_2P)(I_p + P^{\mathsf{H}}P)^{-\frac{1}{2}},\\ \hat{U}_2 &= (U_2 - U_1P^{\mathsf{H}})(I_{n-p} + PP^{\mathsf{H}})^{\frac{1}{2}}.\end{aligned}$$

Proof. By Lemma 9.5, the matrix $\hat{U} = (\hat{U}_1 \ \hat{U}_2)$ is unitary. Furthermore, by Lemma 9.3, $\mathsf{range}(\hat{U}_1)$ and $\mathsf{range}(\hat{U}_2)$ are invariant subspaces of A and A^{H}, respectively, if and only if $\hat{U}_2^{\mathsf{H}} A \hat{U}_1 = O_{n-p,p}$, which is equivalent to

$$\begin{aligned}
&(I_{n-p} + PP^{\mathsf{H}})^{\frac{1}{2}}(U_2^{\mathsf{H}} - PU_1^{\mathsf{H}})A(U_1 + U_2P)(I_p + P^{\mathsf{H}}P)^{-\frac{1}{2}} \\
&\quad = (I_{n-p} + PP^{\mathsf{H}})^{\frac{1}{2}}(U_2^{\mathsf{H}} - PU_1^{\mathsf{H}})A(U_1 + U_2P)(I_p + P^{\mathsf{H}}P)^{-\frac{1}{2}} \\
&\quad = (I_{n-p} + PP^{\mathsf{H}})^{\frac{1}{2}}(G - PK + MP - PHP)(I_p + P^{\mathsf{H}}P)^{-\frac{1}{2}} \\
&\quad = O_{n-p,p},
\end{aligned}$$

because P satisfies the condition stated in the theorem. □

The angles between the subspace $\mathsf{range}(U_1)$ and the invariant subspace $\mathsf{range}(\hat{U}_1)$ are determined by the singular values of the matrix $U_2^{\mathsf{H}}\hat{U}_1$. By the definition of $\hat{U}_1$, we have

$$\begin{aligned}
U_2^{\mathsf{H}}\hat{U}_1 &= U_2^{\mathsf{H}}(U_1 + U_2P)(I_p + P^{\mathsf{H}}P)^{-\frac{1}{2}}, \\
&= P(I_p + P^{\mathsf{H}}P)^{-\frac{1}{2}}.
\end{aligned}$$

If $\pi_1, \pi_2, \ldots$ are the singular values of P, then the singular values of $P(I_p + P^{\mathsf{H}}P)^{-\frac{1}{2}}$ are $\frac{\pi_1}{\sqrt{1+\pi_1^2}}, \frac{\pi_2}{\sqrt{1+\pi_2^2}}, \ldots$, so we have $\sin\theta_i = \frac{\pi_i}{\sqrt{1+\pi_i^2}}$. Thus, $\sin^2\theta_i = \frac{\pi_i^2}{1+\pi_i^2}$, which implies $\pi_i = \tan\theta_i$.

Theorem 9.23. *Let $A \in \mathbb{C}^{n\times n}$ and let $U \in \mathbb{C}^{n\times n}$ be a unitary matrix such that*

$$U^{\mathsf{H}}AU = \begin{pmatrix} K & H \\ G & M \end{pmatrix},$$

where $K \in \mathbb{C}^{p\times p}$, $H \in \mathbb{C}^{p\times(n-p)}$, $G \in \mathbb{C}^{(n-p)\times p}$, $M \in \mathbb{C}^{(n-p)\times(n-p)}$, and $\mathsf{spec}(K) \cap \mathsf{spec}(M) = \emptyset$. Define $\gamma = \|G\|$, $\eta = \|H\|$, and $\delta = \mathsf{sep}(K, M)$ and assume that $\gamma\eta < \frac{1}{4}\delta^2$.

Let $P \in \mathbb{C}^{(n-p)\times p}$ be the unique solution of the equation $\boldsymbol{S}_{K,M}(X) = XHX - G$ such that

$$\|P\| \leqslant \frac{2\gamma}{\delta + \sqrt{\delta^2 - 4\gamma\eta}} < 2\frac{\gamma}{\delta},$$

and let $U = (U_1 \ U_2)$, where $U_1 \in \mathbb{C}^{n\times p}$ and $U_2 \in \mathbb{C}^{n\times(n-p)}$.

If $\hat{U}_1$ and $\hat{U}_2$ are defined as

$$\hat{U}_1 = (U_1 + U_2 P)(I + P^{\mathsf{H}} P)^{-\frac{1}{2}},$$
$$\hat{U}_2 = (U_2 - U_1 P^{\mathsf{H}})(I + P^{\mathsf{H}} P)^{-\frac{1}{2}},$$

then range$(\hat{U}_1)$ *and* range$(\hat{U}_2)$ *are right and left invariant spaces of A, respectively. Furthermore, the matrices that represent A relative to these subspaces are*

$$L_1 = (I + P^{\mathsf{H}} P)^{\frac{1}{2}}(K + HP)(I + P^{\mathsf{H}} P)^{-\frac{1}{2}},$$
$$L_2 = (I + PP^{\mathsf{H}})^{-\frac{1}{2}}(M - PH)(I + PP^{\mathsf{H}})^{\frac{1}{2}},$$

respectively.

Proof. The existence of the matrix P follows immediately from Lemma 9.4. By Lemma 9.5, the matrix $\hat{U} = (\hat{U}_1\ \hat{U}_2)$ is a unitary matrix, the subspace range$(\hat{U}_1)$ is a right invariant subspace of A, and range$(\hat{U}_2)$ are left invariant subspaces of A. We have

$$\hat{U}_1^{\mathsf{H}} A \hat{U}_1 = (I + P^{\mathsf{H}} P)^{-\frac{1}{2}}(U_1^{\mathsf{H}} + P^{\mathsf{H}} U_2^{\mathsf{H}}) A (U_1 + U_2 P)(I + P^{\mathsf{H}} P)^{-\frac{1}{2}}.$$

and

$$\begin{aligned}
&(U_1^{\mathsf{H}} + P^{\mathsf{H}} U_2^{\mathsf{H}}) A (U_1 + U_2 P)\\
&\quad = (U_1^{\mathsf{H}} + P^{\mathsf{H}} U_2^{\mathsf{H}}) U \begin{pmatrix} K & H \\ G & M \end{pmatrix} U^{\mathsf{H}} (U_1 + U_2 P)\\
&\quad = (U_1^{\mathsf{H}} + P^{\mathsf{H}} U_2^{\mathsf{H}})(U_1\ U_2) \begin{pmatrix} K & H \\ G & M \end{pmatrix} \begin{pmatrix} U_1^{\mathsf{H}} \\ U_2^{\mathsf{H}} \end{pmatrix} (U_1 + U_2 P)\\
&\quad = (I_p\ P^{\mathsf{H}}) \begin{pmatrix} K & H \\ G & M \end{pmatrix} \begin{pmatrix} I_p \\ P \end{pmatrix}\\
&\quad = K + HP + P^{\mathsf{H}} G + P^{\mathsf{H}} M P.
\end{aligned}$$

In view of the definition of P, we have $MP = PK + PHP - G$ and this implies

$$\hat{U}_1^{\mathsf{H}} A \hat{U}_1 = (I + P^{\mathsf{H}} P)^{-\frac{1}{2}}(U_1^{\mathsf{H}} + P^{\mathsf{H}} U_2^{\mathsf{H}}) A (U_1 + U_2 P)(I + P^{\mathsf{H}} P)^{-\frac{1}{2}}$$

$$= (I + P^{\mathsf{H}}P)^{-\frac{1}{2}}(K + HP + P^{\mathsf{H}}G + P^{\mathsf{H}}PK$$
$$+ P^{\mathsf{H}}PHP - P^{\mathsf{H}}G)(I + P^{\mathsf{H}}P)^{-\frac{1}{2}}$$
$$= (I + P^{\mathsf{H}}P)^{\frac{1}{2}}(K + HP)(I + P^{\mathsf{H}}P)^{-\frac{1}{2}}.$$

The proof of the second equality is similar. □

Let S and T be two subspaces of $\mathbb{R}^n$ such that $\dim(S) = \dim(T) = p$ and $S = \mathsf{range}(X_1)$, $T = \mathsf{range}(Z)$, where $X_1, Z \in \mathbb{C}^{n\times p}$ are matrices of rank p having orthonormal columns. Next, we give an upper bound for the norm of $\mathsf{SIN}(S, T)$.

Theorem 9.24 (Davis–Kahan SIN Theorem). Let $U = (X_1\ X_2)$ be a unitary matrix such that $X_1 \in \mathbb{C}^{n\times p}$ and $X_2 \in \mathbb{C}^{n\times(n-p)}$, and let $A \in \mathbb{C}^{n\times n}$ be a Hermitian matrix defined by

$$A = (X_1\ X_2)\begin{pmatrix} L_1 & O_{p,n-p} \\ O_{n-p,p} & L_2 \end{pmatrix}\begin{pmatrix} X_1^{\mathsf{H}} \\ X_2^{\mathsf{H}} \end{pmatrix},$$

where $L_1 \in \mathbb{C}^{p\times p}$ and $L_2 \in \mathbb{C}^{(n-p)\times(n-p)}$.

Let $Z \in \mathbb{C}^{n\times p}$ be a matrix with orthonormal columns (that is, $Z^{\mathsf{H}}Z = I_p$) and let $\theta_1, \ldots, \theta_p$ be the angles between the p-dimensional subspaces $S = \mathsf{range}(X_1)$ and $T = \mathsf{range}(Z)$.

Define the function $\zeta_{A,Z}$ on the set of Hermitian matrices of order p as $\zeta_{A,Z}(M) = \boldsymbol{S}_{A,M}(Z) = AZ - ZM$ for every Hermitian matrix $M \in \mathbb{C}^{p\times p}$. Clearly, we have $\zeta_{A,Z}(M) \in \mathbb{C}^{n\times p}$.

Let $\delta(L_2, M) = \min\{|\lambda - \mu|, \lambda \in \mathsf{spec}(L_2), \mu \in \mathsf{spec}(M)\}$. If $\delta(L_2, M) > 0$, then

$$\|\mathsf{SIN}(S, T)\|_F = \|X_2^{\mathsf{H}}Z\|_F \leqslant \frac{\|\zeta_{A,Z}(M)\|_F}{\delta(L_2, M)}.$$

Proof. Since $U = (X_1\ X_2)$ is a unitary matrix, the subspaces $S = \mathsf{range}(X_1)$ and $S^{\perp} = \mathsf{range}(X_2)$ are orthogonal. The equality

$$A = X_1L_1X_1^{\mathsf{H}} + X_2L_2X_2^{\mathsf{H}}.$$

implies

$$\zeta_{A,Z}(M) = AZ - ZM = X_1L_1X_1^{\mathsf{H}}Z + X_2L_2X_2^{\mathsf{H}}Z - ZM.$$

Therefore,

$$\begin{aligned} X_2^{\mathsf{H}}\zeta_{A,Z}(M) &= X_2^{\mathsf{H}}X_1L_1X_1^{\mathsf{H}}Z + X_2^{\mathsf{H}}X_2L_2X_2^{\mathsf{H}}Z - X_2^{\mathsf{H}}ZM \\ &= L_2X_2^{\mathsf{H}}Z - X_2^{\mathsf{H}}ZM = \boldsymbol{S}_{L_2,M}(X_2^{\mathsf{H}}Z), \end{aligned}$$

because $X_2^{\mathsf{H}}X_1 = O_{n-p,p}$ and $X_2^{\mathsf{H}}X_2 = I_{n-p}$.

By Supplement 9.9, we have

$$\begin{aligned} \delta(L_2, M) &= \min\{|\lambda - \mu|, \lambda \in \mathsf{spec}(L_2), \mu \in \mathsf{spec}(M)\} \\ &= \mathsf{sep}_F(L_2, M) = \min\{\|\boldsymbol{S}_{L_2,M}(X)\| \mid \|X\| = 1\}, \end{aligned}$$

for every X such that $\|X\|_F = 1$. Choosing $X = \frac{1}{\|X_2^{\mathsf{H}}Z\|_F}X_2^{\mathsf{H}}Z$, it follows that

$$\|X_2^{\mathsf{H}}Z\|_F \leqslant \frac{\|\boldsymbol{S}_{L_2,M}(X_2^{\mathsf{H}}Z)\|_F}{\delta(L_2, M)} = \frac{\|X_2^{\mathsf{H}}\zeta_{A,Z}(M)\|_F}{\delta(L_2, M)} \leqslant \frac{\|\zeta_{A,Z}(M)\|_F}{\delta(L_2, M)},$$

which gives the desired inequality. □

9.9 MATLAB Computations

The singular value decomposition of a matrix can be computed in MATLAB using the function **svd**. To illustrate its use in the simplest form, consider the matrix

$$A = \begin{pmatrix} \frac{1}{3} & \frac{4}{3} & \frac{5}{3} \\ \frac{11}{5} & \frac{4}{5} & 3 \\ \frac{1}{7} & \frac{4}{7} & \frac{5}{7} \\ \frac{1}{9} & \frac{4}{9} & \frac{5}{9} \end{pmatrix},$$

defined in MATLAB by

```
A = [1/3 4/3 5/3 ; 11/5 4/5 3 ; 1/7 4/7 5/7 ; 1/9 4/9 5/9]
```

A call the `svd(A)` function yields a vector containing the singular values of A, as in

```
>> svd(A)
ans =
     4.3674
     1.2034
     0.0000
```

It is interesting to note that $\mathsf{rank}(A) = 2$ since the last column of A is the sum of the first two columns. Thus, we would expect to see two non-zero singular values.

To compute $\|A\|_2$, which equals the largest singular value of A, we can use `max(svd(A))`.

MATLAB has the function `rank` that computes the numerical rank of a matrix $\mathsf{numrank}_A(d)$. By default, the value of d is set to a value known as *tolerance*, which is `max(size(A)*eps(max(s))`, where s is the vector of the singular values of A. Otherwise, we can enter our own tolerance `tol` level using a call `rank(A,tol)`.

Small variation in the values of the elements of a matrix can trigger sudden changes in rank. These variations may occur due to rounding errors as rational numbers (such as $\frac{1}{3}$) may be affected by these rounding errors.

Example 9.4. When we enter the matrix A considered above, the system responds with

```
A =
    0.3333    1.3333    1.6667
    2.2000    0.8000    3.0000
    0.1429    0.5714    0.7143
    0.1111    0.4444    0.5556
```

Calls of the rank function with several levels of tolerance return results such as

```
>> rank(A,10^(-9))
ans =
     2
>> rank(A,10^(-15))
ans =
     2
>> rank(A,10^(-16))
ans =
     2
>> rank(A,10^(-17))
ans =
     3
```

If d of $\mathsf{numrank}_A(d)$ is sufficiently low, the numerical rank of the representation of A (affected by rounding errors) becomes 3.

Another variant of the **svd** function, `[U,S,V] = svd(A)`, yields a diagonal matrix S, of the same format as A and with nonnegative diagonal elements in decreasing order, and unitary matrices U and V so that $A = USV^{\mathsf{H}}$. For the matrix A shown above, we obtain

```
>> [U,S,V] = svd(A)
U =
   -0.4487    0.7557    0.4769    0.0161
   -0.8599   -0.5105    0.0000   -0.0000
   -0.1923    0.3239   -0.6726   -0.6370
   -0.1496    0.2519   -0.5658    0.7707
S =
    4.3674         0         0
         0    1.2034         0
         0         0    0.0000
         0         0         0
V =

   -0.4775   -0.6623    0.5774
   -0.3349    0.7447    0.5774
   -0.8123    0.0823   -0.5774
```

The "economical form" of the **svd** function is

`[U,S,V] = svd(A,'econ')`

If $A \in \mathbb{R}^{m\times n}$ and $m > n$, only the first n columns of U are computed and $S \in \mathbb{R}^{n\times n}$. If $m < n$, only the first m columns of V are computed.

Example 9.5. Starting from the matrix

$$A = \begin{pmatrix} 18 & 8 & 20 \\ -4 & 20 & 1 \\ 25 & 8 & 27 \\ 9 & 4 & 10 \end{pmatrix} \in \mathbb{R}^{4\times 3}$$

a call to the economical variant of the **svd** function yields

```
>> [U,D,V] = svd(A,'econ')
U =
   -0.5717   -0.0211    0.8095
   -0.0721   -0.9933   -0.0669
```

```
    -0.7656    0.1133   -0.4685
    -0.2859   -0.0105   -0.3474


S =
    49.0923         0         0
          0   20.2471         0
          0         0    0.0000


V =
    -0.6461    0.3127    0.6963
    -0.2706   -0.9468    0.1741
    -0.7137    0.0760   -0.6963
```

Example 9.6. The function svapprox given in what follows computes the successive approximations $B(k) = \sum_{i=1}^{k} \sigma_i \boldsymbol{u}_i \boldsymbol{v}_i^{\mathsf{H}}$ of a matrix $A \in \mathbb{R}^{m \times n}$ having the SVD $A = UDV^{\mathsf{H}}$ and produces a three-dimensional array $C \in \mathbb{R}^{m \times n \times r}$, where r is the numerical rank of A and $C(:,:,k) = B(k)$ for $1 \leqslant k \leqslant r$.

```
function [C] = svapprox(A)
%SVAPPROX computes the successive approximations
%      of A using the singular component decomposition.
%      The number of approximations equals the
%          numerical rank of A.

% determine the format of A and its numerical rank
[m, n] = size(A);
r = rank(A,10^(-5));
% compute the SVD of A
[U,D,V] = svd(A);
C = zeros(m,n,r);
C(:,:,1) = D(1,1) * U(:,1) * (V(:,1))';
for k=2:r
    C(:,:,k) = D(k,k) * U(:,k) * (V(:,k))' + C(:,:,k-1);
end;
```

This function is used in Example 9.7.

Example 9.7. In Figure 4.10(a), we have an image of the digit 4 created from a pgm file that contains the representation of this digit, as discussed in Example 4.56.

The numerical rank of the matrix A introduced in the example mentioned above is 8. Therefore, the array C computed by

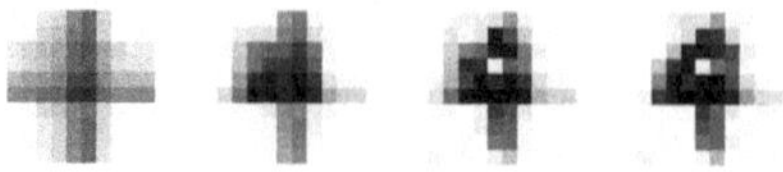

(a) (b) (c) (d)

Fig. 9.1 Successive approximations of A.

`C = svapprox(A)` consists of 8 matrices. To represent these matrices in the `pgm` format, we cast the components of C to integers of the type `uint8` using `D = min(16,uint8(C))`. Thus, $D(:,:,j)$ contains the rounded jth approximation of A and we represent the images for the first four approximations in Figure 9.1. Note that the digit four is easily recognizable beginning with the second approximation. The use of SVD in approximative recognition of digits is discussed extensively in [44].

Exercises and Supplements

(1) Prove that if all singular values of a matrix $A \in \mathbb{C}^{n\times n}$ are equal, then there exists a unitary matrix $U \in \mathbb{C}^{n\times n}$ and a real number a such that $A = aU$.

(2) Prove that if $A \in \mathbb{C}^{m\times n}$ has an orthonormal set of columns (an orthonormal set of rows), then its singular values equal 0 or 1.

(3) Let $c_1, \ldots, c_\ell, s_1, \ldots, s_\ell$ be 2ℓ real numbers such that $s_i^2 + c_i^2 = 1$ and $s_i \geqslant 0$ for $1 \leqslant i \leqslant n$. Prove that the singular values of the matrix

$$X = \begin{pmatrix} \mathsf{diag}(s_1^2, \ldots, s_\ell^2) & \mathsf{diag}(-c_1 s_1, \ldots, -c_\ell s_\ell) \\ \mathsf{diag}(-c_1 s_1, \ldots, -c_\ell s_\ell) & -\mathsf{diag}(s_1^2, \ldots, s_\ell^2) \end{pmatrix}$$

are $s_1, s_1, \ldots, s_\ell, s_\ell$.

Solution: Note that

$$XX' = \begin{pmatrix} E & O \\ O & E \end{pmatrix},$$

where

$$\begin{aligned} E &= \mathsf{diag}(s_1^2, \ldots, s_\ell^2)^2 + \mathsf{diag}(-c_1 s_1, \ldots, -c_\ell s_\ell)^2 \\ &= \mathsf{diag}(s_1^4 + c_1^2 s_1^2, \ldots, s_\ell^4 + c_\ell^2 s_\ell^2) \\ &= \mathsf{diag}(s_1^2, \ldots, s_\ell^2). \end{aligned}$$

Thus, XX' has the eigenvalues $s_1^2, \ldots, s_\ell^2$ each with algebraic multiplicity 2, which implies that X has the eigenvalues $s_1, s_1, \ldots, s_\ell, s_\ell$.

(4) Prove that if P_ϕ is a permutation matrix, where $\phi \in PERM_m$, then the set of singular values of the matrix $P_\phi A$ is the same as the set of singular values of the matrix A for every $A \in \mathbb{C}^{m\times n}$.

(5) Prove that $0 \in \mathsf{spec}(A)$, where $A \in \mathbb{C}^{n\times n}$ if and only if 0 is a singular value of A.

(6) Compute the singular values of the matrices

$$A = \begin{pmatrix} 1 & 0 \\ 0 & 0 \end{pmatrix}, B = \begin{pmatrix} 0 & 1 \\ 0 & 0 \end{pmatrix}, C = \begin{pmatrix} 0 & 0 \\ 0 & 1 \end{pmatrix}, D = \begin{pmatrix} 0 & 0 \\ 1 & 0 \end{pmatrix}.$$

Let X, Y be any two of these four matrices. Verify that XY and YX have 0 as an eigenvalue with multiplicity 2. However, the sets of singular values of each of the matrices in the pairs $(AB, BA), (AD, DA), (BC, CB)$, and (CD, DC) are distinct.

(7) Let $A \in \mathbb{C}^{n\times n}$ be a Hermitian matrix, and let $A = U^{\mathsf{H}}DU$ be the decomposition given by the spectral theorem for Hermitian matrices.

   (a) Prove that if $\mathsf{spec}(A) \subseteq \mathbb{R}_{\geqslant 0}$, then $A = U^{\mathsf{H}}DU$ is an SVD of A.

   (b) If $\mathsf{spec}(A) = \{\lambda_1, \ldots, \lambda_n\} \not\subseteq \mathbb{R}_{\geqslant 0}$, let S be diagonal matrix such that $s_{ii} = 1$ if $\lambda_i \geqslant 0$ and $s_{ii} = -1$ if $\lambda_i < 0$. Prove that $A = (U^{\mathsf{H}}S)(SD)U$ is an SVD of A.

(8) Let $A \in \mathbb{C}^{n\times n}$ be a Hermitian matrix such that $\mathsf{spec}(A) = \{\lambda_1, \ldots, \lambda_n\} \subseteq \mathbb{R}$. Prove that the singular values of A are $|\lambda_1|, \ldots, |\lambda_n|$.

   Solution: The singular values of A equal the square roots of the non-zero eigenvalues of the matrix $A^{\mathsf{H}}A$. Since A is Hermitian, these eigenvalues equal the non-zero eigenvalues of A^2, which means that the singular values of A are $\sqrt{\lambda_1^2}, \ldots, \sqrt{\lambda_n^2}$, that is, $|\lambda_1|, \ldots, |\lambda_n|$.

(9) Next we give a stronger form of Corollary 8.20 that holds for Hermitian matrices. Let $A \in \mathbb{C}^{n\times n}$ and $B \in \mathbb{C}^{m\times m}$ be two matrices. Prove that if A and B are Hermitian, then $\mathsf{sep}_F(A, B) = \min\{|\lambda - \mu| \mid \lambda \in \mathsf{spec}(A), \mu \in \mathsf{spec}(B)\}$.

Solution: The Sylvester operator $\boldsymbol{S}_{A,B}(X) = AX - XB$ can be written as Kronecker operations as

$$\begin{aligned}\mathsf{vec}(\boldsymbol{S}_{A,B}(X)) &= \mathsf{vec}(AX) - \mathsf{vec}(XB)\\ &= (I_n \otimes A)\mathsf{vec}(X) - (B \otimes I_m)\mathsf{vec}(X)\\ &= (I_n \otimes A - B \otimes I_m)\mathsf{vec}(X) = (A \ominus B)\mathsf{vec}(X)\end{aligned}$$

by applying the identities of Supplement 90 of Chapter 3. Since both A and B are Hermitian, by Supplement 83 of the same chapter, the matrix $A \ominus B$ is Hermitian and, by Theorem 7.25, $\mathsf{spec}(A \ominus B) = \{\lambda - \mu \mid \lambda \in \mathsf{spec}(A), \mu \in \mathsf{spec}(B)\}$. Also, note that $\|\mathsf{vec}(X)\|_2 = \|X\|_F$, so $\|\boldsymbol{S}_{A,B}(X)\|_F = \|(A \ominus B)\mathsf{vec}(X)\|_2$. Since $\mathsf{sep}_F(A, B) = \min\{\|\boldsymbol{S}_{A,B}(X)\|_F \mid \|X\|_F = 1\} = \min\{\|(A \ominus B)\mathsf{vec}(X)\|_2 \mid \|\mathsf{vec}X\|_F = 1$, it follows that $\mathsf{sep}_F(A, B)$ equals the smallest singular value of the Hermitian matrix $A \ominus B$. By Corollary 9.2 and by supplement 9.9, this equals $\min\{|\lambda - \mu| \mid \lambda \in \mathsf{spec}(A), \mu \in \mathsf{spec}(B)\}$.

(10) Let $A \in \mathbb{C}^{m\times n}$ be a matrix having full rank r and let $\sigma_1 \geqslant \cdots \geqslant \sigma_r > 0$ be its non-zero singular values. Prove that if $B \in \mathbb{C}^{m\times n}$ is a matrix such that $\|A - B\|_2 < \sigma_r$, then B has full rank.

(11) Let $A \in \mathbb{C}^{m\times n}$ be a matrix. Prove that
   (a) for every $\epsilon > 0$ there exists a matrix B having distinct singular values such that $\|A - B\|_F < \epsilon$;
   (b) if $\mathsf{rank}(A) = r < \min\{m, n\}$, then for every $\epsilon > 0$ there exists a matrix $B \in \mathbb{R}^{m\times n}$ having full rank such that $\|A - B\|_F \leqslant \epsilon$.

Solution: We discuss only the first part of this Supplement. Let $A = UDV^{\mathsf{H}}$ be an SVD of the matrix A, where $D = \mathsf{diag}(\sigma_1, \ldots, \sigma_p) \in \mathbb{C}^{m\times n}$ and $\sigma_1 \geqslant \sigma_2 \geqslant \cdots \geqslant \sigma_p$. Suppose that $m \geqslant n$ and consider the matrix

$$D_\delta = \mathsf{diag}(d_{11} + \delta, d_{22} + 2\delta, \ldots, d_{nn} + n\delta) \in \mathbb{C}^{m\times n},$$

where $\delta > 0$. Clearly, $d_{ii} = \sigma_i$ for $1 \leqslant i \leqslant p$ and $d_{ii} = 0$ for $p+1 \leqslant i \leqslant n$.

If $\sigma_1 = \cdots = \sigma_p$, then the diagonal elements of D_δ are all distinct. Otherwise, choose δ such that $0 < \delta < \min_{1\leqslant i\leqslant p-1}(\sigma_i - \sigma_{i+1})$; this results again in a matrix D_δ with distinct diagonal entries. Define $A_\delta = UD_\delta V^{\mathsf{H}}$. Since U and V are unitary matrices, it follows that $\|A - A_\delta\|_F = \|D - D_\delta\|_F = \delta\sqrt{\frac{n(n+1)}{2}}$. Choosing δ such as

$$\delta < \min\left\{\min_{1 \leqslant i \leqslant p-1}(\sigma_i - \sigma_{i+1}), \frac{\epsilon\sqrt{2}}{\sqrt{n(n+1)}}\right\},$$

implies $\|A - A_\delta\|_F < \epsilon$.

(12) Let $A \in \mathbb{C}^{n\times n}$ be a non-singular matrix such that the matrix $A^{\mathsf{H}}A$ can be unitarily diagonalized as $A^{\mathsf{H}}A = US^2U^{\mathsf{H}}$, where U is a unitary matrix and $S = \mathsf{diag}(\sigma_1, \ldots, \sigma_p)$. Prove that the matrix $V = A^{\mathsf{H}}US^{-1}$ is unitary and that A has the singular value decomposition $A = USV^{\mathsf{H}}$.

Let $A \in \mathbb{C}^{m\times n}$ be a matrix of rank r. By Corollary 6.21, we have

$$\mathbb{C}^m = \mathsf{range}(A) \boxplus \mathsf{null}(A^{\mathsf{H}}),$$
$$\mathbb{C}^n = \mathsf{range}(A^{\mathsf{H}}) \boxplus \mathsf{null}(A).$$

Let

$$B_{\mathsf{range}(A)} = \{\boldsymbol{u}_1, \ldots, \boldsymbol{u}_r\},\ B_{\mathsf{null}(A^{\mathsf{H}})} = \{\boldsymbol{u}_{r+1}, \ldots, \boldsymbol{u}_m\},$$
$$B_{\mathsf{range}(A^{\mathsf{H}})} = \{\boldsymbol{v}_1, \ldots, \boldsymbol{v}_r\},\quad B_{\mathsf{null}(A)} = \{\boldsymbol{v}_r, \ldots, \boldsymbol{r}_n\},$$

be orthonormal bases of the subspaces $\mathsf{range}(A)$, $\mathsf{null}(A^{\mathsf{H}})$, $\mathsf{range}(A^{\mathsf{H}})$, and $\mathsf{null}(A)$, respectively.

(13) Let $A \in \mathbb{C}^{m\times n}$ be a matrix of rank r. Define the unitary matrices

$$U = (\boldsymbol{u}_1 \ \cdots \ \boldsymbol{u}_r \ \boldsymbol{u}_{r+1} \ \cdots \ \boldsymbol{u}_m) \in \mathbb{C}^{m\times m},$$
$$V = (\boldsymbol{v}_1 \ \cdots \ \boldsymbol{v}_r \ \boldsymbol{v}_{r+1} \ \cdots \ \boldsymbol{v}_n) \in \mathbb{C}^{n\times n}.$$

An alternative method for obtaining the singular value decomposition of a matrix is as follows.

(a) Prove that

$$UAV^{\mathsf{H}} = \begin{pmatrix} C & O_{r,n-r} \\ O_{m-r,r} & O_{m-r,n-r} \end{pmatrix},$$

where $C \in \mathbb{C}^{r\times r}$ is a matrix of rank r.

(b) Prove that there are two unitary matrices $U_1 \in \mathbb{C}^{m\times m}$ and $V_1 \in \mathbb{C}^{n\times n}$ such that

$$U_1 \begin{pmatrix} C & O_{r,n-r} \\ O_{m-r,r} & O_{m-r,n-r} \end{pmatrix} V_1^{\mathsf{H}} = \begin{pmatrix} D & O_{r,n-r} \\ O_{m-r,r} & O_{m-r,n-r} \end{pmatrix},$$

where $D \in \mathbb{C}^{r\times r}$ is a diagonal matrix.

(14) Let $A \in \mathbb{C}^{n\times n}$ be a matrix whose singular values are $\sigma_1, \ldots, \sigma_n$. Define the matrix $A^{[k]} = A \otimes A \otimes \cdots \otimes A \in \mathbb{C}^{n^k \times n^k}$ as in Exercise 32. Prove that the singular values of $A^{[k]}$ have the form $\sigma_{i_1}\sigma_{i_2}\cdots\sigma_{i_k}$, where $1 \leqslant i_1, \ldots, i_k \leqslant n$.

(15) Let $A \in \mathbb{C}^{n\times n}$ be a square matrix having the eigenvalues $\lambda_1, \lambda_2, \ldots, \lambda_n$ and the singular values $\sigma_1, \sigma_2, \ldots, \sigma_n$, where $|\lambda_1| \geqslant |\lambda_2| \geqslant \cdots \geqslant |\lambda_n|$ and $\sigma_1 \leqslant \sigma_2 \geqslant \cdots \geqslant \sigma_n$. Prove that $|\lambda_1 \cdots \lambda_k| \geqslant \sigma_1 \cdots \sigma_k$ for every k, $1 \leqslant k \leqslant n-1$ and that $|\lambda_1 \cdots \lambda_n| = \sigma_1 \cdots \sigma_n$.

(16) Let $A = PY$ be the polar decomposition of the matrix A, where $P \in \mathbb{C}^{m\times m}$ is a positive semidefinite matrix such that $P^2 = AA^{\mathsf{H}}$, and $Y \in \mathbb{C}^{m\times n}$ is such that $YY^{\mathsf{H}} = I_m$. Obtain the SVD of A by applying Theorem 8.14 to the Hermitian matrix P^2.

(17) Prove that if $A \in \mathbb{C}^{n\times n}$, then $|\det(A)| = \prod_{i=1}^{n} \sigma_i$, where $\sigma_1, \ldots, \sigma_n$ are the singular values of A.

(18) Prove that $A \in \mathbb{C}^{n\times n}$, then the singular values of the matrix $\mathsf{adj}(A)$ are $\prod\{\sigma_j \mid j \in \{1, \ldots, n\} - \{i\}\}$ for $1 \leqslant i \leqslant n$.

(19) Let $S = \{\boldsymbol{x} \in \mathbb{R}^n \mid \|\boldsymbol{x}\|_2 = 1\}$ be the surface of the unit sphere in $\mathbb{R}^n$ and let $A \in \mathbb{R}^{n\times n}$ be an invertible matrix whose SVD is $A = UDV'$, where U, V are orthogonal matrices and $D = \mathsf{diag}(\sigma_1, \ldots, \sigma_n)$. Prove that

(a) the image of S under left multiplication by A, that is, the set $\{A\boldsymbol{x} \mid \boldsymbol{x} \in S\}$, is an ellipsoid having the semiaxes $\sigma_1, \ldots, \sigma_n$;

(b) The columns of U give the directions of the semiaxes of the ellipsoid.

Solution: Since $A = UDV'$, it follows immediately that $A^{-1} = VD^{-1}U'$. Thus, if $\boldsymbol{y} = A\boldsymbol{x}$ for some $\boldsymbol{x} \in S$, then, by defining $\boldsymbol{w} = U'\boldsymbol{y}$, we have

$$\begin{aligned}\frac{w_1^2}{\sigma_1^2} + \frac{w_2^2}{\sigma_2^2} + \cdots + \frac{w_n^2}{\sigma_n^2} &= \|D^{-1}\boldsymbol{w}\|_2^2 = \|D^{-1}U'\boldsymbol{y}\|_2^2 \\ &= \|VD^{-1}U'\boldsymbol{u}\|_2^2 = \|A^{-1}\boldsymbol{y}\|_2^2 \\ &= \|A^{-1}A\boldsymbol{x}\|_2^2 = \|\boldsymbol{x}\|_2^2 = 1.\end{aligned}$$

Thus, $\{U'A\boldsymbol{x} \mid \boldsymbol{x} \in S\}$ is an ellipsoid having $\sigma_1, \ldots, \sigma_n$ as semiaxes. Since multiplication by an orthogonal matrices is isometric, it follows that the set $\{A\boldsymbol{x} \mid \boldsymbol{x} \in S\}$ is obtained by rotating the previous ellipsoid, so it is again an ellipsoid having the same semiaxes.

(20) Let $A = \sigma_1\boldsymbol{u}_1\boldsymbol{v}_1^{\mathsf{H}} + \cdots + \sigma_r\boldsymbol{u}_r\boldsymbol{v}_r^{\mathsf{H}}$ be the SVD of a matrix $A \in \mathbb{R}^{m\times n}$, where $\sigma_1 \geqslant \cdots \geqslant \sigma_r > 0$. Prove that the matrix $B(k) = \sum_{i=1}^{k}\sigma_i\boldsymbol{u}_i\boldsymbol{v}_i^{\mathsf{H}}$ from Lemma 9.2 can be written as

$$B(k) = \sum_{i=1}^{k}\boldsymbol{u}_i(\boldsymbol{u}_i)^{\mathsf{H}}A = \sum_{i=1}^{k}A\boldsymbol{v}_i(\boldsymbol{v}_i)^{\mathsf{H}}.$$

In other words, $B(k)$ is the projection of A on the subspace generated by the first k left singular vectors of A.

Solution: The equalities follow immediately from Theorem 9.7.

(21) Let $A \in \mathbb{C}^{m\times n}$ be a matrix. Define the square matrix $\tilde{A} \in \mathbb{C}^{(m+n)\times(m+n)}$ as

$$\tilde{A} = \begin{pmatrix} O_{m,m} & A \\ A^{\mathsf{H}} & O_{n,n}\end{pmatrix}.$$

Prove that:

(a) $\tilde{A}$ is a Hermitian matrix;

(b) the matrix A has the singular values $\sigma_1 \geqslant \cdots \geqslant \sigma_k$, where $k = \min\{m, n\}$ if and only if $\mathsf{spec}(\tilde{A})$ consists of the numbers $-\sigma_1 \leqslant \cdots \leqslant -\sigma_k \leqslant \sigma_k \leqslant \ldots \leqslant \sigma_1$ and an additional number of 0s.

Solution: The first part is immediate.

Suppose that $m \geqslant n$ and let $A = UDV^{\mathsf{H}}$ be the singular value decomposition of A, where

$$D = \begin{pmatrix} S \\ O_{m-n,n} \end{pmatrix},$$

$U \in \mathbb{C}^{m\times m}$, $V \in \mathbb{C}^{n\times n}$, and $S = \mathsf{diag}(\sigma_1, \ldots, \sigma_n)$. Since $m \geqslant n$, we can write $U = (Y\ Z)$, where $Y \in \mathbb{C}^{m\times n}$ and $Z \in \mathbb{C}^{m\times(m-n)}$. Define the matrix

$$W = \begin{pmatrix} \frac{1}{\sqrt{2}}Y & -\frac{1}{\sqrt{2}}Y & \frac{1}{\sqrt{2}}Z \\ \frac{1}{\sqrt{2}}Z & \frac{1}{\sqrt{2}}Z & O_{n,m-n} \end{pmatrix} \in \mathbb{C}^{(m+n)\times(m+n)}.$$

This matrix is unitary because

$$\begin{aligned} WW^{\mathsf{H}} &= \begin{pmatrix} \frac{1}{\sqrt{2}}Y & -\frac{1}{\sqrt{2}}Y & \frac{1}{\sqrt{2}}Z \\ \frac{1}{\sqrt{2}}Z & \frac{1}{\sqrt{2}}Z & O_{n,m-n} \end{pmatrix} \begin{pmatrix} \frac{1}{\sqrt{2}}Y^{\mathsf{H}} & \frac{1}{\sqrt{2}}Z^{\mathsf{H}} \\ -\frac{1}{\sqrt{2}}Y^{\mathsf{H}} & \frac{1}{\sqrt{2}}Z^{\mathsf{H}} \\ \frac{1}{\sqrt{2}}Z^{\mathsf{H}} & O_{m-n,n} \end{pmatrix} \\ &= \begin{pmatrix} I_m & O_{m,n} \\ O_{n,m} & I_n \end{pmatrix} = I_{m+n}. \end{aligned}$$

Then, $\tilde{A}$ can be factored as

$$\tilde{A} = W \begin{pmatrix} S & O_{n,n} & O_{n,m-n} \\ O_{n,n} & -S & O_{n,m-n} \\ O_{m-n,n} & O_{m-n,n} & O_{m-n,m-n} \end{pmatrix} W^{\mathsf{H}},$$

which shows that the spectrum of $\tilde{A}$ has the form claimed above.

(22) Let A and E be two matrices in $\mathbb{C}^{m\times n}$ and let $p = \min\{m, n\}$. Denote by $\sigma_1(A) \geqslant \cdots \geqslant \sigma_p(A)$ the singular values of A arranged in decreasing order and by $\sigma_1(E) \geqslant \cdots \geqslant \sigma_p(E)$ and $\sigma_1(A+E) \geqslant \cdots \geqslant \sigma_p(A+E)$ the similar sequences for E and $A+E$, respectively, where $p = \min\{m, n\}$. Prove that $|\sigma_i(A+E) - \sigma_i(A)| \leqslant \|E\|_2$ for $1 \leqslant i \leqslant p$ and $\sum_{i=1}^{p}(\sigma_i(A+E) - \sigma_i(A))^2 \leqslant \|E\|_F^2$.

Solution: Consider the matrices $\tilde{A}, \tilde{E}$ and $\widetilde{A+E}$ defined in Supplement 9.9 . It is immediate to verify that $\widetilde{A+E} = \tilde{A} + \tilde{E}$.

The eigenvalues of the matrix $\tilde{A}$ arranged in increasing order are

$$-\sigma_1(A) \leqslant \cdots \leqslant -\sigma_k(A) \leqslant \sigma_k(A) \leqslant \cdots \leqslant \sigma_1(A),$$

and the eigenvalues of $\tilde{E}$ in the same order are

$$-\sigma_1(E) \leqslant \cdots \leqslant -\sigma_k(E) \leqslant \sigma_k(E) \leqslant \cdots \leqslant \sigma_1(E).$$

By Weyl's Theorem (Supplement 30 of Chapter 8), we have

$$-\sigma_1(E) \leqslant \sigma_i(A+E) - \sigma_i(A) \leqslant \sigma_1(E) = \|E\|_2,$$

by Theorem 9.5, so $|\sigma_i(A+E) - \sigma_i(A)| \leqslant \|E\|_2$ for $1 \leqslant i \leqslant p$.

(23) Let S, T be two subspaces of $\mathbb{R}^n$ such that $\dim(S) = \dim(T) = r$ and let $\{\boldsymbol{u}_1, \ldots, \boldsymbol{u}_r\}$ and $\{\boldsymbol{v}_1, \ldots, \boldsymbol{v}_r\}$ be two orthonormal bases of S and of T represented by the matrices $B_S = (\boldsymbol{u}_1 \ \ldots \ \boldsymbol{u}_r) \in \mathbb{C}^{n\times r}$ and $B_T = (\boldsymbol{v}_1 \ \ldots \ \boldsymbol{v}_r) \in \mathbb{C}^{n\times r}$, respectively. Prove that the singular values of the matrix $B_T' B_S$ are located in the interval $[0, 1]$.

(24) Let $\|\cdot\|$ be a unitarily invariant norm on $\mathbb{C}^{m\times n}$, where $m \leqslant n$, $\boldsymbol{z} \in \mathbb{C}^m$. Define the matrix $Z = (\mathsf{diag}(z_1, \ldots, z_m) \ O_{m,n-m} \in \mathbb{C}^{m\times n}$. Prove that
   (a) the set of singular values of Z is $\{|z_1|^2, \ldots, |z_m|^2\}$;
   (b) the function $g : \mathbb{C}^m \longrightarrow \mathbb{R}_{\geqslant 0}$ defined by $g(\boldsymbol{z}) = \|Z\|$ is a symmetric gauge function.

(25) Let $g : \mathbb{R}^n \longrightarrow \mathbb{R}_{\geqslant 0}$ be a symmetric gauge function and let $\|A\|_g$ be defined by $\|A\|_g = g(\boldsymbol{s})$, where $A \in \mathbb{C}^{m\times n}$, and $\boldsymbol{z} = \begin{pmatrix} \sigma_1 \\ \vdots \\ \sigma_n \end{pmatrix}$ is a vector whose components are the singular values of A. Prove that $\|\cdot\|_g$ is a unitarily invariant norm.

In Exercise 51 of Chapter 6, we noted that ν_p is a symmetric gauge function on $\mathbb{R}^n$. The norms $\|\cdot\|_{\nu_p}$ for $p \geqslant 1$ are known as the *Schatten norms.* Note that for $p = 2$ the Schatten norm coincides with the Frobenius norm. For $p = 1$, we obtain $\|A\|_{\nu_1} = \sigma_1 + \cdots + \sigma_r$, where $\sigma_1, \ldots, \sigma_r$ are the non-zero singular values of A. Since $\|A\|_{\nu_1} = \mathsf{trace}(D)$, where D is the diagonal matrix that occurs in the SVD of A, this Schatten norm is also known as the *trace norm* or *Ky Fan's norm.*

(26) Let $A \in \mathbb{C}^{m\times n}$ be a matrix of rank r such that there exists a factorization

$$A = U \begin{pmatrix} R & O_{r,n-r} \\ O_{m-r,r} & O_{m-r,n-r} \end{pmatrix} V^{\mathsf{H}},$$

where $R \in \mathbb{C}^{r\times r}$ is a matrix of rank r, and $U \in \mathbb{C}^{m\times m}$ and $V \in \mathbb{C}^{n\times n}$ are unitary matrices. Note that the SVD of A is a special case of this factorization.
Prove that the Moore–Penrose pseudoinverse of A is given by

$$A^{\dagger} = V \begin{pmatrix} R^{-1} & O_{k,m-k} \\ O_{n-k,k} & O_{n-k,m-k} \end{pmatrix} U^{\mathsf{H}}.$$

Solution: By Theorem 3.26, it suffices to verify that $AA^{\dagger}A = A$ and $A^{\dagger}AA^{\dagger} = A^{\dagger}$.

(27) Prove that for the matrices $A^{(1)}, \ldots, A^{(n)}$, we have

$$(A^{(1)} \otimes A^{(2)} \otimes \cdots \otimes A^{(n)})^{\dagger} = A^{(1)\dagger} \otimes A^{(2)\dagger} \otimes \cdots A^{(n)\dagger},$$

where $A^{(k)} \in \mathbb{R}^{m_k \times n_k}$ for $1 \leqslant k \leqslant n$.

Solution: Note that the matrices $A^{(i)}$ can be factored as

$$A^{(i)} = U_i \begin{pmatrix} R_i & O_{r_i,n_i-r_i} \\ O_{m_i-r_i,r_i} & O_{m_i-r_i,n_i-r_i} \end{pmatrix} V_i',$$

where $U_i \in \mathbb{R}^{m_i\times m_i}$, $V_i \in \mathbb{R}^{n_i\times n_i}$ are orthonormal matrices, $R_i \in \mathbb{R}^{r_i\times r_i}$, and $r_i = \mathsf{rank}(A^{(i)})$.
The argument is by induction on n. For the base case $n = 2$, observe that the pseudoinverses $A^{(i)\dagger}$ are

$$A^{(i)\dagger} = V_i \begin{pmatrix} R_i^{-1} & O \\ 0 & O \end{pmatrix} U_i'$$

for $i = 1, 2$. Therefore,

$$\begin{aligned}
&A^{(1)\dagger} \otimes A^{(2)\dagger} \\
&= \left(V_1 \begin{pmatrix} R_1^{-1} & O \\ 0 & O \end{pmatrix} \otimes U_1' \right) \otimes \left(V_2 \begin{pmatrix} R_2^{-1} & O \\ O & O \end{pmatrix} U_2' \right) \\
&= (V_1 \otimes V_2) \begin{pmatrix} R_1^{-1} \otimes R_2^{-1} & O \\ O & O \end{pmatrix} (U_1' \otimes U_2').
\end{aligned}$$

On the other hand, we have

$$\begin{aligned}
&\left(A^{(1)} \otimes A^{(2)}\right)^{\dagger} \\
&= \left(\left(U_1 \begin{pmatrix} R_1 & O \\ O & O \end{pmatrix} V_1'\right) \otimes \left(U_1 \begin{pmatrix} R_1 & O \\ O & O \end{pmatrix} V_2'\right)\right)^{\dagger} \\
&= \left((U_1 \otimes U_2)\left(\begin{pmatrix} R_1 & O \\ O & O \end{pmatrix} \otimes \begin{pmatrix} R_2 & O \\ O & O \end{pmatrix}\right)(V_1' \otimes V_2')\right)^{\dagger} \\
&= \left((V_1 \otimes V_2)\left(\begin{pmatrix} (R_1 \otimes R_2)^{-1} & O \\ O & O \end{pmatrix}\right)(U_1 \otimes U_2)'\right) \\
&= \left((V_1 \otimes V_2)\left(\begin{pmatrix} (R_1^{-1} \otimes R_2^{-1} & O \\ O & O \end{pmatrix}\right)(U_1 \otimes U_2)'\right),
\end{aligned}$$

which concludes the base step. The proof of the induction step is straightforward.

(28) Let $A, B \in \mathbb{R}^{m\times p}$ and let $Q \in \mathbb{R}^{p\times p}$ be an orthogonal matrix. Prove that

(a) $\|A - BQ\|_F$ is minimal if and only if $\mathsf{trace}(Q'B'A)$ is maximal;

(b) if the SVD of the matrix $B'A$ is $B'A = U\mathsf{diag}(\sigma_1, \ldots, \sigma_p)V'$. The number $\mathsf{trace}(Q'B'A)$ is maximal if $Q = UV'$.

Solution: As we saw in Example 6.21,

$$\begin{aligned}
&\|A - BQ\|_F^2 \\
&\quad = \mathsf{trace}((A - BQ)'(A - BQ)) \\
&\quad = \mathsf{trace}((A' - Q'B')(A - BQ)) \\
&\quad = \mathsf{trace}(A'A - Q'B'A - A'BQ + Q'B'BQ) \\
&\quad = \mathsf{trace}(A'A) - \mathsf{trace}(Q'B'A) \\
&\qquad - \mathsf{trace}(A'BQ) + \mathsf{trace}(Q'B'BQ) \\
&\quad = \|A\|_F^2 - 2\mathsf{trace}(Q'B'A) + \mathsf{trace}(BQQ'B') \\
&\quad = \|A\|_F^2 - 2\mathsf{trace}(Q'B'A) + \mathsf{trace}(BB')
\end{aligned}$$

$$\text{(because the orthogonality of } Q \text{ means that } QQ' = I)$$
$$= \|A\|_F^2 + \|B\|_F^2 - 2\mathsf{trace}(Q'B'A),$$

which justifies the claim made in the first part.

For the second part, we start from the equality

$$\begin{aligned}\mathsf{trace}(Q'B'A) &= \mathsf{trace}(Q'U\mathsf{diag}(\sigma_1, \ldots, \sigma_p)V') \\ &= \mathsf{trace}(V'Q'U\mathsf{diag}(\sigma_1, \ldots, \sigma_p)).\end{aligned}$$

Note that the matrix $T = V'Q'U$ is orthogonal and

$$\mathsf{trace}(T\mathsf{diag}(\sigma_1, \ldots, \sigma_p) = \sum_{i=1}^{p} t_{ii}\sigma_i \leqslant \sum_{i=1}^{p} \sigma_i \leqslant .$$

When $Q = UV'$, we have $T = I$, so the maximum is achieved.

(29) Let $A \in \mathbb{C}^{m\times n}$ be a matrix having $\sigma_1, \ldots, \sigma_r$ as its non-zero singular values. The volume of A is the number $vol(A) = \sigma_1 \cdots \sigma_r$. Using the notations of Supplement 35 of Chapter 5, prove that $vol(A) = \sqrt{D_r^2(A)}$.

Solution: Let $A = UDV^{\mathsf{H}}$ be the thin SVD decomposition of A (see Corollary 9.3), where $U \in \mathbb{C}^{m\times r}$ and $V \in \mathbb{C}^{n\times r}$ are matrices having orthonormal sets of columns and $D = \mathsf{diag}(\sigma_1, \ldots, \sigma_r)$. Then, $vol(D) = \sigma_1 \cdots \sigma_r$. Note that matrices U and DV^{H} constitute a full-rank decomposition of A, so by Supplement 35 of Chapter 5,

$$D_r^2(A) = D_r^2(U)D_r^2(DV^{\mathsf{H}}),$$

In turn, D and V^{H} are full-rank matrices, so $D_r^2(DV^{\mathsf{H}}) = D_r^2 = \sigma_1^2 \cdots \sigma_r^2$, which concludes the argument.

(30) Let $A \in \mathbb{C}^{m\times n}$ be a matrix and let $B = A\begin{bmatrix}1, \ldots, p \\ 1, \ldots, n\end{bmatrix}$. If $\sigma_1 \geqslant \cdots \geqslant \sigma_n$ are the singular values of A and $\tau_1 \geqslant \cdots \geqslant \tau_n$ are the singular values of B, prove that $\sigma_i \geqslant \tau_i$ for $1 \leqslant i \leqslant n$.

Solution: We can write

$$A = \begin{pmatrix} B \\ C \end{pmatrix},$$

where $C \in \mathbb{C}^{(m-p)\times n}$. We have $A^{\mathsf{H}}A = B^{\mathsf{H}}B + C^{\mathsf{H}}C$. The singular values of A are the eigenvalues of $A^{\mathsf{H}}A$, $\tau_1, \ldots, \tau_n$ are the singular values $B^{\mathsf{H}}B$. The singular values of C, $\gamma_1, \ldots, \gamma_n$ are non-negative. By applying Theorem 8.22, we obtain $\sigma_i \geqslant \tau_i$ for $1 \leqslant i \leqslant n$.

(31) Let $A \in \mathbb{C}^{m\times n}$ be a matrix such that $\mathsf{rank}(A) = r$. Prove that there exists a set of $r(m+n+1)$ numbers that determines the matrix A.

Solution: The matrix A has mn components and a thin SVD decomposition of the form $A = \sigma_1 \boldsymbol{u}_1 \boldsymbol{v}_1' + \cdots + \sigma_r \boldsymbol{u}_r \boldsymbol{v}_r'$, where $\boldsymbol{u}_i \in \mathbb{C}^m$ and $\boldsymbol{v}_i \in \mathbb{C}^n$. Thus, if we select r singular values, mr components for the vectors $\boldsymbol{u}_i$ and nr components for the vectors $\boldsymbol{v}_i$, the matrix A is determined by $r + mr + nr = r(m + n + 1)$ numbers.

(32) Let $A = UDV^{\mathsf{H}}$ be the thin SVD decomposition of the matrix $A \in \mathbb{C}^{m\times n}$. Prove that $A(A^{\mathsf{H}}A)^{-1/2} = UV^{\mathsf{H}}$.

Solution: Note that $A^{\mathsf{H}}A = VDU^{\mathsf{H}}UDV^{\mathsf{H}} = VD^2V^{\mathsf{H}}$ because U is a Hermitian matrix. Thus, $(A^{\mathsf{H}}A)^{-1/2} = VD^{-1}V^{\mathsf{H}}$, which yields $A(A^{\mathsf{H}}A)^{-1/2} = UDV^{\mathsf{H}}VD^{-1}V^{\mathsf{H}} = UV^{\mathsf{H}}$.

(33) Let $A \in \mathbb{C}^{m\times n}$ be a matrix with $m \geqslant n$ and $\mathsf{rank}(A) = r \leqslant n$. Prove that A can be factored as $A = U_1 D_1 V^{\mathsf{H}}$, where $U_1 \in \mathbb{C}^{m\times n}$ matrix, $D \in \mathbb{C}^{n\times n}$ is a diagonal matrix, and $V \in \mathbb{C}^{n\times n}$ such that $U^{\mathsf{H}}U = VV^{\mathsf{H}} = V^{\mathsf{H}}V = I_n$.

Solution: Starting from the full SVD of A, $A = UDV^{\mathsf{H}}$ let $U_1 \in \mathbb{C}^{m\times n}$ be the matrix that consists of the first n columns of the matrix U, and let D_1 be the matrix that consists of the first n rows of D. Then, $A = U_1 D_1 V^{\mathsf{H}}$ is the desired decomposition.

(34) Prove that a matrix A is subunitary if and only if its singular values belong to $[0, 1]$.

Solution: Let $A \in \mathbb{C}^{n\times k}$ with $n \geqslant k$ and let $A = U_1 D_1 V^{\mathsf{H}}$ be the SVD decomposition of A as established in Supplement 9.9. Suppose that for the largest singular value we have $\sigma_1 \leqslant 1$ and let Z be the unitary completion of U_1 to a unitary matrix $(U_1 \;\; Z) \in \mathbb{C}^{m\times m}$. For the matrix $Y = (A \;\; U_1(I_n - D_1^2)^{1/2} \;\; Z) \in \mathbb{C}^{m\times(m+n)}$, we have $YY^{\mathsf{H}} = AA^{\mathsf{H}} + ZZ^{\mathsf{H}} = I_m$, so Y is a semi-unitary matrix, which implies that A is a subunitary matrix. Conversely, suppose that A is a subunitary matrix and let W be a semiunitary matrix such that $W = (A \;\; T)$ and whose rows are orthonormal. If $\boldsymbol{w} \in \mathbb{C}^m$ is a vector with $\boldsymbol{w}^{\mathsf{H}}\boldsymbol{w} = 1_m$, we have

$$\boldsymbol{w}^{\mathsf{H}}\boldsymbol{w} = \boldsymbol{w}^{\mathsf{H}} I_m \boldsymbol{w} = \boldsymbol{w}^{\mathsf{H}}(AA^{\mathsf{H}} + TT^{\mathsf{H}})\boldsymbol{w} \geqslant \boldsymbol{w}^{\mathsf{H}} AA^{\mathsf{H}}\boldsymbol{w},$$

which implies that A has no singular value greater than 1.

(35) The product of two subunitary (suborthogonal) matrices is a subunitary (suborthogonal) matrix.

Solution: Let $A \in \mathbb{C}^{m\times k}$ and $B \in \mathbb{C}^{k\times n}$ be two subunitary matrices. Let A_+ be the expansion of A to a matrix with orthonormal columns and B_+ be the expansion of B to a matrix with orthonormal rows. Since the non-zero singular values of the matrix $A_+B_+B_+^{\mathsf{H}}A^{\mathsf{H}} = A_+A_+^{\mathsf{H}}$ are the same as the eigenvalues of $A_+^{\mathsf{H}}A_+ = I$, it follows that A_+B_+ is subunitary. Since AB is a submatrix of A_+B_+, it follows that AB is subunitary.

(36) Let A and B be two matrices in $\mathbb{C}^{n\times n}$ having the singular values $\sigma_1 \geqslant \cdots \geqslant \sigma_n$ and $\tau_1 \geqslant \cdots \geqslant \tau_n$, respectively. Prove that

$$|\mathsf{trace}(AB)| \leqslant \sigma_{k=1}^{n}\sigma_k\tau_k.$$

Solution: Let $A = UDV^{\mathsf{H}}$ and $B = WCZ^{\mathsf{H}}$ be the singular value decompositions of A and B, where U, V, W, Z are unitary matrices and $D = \mathsf{diag}(\sigma_1, \ldots, \sigma_n)$ and $C = \mathsf{diag}(\tau_1, \ldots, \tau_n)$. We have

$$\begin{aligned}\mathsf{trace}(AB) &= \mathsf{trace}(UDV^{\mathsf{H}}WCZ^{\mathsf{H}}) = \mathsf{trace}(Z^{\mathsf{H}}UDV^{\mathsf{H}}WC)\\ &= \mathsf{trace}(SDTC),\end{aligned}$$

where $S = Z^{\mathsf{H}}U$ and $T = V^{\mathsf{H}}W$ are unitary matrices. Therefore,

$$\begin{aligned}\mathsf{trace}(AB) &= \sum_{p,q=1}^{n} s_{pq}t_{pq}\sigma_p\tau_q\\ &\leqslant \frac{1}{2}\sum_{p,q=1}^{n} s_{pq}^2\sigma_p\tau_q + \frac{1}{2}\sum_{p,q=1}^{n} t_{pq}^2\sigma_p\tau_q.\end{aligned}$$

Since S and T are unitary matrices, it follows that the matrices $S_1 = (|s_{pq}|^2)$ and $T_1 = (|t_{pq}^2|)$ are doubly stochastic, and the desired conclusion follows immediately from Supplement 106 of Chapter 3.

(37) Let $A \in \mathbb{R}^{n\times n}$ be a suborthogonal matrix with $\mathsf{rank}(A) = r$ and let $C \in \mathbb{R}^{n\times n}$ be a diagonal matrix, $C = \mathsf{diag}(c_1, \ldots, c_n)$, where $c_1 \geqslant \cdots \geqslant c_n \geqslant 0$. Prove that $\mathsf{trace}(AC) \leqslant c_1 + \cdots + c_r$.

Hint: Apply Supplement 9.9.

(38) Let $A \in \mathbb{R}^{m\times n}$ be a rectangular matrix. Extend the Rayleigh–Ritz function ral_A defined for square matrices in Theorem 8.17 to rectangular matrices as

$$\mathsf{ral}_A = \frac{\boldsymbol{x}'A\boldsymbol{y}}{\|\boldsymbol{x}\|\|\boldsymbol{y}\|}$$

for $\boldsymbol{x} \in \mathbb{R}^m$ and $\boldsymbol{y} \in \mathbb{R}^n$. Prove that σ is a singular value corresponding to singular vectors $\boldsymbol{x}$ and $\boldsymbol{y}$ when $(\boldsymbol{x}, \boldsymbol{y}, \sigma)$ is a critical point of ral_A.

Solution: Consider the Lagrangian function $L : \mathbb{R}^m \times \mathbb{R}^n \times \mathbb{R} \longrightarrow \mathbb{R}$ given by

$$L(\boldsymbol{x}, \boldsymbol{y}, \sigma) = \boldsymbol{x}'A\boldsymbol{y} - \sigma(\|\boldsymbol{x}\|\ \|\boldsymbol{y}\| - 1)$$

that is continuously differentiable for $\boldsymbol{x} \neq \mathbf{0}_m$ and $\boldsymbol{y} \neq \mathbf{0}_n$. The first-order optimality conditions yield

$$\frac{A\tilde{\boldsymbol{y}}}{\|\tilde{\boldsymbol{y}}\|} = \sigma\frac{\tilde{\boldsymbol{x}}}{\|\tilde{\boldsymbol{x}}\|} \text{ and } \frac{A'\tilde{\boldsymbol{x}}}{\|\tilde{\boldsymbol{x}}\|} = \sigma\frac{\tilde{\boldsymbol{y}}}{\|\tilde{\boldsymbol{y}}\|}.$$

By defining $\boldsymbol{u} = \frac{1}{\|\tilde{\boldsymbol{x}}\|}\tilde{\boldsymbol{x}}$ and $\boldsymbol{v} = \frac{1}{\|\tilde{\boldsymbol{y}}\|}\tilde{\boldsymbol{y}}$, we obtain $A\boldsymbol{v} = \lambda\boldsymbol{u}$ and $A'\boldsymbol{u} = \lambda\boldsymbol{v}$.

Bibliographical Comments

Theorem 9.9 was obtained in [41]. An important historical presentation of the singular value decomposition can be found in [159]. The proof of Theorem 9.10 is given in this article.

The proof of the CS Decomposition Theorem was given in [129] and [128]. Theorem 9.22 is a result of [160].

Supplements 33 and 35 are results of ten Berge that appear in [12]. In Supplement 36, we give a result of von Neumann [85]; the solution belongs to Mirsky [114]. For Supplement 37, see [12]. The proof of Supplement 27 appears in [99].

Chapter 10

The k-Means Clustering

10.1 Introduction

The *k-means algorithm* is one of the best-known clustering algorithms and has been in existence for a long time [73]. In a recent publication [173], the k-means algorithm was listed among the top ten algorithms in data mining. This algorithm computes a partition of a set of points in $\mathbb{R}^n$ that consists of k blocks (clusters) such that the objects that belong to the same block have a high degree of similarity, and the objects that belong to distinct blocks are dissimilar.

The algorithm requires the specification of the number of clusters k as an input. The set of objects to be clustered $S = \{\boldsymbol{u}_1, \ldots, \boldsymbol{u}_m\}$ is a subset of $\mathbb{R}^n$. Due to its simplicity and to its many implementations, it is a very popular algorithm despite this requirement.

10.2 The k-Means Algorithm and Convexity

The k-means algorithm begins with a randomly chosen set of k points $\boldsymbol{c}_1, \ldots, \boldsymbol{c}_k$ in $\mathbb{R}^n$ called *centroids*. An initial partition of the set S of objects is computed by assigning each object $\boldsymbol{u}_i$ to its closest centroid $\boldsymbol{c}_j$ and adopting a rule for breaking ties when there are several centroids that are equally distanced from $\boldsymbol{u}_i$ (e.g., assigning $\boldsymbol{u}_i$ to the centroid with the lowest index). As we shall see, the algorithm alternates between assigning cluster membership for each object and computing the center of each cluster.

Let C_j be the set of points assigned to the centroid $\boldsymbol{c}_j$. The assignments of objects to centroids are expressed by a matrix $B = (b_{ij}) \in \mathbb{R}^{m\times k}$, where

$$b_{ij} = \begin{cases} 1 & \text{if } \boldsymbol{u}_i \in C_j, \\ 0 & \text{otherwise.} \end{cases}$$

Since each object is assigned to exactly one cluster, we have $\sum_{j=1}^{k} b_{ij} = 1$. On the other hand, $\sum_{i=1}^{m} b_{ij}$ equals the number of objects assigned to the centroid $\boldsymbol{c}_j$. After these assignments, expressed by the matrix B, the centroid $\boldsymbol{c}_j$ is recomputed using the following formula:

$$\boldsymbol{c}_j = \frac{\sum_{i=1}^{m} b_{ij}\boldsymbol{u}_i}{\sum_{i=1}^{m} b_{ij}} \tag{10.1}$$

for $1 \leq j \leq k$.

The matrix of the centroids,

$$C = (\boldsymbol{c}_1 \ \cdots \ \boldsymbol{c}_k) \in \mathbb{R}^{n\times k}$$

can be written as

$$C = (\boldsymbol{u}_1 \ \cdots \ \boldsymbol{u}_m) B \mathsf{diag}\left(\frac{1}{m_1}, \ldots, \frac{1}{m_k}\right),$$

where $m_j = \sum_{i=1}^{m} b_{ij}$ is the number of objects of cluster C_j.

The sum of squared errors of a partition $\pi = \{C_1, \ldots, C_k\}$ of a set of objects S is intended to measure the quality of the assignment of objects to centroids and is defined as

$$sse(\pi) = \sum_{j=1}^{k} \sum_{\boldsymbol{u}\in C_j} d^2(\boldsymbol{u}, \boldsymbol{c}_j), \tag{10.2}$$

where $\boldsymbol{c}_j$ is the centroid of C_j for $1 \leq j \leq k$. It is clear that $sse(\pi)$ is unaffected if the data are centered.

We can write $sse(\pi)$ as

$$\begin{aligned} sse(\pi) &= \sum_{i=1}^{m} \sum_{j=1}^{k} b_{ij} \|\boldsymbol{x}_i - \boldsymbol{c}_j\|_2^2 \\ &= \sum_{i=1}^{m} \sum_{j=1}^{k} b_{ij} \sum_{p=1}^{n} (x_{ip} - c_{jp})^2 . \end{aligned}$$

The nk necessary conditions for a local minimum of this function,

$$\frac{\partial sse(\pi)}{\partial c_{jp}} = \sum_{i=1}^{m} b_{ij} \left(-2(x_{ip} - c_{jp})\right) = 0$$

for $1 \leq p \leq n$ and $1 \leq j \leq k$, can be written as

$$\sum_{i=1}^{m} b_{ij} x_{ip} = \sum_{i=1}^{m} b_{ij} c_{jp} = c_{jp} \sum_{i=1}^{m} b_{ij},$$

or, as

$$c_{jp} = \frac{\sum_{i=1}^{m} b_{ij} x_{ip}}{\sum_{i=1}^{m} b_{ij}}$$

for $1 \leq p \leq n$. In vectorial form, these conditions amount to

$$\boldsymbol{c}_j = \frac{\sum_{i=1}^{m} b_{ij} \boldsymbol{x}_i}{\sum_{i=1}^{m} b_{ij}},$$

which is exactly the formula (10.1) that is used to update the centroids. Thus, the choice of the centroids can be justified by the goal of obtaining local minima of the sum of squared errors of the clusterings.

Since we have new centroids, objects must be reassigned, which means that the values of b_{ij} must be re-computed, which, in turn, will affect the values of the centroids, etc.

The halting criterion of the algorithm depends on particular implementations and it may involve

(i) performing a certain number of iterations;
(ii) lowering the sum of squared errors $sse(\pi)$ below a certain limit;
(iii) the current partition coincides with the previous partition.

This variant of the k-means algorithm is known as *Forgy–Lloyd* algorithm (Algorithm 10.2.1) [56, 107]:

Algorithm 10.2.1: Forgy–Lloyd Algorithm

Data: set of objects to be clustered $S = \{\boldsymbol{x}_1, \ldots, \boldsymbol{x}_m\} \subseteq \mathbb{R}^n$ and the number of clusters k.
Result: a collection of k clusters

generate a randomly chosen collection of k points $\boldsymbol{c}_1, \ldots, \boldsymbol{c}_k$ in $\mathbb{R}^n$;
assign each object $\boldsymbol{x}_i$ to the closest centroid $\boldsymbol{c}_j$;
let $\pi = \{C_1, \ldots, C_k\}$ be the partition defined by $\boldsymbol{c}_1, \ldots, \boldsymbol{c}_k$;
recompute the centroids of the clusters $C_1, \ldots, C_k$;
while *halting criterion is not met* **do**
    compute the new value of the partition π using the current centroids;
    recompute the centroids of the blocks of π;
end

The popularity of the k-means algorithm stems from its simplicity and its low time complexity which is $O(km\ell)$, where m is the number of objects to be clustered and ℓ is the number of iterations that the algorithm is performing.

Another variant of the k-means algorithm redistributes objects to clusters based on the effect of such a re-assignment on the objective function. If $sse(\pi)$ decreases, the object is moved and the two centroids of the affected clusters are recomputed. This variant is carefully analyzed in [16].

As the iterations of the k-means algorithm develop, the algorithm may be trapped into a local minimum, which is the serious limitation of this clustering.

The next theorem shows another limitation of the k-means algorithm because this algorithm produces only clusters whose convex closures may intersect only at the points of S.

Theorem 10.1. *Let $S = \{\boldsymbol{x}_1, \ldots, \boldsymbol{x}_m\} \subseteq \mathbb{R}^n$ be a set of m objects. If $C_1, \ldots, C_k$ is the set of clusters computed by the k-means algorithm in any step, then the convex closure of each cluster C_i, $\boldsymbol{K}_{conv}(C_i)$ is included in a polytope P_i that contains $\mathbf{c}_i$ for $1 \leq i \leq k$.*

Proof. Suppose that the centroids of the partition $\{C_1, \ldots, C_k\}$ are $\boldsymbol{c}_1, \ldots, \boldsymbol{c}_k$. Let $\boldsymbol{m}_{ij} = \frac{1}{2}(\boldsymbol{c}_i + \boldsymbol{c}_j)$ be the midpoint of the segment $\overline{\boldsymbol{c}_i\boldsymbol{c}_j}$ and let H_{ij} be the hyperplane $(\boldsymbol{c}_i - \boldsymbol{c}_j)'(\boldsymbol{x} - \boldsymbol{m}_{ij}) = 0$ that is the perpendicular bisector of the segment $\overline{\boldsymbol{c}_i\boldsymbol{c}_j}$. Equivalently,

$$H_{ij} = \left\{\boldsymbol{x} \in \mathbb{R}^m \mid (\boldsymbol{c}_i - \boldsymbol{c}_j)'\boldsymbol{x} = \frac{1}{2}(\boldsymbol{c}_i - \boldsymbol{c}_j)'(\boldsymbol{c}_i + \boldsymbol{c}_j)\right\}.$$

The halfspaces determined by H_{ij} are described by the following inequalities:

$$H_{ij}^+ : (\boldsymbol{c}_i - \boldsymbol{c}_j)'\boldsymbol{x} \leq \frac{1}{2}(\|\boldsymbol{c}_i\|_2^2 - \|\boldsymbol{c}_j\|_2^2),$$

$$H_{ij}^- : (\boldsymbol{c}_i - \boldsymbol{c}_j)'\boldsymbol{x} \geq \frac{1}{2}(\|\boldsymbol{c}_i\|_2^2 - \|\boldsymbol{c}_j\|_2^2).$$

It is easy to see that $\boldsymbol{c}_i \in H_{ij}^+$ and $\boldsymbol{c}_j \in H_{ij}^-$. Moreover, if $d_2(\boldsymbol{c}_i, \boldsymbol{x}) < d_2(\boldsymbol{c}_j, \boldsymbol{x})$, then $\boldsymbol{x} \in H_{ij}^+$, and if $d_2(\boldsymbol{c}_i, \boldsymbol{x}) > d_2(\boldsymbol{c}_j, \boldsymbol{x})$, then $\boldsymbol{x} \in H_{ij}^-$. Indeed, suppose that $d_2(\boldsymbol{c}_i, \boldsymbol{x}) < d_2(\boldsymbol{c}_j, \boldsymbol{x})$, which amounts to $\|\boldsymbol{c}_i - \boldsymbol{x}\|_2^2 < \|\boldsymbol{c}_j - \boldsymbol{x}\|_2^2$. This is equivalent to

$$(\boldsymbol{c}_i - \boldsymbol{x})'(\boldsymbol{c}_i - \boldsymbol{x}) < (\boldsymbol{c}_j - \boldsymbol{x})'(\boldsymbol{c}_j - \boldsymbol{x}).$$

The last inequality is equivalent to

$$\|\boldsymbol{c}_i\|_2^2 - 2\boldsymbol{c}_i'\boldsymbol{x} < \|\boldsymbol{c}_j\|_2^2 - 2\boldsymbol{c}_j'\boldsymbol{x},$$

which implies that $\boldsymbol{x} \in H_{ij}^+$. In other words, $\boldsymbol{x}$ is located in the same half-space as the closest centroid of the set $\{\boldsymbol{c}_i, \boldsymbol{c}_j\}$. Note also that if $d_2(\boldsymbol{c}_i, \boldsymbol{x}) = d_2(\boldsymbol{c}_j, \boldsymbol{x})$, then $\boldsymbol{x}$ is located in $H_{ij}^+ \cap H_{ij}^- = H_{ij}$, that is, on the hyperplane shared by P_i and P_j.

Let P_i be the closed polytope defined by

$$P_i = \bigcap\{H_{ij}^+ \mid j \in \{1, \ldots, k\} - \{i\}\}.$$

Objects that are closer to $\boldsymbol{c}_i$ than to any other centroid $\boldsymbol{c}_j$ are located in the closed polytope P_i. Thus, $C_i \subseteq P_i$ and this implies $\boldsymbol{K}_{\mathrm{conv}}(C_i) \subseteq P_i$. □

10.3 Relaxation of the k-Means Problem

The idea of relaxing the requirements of the k-means algorithm in order to apply ideas that originate in principal component analysis belongs to Ding and He [35].

Let $X \in \mathbb{R}^{m\times n}$ be the data matrix of the set $S = \{\boldsymbol{x}_1,\ldots,\boldsymbol{x}_m\} \subseteq \mathbb{R}^n$ of m objects to be clustered,

$$X = \begin{pmatrix} \boldsymbol{x}_1' \\ \vdots \\ \boldsymbol{x}_m' \end{pmatrix}.$$

Note that

$$XX' = \begin{pmatrix} \boldsymbol{x}_1' \\ \vdots \\ \boldsymbol{x}_m' \end{pmatrix} (\boldsymbol{x}_1 \ \cdots \ \boldsymbol{x}_m) = \begin{pmatrix} \boldsymbol{x}_1'\boldsymbol{x}_1 & \cdots & \boldsymbol{x}_1'\boldsymbol{x}_m \\ \vdots & \cdots & \vdots \\ \boldsymbol{x}_m'\boldsymbol{x}_1 & \cdots & \boldsymbol{x}_m'\boldsymbol{x}_m \end{pmatrix},$$

which implies $\mathsf{trace}(XX') = \sum_{j=1} \boldsymbol{x}_j\boldsymbol{x}_j'$. To simplify the arguments, we shall assume that X is a centered matrix.

Let $\pi = \{C_1,\ldots,C_k\}$ be a clustering of the set S and let $m_j = |C_j|$ for $1 \leq j \leq k$. Clearly, we have $\sum_{j=1}^k m_j = m$. There exists a permutation matrix P that allows us to rearrange the rows of the matrix X such that the rows that belong to a cluster C_j are located in contiguous columns. In other words, we can write PX as a block matrix

$$PX = \begin{pmatrix} X(1) \\ \vdots \\ X(k)) \end{pmatrix},$$

where $X(j) \in \mathbb{R}^{m_j\times n}$. The centroid $\boldsymbol{c}_j$ of C_j can be written as $\boldsymbol{c}_j' = \frac{1}{m_j}\boldsymbol{1}_{m_j}' X(j)$. Furthermore, we have

$$sse(\pi) = \sum_{j=1}^k \sum_{\boldsymbol{x}\in C_j} d^2(\boldsymbol{x},\boldsymbol{c}_j) = \sum_{j=1}^k \sum_{\boldsymbol{x}\in C_j} \|\boldsymbol{x} - \boldsymbol{c}_j\|_2^2$$

$$= \sum_{j=1}^{k} \sum_{\boldsymbol{x} \in C_j} (\boldsymbol{x} - \mathbf{c}_j)'(\boldsymbol{x} - \mathbf{c}_j) = \sum_{j=1}^{k} \sum_{\boldsymbol{x} \in C_j} (\boldsymbol{x}'\boldsymbol{x} - 2\mathbf{c}_j'\boldsymbol{x} + \mathbf{c}_j'\mathbf{c}_j)$$

$$= \sum_{\boldsymbol{x} \in S} \boldsymbol{x}'\boldsymbol{x} - 2\sum_{j=1}^{k} \sum_{\boldsymbol{x} \in C_j} \mathbf{c}_j'\boldsymbol{x} + \sum_{j=1}^{k} \sum_{\boldsymbol{x} \in C_j} \mathbf{c}_j'\mathbf{c}_j$$

$$= \sum_{\boldsymbol{x} \in S} \boldsymbol{x}'\boldsymbol{x} - \sum_{j=1}^{k} m_j \mathbf{c}_j'\mathbf{c}_j$$

$$= \sum_{\boldsymbol{x} \in S} \boldsymbol{x}'\boldsymbol{x} - \sum_{j=1}^{k} \frac{1}{m_j} \mathbf{1}_{m_j}' X(j) X(j)' \mathbf{1}_{m_j}.$$

Let $R_k = (\boldsymbol{r}_1 \ \cdots \ \boldsymbol{r}_k) \in \mathbb{R}^{m \times k}$ be the matrix defined by

$$\boldsymbol{r}_j = \frac{1}{\sqrt{m_j}} \begin{pmatrix} \mathbf{0}_{m_1} \\ \vdots \\ \mathbf{0}_{m_{j-1}} \\ \mathbf{1}_{m_j} \\ \mathbf{0}_{m_{j+1}} \\ \vdots \\ \mathbf{0}_{m_k} \end{pmatrix}$$

for $1 \leq j \leq k$. The columns of R_k are multiples of the characteristic vectors of the clusters, we have $\sum_{j=1}^{k} \sqrt{m_j}\, \boldsymbol{r}_j = \mathbf{1}_m$ and the set of columns of R_k is orthonormal, that is, $R_k' R_k = I_k$. Note that the centroids of the clusters can be written now as $\mathbf{c}_j = \frac{1}{\sqrt{m_j}} X' \boldsymbol{r}_j$.

Now $sse(\pi)$ can be written as $sse(\pi) = \mathsf{trace}(XX') - \mathsf{trace}(R_k' X X' R_k)$. The first term of $sse(\pi)$ is constant; therefore, to minimize $sse(\pi)$, we need to maximize $\mathsf{trace}(R_k' X X' R_k)$.

Let $T \in \mathbb{R}^{k \times k}$ be an orthonormal matrix ($T'T = TT' = I_k$) having the last column

$$\boldsymbol{t}_k = \begin{pmatrix} \sqrt{\frac{m_1}{m}} \\ \vdots \\ \sqrt{\frac{m_k}{m}} \end{pmatrix}. \tag{10.3}$$

Define the matrix $Q_k = (\boldsymbol{q}_1 \ \cdots \ \boldsymbol{q}_k) \in \mathbb{R}^{m\times k}$ as $Q_k = R_k T$. We have $\boldsymbol{q}_i = \boldsymbol{r}_1 t_{1i} + \cdots + \boldsymbol{r}_k t_{ki}$ for $1 \leq i \leq k$, so the last column of Q_k is

$$\boldsymbol{q}_k = \boldsymbol{r}_1\sqrt{\frac{m_1}{m}} + \cdots + \boldsymbol{r}_k\sqrt{\frac{m_k}{m}} = \frac{1}{\sqrt{m}}\mathbf{1}_k.$$

Also, $Q_k' Q_k = T' R_k' R_k T = T'T = I_k$. Let Q_{k-1} be the matrix that consists of the first $k-1$ columns of Q_k. Clearly, Q_{k-1} has orthogonal columns and $q_i'\mathbf{1}_m = 0$ for $1 \leq i \leq k-1$. Since $Q_k' X X' Q_k = T' R_k' X X' R_k T$, we have

$$\begin{aligned}\mathsf{trace}(Q_k' X X' Q_k) &= \mathsf{trace}(T' R_k' X X' R_k T) = \|X' R_k T\|_F^2\\ &= \|X' R_k\|_F^2 = \mathsf{trace}(R_k' X X' R_k).\end{aligned}$$

Therefore,

$$sse(\pi) = \mathsf{trace}(XX') - \mathsf{trace}(Q_k' X X' Q_k).$$

This allows us to write

$$\begin{aligned}sse(\pi) &= \mathsf{trace}(XX') - \mathsf{trace}(Q_k' X X' Q_k)\\ &= \mathsf{trace}(XX') - \frac{1}{\sqrt{m}}\mathbf{1}_m' X X' \mathbf{1}_m - \mathsf{trace}(Q_{k-1}' X X' Q_{k-1})\\ &= \mathsf{trace}(XX') - \mathsf{trace}(Q_{k-1}' X X' Q_{k-1}),\end{aligned}$$

because X being a centered matrix, we have $X'\mathbf{1}_m = \mathbf{0}_m$.

The above argument reduces the k-means clustering to determining a matrix Q_{k-1} such that $\mathsf{trace}(Q_{k-1}' X X' Q_{k-1})$ is maximal, subjected to the following restrictions:

(i) $Q_{k-1}' Q_{k-1} = I_{k-1}$;

(ii) $\boldsymbol{q}_j'\mathbf{1}_m = 0$ for $1 \leq j \leq k$;

(iii) vectors of the form $\boldsymbol{q}_j$ are obtained from vectors of the form $\boldsymbol{r}_j$ via a linear transformation because Q_k is defined as $Q_k = R_k T$.

Recall that directions and the principal components of the centered data matrix X are the vectors $\boldsymbol{d}_j$ and $\boldsymbol{q}_j$, respectively, such that $X\boldsymbol{d}_j = \sigma_j \boldsymbol{q}_j$ and $X'\boldsymbol{q}_j = \sigma_j \boldsymbol{d}_j$ for $1 \leq j \leq k$.

In [35], the idea of relaxing this formulation of k-means by dropping the last requirement is introduced. By Ky Fan's Theorem (Theorem 7.14), we have

$$\max\{\mathsf{trace}(Q'_{k-1}XX'Q_{k-1}) \mid Q'_{k-1}Q_{k-1} = I_{k-1}\} = \lambda_1 + \cdots + \lambda_{k-1},$$

where $\lambda_1 \geq \lambda_2 \geq \cdots$ are the eigenvalues of the matrix XX', that is, the squares of the singular values of X. Moreover, the columns of the optimal Q_{k-1} are the $k-1$ eigenvectors that correspond to the top $k-1$ eigenvalues of the matrix XX', that is, the top $q-1$ principal components of X. These columns represent the continuous solutions for the transformed discrete cluster membership problem.

Theorem 10.2. *Let $\mathbf{c}_1, \ldots, \mathbf{c}_k$ be the centroids of the clusters obtained by applying the k-means algorithm to the centered data matrix $X \in \mathbb{R}^{m\times n}$.*

For the optimal solution of the relaxed problem, the subspace $\langle \mathbf{c}_1, \ldots, \mathbf{c}_k\rangle$ is spanned by the first $k-1$ principal directions of the centered data matrix X.

Proof. If $\boldsymbol{q}_1, \ldots, \boldsymbol{q}_{k-1}$ be the vectors that correspond to the optimal solution, that is, the principal components of the centered matrix X. Since $R_k = Q_kT'$, we can write

$$\begin{aligned}
\mathbf{c}_j &= \frac{1}{\sqrt{m_j}}X'\boldsymbol{r}_j = \frac{1}{\sqrt{m_j}}X'(\boldsymbol{q}_1 t'_{1j} + \cdots + \boldsymbol{q}_k t'_{kj})\\
&= \frac{1}{\sqrt{m_j}}X'(\boldsymbol{q}_1 t_{j1} + \cdots + \boldsymbol{q}_k t_{jk})\\
&= \frac{1}{\sqrt{m_j}}\sigma_1 \boldsymbol{d}_1 t_{j1} + \cdots + \sigma_k \boldsymbol{d}_k t_{jk})
\end{aligned}$$

for $1 \leq j \leq k$. Thus, every centroid belongs to the subspace generated by the principal directions of the matrix X.

Conversely, let $\boldsymbol{d}_j$ be a principal direction of X. Since

$$\begin{aligned}
\boldsymbol{d}_j &= \frac{1}{\sigma_j}X'\boldsymbol{q}_j = \frac{1}{\sigma_j}X'(\boldsymbol{r}_1 t_{1j} + \cdots + \boldsymbol{r}_k t_{kj})\\
&= \frac{1}{\sigma_j}(\sqrt{m_1}\mathbf{c}_1 t_{1j} + \cdots + \sqrt{m_k}\mathbf{c}_k t_{kj}),
\end{aligned}$$

which shows that every principal direction of X belongs to the subspace generated by an optimal set of centroids. □

Note that the vectors $\boldsymbol{q}_j$ may have both positive and negative components. The recovery of the characteristic vectors $\boldsymbol{r}_1, \ldots, \boldsymbol{r}_{k-1}, \boldsymbol{r}_k$ of the clusters cannot proceed from the equality $Q_k = R_k T$ because the matrix T itself can be determined only after obtaining the results of the clustering. Instead, the idea proposed in [35] takes advantage of Theorem10.2.

Let $C \in \mathbb{R}^{m \times m}$ be the matrix $C = \sum_{i=1}^{k} \boldsymbol{q}_j \boldsymbol{q}_j'$. Recall that $\boldsymbol{q}_j = \boldsymbol{r}_j T$, so $\boldsymbol{q}_j \boldsymbol{q}_j' = \boldsymbol{r}_j T T' \boldsymbol{r}_j = \boldsymbol{r}_j \boldsymbol{r}_j'$. The rank-1 matrix $\boldsymbol{r}_j \boldsymbol{r}_j'$ consists of a block surrounded by 0s, so C has a block-diagonal structure. Thus, if $c_{ij} \neq 0$, $\boldsymbol{x}_i$ and $\boldsymbol{x}_j$ belong to the same cluster. The matrix C computed starting from $\boldsymbol{q}_1, \ldots, \boldsymbol{q}_k$ may have negative elements. In general, entries of C below a certain value are ignored to eliminate noise.

10.4 SVD and Clustering

We refer to the current approach for seeking a k-means clustering as the *discrete clustering problem (DCP)*. The DCP problem can be relaxed by seeking a subspace V of $\mathbb{R}^n$ so that $\dim(V) \leq k$ and $\sum_{i=1}^{m} d(\boldsymbol{x}_i, V)^2$ is minimal. This new problem will be referred to as the *continuous clustering problem* (CCP). It was shown in [39], using the SVD, that the relaxation of the problem because it can be used to obtain a 2-approximation of the optimum for the original problem and the approximative solution is interesting in its own right. Furthermore, the CCP problem can be solved in polynomial time because it can be obtained from the SVD of a matrix constructed on the data points. It is easy to see that $sse(\pi) \geq g_S(V)$ for any k-clustering π. Indeed, if $U = \langle \mathbf{c}_1, \ldots, \mathbf{c}_k \rangle$, then $\dim(U) \leq k$ and $sse(\pi) \geq g_S(V)$.

Observe that for each cluster center $\mathbf{c}_j$, the set

$$C_j = \{\boldsymbol{x} \in S \mid d(\boldsymbol{x}, \mathbf{c}_j) \leq d(\boldsymbol{x}, \mathbf{c}_k), \text{ where } k \neq j\}$$

that consists of points whose closest point in C is $\mathbf{c}_j$, which is a polyhedron. The total number of faces of the polyhedra $C_1, \ldots, C_k$ does not exceed $\binom{k}{2}$ because each face is determined by at least one pair of points.

The hyperplanes that define the faces of the polyhedra can be moved without modifying the partition κ such that each face contains at least n points of the set S. If the points of S are in general position and $\mathbf{0}_n \notin S$, each face contains n affinely independent points of S. Observe that there are

$$N(k,m,n) = \sum \left\{ \binom{\binom{m}{n}}{t} \,\middle|\, k \leq t \leq \binom{k}{2} \right\} \tag{10.4}$$

hyperplanes each of which contains n affinely independent points of M.

The following is an enumerative algorithm for resolving the discrete clustering problem (Algorithm 10.4.1).

Algorithm 10.4.1: Enumerative Algorithm

Data: set of objects to be clustered $S = \{\boldsymbol{x}_1, \ldots, \boldsymbol{x}_m\} \subseteq \mathbb{R}^n$ and the number of clusters k.

Result: a collection of k clusters

1. enumerate all $N(k,m,n)$ sets of hyperplanes each of which contains n affinely independent points;
2. retain only family of hyperplanes that partition S into k cells;
3. make one of the 2^{tn} choices as to which cell to assign each point of S located on a hyperplane;
4. find the centroid of each set in the partition π and compute $sse(\pi)$;

Let V be a k-dimensional subspace of $\mathbb{R}^n$ and let $\boldsymbol{z}_1, \ldots, \boldsymbol{z}_m$ be the orthogonal projections of the vectors $\boldsymbol{x}_1, \ldots, \boldsymbol{x}_m$ on V. Consider the matrices

$$A = \begin{pmatrix} \boldsymbol{x}_1' \\ \vdots \\ \boldsymbol{x}_m' \end{pmatrix} \text{ and } B = \begin{pmatrix} \boldsymbol{z}_1' \\ \vdots \\ \boldsymbol{z}_m' \end{pmatrix}.$$

Clearly, we have $\mathsf{rank}(B) \leq k$. We have

$$\|A - B\|_F^2 = \sum_{i=1}^{m} \|\boldsymbol{x}_i - \boldsymbol{z}_i\|_2^2 = \sum_{i=1}^{m} d(\boldsymbol{x}_i, V)^2.$$

Let $(\sigma_1, \ldots, \sigma_r)$ be the sequence of non-zero singular values of A, where $\sigma_1 \geq \cdots \geq \sigma_r > 0$. The SVD Theorem implies that A can be written as

$$A = \sigma_1 \boldsymbol{u}_1 \boldsymbol{v}_1^{\mathsf{H}} + \cdots + \sigma_r \boldsymbol{u}_r \boldsymbol{v}_r^{\mathsf{H}}.$$

Let $B(k) \in \mathbb{C}^{m \times n}$ be the matrix defined by

$$B(k) = \sum_{i=1}^{k} \sigma_i \boldsymbol{u}_i \boldsymbol{v}_i^{\mathsf{H}}.$$

By Theorem 9.10, $B(k)$ is the best approximation of A among the matrices of rank no larger than k in the sense of Frobenius norm and

$$\|A - B(k)\|_F^2 = \|A\|_F^2 - \sum_{i=1}^{k} \sigma_i^2.$$

To solve the CCP, we choose the subspace V as the subspace generated by the first k left singular vectors, $\boldsymbol{u}_1, \ldots, \boldsymbol{u}_k$ of the matrix A. We saw in Supplement 20 of Chapter 9 that $B(k)$ is the projection of A on this subspace.

Combining the DCP Algorithm 10.4.1 with the CCP approach produces a 2-approximation of the DCP (Algorithm 10.4.2):

Algorithm 10.4.2: 2-Approximation Algorithm

Data: set of objects to be clustered $S = \{\boldsymbol{x}_1, \ldots, \boldsymbol{x}_m\} \subseteq \mathbb{R}^n$ and the number of clusters k.

Result: a set D of cluster centers

compute the k-dimensional subspace generated by the first k left singular vectors, $\boldsymbol{u}_1, \ldots, \boldsymbol{u}_k$, of the matrix A;

solve the DCP with input A to obtain the set of centers $D = \{\boldsymbol{d}_1, \ldots, \boldsymbol{d}_k\}$;

output the set of centers $\{\boldsymbol{d}_1, \ldots, \boldsymbol{d}_k\}$.

The optimal value Z obtained from the DCP algorithm satisfies the inequality

$$Z \geq \sum_{i=1}^{m} d(\boldsymbol{x}_i, \boldsymbol{z}_i)^2. \tag{10.5}$$

If $C = \{\boldsymbol{c}_1, \ldots, \boldsymbol{c}_k\}$ is an optimal center set for the DCP, and $\boldsymbol{d}_1, \ldots, \boldsymbol{d}_k$ are the projections of these centers on W, then

$$Z = \sum_{i=1}^{m} d(\boldsymbol{x}_i, C)^2 \geq \sum_{i=1}^{m} d(\boldsymbol{z}_i, D)^2. \tag{10.6}$$

The Inequalities (10.5) and (10.6) imply

$$\begin{aligned} 2Z &\geq \sum_{i=1}^{m} d(\boldsymbol{x}_i, \boldsymbol{z}_i)^2 + \sum_{i=1}^{m} d(\boldsymbol{z}_i, D)^2 \\ &= \sum i = 1^m d(\boldsymbol{x}_i, D)^2, \end{aligned}$$

which shows that the Algorithm 10.4.2 produces indeed a 2-approximative solution of the DCP.

10.5 Evaluation of Clusterings

The *silhouette method* is an unsupervised method for evaluation of clusterings that computes certain coefficients for each object. The set of these coefficients allows an evaluation of the quality of the clustering.

Let $S = \{\boldsymbol{x}_1, \ldots, \boldsymbol{x}_n\}$ be a collection of objects, $d : S \times S \longrightarrow \mathbb{R}_{\geq 0}$ a dissimilarity on S, and let $\{C_1, \ldots, C_k\}$ be a clustering of S.

Suppose that $\boldsymbol{x}_i \in C_\ell$. The *average dissimilarity* of $\boldsymbol{x}_i$ is given by

$$a(\boldsymbol{x}_i) = \frac{\sum\{d(\boldsymbol{x}_i, \boldsymbol{u}) \mid u \in C_\ell - \{\boldsymbol{x}_i\}\}}{|C_\ell|},$$

that is, the average dissimilarity between $\boldsymbol{x}_i$ to all other objects of C_ℓ, the cluster to which $\boldsymbol{x}_i$ is assigned.

For $\boldsymbol{x}_i$ and a cluster $C \neq C_\ell$, let

$$d(\boldsymbol{x}_i, C) = \frac{\sum\{d(\boldsymbol{x}_i, u) \mid f(u) = C\}}{|C|},$$

be the average dissimilarity between $\boldsymbol{x}_i$ and the objects of the cluster C.

Definition 10.1. Let $\{C_1, \ldots, C_k\}$ be a clustering. A *neighbor of* $\boldsymbol{x}_i$ is a cluster $C \neq C_\ell$ for which $d(\boldsymbol{x}_i, C)$ is minimal.

In other words, a neighbor of an object $\boldsymbol{x}_i$ is "the second best choice" for a cluster for $\boldsymbol{x}_i$. Let $b : S \longrightarrow \mathbb{R}_{\geq 0}$ be the function defined by

$$b(\boldsymbol{x}_i) = \min\{d(\boldsymbol{x}_i, C) \mid C \neq C_\ell\}.$$

Definition 10.2. The *silhouette* of the object $\boldsymbol{x}_i$ for which $|C_\ell| \geq 2$, is the number $sil(\boldsymbol{x}_i)$ given by

$$sil(\boldsymbol{x}_i) = \begin{cases} 1 - \frac{a(\boldsymbol{x}_i)}{b(\boldsymbol{x}_i)} & \text{if } a(\boldsymbol{x}_i) < b(\boldsymbol{x}_i) \\ 0 & \text{if } a(\boldsymbol{x}_i) = b(\boldsymbol{x}_i) \\ \frac{b(\boldsymbol{x}_i)}{a(\boldsymbol{x}_i)} - 1 & \text{if } a(\boldsymbol{x}_i) > b(\boldsymbol{x}_i). \end{cases}$$

Equivalently, we have

$$sil(\boldsymbol{x}_i) = \frac{b(\boldsymbol{x}_i) - a(\boldsymbol{x}_i)}{\max\{a(\boldsymbol{x}_i), b(\boldsymbol{x}_i)\}}$$

for $\boldsymbol{x}_i \in O$.

Observe that $-1 \leq sil(\boldsymbol{x}_i) \leq 1$ (see Exercise 10.6). When $sil(\boldsymbol{x}_i)$ is close to 1, this means that $a(\boldsymbol{x}_i)$ is much smaller than $b(\boldsymbol{x}_i)$ and we may conclude that $\boldsymbol{x}_i$ is well-classified. When $sil(\boldsymbol{x}_i)$ is near 0, it is not clear which is the best cluster for $\boldsymbol{x}_i$. Finally, if $sil(\boldsymbol{x}_i)$ is close to -1, the average distance from u to its neighbor(s) is much smaller than the average distance between $\boldsymbol{x}_i$ and other objects that belong to the same cluster $f(\boldsymbol{x}_i)$. In this case, it is clear that $\boldsymbol{x}_i$ is poorly classified.

Definition 10.3. The *average silhouette width of a cluster C* is

$$sil(C) = \frac{\sum\{sil(u) \mid u \in C\}}{|C|}.$$

The *average silhouette width of a clustering κ* is

$$sil(\kappa) = \frac{\sum\{sil(u) \mid u \in O\}}{|O|}.$$

The silhouette of a clustering can be used for determining the "optimal" number of clusters. If the average silhouette of the clustering is above 0.7, we have a strong clustering.

10.6 MATLAB Computations

MATLAB implements the k-means algorithm using the function `kmeans(T,k)`, where `T` is a data matrix in $\mathbb{R}^{m \times n}$ and `k` is the number of clusters. The function considers the *rows* of T as the points to be clustered. A vector $\texttt{idx} \in \mathbb{R}^m$ is returned indicating that the object $\boldsymbol{x}_i$ belongs to the cluster $\texttt{idx}_i$.

Non-numeric values in data are treated by `kmeans` as missing data and rows that contain such data are ignored. By default, `kmeans` uses the Euclidean distance between vectors.

The function `kmeans(S,k)` has several return formats as indicated in the following table:

Return format	Effect
[IDX,C]	Returns the k cluster centroid locations in the $k \times n$ matrix C
[IDX, C, SUMD]	Returns, additionally, the within-cluster sums of point-to-centroid distances in the row vector `sumD`
[IDX, C, SUMD, D]	`D` is an $m \times k$ matrix that contains the distances from each point to every centroid

A more general format

```
[ ... ] = kmeans(..., 'par1',val1, 'par2',val2, ...)
```

specifies optional parameter/value pairs to direct the algorithm. The following is a partial list of these choices:

Parameter	Value	Meaning
'Distance'		Distance between objects
	'sqEuclidean'	Squared Euclidean distance (default)
	'cityblock'	d_1 distance
	'cosine'	$1 - \cos\alpha$, where $\alpha = \angle(\boldsymbol{u}, \boldsymbol{v})$
	'correlation'	$1 - corr(\boldsymbol{u}, \boldsymbol{v})$
'Start'		method used to choose initial cluster centroid positions;
	'sample'	choose k vectors from S at random (default);
	'uniform'	choose k vectors uniformly from $\langle S \rangle$;
	'cluster'	perform preliminary clustering phase on a random 10% subsample of S; this first phase is itself initialized using 'sample'.
'Replicates'	ℓ	number of times to repeat the clustering, each with a new set of initial centroids; ℓ must be a positive integer, default is 1.

'EmptyAction'		Action to take if a cluster loses all of its members;
	'error'	treat an empty cluster as an error (default);
	'drop'	remove any clusters that become empty;
	'singleton'	create a new cluster consisting of the one observation furthest from its centroid.
'Options'		Options for the iterative algorithm used to minimize the fitting criterion;
	'Display'	level of display output. Choices are 'off', (default), 'iter', and 'final'.
	'MaxIter'	maximum number of iterations allowed; default is 100.

Example 10.1. We begin by generating a dataset containing three clusters using the function `datagen` introduced in Example 4.50.

```
>> opts=statset('Display','final');
>> [idclust,centers]=kmeans(T,3,'Replicates',4,'Options',opts);
>>  plot(T(idclust==1,1),T(idclust==1,2),'+',...
         T(idclust==2,1),T(idclust==2,2),'*',...
         T(idclust==3,1),T(idclust==3,2),'x',...
                              centers(:,1),centers(:,2),'o');
```

The cluster centers are represented by small circles in Figure 10.1. Note the use of the MATLAB function `statset` that creates an option structure such that the named parameters have the specified values.

If $X \in \mathbb{R}^{m\times n}$ is a data matrix whose m rows are the $\boldsymbol{x}'_1, \ldots, \boldsymbol{x}'_m$, then `silhouette(X, CLUST)` plots cluster silhouettes for the matrix X, with clusters defined by `CLUST`. By default, `silhouette` uses the squared Euclidean distance between points; the function call `[S,H] = silhouette(X,CLUST)` plots the silhouettes, and returns the silhouette values in the vector `S` and the figure handle in `H`. Other inter-point distances can be used by calling `silhouette(X,CLUST,d)`, where `d` specifies the distance ('Euclidean','cityblock','cosine', etc). The silhouettes of the 140 points in $\mathbb{R}^2$ generated above using

```
[S,H]=silhouette(T,idclust)
```

are shown in Figure 10.2. As expected, the values of these silhouettes are quite high, since the points were grouped tightly around the centers and we used the same number of clusters ($k = 3$) as the number used in the generation process.

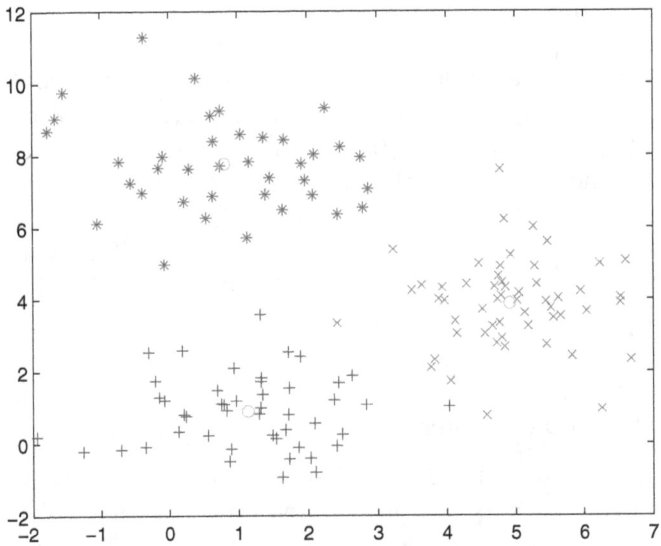

Fig. 10.1 Clustering of a dataset.

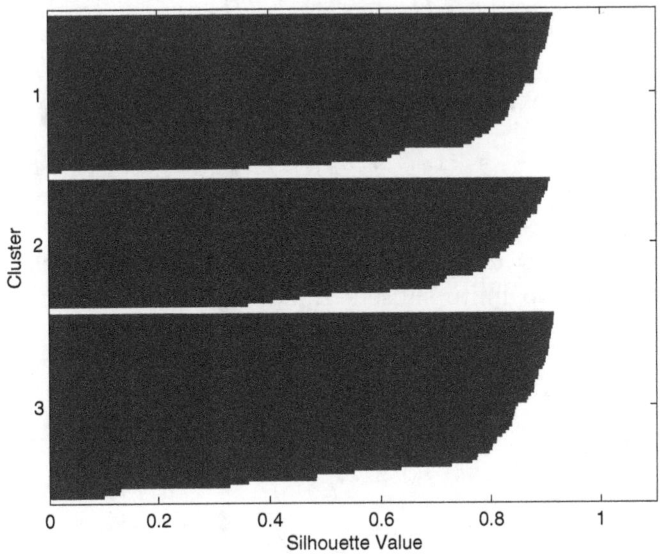

Fig. 10.2 Silhouettes of the 3-clustering.

Exercises and Supplements

(1) Let $\mathbf{c}_0, \mathbf{c}_1$, and $\boldsymbol{x}$ be three vectors in $\mathbb{R}^n$. Prove the following:
   (a) if $d(\mathbf{c}_0, \mathbf{c}_1) > 2d(\boldsymbol{x}, \mathbf{c}_1)$, then $d(\boldsymbol{x}, \mathbf{c}_0) \geq d(\boldsymbol{x}, \mathbf{c}_1)$;
   (b) $d(\boldsymbol{x}, \mathbf{c}_1) \geq \max\{0, d(\boldsymbol{x}, \mathbf{c}_0) - d(\mathbf{c}_0, \mathbf{c}_1)\}$.

   Examine ways to use the above inequalities to improve the performance of the k-means algorithm.

(2) Let $X = \{x_1, \ldots, x_{2n}\}$ be a set that consists of $2n$ real numbers. A *balanced clustering* of X is a partition $\pi = \{C, D\}$ of X such that $|C| = |D| = n$. Prove that if π is such that $\sum_{x,y \in C} d(x, y) + \sum_{x,y \in D} d(x, y)$ is minimal, then one of the clusters consists of the first n points detected by scanning the line from left to right and the other cluster consists of the remaining points.

   Solution: Let $z \in \mathbb{R} - X$ be such that there are n points of X at its left and an equal number of points at its right. For an arbitrary balanced clustering $\{C, D\}$ of X, define the sets

$$C_{l,z} = \{x \in C \mid x < z\}, \quad C_{r,z} = \{x \in C \mid x > z\},$$
$$D_{l,z} = \{x \in D \mid x < z\}, \quad D_{r,z} = \{x \in D \mid x > z\}.$$

   Note that $|C_{l,z}|+|D_{l,z}| = |C_{r,z}|+|D_{r,z}| = n$, so $\{C_{l,z} \cup D_{l,z}, C_{r,z} \cup D_{r,z}\}$ is again a balanced clustering. Suppose that

$$C_{l,z} = \{u_1, \ldots, u_k\}, \quad C_{r,z} = \{u_{k+1}, \ldots, y_n\},$$
$$D_{l,z} = \{v_{k+1}, \ldots, v_n\}, \quad D_{r,z} = \{v_1, \ldots, v_k\}.$$

   Clearly, $C = C_{l,z} \cup C_{r,z}$ and $D = D_{l,z} \cup D_{r,z}$. If the distance between two finite subsets A, B of $\mathbb{R}$ is defined as $d(A, B) = \sum\{|a - b| \mid a \in A, b \in B\}$, then we claim that

$$d(C_{l,z}, C_{r,z}) + d(D_{l,z}, D_{r,z}) \geq d(C_{l,z}, D_{l,z}) + d(C_{r,z}, D_{r,z}).$$

   We have

$$\begin{aligned} d(C_{l,z}, C_{r,z}) &= \sum_{i=1}^{k} \sum_{j=k+1}^{n} (u_j - u_i) \\ &= \sum_{i=1}^{k} \left(\sum_{j=k+1}^{n} u_j \right) - (n - k)u_i \end{aligned}$$

$$= k \sum_{j=k+1}^{n} u_j - (n-k)\sum_{i=1}^{k} u_i,$$

$$\begin{aligned} d(D_{l,z}, D_{r,z}) &= \sum_{j=k+1}^{n} \sum_{i=1}^{k} (v_i - v_j) \\ &= \sum_{j=k+1}^{n} \left(\sum_{i=1}^{k} v_i\right) - kv_j \\ &= (n-k)\sum_{i=1}^{k} v_i - k \sum_{j=k+1}^{n} v_j, \end{aligned}$$

and

$$\begin{aligned} d(C_{l,z}, D_{l,z}) &= \sum_{i=1}^{k} \sum_{j=k+1}^{n} |u_i - v_j|, \\ &\leq \sum_{i=1}^{k} \sum_{j=k+1}^{n} (|z - u_i| + |v_j - z|), \\ &= (n-k)\sum_{i=1}^{k} |z - u_i| + k \sum_{j=k+1}^{n} |v_j - z|, \\ d(C_{r,z}, D_{r,z}) &= \sum_{j=k+1}^{n} \sum_{i=1}^{k} |u_j - v_i| \\ &\leq \sum_{j=k+1}^{n} \sum_{i=1}^{k} |u_j - z| + |z - v_i| \\ &= k \sum_{j=k+1}^{n} |u_j - z| + (n-k)\sum_{i=1}^{k} |z - v_i|. \end{aligned}$$

The desired equality follows after some elementary algebra.

(3) Let $\boldsymbol{g}$ be the centroid of a finite, non-empty subset S of $\mathbb{R}^n$. Define the set of centered vectors $\hat{S}=\{\boldsymbol{y} \in \mathbb{R}^n \mid \boldsymbol{y}=\boldsymbol{x}-\boldsymbol{g}, \boldsymbol{x} \in S\}$.

Prove that

$$\sum\{\|\boldsymbol{y}\|_2^2 \mid \boldsymbol{y} \in \hat{S}\} = \sum\{\|\boldsymbol{x}\|_2^2 \mid \boldsymbol{x} \in S\} - |S|\|\boldsymbol{g}\|_2^2.$$

(4) Let S be a finite, non-empty subset of $\mathbb{R}^n$, $\pi = \{C_1, \ldots, C_k\}$ be a partition of S, $\boldsymbol{g}$ be the centroid of S, and $\boldsymbol{c}_i$ be the centroid of C_i for $1 \leq i \leq k$. Prove that $\boldsymbol{g}$ is the convex combination

$$\boldsymbol{g} = \sum_{i=1}^{k} \frac{|C_i|}{|S|} \boldsymbol{c}_i.$$

(5) This exercise, based on an example given in [88], shows that the Forgy–Lloyd k-means algorithm may converge to a locally minimal solution that is arbitrarily bad compared to the optimal solution.
Let $S = \{x_1, x_2, x_3, x_4\} \subseteq \mathbb{R}$ be the set to be clustered using the Forgy–Lloyd k-means algorithm with $k = 3$. Suppose that $x_1 < x_2 < x_3 < x_4$ and that $x_4 - x_3 < x_2 - x_1 < x_3 - x_2$. Prove that:
   (a) there are only three possible partition outcomes of the Forgy–Lloyd algorithm: $\pi_1 = \{\{x_1\}, \{x_2\}, \{x_3, x_4\}\}$, $\pi_2 = \{\{x_1, x_2\}, \{x_3\}, \{x_4\}\}$, and $\pi_3 = \{\{x_1\}, \{x_2, x_3\}, \{x_4\}\}$;
   (b) $sse(\pi_1) = \frac{(x_4 - x_3)^2}{2}$, $sse(\pi_2) = \frac{(x_2 - x_1)^2}{2}$, and $sse(\pi_3) = \frac{(x_3 - x_2)^2}{2}$.
   (c) Show that π_1 is the optimal partition and the ratio $\frac{sse(\pi_2)}{sse(\pi_1)}$ (where π_2 is a locally minimal partition and, therefore, a possible outcome of the Forgy–Lloyd algorithm) can be arbitrarily large if the distance between x_2 and x_3 is sufficiently large.

(6) Let $\pi = \{C_1, \ldots, C_k\}$ be a clustering of a finite set of objects $S = \{\boldsymbol{x}_1, \ldots, \boldsymbol{x}_m\}$. Prove that

$$sse(\pi) = \sum_{j=1}^{k} \frac{1}{2m_j} \sum_{\boldsymbol{x}_p, \boldsymbol{x}_q \in C_j} \|\boldsymbol{x}_p - \boldsymbol{x}_q\|_2^2,$$

where $m_j = |C_j|$ for $1 \leq j \leq k$.

Solution: To prove this equality, it suffices to show that

$$\sum_{\boldsymbol{x}\in C_j} \|\boldsymbol{x} - \mathbf{c}_j\|_2^2 = \frac{1}{2m_j} \sum_{\boldsymbol{x}_p,\boldsymbol{x}_q\in C_j} \|\boldsymbol{x}_p - \boldsymbol{x}_q\|_2^2.$$

This follows by writing

$$\begin{aligned}
&\frac{1}{2m_j} \sum_{\boldsymbol{x}_p,\boldsymbol{x}_q\in C_j} \|\boldsymbol{x}_p - \boldsymbol{x}_q\|_2^2 \\
&\quad = \frac{1}{2m_j} \sum_{\boldsymbol{x}_p,\boldsymbol{x}_q\in C_j} (\boldsymbol{x}_p'\boldsymbol{x}_p + \boldsymbol{x}_q'\boldsymbol{x}_q - 2\boldsymbol{x}_p'\boldsymbol{x}_q) \\
&\quad = \frac{1}{2m_j} \left(m_j \sum_{\boldsymbol{x}_p\in C_j} \boldsymbol{x}_p'\boldsymbol{x}_p + m_j \sum_{\boldsymbol{x}_q\in C_j} \boldsymbol{x}_q'\boldsymbol{x}_q \right. \\
&\qquad \left. - 2 \sum_{\boldsymbol{x}_p\in C_j} \sum_{\boldsymbol{x}_q\in C_j} \boldsymbol{x}_p'\boldsymbol{x}_q \right) \\
&\quad = \sum_{\boldsymbol{x}\in C_j} \boldsymbol{x}'\boldsymbol{x} - \frac{1}{m_j} \sum_{\boldsymbol{x}_p\in C_j} \sum_{\boldsymbol{x}_q\in C_j} \boldsymbol{x}_p'\boldsymbol{x}_q \\
&\quad = \sum_{\boldsymbol{x}\in C_j} \boldsymbol{x}'\boldsymbol{x} - m_j\mathbf{c}_j'\mathbf{c}_j,
\end{aligned}$$

and

$$\begin{aligned}
\sum_{\boldsymbol{x}\in C_j} \|\boldsymbol{x} - \mathbf{c}_j\|_2^2 &= \sum_{\boldsymbol{x}\in C_j} (\boldsymbol{x}'\boldsymbol{x} - 2\boldsymbol{x}'\mathbf{c}_j + \mathbf{c}_j'\mathbf{c}_j) \\
&= \sum_{\boldsymbol{x}\in C_j} \boldsymbol{x}'\boldsymbol{x} - m_j\mathbf{c}'\mathbf{c}_j,
\end{aligned}$$

which confirms the needed equality.

Let $\pi = \{C_1, \ldots, C_k\}$ be a partition of a finite, non-empty set $S \subseteq \mathbb{R}^n$ and let α be a non-negative number. Define $sse_\alpha(\pi)$ as $sse_\alpha(\pi) = \sum_{j=1}^{k} |C_j|^\alpha \sum_{\boldsymbol{x}\in C_j} d^2(\boldsymbol{x}, \mathbf{c}_j)$. Note that for $\alpha = 0$, we have $sse_0(\pi) = sse(\pi)$. This idea was introduced in [81].

(7) Examine the effect of using sse_α with $\alpha > 0$ instead of sse_0 on the relative size of the clusters resulting from an application of the k-means algorithm.

(8) Prove that the number of hyperplanes given in Formula (10.4)

$$\sum\left\{\binom{\binom{m}{n}}{t} \,\middle|\, k \leq t \leq \binom{k}{2}\right\}$$

is in $O\left(m^{\frac{nk^2}{2}}\right)$.

(9) Let $C, D \subseteq \mathbb{R}^n$ be two finite, non-empty, and disjoint subsets of $\mathbb{R}^n$ and let

$$\delta(C, D) = \sum\{\|\boldsymbol{x} - \boldsymbol{y}\|_2^2 \mid \boldsymbol{x} \in C, \boldsymbol{y} \in D\}.$$

Prove that

(a) $\frac{\delta(C,D)}{mp} = \frac{\delta(C,C)}{m^2} + \frac{\delta(D,D)}{p^2} + \|\mathbf{c} - \mathbf{d}\|_2^2$, where $m = |C|$, $p = |D|$ and $\mathbf{c}$,$\mathbf{d}$ are the centroids of C and D, respectively.

(b) $\delta(C, D) + \delta(C, C) + \delta(D, D) = |S| \sum_{y \in \hat{S}} \|\boldsymbol{y}\|_2^2$.

Solution: For the first part, we have

$$\begin{aligned}\delta(C, C) &= \sum\{\|\boldsymbol{x}_i - \boldsymbol{x}_j\|_2^2 \mid \boldsymbol{x}_i, \boldsymbol{x}_j \in C\}\\ &= \sum\{\boldsymbol{x}_i'\boldsymbol{x}_i - 2\boldsymbol{x}_i'\boldsymbol{x}_j + \boldsymbol{x}_j'\boldsymbol{x}_j \mid \boldsymbol{x}_i, \boldsymbol{x}_j \in C\}\\ &= (m-1)\sum\{\boldsymbol{x}_i'\boldsymbol{x}_i \mid \boldsymbol{x}_i \in C\}\\ &\quad -2\sum_{\boldsymbol{x}_i \in C}\sum_{\boldsymbol{x}_j \in C}\{\boldsymbol{x}_i'\boldsymbol{x}_j \mid i < j\}.\end{aligned}$$

Since $\sum_{i=1}^m \boldsymbol{x}_i = m\mathbf{c}$, the norm of the centroid of C is

$$m^2\mathbf{c}'\mathbf{c} = \sum\{\boldsymbol{x}_i'\boldsymbol{x}_i \mid \boldsymbol{x}_i \in C\} + 2\sum_{\boldsymbol{x}_i \in C}\sum_{\boldsymbol{x}_j \in C}\{\boldsymbol{x}_i'\boldsymbol{x}_j \mid i < j\},$$

so

$$\delta(C, C) + m^2\mathbf{c}'\mathbf{c} = m\sum\{\boldsymbol{x}_i'\boldsymbol{x}_i \mid \boldsymbol{x}_i \in C\}.$$

Similarly,

$$\delta(D, D) + p^2\mathbf{d}'\mathbf{d} = p\sum\{\boldsymbol{x}_j'\boldsymbol{x}_j \mid \boldsymbol{x}_j \in D\}.$$

For $\delta(C, D)$, we have

$$\delta(C, D) = \sum_{\boldsymbol{x}_i \in C, \boldsymbol{x}_j \in D}(\boldsymbol{x}_i'\boldsymbol{x}_i + \boldsymbol{x}_j'\boldsymbol{x}_j - 2\boldsymbol{x}_i'\boldsymbol{x}_j)$$

$$= p \sum_{\boldsymbol{x}_i \in C} \boldsymbol{x}_i' \boldsymbol{x}_i + m \sum_{\boldsymbol{x}_j \in D} \boldsymbol{x}_j' \boldsymbol{x}_j - 2 \sum_{\boldsymbol{x}_i \in C} \sum_{\boldsymbol{x}_j \in D} \boldsymbol{x}_i' \boldsymbol{x}_j$$

$$= p \sum_{\boldsymbol{x}_i \in C} \boldsymbol{x}_i' \boldsymbol{x}_i + m \sum_{\boldsymbol{x}_j \in D} \boldsymbol{x}_j' \boldsymbol{x}_j - 2mp\, \boldsymbol{c}' \boldsymbol{d}.$$

These equalities yield

$$\begin{aligned}
&\frac{\delta(C,D)}{mp} - \frac{\delta(C,C)}{m^2} - \frac{\delta(D,D)}{p^2} \\
&\quad = \frac{1}{m} \sum_{\boldsymbol{x}_i \in C} \boldsymbol{x}_i' \boldsymbol{x}_i + \frac{1}{p} \sum_{\boldsymbol{x}_j \in D} \boldsymbol{x}_j' \boldsymbol{x}_j - 2\, \boldsymbol{c}' \boldsymbol{d} \\
&\qquad + \boldsymbol{c}' \boldsymbol{c} - \frac{1}{m} \sum \{\boldsymbol{x}_i' \boldsymbol{x}_i \mid \boldsymbol{x}_i \in C\} + \boldsymbol{d}' \boldsymbol{d} \\
&\qquad - \frac{1}{p} \sum \{\boldsymbol{x}_j' \boldsymbol{x}_j \mid \boldsymbol{x}_j \in D\} \\
&\qquad + \boldsymbol{c}' \boldsymbol{c}' + \boldsymbol{d}' \boldsymbol{d} - 2\boldsymbol{c}' \boldsymbol{d} = \| \boldsymbol{c} - \boldsymbol{d} \|_2^2,
\end{aligned}$$

which concludes the argument for the first part.

For the second part, using the expressions derived above, we obtain

$$\begin{aligned}
&\delta(C,D) + \delta(C,C) + \delta(D,D) \\
&\quad = p \sum_{\boldsymbol{x}_i \in C} \boldsymbol{x}_i' \boldsymbol{x}_i + m \sum_{\boldsymbol{x}_j \in D} \boldsymbol{x}_j' \boldsymbol{x}_j - 2mp\, \boldsymbol{c}' \boldsymbol{d} \\
&\qquad + m \sum \{\boldsymbol{x}_i' \boldsymbol{x}_i \mid \boldsymbol{x}_i \in C\} - m^2 \boldsymbol{c}' \boldsymbol{c} \\
&\qquad + p \sum \{\boldsymbol{x}_j' \boldsymbol{x}_j \mid \boldsymbol{x}_j \in D\} - p^2 \boldsymbol{d}' \boldsymbol{d} \\
&\quad = |S| \sum_{\boldsymbol{x} \in S} \boldsymbol{x}' \boldsymbol{x} - 2mp\, \boldsymbol{c}' \boldsymbol{d} - m^2 \boldsymbol{c}' \boldsymbol{c} - p^2 \boldsymbol{d}' \boldsymbol{d} \\
&\quad = |S| \sum_{\boldsymbol{x} \in S} \boldsymbol{x}' \boldsymbol{x} - \| m\boldsymbol{c} + p\boldsymbol{d} \|_2^2 \\
&\quad = |S| \sum_{\boldsymbol{x} \in S} \boldsymbol{x}' \boldsymbol{x} - |S|^2 \| \boldsymbol{g} \|_2^2
\end{aligned}$$

$$= |S| \sum_{y \in \hat{S}} \|\boldsymbol{y}\|_2^2,$$

taking into account Exercise 3.

(10) Using the notations introduced in Supplement 9, prove that for any two finite, non-empty, and disjoint subsets C, D of $\mathbb{R}^n$, we have

$$2\frac{\delta(C,D)}{mp} \geq \frac{\delta(C,C)}{m^2} + \frac{\delta(D,D)}{p^2}.$$

Hint: The inequality follows immediately from Supplement 9.

(11) Let $\pi = \{C, D\}$ be a clustering of the set $S \subseteq \mathbb{R}^n$, where $C = \{\boldsymbol{x}_1, \ldots, \boldsymbol{x}_m\}$ and $D = \{\boldsymbol{x}_{m+1}, \ldots, \boldsymbol{x}_{m+p}\}$, $m + p = |S|$, and let $T = (t_{ij}) \in \mathbb{R}^{s \times s}$ be the matrix defined by $t_{ij} = \|\boldsymbol{x}_i - \boldsymbol{x}_j\|_2^2$ for $1 \leq i, j \leq s$, where $s = |S| = m + p$. Define

$$J(\pi) = \frac{mp}{|S|}\left(2\frac{\delta(C,D)}{mp} - \frac{\delta(C,C)}{m^2} - \frac{\delta(D,D)}{p^2}\right).$$

Let $\boldsymbol{q} \in \mathbb{R}^s$ be the vector given by

$$q_i = \begin{cases} \sqrt{\frac{p}{ms}} & \text{if } 1 \leq i \leq m, \\ \sqrt{-\frac{m}{ps}} & \text{if } m+1 \leq i \leq m+p. \end{cases}$$

Prove that

(a) $\boldsymbol{1}_|S|'\boldsymbol{q} = 0$ and $\|\boldsymbol{q}\|_2 = 1$;

(b) $\boldsymbol{q}'T\boldsymbol{q} = -J(\pi)$;

(c) the previous two parts hold if the vectors of S are not arranged in any particular order.

Solution: The first part is immediate. Applying the definition of T, we obtain

$$\boldsymbol{q}'T\boldsymbol{q} = \sum_{i=1}^{m+p}\sum_{j=1}^{m+p} \boldsymbol{q}_i t_{ij} q_j$$

$$= \sum_{i=1}^{m}\sum_{j=1}^{m} \boldsymbol{q}_i t_{ij} q_j + \sum_{i=1}^{m}\sum_{j=m+1}^{m+p} \boldsymbol{q}_i t_{ij} q_j$$

$$\sum_{i=m+1}^{m+p}\sum_{j=1}^{m}\boldsymbol{q}_i t_{ij} q_j + \sum_{i=m+1}^{m+p}\sum_{j=m+1}^{m+p}\boldsymbol{q}_i t_{ij} q_j$$

$$= \frac{p}{ms}\sum_{i=1}^{m}\sum_{j=1}^{m}\|\boldsymbol{x}_i - \boldsymbol{x}_j\|_2^2 - \frac{1}{s}\sum_{i=1}^{m}\sum_{j=m+1}^{m+p}\|\boldsymbol{x}_i - \boldsymbol{x}_j\|_2^2$$

$$- \frac{1}{s}\sum_{i=m+1}^{m+p}\sum_{j=1}^{m}\|\boldsymbol{x}_i - \boldsymbol{x}_j\|_2^2$$

$$+ \frac{m}{ps}\sum_{i=m+1}^{m+p}\sum_{j=m+1}^{m+p}\|\boldsymbol{x}_i - \boldsymbol{x}_j\|_2^2$$

$$= \frac{p}{ms}\delta(C,C) - \frac{2}{s}\delta(C,D) + \frac{m}{ps}\delta(D,D)$$

$$= -\frac{mp}{s}\left(\frac{2\delta(C,D)}{mp} - \frac{\delta(C,C)}{m^2} - \frac{\delta(D,D)}{p^2}\right) = -J(\pi).$$

(12) Prove that for 2-means clustering $\pi = \{C, D\}$ of a finite, nonempty set $S \subseteq \mathbb{R}^n$, we have $sse(\pi) + \frac{1}{2}J(\pi) = \frac{1}{2}\sum_{\boldsymbol{y}\in\hat{S}} \boldsymbol{y}'\boldsymbol{y}$, where $\hat{S}$ is the set of centered vectors that correspond to the vectors of S.

Solution: The sum $sse(\pi) + \frac{1}{2}J(\pi)$ can be expressed using the function δ introduced earlier as follows:

$$sse(\pi) + \frac{1}{2}J(\pi)$$

$$= \frac{1}{2m}\delta(C,C) + \frac{1}{2p}\delta(D,D) + \frac{1}{2|S|}\delta(C,D)$$

$$- \frac{p\delta(C,C)}{2m|S|} - \frac{m\delta(D,D)}{2p|S|}$$

(by Supplements 10.6 and 10.6)

$$= \frac{1}{2|S|}(\delta(C,D) + \delta(C,C) + \delta(D,D)) = \frac{1}{2}\sum_{y\in\hat{S}}\|\boldsymbol{y}\|_2^2,$$

(by the second part of Supplement 9),

where $m = |C|$ and $p = |D|$.

Supplement 12 suggests a way for obtaining optimal 2-means clustering. Since $sse(\pi)+\frac{1}{2}J(\pi)$ does not depend on the partition π, minimizing $sse(\pi)$ is tantamount to maximizing $J(\pi)$. Note that the effect of maximizing $J(\pi)$ is to ensure that the average inter-cluster distance $\frac{\delta(C,D)}{mp}$ is maximal, while the average intra-cluster distances $\frac{\delta(C,C)}{m^2}$ and $\frac{\delta(D,D)}{p^2}$ are minimal. This ensures well-separated, compact clusters.

In [35], a relaxation of this problem is considered by allowing the components of $\boldsymbol{q}$ to assume values in the interval $[-1,1]$ rather than two discrete values, but still requires that $\boldsymbol{q}'\boldsymbol{q} = 1$. Then, to maximize $J(\pi)$, that is, to minimize $-J(\pi)$ amounts to finding $\boldsymbol{q}$ as the unit eigenvector that corresponds to the least eigenvalue of T. An alternative relaxation method is discussed next.

(13) Let S be a finite, non-empty subset of $\mathbb{R}^n$ with $|S| = m$, and let $\hat{T} = H_m T H_m \in \mathbb{R}^{m\times m}$ be the centered distance matrix, where T is the matrix of squared distances between the vectors of S. Prove that

(a) for every vector $\boldsymbol{q}$ such that $\boldsymbol{1}_m'\boldsymbol{q} = 0$, we have $\boldsymbol{q}'\hat{T}\boldsymbol{q} = \boldsymbol{q}'T\boldsymbol{q}$;

(b) every eigenvector of $\hat{T}$ that corresponds to a non-zero eigenvalue is orthogonal on $\boldsymbol{1}_m$.

Solution: Observe that $\boldsymbol{q}'\hat{T}\boldsymbol{q} = \boldsymbol{q}'H_mTH_m\boldsymbol{q} = \boldsymbol{q}'T\boldsymbol{q}$ because $H_m\boldsymbol{q} = \boldsymbol{q}'H_m = \boldsymbol{q}$. Since $\boldsymbol{1}_m$ is an eigenvector of $\hat{T}$ that corresponds to the eigenvalue 0, every eigenvector of $\hat{T}$ that belongs to a different eigenvalue is orthogonal on $\boldsymbol{1}_m$.

(14) Prove that $-1 \leq sil(\boldsymbol{x}) \leq 1$ for every object $\boldsymbol{x} \in S$, where S is a set equipped with a clustering.

Bibliographical Comments

The spectral relaxation of the k-means algorithm was obtained in [176]. The link between k-means clustering and non-negative matrix factorization was established in [36]. For a solution to Exercise 1, see [45]. Supplement 2 is a result obtained in [132]. Supplements 9–13 are based on Ding and He's paper where the *PCA-guided clustering* is introduced.

Chapter 11

Data Sample Matrices

11.1 Introduction

Matrices are natural tools for organizing datasets. Let such a dataset consist of a sequence $\mathcal{E}$ of m vectors of $\mathbb{R}^n$, $(\boldsymbol{u}_1, \ldots, \boldsymbol{u}_m)$. The jth components $(\boldsymbol{u}_i)_j$ of these vectors correspond to the values of a random variable $\mathcal{V}_j$, where $1 \leq j \leq n$. This data series will be represented as a matrix having m rows $\boldsymbol{u}'_1, \ldots, \boldsymbol{u}'_m$ and n columns $\boldsymbol{v}_1, \ldots, \boldsymbol{v}_n$. We refer to matrices obtained in this manner as *sample matrices*. The number m is the *size* of the sample.

In this chapter, we present algebraic properties of vectors and matrices associated with a sample matrix: the mean vector and the covariance matrix. Biplots, a technique for exploring and visualizing sample matrices, are also introduced.

11.2 The Sample Matrix

Each row vector $\boldsymbol{u}'_i$ corresponds to an experiment E_i in the series of experiments $\mathcal{E} = (E_1, \ldots, E_m)$; the experiment E_i consists of measuring the n components of $\boldsymbol{u}'_i = (x_{i1}, \ldots, x_{in})$, as follows.

$$\begin{array}{c|ccc} & \boldsymbol{v}_1 & \cdots & \boldsymbol{v}_n \\ \hline \boldsymbol{u}'_1 & x_{11} & \cdots & x_{1n} \\ \boldsymbol{u}'_2 & x_{21} & \cdots & x_{2n} \\ \vdots & \vdots & \vdots & \vdots \\ \boldsymbol{u}'_m & x_{m1} & \cdots & x_{mn} \\ \hline \end{array}.$$

The column vector

$$\boldsymbol{v}_j = \begin{pmatrix} x_{1j} \\ x_{2j} \\ \vdots \\ x_{mj} \end{pmatrix}$$

represents the measurements of the jth variable $\mathcal{V}_j$ of the experiment, for $1 \leq j \leq n$, as shown in the following. These variables are usually referred to as *attributes* or *features* of the series $\mathcal{E}$.

Definition 11.1. The *sample matrix* of $\mathcal{E}$ is the matrix $X \in \mathbb{C}^{m \times n}$ given by

$$X = \begin{pmatrix} \boldsymbol{u}'_1 \\ \vdots \\ \boldsymbol{u}'_m \end{pmatrix} = (\boldsymbol{v}_1 \ \cdots \ \boldsymbol{v}_n).$$

Clearly, we have $(\boldsymbol{v}_j)_i = (\boldsymbol{u}'_i)_j = x_{ij}$ for $1 \leq i \leq m$ and $1 \leq j \leq n$. If $\mathcal{E}$ is clear from the context, the subscript $\mathcal{E}$ is omitted. We will use both representations of the sample matrix and will write

$$X = \begin{pmatrix} \boldsymbol{u}'_1 \\ \vdots \\ \boldsymbol{u}'_m \end{pmatrix} = (\boldsymbol{v}_1 \ \cdots \ \boldsymbol{v}_n),$$

when we are interested in the vectors that represent results of experiments and

$$X = (\boldsymbol{v}_1, \ldots, \boldsymbol{v}_n),$$

when we need to work with vectors that represent the values of variables.

Pairwise distances between the row vectors of the sample matrix $X \in \mathbb{R}^{m \times n}$ can be computed with the MATLAB function `pdist(X)`. This form of the function returns a vector $\boldsymbol{D}$ having $\frac{m(m-1)}{2}$ components corresponding to $\binom{m}{2}$ pairs of observations arranged in the order $d_2(\boldsymbol{u}'_2, \boldsymbol{u}'_1), d_2(\boldsymbol{u}'_3, \boldsymbol{u}'_1), d_2(\boldsymbol{u}'_3, \boldsymbol{u}'_2), \ldots$, that is the order of the lower triangle of the distance matrix.

Example 11.1. Let X be the data matrix

$$X = \begin{pmatrix} 1 & 4 & 5 \\ 2 & 3 & 7 \\ 5 & 1 & 4 \\ 6 & 2 & 4 \end{pmatrix}.$$

The function call `D = pdist(X)` returns

```
D =
    2.4495     6.0000     5.4772     7.3485     5.0990     5.0990
```

Equivalently, a distance matrix can be obtained using the auxiliary function `squareform`, by writing `E = squareform(D)`, which yields

```
E =
         0    2.4495    6.0000    5.4772
    2.4495         0    7.3485    5.0990
    6.0000    7.3485         0    5.0990
    5.4772    5.0990    5.0990         0
```

There are versions of `pdist` that can return other distances by using a second string parameter. For instance, `pdist(X, 'cityblock')` computes $d_1(\boldsymbol{x}_i, \boldsymbol{x}_j)$ and `pdist(X,'cebyshev')` computes $d_\infty(\boldsymbol{x}_i, \boldsymbol{x}_j)$. In general, Minkowski's distance d_p can be computed using `D = pdist(X,'minkowski',p)`.

A *linear data mapping* for a data sequence $(\boldsymbol{u}_1, \ldots, \boldsymbol{u}_m) \in \mathbf{Seq}_m(\mathbb{R}^n)$ is the morphism $\mathsf{r} : \mathbb{R}^n \longrightarrow \mathbb{R}^q$. If $R \in \mathbb{R}^{n \times q}$ is the matrix that represents this mapping, then $\mathsf{r}(\boldsymbol{u}_i) = R\boldsymbol{u}_i$ for $1 \leq i \leq m$.

If $q < n$, we refer to r as a *linear dimensionality-reduction mapping*. The *reduced data matrix* is given by

$$\mathsf{r}(X_{\mathcal{E}}) = \begin{pmatrix} \mathsf{r}(\boldsymbol{u}_1)' \\ \vdots \\ \mathsf{r}(\boldsymbol{u}_m)' \end{pmatrix} = \begin{pmatrix} (R\boldsymbol{u}_1)' \\ \vdots \\ (R\boldsymbol{u}_m)' \end{pmatrix} = X_{\mathcal{E}} R \in \mathbb{R}^{m \times q}.$$

The reduced dataset $\mathsf{r}(X_{\mathcal{E}})$ has new variables $\mathcal{Y}_1, \ldots, \mathcal{Y}_q$. We denote this by writing

$$(\mathcal{Y}_1, \ldots, \mathcal{Y}_q) = \mathsf{r}(\mathcal{V}_1, \ldots, \mathcal{V}_n).$$

The mapping r is a *linear feature selection mapping* if $R \in \{0,1\}^{q \times n}$ is a 0/1-matrix having exactly one unit in every row and at most one unit in every column.

Definition 11.2. Let $(\boldsymbol{u}_1, \ldots, \boldsymbol{u}_m)$ be a series of observations in $\mathbb{R}^n$. The *sample mean* of this sequence is the vector

$$\tilde{\boldsymbol{u}} = \frac{1}{m} \sum_{i=1}^{m} \boldsymbol{u}_i \in \mathbb{R}^n. \tag{11.1}$$

The series is *centered* if $\tilde{\boldsymbol{u}} = \mathbf{0}_n$.

Note that the series $(\boldsymbol{u}_1 - \tilde{\boldsymbol{u}}, \ldots, \boldsymbol{u}_m - \tilde{\boldsymbol{u}})$ is always centered. Also, observe that

$$\tilde{\boldsymbol{u}} = \frac{1}{m}(\boldsymbol{u}_1 \ \cdots \ \boldsymbol{u}_m)\mathbf{1}_m. \tag{11.2}$$

If $n = 1$, the series of observations is reduced to a vector $\boldsymbol{v} \in \mathbb{R}^m$.

Definition 11.3. The *standard deviation of a vector* $\boldsymbol{v} \in \mathbb{R}^m$ is the number

$$s_{\boldsymbol{v}} = \sqrt{\frac{1}{m-1} \sum_{i=1}^{m} (v_i - v)^2},$$

where v is the mean of the components of $\boldsymbol{v}$.

The *standard deviation of sample matrix* $X \in \mathbb{R}^{m \times n}$, where $X = (\boldsymbol{v}_1 \ \cdots \ \boldsymbol{v}_n)$, is the row $\boldsymbol{s} = (s_{\boldsymbol{v}_1}, \ldots, s_{\boldsymbol{v}_n})$.

If the measurement scale for the variables $\mathcal{V}_1, \ldots, \mathcal{V}_n$ involved in the experiment are very different due to different measurement units, some variables may inappropriately influence the analysis process. Therefore, the columns of the data sample matrix need to be *scaled* in order to make their values comparable. To scale a matrix, we need to replace each column $\boldsymbol{v}_i$ by $\frac{1}{s_{\boldsymbol{v}_i}}\boldsymbol{v}_i$. This will yield a matrix having the standard deviation of each column equal to 1.

Next, we examine the effect of centering on a sample matrix.

Theorem 11.1. *Let* $X \in \mathbb{R}^{m \times n}$ *be a sample matrix*

$$X = \begin{pmatrix} \boldsymbol{u}'_1 \\ \vdots \\ \boldsymbol{u}'_m \end{pmatrix}.$$

The sample matrix that corresponds to the centered sequence is

$$\hat{X} = \left(I_m - \frac{1}{m}\mathbf{1}_m\mathbf{1}'_m \right) X.$$

Proof. The matrix that corresponds to the centered sequence is

$$\hat{X} = \begin{pmatrix} \boldsymbol{u}_1' - \tilde{\boldsymbol{u}}' \\ \vdots \\ \boldsymbol{u}_m' - \tilde{\boldsymbol{u}}' \end{pmatrix} = X - \mathbf{1}_m \tilde{\boldsymbol{u}}'.$$

By Equality (11.2), it follows that

$$\hat{X} = X - \mathbf{1}_m \tilde{\boldsymbol{u}}' = X - \frac{1}{m}\mathbf{1}_m \mathbf{1}_m' X = \left(I_m - \frac{1}{m}\mathbf{1}_m \mathbf{1}_m'\right) X,$$

which yields the desired equality. □

Theorem 11.1 shows that to center a data matrix $X \in \mathbb{R}^{m\times n}$, we need to multiply it at the left by the *centering matrix*

$$H_m = I_m - \frac{1}{m}\mathbf{1}_m \mathbf{1}_m' \in \mathbb{R}^{m\times m},$$

that is, $\hat{X} = H_m X$. Note that $H_m = I_m - \frac{1}{m} J_m$. It is easy to see that H_m is both symmetric and idempotent. Since

$$H_m \mathbf{1}_m = \mathbf{1}_m - \frac{1}{m}\mathbf{1}_m \mathbf{1}_m' \mathbf{1}_m = \mathbf{0},$$

it follows that H_m has the eigenvalue 0.

If $X \in \mathbb{R}^{m\times n}$ is a matrix, the standard deviations are computed in MATLAB using the function `std(X)`, which returns an n-dimensional row $\boldsymbol{s}$ containing the square roots of the sample variances of the columns of U, that is, their standard deviations. The means of the columns of X is computed in MATLAB using the function `mean(X)`.

The MATLAB function `Z = zscore(X)` computes a centered and scaled version of a data sample matrix having the same format as X.

If X is a matrix, then z-scores are computed using the mean and standard deviation along each column of X. The columns of Z have sample mean zero and sample standard deviation one (unless a column of X is constant, in which case that column of Z is constant at 0). If we use the format

```
[Z,mu,sigma] = zscore(X),
```

the mean vector is returned to `mu` and the vector of standard deviations, to `sigma`.

Example 11.2. Let X be the matrix

```
X =
        1    12    77
        3    15    80
        2    15    75
        5    18    98
```

The means and the standard deviations of the columns of X are obtained as follows.

```
>> m = mean(X)

m =
    2.7500   15.0000   82.5000

>> s=std(X)

s =
    1.7078    2.4495   10.5357
```

Finally, to compute together the mean, the standard deviation, and the matrix Z, we write

```
>> [Z,m,s]=zscore(A)

Z =
   -1.0247   -1.2247   -0.5220
    0.1464         0   -0.2373
   -0.4392         0   -0.7119
    1.3175    1.2247    1.4712


m =
    2.7500   15.0000   82.5000


s =
    1.7078    2.4495   10.5357
```

Definition 11.4. Let $\boldsymbol{u} = (\boldsymbol{u}_1, \ldots, \boldsymbol{u}_m)$ be a sequence of vectors in $\mathbb{R}^n$. The *inertia* of this sequence relative to a vector $\boldsymbol{z} \in \mathbb{R}^n$ is the number

$$I_{\boldsymbol{z}}(\boldsymbol{u}) = \sum_{j=1}^{m} \|\boldsymbol{u}_j - \boldsymbol{z}\|_2^2.$$

Theorem 11.2 (Huygens's Inertia Theorem). Let $\boldsymbol{u} = (\boldsymbol{u}_1, \ldots, \boldsymbol{u}_m) \in \mathbf{Seq}_m(\mathbb{R}^n)$. We have

$$I_{\boldsymbol{z}}(\boldsymbol{u}) - I_{\tilde{\boldsymbol{u}}}(\boldsymbol{u}) = m\|\tilde{\boldsymbol{u}} - \boldsymbol{z}\|_2^2,$$

for every $\boldsymbol{z} \in \mathbb{R}^n$.

Proof. The inertia of $\boldsymbol{u}$ relative to $\tilde{\boldsymbol{u}}$ is

$$\begin{aligned} I_{\tilde{\boldsymbol{u}}}(\boldsymbol{u}) &= \sum_{j=1}^{m} \|\boldsymbol{u}_j - \tilde{\boldsymbol{u}}\|_2^2 \\ &= \sum_{j=1}^{m} (\boldsymbol{u}_j - \tilde{\boldsymbol{u}})'(\boldsymbol{u}_j - \tilde{\boldsymbol{u}}) \\ &= \sum_{j=1}^{m} (\boldsymbol{u}_j'\boldsymbol{u}_j - \tilde{\boldsymbol{u}}'\boldsymbol{u}_j - \boldsymbol{u}_j'\tilde{\boldsymbol{u}} + \tilde{\boldsymbol{u}}'\tilde{\boldsymbol{u}}). \end{aligned}$$

Similarly, we have

$$I_{\boldsymbol{z}}(\boldsymbol{u}) = \sum_{j=1}^{m} (\boldsymbol{u}_j'\boldsymbol{u}_j - \boldsymbol{z}'\boldsymbol{u}_j - \boldsymbol{u}_j'\boldsymbol{z} + \boldsymbol{z}'\boldsymbol{z}).$$

This allows us to write

$$\begin{aligned} I_{\boldsymbol{z}}(\boldsymbol{u}) - I_{\tilde{\boldsymbol{u}}}(\boldsymbol{u}) &= \sum_{j=1}^{m} (\tilde{\boldsymbol{u}} - \boldsymbol{z})'\boldsymbol{u}_j + \sum_{j=1}^{m} \boldsymbol{u}_j'(\tilde{\boldsymbol{u}} - \boldsymbol{z}) + \boldsymbol{z}'\boldsymbol{z} - \tilde{\boldsymbol{u}}'\tilde{\boldsymbol{u}} \\ &= (\tilde{\boldsymbol{u}} - \boldsymbol{z})' \sum_{i=1}^{m} \boldsymbol{u}_j + \left(\sum_{j=1}^{m} \boldsymbol{u}_j \right)' (\tilde{\boldsymbol{u}} - \boldsymbol{z}) + m(\boldsymbol{z}'\boldsymbol{z} - \tilde{\boldsymbol{u}}'\tilde{\boldsymbol{u}}) \\ &= m(\tilde{\boldsymbol{u}} - \boldsymbol{z})'\tilde{\boldsymbol{u}} + m\tilde{\boldsymbol{u}}'(\tilde{\boldsymbol{u}} - \boldsymbol{z}) + m(\boldsymbol{z}'\boldsymbol{z} - \tilde{\boldsymbol{u}}'\tilde{\boldsymbol{u}}) \\ &= m\|\tilde{\boldsymbol{u}} - \boldsymbol{z}\|_2^2, \end{aligned}$$

which is the equality of the theorem. □

Corollary 11.1. *Let* $\boldsymbol{u} = (\boldsymbol{u}_1, \ldots, \boldsymbol{u}_m) \in \mathbf{Seq}_m(\mathbb{R}^n)$. *The minimal value of the inertia* $I_{\boldsymbol{z}}(\boldsymbol{u})$ *is achieved for* $\boldsymbol{z} = \tilde{\boldsymbol{u}}$.

Proof. This statement follows immediately from Theorem 11.2. □

Let $\boldsymbol{u}$ and $\boldsymbol{w}$ be two vectors in $\mathbb{R}^m$, where $m > 1$, having the means u and w, and the standard deviations s_u and s_v, respectively.

Definition 11.5. The *covariance coefficient* of $\boldsymbol{u}$ and $\boldsymbol{w}$ is the number

$$cov(\boldsymbol{u}, \boldsymbol{w}) = \frac{1}{m-1} \sum_{i=1}^{m-1} (u_i - u)(w_i - w).$$

The *correlation coefficient* of $\boldsymbol{u}$ and $\boldsymbol{w}$ is the number

$$\rho(\boldsymbol{u}, \boldsymbol{w}) = \frac{cov(\boldsymbol{u}, \boldsymbol{w})}{s_u s_w}.$$

By the Cauchy–Schwarz Inequality (Corollary 6.1), we have

$$\left| \sum_{i=1}^{m} (u_i - u)(w_i - w) \right| \leq \sqrt{\sum_{i=1}^{m} (u_i - u)^2} \cdot \sqrt{\sum_{i=1}^{m} (w_i - w)^2},$$

which implies

$$-1 \leq \rho(\boldsymbol{u}, \boldsymbol{w}) \leq 1.$$

Definition 11.6. Let $X \in \mathbb{R}^{m \times n}$ be a sample matrix and let $\hat{X}$ be the centered sample matrix corresponding to X. The *sample covariance matrix* is the matrix

$$cov(X) = \frac{1}{m-1} \hat{X}' \hat{X} \in \mathbb{R}^{n \times n}.$$

Note that if X is centered, $cov(X) = \frac{1}{m-1} X'X$.

If $n = 1$, the matrix is reduced to one column $X = (\boldsymbol{v})$ and

$$cov(\boldsymbol{v}) = \frac{1}{m-1} \boldsymbol{v}'\boldsymbol{v} \in \mathbb{R}.$$

In this case, we refer to $cov(\boldsymbol{v})$ as the *variance* of $\boldsymbol{v}$; this number is denoted by $var(\boldsymbol{v})$.

If $X = (\boldsymbol{v}_1 \ \cdots \ \boldsymbol{v}_n)$, then $(cov(X))_{ij} = cov(\boldsymbol{v}_i, \boldsymbol{v}_j)$ for $1 \leq i, j \leq n$. The covariance matrix can be written also as

$$cov(X) = \frac{1}{m-1} X' H_m H_m X = \frac{1}{m-1} X' H_m X.$$

The *sample correlation matrix* is the matrix $corr(X)$ given by $(corr(X))_{ij} = \rho(\boldsymbol{v}_i, \boldsymbol{v}_j)$ for $1 \leq i, j \leq n$.

If X is centered, then $cov(X) = \frac{1}{m-1} X'X$. Clearly, the covariance matrix is a symmetric, positive semidefinite matrix. Furthermore, by Theorem 3.33, the rank of $cov(X)$ is the same as the rank of $\hat{X}$ and, since m, the size of the sample, is usually much larger than n, we are often justified in assuming that $\mathsf{rank}(cov(X)) = n$.

Let $X = (\boldsymbol{v}_1 \ \cdots \ \boldsymbol{v}_n) \in \mathbb{R}^{m \times n}$ be a sample matrix. Note that

$$H_m \boldsymbol{v}_p = \left(I_m - \frac{1}{m} \mathbf{1}_m \mathbf{1}_m' \right) \boldsymbol{v}_p = \boldsymbol{v}_p - \frac{1}{m} \mathbf{1}_m \mathbf{1}_m' \boldsymbol{v}_p = \boldsymbol{v}_p - a_p \mathbf{1}_m,$$

because $\frac{1}{m} \mathbf{1}_m' \boldsymbol{v}_p = a_p$ for $1 \leq p \leq n$, where $\tilde{\boldsymbol{u}}' = (a_1, \ldots, a_n)$.

The covariance matrix can be written as

$$\begin{aligned} cov(X) &= \frac{1}{m-1} (\boldsymbol{v}_1 \ \cdots \ \boldsymbol{v}_n)' H_m' H_m (\boldsymbol{v}_1 \ \cdots \ \boldsymbol{v}_n) \\ &= \frac{1}{m-1} (H_m \boldsymbol{v}_1 \ \cdots \ H_m \boldsymbol{v}_n)' (H_m \boldsymbol{v}_1 \ \cdots \ H_m \boldsymbol{v}_n), \end{aligned}$$

which implies that the (p, q)-entry of this matrix is

$$cov(X)_{pq} = \frac{1}{m-1} (H_m \boldsymbol{v}_p)' (H_m \boldsymbol{v}_q) = \frac{1}{m-1} (\boldsymbol{v}_p - a_p \mathbf{1}_m)' (\boldsymbol{v}_q - a_q \mathbf{1}_m).$$

For a diagonal element, we have

$$cov(X)_{pp} = \frac{1}{m-1} \sum_{i=1}^{m} (\boldsymbol{v}_q - a_q \mathbf{1}_m)_i^2,$$

which shows that $cov(X)_{pp}$ measures the scattering of the values of the pth variable around the corresponding component a_i of the mean sample. This quantity is known as the pth *variance* and is denoted by σ_p^2 for $1 \leq p \leq n$. The *total variance* $\mathsf{tvar}(X)$ of X is $\mathsf{trace}(cov(X))$.

For $p \neq q$, the element c_{pq} of the matrix $C = cov(X)$ is referred to as the (p,q)-*covariance*. We have

$$\begin{aligned}(cov(X))_{pq} &= \frac{1}{m-1}(\boldsymbol{v}_p - a_p\mathbf{1}_m)'(\boldsymbol{v}_q - a_q\mathbf{1}_m)\\ &= \frac{1}{m}\left(\boldsymbol{v}_p'\boldsymbol{v}_q - a_p\mathbf{1}_m'\boldsymbol{v}_q - a_q\boldsymbol{v}_p'\mathbf{1}_m + ma_pa_q\right)\\ &= \boldsymbol{v}_p'\boldsymbol{v}_q - a_pa_q.\end{aligned}$$

If $cov(X)_{pq} = 0$, then we say that the variables $\mathcal{V}_p$ and $\mathcal{V}_q$ are *uncorrelated.*

The behavior of the covariance matrix with respect to multiplication by orthogonal matrices is discussed next.

Theorem 11.3. *Let*

$$X = \begin{pmatrix} \boldsymbol{x}_1 \\ \vdots \\ \boldsymbol{x}_m \end{pmatrix}$$

be a centered sample matrix and let $R \in \mathbb{R}^{n\times n}$ be an orthogonal matrix. If $Z \in \mathbb{R}^{m\times n}$ is a matrix such that $Z = XR$, then Z is centered, $cov(Z) = R'cov(X)R$ and $\mathsf{tvar}(Z) = \mathsf{tvar}(X)$.

Proof. By writing explicitly the rows of the matrix Z,

$$Z = \begin{pmatrix} \boldsymbol{z}_1 \\ \vdots \\ \boldsymbol{z}_m \end{pmatrix},$$

we have $\boldsymbol{z}_i = \boldsymbol{x}_iR$ for $1 \leq i \leq m$ because $Z = XR$. Note that the sample mean of Z is

$$\tilde{Z} = \frac{1}{m}\mathbf{1}_m'Z = \frac{1}{m}\mathbf{1}_m'XR = \tilde{X}R,$$

where $\tilde{X}$ is the sample mean of X. Since X is centered, we have $\tilde{Z} = \tilde{X} = \mathbf{0}_n'$, so Z is centered as well.

The covariance matrix of Z is

$$cov(Z) = \frac{1}{m-1} Z'Z = \frac{1}{m-1} R'X'XR = R' cov(X) R.$$

Since the trace of two similar matrices are equal (by Theorem 8.5) and $cov(Z)$ is similar to $cov(X)$, the total variance of Z equals the total variance of X, that is,

$$\mathsf{tvar}(Z) = \mathsf{trace}(cov(Z)) = \mathsf{trace}(cov(X)) = \mathsf{tvar}(X).$$

□

Since the covariance matrix of a centered matrix X, $cov(X) = \frac{1}{m-1} X'X \in \mathbb{R}^{n \times n}$ is symmetric, by Corollary 8.8, $cov(X)$ is orthonormally diagonalizable, so there exists an orthogonal matrix $R \in \mathbb{R}^{n \times n}$ such that $R' cov(X) R = D$, which corresponds to a sample matrix $Z = XR$. Let $cov(Z) = D = \mathsf{diag}(d_1, \ldots, d_n)$. The number d_p is the sample variance of the pth variable of the data matrix, and the covariances of the form $cov(Z)_{pq}$ with $p \neq q$ are 0. From a statistical point of view, this means that the components p and q are uncorrelated. Without loss of generality we can assume that $d_1 \geq \cdots \geq d_n$.

The columns of the matrix Z correspond to the new variables $\mathcal{Z}_1, \ldots, \mathcal{Z}_n$.

Often the variables of a data sample matrix are not expressed using different units. In this case, the components of the covariance have no meaning because variables that have large numerical values have a disproportionate influence compared to variables that have small numerical value. For example, if a spatial variable is measured in millimeters, its values are three orders of magnitude larger than the values of a variable expressed in meters.

11.3 Biplots

Biplots introduced by Gabriel in [61] offer a succinct and powerful way of representing graphically the elements of a matrix using two sets of vectors (hence, the term *biplot*).

Let $A \in \mathbb{R}^{m \times n}$ be a matrix that can be written as a product, $A = LR$, where $L \in \mathbb{R}^{m \times r}$, $R \in \mathbb{R}^{r \times n}$ are the *left* and the *right* factors, respectively. Suppose that

$$L = \begin{pmatrix} \boldsymbol{l}'_1 \\ \vdots \\ \boldsymbol{l}'_m \end{pmatrix} \text{ and } R = (\boldsymbol{r}_1 \ \cdots \ \boldsymbol{r}_m),$$

where $\boldsymbol{l}_1, \ldots, \boldsymbol{l}_m, \boldsymbol{r}_1, \ldots, \boldsymbol{r}_n$ are $m+n$ vectors in $\mathbb{R}^r$. Then, each element a_{ij} of A can be regarded as a inner product of two vectors

$$a_{ij} = \boldsymbol{l}'_i \boldsymbol{r}_j \tag{11.3}$$

for $1 \leq i \leq m$ and $1 \leq j \leq n$. Such matrix factorizations are common in linear algebra and we have already discussed a number of factorization techniques (full-rank decompositions, QR decompositions, etc.) Starting from the factorization $A = LR$, new factorizations of A can be built as $A = (LK')(R'K^{-1})'$ for every invertible matrix $K \in \mathbb{R}^{r \times r}$. Therefore, the above representation for A is not unique in general.

Thus, to use the biplot for a representation of the relations between the rows $\boldsymbol{w}_1, \ldots, \boldsymbol{w}_n$ of A, one could choose R such that $RR' = I_r$, which yields $AA' = LL'$. This implies $\boldsymbol{w}'_i \boldsymbol{w}_j = \boldsymbol{l}'_i \boldsymbol{l}_j$ for $1 \leq i, j \leq n$. Taking $i = j$, we have $\|\boldsymbol{w}'_i\| = \|\boldsymbol{l}'_i\|$, which , in turn, implies $\angle(\boldsymbol{w}'_i, \boldsymbol{w}'_j) = \angle(\boldsymbol{l}'_i, \boldsymbol{l}'_j)$. A similar choice can be made for the columns of A by imposing the requirement $L'L = I_r$, which implies $A'A = R'R$.

The case when the rank r of the matrix A is 2 is especially interesting because we can draw the vectors $\boldsymbol{l}_1, \ldots, \boldsymbol{l}_m, \boldsymbol{r}_1, \ldots, \boldsymbol{r}_n$ to obtain an exact two-dimensional representation of A, as we show in the next example.

Example 11.3. Let

$$A = \begin{pmatrix} 18 & 8 & 20 \\ -4 & 20 & 1 \\ 25 & 8 & 27 \\ 9 & 4 & 10 \end{pmatrix}$$

be a matrix of rank 2 in $\mathbb{R}^{4 \times 3}$ that can be written as $A = LR$, where

$$L = \begin{pmatrix} 2 & 4 \\ -2 & 3 \\ 3 & 5 \\ 1 & 2 \end{pmatrix} \text{ and } R = \begin{pmatrix} 5 & -4 & 4 \\ 2 & 4 & 3 \end{pmatrix}.$$

The vectors that help us with the representation of A are

$$\boldsymbol{l}_1 = \begin{pmatrix} 2 \\ 4 \end{pmatrix}, \boldsymbol{l}_2 = \begin{pmatrix} -2 \\ 3 \end{pmatrix}, \boldsymbol{l}_3 = \begin{pmatrix} 3 \\ 5 \end{pmatrix}, \boldsymbol{l}_4 = \begin{pmatrix} 1 \\ 2 \end{pmatrix}$$

and

$$\boldsymbol{r}_1 = \begin{pmatrix} 5 \\ 2 \end{pmatrix}, \boldsymbol{r}_2 = \begin{pmatrix} -4 \\ 4 \end{pmatrix}, \boldsymbol{r}_3 = \begin{pmatrix} 4 \\ 3 \end{pmatrix}.$$

For example, the a_{32} element of A can be written as

$$a_{32} = \boldsymbol{l}_3'\boldsymbol{r}_2 = (3\ 5)\begin{pmatrix} -4 \\ 4 \end{pmatrix} = 8.$$

Equality (11.3) shows that each vector $\boldsymbol{l}_i$ corresponds to a row of A and each vector $\boldsymbol{r}_j$, to a column of A.

When we can factor a sample data matrix X as $X = LR$, a column of the right factor $\boldsymbol{r}_j$ is referred to as the *biplot axis* and corresponds to a variable $\mathcal{V}_j$ (Figure 11.1). Each vector $\boldsymbol{l}_i'$ represents an observation in the sample matrix. It is interesting to observe that Equality (11.3) implies that the magnitude of projection of $\boldsymbol{l}_i$ on the biplot axis $\boldsymbol{r}_j$ is

$$\|\boldsymbol{l}_i\|_2 \cos \angle(\boldsymbol{l}_i, \boldsymbol{r}_j) = \frac{\boldsymbol{l}_i'\boldsymbol{r}_j}{\|\boldsymbol{r}_j\|_2} = \frac{a_{ij}}{\|\boldsymbol{r}_j\|_2}.$$

Therefore, if we choose the unit of measure on the axis $\boldsymbol{r}_j$ as the number $\frac{1}{\|\boldsymbol{r}_j\|_2}$, we can read the values of the entries a_{ij} directly on

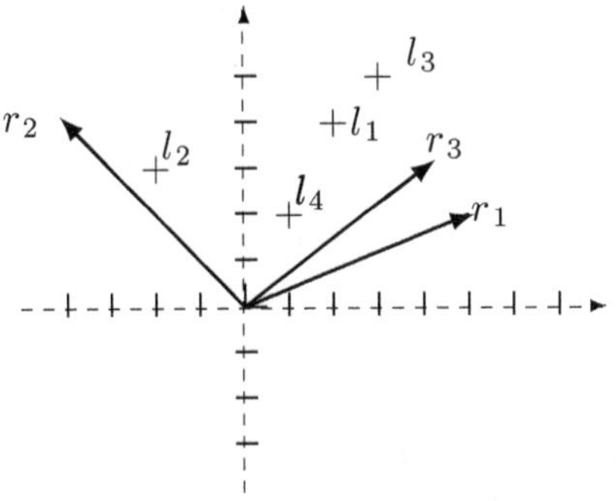

Fig. 11.1 Representation of the vectors $\boldsymbol{l}_i$ and $\boldsymbol{r}_j$.

the axis $\boldsymbol{r}_j$. For instance, the unit along the biplot axis is $\frac{1}{\|\boldsymbol{r}_3\|_2} = 0.2$. It is also clear that if two axis of the biplot point roughly in the same direction, the corresponding variables will show a strong correlation.

In general, the rank of the data matrix A is larger than 2. In this case, approximative representations of A can be obtained by using the thin singular value decomposition of matrices (Corollary 9.3).

Let A be a matrix of rank r and let

$$A = UDV' = \sum_{i=1}^{r} \sigma_i \boldsymbol{u}_i \boldsymbol{v}_i'$$

be the thin SVD, where $U \in \mathbb{R}^{m\times r}$ and $V \in \mathbb{R}^{n\times r}$ are matrices of rank r (and, therefore, full-rank matrices) having orthonormal sets of columns. Here $U = (\boldsymbol{u}_1 \ \cdots \ \boldsymbol{u}_r)$ and $V = (\boldsymbol{v}_1 \ \cdots \ \boldsymbol{v}_r)$.

The matrix D containing singular values can be split between U and V by defining $L = U\sqrt{D}$ and $R = \sqrt{D}V'$. The usefulness of the SVD for biplots is based on the Eckhart–Young Theorem (Theorem 9.9), which stipulates that the best approximation of A in the sense of the matrix norm $\|\!|\cdot|\!\|_2$ in the class of matrix of rank k is the matrix defined by

$$B(k) = \sum_{i=1}^{k} \sigma_i \boldsymbol{u}_i \boldsymbol{v}_i'.$$

According to Theorem 9.10, the same matrix $B(k)$ is the best approximation of A in the sense of Frobenius norm. The extent of the deficiency of this approximation is measured by $\|A - B(k)\|_F^2 = \sigma_{k+1}^2 + \cdots + \sigma_r^2$. Since $\|A\|_F^2 = \sigma_1^2 + \cdots + \sigma_r^2$, an absolute measure of the quality of the approximation of A by $B(k)$ is

$$q_k = 1 - \frac{\|A - B(k)\|_F^2}{\|A\|_F^2} = \frac{\sigma_1^2 + \cdots + \sigma_k^2}{\sigma_1^2 + \cdots + \sigma_r^2}.$$

In the special case, $k = 2$, the quality of the approximation is

$$q_2 = \frac{\sigma_1^2 + \sigma_2^2}{\sigma_1^2 + \cdots + \sigma_r^2}$$

and it is desirable that this number be as close to one as possible. The rank-2 approximation of A is useful because we can apply biplots to the visualization of A.

Example 11.4. Let $A \in \mathbb{R}^{5\times 3}$ be the matrix defined by

$$A = \begin{pmatrix} 1 & 0 & 0 \\ 0 & 1 & 0 \\ 1 & 1 & 1 \\ 1 & 1 & 0 \\ 0 & 0 & 1 \end{pmatrix}.$$

It is easy to see that the rank of this matrix is 3 and, using MATLAB, a singular value decomposition can be obtained as

```
U =
      0.2787   -0.2176   -0.7071   -0.2996   -0.5341
      0.2787   -0.2176    0.7071   -0.2996   -0.5341
      0.7138    0.3398   -0.0000   -0.4037    0.4605
      0.5573   -0.4352   -0.0000    0.7033    0.0736
[       0.1565    0.7749    0.0000    0.4037   -0.4605

S =
      2.3583         0         0
           0    1.1994         0
           0         0    1.0000
           0         0         0
           0         0         0

V =
      0.6572   -0.2610   -0.7071
      0.6572   -0.2610    0.7071
      0.3690    0.9294    0.0000
```

The rank-2 approximation of this matrix is

$$B(2) = \sigma_1 \boldsymbol{u}_1 \boldsymbol{v}_1^{\mathsf{H}} + \sigma_2 \boldsymbol{u}_2 \boldsymbol{v}_2^{\mathsf{H}},$$

and is computed in MATLAB using

```
>> B2 = 2.3583* U(:,1) * V(:,1)' + 1.1994 * U(:,2) * V(:,2)'

B2 =
      0.5000    0.5000   -0.0000
      0.5000    0.5000   -0.0000
      1.0000    1.0000    1.0000
      1.0000    1.0000   -0.0000
     -0.0000   -0.0000    1.0000
```

If we split the singular values as

$$B(2) = (\sqrt{\sigma_1}\boldsymbol{u}_1)(\sqrt{\sigma_1}\boldsymbol{v}_1)^{\mathsf{H}} + (\sqrt{\sigma_2}\boldsymbol{u}_2)(\sqrt{\sigma_2}\boldsymbol{v}_2)^{\mathsf{H}},$$

then $B(2)$ can be written as

$$B(2) = \begin{pmatrix} 0.4280 & -0.2383 \\ 0.4280 & -0.2383 \\ 1.0962 & 0.3721 \\ 0.8559 & -0.4766 \\ 0.2403 & 0.8487 \end{pmatrix} \begin{pmatrix} 1.0092 & 1.0092 & 0.5667 \\ -0.2858 & -0.2858 & 1.0179 \end{pmatrix}.$$

The biplot that represents matrix A is shown in Figure 11.2. The quality of the approximation of A is

$$q_2 = \frac{2.3583^2 + 1.1994^2}{2.3583^2 + 1.1994^2 + 1} = 0.875.$$

The "allocation" of singular values among the columns of the matrices U and V may lead to biplots that have distinct properties.

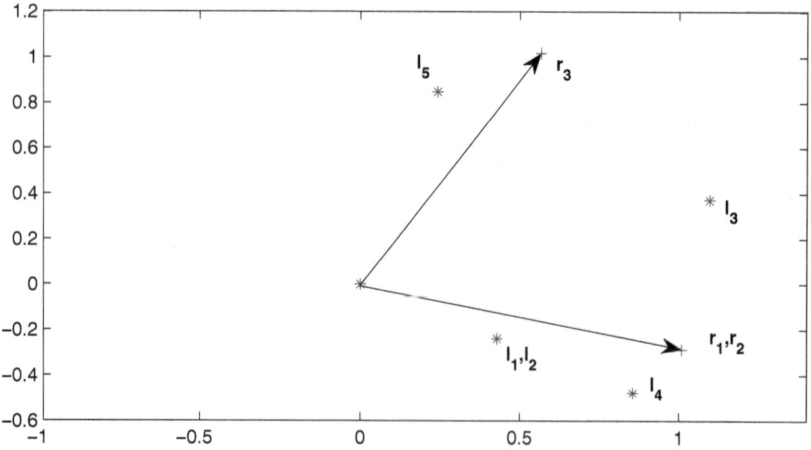

Fig. 11.2 Biplot of the rank-2 approximation of A.

For example, we could write

$$B(2) = (\sigma_1 \boldsymbol{u}_1)\boldsymbol{v}_1^{\mathsf{H}} + (\sigma_2 \boldsymbol{u}_2)\boldsymbol{v}_2^{\mathsf{H}}, \tag{11.4}$$

or

$$B(2) = \boldsymbol{u}_1(\sigma_1 \boldsymbol{v}_1)^{\mathsf{H}} + \boldsymbol{u}_2(\sigma_2 \boldsymbol{v}_2)^{\mathsf{H}}. \tag{11.5}$$

The first allocation leads to the factorization $B(2) = LR$, where

$$L = \begin{pmatrix} 0.6572 & -0.2610 \\ 0.6572 & -0.2610 \\ 1.6834 & 0.4075 \\ 1.3144 & -0.5219 \\ 0.3690 & 0.9294 \end{pmatrix}$$

and

$$R = \left(0.65720.65720.3690 - 0.2610 - 0.26100.9294\right),$$

while the second yields the factors

$$L = (0.2787 \quad -0.21760.2787 \quad -0.21760.7138 \quad 0.33980.5573 \quad -0.43520.1565 \quad 0.7749)$$

and

$$R = \begin{pmatrix} 1.5499 & 1.5499 & 0.8703 \\ -0.3130 & -0.3130 & 1.1147 \end{pmatrix}.$$

The first variant (Equality 11.4) leads to a representation, where the distances between the vectors $\boldsymbol{l}_i$ approximates the Euclidean distances between rows, while for the second variant (Equality 11.5), the cosine of angles between the vectors $\boldsymbol{r}_j$ approximates the correlations between variables.

Exercises and Supplements

(1) Verify that the centering matrix H_m is both symmetric and idempotent.
(2) Compute the spectrum of a centering matrix H_m.

(3) Let $X \in \mathbb{R}^{m\times n}$ be a matrix and let $Q_{\boldsymbol{c}} \in \mathbb{R}^{m\times m}$ be the matrix introduced in Supplement 10 of Chapter 3. Prove that
   (a) if $\boldsymbol{c} \in \mathbb{R}^m$, then $Q_{\boldsymbol{c}}X = X - \boldsymbol{1}_m\boldsymbol{c}'X$; if $\boldsymbol{c}$ is the mean of the rows of X, that is if $\boldsymbol{c} = \frac{1}{m}X'\boldsymbol{1}_m$, then $Q_{\boldsymbol{c}}X = H_mX$;
   (b) if $\boldsymbol{c} \in \mathbb{R}^n$, then $XQ_{\boldsymbol{c}} = X - X\boldsymbol{1}_m\boldsymbol{c}'$; if $\boldsymbol{c}$ is the mean of the columns of X, that is if $\boldsymbol{c} = \frac{1}{n}X\boldsymbol{1}_n$, then $XQ_{\boldsymbol{c}} = XH_n$.

(4) Usually, for a data sample matrix $X \in \mathbb{R}^{m\times n}$, we can assume that $m \geq n$. Examine the possibility that $\mathsf{rank}(X) < n$.

(5) Let $(\boldsymbol{u}_1, \ldots, \boldsymbol{u}_m)$ be a sequence of m vectors in $\mathbb{R}^n$. Prove that the covariance matrix C of this sequence can be written as

$$C = \frac{1}{m-1}(\boldsymbol{u}_1 - \tilde{\boldsymbol{u}} \ \cdots \ \boldsymbol{u}_n - \tilde{\boldsymbol{u}})\begin{pmatrix} \boldsymbol{u}_1' - \tilde{\boldsymbol{u}}' \\ \vdots \\ \boldsymbol{u}_n' - \tilde{\boldsymbol{u}}' \end{pmatrix}.$$

(6) Let $\boldsymbol{q} \in \mathbb{R}^n$ be a vector such that $\sum i = 1^n q_i = 0$. Prove that $H_n\boldsymbol{q} = \boldsymbol{q}'H_n = \boldsymbol{q}$.

(7) Let $A \in \mathbb{R}^{n\times n}$ and $\boldsymbol{c} \in \mathbb{R}^n$. Define the mapping $f : \mathbb{R}^n \longrightarrow \boldsymbol{r}_n$ as $f(\boldsymbol{u}) = A\boldsymbol{u} + \boldsymbol{c}$ for $\boldsymbol{u} \in \mathbb{R}^n$ and let $\boldsymbol{w}_i = f(\boldsymbol{u}_i)$ for $1 \leq i \leq m$. Prove that the covariance matrix D of the sequence $(\boldsymbol{w}_1, \ldots, \boldsymbol{w}_m)$ is $D = ACA'$.

(8) Prove that the function $d_C : \{\boldsymbol{u}_1, \ldots, \boldsymbol{u}_m\}^2 \longrightarrow \mathbb{R}$ defined by $d_C(\boldsymbol{u}_i, \boldsymbol{u}_j) = (\boldsymbol{u}_i - \tilde{\boldsymbol{u}})'C^{-1}(\boldsymbol{u}_j - \tilde{\boldsymbol{u}})$ for $1 \leq i, j \leq m$ is a metric on $\{\boldsymbol{u}_1, \ldots, \boldsymbol{u}_m\}$. Furthermore, prove that $d_D(f(\boldsymbol{u}_i), f(\boldsymbol{u}_j)) = d(\boldsymbol{u}_i, \boldsymbol{u}_j)$ for $1 \leq i, j \leq m$.
The function d_C is known as the *Mahalanobis metric*.

(9) Let $X \in \mathbb{R}^{n\times p}$ be a data sample matrix. Prove that

$$\hat{X} = X - \left(\frac{1}{n} * \boldsymbol{1}_n'X\right) \otimes \boldsymbol{1}_n.$$

Write a MATLAB function that computes $\hat{X}$ starting from X.

(10) Prove that if $\boldsymbol{x} \in \mathbb{R}^m$, then $H_m\boldsymbol{x} = \boldsymbol{x} - x\boldsymbol{1}_m$, where x is the mean of $\boldsymbol{x}$.

(11) Prove that $\boldsymbol{x}'H_m\boldsymbol{x} = \frac{1}{m}\sum_{i=1}^m (x_i - x)$, where $\boldsymbol{x} = \begin{pmatrix} x_1 \\ \vdots \\ x_m \end{pmatrix}$ and x is the mean of $\boldsymbol{x}$.

Let $X \in \mathbb{R}^{m \times n}$ be a sample matrix. Its correlation matrix, $corr(X) \in \mathbb{R}^{n \times n}$, is defined by

$$(corr(X))_{ij} = \rho(\boldsymbol{v}_i, \boldsymbol{v}_j)$$

for $1 \leq i, j \leq n$, where $X = (\boldsymbol{v}_1 \ \cdots \ \boldsymbol{v}_n)$.

(12) Prove that $corr(X) = D^{-1} cov(X) D^{-1}$, where $D = \mathsf{diag}(s_{\boldsymbol{v}_1}, \ldots, s_{\boldsymbol{v}_n})$.

(13) Justify the claims made at the end of Example 11.4 that refer to the allocation variants for singular values.

The *generalized sample variance* of a data sample matrix X is the determinant $\det(cov(X))$, which offers a succinct summary of the variances and covariances of X.

(14) Let $X \in \mathbb{R}^{m \times n}$ be a centered data sample matrix, $X = (\boldsymbol{v}_1, \ldots, \boldsymbol{v}_n)$. Prove that the generalized sample variance of X equals $\frac{1}{n-1} V_n(X)^2$, where $V_n(X)$ is the volume of the parallelepiped constructed on the vectors $\boldsymbol{v}_1, \ldots, \boldsymbol{v}_n$.

Hint: See Supplement 107 of Chapter 6.

Bibliographical Comments

There are several excellent sources for biplots [67, 68, 96]. A recent readable introduction to biplots is [69].

Chapter 12

Least Squares Approximations and Data Mining

12.1 Introduction

The least square method is used in data mining as a method of estimating the parameters of a model by adopting the values that minimize the sum of the squared differences between the predicted and the observed values of data. This estimation process is also known as *regression*, and several types of regression exist depending on the nature of the assumed model of dependency between the predicted and the observed data.

12.2 Linear Regression

The aim of *linear regression* is to explore the existence of a linear relationship between the outcome of an experiment and values of variables that are measured during the experiment. As we saw in Chapter 11, experimental data often are presented as a data sample matrix $B \in \mathbb{R}^{m \times n}$, where m is the number of experiments and n is the number of variables measured. The results of the experiments are

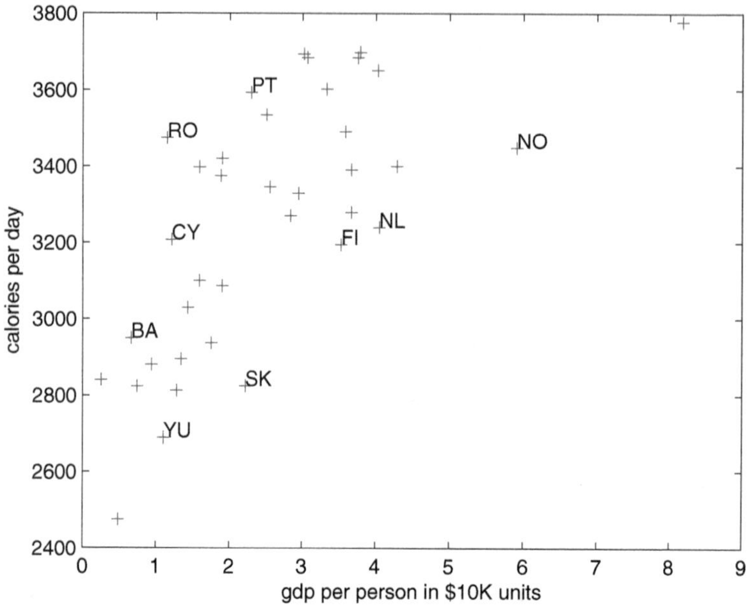

Fig. 12.1 Calories vs. GDP in 10K units per person in Europe.

the components of a vector

$$\boldsymbol{b} = \begin{pmatrix} b_1 \\ \vdots \\ b_m \end{pmatrix}.$$

Linear regression amounts to determining $\boldsymbol{r} \in \mathbb{R}^n$ such that $B\boldsymbol{r} = \boldsymbol{b}$. Knowing the components of $\boldsymbol{r}$ allows us to express the value of the result as a linear combination of the values of the variables. Unfortunately, since m is usually much larger than n, this system is overdetermined and, in general, is inconsistent.

The columns $\boldsymbol{v}_1, \ldots, \boldsymbol{v}_n$ of the matrix B are referred to as the *regressors*; the linear combination $r_1\boldsymbol{v}_1 + \cdots + r_n\boldsymbol{v}_n$ is the *regression of $\boldsymbol{b}$ onto the regressors $\boldsymbol{v}_1, \ldots, \boldsymbol{v}_n$*.

Example 12.1. In Figure 12.1, we represent (using the function `plot` of MATLAB), the number of calories consumed by a person per day vs. the gross national product per person in European countries starting from the following table.

ccode	gdp	cal	ccode	gdp	cal
'AL'	0.74	2824.00	'IT'	3.07	3685.00
'AT'	4.03	3651.00	'LV'	1.43	3029.00
'BY'	1.34	2895.00	'LT'	1.59	3397.00
'BE'	3.79	3698.00	'LU'	8.18	3778.00
'BA'	0.66	2950.00	'MK'	0.94	2881.00
'BG'	1.28	2813.00	'MT'	2.51	3535.00
'HR'	1.75	2937.00	'MD'	0.25	2841.00
'CY'	1.21	3208.00	'NL'	4.05	3240.00
'CZ'	2.56	3346.00	'NO'	5.91	3448.00
'DK'	3.67	3391.00	'PL'	1.88	3375.00
'EE'	1.90	3086.00	'PT'	2.30	3593.00
'FI'	3.53	3195.00	'RO'	1.15	3474.00
'FR'	3.33	3602.00	'RU'	1.59	3100.00
'GE'	0.48	2475.00	'YU'	1.10	2689.00
'DE'	3.59	3491.00	'SK'	2.22	2825.00
'GR'	3.02	3694.00	'SI'	2.84	3271.00
'HU'	1.90	3420.00	'ES'	2.95	3329.00
'IS'	3.67	3279.00	'CH'	4.29	3400.00
'IE'	3.76	3685.00			

This data set was extracted from [55].

We seek to approximate the calorie intake as a linear function of the gdp of the form

$$\text{cal} = r_1 + r_2 \,\text{gdp}.$$

This amounts to solving a linear system that consists of 37 equations and two unknowns:

$$\begin{aligned} r_1 + 0.74 r_2 &= 2824 \\ &\vdots \\ r_1 + 4.29 r_2 &= 3400 \end{aligned}$$

and, clearly such a system is inconsistent.

If the linear system $B\boldsymbol{r} = \boldsymbol{b}$ has no solution, the "next best thing" is to find a vector $\boldsymbol{c} \in \mathbb{R}^n$ such that $\|B\boldsymbol{c} - \boldsymbol{b}\|_2 \leq \|B\boldsymbol{w} - \boldsymbol{b}\|_2$ for every $\boldsymbol{w} \in \mathbb{R}^n$, an approach known as *the least square method.* We will refer to the triple $(B, \boldsymbol{r}, \boldsymbol{b})$ as an *instance of the least square problem.*

Note that $B\boldsymbol{r} \in \mathsf{range}(B)$ for any $\boldsymbol{r} \in \mathbb{R}^n$. Thus, solving this problem amounts to finding a vector $B\boldsymbol{r}$ in the subspace $\mathsf{range}(B)$ such that $B\boldsymbol{r}$ is as close to $\boldsymbol{b}$ as possible.

Let $B \in \mathbb{R}^{m\times n}$ be a full-rank matrix such that $m > n$, so $\mathsf{rank}(B) = n$. The symmetric square matrix $B'B \in \mathbb{R}^{n\times n}$ has the same rank n as the matrix B, as we saw in Theorem 3.33. Therefore, the system $(B'B)\boldsymbol{r} = B'\boldsymbol{b}$ has a unique solution $\boldsymbol{s}$. Moreover, $B'B$ is positive definite because $\boldsymbol{r}'B'B\boldsymbol{r} = (B\boldsymbol{r})'B\boldsymbol{r} = \|B\boldsymbol{r}\|_2^2 > 0$ for $\boldsymbol{r} \neq \boldsymbol{0}$.

Theorem 12.1. *Let $B \in \mathbb{R}^{m\times n}$ be a full-rank matrix such that $m > n$ and let $\boldsymbol{b} \in \mathbb{R}^m$. The unique solution of the system $(B'B)\boldsymbol{r} = B'\boldsymbol{b}$ equals the projection of the vector $\boldsymbol{b}$ on the subspace* $\mathsf{range}(B)$.

Proof. The n columns of the matrix $B = (\boldsymbol{v}_1 \ \cdots \ \boldsymbol{v}_n)$ constitute a basis of the subspace $\mathsf{range}(B)$. Therefore, we seek the projection $\boldsymbol{c}$ of $\boldsymbol{b}$ on $\mathsf{range}(B)$ as a linear combination $\boldsymbol{c} = B\boldsymbol{t}$, which allows us to reduce this problem to a minimization of the function

$$\begin{aligned} f(\boldsymbol{t}) &= \|B\boldsymbol{t} - \boldsymbol{b}\|_2^2 \\ &= (B\boldsymbol{t} - \boldsymbol{b})'(B\boldsymbol{t} - \boldsymbol{b}) = (\boldsymbol{t}'B' - \boldsymbol{b}')(B\boldsymbol{t} - \boldsymbol{b}) \\ &= \boldsymbol{t}'B'B\boldsymbol{t} - \boldsymbol{b}'B\boldsymbol{t} - \boldsymbol{t}'B'\boldsymbol{b} + \boldsymbol{b}'\boldsymbol{b}. \end{aligned}$$

The necessary condition for the minimum is

$$(\nabla f)(\boldsymbol{t}) = 2B'B\boldsymbol{t} - 2B'\boldsymbol{b} = 0,$$

which implies $B'B\boldsymbol{t} = B'\boldsymbol{b}$. □

The linear system $(B'B)\boldsymbol{t} = B'\boldsymbol{b}$ is known as the *system of normal equations* of B and $\boldsymbol{b}$.

Example 12.2. We augment the data sample matrix by a column that consists of 1s to accommodate a constant term r_1; thus, we work with the data sample matrix $B \in \mathbb{R}^{37\times 2}$ given by

$$B = \begin{pmatrix} 1 & 0.74 \\ \vdots & \vdots \\ 1 & 4.29 \end{pmatrix}$$

whose second column consists of the countries' gross domestic products in \$10K units. The matrix $C = B'B$ is

$$C = \begin{pmatrix} 37.0000 & 94.4600 \\ 94.4600 & 333.6592 \end{pmatrix}.$$

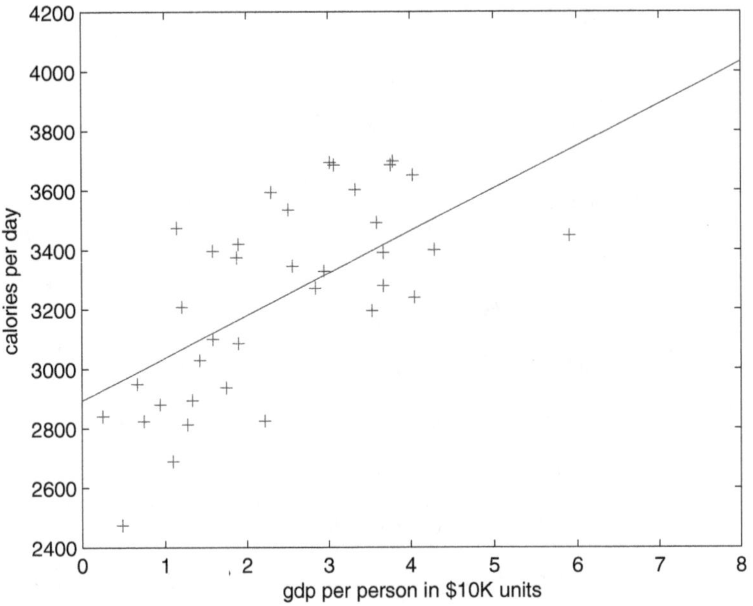

Fig. 12.2 Regression line.

Solving the normal system using the MATLAB statement $\boldsymbol{r} = C\backslash(B' * b)$ yields

$$\boldsymbol{r} = \begin{pmatrix} 2894.2 \\ 142.3 \end{pmatrix},$$

so the regression line is $\mathsf{cal} = 142.3 * \mathsf{gdp} + 2894.2$, shown in Figure 12.2.

Suppose now that $B \in \mathbb{R}^{m \times n}$ has rank k, where $k < \min\{m, n\}$, and $U \in \mathbb{R}^{m \times m}$, $V \in \mathbb{R}^{n \times n}$ are orthonormal matrices such that B can be factored as $B = UMV'$, where

$$M = \begin{pmatrix} R & O_{k,n-k} \\ O_{m-k,k} & O_{m-k,n-k} \end{pmatrix} \in \mathbb{R}^{m \times n},$$

$R \in \mathbb{R}^{k \times k}$, and $\mathsf{rank}(R) = k$.

For $\boldsymbol{b} \in \mathbb{R}^m$ define $\boldsymbol{c} = U'\boldsymbol{b} \in \mathbb{R}^m$ and let $\boldsymbol{c} = \begin{pmatrix} \boldsymbol{c}_1 \\ \boldsymbol{c}_2 \end{pmatrix}$, where $\boldsymbol{c}_1 \in \mathbb{R}^k$ and $\boldsymbol{c}_2 \in \mathbb{R}^{m-k}$. Since $\mathsf{rank}(R) = k$, the linear system $R\boldsymbol{z} = \boldsymbol{c}_1$ has a unique solution $\boldsymbol{z}_1$.

Theorem 12.2. *All vectors $\boldsymbol{r}$ that minimize $\|B\boldsymbol{r} - \boldsymbol{b}\|_2$ have the form*

$$\boldsymbol{r} = V\begin{pmatrix}\boldsymbol{z}\\ \boldsymbol{w}\end{pmatrix}$$

for an arbitrary $\boldsymbol{w}$.

Proof. We have

$$\begin{aligned}\|B\boldsymbol{r} - \boldsymbol{b}\|_2^2 &= \|UMV'\boldsymbol{r} - UU'\boldsymbol{b}\|_2^2\\ &= \|U(MV'\boldsymbol{r} - U'\boldsymbol{b})\|_2^2 = \|MV'\boldsymbol{r} - U'\boldsymbol{b}\|_2^2\\ &\quad \text{(because multiplication by an orthonormal matrix}\\ &\quad \text{is norm-preserving)}\\ &= \|MV'\boldsymbol{r} - \boldsymbol{c}\|_2^2 = \|M\boldsymbol{y} - \boldsymbol{c}\|_2^2\\ &= \|R\boldsymbol{z} - \boldsymbol{c}_1\|_2^2 + \|\boldsymbol{c}_2\|_2^2,\end{aligned}$$

where $\boldsymbol{z}$ consists of the first r components of $\boldsymbol{y}$. This shows that the minimal value of $\|B\boldsymbol{r} - \boldsymbol{b}\|_2^2$ is achieved by the solution of the system $R\boldsymbol{z} = \boldsymbol{c}_1$ and is equal to $\|\boldsymbol{c}_2\|_2^2$. Therefore, the vectors $\boldsymbol{r}$ that minimize $\|B\boldsymbol{r} - \boldsymbol{b}\|_2^2$ have the form $\begin{pmatrix}\boldsymbol{z}\\ \boldsymbol{w}\end{pmatrix}$ for an arbitrary $\boldsymbol{w} \in \mathbb{R}^{n-r}$. □

Instead of the Euclidean norm we can use the $\|\cdot\|_\infty$. Note that we have $t = \|B\boldsymbol{r}-\boldsymbol{b}\|_\infty$ if and only if $-t\boldsymbol{1} \leq B\boldsymbol{r}-\boldsymbol{b} \leq t\boldsymbol{1}$, so finding $\boldsymbol{r}$ that minimizes $\|\cdot\|_\infty$ amounts to solving a linear programming problem: minimize t subjected to the restrictions $-t\boldsymbol{1} \leq B\boldsymbol{r} - \boldsymbol{b} \leq t\boldsymbol{1}$.

Similarly, we can use the norm $\|\cdot\|_p$. If $\boldsymbol{y} = B\boldsymbol{r} - \boldsymbol{b}$, then we need to minimize $\|\boldsymbol{y}\|_p^p = |y_1|^p + \cdots + |y_m|^p$, subjected to the restrictions $-\boldsymbol{y} \leq A\boldsymbol{r} - \boldsymbol{b} \leq \boldsymbol{y}$.

12.3 The Least Square Approximation and QR Decomposition

Solving the system of normal equation presents numeric difficulties because, by Corollary 9.5 the condition number of the matrix $B'B$ is the square of the condition number of B. An alternative approach to finding $\boldsymbol{r} \in \mathbb{R}^n$ that minimizes $f(\boldsymbol{u}) = \|B\boldsymbol{u} - \boldsymbol{b}\|_2^2$ is to use a full QR decomposition of the matrix B, where $B \in \mathbb{R}^{m\times n}$ is a full-rank matrix and $m > n$, as described in Theorem 6.67.

Suppose that

$$B = Q\begin{pmatrix} R \\ O_{m-n,n}\end{pmatrix},$$

where $Q \in \mathbb{R}^{m\times m}$ is an orthonormal matrix and $R \in \mathbb{R}^{n\times n}$ is an upper triangular matrix such that

$$\begin{pmatrix} R \\ O_{m-n,n}\end{pmatrix} \in \mathbb{R}^{m\times n}.$$

We have

$$\begin{aligned} B\boldsymbol{u} - \boldsymbol{b} &= Q\begin{pmatrix} R \\ O_{m-n,n}\end{pmatrix}\boldsymbol{u} - \boldsymbol{b} \\ &= Q\begin{pmatrix} R \\ O_{m-n,n}\end{pmatrix}\boldsymbol{u} - QQ'\boldsymbol{b} \\ &\qquad \text{(because } Q \text{ is orthonormal and therefore } QQ' = I_m\text{)} \\ &= Q\left(\begin{pmatrix} R \\ O_{m-n,n}\end{pmatrix}\boldsymbol{u} - Q'\boldsymbol{b}\right). \end{aligned}$$

By Theorem 6.24, multiplication by an orthogonal matrix preserves the Euclidean norm of vectors. Thus,

$$\|B\boldsymbol{u} - \boldsymbol{b}\|_2^2 = \left\|\begin{pmatrix} R \\ O_{m-n,n}\end{pmatrix}\boldsymbol{u} - Q'\boldsymbol{b}\right\|_2^2.$$

If we write $Q = (L_1\ L_2)$, where $L_1 \in \mathbb{R}^{m\times n}$ and $L_2 \in \mathbb{R}^{m\times(m-n)}$, then

$$\begin{aligned} \|B\boldsymbol{u} - \boldsymbol{b}\|_2^2 &= \left\|\begin{pmatrix} R \\ O_{m-n,n}\end{pmatrix}\boldsymbol{u} - \begin{pmatrix} L_1'\boldsymbol{b} \\ L_2'\boldsymbol{b}\end{pmatrix}\right\|_2^2 \\ &= \left\|\begin{pmatrix} R\boldsymbol{u} - L_1'\boldsymbol{b} \\ -L_2'\boldsymbol{b}\end{pmatrix}\right\|_2^2 \\ &= \|R\boldsymbol{u} - L_1'\boldsymbol{b}\|_2^2 + \|L_2'\boldsymbol{b}\|_2^2. \end{aligned}$$

Observe that the system $R\boldsymbol{u} = L_1'\boldsymbol{b}$ can be solved and its solution minimizes $\|B\boldsymbol{u} - \boldsymbol{b}\|_2$.

12.4 Partial Least Square Regression

When the number of variables of an experiment is large, multiple output variables exist, and some or all these variables are correlated, the least square regression is replaced by a dimensionality reduction technique known as *partial least square regression* (PLS). Its inventor, the Swedish mathematician Herman Wold, observed that the acronym PLS is also consistent with "projection on latent structures" and this is a more accurate description of this data analysis method.

In this section, we assume that the set of variables of an experiment $\mathcal{E}$ is partitioned into two disjoint sets $\{\mathcal{X}_1, \ldots, \mathcal{X}_p\}$ referred to as the set of *predictor variables* and $\{\mathcal{Y}_1, \ldots, \mathcal{Y}_r\}$ named the set of *response variables*. Thus, the sample matrix $X_{\mathcal{E}} \in \mathbb{R}^{m\times(p+r)}$ can be written as

$$X_{\mathcal{E}} = (X, Y),$$

where $X \in \mathbb{R}^{m\times p}$ and $Y \in \mathbb{R}^{m\times r}$.

The basic model of PLS starts from the assumption that the matrices X and Y can be written as

$$X = TP' + E,$$
$$Y = TQ' + F,$$

where the matrices T, U, P, Q, E, F are described in the following table.

Matrix Notation	Format	Matrix Designation
T	$\mathbb{R}^{n\times s}$	Matrix of score vectors
P	$\mathbb{R}^{p\times s}$	Matrix of loadings
Q	$\mathbb{R}^{r\times s}$	Matrix of loadings
E	$\mathbb{R}^{n\times p}$	Matrix of residuals
F	$\mathbb{R}^{n\times r}$	Matrix of residuals

The iterative process introduced in Algorithm 12.4.1 begins with the matrices X and Y and proceeds to construct two sequences of matrices $X_0, \ldots, X_k$ and $Y_0, \ldots, Y_k$, where $X_0 = X$ and $Y_0 = Y$, as follows.

In the **repeat** loop that extends between lines 4 and 12, we compute four sequence of vectors $\boldsymbol{w}_\ell \in \mathbb{R}^p$, $\boldsymbol{t}_\ell \in \mathbb{R}^n$, $\boldsymbol{c}_\ell \in \mathbb{R}^r$, and $\boldsymbol{u}_\ell \in \mathbb{R}^n$.

Algorithm 12.4.1: Iterative Algorithm for PLS

Data: Matrices $X \in \mathbb{R}^{n\times p}$ and $Y \in \mathbb{R}^{n\times r}$

Result: Sequences of matrices $X_0, \ldots, X_k$ and $Y_0, \ldots, Y_k$

for $i = 1$ *to* k **do**

  $\ell = 0$;

  initialize $\boldsymbol{u}_0$ as the first column of Y_{i-1};

  repeat

    $\boldsymbol{w}_\ell = \frac{X'\boldsymbol{u}_\ell}{\boldsymbol{u}_\ell'\boldsymbol{u}_\ell}$;

    $\boldsymbol{w}_\ell \leftarrow \frac{1}{\|\boldsymbol{w}_\ell\|}\boldsymbol{w}_\ell$;

    $\boldsymbol{t}_\ell = \frac{X\boldsymbol{w}_\ell}{\boldsymbol{w}_\ell'\boldsymbol{w}_\ell}$;

    $\boldsymbol{c}_\ell = \frac{Y'\boldsymbol{t}_\ell}{\boldsymbol{t}_\ell'\boldsymbol{t}_\ell}$;

    $\boldsymbol{c}_\ell \leftarrow \frac{1}{\|\boldsymbol{c}_\ell\|}\boldsymbol{c}_\ell$;

    $\boldsymbol{u}_{\ell+1} = \frac{Y\boldsymbol{c}_\ell}{\boldsymbol{c}_\ell'\boldsymbol{c}_\ell}$;

    $\ell \leftarrow \ell + 1$;

  until *until convergence condition is met*;

  $\boldsymbol{t}^{(i)} = \boldsymbol{t}_\ell$;

  $\boldsymbol{u}^{(i)} = \boldsymbol{u}_\ell$;

  $\boldsymbol{c}^{(i)} = \boldsymbol{c}_\ell$;

  $\boldsymbol{p} = \frac{1}{(\boldsymbol{t}^{(i)})'\boldsymbol{t}^{(i)}} X_{i-1}'\boldsymbol{t}^{(i)}$;

  $\boldsymbol{q} = \frac{1}{(\boldsymbol{u}^{(i)})'\boldsymbol{u}^{(i)}} Y_{i-1}'\boldsymbol{u}^{(i)}$;

  $b_i = \frac{(\boldsymbol{t}^{(i)})'\boldsymbol{t}^{(i)}}{(\boldsymbol{u}^{(i)})'\boldsymbol{t}^{(i)}}$;

  $X_i \leftarrow X_{i-1} - \boldsymbol{t}^{(i)}(\boldsymbol{p}^{(i)})'$;

  $Y_i \leftarrow Y_{i-1} - b_i\boldsymbol{t}^{(i)}(\boldsymbol{c}^{(i)})'$;

end

Observe that

$$\begin{aligned}
\boldsymbol{u}_{\ell+1} &= Y\boldsymbol{c}_\ell/(\boldsymbol{c}_\ell'\boldsymbol{c}_\ell)\\
&= YY'\boldsymbol{t}_\ell/(\boldsymbol{c}_\ell'\boldsymbol{c}_\ell)(\boldsymbol{t}_\ell'\boldsymbol{t}_\ell)\\
&= YY'X\boldsymbol{w}_\ell/(\boldsymbol{c}_\ell'\boldsymbol{c}_\ell)(\boldsymbol{t}_\ell'\boldsymbol{t}_\ell)(\boldsymbol{w}_\ell'\boldsymbol{w}_\ell)\\
&= YY'XX'\boldsymbol{u}_\ell/(\boldsymbol{c}_\ell'\boldsymbol{c}_\ell)(\boldsymbol{t}_\ell'\boldsymbol{t}_\ell)(\boldsymbol{w}_\ell'\boldsymbol{w}_\ell)(\boldsymbol{u}_\ell'\boldsymbol{u}_\ell).
\end{aligned}$$

Similarly, we have

$$\begin{aligned}
\boldsymbol{w}_{\ell+1} &= X'\boldsymbol{u}_{\ell+1}/(\boldsymbol{u}'_{\ell+1}\boldsymbol{u}_{\ell+1}) \\
&= X'YY'XX'\boldsymbol{u}_\ell/(\boldsymbol{c}'_\ell\boldsymbol{c}_\ell)(\boldsymbol{t}'_\ell\boldsymbol{t}_\ell)(\boldsymbol{w}'_\ell\boldsymbol{w}_\ell)(\boldsymbol{u}'_\ell\boldsymbol{u}_\ell)(\boldsymbol{u}'_{\ell+1}\boldsymbol{u}_{\ell+1}) \\
&= X'YY'X\boldsymbol{w}_\ell/(\boldsymbol{c}'_\ell\boldsymbol{c}_\ell)(\boldsymbol{t}'_\ell\boldsymbol{t}_\ell)(\boldsymbol{w}'_\ell\boldsymbol{w}_\ell)(\boldsymbol{u}'_{\ell+1}\boldsymbol{u}_{\ell+1}), \\
\boldsymbol{t}_{\ell+1} &= X\boldsymbol{w}_{\ell+1}/(\boldsymbol{w}'_{\ell+1}\boldsymbol{w}_{\ell+1} \\
&= XX'YY'X\boldsymbol{w}_\ell/(\boldsymbol{c}'_\ell\boldsymbol{c}_\ell)(\boldsymbol{t}'_\ell\boldsymbol{t}_\ell)(\boldsymbol{w}'_\ell\boldsymbol{w}_\ell)(\boldsymbol{u}'_{\ell+1}\boldsymbol{u}_{\ell+1})(\boldsymbol{w}'_{\ell+1}\boldsymbol{w}_{\ell+1} \\
&= XX'YY'\boldsymbol{t}_\ell/(\boldsymbol{c}'_\ell\boldsymbol{c}_\ell)(\boldsymbol{t}'_\ell\boldsymbol{t}_\ell)(\boldsymbol{u}'_{\ell+1}\boldsymbol{u}_{\ell+1})(\boldsymbol{w}'_{\ell+1}\boldsymbol{w}_{\ell+1}),
\end{aligned}$$

and

$$\begin{aligned}
\boldsymbol{c}_{\ell+1} &= Y'\boldsymbol{t}_{\ell+1}/(\boldsymbol{t}'_{\ell+1}\boldsymbol{t}_{\ell+1}) \\
&= Y'XX'YY'\boldsymbol{t}_\ell/(\boldsymbol{c}'_\ell\boldsymbol{c}_\ell)(\boldsymbol{t}'_\ell\boldsymbol{t}_\ell)(\boldsymbol{u}'_{\ell+1}\boldsymbol{u}_{\ell+1})(\boldsymbol{w}'_{\ell+1}\boldsymbol{w}_{\ell+1})(\boldsymbol{t}'_{\ell+1}\boldsymbol{t}_{\ell+1}) \\
&= Y'XX'Y\boldsymbol{c}_\ell/(\boldsymbol{c}'_\ell\boldsymbol{c}_\ell)(\boldsymbol{u}'_{\ell+1}\boldsymbol{u}_{\ell+1})(\boldsymbol{w}'_{\ell+1}\boldsymbol{w}_{\ell+1})(\boldsymbol{t}'_{\ell+1}\boldsymbol{t}_{\ell+1}).
\end{aligned}$$

12.5 Locally Linear Embedding

Often data that results from observations obey mathematical relationships that are not immediately apparent because of the high dimensionality of the space where data resides. These relationships, once identified, could point to the location of the data vectors on a manifold whose intrinsic dimensionality could be much lower than the dimension of the ambient space.

The *locally linear embedding* (LLE) introduced by Roweis and Saul in [141] is based on the basic idea that a point and its immediate neighbors are approximatively located on a linear manifold that is locally close to the real underlying manifold.

Suppose that we have a sample data set $\{\boldsymbol{x}_1 \cdots \boldsymbol{x}_n\} \subseteq R^m$ that consists of n vectors in a high-dimension space $\mathbb{R}^m$. The matrix $X \in \mathbb{R}^{m\times n}$ is defined by $X = (\boldsymbol{x}_1 \cdots \boldsymbol{x}_n)$.

The LLE algorithm computes the neighbors $N_i = \{\boldsymbol{x}_j \mid j \in J_i\}$ of each data vector $\boldsymbol{x}_i$ (as, for example, the nearest k data points in the sense of Euclidean distance). Then, it seeks to compute a *matrix of weights* $W \in \mathbb{R}^{n\times n}$ such that

(i) if $j \notin J_i$, then $w_{ij} = 0$;

(ii) $\sum_{i=1}^n w_{ij} = 1$ for every i, $1 \leq i \leq n$;

(iii) the error $\text{err} = \sum_{i=1}^{n} \left\| \boldsymbol{x}_i - \sum_{j=1}^{n} w_{ij}\boldsymbol{x}_j \right\|_2^2$ is minimal for $1 \leq i \leq n$.

Observe that the second condition can be equivalently written as $W\mathbf{1}_n = \mathbf{1}_n$.

Using the weights obtained in the second phase, we seek to determine a set of low-dimensional vectors $\boldsymbol{y}_1, \ldots, \boldsymbol{y}_n$ in $\mathbb{R}^d$, where d is much lower than m such that the local linear structure defined by the matrix W is preserved.

An important property of the weights w_{ij} is their invariance with respect to rotations, rescalings, and translations of the data points and of their neighbors. Indeed, if a rotation is applied to $\boldsymbol{x}_i$ and to its neighbors in N_i, the value of err remains the same. Also, if both $\boldsymbol{x}_i$ and the vectors in N_i are rescaled by the same factor, the values of the coefficients remain unchanged. Finally, if a translation by a vector $\boldsymbol{t}$ is applied to $\boldsymbol{x}_i$ and to all $\boldsymbol{x}_j \in J_i$, we have

$$\boldsymbol{x}_i + \boldsymbol{t} - \sum_{j=1}^{n} w_{ij}(\boldsymbol{x}_j + \boldsymbol{t}) = \boldsymbol{x}_i - \sum_{j=1}^{n} w_{ij}\boldsymbol{x}_j,$$

because $\sum_{j=1}^{n} w_{ij} = 1$. Thus, the weights w_{ij} characterize the intrinsic geometric properties of the neighborhood of $\boldsymbol{x}_i$.

In its first phase, the Saul–Roweis algorithm identifies the neighbors of each data vector either by seeking the k closest vectors or by identifying the points within a closed sphere of fixed radius. The number k of neighbors of a point $\boldsymbol{x}_i$ is, in general, not larger than the input dimensionality m; if this is not the case, special terms need to be added to the reconstruction costs (see Supplement 6).

The second phase of the algorithm involves computing the weights w_{ij} such that err is minimal and $\sum_{j=1}^{n} w_{ij} = 1$. Note that

$$\begin{aligned}
\text{err} &= \sum_{i=1}^{n} \left\| \boldsymbol{x}_i - \sum_{j=1}^{n} w_{ij}\boldsymbol{x}_j \right\|_2^2 = \sum_{i=1}^{n} \left\| \sum_{j=1}^{n} w_{ij}(\boldsymbol{x}_i - \boldsymbol{x}_j) \right\|_2^2 \\
&= \sum_{i=1}^{n} \left(\sum_{j=1}^{n} w_{ij}(\boldsymbol{x}_i - \boldsymbol{x}_j) \right)' \left(\sum_{k=1}^{n} w_{ik}(\boldsymbol{x}_i - \boldsymbol{x}_k) \right) \\
&= \sum_{i=1}^{n} \sum_{j=1}^{n} \sum_{k=1}^{n} w_{ij}(\boldsymbol{x}_i - \boldsymbol{x}_j)' w_{ik}(\boldsymbol{x}_i - \boldsymbol{x}_k).
\end{aligned}$$

The optimal weights are determined using the Lagrangian

$$\begin{aligned}
L(W, \lambda_1, \ldots, \lambda_n) &\\
&= \sum_{i=1}^{n} \sum_{j=1}^{n} \sum_{k=1}^{n} w_{ij}(\boldsymbol{x}_i - \boldsymbol{x}_j)' w_{ik}(\boldsymbol{x}_i - \boldsymbol{x}_k) \\
&\quad - \sum_{i=1}^{n} \lambda_i \left(\sum_{j=1}^{n} w_{ij} - 1 \right).
\end{aligned}$$

The necessary extremal conditions $\frac{\partial L}{\partial w_{pq}}(W, \lambda_1, \ldots, \lambda_n) = 0$ imply

$$2 \sum_{k=1}^{n} w_{pk}(\boldsymbol{x}_p - \boldsymbol{x}_q)'(\boldsymbol{x}_p - \boldsymbol{x}_k) - \lambda_p = 0$$

for $1 \leq p, q \leq n$.

Let k be the chosen number of neighbors of points and let $G(p) \in \mathbb{R}^{k \times k}$ be the *local Gram matrix in* $\boldsymbol{x}_p$, $(G(p))_{qr} = (\boldsymbol{x}_p - \boldsymbol{x}_q)'(\boldsymbol{x}_p - \boldsymbol{x}_r)$ for $1 \leq q, r \leq k$. The matrix $G(p)$ is symmetric and positive semidefinite. The extremum condition can be written as

$$2 \sum_{\ell=1}^{k} w_{p\ell} G(p)_{q\ell} = \lambda_p$$

or, as $G(p)\boldsymbol{w}_p' = \frac{1}{2}\lambda_p \mathbf{1}_n$, where $\boldsymbol{w}_p$ is the pth row of W (involving the coefficients that correspond to $\boldsymbol{x}_p$). Therefore, $\boldsymbol{w}_p' = \frac{1}{2}\lambda_p G(p)^{-1}\mathbf{1}_k$. Using the condition $\mathbf{1}_k' \boldsymbol{w}_p' = 1$, we have $\lambda_p \mathbf{1}_k' G(p)^{-1}\mathbf{1}_k = 2$, so

$$\lambda_p = \frac{2}{\mathbf{1}_k' G(p)^{-1} \mathbf{1}_k}.$$

Consequently, $\boldsymbol{w}_i$ is given by

$$\boldsymbol{w}_p = \frac{1}{\mathbf{1}_k' G(p)^{-1} \mathbf{1}_k} G(p)^{-1} \mathbf{1}_k.$$

For the final phase of the algorithm we seek a set of vectors $\{\boldsymbol{y}_1, \ldots, \boldsymbol{y}_n\}$ in the lower-dimensional space $\mathbb{R}^d$, where $d < m$, such that the matrix $Y = (\boldsymbol{y}_1 \; \cdots \; \boldsymbol{y}_n) \in \mathbb{R}^{d \times n}$ minimizes the cost function $\Phi(Y) = \sum_{i=1}^{n} \left\| \boldsymbol{y}_i - \sum_{j=1}^{n} w_{ij} \boldsymbol{y}_j \right\|_2^2$. Observe that in the absence of other conditions $\Phi(Y)$ is optimal for $Y = O_{d,n}$. Therefore, to make this problem well-posed it is necessary to require that the covariance matrix $cov(Y)$ equals I_n. This will compel the vectors $\boldsymbol{y}_i$ to be different from $\mathbf{0}_d$ and pairwise orthogonal.

Let $M \in \mathbb{R}^{n\times n}$ be the symmetric and positive semidefinite matrix defined by $M = (I_n - W)'(I_n - W)$. The cost function can be written as

$$\begin{aligned}\Phi(Y) &= \sum_{i=1}^{n}(\boldsymbol{y}_i - \sum_{j=1}^{n} w_{ij}\boldsymbol{y}_j)'(\boldsymbol{y}_i - \sum_{h=1}^{n} w_{ih}\boldsymbol{y}_h)\\ &= \sum_{i=1}^{n} \boldsymbol{y}_i'\boldsymbol{y}_i - \sum_{i=1}^{n}\sum_{j=1}^{n} w_{ij}\boldsymbol{y}_j'\boldsymbol{y}_i - \sum_{i=1}^{n}\sum_{h=1}^{n} w_{ih}\boldsymbol{y}_i'\boldsymbol{y}_h\\ &\quad+\sum_{i=1}^{n}\sum_{j=1}^{n}\sum_{h=1}^{n} w_{ij}w_{ih}\boldsymbol{y}_j'\boldsymbol{y}_h.\end{aligned}$$

Note that

$$\begin{aligned}YMY' &= Y(I_n - W' - W + W'W)Y'\\ &= YY' - YW'Y' - YWY' + YW'WY'\end{aligned}$$

and we have the obvious equalities

$$\begin{aligned}\mathsf{trace}(YY') &= \sum_{i=1}^{n} \boldsymbol{y}_i\boldsymbol{y}_i'\\ \mathsf{trace}(YWY') &= \sum_{i=1}^{n}\sum_{j=1}^{n} \boldsymbol{y}_i w_{ij}\boldsymbol{y}_j'\\ \mathsf{trace}(YW'Y') &= \sum_{i=1}^{n}\sum_{j=1}^{n} \boldsymbol{y}_i (W')_{ij}\boldsymbol{y}_j' = \sum_{i=1}^{n}\sum_{j=1}^{n} \boldsymbol{y}_i w_{ji}\boldsymbol{y}_j'\\ \mathsf{trace}(YW'WY') &= \sum_{i=1}^{n}\sum_{j=1}^{n}\sum_{h=1}^{n} \boldsymbol{y}_i (W')_{ij} W_{jh}\boldsymbol{y}_h'\\ &= \sum_{i=1}^{n}\sum_{j=1}^{n}\sum_{h=1}^{n} \boldsymbol{y}_i w_{ji} w_{jh}\boldsymbol{y}_h',\end{aligned}$$

so $\Phi(Y) = \mathsf{trace}(YMY')$.

It is important to observe that the matrix M has $(\mathbf{1}_n, 0)$ as an eigenpair since

$$M\mathbf{1}_n = (I_n - W)'(I_n - W)\mathbf{1}_n = (I_n - W)'\mathbf{0}_n = \mathbf{0}_n,$$

and all its eigenvalues are non-negative.

The cost function is invariant under translations. This makes it necessary to add the second condition that fixes the center of the set of vectors $\{\boldsymbol{y}_1, \ldots, \boldsymbol{y}_n\}$ in $\mathbf{0}_n$, which is equivalent to asking $M\mathbf{1}_n = \mathbf{0}$.

For the case of dimension d, we can write the matrix Y as a collection of n-dimensional rows as follows:

$$Y = \begin{pmatrix} \boldsymbol{z}_1 \\ \cdots \\ \boldsymbol{z}_d \end{pmatrix}.$$

Now the cost function becomes

$$\begin{aligned}\Phi(Y) &= \mathsf{trace}\left(\begin{pmatrix} \boldsymbol{z}_1 \\ \cdots \\ \boldsymbol{z}_d \end{pmatrix} M(\boldsymbol{z}_1' \ \cdots \ \boldsymbol{z}_d')\right) \\ &= \sum_{p=1}^{d} \mathsf{trace}(\boldsymbol{z}_p M \boldsymbol{z}_p') = \sum_{p=1}^{d} \boldsymbol{z}_p M \boldsymbol{z}_p'.\end{aligned}$$

Let $\lambda_1 \geq \cdots \geq \lambda_d \geq \lambda_{d+1} = 0$ be the least $d+1$ eigenvalues of M. To minimize $\Phi(Y)$ we need to minimize each of the non-negative numbers $\boldsymbol{z}_p M \boldsymbol{z}_p'$, which means that we need to adopt for $\boldsymbol{z}_1', \ldots, \boldsymbol{z}_d'$ the unit eigenvectors that correspond to the eigenvalues $\lambda_1, \ldots, \lambda_d$. The matrix Y is given by

$$Y = \begin{pmatrix} \boldsymbol{z}_1 \\ \vdots \\ \boldsymbol{z}_d \end{pmatrix} = (\boldsymbol{y}_1 \ \cdots \ \boldsymbol{y}_n).$$

The eigenvectors that correspond to the smallest $d+1$ eigenvalues provide a matrix $(\boldsymbol{y}_1, \ldots, \boldsymbol{y}_d, \boldsymbol{y}_{d+1}) \in \mathbb{R}^{n \times (d+1)}$. The last eigenvector $\boldsymbol{y}_{d+1} = \mathbf{1}_n$ corresponds to the eigenvalue 0. The d-dimensional rows $\boldsymbol{z}_1, \ldots, \boldsymbol{z}_n$ of the matrix $S = (\boldsymbol{y}_1, \ldots, \boldsymbol{y}_d) \in \mathbb{R}^{n \times d}$ yield the low-dimensional representation $\boldsymbol{z}_1', \ldots, \boldsymbol{z}_n'$ of $\{\boldsymbol{x}_1, \ldots, \boldsymbol{x}_n\}$.

Example 12.3. We discuss the MATLAB algorithm developed for LLE in [140]. We begin by generating a data set in $\mathbb{R}^2$ using the MATLAB code

```
i = 1;
for x = -pi:0.1:pi
    X(1,i) = x*sqrt(3)/2 -1 -0.5*sin(x - pi);
    X(2,i) = 0.5*x + sqrt(3) + sqrt(3)*sin(x-pi)/2;
    i=i+1;
end
```

The 63 columns of matrix $X \in \mathbb{R}^{2\times 63}$ have been plotted in Figure 12.3 and they are located on a curve in $\mathbb{R}^2$ which is a one-dimensional manifold.

Next we determined the nearest two points for each point. For this computation we need to compute the matrix of distances $\texttt{dist} \in \mathbb{R}^{63\times 63}$ between points

```
for i=1:1:63
    for j=1:1:63
    dist(i,j)= norm(X(:,i)-X(:,j));
    end
end
[sorted,index] = sort(dist);
neighborhood = index(2:3,:);
```

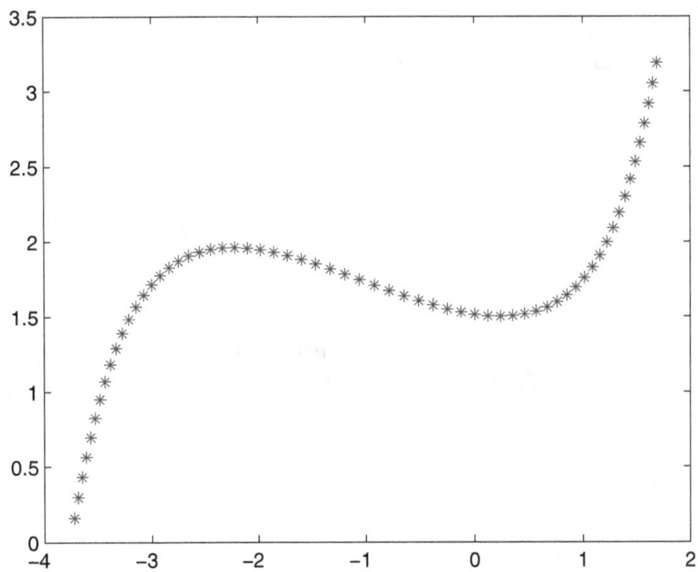

Fig. 12.3 Representation of the data set in $\mathbb{R}^2$.

The matrix **dist** is sorted yielding the matrix **sorted** having each of its columns arranged in ascending order. Thus, the jth column of **sorted** contains the distances between $\boldsymbol{x}_j$ and the remaining points. The matrix **index** contains in its jth column the indices of the points that correspond to the sorted distances. This allows us to find the next two neighbors of $\boldsymbol{x}_j$ by simply extracting the second and the third elements of **dist**.

```
W = zeros(2,63);
for ii = 1:63
   z = X(:,neighborhood(:,ii))-repmat(X(:,ii),1,2);
   % next we compute the local Gram matrix
   G = z'*z;
   % we solve Gw=1
   W(:,ii) = G\ones(2,1);
   % enforce sum(w)=1
   W(:,ii) = W(:,ii)/sum(W(:,ii));
end;
```

In the next phase the eigenvectors of the cost matrix are computed.

```
% M is a sparse matrix with storage for 4*2*63 nonzero
    elements
M = sparse(1:63,1:63,ones(1,63),63,63,4*2*63);
for ii=1:63
   w = W(:,ii);
   jj = neighborhood(:,ii);
   M(ii,jj) = M(ii,jj) - w';
   M(jj,ii) = M(jj,ii) - w;
   M(jj,jj) = M(jj,jj) + w*w';
end;
options.disp = 0; options.isreal = 1; options.issym = 1;
[Y,eigenvals] = eigs(M,2,0,options);
Y = Y(:,2:2)'*sqrt(63);
```

The resulting matrix Y is a row vector containing the representation of the 63 points located on the curve.

12.6 MATLAB Computations

The MATLAB function **lsqr** attempts to compute the least square solution $\boldsymbol{x}$ to the linear system of equations $A\boldsymbol{x} = \boldsymbol{b}$ by finding $\boldsymbol{x}$ that minimizes $\|A\boldsymbol{x} - \boldsymbol{b}\|$. If the system is consistent, the linear squares solution is also a solution of the linear system and a corresponding message is displayed.

Example 12.4. Let A and $\boldsymbol{b}$ be given by

$$A = \begin{pmatrix} 1 & 1 & -2 \\ 2 & -1 & 0 \\ 3 & 1 & -1 \end{pmatrix} \text{ and } \boldsymbol{b} = \begin{pmatrix} -2 \\ 1 \\ 2 \end{pmatrix}.$$

The following MATLAB code solves the system $A\boldsymbol{x} = \boldsymbol{b}$:

```
>> A=[1 1 -2;2 -1 0;3 1 -1]
A =

     1     1    -2
     2    -1     0
     3     1    -1
>> b=[-2;1;2]
b =
    -2
     1
     2
>> x=lsqr(A,b)
x =
    1.0000
    1.0000
    2.0000
```

Example 12.5. Let us expand the matrix A by adding one more row:

```
>> A=[1 1 -2;2 -1 0;3 1 -1;1 -1 3]
```

and setting

```
\bfb = [-2; 1; 2; 8]
```

The system $A\boldsymbol{x} = \boldsymbol{b}$ is now incompatible and the result returned is

```
>> x=lsqr(A,b)
lsqr converged at iteration 3 to a solution with relative
       residual 0.096.

x =
    1.1905
    1.3810
    2.6190
```

The result includes the relative residual error $\frac{\|\|}{\boldsymbol{b}-A\boldsymbol{x}}\|x\|$ and the iteration number at which the algorithm halted.

If $A \in \mathbb{R}^{n\times n}$ and B is a column vector with n components, or a matrix with several such columns, then `X = A\B` is the solution to the equation $AX = B$ in the least square sense. A warning message is displayed if A is badly scaled or nearly singular.

Exercises and Supplements

(1) Let $X \in \mathbb{R}^{m\times n}$, where $m \geq n$, be a full-rank matrix, $\boldsymbol{b}$ be a vector in $\mathbb{R}^m$, and let $\boldsymbol{r} = X\boldsymbol{x} - \boldsymbol{b}$ be the *residual vector* corresponding to $\boldsymbol{x}$.
Prove that $\boldsymbol{r}$ is the solution of the normal system $X'X\boldsymbol{r} = X'\boldsymbol{b}$ if and only if $\boldsymbol{r}$ is orthogonal to the columns of X.

(2) Let $\boldsymbol{a}, \boldsymbol{b} \in \mathbb{R}^m$ be two vectors and let $\boldsymbol{a}x = \boldsymbol{b}$ be a one-indeterminate linear system. The system is, in general, incompatible, if $m > 1$.
   (a) Prove that the best approximation of the solution in the sense defined in Section 12.1 is

$$x = \frac{(\boldsymbol{a}, \boldsymbol{b})}{\|\boldsymbol{a}\|_2^2}.$$

   Replace the norm $\|\cdot\|_2$ with $\|\cdot\|_1$; in other words, define the best approximation of the solution as x such that $\|Ax - \boldsymbol{b}\|_1$ is minimal. Prove that if $\boldsymbol{a} = (a, a, a) \in \mathbb{R}^3$ with $a > 0$, then the optimum is achieved for $x = b_2$.

(3) Let $\boldsymbol{b}_1, \ldots, \boldsymbol{b}_n$ be n vectors in $\mathbb{R}^m$ such that $\sum_{i=1}^n \boldsymbol{b}_i = \boldsymbol{0}_m$ and let $H_{\boldsymbol{w},a}$ be a hyperplane in $\mathbb{R}^m$, where $\boldsymbol{w}$ is a unit vector. Let $B = (\boldsymbol{b}_1 \ \cdots \ \boldsymbol{b}_n)$.
Prove that if $\sum_{i=1}^n d(H_{\boldsymbol{w},a}, \boldsymbol{b}_i)^2$ is minimal (where $d(H_{\boldsymbol{w},a}, \boldsymbol{b}_i)$ is the distance from $\boldsymbol{b}_i$ to $H_{\boldsymbol{w},a}$), then $\boldsymbol{w}$ is an eigenvector of the matrix BB' that corresponds to the least eigenvalue of this matrix and $a = 0$.

   Solution: We need to minimize $\sum_{i=1}^n (\boldsymbol{w}'\boldsymbol{b}_i - a)^2 = \|\boldsymbol{w}'B - a\boldsymbol{1}'\|_2^2$ subjected to the restriction $\|\boldsymbol{w}\|_2 = 1$. Consider the

Lagrangian

$$\begin{aligned}F(w_1,\ldots,w_m,\lambda) &= \sum_{i=1}^{n}(\boldsymbol{w}'\boldsymbol{b}_i - a)^2 - \lambda\left(\sum_{\ell=1}^{n} w_\ell^2 - 1\right)\\ &= \sum_{i=1}^{n}\left(\sum_{j=1}^{m}\boldsymbol{w}_j b_{ji} - a\right)^2 - \lambda\left(\sum_{\ell=1}^{n} w_\ell^2 - 1\right).\end{aligned}$$

Then, we can write

$$\frac{\partial F}{\partial w_k} = \sum_{i=1}^{n} 2\left(\sum_{j=1}^{m}\boldsymbol{w}_j b_{ji} - a\right) b_{ki} - 2\lambda w_k = 0,$$

which is equivalent to

$$\sum_{i=1}^{n}\sum_{j=1}^{m}\boldsymbol{w}_j b_{ji} b_{ki} - a\sum_{j=1}^{m} b_{kj} - \lambda w_k = 0.$$

Since $\sum_{i=1}^{n}\boldsymbol{b}_i = \boldsymbol{0}_m$, it follows that $\sum_{j=1}^{n} b_{kj} = 0$ for every k so the necessary extremal conditions are

$$\sum_{i=1}^{n}\sum_{j=1}^{m}\boldsymbol{w}_j b_{ji} b_{ki} = \lambda w_k = 0$$

for $1 \leq k \leq m$. This equality amounts to $\boldsymbol{w}'BB' = \lambda\boldsymbol{w}'$, or $BB'\boldsymbol{w} = \lambda\boldsymbol{w}$, which shows that $\boldsymbol{w}$ must be an eigenvector of BB'.

The total sum of the squared distances is

$$\begin{aligned}D &= \|\boldsymbol{w}'B - a\mathbf{1}'\|_2^2 = (\boldsymbol{w}'B - a\mathbf{1}')(\boldsymbol{w}'B - a\mathbf{1}')'\\ &= (\boldsymbol{w}'B - a\mathbf{1}')(B'\boldsymbol{w} - a\mathbf{1})\\ &= \boldsymbol{w}'BB'\boldsymbol{w} - 2a\mathbf{1}'B'\boldsymbol{w} + a^2\mathbf{1}'\mathbf{1} = \lambda\boldsymbol{w}'\boldsymbol{w} + ma^2 = \lambda + ma^2.\end{aligned}$$

Thus, $\boldsymbol{w}$ must be a unit eigenvector of BB' that corresponds to the least eigenvalue. The minimal value of D is obtained when $a = 0$.

(4) Formulate and prove a result similar to the one in Supplement 3 without the condition $\sum_{i=1}^{n} \boldsymbol{b}_i = \mathbf{0}_m$.

Hint: Consider the vectors $\boldsymbol{c}_i = \boldsymbol{b}_i - \boldsymbol{b}_0$, where $\boldsymbol{b}_0 = \frac{1}{n}\,\mathrm{sum}_{i=1}^{n} \boldsymbol{b}_i$ and observe that $\sum_{i=1}^{n} \boldsymbol{c}_i = \mathbf{0}$.

(5) Let $W \in \mathbb{R}^{n\times n}$ be a matrix. Prove that $(I_n - W)'(I_n - W)\mathbf{1}_n = \mathbf{0}_n$ if and only if $W\mathbf{1}_n = \mathbf{0}_n$.

(6) Prove that when k, the number of nearest k neighbors exceeds the dimension m of the input space of the locally linear embedding algorithm, the local Gram matrix can be non-invertible. Prove that this problem can be corrected by replacing $G(p)$ by $G(p) + c/kI_n$, where $c > 0$ is a small number.

(7) Let $A \in \mathbb{C}^{m\times n}$ be a matrix with $\mathsf{rank}(A) = k$ such that $A = URV^{\mathsf{H}}$. Here we assume that $U \in \mathbb{C}^{m\times n}$, $V \in \mathbb{C}^{n\times n}$ are unitary matrices, and $R \in \mathbb{C}^{m\times n}$ can be written as

$$R = \begin{pmatrix} Q & O_{k,n-k} \\ O_{m-k,k} & O_{m-k,n-k} \end{pmatrix},$$

where $Q \in \mathbb{C}^{k\times k}$ and $\mathsf{rank}(Q) = k$. Let $\boldsymbol{b} \in \mathbb{C}^m$, $\boldsymbol{x} \in \mathbb{C}^n$, $\boldsymbol{g} = U^{\mathsf{H}}\boldsymbol{b} \in \mathbb{C}^m$, and $\boldsymbol{y} = V^{\mathsf{H}}\boldsymbol{x}$. Write

$$\boldsymbol{g} = \begin{pmatrix} \boldsymbol{g}_1 \\ \boldsymbol{g}_2 \end{pmatrix} \text{ and } \boldsymbol{y} = \begin{pmatrix} \boldsymbol{y}_1 \\ \boldsymbol{y}_2 \end{pmatrix},$$

where $\boldsymbol{g}_1 \in \mathbb{C}^k$, $\boldsymbol{g}_2 \in \mathbb{C}^{m-k}$, $\boldsymbol{y}_1 \in \mathbb{C}^k$, and $\boldsymbol{y}_2 \in \mathbb{C}^{n-k}$.
Prove that

(a) if $\boldsymbol{z}$ is the unique solution of $Q\boldsymbol{z} = \boldsymbol{g}_1$, then any vector $\boldsymbol{x}$ that minimizes $\|A\boldsymbol{x} - \boldsymbol{b}\|$ has the form

$$\hat{\boldsymbol{x}} = V\begin{pmatrix} \boldsymbol{z} \\ \boldsymbol{y}_2 \end{pmatrix};$$

(b) any optimal solution gives the residual vector

$$\boldsymbol{r} = \boldsymbol{b} - A\hat{\boldsymbol{x}} = U\begin{pmatrix} \mathbf{0}_k \\ \boldsymbol{g}_2 \end{pmatrix},$$

where $\|\boldsymbol{r}\| = \|\boldsymbol{g}_2\|$;

(c) the unique solution of minimal Euclidean norm is

$$\hat{\boldsymbol{x}} = V \begin{pmatrix} \boldsymbol{z} \\ \boldsymbol{0}_{n-k} \end{pmatrix},$$

where $\boldsymbol{z}$ was defined above;

(d) the solution of minimum Euclidean length, the minimal value of $\|\boldsymbol{b} - A\boldsymbol{x}\|$, and set of all solutions are unique.

Solution: The hypothesis implies

$$\begin{aligned}\|A\boldsymbol{x} - \boldsymbol{b}\|^2 &= \|URV^{\mathsf{H}}\boldsymbol{x} - \boldsymbol{b}\|^2 \\ &= \|URV^{\mathsf{H}}\boldsymbol{x} - UU^{\mathsf{H}}\boldsymbol{b}\|^2 = \|U(RV^{\mathsf{H}}\boldsymbol{x} - U^{\mathsf{H}}\boldsymbol{b})\|^2 \\ &= \|RV^{\mathsf{H}}\boldsymbol{x} - U^{\mathsf{H}}\boldsymbol{b}\|^2 \\ &\quad \text{(because } U \text{ is a unitary matrix)} \\ &= \|R\boldsymbol{y} - \boldsymbol{g}\|^2 = \left\| \begin{pmatrix} Q\boldsymbol{y}_1 - \boldsymbol{g}_1 \\ \boldsymbol{g}_2 \end{pmatrix} \right\| \\ &= \|Q\boldsymbol{y}_1 - \boldsymbol{g}_1\|^2 + \|\boldsymbol{g}_2\|^2.\end{aligned}$$

The minimum value of $\|A\boldsymbol{x} - \boldsymbol{b}\|^2$ is achieved when $Q\boldsymbol{y}_1 = \boldsymbol{g}_1$ and it equals $\|\boldsymbol{g}_2\|^2$. Thus, if $\boldsymbol{z}$ is the unique solution of the equation $Q\boldsymbol{z} = \boldsymbol{g}_1$, then any solution that minimizes $\|A\boldsymbol{x} - \boldsymbol{b}\|$ has the form

$$\hat{\boldsymbol{x}} = V \begin{pmatrix} \boldsymbol{z} \\ \boldsymbol{y}_2 \end{pmatrix} = V \begin{pmatrix} Q^{-1}\boldsymbol{g}_1 \\ \boldsymbol{g}_2 \end{pmatrix},$$

where $\boldsymbol{y}_2$ is arbitrary. The solution that minimizes $\|A\boldsymbol{x} - \boldsymbol{b}\|$ and has the minimum Euclidean norm is

$$\hat{\boldsymbol{x}} = V \begin{pmatrix} \boldsymbol{z} \\ \boldsymbol{0}_{n-k} \end{pmatrix}.$$

Let $\hat{\boldsymbol{y}} = V^{\mathsf{H}}\hat{\boldsymbol{x}} = \begin{pmatrix} \hat{\boldsymbol{y}}_1 \\ \hat{\boldsymbol{y}}_2 \end{pmatrix}$. The residual vector of an optimal solution is

$$\begin{aligned}\boldsymbol{r} &= \boldsymbol{b} - A\hat{\boldsymbol{x}} = UU^{\mathsf{H}}\boldsymbol{b} - URV^{\mathsf{H}}\hat{\boldsymbol{x}} \\ &= U(U^{\mathsf{H}}\boldsymbol{b} - RV^{\mathsf{H}}\hat{\boldsymbol{x}}) = U(U^{\mathsf{H}}\boldsymbol{b} - R\hat{\boldsymbol{y}}) = U \begin{pmatrix} \boldsymbol{0}_k \\ \boldsymbol{g}_2 \end{pmatrix}.\end{aligned}$$

The last part follows immediately.

(8) Prove that the unique minimum Euclidean length solution $\boldsymbol{x}$ of the minimization problem has the form

$$\hat{\boldsymbol{x}} = V \begin{pmatrix} Q^{-1} & O_{k,m-k} \\ O_{n-k,k} & O_{n-k,m-k} \end{pmatrix} U^{\mathsf{H}}\boldsymbol{b},$$

where the notations are the same as in Supplement 7.

Solution: The conclusion is immediate from the equalities

$$\begin{aligned} \hat{\boldsymbol{x}} &= V \begin{pmatrix} Q^{-1}\boldsymbol{g}_1 \\ \boldsymbol{0}_{n-k} \end{pmatrix} V \begin{pmatrix} Q^{-1} & O_{k,m-k} \\ O_{n-k,k} & O_{n-k,m-k} \end{pmatrix} \begin{pmatrix} \boldsymbol{g}_1 \\ \boldsymbol{g}_2 \end{pmatrix} \\ &= V \begin{pmatrix} Q^{-1} & O_{k,m-k} \\ O_{n-k,k} & O_{n-k,m-k} \end{pmatrix} U^{\mathsf{H}}\boldsymbol{b}. \end{aligned}$$

(9) Let $A \in \mathbb{C}^{m\times n}$ be a matrix and let $\boldsymbol{z}_i \in \mathbb{R}^m$ be the solution of minimal Euclidean length of the least square problem instance $(A, \boldsymbol{z}, \boldsymbol{e}_i)$, where $\boldsymbol{e}_i \in \mathbb{C}^m$ and $\boldsymbol{z}_i \in \mathbb{C}^n$ for $1 \leq i \leq m$. Prove that the Moore–Penrose pseudoinverse of A is $A^\dagger = (\boldsymbol{z}_1 \ \cdots \ \boldsymbol{z}_m)$.

Solution: In Supplement 26 of Chapter 9, we have shown that the pseudoinverse of A has the form

$$A^\dagger = V \begin{pmatrix} R^{-1} & O_{k,m-k} \\ O_{n-k,k} & O_{n-k,m-k} \end{pmatrix} U^{\mathsf{H}},$$

where

$$A = U \begin{pmatrix} R & O_{k,n-k} \\ O_{m-k,k} & O_{m-k,n-k} \end{pmatrix} V^{\mathsf{H}},$$

$R \in \mathbb{C}^{k\times k}$ is a matrix of rank k, and $U \in \mathbb{C}^{m\times m}$ and $V \in \mathbb{C}^{n\times n}$ are unitary matrices. The columns of $A^\dagger$ are given by

$$\boldsymbol{z}_i = A^\dagger \boldsymbol{e}_i = V \begin{pmatrix} R^{-1} & O_{k,m-k} \\ O_{n-k,k} & O_{n-k,m-k} \end{pmatrix} U^{\mathsf{H}}\boldsymbol{e}_i,$$

and $\boldsymbol{z}_i$ is precisely the solution of minimal Euclidean length of the least square problem $\min\{\boldsymbol{e}_i - A\boldsymbol{z}\}$.

(10) Let $A \in \mathbb{R}^{m\times n}$ and $\boldsymbol{b} \in \mathbb{R}^n$. Define $\boldsymbol{d} = A'\boldsymbol{b}$, $B = (A\ \boldsymbol{b})$, and

$$H = B'B = \begin{pmatrix} A'A & \boldsymbol{d}' \\ \boldsymbol{d} & a^2 \end{pmatrix},$$

where $a^2 = \boldsymbol{b}'\boldsymbol{b}$. Prove that if H has the Cholesky factorization $H = U'U$, where

$$U = \begin{pmatrix} W & \boldsymbol{y} \\ \boldsymbol{0} & \rho \end{pmatrix},$$

then $|\rho| = \|\boldsymbol{b} - A\hat{\boldsymbol{x}}\|$ for every $\hat{\boldsymbol{x}}$ that minimizes $\|\boldsymbol{b} - A\boldsymbol{x}\|$.

(11) Let $A \in \mathbb{R}^{m\times n}$ and $C \in \mathbb{R}^{m\times p}$ be two matrices and let $X \in \mathbb{R}^{n\times p}$ be a matrix such that $\|AX - C\|_F$ is minimal. Prove that the equation $A'AX = A'C$ is always solvable in X and that its solution minimizes $\|AX - C\|_F$.

Solution: Let $C = \{\boldsymbol{c}_1\ \cdots\ \boldsymbol{c}_p\}$. We saw that the equation $A'A\boldsymbol{x} = A'\boldsymbol{c}_i$ is always solvable with respect to $\boldsymbol{x}$; let $\boldsymbol{x}_i$ be its solution, where $1 \leq i \leq p$. Then, we have

$$A'A(\boldsymbol{x}_1\ \cdots\ \boldsymbol{x}_p) = (\boldsymbol{c}_1\ \ldots\ \boldsymbol{c}_p) = C,$$

so $X = (\boldsymbol{x}_1\ \cdots\ \boldsymbol{x}_p)$ is a solution of $A'AX = A'C$. Note that

$$(X'A' - C')(AX - C) = \begin{pmatrix} \boldsymbol{x}_1'A - \boldsymbol{c}_1' \\ \vdots \\ \boldsymbol{x}_p' - \boldsymbol{c}_p' \end{pmatrix} (A\boldsymbol{x}_1 - \boldsymbol{c}_1\ \cdots\ A\boldsymbol{x}_p - \boldsymbol{c}_p),$$

which implies

$$\begin{aligned} \|AX - C\|_F^2 &= \mathsf{trace}(X'A' - C')(AX - C) \\ &= \sum_{i=1}^p (\boldsymbol{x}_i'A - \boldsymbol{c}_i')(A\boldsymbol{x}_i - \boldsymbol{c}_i) \\ &= \sum_{i=1}^p \|A\boldsymbol{x}_i - \boldsymbol{c}_i\|_2^2. \end{aligned}$$

This implies the second part of this supplement.

(12) Let $A \in \mathbb{R}^{m\times n}$ be a matrix whose columns form an orthonormal set of vectors and let $\boldsymbol{c} \in \mathbb{R}^m$. Prove that $\boldsymbol{x} = A'\boldsymbol{c}$ minimizes $\|A\boldsymbol{x} - \mathbf{c}\|$. Note that in this case $A' = A^\dagger$ by Supplement 45 of Chapter 3.

(13) Let $A \in \mathbb{R}^{m\times n}$ be a matrix with $\mathsf{rank}(A) = r < n$. If $X \subseteq \mathbb{R}^n$ is the set of vectors $\boldsymbol{x}$ that minimize $\|A\boldsymbol{x} - \boldsymbol{b}\|$, prove that this set is convex. Therefore, X contains a *unique* element having minimum 2-norm, denoted by $\boldsymbol{x}_{LS}$ (the subscript suggests the words "least square").

Solution: Suppose that both $\boldsymbol{u}$ and $\boldsymbol{v}$ minimize $\|A\boldsymbol{x}-\boldsymbol{b}\|$. Then, for $\lambda \in [0,1]$, we have

$$\begin{aligned}\|A(\lambda\boldsymbol{u} + (1-\lambda)\boldsymbol{v} - \boldsymbol{b}\| &\leqslant \lambda\|A\boldsymbol{u} - \boldsymbol{b}\| + (1-\lambda)\|A\boldsymbol{v} - \boldsymbol{b}\|\\ &= \|A\boldsymbol{x} - \boldsymbol{b}\|.\end{aligned}$$

(14) Let $A \in \mathbb{R}^{m\times n}$ be a matrix and $\mathsf{rank}(A) = r$, and let $A = UDV'$ be the SVD decomposition of A, where $U = (\boldsymbol{u}_1 \ \cdots \ \boldsymbol{u}_m) \in \mathbb{R}^{m\times m}$, $V = (\boldsymbol{v}_1 \ \cdots \ \boldsymbol{v}_n) \in \mathbb{R}^{n\times n}$ are orthogonal matrices and $D = \mathsf{diag}(\sigma_1, \ldots, \sigma_r, 0, \ldots, 0)$.
If $\boldsymbol{b} \in \mathbb{R}^m$, prove that $\boldsymbol{x}_{LS} = \sum_{i=1}^{r} \frac{\boldsymbol{u}_i'\boldsymbol{b}}{\sigma_i}\boldsymbol{v}_i$ minimizes $\|A\boldsymbol{x} - \boldsymbol{b}\|$ has the smallest norm of all minimizers.

Solution: We have

$$\begin{aligned}\|A\boldsymbol{x} - \boldsymbol{b}\|^2 &= \|(U'AV)(V'\boldsymbol{x}) - U'\boldsymbol{b}\|^2\\ &\qquad\text{(because } U \text{ and } V \text{ are orthogonal)}\\ &= \sum_{i=1}^{r} \left(\sigma_i(V'\boldsymbol{x})_i - \boldsymbol{u}_i'\boldsymbol{b}\right)^2 + \sum_{i=r+1}^{m} (\boldsymbol{u}_i\boldsymbol{b})^2.\end{aligned}$$

If $\boldsymbol{x}_{LS}$ solves the least square problem, this implies $(V'\boldsymbol{x})_i = \frac{\boldsymbol{u}_i'\boldsymbol{b}}{\sigma_i}$.

(15) Let $A \in \mathbb{R}^{m\times n}$ be a matrix and $\mathsf{rank}(A) = r$, and let $A = UDV'$ be the SVD decomposition of A. Prove that the matrix

$$A^\dagger = V\mathsf{diag}\left(\frac{1}{\sigma_1}, \ldots, \frac{1}{\sigma_r}, 0, \ldots, 0\right) U'$$

is the pseudoinverse of A and $\boldsymbol{x}_{LS} = A^\dagger\boldsymbol{b}$ is the least square solution.

Bibliographical Comments

Exercise 2 is based on an observation made in [65]. Monographs dedicated to the least squares problems are [21] and [103].

Partial square regression is discussed in [1]. Locally linear embedding is due to S. T. Roweis and L. K. Saul [141]. Further improvements and variations on the LLE theme can be found in [9, 38, 144].

Supplement 6 is discussed in [143]. Supplements 7–10 are results of [103].

Chapter 13

Dimensionality Reduction Techniques

13.1 Introduction

Physical and biological data as well as economic and demographic data often have high dimensionality. Intelligent data-mining algorithms work best in interpretation and decision-making based on this data when we are able to simplify their tasks by reducing the high dimensionality of the data.

Dimensionality reduction refers to the extraction of the relevant information for a specific objective, while ignoring the unnecessary information and is a key concept in pattern recognition, data mining, feature processing, and machine learning. Dimensionality reduction requires tuning in terms of the expected number of dimensions, or the parameters of the learning algorithms.

13.2 Principal Component Analysis

Principal component analysis (PCA) is a dimensionality reduction technique that aims to create a few new, uncorrelated linear combinations of the variables of experiments that "explain" the major parts of the data variability.

Let $W \in \mathbb{R}^{m \times n}$ be a data matrix given by

$$W = \begin{pmatrix} \boldsymbol{u}_1' \\ \vdots \\ \boldsymbol{u}_m' \end{pmatrix}.$$

Definition 13.1. Let $\boldsymbol{w} \in \mathbb{R}^n$ be a unit vector. The *residual* of $\boldsymbol{u}_i$ relative to $\boldsymbol{w}$ is the number $r(\boldsymbol{u}_i) = \|\boldsymbol{u}_i - (\boldsymbol{u}_i'\boldsymbol{w})\boldsymbol{w}\|^2$ and it represents the error committed when the vector $\boldsymbol{u}_i$ is replaced by its projection on $\boldsymbol{w}$.

Theorem 13.1. *If $r(\boldsymbol{u}_i)$ is the residual of the vector $\boldsymbol{u}_i$ of a data matrix W relative to the vector $\boldsymbol{w}$ with $\|\boldsymbol{w}\| = 1$, then $r(\boldsymbol{u}_i) = \|\boldsymbol{u}_i\|^2 - (\boldsymbol{u}_i'\boldsymbol{w})^2$.*

Proof. We have

$$\begin{aligned}
r(\boldsymbol{u}_i) &= \|\boldsymbol{u}_i - (\boldsymbol{u}_i'\boldsymbol{w})\boldsymbol{w}\|^2 \\
&= (\boldsymbol{u}_i - (\boldsymbol{u}_i'\boldsymbol{w})\boldsymbol{w})'(\boldsymbol{u}_i - (\boldsymbol{u}_i'\boldsymbol{w})\boldsymbol{w}) \\
&= (\boldsymbol{u}_i' - (\boldsymbol{u}_i'\boldsymbol{w})\boldsymbol{w}')(\boldsymbol{u}_i - (\boldsymbol{u}_i'\boldsymbol{w})\boldsymbol{w}) \\
&= \boldsymbol{u}_i'\boldsymbol{u}_i - (\boldsymbol{u}_i'\boldsymbol{w})\boldsymbol{w}'\boldsymbol{u}_i - \boldsymbol{u}_i'(\boldsymbol{u}_i'\boldsymbol{w})\boldsymbol{w} + (\boldsymbol{u}_i'\boldsymbol{w})\boldsymbol{w}'(\boldsymbol{u}_i'\boldsymbol{w})\boldsymbol{w} \\
&= \|\boldsymbol{u}_i\|^2 - 2(\boldsymbol{u}_i'\boldsymbol{w})^2 + (\boldsymbol{u}_i'\boldsymbol{w})^2 \\
&= \|\boldsymbol{u}_i\|^2 - (\boldsymbol{u}_i'\boldsymbol{w})^2
\end{aligned}$$

because $\boldsymbol{w}'\boldsymbol{w} = 1$. □

Definition 13.2. The *mean square error* $\mathsf{MSE}(W, \boldsymbol{w})$ of the projections of the experiments $\boldsymbol{u}_1, \ldots, \boldsymbol{u}_m$ of the data matrix $W \in \mathbb{R}^{m\times n}$ on the unit vector $\boldsymbol{w} \in \mathbb{R}^n$ is the sum of the residuals

$$\mathsf{MSE}(W, \boldsymbol{w}) = \frac{1}{m}\sum_{i=1}^{m} r(\boldsymbol{u}_i).$$

The *average of the projections of the experiment vectors on the unit vector* $\boldsymbol{w}$ is

$$u_{\boldsymbol{w}} = \frac{1}{m}\sum_{i=1}^{m} \boldsymbol{u}_i'\boldsymbol{w} = \tilde{\boldsymbol{u}}'\boldsymbol{w},$$

where $\tilde{\boldsymbol{u}}$ is the sample mean defined by Equality (11.1).

The *variance* of the projections of the experiment vectors on $\boldsymbol{w}$ is

$$V(W, \boldsymbol{w}) = \frac{1}{m}\sum_{i=1}^{m}(\boldsymbol{u}_i'\boldsymbol{w} - u_{\boldsymbol{w}})^2.$$

Note that if W is centered, we have $\tilde{\boldsymbol{u}} = \mathbf{0}_n$ and, therefore, we have $u_{\boldsymbol{w}} = 0$.

Theorem 13.2. *We have the following equalities:*

$$V(W, \boldsymbol{w}) = \frac{1}{m} \sum_{i=1}^{m} (\boldsymbol{u}_i' \boldsymbol{w})^2 - u_{\boldsymbol{w}}^2,$$

and

$$\mathsf{MSE}(W, \boldsymbol{w}) = \frac{1}{m} \sum_{i=1}^{m} \|\boldsymbol{u}_i\|^2 - u_{\boldsymbol{w}}^2 - V(W, \boldsymbol{w}).$$

Proof. The first equality is well known from elementary statistics. For the second equality, we can write

$$\begin{aligned}
\mathsf{MSE}(W, \boldsymbol{w}) &= \frac{1}{m} \sum_{i=1}^{m} r(\boldsymbol{u}_i) \\
&= \frac{1}{m} \sum_{i=1}^{m} \left(\|\boldsymbol{u}_i\|^2 - (\boldsymbol{u}_i' \boldsymbol{w})^2\right) \\
&= \frac{1}{m} \sum_{i=1}^{m} \|\boldsymbol{u}_i\|^2 - \frac{1}{m} \sum_{i=1}^{m} (\boldsymbol{u}_i' \boldsymbol{w})^2 \\
&= \frac{1}{m} \sum_{i=1}^{m} \|\boldsymbol{u}_i\|^2 - u_{\boldsymbol{w}}^2 - V(W, \boldsymbol{w}).
\end{aligned}$$

□

Corollary 13.1. *If W is a centered data matrix, then minimizing $\mathsf{MSE}(W, \boldsymbol{w})$ amounts to maximizing the variance of the projections of the vectors of the experiments.*

Proof. Since W is centered, we have $u_{\boldsymbol{w}} = 0$. Therefore, the equality involving $\mathsf{MSE}(W, \boldsymbol{w})$ from Theorem 13.2 becomes

$$\mathsf{MSE}(W, \boldsymbol{w}) = \frac{1}{m} \sum_{i=1}^{m} \|\boldsymbol{u}_i\|^2 - V(W, \boldsymbol{w}).$$

The first term does not depend on $\boldsymbol{w}$. Therefore, to minimize the mean square error, we need to maximize the variance of the projections of the vectors of the experiments. □

If W is a centered data matrix, we have $u_{\boldsymbol{w}} = 0$ and the variance of the data matrix reduces to

$$V(W, \boldsymbol{w}) = \frac{1}{m} \sum_{i=1}^{m} (\boldsymbol{u}_i' \boldsymbol{w})^2.$$

This expression can be transformed as

$$V(W, \boldsymbol{w}) = \frac{1}{m}(W\boldsymbol{w})'(W\boldsymbol{w}) = \frac{1}{m}\boldsymbol{w}'W'W\boldsymbol{w} = \boldsymbol{w}'Z\boldsymbol{w},$$

where $Z = \frac{1}{m}W'W$.

We need to choose the unit vector $\boldsymbol{w}$ to maximize $V(W, \boldsymbol{w})$. In other words, we need to maximize $V(W, \boldsymbol{w})$ subjected to the restriction $\boldsymbol{w}'\boldsymbol{w} - 1 = 0$. This can be resolved using a Lagrange multiplier λ to optimize the function

$$L(\boldsymbol{w}, \lambda) = \frac{1}{m}\boldsymbol{w}'W'W\boldsymbol{w} - \lambda(\boldsymbol{w}'\boldsymbol{w} - 1).$$

Since

$$\frac{\partial L}{\partial \lambda} = \boldsymbol{w}'\boldsymbol{w} - 1,$$

$$\frac{\partial L}{\partial \boldsymbol{w}} = 2Z'\boldsymbol{w} - 2\lambda\boldsymbol{v},$$

which implies $\boldsymbol{w}'\boldsymbol{w} = 1$ and $Z\boldsymbol{w} = \lambda\boldsymbol{w}$. The last equality amounts to

$$\frac{1}{m}W'W\boldsymbol{w} = \lambda\boldsymbol{w},$$

which means that $\boldsymbol{w}$ must be an eigenvector of the covariance matrix $cov(W)$. This is an $n \times n$ symmetric matrix, so its eigenvectors are mutually orthogonal and all its eigenvalues are non-negative. These eigenvectors are the *principal components* of the data.

Definition 13.3. Let

$$W = \begin{pmatrix} \boldsymbol{u}_1' \\ \vdots \\ \boldsymbol{u}_m' \end{pmatrix} = (\boldsymbol{v}_1, \ldots, \boldsymbol{v}_n) \in \mathbb{R}^{m \times n}$$

be a data sample matrix and let $\hat{W} = H_m W$ be the corresponding centered data matrix.

The *principal directions* of W are the eigenvectors of the covariance matrix

$$cov(W) = \frac{1}{m-1}\hat{W}'\hat{W} = \frac{1}{m-1}W'H_m'H_mW\frac{1}{m-1}W'H_mW \in \mathbb{R}^{n\times n}.$$

The *principal components* of W are the eigenvectors of the matrix $\hat{W}\hat{W}'$.

Note that the covariance matrix $cov(W)$ is a scalar multiple of the Gram matrix $\hat{W}'\hat{W}$ of the columns $\hat{\boldsymbol{v}}_1, \ldots, \hat{\boldsymbol{v}}_n$ of the centered data matrix $\hat{W}$.

If $R \in \mathbb{R}^{n\times n}$ is the orthogonal matrix that diagonalizes $cov(W)$, then the principal directions of W are the columns of R because $R'cov(W)R = D$, or equivalently, $cov(W)R = RD$. Without loss of generality, we assume in this section that $D = \mathsf{diag}(d_1, d_2, \ldots, d_n)$ and that $d_1 \geqslant d_2 \geqslant \cdots \geqslant d_n$.

The first eigenvector of $cov(W)$ (which corresponds to d_1) is the *first principal direction* of the data matrix W; in general, the kth eigenvector $\boldsymbol{r}_k$ is called the kth *principal direction* of W.

There exists an immediate link between PCA and the SVD decomposition of a centered data matrix $\hat{W}$. Namely, if $\hat{W}\boldsymbol{s} = \sigma\boldsymbol{r}$ and $\hat{W}'\boldsymbol{r} = \sigma\boldsymbol{s}$, then $\boldsymbol{r}$ is a principal component of $\hat{W}$ and $\boldsymbol{s}$ is the corresponding principal direction of $\hat{W}$.

If

$$\hat{W} = \begin{pmatrix} \hat{\boldsymbol{u}}_1' \\ \vdots \\ \hat{\boldsymbol{u}}_m' \end{pmatrix},$$

we have

$$\hat{\boldsymbol{u}}_1'\boldsymbol{r} = \sigma s_1, \ldots, \hat{\boldsymbol{u}}_m'\boldsymbol{r} = s_m,$$

which shows that the principal components are the projections of the centered data points on the principal directions.

As we saw in Theorem 11.3, the sum of the elements of D's main diagonal equals the total variance $\mathsf{tvar}(W)$. The principal directions "explain" the sources of the total variance: sample vectors grouped around $\boldsymbol{r}_1$ explain the largest portion of the variance; sample vectors grouped around $\boldsymbol{r}_2$ explain the second largest portion of the variance, etc.

Let $Q \in \mathbb{R}^{n\times\ell}$ be a matrix having orthogonal columns. Starting from a sample matrix $X \in \mathbb{R}^{m\times n}$, we can construct a new sample matrix $W \in \mathbb{R}^{m\times\ell}$ having ℓ variables. Each experiment E_i is represented now by a row $\boldsymbol{w}_i'$ that is linked by $\boldsymbol{u}_i'$ by the equality $\boldsymbol{w}_i' = \boldsymbol{u}_i'Q$. This means that the component $(\boldsymbol{w}_i')_k$ that corresponds to the new variable $\mathcal{W}_k$ is obtained as

$$(\boldsymbol{w}_i')_k = \sum_{p=1}^{n} (\boldsymbol{u}_i')_p q_{pk},$$

a linear combination of the values that correspond to the previous variables.

Theorem 13.3. *Let $W \in \mathbb{R}^{m\times n}$ be a centered sample matrix and let $R \in \mathbb{R}^{n\times n}$ be an orthogonal matrix such that $R'\mathit{cov}(W)R = D$, where $D \in \mathbb{R}^{n\times n}$ is a diagonal matrix $D = \mathsf{diag}(d_1,\ldots,d_n)$ and $d_1 \geqslant \cdots \geqslant d_n$.*

Let $Q \in \mathbb{R}^{n\times\ell}$ be a matrix having orthogonal columns and let $X = WQ \in \mathbb{R}^{m\times\ell}$. Then, $\mathsf{trace}(\mathit{cov}(X))$ is maximized when Q consists of the first ℓ columns of R and is minimized when Q consists of the last ℓ columns of R.

Proof. This result follows from Ky Fan's Theorem (Theorem 7.14) applied to the symmetric covariance matrix of the transformed dataset. □

Let $\hat{W} = UDV'$ be the thin SVD of the centered data matrix $\hat{W} \in \mathbb{R}^{m\times n}$, where $U \in \mathbb{R}^{m\times r}$ and $V \in \mathbb{R}^{r\times n}$ are matrices having orthogonal columns and

$$D = \begin{pmatrix} \sigma_1 & 0 & \cdots & 0 \\ 0 & \sigma_2 & \cdots & 0 \\ \vdots & \vdots & \cdots & \vdots \\ 0 & 0 & \cdots & \sigma_r \end{pmatrix},$$

where $\sigma_1 \geqslant \cdots \geqslant \sigma_r > 0$ are the singular values of A. For the covariance matrix of $\mathit{cov}(W)$, we have

$$\begin{aligned} \mathit{cov}(W) &= \frac{1}{m-1}\hat{W}'\hat{W} = \frac{1}{m-1}VD'U'UDV' \\ &= \frac{1}{m-1}VD'DV' = \frac{1}{m-1}VD^2V', \end{aligned}$$

due to the orthogonality of the columns of U. As we saw before, the columns of V are the eigenvectors of $cov(W)$. The matrix V is known as the *matrix of loadings*.

The matrix $S = UD \in \mathbb{R}^{m\times r}$ is known as the *matrix of scores*. It is clear that $\sigma_1^2, \ldots, \sigma_r^2$ coincide with the eigenvalues of $\hat{W}'\hat{W}$.

Observe that $\hat{W} = SV$, where S is the scores matrix and V is the loadings matrix. Since the columns of V are orthogonal, we also have $S = \hat{W}V$.

The SVD of $\hat{W}$ can be written as

$$\hat{W} = \sum_{i=1}^{r} \sigma_i \boldsymbol{u}_i \boldsymbol{v}_i'.$$

This implies $\hat{W}'\hat{W}\boldsymbol{v}_i = \sigma_i^2 \boldsymbol{v}_i$. Since $\boldsymbol{u}_i'\hat{W} = \sigma_i \boldsymbol{v}_i'$, it follows that $\boldsymbol{v}_i'$ is a weighted sum of the rows of the matrix $\hat{W}$. Similarly, $\boldsymbol{u}_i$ are weighted sums of the columns of $\hat{W}$.

As observed in [61], if

$$\hat{W} = (\boldsymbol{u}_1 \ \cdots \ \boldsymbol{u}_r)(\sigma_1 \boldsymbol{v}_1' \ \ldots \ \sigma_r \boldsymbol{v}_r)',$$

then

$$\begin{aligned} I_r &= (\boldsymbol{u}_1 \ \cdots \ \boldsymbol{u}_r)'(\boldsymbol{u}_1 \ \cdots \ \boldsymbol{u}_r), \\ (n-1)\hat{X}'\hat{X} &= (\boldsymbol{u}_1 \ \cdots \ \boldsymbol{u}_r)(\boldsymbol{u}_1 \ \cdots \ \boldsymbol{u}_r)'. \end{aligned}$$

Example 13.1. We use the FAO dataset introduced in Example 12.1 showing the protein and fat consumption for 37 European countries.

The sample matrix $X \in \mathbb{R}^{37\times 2}$ is obtained from the second and third columns of this table that correspond to the variables *prot* and *fat*. The vector of the sample variances of the two columns is $\boldsymbol{s} = (15.5213\ 28.9541)$. Since the magnitudes of the sample variances are substantial and quite distinct, we normalize the data by dividing the columns of X by their respective sample variances. The normalization is done by using the function `zscore`; namely, `zscore(X)` returns a centered and scaled version of X having the same format as X such that the columns of the result have sample mean 0 and sample variance 1.

The loading matrix or the coefficient matrix is given by

$$\begin{pmatrix} 0.7071 & -0.7071 \\ 0.7071 & 0.7071 \end{pmatrix}.$$

Both coefficients in the first column (which represents the first principal component) are equal and positive, which means that the first principal component is a weighted average of the two variables. The second principal component corresponds to a weighted difference of the original variables. The coordinates of the data in the new coordinate system is defined by the matrix scores. These scores have been plotted in Figure 13.1.

Principal component analysis in MATLAB is done using the function `pca` of the statistics toolbox. There are several signatures of this function which we review next.

The statement `coeff = pca(A)` performs principal components analysis (PCA) on the matrix $A \in \mathbb{R}^{m \times n}$, and returns the principal component coefficients, also known as *loadings*. Rows of A correspond to observations, and columns to variables. The columns of the matrix `coeff` (an $n \times n$ matrix) contain coefficients for one principal

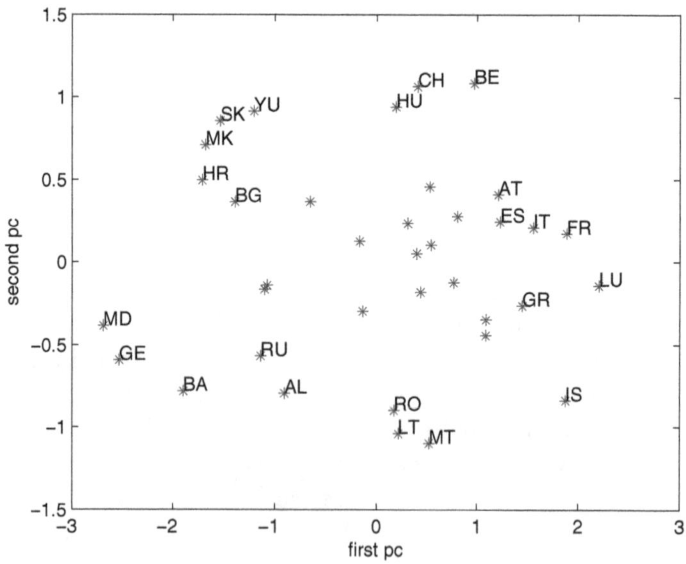

Fig. 13.1 The first two principal components of the FAO dataset.

component and these columns are in order of decreasing component variance.

The function `pca` computes the principal components of a sample matrix X. The are several incarnations of the function `pca`, described as follows.

(i) `[coeff,score] = pca(X)` returns the matrix `score`, the principal component scores, that is, the representation of X in the principal component space. The rows of `score` correspond to observations, columns to components.

(ii) `[coeff,score,latent] = pca(X)` returns the vector `latent` which contains the eigenvalues of the covariance matrix of X.

The matrix `score` contains the data formed by transforming the original data into the space of the principal components. The values of the vector `latent` are the variance of the columns of `score`.

The function `pca` centers X by subtracting off column variance means, but does not rescale the columns of X. To perform principal components analysis with standardized variables, we need to use `pca(zscore(X))`.

Example 13.2. The dataset that we are about to analyze originates in a study of the health condition of Boston neighborhoods [28] produced by the Health Department of the City of Boston. The data include incidence of various diseases and health events that occur in the 16 neighborhoods of the city identified as

Neighborhood	**Code**	**Neighborhood**	**Code**
Allston/Brighton	AB	North Dorchester	ND
Back Bay	BB	North End	NE
Charlestown	CH	Roslindale	RO
East Boston	EB	Roxbury	RX
Fenway	FW	South Boston	SB
Hyde Park	HP	South End	SE
Jamaica Plain	JP	South Dorchester	SD
Mattapan	MT	West Roxbury	WR

This is entered in MATLAB as

```
neighborhoods = ['AB';'BB';'CH';'EB';'FW';'HP';'JP';...
'MT';'ND';'NE';'RS';'RX';'SB';'SD';'SE';'WR']
```

The diseases and the health conditions are listed as the vector `categories`:

Category	**Code**	**Category**	**Code**
Hepatitis B	HepB	Tuberculosis	TBCD
Hepatitis C	HepC	Live Births	B154
HIV/AIDS	HIVA	Low weight at birth	LBWE
Chlamydia	CHLA	Infant Mortality	INFM
Syphilis	SYPH	Children with Elevated Lead	CELL
Gonorrhea	GONO	Subst. Abuse Treat. Admissions	SATA

This is entered in MATLAB as

```
categories = ['HepB';'HepC';'HIVA';'CHLA';'SYPH';'GONO';...
              'TBCD';'B154';'LBWE';'INFM';'CELL';'SATA']
```

The data itself are contained in the matrix `diseaseinc` in $\mathbb{R}^{16\times 12}$ and have the form

$$\begin{matrix}
38 & 57 & 13 & 168 & 5 & 22 & 20 & 607 & 40 & 7 & 13 & 624\\
17 & 24 & 16 & 179 & 11 & 52 & 8 & 306 & 25 & 0 & 0 & 497\\
10 & 13 & 0 & 46 & 0 & 8 & 0 & 284 & 25 & 0 & 0 & 489\\
12 & 46 & 11 & 150 & 10 & 16 & 17 & 718 & 43 & 10 & 43 & 1009\\
18 & 19 & 8 & 163 & 9 & 44 & 7 & 125 & 10 & 0 & 0 & 272\\
11 & 18 & 8 & 179 & 9 & 25 & 9 & 487 & 46 & 12 & 19 & 2781\\
6 & 32 & 18 & 213 & 10 & 46 & 6 & 420 & 35 & 5 & 10 & 1071\\
15 & 24 & 11 & 264 & 14 & 56 & 8 & 285 & 26 & 7 & 25 & 390\\
42 & 76 & 22 & 611 & 15 & 135 & 29 & 1350 & 168 & 28 & 88 & 1492\\
0 & 0 & 0 & 0 & 0 & 0 & 0 & 89 & 5 & 0 & 0 & 130\\
13 & 22 & 0 & 115 & 6 & 31 & 0 & 488 & 39 & 6 & 28 & 330\\
21 & 50 & 17 & 477 & 8 & 72 & 8 & 829 & 87 & 27 & 30 & 2075\\
9 & 52 & 0 & 85 & 7 & 25 & 5 & 403 & 25 & 0 & 18 & 1335\\
68 & 78 & 23 & 760 & 24 & 176 & 24 & 656 & 67 & 9 & 63 & 1464\\
51 & 35 & 35 & 124 & 31 & 61 & 11 & 439 & 34 & 0 & 0 & 6064\\
9 & 10 & 0 & 17 & 0 & 0 & 0 & 419 & 34 & 0 & 11 & 179
\end{matrix}$$

The array containing the sample variances of columns is computed by applying the function `std`:

```
stdinc = std(diseaseinc)
```

Next, by using the function `repmat` as in

```
si = diseaseinc./repmat(stdinc,16,1)
```

we create a 16×12 matrix consisting of 16 copies of `stdinc` and compute the normalized matrix `si` that is subjected to PCA in

```
[loadings,scores,variances]=pca(si)
```

This is one of several formats of the function `pca`. This function is applied to a data matrix and it centers the matrix by subtracting off column means. In the format that we use here, the function returns the matrices `loadings`, `scores`, and `variances` that contain the following data:

(i) The columns of the matrix `loadings` contain the principal components. The entries of this matrix are, of course, known as *loadings*. In our case, `loadings` is a 12×12 matrix, where each column represents one principal component. The columns are in order of decreasing component variance. We reproduce the first three columns of this matrix as follows:

$$\begin{matrix} 0.2914 & 0.2732 & -0.2641 \\ 0.3207 & -0.0568 & -0.1593 \\ 0.2666 & 0.3848 & 0.1427 \\ 0.3267 & -0.0800 & -0.2671 \\ 0.2426 & 0.4650 & -0.0270 \\ 0.3209 & 0.0668 & -0.3463 \\ 0.3215 & -0.0161 & -0.1444 \\ 0.3055 & -0.2594 & 0.3184 \\ 0.3061 & -0.2659 & 0.2735 \\ 0.2655 & -0.3215 & 0.3832 \\ 0.3026 & -0.2903 & -0.0815 \\ 0.1423 & 0.4702 & 0.5848 \end{matrix}$$

(ii) The matrix `scores` in $\mathbb{R}^{16\times 12}$ contain the principal component scores, that is, the representation of `si` in the principal component space. Rows of `scores` correspond to neighborhoods and columns, to components.

(iii) The matrix `variances` contain the principal component variances, that is, the eigenvalues of the covariance matrix of `si`.

The first two columns of the matrix `scores` contain the projections of data on the first two principal components. This is done by running

```
plot(scores(:,1),scores(:,2),'*')
```

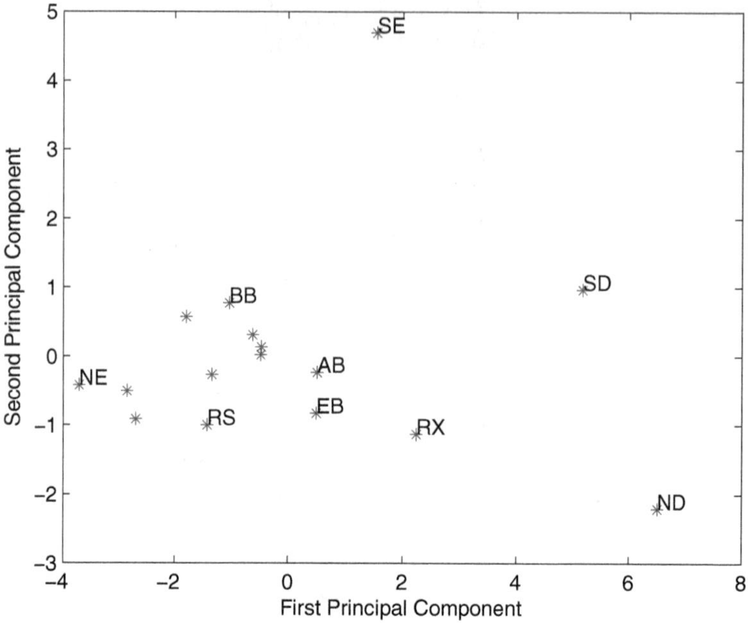

Fig. 13.2 Projections on the first two principal components.

After the plot is created, labels can be added to the axes using

```
xlabel('First Principal Component')
ylabel('Second Principal Component')
```

The resulting plot is shown in Figure 13.2. The neighborhood codes are applied to this plot by running `gname(neighborhoods)`. An inspection of the figure shows that the health issues are different for neighborhoods like South End (SE), South Dorchester (SD), and North Dorchester (ND).

The matrix `variances` allows us to examine the percentage of the total variability explained by each principal component. Initially, we compute the matrix `percent_explained` as

```
percent_explained=100*variances/sum(variances)
```

and using the function `pareto` we write

```
pareto(percent_explained)
xlabel('Principal Component')
ylabel('Variance Explained')
```

This code produces the histogram as shown in Figure 13.3.

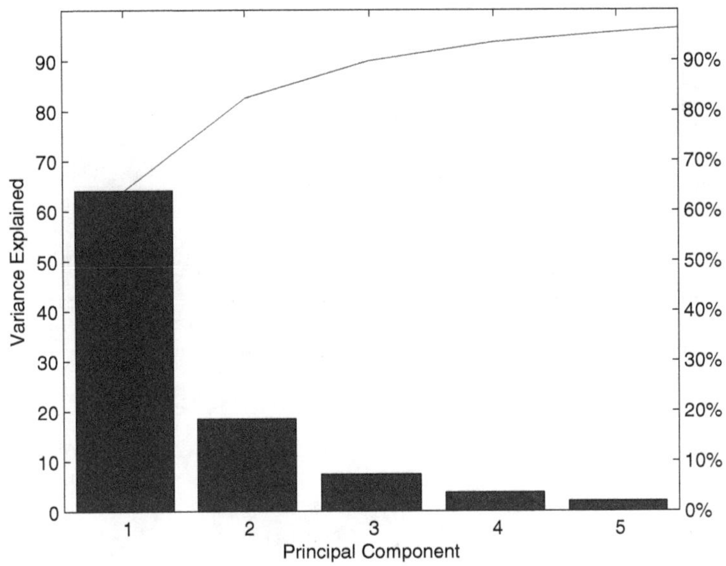

Fig. 13.3 Percentage of variability explained by principal components.

To visualize the results, one can use the **biplot** function as in

```
biplot(loadings(:,1:2),'scores',scores(:,1:2),'varlabels',
categories)
```

resulting in Figure 13.4. Each of the 12 variables is represented by a vector in this figure. Since the first principal component has positive coefficients, all vectors are located in the right half-plane. On the other hand, the signs of the coefficients of the second principal component are varying. These components distinguish between neighborhoods where there is a high incidence of substance abuse treatment admissions (SATA), syphilis (SYPH), HIV/Aids (HIVA), Hepatitis B (HepB), and Gonorrhea (GONO) and low incidence of the others and neighborhoods where the opposite situation occurs.

As observed in [84], the conclusions of a PCA analysis of data are mainly qualitative. The numerical precision (4 decimal digits) is not especially relevant for the PCA.

Next, we present a geometric point of view of principal component analysis.

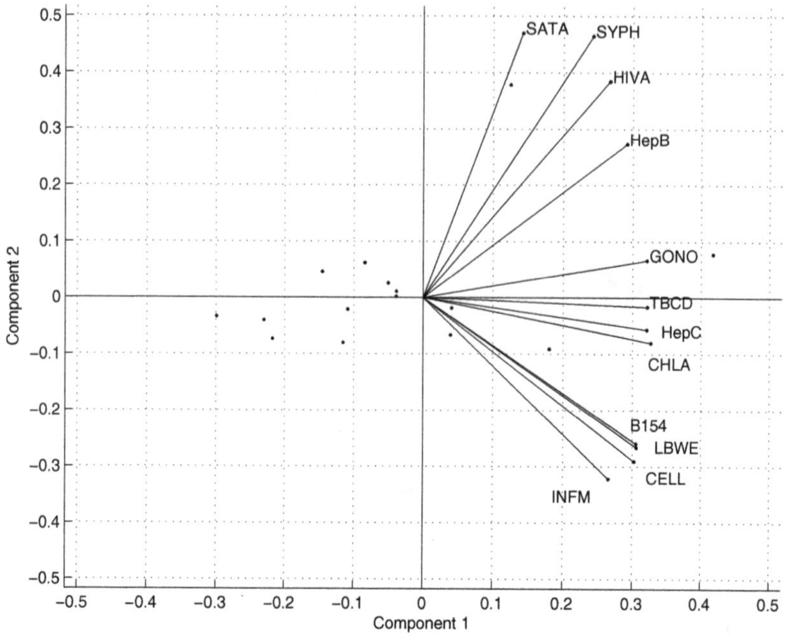

Fig. 13.4 Representation of the 12 variables.

Let $\boldsymbol{t} \in \mathbb{R}^n$ be a unit vector. The projection of a vector $\boldsymbol{w} \in \mathbb{R}^n$ on the subspace $\langle \boldsymbol{t} \rangle$ generated by $\boldsymbol{t}$ is given by

$$\mathsf{proj}_{\langle \boldsymbol{t} \rangle}(\boldsymbol{w}) = \boldsymbol{t}\boldsymbol{t}'\boldsymbol{w}.$$

To simplify the notation, we shall write $\mathsf{proj}_{\boldsymbol{t}}$ instead of $\mathsf{proj}_{\langle \boldsymbol{t} \rangle}$. Let $\hat{W} \in \mathbb{R}^{m \times n}$ be a centered sample matrix that corresponds to a sequence of experiments $(\boldsymbol{u}_1, \ldots, \boldsymbol{u}_m)$, that is

$$\hat{W} = \begin{pmatrix} \boldsymbol{u}_1' \\ \vdots \\ \boldsymbol{u}_m' \end{pmatrix}.$$

We seek to evaluate the inertia $I_0(\mathsf{proj}_{\boldsymbol{t}}(\hat{W}'))$ on the subspace generated by the unit vector $\boldsymbol{t} \in \mathbb{R}^n$. Since $\hat{W}' = (\boldsymbol{u}_1 \ \cdots \ \boldsymbol{u}_m)$, by the definition of inertia, we have

$$\begin{aligned}
I_0(\mathsf{proj}_{\boldsymbol{t}}(\hat{W}')) &= \sum_{j=1}^{m} \|\boldsymbol{t}\boldsymbol{t}'\boldsymbol{u}_j\|_2^2 \\
&= \sum_{j=1}^{m} \boldsymbol{u}_j'\boldsymbol{t}\boldsymbol{t}'\boldsymbol{t}\boldsymbol{t}'\boldsymbol{u}_j \\
&= \sum_{j=1}^{m} \boldsymbol{u}_j'\boldsymbol{t}\boldsymbol{t}'\boldsymbol{u}_j \\
&\quad \text{(because } \boldsymbol{t}'\boldsymbol{t} = 1\text{)} \\
&= \sum_{j=1}^{m} \boldsymbol{t}'\boldsymbol{u}_j\boldsymbol{u}_j'\boldsymbol{t} \\
&\quad \text{(because both } \boldsymbol{u}_j'\boldsymbol{t} \text{ and } \boldsymbol{t}'\boldsymbol{u}_j \text{ are scalars)} \\
&= \boldsymbol{t}'X'X\boldsymbol{t}.
\end{aligned}$$

The necessary condition for the existence of extreme values of this inertia as a function of $\boldsymbol{t}$ is

$$\begin{aligned}
\mathsf{grad}\ \left(I_0(\mathsf{proj}_{\boldsymbol{t}}(\hat{W}')) + \lambda(1 - \boldsymbol{t}'\boldsymbol{t})\right) &= \mathsf{grad}\ \left(\boldsymbol{t}'\hat{W}'\hat{W}\boldsymbol{t} + \lambda(1 - \boldsymbol{t}'\boldsymbol{t})\right) \\
&= 2\hat{W}'\hat{W}\boldsymbol{u} - 2\lambda\boldsymbol{t} = \mathbf{0},
\end{aligned}$$

where λ is a Lagrange multiplier. This implies $\hat{W}'\hat{W}\boldsymbol{t} = \lambda\boldsymbol{t}$. In other words, to achieve extreme values of the inertia $I_0(\mathsf{proj}_{\boldsymbol{t}}(\hat{W}'))$, $\boldsymbol{t}$ must be chosen as an eigenvector of the covariance matrix of $\hat{W}$, that is, as a principal direction of $\hat{W}$.

The principal directions of a data sample matrix $\hat{W} \in \mathbb{R}^{m\times n}$ can be obtained directly from the data sample matrix W by applying Corollary 8.16, a consequence of the Courant–Fisher theorem.

Suppose that the eigenvalues of $\hat{W}'\hat{W}$ are the numbers $\lambda_1 \geqslant \cdots \geqslant \lambda_n$. The first principal direction $\boldsymbol{t}_1$ of W, which corresponds to the largest eigenvalue of $\hat{W}'\hat{W}$, is

$$\begin{aligned}
\boldsymbol{t}_1 &= \arg\max_{\boldsymbol{t}} \left\{\boldsymbol{t}'\hat{W}\hat{W}'\boldsymbol{t} \mid \boldsymbol{t} \in \mathbb{R}^n, \|\boldsymbol{t}\|_2 = 1\right\} \\
&= \arg\max_{\boldsymbol{t}} \left\{\|\hat{W}'\boldsymbol{t}\|_2^2 \mid \|\boldsymbol{t}\|_2 = 1\right\}.
\end{aligned}$$

Suppose that we computed the principal directions $\boldsymbol{t}_1, \ldots, \boldsymbol{t}_k$ of $\hat{W}$. Then, by Corollary 8.16, $\boldsymbol{t}_{k+1} \in \mathbb{R}^n$ is a unit vector $\boldsymbol{t}$ that maximizes

$$\boldsymbol{t}'\hat{W}\hat{W}'\boldsymbol{t} = \|\hat{W}'\boldsymbol{t}\|_2^2$$

and belongs to the subspace orthogonal to the subspace generated by the first k principal directions of $\hat{W}$, that is,

$$\boldsymbol{t}_{k+1} = \arg\max_{\boldsymbol{t}} \left\{ \|\hat{W}'\boldsymbol{t}\|_2^2 \mid \boldsymbol{t} \in \mathbb{R}^n, \|\boldsymbol{t}\|_2 = 1, \boldsymbol{t} \in \langle \boldsymbol{t}_1, \ldots, \boldsymbol{t}_k \rangle^{\perp} \right\}.$$

Note that for every vector $\boldsymbol{z} \in \mathbb{R}^n$, we have

$$\left(I - \sum_{j=1}^{k} \boldsymbol{t}_j \boldsymbol{t}_j' \right) \boldsymbol{z} = \boldsymbol{z} - \mathsf{proj}_{\langle \boldsymbol{t}_1, \ldots, \boldsymbol{t}_k \rangle} \boldsymbol{z} \in \langle \boldsymbol{t}_1, \ldots, \boldsymbol{t}_k \rangle^{\perp}.$$

Therefore, $\boldsymbol{x} \in \langle \boldsymbol{t}_1, \ldots, \boldsymbol{t}_k \rangle^{\perp}$ is equivalent to $\boldsymbol{x} = (I - \sum_{j=1}^{k} \boldsymbol{t}_j \boldsymbol{t}_j')\boldsymbol{x}$. Thus, we can write

$$\boldsymbol{t}_{k+1} = \arg\max_{\boldsymbol{t}} \left\{ \left\| \hat{W} \left(I - \sum_{j=1}^{k} \boldsymbol{t}_j \boldsymbol{t}_j' \right) \boldsymbol{t} \right\|_2 \mid \|\boldsymbol{t}\|_2 = 1 \right\},$$

for $0 \leqslant k \leqslant n-1$. This technique allows finding the principal directions of $\hat{W}$ by solving a sequence of optimization problems involving the matrix $\hat{W}$.

Next, we discuss an algorithm known as the *Nonlinear Iterative Partial Least Squares Algorithm* (*NIPALS*), which can be used for computing the principal components of a centered data matrix (Algorithm 13.2.1). The algorithm computes a sequence of matrices $X_1, X_2, \ldots$ beginning with the matrix $X_1 = \hat{W}$ and needs a parameter ϵ to impose an iteration limitation.

If $c = \mathsf{rank}(\hat{W})$, we found the principal directions of W. In this case, $\hat{W}$ can be written as $\hat{W} = TV$, where $T = (\boldsymbol{t}_1 \ \cdots \ \boldsymbol{t}_c)$ and $V = (\boldsymbol{v}_1 \ \cdots \ \boldsymbol{v}_c)$.

Let $\ell_j = \|X_j'\boldsymbol{t}\|_2$. When the algorithm completes the **repeat** loop, we have $\breve{\boldsymbol{t}} = X_j \boldsymbol{v}_j$ is approximatively equal to $\boldsymbol{t}_j$, so $X_j' X_j \boldsymbol{v}_j = \ell_j \boldsymbol{v}_j$, which implies that ℓ_j is close to an eigenvalue and $\boldsymbol{v}_j$ is close to an eigenvector of $X_j' X_j$. Also, we have

$$\boldsymbol{t}'\boldsymbol{t} = \boldsymbol{v}_j' X_j' X_j \boldsymbol{v}_j = \boldsymbol{v}_j'(X_j' X_j \boldsymbol{v}_j) = \ell_j \boldsymbol{v}_j' \boldsymbol{v}_j = \ell_j,$$

since $\boldsymbol{v}_j$ is a unit vector.

Algorithm 13.2.1: The NIPALS Algorithm

Data: A centered sample matrix $\hat{W} \in R^{m \times n}$
Result: The first c principal directions of $\hat{W}$, where $1 \leqslant c \leqslant \mathsf{rank}(A)$

$X_1 = \hat{W}$;
for $j = 1$ *to* c **do**
  choose $\boldsymbol{t}_j$ as any column of X_j;
  repeat
    $\boldsymbol{v}_j = \frac{X_j'\boldsymbol{t}}{\|X_j'\boldsymbol{t}\|_2}$;
    $\breve{\boldsymbol{t}} = X_j \boldsymbol{v}_j$
  until $\|\breve{\boldsymbol{t}} - \boldsymbol{t}_j\|_2 < \epsilon$;
  $X_{j+1} = X_j - \boldsymbol{t}_j \boldsymbol{v}_j'$;
end

After the first run, we have $\hat{W} = X_1 = \boldsymbol{t}_1\boldsymbol{v}_1' + X_2$. We have

$$(\hat{W} - \boldsymbol{t}_1\boldsymbol{v}_1')'\boldsymbol{t}_1 = \hat{W}'\boldsymbol{t}_1 - \boldsymbol{v}_1\boldsymbol{t}_1'\boldsymbol{t}_1 = \hat{W}'\hat{W}\boldsymbol{v}_1 - \boldsymbol{v}_1\lambda_1 = 0.$$

Since $\boldsymbol{t}_2$ is initially a column of X_2, $\boldsymbol{t}_2$ it is orthogonal to $\boldsymbol{t}_1$ and remains so to the end of the loop.

After the second run through the **for** loop, we have $\hat{W} = \boldsymbol{t}_1\boldsymbol{v}_1' + \boldsymbol{t}_2\boldsymbol{v}_2' + X_3$, etc.

When the loop is completed, we have

$$\hat{W} = \boldsymbol{t}_1\boldsymbol{v}_1' + \boldsymbol{t}_2\boldsymbol{v}_2' + \cdots + \boldsymbol{t}_c\boldsymbol{v}_c + X_{c+1}.$$

If $c = \mathsf{rank}(\hat{W})$, then $X_{c+1} = O$.

13.3 Linear Discriminant Analysis

Linear discriminant analysis (LDA) is a supervised dimension-reduction technique. If a dataset contains data that belong to k distinct classes, LDA aims to find a low-dimensional subspace such that the projections of the data vectors on this subspace are well-separated clusters.

Let $(\boldsymbol{u}_1, \ldots, \boldsymbol{u}_m) \in \mathbf{Seq}(\mathbb{R}^n)$ be a sequence of vectors that belong to two classes C_1 and C_2. Denote by $\boldsymbol{m}_i$ the mean of the vectors that belong to the class C_i for $i = 1, 2$ and by $\boldsymbol{m}$ the global mean

$$\boldsymbol{m} = \frac{1}{m}\sum_{i=1}^{m} \boldsymbol{u}_i = \frac{|C_1|}{m}\boldsymbol{m}_1 + \frac{|C_2|}{m}\boldsymbol{m}_2.$$

We seek a unit vector $\boldsymbol{w}$ such that the projections on the subspace $\langle \boldsymbol{w} \rangle$ of the vectors that belong to the two classes are as separated as possible. The function $\phi : \mathbb{R}^n \longrightarrow \mathbb{R}$ defined by $\phi(\boldsymbol{u}) = \boldsymbol{w}'\boldsymbol{u}$ is a *discriminant function* and the set of projections consists of $\{\phi(\boldsymbol{u}_i) = \boldsymbol{w}'\boldsymbol{u}_i \mid 1 \leqslant i \leqslant m\}$.

The means of the projections on $\langle \boldsymbol{w} \rangle$ are

$$m_{i,\boldsymbol{w}} = \frac{1}{|C_i|}\sum\{\boldsymbol{w}'\boldsymbol{u} \in C_i\} = \boldsymbol{w}'\boldsymbol{m}_i$$

for $i = 1, 2$.

The *class scatter matrix* for C_i is the matrix $S_i \in \mathbb{R}^{n \times n}$ defined by

$$S_i = \sum\{(\boldsymbol{u} - \boldsymbol{m}_i)(\boldsymbol{u} - \boldsymbol{m}_i)' \mid \boldsymbol{u} \in C_i\}$$

for $i = 1, 2$.

The *intra-class scatter matrix* is $S_{\mathsf{intra}} = S_1 + S_2$. The *scatter of C_i relative to $\boldsymbol{w}$* is defined as the number

$$s_{i,\boldsymbol{w}}^2 = \boldsymbol{w}' S_{\mathsf{intra}} \boldsymbol{w}$$

and the *total intra-scatter relative to $\boldsymbol{w}$* is

$$s_{\boldsymbol{w}}^2 = s_{1,\boldsymbol{w}}^2 + s_{2,\boldsymbol{w}}^2 = \boldsymbol{w}' S_{\mathsf{intra}} \boldsymbol{w}. \tag{13.1}$$

Note that the matrix S_{intra} is symmetric and positive semidefinite.

Let S_{inter} be the *inter-class scatter matrix* defined by

$$S_{\mathsf{inter}} = |C_1|(\boldsymbol{m}_1 - \boldsymbol{m})(\boldsymbol{m}_1 - \boldsymbol{m})' + |C_2|(\boldsymbol{m}_2 - \boldsymbol{m})(\boldsymbol{m}_2 - \boldsymbol{m})'.$$

By substituting the value of $\boldsymbol{m}$ in the definition of S_{inter}, we obtain

$$S_{\mathsf{inter}} = \frac{|C_1||C_2|}{m}(\boldsymbol{m}_1 - \boldsymbol{m}_2)(\boldsymbol{m}_1 - \boldsymbol{m}_2)'. \tag{13.2}$$

Thus, the inter-class scatter matrix is a matrix of rank at most 1 (when $\boldsymbol{m}_1 \neq \boldsymbol{m}_2$).

Lemma 13.1. *The* separation between the means of the classes *is given by*

$$|m_{1,\boldsymbol{w}} - m_{2,\boldsymbol{w}}|^2 = \boldsymbol{w}' S_{\mathit{inter}} \boldsymbol{w}. \tag{13.3}$$

Proof. By applying the definitions of $m_{1,\boldsymbol{w}}$ and $m_{2,\boldsymbol{w}}$, we have

$$\begin{aligned}|m_{1,\boldsymbol{w}} - m_{2,\boldsymbol{w}}|^2 &= (\boldsymbol{w}'\boldsymbol{m}_1 - \boldsymbol{w}'\boldsymbol{m}_2)^2\\ &= \boldsymbol{w}'(\boldsymbol{m}_1 - \boldsymbol{m}_2)(\boldsymbol{m}_1 - \boldsymbol{m}_2)'\boldsymbol{w}\\ &= \frac{m}{|C_1|\,|C_2|}\boldsymbol{w}'S_{\mathsf{inter}}\boldsymbol{w}.\end{aligned}$$

□

Definition 13.4. The *Fisher linear discriminant* is the generalized Rayleigh–Ritz quotient of the inter-class scattering matrix and the intra-class scattering matrix:

$$\mathsf{F}(\boldsymbol{w}) = \frac{\boldsymbol{w}'S_{\mathsf{inter}}\boldsymbol{w}}{\boldsymbol{w}'S_{\mathsf{intra}}\boldsymbol{w}}$$

for $\boldsymbol{w} \in \mathbb{R}^n - \{\mathbf{0}_n\}$.

Equalities (13.1) and (13.1) imply the equality

$$\mathsf{F}(\boldsymbol{w}) = \frac{|C_1|\,|C_2|}{m}\frac{|m_{1,\boldsymbol{w}} - m_{2,\boldsymbol{w}}|^2}{s^2_{\boldsymbol{w}}}. \tag{13.4}$$

To obtain well-separated projections of the vectors of the classes, we need to find $\boldsymbol{w}$ that maximizes Fisher's linear discriminant. As observed in Section 8.9, the largest value of the generalized Rayleigh–Ritz quotient (and, therefore the largest value of the Fisher linear discriminant) is obtained when $\boldsymbol{w}$ is a generalized eigenvector that corresponds to the largest generalized eigenvalue of the matrix pencil $(S_{\mathsf{inter}}, S_{\mathsf{intra}})$.

When the vectors of the sequence $(\boldsymbol{u}_1, \ldots, \boldsymbol{u}_m) \in \mathbf{Seq}(\mathbb{R}^n)$ belong to several classes $C_1, \ldots, C_k$, we need to consider the projection on a $(k-1)$-dimensional subspace.

If $\boldsymbol{m}_i = \frac{1}{|C_i|}\sum\{\boldsymbol{u} \mid \boldsymbol{u} \in C_i\}$, then, as before, the intra-class scatter matrix is $S_{\mathsf{intra}} = \sum_{i=1}^k S_i$, where

$$S_i = \sum\{(\boldsymbol{u} - \boldsymbol{m}_i)'(\boldsymbol{u} - \boldsymbol{m}_i) \mid \boldsymbol{u} \in C_i\} \in \mathbb{R}^{n\times n}$$

for $1 \leqslant i \leqslant k$. The inter-class scatter matrix is now given by

$$S_{\mathsf{inter}} = \sum_{j=1}^{k} |C_i|(\boldsymbol{m}_i - \boldsymbol{m})(\boldsymbol{m}_i - \boldsymbol{m})',$$

where $\boldsymbol{m}$ is the global mean of the sequence of vectors. Thus, S_{inter} is the sum of k matrices of rank 1. Since $\sum_{i=1}^k(\boldsymbol{m}_i - \boldsymbol{m}) = 0$, the actual rank of S_{inter} is $k-1$.

If S_{intra} is a non-singular matrix, then the eigenvalues of the pencil $(S_{\mathsf{inter}}, S_{\mathsf{intra}})$ coincide with the eigenvalues of the matrix $S_{\mathsf{intra}}^{-1} S_{\mathsf{inter}}$.

Let $P \in \mathbb{R}^{n \times n}$ be the projection matrix on the desired subspace. The projection of the mean of the class C_i is

$$\breve{\boldsymbol{m}}_i = \frac{1}{|C_i|}\{P\boldsymbol{u} \mid \boldsymbol{u} \in C_i\} = P\boldsymbol{m}_i$$

and the projection of the global mean is

$$\breve{\boldsymbol{m}} = \sum \frac{|C_i|}{\sum_{i=1}^{k} |C_i|}\tilde{\boldsymbol{m}}_i = P\boldsymbol{m}.$$

Thus, the intra-scatter matrix of the projections of class C_i is

$$\begin{aligned}\breve{S}_i &= \sum\{(P\boldsymbol{u} - P\boldsymbol{m}_i)(P\boldsymbol{u} - P\boldsymbol{m}_i)' \mid \boldsymbol{u} \in C_i\} \\ &= P\sum\{(\boldsymbol{u} - \boldsymbol{m}_i)(\boldsymbol{u} - \boldsymbol{m}_i)'P' \mid \boldsymbol{u} \in C_i\} = PS_iP',\end{aligned}$$

which implies that the intra-scatter matrix of the projections is

$$\breve{S}_{\mathsf{intra}} = PS_{\mathsf{intra}}P'.$$

Similarly, the inter-scatter matrix of the projections is

$$\breve{S}_{\mathsf{inter}} = PS_{\mathsf{inter}}P'.$$

The projection matrix P is chosen in this case such that $\mathsf{trace}(\breve{S}_{\mathsf{intra}}^{-1}\breve{S}_{\mathsf{inter}})$ is maximal.

An interesting connection between LDA and the least square regression is shown in [174].

13.4 Latent Semantic Indexing

Latent semantic indexing (LSI) is an information-retrieval technique presented in [34, 40] based on SVD.

The central problem in information retrieval (IR) is the computation of sets of documents that contain terms specified by queries submitted by users of collections of documents. The problem is quite challenging because any such retrieval needs to take into account that a concept can be expressed by many equivalent words (*synonimy*) and the same word may mean different things in various contexts (*polysemy*). This can lead the retrieval technique to return documents that are irrelevant to the query (*false positive*) or to omit documents that may be relevant (*false negatives*).

Several models of information retrieval exist. We refer the interested reader to [6] for a comprehensive presentation. We are discussing here the *vector model* and examine an application of SVD to this model.

The next definition establishes the framework for discussing the LSI.

Definition 13.5. A *corpus* is a pair $\mathcal{K} = (T, \mathcal{D})$, where $T = \{t_1, \ldots, t_m\}$ is a finite set whose elements are referred to as *terms*, and $\mathcal{D} = (D_1, \ldots, D_n)$ is a set of *documents*. Each document D_i is a finite sequence of terms, $D_j = (t_{j1}, \ldots, t_{jk}, \ldots, t_{j\ell_j})$.

Let $\mathcal{K}$ be a corpus such that $|T| = m$ terms and $\mathcal{D}$ contains n documents. If t_i is a term and D_j is a document of $\mathcal{K}$, the *frequency* of t_i in D_j is the number of occurrences of t_i in d_j, that is,

$$a_{ij} = |\{p \mid t_{jp} = t_i\}|.$$

The *frequency matrix of the corpus* $\mathcal{K}$ is the matrix $A \in \mathbb{R}^{m \times n}$ defined by $A = (a_{ij})$.

Each term t_i generates a row vector $(a_{i1}, a_{i2}, \ldots, a_{in})$ referred to as a *term vector* and each document d_j generates a column vector

$$\boldsymbol{d}_j = \begin{pmatrix} a_{1j} \\ \vdots \\ a_{mj} \end{pmatrix}.$$

A *query* is a sequence of terms $q \in \mathbf{Seq}(T)$ and it is also represented as a vector

$$\boldsymbol{q} = \begin{pmatrix} q_1 \\ \vdots \\ q_m \end{pmatrix},$$

where $q_i = 1$ if the term t_i occurs in q, and 0 otherwise.

When a query q is applied to a corpus $\mathcal{K}$, an IR system that is based on the vector model computes the similarity between the query and the documents of the corpus by evaluating the cosine of the angle between the query vector $\boldsymbol{q}$ and the vectors of the documents of the corpus. For the angle α_j between $\boldsymbol{q}$ and $\boldsymbol{d}_j$, we have

$$\cos \alpha_j = \frac{(\boldsymbol{q}, \boldsymbol{d}_j)}{\|\boldsymbol{q}\|_2 \|\boldsymbol{d}_j\|_2}.$$

The IR system returns those documents D_j for which this angle is small, that is $\cos \alpha_j \geqslant t$, where t is a parameter provided by the user.

The LSI method aims to capture relationships between documents motivated by the underlying structure of the documents. This structure is obscured by synonimy, polysemy, the use of insignificant syntactic-sugar words, and plain noise, which is caused my misspelled words or counting errors.

By Theorem 9.8, if $A \in \mathbb{R}^{m \times n}$ is the matrix of a corpus $\mathcal{K}$, $A = UDV^{\mathsf{H}}$ is an SVD of A, and $\mathsf{rank}(A) = p$, then the first p columns of U form an orthonormal basis for $\mathsf{range}(A)$, the subspace generated by the vector documents of $\mathcal{K}$; and the last $n-p$ columns of V constitute an orthonormal basis for $\mathsf{null}(A)$. Also, by Corollary 9.6, the first p transposed columns of V form an orthonormal basis for the subspace of $\mathbb{R}^n$ generated by the term vectors of $\mathcal{K}$.

Example 13.3. Consider a tiny corpus $\mathcal{K} = (T, \mathcal{D})$, where $T = \{t_1, \ldots, t_5\}$ and $\mathcal{D} = \{D_1, D_2, D_3\}$. Suppose that the matrix of the corpus is

$A =$

	D_1	D_2	D_3
t_1	1	0	0
t_2	0	1	0
t_3	1	1	1
t_4	1	1	0
t_5	0	0	1

A matrix of this type may occur when documents D_1 and D_2 are fairly similar (they contain two common terms, t_3 and t_4) and when t_1 and t_2 are synonyms.

A query that seeks documents that contain t_1 returns the single-document D_1. However, since t_1 and t_2 are synonyms, it would be desirable to have both D_1 and D_2 returned. Of course, the matrix representation cannot directly account for the equivalence of the terms t_1 and t_2. The fact that t_1 and t_2 are synonymous is consistent with the fact that both these terms appear in the common context $\{t_3, t_4\}$ in D_1 and D_2.

The successive approximations of A are

$$B(1) = \sigma_1 * \boldsymbol{u}_1 * \boldsymbol{v}_1^{\mathsf{H}} = \begin{pmatrix} 0.4319 & 0.4319 & 0.2425 \\ 0.4319 & 0.4319 & 0.2425 \\ 1.1063 & 1.1063 & 0.6213 \\ 0.8638 & 0.8638 & 0.4851 \\ 0.2425 & 0.2425 & 0.1362 \end{pmatrix},$$

$$B(2) = \sigma_1 * \boldsymbol{u}_1 * \boldsymbol{v}_1^{\mathsf{H}} + \sigma_2 * \boldsymbol{u}_2 * \boldsymbol{v}_2^{\mathsf{H}} = \begin{pmatrix} 0.5000 & 0.5000 & 0.0000 \\ 0.5000 & 0.5000 & -0.0000 \\ 1.0000 & 1.0000 & 1.0000 \\ 1.0000 & 1.0000 & 0.0000 \\ 0.0000 & -0.0000 & 1.0000 \end{pmatrix}.$$

Another property of SVDs that is useful for LSI was shown in Theorem 9.7, namely that the rank-1 matrices $\boldsymbol{u}_i \boldsymbol{v}_i^{\mathsf{H}}$ are pairwise orthogonal, and their Frobenius norms are all equal to 1. If we regard the noise as distributed with relative uniformity with respect to the p orthogonal components of the SVD, then by omitting several such components that correspond to relatively small singular values, we eliminate a substantial part of the noise and we obtain a matrix that better reflects the underlying hidden structure of the corpus.

Example 13.4. Suppose that we apply to the miniature corpus described in Example 13.3 a query whose vector is

$$\boldsymbol{q} = \begin{pmatrix} 1 \\ 0 \\ 0 \\ 1 \\ 0 \end{pmatrix}.$$

The similarity between $\boldsymbol{q}$ and the document vectors $\boldsymbol{d}_i$, $1 \leqslant i \leqslant 3$, that constitute the columns of $\mathcal{A}$ is

$$\cos(\boldsymbol{q}, \boldsymbol{d}_1) = 0.8165, \cos(\boldsymbol{q}, \boldsymbol{d}_2) = 0.482, \text{ and } \cos(\boldsymbol{q}, \boldsymbol{d}_3) = 0,$$

suggesting that d_1 is by far the most relevant document for q.

However, if we compute the same value of cosine for q and the columns $\boldsymbol{b}_1, \boldsymbol{b}_2$, and $\boldsymbol{b}_3$ of the matrix $B(2)$, we have

$$\cos(\boldsymbol{q}, \boldsymbol{b}_1) = \cos(\boldsymbol{q}, \boldsymbol{b}_2) = 0.6708, \text{ and } \cos(\boldsymbol{q}, \boldsymbol{b}_3) = 0.$$

This approximation of A uncovers the hidden similarity of d_1 and d_2, a fact that is quite apparent from the structure of the matrix $B(2)$.

13.5 Recommender Systems and SVD

Recommender systems use data analysis techniques to help customers find products they would like to buy, visit websites they would be interested to see, choose entertainment that is appropriate for their tastes, etc. Frequently, these systems are based on collaborative filtering, a process that involves collaboration among multiple agents having different viewpoints. Collaborative filtering requires collecting data concerning the interests or tests from many users and is based on the assumption that users who made similar choices in the past will agree in the future. Thus, collaborative recommender systems recommend items to users based on ratings of these items awarded by other users.

Formally, we can regard a recommender system as a bipartite weighted graph $\mathcal{G} = (B \cup T, E, r)$, where $\{B, T\}$ is a partition of the set of vertices of the graph. The sets $B = \{b_1, \ldots, b_m\}$ and $T = \{t_1, \ldots, t_n\}$ are the set of buyers and the set of items. An edge $(b_i, t_j) \in B \times T$ exists in E if buyer b_i has rated item t_j with the rating a_{ij}. In this case, $r(b_i, t_j) = a_{ij}$. For large recommender systems, the sizes of the sets B and C can be of the order of millions, an aspect that raises issues of performance and scalability for RS designers.

SVDs can be used for recommender systems in the same way that they are used for latent semantic indexing (see Section 13.4), that is, by using lower-rank approximations of the recommender system matrix, that are likely to filter out the noise effect of smaller singular value components.

Generally, each buyer rates a rather small number of items. Thus, the set of ratings is sparse. To remedy this sparsity, one replaces each missing rating of an item by the average of the existing ratings for that item as proposed in [142]. We obtain a matrix $A \in \mathbb{R}^{m \times n}$. The vector $\boldsymbol{r} \in \mathbb{R}^m$ of average ratings for each buyer is $\boldsymbol{r} = \frac{1}{n} A \mathbf{1}_n$ and serves the normalization of A. This normalization process consists in subtracting the average rating of each buyer from each rating generated by the buyer. The normalized matrix N is given by

$$N = A - \mathsf{diag}(r_1, \ldots, r_m) J_{m,n}.$$

The SVD decomposition of the resulting matrix is computed and a low rank approximation of the rating matrix is considered (Algorithm 13.5.1). The new rating matrix $\hat{A}$ is obtained by adding the average ratings of customers to the rank k approximation of the normalized rating matrix, that is

$$\begin{aligned}\hat{A} = &\ \mathsf{diag}(r_1, \cdots, r_m) J_{m,m} \\ &+ U(:, 1:k) D(1:k, 1:k) V(:, 1:k))'.\end{aligned}$$

Algorithm 13.5.1: Algorithm for Computing a Ratings Matrix

Data: The bipartite graph $\mathcal{G}$ of the recommender system
Result: A matrix of ratings

1. compute vector of average ratings $\boldsymbol{r}$ for items;
2. construct the matrix A by adopting average ratings of items for missing ratings;
3. compute the normalized matrix N by subtracting the average ratings of buyers from each rating in A;
4. compute the SVD, (U, D, V) of the normalized matrix N;
5. adopt a rank k;
6. compute a new ratings matrix as

$$\begin{aligned}&\mathsf{diag}(r_1, \ldots, r_m) J_{m,m} \\ &+ U(:, 1:k) D(1:k, 1:k) V(:, 1:k))'\end{aligned}$$

Example 13.5. Suppose we have a collaborative system having the graph shown in Figure 13.5. The average ratings of the items are as

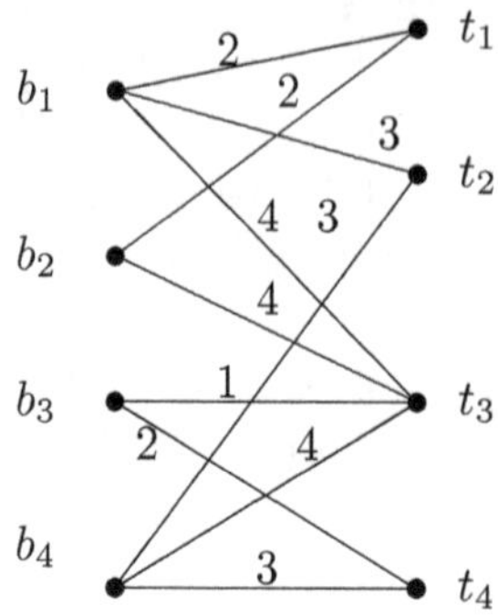

Fig. 13.5 Bipartite graph of a recommender system.

follows:

Item	t_1	t_2	t_3	t_4
Average rating	2	3	3.25	2.5

Thus, the rating matrix is

$$A = \begin{pmatrix} 2 & 2 & 4 & 2.5 \\ 2 & 3 & 4 & 2.5 \\ 2 & 3 & 1 & 2 \\ 2 & 3 & 4 & 3 \end{pmatrix}.$$

The average ratings are given by $\boldsymbol{r} = \frac{1}{4}A\mathbf{1}_4$ and are obtained in MATLAB using the expression `0.25*A*ones(4,1)`. We have the vector of the averages

```
2.6250
2.8750
2.0000
3.0000
```

The normalized matrix N is

```
-0.6250   -0.6250    1.3750   -0.1250
-0.8750    0.1250    1.1250   -0.3750
      0    1.0000   -1.0000         0
-1.0000         0    1.0000         0
```

The SVD decomposition $N = UDV'$ is computed by `[U,D,V]=svd(N)`:

```
U =
   -0.5962    0.2111    0.0004   -0.7746
   -0.4995   -0.4647    0.6840    0.2582
    0.4036   -0.7549   -0.0232   -0.5164
   -0.4818   -0.4118   -0.7291    0.2582
D =
    2.7177         0         0         0
         0    1.1825         0         0
         0         0    0.3013         0
         0         0         0    0.0000
V =
    0.4752    0.5805    0.4326    0.5000
    0.2626   -0.7991    0.2060    0.5000
   -0.8342    0.0936    0.2129    0.5000
    0.0963    0.1250   -0.8515    0.5000
```

The approximation $B(2)$ of rank 2 of A is computed as the product of the matrices $U_2\sqrt{D_2}$ and $\sqrt{D_2}V_2'$, where U_2 consists of the first two columns of U, V_2 consists of the first two column of V, and D_2 is the matrix

$$D_2 = \begin{pmatrix} 2.7177 & 0 \\ 0 & 1.1825 \end{pmatrix},$$

which gives

$$\sqrt{D_2} = \begin{pmatrix} 1.6485 & 0 \\ 0 & 1.0874 \end{pmatrix}.$$

13.6 Metric Multidimensional Scaling

Multidimensional scaling (MDS) is a process that allows us to represent a dissimilarity space using a low-dimensional Euclidean space. Scaling is important for visualizing the result of data explorations.

Two basic types of scaling algorithms exist. The *metric multidimensional scaling* starts with a finite dissimilarity space and produces a set of vectors that optimizes a function known as *strain*. The *non-metric multidimensional scaling* seeks a monotonic relationship between the values of a finite dissimilarity and the distances between the vectors that represent the elements of the dissimilarity space. We discuss only the metric multidimensional scaling.

Let $(\boldsymbol{x}_1, \ldots, \boldsymbol{x}_m)$ be a sequence of m vectors in $\mathbb{R}^n$. The corresponding matrix is

$$X = (\boldsymbol{x}_1, \ldots, \boldsymbol{x}_m) \in \mathbb{R}^{n \times m}.$$

Note that X is the transpose of the sample data matrix previously considered (which had $\boldsymbol{x}_1', \ldots, \boldsymbol{x}_m'$ as its rows).

Given the matrix of Euclidean distances $D = (d_{ij}^2) \in \mathbb{R}^{m \times m}$, where

$$d_{ij}^2 = \|\boldsymbol{x}_i - \boldsymbol{x}_j\|_2^2 = (\boldsymbol{x}_i - \boldsymbol{x}_j)'(\boldsymbol{x}_i - \boldsymbol{x}_j)$$

for $1 \leqslant i, j \leqslant m$, we need to retrieve the vectors $\boldsymbol{x}_1, \ldots, \boldsymbol{x}_m$.

Clearly, this problem does not have a unique solution because the matrix D is the same for $(\boldsymbol{x}_1, \ldots, \boldsymbol{x}_m)$ and for $(\boldsymbol{x}_1 + \boldsymbol{c}, \ldots, \boldsymbol{x}_m + \boldsymbol{c})$ for every $\boldsymbol{c} \in \mathbb{R}^n$.

Let $G \in \mathbb{R}^{m \times m}$ be the Gram matrix of X, $G = G_X = X'X$. Since $g_{pq} = \boldsymbol{x}_p'\boldsymbol{x}_q$, we have

$$d_{ij}^2 = (\boldsymbol{x}_i - \boldsymbol{x}_j)'(\boldsymbol{x}_i - \boldsymbol{x}_j) = g_{ii} + g_{jj} - 2g_{ij} \tag{13.5}$$

for $1 \leqslant i, j \leqslant m$.

Suppose now that $F \in \mathbb{R}^{m \times m}$ is the Gram matrix of another sequence of vectors $(\boldsymbol{y}_1, \ldots, \boldsymbol{y}_m)$, that is, $F = G_Y = Y'Y$, where $Y = (\boldsymbol{y}_1, \ldots, \boldsymbol{y}_m) \in \mathbb{R}^{n \times m}$ such that $d_{ij}^2 = f_{ii} + f_{jj} - 2f_{ij}$. Then $g_{ii} + g_{jj} - 2g_{ij} = f_{ii} + f_{jj} - 2f_{ij}$ for $1 \leqslant i, j \leq m$. Let $W = G - F$. Then W is a symmetric matrix and $w_{ii} + w_{jj} - 2w_{ij} = 0$, so $w_{ij} = \frac{1}{2}(w_{ii} + w_{jj})$. Let

$$\boldsymbol{w} = \frac{1}{2}\begin{pmatrix} w_{11} \\ \vdots \\ w_{mm} \end{pmatrix}$$

and note that the matrix W can now be written as $W = \boldsymbol{w}\mathbf{1}_m' + \mathbf{1}_m\boldsymbol{w}'$, which proves that W is a special, rank-2 matrix. Consequently,

$$G = F + \boldsymbol{w}\mathbf{1}_m' + \mathbf{1}_m\boldsymbol{w}'. \tag{13.6}$$

Thus, the set of vectors that correspond to a distance matrix is not unique, and the Gram matrices of any two such sequences differ by a symmetric matrix of rank 2.

It is possible to construct X starting from D if we assume that the centroid of the vectors of X is $\mathbf{0}_n$, that is, if $\sum_{i=1}^m \boldsymbol{x}_i = \mathbf{0}_n$.

Let $A \in \mathbb{R}^{m\times m}$ be the matrix defined by $A = -\frac{1}{2}D$. Elementwise, this means that $a_{ij} = -\frac{1}{2}d_{ij}^2$ for $1 \leqslant i, j \leqslant m$. Consider the averages defined by

$$a_{i\cdot} = \frac{1}{m}\sum_{j=1}^m a_{ij},$$

$$a_{\cdot j} = \frac{1}{m}\sum_{i=1}^m a_{ij},$$

$$a_{\cdot\cdot} = \frac{1}{m^2}\sum_{i=1}^m\sum_{j=1}^m a_{ij}.$$

The components of the Gram matrix $G \in \mathbb{C}^{m\times m}$, $g_{ij} = \boldsymbol{x}_i'\boldsymbol{x}_j$ for $1 \leqslant i, j \leqslant m$, can be expressed using these averages, assuming that the set of columns of X is centered in $\mathbf{0}_n$.

Theorem 13.4. *Let $X = (\boldsymbol{x}_1, \ldots, \boldsymbol{x}_m) \in \mathbb{R}^{n\times m}$ be a matrix such that $\sum_{i=1}^m \boldsymbol{x}_i = \mathbf{0}_n$ and let A be the matrix defined by $a_{ij} = -\frac{1}{2}\|\boldsymbol{x}_i - \boldsymbol{x}_j\|_2^2$ for $1 \leqslant i, j \leqslant m$. The components of the Gram matrix G, $g_{ij} = \boldsymbol{x}_i'\boldsymbol{x}_j$ are given by*

$$g_{ij} = a_{ij} - a_{i\cdot} - a_{\cdot j} + a_{\cdot\cdot}$$

for $1 \leqslant i, j \leqslant m$.

Proof. By Equality (13.5), we have

$$-2a_{ij} = g_{ii} + g_{jj} - 2g_{ij}$$

for $1 \leqslant i, j \leqslant m$. Note that

$$\sum_{i=1}^m g_{ij} = \sum_{j=1}^m g_{ij} = 0.$$

The averages introduced earlier can be written as

$$
\begin{aligned}
-2a_{\cdot j} &= \frac{1}{m}\sum_{i=1}^{m} d_{ij}^2 = \frac{1}{m}\sum_{i=1}^{m}(g_{ii} + g_{jj} - 2g_{ij}) \\
&= \frac{1}{m}\sum_{i=1}^{m} g_{ii} + g_{jj} - 2\left(\frac{1}{m}\sum_{i=1}^{m}\boldsymbol{x}_i'\right)\boldsymbol{x}_j \\
&= g_{jj} + \frac{1}{m}\sum_{i=1}^{m} g_{ii},
\end{aligned}
$$

because $\sum_{i=1}^{m}\boldsymbol{x}_i = \mathbf{0}_n$. Similarly, we have

$$
-2a_{i\cdot} = g_{ii} + \frac{1}{m}\sum_{j=1}^{m} g_{jj}. \tag{13.7}
$$

Therefore,

$$
\frac{1}{n^2}\sum_{i=1}^{m}\sum_{j=1}^{m} d_{ij}^2 = \frac{2}{m}\sum_{i=1}^{m} g_{ii}. \tag{13.8}
$$

Thus, we have

$$
\begin{aligned}
g_{ii} &= \frac{1}{m}\sum_{j=1}^{m} d_{ij}^2 - \frac{1}{m}\sum_{j=1}^{m} g_{jj}, \\
g_{jj} &= \frac{1}{m}\sum_{j=1}^{m} d_{ij}^2 - \frac{1}{m}\sum_{i=1}^{m} g_{ii}.
\end{aligned}
$$

Equality (13.5) yields

$$
\begin{aligned}
g_{ij} &= -\frac{1}{2}\left(d_{ij}^2 - g_{ii} - g_{jj}\right) \\
&= -\frac{1}{2}\left(d_{ij}^2 - \frac{1}{m}\sum_{j=1}^{m} d_{ij}^2 + \frac{1}{m}\sum_{j=1}^{m}\boldsymbol{x}_j\boldsymbol{x}_j' - \frac{1}{m}\sum_{j=1}^{m} d_{ij}^2 + \frac{1}{m}\sum_{i=1}^{m}\boldsymbol{x}_i\boldsymbol{x}_i'\right) \\
&= -\frac{1}{2}\left(d_{ij}^2 - \frac{1}{m}\sum_{j=1}^{m} d_{ij}^2 - \frac{1}{m}\sum_{j=1}^{m} d_{ij}^2 + \frac{1}{r^2}\sum_{i=1}^{m}\sum_{j=1}^{m} d_{ij}^2\right) \\
&= a_{ij} - a_{i\cdot} - a_{\cdot j} + a_{\cdot\cdot},
\end{aligned}
$$

which completes the proof. □

Corollary 13.2. *The Gram matrix $G = X'X$ of the sequence of $\mathbb{R}^n$ vectors $X = (\boldsymbol{x}_1, \ldots, \boldsymbol{x}_m)$ can be obtained from the matrix $A = \left(-\frac{1}{2}\|\boldsymbol{x}_i - \boldsymbol{x}_j\|_2^2\right)$ as $G = H_m A H_m$, where H_m is the centering matrix $H_m = I_m - \frac{1}{m}\mathbf{1}_m\mathbf{1}_m'$.*

Proof. The matrix $H_m A H_m$ can be written as

$$\begin{aligned}
H_m A H_m &= \left(I_m - \frac{1}{m}\mathbf{1}_m\mathbf{1}_m'\right) A \left(I_m - \frac{1}{m}\mathbf{1}_m\mathbf{1}_m'\right) \\
&= \left(I_m - \frac{1}{m}\mathbf{1}_m\mathbf{1}_m'\right)\left(A - \frac{1}{m}A\mathbf{1}_m\mathbf{1}_m'\right) \\
&= A - \mathbf{1}_m\left(\frac{1}{m}\mathbf{1}_m' A\right) - \left(\frac{1}{m}A\mathbf{1}_m\right)\mathbf{1}' + \frac{1}{m^2}\mathbf{1}_m\left(\mathbf{1}_m' A \mathbf{1}_m\right)\mathbf{1}_m'.
\end{aligned}$$

The terms of the above sum correspond to $a_{ij}, a_{\cdot j}, a_{i\cdot}$, and $a_{\cdot\cdot}$, respectively. The desired conclusion then follows from Theorem 13.4. □

The rank of the matrix $G = X'X \in \mathbb{R}^{m\times m}$ is equal to the rank of X, namely $\mathsf{rank}(G) = n$, as we saw in Theorem 3.33. Since G is symmetric, positive semidefinite, and of rank n, it follows that G has n non-negative eigenvalues and $m - n$ zero eigenvalues. By the Spectral Theorem for Hermitian matrices (Theorem 8.14), we have $G = U'DU$, where U is an orthogonal matrix, $U = (\boldsymbol{u}_1 \ \cdots \ \boldsymbol{u}_m)$, $D = (\lambda_1, \ldots, \lambda_n, 0, \ldots, 0)$, and $\lambda_1 \geqslant \cdots \geqslant \lambda_n > 0$.

Taking into account that the last $m-n$ elements of the diagonal of G are 0, we can write $G = V'\mathsf{diag}(\lambda_1, \ldots, \lambda_n)V$, where $V \in \mathbb{R}^{n\times m}$. By defining X as $X = \mathsf{diag}(\sqrt{\lambda_1}, \ldots, \sqrt{\lambda_n})V \in \mathbb{R}^{n\times m}$, we have $G = X'X$ and the m columns of X yield the desired vectors in $\mathbb{R}^n$.

A more general problem begins with a matrix of dissimilarities $\Delta = (\delta_{ij}) \in \mathbb{R}^{m\times m}$ and seeks to determine whether there exists a sequence of vectors $(\boldsymbol{x}_1, \ldots, \boldsymbol{x}_m)$ in $\mathbb{R}^n$ such that $d(\boldsymbol{x}_i, \boldsymbol{x}_j) = \delta_{ij}$ for $1 \leqslant i, j \leqslant m$.

Lemma 13.2. *Let $A, G \in \mathbb{R}^{m\times m}$ be two matrices such that $G = H_m A H_m$, where H_m is the centering matrix $H_m = I_m - \frac{1}{n}\mathbf{1}_m\mathbf{1}_m'$. Then $g_{ii} + g_{jj} - 2g_{ij} = a_{ii} + a_{jj} - 2a_{ij}$ for $1 \leqslant i, j \leqslant m$.*

Proof. We saw that if $G = H_m A H_m$, then $g_{ij} = a_{ij} - a_{i\cdot} - a_{\cdot j} + a_{\cdot\cdot}$. Therefore, we have

$$g_{ii} = a_{ii} - 2a_{i\cdot} + a_{\cdot\cdot},$$
$$g_{jj} = a_{jj} - 2a_{\cdot j} + a_{\cdot\cdot}.$$

This allows us to write

$$\begin{aligned} g_{ii} + g_{jj} - 2g_{ij} &= a_{ii} - 2a_{i\cdot} + a_{\cdot\cdot} + a_{jj} - 2a_{\cdot j} + a_{\cdot\cdot} \\ &\quad -2(a_{ij} - a_{i\cdot} - a_{\cdot j} + a_{\cdot\cdot}) \\ &= a_{ii} + a_{jj} - 2a_{ij}. \end{aligned}$$

□

Theorem 13.5. *Let $\Delta \in \mathbb{R}^{m\times m}$ be a matrix of dissimilarities, $A \in \mathbb{R}^{m\times m}$ be the matrix defined by $a_{ij} = -\frac{1}{2}\delta_{ij}^2$ for $1 \leqslant i,j \leqslant m$, and let G be the centered matrix $G = H_m A H_m$. If G is a positive semidefinite matrix and $\mathsf{rank}(G) = n$, then there exists a sequence $(\boldsymbol{x}_1, \ldots, \boldsymbol{x}_m)$ of vectors in $\mathbb{R}^n$ such that $d(\boldsymbol{x}_i, \boldsymbol{x}_j) = \delta_{ij}$ for $1 \leqslant i,j \leqslant m$.*

Proof. Since G is a symmetric, positive semidefinite matrix having rank n, by Theorem 8.14, it is possible to write $G = V'DV$, where $V = (\boldsymbol{v}_1 \cdots \boldsymbol{v}_m) \in \mathbb{R}^{n\times m}$, $D = (\lambda_1, \ldots, \lambda_n)$, and $\lambda_1 \geqslant \cdots \geqslant \lambda_n > 0$.

Let $X = \mathsf{diag}(\sqrt{\lambda_1}, \ldots, \sqrt{\lambda_n})V \in \mathbb{R}^{n\times m}$. We claim that the distances between $\boldsymbol{x}_1, \ldots, \boldsymbol{x}_m$ equal the prescribed dissimilarities. Indeed, since $\boldsymbol{x}_i = \sqrt{\lambda_i}\boldsymbol{v}_i$, we have

$$\begin{aligned} d(\boldsymbol{x}_i, \boldsymbol{x}_j)^2 &= (\boldsymbol{x}_i - \boldsymbol{x}_j)'(\boldsymbol{x}_i - \boldsymbol{x}_j) \\ &= \boldsymbol{x}_i'\boldsymbol{x}_i + \boldsymbol{x}_j'\boldsymbol{x}_j - 2\boldsymbol{x}_i'\boldsymbol{x}_j \\ &= \lambda_i \boldsymbol{v}_i'\boldsymbol{v}_i + \lambda_j \boldsymbol{v}_j'\boldsymbol{v}_j - 2\lambda_i\lambda_j \boldsymbol{v}_i'\boldsymbol{v}_j \\ &= g_{ii} + g_{jj} - 2g_{ij} \\ &\qquad \text{(by Lemma 13.2)} \\ &= a_{ii} + a_{jj} - 2a_{ij} = -2a_{ij} = \delta_{ij}^2, \end{aligned}$$

which is the desired conclusion. □

Since $G = H_m A H_m$ and H_m has an eigenvalue equal to 0, it is clear that G also has such an eigenvalue. Therefore, $\mathsf{rank}(G) \leq m-1$, so there exist m vectors of dimensionality not larger than $m-1$ such that their distances are equal to the given dissimilarities.

We saw that the matrices XX' and $G = X'X$ have the same rank and their non-zero eigenvalues are positive numbers and have the same algebraic multiplicities for both matrices (by Corollary 7.6).

Let $\boldsymbol{w}$ be a principal component of the matrix $X' \in \mathbb{R}^{n\times m}$, that is, an eigenvector of the matrix XX'. Suppose that $\mathsf{rank}(G) = r$ and let $X' = UDV'$ be the thin SVD decomposition of the matrix X', where $D = (\sigma_1, \ldots, \sigma_r)$ and $U, V \in \mathbb{R}^{m\times r}$ (see Corollary 9.3). The matrices U and V have orthogonal columns, so $U'U = V'V = I_r$. Since the numbers $\sigma_1, \ldots, \sigma_r$ are positive, D is invertible and we obtain $U = X'VD^{-1}$. Thus, MDS involves a process that is dual to the usual PCA; some authors refer to it as the *dual PCA*.

MATLAB deals with metric MDS using the function `cmdscale`. The function call `X = cmdscale(D)` is applied to a distance matrix $D \in \mathbb{R}^{m\times n}$, and returns a matrix $X \in \mathbb{R}^{m\times n}$. The *rows* of X are the coordinates of m points in n-dimensional space for some n, where $n < m$. When D is a Euclidean distance matrix, the distances between those points are given by D. The number n is the smallest dimension of the subspace in which the m points whose inter-point distances are given by D can be embedded.

`[X,e] = cmdscale(D)` also returns the eigenvalues of XX' as components of the vector e. When D is Euclidean, the first n elements of e are positive, the rest zero. If the first k elements of e are much larger than the remaining $n - k$, then it is possible to use the first k columns of X to produce k-dimensional vectors whose inter-point distances approximate D. This can provide a useful dimension reduction for visualization, e.g., for $k = 2$.

D need not be a Euclidean distance matrix. If it is non-Euclidean or a more general dissimilarity matrix, then some elements of e are negative, and `cmdscale` chooses n as the number of positive eigenvalues. In this case, the reduction to n or fewer dimensions provides a reasonable approximation to D only if the negative elements of e are small in magnitude.

D can be specified as either a full dissimilarity matrix, or in upper-triangle vector form such as is output by `pdist`. A full dissimilarity matrix must be real and symmetric, and have zeros along the diagonal and positive elements everywhere else. A dissimilarity matrix in upper triangle form must have real, positive entries. D can be specified as a full similarity matrix, with ones along the diagonal and all other elements less than one. The function `cmdscale` transforms a similarity matrix to a dissimilarity matrix in such a way that distances between the points returned in Y are equal to or approximate the distances given by $\sqrt{I - D}$.

Example 13.6. We start from a matrix of driving distances between five northeastern cities: Boston, Providence, Hartford, New York, and Concord.

```
dist =

         0   41.9000   92.8800  189.9000   63.4700
   41.9000         0   65.3600  154.8400   95.7800
   92.8800   65.3600         0   99.7600  115.5900
  189.9000  154.8400   99.7600         0  213.7800
   63.4700   95.7800  115.5900  213.7800         0
```

Applying `cmdscale` to this matrix, `[X,e]=cmdscale(dist)` produces the matrix

```
X =
   58.1439  -20.4773   -4.2664
   19.3304  -34.2586    3.4664
  -29.8485    8.8070    1.1787
 -129.6169    7.7975   -1.1686
   81.9911   38.1313    0.7899
```

and the matrix of eigenvalues

```
e =
  1.0e+004 *
    2.8168
    0.3185
    0.0034
   -0.0000
   -0.0006
```

Next, by defining the matrix

```
>> cities = ['BOS';'PRO';'HAR';'NYC';'CON']
```

the results are displayed using

```
>> plot(X(:,1),X(:,2),'+')
>> gname(cities)
```

which results in the representation contained in Figure 13.6.

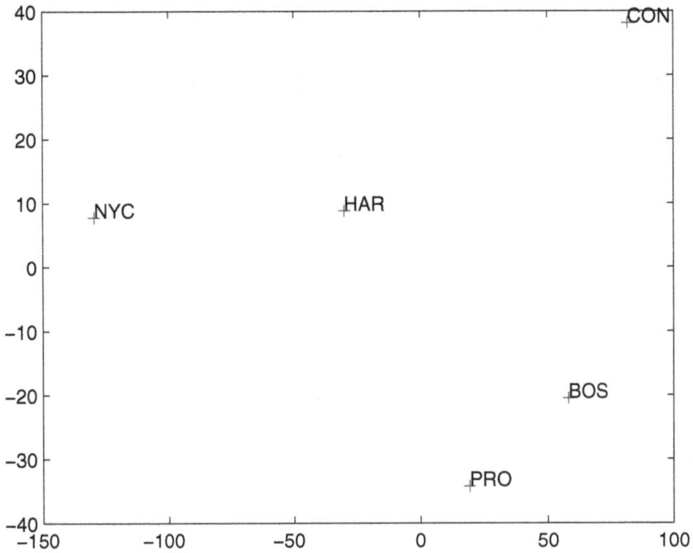

Fig. 13.6 Representation of the five cities.

Of course, the representation is approximative, but the relative positions of the cities is reasonably close to their real placement.

13.7 Procrustes Analysis

Procrustes[1] analysis (PRA) starts with two sets of vectors $U = \{\boldsymbol{u}_1, \ldots, \boldsymbol{u}_m\} \subseteq \mathbb{R}^n$, and $V = \{\boldsymbol{v}_1, \ldots, \boldsymbol{v}_m\} \subseteq \mathbb{R}^p$, where $p < n$, and tries to evaluate and improve the quality of the correspondence between these sets, as specified by the subscripts of the elements of the two sets. In general, we have $p \leqslant n$. The dimensionality of the vectors of the two sets can be equalized by appending $n - p$ zeroes as the last components of the vectors of V.

[1]Damastes, known as Procrustes (the stretcher), was a mythological Greek character who had the annoying habit of fitting his guests to the bed he offered them. If the guest was too short, he stretched the guest; if the guest was too long, he cut off a part of the body to make the guest fit the bed. He was killed by hero Theseus, the founder of Athens, who treated him to his own bed.

PRA tries to construct a linear transformation (through the application of dilations, rotations, reflections, and translations) that maps the point of V to be as close as possible to the points of U. The evaluation criterion for the quality of the matching is

$$d(V, U) = \sum_{i=1}^{m} (\boldsymbol{v}_i - \boldsymbol{u}_i)'(\boldsymbol{v}_i - \boldsymbol{u}_i),$$

which we need to minimize.

Define the transformation $h : \mathbb{R}^n \longrightarrow \mathbb{R}^n$ as $h(\boldsymbol{v}) = bT'\boldsymbol{v} + \boldsymbol{c}$, where $b \in \mathbb{R}$, $T \in \mathbb{R}^{n \times n}$ is an orthogonal matrix and $\boldsymbol{c} \in \mathbb{R}^n$. Clearly, h is the composition of a rotation (or a reflection), a dilation, and a translation. The quality of the matching between the displaced set of vectors $h(V)$ and U is

$$d(h(V), U) = \sum_{i=1}^{m} (aA'\boldsymbol{v}_i + \boldsymbol{b} - \boldsymbol{u}_i)'(aA'\boldsymbol{v}_i + \boldsymbol{b} - \boldsymbol{u}_i).$$

Let $\tilde{\boldsymbol{u}}$ and $\tilde{\boldsymbol{v}}$ be the means of the vectors of U and V, respectively. Since

$$bT'\boldsymbol{v}_i + \boldsymbol{c} - \boldsymbol{u}_i = bT'(\boldsymbol{v}_i - \tilde{\boldsymbol{v}}) - \tilde{\boldsymbol{u}} + bT'\tilde{\boldsymbol{v}} + \boldsymbol{c} - (\boldsymbol{u}_i - \tilde{\boldsymbol{u}}),$$

and

$$\sum_{i=1}^{m} (bT'(\boldsymbol{v}_i - \tilde{\boldsymbol{v}}) - (\boldsymbol{u}_i - \tilde{\boldsymbol{u}})) = \boldsymbol{0}_n,$$

it follows that

$$\begin{aligned} d(h(V), U) = & \sum_{i=1}^{m} (bT'(\boldsymbol{v}_i - \tilde{\boldsymbol{v}}) - \boldsymbol{u}_i + \tilde{\boldsymbol{u}})'(bT'(\boldsymbol{v}_i - \tilde{\boldsymbol{v}}) - \boldsymbol{u}_i + \tilde{\boldsymbol{u}}) \\ & + m(bT'\tilde{\boldsymbol{v}} + \boldsymbol{c} - \tilde{\boldsymbol{u}})'(bT'\tilde{\boldsymbol{v}} + \boldsymbol{c} - \tilde{\boldsymbol{u}}). \end{aligned}$$

To ensure the minimum of $d(h(V), U)$, we need to take $\boldsymbol{c} = \tilde{\boldsymbol{u}} - bT'\tilde{\boldsymbol{v}}$. Thus, the new set of vectors is defined by $h(\boldsymbol{v}_i) = bT'(\boldsymbol{v}_i - \tilde{\boldsymbol{v}}) + \tilde{\boldsymbol{u}}$. This implies that the mean of the set $\{h(\boldsymbol{v}_i) \mid 1 \leqslant i \leqslant m\}$ coincides with $\tilde{\boldsymbol{u}}$, that is, the set of vectors $\{\boldsymbol{v}_i \mid 1 \leq i \leqslant m\}$ is transformed such that the means of the two groups of vectors, $h(V)$ and U, coincide.

Suppose now that both means of the sets U and V coincide with $\mathbf{0}_n$. In this case,

$$\begin{aligned} d(h(V), U) &= \sum_{i=1}^{m} (bT'\boldsymbol{v}_i - \boldsymbol{u}_i)'(bT'\boldsymbol{v}_i - \boldsymbol{u}_i) \\ &= b^2 \sum_{i=1}^{m} \boldsymbol{v}_i' TT' \boldsymbol{v}_i + \sum_{i=1}^{m} \boldsymbol{u}_i' \boldsymbol{u}_i - 2b \sum_{i=1}^{m} \boldsymbol{u}_i' T \boldsymbol{v}_i \\ &= b^2 \sum_{i=1}^{m} \boldsymbol{v}_i' \boldsymbol{v}_i + \sum_{i=1}^{m} \boldsymbol{u}_i' \boldsymbol{u}_i - 2b \sum_{i=1}^{m} \boldsymbol{u}_i' T \boldsymbol{v}_i \\ &= b^2 \mathsf{trace}(VV') + \mathsf{trace}(UU') - 2b\mathsf{trace}(VTU'), \end{aligned} \tag{13.9}$$

where

$$U = \begin{pmatrix} \boldsymbol{u}_1' \\ \vdots \\ \boldsymbol{u}_m' \end{pmatrix} \text{ and } V = \begin{pmatrix} \boldsymbol{v}_1' \\ \vdots \\ \boldsymbol{v}_m' \end{pmatrix}.$$

The ambiguity introduced by denoting the above matrices by the same letter as the set of vectors is harmless. The minimum of this quadratic function of b is achieved when

$$b = \frac{\mathsf{trace}(VTU')}{\mathsf{trace}(VV')} = \frac{\mathsf{trace}(TU'V)}{\mathsf{trace}(VV')}$$

and it equals

$$d_{min} = \frac{\mathsf{trace}(VV')\mathsf{trace}(UU') - \mathsf{trace}(VTU')^2}{\mathsf{trace}(VV')}.$$

Once the translation ($\boldsymbol{c}$) and the dilation ratio (b) have been chosen, we need to focus on the rotation matrix T.

Theorem 13.6 (Berge–Sibson Theorem). *Let $M \in \mathbb{R}^{n\times n}$ be a matrix having the singular value decomposition $M = PDQ'$, where $P, Q \in \mathbb{R}^{n\times n}$ are orthogonal matrices and $D = \mathsf{diag}(\sigma_1, \ldots, \sigma_n)$, where $\sigma_i \geqslant 0$ for $1 \leqslant i \leqslant n$. Then, for every orthogonal matrix $T \in \mathbb{R}^{n\times n}$, we have $\mathsf{trace}(TM) \leqslant \mathsf{trace}((M'M)^{1/2})$ and the equality takes place when $T = QP'$.*

Proof. Starting from the SVD $M = PDQ'$, we have

$$\mathsf{trace}(TM) = \mathsf{trace}(TPDQ') = \mathsf{trace}(Q'TPD) = \mathsf{trace}(LD),$$

where $L = Q'TP \in \mathbb{R}^{n\times n}$ is an orthogonal matrix (as a product of orthogonal matrices). Since the elements of an orthogonal matrix cannot exceed 1, it follows that

$$\mathsf{trace}(TM) \leqslant \sum_{i=1}^{n} \sigma_i = \mathsf{trace}(D) = \mathsf{trace}((D'D)^{\frac{1}{2}}).$$

Taking into account that $D = P'MQ$ (and $D' = Q'M'P$), we have

$$\begin{aligned}
\mathsf{trace}((D'D)^{\frac{1}{2}}) &= \mathsf{trace}((Q'M'PP'MQ)^{\frac{1}{2}}) \\
&= \mathsf{trace}((Q'M'MQ)^{\frac{1}{2}}) = \mathsf{trace}((MQQ'M)^{\frac{1}{2}}) \\
&\quad \text{(by Corollary 7.6)} \\
&= \mathsf{trace}((MM)^{\frac{1}{2}}),
\end{aligned}$$

which is the desired inequality.

If $T = QP'$, we can write

$$\begin{aligned}
TM = QP'M = QP'PDQ' &= (QDQ'QDQ')^{\frac{1}{2}} \\
&= (QDDQ')^{\frac{1}{2}} = (QDP'PDQ')^{\frac{1}{2}} \\
&= (A'A)^{\frac{1}{2}},
\end{aligned}$$

which concludes the argument. □

Berge–Sibson Theorem allows us to choose the optimal orthogonal matrix in Equality (13.9) by maximizing $\mathsf{trace}(VTU') = \mathsf{trace}(TU'V)$. To this end, we need to take $T = QP'$, where $U'V$ has the SVD $U'V = PDQ'$.

Since $T = QP'$, it follows that $Q'TPD = D$ because P and Q are orthogonal. In turn, this implies $QQ'TPDQ' = QDQ'$, so $TPDQ' = QDQ'$. Therefore,

$$\begin{aligned} TU'V &= TPDQ' = QDQ' = (QD^2Q')^{\frac{1}{2}} = (QDPP'DQ')^{\frac{1}{2}} \\ &= ((U'V)'(U'V))^{\frac{1}{2}}. \end{aligned}$$

The optimal dilation ratio can be written now as

$$b = \frac{\mathsf{trace}((U'V)'(U'V))^{\frac{1}{2}}}{\mathsf{trace}(VV')}.$$

Since $\mathsf{trace}(TU'V) = \mathsf{trace}(VTU')$, the minimal value of $d(h(V), U)$ is

$$\begin{aligned} d_{min} &= \frac{\mathsf{trace}(VV')\mathsf{trace}(UU') - \mathsf{trace}(VTU')^2}{\mathsf{trace}(VV')} \\ &= \mathsf{trace}(UU') - \frac{\mathsf{trace}(VTU')^2}{\mathsf{trace}(VV')} \\ &= \mathsf{trace}(UU') - \frac{\mathsf{trace}(((U'V)'(U'V))^{\frac{1}{2}})^2}{\mathsf{trace}(VV')}. \end{aligned}$$

A normalized version of this criterion is

$$\frac{d_{min}}{\mathsf{trace}(U'U)} = 1 - \frac{\mathsf{trace}(((U'V)'(U'V))^{\frac{1}{2}})^2}{\mathsf{trace}(VV')\mathsf{trace}(UU')},$$

known as the *Procrustes criterion.*

MATLAB performs PRA starting from two sample data matrices U and V, which have the same number of rows. The rows of U and V are obtained by transposing the vectors of U and V, respectively. The PRA matches the ith row in V to the ith row in U. Rows in U can have smaller dimensions (number of columns) than those in V; in this case columns of zeros are added to U to equalize the number of columns of the two matrices. Calling the function

```
[d, Z] = procrustes(V,U)
```

also returns

$$Z = (h(\boldsymbol{u}_1)' \ \cdots \ h(\boldsymbol{u}_m)') = bUT + \boldsymbol{c}'$$

containing the transformed U vectors.

The transformation that maps U into Z is returned by the function call

```
[d, Z, transform] = procrustes(V,U)
```

The variable **transform** is a structure that consists of the following components:

- c', that is the translation component;
- T, the orthogonal matrix involved;
- b, the dilation factor.

Other further interesting variants of **procrustes** are described as follows:

procrustes(..., 'Scaling',false)	no scale component ($b = 1$)
procrustes(..., 'Scaling',true)	includes a scale component (default)
procrustes(..., 'Reflection',false)	no reflection component ($\det(T) = 1$)
procrustes(..., 'Reflection','best')	best fit solution (default)
procrustes(..., 'Reflection',true)	solution includes a reflection ($\det(T) = -1$)

Example 13.7. We apply the Procrustes analysis to the sets of points X and Y in $\mathbb{R}^2$ produced as in Example 4.18:

```
U =                          V =
    0.5377   -0.2050             2.2260    1.7753
    1.8339   -0.1241             2.7057    1.4795
   -2.2588    1.4897             1.3409    3.2412
    0.8622    1.4090             2.6851    2.4492
    0.3188    1.4172             2.3451    2.5894
   -1.3077    0.6715             1.6735    2.5745
   -0.4336   -1.2075             1.5266    1.5894
    0.3426    0.7172             2.2899    2.1642
    3.5784    1.6302             4.0256    1.7556
    2.7694    0.4889             3.2358    1.5190
   -1.3499    1.0347             1.6690    2.8621
    3.0349    0.7269             3.4838    1.5175
    0.7254   -0.3034             2.2542    1.7058
   -0.0631    0.2939             2.0618    2.1317
    0.7147   -0.7873             2.0694    1.5363
```

The function call

```
>> [d,Z,tr] = procrustes(U,V)
```

returns

```
d =
    0.0059

Z =
    0.6563   -0.1488
    1.7723   -0.1696
   -2.3302    1.4579
    0.7597    1.4628
    0.0388    1.3591
   -1.0910    0.6567
   -0.3485   -1.1707
    0.3733    0.5786
    3.7441    1.6315
    2.6363    0.4321
   -1.3887    1.1426
    3.0605    0.6795
    0.7745   -0.2389
    0.0172    0.2933
    0.6304   -0.7142

tr =
    T: [2x2 double]
    b: 1.9805
    c: [15x2 double]
```

The results can be represented graphically using

```
>> plot(U(:,1),X(:,2),'bx',V(:,1),Y(:,2),'b+',Z(:,1),Z(:,2),
'bo')
```

and as shown in Figure 13.7. A summary inspection of Figure 13.7 shows that the transformation of the points of V (marked by a symbol '+') into the points of Z (marked by 'o') yields points that are quite close to the points of X (marked by 'x').

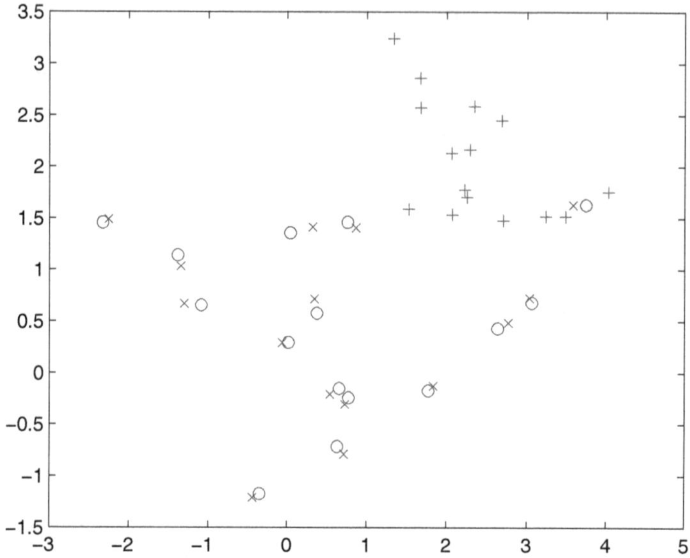

Fig. 13.7 Procrustes analysis of the sets X and Y.

13.8 Non-negative Matrix Factorization

Throughout this book we presented several factorization techniques for matrices that satisfy specific conditions involving the matrix to be factored or the factor matrices. In this chapter, we are concerned with a special type of *approximative factoring* of a matrix. Namely, starting from a matrix $A \in \mathbb{R}^{n\times m}$ and a positive integer $p < \min\{m, n\}$, we seek to find a pair of matrices (U, V) such that $U \in \mathbb{R}^{n\times p}$, $V \in \mathbb{R}^{p\times m}$, and $\|A - UV\|_F$ is minimal. If A is a non-negative matrix and we impose the non-negativity condition on the factors U, V, then we have the *non-negative factorization* of A. Note that $\mathsf{rank}(UV) \leqslant p$.

We begin by examining the existence of an approximative factorization for an arbitrary matrix $A \in \mathbb{R}^{m\times n}$. The minimization of $\|A - UV\|_F$ is equivalent to minimizing the function $f(U, V) = \frac{1}{2}\|A - UV\|_F^2$. We have

$$\mathsf{grad}_U \|A - UV\|_F^2 = UVV' - AV' \tag{13.10}$$

and

$$\text{grad}_V \|A - UV\|_F^2 = U'UV - U'A. \qquad (13.11)$$

Thus, the necessary conditions for finding a stationary point are

$$UVV' = AV' \text{ and } U'UV = U'A.$$

Also, if (U, V) is a solution of the problem, then so is $(aD, \frac{1}{a}V)$, where a is a positive number. Thus, if the problem has a solution, this solution is not unique, in general.

The minimization of $\|A - UV\|_F$ when we seek non-negative factors U and V is equivalent to minimizing the function $f(U, V) = \frac{1}{2}\|A - UV\|_F^2$ subjected to $np + pm$ restrictions: $u_{ij} \geqslant 0$ for $1 \leqslant i \leqslant n$ and $1 \leqslant j \leqslant p$ and $v_{jk} \geqslant 0$ for $1 \leqslant j \leqslant p$ and $1 \leqslant k \leqslant m$, respectively.

We apply the Karush–Kuhn–Tucker Theorem to the function $f(U, V)$ that depends on $np + pm$ components u_{ij} and v_{jk} of the matrices $U \in \mathbb{R}^{n\times p}$ and $V \in \mathbb{R}^{p\times m}$.

The functions g and h that capture the constraints are $g : \mathbb{R}^{n\times p} \times \mathbb{R}^{n\times p} \longrightarrow \mathbb{R}^{n\times p}$ and $h : \mathbb{R}^{p\times m} \times \mathbb{R}^{n\times p} \longrightarrow \mathbb{R}^{p\times m}$, which are defined by

$$g\begin{pmatrix} U \\ V \end{pmatrix} = -U \text{ and } h\begin{pmatrix} U \\ V \end{pmatrix} = -V.$$

Let $E_{qr}^{a\times b}$ be an $a \times b$ matrix whose single non-zero component is equal to 1 and is located at the intersection of line q and column r.

For the constraint functions, we have

$$\text{grad}_U g_{ij} = -E_{ij}^{n\times p}, \text{grad}_V g_{ij} = O_{p,m},$$
$$\text{grad}_U h_{jk} = O_{n,p}, \text{grad}_V h_{jk} = -E_{jk}^{p\times m}$$

for $1 \leqslant i \leqslant n$, $1 \leqslant j \leqslant p$, and $1 \leqslant k \leqslant m$. We have

$$g\begin{pmatrix} U \\ V \end{pmatrix} = -U \leqslant O_{n,p} \text{ and } h\begin{pmatrix} U \\ V \end{pmatrix} = -V \leqslant O_{p,m}.$$

We introduce two matrices of Lagrange multipliers, $B \in \mathbb{R}^{n\times p}$ and $C \in \mathbb{R}^{p\times m}$.

The first Karush–Kuhn–Tucker condition amounts to the equalities

$$\mathsf{grad}_{\ U}\|A-UV\|_F^2+\sum_{i=1}^{n}\sum_{j=1}^{p}b_{ij}\mathsf{grad}_{\ U}g_{ij}+\sum_{j=1}^{p}\sum_{k=1}^{m}c_{jk}\mathsf{grad}_{\ U}h_{jk}=O_{n,p},$$

$$\mathsf{grad}_{\ V}\|A-UV\|_F^2+\sum_{i=1}^{n}\sum_{j=p}^{m}b_{ij}\mathsf{grad}_{\ V}g_{ij}+\sum_{j=1}^{p}\sum_{k=1}^{m}c_{jk}\mathsf{grad}_{\ V}g_{jk}=O_{p,m},$$

which are equivalent to

$$UVV'-AV'-B=O_{n,p},$$
$$U'UV-U'A-C=O_{p,m}.$$

The remaining conditions can be written as

$$b_{ij}u_{ij}=0 \text{ and } c_{jk}v_{jk}=0$$

for $1\leqslant i\leqslant n$, $1\leqslant j\leqslant p$, and $1\leqslant k\leqslant m$, and $B\geqslant O_{n,p}$, $C\geqslant O_{p,m}$.

If both $U>O_{n,p}$ and $V>O_{p,m}$, then we have $B=O_{n,p}$ and $C=O_{p,m}$, and the Karush–Kuhn–Tucker conditions imply

$$UVV'=AV' \text{ and } U'UV=U'A.$$

If we seek matrices U and V of full rank, then, by Corollary 3.3, VV' and $U'U$ are non-singular matrices and we obtain

$$U=AV'(VV')^{-1} \text{ and } V=(U'U)^{-1}U'A.$$

These arguments suggest a nonnegative matrix factorization known as the alternate least-square (ALS) algorithm (see [17, 127]). This algorithm takes advantage of the fact that although the function $F:\mathbb{R}^{m\times p}\times\mathbb{R}^{p\times n}\longrightarrow\mathbb{R}$ defined by $F(U,V)=\|A-UV\|_F$ is not convex with respect to the pair (U,V), it is convex on either U or V. Its basic form is Algorithm 13.8.1.

The factorization technique for non-negative matrices that we discuss next was developed by Lee and Seung [105] and consists of constructing iteratively a sequence of pairs of matrices $(U^0,V^0),(U^1,V^1),\ldots$ such that the products U^tV^t offer increasingly better approximations of a non-negative matrix A. We need the following preliminary notion.

Definition 13.6. Let $G:(\mathbb{R}^k)^2\longrightarrow\mathbb{R}$ and $F:\mathbb{R}^k\longrightarrow\mathbb{R}$ be two functions. G is an *auxiliary function* for F if $G(\boldsymbol{x},\boldsymbol{y})\geq F(\boldsymbol{x})$ and $G(\boldsymbol{x},\boldsymbol{x})=F(\boldsymbol{x})$ for $\boldsymbol{x},\boldsymbol{y}\in\mathbb{R}^k$.

Algorithm 13.8.1: Alternate Least-Square Factorization Algorithm

Data: A nonnegative matrix $A \in \mathbb{R}^{m\times n}$ and a maximum number of iterations r
Result: An approximate factorization $A = UV$
initialize U^1 to a random matrix;
for $t = 1$ *to* r **do**
  solve for V^t the equation $(U^t)'U^tV^t = (U^t)'A$;
  set all negative elements in V^t to 0;
  if $t < r$ **then**
    solve for U^{t+1} the equation $V^t(V^t)'U^{t+1} = (V^t)'A^t$;
    set all negative elements in U^{t+1} to 0;
  end
end
return U^r and V^r;

Consider a sequence of vectors $(\boldsymbol{u}^t)_{t\in\mathbb{N}}$ in $\mathbb{R}^k$ defined by

$$\boldsymbol{u}^{t+1} = \arg\min_{\boldsymbol{u}} G(\boldsymbol{u}, \boldsymbol{u}^t) \tag{13.12}$$

for $t \in \mathbb{N}$. Then,

$$\begin{aligned} F(\boldsymbol{u}^{t+1}) &\leqslant G(\boldsymbol{u}^{t+1}, \boldsymbol{u}^t) && \text{(by the first condition of Definition 13.6)}\\ &\leqslant G(\boldsymbol{u}^t, \boldsymbol{u}^t) && \text{(by Equality (13.12))}\\ &= F(\boldsymbol{u}^t) && \text{(by the second condition of Definition 13.6)} \end{aligned}$$

for $t \in \mathbb{N}$. Thus, the sequence $(F(\boldsymbol{u}^t))_{t\in\mathbb{N}}$ is non-increasing.

The inequalities

$$F(\boldsymbol{u}^{t+1}) \leqslant G(\boldsymbol{u}^t, \boldsymbol{u}^t) \leqslant F(\boldsymbol{u}^t)$$

show that if $F(\boldsymbol{u}^{t+1}) = F(\boldsymbol{u}^t)$, then $G(\boldsymbol{u}^t, \boldsymbol{u}^t) = F(\boldsymbol{u}^t)$ so $\boldsymbol{u}^t$ is a local minimum of $G(\boldsymbol{u}, \boldsymbol{u}^t)$. If the derivatives of F exist and are continuous in a neighborhood of $\boldsymbol{u}^t$, then $\nabla F(\boldsymbol{u}^t) = 0$. Thus, the sequence $(\boldsymbol{u}^t)_{t\in\mathbb{N}}$ converges to a local minimum $\boldsymbol{u}_{min}$ of F.

Let $A = (\boldsymbol{a}_1 \ \cdots \ \boldsymbol{a}_m)$ be the matrix to be factored, where $\boldsymbol{a}_j \in \mathbb{R}^n$ for $1 \leqslant j \leqslant m$, and let $F_A : \mathbb{R}^{n\times p} \times \mathbb{R}^{p\times m} \longrightarrow \mathbb{R}_{\geqslant 0}$ be the objective function that we seek to minimize in NNMF, as follows:

$$F_A(U,V) = \frac{1}{2}\|A - UV\|_F^2.$$

If we write $A = (\boldsymbol{a}_1 \ \cdots \ \boldsymbol{a}_m)$ and $V = (\boldsymbol{v}_1 \ \cdots \ \boldsymbol{v}_m)$, then the matrix $A - UV$ can be written as

$$A - UV = (\boldsymbol{a}_1 - U\boldsymbol{v}_1 \ \cdots \ \boldsymbol{a}_m - U\boldsymbol{v}_m),$$

so

$$\|A - UV\|_F^2 = \sum_{j=1}^{n} \|\boldsymbol{a}_j - U\boldsymbol{v}_j\|_F^2.$$

Let $F_{\boldsymbol{a}}(U,\boldsymbol{v}) = \|\boldsymbol{a} - U\boldsymbol{v}\|_F^2$. The last equality can be written as

$$F_A(U,V) = \sum_{j=1}^{n} F_{\boldsymbol{a}_j}(U,\boldsymbol{v}_j).$$

Lemma 13.3. *Let $\boldsymbol{v} \in \mathbb{R}^k$ and let $d_i(\boldsymbol{v})$ be defined by $d_i(\boldsymbol{v}) = \frac{(U'U\boldsymbol{v})_i}{v_i}$ for $1 \leqslant i \leqslant k$ and $K(\boldsymbol{v}) = \mathsf{diag}(d_1(\boldsymbol{v}), \ldots, d_k(\boldsymbol{v}))$. The matrix $K(\boldsymbol{v}) - U'U$ is positive semidefinite.*

Proof. Define the matrix $M(\boldsymbol{v}) \in \mathbb{R}^{k\times k}$ as $M(\boldsymbol{v})_{ij} = v_i(K(\boldsymbol{v}) - U'U)_{ij}v_j$ for $1 \leqslant i,j \leqslant k$. In Supplement 96 of Chapter 6, we proved that M is positive semidefinite if and only if $K(\boldsymbol{v}) - U'U$ is positive semidefinite. Note that

$$M(\boldsymbol{v})_{ii} = (K(\boldsymbol{v}) - U'U)_{ii}v_i^2 = (U'U\boldsymbol{v})_i v_i = \sum_j (U'U)_{ij}v_j v_i$$

for $1 \leqslant i \leqslant k$. Also, if $i \neq j$, we have

$$M(\boldsymbol{v})_{ij} = -v_i(U'U)_{ij}v_j.$$

We have

$$\begin{aligned}
\boldsymbol{z}'M(\boldsymbol{v})\boldsymbol{z} &= \sum_i \sum_j z_i M(\boldsymbol{v})_{ij} z_j \\
&= \sum_i M(\boldsymbol{v})_{ii} z_i^2 + \sum_{i\neq j} z_i M(\boldsymbol{v})_{ij} z_j
\end{aligned}$$

$$= \sum_i \sum_j (U'U)_{ij} v_j v_i - \sum_{i \neq j} z_i v_i (U'U)_{ij} v_j z_j$$

$$= \sum_{i,j} (U'U)_{ij} v_i v_j \left(\frac{1}{2} z_i^2 + \frac{1}{2} z_j^2 - z_i z_j \right)$$

$$= \frac{1}{2} \sum_{i,j} (U'U)_{ij} v_i v_j (z_i - z_j)^2,$$

which shows the positive semidefiniteness of $M(\boldsymbol{v})$. □

Lemma 13.4. *The function* $G : (\mathbb{R}^k)^2 \longrightarrow \mathbb{R}$ *defined by*

$$G(\boldsymbol{s}, \boldsymbol{v}) = F(\boldsymbol{v}) + (\boldsymbol{s} - \boldsymbol{v})' \nabla F(\boldsymbol{v}) + \frac{1}{2} (\boldsymbol{s} - \boldsymbol{v})' K(\boldsymbol{v}) (\boldsymbol{s} - \boldsymbol{v})$$

is an auxiliary function for the quadratic function $F : \mathbb{R}^k \longrightarrow \mathbb{R}$ *given by*

$$F(\boldsymbol{s}) = F(\boldsymbol{v}) + (\boldsymbol{s} - \boldsymbol{v})' \nabla F(\boldsymbol{v}) + \frac{1}{2} (\boldsymbol{s} - \boldsymbol{v})' U'U (\boldsymbol{s} - \boldsymbol{v})$$

Proof. The inequality $G(\boldsymbol{s}, \boldsymbol{v}) \geqslant F(\boldsymbol{s})$ amounts to

$$(\boldsymbol{s} - \boldsymbol{v})' K(\boldsymbol{v}) (\boldsymbol{s} - \boldsymbol{v}) \geqslant (\boldsymbol{s} - \boldsymbol{v})' U'U (\boldsymbol{s} - \boldsymbol{v}),$$

which follows immediately from Lemma 13.3. Thus, the first condition of Definition 13.6 is satisfied. It is immediate that $G(\boldsymbol{v}, \boldsymbol{v}) = F(\boldsymbol{v})$. □

Choose $F = F_{\boldsymbol{a}}$. We have

$$F(\boldsymbol{s}) = F_{\boldsymbol{a}}(\boldsymbol{s}) = \frac{1}{2} \| \boldsymbol{a} - U\boldsymbol{s} \|_F^2$$

$$= \frac{1}{2} \sum_i (\boldsymbol{a} - U\boldsymbol{s})_i^2 = \frac{1}{2} \sum_i \left(a_i - \sum_j U_{ij} s_j \right)^2$$

Therefore,

$$\frac{\partial F}{\partial s_\ell} = - \sum_i \left(a_i - \sum_j U_{ij} s_j \right) U_{i\ell} = - \sum_i a_i U_{i\ell} + \sum_i \sum_j U_{ij} U_{i\ell} s_j$$

for $1 \leqslant i \leqslant k$. Thus,

$$\nabla F(\boldsymbol{s}) = -U'\boldsymbol{a} + U'U\boldsymbol{s}.$$

The corresponding auxiliary function is now

$$G(\boldsymbol{s},\boldsymbol{v}) = F(\boldsymbol{v}) + (\boldsymbol{s}-\boldsymbol{v})'(-U'\boldsymbol{a} + U'U\boldsymbol{v})) + \frac{1}{2}(\boldsymbol{s}-\boldsymbol{v})'K(\boldsymbol{v})(\boldsymbol{s}-\boldsymbol{v}).$$

To obtain the update rule, we need to find $\boldsymbol{v}$ such that

$$G(\boldsymbol{v},\boldsymbol{v}^t) = F(\boldsymbol{v}^t) + (\boldsymbol{v}-\boldsymbol{v}^t)'(-U'\boldsymbol{a} + U'U\boldsymbol{v}^t) + \frac{1}{2}(\boldsymbol{v}-\boldsymbol{v}^t)'K(\boldsymbol{v}^t)(\boldsymbol{v}-\boldsymbol{v}^t)$$

is minimal. Regarding $G(\boldsymbol{v},\boldsymbol{v}^t)$ as a function of $\boldsymbol{v}$, we obtain

$$\nabla_{\boldsymbol{v}} G(\boldsymbol{v},\boldsymbol{v}^t) = -U'\boldsymbol{a} + U'U\boldsymbol{v}^t + K(\boldsymbol{v}^t)(\boldsymbol{v}-\boldsymbol{v}^t),$$

by applying the differentiation rules $\nabla(\boldsymbol{a}'\boldsymbol{x}) = \nabla(\boldsymbol{x}'\boldsymbol{a})$ and $\nabla(\boldsymbol{x}'A\boldsymbol{x}) = 2A\boldsymbol{x}$ (for a symmetric matrix).

Therefore, the necessary extremal condition $\nabla_{\boldsymbol{v}} G(\boldsymbol{v},\boldsymbol{v}^t) = \mathbf{0}$ implies

$$\boldsymbol{v} = \boldsymbol{v}^t - K(\boldsymbol{v}^t)^{-1}(U'U\boldsymbol{v}^t - U'\boldsymbol{a}),$$

which is the update rule for the columns of the matrix V if U is kept constant.

Note that

$$\begin{aligned} K(\boldsymbol{v}^t)^{-1} &= \mathsf{diag}\left(\frac{1}{d_1(\boldsymbol{v}^t)}, \ldots, \frac{1}{d_k(\boldsymbol{v}^t)}\right) \\ &= \mathsf{diag}\left(\frac{v_1^t}{(U'U\boldsymbol{v}^t)_1}, \ldots, \frac{v_k^t}{(U'U\boldsymbol{v}^t)_k}\right). \end{aligned}$$

Thus, we obtain the update rule for the ith column of the matrix V for a fixed matrix U

$$\begin{aligned} v_i^{t+1} &= v_i^t - \frac{v_i^t}{(U'U\boldsymbol{v}^t)_i}((U'U\boldsymbol{v}^t)_i - (U'\boldsymbol{a})_i) \\ &= \frac{v_i^t(U'\boldsymbol{a})_i}{(U'U\boldsymbol{v}^t)_i}. \end{aligned}$$

The components of the updated matrix V can be written as

$$v_{ij}^{t+1} = \frac{v_{ij}^t (U'A)_{ij}}{(U'UV^t)_{ij}}.$$

Similarly, the updated components of the matrix U, when V is fixed, are given by

$$u_{ij}^{t+1} = \frac{u_{ij}^t (AV')_{ij}}{(U^t VV')_{ij}}.$$

Lee and Seung algorithm [105] consists of applying alternating updates of the matrices U and V (Algorithm 13.8.2).

Algorithm 13.8.2: Lee–Seung Factorization Algorithm

Data: A nonnegative matrix $A \in \mathbb{R}^{m \times n}$ and a maximum number of iterations r

Result: An approximate factorization $A = UV$

```
initialize U^1 to a random matrix;
initialize V^1 to a random matrix;
for t = 1 to r do
    V^{t+1} = (V^t ⊙ ((U^t)'A)) ⊘ ((U^t)'U^t V^t + 10^{-10} J_{p,m};
    U^{t+1} = (U^t ⊙ (A(V^t)')) ⊘ (U^t V^t (V^t)' + 10^{-10} J_{n,p};
end
return U^t and V^t;
```

Using the Hadamard product and quotient of matrices introduced in Definition 3.45, the computations involving the sequences $V^1, \ldots, V^t, \ldots$ and $U^1, \ldots, U^t, \ldots$ can be written in a more succinct form as

$$V^{t+1} = (V^t \odot ((U^t)'A)) \oslash ((U^t)'U^t V^t), \tag{13.13}$$

$$U^{t+1} = (U^t \odot (A(V^t)')) \oslash (U^t V^t (V^t)'). \tag{13.14}$$

The algorithm is given next. The terms $10^{-10} J_{p,m}$ and $10^{-10} J_{n,p}$ are added to avoid division by 0.

The Equalities (13.13) and (13.14) show that if U^1 and V^1 are positive, the matrices U^t and V^t remain positive throughout the iterations.

MATLAB uses the function **nnmf** of the Statistics toolbox to compute a factorization of a nonnegative matrix. When using its simplest syntax, `[U,V] = nnmf(A,k)`, this function computes an approximate decomposition of the non-negative matrix $A \in \mathbb{R}^{m\times n}$ into the non-negative factors $U \in \mathbb{R}^{m\times k}$ and $V \in \mathbb{R}^{k\times n}$, where the rows of V have unit length. Actually, the product UV is an approximation of A that minimizes the quantity $d = \sqrt{\|A - UV\|_F/mn}$ known as the *root-mean-squared residual*. To add d to the results, we can write `[U,V,d] = nnmf(A,k)`.

The function **nnmf** can be called with the number of options detailed in the documentation of MATLAB . Such a call has the form

```
[U,V] = nnmf(A,k,param1,val1,param2,val2,...)
```

For example, for the option **'algorithm'** the possible values are **'mult'** for the multiplicative algorithm or **'als'** for an alternating least-squares algorithm.

Example 13.8. The IRIS dataset from the UCI machine learning repository consists of 150 records that describe characteristics of three varieties of the iris flower: *Iris Setosa*, *Iris Versicolor*, and *Iris Virginica*. Each record has four components that correspond to the sepal length, sepal width, petal length, and petal width (in centimeters).

To compute a nonnegative matrix factorization of the matrix $D \in \mathbb{R}^{150\times 4}$ as $D = UV$, where $U \in \mathbb{R}^{150,2}$ and $V \in \mathbb{R}^{2,4}$, we write

```
>> [U,V] = nnmf(D,2)
```

This produces the matrix U (which we omit) and the matrix V given by

```
V =

    0.6942    0.2855    0.6223    0.2221
    0.8027    0.5675    0.1829    0.0142
```

corresponding to the four variables mentioned above. The factorization of D allows us to express each row $\boldsymbol{d}_i = (\text{sl}_i, \text{sw}_i, \text{pl}_i, \text{pw}_i)$ as

$$\text{sl}_i = 0.6942u_{i1} + 0.807u_{i2} \quad \text{sw}_i = 0.2855u_{i1} + 0.5675u_{i2},$$
$$\text{pl}_i = 0.6223u_{i1} + 0.1829u_{i2} \; \text{pw}_i = 0.2221u_{i1} + 0.0142u_{i2}.$$

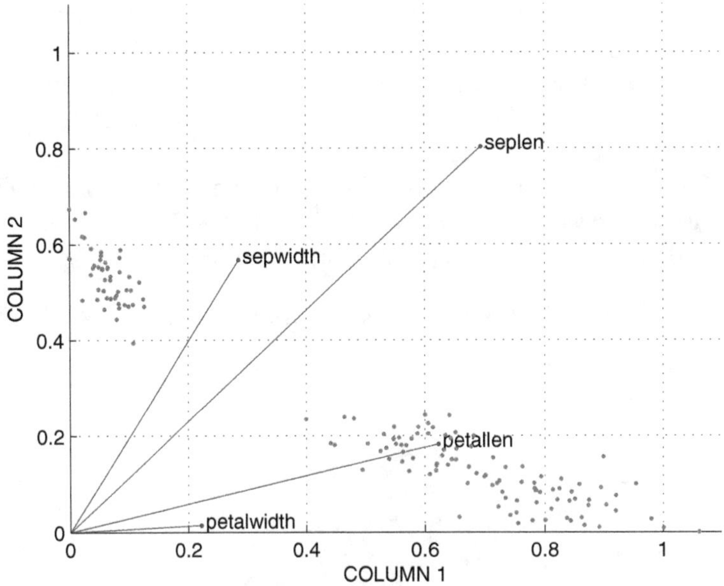

Fig. 13.8 Biplot of the IRIS dataset.

The first column of U has a strong influence on the sepal length sl_i and petal length pl_i; in contrast, the second column of U has little influence on the petal width pw_i. To visualize these influences, a biplot is constructed.

```
>> biplot(V','scores',U,'varlabels',{'seplen','sepwidth',...
                                    'petallen','petalwidth'});
>> axis([0 1.1 0 1.1])
>> xlabel('COLUMN 1')
>> ylabel('COLUMN 2')
```

The result is shown in Figure 13.8.

Exercises and Supplements

(1) Let $X = (\boldsymbol{x}_1 \ \cdots \ \boldsymbol{x}_n) \in \mathbb{R}^{m\times n}$ be a matrix. A *Karhunen–Loeve basis for* X is a sequence of pairwise orthogonal k unit vectors $(k < n)$ $\boldsymbol{u}_1, \ldots, \boldsymbol{u}_k$ in $\mathbb{R}^m$ such that for the orthogonal

projections $\mathsf{proj}_S(\boldsymbol{u}_1), \ldots, \mathsf{proj}_S(\boldsymbol{u}_n)$ on the subspace S generated by $\boldsymbol{u}_1, \ldots, \boldsymbol{u}_k$, $\sum_{i=1}^{n} \|\boldsymbol{x}_i - \mathsf{proj}_S(\boldsymbol{x}_i)\|^2$ is minimal. Prove that the first k columns of the orthogonal matrix U involved in the SVD of the matrix X, $X = UDV'$ is a Karhunen–Loeve basis for X.

(2) Let $X \in \mathbb{R}^{m\times n}$ be a centered data matrix and let $X = UDV'$ be the thin SVD of X, where $U \in \mathbb{R}^{m\times r}$, $D \in \mathbb{R}^{r\times r}$, and $V \in \mathbb{R}^{n\times r}$ and each U and V has orthonormal columns. If

$$S = UD = (\boldsymbol{s}_1 \ \cdots \ \boldsymbol{s}_r) = (d_1\boldsymbol{u}_1 \ \cdots \ d_r\boldsymbol{u}_r) \in \mathbb{R}^{m\times r}$$

is the matrix of scores and $V \in \mathbb{R}^{n\times r}$ is the matrix of loadings, prove that

(a) the variance of a score vector $\boldsymbol{s}_i$ is $var(\boldsymbol{s}_i) = \frac{1}{m-1}d_i^2$;

(b) $X = \boldsymbol{s}_1\boldsymbol{v}_1' + \cdots + \boldsymbol{s}_r\boldsymbol{v}_r'$;

(c) if $X_k = \boldsymbol{s}_1\boldsymbol{v}_1' + \cdots + \boldsymbol{s}_k\boldsymbol{v}_k'$, where $k \leqslant r$, then

$$\frac{\mathsf{tvar}(X_k)}{\mathsf{tvar}(X)} = \frac{\sum_{i=1}^{k} d_i^2}{\sum_{i=1}^{r} d_i^2}.$$

In other words, $\frac{\sum_{i=1}^{k} d_i^2}{\sum_{i=1}^{r} d_i^2}$ indicates the portion of the total variance of X explained by the first k scores.

(3) Let $X \in \mathbb{R}^{m\times n}$ be a centered data sample matrix whose covariance matrix is $cov(X) = aI_n + bJ_{n,n}$. Prove that $a \geqslant 0$ and $b \geqslant 0$.

(4) The orthogonal factor model [82] seeks to express a centered data sample matrix $X \in \mathbb{R}^{m\times n} = (\boldsymbol{v}_1 \ \cdots \ \boldsymbol{v}_n)$ as $X = FL + S$, where $L \in \mathbb{R}^{q\times n}$ is the matrix of factor loadings, $F \in \mathbb{R}^{m\times q}$ is the matrix of factors, and $S \in \mathbb{R}^{m\times n}$ is the matrix of specific factors. The model assumes that F is a centered matrix, $cov(F) = I_n$, S is a centered matrix, $cov(S)$ is a diagonal matrix, and $F'S = O_{n,n}$.

Prove that

(a) $cov(X) = (q-1)cov(L) + cov(S)$ and $F'X = L$;

(b) $var(\boldsymbol{v}_i) = (q-1)var(\boldsymbol{l}_i) + var(\boldsymbol{s}_i)$ for $1 \leqslant i \leqslant n$.

(5) Let $X \in \mathbb{R}^{m\times n}$ be a centered data sample matrix and let $p, q \in \mathbb{N}$ be such that $p + q = n$ and $X = (UV)$, where $U \in \mathbb{R}^{m\times p}$ and $V \in \mathbb{R}^{m\times q}$. Define the *mixed covariance* of U and V as $cov(U, V) = \frac{1}{m-1}U'V \in \mathbb{R}^{p\times q}$. Prove that

$$cov(X) = \begin{pmatrix} cov(U) & cov(U,V) \\ cov(U,V)' & cov(V) \end{pmatrix}.$$

(6) Using the same notations as in Exercise 5, let $\boldsymbol{u} = U\boldsymbol{a}$ and $\boldsymbol{v} = V\boldsymbol{b}$ be two vectors in $\mathbb{R}^m$ that aggregate the columns of U and V, respectively. We assume here that $\boldsymbol{a} \in \mathbb{R}^p$ and $\boldsymbol{v} \in \mathbb{R}^q$. Prove that

(a) $var(\boldsymbol{u}) = \boldsymbol{a}'cov(U)\boldsymbol{a}$, $var(\boldsymbol{v}) = \boldsymbol{b}'cov(V)\boldsymbol{b}$, and $cov(\boldsymbol{u}, \boldsymbol{v}) = \boldsymbol{a}'cov(U, V)\boldsymbol{b}$;

(b) if the correlation of $\boldsymbol{u}$ and $\boldsymbol{v}$ is defined as

$$corr(\boldsymbol{u}, \boldsymbol{v}) = \frac{cov(\boldsymbol{u}, \boldsymbol{v})}{\sqrt{var(\boldsymbol{u})}\sqrt{var(\boldsymbol{v})}},$$

prove that the largest value of $corr(\boldsymbol{u}, \boldsymbol{v})$ is the largest eigenvalue of

$$cov(U)^{-1/2}cov(U, V)cov(V)^{-1}cov(V, U)cov(U)^{-1/2}.$$

(7) Let $A \in \mathbb{R}^{n\times m}$ be a positive matrix. Define $F_A : \mathbb{R}^{n\times p}_{>0} \times \mathbb{R}^{p\times m}_{>0} \longrightarrow \mathbb{R}$ by $F_A(U, V) = D(A, UV)$, where D is the matrix divergence introduced in Supplement 108 of Chapter 3. If $A = (\boldsymbol{a}_1 \cdots \boldsymbol{a}_m)$ and $V = (\boldsymbol{v}_1 \cdots \boldsymbol{v}_m)$, prove that

$$D(A, UV) = \sum_{j=1}^{n} D(\boldsymbol{a}_j, U\boldsymbol{v}_j).$$

(8) Let $F_{\boldsymbol{a}} : \mathbb{R}^k_{>0} \longrightarrow \mathbb{R}$ be the function defined by

$$F_{\boldsymbol{a}}(\boldsymbol{s}) = \sum_i \left(\sum_j u_{ij}s_j - a_i + a_i \ln \frac{a_i}{\sum_j u_{ij}s_j} \right)$$

for $\boldsymbol{s} \in \mathbb{R}^k_{>0}$, where $\boldsymbol{a} \in \mathbb{R}^k_{>0}$. Prove that the function $G_{\boldsymbol{a}} : (\mathbb{R}^k_{>0})^2 \longrightarrow \mathbb{R}$ given by

$$G_{\boldsymbol{a}}(\boldsymbol{s}, \boldsymbol{v}) = \sum_i (a_i \ln a_i - a_i) + \sum_i \sum_j u_{ij} s_i - \sum_i \sum_j a_i \frac{u_{ij} v_j}{\sum_p u_{ip} s_p} \left(\ln u_{ij} s_j - \ln \frac{u_{ij} v_j}{\sum_p u_{ip} s_p} \right)$$

is an auxiliary function for $F_{\boldsymbol{a}}$.

Solution: We have

$$\begin{aligned}
G_{\boldsymbol{a}}(\boldsymbol{s}, \boldsymbol{s}) &= \sum_i (a_i \ln a_i - a_i) + \sum_i \sum_j u_{ij} s_i \\
&\quad - \sum_i \sum_j a_i \frac{u_{ij} s_j}{\sum_p u_{ip} s_p} \left(\ln u_{ij} s_j - \ln \frac{u_{ij} s_j}{\sum_p u_{ip} s_p} \right) \\
&= \sum_i (a_i \ln a_i - a_i) + \sum_i \sum_j u_{ij} s_i \\
&\quad - \sum_i \sum_j a_i \frac{u_{ij} s_j}{\sum_p u_{ip} s_p} \ln \sum_p u_{ip} s_p \\
&= \sum_i \left(a_i \ln a_i - a_i + \sum_j u_{ij} s_j - a_i \ln \sum_p u_{ip} s_p \right) \\
&= F_{\boldsymbol{a}}(\boldsymbol{s}).
\end{aligned}$$

The second condition of Definition 13.6, $G_{\boldsymbol{a}}(\boldsymbol{s}, \boldsymbol{v}) \geqslant F_{\boldsymbol{a}}(\boldsymbol{s})$, can be proven starting from the convexity of the function $\ell(x) = -\ln x$. By Jensen's Theorem, we have

$$-\ln \left(\sum_{j=1}^k u_{ij} s_j \right) \leqslant -\sum_{j=1}^k t_j \ln \frac{u_{ij} s_j}{t_j}$$

for any set of non-negative numbers $t_1, \ldots, t_k$ such that $\sum_{j=1}^{k} t_j = 1$. Choosing $t_j = \frac{u_{ij}v_j}{\sum_{p=1}^{k} u_{ip}v_p}$ yields the inequality

$$
\begin{aligned}
-\ln\left(\sum_{j=1}^{k} u_{ij}s_j\right) &\leq -\sum_{j=1}^{k} \frac{u_{ij}v_j}{\sum_{p=1}^{k} u_{ip}v_p} \ln \frac{u_{ij}s_j}{\frac{u_{ij}v_j}{\sum_{p=1}^{k} u_{ip}v_p}} \\
&= -\sum_{j=1}^{k} \frac{u_{ij}v_j}{\sum_{p=1}^{k} u_{ip}v_p} \\
&\quad \times \left(\ln u_{ij}s_j - \ln \frac{u_{ij}v_j}{\sum_{p=1}^{k} u_{ip}v_p}\right).
\end{aligned}
$$

This inequality implies

$$
\begin{aligned}
G_{\boldsymbol{a}}(\boldsymbol{s}, \boldsymbol{v}) &\geqslant \sum_i (a_i \ln a_i - a_i) + \sum_i \sum_j u_{ij}s_i \\
&\quad - \sum_i a_i \ln\left(\sum_{j=1}^{k} u_{ij}s_j\right) = F_{\mathcal{A}}(\boldsymbol{s}).
\end{aligned}
$$

Thus, $G_{\boldsymbol{a}}$ is indeed an auxiliary function for $F_{\boldsymbol{a}}$.

(9) Let X be a non-negative matrix and let $X = UV$ be a non-negative matrix factorization of X. Prove that all rows of X are non-negative combinations of the rows of U, that is, they lie in the simplicial cone generated by the rows of U.

(10) Let $X = (\boldsymbol{x}_1, \ldots, \boldsymbol{x}_p)$, $Y = (\boldsymbol{y}_1, \ldots, \boldsymbol{y}_q)$ be two sequences of vectors in $\mathbb{R}^n$. The matrix $D \in \mathbb{R}^{p \times q}$ of squared distances between these sequences is $D_{X,Y} = (d_{ij})$, where $d_{ij} = \|\boldsymbol{x}_i - \boldsymbol{y}_j\|_2^2$ for $1 \leqslant i \leqslant p$ and $1 \leqslant j \leqslant q$. Prove that

$$
D_{X,Y} = \begin{pmatrix} \boldsymbol{x}_1'\boldsymbol{x}_1 \\ \vdots \\ \boldsymbol{x}_p'\boldsymbol{x}_p \end{pmatrix} \mathbf{1}_q' + \mathbf{1}_p(\boldsymbol{y}_1'\boldsymbol{y}_1, \ldots, \boldsymbol{y}_q'\boldsymbol{y}_q) - 2\, X'Y.
$$

The analogue of multidimensional metric scaling for two sets of vectors has been introduced and studied as multidimensional unfolding (MDU) by Schönemann in [147]. For MDU, the issue is to reconstitute two finite vector sets given their mutual distances.

Let $X = (\boldsymbol{x}_1 \ \cdots \ \boldsymbol{x}_p) \in \mathbb{R}^{n\times p}$ and let $\mathsf{t}_{\boldsymbol{z}}$ be a translation of $\mathbb{R}^n$. Denote by $\mathsf{t}_{\boldsymbol{z}}(X)$ the matrix $(\mathsf{t}_{\boldsymbol{z}}(\boldsymbol{x}_1) \ \cdots \ \mathsf{t}_{\boldsymbol{z}}(\boldsymbol{x}_p))$.

(11) Let $X = (\boldsymbol{x}_1 \ \cdots \ \boldsymbol{x}_p) \in \mathbb{R}^{n\times p}$, $\boldsymbol{x}_0 \in \mathbb{R}^n$, and let $\boldsymbol{c} \in \mathbb{R}^p$ be a vector such that $\boldsymbol{x}_0 = X\boldsymbol{c}$. Prove that if $\tilde{X} = \mathsf{t}_{-\boldsymbol{x}_0}(X)$, then $\tilde{X} = XQ'_{\boldsymbol{c}}$, where $Q_{\boldsymbol{c}} = I_p - \mathbf{1}_p\boldsymbol{c}'$ is the projection matrix introduced in Supplement 10 of Chapter 3.

Solution: We have

$$\begin{aligned}\mathsf{t}_{-\boldsymbol{x}_0}(X') &= \begin{pmatrix} \boldsymbol{x}'_1 - \boldsymbol{x}'_0 \\ \vdots \\ \boldsymbol{x}'_p - \boldsymbol{x}'_0 \end{pmatrix} \\ &= X' - \mathbf{1}_p\boldsymbol{x}'_0 = X' - \mathbf{1}_p\boldsymbol{c}'X' \\ &= (I_p - \mathbf{1}_p\boldsymbol{c}')X' = Q_{\boldsymbol{c}}X',\end{aligned}$$

which is equivalent to $\mathsf{t}_{-\boldsymbol{x}_0}(X) = XQ'_{\boldsymbol{c}}$.

(12) Let $X = (\boldsymbol{x}_1 \ \cdots \ \boldsymbol{x}_p) \in \mathbb{R}^{n\times p}$, $Y = (\boldsymbol{y}_1 \ \cdots \ \boldsymbol{y}_q) \in \mathbb{R}^{n\times q}$, $\boldsymbol{x}_0, \boldsymbol{y}_0 \in \mathbb{R}^n$. Let $\tilde{X}$ and $\tilde{Y}$ be matrices obtained by a translation by $-\boldsymbol{x}_0$ and $-\boldsymbol{y}_0$ of the columns of X and Y, respectively. If $\boldsymbol{c} \in \mathbb{R}^p$ and $\boldsymbol{d} \in \mathbb{R}^q$ are such that $\boldsymbol{x}_0 = X\boldsymbol{c}$ and $\boldsymbol{y}_0 = Y\boldsymbol{d}$, prove that $\tilde{X}'\tilde{Y} = -\frac{1}{2}Q_{\boldsymbol{c}}D_{X,Y}Q'_{\boldsymbol{d}}$.

Solution: Starting from the equality given in Exercise 10, we have

$$\begin{aligned}&Q_{\boldsymbol{c}}D_{X,Y}Q'_{\boldsymbol{d}} \\ &= Q_{\boldsymbol{c}}\begin{pmatrix} \boldsymbol{x}'_1\boldsymbol{x}_1 \\ \vdots \\ \boldsymbol{x}'_p\boldsymbol{x}_p \end{pmatrix}\mathbf{1}'_qQ'_{\boldsymbol{d}} + Q_{\boldsymbol{c}}\mathbf{1}_p(\boldsymbol{y}'_1\boldsymbol{y}_1, \ldots, \boldsymbol{y}'_q\boldsymbol{y}_q)Q'_{\boldsymbol{d}} \\ &\quad -2\,Q_{\boldsymbol{c}}X'YQ'_{\boldsymbol{d}} \\ &= -2\,Q_{\boldsymbol{c}}X'YQ'_{\boldsymbol{d}} = -2\tilde{X}'\tilde{Y},\end{aligned}$$

because $Q_{\boldsymbol{d}}\mathbf{1}_q = \mathbf{0}_q$ and $Q_{\boldsymbol{c}}\mathbf{1}_p = \mathbf{0}_p$.

Let $C_{X,Y} = \tilde{X}'\tilde{Y}$ and let $C_{X,Y} = GH$ be a full-rank decomposition of $C_{X,Y}$ (see, Theorem 3.35). Note that such a decomposition is not unique for, if T is a non-singular matrix, $C_{X,Y}$ can be written as $C_{X,Y} = (\tilde{X}'T^{-1})(T\tilde{Y})$ is yet another full-rank factorization of $C_{X,Y}$.

(13) Let $C = GH$ be a full-rank decomposition of the matrix $C_{X,Y}$. Prove that

(a) starting from $G = (\boldsymbol{g}_1 \ \cdots \ \boldsymbol{g}_p)$ and $H = (\boldsymbol{h}_1 \ \cdots \ \boldsymbol{h}_q)$, if

$$\boldsymbol{x}_i = T'\boldsymbol{g}_i + \boldsymbol{x}_0 \text{ and } \boldsymbol{y}_j = T^{-1}\boldsymbol{h}_j + \boldsymbol{y}_0$$

for $1 \leqslant i \leqslant p$ and $1 \leqslant j \leqslant q$, then the matrices $X = (\boldsymbol{x}_1 \ \cdots \ \boldsymbol{x}_p)$ and $Y = (\boldsymbol{y}_1 \ \cdots \ \boldsymbol{y}_q)$ give a solution of the metric unfolding problem;

(b) with the previous choice for X and Y and with $M = TT'$, we have

$$\begin{aligned} d_{ij}^2 &= \boldsymbol{g}_i'M\boldsymbol{g}_i + \boldsymbol{h}_j'M^{-1}\boldsymbol{h}_j + (\boldsymbol{x}_0 - \boldsymbol{y}_0)'(\boldsymbol{x}_0 - \boldsymbol{y}_0) \\ &\quad + 2\boldsymbol{g}_i'T(\boldsymbol{x}_0 - \boldsymbol{y}_0) - 2\boldsymbol{h}_j'(T^{-1})'(\boldsymbol{x}_0 - \boldsymbol{y}_0) - 2\boldsymbol{g}_i'\boldsymbol{h}_j, \end{aligned}$$

for $1 \leqslant i \leqslant p$ and $1 \leqslant j \leqslant q$.

Let S be the finite set of numbers, $S = \{-1, 0, 1\}$. A *semidiscrete decomposition* (SDD) of a matrix $A \in \mathbb{R}^{m\times n}$ is a sum of the form

$$A = \sum_{p=1}^{k} d_p \boldsymbol{x}_p \boldsymbol{y}_p',$$

where $\boldsymbol{x}_p \in S^m$, $\boldsymbol{y}_p \in S^n$, and $d_1, \ldots, d_k$ are real numbers.

Note that every matrix A can be expressed as a sum of mn rank-1 matrices $A = \sum_{i=1}^m \sum_{j=1}^n a_{ij}\boldsymbol{e}_i\boldsymbol{e}_j'$; the purpose of the SDD is to develop sum of fewer terms that approximate the matrix A.

(14) Prove that for a fixed $A, \boldsymbol{x}$, and $\boldsymbol{y}$, the least value of $\| A - d\boldsymbol{x}\boldsymbol{y}' \|_F^2$ is achieved for $d_* = \frac{(\boldsymbol{x}'A\boldsymbol{y})^2}{\|\boldsymbol{x}\|_F^2 \|\boldsymbol{y}\|_F^2}$ and it equals $\|A\|_F^2 - \frac{(\boldsymbol{x}'A\boldsymbol{y})^2}{\|\boldsymbol{x}\|_F^2\|\boldsymbol{y}\|_F^2}$.

Solution: Let R be the *residual* $R = A - d\boldsymbol{x}\boldsymbol{y}'$ obtained by approximating A by $d\boldsymbol{x}\boldsymbol{y}'$ and let $r_A(d, \boldsymbol{x}, \boldsymbol{y}) = \|R\|_F^2$. We have

$$\begin{aligned} r_A(d, \boldsymbol{x}, \boldsymbol{y}) &= (A - d\boldsymbol{x}\boldsymbol{y}')'(A - d\boldsymbol{x}\boldsymbol{y}') = (A' - d\boldsymbol{y}\boldsymbol{x}')(A - d\boldsymbol{x}\boldsymbol{y}') \\ &= A'A - d\boldsymbol{y}\boldsymbol{x}'A - dA'\boldsymbol{x}\boldsymbol{y}' + d^2\boldsymbol{y}\boldsymbol{x}'\boldsymbol{x}\boldsymbol{y}' \\ &= \|A\|_F^2 - 2d\boldsymbol{x}'A\boldsymbol{y} + d^2\|\boldsymbol{x}\|_F^2\|\boldsymbol{y}\|_F^2. \end{aligned}$$

The optimal solution for d is therefore $d = \frac{\boldsymbol{x}'A\boldsymbol{y}}{\|\boldsymbol{x}\|_F^2\|\boldsymbol{y}\|_F^2}$. The final part follows from the substitution of d in $r_A(d, \boldsymbol{x}, \boldsymbol{y})$.

(15) Suppose that $A \in \mathbb{R}^{m\times n}$ is a matrix and let $y \in S^n$. Determine d and $\boldsymbol{x} \in S^m$ such that $\|A - d\boldsymbol{x}\boldsymbol{y}'\|_F^2$ is minimal.

Solution: In principle, we need to examine 3^m possible value for $\boldsymbol{x}$, but we will show that it is possible to limit the search to m values. Let $z_i = x_i \sum_{j=1}^n a_{ij}y_j$. We have

$$\begin{aligned} r_A(d, \boldsymbol{x}, \boldsymbol{y}) &= \sum_{i=1}^m \sum_{j=1}^n (a_{ij} - dx_iy_j)^2 \\ &= \sum_{i=1}^m \sum_{j=1}^n (a_{ij}^2 - 2da_{ij}x_iy_j + d^2x_i^2y_j^2) \\ &= \sum_{i=1}^m \sum_{j=1}^n a_{ij}^2 - 2d\sum_{i=1}^m \sum_{j=1}^n a_{ij}x_iy_j + d^2\sum_{i=1}^m \sum_{j=1}^n x_i^2y_j^2 \\ &= \sum_{i=1}^m \sum_{j=1}^n a_{ij}^2 - 2d\sum_{i=1}^m z_i + d^2\sum_{i=1}^m x_i^2 \sum_{j=1}^n y_j^2. \end{aligned}$$

The signs of the components of $\boldsymbol{x}$ do not affect the last term. Thus, to minimize $r_A(d, \boldsymbol{x}, \boldsymbol{y})$, the signs of x_i must be chosen such that all z_i have the same sign: positive if $d > 0$ and negative, otherwise. Let J be the number of non-zero components of $\boldsymbol{x}$. Clearly, we have $1 \leqslant J \leqslant n$. Then,

$$r_A(d, \boldsymbol{x}, \boldsymbol{y}) = \sum_{i=1}^m \sum_{j=1}^n a_{ij}^2 - 2d\sum_{i=1}^m z_i + d^2J\sum_{j=1}^n y_j^2.$$

For a particular choice of $\boldsymbol{x}$, the d that minimizes $r_A(d, \boldsymbol{x}, \boldsymbol{y})$ is

$$d_* = \frac{\sum_{i=1}^m z_i}{J \sum_{j=1}^n y_j^2}, \tag{13.15}$$

and the minimum value of $r_A(d, \boldsymbol{x}, \boldsymbol{y})$ is

$$r_A(d_*, \boldsymbol{x}, \boldsymbol{y}) = \sum_{i=1}^m \sum_{j=1}^n a_{ij}^2 - d_*^2 J \sum_{j=1}^n y_j^2.$$

Thus, minimizing $r_A(d, \boldsymbol{x}, \boldsymbol{y})$ amounts to maximizing $d^2 J$. For a given J, x_i must be chosen equal to *sign* $(|\sum_{j=1} a_{ij} y_j|)$ for those i that correspond to the largest sums $|\sum_{j=1} a_{ij} y_j|$. This shows that there are m choices of $\boldsymbol{x}$ to check: set the x_i's corresponding to the J largest sums $|\sum_{j=1} a_{ij} y_j|$ to 1 or to -1 such that all z_i are positive, find d from Equality 13.15, and choose $d^2 J$ among the n choices for J.

The O'Leary–Peleg algorithm [95, 122] for SDD starts with a matrix $A \in \mathbb{R}^{m \times n}$ and with a number k and consists of alternative steps that determine two sequences of vectors $\boldsymbol{x}_1, \ldots, \boldsymbol{x}_k$ and $\boldsymbol{y}_1, \ldots, \boldsymbol{y}_k$ that allow the construction of a sequence of approximations $A_0, A_1, \ldots, A_k, \ldots$ of A.

The squares of the norms of the residuals of successive approximations are $r_1, r_2, \ldots$. The desired accuracy of the approximation is r_{min}. Also, ℓ_{max} is the maximum allowable number of inner iterations, and α_{min} is the minimum relative improvement.

(16) Let $A \in \mathbb{R}^{m \times n}$ and let $A_0, A_1, \ldots$ and $R_1, R_2, \ldots$ be two sequences of matrices defined by $A_0 = O_{m,n}$, $R_1 = A$, $A_k = A_{k-1} + d_k \boldsymbol{x}_k \boldsymbol{y}_k'$, and $R_{k+1} = R_k - d_k \boldsymbol{x}_k \boldsymbol{y}_k$, where the sequences $\boldsymbol{x}_1, \boldsymbol{x}_2, \ldots$ and $\boldsymbol{y}_1, \boldsymbol{y}_2, \ldots$ are defined as in Algorithm 13.8.3. Prove that
   (a) $\| R_{k+1} \|_F < \| R_k \|_F$, which ensures the convergence of the O'Leary–Peleg algorithm;
   (b) $\|_{\|} R_{k+1}{}_F^2 \leq \left(1 - \frac{1}{mn}\right)^k \| R_0 \|_F^2$, which proves that $\lim_{k \to \infty} \| R_k \|_F = 0$ and the rate of convergence is at least linear.

Algorithm 13.8.3: O'Leary–Peleg Factorization Algorithm

Data: A matrix $A \in \mathbb{R}^{m \times n}$ and a number k_{max} of approximating terms

Result: An SDD approximate factorization of A

initialize $R_1 = A$; initialize $r_1 = \|R_1\|_F^2$;
for $k = 1$ *to* k_{max} **do**
    while $r_k > r_{min}$ **do**
        choose $\boldsymbol{y} = \boldsymbol{e}_j$ if the jth column of R_k contains the largest magnitude entry in R_k;
        for $\ell = 1$ *to* ℓ_{max} **do**
            while $\alpha > \alpha_{min}$ **do**
                $\boldsymbol{z} = \frac{1}{\|\boldsymbol{y}\|_F^2} R_k \boldsymbol{y}$;
                find $\boldsymbol{x} \in S^m$ that maximizes $\frac{1}{\|\boldsymbol{x}\|_F^2}(\boldsymbol{x}'\boldsymbol{z})^2$;
                $\boldsymbol{z} = \frac{1}{\|\boldsymbol{x}\|_F^2} R_k \boldsymbol{x}$;
                find $\boldsymbol{y} \in S^n$ that maximizes $\frac{1}{\|\boldsymbol{y}\|_F^2}(\boldsymbol{y}'\boldsymbol{z})^2$;
                $\beta = \frac{(\boldsymbol{x}' R_k \boldsymbol{y})^2}{\|\boldsymbol{x}\|_F^2 \|\boldsymbol{y}\|_F^2}$;
                if $\ell > 1$ **then**
                    $\alpha = \frac{\beta - \bar{\beta}}{\bar{\beta}}$
                end
                $\bar{\beta} = \beta$;
            end
        end
        $\boldsymbol{x}_k = \boldsymbol{x}$;
        $\boldsymbol{y}_k = \boldsymbol{y}$;
        $d_k = \frac{\boldsymbol{x}_k' R_k \boldsymbol{y}_k}{\|\boldsymbol{x}_k\|_F^2 \|\boldsymbol{y}_k\|_F^2}$;
        $A_k = A_{k-1} + d_k \boldsymbol{x}_k \boldsymbol{y}_k'$;
        $R_{k+1} = R_k - d_k \boldsymbol{x}_k \boldsymbol{y}_k'$;
        $r_{k+1} = r_k - \beta$;
    end
end

Bibliographical Comments

Spectral methods for dimensionality reduction are surveyed in [145]. A main reference for multidimensional scaling is [32].

The application of SVDs to the recommender system was initiated in [19].

Schoenberg's paper on metric spaces [146] in one of the earliest contributions to metric scaling. Theorem 13.6 was obtained by Berge in [15] and Sibson in [150]. The connection between multidimensional metric scaling and PCA was studied by Gower in [66].

The nonnegative factorization problem attracted a huge amount of interest after Lee and Seung's paper [105] was published in 1999. The research in this type of problems was initiated earlier by Paatero in [127] followed by [125, 126]. Applications in text mining and document clustering are discussed in several important publications [17, 26, 131, 149] on which this chapter is based.

The semidiscrete matrix decomposition of matrices was introduced in [122] and further explored in [94, 95].

Chapter 14

Tensors and Exterior Algebras

14.1 Introduction

In many machine learning-oriented literature sources, tensors are regarded as multidimensional arrays. While the sets of values of tensors over finite-dimensional spaces are indeed such arrays, this point of view misses fundamental properties of tensors, especially their behaviors relative to basis changes in the underlying linear spaces.

Recall that the elements of a linear space V are referred to as contravariant vectors while the elements of the dual space V^* are referred to as covariant vectors (see Section 2.8). The components of contravariant vectors will be denoted with letters with superscripts, while the components of covariant vectors will be designated by letters with subscripts. The opposite convention is applied to vectors themselves: vectors in V are denoted by letters with subscripts, while vectors in V^* are denoted by letters with superscripts.

14.2 The Summation Convention

The *summation convention*, a modality of simplifying expression with multiple indices that is widely used in presenting tensors, was introduced by Albert Einstein. This convention stipulates that *when an index variable appears twice in a single term and is not otherwise defined, it implies summation of that term over all the values of the index.*

Indices that are not summation indices are referred to as *free indices.*

Example 14.1. If i ranges over the set $\{1,2,3,4\}$, the expression $x^i y_i$ stands for $x^1 y_1 + x^2 y_2 + x^3 y_3 + x^4 y_4$. The summation index is a dummy index, in the sense that it can be replaced by any other index ranging over the same set without modifying the value of the expression. For instance, we have $x^k y_k = x^i y_i$, if k ranges over the set $\{1,2,3,4\}$.

Example 14.2. Using the summation convention, the definition of a determinant (Definition 5.1) can be written using the summation convention and the Levi-Civita symbols (introduced in Definition 1.11) as

$$\det(A) = \epsilon_{i_1 \cdots i_n} a_{1i_1} \cdots a_{ni_n}.$$

Example 14.3. The inner product of two vectors $\boldsymbol{x}, \boldsymbol{y} \in \mathbb{R}^n$ can be written as

$$\boldsymbol{x}'\boldsymbol{y} = x_i y_i.$$

Example 14.4. Let $y_i = y_i(x_1, \ldots, x_n)$ be n functions which have continuous partial derivatives relative to $x_1, \ldots, x_n$. The *Jacobian matrix* J is

$$J = \begin{pmatrix} \frac{\partial y_1}{\partial x_1} & \frac{\partial y_1}{\partial x_2} & \cdots & \frac{\partial y_1}{\partial x_n} \\ \frac{\partial y_2}{\partial x_1} & \frac{\partial y_2}{\partial x_2} & \cdots & \frac{\partial y_2}{\partial x_n} \\ \vdots & \vdots & \vdots & \vdots \\ \frac{\partial y_n}{\partial x_1} & \frac{\partial y_n}{\partial x_2} & \cdots & \frac{\partial y_n}{\partial x_n} \end{pmatrix}.$$

The determinant $\det(J)$ is the *Jacobian* of the transformation of $\mathbb{R}^n$ defined by the functions $y_1, \ldots, y_n$. The transformation is locally bijective on an open subset U of $\mathbb{R}^n$ if and only if $\det(J) \neq 0$ at each point of U. When $\det(J) \neq 0$ and the functions $y_1, \ldots, y_n$ have continuous partial derivatives of second order in U, then the transformation is called an *admissible change of coordinates.*

If $z_k = z_k(y_1, \ldots, y_n)$ are n functions (for $1 \leqslant k \leqslant n$) which have continuous partial derivatives relative to $y_1, \ldots, y_n$, we derive the following link between partial derivatives written by using the summation convention:

$$\frac{\partial z_k}{\partial x_i} = \frac{\partial z_k}{\partial y_j}\frac{\partial y_j}{\partial x_i}.$$

14.3 Tensor Products of Linear Spaces

Recall that multilinear functions were introduced in Definition 2.27.

Definition 14.1. Let V and W be finite-dimensional linear spaces. Their *tensor product* is a linear space $V \otimes W$ equipped with a bilinear map $\otimes : V \times W \longrightarrow V \otimes W$ such that for any linear space U and bilinear mapping $f : V \times W \longrightarrow U$, there exists a unique linear mapping $g : V \otimes W \longrightarrow U$ such that the diagram

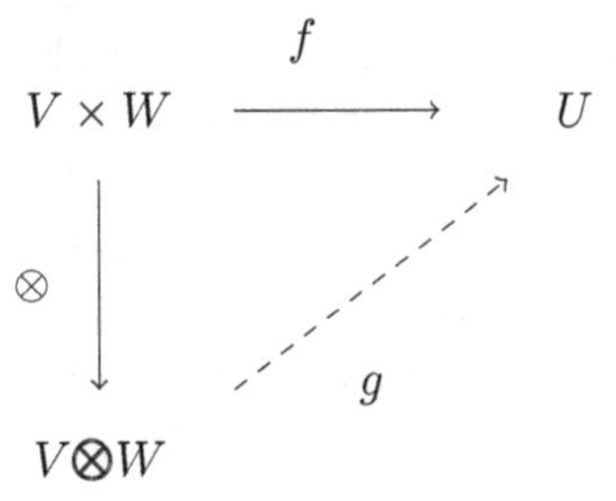

is commutative.

The linear space $V \otimes W$ is the *tensor product* of the linear spaces V and W, its elements are referred to as *tensors*, and the elements in $\mathsf{Im}(\otimes)$, the image of the bilinear function $\otimes$, are said to be the *decomposable* tensors.

The existence of the mapping g is the *universal property* of the tensor product. It means that any bilinear function $f : V \times W \longrightarrow U$ can be obtained by applying a linear function g to the special bilinear function $\otimes$.

We defined the tensor product of two linear spaces as a new linear space that satisfies a certain property. As we shall see, if such a linear space exists, then it is unique. However, it is incumbent on us to show the *existence* of the tensor product. We will do that later in this section (see Theorem 14.1).

The bilinearity of the $\otimes$ mapping implies the following *axioms* for the tensor product:

- *Distributivity of tensor product*: for any vectors $\boldsymbol{v}, \boldsymbol{v}_1, \boldsymbol{v}_2$ in V and any vectors $\boldsymbol{w}, \boldsymbol{w}_1, \boldsymbol{w}_2$ in W, we have

$$\boldsymbol{v} \otimes (a\boldsymbol{w}_1 + b\boldsymbol{w}_2) = a(\boldsymbol{v} \otimes \boldsymbol{w}_1) + b(\boldsymbol{v} \otimes \boldsymbol{w}_2), \quad (14.1)$$

$$(a\boldsymbol{v}_1 + b\boldsymbol{v}_2) \otimes \boldsymbol{w} = a(\boldsymbol{v}_1 \otimes \boldsymbol{w}) + b(\boldsymbol{v}_2 \otimes \boldsymbol{w}); \quad (14.2)$$

- *Existence of basis in tensor product*: if $\{\boldsymbol{v}_1, \ldots, \boldsymbol{v}_m\}$ is a basis in V and $\{\boldsymbol{w}_1, \ldots, \boldsymbol{w}_m\}$ is a basis in W, then $\{\boldsymbol{u}_{ij} = \boldsymbol{v}_i \otimes \boldsymbol{w}_j \mid 1 \leqslant i \leqslant m, 1 \leqslant j \leqslant n\}$ is a basis in $U = V \bigotimes W$.

If $\boldsymbol{v} = \boldsymbol{0}_V$ or $\boldsymbol{w} = \boldsymbol{0}_W$, then $\boldsymbol{v} \otimes \boldsymbol{w} = \boldsymbol{0}_{V \bigotimes W}$.

By taking $a = 1$, $b = -1$, and $\boldsymbol{w}_1 = \boldsymbol{w}_2$ in Equality (14.1), we obtain $\boldsymbol{v} \otimes \boldsymbol{0}_W = \boldsymbol{0}_{V \bigotimes W}$ for every $\boldsymbol{v} \in V$. Similarly, $\boldsymbol{0}_V \otimes \boldsymbol{w} = \boldsymbol{0}_{V \bigotimes W}$.

Thus, if V, W are finite-dimensional vector spaces having the bases $\{\boldsymbol{v}_1, \ldots, \boldsymbol{v}_m\}$ and $\{\boldsymbol{w}_1, \ldots, \boldsymbol{w}_m\}$, respectively, and $\{\boldsymbol{u}^{ij} = \boldsymbol{v}^i \otimes \boldsymbol{w}^j \mid 1 \leqslant i \leqslant m, 1 \leqslant j \leqslant n\}$ is a basis in $U = V \bigotimes W$, then, if $\boldsymbol{x} = x^i \boldsymbol{v}_i \in V$ and $\boldsymbol{y} = y^j \boldsymbol{w}_j$, the unique tensor product that satisfies the axioms has the components $x^i y^j$ in the basis $\{\boldsymbol{u}_{ij} \mid 1 \leqslant i \leqslant m, 1 \leqslant j \leqslant n\}$.

The tensor space $V \bigotimes W$ is the set of all sums of the form $\sum_i \sum_j a_{ij} \boldsymbol{v}_i \otimes \boldsymbol{w}_j$ which have a finite number of non-zero coefficients. We will refer to the set of coefficients a_{ij} as a *tensor*.

Example 14.5. Suppose that V and W are the unidimensional linear space $\mathbb{R}$. The direct product $\mathbb{R} \times \mathbb{R}$ consists of all pairs $(u, v) \in \mathbb{R} \times \mathbb{R}$. Addition of (u_1, v_1) and (u_2, v_2) in $\mathbb{R} \times \mathbb{R}$ is defined by $(u_1, v_1) + (u_2, v_2) = (u_1 + u_2, v_1 + v_2)$ and scalar multiplication is defined as $a(u, v) = (au, av)$ for $a \in \mathbb{R}$ and $(u, v) \in \mathbb{R} \times \mathbb{R}$.

The tensor product $\mathbb{R} \bigotimes \mathbb{R}$ also consists of pairs (u, v), where $u, v \in \mathbb{R}$ are denoted as $u \otimes v$. Scalar multiplication in the tensor space is defined as

$$a(u \otimes v) = (au) \otimes v = u \otimes (av)$$

for $a, u, v \in \mathbb{R}$.

Addition of two pairs $u_1 \otimes v_1$ and $u_2 \otimes v_2$ in $\mathbb{R} \bigotimes \mathbb{R}$ is denoted as $u_1 \otimes v_1 + u_2 \otimes v_2$ and these pairs interact only when one of their components are the same, that is,

$$u_1 \otimes v + u_2 \otimes v = (u_1 + u_2) \otimes v,$$

$$u \otimes v_1 + u \otimes v_2 = u \otimes (v_1 + v_2).$$

In other words, addition in $\mathbb{R} \otimes \mathbb{R}$ is symbolic unless one of the components is the same in both terms. For instance, we can write

$$(\boldsymbol{u} \otimes 5\boldsymbol{v}) + (2\boldsymbol{u} \otimes \boldsymbol{v}) = 5(\boldsymbol{u} \otimes \boldsymbol{v}) + 2(\boldsymbol{u} \otimes \boldsymbol{v}) = 7(\boldsymbol{u} \otimes \boldsymbol{v}).$$

Next, we prove that the tensor product of linear spaces introduced above exists and is unique up to an isomorphism and this will be done in several steps.

Theorem 14.1. *Let V and W be two finite-dimensional linear spaces. If $V \otimes W$ exists, then the linear spaces $\mathsf{Hom}(V \otimes W, U)$ and $\mathfrak{M}(V \times W, U)$ are isomorphic.*

Proof. Let $h : V \otimes W \longrightarrow U$ be a linear mapping in $\mathsf{Hom}(V \otimes W, U)$. Then $h\otimes : V \times W \longrightarrow U$ is bilinear.

Conversely, if $f \in \mathfrak{M}(V \times W, U)$, there is a unique linear map $g : V \otimes W \longrightarrow U$ such that $g(\boldsymbol{v} \otimes \boldsymbol{w}) = f(\boldsymbol{v}, \boldsymbol{w})$ for all $(\boldsymbol{v}, \boldsymbol{w}) \in V \times W$. □

Theorem 14.2. *If the tensor products $V \otimes_1 W$ and $V \otimes_2 W$ of two linear spaces V and W exist, then they are isomorphic.*

Proof. Suppose that $V \otimes_1 W$ and $V \otimes_2 W$ are both tensor products of the linear spaces V and W. By the universal property of the tensor product of linear spaces, there exist unique linear maps $g_1 : V \otimes_2 W \longrightarrow V \otimes_1 W$ and $g_2 : V \otimes_1 W \longrightarrow V \otimes_2 W$ that make the diagrams

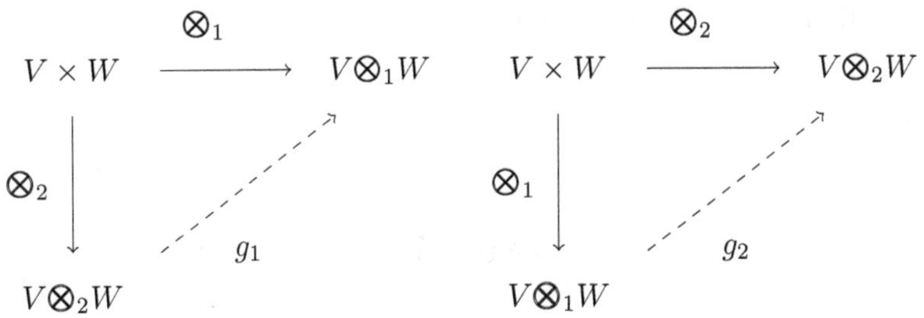

commutative. Define the linear mappings $h_1 = g_1 g_2 : V \otimes_1 W \longrightarrow V \otimes V_1$ and $h_2 = g_2 g_1 : V \otimes_2 W \longrightarrow V \otimes V_2$ which make the diagrams

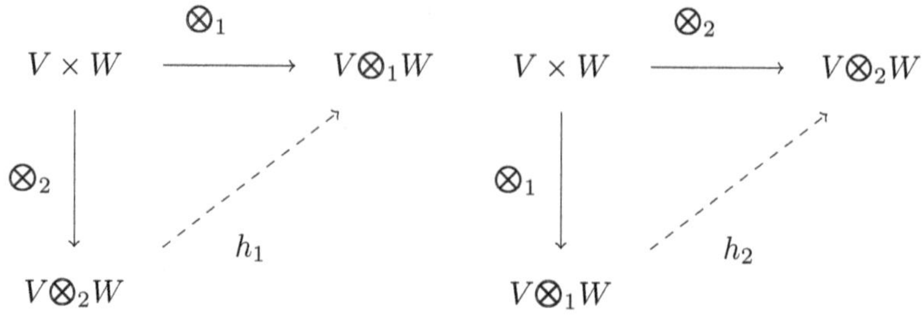

commute. The identity maps $i_1 : V \otimes_1 W \longrightarrow V \otimes_1 W$ and $i_2 : V \otimes_2 W \longrightarrow V \otimes_2 W$ are linear and also make the diagrams commute. By the uniqueness hypothesis, we have $h_1 = i_1$ and $h_2 = i_2$, so g_1 and g_2 are inverses to each other and this proves that $V \otimes_1 W$ and $V \otimes_2 W$ are isomorphic. □

Theorem 14.3. *Let V, W, and U be real finite-dimensional linear spaces. There exists a finite-dimensional real linear space T and a bilinear mapping $t : V \times W \longrightarrow T$ denoted by $t(\boldsymbol{u}, \boldsymbol{v}) = \boldsymbol{u} \otimes \boldsymbol{v}$ satisfying the following properties:*

(i) *for every bilinear mapping $f : V \times W \longrightarrow U$, there exists a unique linear mapping $g : T \longrightarrow U$ such that*

$$f(\boldsymbol{v}, \boldsymbol{w}) = g(t(\boldsymbol{v}, \boldsymbol{w}));$$

(ii) *if $\{\boldsymbol{v}_1, \ldots, \boldsymbol{v}_m\}$ is a basis in V and $\{\boldsymbol{w}_1, \ldots, \boldsymbol{w}_n\}$ is a basis in W, then $\{(\boldsymbol{v}_i \otimes \boldsymbol{w}_j) \mid 1 \leqslant i \leqslant m, 1 \leqslant j \leqslant n\}$ is a basis in T, hence* $\dim(T) = \dim(V)\dim(W)$.

Proof. In this proof, we will use the summation convention. For each pair $(i, j) \in \{1, \ldots, m\} \times \{1, \ldots, n\}$, let t_{ij} be a symbol. Define T to be the real linear space that consists of all formal linear combinations with real coefficients $a^{ij} t_{ij}$ of the symbols t_{ij}.

Define the bilinear mapping $t : V \times W \longrightarrow T$ by $t(\boldsymbol{v}_i, \boldsymbol{w}_j) = t_{ij}$ and denote $t(\boldsymbol{v}_i, \boldsymbol{w}_j)$ as $\boldsymbol{v}_i \otimes \boldsymbol{w}_j$. This mapping is extended to $V \times W$ as a bilinear mapping. Thus, if $\boldsymbol{v} = a^i \boldsymbol{v}_i$ and $\boldsymbol{w} = b^j \boldsymbol{w}_j$, we define $\boldsymbol{v} \otimes \boldsymbol{w}$ as being the element of T given by $\boldsymbol{v} \otimes \boldsymbol{w} = t(\boldsymbol{v}, \boldsymbol{w}) = a^i b^j t_{ij}$.

Suppose now that $f : V \times W \longrightarrow U$ is an arbitrary bilinear map. Since every element of T is a linear combination of t_{ij}, we can define a unique linear transformation $g : T \longrightarrow U$ as $g(t_{ij}) = f(\boldsymbol{v}_i, \boldsymbol{w}_j)$.

The bilinearity of f and the linearity of g imply

$$f(\boldsymbol{v}, \boldsymbol{w}) = f\left(a^i \boldsymbol{v}_j, b^j \boldsymbol{w}_j\right) = a^i b^j f(\boldsymbol{v}_i, \boldsymbol{w}_j) = a^i b^j g(t_{ij})$$
$$= g(a^i b^j t_{ij}) = g(\boldsymbol{u} \otimes \boldsymbol{v}) = g(t(u, v)).$$

This proves the existence and uniqueness of g such that $f = gt$ as claimed in the first part of the statement.

We show now that for arbitrary bases in V and W, $\{\hat{\boldsymbol{v}}_1, \ldots, \hat{\boldsymbol{v}}_m\}$ and $\{\hat{\boldsymbol{w}}_1, \ldots, \hat{\boldsymbol{w}}_n\}$, the set $\{\hat{\boldsymbol{v}}_i \otimes \hat{\boldsymbol{w}}_j \mid 1 \leqslant i \leqslant m, 1 \leqslant j \leqslant n\}$ is a basis in T.

For $\boldsymbol{v} = \hat{a}^i \hat{\boldsymbol{v}}_i$ and $\boldsymbol{w} = \hat{b}^j \hat{\boldsymbol{w}}_j$, by the bilinearity of $\otimes$, we have

$$\boldsymbol{v} \otimes \boldsymbol{w} = \hat{a}^i \hat{b}^j (\hat{\boldsymbol{v}}_i \otimes \hat{\boldsymbol{w}}_j),$$

which shows that the mn elements $\hat{\boldsymbol{v}}_i \otimes \hat{\boldsymbol{w}}_j$ span T. If these elements were linearly dependent, we would have $\dim(T) < mn$, which is a contradiction. Thus, $\{\hat{\boldsymbol{v}}_i \otimes \hat{\boldsymbol{w}}_j \mid 1 \leqslant i \leqslant m, 1 \leqslant j \leqslant n\}$ is a basis for T. □

The space T defined above is the tensor product of V and W and is denoted by $V \otimes W$.

Corollary 14.1. *Let V, W be two real linear spaces. If $\boldsymbol{v} \neq \boldsymbol{0}_V$ and $\boldsymbol{w} \neq \boldsymbol{0}_W$, then $\boldsymbol{v} \otimes \boldsymbol{w} \neq \boldsymbol{0}_{V \otimes W}$.*

Proof. Let B_V and B_W be bases of V and W that contain $\boldsymbol{v}$ and $\boldsymbol{w}$, respectively. Then $\boldsymbol{v} \otimes \boldsymbol{w}$ is a member of a basis of $V \otimes W$, hence $\boldsymbol{v} \otimes \boldsymbol{w} \neq \boldsymbol{0}_{V \otimes W}$. □

Let V, W be two real linear spaces with $\dim(V) = m$ and $\dim(W) = n$ having the bases $\{\boldsymbol{v}_i \mid 1 \leqslant i \leqslant m\}$ and $\{\boldsymbol{w}_j \mid 1 \leqslant j \leqslant n\}$, respectively. The equalities

$$\sum_{i=1}^{n} \sum_{j=1}^{m} a_{ij} (\boldsymbol{v}_i \otimes \boldsymbol{w}_j)$$
$$= \sum_{i=1}^{n} \left(\boldsymbol{v}_i \otimes \left(\sum_{j=1}^{m} a_{ij} \boldsymbol{w}_j \right) \right)$$
$$= \sum_{j=1}^{m} \left(\left(\sum_{i=1}^{n} a_{ij} \boldsymbol{v}_i \right) \otimes \boldsymbol{w}_j \right),$$

imply that an element of $V \otimes W$ can be expressed in many ways as sums of tensor products of vectors in V and W.

Theorem 14.4. *Let V, W be two real linear spaces. The linear spaces $V \otimes W$ and $W \otimes V$ are isomorphic.*

Proof. Define the bilinear mapping $h : V \otimes W \longrightarrow W \otimes V$ as $h(\boldsymbol{v}, \boldsymbol{w}) = \boldsymbol{w} \otimes \boldsymbol{v}$ for $\boldsymbol{v} \in V$ and $\boldsymbol{w} in W$. By the universal property, there exists a unique linear mapping $g : V \otimes W \longrightarrow W \otimes V$ that makes the diagram

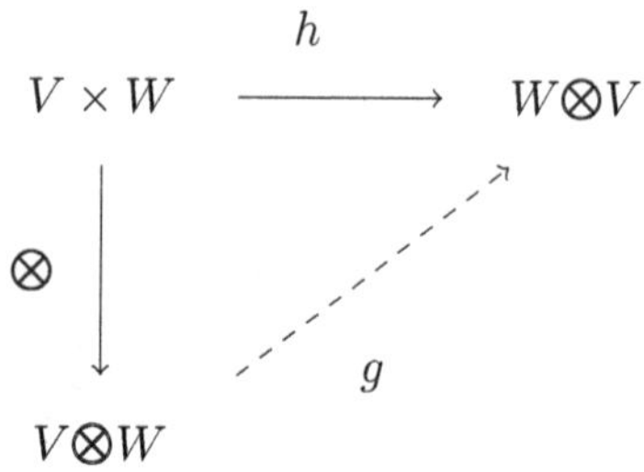

commutative. Since the set $\{\boldsymbol{w} \otimes \boldsymbol{v} \mid \boldsymbol{v} \in V, \boldsymbol{w} \in W\}$ spans the space $W \otimes V$, the function g is surjective and, therefore, it is an isomorphism. □

Theorem 14.5. *Let V, W be two real linear spaces. The linear spaces $V^* \otimes W$ and $\mathsf{Hom}(V, W)$ are isomorphic.*

Proof. Let $h : V^* \times W \longrightarrow \mathsf{Hom}(V, W)$ be defined as $h(\boldsymbol{f}, \boldsymbol{w})(\boldsymbol{v}) = \boldsymbol{f}(\boldsymbol{v})\boldsymbol{w}$ for $\boldsymbol{f} \in V^*$, $\boldsymbol{v} \in V$, and $\boldsymbol{w} \in W$. Since h is bilinear, it induces a linear mapping $g : V^* \otimes W \longrightarrow \mathsf{Hom}(V, W)$ by the universal property such that $g(\boldsymbol{f} \otimes \boldsymbol{w}) = h(\boldsymbol{f}, \boldsymbol{w})$ for $\boldsymbol{f} \in V^*$, and $\boldsymbol{w} \in W$. In other words, the diagram

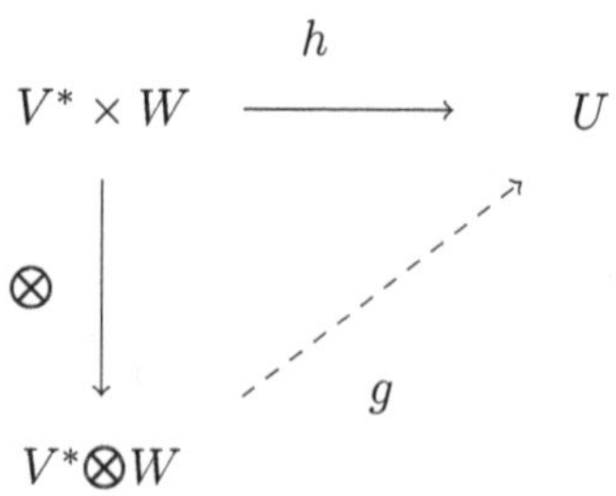

is commutative.

By Supplement 23 of Chapter 2, mappings of the form $h(\boldsymbol{f}, \boldsymbol{w})$ span $\mathsf{Hom}(V, W)$, which implies that g is an isomorphism. $\square$

Theorem 14.6. *Let V, W, X, Y be $\mathbb{R}$-linear space and let $f \in Hom(V, W)$ and $h \in \mathsf{Hom}(X, Y)$. There is a unique linear map $g = f \otimes h : V \otimes X \longrightarrow W \otimes Y$ such that*

$$(f \otimes h)(\boldsymbol{v} \otimes \boldsymbol{x}) = f(\boldsymbol{v}) \otimes h(\boldsymbol{x})$$

for $\boldsymbol{v} \in V$ and $\boldsymbol{x} \in X$.

Proof. Let $\ell : V \times X \longrightarrow W \otimes Y$ be the bilinear mapping defined by $\ell(\boldsymbol{v}, \boldsymbol{x}) = f(\boldsymbol{v}) \otimes h(\boldsymbol{x})$ for $\boldsymbol{v} \in V$ and $\boldsymbol{x} \in X$. By the universal property of the tensor product, there is a unique linear mapping $g : V \otimes X \longrightarrow W \otimes Y$ such that $g(\boldsymbol{v} \otimes \boldsymbol{x}) = \ell(\boldsymbol{v}, \boldsymbol{x}) = f(\boldsymbol{v}) \otimes h(\boldsymbol{x})$ for $\boldsymbol{v} \in V$ and $\boldsymbol{x} \in X$. Thus, g is the linear mapping $f \otimes h$. $\square$

Example 14.6. Let V, W, X, Y be four $\mathbb{R}$-linear spaces such that $\dim(V) = m$, $\dim(W) = n$, $\dim(X) = p$, $\dim(Y) = q$.

If $f \in Hom(V, W)$ and $h \in \mathsf{Hom}(X, Y)$, let $A_f \in \mathbb{R}^{n \times m}$ and $A_h \in \mathbb{R}^{q \times p}$ be the matrices associated to these linear transformations and the bases $\{\boldsymbol{v}_1, \ldots, \boldsymbol{v}_m\}$, $\{\boldsymbol{w}_1, \ldots, \boldsymbol{w}_n\}$, $\{\boldsymbol{x}_1, \ldots, \boldsymbol{x}_p\}$, and $\{\boldsymbol{y}_1, \ldots, \boldsymbol{y}_q\}$, in V, W, X, and Y, respectively.

The matrices A_f and A_h associated with the linear transformations f and h, respectively, are given by $(A_f)_{ij} = (f(\boldsymbol{v}_i))_j$ for $1 \leqslant i \leqslant m$ and $1 \leqslant j \leqslant n$, and $(A_h)_{rs} = (h(\boldsymbol{x}_r)_s$ for $1 \leqslant r \leqslant p$ and $1 \leqslant s \leqslant q$.

By Theorem 14.3, a basis in the linear space $V \otimes X$ consists of mp tensors of the form $\boldsymbol{v}_i \otimes \boldsymbol{x}_j$, while a basis in $W \otimes Y$ consists of nq tensors of the form $\boldsymbol{w}_r \otimes \boldsymbol{y}_s$.

The matrix $A_{f \otimes h} \in \mathbb{R}^{nq \times mp}$ associated with the linear transformation $g = f \otimes h : V \otimes X \longrightarrow W \otimes Y$ is given by $(A_{f \otimes h})_{(ij),(rs)} = (((f \otimes h)(\boldsymbol{v}_i \otimes \boldsymbol{x}_j))_{rs}$, which is the Kronecker product $A_f \otimes A_h$ of the matrices A_f and A_h.

The next theorem establishes the associativity of the tensor product of linear spaces.

Theorem 14.7. *Let V, W, and U be linear spaces. The tensor spaces $V \otimes (W \otimes U)$ and $(V \otimes W) \otimes U$ are isomorphic.*

Proof. For a fixed $\boldsymbol{u} \in U$, the mapping $f_{\boldsymbol{u}} : V \times W \longrightarrow V \otimes (W \otimes U)$ given by $f_{\boldsymbol{u}}(\boldsymbol{v}, \boldsymbol{w}) = \boldsymbol{v} \otimes (\boldsymbol{w} \otimes \boldsymbol{u})$ is bilinear because

$$\begin{aligned} f_{\boldsymbol{u}}(\boldsymbol{v}, \boldsymbol{w}_1 + \boldsymbol{w}_2) &= \boldsymbol{v} \otimes (\boldsymbol{w}_1 \otimes \boldsymbol{u} + \boldsymbol{w}_2 \otimes \boldsymbol{u}) \\ &= \boldsymbol{v} \otimes (\boldsymbol{w}_1 \otimes \boldsymbol{u}) + \boldsymbol{v} \otimes (\boldsymbol{w}_2 \otimes \boldsymbol{u}) \\ &= f_{\boldsymbol{u}}(\boldsymbol{v}, \boldsymbol{w}_1) + f_{\boldsymbol{u}}(\boldsymbol{v}, \boldsymbol{w}_2). \end{aligned}$$

By the universal property of tensor products, there is a unique linear mapping $f'_{\boldsymbol{u}} : V \otimes W \longrightarrow V \otimes (W \otimes U)$ such that $f_{\boldsymbol{u}} = f'_{\boldsymbol{u}} h$, where $h : V \times W \longrightarrow V \otimes W$ is the bilinear mapping $h(\boldsymbol{v}, \boldsymbol{w}) = \boldsymbol{v} \otimes \boldsymbol{w}$ for $\boldsymbol{v} \in V$ and $\boldsymbol{w} \in W$. In other words, the diagram

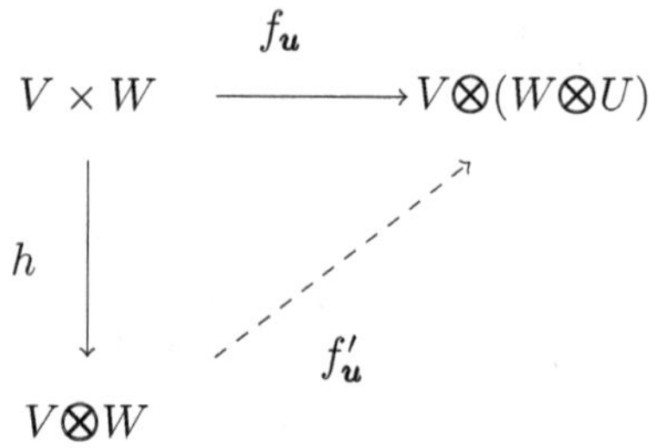

is commutative.

For $(\boldsymbol{v}, \boldsymbol{w}) \in V \times W$, we have

$$f_{\boldsymbol{u}}(\boldsymbol{v}, \boldsymbol{w}) = f'_{\boldsymbol{u}} h(\boldsymbol{v}, \boldsymbol{w}) = f'_{\boldsymbol{u}}(\boldsymbol{v} \otimes \boldsymbol{w}),$$

which implies $f'_{\boldsymbol{u}}(\boldsymbol{v}, \boldsymbol{w}) = \boldsymbol{v} \otimes (\boldsymbol{w} \otimes \boldsymbol{u})$ due to the definition of $f_{\boldsymbol{u}}$. Since $\boldsymbol{u}$ is arbitrary, there are morphisms $f'_{\boldsymbol{u}+\boldsymbol{t}}$ and $f'_{a\boldsymbol{u}}$ such that for any $(\boldsymbol{v}, \boldsymbol{w}) \in V \times W$, we have

$$\begin{aligned} f'_{\boldsymbol{u}+\boldsymbol{t}}(\boldsymbol{v}, \boldsymbol{w}) &= \boldsymbol{v} \otimes (\boldsymbol{w} \otimes (\boldsymbol{u} + \boldsymbol{t})) \\ &= \boldsymbol{v} \otimes (\boldsymbol{w} \otimes \boldsymbol{u} + \boldsymbol{w} \otimes \boldsymbol{t}) \\ &= \boldsymbol{v} \otimes (\boldsymbol{w} \otimes \boldsymbol{u}) + \boldsymbol{v} \otimes (\boldsymbol{w} \otimes \boldsymbol{t}), \end{aligned}$$

which amounts to

$$f'_{\boldsymbol{u}+\boldsymbol{t}}(\boldsymbol{v}, \boldsymbol{w}) = f_{\boldsymbol{u}}(\boldsymbol{v}, \boldsymbol{w}) + f'_{\boldsymbol{t}}(\boldsymbol{v}, \boldsymbol{w})$$

and

$$\begin{aligned} f'_{a\boldsymbol{u}}(\boldsymbol{v}, \boldsymbol{w}) &= \boldsymbol{v} \otimes (\boldsymbol{w} \otimes a\boldsymbol{u}) \\ &= \boldsymbol{v} \otimes (a(\boldsymbol{w} \otimes \boldsymbol{u})) \\ &= a(\boldsymbol{v} \otimes (\boldsymbol{w} \otimes \boldsymbol{u})), \end{aligned}$$

that is

$$f'_{a\boldsymbol{u}}(\boldsymbol{v}, \boldsymbol{w}) = a f'_{\boldsymbol{u}}(\boldsymbol{v}, \boldsymbol{w}).$$

Therefore, the function $\phi : (V \otimes W) \times U \longrightarrow V \otimes (W \otimes U)$ given by $\phi((\boldsymbol{v}, \boldsymbol{w}), \boldsymbol{u}) = f_{\boldsymbol{u}}(\boldsymbol{v}, \boldsymbol{w})$ is bilinear. By the universal property of the tensor product $(V \otimes W) \otimes U$, ϕ induces a unique morphism $\phi' : (V \otimes W) \otimes U \longrightarrow V \otimes (W \otimes U)$ such that

$$\phi'((\boldsymbol{v} \otimes \boldsymbol{w}) \otimes \boldsymbol{u}) = \boldsymbol{v} \otimes (\boldsymbol{w} \otimes \boldsymbol{u}),$$

which is reflected in the following commutative diagram:

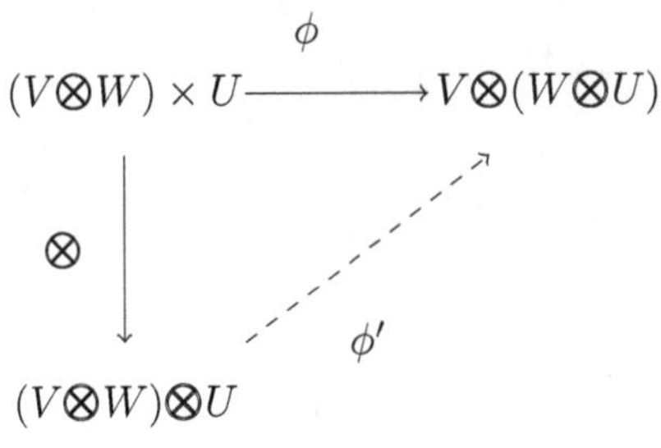

Similarly, there is a unique morphism $\psi' : V \otimes (W \otimes U) \longrightarrow (V \otimes W) \otimes U$ such that

$$\psi'(\boldsymbol{v} \otimes (\boldsymbol{w} \otimes \boldsymbol{u})) = (\boldsymbol{v} \otimes \boldsymbol{w}) \otimes \boldsymbol{u}.$$

It is immediate that $\psi'\phi'$ and $1_{((V\otimes W)\otimes U)}$ coincide on the generators of $(V \otimes W) \otimes U$, that is $\psi'\phi' = 1_{((V\otimes W)\otimes U}$. Similarly, we have $\phi'\psi' = 1_{(V\otimes(W\otimes U)}$. Thus, ϕ' and ψ' are inverse of each other, which means that they are isomorphisms. □

The associativity of the tensor product of linear spaces allows us to write $V \otimes W \otimes U$ instead of $V \otimes (W \otimes U)$ or $(V \otimes W) \otimes U$ for any linear spaces V, W, U. Also, taking into account Theorem 14.4, we could freely change the order in which we list the linear spaces in any tensor product.

Example 14.7. Let $\{\boldsymbol{e}_1, \boldsymbol{e}_2, \boldsymbol{e}_3\}$ be the standard basis for $\mathbb{R}^3$ and let $\{\boldsymbol{e}_1, \boldsymbol{e}_2\}$ be the standard basis for $\mathbb{R}^2$. A basis for $\mathbb{R}^3 \otimes \mathbb{R}^2$ consists of the tensors

$$\boldsymbol{e}_1 \otimes \boldsymbol{e}_1, \boldsymbol{e}_1 \otimes \boldsymbol{e}_2, \boldsymbol{e}_2 \otimes \boldsymbol{e}_1,$$
$$\boldsymbol{e}_2 \otimes \boldsymbol{e}_2, \boldsymbol{e}_3 \otimes \boldsymbol{e}_1, \boldsymbol{e}_3 \otimes \boldsymbol{e}_2.$$

Let U be an $\mathbb{R}$-linear space and let $f : \mathbb{R}^3 \times \mathbb{R}^2 \longrightarrow U$ be a bilinear mapping. Define the mapping $g : \mathbb{R}^3 \otimes \mathbb{R}^2 \longrightarrow U$ by $g(\boldsymbol{e}_i \otimes \boldsymbol{e}_j) = f(\boldsymbol{e}_i, \boldsymbol{e}_j)$ for $i \in \{1, 2, 3\}$ and $j \in \{1, 2\}$. Since f is a bilinear mapping, g is well-defined and $g(\boldsymbol{u} \otimes \boldsymbol{v}) = f(\boldsymbol{u}, \boldsymbol{v})$.

Theorem 14.3 can be extended to the case of m linear spaces as follows. If $V_1, \ldots, V_m$ are m real and finite-dimensional spaces, there exists a linear space $V_1 \otimes \cdots \otimes V_m$ and a mapping $\otimes : V_1 \times \cdots \times V_m \longrightarrow V_1 \otimes \cdots \otimes V_m$ such that for any linear space U and multilinear mapping $f : V_1 \times \cdots \times V_m \longrightarrow U$, there exists a unique linear mapping $g : V_1 \otimes \cdots \otimes V_m \longrightarrow U$ such that the diagram

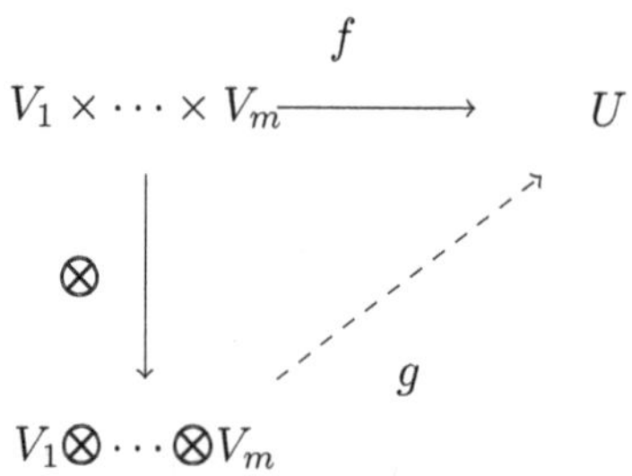

is commutative.

Recall the definition of tensors that we introduced in Chapter 2:

Definition 14.2. A *tensor* t *over the linear space* V is a multilinear function

$$\mathsf{t} : V^{\otimes p_1} \otimes V^{*\otimes q_1} \otimes \cdots \otimes V^{\otimes p_k} \otimes V^{*\otimes q_\ell} \longrightarrow \mathbb{R}.$$

In this case, we say that the pair $(p, q) = (p_1 + \cdots + p_k, q_1 + \cdots + q_\ell)$ is the *type* t while $p + q$ is the *valence* of t.

Furthermore, we say that $\mathfrak{t}$ is contravariant of order p and covariant of order q.

The valence of a tensor is the total number of arguments of a tensor regarded as a multilinear function.

Example 14.8. The tensors in $V \otimes V^* \otimes V^* \longrightarrow \mathbb{R}$ have valence 3 and type $(1,2)$. In other words, such tensors are 1-contravariant and 2-covariant. If $\dim(V) = m$, a tensor $\mathfrak{t}$ in this space can be written as $\mathfrak{t} = t^{i}{}_{j\,k}\, \boldsymbol{e}_i \otimes \boldsymbol{f}^j \otimes \boldsymbol{f}^k$ and each of the indices i, j, k varies between 1 and m. This tensor is once contravariant and twice covariant.

It is desirable to avoid writing the tensor indices directly underneath each other as in t^i_{jk} because when these indices are lowered or raised (as we will see that it is necessary sometimes), then it is difficult to put back these indices in the places they left.

Note that

$$\begin{aligned}\mathfrak{t} = t^{i_1 \cdots i_{p_1}}{}_{j_1 \cdots j_q}{}^{h_1 \cdots h_{p_r}}{}_{\ell_1 \ldots \ell_k}\, & \boldsymbol{e}_{i_1} \otimes \cdots \otimes \boldsymbol{e}_{i_{p_1}} \\ & \otimes \boldsymbol{f}^{j_1} \otimes \cdots \otimes \boldsymbol{f}^{j_{q_1}} \\ & \otimes \cdots \otimes \boldsymbol{f}^{\ell_1} \otimes \cdots \otimes \boldsymbol{f}^{\ell_{q_k}} \otimes \boldsymbol{e}_{h_1} \otimes \cdots \otimes \boldsymbol{e}_{h_{p_r}}.\end{aligned}$$

is a tensor in the product space $V^{\otimes p_1} \otimes V^{*\otimes q_1} \otimes V^{*\otimes q_k} \otimes V^{\otimes p_r}$.

Taking into account the commutativity and associativity of tensor products of linear spaces, we can reformulate the definition of tensors.

Definition 14.3. Let V be a real linear space and let $p, q \in \mathbb{N}$. A *tensor* of order (p,q) on V is a multilinear mapping $\mathfrak{t}$, where

$$\mathfrak{t} : \underbrace{V \times \cdots \times V}_{p} \times \underbrace{V^* \times \cdots \times V^*}_{q} \longrightarrow \mathbb{R}.$$

Also, we refer to a tensor $\mathfrak{t}$ defined as above as a *p-contravariant* and *q-covariant* tensor.

If $p = q = 0$, $\mathfrak{t}$ is a member of $\mathbb{R}$. If $q = 0$, $\mathfrak{t}$ is a *p-contravariant tensor*, and if $p = 0$, $\mathfrak{t}$ is a *q-covariant tensor*. If $(p,q) \neq (0,0)$, then $\mathfrak{t}$ is a *mixed tensor*.

The set of tensors of order (p,q) on V is denoted as $\mathfrak{T}^p_q(V)$. Similarly, the set of p-contravariant tensors will denoted by $\mathfrak{T}^p(V)$ and the set of q-covariant tensors will be denoted by $\mathfrak{T}_q(V)$.

If $\dim(V) = n$, then $\dim(\mathfrak{T}^p_q) = n^{p+q}$ because the dimension of a tensor product is the product of the dimensions of the factors. If $B = \{\boldsymbol{e}_1, \ldots, \boldsymbol{e}_p\}$ is a basis of V and $\{\boldsymbol{f}^1, \ldots, \boldsymbol{f}^n\}$ is its dual basis in V^*, then the set

$$\{\boldsymbol{e}_{i_1} \otimes \cdots \otimes \boldsymbol{e}_{i_p} \otimes \boldsymbol{f}^{j_1} \otimes \cdots \otimes \boldsymbol{f}^{j_q} \mid 1 \leqslant i_\ell \leqslant n, 1 \leqslant j_m \leqslant n\}$$

is a basis of $\mathfrak{T}^p_q$ called the *standard basis* corresponding to B.

If the element $\boldsymbol{e}_{i_1} \otimes \cdots \otimes \boldsymbol{e}_{i_p} \otimes \boldsymbol{f}^{j_1} \otimes \cdots \otimes \boldsymbol{f}^{j_q}$ of the standard basis is denoted by $t^{i_1 \cdots i_p}{}_{j_1 \cdots j_q}$, then a tensor $\mathfrak{z} \in \mathfrak{T}^p_q$ can be expressed uniquely as

$$\mathfrak{z} = \sum z^{j_1 \cdots j_q}{}_{i_1 \cdots i_p} t^{i_1 \cdots i_p}{}_{j_1 \cdots j_q}.$$

The n^{p+q} numbers $z^{j_1 \cdots j_q}{}_{i_1 \cdots i_p}$ are the *components* of $\mathfrak{z}$ relative to the standard basis. The upper indices of the components z correspond to V and the lower indices correspond to V^*.

Example 14.9. Let V be a real linear space and let $\boldsymbol{f}^1, \ldots, \boldsymbol{f}^n \in V^*$. Define $g_{\boldsymbol{f}^1,\ldots,\boldsymbol{f}^n} : V^n \longrightarrow \mathbb{R}$ as the multilinear function

$$g_{\boldsymbol{f}^1,\ldots,\boldsymbol{f}^n}(\boldsymbol{x}_1, \ldots, \boldsymbol{x}_n) = \boldsymbol{f}^1(\boldsymbol{x}_1) \cdots \boldsymbol{f}^n(\boldsymbol{x}_n)$$

for $\boldsymbol{x}_i \in V$ and $1 \leqslant i \leqslant n$. The multilinear function $g_{\boldsymbol{f}^1,\ldots,\boldsymbol{f}^n} : V^{\times n} \longrightarrow \mathbb{R}$ is denoted by $\boldsymbol{f}^1 \otimes \cdots \otimes \boldsymbol{f}^n$, and is an n-contravariant tensor.

Example 14.10. If $\boldsymbol{x}_i \in V_i$ for $1 \leqslant i \leqslant n$, then the multilinear function $\ell^{\boldsymbol{x}_1,\ldots,\boldsymbol{x}_n} : (V^*)^n \longrightarrow \mathbb{R}$, defined by

$$\ell^{\boldsymbol{x}_1,\ldots,\boldsymbol{x}_n}(\boldsymbol{f}^1, \ldots, \boldsymbol{f}^n) = \boldsymbol{f}^1(\boldsymbol{x}_1) \cdots \boldsymbol{f}^n(\boldsymbol{x}_n)$$

for $\boldsymbol{f}^i \in V^*$ for $1 \leqslant i \leqslant n$ belonging to $\mathcal{M}(V^*, \ldots, V^*; \mathbb{R})$, is denoted by $\boldsymbol{x}_1 \otimes \cdots \otimes \boldsymbol{x}_n$, and is a n-covariant tensor.

Let $B = \{\boldsymbol{e}_1, \ldots, \boldsymbol{e}_n\}$ be a basis in the linear space V and let $\tilde{B} = \{\boldsymbol{f}^1, \ldots, \boldsymbol{f}^n\}$ be its dual basis in V^*. Define the new bases $B' =$

$\{e_{1'},\ldots,e_{n'}\}$ and $\tilde{B}' = \{f^{1'},\ldots,f^{n'}\}$ in V and V^*, respectively, using the equalities $e_i = a_i^{j'} e_{j'}$ and $f^{i'} = a_j^{i'} f^j$ as follows:

$$e_{i'} = a_{i'}^{i} e_i, e_i = a_i^{i'} e_{i'}$$
$$f^{j'} = a_j^{j'} f^j, f^j = a_{j'}^{j} f^{j'}.$$

Of course, summation indices can be changed consistently without affecting the correctness of these formulas.

Recall that if V is a real linear space and V^* is its dual, the vectors of V are said to be contravariant, while those of V^* are said to be covariant for reasons that were discussed in Section 3.7

If $\{e_1,\ldots,e_m\}$ is a basis in the linear space V, a contravariant vector $x \in V$ can be written (with the summation convention) as $x = x^i e_i$. If $\{f^1,\ldots,f^n\}$ is a basis in V^*, a covariant vector $y \in V^*$ can be written as $y = y_j f^j$.

Let $\mathfrak{t}$ be the tensor

$$\mathfrak{t} = t^{i_1 \cdots i_h}{}_{j_1 \cdots j_k} e_{i_1} \otimes \cdots \otimes e_{i_h} \otimes f^{j_1} \otimes \cdots f^{j_k}.$$

Applying the change of bases, we have

$$\mathfrak{t} = t^{i_1 \cdots i_h}{}_{j_1 \cdots j_k} a_{i_1}^{i'_1} \cdots a_{i_h}^{i'_h} a_{j'_1}^{j_1} \cdots a_{j'_k}^{j_k} e_{i'_1} \otimes \cdots \otimes e_{i'_h} \otimes f^{j'_1} \otimes \cdots \otimes f^{j'_k}.$$

Therefore, the components $\tilde{t}^{i'_1 \cdots i'_h}{}_{j'_1 \cdots j'_k}$ of $\mathfrak{t}$ relative to the bases $\{e_{1'},\ldots,e_{n'}\}$ and $\{f^{1'},\ldots,f^{n'}\}$ are given by

$$\tilde{t}^{i'_1 \cdots i'_h}{}_{j'_1 \cdots j'_k} = t^{i_1 \cdots i_h}{}_{j_1 \cdots j_k} a_{i_1}^{i'_1} \cdots a_{i_h}^{i'_h} a_{j'_1}^{j_1} \cdots a_{j'_k}^{j_k}. \tag{14.3}$$

Similarly, we obtain the formula

$$t^{i_1 \cdots i_h}{}_{j_1 \cdots j_k} = a_{i'_1}^{i_1} \cdots a_{i'_h}^{i_h} a_{j_1}^{j'_1} \cdots a_{j_k}^{j'_k} \tilde{t}^{i'_1 \cdots i'_h}{}_{j'_1 \cdots j'_k}. \tag{14.4}$$

The operation of tensor products can be extended from vectors and covectors to tensors.

Definition 14.4. Let $\mathfrak{u} \in \mathfrak{T}^p_q(V)$ and $\mathfrak{w} \in \mathfrak{T}^r_s(V)$ be two tensors. Their product is the tensor $\mathfrak{u} \otimes \mathfrak{w} \in \mathfrak{T}^{p+r}_{q+s}(V)$ defined as

$$
\begin{aligned}
&(\mathfrak{u} \otimes \mathfrak{w})(\boldsymbol{v}^1, \ldots, \boldsymbol{v}^{p+r}, \boldsymbol{v}_1, \ldots, \boldsymbol{v}_{q+s}) \\
&= \mathfrak{u}(\boldsymbol{v}^1, \ldots, \boldsymbol{v}^p, \boldsymbol{v}_1, \ldots, \boldsymbol{v}_q)\mathfrak{w}(\boldsymbol{v}^{p+1}, \ldots, \boldsymbol{v}^{p+r}, \boldsymbol{v}_{q+1}, \ldots, \boldsymbol{v}_{q+s})
\end{aligned}
$$

for $\boldsymbol{v}^1, \ldots, \boldsymbol{v}^{p+r} \in V^*$ and $\boldsymbol{v}_1, \ldots, \boldsymbol{v}_{q+s} \in V$.

Thus, the tensor product is an operation $\otimes : \mathfrak{T}^p_q(V) \times \mathfrak{T}^r_s(V) \longrightarrow \mathfrak{T}^{p+r}_{q+s}(V)$.

The tensor product can be extended to any finite number of tensor spaces,

$$\otimes : \mathfrak{T}^{p_1}_{q_1}(V) \times \cdots \times \mathfrak{T}^{p_1+\cdots+p_k}_{q_1+\cdots+q_k}(V),$$

such that $\otimes(\mathfrak{t}_1, \ldots, \mathfrak{t}_k) = \mathfrak{t}_1 \otimes \cdots \otimes \mathfrak{t}_k$, where $\mathfrak{t}_\ell \in \mathfrak{T}^{p_\ell}_{q_\ell}$ for $1 \leqslant \ell \leqslant k$. The extended operation is multilinear, associative, and distributive.

14.4 Tensors on Inner Product Spaces

Let V be an n-dimensional real inner product space and let $B = \{\boldsymbol{e}_1, \ldots, \boldsymbol{e}_n\}$ be a basis in V. In Section 6.15, we examined the link between the contravariant components x^i of a vector $\boldsymbol{x} \in V$ relative to a basis B and the covariant components x^i of the same vector $\boldsymbol{x}$ using the fundamental matrix G of the basis B.

Let $\mathfrak{t}$ be a tensor $\mathfrak{t} = \boldsymbol{x}_1 \otimes \cdots \otimes \boldsymbol{x}_q$. If $x_r^{i_r}$ is a contravariant component of $\boldsymbol{x}_r$, then we have $t^{i_1 \cdots i_q} = x_1^{i_1} \cdots x_q^{i_q}$. Taking into account the links between contravariant and covariant components of vectors, we can write

$$
\begin{aligned}
t_{i_1}{}^{i_2 \cdots i_q} &= x_1^{i_1} x_2^{i_2} \cdots x_q^{i_q} \\
&= g_{i_1 j} x_1^j x_2^{i_2} \cdots x_q^{i_q} \\
&\quad (\text{because } x_1^{i_1} = g_{i_1 j} x_1^j) \\
&= g_{i_1 j} t^{j i_2 \cdots i_q}.
\end{aligned}
$$

Similarly, one obtains

$$t^{i_1\cdots i_q} = g^{i_1 j} t_j^{\ \ i_2\cdots i_q}.$$

Repeated applications of the previous transformations allow us to write the following:

$$t_{i_1 i_2\cdots i_q} = g_{i_1 j_1} g_{i_2 j_2}\cdots g_{i_q j_q} t^{j_1 j_2\cdots j_q},$$

$$t^{i_1 i_2\cdots i_q} = g^{i_1 j_1} g^{i_2 j_2}\cdots g^{i_q j_q} t_{j_1 j_2\cdots j_q}.$$

We can formulate now a tensoriality criterion for tensors in Euclidean spaces.

Theorem 14.8. *A collection of numbers $t_{i_1 i_2\cdots i_p}$ depending on the choice of basis forms a tensor if and only if*

$$t_{i'_1 i'_2\cdots i'_p} = g_{i'_1 i_1} g_{i'_2 i_2}\cdots g_{i'_p i_p} t_{i_1 i_2\cdots i_p},$$

where $B = \{\mathbf{e}_1, \ldots, \mathbf{e}_n\}$, $g_{ij} = (\mathbf{e}_i, \mathbf{e}_j)$ for $1 \leqslant i, j \leqslant n$.

Proof. Suppose that $t_{i_1 i_2\cdots i_p}$ are the components of a tensor $\mathfrak{t}$. Then, the numbers $t_{i_1 i_2\cdots i_p}$ are the coefficients of some multilinear form f, that is,

$$t_{i_1 i_2\cdots i_p} = f(\mathbf{e}_{i_1}, \mathbf{e}_{i_2}, \ldots, \mathbf{e}_{i_p}).$$

The coefficients $t_{i'_1 i'_2\cdots i'_p}$ in the new basis $\mathbf{e}'_1, \ldots, \mathbf{e}'_m$ are

$$t_{i'_1 i'_2\cdots i'_p} = f(\mathbf{e}'_{i_1}, \mathbf{e}'_{i_2}, \ldots, \mathbf{e}'_{i_p}).$$

Since $\mathbf{e}'_{i_1} = g_{i'_1 i_1}\mathbf{e}_{i_1}, \ldots, \mathbf{e}'_{i_m} = g_{i'_m i_m}\mathbf{e}_{i_m}$, we can write

$$\begin{aligned} t_{i'_1 i'_2\cdots i'_p} &= f(g_{i'_1 i_1}\mathbf{e}_{i_1}, g_{i'_2 i_2}\mathbf{e}_{i_2}, \ldots, g_{i'_p, i_p}\mathbf{e}_p) \\ &= g_{i'_1 i_1} g_{i'_2 i_2}\cdots g_{i'_p i_p} t_{i_1 i_2\cdots i_p}, \end{aligned}$$

because f is a multilinear form.

Conversely, suppose that $t_{i_1 i_2 \cdots i_p}$ transform as indicated when switching to a new basis. Let $\boldsymbol{x}_1, \ldots, \boldsymbol{x}_p$ be p vectors such that $\boldsymbol{x}_i = x_{ij}\boldsymbol{e}_j$ for $1 \leqslant i \leqslant p$. We must show that

$$t_{i_1 i_2 \cdots i_p} x_{i_1 j_1} \cdots x_{i_p j_p}$$

is a multilinear form in $\boldsymbol{x}_1, \ldots, \boldsymbol{x}_p$, which means that it depends only on the vectors $\boldsymbol{x}_1, \ldots, \boldsymbol{x}_p$ and not on the choice of the basis.

In a basis $B' = \{\boldsymbol{e}'_1, \ldots, \boldsymbol{e}'_n\}$, the previous expression is

$$t_{i'_1 i'_2 \cdots i'_p} x_{i'_1 j'_1} \cdots x_{i'_p j'_p}.$$

The hypothesis of the theorem allows us to write the following:

$$\begin{aligned}
&t_{i'_1 i'_2 \cdots i'_p} x_{i'_1 j'_1} \cdots x_{i'_p j'_p} \\
&= g_{i'_1 i_1} g_{i'_2 i_2} \cdots g_{i'_p i_p} t_{i_1 i_2 \cdots i_p} x_{i'_1 j'_1} \cdots x_{i'_p j'_p} \\
&= g_{i'_1 i_1} g_{i'_2 i_2} \cdots g_{i'_p i_p} t_{i_1 i_2 \cdots i_p} x_{i'_1 j'_1} \cdots x_{i'_p j'_p} \\
&= g_{i'_1 i_1} g_{i'_2 i_2} \cdots g_{i'_p i_p} t_{i_1 i_2 \cdots i_p} g_{i'_1 k_1} x_{k_1 j'_1} \cdots g_{i'_p k_p} x_{k_p j'_p} \\
&= g_{i'_1 i_1} g_{i'_1 k_1} g_{i'_2 i_2} g_{i'_2 k_2} \cdots g_{i'_p i_p} g_{i'_p k_p} t_{i_1 i_2 \cdots i_p} x_{k_1 j'_1} \cdots x_{k_p j'_p} \\
&= \delta_{i_1 k_1} \delta_{i_2 k_2} \cdots \delta_{i_p k_p} t_{i_1 i_2 \cdots i_p} x_{k_1 j'_1} \cdots x_{k_p j'_p} \\
&= t_{i_1 i_2 \cdots i_p} x_{i_1 j_1} \cdots x_{i_p j_p}
\end{aligned}$$

because

$$x_{i'_1 j'_1} = g_{i'_1 k_1} x_{k_1 j'_1}, \ldots, x_{i'_p j'_p} = g_{i'_p k_p} x_{k_p j'_p}.$$

□

14.5 Contractions

Theorem 14.9. *Let* $\mathfrak{t} = \boldsymbol{v}_1 \otimes \cdots \otimes \boldsymbol{v}_p \otimes \boldsymbol{f}^1 \otimes \cdots \otimes \boldsymbol{f}^q$ *be a simple tensor in* $\boldsymbol{T}^p_q$*, where* $p, q \geqslant 1$*. There exists a unique linear mapping* $c : \mathfrak{T}^p_q \longrightarrow \mathfrak{T}^{p-1}_{q-1}$ *(referred to as* contraction*) such that*

$$\begin{aligned}
c(\mathfrak{t}) = \boldsymbol{f}^j(\boldsymbol{v}_i)\boldsymbol{v}_1 \otimes \cdots \otimes \boldsymbol{v}_{i-1} \otimes \boldsymbol{v}_{i+1} \cdots \otimes \boldsymbol{v}_p \\
\otimes \boldsymbol{f}^1 \otimes \cdots \otimes \boldsymbol{f}_{j-1} \cdots \boldsymbol{f}^{j+1} \otimes \cdots \otimes \boldsymbol{f}^q.
\end{aligned}$$

Proof. Define the multilinear mapping $f : V^p \times (V^*)^q \longrightarrow \mathfrak{T}^{p-1}_{q-1}$ as

$$
\begin{aligned}
& f(\boldsymbol{v}_1, \ldots, \boldsymbol{v}_p, \boldsymbol{f}^1, \ldots, \boldsymbol{f}^q) \\
& = \boldsymbol{f}^j(\boldsymbol{v}_i)\boldsymbol{v}_1 \otimes \cdots \otimes \boldsymbol{v}_{i-1} \otimes \boldsymbol{v}_{i+1} \cdots \otimes \boldsymbol{v}_p \\
& \quad \otimes \boldsymbol{f}^1 \otimes \cdots \otimes \boldsymbol{f}_{j-1} \cdots \boldsymbol{f}^{j+1} \otimes \cdots \otimes \boldsymbol{f}^q.
\end{aligned}
$$

Let $c : \mathfrak{T}^p_q \longrightarrow \mathfrak{T}^{p-1}_{q-1}$ be a linear mapping defined by its values on the basis of $\mathfrak{T}^p$ as

$$
c(\boldsymbol{e}_{i_1} \otimes \cdots \otimes \boldsymbol{e}_{i_p} \otimes \boldsymbol{f}^{j_1} \otimes \boldsymbol{f}^{j_q}) = f(\boldsymbol{e}_1, \ldots, \boldsymbol{e}_p, \boldsymbol{f}^1, \ldots, \boldsymbol{f}^q),
$$

and denote its extension to a linear mapping defined on $\mathfrak{T}^p_q$ also by c.

Since $c\otimes$ and f agree on the basis of $V^p \times (V^*)^q$, the diagram

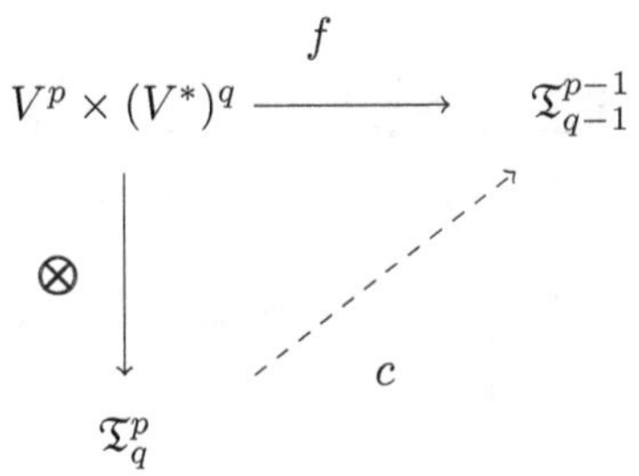

is commutative.

The universal property of the tensor product implies the uniqueness of this extension. □

Example 14.11. Let V be a real linear space and let $\boldsymbol{v} \in V$ and $\boldsymbol{f} \in V^*$. Define the bilinear mapping $\phi : V \times V^* \longrightarrow \mathbb{R}$ as $\phi(\boldsymbol{v}, \boldsymbol{f}) = \boldsymbol{f}(\boldsymbol{v})$. By the universal property of a tensor product, ϕ can be factored as $\phi = c\otimes$, where $c : \mathfrak{T}^1_1 \longrightarrow \mathbb{R} = \mathfrak{T}^0_0$ is a linear mapping. In other words, we have $\phi(\boldsymbol{v}, \boldsymbol{f}) = \boldsymbol{f}(\boldsymbol{v})\boldsymbol{v} \otimes \boldsymbol{f}$ for all $\boldsymbol{v} \in V$ and $\boldsymbol{f} \in V^*$.

14.6 Symmetric and Skew-Symmetric Tensors

Theorem 14.10. *Let ϕ be a permutation in $PERM_r$, and let V be a linear space.*

There exists a unique linear mapping $\boldsymbol{P}_\phi : V^{\otimes r} \longrightarrow V^{\otimes r}$ *such that*

$$\boldsymbol{P}_\phi(\boldsymbol{v}^1 \otimes \cdots \otimes \boldsymbol{v}^r) = \boldsymbol{v}^{\phi^{-1}(1)} \otimes \cdots \otimes \boldsymbol{v}^{\phi^{-1}(r)}.$$

The mapping $\boldsymbol{P}_\phi$ *is an isomorphism of* $V^{\otimes r}$ *and* $\boldsymbol{P}_\psi \boldsymbol{P}_\phi = \boldsymbol{P}_{\psi\phi}$ *for every* $\psi, \phi \in PERM_r$.

Proof. The mapping $f : \underbrace{V \times \cdots \times V}_{r} \longrightarrow V^{\otimes r}$ defined by

$$f(\boldsymbol{v}^1, \ldots, \boldsymbol{v}^r) = \boldsymbol{v}^{\phi^{-1}(1)} \otimes \cdots \otimes \boldsymbol{v}^{\phi^{-1}(r)}$$

is an r-multilinear mapping. By the universal property of the tensor products, there exists a unique linear mapping $\boldsymbol{P}_\phi$ that makes the diagram

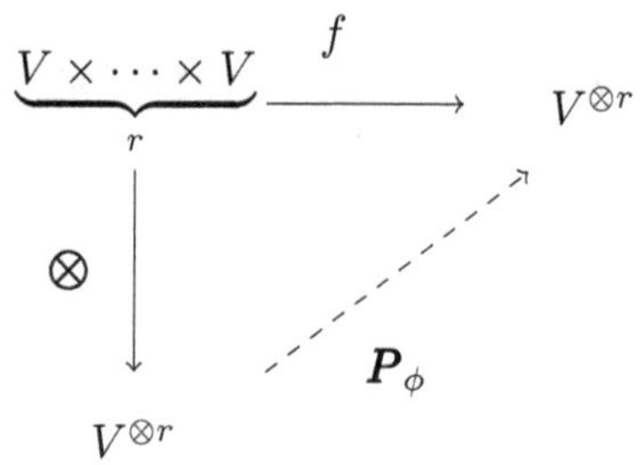

commutative, that is,

$$\boldsymbol{P}_\phi(\boldsymbol{v}^1 \otimes \cdots \otimes \boldsymbol{v}^r) = \boldsymbol{v}^{\phi^{-1}(1)} \otimes \cdots \boldsymbol{v}^{\phi^{-1}(r)}$$

for $\boldsymbol{v}^1, \ldots, \boldsymbol{v}^r$. Since $\boldsymbol{P}_\phi$ induces a permutation of the standard basis of $V^{\otimes r}$, it is immediate that $\boldsymbol{P}_\phi$ is an isomorphism.

We have

$$\begin{aligned}
&\boldsymbol{P}_\psi \boldsymbol{P}_\phi(\boldsymbol{v}^1 \otimes \cdots \otimes \boldsymbol{v}^r)\\
&= \boldsymbol{P}_\psi(\boldsymbol{v}^{\phi^{-1}(1)} \otimes \cdots \otimes \boldsymbol{v}^{\phi^{-1}(r)})\\
&= (\boldsymbol{v}^{\phi^{-1}\psi^{-1}(1)} \otimes \cdots \otimes \boldsymbol{v}^{\phi^{-1}\psi^{-1}(r)})\\
&= \boldsymbol{P}_{\psi\phi}(\boldsymbol{v}^1 \otimes \cdots \otimes \boldsymbol{v}^r).
\end{aligned}$$

□

Example 14.12. Let $\phi, \psi \in PERM_4$ be the permutations given by

$$\phi : \begin{pmatrix} 1 & 2 & 3 & 4 \\ 2 & 4 & 1 & 3 \end{pmatrix} \text{ and } \psi : \begin{pmatrix} 1 & 2 & 3 & 4 \\ 4 & 2 & 1 & 3 \end{pmatrix}.$$

This implies

$$\phi^{-1} : \begin{pmatrix} 1 & 2 & 3 & 4 \\ 3 & 1 & 4 & 2 \end{pmatrix} \text{ and } \psi^{-1} : \begin{pmatrix} 1 & 2 & 3 & 4 \\ 3 & 2 & 4 & 1 \end{pmatrix}$$

and

$$\begin{aligned} \boldsymbol{P}_\phi(\boldsymbol{v}^1 \otimes \boldsymbol{v}^2 \otimes \boldsymbol{v}^3 \otimes \boldsymbol{v}^4) &= \boldsymbol{v}^3 \otimes \boldsymbol{v}^1 \otimes \boldsymbol{v}^4 \otimes \boldsymbol{v}^2, \\ \boldsymbol{P}_\psi(\boldsymbol{v}^1 \otimes \boldsymbol{v}^2 \otimes \boldsymbol{v}^3 \otimes \boldsymbol{v}^4) &= \boldsymbol{v}^3 \otimes \boldsymbol{v}^2 \otimes \boldsymbol{v}^4 \otimes \boldsymbol{v}^1. \end{aligned}$$

Therefore,

$$\begin{aligned} \boldsymbol{P}_\psi \boldsymbol{P}_\phi(\boldsymbol{v}^1 \otimes \boldsymbol{v}^2 \otimes \boldsymbol{v}^3 \otimes \boldsymbol{v}^4) &= \boldsymbol{P}_\psi(\boldsymbol{v}^3 \otimes \boldsymbol{v}^1 \otimes \boldsymbol{v}^4 \otimes \boldsymbol{v}^2) \\ &= \boldsymbol{v}^4 \otimes \boldsymbol{v}^1 \otimes \boldsymbol{v}^2 \otimes \boldsymbol{v}^3. \end{aligned}$$

Since

$$\psi\phi : \begin{pmatrix} 1 & 2 & 3 & 4 \\ 2 & 3 & 4 & 1 \end{pmatrix}, \text{ and } (\psi\phi)^{-1} : \begin{pmatrix} 1 & 2 & 3 & 4 \\ 4 & 1 & 2 & 3 \end{pmatrix},$$

it follows that

$$\begin{aligned} \boldsymbol{P}_{\psi\phi}(\boldsymbol{v}^1 \otimes \boldsymbol{v}^2 \otimes \boldsymbol{v}^3 \otimes \boldsymbol{v}^4) &= (\boldsymbol{v}^4 \otimes \boldsymbol{v}^1 \otimes \boldsymbol{v}^2 \otimes \boldsymbol{v}^3) \\ &= \boldsymbol{P}_\psi \boldsymbol{P}_\phi(\boldsymbol{v}^1 \otimes \boldsymbol{v}^2 \otimes \boldsymbol{v}^3 \otimes \boldsymbol{v}^4). \end{aligned}$$

Definition 14.5. A tensor $\mathsf{t} \in V^{\otimes r}$ is *symmetric* if $\boldsymbol{P}_\phi(\mathsf{t}) = \mathsf{t}$ for every $\phi \in PERM_r$.

A tensor $\mathsf{t} \in V^{\otimes r}$ is *skew-symmetric* or *alternating* if $\boldsymbol{P}_\phi(\mathsf{t}) = \text{sign}(\phi)\mathsf{t}$ for every permutation $\phi \in PERM_r$.

The set of symmetric tensors in $V^{\otimes r}$ is denoted as $\mathsf{SYM}_{V,r}$. The set of skew-symmetric tensors in $V^{\otimes r}$ is denoted as $\mathsf{SKS}_{V,r}$. Both $\mathsf{SYM}_{V,r}$ and $\mathsf{SKS}_{V,r}$ are subspaces of $V^{\otimes r}$.

Definition 14.6. Let V, W be $\mathbb{F}$-linear spaces and k be a positive integer. A multilinear map $f : V^k \longrightarrow W$ is *alternating* if it vanishes whenever two arguments are equal, that is, $f(\ldots, \boldsymbol{x}, \ldots, \boldsymbol{x}, \ldots) = \mathbf{0}_W$ for every $\boldsymbol{x} \in V$.

A multilinear mapping $f : V^k \longrightarrow W$ is *skew-symmetric* if the sign of f changes when two arguments are permuted, that is,

$$f(\ldots, \boldsymbol{x}, \ldots, \boldsymbol{y}, \ldots) = -f(\ldots, \boldsymbol{y}, \ldots, \boldsymbol{x}, \ldots)$$

for $\boldsymbol{x}, \boldsymbol{y} \in V$.

The notions of alternating multilinear function and skew-symmetric function are identical, as the next statement shows.

Theorem 14.11. *Let V, W be $\mathbb{F}$-linear spaces and k be a positive integer such that $k \geqslant 2$. A multilinear mapping $f : V^k \longrightarrow W$ is alternating if and only if it is skew-symmetric.*

Proof. Let f be an alternating multilinear mapping. By the multilinearity of f, we can write

$$\begin{aligned} & f(\ldots, \boldsymbol{x}+\boldsymbol{y}, \ldots, \boldsymbol{x}+\boldsymbol{y}, \ldots) \\ & = f(\ldots, \boldsymbol{x}, \ldots, \boldsymbol{x}, \ldots) + f(\ldots, \boldsymbol{x}, \ldots, \boldsymbol{y}, \ldots) \\ & \quad + f(\ldots, \boldsymbol{y}, \ldots, \boldsymbol{x}, \ldots) + f(\ldots, \boldsymbol{y}, \ldots, \boldsymbol{y}, \ldots) \\ & = f(\ldots, \boldsymbol{x}, \ldots, \boldsymbol{y}, \ldots) + f(\ldots, \boldsymbol{y}, \ldots, \boldsymbol{x}, \ldots) = \mathbf{0}_W, \end{aligned}$$

hence f is skew-symmetric.

Conversely, if f is skew-symmetric, we have

$$f(\ldots, \boldsymbol{x}, \ldots, \boldsymbol{x}, \ldots) = -f(\ldots, \boldsymbol{x}, \ldots, \boldsymbol{x}, \ldots),$$

hence $f(\ldots, \boldsymbol{x}, \ldots, \boldsymbol{x}, \ldots) = \mathbf{0}_W$, which shows that f is alternating. □

The set of alternating multilinear mappings from V^k to W is a subspace of $\mathfrak{M}(V, \ldots, V; W)$.

Theorem 14.12. *Let V, W be $\mathbb{F}$-linear spaces. A multilinear mapping $f : V^k \longrightarrow W$ is skew-symmetric if and only if*

$$f(\boldsymbol{x}_{\phi(1)}, \ldots, \boldsymbol{x}_{\phi(k)}) = \operatorname{sign}(\phi) f(\boldsymbol{x}_1, \ldots, \boldsymbol{x}_k)$$

for every $\phi \in PERM_k$ and $\boldsymbol{x}_1, \ldots, \boldsymbol{x}_k \in V$.

Proof. Suppose that the condition of the theorem is satisfied. If ϕ is a transposition,

$$\phi : \begin{pmatrix} 1 & \cdots & i & \cdots & j \cdots & k \\ 1 & \cdots & j & \cdots & i \cdots & k \end{pmatrix},$$

the equality of the theorem amounts to

$$f(\boldsymbol{x}_1, \ldots, \boldsymbol{x}_i, \ldots, x_j, \ldots, x_k) = -f(\boldsymbol{x}_1, \ldots, \boldsymbol{x}_j, \ldots, x_i, \ldots, x_k),$$

which shows that f is skew-symmetric.

Conversely, if f is skew-symmetric, for each transposition ψ we have

$$f(\boldsymbol{x}_{\psi(1)}, \ldots, \boldsymbol{x}_{\psi(k)}) = -f(\boldsymbol{x}_1, \ldots, \boldsymbol{x}_k).$$

If ϕ is a product of transpositions $\phi = \psi_1 \cdots \psi_r$, then the k-tuple $(\boldsymbol{x}_{\phi(1)}, \ldots, \boldsymbol{x}_{\phi(k)})$ can be obtained from the k-tuple $(\boldsymbol{x}_1, \ldots, \boldsymbol{x}_k)$ by applying successively the transpositions $\psi_1, \ldots \psi_r$. This implies

$$\begin{aligned} f(\boldsymbol{x}_{\phi(1)}, \ldots, \boldsymbol{x}_{\phi(k)}) &= \text{sign}(\psi_1) \cdots \text{sign}(\psi_r) f(\boldsymbol{x}_1, \ldots, \boldsymbol{x}_k) \\ &= \text{sign}(\phi) f(\boldsymbol{x}_1, \ldots, \boldsymbol{x}_k). \end{aligned}$$

□

Theorem 14.13. *If $\boldsymbol{v}^1 \otimes \cdots \otimes \boldsymbol{v}^r \in \mathsf{SKS}_{V,r}$, then $\boldsymbol{v}^1 \otimes \cdots \otimes \boldsymbol{v}^r = \mathbf{0}_{V,r}$ if any two of the vectors $\boldsymbol{v}^1, \ldots, \boldsymbol{v}^r$ are equal.*

Proof. For the transposition

$$\phi : \begin{pmatrix} 1 & \cdots & i & \cdots & j & \cdots & r \\ 1 & \cdots & j & \cdots & i & \cdots & r. \end{pmatrix},$$

we have $inv(\phi) = 1$. Suppose that the ith argument and the jth argument of $\boldsymbol{t}$ are both equal to $\boldsymbol{v}$. Then, by the skew-symmetry of t, we have

$$\boldsymbol{v}^1 \otimes \cdots \otimes \boldsymbol{v} \otimes \cdots \otimes \boldsymbol{v}, \cdots, \boldsymbol{v}^r) = -\boldsymbol{v}^1 \otimes \cdots \otimes \boldsymbol{v} \otimes \cdots \otimes \boldsymbol{v} \otimes \cdots \otimes \boldsymbol{v}^r,$$

hence $\boldsymbol{v}^1 \otimes \cdots \otimes \boldsymbol{v} \otimes \cdots \otimes \boldsymbol{v} \otimes \cdots \otimes \boldsymbol{v}^r = \mathbf{0}_{V,r}$. □

Corollary 14.2. *If $\{\boldsymbol{v}^1, \ldots, \boldsymbol{v}^r\}$ is a linearly dependent set of vectors in $\mathsf{SKS}_{V,r}$, then $\boldsymbol{v}^1 \otimes \cdots \otimes \boldsymbol{v}^r = \mathbf{0}_{V,r}$.*

Proof. Since $\{\boldsymbol{v}^1, \ldots, \boldsymbol{v}^r\}$ is a linearly dependent set, there exists a vector $\boldsymbol{v}^i$ that can be expressed as a linear combination of the other vectors. Without loss of generality, we may assume that $\boldsymbol{v}^1 = a_2\boldsymbol{v}^2 + \cdots + a_r\boldsymbol{v}^r$. This implies

$$\begin{aligned}\boldsymbol{v}^1 \otimes \cdots \otimes \boldsymbol{v}^r &= (a_2\boldsymbol{v}^2 + \cdots + a_r\boldsymbol{v}^r) \otimes \boldsymbol{v}^2 \otimes \cdots \otimes \boldsymbol{v}^r \\ &= \sum_{i=2}^{r} a_i(\boldsymbol{v}^i \otimes \boldsymbol{v}^2 \otimes \cdots \otimes \boldsymbol{v}^r) = \boldsymbol{0}_{V,r}.\end{aligned}$$

□

Corollary 14.3. *Let V be a real linear space with $\dim(V) = n$. If $r > n$, then $V^{\otimes r} = \boldsymbol{0}_{V,r}$.*

Proof. Since every subset of V that contains more than $n = \dim(V)$ vectors is linearly dependent, the statement follows from Corollary 14.2. □

Define the linear mappings $\boldsymbol{S}_{V,r} : V^{\otimes r} \longrightarrow V^{\otimes r}$ and $\boldsymbol{A}_{V,r} : V^{\otimes r} \longrightarrow V^{\otimes r}$ as

$$\boldsymbol{S}_{V,r}(\mathrm{t}) = \frac{1}{r!} \sum_{\phi \in PERM_r} \boldsymbol{P}_\phi(\mathrm{t}), \tag{14.5}$$

$$\boldsymbol{A}_{V,r}(\mathrm{t}) = \frac{1}{r!} \sum_{\phi \in PERM_r} \mathrm{sign}(\phi)\boldsymbol{P}_\phi(\mathrm{t}). \tag{14.6}$$

These mappings are referred to as the *symmetrizer* and the *alternator* on $V^{\otimes r}$ for reasons that will become apparent immediately. If V is clear from the context, these mappings will be denoted as $\boldsymbol{S}_r$ and $\boldsymbol{A}_r$, respectively.

Theorem 14.14. *For every permutation $\psi \in PERM_r$ we have $\boldsymbol{P}_\psi \boldsymbol{S}_r = \boldsymbol{S}_r \boldsymbol{P}_\psi = \boldsymbol{S}_r$ and $\boldsymbol{P}_\psi \boldsymbol{A}_r = \boldsymbol{A}_r \boldsymbol{P}_\psi = \mathrm{sign}(\psi)\boldsymbol{A}_r$.*

Proof. We have

$$\begin{aligned}\boldsymbol{P}_\psi \boldsymbol{S}_r(\mathrm{t}) &= \frac{1}{r!}\boldsymbol{P}_\psi \left(\sum_{\phi \in PERM_r} \boldsymbol{P}_\phi(\mathrm{t}) \right) \\ &= \frac{1}{r!} \left(\sum_{\phi \in PERM_r} \boldsymbol{P}_\psi \boldsymbol{P}_\phi(\mathrm{t}) \right)\end{aligned}$$

$$= \frac{1}{r!}\left(\sum_{\phi\in PERM_r} \boldsymbol{P}_{\psi\phi}(\mathrm{t})\right).$$

As ϕ runs through $PERM_r$, the same holds for $\tau = \psi\phi$, hence,

$$\boldsymbol{P}_{\psi}\boldsymbol{S}_r(\mathrm{t}) = \frac{1}{r!}\sum_{\tau\in PERM_r} \boldsymbol{P}_{\tau}(\mathrm{t}) = \boldsymbol{S}_{\mathbf{r}}(\mathrm{t}).$$

For the second equality of the theorem, we have

$$\begin{aligned}
\boldsymbol{P}_{\psi}\boldsymbol{A}_r(\boldsymbol{t}) &= \frac{1}{r!}\boldsymbol{P}_{\psi}\left(\sum_{\phi\in PERM_r} \mathrm{sign}(\phi)\boldsymbol{P}_{\phi}(\mathrm{t})\right)\\
&= \frac{1}{r!}\sum_{\phi\in PERM_r} \mathrm{sign}(\phi)\boldsymbol{P}_{\psi\phi}(\mathrm{t})\\
&= \frac{1}{r!}\sum_{\phi\in PERM_r} \mathrm{sign}(\phi)\boldsymbol{P}_{\psi\phi}(\mathrm{t})\\
&= \frac{1}{r!}\sum_{\phi\in PERM_r} \mathrm{sign}(\psi)\mathrm{sign}(\psi\phi)\boldsymbol{P}_{\psi\phi}(\mathrm{t})\\
&= \mathrm{sign}(\psi)\frac{1}{r!}\sum_{\tau\in PERM_r} \mathrm{sign}(\tau)\boldsymbol{P}_{\tau}(\mathrm{t})
\end{aligned}$$

because

$$\mathrm{sign}(\phi) = \mathrm{sign}(\psi)\mathrm{sign}(\psi\phi)$$

and $\tau = \psi\phi$ runs through $PERM_r$.

The remaining equalities, $\boldsymbol{S}_r\boldsymbol{P}_{\psi} = \boldsymbol{S}_r$ and $\boldsymbol{A}_r\boldsymbol{P}_{\psi} = \mathrm{sign}(\psi)\boldsymbol{A}_r$, have similar arguments. □

Theorem 14.15. *If* $\mathrm{t} \in \mathsf{SKS}_{V,r}$, *then* $\boldsymbol{A}_r(\mathrm{t}) = \mathrm{t}$.

Proof. Since $\mathsf{t} \in \mathsf{SKS}_{V,r}$, we have $\boldsymbol{P}_\phi(\mathsf{t}) = \text{sign}(\phi)\mathsf{t}$ for every $\phi \in PERM_r$. This allows us to write

$$\begin{aligned}\boldsymbol{A}_r(\mathsf{t}) &= \frac{1}{r!} \sum_{\phi\in PERM_r} \text{sign}(\phi)\boldsymbol{P}_\phi(\mathsf{t}) \\ &= \frac{1}{r!} \sum_{\phi\in PERM_r} \text{sign}(\phi)^2\mathsf{t} = \mathsf{t}\end{aligned}$$

because the last sum contains $r!$ terms and $\text{sign}(\phi)^2 = 1$. □

Theorem 14.16. *$\boldsymbol{A}_r(\mathsf{t})$ is a skew-symmetric tensor for every tensor $\mathsf{t} \in V^{\otimes r}$.*

Proof. We need to show that $\boldsymbol{P}_\psi(\boldsymbol{A}_r(\mathsf{t})) = \text{sign}(\psi)\boldsymbol{A}_r(\mathsf{t})$ for every permutation $\psi \in PERM_r$.

By the definition of $\boldsymbol{A}_r$, we have

$$\begin{aligned}\boldsymbol{P}_\psi(\boldsymbol{A}_r(\mathsf{t})) &= \frac{1}{r!} \sum_{\phi\in PERM_r} \text{sign}(\phi)P_\psi P_\phi(\mathsf{t}) \\ &= \frac{1}{r!} \sum_{\phi\in PERM_r} \text{sign}(\phi)P_{\psi\phi}(\mathsf{t}) \\ &= \frac{1}{r!} \sum_{\phi\in PERM_r} \text{sign}(\phi)P_{\psi\phi}(\mathsf{t}).\end{aligned}$$

Since $\text{sign}(\psi)^2 = 1$, the last equality can be written as

$$\begin{aligned}\boldsymbol{P}_\psi(\boldsymbol{A}_r(\mathsf{t})) &= \frac{1}{r!} \sum_{\phi\in PERM_r} \text{sign}(\psi)\text{sign}(\psi\phi)\boldsymbol{P}_{\psi\phi}(\mathsf{t}) \\ &= \text{sign}(\psi)\frac{1}{r!} \sum_{\phi\in PERM_r} \text{sign}(\psi\phi)\boldsymbol{P}_{\psi\phi}(\mathsf{t}).\end{aligned}$$

By Theorem 1.4, for a fixed ψ we have the equality $\{\psi\phi \mid \phi \in PERM_r\} = PERM_r$, which allows us to write

$$\boldsymbol{P}_\psi(\boldsymbol{A}_r(\mathsf{t})) = \text{sign}(\psi)\frac{1}{r!} \sum_{\phi\in PERM_r} \text{sign}(\phi)\boldsymbol{P}_\phi(\mathsf{t}) = \text{sign}(\psi)\boldsymbol{A}_r(\mathsf{t}).$$

□

Theorem 14.17. *The mappings $\boldsymbol{S}_{V,r}$ and $\boldsymbol{A}_{V,r}$ defined on $V^{\otimes r}$ are projections of $V^{\otimes r}$ onto the subspaces $\mathsf{SYM}(V,r)$ and $\mathsf{SKS}_{V,r}$ of $V^{\otimes r}$, respectively.*

Proof. We begin by proving that both $\boldsymbol{S}_{V,r}$ and $\boldsymbol{A}_{V,r}$ are idempotent. We have

$$\begin{aligned}\boldsymbol{S}_{V,r}(\boldsymbol{S}_{V,r}(\mathrm{t})) &= \frac{1}{r!}\sum_{\phi\in PERM_r} \boldsymbol{P}_\phi \boldsymbol{S}_{V,r}(\mathrm{t}) \\ &= \frac{1}{r!}\sum_{\phi\in PERM_r} \boldsymbol{S}_{V,r}(\mathrm{t}) \\ &\quad \text{(by Theorem 14.14)} \\ &= \boldsymbol{S}_{V,r}(\mathrm{t}) \\ &\quad \text{(because the previous sum contains } r! \text{ terms),}\end{aligned}$$

hence $\boldsymbol{S}_{V,r}\boldsymbol{S}_{V,r} = \boldsymbol{S}_{V,r}$.

A similar computation leads to the same conclusion for $\boldsymbol{A}_{V,r}$:

$$\begin{aligned}\boldsymbol{A}_{V,r}(\boldsymbol{A}_{V,r}(\mathrm{t})) &= \frac{1}{r!}\sum_{\phi\in PERM_r} \mathrm{sign}(\phi)\boldsymbol{P}_\phi \boldsymbol{A}_{V,r}(\mathrm{t}) \\ &= \frac{1}{r!}\sum_{\phi\in PERM_r} \mathrm{sign}(\phi)^2 \boldsymbol{A}_{V,r}(\mathrm{t}) \\ &\quad \text{(by Theorem 14.14)} \\ &= \boldsymbol{A}_{V,r}(\mathrm{t}) \\ &\quad \text{(because the previous sum} \\ &\quad \text{contains } r! \text{ terms and } \mathrm{sign}(\phi)^2 = 1).\end{aligned}$$

Note that $\mathrm{t} \in \mathsf{SYM}_{V,r}$ implies $\boldsymbol{P}_\phi(\mathrm{t}) = \mathrm{t}$ for all $\phi \in PERM_r$. Therefore, if $\mathrm{t} \in \mathsf{SYM}_{V,r}$, we have $\boldsymbol{S}_{V,r}(\mathrm{t}) = \frac{1}{r!}\sum_{\phi\in PERM_r} \boldsymbol{P}_\phi(\mathrm{t}) = \mathrm{t}$. Conversely, if $\boldsymbol{S}_{V,r}(\mathrm{t}) = \mathrm{t}$, we have $\boldsymbol{P}_\phi(\mathrm{t}) = \boldsymbol{P}_\phi(\boldsymbol{S}_{V,r}(\mathrm{t})) = \boldsymbol{S}_{V,r}(\mathrm{t}) = \mathrm{t}$ for all $\sigma \in PERM_r$ and $\mathrm{t} \in \mathsf{SYM}_{V,r}$.

The membership $\mathsf{t} \in \mathsf{SKS}_{V,r}$ is equivalent to $\boldsymbol{P}_\phi(\mathsf{t}) = \text{sign}(\phi)\mathsf{t}$ for every $\phi \in PERM_r$. Thus, if $\mathsf{t} \in \mathsf{SKS}_{V,r}$, we have

$$\boldsymbol{A}_{V,r}(\mathsf{t}) = \frac{1}{r!} \sum_{\phi \in PERM_r} \text{sign}(\phi)\boldsymbol{P}_\phi(\mathsf{t}) = \frac{1}{r!} \sum_{\phi \in PERM_r} \text{sign}(\phi)^2 \mathsf{t} = \mathsf{t}.$$

Conversely, if $\boldsymbol{A}_{V,r}(\mathsf{t}) = \mathsf{t}$, we have $\boldsymbol{P}_\phi(\mathsf{t}) = \boldsymbol{P}_\phi \boldsymbol{A}_{V,r}(\mathsf{t}) = \text{sign}(\phi)\boldsymbol{A}_{V,r}(\mathsf{t}) = \text{sign}(\phi)\mathsf{t}$ for all $\phi \in PERM_r$ and $\boldsymbol{t} \in \mathsf{SKS}_{V,r}$.

Furthermore, we have

$$\begin{aligned}\boldsymbol{S}_{V,r}(\boldsymbol{A}_{V,r}(\mathsf{t})) &= \frac{1}{r!} \sum_{\sigma \in PERM_r} \boldsymbol{P}_\sigma(\boldsymbol{A}_{V,r}(\mathsf{t})) \\ &= \frac{1}{r!} \left(\sum_{\sigma \in PERM_r} \text{sign}(\phi) \right) \boldsymbol{A}_{V,r}(\mathsf{t}) = \mathbf{0}_{V^{\otimes r}}\end{aligned}$$

because $\sum_{\sigma \in PERM_r} \text{sign}(\phi) = 0$. A similar computation yields $\boldsymbol{A}_{V,r}(\boldsymbol{S}_{V,r}(\mathsf{t})) = \mathbf{0}_{V^{\otimes r}}$.

Let $\mathsf{t} \in \mathsf{SYM}_{V,r} \cap \mathsf{SKS}_{V,r}$, we have $\mathsf{t} = \boldsymbol{S}_{V,r}(\mathsf{t})$, hence $\boldsymbol{A}_{V,r}(\boldsymbol{k}) = \boldsymbol{A}_{V,r}(\boldsymbol{S}_{V,r}(\boldsymbol{t})) = \mathbf{0}_{V^{\otimes r}}$, hence $\boldsymbol{A}_{V,r}\boldsymbol{S}_{V,r} = \mathbf{0}$. Similarly, $\boldsymbol{S}_{V,r}\boldsymbol{A}_{V,r} = \mathbf{0}_{V^{\otimes r}}$. □

Example 14.13. Let V be a linear space and let t be a tensor in V^2. We have

$$PERM_2 = \left\{ \phi_0 : \begin{pmatrix} 1 & 2 \\ 1 & 2 \end{pmatrix}, \phi_1 : \begin{pmatrix} 1 & 2 \\ 2 & 1 \end{pmatrix} \right\}$$

and, therefore, we can write

$$\boldsymbol{A}_{V,2}(\boldsymbol{t}) = \frac{1}{2}\left(\boldsymbol{P}_{\phi_0}(\mathsf{t}) - \boldsymbol{P}_{\phi_1}(\mathsf{t})\right).$$

Thus, if $\boldsymbol{t}$ is a simple tensor $\mathsf{t} = \boldsymbol{u} \otimes \boldsymbol{w} \in V^{*2}$, we have

$$\boldsymbol{A}_{V,2} = \frac{1}{2}(\boldsymbol{u} \otimes \boldsymbol{w} - \boldsymbol{w} \otimes \boldsymbol{u}).$$

Let V be a linear space with $\dim(V) = n$. For $\boldsymbol{e}_{i_1}, \ldots, \boldsymbol{e}_{i_k} \in V$ with $i_1, \ldots, i_k \in \{1, \ldots, n\}$, following [30], we denote by $\boldsymbol{e}_{i_1} \cdots \boldsymbol{e}_{i_k}$

the tensor

$$\boldsymbol{S}_{V,k}(\boldsymbol{e}_{i_1} \otimes \cdots \otimes \boldsymbol{e}_{i_k}) = \frac{1}{k!} \sum_{\phi \in PERM_k} \boldsymbol{e}_{i_{\phi(1)}} \otimes \cdots \otimes \boldsymbol{e}_{i_{\phi(k)}}.$$

The factor $\boldsymbol{e}_{i_1} \cdots \boldsymbol{e}_{i_k}$ depends only on the number of times each $\boldsymbol{e}_i$ enters this product, and we may write

$$\boldsymbol{e}_{i_1} \cdots \boldsymbol{e}_{i_k} = \boldsymbol{e}_1^{p_1} \cdots \boldsymbol{e}_n^{p_n},$$

where p_i is the multiplicity of occurrence of $\boldsymbol{e}_i$ in $\boldsymbol{e}_{i_1} \cdots \boldsymbol{e}_{i_k}$ (which may also be 0). Thus, the numbers $p_1, \ldots, p_n$ are non-negative integers and $p_1 + \cdots + p_n = k$.

Theorem 14.18. *Let $\{\boldsymbol{e}_1, \ldots, \boldsymbol{e}_n\}$ be a basis of a linear space V. Then*

$$\{\boldsymbol{S}_{V,k}(\boldsymbol{e}_{i_1} \otimes \cdots \otimes \boldsymbol{e}_{i_k}) \mid 1 \leqslant i_1 \leqslant \cdots \leqslant i_k \leqslant n\}$$

is a basis of $S_{V,k}$. Furthermore, $\dim(S_{V,k}) = \binom{n+k-1}{k}$.

Proof. Since $B = \{\boldsymbol{e}_{i_1} \otimes \cdots \otimes \boldsymbol{e}_{i_k} \mid 1 \leqslant i_1 \leqslant n, \ldots, 1 \leqslant i_k \leqslant n\}$ is a basis for $\mathfrak{T}^k(V)$ and $\boldsymbol{S}_{V^n,k}$ maps $\mathfrak{T}^k(V^n)$ into $S_{V,k}$, the set

$$\begin{aligned} S(\mathcal{B}) &= \{\boldsymbol{e}_{i_1} \cdots \boldsymbol{e}_{i_k} \mid 1 \leqslant i_1 \leqslant \cdots \leqslant i_k \leqslant n\} \\ &= \{\boldsymbol{e}_1^{p_1} \cdots \boldsymbol{e}_n^{p_n} \mid p_1 + \cdots + p_n = k\} \end{aligned}$$

spans $S_{V,k}$. Vectors in $S_{V,k}$ are linearly independent because, if $(p_1, \ldots, p_n) \neq (q_1, \ldots, q_n)$, then the tensors $\boldsymbol{e}_1^{p_1} \cdots \boldsymbol{e}_n^{p_n}$ and $\boldsymbol{e}_1^{q_1} \cdots \boldsymbol{e}_n^{q_n}$ are linear combinations of two non-intersecting subsets of the basic elements of $\mathfrak{T}^k(V)$. By Supplement 12 of Chapter 1, the cardinality of $S(\mathcal{B})$ is the number of combinations with repetition of n objects taken k at a time. □

Theorem 14.19. *Let $t_{i_1,\ldots,i_p}$ be a standard basis element of a tensor space $\mathfrak{T}^p = V^{\otimes p}$. If $i_k = i_\ell$ for some $1 \leqslant k < \ell \leqslant p$, then* $\boldsymbol{A}_{V,p}(t_{i_1,\ldots,i_p}) = \boldsymbol{0}_{V^{\otimes p}}$.

Proof. Let $\phi \in PERM_p$ be the transposition that inverts the places of k and ℓ. Since $i_k = i_\ell$, we have $\boldsymbol{P}_\phi(t_{i_1,\ldots,i_p}) = t_{i_1,\ldots,i_p}$. This implies

$$\begin{aligned} \boldsymbol{A}_{V,p}(t_{i_1,\ldots,i_p}) &= \boldsymbol{A}_{V,p}(P_\phi(t_{i_1,\ldots,i_p})) \\ &= \operatorname{sign}(\phi)\boldsymbol{A}_{V,p}(t_{i_1,\ldots,i_p}) \\ &= -\boldsymbol{A}_{V,p}(t_{i_1,\ldots,i_p}), \end{aligned}$$

hence $\boldsymbol{A}_{V,p}(t_{i_1,\ldots,i_p}) = \boldsymbol{0}_{V^{\otimes p}}$. □

Corollary 14.4. *If $p > n$, where $n = \dim(V)$, then $\dim(V^{\otimes p}) = 0$.*

Proof. This follows from Theorem 14.19. □

Theorem 14.20. *Let V be an n-dimensional linear space. If $p \leqslant n$, then $\{\boldsymbol{A}_{V,p}(t_{i_1,\ldots,i_p}) \mid i_1 < i_2 < \cdots < i_p\}$ is a basis of the linear space $\mathsf{SKS}_{V,p}$ and $\dim(\mathsf{SKS}_{V,p}) = \binom{n}{p}$.*

Proof. By Theorem 14.19, we need to consider only those $t_{i_1,\ldots,i_p}$ having all indices distinct. For $\phi, \psi PERM_p$, we have $t_{i_{\phi(1)},\ldots,i_{\phi(p)}} \neq t_{i_{\psi(1)},\ldots,i_{\psi(p)}}$ if and only if $\phi \neq \psi$. Thus, we have $\boldsymbol{A}_{V,p}(t_{i_1,\ldots,i_p}) \neq \mathbf{0}_{V^{\otimes p}}$.

If the sets $\{i_1, \ldots, i_p\}$ and $\{k_1, \ldots, k_p\}$ contain the same elements, then $\boldsymbol{A}_{V,p}(t_{i_1,\ldots,i_p}) = \pm \boldsymbol{A}_{V,p}(t_{k_1,\ldots,k_p})$.

If the sets $\{i_1, \ldots, i_p\}$ and $\{k_1, \ldots, k_p\}$ are distinct, there are no common elements of the basis when we expend $\boldsymbol{A}_{V,p}(t_{i_1,\ldots,i_p})$ and $\boldsymbol{A}_{V,p}(t_{k_1,\ldots,k_p})$ as linear combinations of the elements of the basis. Therefore, the set of elements of the form $\boldsymbol{A}_{V,p}(t_{i_1,\ldots,i_p})$ is linearly independent. □

Corollary 14.5. *We have*

$$\dim(\mathsf{SKS}_{V,n}) = 1 \text{ and } \dim(\mathsf{SKS}_{V,p}) = \dim(\mathsf{SKS}_{V,n-p}).$$

Proof. These equalities follow from Theorem 14.20. □

Let $\mathsf{t} \in V^{\otimes p}$. We have

$$\mathsf{t} - \boldsymbol{S}_{V,p}(\mathsf{t}) = \frac{1}{p!}\left(p!\mathsf{t} - \sum_{\phi \in PERM_p} \boldsymbol{P}_\phi(\mathsf{t})\right) = \frac{1}{p!}\sum_{\phi \in PERM_p} (\mathsf{t} - \boldsymbol{P}_\phi(\mathsf{t})).$$

If $\mathsf{t} \in \ker \boldsymbol{S}_{V,p}$, this implies

$$\mathsf{t} = \frac{1}{p!}\sum_{\phi \in PERM_p} (\mathsf{t} - \boldsymbol{P}_\phi(\mathsf{t})).$$

Let $\mathfrak{z} = \mathsf{t} - \boldsymbol{P}_\phi(\mathsf{t})$. Then $\boldsymbol{S}_{V,p}(\mathfrak{z}) = \boldsymbol{S}_{V,p}(\mathsf{t}) - \boldsymbol{S}_{V,p}(\mathsf{t}) = \mathbf{0}_{V^{\otimes p}}$, so $\mathfrak{z} \in \ker(\boldsymbol{S}_{V,p})$.

For $\phi \in PERM_p$ and $\mathsf{t} \in V^{\otimes p}$, let $\mathfrak{w}(\boldsymbol{t}, \phi) = \mathsf{t} - \boldsymbol{P}_\phi(\mathsf{t})$. We have

$$\boldsymbol{S}_{V,p}(\mathfrak{w}(\boldsymbol{t}, \phi)) = \boldsymbol{S}_{V,p}(\boldsymbol{t}) - \boldsymbol{S}_{V,p}(\boldsymbol{P}_\phi(\boldsymbol{t})) = \mathbf{0}_{V^{\otimes p}},$$

hence $\mathfrak{w}(\boldsymbol{t}, \phi) \in \ker(\boldsymbol{S}_{V,p})$. Thus, if W is the subspace generated by $\{\mathsf{t} - \boldsymbol{P}_\phi \mid \mathsf{t} \in V^{\otimes p} \text{ and } \phi \in PERM_p\}$, we have $\ker(\boldsymbol{S}_{V,p}) = W$.

14.7 Exterior Algebras

In this section, we introduce exterior algebra also known as Grassmann[1] algebra, a construct that makes use of a new concept known as *wedge product.*

Definition 14.7. Let V be a linear space and let $\mathfrak{t} = \boldsymbol{v}^1 \otimes \cdots \otimes \boldsymbol{v}^r$ be a simple tensor in $\mathfrak{T}_r(V)$. The *wedge product* of the vectors $\boldsymbol{v}^1, \ldots, \boldsymbol{v}^r$ is the tensor $r!\boldsymbol{A}_{V,r}(\mathfrak{t})$ denoted as $\boldsymbol{v}^1 \wedge \cdots \wedge \boldsymbol{v}^r$.

Example 14.14. The wedge product of the vectors $\boldsymbol{v}^1, \boldsymbol{v}^2 \in V$ is the tensor

$$\begin{aligned}\boldsymbol{v}^1 \wedge \boldsymbol{v}^2 &= 2!\boldsymbol{A}_{V,r}(\boldsymbol{v}^1 \otimes \boldsymbol{v}^2)\\ &= \boldsymbol{v}_1 \otimes \boldsymbol{v}_2 - \boldsymbol{v}^2 \otimes \boldsymbol{v}^1.\end{aligned}$$

The following properties are immediate:

(i) $\boldsymbol{v}^1 \wedge \boldsymbol{v}^2 = -\boldsymbol{v}^2 \wedge \boldsymbol{v}^1$,

(ii) $(a\boldsymbol{v}^1) \wedge \boldsymbol{v}^2 = a(\boldsymbol{v}^1 \wedge \boldsymbol{v}^2)$,

(iii) $(\boldsymbol{u}_1 + \boldsymbol{u}_2) \wedge \boldsymbol{v} = (\boldsymbol{u}_1 \wedge \boldsymbol{v}) + (\boldsymbol{u}_2 \wedge \boldsymbol{v})$,

for every $\boldsymbol{u}^1, \boldsymbol{u}^2, \boldsymbol{v}^1, \boldsymbol{v}^2$, and $\boldsymbol{v} \in V$.

By Property (i), we have $\boldsymbol{u} \wedge \boldsymbol{u} = \boldsymbol{0}_{V^{\otimes 2}}$, for every $\boldsymbol{u} \in L$.

A tensor of the form $\boldsymbol{u} \wedge \boldsymbol{v}$ is also referred to as a *bivector.*

In general, we have

$$\begin{aligned}\boldsymbol{v}^1 \wedge \cdots \wedge \boldsymbol{v}^r &= \sum_{\phi \in PERM_r} \mathrm{sign}(\phi)\boldsymbol{P}_\phi(\boldsymbol{v}^{\phi^{-1}(1)} \otimes \cdots \otimes \boldsymbol{v}^{\phi^{-1}(r)})\\ &= \sum_{\phi \in PERM_r} \mathrm{sign}(\phi)\boldsymbol{P}_\phi(\boldsymbol{v}^{\phi(1)} \otimes \cdots \otimes \boldsymbol{v}^{\phi(r)}),\end{aligned}$$

because $\mathrm{sign}(\phi) = \mathrm{sign}(\phi^{-1})$.

The next statement presents a universal property of the wedge product.

[1]Hermann Günter Grassmann was born in Stettin in 1809 and died in the same place in 1877. He was a mathematician and linguist and he made important contributions to an algebraic approach to geometry.

Theorem 14.21. *Let V and W be two linear spaces and let $f : \underbrace{V \times \cdots \times V}_{r} \longrightarrow W$ be a skew-symmetric multilinear function. There exists a unique linear transformation $d : \mathsf{SKS}_{V,r} \longrightarrow W$ such that the diagram*

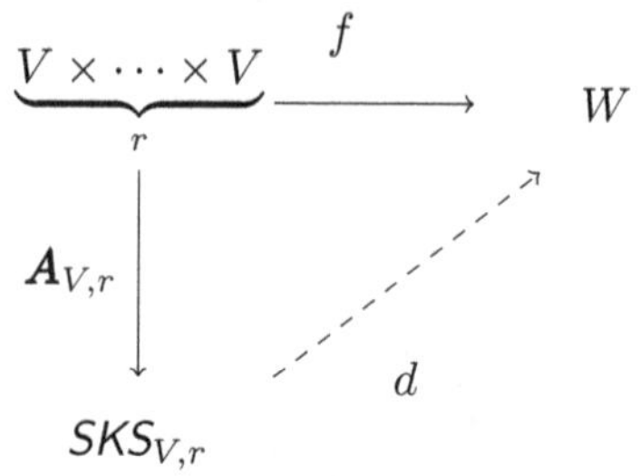

is commutative.

Proof. By the universal property of tensor products, there exists a linear mapping $c : V^{\otimes r} \longrightarrow W$ such that $f = c\otimes$. Since f is skew-symmetric, it follows that

$$\begin{aligned} c(\boldsymbol{v}^{\phi(1)} \otimes \cdots \otimes \boldsymbol{v}^{\phi(r)}) &= f(\boldsymbol{v}^{\phi(1)}, \ldots, \boldsymbol{v}^{\phi(r)}) \\ &= \operatorname{sign}(\phi) f(\boldsymbol{v}_1, \ldots, \boldsymbol{v}_k) \\ &\quad \text{(by Theorem 14.12)} \\ &= \operatorname{sign}(\phi) c(\boldsymbol{v}^1 \otimes \cdots \otimes \boldsymbol{v}^r) \end{aligned}$$

for all simple tensors $\boldsymbol{v}^1 \otimes \cdots \otimes \boldsymbol{v}^r \in V^{\otimes r}$ and all $\phi \in PERM_r$. Therefore, $c(\boldsymbol{v}^1 \otimes \cdots \otimes \boldsymbol{v}^r) = \operatorname{sign}(\phi) c(\boldsymbol{v}^{\phi(1)} \otimes \cdots \otimes \boldsymbol{v}^{\phi(r)})$. Summing up over all permutations in $PERM_r$, we obtain

$$\begin{aligned} r! c(\boldsymbol{v}^1 \otimes \cdots \otimes \boldsymbol{v}^r) &= \sum_{\phi \in PERM_r} \operatorname{sign}(\phi) c(\boldsymbol{v}^{\phi(1)} \otimes \cdots \otimes \boldsymbol{v}^{\phi(r)}) \\ &= r! f(\boldsymbol{v}^1, \ldots, \boldsymbol{v}^r). \end{aligned}$$

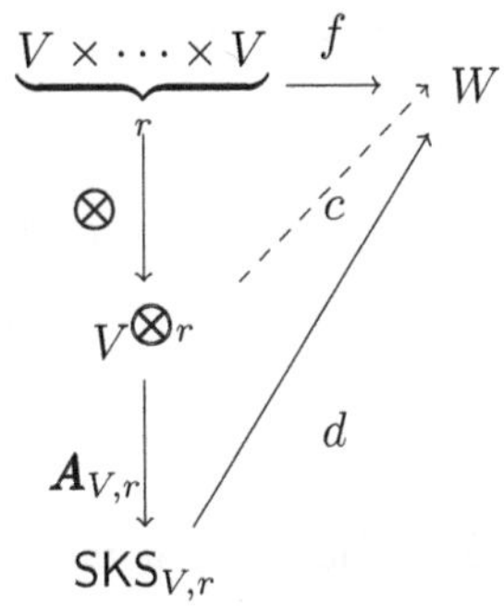

Define the mapping $d : \mathsf{SKS}_{V,r} \longrightarrow W$ as $d(\mathrm{t}) = c(\mathrm{t})$ for $\mathrm{t} \in \mathsf{SKS}_{V,r}$, that is, the restriction of c to the subspace $\mathsf{SKS}_{V,r}$, that is, $d = \frac{1}{r!} c \; 1_{\mathsf{SKS}_{V,r}}$. It is clear that d makes commutative the above diagram.

Simple tensors of the form $\boldsymbol{v}^1 \wedge \cdots \wedge \boldsymbol{v}^r \in \mathsf{SKS}_{V,r}$ generate the subspace $\mathsf{SKS}_{V,r}$ due to the fact that $\mathfrak{T}_r(V)$ is generated by tensors of the form $\boldsymbol{v}^1 \otimes \cdots \otimes \boldsymbol{v}_r$ and $\boldsymbol{A}_{V,r}$ is the projection of $\mathfrak{T}_r(V)$ on $\mathsf{SKS}_{V,r}$. Thus, d is unique because its values on $\mathsf{SKS}_{V,r}$ are uniquely determined by the condition $d(\boldsymbol{v}^1 \wedge \cdots \wedge \boldsymbol{v}^r) = f(\boldsymbol{v}_1, \ldots, \boldsymbol{v}_r)$. □

Let $\mathrm{t}_1 \in \mathsf{SKS}_{V,p_1}$ and $\mathrm{t}_2 \in \mathsf{SKS}_{V,p_2}$, the tensor $\mathrm{t}_1 \wedge \mathrm{t}_2$ is defined as

$$\mathrm{t}_1 \wedge \mathrm{t}_2 = \binom{p_1 + p_2}{p_1} \boldsymbol{A}_{V,p_1+p_2}(\mathrm{t}_1 \otimes \mathrm{t}_2). \tag{14.7}$$

Equivalently, we have

$$\begin{aligned} \mathrm{t}_1 \wedge \mathrm{t}_2 &= \binom{p_1 + p_2}{p_1} \frac{1}{(p_1+p_2)!} \sum_{\phi \in PERM_{p_1+p_2}} \operatorname{sign}(\phi) \boldsymbol{P}_\phi(\mathrm{t}_1 \otimes \mathrm{t}_2) \\ &= \frac{1}{p_1! p_2!} \sum_{\phi \in PERM_{p_1+p_2}} \operatorname{sign}(\phi) \boldsymbol{P}_\phi(\mathrm{t}_1 \otimes \mathrm{t}_2). \end{aligned}$$

Definition 14.8. The *Grassmann algebra* or the *exterior algebra* of order n of a linear space V is the algebra $\bigwedge(V)$ defined as a direct sum of the subspaces $\mathsf{SKS}_{V,p}$ for $0 \leqslant p \leqslant n$, where the product $\mathrm{t}_1 \wedge \mathrm{t}_2$ of $\mathrm{t}_1 \in \mathsf{SKS}_{V,p_1}$ and $\mathrm{t}_2 \in \mathsf{SKS}_{V,p_2}$ is defined as $\boldsymbol{A}_{V,p_1+p_2}$ and this operation is extended to $\bigwedge(V)$ by linearity.

If $\mathfrak{t}_1 = \boldsymbol{u}_1 \otimes \cdots \otimes \boldsymbol{u}_{p_1}$ and $\mathfrak{t}_2 = \boldsymbol{v}_1 \otimes \cdots \otimes \boldsymbol{v}_{p_2}$, we have

$$\mathfrak{t}_1 \otimes \mathfrak{t}_2 = \boldsymbol{u}_1 \otimes \cdots \otimes \boldsymbol{u}_{p_1} \otimes \boldsymbol{v}_1 \otimes \cdots \otimes \boldsymbol{v}_{p_2},$$
$$\mathfrak{t}_2 \otimes \mathfrak{t}_1 = \boldsymbol{v}_1 \otimes \cdots \otimes \boldsymbol{v}_{p_2} \otimes \boldsymbol{u}_1 \otimes \cdots \otimes \boldsymbol{u}_{p_1}.$$

Theorem 14.22. *Let* $\mathfrak{t} \in \mathsf{SKS}_{V,p}$ *and* $\mathfrak{s} \in \mathsf{SKS}_{V,q}$. *We have:*

$$\mathfrak{s} \wedge \mathfrak{t} = (-1)^{pq} \mathfrak{t} \wedge \mathfrak{s}.$$

Proof. Let $\theta \in PERM_{p+q}$ be the permutation introduced in Exercise 7 of Chapter 1,

$$\theta : \begin{pmatrix} 1 & 2 & \cdots & p & p+1 & \cdots & p+q \\ p+1 & p+2 & \cdots & p+q & 1 & \cdots & q \end{pmatrix}.$$

Recall that $inv(\theta) = (-1)^{pq}$.

Let $\mathfrak{t} = t^{i_1 \cdots i_p} \boldsymbol{e}_{i_1} \otimes \cdots \otimes \boldsymbol{e}_{i_p}$ and $\mathfrak{s} = s^{j_1 \cdots j_q} \boldsymbol{e}_{j_1} \otimes \cdots \otimes \boldsymbol{e}_{j_q}$. We have

$$\boldsymbol{P}_\theta(\mathfrak{t} \otimes \mathfrak{s}) = \mathfrak{s} \otimes \mathfrak{t}.$$

Therefore,

$$\mathfrak{t} \wedge \mathfrak{s} = \frac{1}{p_1! p_2!} \sum_{\phi \in PERM_{p+q}} \text{sign}(\phi) \boldsymbol{P}_\phi(\mathfrak{t} \otimes \mathfrak{s})$$
$$\mathfrak{s} \wedge \mathfrak{t} = \frac{1}{p_1! p_2!} \sum_{\phi \in PERM_{p+q}} \text{sign}(\phi) \boldsymbol{P}_\phi(\mathfrak{s} \otimes \mathfrak{t})$$
$$= \frac{1}{p_1! p_2!} \sum_{\phi \in PERM_{p+q}} \text{sign}(\phi) \boldsymbol{P}_\phi \boldsymbol{P}_\theta(\mathfrak{t} \otimes \mathfrak{s})$$
$$= \frac{1}{p_1! p_2!} \sum_{\phi \in PERM_{p+q}} \text{sign}(\phi) \text{sign}\theta^2 \boldsymbol{P}_{\phi\theta}(\mathfrak{t} \otimes \mathfrak{s})$$
$$= \text{sign}\theta \frac{1}{p_1! p_2!} \sum_{\phi \in PERM_{p+q}} \text{sign}(\phi\theta) \boldsymbol{P}_{\phi\theta}(\mathfrak{t} \otimes \mathfrak{s})$$
$$= \text{sign}\theta \frac{1}{p_1! p_2!} \sum_{\psi \in PERM_{p+q}} \text{sign}(\psi) \boldsymbol{P}_\psi(\mathfrak{t} \otimes \mathfrak{s})$$
$$= (-1)^{pq} \mathfrak{t} \wedge \mathfrak{s}.$$

□

Corollary 14.6. *If* $\mathsf{t} \in \mathsf{SKS}_{V,p}$ *and* p *is an odd number, then* $\mathsf{t} \wedge \mathsf{t} = \mathbf{0}_{V,p}$.

Proof. By Theorem 14.22, we have $\mathsf{t} \wedge \mathsf{t} = (-1)^{p^2} \mathsf{t} \wedge \mathsf{t}$. Since p is an odd number, so is p^2, hence $\mathsf{t} \wedge \mathsf{t} = \mathbf{0}_{V,p}$. □

Let $\mathsf{t}_i \in \mathsf{SKS}_{V,p_i}$ for $1 \leqslant i \leqslant n$ be n tensors. Define the tensor $\mathsf{t}_1 \wedge \mathsf{t}_2 \wedge \cdots \wedge \mathsf{t}_n$ as

$$\begin{aligned}
&\mathsf{t}_1 \wedge \mathsf{t}_2 \wedge \cdots \wedge \mathsf{t}_n \\
&= \frac{(p_1 + p_2 + \cdots + p_n)!}{p_1! p_2! \cdots p_n!} \boldsymbol{A}_{V,p_1+p_2+\cdots p_n}(\mathsf{t}_1 \otimes \mathsf{t}_2 \otimes \cdots \otimes \mathsf{t}_n).
\end{aligned}$$

The wedge product of tensors is multilinear. For simple tensors, we have the associativity property:

$$\begin{aligned}
&(\boldsymbol{v}^1 \wedge \cdots \wedge \boldsymbol{v}^p) \wedge ((\boldsymbol{u}^1 \wedge \cdots \wedge \boldsymbol{u}^q) \wedge (\boldsymbol{w}^1 \wedge \cdots \boldsymbol{w}^r)) \\
&= ((\boldsymbol{v}^1 \wedge \cdots \wedge \boldsymbol{v}^p) \wedge (\boldsymbol{u}^1 \wedge \cdots \wedge \boldsymbol{u}^q)) \wedge (\boldsymbol{w}^1 \wedge \cdots \boldsymbol{w}^r),
\end{aligned}$$

Since $\mathsf{SKS}_{V,r}$ is a subspace of $\mathfrak{T}_r$ and the simple tensors $\boldsymbol{e}^{i_1} \cdots \otimes \cdots \boldsymbol{e}^{i_r}$ form a base of $\mathfrak{T}_r$, it follows that every $\mathsf{t} \in \mathsf{SKS}_{V,r}$ can be written as $\mathsf{t} = t_{i_1 \cdots i_r} \boldsymbol{e}^{i_1} \cdots \otimes \cdots \boldsymbol{e}^{i_r}$. This implies

$$\begin{aligned}
\mathsf{t} &= \boldsymbol{A}_{V,r}(\mathsf{t}) = t_{i_1 \cdots i_r} \boldsymbol{A}_{V,r}(\boldsymbol{e}^{i_1} \otimes \cdots \boldsymbol{e}^{i_r}) \\
&= \frac{1}{r!} t_{i_1 \cdots i_r} \boldsymbol{e}^{i_1} \wedge \cdots \wedge \boldsymbol{e}^{i_r}.
\end{aligned}$$

These equalities show that the set of simple skew-symmetric tensors of the form $\boldsymbol{e}^{i_1} \wedge \cdots \wedge \boldsymbol{e}^{i_r}$ span the subspace $\mathsf{SKS}_{V,r}$. However, the set of simple skew-symmetric tensors of $\mathsf{SKS}_{V,r}$ does not form a basis of this space because it is not linearly independent. Indeed, if $\phi \in PERM_r$, we have $\boldsymbol{e}^{i_{\phi(1)}} \wedge \cdots \wedge \boldsymbol{e}^{i_{\phi(1)}} = \text{sign}(\phi) \boldsymbol{e}^{i_1} \wedge \cdots \wedge \boldsymbol{e}^{i_r}$.

To identify a basis of $\mathsf{SKS}_{V,r}$, we introduce the notation $\mathsf{t}^{[i_1 \cdots i_r]}$ for the component $\mathsf{t}^{i_1 \cdots i_r}$ that satisfies the restriction $i_1 < \cdots < i_r$; similarly, we denote by $c_{[i_1 \cdots i_r]}$ the fact that $i_1 < \cdots < i_r$.

Theorem 14.23. *If* V *is a linear space that has a* d*-element spanning subset, then* $\mathsf{SKS}_{V,k} = \{\mathbf{0}_{V,k}\}$ *for* $k > d$.

Proof. Indeed, suppose that $\{\boldsymbol{x}_1, \ldots, \boldsymbol{x}_d\}$ spans V. When $k > d$ and $\boldsymbol{x}_{i_1} \wedge \cdots \wedge \boldsymbol{x}_{i_k}$ contains two equal factors, so it is zero. Thus, $\mathsf{SKS}_{V,k}$ is spanned by zero, so $\mathsf{SKS}_{V,k} = \{0_{\mathsf{SKS}_{V,k}}\}$. □

Next, we present a universal property for $\mathsf{SKS}_{V,2}$, where V is a linear space.

Theorem 14.24. *Let V be a real linear space. There exists a linear space W and an alternating multilinear mapping $f : V \times V \longrightarrow W$ such that if ψ is the exterior multiplication $\psi : V \times V \longrightarrow \mathsf{SKS}_{V,2}$, then f factors uniquely as a composition $f = g\psi$, where $g : \mathsf{SKS}_{V,2} \longrightarrow W$ is a linear mapping. In other words, there exists a unique linear mapping g that makes the diagram*

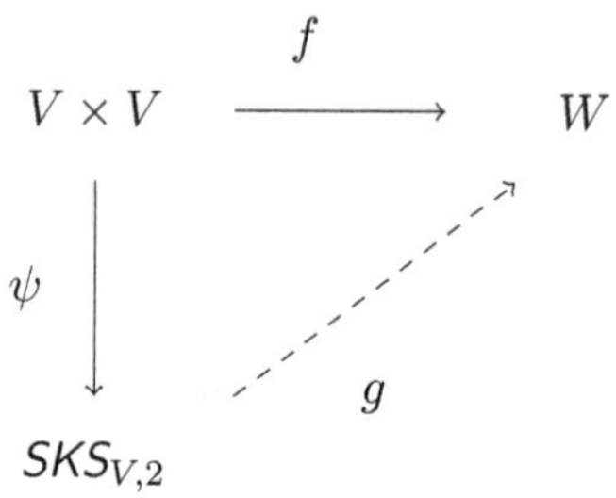

commutative.

Proof. Recall that $\mathsf{FREE}(V \times V)$ is the set of all maps over $V \times V$ with values in $\mathbb{R}$ that are 0 everywhere with the exception of a finite number of pairs in $V \times V$.

For $\boldsymbol{x}, \boldsymbol{y} \in V$, define $\lambda_{\boldsymbol{x},\boldsymbol{y}} : V \times V \longrightarrow \mathbb{R}$ as

$$\lambda_{\boldsymbol{x},\boldsymbol{y}}(\boldsymbol{r}, \boldsymbol{s}) = \begin{cases} 1 & \text{if } (\boldsymbol{x}, \boldsymbol{y}) = (\boldsymbol{r}, \boldsymbol{s}) \\ 0 & \text{if } (\boldsymbol{x}, \boldsymbol{y}) \neq (\boldsymbol{r}, \boldsymbol{s}). \end{cases}$$

The set $\{\lambda_{\boldsymbol{x},\boldsymbol{y}} \mid \boldsymbol{x}, \boldsymbol{y} \in V\}$ is a basis in the linear space $\mathcal{F}(V \times V \longrightarrow \mathbb{R})$ of real-valued functions defined on $\mathsf{FREE}(V \times V)$ which have finite supports.

Define the set of functions $\mathcal{G}$ in $\mathcal{F}(V \times V \longrightarrow \mathbb{R})$ as consisting of the functions of the form

$$\begin{aligned} &\lambda_{a\boldsymbol{x}+b\boldsymbol{y},c\boldsymbol{z}+d\boldsymbol{t}} + \tfrac{1}{2}\left(ac\lambda_{\boldsymbol{z},\boldsymbol{x}} + ad\lambda_{\boldsymbol{t},\boldsymbol{x}} + bc\lambda_{\boldsymbol{z},\boldsymbol{y}} + bd\lambda_{\boldsymbol{t},\boldsymbol{y}}\right) \\ &-\tfrac{1}{2}\left(ac\lambda_{\boldsymbol{x},\boldsymbol{z}} + ad\lambda_{\boldsymbol{x},\boldsymbol{t}} + bc\lambda_{\boldsymbol{y},\boldsymbol{z}} + bd\lambda_{\boldsymbol{y},\boldsymbol{t}}\right), \end{aligned}$$

for $\boldsymbol{x}, \boldsymbol{y}, \boldsymbol{z}, \boldsymbol{t} \in V$ and $a, b, c, d \in \mathbb{R}$. The subspace of $\mathcal{F}(V \times V \longrightarrow \mathbb{R})$ generated by $\mathcal{G}$ is denoted by $\langle \mathcal{F} \rangle$. The relation ρ in $\mathsf{FREE}(V \times V)$ consists of those pairs f, g such that $f \rho g$ if $f - g \in \langle \mathcal{G} \rangle$. The linear space W is defined as $W = \mathsf{FREE}(V \times V)/\rho$.

Let J be the canonical surjection of $\mathsf{FREE}(V \times V)$ in $\mathsf{FREE}(V \times V)/\rho$, which has $\langle \mathcal{G} \rangle$ as a kernel. Then the map $f = J\lambda$ defined on $\mathsf{FREE}(V \times V)$ is an alternate bilinear map.

Indeed, we have

$$\begin{aligned} &J(\lambda_{\boldsymbol{x},\boldsymbol{y}}) + J\left(\frac{1}{2}\lambda_{\boldsymbol{y},\boldsymbol{x}}\right) - J\left(\frac{1}{2}\lambda_{\boldsymbol{x},\boldsymbol{y}}\right) \\ &= J\left(\lambda_{\boldsymbol{x},\boldsymbol{y}} + \frac{1}{2}\lambda_{\boldsymbol{y},\boldsymbol{x}} - \frac{1}{2}\lambda_{\boldsymbol{x},\boldsymbol{y}}\right) \\ &\qquad \text{(due to the linearity of } J) \\ &= 0, \end{aligned}$$

because J is linear with kernel $\langle \mathcal{G} \rangle$. Thus, the function

$$\lambda_{\boldsymbol{x},\boldsymbol{y}} + \frac{1}{2}\lambda_{\boldsymbol{y},\boldsymbol{x}} - \frac{1}{2}\lambda_{\boldsymbol{x},\boldsymbol{y}}$$

belongs to $\mathcal{G}$ (with $(\boldsymbol{x}, \boldsymbol{y}, \boldsymbol{z}, \boldsymbol{t}) = (\boldsymbol{x}, \boldsymbol{0}, \boldsymbol{y}, \boldsymbol{0})$ and $(a, b, c, d) = (1, 0, 1, 0)$). Consequently, we have $J(\lambda_{\boldsymbol{x},\boldsymbol{y}} + \lambda_{\boldsymbol{y},\boldsymbol{x}}) = 0$, hence $f(\boldsymbol{x}, \boldsymbol{y}) = -f(\boldsymbol{y}, \boldsymbol{x})$, which shows that the function f, is skew-symmetric.

To prove the bilinearity of f observe that

$$J\left(\lambda_{\boldsymbol{x}+a\boldsymbol{y},\boldsymbol{z}} + \frac{1}{2}\left(\lambda_{\boldsymbol{z},\boldsymbol{x}} + a\lambda_{\boldsymbol{z},\boldsymbol{y}}\right) - \frac{1}{2}\left(\lambda_{\boldsymbol{x},\boldsymbol{z}} + a\lambda_{\boldsymbol{x},\boldsymbol{z}}\right)\right) = 0$$

because the argument of J has the form prescribed for the subspace $\mathsf{FREE}(V \times V)$ with $(1, a, 1, 0)$ taken for (a, b, c, d). Thus, $f(\boldsymbol{x}+a\boldsymbol{y}, \boldsymbol{z}) + f(\boldsymbol{z}, \boldsymbol{x}) + af(\boldsymbol{z}, \boldsymbol{y}) = 0$ amounts to $f(\boldsymbol{x}+a\boldsymbol{y}, \boldsymbol{z}) = -f(\boldsymbol{z}, \boldsymbol{x}) - af(\boldsymbol{z}, \boldsymbol{y})$, which implies the bilinearity of f by applying the skew-symmetry previously shown.

We show now that the space $W = \mathsf{FREE}(V \times V)/\rho$ and the mapping f satisfy the conditions of the theorem. Let ψ be a skew-symmetric map of $V \times V$ into $\mathsf{SKS}_{V,2}$.

Consider a basis that consists of elements $\lambda_{\boldsymbol{x},\boldsymbol{y}}$, where $\boldsymbol{x}, \boldsymbol{y} \in V$, and consider in $\mathsf{SKS}_{V,2}$ the family $\{\psi(\boldsymbol{x}, \boldsymbol{y}) \mid \boldsymbol{x}, \boldsymbol{y} \in V\}$. There exists a unique linear map $\Lambda : \mathsf{FREE}(V \times V) \longrightarrow W$ such that $\Lambda(\lambda_{\boldsymbol{x},\boldsymbol{y}}) = \psi(\boldsymbol{x}, \boldsymbol{y})$.

If $f, g \in \mathsf{FREE}(V \times V)$ are such that $f \rho g$, then $\Lambda(f) = \Lambda(g)$. Indeed, note that any function in $\langle \mathcal{G} \rangle$ is null for Λ. Now, we have

$$\Lambda\left(\sum\left(\lambda_{a\boldsymbol{x}+b\boldsymbol{y},c\boldsymbol{z}+d\boldsymbol{t}} + \frac{1}{2}(ac\lambda_{\boldsymbol{z},\boldsymbol{x}} + ad\lambda_{\boldsymbol{t},\boldsymbol{x}} + bc\lambda_{\boldsymbol{z},\boldsymbol{y}} + bd\lambda_{\boldsymbol{t},\boldsymbol{y}})\right.\right.$$
$$\left.\left.-\frac{1}{2}(ac\lambda_{\boldsymbol{x},\boldsymbol{z}} + ad\lambda_{\boldsymbol{x},\boldsymbol{t}} + bc\lambda_{\boldsymbol{x},\boldsymbol{z}} + bd\lambda_{\boldsymbol{y},\boldsymbol{t}})\right)\right)$$
$$= \sum \psi(a\boldsymbol{x} + b\boldsymbol{y}, c\boldsymbol{z}, d\boldsymbol{t})$$
$$+\frac{1}{2}\left(ac\psi(\boldsymbol{z}, \boldsymbol{x}) + ad\psi(\boldsymbol{t}, \boldsymbol{x}(+bc\psi(\boldsymbol{z}, \boldsymbol{y}) + bd\psi(\boldsymbol{t}, \boldsymbol{y})\right)$$
$$-\frac{1}{2}\left(ac\psi(\boldsymbol{x}, \boldsymbol{z}) + ad\psi(\boldsymbol{x}, \boldsymbol{t}) + bc\psi(\boldsymbol{y}, \boldsymbol{z}) + bd\psi(\boldsymbol{y}, \boldsymbol{t})\right),$$

which is 0 due to the bilinearity and skew symmetry of ψ.

Starting from Λ, define a unique mapping $g : \mathsf{SKS}_{V,2} \longrightarrow W$ such that for every $\boldsymbol{v} \in \mathsf{SKS}_{V,2}$ we have $g(\boldsymbol{v}) = \Lambda(\lambda_{\boldsymbol{x},\boldsymbol{y}})$, where $\lambda_{\boldsymbol{x},\boldsymbol{y}}$ is a representative of $\boldsymbol{v}$ in $\mathsf{FREE}(V \times V)$. Then, we have $g(J(\lambda_{\boldsymbol{x},\boldsymbol{y}})) = \psi(\boldsymbol{x}, \boldsymbol{y})$, which shows that $gJ = \psi$.

The map g is unique. Indeed, the family $\{J(\lambda_{\boldsymbol{x},\boldsymbol{y}}) \mid \boldsymbol{x}, \boldsymbol{y} \in L\}$ is a generating set of W since J is a surjective mapping from $\mathsf{FREE}(V \times V)$ into W and $\{\lambda_{\boldsymbol{x},\boldsymbol{y}} \mid \boldsymbol{x}, \boldsymbol{y} \in L\}$ is a basis of $\mathsf{FREE}(V \times V)$. This family contains a basis $\{J(\lambda_{\boldsymbol{x}_i,\boldsymbol{y}_i}) \mid i \in I\}$ of W. Then, we have

$$g(f(\boldsymbol{x}, \boldsymbol{y})) = \psi(\boldsymbol{x}, \boldsymbol{y})$$

for $\boldsymbol{x}, \boldsymbol{y} \in V$, which implies $g(f(\boldsymbol{x}_i, \boldsymbol{y}_i)) = \psi(\boldsymbol{x}_i, \boldsymbol{y}_i)$. Thus, there exists a unique mapping $g : \mathsf{SKS}_{V,2} \longrightarrow W$ such that $f = g\psi$, where $W = \mathsf{FREE}(V \times V)/\rho$. □

A more general result having a similar argument is

Theorem 14.25. *Let V be a real linear space. There exists a linear space W and an alternating multilinear mapping $f : \underbrace{V \times \cdots \times V}_{k} \longrightarrow W$ such that if ψ is the exterior multiplication $\psi : \mathsf{SKS}_{V,k} \longrightarrow W$, then f factors uniquely as a composition $f = g\psi$, where $g : \mathsf{SKS}_{V,k} \longrightarrow M$ is a linear map.*

Corollary 14.7. *Let V be a linear space with $\dim(V) = n$ and let $\psi : \underbrace{V \times \cdots \times V}_{k} \longrightarrow \mathsf{SKS}_{V,k}$. If $k > n$, we have $\mathsf{SKS}_{V,k} = \{\mathbf{0}_V\}$.*

Proof. This follows from the fact that if $k > \dim(V)$, every k-blade must contain two identical factors and, therefore, must equal $\mathbf{0}_V$. □

Also, observe that if $\dim(V) = n$, we have $\dim(\mathsf{SKS}_{V,r}) = 1$.

Theorem 14.26. *Let V be a linear space. We have $\boldsymbol{v}_1 \wedge \boldsymbol{v}_2 \wedge \cdots \wedge \boldsymbol{v}_k = \mathbf{0}_{V,k}$ if and only if $\{\boldsymbol{v}_1, \boldsymbol{v}_2, \ldots, \boldsymbol{v}_k\}$ is a linearly dependent set in V.*

Proof. Suppose that $\{\boldsymbol{v}_1, \boldsymbol{v}_2, \ldots, \boldsymbol{v}_k\}$ is a linearly dependent set in V. Without loss of generality, assume that $\boldsymbol{v}_1 = a_2\boldsymbol{v}_2 + \cdots + a_k\boldsymbol{v}_k$. Then,

$$\begin{aligned}
&\boldsymbol{v}_1 \wedge \boldsymbol{v}_2 \wedge \cdots \wedge \boldsymbol{v}_k\\
&= (a_2\boldsymbol{v}_2 + \cdots + a_k\boldsymbol{v}_k) \wedge \boldsymbol{v}_2 \wedge \cdots \wedge \boldsymbol{v}_k\\
&= \sum_{i=2}^{k} (a_i\boldsymbol{v}_i \wedge \boldsymbol{v}_2 \cdots \wedge \boldsymbol{v}_i \wedge \cdots \wedge \boldsymbol{v}_k) = \mathbf{0}_{V,k}
\end{aligned}$$

because each wedge in the last sum has a repeated vector.

Conversely, suppose that $\{\boldsymbol{v}_1, \boldsymbol{v}_2, \ldots, \boldsymbol{v}_k\}$ is linearly independent and let $\{\boldsymbol{v}_1, \boldsymbol{v}_2, \ldots, \boldsymbol{v}_k, \ldots, \boldsymbol{v}_n\}$ be its extension to a basis of V. As we saw above, the collection $\{\boldsymbol{v}_{i_1} \wedge \cdots \wedge \boldsymbol{v}_{i_k} \mid 1 \leqslant i_1 < \cdots < i_k \leqslant n\}$ is a basis of $\mathsf{SKS}_{V,k}$. Since $\boldsymbol{v}_1 \wedge \cdots \wedge \boldsymbol{v}_k$ belongs to this basis, it cannot be $\mathbf{0}_{V,k}$. □

Theorem 14.27. *Let V be a linear space and let $\boldsymbol{v}_1 \wedge \cdots \wedge \boldsymbol{v}_k$ and $\boldsymbol{w}_1 \wedge \cdots \wedge \boldsymbol{w}_k$ be two wedges in $\mathsf{SKS}_{V,k}$. There exists $c \in \mathbb{R} - \{0\}$ such that*

$$\boldsymbol{v}_1 \wedge \cdots \wedge \boldsymbol{v}_k = c\, \boldsymbol{w}_1 \wedge \cdots \wedge \boldsymbol{w}_k$$

if and only if $\langle \boldsymbol{v}_1, \ldots, \boldsymbol{v}_k \rangle = \langle \boldsymbol{w}_1, \ldots, \boldsymbol{w}_k \rangle$.

Proof. Let $R = \langle \boldsymbol{v}_1, \ldots, \boldsymbol{v}_k \rangle$ and $S = \langle \boldsymbol{w}_1, \ldots, \boldsymbol{w}_k \rangle$.

If $R = S$, every $\boldsymbol{v}_i$ is a linear combination of $\{\boldsymbol{w}_1, \ldots, \boldsymbol{w}_k\}$, $\boldsymbol{v}_i = c_{i1}\boldsymbol{w}_1 + \ldots + c_{ik}\boldsymbol{w}_k$. After substituting the expressions of $\boldsymbol{v}_i$ in $\boldsymbol{v}_1 \wedge \cdots \wedge \boldsymbol{v}_k$, applying multilinearity and the alternating property, we are left with an expression of the form $c\boldsymbol{w}_1 \wedge \cdots \wedge \boldsymbol{w}_k$ for some non-zero c. Note that $c \neq 0$ because a wedge cannot be $\mathbf{0}_V$.

If $R \neq S$, then let $\ell = \dim(R \cap S) < k$. Without loss of generality, we may assume that the first elements $(\boldsymbol{u}_1, \ldots, \boldsymbol{u}_\ell)$ of both lists $(\boldsymbol{v}_1, \ldots, \boldsymbol{v}_k)$ and $(\boldsymbol{w}_1, \ldots, \boldsymbol{w}_k)$ form a basis for $R \cap S$. Thus, the previous lists can be written as $(\boldsymbol{u}_1, \ldots, \boldsymbol{u}_\ell, \boldsymbol{v}_1, \ldots, \boldsymbol{v}_{k-\ell})$

and $(\boldsymbol{u}_1, \ldots, \boldsymbol{u}_\ell, \boldsymbol{w}_1, \ldots, \boldsymbol{w}_{k-\ell})$. Then, the members of the list $(\boldsymbol{u}_1, \ldots, \boldsymbol{u}_\ell, \boldsymbol{v}_1, \ldots, \boldsymbol{v}_{k-\ell}, \boldsymbol{w}_1, \ldots, \boldsymbol{w}_{k-\ell})$ form a linearly independent set, which can be extended to a basis of V. This implies that the vectors $\boldsymbol{u}_1 \wedge \cdots \wedge \boldsymbol{u}_\ell \wedge \boldsymbol{v}_1 \wedge \cdots \wedge \boldsymbol{v}_{k-\ell}$ and $\boldsymbol{u}_1 \wedge \cdots \wedge \boldsymbol{u}_\ell \wedge \boldsymbol{w}_1 \wedge \cdots \wedge \boldsymbol{w}_{k-\ell}$ belong to the same basis, which means that they cannot differ by a constant multiple. □

Lemma 14.1. *Let V be an n-dimensional linear space and let $B = \{\boldsymbol{e}_1, \ldots, \boldsymbol{e}_n\}$ be a basis for V. The set of wedge products $\boldsymbol{e}_{i_1} \wedge \cdots \wedge \boldsymbol{e}_{i_k}$ with $1 \leqslant i_1 < \cdots < i_k \leqslant n$ spans $\mathsf{SKS}_{V,k}$.*

Proof. We start from the fact that $\mathsf{SKS}_{V,k}$ is spanned by k-wedges $\boldsymbol{w}_1 \wedge \cdots \wedge \boldsymbol{w}_k$. Therefore, it suffices to prove that any such wedge is in the span of B_k.

Since B is a basis for V, each $\boldsymbol{w}_i$ is a linear combination of the basis vectors $\boldsymbol{e}_1, \ldots, \boldsymbol{e}_k$. If $\boldsymbol{w}_1 \wedge \cdots \wedge \boldsymbol{w}_k$ is expanded using distributivity, each term in the resulting sum is a multiple of a k-wedge of the form $\boldsymbol{e}_{i_1} \wedge \cdots \wedge \boldsymbol{e}_{i_k}$. This suffices to conclude that $\mathsf{SKS}_{V,k}$ is generated by the n^k blades of the form $\boldsymbol{e}_{i_1} \wedge \boldsymbol{e}_{i_2} \wedge \cdots \wedge \boldsymbol{e}_{i_k}$.

Among the blades of the form $\boldsymbol{e}_{i_1} \wedge \boldsymbol{e}_{i_2} \wedge \cdots \wedge \boldsymbol{e}_{i_k}$ those which contain two equal indices are $\boldsymbol{0}_{V,k}$, and the blades can always be rearranged in the increasing order of the indices. Therefore, $\mathcal{B}_k$ spans $\mathsf{SKS}_{V,k}$. □

Theorem 14.28. *Let $B = \{\boldsymbol{e}_1, \ldots, \boldsymbol{e}_n\}$ be a basis in the linear space V. Then the set*

$$\mathcal{B}_k = \{\boldsymbol{e}_{i_1} \wedge \boldsymbol{e}_{i_2} \wedge \cdots \wedge \boldsymbol{e}_{i_k} \mid 1 \leqslant i_1 < i_3 < \cdots < i_k \leqslant n\}.\}$$

is a basis of $\mathsf{SKS}_{V,k}$.

Proof. For $k = 0$, there is nothing to prove, so we may assume that $k \geqslant 1$.

The idea is to embed $\mathsf{SKS}_{V,k}$ as a subspace of the tensor power $V^{\otimes k}$. Let $f : V^k \longrightarrow V^{\otimes k}$ be the multilinear function defined as

$$f(\boldsymbol{e}_1, \ldots, \boldsymbol{e}_k) = \sum_{\phi \in PERM_k} (-1)^{inv(\phi)} \boldsymbol{e}_{\phi(1)} \otimes \cdots \otimes \boldsymbol{e}_{\phi(k)}.$$

The multilinearity of f follows from the fact that each term of the sum contains each $\boldsymbol{e}_i$ only once. It is easy to see that f is also alternating. By the universal property of the exterior power, there is a linear map $g_{k,V} : \mathsf{SKS}_{V,k} \longrightarrow V^{\otimes k}$ such that

$$g_{k,V}(\boldsymbol{e}_1 \wedge \cdots \boldsymbol{e}_k) = \sum_{\phi \in PERM_k} (-1)^{inv(\phi)} \boldsymbol{e}_{\phi(1)} \otimes \cdots \otimes \boldsymbol{e}_{\phi(k)}.$$

Injectivity of $g_{k,V}$ is clear for $k > n$. Therefore, suppose that $1 \leqslant k \leqslant n$. We may assume $k \geqslant 2$ because the statement is immediate for $k = 1$.

Since $B = \{\boldsymbol{e}_1, \ldots, \boldsymbol{e}_n\}$ is a basis for V, the wedge products $\boldsymbol{e}_{i_1} \wedge \cdots \wedge \boldsymbol{e}_{i_k}$ with $1 \leqslant i_1 < \cdots < i_k \leqslant n$ spans $\mathsf{SKS}_{V,k}$ by Lemma 14.1. Since $V^{\otimes k}$ has a basis, $\boldsymbol{e}_{i_1} \otimes \cdots \otimes \boldsymbol{e}_{i_k}$, where $1 \leqslant i_1, \ldots, i_k \leqslant n$.

Suppose $\mathsf{t} \in \mathsf{SKS}_{V,k}$ satisfies $g_{k,V}(\mathsf{t}) = 0$, where $\mathsf{t} = \sum\{c_{i_1 \cdots i_k} \boldsymbol{e}_{i_1} \wedge \cdots \wedge \boldsymbol{e}_{i_k} | 1 \leqslant i_1 < \cdots < i_k \leqslant n\}$, where $c_{i_1 \cdots i_k} \in \mathbb{R}$. Then $g_{k,V} = 0$ implies

$$\sum_{1 \leqslant i_1 < \cdots < i_k \leqslant n} c_{i_1 \cdots i_k} \sum_{\phi \in PERM_k} (-1)^{inv(\phi)} \boldsymbol{e}_{\phi(1)} \otimes \cdots \otimes \boldsymbol{e}_{\phi(k)} = 0,$$

which amounts to

$$\sum_{\phi \in PERM_k} (-1)^{inv(\phi)} \sum_{1 \leqslant i_1 < \cdots < i_k \leqslant n} c_{i_1 \cdots i_k} \boldsymbol{e}_{\phi(1)} \otimes \cdots \otimes \boldsymbol{e}_{\phi(k)} = 0.$$

The tensors $\boldsymbol{e}_{\phi(1)} \otimes \cdots \otimes \boldsymbol{e}_{\phi(k)}$ are a basis of $V^{\otimes k}$, so all coefficients $c_{i_1 \cdots i_k}$ are 0. Thus, $g_{k,V}$ is injective and this implies that the set $\{\boldsymbol{e}_{i_1} \wedge \cdots \wedge \boldsymbol{x}_{i_k} \mid 1 \leqslant i_1 < \cdots < i_k \leqslant n\}$ is a linearly independent set (by Supplement 3 of Chapter 2) and, therefore, a basis of $\mathsf{SKS}_{V,k}$. □

Corollary 14.8. *If V is a linear space with* $\dim(V) = n$, *we have*

$$\dim(\mathsf{SKS}_{V,k}) = \binom{n}{k}.$$

Proof. This follows from the fact that the basis of $\mathsf{SKS}_{V,k}$ that consists of the tensors $\boldsymbol{e}_{i_1} \wedge \cdots \wedge \boldsymbol{e}_{i_k}$ with $i_1 < \cdots < i_k$ contains $\binom{n}{k}$ vectors. □

If $\mathsf{t} \in \mathsf{SKS}_{V,k}$ is an antisymmetric tensor,

$$\mathsf{t} = \sum t_{i_1 \cdots i_r} \boldsymbol{e}^{i_1} \wedge \cdots \wedge \boldsymbol{e}^{i_r},$$

then $t_{i_1 i_2 \cdots i_k}$ with $i_1 < i_2 < \cdots < i_k$ is referred to as a *strict component* and is denoted as $t_{(i_1 i_2 \cdots i_k)}$. Thus, a component $t_{\ell_1 \ell_2 \cdots \ell_k}$ of t

whose indices are pairwise distinct can be written in terms of strict components as

$$t_{\ell_1\ell_2\cdots\ell_k} = (-1)^{inv(\phi)} t_{(i_1 i_2 \cdots i_k)},$$

where ϕ is the permutation

$$\phi : \begin{pmatrix} i_1 & i_2 & \cdots & i_k \\ \ell_1 & \ell_2 & \cdots & \ell_k \end{pmatrix}.$$

Let $\boldsymbol{v}^1, \ldots, \boldsymbol{v}^r$ be r vectors in the linear space V having the basis $\{\boldsymbol{e}^1, \ldots, \boldsymbol{e}^r\}$ and let $\boldsymbol{v}^1 \wedge \cdots \wedge \boldsymbol{v}^r \in \mathsf{SKS}_{V,r}$. If and $\boldsymbol{v}^i = a^i_j \boldsymbol{e}^j$, then

$$\begin{aligned} \boldsymbol{v}^1 \wedge \cdots \wedge \boldsymbol{v}^r &= a^1_{j_1} \cdots a^r_{j_r} \boldsymbol{e}^{j_1} \wedge \cdots \wedge \boldsymbol{e}^{j_r} \\ &= a^1_{j_1} \cdots a^r_{j_r} \delta^{i_1 \cdots i^r}_{j_1 \cdots j_r} \boldsymbol{e}^{i_1} \wedge \cdots \wedge \boldsymbol{e}^{i_r}. \end{aligned}$$

Therefore, the strict components of $\boldsymbol{v}^1 \wedge \cdots \wedge \boldsymbol{v}^r$ are $a^1_{j_1} \cdots a^r_{j_r} \delta^{i_1 \cdots i^r}_{j_1 \cdots j_r}$ for $i_1 < \cdots < i_r$.

If $\mathfrak{t} \in \mathsf{SKS}_{V,p}$ and $\mathfrak{s} \in \mathsf{SKS}_{V,q}$, we define their *exterior product* $\mathfrak{t} \wedge \mathfrak{s}$ as an element of $\mathsf{SKS}_{V,p+q}$. This product satisfies the following properties:

(i) distributivity of exterior product: if $\mathfrak{t}, \mathfrak{s} \in \mathsf{SKS}_{V,p}$ and $\mathfrak{r} \in \mathsf{SKS}_{V,q}$, then

$$(\mathfrak{t} + \mathfrak{s}) \wedge \boldsymbol{r} = \mathfrak{t} \wedge \boldsymbol{r} + \mathfrak{s} \wedge \boldsymbol{r};$$

if $\mathfrak{t} \in \mathsf{SKS}(V, p)$ and $\boldsymbol{s}, \mathfrak{r} \in \mathsf{SKS}_{V,q}$, then

$$\mathfrak{t} \wedge (\mathfrak{s} + \boldsymbol{r}) = \mathfrak{t} \wedge \boldsymbol{s} + \mathfrak{t} \wedge \boldsymbol{r};$$

(ii) if $a \in \mathbb{R}$, then

$$a\, \mathfrak{t} \wedge \mathfrak{s} = (a\mathfrak{t}) \wedge \mathfrak{s} = \mathfrak{t} \wedge (a\mathfrak{s});$$

(iii) if $\dim(V) = n$, $\{\boldsymbol{e}_{i_1} \wedge \boldsymbol{e}_{i_2} \wedge \cdots \wedge \boldsymbol{e}_{i_k} \mid\mid 1 \leqslant i_1 < i_2 < \cdots < i_k \leqslant n\}$ is a basis of $\mathsf{SKS}_{V,k}$, $\{\boldsymbol{e}_{j_1} \wedge \boldsymbol{e}_{j_2} \wedge \cdots \wedge \boldsymbol{e}_{j_\ell} \mid 1 \leqslant j_1 < j_2 < \cdots < j_\ell \leqslant n\}$ is a basis of $\bigwedge^\ell V$, then

$$(\boldsymbol{e}_{i_1} \wedge \boldsymbol{e}_{i_2} \wedge \cdots \wedge \boldsymbol{e}_{i_k}) \wedge (\boldsymbol{e}_{j_1} \wedge \boldsymbol{e}_{j_2} \wedge \cdots \wedge \boldsymbol{e}_{j_\ell}) = \mathbf{0}_{V,k+\ell}$$

if the sequences $(\boldsymbol{e}_{i_1}, \boldsymbol{v}_{i_2}, \ldots, \boldsymbol{e}_{i_k})$ and $(\boldsymbol{e}_{j_1}, \boldsymbol{e}_{j_2}, \ldots, \boldsymbol{e}_{j_\ell})$ have any common components; otherwise,

$$\begin{aligned} &(\boldsymbol{e}_{i_1} \wedge \boldsymbol{e}_{i_2} \wedge \cdots \wedge \boldsymbol{e}_{i_k}) \wedge (\boldsymbol{e}_{j_1} \wedge \boldsymbol{e}_{j_2} \wedge \cdots \wedge \boldsymbol{e}_{j_\ell}) \\ &= (-1)^v \boldsymbol{e}_{i_1} \wedge \boldsymbol{e}_{i_2} \wedge \cdots \wedge \boldsymbol{e}_{i_k}) \wedge \boldsymbol{e}_{j_1} \wedge \boldsymbol{e}_{j_2} \wedge \cdots \wedge \boldsymbol{e}_{j_\ell}, \end{aligned}$$

where $v = inv(i_1, \ldots, i_k, j_1, \ldots, j_\ell)$ is the number of inversions of this sequence.

It is immediate that this multiplication satisfies conditions (i)–(iii) considered above.

Let V be a linear space with $\dim(V) = n$ having the basis $\boldsymbol{e}_1, \ldots, \boldsymbol{e}_n$. An antisymmetric tensor t in $V^{\otimes k}$ can be written as

$$\mathsf{t} = \sum_{a_1 \cdots a_k} t^{a_1 \cdots a_k} \boldsymbol{e}_{a_1} \otimes \cdots \otimes \boldsymbol{e}_{a_k}.$$

If any two indices a_i, a_j are equal, the term that contains these indices vanishes due to the antisymmetry of t. Thus, the number of non zero terms in this sum is $\binom{n}{k}$.

A component $t_{a_1 \cdots a_k}$ of the antisymmetric tensor $\boldsymbol{k}$ having distinct indices is related to a component $t_{i_1 \cdots i_k}$ such that $i_1 < i_2 < \cdots < i_k$ and $\{a_1, \ldots, a_k\} = \{i_1, \ldots, i_k\}$ by the formula

$$t_{a_1 \cdots a_k} = (-1)^{inv(a_1, \ldots, a_k)} t_{i_1 \cdots i_k},$$

where $inv(a_1, \ldots, a_k)$ is the number of inversions of the sequence $(a_1, \ldots, a_k)$ due to the antisymmetry of t.

14.8 Linear Mappings between Spaces $\mathsf{SKS}_{V,k}$

Tensors in $\mathsf{SKS}_{V,k}$ are representable as linear combinations of blades, but these representations are not unique.

Example 14.15. Let $\boldsymbol{v}_1, \boldsymbol{v}_2$, and $\boldsymbol{v}_3$ be three vectors in the linear space V. Then the 2-blade $\mathsf{t} = \boldsymbol{v}_1 \wedge \boldsymbol{v}_2 + \boldsymbol{v}_1 \wedge \boldsymbol{v}_3 + \boldsymbol{v}_2 \wedge \boldsymbol{v}_3$ has the alternative representation

$$\mathsf{t} = (\boldsymbol{v}_1 + \boldsymbol{v}_2) \wedge (\boldsymbol{v}_2 + \boldsymbol{v}_3)$$

because $\boldsymbol{v}_2 \wedge \boldsymbol{v}_2 = \mathbf{0}_{V,2}$.

Definition 14.9. Let $h : V \longrightarrow W$ be a linear mapping between the linear spaces V and W. The *mapping induced* by h between $\mathsf{SKS}_{V,k}$ and $\mathsf{SKS}_{W,k}$ is defined on k-blades as

$$h_*(\boldsymbol{v}_1 \wedge \cdots \wedge \boldsymbol{v}_k) = h(\boldsymbol{v}_1) \wedge \cdots \wedge h(\boldsymbol{v}_k)$$

for $\boldsymbol{v}_1, \ldots, \boldsymbol{v}_k \in V$.

Theorem 14.29. *Let V be a non-zero linear space. The set $S = \{\boldsymbol{v}_1, \ldots, \boldsymbol{v}_k\}$ is linearly independent if and only if $\boldsymbol{v}_1 \wedge \cdots \wedge \boldsymbol{v}_k \neq \boldsymbol{0}_{V,k}$.*

Proof. We assume that $k \leqslant \dim(V)$ because for $k > \dim(V)$ the set $\{\boldsymbol{v}_1, \ldots, \boldsymbol{v}_k\}$ cannot be linearly independent. Also, the case $k = 1$ is trivial, so assume that $k > 2$.

Suppose that S is linearly independent and let B be the extension of S to a basis B for V, so $\boldsymbol{v}_1 \wedge \cdots \wedge \boldsymbol{v}_k$ is part of a basis of $\mathsf{SKS}_{V,k}$. Therefore, $\boldsymbol{v}_1 \wedge \cdots \boldsymbol{v}_k \neq \boldsymbol{0}_{V,k}$.

Suppose now that S is linearly dependent, so there exists $\boldsymbol{v}_i \in S$ that is a linear combination of vectors in $S - \{\boldsymbol{v}_i\}$, say $\boldsymbol{v}_k = a_1\boldsymbol{v}_1 + \cdots + a_{k-1}\boldsymbol{v}_{k-1}$. This implies

$$\boldsymbol{v}_1 \wedge \cdots \wedge \boldsymbol{v}_{k-1} \wedge \left(\sum_{i=1}^{k-1} a_i \boldsymbol{v}_i \right).$$

This is sum of $k-1$ wedges, each containing a repeated factor, hence every term of this sum vanishes. □

14.9 Determinants and Exterior Algebra

Next, we redefine the notion of determinant using exterior algebra.

Definition 14.10. Let V be a linear space with $\dim(V) = n$, having a basis $B = \{\boldsymbol{e}_1, \ldots, \boldsymbol{e}_n\}$. The *determinant* of the family of vectors $\{\boldsymbol{v}_1, \ldots, \boldsymbol{v}_n\}$ in the basis B is the unique real number δ such that

$$\boldsymbol{v}_1 \wedge \cdots \wedge \boldsymbol{v}_n = \delta \boldsymbol{e}_1 \wedge \cdots \wedge \boldsymbol{e}_n.$$

This definition is correct because if $\dim(V) = n$, then $\dim(\mathsf{SKS}_{V,n}) = \binom{n}{n} = 1$ having $\{\boldsymbol{e}_1 \wedge \cdots \wedge \boldsymbol{e}_n\}$ as a basis.

Of course, we will use the familiar notation $\delta = \det(\boldsymbol{v}_1, \ldots, \boldsymbol{v}_n)$.

Classical properties of determinants can be retrieved using exterior algebra. For instance, we can reprove in this context the Leibniz formula (Equality (5.1)). Indeed, note that

$$\boldsymbol{v}_1 \wedge \cdots \wedge \boldsymbol{v}_n = \sum_{i_1=1}^{n} \cdots \sum_{i_n=1}^{n} v_{i_1 1} \cdots v_{i_n n} \boldsymbol{e}_{i_1} \wedge \cdots \wedge \boldsymbol{e}_{i_n}.$$

We may assume that in the right-member summations, all terms involve indices $i_1, \ldots, i_n$ having distinct values. In other words, there exists a permutation $\phi \in PERM_n$ such that $i_k = \phi(k)$ for $1 \leqslant k \leqslant n$. Therefore,

$$\boldsymbol{v}_1 \wedge \cdots \wedge \boldsymbol{v}_n = \sum_{\phi \in PERM_n} v_{\phi(1)1} \cdots v_{\phi(n)n} \boldsymbol{e}_{\phi(1)} \wedge \cdots \wedge \boldsymbol{e}_{\phi(n)}.$$

This means that

$$\boldsymbol{v}_1 \wedge \cdots \wedge \boldsymbol{v}_n = \left(\sum_{\phi \in PERM_n} (-1)^{inv(\phi)} v_{\phi(1)1} \cdots v_{\phi(n)n} \right) \boldsymbol{e}_1 \wedge \cdots \wedge \boldsymbol{e}_n,$$

which is equivalent to the Leibniz formula.

Definition 14.11. Let V be a linear space and let $\boldsymbol{v}_1, \ldots, \boldsymbol{v}_k \in V$. The k-dimensional parallelepiped $P(\boldsymbol{v}_1, \ldots, \boldsymbol{v}_k)$ determined by $\boldsymbol{v}_1, \ldots, \boldsymbol{v}_k$ is the subset of V defined as

$$P(\boldsymbol{v}_1, \ldots, \boldsymbol{v}_k) = \{a_1 \boldsymbol{v}_1 + \cdots + a_k \boldsymbol{v}_k \mid a_i \in [0,1] \text{ for } 1 \leqslant i \leqslant k\}.$$

If $k = 2$, the bidimensional parallelepiped is referred to as the *parallelogram* determined by the vectors $\boldsymbol{v}_1, \boldsymbol{v}_2 \in V$,

$$P(\boldsymbol{v}_1, \boldsymbol{v}_2) = \{a_1 \boldsymbol{v}_1 + a_2 \boldsymbol{v}_2 \mid 0 \leqslant a_1, a_2 \leqslant 1\}.$$

Let $\boldsymbol{v}_1, \boldsymbol{v}_2$ be two vectors in $\mathbb{R}^n$, U be the bidimensional subspace of $\mathbb{R}^n$ determined by $\boldsymbol{v}_1$ and $\boldsymbol{v}_2$, and let $\boldsymbol{e}_1, \boldsymbol{e}_2$ be a basis in U. We can write

$$\boldsymbol{v}_1 = a_{11}\boldsymbol{e}_1 + a_{21}\boldsymbol{e}_2 \text{ and } \boldsymbol{v}_2 = a_{12}\boldsymbol{e}_1 + a_{22}\boldsymbol{e}_2,$$

which allows us to write

$$\begin{aligned}
\boldsymbol{v}_1 \wedge \boldsymbol{v}_2 &= (a_{11}\boldsymbol{e}_1 + a_{21}\boldsymbol{e}_2) \wedge (a_{12}\boldsymbol{e}_1 + a_{22}\boldsymbol{e}_2) \\
&= (a_{11}a_{22} - a_{21}a_{12}) \boldsymbol{e}_1 \wedge \boldsymbol{e}_2 \\
&= \begin{vmatrix} a_{11} & a_{12} \\ a_{21} & a_{22} \end{vmatrix} \boldsymbol{e}_1 \wedge \boldsymbol{e}_2.
\end{aligned}$$

If we interpret $\boldsymbol{e}_1 \wedge \boldsymbol{e}_2$ as the unit area of a square in the subspace U, then $\boldsymbol{v}_1 \wedge \boldsymbol{v}_2$ is the measure of the *oriented* area of $P(\boldsymbol{v}_1, \boldsymbol{v}_2)$.

In a similar manner, one can consider oriented volumes, starting from a three-dimensional subspace U of $\mathbb{R}^n$. If $\{\boldsymbol{e}_1, \boldsymbol{e}_2, \boldsymbol{e}_3\}$ is a basis in U and $\boldsymbol{v}_i = \begin{pmatrix} v_{1i} \\ v_{2i} \\ v_{3i} \end{pmatrix}$ for $i \in \{1,2,3\}$ are the expressions of these vectors in the basis $\{\boldsymbol{e}_1, \boldsymbol{e}_2, \boldsymbol{e}_3\}$, then, by taking into account the associativity of $\wedge$, we can write

$$\begin{aligned}
&\boldsymbol{v}_1 \wedge \boldsymbol{v}_2 \wedge \boldsymbol{v}_3 \\
&= (v_{11}\boldsymbol{e}_1 + v_{21}\boldsymbol{e}_2 + v_{31}\boldsymbol{e}_3) \wedge (v_{12}\boldsymbol{e}_1 + v_{22}\boldsymbol{e}_2 + v_{32}\boldsymbol{e}_3) \\
&\quad \wedge (v_{13}\boldsymbol{e}_1 + v_{23}\boldsymbol{e}_2 + v_{33}\boldsymbol{e}_3) \\
&= (v_{11}v_{22}\ \boldsymbol{e}_1 \wedge \boldsymbol{e}_2 + v_{11}v_{32}\ \boldsymbol{e}_1 \wedge \boldsymbol{e}_3 + v_{21}v_{12}\ \boldsymbol{e}_2 \wedge \boldsymbol{e}_1 + v_{21}v_{32}\ \boldsymbol{e}_2 \wedge \boldsymbol{e}_3 \\
&\quad + v_{31}v_{12}\ \boldsymbol{e}_3 \wedge \boldsymbol{e}_1 + v_{31}v_{22}\ \boldsymbol{e}_3 \wedge \boldsymbol{e}_2) \wedge (v_{13}\boldsymbol{e}_1 + v_{23}\boldsymbol{e}_2 + v_{33}\boldsymbol{e}_3) \\
&= ((v_{11}v_{22} - v_{21}v_{12})\ \boldsymbol{e}_1 \wedge \boldsymbol{e}_2 + (v_{11}v_{32} - v_{31}v_{12})\boldsymbol{e}_1 \wedge \boldsymbol{e}_3 \\
&\quad + (v_{21}v_{32} - v_{31}v_{22})\ \boldsymbol{e}_2 \wedge \boldsymbol{e}_3) \wedge (v_{13}\boldsymbol{e}_1 + v_{23}\boldsymbol{e}_2 + v_{33}\boldsymbol{e}_3) \\
&= (v_{11}v_{22} - v_{21}v_{12})v_{33}\ \boldsymbol{e}_1 \wedge \boldsymbol{e}_2 \wedge \boldsymbol{e}_3 \\
&\quad + (v_{11}v_{32} - v_{31}v_{12})v_{23}\ \boldsymbol{e}_1 \wedge \boldsymbol{e}_3 \wedge \boldsymbol{e}_2 \\
&\quad + (v_{21}v_{32} - v_{31}v_{22})v_{13}\ \boldsymbol{e}_2 \wedge \boldsymbol{e}_3 \wedge \boldsymbol{e}_1 \\
&= \begin{vmatrix} v_{11} & v_{12} & v_{13} \\ v_{21} & v_{22} & v_{23} \\ v_{31} & v_{32} & v_{33} \end{vmatrix}\ \boldsymbol{e}_1 \wedge \boldsymbol{e}_2 \wedge \boldsymbol{e}_3 \\
&= \det[\boldsymbol{v}_1\ \boldsymbol{v}_2\ \boldsymbol{v}_3]\ \boldsymbol{e}_1 \wedge \boldsymbol{e}_2 \wedge \boldsymbol{e}_3.
\end{aligned}$$

Exercises and Supplements

(1) Let V, W be real linear spaces and let $\boldsymbol{v} \in V$ and $\boldsymbol{w} \in W$. Prove that if $\boldsymbol{v} \neq \boldsymbol{0}_V$ and $\boldsymbol{w} \neq \boldsymbol{0}_W$, then $\boldsymbol{v} \otimes \boldsymbol{w} \neq \boldsymbol{0}_{V \otimes M}$.

Solution: Since $\boldsymbol{v} \neq \boldsymbol{0}_V$ and $\boldsymbol{w} \neq \boldsymbol{0}_W$, there exist $\boldsymbol{f} \in V^*$ and $\boldsymbol{g} \in W^*$ such that $\boldsymbol{f}(\boldsymbol{v}) \neq 0$ and $\boldsymbol{g}(\boldsymbol{w}) \neq 0$. Consider the bilinear mapping $\psi : V \times W \longrightarrow \mathbb{R}$ defined by $\psi(\boldsymbol{v}, \boldsymbol{w}) = \boldsymbol{f}(\boldsymbol{v})\boldsymbol{g}(\boldsymbol{w})$. There exists a linear application $h : V \otimes W \longrightarrow \mathbb{R}$ such that $\boldsymbol{f}(\boldsymbol{v})\boldsymbol{g}(\boldsymbol{w}) = \psi(\boldsymbol{v}, \boldsymbol{w}) = h(\boldsymbol{v} \otimes \boldsymbol{w})$. Therefore, $\boldsymbol{v} \otimes \boldsymbol{w} \neq \boldsymbol{0}_{V \otimes W}$.

(2) Let V, W be real linear spaces and let $\boldsymbol{v}_1, \tilde{\boldsymbol{v}} \in V$ and $\boldsymbol{w}_1, \tilde{\boldsymbol{w}} \in W$. Prove that $\boldsymbol{v}_1 \otimes \boldsymbol{w}_1 = \tilde{\boldsymbol{v}} \otimes \tilde{\boldsymbol{w}} \neq \mathbf{0}_{V \otimes W}$ if and only if $\boldsymbol{v}_1 = a\tilde{\boldsymbol{v}}$ and $\tilde{\boldsymbol{w}} = a\boldsymbol{w}_1$ for some $a \in \mathbb{R}$.

Solution: Observe that $\boldsymbol{v}_1 \neq \mathbf{0}_V$ and $\boldsymbol{w}_1 \neq \mathbf{0}_W$ because $\boldsymbol{v}_1 \otimes \boldsymbol{w}_1 \neq \mathbf{0}_{V \otimes W}$. Let $\{\boldsymbol{v}_1, \boldsymbol{v}_2, \ldots, \boldsymbol{v}_n\}$ be a basis of V and let $\{\boldsymbol{w}_1, \boldsymbol{w}_2, \ldots, \boldsymbol{w}_m\}$ be a basis of W obtained by applying the independent set extension corollary to the sets $\{\boldsymbol{v}_1\}$ and $\{\boldsymbol{w}_1\}$ (Corollary 2.3). If $\tilde{\boldsymbol{v}} \otimes \tilde{w} = \boldsymbol{v}_1 \otimes \boldsymbol{w}_1$, then $\tilde{\boldsymbol{v}} = b\boldsymbol{v}_1$ and $\tilde{\boldsymbol{w}} = a\boldsymbol{w}_1$. Therefore, $ab\boldsymbol{v}_1 \otimes \boldsymbol{w}_1 = \boldsymbol{v}_1 \otimes \boldsymbol{w}_1$, so $ab = 1$. It follows that $\tilde{\boldsymbol{w}} = a\boldsymbol{w}_1$ and $\boldsymbol{v}_1 = \frac{1}{b}\tilde{\boldsymbol{v}} = a\tilde{\boldsymbol{v}}$.

(3) Let V, W, and U be three real linear spaces. Show that the linear spaces $\mathsf{Hom}(V \otimes W, U)$ and $\mathsf{Hom}(V, Hom(W, U))$ are isomorphic.

(4) Let V, W be real linear spaces. Prove that the map $h : V \otimes W \longrightarrow W \otimes V$ defined as $h(\boldsymbol{v} \otimes \boldsymbol{w}) = \boldsymbol{w} \otimes \boldsymbol{v}$ for $\boldsymbol{v} \in V$ and $\boldsymbol{w} \in W$ is an isomorphism.

(5) Let V, W, U be real linear spaces and let $f : V \longrightarrow W$ be a linear mapping. Prove the following:
   (a) the mapping $g : V \times U \longrightarrow W \otimes U$ defined as $g(\boldsymbol{v}, \boldsymbol{u}) = f(\boldsymbol{v}) \otimes \boldsymbol{u}$ is bilinear;
   (b) by the universal property, g induces a linear mapping $f \otimes 1_U : V \times U \longrightarrow W \times U$;
   (c) if f is surjective, then $f \otimes 1_U$ is surjective;
   (d) if f is injective, then $f \otimes 1_U$ is injective.

(6) Let V, W be real linear spaces, V_1, V_2 be two subspaces of V, and W_1, W_2 be two subspaces of W. Prove that

$$(V_1 \otimes W_1) \cap (V_2 \otimes W_2) = (V_2 \cap V_2) \otimes (W_1 \cap W_2).$$

(7) Prove that the tensor t^{ij} is the tensor product of two vectors if and only if $t^{ai}t^{bj} - t^{aj}t^{bi} = 0$ for $1 \leqslant a, b, i, j \leqslant n$.

(8) Let $x^1, \ldots, x^n$ be the components of a vector $\boldsymbol{x} \in \mathbb{R}^n$ relative to a basis $\mathbf{e}_1, \ldots, \mathbf{e}_n$ of $\mathbb{R}^n$. Show that the function that maps $\boldsymbol{x}$ into $x^i x^j$ is a contravariant tensor.

(9) Show that every tensor t of valence 2 whose components are the same in all bases has the form $a\delta_i^j$.

(10) Let V and W be two real linear spaces, where $\dim(V) = n$, $\dim(W) = m$, $R = \{\boldsymbol{v}_1, \ldots, \boldsymbol{v}_r\}$ is a subset of V, and $S = \{\boldsymbol{w}_1, \ldots, \boldsymbol{w}_r\}$ is a subset of W. If $\mathsf{t} = \sum_{i=1}^r \boldsymbol{v}_i \otimes \boldsymbol{w}_i$ and R is

linearly independent, prove that $\boldsymbol{w}_1, \ldots, \boldsymbol{w}_r$ are uniquely determined.

Solution: Let $B = \{\boldsymbol{v}_1, \ldots, \boldsymbol{v}_r, \boldsymbol{v}_{r+1}, \ldots, \boldsymbol{v}_n\}$ be a basis for V obtained by extending the linearly independent set R. Suppose that $\{\tilde{\boldsymbol{w}}_1, \ldots, \tilde{\boldsymbol{w}}_r\}$ is an alternate collection of elements such that $\mathsf{t} = \sum_{i=1}^{r} \boldsymbol{v}_i \otimes \tilde{\boldsymbol{w}}_i$. Then, we have $\sum_{i=1}^{r} \boldsymbol{v}_i \otimes (\boldsymbol{w}_i - \tilde{\boldsymbol{w}}_i) = \mathbf{0}_{V\otimes W}$. Suppose that $B' = \{\boldsymbol{y}_1, \ldots, \boldsymbol{y}_m\}$ is a basis for W and $\boldsymbol{w}_i - \tilde{\boldsymbol{w}}_i = \sum_{j=1}^{m} a_{ij}\boldsymbol{y}_j$. Then we have

$$\begin{aligned}\sum_{i=1}^{r} \boldsymbol{v}_i \otimes (\boldsymbol{w}_i - \tilde{\boldsymbol{w}}_i) &= \sum_{i=1}^{r} \left(\boldsymbol{v}_i \otimes \sum_{j=1}^{m} a_{ij}\boldsymbol{y}_j \right) \\ &= \sum_{i=1}^{r} \sum_{j=1}^{m} a_{ij}(\boldsymbol{v}_i \otimes \boldsymbol{y}_j) = \mathbf{0}_{V\otimes W}.\end{aligned}$$

Since $\boldsymbol{v}_i \otimes \boldsymbol{y}_j$ is a basis for $V \otimes W$, it follows that $a_{ij} = 0$ for $1 \leqslant i \leqslant r$ and $1 \leqslant j \leqslant m$, which means that $\boldsymbol{w}_i = \tilde{\boldsymbol{w}}_i$ for $1 \leqslant i \leqslant r$.

(11) Let V, W, X, Y be four real, finite-dimensional linear spaces and let $h \in \mathsf{Hom}(V, W)$ and $g \in \mathsf{Hom}(X, Y)$. Denote by $\Phi : V \times X \longrightarrow W \otimes Y$ the bilinear mapping defined as $\Phi(\boldsymbol{v}, \boldsymbol{x}) = h(\boldsymbol{v}) \otimes g(\boldsymbol{x})$ and let $f : V \otimes X \longrightarrow W \otimes Y$ be the linear operator that exists by Theorem 14.3 such that $f(\boldsymbol{v} \otimes \boldsymbol{x}) = h(\boldsymbol{v}) \otimes g(\boldsymbol{x})$. Prove that
   (a) the mapping
   $$\phi : \mathsf{Hom}(V, W) \times \mathsf{Hom}(X, Y) \longrightarrow \mathsf{Hom}(V \otimes X, W \otimes Y)$$
   given by $\phi(h, g) = f$ is linear;
   (b) there exists an injective linear application:
   $$\chi : \mathsf{Hom}(V, W) \otimes \mathsf{Hom}(X, Y) \longrightarrow \mathsf{Hom}(V \otimes X, W \otimes Y).$$

(12) Let V be a linear space with $\dim(V) = 3$. Prove that every vector in $\mathsf{SKS}_{V,2}$ can be represented by a single blade.

Solution: Let $B = \{\boldsymbol{e}_1, \boldsymbol{e}_2, \boldsymbol{e}_3\}$ be a basis in V. Note that a tensor $\mathfrak{t}$ of the form

$$\mathfrak{t} = a\, \boldsymbol{e}_1 \wedge \boldsymbol{e}_2 + b\, \boldsymbol{e}_1 \wedge \boldsymbol{e}_3 + c\, \boldsymbol{e}_2 \wedge \boldsymbol{e}_3$$

can be written as

$$\mathfrak{t} = (p\, \boldsymbol{e}_1 + q\, \boldsymbol{e}_2) \wedge (r\, \boldsymbol{e}_2 + s\, \boldsymbol{e}_3),$$

where $pr = a$, $ps = b$, and $qs = c$. These equalities are satisfied by taking $p = \frac{a}{r}$, $s = \frac{br}{a}$, and $q = \frac{ac}{br}$ for an arbitrary $r \neq 0$.

(13) Let $\boldsymbol{x}, \boldsymbol{y} \in \mathbb{R}^n$. Define $t_{ij} = x_i y_j - x_j y_i$. Prove that
 (a) We have

$$\begin{vmatrix} t_{ij} & t_{ik} & t_{i\ell} \\ x_j & x_k & x_\ell \\ y_j & y_k & y_\ell \end{vmatrix} = 0.$$

 (b) The following equality holds:

$$t_{ij}t_{kl} + t_{ik}t_{lj} + t_{i\ell}t_{jk} = 0,$$

 for every four distinct indices i, j, k, ℓ.

(14) Let V, W be two linear spaces and let $\mathfrak{t} \in V \otimes W$ be a non-zero tensor given by $\mathfrak{t} = \sum_{i=1}^{n} \boldsymbol{v}_i \otimes \boldsymbol{w}_i$, where $\boldsymbol{v}_i \in V$ and $\boldsymbol{w}_i \in W$ for $1 \leqslant i \leqslant n$. Prove that if n is the least number with this property, the sets $\{\boldsymbol{v}_1, \ldots, \boldsymbol{v}_n\}$ and $\{\boldsymbol{w}_1, \ldots, \boldsymbol{w}_n\}$ are linearly independent.

(15) Verify that mappings $\boldsymbol{S}_{V,r} : V^{\otimes r} \longrightarrow V^{\otimes r}$ and $\mathsf{SKS}_{\mathbf{V,r}} : \mathbf{V}^{\otimes \mathbf{r}} \longrightarrow \mathbf{V}^{\otimes \mathbf{r}}$ defined by Equalities (14.5) and (14.6) are indeed linear.

(16) Let $T = \{t_{i_1 \cdots i_q} \mid 1 \leqslant i_1, \ldots, i_q \leqslant n\}$ be a set of n^q numbers. Prove that there exists an n-dimensional linear space V such that T is the set of components of a covariant tensor if and only if for every set of contravariant vectors $\{\boldsymbol{v}_1, \ldots, \boldsymbol{v}_q\}$, where $\boldsymbol{v}_j = (v_j^i)$ for $1 \leqslant j \leqslant q$ and $1 \leqslant i \leqslant n$, the quantity $t_{i_1 \cdots i_q} v_1^{i_1} \cdots v_q^{i_q}$ is invariant with respect to all basis changes.

(17) Let V, W, X, Y be four finite-dimensional real linear spaces having the bases $B_V = \{\boldsymbol{v}_1, \ldots, \boldsymbol{v}_m\}$, $B_W = \{\boldsymbol{w}_1, \ldots, \boldsymbol{w}_n\}$, $B_X = \{\boldsymbol{x}_1, \ldots, \boldsymbol{x}_p\}$, and $B_Y = \{\boldsymbol{y}_1, \ldots, \boldsymbol{y}_q\}$, respectively. Also, let $f_1 \in \mathsf{Hom}(V, X)$ and $f_2 \in \mathsf{Hom}(W, Y)$ be two linear mappings.

(a) Let $\ell : V \times W \longrightarrow X \times Y$ be the mapping defined as $\ell(\boldsymbol{v}, \boldsymbol{w}) = f_1(\boldsymbol{v}) \otimes f_2(\boldsymbol{w})$ for $\boldsymbol{v} \in V$ and $\boldsymbol{w} \in W$. Prove that ℓ is a bilinear mapping.

(b) Prove that there exists a unique linear mapping $g : V \otimes W \longrightarrow X \otimes Y$ such that $g(\boldsymbol{v} \otimes \boldsymbol{w}) = f_1(\boldsymbol{v}) \otimes f_2(\boldsymbol{w})$. The linear mapping g is referred to as the *tensor product of the mappings* f_1 *and* f_2 and is denoted by $f_1 \otimes f_2$.

(18) Using the same notations as in Exercise 17, suppose that the matrices representing the linear functions f_1 and f_2 relative to the bases B_V, B_X and B_W, B_Y are $A \in \mathbb{R}^{m\times n}$ and $B \in \mathbb{R}^{p\times q}$, respectively. Prove that the matrix representing the tensor product of the linear mappings f_1 and f_2, $f_1 \otimes f_2$ is the Kronecker product $A \otimes B$.

Solution: The matrices A and B can be written as $A = (f_1(\boldsymbol{v}_i, \boldsymbol{w}_j))$ and $B = (f_2(\boldsymbol{x}_j, \boldsymbol{x}_k))$. The rows of the matrix $C \in \mathbb{R}^{mn\times pq}$ that represent g correspond to the pairs $(\boldsymbol{v}_i, \boldsymbol{w}_j)$ and the columns of C correspond to the pairs $(\boldsymbol{x}_p, \boldsymbol{y}_q)$. Thus, at the intersection of a row $(\boldsymbol{v}_i, \boldsymbol{w}_j)$ and a column $(\boldsymbol{x}_p, \boldsymbol{y}_q)$ we have $f_1(\boldsymbol{v}_i, \boldsymbol{w}_j) f_2(\boldsymbol{x}_p, \boldsymbol{y}_q)$. With an appropriate numbering of the rows and columns of C, this matrix equals $A \otimes B$.

(19) Let $U_n = \mathbb{R}^{I_n}$ and $V_n = \mathbb{R}^{J_n}$ be linear spaces, and let $f_n : U_n \longrightarrow V_n$ be linear mappings, for $1 \leqslant n \leqslant N$. Define $f : U_1 \times \cdots \times U_N \longrightarrow V$, where $V = V_1 \otimes \cdots \otimes V_N$ as

$$f(\boldsymbol{u}_1, \ldots, \boldsymbol{u}_N) = f_1(\boldsymbol{u}_1) \otimes \cdots \otimes f_N(\boldsymbol{u}_N).$$

Prove the following:

(a) f is a multilinear function;

(b) there exists a unique linear mapping $g : U_1 \otimes \cdots \otimes U_n \longrightarrow V_1 \otimes \cdots \otimes V_N$ such that

$$g(\boldsymbol{u}_1 \otimes \cdots \otimes \boldsymbol{u}_N) = f_1(\boldsymbol{u}_1) \otimes \cdots \otimes f_N(\boldsymbol{u}_N).$$

The mapping g is referred to as the *tensor product of the linear mappings* $f_1, \ldots, f_N$;

(c) if A_n is the matrix of the linear mapping f_n for $1 \leqslant n \leqslant N$, then the Kronecker product $A_1 \otimes \cdots \otimes A_N$ is the matrix associated with linear mapping g.

(20) Let $\mathfrak{t}, \mathfrak{s} \in V^{\otimes 2}$ be two tensors such that $\mathfrak{t}$ is symmetric and $\mathfrak{s}$ is antisymmetric. Prove that $\sum_{i,j} t_{ij} s_{ij} = 0$.

(21) Let $\mathfrak{s}$ and $\mathfrak{z}$ be two skew-symmetric tensors in $\mathsf{SKS}_{V,n}$. Prove that we have for their contraction

$$s_{i_1 \cdots i_n} z^{i_1 \cdots i_n} = n! s_{[i_1 \cdots i_n]} z^{i_1 \cdots i_n},$$

where $s_{[i_1 \cdots i_n]}$ denotes that we are summing over increasing sequences of indices only.

The notion of isotropic vector introduced in Supplement 99 of Chapter 6 can be extended to tensors. An *isotropic* tensor is a tensor whose components are the same in all rotated bases.

(22) Prove that the only isotropic tensor of rank 2 is a multiple of Kronecker's tensor δ_{ij}.

Solution: Let $R \in \mathbb{R}^{2\times 2}$ be an orthogonal matrix. Since $\sum_{k,\ell} r_{ik} r_{j\ell} \delta_{kl} = \delta_{ij}$, the isotropy of δ_{ij} is immediate.
Conversely, suppose that t_{ij} is an isotropic tensor, that is, $\sum_{k,\ell} r_{ik} r_{j\ell} t_{kl} = t_{ij}$ for every orthogonal matrix R and $i, j \in \{1, 2\}$. In matrix terms this can be expressed as $R\Delta R' = \Delta$, where $R = (r_i j)$ and $\Delta = (\delta_{ij})$.
Choosing

$$R = \begin{pmatrix} 0 & 1 & 0 \\ -1 & 0 & 0 \\ 0 & 0 & 1 \end{pmatrix},$$

we obtain $\delta_{11} = \delta_{22}$, $\delta_{13} = \delta_{23} = 0$, and $\delta_{31} = \delta_{32} = 0$.
Similarly, considering the rotation

$$R = \begin{pmatrix} 1 & 0 & 0 \\ 0 & 0 & 1 \\ 0 & -1 & 0 \end{pmatrix},$$

we obtain $\delta_{22} = \delta_{33}$, and $\delta_{21} = \delta_{31} = -\delta_{21}$, $\delta_{13} = \delta_{12} = -\delta_{13}$, $\delta_{23} = -\delta_{32} = -\delta_{23}$. This means that $\delta_{21} = \delta_{31} = \delta_{13} = \delta_{12} = \delta_{23} = \delta_{32} = 0$. Thus, all off-diagonal elements of δ are 0 and all diagonal elements are equal to constant, say a. This means that t has the form $a\delta_{ij}$.

(23) Prove that in three dimensions we have the following equalities involving Levi-Civita symbols:

$$\sum_{i=1}^{3} \epsilon_{ijk}\epsilon_{imn} = \delta_{jm}\delta_{kn} - \delta_{jn}\delta_{km}$$

for $1 \leqslant j, k, m, n \leqslant 3$.

(24) Prove that

$$\epsilon_{i_1 i_2 \cdots i_n}\epsilon_{j_1 j_2 \cdots j_n} = \begin{vmatrix} \delta_{i_1 j_1} & \delta_{i_1 j_2} & \cdots & \delta_{i_1 j_n} \\ \delta_{i_2 j_1} & \delta_{i_2 j_2} & \cdots & \delta_{i_2 j_n} \\ \vdots & \vdots & \vdots & \vdots \\ \delta_{i_n j_1} & \delta_{i_n j_2} & \cdots & \delta_{i_n j_n} \end{vmatrix}.$$

(25) Prove that the determinant of a matrix $A \in \mathbb{C}^{n\times n}$ can be written as

$$\det(A) = \sum_{i_1 \cdots i_n} \epsilon_{i_1 \cdots i_n} a_{1i_1} \cdots a_{ni_n}.$$

(26) Let $\boldsymbol{e}_2, \ldots, \boldsymbol{e}_{2n}$ be a basis of a linear space V. If $\boldsymbol{w} = \boldsymbol{e}_1 \wedge \boldsymbol{e}_2 + \boldsymbol{e}_3 \wedge \boldsymbol{e}_4 + \cdots + \boldsymbol{e}_{2n-1} \wedge \boldsymbol{e}_{2n}$, prove that

$$\mathsf{SKS}_{V,n}(\boldsymbol{w}) = n!\boldsymbol{e}_1 \wedge \boldsymbol{e}_2 \wedge \cdots \wedge \boldsymbol{e}_{2n}.$$

Solution: Let $\boldsymbol{u}_i = \boldsymbol{e}_{2i-1} \wedge \boldsymbol{e}_{2i} \in \mathsf{SKS}_{V,2}$. We have $\boldsymbol{u}_i \in \mathsf{SKS}_{V,2}$, hence, $\boldsymbol{u}_i \wedge \boldsymbol{u}_j = \boldsymbol{u}_j \wedge \boldsymbol{u}_i$ and $\boldsymbol{u}_i \wedge \boldsymbol{u}_i = \boldsymbol{0}$. Therefore,

$$\begin{aligned} \mathsf{SKS}_{V,n}(\boldsymbol{w}) &= \sum_{i_1 \cdots i_n} \boldsymbol{u}_{i_1} \wedge \cdots \wedge \boldsymbol{u}_{i_n} \\ &= n!\boldsymbol{u}_1 \wedge \cdots \boldsymbol{u}_n = n!\boldsymbol{e}_1 \wedge \boldsymbol{e}_2 \wedge \cdots \wedge \boldsymbol{e}_{2n}. \end{aligned}$$

(27) Let $\mathfrak{t} = \boldsymbol{x}_1 \otimes \cdots \otimes \boldsymbol{x}_p \in \mathsf{SKS}_{V,p}$ and $\mathfrak{s} = \boldsymbol{x}_{p+1} \otimes \cdots \otimes \boldsymbol{x}_{p+q} \in \mathsf{SKS}_{V,q}$. Prove that

$$\begin{aligned} \boldsymbol{A}_{V,p+q}(\mathfrak{t} \otimes \mathfrak{s}) &= \boldsymbol{A}_{V,p+q}(\boldsymbol{A}_{V,p}(\boldsymbol{t}) \otimes \boldsymbol{s}), \\ \boldsymbol{A}_{V,p+q}(\mathfrak{t} \otimes \mathfrak{s}) &= \boldsymbol{A}_{V,p+q}(\boldsymbol{t} \otimes \boldsymbol{A}_{V,q}(\boldsymbol{s}). \end{aligned}$$

Extend these equalities to tensors in $\Lambda(V)$.

Solution: Since

$$\begin{aligned}
&\boldsymbol{A}_{V,p}(\boldsymbol{x}_1 \otimes \cdots \otimes \boldsymbol{x}_p)\\
&= \frac{1}{p!} \sum_{\phi \in PERM_p} \text{sign}(\phi) \boldsymbol{P}_\phi(\boldsymbol{x}_1 \otimes \cdots \otimes \boldsymbol{x}_p)\\
&= \frac{1}{p!} \sum_{\phi \in PERM_p} \text{sign}(\phi) \boldsymbol{x}_{\phi^{-1}(1)} \otimes \cdots \otimes \boldsymbol{x}_{\phi^{-1}(p)},
\end{aligned}$$

it follows that

$$\begin{aligned}
&\boldsymbol{A}_{V,p+q}(\boldsymbol{A}_{V,p}(\boldsymbol{t}) \otimes \boldsymbol{s})\\
&= \boldsymbol{A}_{V,p+q}\left(\frac{1}{p!} \sum_{\phi \in PERM_p} \text{sign}(\phi) \boldsymbol{x}_{\phi^{-1}(1)} \otimes \cdots \otimes \boldsymbol{x}_{\phi^{-1}(p)}\right.\\
&\qquad \left. \otimes\, \boldsymbol{x}_{p+1} \otimes \cdots \otimes \boldsymbol{x}_{p+q}\right)\\
&= \frac{1}{p!(p+q)!} \sum_{\phi \in PERM_p} \sum_{\psi \in PERM_{p+q}}\\
&\qquad \times \text{sign}(\psi^{-1}\phi^{-1}) \boldsymbol{x}_{\psi^{-1}(\phi^{-1}(1))} \otimes \cdots \otimes \boldsymbol{x}_{\psi^{-1}(p+q)}.
\end{aligned}$$

The desired equality follows from the fact that

$$\begin{aligned}
&\sum_{\phi \in PERM_p} (-1)^{\text{sign}(\psi^{-1}\phi^{-1})} \boldsymbol{x}_{\psi^{-1}(\phi^{-1}(1))} \otimes \cdots \otimes \boldsymbol{x}_{\psi^{-1}(p+q)}\\
&\quad = p! \sum_{\text{sign}(\tau)} \boldsymbol{x}_{\tau(1)} \cdots \boldsymbol{x}_{\tau(p+q)},
\end{aligned}$$

where $\tau = (\psi^{-1}(\phi^{-1}(1)), \ldots, \psi^{-1}(\phi^{-1}(p)), \psi^{-1}(p+1), \ldots, \psi^{-1}(p+q))$.

(28) Prove that if $\mathfrak{t} \in \mathsf{SKS}_{V,p}$ and $\mathfrak{s} \in \mathsf{SKS}(V,q)$, then $\mathfrak{t} \wedge \mathfrak{s} = (-1)^{pq}$ $\boldsymbol{s} \wedge \mathfrak{t}$.

(29) Prove that if $\mathfrak{t} \in \mathfrak{T}^2(V)$ with $\dim(V) = n$, then $\mathsf{SYM}_{V,2}(\mathfrak{t}) = \frac{t^{ij}+t^{ji}}{2} \boldsymbol{e}_i \otimes \boldsymbol{e}_j$ and $\mathsf{SKS}_{V,2}(\mathfrak{t}) = \frac{t^{ij}-t^{ji}}{2} \boldsymbol{e}_i \otimes \boldsymbol{e}_j$.

(30) Let $B = \{\boldsymbol{e}_1, \ldots, \boldsymbol{e}_n\}$ be a basis of the linear space V and let $\boldsymbol{g}^1, \ldots, \boldsymbol{g}^r \in V^*$. Prove that

$$(\boldsymbol{e}_{i_1} \wedge \cdots \wedge \boldsymbol{e}_{i_r})(\boldsymbol{g}^1, \ldots, \boldsymbol{g}^r) = \frac{1}{r!} \begin{vmatrix} \boldsymbol{g}^1(\boldsymbol{e}_{i_1}) & \cdots & \boldsymbol{g}^r(\boldsymbol{e}_{i_1}) \\ \vdots & \vdots & \vdots \\ \boldsymbol{g}^1(\boldsymbol{e}_{i_r}) & \cdots & \boldsymbol{g}^r(\boldsymbol{e}_{i_r}). \end{vmatrix}.$$

(31) Let $\{\boldsymbol{x}_1, \ldots, \boldsymbol{x}_p\}$ be a linearly independent set in the real linear space V, where $p \leqslant n = \dim(V)$. Suppose that $\{\boldsymbol{y}_1, \ldots, \boldsymbol{y}_p\}$ is a set of vectors in V such that

$$\boldsymbol{x}_1 \wedge \boldsymbol{y}_1 + \cdots + \boldsymbol{x}_p \wedge \boldsymbol{y}_p = \boldsymbol{0}_{\mathsf{SKS}_{V,2}}.$$

Prove that $\boldsymbol{y}_i = \sum_{j=1}^p a_{ij}\boldsymbol{x}_j$ and $a_{ij} = a_{ji}$ for $1 \leqslant i, j \leqslant$. This statement is known as Cartan's Lemma.[2]

Solution: Extend the set $\{\boldsymbol{x}_1, \ldots, \boldsymbol{x}_p\}$ to a basis $\{\boldsymbol{x}_1, \ldots, \boldsymbol{x}_p, \ldots, \boldsymbol{x}_n\}$ of V. Then we have $\boldsymbol{y}_i = \sum_{j=1}^p a_{ij}\boldsymbol{x}_j + \sum_{j=p+1}^n b_{ij}\boldsymbol{x}_j$. Substituting the values of $\boldsymbol{y}_i$ in

$$\boldsymbol{x}_1 \wedge \boldsymbol{y}_1 + \cdots + \boldsymbol{x}_p \wedge \boldsymbol{y}_p = \boldsymbol{0}_{\mathsf{SKS}_{V,2}},$$

we obtain

$$\sum_{i<j\leqslant p} (a_{ij} - a_{ji})\boldsymbol{x}_i \wedge \boldsymbol{x}_j + \sum_{i\leqslant p, j>p} b_{ij}\boldsymbol{x}_i \wedge \boldsymbol{x}_j = \boldsymbol{0}_{\mathsf{SKS}_{V,2}}.$$

Since $\boldsymbol{x}_i \wedge \boldsymbol{x}_j$ with $i < j$ are linearly independent, it follows that $b_{ij} = 0$ and $a_{ij} = a_{ji}$.

[2]Elie Joseph Cartan, one of the foremost mathematicians of the 20th century, was born in Dolomieu in Savoie on April 9, 1868 and died on May 6, 1951 in Paris. Cartan taught at the universities of Montpellier, Lyon, and Nancy and became a professor at Sorbonne in 1912. His main contributions are in differential geometry, theory of Lie groups, and general relativity.

Bibliographical Comments

There are many sources for studying multilinear algebras and tensors. We recommend [37, 70, 72, 106] and [153]. Important references that focus on specific themes are [93, 130, 151] and [5].

Accessible and lucid introductions to tensors can be found in [2] and [24]. Exercise 14.9 appears in [135]. A terse but useful introduction to tensors and exterior algebra is given in[175].

The proof of Cartan's Lemma follows [156].

Chapter 15

Multidimensional Array and Tensors

15.1 Introduction

In many machine learning-oriented literature sources, tensors are regarded as multidimensional arrays known also as *hypermatrices* (*cf.* [50]). While the sets of values of tensors over finite-dimensional spaces are indeed such arrays, this point of view misses fundamental properties of tensors, especially their behaviors relative to base changes in the underlying linear spaces.

Recall that the elements of a linear space V are referred to as contravariant vectors while the elements of the dual space V^* are referred to as covariant vectors (see Section 2.8). The components of contravariant vectors will be denoted with letters with superscripts, while the components of covariant vectors will be designated by letters with subscripts. The opposite convention is applied to vectors themselves: vectors in V are denoted by letters with subscripts, while vectors in V^* are denoted by letters with superscripts.

15.2 Multidimensional Arrays

The notion of multidimensional array (mda) is a generalization of the notion of a matrix. An alternative term used by some authors (*cf.* [50]) is that of *hypermatrix*.

The notion of mda is closely related to the notion of tensor. This section discusses mdas following [4, 5, 92] and [51]. Multidimensional arrays are the focus when we study the sets of values of tensors in fixed bases, rather than the behavior of tensors relative to base changes.

The components of a tensor are denoted by indices giving their position in the mda, as subscripts and superscripts, following the symbolic name of the tensor. Whether an index is displayed as a superscript or subscript depends on the transformation properties of the tensor previously described.

Definition 15.1. A *multidimensional array* (mda) on a field $\mathbb{F}$ is a function

$$T : \{1, \ldots, I_1\} \times \{1, \ldots, I_2\} \times \cdots \times \{1, \ldots, I_m\} \longrightarrow \mathbb{F}.$$

The set $\mathcal{M} = \{I_1, \ldots, I_m\}$ is known as the *set of modes* of T. The field $\mathbb{F}$ is, usually, the real field $\mathbb{R}$ or the complex field $\mathbb{C}$.

The *dimension vector* of T is

$$\dim(T) = (I_1, I_2, \ldots, I_m).$$

The element $T(i_1, \ldots, i_m)$ is also denoted as $T_{i_1 \cdots i_m}$.

The *order* of an mda T is the number of dimensions m. Vectors and matrices are mdas of order one and two, respectively.

When $I_1 = I_2 = \cdots = I_m = n$, T is an *m-order, n-dimensional mda* or a *hypercubic mda*. The set of mdas of order m and dimension n is denoted by $\mathfrak{M}_{m,n}$.

If $T \in \mathfrak{M}_{m,n}$, $T = (t_{i_1 \cdots i_m})$, the entries of the form $t_{i \cdots i}$ are the *diagonal entries*. The other entries are *off-diagonal entries*.

A *diagonal mda* is an mda having all off-diagonal entries equal to 0. The *identity mda* denoted by $I_{m,n}$ is a diagonal mda having all diagonal entries equal to 1.

The set of hypercubic mdas of order 2 corresponds to the set of square matrices.

Recall that the tensor space $V_1 \otimes \cdots \otimes V_m$ is the set of tensors of order m and *size* $\mathsf{size}(T) = I_1 \times \cdots \times I_m$, where $I_n = \dim(V_n)$ for $1 \leqslant n \leqslant m$.

By considering a basis $B_n = \{\boldsymbol{b}_1^{(n)}, \ldots, \boldsymbol{b}_{I_n}^{(n)}\}$ in each vector space V_n, T can be written as

$$T = \sum_{i_1}^{I_1} \cdots \sum_{i_m=1}^{I_m} t^{i_1 \cdots i_m} \boldsymbol{b}_{i_1}^{(1)} \otimes \cdots \otimes \boldsymbol{b}_{i_m}^{(m)},$$

where $t^{i_1 \cdots i_m}$ are the contravariant components of T. These components define an mda relative to the bases $B_1, \ldots, B_m$. If the bases $B_1, \ldots, B_m$ are fixed, we may identify the tensor T with this mda that consists of the entries $t^{i_1 \cdots i_m}$. The components of the mda T will be written using lower indices.

As usual, the ith entry of a vector $\boldsymbol{a}$ is denoted by a_i, and the element (i, j) of a matrix A is denoted by a_{ij}. As stated previously, the element (i, j, k) of a third-order mda T is denoted by T_{ijk} or with $T(i, j, k)$, etc.

Subarrays are formed when a subset of the indices is fixed. For matrices, these are the rows and columns.

A colon is used to indicate all elements of a mode. Thus, the jth column of a matrix A is denoted by $A(:, j)$, and the ith row of A is denoted by $A(i, :)$.

Fibers are the higher-order analogue of matrix rows and columns. A fiber of an mda is defined by fixing every index but one. A matrix column is a mode-1 fiber and a matrix row is a mode-2 fiber.

Third-order mdas have *column*, *row*, and *tube* fibers, denoted by $T(:, j, k)$, $T(i, :, k)$, and $T(i, j, :)$, respectively, as shown in Figure 15.1.

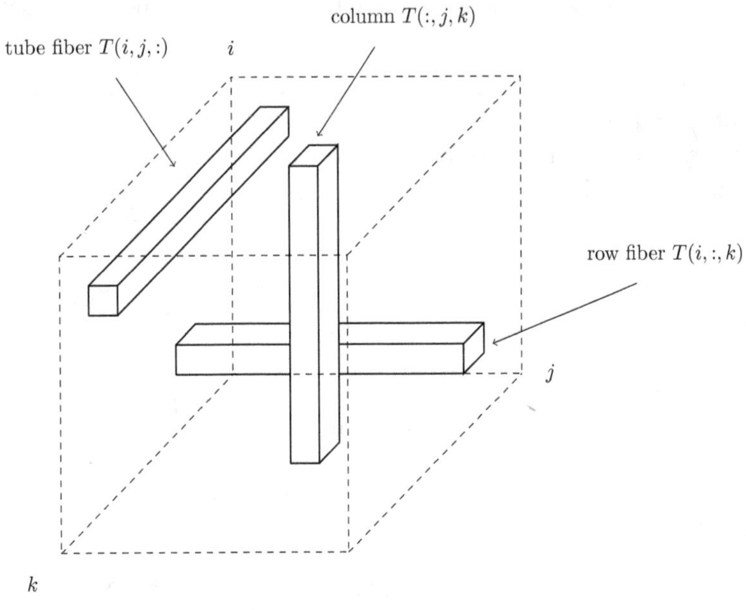

Fig. 15.1 Fibers of a third-order mda T.

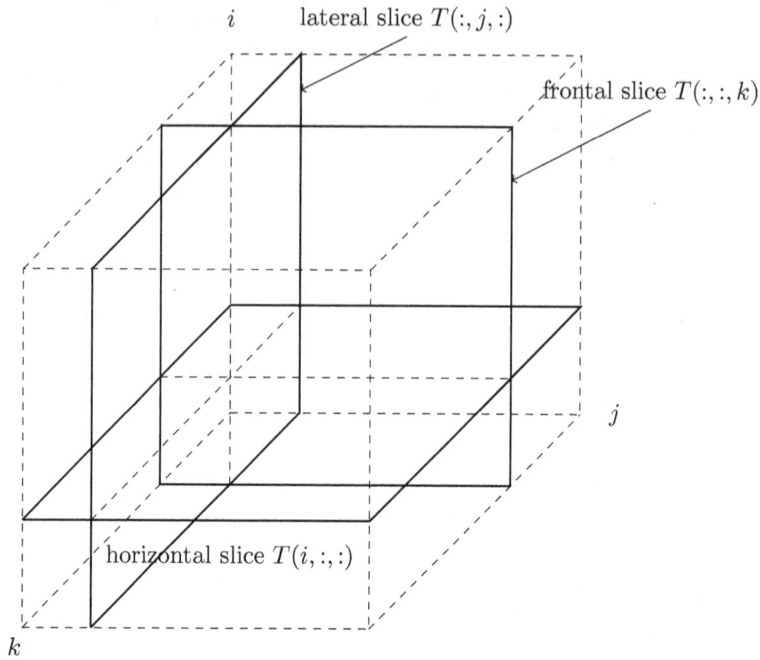

Fig. 15.2 Horizontal, lateral, and frontal slices of a third-order mda T.

When extracted from an mda, fibers are always assumed to be oriented as column vectors.

Slices are two-dimensional sections of an mda, defined by fixing all but two indices. Figure 15.2 shows the horizontal, lateral, and frontal slices of a third-order mda T, denoted by $T(i, :, :)$, $T(:, j, :)$, and $T(:, :, k)$, respectively.

The notion of symmetric mda corresponds to the notion of symmetric tensor introduced in Definition 14.5.

Definition 15.2. An mda is *symmetric* if its elements remain constant under every permutation of the indices.

An alternative term used for symmetric mdas and tensors is the term *supersymmetric*.

Example 15.1. Let T be an mda of order 3 in $\mathbb{R}^{I \times I \times I}$. T is symmetric if $T_{ijk} = T_{ikj} = T_{jik} = T_{jki} = T_{kij} = T_{kji}$ for every $i, j, k \in I$.

An mth order, n-dimensional array T defines a homogeneous polynomial $f_T(\boldsymbol{x})$ as

$$f_T(\boldsymbol{x}) = \sum_{1 \leqslant i_1, i_2, \ldots, i_m \leqslant n} T_{i_1 i_2 \cdots i_m} x_{i_1} x_{i_2} \cdots x_{i_m}.$$

Conversely, a homogeneous polynomial $f(\boldsymbol{x})$ of degree m on $\mathbb{R}^n$ determines a symmetric mda.

15.3 Outer Products

Definition 15.3. Let T and S be two mdas with $\dim(T) = (K_1, \ldots, K_m)$ and $\dim(S) = (L_1, \ldots, L_p)$. Their *outer product* $T \circ S$ is an mda with $\dim(T \circ S) = (K_1, \ldots, K_m, L_1, \ldots, L_p)$ and

$$(T \circ S)(i_1, \ldots, i_m, j_1, \ldots, j_p) = T(i_1, \ldots, i_m) S(j_1, \ldots, j_p).$$

Example 15.2. The outer product of two vectors $\boldsymbol{u} \in \mathbb{R}^m$ and $\boldsymbol{v} \in \mathbb{R}^n$ was introduced in Definition 3.47 as the matrix $\boldsymbol{u} \circ \boldsymbol{v} = \boldsymbol{u}\boldsymbol{v}'$. Under Definition 15.3, we have

$$(\boldsymbol{u} \circ \boldsymbol{v})_{ij} = u_i v_j,$$

which shows that the outer product of mda extends the outer product of vectors.

Example 15.3. If T and S are mdas with $\dim(T) = (2, 7, 5)$ and $\dim(S) = (5, 2, 9, 2)$, then $\dim(T \circ S) = (2, 7, 5, 5, 2, 9, 2)$. Assuming that $T(1, 4, 3) = 10$ and $S(3, 2, 7, 1) = 11$, we have

$$(T \circ S)(1, 4, 3, 3, 2, 7, 1) = 110.$$

The notion of outer product can be extended to an arbitrary finite number of vectors.

Definition 15.4. Let $\boldsymbol{u}_1 \in \mathbb{C}^{I_1}, \ldots, \boldsymbol{u}_k \in \mathbb{C}^{I_k}$. The *outer product* $\boldsymbol{u}_1 * \cdots * \boldsymbol{u}_k$ is the mda $T = \boldsymbol{u}_1 \circ \cdots \circ \boldsymbol{u}_k$, where

$$T_{i_1 i_2 \cdots i_k} = (\boldsymbol{u}_1)_{i_1} (\boldsymbol{u}_2)_{i_2} \cdots (\boldsymbol{u}_k)_{i_k}$$

for $1 \leqslant i_p \leqslant I_p$ and $1 \leqslant p \leqslant k$.

Theorem 15.1. *Let T, S, R be three mdas. The outer product is associative, that is, $(T \circ S) \circ R = T \circ (S \circ R)$.*

Proof. The proof is immediate and it is left to the reader. □

15.4 Tensor Rank

The *n-mode vectors* of an mda $T \subseteq \mathbb{C}^{I_1 \times I_2 \times \cdots \times I_N}$ are the I_n-dimensional vectors obtained from T by varying the index i_n and keeping the other indices fixed. The *n-rank of* T denoted by $R_n = \mathsf{rank}_n(T)$ is the dimension of the vector space spanned by the n-mode vectors.

Note that the n-mode vectors of T are the column vectors of the matrix unfolding $T_{(n)}$ and that $\mathsf{rank}_n(T) = \mathsf{rank}(A_{(n)})$. As we shall see, the different n-ranks of an mda are not necessarily equal.

Definition 15.5. A *simple tensor* or a *tensor of rank* 1 in T^p_q is a tensor of the form $\boldsymbol{v}_1 \otimes \cdots \otimes \boldsymbol{v}_p \otimes \boldsymbol{f}^1 \otimes \cdots \otimes \boldsymbol{f}^q$.

In terms of multidimensional arrays, an m-mode mda $T \subseteq \mathbb{R}^{I_1 \times I_2 \times \cdots \times I_m}$ is of rank 1 if its $(i_1, \ldots, i_m)$-component can be written as a product of n vectors:

$$T(i_1, \ldots, i_n) = \boldsymbol{v}_1(i_1) \cdots \boldsymbol{v}_m(i_m)$$

for $1 \leqslant i_j \leqslant I_j$ and $1 \leqslant j \leqslant n$.

The representation of a simple tensor product of vectors is not unique in general. For example, we can easily verify that

$$\begin{aligned}&(a\boldsymbol{v}_1) \otimes \boldsymbol{v}_2 \cdots \otimes \boldsymbol{v}_p \otimes \boldsymbol{v}^1 \otimes \cdots \otimes \boldsymbol{v}^q\\&= \boldsymbol{v}_1 \otimes (a\boldsymbol{v}_2) \otimes \cdots \otimes \boldsymbol{v}_p \otimes \boldsymbol{v}^1 \otimes \cdots \otimes \boldsymbol{v}^q,\end{aligned}$$

which justifies the above claim.

Definition 15.6. The *rank* of a tensor $T \in T^p_q$ is the minimum number of simple tensors (tensors of rank 1) that sum up to T.

If $T = T_1 + \cdots + T_r$, where $T_1, \ldots, T_r$ are simple tensors, we say that $(T_1, \ldots, T_r)$ is a *realization* of the T.

Example 15.4. Let V_1, V_2, V_3 be three real vector spaces and let $\boldsymbol{v}_i \in V_i$ for $1 \leqslant i \leqslant 3$. The tensor $2\boldsymbol{v}_1 \otimes 3\boldsymbol{v}_2 \otimes \boldsymbol{v}_3$ is a simple tensor.

A simple tensor can be written in several equivalent ways. Note that t is the same as the tensor $\boldsymbol{v}_1 \otimes 6\boldsymbol{v}_2 \otimes \boldsymbol{v}_3$; however, $(2\boldsymbol{v}_1, 3\boldsymbol{v}_2, \boldsymbol{v}_3)$ and $(\boldsymbol{v}_1, 6\boldsymbol{v}_2, \boldsymbol{v}_3)$ are distinct when considered as elements of the direct product $V_1 \times V_2 \times V_3$.

Example 15.5. A matrix $A \in \mathbb{C}^{p\times q}$ can be regarded as a tensor of type $(1,1)$. As we saw in Lemma 9.2, a matrix A of rank r can be written as a sum of matrices of rank 1,

$$A = \sum_{i=1}^{r} \sigma_i \boldsymbol{u}_i \boldsymbol{v}_i^{\mathsf{H}},$$

and the rank r gives the minimum number of terms in the right member of the equality.

If we regard the rank-1 matrices $\sigma_i \boldsymbol{u}_i \boldsymbol{v}_i^{\mathsf{H}}$ as corresponding to simple tensors, it follows that the notion of tensor rank is a generalization of the notion of matrix rank.

The terms *three-way array* and *triad* were introduced in [97]. A three-way array is an array of numbers x_{ijk} subscripted by three indices. A *triad* is a three-way array that has multiplicative form, that is an array $X = (x_{ijk})$, where $x_{ijk} = a_i b_j c_k$, $1 \leqslant i \leqslant I$, $1 \leqslant j \leqslant J$, and $1 \leqslant k \leqslant K$.

Definition 15.7. Let $A \in \mathbb{R}^{I\times R}$, $B \in \mathbb{R}^{J\times R}$, and $C \in \mathbb{R}^{K\times R}$ be three matrices with the same number of columns.

The *triple product* of these matrices is the three-way array $T = [A, B, C]$ whose (i, j, k)-element is $\sum_{r\in R} a_{ir} b_{jr} c_{kr}$.

A triple product exists only if A, B, and C have the same number of columns R.

If $T = [A, B, C]$, the triplet $[A, B, C]$ is a *decomposition* of the mda T. The *rank* of $[A, B, C]$ is the number of columns R in each of the matrices A, B, C. The *rank of a three-way array* T, $\mathsf{rank}(T)$, is the smallest rank of any decomposition of T as $T = [A, B, C]$.

Definition 15.8. Let T be a three-way array. A *k-slab* of T is a matrix obtained from T by fixing the kth index for $1 \leqslant k \leqslant 3$. The dimension of the linear space generated by the k-slabs of T is denoted by $\dim_k(T)$.

T is said to be *k-nondegenerate* if $\dim_k(T)$ equals the number of slabs T_k, where $k \in \{1, 2, 3\}$.

If a three-way $I \times J \times K$-array T degenerates to a $I \times J$-matrix because $K = 1$, then $\dim_1(T)$ and $\dim_2(T)$ correspond to the row rank and column rank of the $I \times J$ matrix.

These notions generalize corresponding notions from matrices. Indeed, if $[A, B, C]$ is a decomposition with $K = 1$, then the array degenerates to an $I \times J$-matrix, $\dim_1(T)$ is the row rank, $\dim_2(T)$ is the column rank, and $\dim_3(T) \in \{1, 0\}$.

Note that $\mathsf{rank}(T) \geqslant \dim_\nu(T)$ for $\nu \in \{1, 2, 3\}$.

Let $T = (t_{ijk})$ be a third-order array where $1 \leqslant i \leqslant I$, $1 \leqslant j \leqslant J$, $1 \leqslant k \leqslant K$. If U, V, W are matrices, define the third-order arrays $UT = \sum_{i=1}^{I} u_{\ell i} t_{ijk}$, $\overset{V}{T} = \sum_{j=1}^{J} v_{\ell j} t_{ijk}$, and $TW = \sum_{i=1}^{K} w_{\ell k} t_{ijk}$.

Lemma 15.1. *Let $T = (t_{ijk}) \in \mathbb{R}^{I \times J \times K}$ be a three-way* mda *and let $U \in \mathbb{R}^{L \times I}, W \in \mathbb{R}^{K \times G}$ be two matrices. We have the associativity property $(UT)W = U(TW)$.*

Proof. Note that

$$(UT)_{\ell jk} = \sum_{i=1}^{I} u_{\ell i} t_{ijk}$$

and

$$((UT)W)_{\ell jg} = \sum_{k=1}^{K} (UT)_{\ell jk} w_{kg} = \sum_{k=1}^{K} \sum_{i=1}^{I} \sum_{i=1}^{I} u_{\ell i} t_{ijk} w_{kg}.$$

On the other hand,

$$(TW)_{ijg} = \sum_{k=1}^{K} t_{ijk} w_{kg}$$

and

$$(U(TW))_{\ell jg} = \sum_{i=1}^{I} u_{\ell i} (TW)_{ijg} = \sum_{i=1}^{I} u_{\ell i} \sum_{k=1}^{K} t_{ijk} w_{kg}.$$

A change of the summation signs yields the associativity property. □

Theorem 15.2. *Let $A \in \mathbb{R}^{I \times R}$, $B \in \mathbb{R}^{J \times R}$, $C \in \mathbb{R}^{K \times R}$ and let $T = [A, B, C]$ be a third-order array. If the matrix products UA, BV, and WC are defined, then*

$$U \overset{V}{T} W = [UA, VB, WC].$$

Proof. Indeed, we have

$$[UA, VB, WC]_{ijk} = \left(\sum_r (UA)_{ir}(VB)_{jr}(WC)_{kr})\right)_{ijk}$$
$$= \sum_r \left(\sum_p u_{ip}a_{pr}\right)\left(\sum_q v_{jq}b_{qr}\right)\left(\sum_s w_{ks}c_{sr}\right).$$

□

Let $X \in \mathbb{R}^{n\times p}$, $U \in \mathbb{C}^{m\times n}$, and $V \in \mathbb{R}^{p\times q}$. The inequality of Frobenius (Corollary 3.5) for these matrices:

$$\mathsf{rank}(X) \geqslant \mathsf{rank}(UX) + \mathsf{rank}(XW) - \mathsf{rank}(UXW)$$

can be extended as follows.

Theorem 15.3 (Kruskal's theorem for three-way arrays). *Let X be a three-way* mda *and let U, V, W be matrices. We have*

$$\mathsf{rank}(X) \geqslant \dim_1(UX) + \mathsf{rank}(\overset{V}{X}\, W) - \dim_1(U\,\overset{V}{X}\, W).$$

Proof. Let $R = \mathsf{rank}(X)$ and let $X = [A, B, C]$ be a minimal representation of X, so (A, B, C) has rank R.

Note that the 1-slabs of the three-way array $[I_R, B, C]$ consist of R matrices of rank 1, which are outer products of the columns of B and the columns of C. Furthermore, each of the I_R slabs of $[A, B, C]$ is a linear combination of these dyads and the entries of A are the coefficients because $[A, B, C] = A[I_R, B, C]$.

The slabs of $[I_R, B, C]$ are linearly independent for otherwise there would be a linear relationship between these slabs, which would allow one of these slabs to be expressed as a linear combination of the others. Using this, the slabs of X could be expressed as a linear combination of $R-1$ dyads, which would contradict the fact that $\mathsf{rank}(X) = R$.

Define $R_1 = \dim_1(UX) = \dim_1(UA[I_R, B, C]) = \mathsf{rank}(UA)$, where the last equality follows from the fact that $[I_R, B, C]$ is non-degenerate.

Let U_1 be the matrix formed by R_1 independent rows of UA. There exists a matrix P_1 formed by a set of distinct rows of an identity matrix and a matrix Q_1 such that

$$U_1 = P_1(UA) \text{ and } UA = Q_1 U_1.$$

Let $R_2 = \dim_1(U \overset{V}{X} W)$. We have

$$R_2 = \dim_1(UA[I, VB, Wc]) = \dim_1(U_1[I, VB, WC]).$$

Since $R_2 \leqslant R_1$, $U_1[I, VB, WC]$ has R_1 slabs which span a space of dimension R_2. There are $R_1 - R_2$ independent linear combinations of slabs that are 0, that is, there is a $(R_1 - R_2) \times R_1$ matrix T of full row rank such that $TU_1[I, VB, WC] = 0$. Therefore,

$$\dim_1[I, VB, WC] \leqslant R - \mathsf{rank}(TU_1) = R - (R_1 - R_2),$$

so we can select some set of $R - (R_1 - R_2)$ slabs of $[I, VB, WC]$ (all dyads) which span the entire space. Then

$$X \overset{V}{W} = A[I, VB, WC]$$

can be represented using these dyads, so

$$\mathsf{rank}(X \overset{V}{W}) \leqslant R - R_1 + R_2.$$

□

Of special interest are sparse mdas, that is, mdas having a large majority of entries equal to 0.

Let $T \subseteq \mathbb{R}^{I \times J \times K}$ be a three-mode mda having P non-zero entries. If we were to store all components of T, we would need IJK locations. If P is the number of non-zero components of T and P is much smaller than IJK, then we need to store P non-zero values and $3P$ index values for a total of $4P$ values.

Example 15.6. The following example is given in [13]. Let

$$\boldsymbol{a} = \begin{pmatrix} 2 \\ 1 \end{pmatrix}, \boldsymbol{b} = \begin{pmatrix} 5 \\ -1 \end{pmatrix}, \text{ and } \boldsymbol{c} = \begin{pmatrix} 3 \\ 1 \\ 4 \end{pmatrix}.$$

The triad $[\boldsymbol{a}, \boldsymbol{b}, \boldsymbol{c}]$ consists of two lateral slices $5\boldsymbol{a}\boldsymbol{c}'$ and $-\boldsymbol{a}\boldsymbol{c}'$. This triad has the frontal slices $3\boldsymbol{a}\boldsymbol{b}', \boldsymbol{a}\boldsymbol{b}'$, and $4\boldsymbol{a}\boldsymbol{b}'$, and the horizontal slices $2\boldsymbol{b}\boldsymbol{c}'$ and $\boldsymbol{b}\boldsymbol{c}'$ (Figure 15.3).

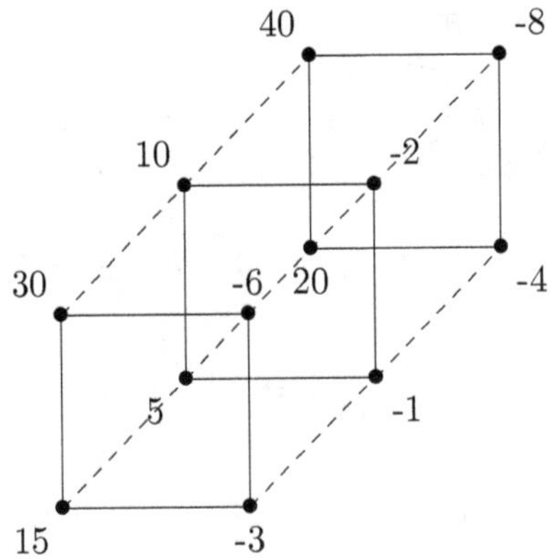

Fig. 15.3 Triad representation.

If an mda T can be written as $T = \sum_{i=1}^{r} \boldsymbol{a}_i \otimes \boldsymbol{b}_i \otimes \boldsymbol{c}_i$, then a jth frontal slice X_j of T can be written as $\sum_{i=1}^{r} \boldsymbol{a}_i \times \boldsymbol{b}_i \times c_{ij}$, or as AC_jB', where $A = (\boldsymbol{a}_1 \ \cdots \ \boldsymbol{a}_r)$, $B = (\boldsymbol{b}_1 \ \cdots \ \boldsymbol{b}_r)$, and $C_j = \mathsf{diag}(c_1, \ldots, c_r)$ is a diagonal matrix.

Conversely, if each frontal slice X_j of an mda T can be written as AC_jB' for some diagonal matrix $C_j \in \mathbb{R}^{r \times r}$, then the mda can be written as $T = \sum_{i=1}^{r} \boldsymbol{a}_i \otimes \boldsymbol{b}_i \otimes \boldsymbol{c}_i$.

It follows that the rank of a $2 \times n \times n$ mda, having the frontal slices $X_1, X_2 \in \mathbb{R}^{n \times n}$ equals the least r for which X_1 and X_2 can be decomposed as $X_1 = ADB'$ and $X_2 = AEB'$ for some matrices $A, B \in \mathbb{R}^{n \times r}$ and diagonal matrices $D, E \in \mathbb{R}^{r \times r}$.

For a $2 \times 2 \times 2$ mda T, the rank of this mda is 0 if and only if T is an mda of zeros. Next, we assume that T is a non-zero mda.

A $2 \times 2 \times 2$ mda containing a non-singular 2×2 slice cannot have a rank less than 2 because we can rearrange the array such that this slice becomes one of the frontal slices, and the previous considerations apply. Conversely, every non-zero $2 \times 2 \times 2$ mda that does not have a non-singular slice in any direction has either rank 1, or it is superdiagonal, which means that

$$X_1 = \begin{pmatrix} \alpha & 0 \\ 0 & 0 \end{pmatrix} \text{ and } X_2 = \begin{pmatrix} 0 & 0 \\ 0 & \beta \end{pmatrix}$$

for some $\alpha \neq 0$ and $\beta \neq 0$.

Indeed, consider the case when some slice is the zero matrix. The assumed singularity of the other slice implies that the mda has rank 1.

If we ignore this case, and assume that $X_1 = (\boldsymbol{x}|\lambda\boldsymbol{x})$, where $\boldsymbol{x} \neq \mathbf{0}_2$ and $\lambda \in \mathbb{R}$, the singularity of the lateral slices yields $X_2 = (\delta\boldsymbol{x}|\epsilon\boldsymbol{x})$ for some $\delta, \epsilon \in \mathbb{R}$. If $\lambda = 0$, then either this expression is still valid for a $\delta \neq 0$, or we have $X_2 = (0|\boldsymbol{y})$ for an arbitrary $\boldsymbol{y}$ (a special case to be treated later).

The singularity of the horizontal slices implies

$$\begin{pmatrix} 1 \\ \lambda \end{pmatrix} = \mu \begin{pmatrix} \delta \\ \epsilon \end{pmatrix}$$

for some $\mu \neq 0$. So, $X_1 = \boldsymbol{x}(1\ \lambda) = \boldsymbol{x}(\mu\delta, \mu\epsilon)$ and $X_2 = \boldsymbol{x}(\delta\ \epsilon)$, and a rank 1 solution has been constructed.

Suppose now that $X_1 = (\boldsymbol{x}\ \mathbf{0}_2)$ and $X_2 = (\mathbf{0}_2\ \boldsymbol{y})$ for arbitrary $\boldsymbol{x}$ and $\boldsymbol{y}$. Since the horizontal slices are of rank 1, X has the superdiagonal form, defined by having precisely two non-zero elements that never occur in the same slice. This array has rank 2.

We conclude that for a $2 \times 2 \times 2$ array the rank is 0 if and only if the array has only 0 entries, the rank is 1 if and only if each of the six slices is singular, except the superdiagonal matrix which has rank two.

In the next theorem proved initially by Kruskal [97] and reproved (with a new argument by ten Berge in [13]), we deal with $2 \times 2 \times 2$ mdas for which at least one slice is non-singular, and therefore, has rank at least 2.

Theorem 15.4. *Let $T \in \mathbb{R}^{2\times2\times2}$ be an mda which has a slice X_1 that is nonsingular. The rank of T is at most* 3.

Proof. Let $X_1, X_2 \in \mathbb{R}^{2\times2}$, where X_1 is non-singular. The system

$$X_1 = ADB' \text{ and } X_2 = AEB'$$

has a solution if and only if a solution for

$$X_1X_1^{-1} = ADB'X_1^{-1} \text{ and } X_2X_1^{-1} = AEB'X_1^{-1}.$$

Thus, we consider solutions for the system

$$Y_1 = I_2 = ADB'X_1^{-1} \text{ and } Y_2 = X_2X_1^{-1} = AEB'X_1^{-1}. \qquad (15.1)$$

Let

$$Y_2 = \begin{pmatrix} y_{11} & y_{12} \\ y_{21} & y_{22} \end{pmatrix}.$$

It is easy to verify that the matrices

$$A = \begin{pmatrix} 1 & 0 & y_{12} \\ 0 & 1 & y_{21} \end{pmatrix}, \; (X_1')^{-1}B = \begin{pmatrix} 1 & 0 & 1 \\ 0 & 1 & 1 \end{pmatrix}$$
$$D = \begin{pmatrix} 1 & 0 & 0 \\ 0 & 1 & 0 \\ 0 & 0 & 0 \end{pmatrix}, \; E = \begin{pmatrix} y_{11} - y_{12} & 0 & 0 \\ 0 & y_{22} - y_{21} & 0 \\ 0 & 0 & 1 \end{pmatrix}$$

form a solution of the system, where B is derived from $(X_1')^{-1}B$. Thus, a $2 \times 2 \times 2$ mda cannot have a rank greater than 3. □

Suppose that X_1, the first frontal slice of the mda T, is nonsingular. If $X_2 = \lambda X_1$, then T has rank 2. Consider therefore the case when such a λ does not exist and suppose that the mda has rank 2. Then, it follows from the equalities $X_1 = ADB'$ and $X_2 = AEB'$ that if $A = (\boldsymbol{a}_1\ \boldsymbol{a}_2)$ and $B = (\boldsymbol{b}_1\ \boldsymbol{b}_2)$, then there are scalars d_1, d_2, e_1, e_2 such that

$$X_1 = d_1 \boldsymbol{a}_1 \boldsymbol{b}_1' + d_2 \boldsymbol{a}_2 \boldsymbol{b}_2'$$

and

$$X_2 = e_1 \boldsymbol{a}_1 \boldsymbol{b}_1' + e_2 \boldsymbol{a}_2 \boldsymbol{b}_2',$$

with the nonzero diagonal entries d_1, d_2 of D, and

$$\frac{e_1}{d_1} \neq \frac{e_2}{d_2}.$$

Both $X_2 - \frac{e_1}{d_1} X_1$ and $X_2 - \frac{e_2}{d_2} X_1$ are singular, hence,

$$\det(X_2 - \lambda X_1) = 0 \tag{15.2}$$

must have two distinct real roots. The last equality can be written as

$$\lambda^2 \det(X_1) + \lambda \left(x_{21}^{(1)} x_{12}^{(2)} + x_{12}^{(1)} x_{21}^{(2)} - x_{11}^{(1)} x_{22}^{(2)} - x_{22}^{(1)} x_{22}^{(2)} \right) + \det(X_2) = 0,$$

where $x_{ij}^{(k)}$ is the ijth element of X_k. Replacing X_1 and X_2 with $Y_1 = I_2$ and $Y_2 = X_2 X_1^{-1}$, this equality becomes

$$\lambda^2 - \lambda(\mathsf{trace}(Y_2)) + \det(Y_2) = 0, \tag{15.3}$$

which has a positive discriminant if and only if

$$(\mathsf{trace}(Y_2))^2 > 4\det(Y_2). \tag{15.4}$$

Thus, Inequality (15.4) is necessary for a rank-2 solution. Conversely, if Inequality (15.4) is satisfied, then Equation (15.2) has two real distinct solutions λ_1 and λ_2, which allows us to define the rank-1 matrices as follows:

$$\begin{aligned}\boldsymbol{u}_1\boldsymbol{v}_1' &= (\lambda_1 - \lambda_2)^{-1}(X_2 - \lambda_2 X_1),\\ \boldsymbol{u}_2\boldsymbol{v}_2' &= -(\lambda_1 - \lambda_2)^{-1}(X_2 - \lambda_1 X_1).\end{aligned}$$

Note that $\boldsymbol{u}_1\boldsymbol{v}_1' + \boldsymbol{u}_2\boldsymbol{v}_2' = X_1$ and $\lambda_1\boldsymbol{u}_1\boldsymbol{v}_1' + \lambda_2\boldsymbol{u}_2\boldsymbol{v}_2' = X_2$, which shows that a rank-2 solution has been constructed, where $A = (\boldsymbol{u}_1\ \boldsymbol{u}_2)$, $B = (\boldsymbol{v}_1\ \boldsymbol{v}_2)$, $D = I_2$, and $E = \mathsf{diag}(\lambda_1, \lambda_2)$. This shows that Equality (15.4) is necessary and sufficient for a $2 \times 2 \times 2$ mda to have rank 2, assuming that X_1 is non-singular, and that X_2 is not proportional to X_1.

Note that ten Berge characterization of mdas of rank 2 is obtained by excluding the case where all slices are singular, and assuming that X_2 is not proportional to X_1.

Example 15.7. Consider the $2 \times 2 \times 2$ mda T having the frontal slices

$$X_1 = \begin{pmatrix} 1 & 0 \\ 0 & 1 \end{pmatrix} \text{ and } X_2 = \begin{pmatrix} 0 & -1 \\ 1 & 0 \end{pmatrix}.$$

The inverse of X_2 is $\begin{pmatrix} 0 & 1 \\ -1 & 0 \end{pmatrix}$, hence $Y_1 = I_2$ and

$$Y_2 = X_2 X_1^{-1} = X_2.$$

By Theorem 15.4, the rank of T is at most 3. We have $\mathsf{trace}(Y_2) = 0$ and $\det(Y_2) = 1$. Thus, Inequality (15.4) fails, hence, T has rank 3.

However, if we regard T as a complex mda, that is as an mda over $\mathbb{C}^{2\times2\times2}$, note that we have

$$T = \frac{1}{2}(1\ \ -i)\otimes(1\ \ i)\otimes(1\ \ -i) + \frac{1}{2}(1\ \ i)\otimes(1\ \ -i)\otimes(1\ \ i),$$

that is, T has rank 2 over the complex field $\mathbb{C}$. Thus, unlike matrices, the rank of a tensor depends on the field of definition.

The rank of an mda is not necessarily equal to any n-rank. Examples 15.8 and 15.9 that explore the relationship between the n-ranks and the rank of an mda originate in [100].

Example 15.8. Consider the $2 \times 2 \times 2$ mda defined in Figure 15.4 given by:

This mda is created by creating a matrix

```
A = [1 0;0 1]
```

expanding it, and transforming it into an mda:

```
A =

     1     0
     0     1

>> A(:,:,2)=[1 0;0 0]

A(:,:,1) =

     1     0
     0     1
```

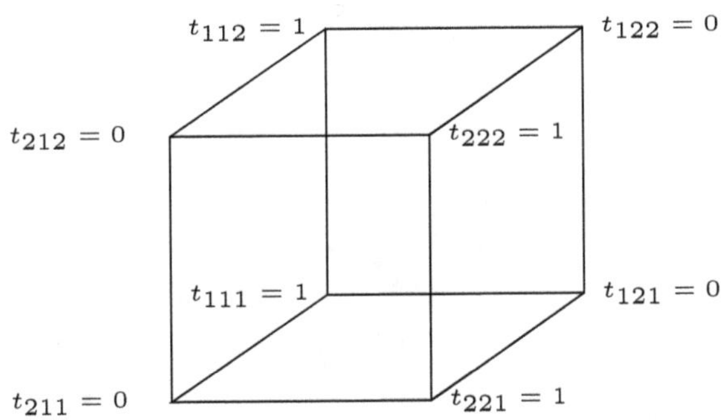

Fig. 15.4 A $2 \times 2 \times 2$-mda.

```
A(:,:,2) =

     1     0
     0     1

>> T=tensor(A)
T is a tensor of size 2 x 2 x 2
    T(:,:,1) =
         1     0
         0     1
    T(:,:,2) =
         1     0
         0     1
```

The three unfoldings of the mda T are obtained using the functions `tensor_as_matrix` of the package 862 described in [4]. For example, to obtain the unfolding after the first dimension, we write `tensor_as_matrix(T,1,'fc')`. The unfoldings produced on the first dimension in this manner are

```
>> tensor_as_matrix(T,1,'fc')

ans is a matrix corresponding to a tensor of size 2 x 2 x 2
ans.rindices = [1] (modes of tensor corresponding to rows)
ans.cindices = [2, 3] (modes of tensor corresponding to columns)

ans.data =
     1     0     1     0
     0     1     0     1
```

Similar unfoldings yield

```
>> tensor_as_matrix(T,2,'fc')

ans.data =
     1     1     0     0
     0     0     1     1

>> tensor_as_matrix(T,3,'fc')

ans.data =
     1     0     0     1
     1     0     0     1
```

Clearly, we have the ranks $R_1 = R_2 = 2$ and $R_3 = 1$, so the ranks can be distinct, and, in turn, can be distinct from the rank of the mda.

Since $T = \boldsymbol{e}_1 \otimes \boldsymbol{e}_2 \otimes (\boldsymbol{e}_1 + \boldsymbol{e}_2) + \boldsymbol{e}_2 \otimes \boldsymbol{e}_2 \otimes (\boldsymbol{e}_1 + \boldsymbol{e}_2)$, the rank of T cannot be larger than 2, which shows that T is indeed of rank 2.

Example 15.9. Let T be the $2 \times 2 \times 2$ mda defined in Figure 15.5. The n-rank is 2 for $1 \leqslant n \leqslant 3$. Since

$$T = \boldsymbol{e}_2 \otimes \boldsymbol{e}_1 \otimes \boldsymbol{e}_1 + \boldsymbol{e}_1 \otimes \boldsymbol{e}_2 \otimes \boldsymbol{e}_1 + \boldsymbol{e}_1 \otimes \boldsymbol{e}_1 \otimes \boldsymbol{e}_2, \tag{15.5}$$

it follows that the rank of T is not larger than 3.

To prove that the rank of T equals 3, we need to show that the decomposition 15.5 is minimal.

Suppose that T would have rank 2, that is,

$$T = \boldsymbol{x}_1 \otimes \boldsymbol{y}_1 \otimes \boldsymbol{z}_1 + \boldsymbol{x}_2 \otimes \boldsymbol{y}_2 \otimes \boldsymbol{z}_2, \tag{15.6}$$

and let $\boldsymbol{X}, \boldsymbol{Y}, \boldsymbol{Z}, D_1$, and D_2 be defined as

$$\boldsymbol{X} = (\boldsymbol{x}_1\ \boldsymbol{x}_2), \boldsymbol{Y} = (\boldsymbol{y}_1\ \boldsymbol{y}_2), \boldsymbol{Z} = (\boldsymbol{z}_1\ \boldsymbol{z}_2),$$

and

$$D_1 = \mathsf{diag}(z_{11}, z_{12}) \text{ and } D_2 = \mathsf{diag}(z_{21}, z_{22}).$$

Note that

$$\begin{aligned} \boldsymbol{X} D_1 \boldsymbol{Y}' &= (\boldsymbol{x}_1\ \boldsymbol{x}_2) \begin{pmatrix} z_{11} & 0 \\ 0 & z_{12} \end{pmatrix} \begin{pmatrix} \boldsymbol{y}_1' \\ \boldsymbol{y}_2' \end{pmatrix} \\ &= \boldsymbol{x}_1 \boldsymbol{y}_1' z_{11} + \boldsymbol{x}_2 \boldsymbol{y}_2' z_{12} = \begin{pmatrix} 0 & 1 \\ 1 & 0 \end{pmatrix}. \end{aligned}$$

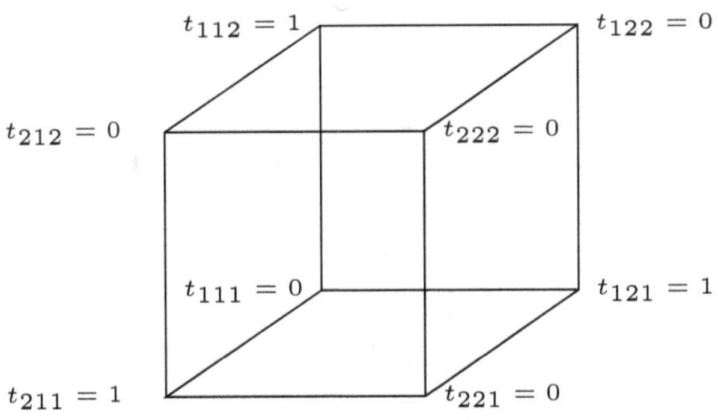

Fig. 15.5 A $2 \times 2 \times 2$-mda.

Similarly,

$$\boldsymbol{x}D_2\boldsymbol{y}' = \boldsymbol{x}_1\boldsymbol{y}_1' z_{21} + \boldsymbol{x}_2\boldsymbol{y}_2' z_{22} = \begin{pmatrix} 1 & 0 \\ 0 & 0 \end{pmatrix},$$

which means that Equality (15.6) is equivalent to

$$XD_1Y' = \begin{pmatrix} 0 & 1 \\ 1 & 0 \end{pmatrix} \text{ and } XD_2Y' = \begin{pmatrix} 1 & 0 \\ 0 & 0 \end{pmatrix}.$$

The last equalities imply that $\mathsf{rank}(D_1) = 2$ and $\mathsf{rank}(D_2) = 1$. The matrices X and Y have full rank. Without loss of generality, we may assume that $z_{22} = 0$.

Observe now that

$$(XD_2Y') \cdot (XD_1Y')^{-1} = X(D_2D_1)^{-1}X^{-1} = \begin{pmatrix} 0 & 1 \\ 0 & 0 \end{pmatrix}.$$

This equality implies that X_1 is proportional to $\begin{pmatrix} 1 \\ 0 \end{pmatrix}$ while the first row is proportional to $(0\ 1)$. This is possible when X is of rank 1, which conflicts with the initial assumption ($R_1 = 2$).

15.5 Matricization and Vectorization

Matricization [92] of an mda is the rearrangement of the elements of an mda into a matrix.

Let $T \in \mathbb{R}^{I_1 \times I_2 \times \cdots \times I_N}$ be an order-N mda whose set of modes is $\mathcal{N} = \{1, 2, \ldots, N\}$.

For a subset $\mathcal{D} = \{d_1, \ldots, d_{|\mathcal{D}|}\}$ of $\mathcal{N}$, define the function $\psi_{\mathcal{D}} : \prod_{\ell=1}^{N}\{1, \ldots, I_\ell\} \longrightarrow \mathbb{N}$ as

$$\psi_{\mathcal{D}}(i_1, \ldots, i_N) = 1 + \sum_{p=1}^{|\mathcal{D}|} \left[(i_{d_p} - 1) \prod_{j=1}^{p-1} I_{d_j} \right],$$

where $1 \leqslant i_j \leqslant I_j$ for $1 \leqslant j \leqslant |\mathcal{D}|$.

Suppose that $\{\mathcal{R}, \mathcal{C} : I_{\mathcal{N}}\}$ is a partition of $\mathcal{N}$, where

$$\mathcal{R} = \{r_1, \ldots, r_p\}, \mathcal{C} = \{s_1, \ldots, s_q\},$$

and $p+q = N$. The set $\mathcal{R}$ contains those indices that will be mapped into row indices of the resulting matrix, while the set $\mathcal{C}$ contains the indices that will be mapped into the column indices of the resulting matrix.

Let $J = \prod_{n \in \mathcal{R}} I_n$ and $K = \prod_{n \in \mathcal{C}} I_n$ and let π be a bijection

$$\pi : \{1, \ldots, I_1\} \times \cdots \times \{1, \ldots, I_N\} \longrightarrow \{1, \ldots, J\} \times \{1, \ldots, K\}$$

defined by

$$\pi(i_1, \ldots, i_N) = (\psi_{\mathcal{R}}(i_1, \ldots, i_N), \psi_{\mathcal{C}}(i_1, \ldots, i_N)).$$

The matrix $T_{(\mathcal{R},\mathcal{C}:I_N)} \in \mathbb{R}^{J \times K}$ is defined as

$$(T_{(\mathcal{R},\mathcal{C}:I_N)})_{jk} = T_{i_1 i_2 \cdots i_N}, \tag{15.7}$$

with $j = \psi_{\mathcal{R}}(i_1, \ldots, i_n)$ and $k = \psi_{\mathcal{C}}(i_1, \ldots, i_n)$.

Definition 15.9. The *matricized mda* is a matrix $T_{(\mathcal{R},\mathcal{C}:I_N)}$ in $\mathbb{R}^{J \times K}$ defined by Equation (15.7).

This form of matricization was introduced in [50–52].

A special case of matricization occurs when $\mathcal{R} = \{n\}$ and $\mathcal{C} = \{1, \ldots, N\} - \{n\}$. In this case, we use the term *n-unfolding*. The *matrix unfolding*

$$T_{(n)} \in \mathbb{R}^{I_n \times (I_{n+1} \cdots I_N I_1 I_2 \cdots I_{n-1})}$$

of T contains the element $T_{i_1 i_2 \cdots i_N}$ at the position with row number i_n and column number $\psi_{\{d_n\}}(i_1, \ldots, i_N)1 + \sum_{m=1, m \neq n}^{N} \left((i_m - 1) \prod_{p=1, p \neq n}^{m-1} I_p\right)$.

Example 15.10. Let T be an mda with the format $3 \times 4 \times 2$ having the following frontal slices:

$$T_1 = \begin{pmatrix} 1 & 4 & 7 & 10 \\ 2 & 5 & 8 & 11 \\ 3 & 6 & 9 & 12 \end{pmatrix} \text{ and } T_2 = \begin{pmatrix} 13 & 16 & 19 & 22 \\ 14 & 17 & 20 & 23 \\ 15 & 18 & 21 & 24 \end{pmatrix}.$$

For $R = \{1, 2\}$ and $C = \{3\}$, the matrix $T_{R;C} \in \mathbb{R}^{12 \times 2}$ is given by

$$T_{12;3} = \sum_{i_1=1}^{3} \sum_{i_2=1}^{4} \sum_{i_3=1}^{2} T_{i_1 i_2 i_3} \mathbf{e}_{i_1}^{3} \otimes \mathbf{e}_{i_2}^{4} \otimes \mathbf{e}_{i_3}^{2'}.$$

Consider the term $T_{231}\boldsymbol{e}_2^3 \otimes \boldsymbol{e}_3^4 \otimes \boldsymbol{e}_1^{2'}$ from the above sum. Since

$$\boldsymbol{e}_2^3 \otimes \boldsymbol{e}_3^4 \otimes \boldsymbol{e}_1^{2'} = \begin{pmatrix} 0 \\ 1 \\ 0 \end{pmatrix} \otimes \begin{pmatrix} 0 \\ 0 \\ 1 \\ 0 \end{pmatrix} \otimes \begin{pmatrix} 1 \\ 0 \end{pmatrix}' = \begin{pmatrix} 0 & 0 \\ 0 & 0 \\ 0 & 0 \\ 0 & 0 \\ 0 & 0 \\ 0 & 0 \\ 1 & 0 \\ 0 & 0 \\ 0 & 0 \\ 0 & 0 \\ 0 & 0 \\ 0 & 0 \end{pmatrix},$$

it follows that T_{231} occupies the position at the 7th row and the first column of the matrix $T_{R,C}$.

Example 15.11. We present an example of the matricization of an mda given in [93, 139].

Let T be an mda with the format $3 \times 4 \times 2$ introduced in Example 15.10 having the following frontal slices:

$$T_1 = \begin{pmatrix} 1 & 4 & 7 & 10 \\ 2 & 5 & 8 & 11 \\ 3 & 6 & 9 & 12 \end{pmatrix} \text{ and } T_2 = \begin{pmatrix} 13 & 16 & 19 & 22 \\ 14 & 17 & 20 & 23 \\ 15 & 18 & 21 & 24 \end{pmatrix}$$

The three unfoldings are

$$T_{(1)} = \begin{pmatrix} 1 & 4 & 7 & 10 & 13 & 16 & 19 & 22 \\ 2 & 5 & 8 & 11 & 14 & 17 & 20 & 23 \\ 3 & 6 & 9 & 12 & 15 & 18 & 21 & 24 \end{pmatrix},$$

$$T_{(2)} = \begin{pmatrix} 1 & 2 & 3 & 13 & 14 & 15 \\ 4 & 5 & 6 & 16 & 17 & 18 \\ 7 & 8 & 9 & 19 & 20 & 21 \\ 10 & 11 & 12 & 22 & 23 & 24 \end{pmatrix},$$

$$T_{(3)} = \begin{pmatrix} 1 & 2 & 3 & 4 & \cdots & 9 & 10 & 11 & 12 \\ 13 & 14 & 15 & 16 & \cdots & 21 & 22 & 23 & 24 \end{pmatrix}.$$

The fibers of n-mode of the mda T can be retrieved as the columns of the unfolding $T_{(n)}$.

Definition 15.10. Let T and S be two mdas having the set of modes $I \cup J$ and $I \cup K$, respectively, where $|I| = M$, $|J| = N$, and $|K| = P$ and I, J, K are pairwise disjoint. We assume that $I = \{i_1, \ldots, i_M\}$.

The *contraction* of T and S over I is the mda $T \bowtie_I S$ given by

$$(T \bowtie_I S)_{j_1 \cdots j_N k_1 \cdots k_P} = \sum_{i_1=1}^{I_1} \cdots \sum_{i_m=1}^{I_m} T_{i_1 \cdots i_m j_1 \cdots j_n} S_{i_1 \cdots i_m k_1 \cdots k_p}.$$

Definition 15.11. Let T be an $(I_1 \times I_2 \times \cdots \times I_N)$-mda and let $A \in \mathbb{R}^{J_n \times I_n}$ be a matrix.

The *n-mode product of T and A* is an mda denoted by $T \times_n A \in \mathbb{R}^{I_1 \times \cdots \times I_{n-1} \times J_n \times I_{n+1} \times I_N}$ given by

$$(T \times_n A)_{i_1 \cdots i_{n-1} j_n i_{n+1} \cdots i_N} = \sum_{i_n=1}^{I_n} T_{i_1 i_2 \cdots i_N} A_{j_n\, i_n}.$$

The result of the n-mode product $T \times_n A$ is an mda of size $I_1 \times I_2 \cdots \times I_{n-1} \times J_n \times I_{n+1} \cdots \times I_N$-mda. Each of the mode-$n$ fibers of T is multiplied by the matrix A.

Example 15.12. Let $A \in \mathbb{R}^{I_1 \times I_2}$ be a matrix, which we regard as an mda. The product $A \times_1 B$ can be considered for a matrix $B \in J \times I_1$ and yields a matrix $A \times_1 B \in \mathbb{R}^{J \times I_2}$ where

$$(A \times_1 B)_{j i_2} = \sum_{i_1=1}^{I_1} A_{i_1 i_2} B_{j i_1} = \sum_{i_1=1}^{I_1} A_{i_1 i_2} B'_{i_1 j}.$$

In terms of matrix products, this is $A \times_1 B = B'A$.

If $C \in \mathbb{R}^{J \times I_2}$, the product $A \times_2 C \in \mathbb{R}^{I_1 \times J}$ is

$$(A \times_2 C)_{i_1 j} = \sum_{i_2=1}^{I_2} A_{i_1 i_2} C_{j\, i_2} = \sum_{i_2=1}^{I_2} A_{i_1 i_2} C'_{i_2 j}.$$

We have $A \times_2 C = AC'$.

Theorem 15.5. *Let $\mathcal{N} = \{1, \ldots, N\}$, and let the sets $\mathcal{R} = \{r_1, \ldots, r_L\}$ and $\mathcal{C} = \{c_1, \ldots, c_M\}$ define a partition of $\mathcal{N} = \{1, \ldots, N\}$. Also, let Y be an $(J_1 \times J_2 \times \cdots \times J_N)$-mda and let $A^{(n)} \in \mathbb{R}^{I_n \times J_n}$ for $n \in \mathcal{N}$ be N matrices.*

The equality

$$X = Y \times_1 A^{(1)} \times_2 A^{(2)} \cdots \times_N A^{(N)} \tag{15.8}$$

holds if and only if we have the following matrix equality:

$$X_{(\mathcal{R}\times\mathcal{C}:I_{\mathcal{N}})} = (A^{(r_L)}\otimes\cdots\otimes A^{(r_1)})Y_{(\mathcal{R}\times\mathcal{C}:I_{\mathcal{N}})}(A^{(c_M)}\otimes\cdots\otimes A^{(c_1)})'. \tag{15.9}$$

Proof. Equality (15.8) is equivalent to

$$X_{i_1 i_2 \cdots i_N} = \sum_{j_1=1}^{I_1} \cdots \sum_{j_N=1}^{I_N} Y_{j_1 \cdots j_n} A^{(1)}_{i_1\, j_1} \cdots A^{(N)}_{i_N\, j_N}.$$

Observe that

$$X_{i_1 i_2 \cdots i_N} = (X_{(\mathcal{R},\mathcal{C}:I_{\mathcal{N}})})_{\psi_{\mathcal{R}}(i_1,i_2,\cdots,i_N),\psi_{\mathcal{C}}(i_1,i_2,\cdots,i_N)},$$

and

$$Y_{i_1 i_2 \cdots i_N} = (Y_{(\mathcal{R},\mathcal{C}:I_{\mathcal{N}})})_{\psi_{\mathcal{R}}(i_1,i_2,\cdots,i_N),\psi_{\mathcal{C}}(i_1,i_2,\cdots,i_N)}.$$

By the definition of matricization, Equality (15.9) implies

$$\begin{aligned}
&X_{(\mathcal{R}\times\mathcal{C}:I_{\mathcal{N}})}\\
&= \sum_{i_1=1}^{I_1} \cdots \sum_{i_N=1}^{I_N} X_{i_1 \cdots i_N} \left(\otimes_{n\in\mathcal{R}} \boldsymbol{e}^{(I_n)}_{i_n}\right) \left(\otimes_{n\in\mathcal{C}} \boldsymbol{e}^{(I_n)}_{i_n}\right)'\\
&= \sum_{i_1=1}^{I_1} \cdots \sum_{i_N=1}^{I_N} \sum_{j_1=1}^{I_1} \cdots \sum_{j_N=1}^{I_N} Y_{j_1 \cdots j_n} A^{(1)}_{i_1\, j_1} \cdots A^{(N)}_{i_N\, j_N}\\
&\quad \left(\otimes_{n\in\mathcal{R}} \boldsymbol{e}^{(I_n)}_{i_n}\right) \left(\otimes_{n\in\mathcal{C}} \boldsymbol{e}^{(I_n)}_{i_n}\right)'\\
&= \sum_{i_1=1}^{I_1} \cdots \sum_{i_N=1}^{I_N} \sum_{j_1=1}^{I_1} \cdots \sum_{j_N=1}^{I_N} (Y_{(\mathcal{R},\mathcal{C}:I_{\mathcal{N}})})_{\psi_{\mathcal{R}}(j_1,j_2,\cdots,j_N),\psi_{\mathcal{C}}(j_1,j_2,\cdots,j_N)}\\
&\quad A^{(1)}_{i_1\, j_1} \cdots A^{(N)}_{i_N\, j_N} \left(\otimes_{n\in\mathcal{R}} \boldsymbol{e}^{(I_n)}_{i_n}\right) \left(\otimes_{n\in\mathcal{C}} \boldsymbol{e}^{(I_n)}_{i_n}\right)'
\end{aligned}$$

$$= \sum_{i_1=1}^{I_1} \cdots \sum_{i_N=1}^{I_N} \sum_{j_1=1}^{I_1} \cdots \sum_{j_N=1}^{I_N} (Y_{(\mathcal{R},\mathcal{C}:I_N)})_{\psi_{\mathcal{R}}(j_1,j_2,\cdots,j_N),\psi_{\mathcal{C}}(j_1,j_2,\cdots,j_N)}$$

$$A^{(1)}_{i_1\,j_1} \cdots A^{(N)}_{i_N\,j_N} e^{\prod_L^{i=1} r_i}_{\psi_{\mathcal{R}}(i_1,\ldots,i_N)} e^{\prod_M^{j=1} c_j}_{\psi_{\mathcal{C}}(j_1,\ldots,j_n)}$$

$$= (A^{(r_L)} \otimes \cdots \otimes A^{(r_1)}) Y_{(\mathcal{R}\times\mathcal{C}:I_N)} (A^{(c_M)} \otimes \cdots \otimes A^{(c_1)})'. \quad \square$$

Corollary 15.1. *Let X be an $I_1 \times \cdots \times I_N$-mda and let $A^{(i)}$ be a $J_i \times I_i$-matrix for $1 \leqslant i \leqslant N$. We have*

$$Y = X \times_1 A^{(1)} \times_2 A^{(2)} \cdots \times_N A^{(N)}$$

if and only if we have the equality

$$Y_{(n)} = A^{(n)} X_{(n)} (A^{(N)} \otimes \cdots \otimes A^{(n+1)} \otimes A^{(n-1)} \otimes \cdots \otimes A^{(1)})'$$

between the unfoldings $Y_{(n)}$ and $X_{(n)}$ for $1 \leqslant n \leqslant N$.

Let X be an $(I_1 \times I_2 \times \cdots \times I_N)$-mda, let A be a matrix, $A \in \mathbb{R}^{I_n \times J_n}$, and let $Y \in \mathbb{R}^{I_1 \times I_2 \times \cdots J_n \times \cdots \times I_n}$.

We have $Y = X \times_n A$ if and only if $Y_{(n)} = AX_{(n)}$.

Proof. These statements follow from Theorem 15.5. □

Theorem 15.6. *Let $T \in \mathbb{R}^{I_1 \times I_2 \times \cdots \times I_N}$ and let $A \in \mathbb{R}^{J_n \times I_n}$ and $B \in \mathbb{R}^{J_m \times I_m}$ be two matrices with $n \neq m$. We have*

$$(T \times_n A) \times_m B = (T \times_m B) \times_n A,$$

and their common value is denoted by $T \times_n A \times_m B$.

Furthermore, if $C \in \mathbb{R}^{J_n \times I_n}$ and $D \in \mathbb{R}^{K_n \times J_n}$, then

$$(T \times_n C) \times_n D = T \times_n (DC). \tag{15.10}$$

Proof. Both equalities follow immediately from the definition of $\times_n$. □

Next we introduce the *n-mode product* $\overline{\times}_n$ of an mda with a vector.

Definition 15.12. Let T be an $(I_1 \times I_2 \times \cdots \times I_N)$-mda and let $\boldsymbol{v} \in \mathbb{R}^{I_n}$ be a matrix. The *n-mode product of T and $\boldsymbol{v}$* is an mda of order $N-1$ of size $I_1 \times \cdots \times I_{n-1} \times I_{n+1} \times \cdots \times I_N$ denoted by $T\overline{\times}_n\boldsymbol{v}$ given by

$$(T\overline{\times}_n\boldsymbol{v})_{i_1\cdots i_{n-1}i_{n+1}\cdots i_N} = \sum_{i_n=1}^{I_n} T_{i_1 i_2 \cdots i_n} v_{i_n}.$$

Clearly, $T\overline{\times}_n \boldsymbol{v}$ computes the inner product of each n-fiber of T with $\boldsymbol{v}$.

Example 15.13. Let T be a tensor of size $4 \times 3 \times 2$ having the 3rd dimension slices

$$T_1 = \begin{pmatrix} 9 & 4 & 6 \\ 5 & 9 & 1 \\ 8 & 8 & 8 \\ 2 & 9 & 9 \end{pmatrix}$$

and

$$T_2 = \begin{pmatrix} 7 & 6 & 3 \\ 7 & 2 & 1 \\ 7 & 7 & 1 \\ 4 & 1 & 8 \end{pmatrix}.$$

If

$$\boldsymbol{u} = \begin{pmatrix} 10 \\ 20 \\ 30 \end{pmatrix},$$

the product $T\overline{\times}_2 u$ is the mda of size 4×2,

$$T\overline{\times}_1 u = \begin{pmatrix} 350 & 280 \\ 260 & 140 \\ 480 & 240 \\ 470 & 300 \end{pmatrix}.$$

The actual implementation of this computation is further discussed in Example 15.25.

Theorem 15.7. *Let T and S be two* mdas *such that*

$$T \in \mathbb{R}^{I_1 \times \cdots \times I_{n-1} \times J \times I_{n+1} \times \cdots \times I_N}$$
$$S \in \mathbb{R}^{I_1 \times \cdots \times I_{n-1} \times K \times I_{n+1} \times \cdots \times I_N}$$

and let A be a matrix, $A \in \mathbb{R}^{J \times K}$. We have $(T, S \times_n A) = (T \times_n A', S)$.

Proof. We have

$$(T \times_n A')_{i_1 \cdots i_{n-1} k i_{n+1} \cdots I_N} = \sum_{j=1}^{J} T_{i_1 \cdots i_{n-1} j i_{n+1} \cdots i_N} a_{kj},$$

$$(S \times_n A)_{i_1 \cdots i_{n-1} j i_{n+1} \cdots i_N} = \sum_{k=1}^{K} S_{i_1 \cdots i_{n-1} k i_{n+1} \cdots i_N} a_{jk},$$

and

$$\begin{aligned}
(T, S \times_n A) &= \sum_{i_1=1}^{I_1} \cdots \sum_{i_n=1}^{I_N} T_{i_1 i_2 \ldots i_N} (S \times_n A)_{i_1 i_2 \cdots i_N} \\
&= \sum_{i_1=1}^{I_1} \cdots \sum_{i_N=1}^{I_N} T_{i_1 i_2 \ldots i_N} \sum_{i=1}^{I_n} S_{i_1 \cdots i_n \cdots i_N} a_{i_n i} \\
&= \sum_{i_1=1}^{I_1} \cdots \sum_{i_n=1}^{I_N} T_{i_1 i_2 \ldots i_n} a'_{i i_n} S_{i_1 \cdots i_n \cdots i_N} \\
&= (T \times_n A', S).
\end{aligned}$$

□

15.6 Inner Product and Norms

Example 15.14. Let T and S be two mdas in $\mathbb{R}^{I_1 \times I_2 \times \cdots \times I_N}$. Their *inner product* is given by

$$(T, S) = \sum_{i_1=1}^{I_1} \sum_{i_2=1}^{I_2} \cdots \sum_{i_N=1}^{I_N} T_{i_1 i_2 \cdots i_N} S_{i_1 i_2 \cdots i_N},$$

which is actually their contraction $T \bowtie_M S$ over the entire set of modes $M = \{1, \ldots, N\}$.

Therefore, it is natural to define the *norm* of an mda T as

$$\|T\|^2 = (T, T) = \sum_{i_1=1}^{I_1} \cdots \sum_{i_N=1}^{I_N} T^2_{i_1 \cdots i_N}.$$

The definition of the norm also implies that for two mdas $T, S \in \mathbb{R}^{I_1\times\cdots\times I_N}$, we have

$$\|T - S\|^2 = \|T\|^2 - 2(T, S) + \|S\|^2.$$

The norm of an mda can be expressed as the norm of a matrix, as the following statement shows.

Theorem 15.8. *Let X be an mda, $X \in \mathbb{R}^{I_1\times I_2\times\cdots\times I_N}$, where the set of modes of X is $\mathcal{N} = \{1,\ldots,N\}$. We have $\|X\| = \|X_{\mathcal{R},calc:I_{\mathcal{N}}}\|_F$.*

Proof. The square of Frobenius norm of the matrix $\|X_{\mathcal{R},calc:I_{\mathcal{N}}}\|_F$ is the sum of all squares of entries of this matrix. Since these entries form a rearrangement of the entries of X, the result follows immediately. □

15.7 Evaluation of a Set of Bilinear Forms

Let X, Y be two $\mathbb{F}$-linear spaces. Following [23], we discuss the evaluation of several bilinear forms $p_i = \sum_j \sum_k h_{ijk} x_j y_k$, where $1 \leqslant i \leqslant m$, $\boldsymbol{x} = (x_j) \in X$, and $\boldsymbol{y} = (y_k) \in Y$.

Let G_i be the matrix $G_i = (h_{ijk})$, where $1 \leqslant i \leqslant m$ that corresponds to the linear form p_i. For $\boldsymbol{s} = \begin{pmatrix} s_1 \\ \vdots \\ s_m \end{pmatrix}$, define $G(s) = \sum_{i=1}^m s_i G_i$. The matrix $G(s)$ is the *characteristic matrix of the problem*; the matrices G_i are called the *basis matrices*.

The scalar

$$H(\boldsymbol{s}, \boldsymbol{x}, \boldsymbol{y}) = \sum_{i=1}^{m}\sum_{j=1}^{p}\sum_{k=1}^{q} h_{ijk} s_i x_j y_k = \boldsymbol{x}' G(s)\boldsymbol{y}$$

is the *defining function* of the problem. The number of linear forms $m = \dim \boldsymbol{s}$ is the *index* of the problem.

When we evaluate the bilinear forms $p_1, \ldots, p_m$ at $(\boldsymbol{x}, \boldsymbol{y})$, we are computing the set of inner products $\boldsymbol{x}'(G_i\boldsymbol{y})$.

Following [23], we assume that $\mathbb{F}$ contains a subset $\mathbb{K}$ such that the computation of $\mathbb{K}$-linear forms is cheap compared with the computations of products of such linear forms. Furthermore, we assume that the additive and multiplicative identities in $\mathbb{F}$ hold in $\mathbb{K}$, and that all elements h_{ijk} belong to $\mathbb{K}$.

The general algorithm for evaluating $\mathbb{K}$-linear forms proposed in [23] has the following form (Algorithm 15.7.1):

Algorithm 15.7.1: Brokett–Dobkin Algorithm

Data: A set of elements h_{ijk} of $\mathbb{K}$

Result: A set of products of the form $p_i = \sum_{j,k} h_{ijk} x_i y_j$

Compute the $\mathbb{K}$-linear forms $\langle \boldsymbol{c}_i, \boldsymbol{x} \rangle$ and $\langle \boldsymbol{b}_i, \boldsymbol{y} \rangle$ for $1 \leqslant i \leqslant d$;

Compute the products $r_i = \langle \boldsymbol{c}_i, \boldsymbol{x} \rangle \langle \boldsymbol{b}_i, \boldsymbol{y} \rangle$ for $1 \leqslant i \leqslant d$;

Compute p_i as $p_i = \sum_{j=1}^{d} a_{ij} r_j$ for $1 \leqslant i \leqslant m$, where $a_{ij} \in \mathbb{K}$;

return p_i for $1 \leqslant i \leqslant m$;

Example 15.15. Let $x = x_1 + ix_2$ and $y = y_1 + iy_2$ be two complex numbers. Their product is

$$xy = x_1y_1 - x_2y_2 + i(x_2y_1 + x_1y_2),$$

which entails the evaluation of the bilinear forms $x_1y_1 - x_2y_2$ and $x_2y_1 + x_1y_2$. These forms correspond to the basis matrices

$$G_1 = \begin{pmatrix} 1 & 0 \\ 0 & -1 \end{pmatrix} \text{ and } G_2 = \begin{pmatrix} 0 & 1 \\ 1 & 0 \end{pmatrix}$$

because

$$x_1y_1 - x_2y_2 = (x_1 \; x_2) \begin{pmatrix} 1 & 0 \\ 0 & -1 \end{pmatrix} \begin{pmatrix} y_1 \\ y_2 \end{pmatrix}$$

and

$$x_2y_1 + x_1y_2 = (x_1 \; x_2) \begin{pmatrix} 0 & 1 \\ 1 & 0 \end{pmatrix} \begin{pmatrix} y_1 \\ y_2 \end{pmatrix}.$$

The characteristic matrix is

$$G(s) = \begin{pmatrix} s_1 & s_2 \\ s_2 & -s_1 \end{pmatrix}$$

and the defining function is

$$\begin{aligned} H(\boldsymbol{s}, \boldsymbol{x}, \boldsymbol{y}) &= (x_1 \; x_2) \begin{pmatrix} s_1 & s_2 \\ s_2 & -s_1 \end{pmatrix} \begin{pmatrix} y_1 \\ y_2 \end{pmatrix} \\ &= x_1y_1s_1 + x_1y_2s_2 + x_2y_1s_2 - x_2y_2s_1. \end{aligned}$$

Example 15.16. The computation of the product of matrices

$$X = \begin{pmatrix} x_1 & x_2 \\ x_3 & x_4 \end{pmatrix} \text{ and } Y = \begin{pmatrix} y_1 & y_2 \\ y_3 & y_4 \end{pmatrix}$$

corresponds to the defining function

$$\begin{aligned} H(\boldsymbol{s}, \boldsymbol{x}, \boldsymbol{y}) &= x_1 y_1 s_1 + x_2 y_3 s_1 + x_1 y_2 s_2 + x_2 y_4 s_2 \\ &\quad + x_3 y_1 s_3 + x_4 y_3 s_3 + x_3 y_2 s_4 + x_4 y_4 s_4 \end{aligned}$$

and the characteristic matrix

$$G(\boldsymbol{s}) = \begin{pmatrix} s_1 & s_2 & 0 & 0 \\ 0 & 0 & s_1 & s_2 \\ s_3 & s_4 & 0 & 0 \\ 0 & 0 & s_3 & s_4 \end{pmatrix}.$$

Example 15.17. Let $x(t)$ and $y(t)$ be two polynomials of degree $m-1$ and $n-1$, respectively,

$$x(t) = \sum_{j=1}^{m} x_j t^{j-1} \text{ and } y(t) = \sum_{k=1}^{n} y_k t^{k-1}.$$

The defining function is

$$H(\boldsymbol{s}, \boldsymbol{x}, \boldsymbol{y}) = \sum_{j=1}^{m} \sum_{k=1}^{n} s_{j+k-1} x_j y_k.$$

The characteristic matrix of this problem is the following Hankel matrix (introduced in Definition 3.22):

$$G_{mn}(\boldsymbol{s}) = \begin{pmatrix} s_1 & s_2 & s_3 & \cdots & s_n \\ s_2 & s_3 & s_4 & \cdots & s_{n+1} \\ s_3 & s_4 & s_5 & \cdots & s_{n+2} \\ \vdots & \vdots & \vdots & \cdots & \vdots \\ s_m & s_{m+1} & s_{m+2} & \cdots & s_{m+n-1} \end{pmatrix}.$$

15.8 Matrix Multiplications and Arrays

The role of bilinear forms is studied in [97] using a special matrix representation. Let $U, V \in \mathbb{R}^{2\times 2}$ be two matrices and let $W = UV$. Index the elements of each matrix by a single index as follows:

$$\begin{pmatrix} u_1 & u_2 \\ u_3 & u_4 \end{pmatrix} \begin{pmatrix} v_1 & v_2 \\ v_3 & v_4 \end{pmatrix} = \begin{pmatrix} w_1 & w_2 \\ w_3 & w_4 \end{pmatrix}.$$

Denote by x_{ijk} the coefficient of $u_i v_j$ in $w_k = \sum_{i=1}^{2} \sum_{j=1}^{2} x_{ijk} u_i v_j$; each x_{ijk} is either 0 or 1. The corresponding matrices (x_{ijk}) for w_k and $1 \leqslant k \leqslant 4$ are

$$\begin{array}{cc} w_1 : \begin{pmatrix} 1 & 0 & 0 & 0 \\ 0 & 0 & 1 & 0 \\ 0 & 0 & 0 & 0 \\ 0 & 0 & 0 & 0 \end{pmatrix} & w_2 : \begin{pmatrix} 0 & 1 & 0 & 0 \\ 0 & 0 & 0 & 1 \\ 0 & 0 & 0 & 0 \\ 0 & 0 & 0 & 0 \end{pmatrix} \\ w_3 : \begin{pmatrix} 0 & 0 & 0 & 0 \\ 0 & 0 & 0 & 0 \\ 1 & 0 & 0 & 0 \\ 0 & 0 & 1 & 0 \end{pmatrix} & w_4 : \begin{pmatrix} 0 & 0 & 0 & 0 \\ 0 & 0 & 0 & 0 \\ 0 & 1 & 0 & 0 \\ 0 & 0 & 0 & 1 \end{pmatrix} \end{array}.$$

A more general situation involves computing a set of K bilinear forms using a three-way array of coefficients x_{ijk} as

$$w_k = \sum_{i=1}^{I} \sum_{j=1}^{J} x_{ijk} u_i v_j,$$

where $1 \leqslant k \leqslant K$.

Example 15.18. Suppose that for a three-mode array X we have the decomposition $x_{ijk} = \sum_{r=1}^{R} a_{ir} b_{jr} c_{kr}$ for $1 \leqslant i \leqslant I$, $1 \leqslant j \leqslant J$, and $1 \leqslant k \leqslant K$. If

$$w_k = \sum_{i=1}^{I} \sum_{j=1}^{J} x_{ijk} u_i v_j$$

for $1 \leqslant k \leqslant K$, then w_k can be written as

$$\begin{aligned} w_k &= \sum_{i=1}^{I} \sum_{j=1}^{J} x_{ijk} u_i v_j \\ &= \sum_{i=1}^{I} \sum_{j=1}^{J} \sum_{r=1}^{R} a_{ir} b_{jr} c_{kr} u_i v_j \\ &= \sum_{r=1}^{R} \left(\sum_{i=1}^{I} a_{ir} u_i \right) \left(\sum_{j=1}^{J} b_{jr} v_j \right) c_{kr} \\ &= \sum_{r=1}^{R} f_r(a_{ir}) g_r(b_{jr}), \end{aligned}$$

where $f_r(a_{ir}) = \sum_{i=1}^{I} a_{ir} u_i$ and $g_r(b_{jr}) = \sum_{j=1}^{J} b_{jr} v_j$ for $1 \leqslant r \leqslant R$. This entails computing $2R$ linear combinations and multiplying these combinations pairwise to form R products. Then, we compute linear combinations of these products with the coefficients c_{kr}:

$$\begin{aligned} w_1 &= u_1 v_1 + u_2 v_3 \; w_2 = u_1 v_2 + u_2 v_4, \\ w_3 &= u_3 v_1 + u_4 v_3 \; w_4 = u_3 v_2 + u_4 v_4. \end{aligned}$$

For the standard matrix multiplication (using Kruskal's one-index notation), we have $R = 8$ bilinear forms, $h_1, \ldots, h_8$:

$$\begin{aligned} h_1 &= u_1 v_1 \; h_2 = u_2 v_3 \; h_3 = u_1 v_2 \; h_4 = u_2 v_4, \\ h_5 &= u_3 v_1 \; h_6 = u_4 v_3 \; h_7 = u_3 v_2 \; h_8 = u_4 v_4, \end{aligned}$$

then

$$\begin{aligned} w_1 &= h_1 + h_2 \; w_2 = h_3 + h_4, \\ w_3 &= h_5 + h_6 \; w_4 = h_7 + h_8. \end{aligned}$$

In tabular form, this algorithm is shown in Figure 15.6.

Example 15.19. Here we used the notation of Strassen (introduced in Exercise 120 of Chapter 3):

$$\begin{aligned} &I = f_1 g_1, \; II = f_2 g_2, \; III = f_3 g_3, \; IV = f_4 g_4, \\ &V = f_5 g_5, \; VI = f_6 g_6, \; VII = f_7 g_7. \end{aligned}$$

	f_1	f_2	f_3	f_4	f_5	f_6	f_7	f_8
u_1	1	0	1	0	0	0	0	1
u_2	0	1	0	1	0	0	0	0
u_3	0	0	0	0	1	0	1	0
u_4	0	0	0	0	0	1	0	1

	g_1	g_2	g_3	g_4	g_5	g_6	g_7	g_8
v_1	1	0	0	0	1	0	0	0
v_2	0	0	1	0	0	0	1	0
v_3	0	1	0	0	0	1	0	0
v_4	0	0	0	1	0	0	0	1

	h_1	h_2	h_3	h_4	h_5	h_6	h_7	h_8
w_1	1	1	0	0	0	0	0	0
w_2	0	0	1	1	0	0	0	0
w_3	0	0	0	0	1	1	0	0
w_4	0	0	0	0	0	0	1	1

Fig. 15.6 Bilinear forms in matrix product computation.

15.9 MATLAB Computations

Following the practice of Computer Science literature, we will use the terms mda and tensor loosely and interchangeably.

The *Tensor Toolbox* for MATLAB consists of a collection of tools for working with mdas. It is an open-source package constructed by researchers at Sandia National Laboratory [4, 5].

Every MATLAB array has at least two dimensions: a scalar is an object of size 1×1, a column vector is an array of size $n \times 1$, *etc.* MATLAB drops trailing singleton dimensions. Thus, a $3 \times 4 \times 1$ object has a reported size of 3×4.

The MATLAB tensor class explicitly stores trailing singleton dimensions.

Example 15.20. To create a tensor T starting from an array A, we write (Figure 15.7)

```
A= rand(3,4,2)

A(:,:,1) =

    0.8147    0.9134    0.2785    0.9649
    0.9058    0.6324    0.5469    0.1576
    0.1270    0.0975    0.9575    0.9706


A(:,:,2) =

    0.9572    0.1419    0.7922    0.0357
    0.4854    0.4218    0.9595    0.8491
    0.8003    0.9157    0.6557    0.9340

>> T=tensor(A)
T is a tensor of size 3 x 4 x 2
    T(:,:,1) =
        0.8147    0.9134    0.2785    0.9649
        0.9058    0.6324    0.5469    0.1576
        0.1270    0.0975    0.9575    0.9706
    T(:,:,2) =
        0.9572    0.1419    0.7922    0.0357
        0.4854    0.4218    0.9595    0.8491
        0.8003    0.9157    0.6557    0.9340
```

	f_1	f_2	f_3	f_4	f_5	f_6	f_7
u_1	1	0	1	0	1	-1	0
u_2	0	0	0	0	1	0	1
u_3	0	1	0	0	0	1	0
u_4	1	1	0	1	0	0	-1

	g_1	g_2	g_3	g_4	g_5	g_6	g_7
v_1	1	1	0	-1	0	1	0
v_2	0	0	1	0	0	1	0
v_3	0	0	0	1	0	0	1
v_4	1	0	-1	0	1	0	0

	I	II	III	IV	V	VI	VII
w_1	1	0	0	1	-1	0	1
w_2	0	1	0	1	0	0	0
w_3	0	0	1	0	1	0	0
w_4	1	-1	1	0	0	0	1

Fig. 15.7 Bilinear forms in Strassen's matrix product computation

The `tensor` class explicitly tracks singleton dimensions.

Example 15.21. Creating an mda of size $4 \times 3 \times 1$ with `A=rand(4,3,1)` ignores trailing singleton dimensions resulting in

```
A =

    0.6787    0.6555    0.2769
    0.7577    0.1712    0.0462
    0.7431    0.7060    0.0971
    0.3922    0.0318    0.8235
```

and

```
>> ndims(A)
ans =
     2
>> size(A)
ans =
     4     3
```

In contrast, using the tensor constructor `T = tensor(A,[4,3,1])` results in

```
T is a tensor of size 4 x 3 x 1
    T(:,:,1) =
        0.6787    0.6555    0.2769
        0.7577    0.1712    0.0462
        0.7431    0.7060    0.0971
        0.3922    0.0318    0.8235
```

and

```
>> ndims(T)
ans =
     3
>> size(T)
ans =
     4     3     1
```

A vector can be stored as a tensor by writing

```
T = tensor(rand(4,1),[4]).
```

Accessors and assignments for tensors work in the same way as for mdas.

Example 15.22. A $2 \times 2 \times 2$ tensor is created with

```
T = tensor(rand(2,2,2))
```

To reassign a 2×2 identity matrix, we write

```
A(:,1,:) = eye(2)
```

Three types of mda multiplications are discussed: multiplication of an mda with a matrix (denoted by `ttm`), with a vector (denoted by `ttv`), and with another mda (denoted by `ttt`). The first two variants can be regarded as special cases of the third.

When mda is multiplied with a matrix, it is necessary to specify the mode of the mda involved in the multiplication.

In [101], the *n-mode multiplication* of an mda by a matrix is discussed. This type of multiplication is a generalization of the matrix product.

Let $A = UBV'$ be a matrix, $A \in \mathbb{R}^{j_1 \times j_2}$, $B \in \mathbb{R}^{i_1 \times i_2}$, $U \in \mathbb{R}^{j_1 \times i_1}$, and $V \in \mathbb{R}^{j_2 \times i_2}$, we adopt the notation proposed in [101] and denote this product by $A = B \times_1 U \times_2 V$ which suggests that the matrix U creates linear combinations of the rows of B (the mode 1 of B), and the matrix V creates linear combinations of the columns of B (the mode 2 of B). More generally, if

$$T \in \mathbb{R}^{J_1 \times J_2 \times \cdots J_n \times \cdots \times J_N}$$

and $A \in \mathbb{R}^{I_n \times J_n}$, the mda

$$T \times_n A \in \mathbb{R}^{J_1 \times \cdots \times J_{n-1} \times I_n \times J_{n+1} \times \cdots J_N}$$

is defined as

$$(T \times_n A)_{j_1 \cdots j_{n-1} i j_{n+1} \cdots j_N} = \sum_{j_n=1}^{J_n} T_{j_1 \cdots j_n \cdots j_N} a_{i j_n},$$

and is obtained by having A act on the nth fibers of T. The MATLAB implementation of the n-mode multiplication is discussed in Example 15.23.

The mda-matrix n-mode product of the mda T and the matrix A is specified by $\texttt{ttm}(T, A, n)$.

Example 15.23. We create a tensor T starting from a $3 \times 4 \times 2$ mda of integers named A. To this end, we define a 3×4-matrix A using

```
> A = [1 4 7 10;2 5 8 11;3 6 9 12]

A =

     1     4     7    10
     2     5     8    11
     3     6     9    12
```

Then, the second page of this array is added using

```
A(:,:,2) = [13 16 19 22;14 17 20 23; 15 18 21 24]
```

which results in

```
A(:,:,1) =

     1     4     7    10
     2     5     8    11
     3     6     9    12


A(:,:,2) =

    13    16    19    22
    14    17    20    23
    15    18    21    24
```

The tensor T is obtained by writing

```
>> T = tensor(A)
T is a tensor of size 3 x 4 x 2
    T(:,:,1) =
         1     4     7    10
         2     5     8    11
         3     6     9    12
    T(:,:,2) =
        13    16    19    22
```

```
    14    17    20    23
    15    18    21    24
```

Let B be the 2×3 matrix defined by

```
>> B=[1 2 3;4 5 6]

B =

     1     2     3
     4     5     6
```

To compute the tensor–matrix product $P = \texttt{ttm}(T, B, 1)$ between the $3 \times 4 \times 2$ tensor T and the matrix $B \in \mathbb{R}^{2\times 3}$ resulting in, a $2 \times 4 \times 2$ tensor, we execute the following code:

```
>> P=ttm(T,B,1)

P is a tensor of size 2 x 4 x 2
     P(:,:,1) =
         14    32    50    68
         32    77   122   167
     P(:,:,2) =
         86   104   122   140
        212   257   302   347
```

corresponding to the following equalities:

$$14 = 1 \cdot 1 + 2 \cdot 2 + 3 \cdot 3$$

$$\vdots$$

$$68 = 1 \cdot 10 + 2 \cdot 11 + 3 \cdot 12$$

$$32 = 4 \cdot 1 + 5 \cdot 2 + 6 \cdot 3$$

$$\vdots$$

$$167 = 4 \cdot 10 + 5 \cdot 11 + 6 \cdot 12$$

and

$$86 = 1 \cdot 13 + 2 \cdot 14 + 3 \cdot 15$$

$$\vdots$$

$$140 = 1 \cdot 22 + 2 \cdot 23 + 3 \cdot 24$$

$$212 = 4 \cdot 13 + 5 \cdot 14 + 6 \cdot 15$$
$$\vdots$$
$$347 = 4 \cdot 22 + 5 \cdot 23 + 6 \cdot 24.$$

If C is the 2×4 matrix,

$$C = \begin{pmatrix} 3 & 2 & 3 & 1 \\ 0 & 3 & 0 & 4 \end{pmatrix},$$

the product of T by C along the second dimension of T is obtained as

```
>> Q=ttm(T,C,2)
Q is a tensor of size 3 x 2 x 2
    Q(:,:,1) =
         42     52
         51     59
         60     66
    Q(:,:,2) =
        150    136
        159    143
        168    150
```

Finally, if we multiply T along its third dimension by the matrix $D \in \mathbb{R}^{2\times 2}$ given by

$$D = \begin{pmatrix} 7 & 3 \\ 5 & 2 \end{pmatrix},$$

the resulting mda $R = \mathtt{ttm}(T, D, 3)$ is

```
>> R=ttm(T,D,3)
R is a tensor of size 3 x 4 x 2
    R(:,:,1) =
         46     76    106    136
         56     86    116    146
         66     96    126    156
    R(:,:,2) =
         31     52     73     94
         38     59     80    101
         45     66     87    108
```

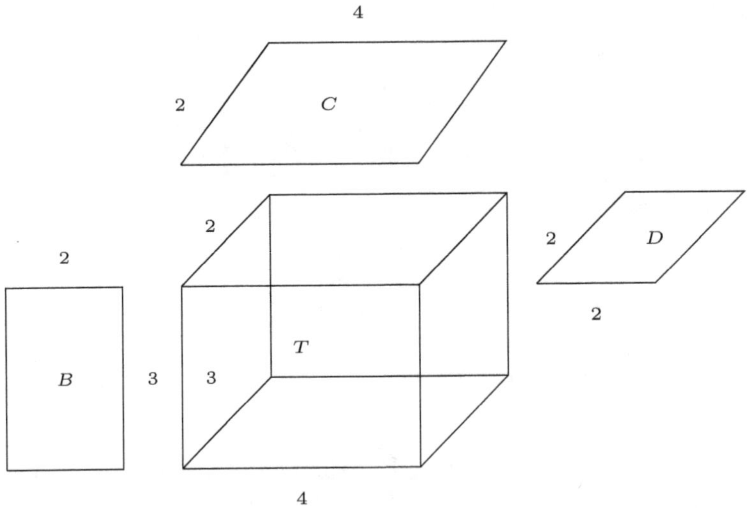

Fig. 15.8 Possible mda-matrix products.

An image of various possible mda-matrix products is shown in Figure 15.8. This image suggests an alternative notation for matrix-mda products proposed by Kruskal in [97], where the products

`ttm(T,B,1), ttm(T,C,2), ttm(T,D,3)`

are denoted as BT, $\overset{C}{T}$, and TD, respectively.

An alternate way of computing n-mode products using cell arrays is shown next.

Example 15.24. Consider the mda $T \in \mathbb{R}^{4\times3\times2}$ produced by

```
>> T = tensor(randi(9,4,3,2))
T is a tensor of size 4 x 3 x 2
    T(:,:,1) =
         9     4     6
         5     9     1
         8     8     8
         2     9     9
    T(:,:,2) =
         7     6     3
         7     2     1
         7     7     1
         4     1     8
```

and the matrices A and B generated by

```
>> A = randi(9,2,4)
A =

     7     9     4     7
     3     1     4     8
```

and

```
>> B=randi(9,3,2)
B =

     2     6
     5     7
     5     7
```

Matrices A and B are stored in the cell array `W` defined next:

```
>> W{1}=A
W =
  1 x 1 cell array

    {2 x 4 double}

>> W{2}=B

W =

  1 x 2 cell array

    {2 x 4 double}    {3 x 2 double}
```

Finally, using the cell array `W` the product $T \times_3 B \times_1 A$ is computed as

```
>> S=ttm(T,W,[1,3])
S is a tensor of size 2 x 3 x 3
    S(:,:,1) =
            1316          978          832
             688          586          714
    S(:,:,2) =
            1946         1685         1360
            1016         1017         1161
    S(:,:,3) =
            1946         1685         1360
            1016         1017         1161
```

If T is an $I_1 \times I_2 \times \cdots \times I_N$ tensor, $A \in \mathbb{R}^{J_m \times I_m}$, and $B \in \mathbb{R}^{J_n \times I_n}$, then

$$(T \times_m A) \times_n B = (T \times_n B) \times_m A.$$

Let T be an $I_1 \times I_2 \times \cdots \times I_N$ tensor and let $(U^{(1)}, \ldots, U^{(N)})$ be a sequence of matrices, where $U^{(i)}$ has the format $J_i \times I_i$. Then $T \times_1 U^{(1)} \times_2 U^{(2)} \times_N U^{(N)}$ is of size $J_1 \times J_2 \times \cdots \times J_N$.

Let T be an mda of format $I_1 \times I_2 \times \cdots I_n \cdots I_N$ and let $\boldsymbol{v}$ be a vector of size I_n. The *contracted n-mode* product of T and $\boldsymbol{v}$ introduced in [4] builds an mda $T\overline{\times}_n\boldsymbol{v}$ of format $I_1 \times \cdots \times I_{n-1} \times I_{n+1} \times \cdots I_n$ given by

$$(T\overline{\times}_n\boldsymbol{v})(i_1, \ldots, i_{n-1}, i_{n+1}, \ldots, i_N) = \sum_{i_n=1}^{I_n} T(i_1, \ldots, i_N)\boldsymbol{v}(i_n).$$

Example 15.25. Let T be the mda introduced in Example 15.13 and let

$$\boldsymbol{u} = \begin{pmatrix} 10 \\ 20 \\ 30 \end{pmatrix}.$$

$T \in \mathbb{R}^{4\times3\times2}$ can be multiplied by the vectors $\boldsymbol{u}$ along its second dimension by writing

```
>> ttv(T,u,2)
ans is a tensor of size 4 x 2
    ans(:,:) =
             350         280
             260         140
             480         240
             470         300
```

Yet another type of mda multiplication involves two tensors. This product comes in three flavors: the outer product, the inner product, and the contracted product.

Definition 15.13. Let T be an mda of size $I_1 \times \cdots \times I_M$ and let S be an mda of size $J_1 \times \cdots \times J_N$. The *outer product* $T \circ S$ is of size

$I_1 \times \cdots \times I_M \times J_1 \times \cdots \times J_N$ and is given by

$$(ToS)(i_1, \ldots, i_M, j_1 \ldots, j_N) = T(i_1, \ldots, i_M)S(j_1, \ldots, j_N).$$

The MATLAB command `ttt` is an abbreviation of "tensor times tensor" and is given by `Z = ttt(T,S)`.

Example 15.26. Let T and S be the mdas produced by

```
>> T = tensor(randi(9,2,2,2))
>> s = tensor(randi(5,2,1,3))
```

and given by

```
T is a tensor of size 2 x 2 x 2
    T(:,:,1) =
           9     1
           6     3
    T(:,:,2) =
           5     9
           9     2
>> S=tensor(randi(5,2,1,3))
S is a tensor of size 2 x 1 x 3
    S(:,:,1) =
           5
           4
    S(:,:,2) =
           5
           4
    S(:,:,3) =
           1
           5
```

Their outer product obtained with `Z = ttt(T,S)` is

```
Z is a tensor of size 2 x 2 x 2 x 2 x 1 x 3
    Z(:,:,1,1,1,1) =
          45     5
          30    15
    Z(:,:,2,1,1,1) =
          25    45
          45    10
    Z(:,:,1,2,1,1) =
          36     4
          24    12
```

```
Z(:,:,2,2,1,1) =
     20    36
     36     8
Z(:,:,1,1,1,2) =
     45     5
     30    15
Z(:,:,2,1,1,2) =
     25    45
     45    10
Z(:,:,1,2,1,2) =
     36     4
     24    12
Z(:,:,2,2,1,2) =
     20    36
     36     8
Z(:,:,1,1,1,3) =
      9     1
      6     3
Z(:,:,2,1,1,3) =
      5     9
      9     2
Z(:,:,1,2,1,3) =
     45     5
     30    15
Z(:,:,2,2,1,3) =
     25    45
     45    10
```

Example 15.27. Let T and S be the mdas

```
T = tensor(randi(9,2,2,2))
S = tensor(randi[10 20],2,2,2)
```

```
>> T= tensor(randi(9,2,2,2))
T is a tensor of size 2 x 2 x 2
    T(:,:,1) =
         8     5
         2     5
    T(:,:,2) =
         6     7
         7     3
>> S= tensor(randi([10 20],2,2,2))
```

```
S is a tensor of size 2 x 2 x 2
     S(:,:,1) =
          14      11
          17      17
     S(:,:,2) =
          10      10
          13      11
```

The inner product of T and S computed with `ttt(T,S)`, or with `[1,2]` is

```
ans =

    540
```

The contracted product of two mdas is a generalization of the mda vector product.

Definition 15.14. Let T be an mda of size $I_1 \times \cdots \times I_M \times J_1 \times \cdots J_N$ and let S be an mda of size $I_1 \times \cdots \times I_M \times K_1 \times \cdots \times K_P$. The *contracted product* along the first M modes is the mda of size $J_1 \times \cdots \times J_N \times K_1 \times K_P$ given by

$$(T,S)_{1,\ldots,M;1\ldots,M}(j_1,\ldots,j_n,k_1,\ldots,k_p)$$
$$= \sum_{i_1=1}^{I_1} \cdots \sum_{i_M=1}^{I_M} T(i_1,\ldots,i_M,j_1,\ldots,j_N)S(i_1,\ldots,i_M,k_1,\ldots,k_p).$$

In MATLAB , the command for mda contracted product is

```
U = ttt(T,S,[1:M],[1:M])
```

The contracted product may involve distinct lists of dimensions of the two mdas of equal lengths provided that the sizes of the involved dimensions are identical. The next example illustrates this point.

Example 15.28. Let T and S be the mdas defined as

```
>> T=tensor(randi(4,3,4,2))
T is a tensor of size 3 x 4 x 2
     T(:,:,1) =
          2      2      3      3
          1      3      2      3
          4      1      3      2
```

```
    T(:,:,2) =
          1     1     4     1
          1     4     1     4
          4     3     2     1
>> S=tensor(randi(5,4,3,2))
S is a tensor of size 4 x 3 x 2
    S(:,:,1) =
          4     2     5
          5     2     1
          5     5     2
          1     3     1
    S(:,:,2) =
          1     1     3
          5     5     3
          3     4     1
          3     2     2
```

The lists of dimensions of T and S are [13] and [23], and the corresponding sizes are 3 and 2. Therefore, the contracted product `ttt(T,S,[1 3],[2 3])` returns

```
>> ttt(T,S,[1 3],[2 3])
ans is a tensor of size 4 x 4
    ans(:,:) =
         44    38    34    22
         33    51    49    29
         42    53    49    30
         36    51    54    27
```

This means that the dimensions involved in the contracted product are the first and the third for the mda T and the second and the third for the mda S. Note that the size of the first dimension of T is 3, the same as the size of the second dimension of S; also, the size of the third dimension of T is 2, which is also the size of the third dimension of S.

The MATLAB function `reshape` given by

```
B = reshape(A,sz)
```

reformats the array A using the size vector `sz` to define `size(B)`. The vector `sz` must contain at least 2 elements, and the product of the components of `sz` must equal `numel(A)`, that gives the number of elements in the array `A`.

Example 15.29. For the array `A=randi([1,10],[2 3 2])` given by

```
A(:,:,1) =

     9     2     7
    10    10     1


A(:,:,2) =

     3    10     2
     6    10    10
```

the effect of the statement `B=reshape(A,[2 6])` is to produce the array

```
B =

     9     2     7     3    10     2
    10    10     1     6    10    10
```

The alternative format `B = reshape(A,sz1,...,szN)` reformats A into an array having `sz1,...,szN` as sizes of each dimension. The vector $A = 1 : 10$ can be reshaped into a 5×2 matrix by

```
B = reshape(A,[5 2])
```

which yields

```
A =

     1     2     3     4     5     6     7     8     9    10

>> B = reshape(A,[5 2])

B =

     1     6
     2     7
     3     8
     4     9
     5    10
```

Finally, to reshape an array of integers A given by

```
A = randi([1 12],[3 2 4])
```

into a 6×4 matrix M, we write

```
M = reshape(A,6,4)
```

and obtain

```
>> A = randi([1 12],[3,2,4])

A(:,:,1) =

    12     2
     6     6
    10    11


A(:,:,2) =

    10     1
    12    11
     8    12


A(:,:,3) =

     9     5
    10     8
     9     3


A(:,:,4) =

     9     1
     1     2
     4    10
```

and

```
M =

    12    10     9     9
     6    12    10     1
```

```
   10    8    9    4
    2    1    5    1
    6   11    8    2
   11   12    3   10
```

The function `permute` applied as `B = permute(A,dimorder)` rearranges the dimensions of an array in the order specified by the vector `dimorder`. For example, `permute(A,[2 1])` switches the row and column dimensions of a matrix A. In general, the ith dimension of the output array is the dimension `dimorder(i)` from the input array.

Example 15.30. To create a $3 \times 4 \times 2$ array and permute it so that the first and third dimensions are switched, resulting in a $2 \times 4 \times 3$ array, we could use the following MATLAB code:

```
A = rand(3,4,2)

A(:,:,1) =

    0.8147    0.9134    0.2785    0.9649
    0.9058    0.6324    0.5469    0.1576
    0.1270    0.0975    0.9575    0.9706


A(:,:,2) =

    0.9572    0.1419    0.7922    0.0357
    0.4854    0.4218    0.9595    0.8491
    0.8003    0.9157    0.6557    0.9340

B = permute(A,[3 2 1])


B(:,:,1) =

    0.8147    0.9134    0.2785    0.9649
    0.9572    0.1419    0.7922    0.0357


B(:,:,2) =

    0.9058    0.6324    0.5469    0.1576
    0.4854    0.4218    0.9595    0.8491
```

```
B(:,:,3) =

    0.1270    0.0975    0.9575    0.9706
    0.8003    0.9157    0.6557    0.9340
```

Example 15.31. We start by creating an mda of random integers using `X=randi(10,4,6,4,2)`:

```
X(:,:,1,1) =

     1     8    10     7     1     3
     3     5     6     7     9     4
     9     6     6     4    10     7
     1     3     3     4     8     2
    10     5     5    10     1     8


X(:,:,2,1) =

     2    10     1     7     2     2
     7     9     8     7     3    10
     5     4     6     9     9     8
     8     7     5     9     1     6
     8     2    10     6     5     5


X(:,:,3,1) =

     1     1     7     5     2     4
     7     9     6     5     4     6
     1     9    10     9     9     5
     1     8     7     1     9     7
     6     2     9     2     1     7


X(:,:,4,1) =

     3     2    10     5     4     7
     5     4    10     6     8     2
     1     2     1    10     7     2
    10     5     8     5     6    10
     2     4     3    10     7     2
```

```
X(:,:,1,2) =

     1     4     7     2     3    10
     6     5     4     6     3     8
     9    10     2     3     7     4
     7     2     5     4     3     6
     2     9     5     6     9     2


X(:,:,2,2) =

    10     1     5     7     6    10
     9     5     1     7     7     6
     9     4     6     1     5     4
     3     2     5     1     9     2
     6     2     7     4     8     7


X(:,:,3,2) =

     8     3     6     7     3     8
     5     5    10     3     3     7
     1     6     7     7     7     1
     3     5    10     7     9     7
     2     9     3     1     4     4


X(:,:,4,2) =

    10     8     2     7     3     6
     1     4     8     2     8     7
     5     8     5     8     2     6
     5     5     2     3     3     5
     5     1     4    10     1     7
```

Next, we create two arrays `R =[2 3]` and `C = [4 1]` that define a partitioning of the set of modes of the array `X`. The array `I = size(X)` defines the sizes of the modes of X; `J = prod(I(R))` and `K = prod(I(C))` define the dimensions of the target matrix `Y`:

```
>> J = prod(I(R))
J =
    24
>> K=prod(I(C))
K =
    10
```

The resulting matrix Y will have 24 rows and 10 columns and can be obtained by

```
Y = reshape(permute(X,[R C]),J,K)
```

as follows:

```
Y =
     1     1     3     6     9     9     1     7    10     2
     8     4     5     5     6    10     3     2     5     9
    10     7     6     4     6     2     3     5     5     5
     7     2     7     6     4     3     4     4    10     6
     1     3     9     3    10     7     8     3     1     9
     3    10     4     8     7     4     2     6     8     2
     2    10     7     9     5     9     8     3     8     6
    10     1     9     5     4     4     7     2     2     2
     1     5     8     1     6     6     5     5    10     7
     7     7     7     7     9     1     9     1     6     4
     2     6     3     7     9     5     1     9     5     8
     2    10    10     6     8     4     6     2     5     7
     1     8     7     5     1     1     1     3     6     2
     1     3     9     5     9     6     8     5     2     9
     7     6     6    10    10     7     7    10     9     3
     5     7     5     3     9     7     1     7     2     1
     2     3     4     3     9     7     9     9     1     4
     4     8     6     7     5     1     7     7     7     4
     3    10     5     1     1     5    10     5     2     5
     2     8     4     4     2     8     5     5     4     1
    10     2    10     8     1     5     8     2     3     4
     5     7     6     2    10     8     5     3    10    10
     4     3     8     8     7     2     6     3     7     1
     7     6     2     7     2     6    10     5     2     7
```

To convert back the matrix Y to an mda, we could use the function `ipermute(B,dimorder)` that rearranges the dimensions of an array `B` in the order specified by the vector `dimorder` as in

```
Z = ipermute(reshape(Y,[I(R) I(C)]),[R C]).
```

Vectorization of an mda is a special case of matricization that converts an mda to a vector. Let $V_1, \ldots, V_p$ be p $\mathbb{F}$-linear spaces, where $\dim(V_i) = n_i$ for $1 \leqslant i \leqslant p$. If $\pi \in PERM_p$, the following linear spaces:

$$\begin{array}{l} V_1 \otimes V_2 \otimes \cdots \otimes V_p \\ \mathbb{F}^{n_1} \otimes \mathbb{F}^{n_2} \otimes \cdots \mathbb{F}^{n_p} \\ \vdots \\ \mathbb{F}^{\pi(n_1)} \otimes \mathbb{F}^{\pi(n_2)} \otimes \cdots \mathbb{F}^{\pi(n_p)} \\ \vdots \\ \mathbb{F}^{n_1 n_2 \cdots n_p} \end{array}$$

are isomorphic. The isomorphisms between these linear spaces make it possible to interpret a p-dimensional array as an mda of order ℓ, where $\ell \leqslant p$.

Let $\{e_1^j, \ldots, e_{n_j}^j\}$ be a basis in $\mathbb{F}^j$ for $1 \leqslant j \leqslant p$. For $m \in \mathbb{N}$, define $[m] = \{1, \ldots, m\}$. A bijective map $\mu : [n_1] \times \cdots \times [n_p] \longrightarrow [n_1 \cdots n_p]$ generates a linear space isomorphism between $\mathbb{F}^{n_1} \otimes \cdots \otimes \mathbb{F}^{n_p}$ and $\mathbb{F}^{n_1 \cdots n_p}$, which maps $\boldsymbol{e}_{j_1}^1 \otimes \cdots \otimes \boldsymbol{e}_{j_p}^p$ into $\boldsymbol{e}_{\mu(j_1, \ldots, j_p)}$.

The isomorphism $V_1 \otimes V_2 \otimes \cdots \otimes V_p \cong \mathbb{F}^{n_1} \otimes \mathbb{F}^{n_2} \otimes \cdots \mathbb{F}^{n_p}$ allows the interpretation of an mda as a vector in $\mathbb{R}^{n_1 n_2 \ldots n_p}$. This interpretation is known as a *vectorization*, and it is a generalization of matrix vectorization that was introduced in Definition 3.16. Namely, the vectorization $\mathsf{vec}(T)$ of the mda T is

$$\mathsf{vec}(T) = \begin{pmatrix} t_{11\cdots 1} \\ \vdots \\ t_{n_1 1 \cdots 1} \\ t_{12\cdots 1} \\ \vdots \\ t_{n_1 n_2 \cdots n_p}. \end{pmatrix}.$$

Example 15.32. The vectorization mapping introduced in Definition 3.16 is implemented in MATLAB using the function **reshape**. If $A \in \mathbb{R}^{3\times 2}$ is the matrix defined as

```
>> A = [1 2;3 4;5 6]

A =
     1 2
     3 4
     5 6
```

then `v = reshape(A,6,1)` will return

```
v =
     1
     3
     5
     2
     4
     6
```

which is $\mathsf{vec}(A)$.

An alternative way to compute A is `v = A(:)`.

To reverse this reshaping, we can write `A = reshape(v,3,2)`.

Example 15.33. Let $T \in \mathbb{R}^{3\times 2\times 3}$ be the mda defined as

$$\begin{aligned}
&a_{111} = a_{112} = a_{211} = -a_{212} = 1,\\
&a_{213} = a_{311} = a_{313} = a_{121} = a_{122} = a_{221} = -a_{222} = 2,\\
&a_{223} = a_{321} = a_{323} = 4,\\
&a_{133} = a_{312} = a_{123} = a_{322} = 0.
\end{aligned}$$

The matrix unfolding $T_{(1)} \in \mathbb{R}^{3\times 6}$ is

$$\begin{matrix} \mathbf{I} & \mathbf{II} & \mathbf{III} & \mathbf{IV} & \mathbf{V} & \mathbf{VI} \end{matrix}$$
$$\begin{pmatrix} 1 & 1 & 0 & 2 & 2 & 0 \\ 1 & -1 & 2 & 2 & -2 & 4 \\ 2 & 0 & 2 & 4 & 0 & 4 \end{pmatrix}.$$

The columns of $T_{(1)}$, numbered with roman numerals from **I** to **VI**, are shown in Figure 15.9.

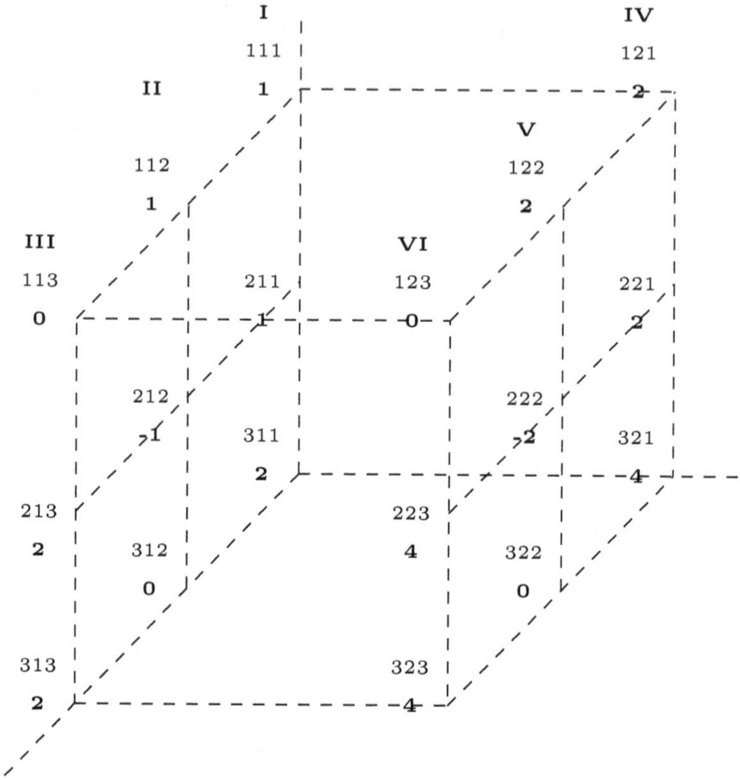

Fig. 15.9 The columns of the unfolding $T_{(1)}$.

15.10 Hyperdeterminants

Hyperdeterminants extend the notion of determinant defined for matrices and offer information about solutions of multilinear systems.

Example 15.34. Let T be an mda of format $3 \times 2 \times 2$. It is possible to slice this array as three 2×2 matrices:

$$\begin{pmatrix} t_{000} & t_{001} \\ t_{010} & t_{011} \end{pmatrix}, \begin{pmatrix} t_{100} & t_{101} \\ t_{110} & t_{111} \end{pmatrix}, \text{ and } \begin{pmatrix} t_{200} & t_{201} \\ t_{210} & t_{211} \end{pmatrix}$$

or as two 3×2 matrices:

$$\begin{pmatrix} t_{000} & t_{001} \\ t_{100} & t_{101} \\ t_{200} & t_{201} \end{pmatrix} \text{ and } \begin{pmatrix} t_{010} & t_{011} \\ t_{110} & t_{111} \\ t_{210} & t_{211} \end{pmatrix}.$$

The homogeneous multilinear system $T(\boldsymbol{x} \otimes \boldsymbol{y}) = 0$, where $\boldsymbol{x}, \boldsymbol{y} \in \mathbb{R}^2$, is

$$t_{000}x_0y_0 + t_{001}x_0y_1 + t_{010}x_1y_0 + t_{011}x_1y_1 = 0$$
$$t_{100}x_0y_0 + t_{101}x_0y_1 + t_{110}x_1y_0 + t_{111}x_1y_1 = 0$$
$$t_{200}x_0y_0 + t_{201}x_0y_1 + t_{210}x_1y_0 + t_{211}x_1y_1 = 0.$$

Note that this system has a solution $x_0 = x_1 = 0$ for any value of $\boldsymbol{y} = \begin{pmatrix} y_0 \\ y_1 \end{pmatrix}$.

Example 15.35. Consider the multilinear system

$$x_0y_0 = z_0, x_0y_1 = z_1, x_1y_0 = z_2, x_1y_1 = z_3.$$

Observe that if the system has a solution, then

$$z_0z_3 - z_1z_2 = x_0y_0x_1y_1 - x_0y_1x_1y_0 = 0.$$

The system has a non-trivial solution if and only if $z_0z_3 - z_1z_2 = 0$ and $(z_0, z_1, x_2, z_3) \neq \mathbf{0}_4$.

Indeed, suppose that at least one number, say z_0, is distinct from 0 and $z_0z_3 - z_1z_2 = 0$. This implies $x_0 \neq 0$ and $y_0 \neq 0$, which, in turn, imply $y_1 = \frac{z_1}{x_0}$ and $x_1 = \frac{z_2}{y_0}$. Therefore, $x_1y_1 = \frac{z_1z_2}{x_0y_0} = \frac{z_1z_2}{z_0} = z_3$, hence the system has a non-trivial solution.

The converse implication is immediate.

Theorem 15.9. *Let T be a $2 \times 2 \times 2$ mda that defines a system in the variables x_0, x_1, y_0, y_1 as follows:*

$$t_{000}x_0y_0 + t_{001}x_0y_1 + t_{010}x_1y_0 + t_{011}x_1y_1 = 0,$$
$$t_{100}x_0y_0 + t_{101}x_0y_1 + t_{110}x_1y_0 + t_{111}x_1y_1 = 0.$$

This system has non-trivial solutions up to a multiplicative constant. Also, the system has a unique solution up to a multiplicative constant if and only if the

$$\begin{vmatrix} t_{000}x_0 + t_{010}x_1 & t_{001}x_0 + t_{011}x_1 \\ t_{100}x_0 + t_{110}X_1 & t_{101}x_0 + t_{111}x_1 \end{vmatrix} = 0.$$

Proof. In matrix form, the system can be written as

$$\begin{pmatrix} t_{000}x_0 + t_{010}x_1 & t_{001}x_0 + t_{011}x_1 \\ t_{100}x_0 + t_{110}X_1 & t_{101}x_0 + t_{111}x_1 \end{pmatrix} \begin{pmatrix} y_0 \\ y_1 \end{pmatrix} = \begin{pmatrix} 0 \\ 0 \end{pmatrix}.$$

Therefore, non-trivial solutions exist for values of x_0, x_1 such that

$$\begin{vmatrix} t_{000}x_0 + t_{010}x_1 & t_{001}x_0 + t_{011}x_1 \\ t_{100}x_0 + t_{110}X_1 & t_{101}x_0 + t_{111}x_1 \end{vmatrix} = 0. \tag{15.11}$$

□

The determinant in the left member of Equality (15.11) equals

$$x_0^2 \begin{vmatrix} t_{000} & t_{001} \\ t_{100} & t_{101} \end{vmatrix} + x_0x_1 \left(\begin{vmatrix} t_{000} & t_{011} \\ t_{100} & t_{111} \end{vmatrix} + \begin{vmatrix} t_{010} & t_{001} \\ t_{110} & t_{101} \end{vmatrix} \right) + x_1^2 \begin{vmatrix} t_{010} & t_{011} \\ t_{110} & t_{111} \end{vmatrix},$$

and is a homogeneous polynomial. Its discriminant,

$$\left(\begin{vmatrix} t_{000} & t_{011} \\ t_{100} & t_{111} \end{vmatrix} + \begin{vmatrix} t_{010} & t_{001} \\ t_{110} & t_{101} \end{vmatrix} \right)^2 - 4 \begin{vmatrix} t_{000} & t_{001} \\ t_{100} & t_{101} \end{vmatrix} \begin{vmatrix} t_{010} & t_{011} \\ t_{110} & t_{111} \end{vmatrix}$$

is referred to as the *hyperdeterminant* of T.

Let T be an mda of format $3 \times 2 \times 2$ and let A_T be the matrix defined as

$$A_T = \begin{pmatrix} t_{000} & t_{001} & t_{010} & t_{011} \\ t_{100} & t_{101} & t_{110} & t_{111} \\ t_{200} & t_{201} & t_{210} & t_{211} \end{pmatrix}. \tag{15.12}$$

Denote by T_{00}, T_{01}, T_{10}, and T_{11} the matrices obtained from A_T by eliminating the first column (00), the second column (01), the third column (10), and the fourth column (11).

If $\mathsf{rank}(A_T) < 3$, then one of the equations of the system is a linear combination of the other two and we obtain a $2 \times 2 \times 2$ system. For example, if the third row of A_T is a linear combination of the first two rows, the system

$$t_{000}x_0y_0 + t_{001}x_0y_1 + t_{010}x_1y_0 + t_{011}x_1y_1 = 0,$$
$$t_{100}x_0y_0 + t_{101}x_0y_1 + t_{110}x_1y_0 + t_{111}x_1y_1 = 0,$$

can be written as

$$(t_{000}y_0 + t_{001}y_1)x_0 + (t_{010}y_0 + t_{011}y_1)x_1 = 0,$$
$$(t_{100}y_0 + t_{101}y_1)x_0 + (t_{110}y_0 + t_{111}y_1)x_1 = 0.$$

This system has a non-trivial solution in x_0, x_1 if

$$\begin{vmatrix} t_{000}y_0 + t_{001}y_1 & t_{010}y_0 + t_{011}y_1 \\ t_{100}y_0 + t_{101}y_1 & t_{110}y_0 + t_{111}y_1 \end{vmatrix} = 0.$$

Theorem 15.10. *Suppose that* $\mathsf{rank}(A_T) = 3$, *where* A_T *is the matrix defined in Equality (15.12). The multilinear system*

$$t_{000}x_0y_0 + t_{001}x_0y_1 + t_{010}x_1y_0 + t_{011}x_1y_1 = 0,$$
$$t_{100}x_0y_0 + t_{101}x_0y_1 + t_{110}x_1y_0 + t_{111}x_1y_1 = 0,$$
$$t_{200}x_0y_0 + t_{201}x_0y_1 + t_{210}x_1y_0 + t_{211}x_1y_1 = 0.$$

has a non-trivial solution if and only if

$$\det(T_{01})\det(T_{10}) - \det(T_{00})\det(T_{11}) = 0.$$

Proof. Let

$$z_0 = x_0y_0, z_1 = x_0y_1, z_2 = x_1y_0, \text{ and } z_3 = x_1y_1,$$

and consider the new multilinear system

$$t_{000}z_0 + t_{001}z_1 + t_{010}z_2 + t_{011}z_3 = 0,$$
$$t_{100}z_0 + t_{101}z_1 + t_{110}z_2 + t_{111}z_3 = 0,$$
$$t_{200}z_0 + t_{201}z_1 + t_{210}z_2 + t_{211}z_3 = 0,$$
$$x_0y_0 = z_0,$$
$$x_0y_1 = z_1,$$
$$x_1y_0 = z_2,$$
$$x_1y_1 = z_3.$$

By Supplement 47 of Chapter 5, if the linear system that consists of the first three equations has a nontrivial solution (z_1, z_2, z_3, z_4), then

$$\frac{z_0}{\det(T_{00})} = \frac{z_1}{-\det(T_{01})} = \frac{z_2}{\det(T_{10})} = \frac{z_3}{-\det(T_{11})},$$

where the matrices T_k are obtained from A_T by eliminating the kth column, where $1 \leqslant k \leqslant 4$. Furthermore, Example 15.35 shows that the full multilinear system having a non-trivial solution implies that $\det(T_{01})\det(T_{10}) - \det(T_{00})\det(T_{11}) = 0$. □

The notion of hyperdeterminant has been extended to mdas having *boundary format*, that is, to mdas having the format $(k_0+1) \times (k_1+1) \times \cdots \times (k_d+1)$, where $k_0 = \sum_{i=1}^{d} k_i$. In this special situation, it is shown in [63] that the hyperdeterminant can be expressed as the determinant of a block matrix. Note that an mda of format $3 \times 2 \times 2$ has this boundary format.

15.11 Eigenvalues and Singular Values

Let T be an mth order, n-dimensional mda. The mth *degree homogeneous polynomial* f_T defined by T is given by

$$f_T(\boldsymbol{x}) = \sum_{i_1 \cdots i_m} T_{i_1 \cdots i_m} x_{i_1} \cdots x_{i_m}.$$

If $\boldsymbol{x}^m$ is the mth order n-dimensional mda with entries $x_{i_1} \cdots x_{i_m}$, then $T\boldsymbol{x}^m = f_T(\boldsymbol{x})$.

The mda T is *positive definite* if f_T is positive definite.

For a vector $\boldsymbol{x} \in \mathbb{R}^n$ and $m \in \mathbb{N}$, $\boldsymbol{x}' = (x_1, \ldots, x_n)$, denote by $\boldsymbol{x}^{[m]}$ the vector in $\mathbb{R}^n$ whose components are $(\boldsymbol{x}[m])_i = x_i^m$ for $1 \leqslant i \leqslant n$.

Starting from the k-mode product of an mda $T = (T_{i_1 \cdots i_m})$ and a matrix $P \in \mathbb{R}^{p \times n}$, we consider the mda $P^m(T)$ given by

$$(P^m(T))_{j_1 \cdots j_m} = \sum_{i_1 \cdots i_m = 1}^{n} T_{i_1 \cdots i_m} p_{j_1 i_1} \cdots p_{j_m i_m}.$$

If P is a row vector, $\boldsymbol{x}' = (x_1, \ldots, x_n)$, the following notations introduced in [138] which involve homogeneous polynomials are used:

$$\begin{aligned}
T\boldsymbol{x}^{m-2} &= T \times_3 \boldsymbol{x}' \times_4 \cdots \times_m \boldsymbol{x}' \\
&= \sum_{i_3,\ldots,i_m=1}^{n} T_{iji_3\cdots i_m} x_{i_3} \cdots x_{i_m} \\
T\boldsymbol{x}^{m-1} &= T \times_2 \boldsymbol{x}' \times_3 \cdots \times_m \boldsymbol{x}' \\
&= \sum_{i_2,\ldots,i_m=1}^{n} T_{ii_2\cdots i_m} x_{i_2} \cdots x_{i_m} \\
T\boldsymbol{x}^{m} &= T \times_1 \boldsymbol{x}' \times_2 \cdots \times_m \boldsymbol{x}' \\
&= \sum_{i_1,\ldots,i_m=1}^{n} T_{ii_1\cdots i_m} x_{i_3} \cdots x_{i_m}.
\end{aligned}$$

The right-hand side of the last equality is the full contraction of T and $\boldsymbol{x}$.

In general, one could define $T\boldsymbol{x}^{m-k}$ as

$$T\boldsymbol{x}^{m-k} = T \times_{k+1} \boldsymbol{x}' \times_{k+2} \cdots \times_m \boldsymbol{x}'$$

as having the components the homogeneous polynomials

$$\sum_{i_{k+1}\cdots i_m=1}^{n} T_{j_1\cdots j_k i_{k+1}\cdots i_m} x_{i_{k+1}} \cdots x_{i_m}.$$

Definition 15.15. Let $T \in \mathfrak{M}_{m,n}$. A complex number λ is an *eigenvalue* of T if there exists $\boldsymbol{x} \in \mathbb{C}^n - \{\mathbf{0}_n\}$ such that the following equalities involving homogeneous polynomials are satisfied:

$$(T\boldsymbol{x}^{m-1})_i = \lambda x_i^{m-1} \tag{15.13}$$

for $1 \leqslant i \leqslant n$.

The vector $\boldsymbol{x}$ is an *eigenvector* of T associated with λ and $(\lambda, \boldsymbol{x})$ is an *eigenpair* of T.

If $\boldsymbol{x}^{[m-1]}$ is a vector in $\mathbb{C}^n$ defined by $(\boldsymbol{x}^{[m-1]})_i = x_i^{m-1}$ for $1 \leqslant i \leqslant n$, then Equality (15.13) can be written as $T\boldsymbol{x}^{m-1} = \lambda\boldsymbol{x}^{[m-1]}$.

The *spectrum* of T is the set of eigenvalues of T. This set is denoted as $\mathsf{spec}(T)$; the spectral radius of T is the number

$$\rho(T) = \max\{|\lambda| \mid \lambda \in \mathsf{spec}(T)\}.$$

The term H-eigenvalue was introduced in [136].

Definition 15.16. An eigenvalue λ of an mda T is an *H-eigenvalue* if λ is real and there is a real eigenvector $\boldsymbol{x}$ for λ. In this case, we refer to $\boldsymbol{x}$ as an H-eigenvector.

If $\lambda \in \mathbb{R}$, $\boldsymbol{x} \in \mathbb{R}^n$, and $A\boldsymbol{x}^{m-1} = \lambda x$ and $\boldsymbol{x}'\boldsymbol{x} = 1$, the $(\lambda, \boldsymbol{x})$ is said to be a Z-pair (*cf.* [136, 138]).

Theorem 15.11. *Let $T \in T_{m,n}$ be an mda, where m is an even number. Then T has H-eigenvalues.*

Proof. Consider the minimization of the continuous function $T\boldsymbol{x}^m$ subjected to the restriction $\sum_{i=1}^{n} x_i^m = 1$. Since m is a positive even integer, the set $\{\boldsymbol{x} \in \mathbb{R}^n \mid \sum_{i=1}^{n} x_i^m = 1\}$ is compact and the minimizer $\boldsymbol{x}^*$ exists.

Using the Lagrangian $L(\boldsymbol{x}, \lambda) = T\boldsymbol{x}^m - \lambda(\|\boldsymbol{x}\|^m - 1)$, the following Karush–Kuhn–Tucker optimality conditions:

$$\frac{\partial L(\boldsymbol{x}, \lambda)}{\partial \boldsymbol{x}} = 0 \text{ and } \frac{\partial L(\boldsymbol{x}, \lambda)}{\partial \lambda} = 0$$

amount to $mT\boldsymbol{x}^{m-1} - \lambda m\boldsymbol{x}^{[m-1]} = 0$ and $1 - \|\boldsymbol{x}\|_m^m = 0$, which imply $T\boldsymbol{x}^{m-1} = \lambda\boldsymbol{x}^{[m-1]}$, which is the equality that defines eigenvalues, where $\boldsymbol{x} = \boldsymbol{x}^*$. □

The notions of positive definiteness and semidefiniteness for matrices can be extended to mdas in $\mathcal{M}_{m,n}$.

Definition 15.17. An mda $T \in \mathcal{M}_{m,n}$ is *positive semidefinite* if $T\boldsymbol{x}^m \geqslant 0$ for all $\boldsymbol{x} \in \mathbb{R}^n$. If $T\boldsymbol{x}^m > 0$ for $\boldsymbol{x} \in \mathbb{R}^n - \{\mathbf{0}_n\}$, then T is *positive definite.*

Let $\lambda_{H\ \min}(T)$ be the smallest eigenvalue of T, where $T \in T_{m,n}$ and m is an even number. Then T is positive definite (positive semidefinite) if and only if $\lambda_{H\ \min}(T) > 0$ ($\lambda_{H\ \min}(T) \geqslant 0$, respectively).

By the definition of $\boldsymbol{x}^*$ given in Theorem 15.11, we have

$$T\left(\frac{\boldsymbol{x}}{(\sum_{i=1}^{n} x_i^m)^{1/m}}\right) \geqslant T(\boldsymbol{x}^*)^m.$$

Since $\sum_{i=1}^{m} x_i^m > 0$ when $\boldsymbol{x} \in \mathbb{R}^n - \{\mathbf{0}_n\}$, it follows that $\lambda_{H\ \min}(T) > 0$.

A generalization of Gershgorin's Theorem (Theorem 7.21) for mdas follows.

Theorem 15.12. *Let $T \in \mathfrak{M}_{m,n}$ be an* mda *and let r_i be the sum of the absolute values of off-diagonal entries of T. The eigenvalues of T are contained in the union of n disks having the diagonal components of T as their centers and the numbers r_i as their radii.*

Proof. Suppose that $(\lambda, \boldsymbol{x})$ is an eigenpair of T and that x_i is

$$|x_i| = \max\{|x_j| \mid 1 \leqslant j \leqslant n\}.$$

We have

$$\lambda x_i^{m-1} = \sum_{i_2,\ldots,i_m=1}^{n} a_{ii_2\cdots i_m} x_{i_2} \cdots x_{i_m},$$

which implies

$$(\lambda - a_{i\cdots i})x_i^{m-1} = \sum_{\substack{i_2,\ldots,i_m=1 \\ \delta_{i_1\cdots i_m}=0}}^{n} a_{ii_2\cdots i_m} x_{i_2} \cdots x_{i_m}.$$

This implies

$$|\lambda - a_{i\cdots i}| \leqslant \sum_{\substack{i_2,\ldots,i_m=1 \\ \delta_{i_1\cdots i_m}=0}}^{n} |a_{ii_2\cdots i_m} \frac{|x_{i_2}|}{|x_i|} \cdots \frac{|x_{i_m}|}{|x_i|}$$

$$\leqslant \sum_{\substack{i_2,\ldots,i_m=1\\ \delta_{i_1\cdots i_m}=0}}^{n} |a_{ii_2\cdots i_m}|,$$

which concludes the argument. □

The notion of submatrix introduced in Definition 3.5 can be extended to mdas.

Definition 15.18. A *submda* (or a subtensor) of an mda $T \in \mathbb{R}^{I_1\times\cdots\times I_N}$ is an mda $T_{i_n=a} \in \mathbb{R}^{I_1\times\cdots\times I_{n-1}\times I_{n+1}\times\cdots\times I_N}$ obtained by fixing the nth index to a.

Example 15.36. Let T be an mda with the format $3\times 4\times 2$ having the following frontal slices:

$$T_1 = \begin{pmatrix} 1 & 4 & 7 & 10\\ 2 & 5 & 8 & 11\\ 3 & 6 & 9 & 12 \end{pmatrix} \text{ and } T_2 = \begin{pmatrix} 13 & 16 & 19 & 22\\ 14 & 17 & 20 & 23\\ 15 & 18 & 21 & 24 \end{pmatrix}.$$

T_1 and T_2 are the subtensors $T_{i_3=1}$ and $T_{i_3=1}$. The subtensors $T_{i_1=1}$, $T_{i_1=2}$, and $T_{i_1=3}$ in $\mathbb{R}^{4\times 2}$ are

$$T_{i_1=1} = \begin{pmatrix} 1 & 4 & 7 & 10\\ 13 & 16 & 19 & 22 \end{pmatrix},$$

$$T_{i_1=2} = \begin{pmatrix} 2 & 5 & 8 & 11\\ 14 & 17 & 20 & 23 \end{pmatrix},$$

$$T_{i_1=3} = \begin{pmatrix} 3 & 6 & 9 & 12\\ 15 & 18 & 21 & 24 \end{pmatrix}.$$

We reformulate the Singular Value Decomposition Theorem for matrices (Theorem 9.1) in preparation for a similar result for mdas. Namely, for a matrix $A \in \mathbb{C}^{I_1\times I_2}$, there is a decomposition

$$A = U^{(1)}SV^{(2)'} = S\times_1 U^{(1)} \times_2 V^{(2)'} = S\times_1 U^{(1)}\times_2 U^{(2)},$$

such that $U^{(1)} = (\boldsymbol{u}_1^{(1)}\boldsymbol{u}_2^{(1)}\cdots\boldsymbol{u}_{I_1}^{(1)}$ is a unitary $(I_1\times I_1)$-matrix, $V^{(2)'} = U^{(2)} = (\boldsymbol{u}_1^{(2)}\boldsymbol{u}_2^{(2)}\cdots\boldsymbol{u}_{I_2}^{(2)}$ is a unitary $(I_2\times I_2)$-matrix, S is an $(I_1\times I_2)$-matrix such that $S = \mathsf{diag}(\sigma_1,\sigma_2,\ldots,\sigma_{\min\{I_1,I_2\}})$, and $\sigma_1 \geqslant \sigma_2 \geqslant \cdots \geqslant \sigma_{\min\{I_1,I_2\}} \geqslant 0$.

The generalization of this result to mdas that follows is known as the Higher Order Singular Value Decomposition (HOSVD). The presentation follows [100, 102].

Theorem 15.13 (HOSVD theorem). *Every mda $T \in \mathbb{R}^{I_1 \times \cdots \times I_N}$ can be written as*

$$T = S \times_1 U^{(1)} \times_2 \cdots \times_N U^{(N)},$$

such that

(i) $U^{(n)} = (\boldsymbol{u}_1^{(n)} \boldsymbol{u}_2^{(n)} \cdots \boldsymbol{u}_{I_n}^{(n)})$ *is an orthogonal* $(I_n \times I_n)$*-matrix;*

(ii) $S \in \mathbb{R}^{I_1 \times \cdots \times I_N}$ *is an mda such that*

  (a) any two distinct mdas $S_{i_n=a}$ *and* $S_{i_n=b}$ *with* $a \neq b$ *are orthogonal (the* general orthogonality *property), and*

  (b)

$$\|S_{i_n=1}\| \geqslant \|S_{i_n=2}\| \geqslant \cdots \geqslant \|S_{i_n=I_n}\|$$

  for all possible values of n, α, β, *with* $\alpha \neq \beta$.

The i*th column* $\boldsymbol{u}_i^{(n)}$ *of the matrix* $U^{(n)}$ *is the* ith n-singular vector. *The numbers* $\|S_{i_n=i}\|$ *denoted by* $\sigma_i^{(n)}$ *are the* n-mode singular values.

Proof. Consider the mdas $T, S \in \mathbb{R}^{I_1 \times \cdots \times I_N}$ such that

$$S = T \times_1 U^{(1)'} \times_2 \cdots \times_N U^{(N)'},$$

where $U^{(1)}, \ldots, U^{(N)}$ are orthogonal matrices. In matrix form, this equality becomes

$$T_{(n)} = U^{(n)} S_{(n)} \left(U^{(n+1)} \otimes \cdots U^{(N)} \otimes U^{(1)} \otimes \cdots \otimes U^{(n-1)}\right)'.$$

If $U^{(n)}$ is obtained from the SVD of $T_{(n)}$ as $T_{(n)} = U^{(n)} \Sigma^{(n)} V^{(n)'}$, where $V^{(n)}$ is orthogonal and $\Sigma^{(n)} = \mathsf{diag}(\sigma_1^{(n)}, \sigma_2^{(n)}, \ldots, \sigma_{I_n}^{(n)})$, where $\sigma_1^{(n)} \geqslant \sigma_2^{(n)} \geqslant \cdots \geqslant \sigma_{I_n}^{(n)} \geqslant 0$, we denote by r_n the highest index for which $\sigma_{r_n}^{(n)} > 0$. Taking into account that the matrix $U^{(n+1)} \otimes \cdots U^{(N)} \otimes U^{(1)} \otimes \cdots \otimes U^{(n-1)}$ is orthogonal (by Supplement 117 of Chapter 6), it follows that

$$S_{(n)} = \Sigma^{(n)} V^{(n)'} \left(U^{(n+1)} \otimes U^{(n+2)} \cdots \otimes U^{(N)} \otimes U^{(1)} \right.$$
$$\left. \otimes U^{(2)} \cdots \otimes U^{(n-1)}\right).$$

This implies for arbitrary orthogonal matrices

$$U^{(1)}, \ldots, U^{(n-1)}, U^{(n+1)}, \ldots, U^{(N)}$$

that if $\alpha \neq \beta$, then $(S_{i_n=\alpha}, S_{i_n=\beta}) = 0$ and

$$\|S_{i_n=1}\| = \sigma_1^{(n)} \geqslant \|S_{i_n=2}\| = \sigma_2^{(n)} \geqslant \cdots \geqslant \|S_{i_n=I_n}\| = \sigma_{I_n}^{(n)} \geqslant 0,$$

and, if $r_n < I_n$,

$$\|S_{i_n=r_n+1}\| = \sigma_{r_n+1}^{(n)} = \cdots = \|S_{i_n=I_n}\| = \sigma_{I_n}^{(n)} = 0.$$

The matrices $U^{(1)}, U^{(n-1)}, U^{(n+1)}, U^{(N)}$ can be constructed in the same manner as $U^{(n)}$ such that S and these matrices satisfy the conditions of the theorem.

On the other hand, all matrices $U^{(1)}, \ldots, U^{(N)}$ and S that satisfy the theorem can be found from the singular value decompositions of $A_{(n)}$. □

Note that the matrix S that occurs in the SVD is replaced by the core mda S and the diagonal character of the matrix S in the SVD Theorem is replaced by the orthogonality of the mdas of the form $S_{i_n=a}$. The role played by the singular values in the SVD theorem is played by the norms of the mdas $S_{i_n=k}$.

Ample details concerning the spectral theory of mdas can be found in the monograph [138].

15.12 Decomposition of Tensors

The notion of singular value defined for matrices was extended to tensors in [101]. This is the basis of the CANDECOMP/PARAFAC technique (abbreviated as CP), which decomposes a tensor as a sum of rank-1 tensors, a concept developed in the psychometric [25, 46] and chemometric publications [89].

A construction introduced by Kruskal [97] starts with the matrices $A \in \mathbb{R}^{I \times R}$, $B \in \mathbb{R}^{J \times R}$, and $C \in \mathbb{R}^{K \times R}$ and defines the mda T as

$$T_{ijk} = \sum_{r=1}^{R} a_{ir} b_{jr} c_{kr}.$$

The mda T is denoted by $[\![A, B, C]\!]$.

A more general decomposition applicable to mdas of order N allows an mda $T \in \mathbb{R}^{I_1 \times \cdots \times I_n}$ to be expressed as

$$T_{i_1 i_2 \cdots i_r} = \sum_{r=1}^{R} \lambda_r a_{i_1 r} a_{i_2 r} \cdots a_{i_n r},$$

where $\lambda \in \mathbb{R}^R$, $A_{i_p} \in \mathbb{R}^{I_p \times R}$ for $1 \leqslant p \leqslant r$. In this case, T is denoted by $[\![\lambda; A, B, C]\!]$.

Another modality of saving space when storing an mda is the *Tucker operator* (see [92] and [164]).

A three-way mda $T \in \mathbb{R}^{I \times J \times K}$ is stored as the product of a *core mda* G of size $R \times S \times Z$ with corresponding matrices $A \in \mathbb{R}^{I \times R}$, $B \in \mathbb{R}^{J \times S}$, and $C \in \mathbb{R}^{K \times Z}$ such that

$$T_{ijk} = \sum_{r=1}^{R} \sum_{s=1}^{S} \sum_{t=1}^{Z} G_{rst} a_{ir} b_{js} c_{ks}.$$

If I, J, K are much larger than R, S, Z, then forming T explicitly requires more memory than it is required to store its components, namely $RSZ + IR + JS + KZ$.

Definition 15.19. Let $T \in \mathbb{R}^{J_1 \times \cdots \times J_N}$, $\mathcal{N} = \{1, \ldots, N\}$, and let N matrices $A^{(n)} \in \mathbb{R}^{I_n \times J_n}$ for $n \in \mathcal{N}$.

The *Tucker operator* applied to T produces the mda $[\![T; A^{(1)}, \ldots, A^{(N)}]\!]$ defined by

$$[\![T; A^{(1)}, \ldots, A^{(N)}]\!] = T \times_1 A^{(1)} \times_2 \cdots \times_n A^{(N)} \in \mathbb{R}^{I_1 \times \cdots \times I_N}.$$

T is referred to as the *core mda*.

Theorem 15.14. *Let $T \in \mathbb{R}^{J_1 \times \cdots \times J_N}$ and let $\mathcal{N} = \{1, \ldots, N\}$.*
If $A^{(n)} \in \mathbb{R}^{I_n \times J_n}$ and $B^{(n)} \in \mathbb{R}^{K_n \times I_n}$ for $1 \leqslant n \leqslant N$, then

$$[\![[\![T; A^{(1)}, \ldots, A^{(N)}]\!]; B^{(1)}, \ldots, B^{(N)}]\!] = [\![T; B^{(1)} A^{(1)}, \ldots, B^{(N)} A^{(N)}]\!].$$

Proof. By the definition of Tucker operator, we have

$$\begin{aligned}
&[\![[\![T; A^{(1)}, \ldots, A^{(N)}]\!]; B^{(1)}, \ldots, B^{(N)}]\!] \\
&= [\![(T \times_1 A^{(1)} \times_2 \cdots \times_N A^{(N)}); B^{(1)}, \ldots, B^{(N)}]\!] \\
&= (T \times_1 A^{(1)} \times_2 \cdots \times_N A^{(N)}) \times_1 B^{(1)} \cdots \times_N B^{(N)}.
\end{aligned}$$

Note that

$$(T \times_1 A^{(1)} \cdots \times_N A^{(N)})_{i_1 \cdots i_N} = \sum_{j_1 \cdots j_n} T_{j_1 \cdots j_N} A^{(1)}_{i_1 j_1} \cdots A^{(N)}_{i_N j_N}.$$

Furthermore,

$$\begin{aligned}
&((T \times_1 A^{(1)} \cdots \times_n A^{(N)}) \times_1 B^{(1)} \cdots \times_N B^{(N)})_{k_1 \ldots k_N} \\
&= \sum_{i_1 \cdots i_N} \sum_{j_1 \cdots j_N} T_{j_1 \cdots j_N} A^{(1)}_{i_1 j_1} \cdots A^{(N)}_{i_N j_N} B^{(1)}_{k_1 i_1} \cdots B^{(N)}_{k_N i_N} \\
&= \sum_{i_1 \cdots i_N} \sum_{j_1 \cdots j_N} T_{j_1 \cdots j_N} B^{(1)}_{k_1 i_1} A^{(1)}_{i_1 j_1} \cdots B^{(N)}_{k_N i_N} A^{(N)}_{i_N j_N} \\
&= \sum_{j_1 \cdots j_n} T_{j_1 \cdots j_N} \left(\sum_{i_1} B^{(1)}_{k_1 i_1} A^{(1)}_{i_1 j_1} \right) \cdots \left(\sum_{i_N} B^{(N)}_{k_N i_N} A^{(N)}_{i_N j_N} \right) \\
&= (T; B^{(1)} A^{(1)}, \ldots, B^{(N)} A^{(N)})_{k_1 \ldots k_N},
\end{aligned}$$

which concludes the argument. □

The next statement allows us to express the Tucker operator in terms of matricized mdas.

Theorem 15.15. *Let $T \in \mathbb{R}^{J_1 \times \cdots \times J_N}$ be an* mda *and let $\mathcal{N} = \{1, \cdots, N\}$.*

If $A^{(n)} \in \mathbb{R}^{I_n \times J_n}$ for $n \in \mathcal{N}$, and $\mathcal{R} = \{r_1, \ldots, r_L\}$, $\mathcal{C} = \{c_1, \ldots, c_M\}$ form a partition of $\mathcal{N}$, then $S = [\![T; A^{(1)}, \ldots, A^{(N)}]\!]$ if and only if

$$S_{(\mathcal{R} \times \mathcal{C}): J_\mathcal{N}} = (A^{(r_L)} \otimes \cdots \otimes A^{(r_1)}) T_{(\mathcal{R} \times \mathcal{C}: I_\mathcal{N})} (A^{(c_M)} \otimes \cdots \otimes A^{(c_1)})'.$$

Here $I_\mathcal{N} = \{I_1, \ldots, I_N\}$ and $J_\mathcal{N} = \{J_1, \ldots, J_N\}$.

Proof. This is a direct consequence of Theorem 15.5. □

The calculation of the norm of a large mda can be reduced to the calculation of the norm of a smaller mda using the thin QR factorization of rectangular matrices discussed in Theorem 6.65.

Theorem 15.16. *Let $T \in \mathbb{R}^{J_1 \times J_2 \times \cdots \times J_N}$ be an* mda *having the set of modes $\mathcal{N} = \{1, \ldots, N\}$, and let $A^{(n)}$ be N rectangular matrices* *having the QR decompositions* $A^{(n)} = Q^{(n)}R^{(n)}$ *for* $n \in \mathcal{N}$, *where the columns of* $Q^{(n)}$ *constitute an orthonormal basis for* $\mathsf{range}(A)$, *and* $R^{(n)}$ *is an upper triangular invertible matrix such that its diagonal elements are real non-negative numbers for* $n \in \mathcal{N}$.

We have

$$\left\| [\![T; A^{(1)}, \ldots, A^{(N)}]\!] \right\| = \left\| [\![T; R^{(1)}, \ldots, R^{(N)}]\!] \right\|.$$

Proof. By the definition of QR decomposition of matrices and Theorem 15.14, we have

$$\begin{aligned} [\![T; A^{(1)}, \ldots, A^{(N)}]\!] &= [\![T; Q^{(1)}R^{(1)}, \ldots, Q^{(N)}R^{(N)}]\!] \\ &= [\![[\![T; R^{(1)}, \ldots, R^{(N)}]\!]; Q^{(1)}, \ldots, Q^{(n)}]\!]. \end{aligned}$$

Supplement 15.13 further implies

$$\left\| [\![T; A^{(1)}, \ldots, A^{(N)}]\!] \right\| = \left\| [\![T; R^{(1)}, \ldots, R^{(N)}]\!] \right\|.$$

□

Definition 15.20. The *Tucker decomposition* of an mda $T \in \mathbb{R}^{I_1 \times I_2 \times \cdots \times I_N}$ is given by

$$T = [\![G; A^{(1)}, A^{(2)}, \ldots, A^{(N)}]\!],$$

where $A^{(n)} \in \mathbb{R}^{I_n \times J_n}$ and $G \in \mathbb{R}^{J_1 \times I_2 \times \cdots \times J_N}$.

The Tucker decomposition allows the generation of an approximation G of T when the ranks of $G_{(n)}$ are smaller than the corresponding ranks of the matrices $T_{(n)}$.

In general, the Tucker decomposition is not unique (see Supplement 15.13).

15.13 Approximation of mdas

The next theorem appears in [92, 102].

Theorem 15.17. *Let* $T \in \mathbb{R}^{I_2 \times I_2 \times \cdots \times I_N}$ *and let* $A^{(n)} \in \mathbb{R}^{I_n \times J_n}$ *be matrices with orthonormal sets of columns for* $1 \leqslant n \leqslant N$.

The optimal mda $G \in \mathbb{R}^{J_1 \times J_2 \times \cdots \times J_N}$ *that minimizes* $\left\| T - [\![G; A^{(1)}, \ldots, A^{(N)}]\!] \right\|$ *for fixed matrices* $A^{(1)}, \ldots, A^{(N)}$ *is given by* $G = [\![T; A^{(1)'}, \ldots, A^{(n)'}]\!]$.

Furthermore, the optimal matrices $A^{(i)}$ *for this problem maximize* $\left\| [\![T; A^{(1)'}, \ldots, A^{(N)'}]\!] \right\|$.

Proof. Choosing $\mathcal{R} = \mathcal{N}$ in Theorem 15.15 allows us to write

$$\begin{aligned}
&\left\| T - [\![G; A^{(1)'}, \ldots, A^{(N)'}]\!] \right\| \\
&= \left\| T - G \times_1 A^{(1)'} \times_2 \cdots \times_n A^{(N)'} \right\| \\
&= \left\| \mathsf{vec}(T) - (A^{(N)} \otimes \ldots \otimes A^{(1)})\mathsf{vec}(G) \right\|.
\end{aligned}$$

The minimization of the norm $\left\| \mathsf{vec}(T) - (A^{(N)} \otimes \ldots \otimes A^{(1)})\mathsf{vec}(G) \right\|$ can be treated as a least square problem and the solution is

$$\mathsf{vec}(G) = \left(A^{(N)} \otimes \cdots \otimes A^{(1)}\right)^{\dagger} \mathsf{vec}(T).$$

The solution can be written as

$$\begin{aligned}
\mathsf{vec}(G) &= (A^{(N)} \otimes \cdots \otimes A^{(1)})^{\dagger}\mathsf{vec}(T) \\
&= (A^{(N)\dagger} \otimes \cdots \otimes A^{(1)\dagger})\mathsf{vec}(T) \\
&\quad \text{(by Supplement 27 of Chapter 9)} \\
&= (A^{(N)'} \otimes \cdots \otimes A^{(1)'})\mathsf{vec}(T) \\
&\quad \text{(because the matrices } A^{(i)} \text{ have orthonormal} \\
&\quad \text{sets of columns).}
\end{aligned}$$

Next, we have

$$\begin{aligned}
&\left\| T - [\![G; A^{(1)}, \ldots, A^{(N)}]\!] \right\| \\
&= \|T\|^2 - 2(T, [\![G; A^{(1)}, \ldots, A^{(N)}]\!]) + \|[\![G; A^{(1)}, \ldots, A^{(n)}]\!]\|^2.
\end{aligned}$$

We have

$$\begin{aligned}
&(T, [\![G; A^{(1)}, \ldots, A^{(N)}]\!]) \\
&= ([\![T; A^{(1)'}, \ldots, A^{(N)'}]\!], G) \\
&\quad \text{(by Theorem 15.7)} \\
&= \|G\|^2.
\end{aligned}$$

By Supplement 17, we have

$$\left\| [\![G; A^{(1)}, \ldots, A^{(N)}]\!] \right\| = \|G\|,$$

which implies

$$\left\| T - [\![G; A^{(1)}, \ldots, A^{(N)}]\!] \right\|^2 = \|T\|^2 - \|G\|^2.$$

Thus, minimizing $\left\| T - [\![G; A^{(1)}, \ldots, A^{(N)}]\!] \right\|$ amounts to maximizing $\|G\|$. □

Next we discuss approximations of mdas by mdas of rank 1. In other words, given an mda $T \in \mathbb{R}^{I_1 \times \cdots \times I_N}$, we seek the determination of $\lambda \in \mathbb{R}$ and of unit-vectors $\boldsymbol{u}^{(1)}, \ldots, \boldsymbol{u}^{(N)}$ that define a rank-1 mda S defined as

$$S_{i_1 \ldots i_N} = \lambda \boldsymbol{u}^{(1)}_{i_1} \cdots \circ \boldsymbol{u}^{(N)}_{i_N}$$

that minimizes the least-square function

$$f(S) = \|T - S\|^2. \tag{15.14}$$

This is a constrained optimization problem solved in [101] by the method of Lagrange multipliers.

Consider the objective function $\tilde{f}$ defined as

$$\sum_{i_1 \cdots i_N} \left(T_{i_1 \cdots i_N} - \lambda u^{(1)}_{i_1} \cdots u^{(N)}_{i_N} \right)^2 + \sum_n \lambda^{(n)} \left(\sum_{i_n} (u^{(n)}_{i_n})^2 - 1 \right).$$

The equality $\frac{\partial \tilde{f}}{\partial u^{(n)}_{i_n}} = 0$ implies

$$\lambda \sum_{i_1 \cdots i_{n-1} i_{n+1} \cdots i_N} T_{i_1 \cdots i_N} u^{(1)}_{i_1} \cdots u^{(n-1)}_{i_{n-1}} u^{(n+1)}_{i_{n+1}} u^{(N)}_{i_N} = \lambda^{(n)} u^{(n)}_{i_n} \tag{15.15}$$

$$+\lambda^2 u^{(n)}_{i_n} \sum_{i_1 \cdots i_{n-1} i_{n+1} \cdots i_N} \left(u^{(1)}_{i_1} \right)^2 \cdots \left(u^{(n-1)}_{i_{n-1}} \right)^2 \left(u^{(n+1)}_{i_{n+1}} \right)^2 \cdots \left(u^{N}_{i_N} \right)^2.$$

The equality $\frac{\partial \tilde{f}}{\partial \lambda^{(n)}} = 0$ implies

$$\sum_{i_n} \left(u_{i_n}^{(n)}\right)^2 = 1. \tag{15.16}$$

Finally, $\frac{\partial \tilde{f}}{\partial \lambda} = 0$ yields

$$\sum_{i_1 \cdots i_N} T_{i_1 \cdots i_N} u_{i_1}^{(1)} \cdots u_{i_N}^{(N)} = \lambda \sum_{i_1 \cdots i_N} \left(u_{i_1}^{(1)}\right)^2 \cdots \left(u_{i_N}^{(N)}\right)^2.$$

The previous equalities imply

$$\sum_{i_1 \cdots i_N} T_{i_1 \cdots i_N} u_{i_1}^{(1)} \cdots u_{i_N}^{(N)} = \lambda, \tag{15.17}$$

and

$$\lambda \sum_{i_1 \cdots i_{n-1} i_{n+1} \cdots i_N} T_{i_1 \cdots i_N} u_{i_1}^{(1)} \cdots u_{i_{n-1}}^{(n-1)} u_{i_{n+1}}^{(n+1)} \cdots U_{i_N}^{(N)} = (\lambda^2 + \lambda^{(n)}) u_{i_n}^{(n)}. \tag{15.18}$$

Taking into account Equality (15.16), we have

$$\sum_{i_1 \cdots i_{n-1} i_{n+1} \cdots i_N} T_{i_1 \cdots i_N} u_{i_1}^{(1)} \cdots u_{i_{n-1}}^{(n-1)} u_{i_{n+1}}^{(n+1)} \cdots u_{i_N}^{(N)} = \lambda u_{i_N}^{(N)}. \tag{15.19}$$

These necessary conditions correspond to

$$T \times_1 \boldsymbol{u}^{(1)'} \cdots \times_{n-1} \boldsymbol{u}^{(n-1)'} \times_{n+1} \boldsymbol{u}^{(n+1)'} \cdots \times_N \boldsymbol{u}^{(N)'} = \lambda \boldsymbol{u}^{(N)}, \tag{15.20}$$

$$T \times_1 \boldsymbol{u}^{(1)'} \times_2 \cdots \times_N \boldsymbol{u}^{(N)'} = \lambda, \tag{15.21}$$

and

$$\|\boldsymbol{u}^{(n)}\| = 1 \tag{15.22}$$

for $1 \leqslant n \leqslant N$.

Minimizing the function f defined by Equality (15.14) is equivalent to maximizing

$$g(\boldsymbol{u}^{(1)}, \boldsymbol{u}^{(2)}, \ldots, \boldsymbol{u}^{(N)}) = \left| T \times_1 \boldsymbol{u}^{(1)'} \times_2 \boldsymbol{u}^{(2)'} \times_3 \cdots \times_N \boldsymbol{u}^{(N)'} \right|^2,$$

over $\boldsymbol{u}^{(1)}, \boldsymbol{u}^{(2)}, \ldots, \boldsymbol{u}^{(N)}$. Moreover, if λ is chosen as in Equality (15.21), then the functions f and g satisfy the equality $f(S) = \|T\|^2 - g(S)$.

Indeed, note that $f(S) = \|T - S\|^2 = \|T\|^2 - 2(T, S) + \|S\|^2$. The definition of λ and Equality (15.17) imply

$$(T, S) = \sum_{i_1 \cdots i_N} T_{i_1 \cdots i_N} \lambda u_{i_1}^{(1)} \cdots u_{i_N}^{(N)} = \lambda^2.$$

Since $\|\boldsymbol{u}^{(1)}\| = \|\boldsymbol{u}^{(2)}\| = \cdots = \|\boldsymbol{u}^{(N)}\| = 1$, we also have $\|S\|^2 = \lambda^2$. This implies $f(S) = \|T\|^2 - g(S)$.

An alternating least squares algorithm for computing the best rank-1 approximation of an mda is presented in [101].

Exercises and Supplements

(1) Let $\boldsymbol{e}_{(i_n)}^{(I_n)}$ be a vector in $\mathbb{R}^{I_n}$ for $1 \leqslant n \leqslant N$. Prove that

$$\boldsymbol{e}_{(i_1)}^{(I_1)} \otimes \cdots \otimes \boldsymbol{e}_{(i_N)}^{(I_N)} = \boldsymbol{e}_i^I,$$

where $I = I_1 \cdot \ldots \cdot I_N$, and

$$i = (i_1 - 1) \prod_{p=2}^{N} I_p + (i_2 - 1) \prod_{p=3}^{N} I_p + \cdots + i_N.$$

(2) Let $T \in \mathbb{R}^{J_1 \times J_2 \times \cdots \times J_N}$ be an mda. Prove that if $A \in \mathbb{R}^{I_m \times J_m}$ and $B \in \mathbb{R}^{I_n \times J_n}$ and $m \neq n$, then $(T \times_m A) \times_n B = (T \times_n B) \times_m A$. The common value of these arrays is denoted by $T \times_m A \times_n B$.

Solution: Without loss of generality, assume that $m < n$. The definition of the mda-matrix mode product allows us to write

$$(T \times_m A)_{j_1,\ldots,j_{m-1},i_m,j_{m+1},\ldots,j_N} = \sum_{i_m=1}^{I_m} T_{j_1 j_2 \cdots j_N} A_{i_m j_m}.$$

Therefore,

$$\begin{aligned}
&((T \times_m A) \times_n B)_{j_1,\ldots,j_{m-1},i_m,j_{m+1},\ldots,j_{n-1},i_n,j_{n+1},\ldots,j_N} \\
&= \textstyle\sum_{i_n=1}^{I_n} (T \times_m A)_{j_1,\ldots,j_{m-1},i_m,j_{m+1},\ldots,j_N} B_{i_n\, j_n} \\
&= \textstyle\sum_{i_n=1}^{I_n} \sum_{i_m=1}^{I_m} T_{j_1 j_2 \cdots j_n \cdots j_m \cdots j_N} A_{i_m j_m} B_{i_n j_n}.
\end{aligned}$$

The same expression is obtained by computing $((T \times_n B) \times_m A)_{j_1 \cdots j_{m-1} i_m j_{m+1} \cdots j_{n-1} i_n j_{n+1} \cdots j_N}$.

(3) Recall that the contracted n-mode product $\overline{\times}_n$ drops the nth singleton dimension. Prove that the contracted n-mode product is not commutative, that is, in general $(T\overline{\times}_n\boldsymbol{x})\overline{\times}_n\boldsymbol{y} \neq (T\overline{\times}_n\boldsymbol{y})\overline{\times}_n\boldsymbol{x}$.

(4) Prove that if T is an mda having the set of modes $\mathcal{M}$, then $\mathsf{vec}(T) = T_{(\mathcal{M}\times\emptyset:\mathcal{M})}$.

(5) Let T, S be two mdas of rank 1, where $T = \boldsymbol{a}^{(1)}\otimes\boldsymbol{a}^{(2)}\otimes\cdots\otimes\boldsymbol{a}^{(n)}$ and $S = \boldsymbol{b}^{(1)} \otimes \boldsymbol{b}^{(2)} \otimes \cdots \otimes \boldsymbol{b}^{(n)}$. Prove that

$$(T, S) = \prod_{i=1}^{n}(\boldsymbol{a}^{(i)} \otimes \boldsymbol{b}^{(i)}).$$

(6) Let $T \in \mathbb{R}^{I_1\times I_2\times\cdots\times I_N}$ be an mda, and let $A \in \mathbb{R}^{J_n\times I_n}$ and $B \in \mathbb{R}^{J_m\times I_m}$ be two matrices, where $m \neq m$. Prove that

$$(T \times_n A) \times_m B = (T \times_m B) \times_n A.$$

(7) Let $A \in \mathbb{R}^{m\times n}$ and $B \in \mathbb{R}^{p\times q}$ be two matrices, and let $\boldsymbol{x} \in \mathbb{R}^{qn}$ be a vector. Compute the vector $\boldsymbol{y} = (A \otimes B)\boldsymbol{x}$ without computing $A\otimes B$. This is important if the size of the Kronecker product is very large.

Solution: The following MATLAB program is a solution:

```
[m,n] = size(A);
[p,q] = size(B);
X = reshape(x,q,n);
Y = B*X*A';
y = reshape(Y,m*p,1);
```

(8) Let

$$A^{(1)}, \ldots, A^{(N)}$$

be N matrices, where $A^{(n)} \in \mathbb{R}^{I_n\times J_n}$ for $n \in \{1, \ldots, N\}$. Prove that $X = [\![Y; A^{(1)}, \ldots, A^{(N)}]\!]$ if and only if $X_{(n)} = A^{(n)}Y_{(n)}(A^{(N)} \otimes \cdots \otimes A^{(n+1)} \otimes A^{(n-1)} \otimes \cdots \otimes A^{(1)})'$.

(9) Let T be an mda in $T_{m,n}$. Prove that if $(\lambda, \boldsymbol{x})$ is an eigenpair of T, then $(a\lambda + b, \boldsymbol{x})$ is an eigenpair of $aT + bI_{m,n}$.

(10) Let $X = [A, B, C]$ be a representation of a three-way array X, and let $\mathcal{M}, \mathcal{N}$ be two sets of matrices such that $\mathcal{M} \subseteq \mathcal{N}$.

Prove that the rank R of X satisfies the following inequality:

$$R \leqslant \min_{N \in \mathcal{N}} \mathsf{rank}(NX) + \max_{M \in \mathcal{M}}(\text{number of zero columns in } MA).$$

Solution: For any $M \in \mathcal{M}$, we have

$$\begin{aligned}&\min_{N \in \mathcal{N}} \mathsf{rank}(NX) \leqslant \mathsf{rank}(MX) = \mathsf{rank}([MA, B, C])\\&\leqslant R - (\text{number of zero columns in } MA).\end{aligned}$$

If we take the minimum over all $M \in \mathcal{M}$, we obtain the statement.

(11) Let X be a three-way array, that is 1-non-degenerate, and let $\mathcal{N} = \{\boldsymbol{u} \mid \boldsymbol{u} \neq \mathbf{0}\}$. Prove that

$$\mathsf{rank}(X) \leqslant \min_{\boldsymbol{u} \in \mathcal{N}} \mathsf{rank}(\boldsymbol{u}X) + \dim_1(X) - 1.$$

Solution: Let $[A, B, C]$ be a representation of X and let $\mathcal{M} = \{\boldsymbol{u} \mid \boldsymbol{u}A \neq \mathbf{0}\}$. Since X is 1-non-degenerate, we have $\mathcal{N} = \{\boldsymbol{u} \mid \boldsymbol{u}X \neq \mathbf{0}\} \subseteq \mathcal{M} \subseteq \mathcal{N}$, hence, $\mathcal{N} = \mathcal{M}$. Then

$$\begin{aligned}&\max(\text{number of zero columns in } \boldsymbol{u}A\\&= \max(\text{number of zero columns in } A\\&\quad\text{which are orthogonal to } \boldsymbol{u})\\&\geqslant \mathsf{rank}(A) - 1\\&\quad(\text{because we can pick some } \mathsf{rank}(A)\\&\quad\text{independent columns of}\\&\quad A \text{ and select } \boldsymbol{u} \text{ orthogonal to } \mathsf{rank}(A) - 1 \text{ of them})\\&\geqslant \dim_1(X) - 1.\end{aligned}$$

Now the statement follows from the previous supplement.

(12) Let X be a 1-non-degenerate array and let $\mathcal{N}$ consist of all $M \times 1$-matrices with full row rank. Prove that

$$\mathsf{rank}(X) \geqslant \min_{A \in \mathcal{N}} \mathsf{rank}(AX) + \dim_1(X) - M.$$

(13) Prove that the following multilinear system:

$$x_0y_0 + x_1y_1 = x_0y_1 - x_1y_0 = 0,$$

has nontrivial solutions in $\mathbb{C}$.

(14) Prove that the following multilinear system:

$$x_0y_0 = x_0y_1 = x_1y_0 = x_1y_1 = 0,$$

has only trivial solutions.

(15) Let $T \in \mathcal{M}_{m,n}$ be a hypercubic mda and let $a, b \in \mathbb{R}$. If $(\lambda, \boldsymbol{x})$ is an eigenpair of T, prove that $(a\lambda + b, \boldsymbol{x})$ is an eigenpair of $aT + bI_{m,n}$.

(16) Consider the odd-order symmetric mda $T \in T_{m,n}$ defined in [27] or [138] as follows:

$$\begin{aligned} T_{111} &= 10 \quad & T_{112} &= -\sqrt{3} \\ T_{121} &= -\sqrt{3} & T_{122} &= \sqrt{3} \\ T_{211} &= -\sqrt{3} & T_{212} &= \sqrt{3} \\ T_{221} &= \sqrt{3} & T_{222} &= 4. \end{aligned}$$

Prove the following:

(a) if $u, v \in \mathbb{R}_{>0}$, then

$$2u^3 + v^3 \geqslant 3u^2v;$$

(b) $T\boldsymbol{x}^3 \geqslant 0$ for $\boldsymbol{x} \in \mathbb{R}^2_{\geqslant 0}$;

(c) there is no H-eigenvalue for T.

Solution: The inequality is equivalent to $2z^3 + 1 \geqslant 3z^2$ for $z > 0$. The function $\phi(z) = 2z^3 - 3x^2 + 1$ has a minimum for $z = 1$, hence, $2z^3 - 3z^2 + 1 \leqslant 0$ for $z \geqslant 0$. Replacing z by $\frac{u}{v}$ gives the desired inequality.
Therefore,

$$\begin{aligned} 10x_1^3 + 4x_2^3 &= \frac{2}{3}(15^{\frac{1}{3}}x_1)^3 + \frac{1}{3}(12^{\frac{1}{3}}x_2)^3 \geqslant (15^{\frac{1}{3}}x_1)^2(12^{\frac{1}{3}}x_2) \\ &\geqslant 3\sqrt{3}x_1^2x_2, \end{aligned}$$

which shows that $T\boldsymbol{x}^3 \geqslant 0$ for $\boldsymbol{x} \in \mathbb{R}^2_{\geqslant 0}$.

Suppose that $(\lambda, \boldsymbol{x}) \in \mathbb{R} \times \mathbb{R}^2_{\neq 0}$ is an eigenpair, that is,

$$\begin{aligned}10x_1^2 - 2\sqrt{3}x_1x_2 + \sqrt{3}x_2^2 &= \lambda x_1^2,\\ -\sqrt{3}x_1^2 + 2\sqrt{3}x_1x_2 + 4x_2^2 &= \lambda x_2^2.\end{aligned}$$

If $x_2 = 0$, then $x_1 = 0$, which is impossible. If $x_2 \neq 0$, let $z = \frac{x_1}{x_2}$. The previous system becomes

$$\begin{aligned}(10-\lambda)z^2 - 2\sqrt{3}z + \sqrt{3} &= 0,\\ -\sqrt{3}z^2 + 2\sqrt{3}z + 4 - \lambda &= 0.\end{aligned}$$

The non-negativity of the discriminants of these trinomials implies

$$\begin{aligned}12 - 4\sqrt{3(10-\lambda)} &\geqslant 0,\\ 12 + 4\sqrt{3}(4-\lambda) &\geqslant 0.\end{aligned}$$

These inequalities lead to $4+\sqrt{3} \geqslant \lambda \geqslant 10-\sqrt{3}$, which is contradictory.

(17) Let $T \in \mathbb{R}^{I_1 \times \cdots \times I_{n-1} \times J \times I_{n+1} \times I_N}$ and let $A \in \mathbb{R}^{J \times I_n}$. If A has orthonormal columns, prove that $\|T\| = \|T \times_n A\|$.

Solution: By Equality (15.10), we have

$$\begin{aligned}\|T \times_n A\|^2 &= (T \times_n A, T \times_n A)\\ &= ((T \times_n A) \times A', T)\\ &\quad \text{(by Exercise 20)}\\ &= (T \times_n (A'A)), X) = (T, T)\\ &\quad \text{(by hypothesis)}\\ &= \|T\|^2,\end{aligned}$$

because $A'A$ is the unit matrix due to the orthonormality of its columns.

(18) Let $\mathcal{N} = \{1, \ldots, N\}$ and let $A^{(n)} \in \mathbb{R}^{I_n \times J_n}$ for $n \in \mathcal{N}$ be N matrices having orthonormal sets of columns. Prove that $T = [\![S; A^{(1)}, \ldots; A^{(N)}]\!]$ implies $S = [\![T; A^{(1)'}, \ldots; A^{(N)'}]\!]$.

Hint: Apply Theorem 15.14.

(19) Let $T = [\![G; A^{(1)}, A^{(2)}, \ldots, A^{(N)}]\!]$ and let $B \in \mathbb{R}^{J_1 \times J_1}$ be an orthogonal matrix. Prove that $T = [\![G \times_1 B; A^{(1)}B, A^{(2)}, \ldots, A^{(N)}]\!]$.

Solution: The hypothesis amounts to $T = G \times_1 A^{(1)} \times_2 \cdots \times_n A^{(N)}$. Therefore, $T = G \times_1 A^{(1)}BB' \times_2 \cdots \times_n A^{(N)} = G \times_1 B; A^{(1)}B \times_2 \cdots \times_n A^{(N)}$, so $T = [\![G \times_1 B; A^{(1)}B, A^{(2)}, \ldots, A^{(N)}]\!]$.

(20) Let

$$T \in \mathbb{R}^{I_1 \times \cdots \times I_{n-1} \times J \times I_{n+1} \times \cdots \times I_N},$$
$$S \in \mathbb{R}^{I_1 \times \cdots \times I_{n-1} \times K \times I_{n+1} \times \cdots \times I_N}$$

be two mdas, and let $A \in \mathbb{R}^{J \times K}$. Prove that

$$(T, S \times_n A) = (T \times_n A', S).$$

Solution: Let $S_{i_1 \cdots i_{n-1} k i_{n+1} \cdots i_N}$ be the components of S. By the definition of $S \times_n A$, we have

$$(S \times_n A)_{i_1 \cdots i_{n-1} j i_{n+1} \ldots i_N} = \sum_{k=1}^{|K|} S_{i_1 \cdots i_{n-1} k i_{n+1} \cdots i_N} A_{jk}$$

and

$$\begin{aligned}
&(T, S \times_n A) \\
&= \sum_{i_1, \ldots, i_{n-1}, j, i_{n+1}, \ldots, i_N} T_{i_1 \cdots i_{n-1} j i_{n+1} \cdots i_N} (S \times A)_{i_1 \cdots i_{n-1} j i_{n+1} \cdots i_N} \\
&= \sum_{i_1, \ldots, i_{n-1}, j, i_{n+1}, \ldots, i_N} T_{i_1 \cdots i_{n-1} j i_{n+1} \cdots i_N} \sum_{k=1}^{|K|} S_{i_1 \cdots i_{n-1} k i_{n+1} \cdots i_N} A_{jk} \\
&= \sum_{i_1, \ldots, i_{n-1}, k, i_{n+1}, \ldots, i_N} T_{i_1 \cdots i_{n-1} j i_{n+1} \cdots i_N} (A)'_{kj} S_{i_1 \cdots i_{n-1} k i_{n+1} \cdots i_N} \\
&= (T \times_n A', S).
\end{aligned}$$

The *Kruskal operator* is a special case of the Tucker operator. For $\mathcal{N} = \{1, \ldots, N\}$ and $A^{(n)} \in \mathbb{R}^{I_n \times R}$ for $n \in \mathcal{N}$, the value of the Kruskal operator applied to matrices $A^{(1)}, \ldots, A^{(N)}$ is defined as

$[\![A^{(1)}, \ldots, A^{(N)}]\!]$. Note that the matrices $A^{(1)}, \ldots, A^{(N)}$ must have the same number of columns. If I is the identity mda (whose diagonal components equal 1, and is 0 elsewhere), the Kruskal operator can be written as $[\![I; A^{(1)}, \ldots, A^{(N)}]\!]$.

If $S = [\![A^{(1)}, \ldots, A^{(N)}]\!]$, then

$$S_{i_1,i_2,\ldots,i_N} = \sum_{j=1}^{R} a^{(1)}_{i_1 j} a^{(2)}_{i_2 j} \cdots a^{(N)}_{i_N j}.$$

The Kruskal operator is used to define the PARAFAC decomposition of an mda T as $T = [\![A^{(1)}, A^{(2)}, \ldots, A^{(n)}]\!]$. The interested reader should consult [47, 92].

(21) Let $\mathcal{N} = \{1, \ldots, N\}$ and $A^{(n)} \in \mathbb{R}^{I_n \times R}$. If $A^{(n)} \in \mathbb{R}^{I_n \times R}$ and $B^{(n)} \in \mathbb{R}^{K_n \times I_n}$, prove that

$$[\![[\![A^{(1)}, \ldots, A^{(N)}]\!]; B^{(1)}, \ldots, B^{(N)}]\!] = [\![B^{(1)}A^{(1)}, \ldots, B^{(N)}A^{(N)}]\!].$$

(22) Let $\mathcal{N} = \{1, \ldots, N\}$ and $A^{(n)} \in \mathbb{R}^{I_n \times R}$ for $n \in \mathcal{N}$. Prove that

$$\left\| [\![A^{(1)}, \ldots, A^{(N)}]\!] \right\|^2 = \sum_{j=1}^{R} \sum_{k=1}^{R} ((A^{(1)'}A^{(1)}) \odot (A^{(2)'}A^{(2)}) \odot \cdots \odot (A^{(N)'}A^{(N)}))_{jk}.$$

Recall that "$\odot$" is the notation for the Hadamard product introduced in Definition 3.45.

Solution: We have

$$\begin{aligned}
&\left\| [\![A^{(1)}, \ldots, A^{(N)}]\!] \right\|^2 \\
&= ([\![A^{(1)}, \ldots, A^{(N)}]\!], [\![A^{(1)}, \ldots, A^{(N)}]\!]) \\
&= \sum_{i_1 \cdots i_N} \left(\sum_{j=1}^{R} a^{(1)}_{i_1 j} a^{(2)}_{i_2 j} \cdots a^{(N)}_{i_N j}, \sum_{k=1}^{R} a^{(1)}_{i_1 k} a^{(2)}_{i_2 k} \cdots a^{(N)}_{i_N k} \right) \\
&= \sum_{j=1}^{R} \sum_{k=1}^{R} \sum_{i_1 \cdots i_N} a^{(1)}_{i_1 j} a^{(2)}_{i_2 j} \cdots a^{(N)}_{i_N j} a^{(1)}_{i_1 k} a^{(2)}_{i_2 k} \cdots a^{(N)}_{i_N k}
\end{aligned}$$

$$= \sum_{j=1}^{R} \sum_{k=1}^{R} \sum_{i_1} a^{(1)}_{i_1 j} a^{(1)}_{i_1 k} \sum_{i_2} a^{(2)}_{i_2 j} a^{(2)}_{i_2 k} \cdots \sum_{i_N} a^{(1)}_{i_N j} a^{(1)}_{i_N k}$$

$$= \sum_{j=1}^{R} \sum_{k=1}^{R} (A^{(1)'} A^{(1)})_{jk} \cdots (A^{(N)'} A^{(N)})_{jk}.$$

A fundamental result of complexity theory [62] is that the *satisfiability problem* stated in what follows is NP-complete.

> **3-SATISFIABILITY** or **3-SAT**:
> *Input:* a collection of m clauses $\mathcal{C} = \{C_1, \ldots, C_m\}$ on a finite set of n variables U such that each clause contains 3 literals.
> *Problem:* is there a truth assignment for U that satisfies all clauses in $\mathcal{C}$?

The NP-completeness of 3-SAT was used in [74] to establish the NP-hardness of determining the rank of a tensor over the field of rational numbers.

This new problem can be formulated as follows:

> **TENSOR RANK**:
> *Input: a three-dimensional tensor T_{ijk}*
> *Problem:* given numbers T_{ijk} where $1 \leqslant i \leqslant n_1$, $1 \leqslant j \leqslant n_2$, $1 \leqslant k \leqslant n_3$, and $r \in \mathbb{N}$ are there r vectors $\boldsymbol{v}^{(\ell)}_e$, $1 \leqslant \ell \leqslant r$, $1 \leqslant e \leqslant 3$ such that $T_{ijk} = \sum_{\ell=1}^{r} v^{\ell}_1(i) v^{\ell}_2(j) v^{\ell}_3(k)$ for all i, j, k?

(23) Give an encoding of an arbitrary **3-SAT** Boolean formula in n variables and m clauses as an mda $T \in \mathbb{Q}^{(n+2m+2)\times 3n\times(3n+m)}$ with the property that the **3-SAT** formula is satisfiable if and only if $\mathsf{rank}(T) \leqslant 4n + 2m$.

Hint: See [74].

Bibliographical Comments

The main sources for this chapter are [4, 5, 91–93, 95]. For tensor ranks in various fields, the reader should consult [58, 115]. A useful survey of tensor decomposition is in [139].

The works of L. Qi and his collaborators [80, 136–138] is a main reference for the spectral theory of tensors. Supplement 10 is a result of Kruskal [97]. The treatment of hyperdeterminants follows [123].

Bibliography

[1] Abdi, H. (2003). Partial least squares regression (PLS-regression), in *Encyclopedia for Research Methods for the Social Sciences* (Sage, Thousand Oaks, CA), pp. 792–795.

[2] Akivis, M. A. and Goldberg, V. V. (1977). *An Introduction to Linear Algebra and Tensors* (Dover Publications, New York).

[3] Artin, M. (1991). *Algebra*, 1st edn. (Prentice-Hall, Englewood Cliffs, NJ), a second edition was published in 2010.

[4] Bader, B. W. and Kolda, T. G. (2006). Algorithm 862: Matlab tensor classes for fast algorithm prototyping, *ACM Transactions on Mathematical Software* **32**, 4, pp. 635–653.

[5] Bader, B. W. and Kolda, T. G. (2008). Efficient Matlab computations with sparse and factored tensors, *SIAM Journal on Scientific Computing* **30**, 1, pp. 205–231.

[6] Baeza-Yates, R. and Ribeiro-Neto, B. (1999). *Modern Information Retrieval* (Addisson-Wesley, Harlow, UK).

[7] Bartle, R. G. and Sherbery, D. R. (1999). *Introduction to Real Analysis*, 3rd edn. (Wiley, New York).

[8] Bauer, F. L. and Fike, C. T. (1960). Norms and exclusion theorems, *Numerische Mathematik* **2**, pp. 137–141.

[9] Belkin, M. and Niyogi, P. (2001). Laplacian eigenmaps and spectral techniques for embedding and clustering, in *Advances in Neural Information Processing Systems 14* (MIT Press), pp. 585–591.

[10] Ben-Israel, A. (1992). A volume associated with $m \times n$ matrices, *Linear Algebra and Its Applications* **167**, pp. 87–111.

[11] Ben-Israel, A. and Greville, T. N. E. (1974). *Generalized Inverses — Theory and Applications* (Wiley-Interscience, New York).

[12] Berge, J. M. F. T. (1983). A generalization of Kristof's theorem on the trace of certain matrix products, *Psychometrika* **48**, pp. 519–523.

[13] Berge, J. M. F. T. (1991). Kruskal's polynomial for 2× 2× 2 arrays and a generalization to 2× n× n arrays, *Psychometrika* **56**, 4, pp. 631–636.

[14] Berge, J. M. F. T. (1993). *Least Squares Optimization in Multivariate Analysis* (DSWO Press, Leiden, The Netherlands).

[15] Berge, T. (1977). Orthogonal Procrustes rotation for two or more matrices, *Psychometrika* **42**, pp. 267–276.

[16] Berkhin, P. and Becher, J. (2002). Learning simple relations: Theory and applications, in *Proceedings of the Second SIAM International Conference on Data Mining* (Arlington, VA).

[17] Berry, M. W., Browne, M., Langville, A. N., Pauca, V. P. and Plemmons, R. J. (2007). Algorithms and applications for approximate nonnegative matrix factorization, *Computational Statistics & Data Analysis* **52**, 1, pp. 155–173.

[18] Bhatia, R. (1997). *Matrix Analysis* (Springer, New York).

[19] Billsus, D. and Pazzani, M. (1998). Learning collaborative information filters, in J. W. Shavlik (ed.), *Proceedings of the 15th International Conference on Machine Learning*, pp. 46–54.

[20] Birkhoff, G. (1973). *Lattice Theory*, 3rd edn. (American Mathematical Society, Providence, RI).

[21] Björk, A. (1996). *Numerical Methods for Least Squares Problems* (SIAM, Philadelphia, PA).

[22] Brand, M. (2003). Fast online SVD revisions for lightweight recommender systems, in D. Barbará and C. Kamath (eds.), *Proceedings of the Third SIAM International Conference on Data Mining* (SIAM, Philadelphia, PA), pp. 37–46.

[23] Brokett, R. W. and Dobkin, D. (1978). On the optimal evaluation of a set of bilinear forms, *Linear Algebra and Its Applications* **19**, pp. 207–235.

[24] Burdick, D. S. (1995). An introduction to tensor product with applications to multiway data analysis, *Chemometrics and Intelligent Laboratory Systems* **28**, pp. 229–237.

[25] Carroll, J. D. and Chang, J. J. (1970). Analysis of individual differences in multidimensional scaling via an n-way generalization of Eckart-Young decomposition, *Psychometrika* **35**, 3, pp. 283–319.

[26] Catral, M., Han, L., Neumann, M. and Plemmons, R. (2004). On reduced rank nonnegative (non-negative) matrix factorizations for symmetric matrices, *Linear Algebra and Applications* **393**, pp. 107–127.

[27] Chen, H., Chen, Y., Li, G. and Qi, L. (2015). Finding the maximum eigenvalue of a class of tensors with applications in copositivity test and hypergraphs, arXiv preprint arXiv:1511.02328.

[28] City of Boston (2008). Public Health Commission Research Office: The health of Boston 2008, Tech. rep., Boston, MA.

[29] Cline, R. E. and Funderlink, R. E. (1979). The rank of a difference of matrices and associated generalized inverses, *Linear Algebra and Its Applications* **24**, pp. 185–215.

[30] Comon, P., Golub, G., Lim, L.-H. and Mourrain, B. (2008). Symmetric tensors and symmetric tensor rank, *SIAM Journal on Matrix Analysis and Applications* **30**, 3, pp. 1254–1279.

[31] Cox, D. A., Little, J. and O'Shea, D. (2005). *Using Algebraic Geometry*, 2nd edn. (Springer, New York).

[32] Cox, T. F. and Cox, M. (2001). *Multidimensional Scaling*, 2nd edn. (Chapman & Hall, Boca Raton).

[33] Davis, C. and Kahan, W. M. (1969). Some new bounds on perturbation of subspaces, *Bulletin of the American Mathematical Society* **75**, pp. 863–868.

[34] Deerwester, S., Dumais, S. T., Furnas, G. W., Landauer, T. K. and Harshman, R. (1990). Indexing by latent semantic analysis, *Journal of the American Society for Information Science* **41**, pp. 391–407.

[35] Ding, C. and He, X. (2004). K-means clustering via principal component analysis, in *Proceedings of the 21st International Conference on Machine Learning* (Banff, Canada), pp. 225–232.

[36] Ding, C., He, X. and Simon, H. (2005). On the equivalence of nonnegative matrix factorization and spectral clustering, in *Proceedings of the SIAM International Conference on Data Mining* (Newport Beach, CA), pp. 606–610.

[37] Dodson, C. T. J. and Poston, T. (1997). *Tensor Geometry*, 2nd edn. (Springer, Berlin).

[38] Donoho, D. and Grimes, G. (2003). Hessian eigenmaps: New locally linear embedding techniques for high-dimensional data, *Proceedings of the National Academy of Sciences of the United States of America* **5**, pp. 5591–5596.

[39] Drineas, P., Frieze, A., Kannan, R., Vampala, S. and Vinay, V. (2004). Clustering large graphs via the singular value decomposition, *Machine Learning* **56**, pp. 9–33.

[40] Dumais, S. T., Furnas, G. W., Landauer, T. K., Deerwester, S. and Harshman, R. (1988). Using latent semantic analysis to improve access to textual information, in *CHI '88: Proceedings of the SIGCHI Conference on Human Factors in Computing Systems* (ACM, New York, NY), ISBN 0-201-14237-6, pp. 281–285.

[41] Eckhart, C. and Young, G. (1936). The approximation of one matrix by another of lower rank, *Psychometrica* **1**, pp. 211–218.

[42] Egerváry, E. (1960). On rank-diminishing operators and their application to the solution of linear systems, *Zeitschrift für Angewandte Mathematik und Physik* **11**, pp. 376–386.

[43] Elad, M. (2010). *Sparse and Redundant Representations* (Springer, New York).

[44] Eldén, L. (2007). *Matrix Methods in Data Mining and Pattern Recognition* (SIAM, Philadelphia, PA).

[45] Elkan, C. (2003). Using the triangle inequality to accelerate k-means, in *Proceedings of the 20th International Conference on Machine Learning*, pp. 147–153.

[46] A. H. et al. (1970). Foundations of the PARAFAC procedure: Models and conditions for an "explanatory" multimodal factor analysis.

[47] Faber, N. M., Bro, R. and Hopke, P. K. (2003). Recent developments in candecomp/parafac algorithms: A critical review, *Chemometrics and Intelligent Laboratory Systems* **65**, pp. 119–137.

[48] Fan, K. (1949). On a theorem of Weil concerning eigenvalues of linear transformations -I, *Proceedings of the National Academy of Sciences of the United States of America* **35**, pp. 652–655.

[49] Fan, K. (1951). Maximum properties and inequalities for the eigenvalues of completely continuous operators, *Proceedings of the National Academy of Sciences of the United States of America* **37**, pp. 760–766.

[50] Favier, G. (2019). *From Algebraic Structure to Tensors*, Vol. 1 (Wiley, Hoboken, NJ).

[51] Favier, G. (2021). *Matrix and Tensor Decompositions in Signal Processing*, Vol. 2 (Wiley, Hoboken, NJ).

[52] Favier, G. and de Almeida, A. (2014). Overview of constrained parafac models, *EURASIP Journal on Advances in Signal Processing* **2014**, 1, pp. 1–25.

[53] Fejer, P. A. and Simovici, D. A. (1991). *Mathematical Foundations of Computer Science*, Vol. 1 (Springer Verlag, New York).

[54] Fletcher, R. and Sorensen, D. (1983). An algorithmic derivation of the Jordan canonical form, *American Mathematical Monthly* **90**, pp. 12–16.

[55] Food and Organization, A. (2009). *FAO Statistical Yearbook* (Statistics Division FAO, Rome, Italy).

[56] Forgy, E. (1965). Cluster analysis of multivariate data: Efficiency vs. interpretabiity of classifications, *Biometrics* **21**, p. 768.

[57] Francis, J. G. F. (1961). The QR transformation — A unitary analogue to the LR transformation — Part I, *The Computer Journal* **4**, pp. 265–271.

[58] Friedland, S. (2012). On the generic and typical rank of 3-tensors, *Linear Algebra and Its Applications* **436**, pp. 478–497.

[59] Frieze, A. and Kannan, R. (1999). Quick approximation to matrices and applications, *Combinatorica* **19**, pp. 175–200.

[60] Fuglede, B. (1950). A commutativity theorem for normal operators, *Proceedings of the National Academy of Sciences of the United States of America* **36**, 1, p. 35.

[61] Gabriel, K. R. (1971). The biplot graph display of matrices with application to principal component analysis, *Biometrika* **58**, pp. 453–467.

[62] Garey, M. R. and Johnson, D. S. (1979). *Computers and Intractability — A Guide to the Theory of NP-Completeness* (W. H. Freeman, New York).

[63] Gelfand, I. M., Kapranov, M. and Zelevinsky, A. (1994). *Discriminants, Resultants, and Multidimensional Determinants* (Birkhäuser, Boston, MA).

[64] Gilat, A. (2011). *MATLAB — An Introduction with Applications*, 4 edn. (John Wiley, New York).

[65] Golub, G. H. and Loan, C. F. V. (1989). *Matrix Computations*, 2nd edn. (The Johns Hopkins University Press, Baltimore, MD).

[66] Gower, J. C. (1966). Some distance properties of latent root and vector methods used in multivariate analysis, *Biometrika* **53**, pp. 325–338.

[67] Gower, J. C. and Hand, D. J. (1995). *Biplots* (Chapman & Hall, London).

[68] Gower, J. C., Lubbe, S. G. and Le Roux, N. (2011). *Understanding Biplots* (Wiley, New York).

[69] Greenacre, M. (2010). *Biplots in Practice* (Fundacion BBVA, Madrid, Spain).

[70] Greub, W. (1981). *Linear Algebra*, 4th edn. (Springer-Verlag, New York).

[71] Greub, W. H. (1978). *Multilinear Algebra*, 2nd edn. (Springer-Verlag, New York).

[72] Hackbusch, W. (2012). *Tensor Spaces and Numerical Tensor Calculus*, 2nd edn. (Springer, Cham, Switzerland).

[73] Hartigan, J. A. and Wong, M. (1979). A k-means clustering algorithm, *Applied Statistics*, pp. 100–108.

[74] Håstad, J. (1990). Tensor rank is NP-complete, *Journal of Algorithms* **4**, 11, pp. 644–654.

[75] Higham, D. J. and Higham, N. J. (2000). *Matlab Guide* (SIAM, Philadelphia, PA).

[76] Higham, N. J. (2002). *Accuracy and Stability of Numerical Algorithms*, 2nd edn. (SIAM, Philadelphia, PA).
[77] Hoffman, A. J. and Wielandt, H. W. (1953). The variation of the spectrum of a normal matrix, *Duke Mathematical Journal* **20**, pp. 37–39.
[78] Horn, R. A. and Johnson, C. R. (1996). *Matrix Analysis* (Cambridge University Press, Cambridge, UK).
[79] Horn, R. A. and Johnson, C. R. (2008). *Topics in Matrix Analysis* (Cambridge University Press, Cambridge, UK).
[80] Hu, S., Huang, Z.-H., Ling, C. and Qi, L. (2013). On determinants and eigenvalue theory of tensors, *Journal of Symbolic Computation* **50**, pp. 508–531.
[81] Inaba, M., Katoh, N. and Imai, H. (1994). Applications of weighted Voronoi diagrams and randomization to variance-based k-clustering, in *10th Computational Geometry* (ACM, New York), pp. 331–339.
[82] Johnson, R. A. and Wichern, D. W. (2002). *Applied Multivaried Statistical Analysis*, 5 edn. (Prentice Hall, Upper Saddle River, NJ).
[83] Johnson, R. K. (2011). *The Elements of MATLAB Style* (Cambridge University Press, Cambridge, UK).
[84] Jolliffe, I. T. (2002). *Principal Component Analysis*, 2nd edn. (Springer, New York).
[85] Jordan, P. and Neumann, J. V. (1935). On inner products in linear, metric spaces, *The Annals of Mathematics* **36**, pp. 719–723.
[86] Juhász, F. (1978). On the spectrum of a random graph, in *Colloquia Mathematica Societatis János Bolyai*, Vol. 25 (Szeged), pp. 313–316.
[87] Kannan, R. (2010). Spectral methods for matrices and tensors, in *Proceedings of the ACM Symposium on the Theory of Computing (STOC)* (ACM, New York), pp. 1–12.
[88] Kanungo, T., Mount, D. M., Netanyahu, N. S. and Piatko, C. D. (2004). A local search approximation algorithm for k-means clustering, *Computational Geometry: Theory and Applications* **28**, pp. 89–112.
[89] Kiers, H. A. (1998). A three-step algorithm for CANDECOMP/PARAFAC analysis of large data sets with multicollinearity, *Journal of Chemometrics: A Journal of the Chemometrics Society* **12**, 3, pp. 155–171.
[90] Knyazev, A. V. and Argentati, M. E. (2006). On proximity of Rayleigh quotients for different vectors and Ritz values generated by different trial subspaces, *Linear Algebra and Its Applications* **415**, pp. 82–95.

[91] Kolda, T. G. (2001). Orthogonal tensor decompositions, *SIAM Journal of Matrix Analysis and Applications* **23**, pp. 243–255.

[92] Kolda, T. G. (2006). Multilinear operators for higher-order decompositions, Tech. Rep. SAND2006-2081, Sandia.

[93] Kolda, T. G. and Bader, B. W. (2009). Tensor decompositions and applications, *SIAM Review* **51**, 3, pp. 455–500.

[94] Kolda, T. G. and O'Leary, D. P. (1998). A semidiscrete matrix decomposition for latent semantic indexing in information retrieval, *ACM Transactions on Information Systems* **16**, pp. 322–346.

[95] Kolda, T. G. and O'Leary, D. P. (1999). Computation and uses of the semidiscrete matrix decomposition, Tech. Rep. ORNL-TM-13766, Oak Ridge National Laboratory, Oak Ridge, TN.

[96] Kroonenberg, P. M. (2008). *Applied Multiway Data Analysis* (Wiley, Hoboken, NJ).

[97] Kruskal, J. B. (1977). Three-way arrays: Rank and uniqueness of trilinear decompositions, with application to arithmetic complexity and statistics, *Linear Algebra and Its Applications* **18**, 2, pp. 95–138.

[98] Lanczos, C. (1958). Linear systems in self-adjoint form, *American Mathematical Monthly* **65**, pp. 665–679.

[99] Langville, A. N. and Stewart, W. J. (2004). The Kronecker product and stochastic automata networks, *Journal of Computational and Applied Mathematics* **167**, 2, pp. 429–447.

[100] Lathauwer, L. D. (1997). Signal processing based on multilinear algebra, Ph.D. thesis, Katholieke Universiteit Leuven.

[101] Lathauwer, L. D., Moor, B. D. and Vandewalle, J. (2000). A multilinear singular value decomposition, *SIAM Journal on Matrix Analysis and Applications* **21**, 4, pp. 1253–1278.

[102] Lathauwer, L. D., Moor, B. D. and Vandewalle, J. (2000). On the best rank-1 and rank-$(r_1, r_2, \ldots, r_n)$ approximation of higher-order tensors, *SIAM Journal on Matrix Analysis and Applications* **21**, 4, pp. 1324–1342.

[103] Lawson, C. L. and Hanson, R. J. (1995). *Solving Least Square Problems* (SIAM, Philadelphia, PA).

[104] Lax, P. D. (2007). *Linear Algebra and Its Applications* (Wiley-International, Hoboken, NJ).

[105] Lee, D. D. and Seung, H. S. (1999). Learning the parts of objects by non-negative matrix factorization, *Nature* **401**, pp. 515–521.

[106] Lichnerowicz, A. (1956). *Algèbre et Analyse Linéaires* (Masson et Co., Paris).

[107] Lloyd, S. P. (1982). Least square quantization in PCM, *IEEE Transactions on Information Theory* **28**, pp. 129–137.

[108] MacLane, S. and Birkhoff, G. (1993). *Algebra*, 3rd edn. (Chelsea Publishing Company, New York).
[109] Magnus, J. R. and Neudecker, H. (1979). The commutation matrix: Some properties and applications, *The Annals of Statistics* **7**, 2, pp. 381–394.
[110] Marcus, M. (1973). *Finite Dimensional Multilinear Algebra — Part I* (Marcel Dekker, New York).
[111] Marcus, M. and Minc, H. (2010). *A Survey of Matrix Theory and Matrix Inequalities* (Dover Publications, New York).
[112] Marsaglia, G. and Styan, G. (1974). Equalities and inequalities for ranks of matrices, *Linear and Multilinear Algebra* **2**, pp. 269–292.
[113] Meyer, C. (2000). *Matrix Analysis and Applied Algebra* (SIAM, Philadelphia, PA).
[114] Mirsky, L. (1975). A trace inequality of john von neumann, *Monatshefte für Mathematik* **79**, pp. 303–306.
[115] Moitra, A. (2018). *Algorithmic Aspects of Machine Learning* (Cambridge University Press, Cambridge, UK).
[116] Moler, C. B. (2004). *Numerical Computing with MATLAB* (SIAM, Philadelphia, PA).
[117] Morozov, A. Y. and Shakirov, S. R. (2010). New and old results in resultant theory, *Theoretical and Mathematical Physics* **163**, 2, pp. 587–617.
[118] Neudecker, H. (1968). The Kronecker matrix product and some of its applications in econometrics, *Statistica Neerlandica* **22**, 1, pp. 69–82.
[119] Ninio, F. (1976). A simple proof of the Perron-Frobenius theorem for positive matrices, *Journal of Physics A* **9**, pp. 1281–1282.
[120] Oldenburger, R. (1940). Infinite powers of matrices and characteristic roots, *Duke Mathematical Journal* **6**, pp. 357–361.
[121] O'Leary, D. P. (2006). Matrix factorization for information retrieval, Tech. Rep., University of Maryland, College Park, MD.
[122] O'Leary, D. P. and Peleg, S. (1983). Digital image compression, *IEEE Transactions on Communications* **31**, pp. 441–444.
[123] Ottaviani, G. (2012). Introduzione all'iperdeterminante, *La Matematica nella Società e nella Cultura. Rivista dell'Unione Matematica Italiana* **5**, 2, pp. 169–195.
[124] Overton, M. L. and Womersley, R. S. (1991). On the sum of the largest eigenvalues of a symmetric matrix, Tech. Rep. 550, Courant Institute of Mathematical Sciences, NYU.
[125] Paatero, P. (1997). Least squares formulation of robust non-negative factor analysis, *Chemometrics and Intelligent Laboratory Systems* **37**, pp. 23–35.

[126] Paatero, P. (1999). The multilinear engine — A table driven least squares program for solving multilinear problems, including the n-way parallel factor analysis model, *Journal of Computational and Graphical Statistics* **8**, pp. 1–35.

[127] Paatero, P. and Tapper, U. (1994). Positive matrix factorization: A non-negative factor model with optimal utilization of error estimates of data values, *Environmetrics* **5**, pp. 111–126.

[128] Paige, C. and Saunders, M. (1981). Toward a generalized singular value decomposition, *SIAM Journal on Numerical Analysis* **18**, pp. 398–405.

[129] Paige, C. C. and Wei, M. (1994). History and generality of the CS decomposition, *Linear Algebra and Its Applications* **208**, pp. 303–326.

[130] Papalexakis, E. E., Faloutsos, C. and Sidiropoulos, N. D. (2016). Tensors for data mining and data fusion: Models, applications, and scalable algorithms, *ACM Transactions on Intelligent Systems and Technology (TIST)* **8**, 2, pp. 1–44.

[131] Pauca, V. P., Shahnaz, F., Berry, M. W. and Plemmons, R. J. (2004). Text mining using non-negative matrix factorizations, in *SIAM Data Mining*, pp. 452–456.

[132] Pferschy, U. and Rudolf, R. (1994). Some geometric clustering problems, *Nordic Journal of Computing* **1**, pp. 246–263.

[133] Pollock, D. S. G. (1979). *The Algebra of Econometrics* (John Wiley & Sons, Chichester).

[134] Pollock, D. S. G. (2021). Multidimensional arrays, indices and Kronecker products, *Econometrics* **9**, p. 18.

[135] Prasolov, V. V. (2000). *Problems and Theorems in Linear Algebra* (American Mathematical Society, Providence, RI).

[136] Qi, L. (2005). Eigenvalues of a real supersymmetric tensor, *Journal of Symbolic Computation* **40**, 6, pp. 1302–1324.

[137] Qi, L. (2007). Eigenvalues and invariants of tensors, *Journal of Mathematical Analysis and Applications* **325**, 2, pp. 1363–1377.

[138] Qi, L. and Luo, Z. (2017). *Tensor Analysis — Spectral Theory and Special Tensors* (SIAM, Philadelphia, PA).

[139] Rabanser, S., Shchur, O. and Günnemann, S. (2017). Introduction to tensor decompositions and their applications in machine learning, arXiv preprint arXiv:1711.10781.

[140] Roweis, S. T. and Saul, L. K. (2000). Locally linear embedding, `http://www.cs.nyu.edu/~roweis/lle/`.

[141] Roweis, S. T. and Saul, L. K. (2000). Nonlinear dimensionality reduction by locally linear embedding, *Science* **290**, pp. 2323–2326.

[142] Sarwar, B., Karypis, G., Konstan, J. and Riedl, J. (2002). Incremental singular value decomposition algorithms for highly scalable recommender systems, in *5th International Conference on Computer and Information Technology (ICCIT)*, pp. 27–28.

[143] Saul, L. K. and Roweis, S. T. (2000). An introduction to locally linear embedding, Tech. rep., AT&T Labs-Research.

[144] Saul, L. K. and Roweis, S. T. (2003). Think globally, fit locally: Unsupervised learning of low dimensional manifolds, *Journal of Machine Learning Research* **4**, pp. 119–155.

[145] Saul, L. K., Weinberger, K., Sha, F., Ham, J. and Lee, D. D. (2006). Spectral methods for dimensionality reduction, in *Semi-Supervised Learning* (The MIT Press, Cambridge, MA), pp. 293–308.

[146] Schoenberg, I. J. (1935). Remarks to Maurice Fréchet's article "sur la définition axiomatique d'une classe d'espaces distanciés vectoriellement applicable sur l'espace de Hilbert", *Annals of Mathematics* **36**, pp. 724–732.

[147] Schönemann, P. H. (1970). On metric multidimensional unfolding, *Psychometrika* **35**, pp. 349–366.

[148] Schwerdtfeger, H. (1960). Direct proof of Lanczos' decomposition theorem, *American Mathematical Monthly* **67**, pp. 855–860.

[149] Shahnaz, F., Berry, M. W., Pauca, V. P. and Plemmons, R. J. (2006). Document clustering using nonnegative matrix factorization, *Information Processing Management* **42**, 2, pp. 373–386.

[150] Sibson, R. (1978). Studies in the robustness of multidimensional scaling: Procrustes statistics, *Journal of the Royal Statistical Society, Series B* **40**, pp. 234–238.

[151] Sidiropoulos, N., Lathauwer, L. D., Fu, X., Huang, K., Papalexakis, E. and Faloutsos, C. (2017). Tensor decomposition for signal processing and machine learning, *IEEE Transactions on Signal Processing* **65**, 13, pp. 3551–3582.

[152] Simovici, D. A. and Djeraba, C. (2015). *Mathematical Tools for Data Mining — Set Theory, Partially Ordered Sets, Combinatorics*, 2nd edn. (Springer-Verlag, London).

[153] Soules, G. W. (1983). Constructing symmetric nonnegative matrices, *Linear and Multilinear Algebra* **13**, pp. 241–251.

[154] Stanley, R. P. (1997). *Enumerative Combinatorics*, Vol. 1 (Cambridge University Press, Cambridge).

[155] Stanley, R. P. (1999). *Enumerative Combinatorics*, Vol. 2 (Cambridge University Press, Cambridge).

[156] Sternberg, S. (1983). *Lectures on Differential Geometry*, 2nd edn. (AMS Chelsea Publishing, Providence, RI).

[157] Stewart, G. W. (1973). Error and perturbation bounds for subspaces associated with certain eigenvalue problems, *SIAM Review* **15**, pp. 727–764.

[158] Stewart, G. W. (1977). On the perturbation of pseudo-inverses, projections, and linear least square problems, *SIAM Review* **19**, pp. 634–662.

[159] Stewart, G. W. (1993). On the early history of the singular value decomposition, *SIAM Review* **35**, pp. 551–566.

[160] Stewart, G. W. and Sun, J.-G. (1990). *Matrix Perturbation Theory* (Academic Press, Boston, MA).

[161] Strassen, V. (1969). Gaussian elimination is not optimal, *Numerische Mathematik* **13**, pp. 354–356.

[162] Trefethen, L. N. and III, D. B. (1997). *Numerical Linear Algebra* (SIAM, Philadelphia, PA).

[163] Trotter, W. T. (1995). Partially ordered sets, in *Handbook of Combinatorics* (The MIT Press, Cambridge, MA), pp. 433–480.

[164] Tucker, L. R. (1966). Some mathematical notes on three-mode factor analysis, *Psychometrica* **31**, pp. 279–331.

[165] van der Waerden, B. L. (1950). *Modern Algebra (Vol. I and II)* (F. Ungar Publ. vo, New York).

[166] van Lint, J. H. and Wilson, R. M. (2002). *A Course in Combinatorics*, 2nd edn. (Cambridge University Press, Cambridge).

[167] Van Loan, C. F. and Fan, K. Y. D. (2010). *Insight through Computing* (SIAM, Philadelphia, PA).

[168] vols.), F. R. G. . (1977). *The Theory of Matrices* (American Mathematical Society, Providence, RI).

[169] Wedderburn, J. (1934). *Lectures on Matrices* (American Mathematical Society, Providence, RI).

[170] Weyr, E. (1887). Note sur la théorie de quantités complexes formées avec n unites principales, *Bulletin des Sciences Mathématiques II* **11**, pp. 205–215.

[171] Wilkerson, J. H. (1965). *The Algebraic Eigenvalue Problem* (Clarendon Press, Oxford, London).

[172] Winograd, S. (1968). A new algorithm for inner product, *IEEE Transactions on Computers* **C-17**, pp. 693–694.

[173] Wu, X. and (editors), V. K. (2009). *The Top Ten Algorithms in Data Mining* (CRC Press, Boca Raton, FL).

[174] Ye, J. (2007). Least square discriminant analysis, in *Proceedings of the 24th International Conference on Machine Learning* (Corvallis, OR), pp. 1087–1094.

[175] Yokonuma, T. (1992). *Tensor Spaces and Exterior Algebra* (American Mathematical Society, Providence, RI).

[176] Zha, H., He, X., Ding, C., Simon, H. and Gu, M. (2001). Spectral relaxation for k-means clustering, *Neural Information Processing Systems* **14**, pp. 1057–1064.

[177] Zhang, F. (1999). *Matrix Theory: Basic Results and Techniques* (Springer Verlag, New York).

[178] Zhang, F. (2005). *The Schur Complement and Its Applications* (Springer, New York).

Index

D

E

F

G

L

M

N

O

P

Q

R

S

W

Z

www.ingramcontent.com/pod-product-compliance
Lightning Source LLC
La Vergne TN
LVHW012052011026
849162LV00018B/183